Physical Effects of Geometric Phases

Physical Effects of Geometric Phases

Qian Niu
The University of Texas at Austin

Ming-Che Chang
National Taiwan Normal University

Biao Wu
Peking University

Di Xiao
Carnegie Mellon University

Ran Cheng
Carnegie Mellon University

World Scientific

NEW JERSEY • LONDON • SINGAPORE • BEIJING • SHANGHAI • HONG KONG • TAIPEI • CHENNAI • TOKYO

Published by

World Scientific Publishing Co. Pte. Ltd.
5 Toh Tuck Link, Singapore 596224
USA office: 27 Warren Street, Suite 401-402, Hackensack, NJ 07601
UK office: 57 Shelton Street, Covent Garden, London WC2H 9HE

British Library Cataloguing-in-Publication Data
A catalogue record for this book is available from the British Library.

PHYSICAL EFFECTS OF GEOMETRIC PHASES

ISBN 978-981-3225-72-5

Desk Editor: Christopher Teo

Typeset by Stallion Press
Email: enquiries@stallionpress.com

Printed in Singapore

To Professor David J. Thouless

Preface

At the finishing stage of this book, my coauthors reminded me of two formalities that books have: a dedication and a preface. We decide to dedicate this book to our academic father and grandfather, David Thouless, because of his profound influence on the subject matter and our career paths. The Nobel Committee apparently agreed with us, for they announced shortly after their 2016 physics prize to him along with Duncan Haldane and Michael Kosterlitz. I will now use this preface to describe briefly my personal journey on the subject of this book and to thank my collaborators along the way.

David gave me the chance to contribute on the development of topological ideas on quantum charge pumping and the quantum Hall effect by addressing disorder and interaction effects. He also led me into the beautiful world of Hofstadter butterfly. My first serious adventure in the study of geometric phases was with Ming-Che Chang to understand the way that the Hofstadter spectrum splits as the magnetic flux approaches an irrational unit through a series of rational approximants. The Chern numbers that David and collaborators (TKNN) calculated for the magnetic Bloch bands turned out to be the key in explaining the anomalies in the spectral splitting rule.

Our approach was a semiclassical one based on wave packet dynamics within each magnetic Bloch band, driven by an external electric field and an extra magnetic field in addition to the rational one defining the magnetic Bloch bands. One striking difference from the textbook equations of motion for Bloch electrons is that the electron velocity acquires an anomalous term proportional to the Berry curvature, which makes up the Chern number for a full band. The Berry phases of cyclotron orbits also affects the quantization rule that ultimately affects how many Landau levels one can fit in a full band. The success in accurately predicting the Landau levels and obtaining the correct number of band splitting gave us the confidence on the semiclassical theory when it is supplemented by the Berry phase effect.

The true power of TKNN was not confined to the quantum Hall effect. There had already existed an elegant explanation by Robert Laughlin based on a gauge-invariance argument, which is much simpler and more pertinent to the particulars of two-dimensional electron system in quantum Hall devices. However, the Chern number not only can stay robust against disorder and many-body effects, as shown in my thesis work, but also does not really require the presence of a magnetic field. With respect to the latter, Duncan Haldane made the first crucial step by proposing a model Hamiltonian for quantum Hall effect in the absence of Landau levels. Experiments of Qikun Xue's group and others more than two decades later have eventually made the case beyond any doubt.

It therefore naturally occurred to us that the Berry phase effect that we studied for magnetic Bloch bands must also exist in ordinary Bloch bands. So enters Ganesh Sundaram, whose thesis work gave a systematic study of Berry phase effects under the general setting of a periodic system influenced by parametric fields that vary slowly in space and time. Six kinds of Berry curvatures enter the equations of motion. Those in real space and time bring in the electromagnetic fields as well as their analog known as artificial gauge fields. Those in momentum space and time underlie the Chern numbers of TKNN as well as Thouless' quantized charge pump. There are also Berry curvatures in the cross planes of positions and momenta, the physical significance of which was to be fully appreciated later in spintronics studies. We are thus led to the revelation that Berry phase and its associated concepts will play out as a central unifying protocol for understanding the ever diverse phenomena in solid state physics.

The usefulness of Berry curvature in transport problems also took some time to be appreciated. The anomalous velocity made its appearance already in the classic work of Karplus and Luttinger in accounting for the phenomenon of anomalous Hall effect, but their results were disputed and displaced by Smit's skew scattering and Berger's side jump mechanisms as the cause for the effect. When I presented our theory and its connection to the classic works in a course on solid state physics, my colleague Allan MacDonald was sitting in the classroom observing my lecture on an official assignment. We discussed afterwards and agreed that the theoretical state of affairs on the anomalous Hall effect was totally contradictory with regard to the Berry curvature nature of the anomalous velocity, because its integral over a full band would be topological and completely insensitive to disorder.

We were thus compelled to confront the situation and got Thomas Jungwirth to evaluate the intrinsic anomalous velocity contribution for a real system. This work and numerous others afterwards clearly established the importance (often the dominance) of the intrinsic Karplus-Luttinger mechanism and thus the momentum-space Berry curvature in transport. In this connection, I am also indebted to Yugui Yao, who performed some of the earliest calculations of Berry curvatures for ferromagnetic crystals from first principles, which ignited the interest of the community of ab-initio calculations on Berry curvature related intrinsic properties. I must also thank Nikolai Sinitsyn, who has put the semiclassical transport theory onto a rigorous basis by including various scattering processes and comparing with diagrammatic calculations.

And I must mention Duncan again, who incorporated the Berry curvature into the Fermi-liquid theoretical framework, making it a robust concept for interacting electrons.

The full appreciation of the power of the Berry-curvature modified semiclassical dynamics had to wait for the arrival of Di Xiao and Junren Shi in my group. We found that the usual phase-space density of states is modified by the Berry curvature in the presence of a magnetic field. This gives rise to a modification of orbital magnetization, which in turn gives rise to anomalous transport currents induced by statistical forces such as the gradient of the chemical potential or temperature. This type of anomalous current is indispensible in the study of thermoelectric effects and thermal Hall effects, and their absence would cause serious issues with the basic transport relations of Einstein, Onsager, and Mott.

The discovery of the modified density of states also led us dive deeper into the non-canonical nature of the physical position and crystal momentum along with the question of re-quantizing the new semiclassical dynamics. With the return of Ming-Che as a visiting professor back to my group, we found a link to the canonical variables. This involves shifts from the physical ones by the vector potential and the Berry connection, with additional terms necessary for the accuracy of the transformation. When we rewrite the wave packet energy in terms of the canonical variables, we were very delighted to see the familiar Pauli Hamiltonian in the non-relativistic limit, including the spin-orbit coupling term as well as an obscure Yafet term. Also, Chih-Piao Chuu led the calculation of the magnetic moment of the Dirac wave packet and found the Bohr magneton. This essentially re-established the rotating charge model of the electron spin, which was originally proposed by Uhlenbeck and Smith but later rejected by Lorentz and others.

This work on the Dirac wave packet was based on a general non-Abelian formulation of semiclassical dynamics, a thesis work of Dimi Culcer. At that time, we were interested in the intrinsic spin Hall effect, an endeavor inspired by the success with the intrinsic anomalous Hall effect. Jairo Sinova led a paper on the Rashba model, and Dimi was able to analyze the results using our semiclassical theory for non-degenerate bands. At the same time, Shuichi Murakami, Naoto Nagaosa and Shou-cheng Zhang addressed a Luttinger model with degenerate bands, which made us to think seriously on the non-Abelian generalization of our semiclassical theory. I am very glad to see that this intellectual exercise eventually led to a new understanding of the Pauli Hamiltonian on which much of solid state physics is based.

In ferromagnets, artificial gauge fields can arise from space-time Berry curvatures in a dynamic texture. Shengyuan Yang led the theoretical analysis in our collaboration with experiment by Geoffrey Beach and others at UT Austin that gave the first measurement of the magnetic Josephson EMF. Although the domain wall is not of the type used by Berger in his original prediction, the result still holds due to the topological nature of the effect. Such artificial gauge fields become non-Abelian in antiferromagnets because of the degeneracy of the bands and, as shown in Ran Chengs thesis work, they also provide an essential coupling between quasiparticles and the order parameter as useful as in ferromagnetic spintronics.

When spin-orbit coupling is also present, Berry curvatures can arise in the cross planes of coordinates and momenta, which were seen to play a crucial role in the inverse spin Hall effect in a work with Junren and Ping Zhang. Phase-space Chern-Simons form appeared in a work with Di, Junren, and Dennis Clougherty, where we analyzed electronic polarization induced by inhomogeneity such as that studied in multiferroics. Interestingly, the magnetoelectric polarizability that characterizes 3D topological insulators can also be obtained this way, if the vector potential of a magnetic field is regarded as an inhomogeneity in the Hamiltonian as done by Andrew Essin, Joel Moore and David Vanderbilt.

For many purposes, it is necessary to extend the semiclassical theory beyond first order in the electromagnetic fields, and this difficult task was taken over by Yang Gao with the help of Shengyuan. The difficulty is measured not only by the size of paper piles on Yangs desk, but also by the number of failed attempts by previous students. In the end, we were very delighted to see that the form of the equations of motion remains intact, and one only needs to correct the Berry curvature to first order in the fields and energy to second order. The former involves a correction to the Berry connection, which turned out to be gauge invariant and has physical significance as a positional shift. There are good reasons to believe that higher order corrections will follow the same structure.

Geometric phase and topological ideas have found good applications in the study of graphene and other two-dimensional materials. I started working in this field with Di and Wang Yao, and predicted the valley Hall effect and valley magnetic moment. Wang led the discovery of valley contrasting optical selection rule, which had a huge impact on optical study of 2D transition metal dichalcogenides years later. I would like to thank Ji Feng for involving me in a theoretical and experimental collaboration validating this effect on MoS_2. And, I must also mention Tianyi Cai for her investigation of magnetic control of valley polarization, and Lifa Zhang for his discovery of chiral phonons in such materials.

We have also investigated the possibility of turning graphene into a topological insulator with quantized charge Hall effect or its spin analog. Zhenghua Qiao led this endeavor by considering proximity Zeeman coupling and the Rashba effect, and James Tse and Hua Jiang joined forces to find the comprehensive phase diagram and disorder effects. I like Zhenhua's work on how Berry curvatures are rearranged in disorder induced transition to Anderson insulator. We also collaborated with Allan's group through Fan Zhang on spontaneous topological phases, with Xiao Li as the major player from my group.

As this preface is getting very long, I still cannot stop but continue to express my gratitude to my collaborators on geometric phase effects in cold atomic systems. I started out collaborating with my colleague Mark Raizen on realizing some classic solid state physics and then rode along the wave of studies of Bose Einstein condensation of cold atoms. With Jie Liu and Biao Wu, we considered the adiabatic theory for the nonlinear Schrödinger equation. The presence of instability and lack of superposition principle posed a great challenge, but we were able to overcome it by drawing analogy with nonlinear classical mechanics and employing the concepts of the Aharonov-Anandan phase and the Hannay angle. Artem Dudarev and Roberto Diener worked on Berry curvatures in atomic Bloch band structures, and wrote one of the earliest papers on realizing effective spin-orbit coupling in optical lattices. Chuanwei Zhang worked on wave packet dynamics of Bogoliubov quasiparticles, and helped clarifying an issue with transverse force on a moving vortex left over from an earlier work of Thouless, Ping Ao, and myself.

I must also express my gratitude to collaborators outside my research group, apart from those already mentioned above. I would like to thank Yong-Shi Wu for his advice on my thesis work, Xiao-Gang Wen for taking me on an early adventure on topological order, Leonard Kleinman for sharing his dream for a first principles theory of spin wave dynamics, Hanno Weitering and Xiaofeng Jin for collaborating on their anomalous Hall experiments, Shun-Qing Shen for keeping me engaged with studies on topological insulators, Hua Chen for recent collaborations on noncollinear antiferromagnets, and my student-time friend Guang-Yu Guo for accompanying me from time to time in this journey.

I especially like to thank my coauthors for their contributions to the subject and for writing the bulk of this book. On their behalf, I would also like to express our gratitude to the International Center for Quantum Materials at Peking University for support. In closing, I would like to use this opportunity to thank my parents for giving me everything, and my family Julie, Bernard, and Lauren for their continued love.

Qian Niu
Austin, Texas
December 2016

Contents

Reprint Papers

1

Introduction

Geometric phase refers to the phase change of a quantum state under a cyclic evolution in addition to the obvious, dynamical phase given by the time integral of the energy of the state. In most cases, the cyclic evolution is realized by a slow periodic change of the system parameters, and this phase generally depends on the geometry of the path in the parameter space, but not on the rate of change as long as it is slow.

In the thirty plus years since Michael Berry's inspiring article, this concept has permeated throughout the physics community, and become a common language across different disciplines. It has become a powerful tool in linking different phenomena, organizing diverse knowledge, and facilitating new ideas and predictions. The generality and usefulness of this sweeping concept compares no less with that of "symmetry" in physics.

The book of *Geometric Phases in Physics* by Shapere and Wilczek [1] has provided us a very early summary of its conceptual development and applications. At that time, one would be very excited to see geometric phases measured experimentally, but the physical effects seemed to be small as well as rare. We have now a dramatically changed landscape of physics research where geometric phase is ubiquitously seen and often with big effects. With a slightly shifted title *Physical Effects of Geometric Phases* for this reprint collection, we hope to emphasize this new outlook [2].

Most of the advances in the applications of geometric phase occur in condensed matter and aggregates of cold atoms. One recurring topic is how geometric phases enter the effective dynamics of the slow degrees of freedom and how they appear in the system response at macroscopic scales. These slow variables are usually in the form of quasiparticles and collective modes, and our view on their behavior and interaction has been fundamentally transformed by systematic application of geometric phase ideas. As a result, we have a much more coherent understanding of the diverse phenomena in condensed matter and atomic systems.

1.1 Geometric phase in momentum space

A key development came from the recognition of geometric phases in crystal momentum space: those across the Brillouin zone and those around small loops in the zone. The former underlies the electrical polarization of crystalline insulators, a most basic property of dielectrics. The latter gives rise to the Berry curvature in momentum space, which modifies the dynamics of Bloch electrons with important consequences in transport as well as thermodynamic properties. Such understanding has, for example, inspired tremendous efforts in the last decade reinvestigating the anomalous Hall effect in magnetized materials, which was discovered about the same time as the ordinary Hall effect but whose theoretical explanation remained controversial. The anomalous velocity induced by the Berry curvature gives rise to an intrinsic contribution to the Hall conductivity. This corresponds to an old finding of Karplus and Luttinger, and it turns out to dominate a broad range of experimental situations.

The integral of the Berry curvature over the entire zone is the topological Chern number that characterizes the quantum Hall states. The original TKNN formulation applied to electrons in a strong magnetic field, and its significance was initially overshadowed by Laughlin's much simpler explanation based on gauge invariance. When Berry's work came along shortly thereafter, Simon immediately found a connection between the two, revealing the topological aspect of geometric phase and also bringing the TKNN formulation into a general perspective. Several years later, Haldane found a concrete model where one can indeed produce non-zero Chern numbers without invoking the Landau levels. After a generation of theoretical proposals based on various mechanisms in different systems, convincing experimental realization of the quantum anomalous Hall effect has appeared only recently.

Topological features of geometric phase and Berry curvature in momentum space have also been the focus of recent studies of topological insulators and superconductors protected by time reversal or other discrete symmetries. Kane and Mele's graphene model with intrinsic spin-orbit coupling consists of two copies of the Haldane model each associated with a spin species and with opposite Chern numbers. Fu and Kane's Z_2 topological number involves an integral of the Berry curvature over half a Brillouin zone under a time reversal symmetric gauge. Qi and Zhang's theta number characterizing three dimensional topological insulators is an integral of the Chern-Simons form in momentum space, a sort of self-linking number of Berry curvature flux lines in the Brillouin zone.

Because of their strong effects on electronic properties, first principles computation of geometric phase, Berry curvature, and associated topological numbers in momentum space has become an important branch of materials research. Apart from electric polarization and the anomalous Hall conductivity mentioned above, we now also have access to orbital magnetization, another basic property of materials, using first principles methods. Systematic searches for realistic topological insulators have been aided by such computations with great success.

Four chapters are devoted to the exposition of the geometric phase concept in momentum space (Chapter 2 by Ming-Che Chang) and its applications. In Chapters 3 and 4, Di Xiao discusses applications in electronic equilibrium and transport properties, including polarization, orbital magnetization, pumping and Hall effects. In Chapter 5, Ming-Che Chang discusses applications on topological insulators.

1.2 Artificial gauge fields in real space

An early impetus of geometric phase studies was the possibility of generating artificial gauge fields in the configuration space of slow variables. The classic example is a molecular system, where geometric phase can arise when atoms move around each other, causing a shift of rotation spectrum of the molecule–just like the Aharonov-Borm effect on the spectrum of an electron moving around a magnetic flux. This can be seen more clearly in the Mead extension of the Born-Oppenheimer theory, where geometric gauge fields enter explicitly into the nuclear equation of motion after separating out the fast electronic degrees of freedom.

With the rise of research on cold atomic systems, this understanding of geometric gauge fields has been put into extensive engineering practice by coupling the atomic internal degrees of freedom, often in the form of spin fine structure, with the polarization texture of an external optical field. Unlike those in molecular systems, the atoms in the dilute gas can be regarded as independent, and the gauge fields act on each atom in exactly the same way that electromagnetic fields act on a charge.

The simplest example is perhaps an electron in a magnetization texture that varies slowly in space and time. With a relatively strong Hund rule coupling, the electron spin will be locked in the direction of the local magnetization orientation. The motion of the electron spin then gives rise to a geometric phase in space-time, producing an effective electromagnetic gauge field depending upon the gradient and time derivative of the magnetization. Experimental confirmation has been done on the physical effects of such gauge fields using magnetization textures with skyrmions and domain walls. Indeed, the Lorentz force from the effective magnetic field can give rise to an anomalous Hall effect, while the effective electric field can give rise to an electromotive force.

An important topic in spintronics research is how electron currents may drive the motion of magnetic order parameters. The gauge coupling discussed above can also provide a back reaction of the quasiparticles on the background magnetization texture. This is the mechanism behind the so-called current driven adiabatic torque, which turns out to be important for moving topological defects such as domain walls and skyrmions and for generating spin waves.

Ran Cheng discusses these topics in Chapter 6, focusing on new developments in the past few decades.

1.3 Non-canonical phase space

In the presence of a magnetic field or an effective one generated by artificial means, the gauge invariant momentum components are no longer canonical variables. What this means is that they do not commute with each other in quantum mechanics, or their Poisson brackets are nonzero in classical mechanics. Similarly, in the presence of Berry curvature in momentum space, the position coordinates become non-canonical, with the nonzero commutator or Poisson bracket given by the Berry curvature.

One peculiar feature of the semiclassical dynamics in such systems is that the conventional measure of phase space volume is no longer conserved along the motion when both the magnetic field and the Berry curvature are present. However, this apparent breakdown of Liouville theorem can be remedied by introducing a correction factor. Consequently, as shown by further analysis, the density of quantum states per phase space volume should also be modified. Because of this, the Berry curvature can generally influence thermodynamic and transport properties through the magnetic field. This is the reason that Berry curvature enters the orbital magnetization, the anomalous Nernst and thermal Hall conductivities, etc.

Geometric phase can also exist in a cross plane of momentum and position. This can be realized by a combined action of a position-dependent Zeeman field and Rashba spin orbit coupling, which can be regarded as a momentum dependent Zeeman field. As the electron moves in phase space, the motion of electron spin naturally introduces a geometric phase. The corresponding Berry curvature in the cross plane can be regarded as a modification of the Planck constant, and is another way of making the phase space non-canonical. The inverse spin Hall effect, charge current driven by a Zeeman field gradient, naturally depends on such cross-plane Berry curvatures due to the combination of this spin force and spin-orbit coupling. As a topic of interest in multiferroics, polarization due to inhomogeneity in the magnetization involves such phase space Berry curvatures through a Chern-Simons form.

These topics are discussed in Chapter 6 jointly by Ming-Che Chang and Di Xiao.

1.4 Non-Abelian generalizations

It often happens that the specification of the slow variables (or external parameters) does not fix the quantum state of fast variables due to degeneracy of the eigenstates. Cyclic motion of the slow variables can in general cause a unitary rotation within the subspace of the degenerate states, not just a geometric phase. The Berry curvature becomes a matrix, and the corresponding gauge structure is known as non-Abelian.

In the Dirac theory for electrons in a vacuum, both the positive and negative energy bands are two-fold degenerate, corresponding to the spin degrees of freedom. In a crystal with both time reversal and spatial inversion symmetry, the Bloch bands are also two fold degenerate at each general point of the Brillouin zone. The Berry curvature in momentum space is spin-dependent, which can cause a spin-dependent Hall effect important for spintronics applications. This spin-dependent Berry curvature also makes the electron position coordinates non-canonical, which is the ultimate reason for spin-orbit coupling.

Artificial non-Abelian gauge field can also be generated in real space and time. A simple example is a collinear antiferromagnet, in which the Bloch bands are also two fold degenerate where time reversal symmetry is replaced by a symmetry of combined time reversal and sublattice swapping. Variation of the antiferromagnetic Néel order parameter in space and time can generate an artificial non-abelian gauge field, which acts on the electron just like the SU(2) Yang-Mills field. This non-abelian gauge coupling between the current carriers and the order parameter field is an important topic of discussion in antiferromagnetic spintronics.

Fractional quantum Hall systems on a multiply-connected closed surface can support multiple ground states, which can be understood in terms of the non-abelian gauge structure with respect to the magnetic fluxes through the holes. On a closed surface of high genus, the ground state degeneracy goes up exponentially, which can be accessed by tuning the fluxes through the various holes. For some more exotic fractional quantum Hall systems, exponentially large degeneracy in the ground states can also occur in a planar geometry with point-like defects (punctures of the plane), and different ground states can be accessed by braiding the defects pairwise. Such non-abelian statistics also exist between Majorana modes localized on point-like defects in some exotic superconductors and quantum spin systems. Extensive efforts have been devoted to this line of research with the tantalizing prospect of realizing topological quantum computing.

These topics are discussed in various places, especially Chapters 2, 6, and 7.

1.5 Effective quantum mechanics

Although it takes great insight to identify the slow degrees of freedom in a given situation, there is a high degree of flexibility in the choice of variables to represent them. The semiclassical equation of motion is covariant with the change of variables as long as the transformation is smooth and invertible. The general structure is not canonical because of the presence of the Berry curvatures, and quantization of the semiclassical theory becomes a nontrivial matter.

Simplification occurs if gauge fields only exist in real space and time. This is exemplified in the Mead extension of the Born-Oppenheimer theory of nuclear dynamics, where one shifts the nuclear momenta by the vector potentials in the kinetic energy. The traditional Peierls substitution does a similar trick, shifting the crystal momenta by the vector potentials to obtain a quantum Hamiltonian for a Bloch band in the presence of a magnetic field.

The situation becomes complicated if Berry curvatures also exist in the momentum space. A first guess of the solution is to also shift the position variables by the Berry connections, which are the momentum-space analog of the vector potentials. However, additional shifts to both momentum and position variables are needed to make the phase space truly canonical. When this process is applied, for example, to the degenerate bands of the Dirac theory or the Bloch bands in a crystal, we obtain spin-orbit terms in the effective Hamiltonian because of the spin-dependent Berry connections. The canonicalization process also explains a previously obscure Yafet term for the orbital magnetic moment.

It must be remembered that the canonical variables are not gauge invariant and therefore unphysical, although they are easier to use in connection with the effective quantum theory. For example, the momentum derivative of the canonical Hamiltonian does not give the full expression for the physical velocity; it must be supplemented with a time derivative of the Berry connection in order to achieve gauge invariance. The reader is also cautioned that canonical variables cannot be defined globally when the bands have a non-trivial topology.

This topic is discussed in Chapter 7 along with an exposition of non-canonical phase space.

1.6 Nonlinear Quantum Systems

Geometric phase ideas are also useful in the study of dynamics of nonlinear quantum systems. One example is the Gross-Pitaevskii equation describing Bose-Einstein condensates. The adiabatic theorem still holds for the ground state, when external parameters are changed slowly in comparison with the energy gap of excitations. The geometric phase, Berry curvatures, etc. are defined and used in the usual manner in terms of the ground state wave function. For other energy eigenstates, there is the possibility of instability because of the nonlinearity of the system–which may ruin the adiabatic theorem.

Another consequence of nonlinearity is the absence of the superposition principle. Therefore, the evolution of non-eigenstates can no longer be inferred from knowledge of the eigenstates. Here, an exact analogy with a classical Hamiltonian system can be of great help. In this analogy, the eigenstates are simply the fixed points in the classical system. The fixed point corresponding to the ground state is always stable, but the others may or may not be stable. Non-eigenstates correspond to orbits beside the fixed points. For a closed orbit around a stable fixed point, the classical adiabatic invariant is the action variable of the orbit. This orbit corresponds to a cyclically varying quantum state, and the action turns out to be exactly its Aharonov-Anandon phase.

One can also define a geometric phase for the evolution of a non-eigenstate accumulated over the course of slow and periodic changes of external parameters. This turns out to be the Hannay's angle in the classical analogy. In the absence of nonlinearity, this geometric phase reduces to a weighted sum of the Berry phases of the eigenstates, with the weights given by the probabilities on the eigenstates.

Excitations in a nonlinear quantum system are more suitably described by the Bogoliubov quasiparticles than the higher eigenstates. The Bogoliubov equation is non-Hermitian, like the equations of motion for other boson quasiparticles such as magnons, phonons, and photons. Geometric phase and Berry curvatures can still be defined if appropriate normalization of the quasiparticle wave function is taken into account. Non-trivial gauge coupling with the velocity field of the condensate can arise this way. For example, in a condensate with a vortex structure in the velocity field, a proper account of equilibrium momentum circulation of quasiparticles shows a deviation from the naïve expectation but agrees with the assumption that normal fluids do not contribute to the circulation.

These topics are discussed in Chapter 8 by Biao Wu.

1.7 Geometric phase in collective coordinates

So far we have been discussing geometric phase or gauge fields for the motion of slow particles. However, it often happens that the slow degrees of freedom may correspond to some collective motion in a many-body system. Goldstone modes and topological defects of order parameters fall into this category, as does the nonlinear Gross-Pitaevskii equation for the Bose-Einstein condensate. The use of geometric phase ideas can lead to precise conclusions about the nature of these collective degrees of freedom.

Wilczek and Shapere gave extensive coverage of vortex-like excitations in fractional Hall states. The fact that they carry fractional charges can be explained by comparing the geometric phase and Aharonov-Bohm phase in traversing such an excitation over a loop. In addition, fractional statistics of these excitations can be found by calculating the geometric phase in circling one around another of such excitations. In this book, we will review two other applications: one on the dynamics of spinwaves, and the other on the Magnus force on a moving vortex.

Spinwaves are often studied with the Heisenberg model, but real materials usually involve itinerant electrons. The spin magnetization at some position and time should be defined as an ensemble average of electron spins which happen to be in the local region at the time. In the adiabatic approximation, spinwave dynamics can be formulated in terms of the constrained ground state, the lowest energy state for a given spin magnetization field. The equations of motion of the spin magnetization involve not only the energy but also the Berry curvature of the constrained ground state in the configuration space of the spin magnetization. Adaptation of this formulation to the spin-density functional theory has been attempted with some success. However, the anticipated non-local Berry curvature effects due to itinerancy of the electrons are yet to be identified in real systems.

Another problem of general interest is the motion of a vortex. In a superfluid, for example, the vortex is defined as the center of circulating velocity field. The Berry curvature of the condensate wave function in the coordinate plane of the vortex gives rise to a Lorentz-like force, the Magnus force, which is proportional and perpendicular to the vortex velocity. Generalizing the argument by using the exact many-body wave functions, one finds that the Magnus force is not affected by the presence of normal fluid at finite temperatures–except for its role in depressing the superfluid density. This is contradictory to an old result of Iordansky, which caused quite some controversy, and its prediction is still waiting to be confirmed experimentally.

These topics are discussed in Chapter 9 by Qian Niu.

Chapter 1

References

1. A. Shapere and F. Wilczek, *Geometric Phases in Physics* (World Scientific, Singapore, 1989).
2. Fourteen years later, and fourteen years ago by now, a monograph on the mathematical foundation and a number of applications of geometric phases was published, see A. Bohm, A. Mostafazadeh, H. Koizumi, Q. Niu, and J. Zwanziger, *The Geometric Phase in Quantum Systems* (Springer-Verlag, Berlin, 2003).

2

Geometric phases in momentum space

Geometric phase (or Berry phase) is an essential ingredient in the theory of quantum Hall effect (QHE). The topological invariant due to the Berry curvature in QHE is first discovered by D. Thouless using linear response analysis. Later, the Berry curvature is incorporated into semiclassical equations of motion. This approach provides the easiest way to understand the Berry curvature-related Hall currents in QHE and anomalous Hall effect. This semiclassical formulation comes with two varieties, depending on whether the Bloch states have a single component (Abelian version) or multiple components (non-Abelian version). The later version is suitable for the study of the spin Hall effect, among other things. This chapter covers essential papers on the semiclassical dynamics, as well as related works on Zak phase (Berry phase in 1D) and semiclassical spin dynamics.

The basics of the Berry phase can be found in, e.g, Chap. 10 of "Quantum mechanics" by Griffiths . Consider a system that varies slowly with a parameter $\boldsymbol{\lambda}(t)$. Then the Hamiltonian $H_{\boldsymbol{\lambda}}$, and its eigenenergies and eigenstates, $\varepsilon_{n,\boldsymbol{\lambda}}$ and $\psi_{n,\boldsymbol{\lambda}}$, would all depend on $\boldsymbol{\lambda}$. Suppose the energy levels do not cross, then according to the adiabatic theorem, if an electron starts in an eigenstate ψ_n, then as long as $\boldsymbol{\lambda}$ changes slowly, it would stay at that eigenstate. In the discussion below, this adiabatic condition is always presumed.

If the system undergoes a cyclic motion, such that after a period of T, $\boldsymbol{\lambda}(0) = \boldsymbol{\lambda}(T)$, then M. Berry finds that, in addition to the usual dynamical phase, there could be an extra phase $\gamma_n(c)$. This Berry phase depends on the level n of the electron, and on the geometric path c it traverses in the *parameter space* of $\boldsymbol{\lambda}$, but it does not depend on the rate of traverse. The Berry phase $\gamma_n(c)$ can be written as a loop integral of the Berry connection $\mathbf{A}_{n,\boldsymbol{\lambda}}$,

$$\gamma_n(c) = \oint_c \mathbf{A}_{n,\boldsymbol{\lambda}} \cdot d\boldsymbol{\lambda}, \tag{2.1}$$

$$\mathbf{A}_{n,\boldsymbol{\lambda}} = i\langle\psi_{n,\boldsymbol{\lambda}}\left|\frac{\partial}{\partial\boldsymbol{\lambda}}\right|\psi_{n,\boldsymbol{\lambda}}\rangle. \tag{2.2}$$

In the Bloch theory of a periodic solid, the crystal momentum **k** plays the role of the slowly varying parameter $\boldsymbol{\lambda}$. The momentum is defined over a Brillouin zone (BZ), which has the topology of a torus. The adiabatic condition demands the electron to stay in one energy band and is not allowed to jump to other bands (one-band approximation). This is a good approximation as long as the energy band under consideration is well isolated from the other bands.

For a *non-degenerate* band, the Berry connection is defined as

$$\mathbf{A}_n(\mathbf{k}) = i\langle u_n \left| \frac{\partial}{\partial \mathbf{k}} \right| u_n\rangle, \tag{2.3}$$

where u_n is the cell-periodic part of the Bloch state, $\psi_{n,\mathbf{k}} = e^{i\mathbf{k}\cdot\mathbf{r}} u_{n,\mathbf{k}}$. The curl of $\mathbf{A}_n$ is called the Berry curvature,

$$\boldsymbol{\Omega}_n(\mathbf{k}) = \nabla \times \mathbf{A}_n = i \left\langle \frac{\partial u_n}{\partial \mathbf{k}} |\times| \frac{\partial u_n}{\partial \mathbf{k}} \right\rangle. \tag{2.4}$$

The appearance of the Berry curvature in a solid would lead to interesting effects. However, if a solid has *both* time reversal and space inversion symmetries, then the Berry curvature would vanish (note: this statement is valid only for a *non-degenerate* band). In order to have non-zero Berry curvature in a non-degenerate band, one needs to break one of the symmetries. For example, when the time-reversal symmetry is broken by a magnetic field, a Bloch band splits to multiple Landau subbands. In each of the subband, the Berry curvature is nonzero.

In a two-dimensional system, the total Berry curvature of a subband, also called the TKNdN number [1] or the first Chern number, is

$$\frac{1}{2\pi} \int_{BZ} d^2\mathbf{k}\, \Omega_z(\mathbf{k}) = C_1 \in Z. \tag{2.5}$$

Mathematically, this integer is a topological invariant; physically, it is equal to the quantum Hall conductance (in units of e^2/h) of that subband. The topological nature of the quantum Hall conductance guarantees the precision of the quantum Hall plateaus.

2.1 Zak phase

In a one-dimensional lattice, an electron moving through the 1D BZ has the Berry phase,

$$\gamma_n = \int_{-\pi/a}^{\pi/a} A_n dk, \tag{2.6}$$

where a is the lattice constant. That this integral could be nonzero is demonstrated by Zak [2], thus this phase is often called the Zak phase.

It worths emphasizing that the Zak phase in 1D would depend on the choice of the origin (or the unit cells). For example, given a translation $x \to x' = x + s$ $(s < a)$, then (see the Supplement of Ref. [3])

$$A_n \to A'_n = A_n + s, \tag{2.7}$$

and

$$\gamma_n \to \gamma'_n = \gamma_n + 2\pi \frac{s}{a}. \tag{2.8}$$

Nevertheless, the difference between the Zak phases of two lattice configurations (with the same origin) is independent of the choice of the origin, and can be measured experimentally [3].

If a system has space-inversion symmetry, then the Zak phase (with a proper choice of the origin) can only be 0 or π : Space inversion flips the sign of the Zak phase, $\gamma_n \to -\gamma_n$ (mod 2π), which is invariant if $\gamma_n = 0, \pi$. In the absence of inversion symmetry (or with a bad choice of the origin), any value of γ_n (mod 2π) is possible.

2.2 Semiclassical dynamics

The Berry curvature contributes to a so-called *anomalous velocity* (the second term in Eq. (2.9)) to the equations of motion of a wavepacket [4, 5],

$$\dot{\mathbf{r}} = \frac{\partial E_n}{\hbar \partial \mathbf{k}} - \dot{\mathbf{k}} \times \boldsymbol{\Omega}_n, \tag{2.9}$$

$$\hbar \dot{\mathbf{k}} = -e\mathbf{E} - e\dot{\mathbf{r}} \times \mathbf{B}, \tag{2.10}$$

where

$$E_n = \varepsilon_n - \mathbf{M}_n \cdot \mathbf{B}. \tag{2.11}$$

E_n is the Bloch energy ε_n shifted due to the magnetic moment $\mathbf{M}_n$ of the wavepacket $W_n(\mathbf{r})$. The magnetic moment is a result of the wavepacket rotation with respect to its own center of mass $\mathbf{r}_c$,

$$\mathbf{M}_n = -\frac{e}{2m} \langle W_n | (\mathbf{r} - \mathbf{r}_c) \times \mathbf{p} | W_n \rangle. \tag{2.12}$$

The same semiclassical formulation can be applied to a quantum Hall system with strong magnetic field, but with one crucial modification: since the vector potential breaks the translation symmetry, the usual Bloch momentum $\mathbf{k}$ is no longer a good quantum number. Nevertheless, it can be shown that if there is p/q flux quantum per plaquette (i.e., the unit cell in the absence of magnetic field), where p and q are co-prime integers, then one can choose an enlarged unit cell with q plaquettes to recover the lattice translation symmetry [4]. As a result, the domain of the *magnetic* Bloch momentum is shrunk by a factor of q compared to the usual BZ. Other than that, the semiclassical equations of motion are the same as those in Eqs. (2.9),(2.10).

2.3 Berry phase and quantization

The classical trajectory in the phase space can be quantized, for example, by the Bohr-Sommerfeld quantization (for an alternative method of quantization, see Chapter 7). An example is the Onsager quantization used in the study of the de Haas-van Alphen effect:

$$\frac{1}{2}\oint_{c_\ell} (\mathbf{k} \times d\mathbf{k}) \cdot \hat{z} = 2\pi \left(\ell + \frac{1}{2} - \frac{\gamma(c_\ell)}{2\pi} \right) \frac{eB}{\hbar}, \tag{2.13}$$

in which a Berry phase correction has been included,

$$\gamma(c_\ell) = \oint_{c_\ell} \mathbf{A} \cdot d\mathbf{k} \tag{2.14}$$

is the Berry phase for an orbit c_ℓ in a particular (say, the n-th) band. If the magnitude of the Berry phase is independent of the electron path c_ℓ (for example, in graphene, any path circling the Dirac point has a Berry phase of π), then this formula immediately gives the allowed area (LHS of Eq. (2.13)) encircled by a cyclotron orbit in momentum space. However, in the general case where $\gamma(c_\ell)$ depends on the size of the orbit on the Fermi surface, Eq. (2.13) needs to be solved self-consistently to determine the actual c_ℓ.

2.4 Berry phase and defect

The semiclassical formulation can be extended to deal with crystal deformation and dislocation. For example, near a (real-space) line defect with Berger vector $\mathbf{b}$, a circling electron would acquire a Berry phase $\gamma_n \propto \mathbf{b}$ [5].

In momentum space, one can also have a point or line defect, which is due to a point or line degeneracy between two Bloch bands. For example, near a point degeneracy at $\mathbf{k} = 0$, one can adopt a two-band approximation and write the Hamiltonian as,

$$H = \alpha \mathbf{f}(\mathbf{k}) \cdot \boldsymbol{\sigma}. \tag{2.15}$$

If $\mathbf{f}(\mathbf{k}) = \mathbf{k}$, then the vector field $\mathbf{f}$ near the degenerate point has a hedgehog texture on the surface S^2 surrounding the origin. When an electron traverses a loop c on S^2, it acquires a Berry phase $\pm\Omega_c/2$ proportional to the solid angle Ω_c of the loop.

If the momentum space is two-dimensional, again with $H = \alpha\mathbf{k} \cdot \boldsymbol{\sigma}$, then a path circling the origin would extend a solid angle of 2π. Therefore the Berry phase $\gamma = \pm\pi$. This is what happens near graphene's Dirac point.

The wrapping number of $\mathbf{f}(\mathbf{k})$ around S^2 is called the *Berry index* of the degenerate point. It determines the change of Chern numbers due to band touching: Suppose the Hamiltonian is controlled by a parameter η. The two bands touch at a point when $\eta = \eta^*$, but depart when $\eta \neq \eta^*$. If a band has

Chern numbers C_- and C_+ before and after the touch, then their difference is equal to the Berry index of the degenerate point [6]. Obviously, at the critical value η^*, the Chern number of a filled band is not well-defined.

The connectedness of the parameter space near a *line* degeneracy in 3D is similar to the point degeneracy in 2D. For a closed path surrounding a line degeneracy in 3D, $\gamma = \pm\pi$ if the level separation away from the line is linear in k [7]. Such a Berry phase enters the Onsager integral (Eq. (2.13)) of a cyclotron orbit. As a result, the cyclotron energy is shifted. Such a shift can be detected through the measurement of the de Haas-van Alphen oscillation [8].

2.5 Non-Abelian generalization

In addition to the point and line degeneracy, two Bloch bands can have *global* degeneracy over the whole BZ. When this happens, the Bloch state becomes a spinor with inner degrees of freedom at each $\mathbf{k}$ point. If the global degeneracy is D-fold, then the Berry connection becomes a $D \times D$ matrix,

$$\mathbf{R}_{mn}(\mathbf{k}) = i\left\langle u_{m\mathbf{k}} \middle| \frac{\partial u_{n\mathbf{k}}}{\partial \mathbf{k}} \right\rangle, \tag{2.16}$$

and the Berry curvature is

$$\boldsymbol{\mathcal{F}}(\mathbf{k}) = \nabla_{\mathbf{k}} \times \boldsymbol{\mathcal{R}} - i\boldsymbol{\mathcal{R}} \times \boldsymbol{\mathcal{R}}, \tag{2.17}$$

where (boldfaced) caligraphic fonts represent (vector-valued) matrices, $(\boldsymbol{\mathcal{R}})_{mn} = \mathbf{R}_{mn}$.

Because of the non-Abelian gauge structure, the line integral for the Berry phase has to be path-ordered (P_c is a path-ordering operator),

$$\mathcal{U}_B = P_c\exp\left(i\oint_c d\mathbf{k}\cdot\boldsymbol{\mathcal{R}}\right). \tag{2.18}$$

This leads to a $SU(D)$ rotation among the degenerate states. $\mathcal{U}_B$ usually cannot be expressed in an analytic form, but has to be computed numerically.

Generalization of the wave packet dynamics, Eqs. (2.9) and (2.10), to non-Abelian case can be found in Refs. [10], [11]. Such a generalization is crucial to the study of the spin Hall effect, which is discussed below.

2.6 Spin Hall effect

In the non-Abelian case, the Berry curvature is spin-dependent. As a result, when one applies an electric field, $\hbar\dot{\mathbf{k}} = -e\mathbf{E}$, the anomalous velocity induced by a Berry curvature is spin-dependent,

$$\mathbf{v}_a = -\dot{\mathbf{k}} \times \mathbf{F} = \frac{e}{\hbar}\mathbf{E} \times \mathbf{F}, \tag{2.19}$$

where $\mathbf{F} \equiv \langle \mathcal{F} \rangle$ is the spinor-averaged Berry curvature (band index n is neglected). In systems with spin-orbit coupling, spin-up and spin-down electrons would have different $\langle \mathcal{F} \rangle$, and thus different anomalous (transverse) velocities. Such a spin separation is the cause of (intrinsic) spin Hall effect (SHE), which is first proposed by Murakami *et al* [12]. (see Chapter 4)

If the electron spin is conserved during the transport, such that spin-up current and spin-down current decouple with each other, and if each of them has a quantized Hall conductivity, $\sigma_H^{\uparrow}$ and $\sigma_H^{\downarrow}$, then the spin Hall conductivity,

$$\sigma_H^s = (\sigma_H^{\uparrow} - \sigma_H^{\downarrow})\frac{\hbar/2}{e}, \tag{2.20}$$

is also quantized. This is the simplest example of the quantum spin Hall effect (QSHE).

However, the electron spin in a solid is often not conserved due to the spin-orbit interaction. As a result, the spin Hall conductivity is no longer quantized. Worse yet, the spin current may not be well defined in the presence of the spin-orbit coupling. Despite these problems, it is possible to have a quantum version of the spin Hall effect in two dimension, in which the *number* of edge channels remains robust. Even though the spin transport is not exactly quantized, it is still called the QSHE. Studying the QSHE in 2D provides an important route to the discovery of topological insulators (see Chapter 5).

2.7 Fermi-Walker transport

There is an interesting connection between the parallel transport in the theory of Berry phase and the Fermi-Walker transport of a triad [13], which we now explain.

Along the trajectory $\mathbf{x}(s)$ of a particle, one can set up a moving local coordinate. A simple choice would be the Frenet-Serret triad $(\mathbf{t}, \mathbf{n}, \mathbf{b})$, where $\mathbf{t}$ is the tangent vector of the trajectory, $\mathbf{n} \parallel d\mathbf{t}/ds$ is the normal vector, and $\mathbf{b} = \mathbf{t} \times \mathbf{n}$ is the bi-normal vector. These three unit vectors are mutually perpendicular to each other. They vary according to the following equations,

$$\frac{d\mathbf{t}}{ds} = \kappa\mathbf{n}, \tag{2.21}$$

$$\frac{d\mathbf{n}}{ds} = -\kappa\mathbf{t} + \tau\mathbf{b}, \tag{2.22}$$

$$\frac{d\mathbf{b}}{ds} = -\tau\mathbf{n}. \tag{2.23}$$

The coefficients κ and τ are called the curvature and the torsion of the curve respectively.

The equations above can also be written as

$$\frac{d\mathbf{v}}{ds} = \boldsymbol{\omega} \times \mathbf{v}, \tag{2.24}$$

where $\boldsymbol{\omega} = \kappa\mathbf{b} + \tau\mathbf{t}$, and $\mathbf{v}$ can be $\mathbf{t}, \mathbf{n}$, or $\mathbf{b}$. Obviously, $\boldsymbol{\omega}$ is the angular velocity of the triad. Because of the torsion, the triad rotates around $\mathbf{t}$ with an angular velocity $\tau\mathbf{t}$. One can rotate the $\mathbf{n}$-$\mathbf{b}$ plane to eliminate the rotation due to the torsion.

After the rotation, $(\mathbf{t}, \mathbf{n}, \mathbf{b})$ become $(\mathbf{t}, \mathbf{n}', \mathbf{b}')$, and the new frame does not rotate with respect to $\mathbf{t}$. Also,

$$\frac{d\mathbf{u}}{ds} - \kappa\mathbf{b} \times \mathbf{u} = 0, \tag{2.25}$$

where $\mathbf{u}$ can be $\mathbf{t}, \mathbf{n}'$, or $\mathbf{b}'$. A vector $\mathbf{u}$ satisfying this equation is said to be Fermi-Walker transported.

An important difference between these two sets of triads is that the former is path independent, while the latter is not. For example, suppose the curve is a helix, then after circling one "rung", the first set of triad would go back to the original orientation, but the second set would have rotated. It is known that for a linearly polarized light in an optical fibre, $(\mathbf{k}, \mathbf{E}, \mathbf{B})$ follows the second set of triad (Rytov, 1938). Therefore, after passing through a coiled optical fibre, the polarization $\mathbf{E}$ rotates by an angle $\theta = \int_c \tau ds$, which is analogous to the Berry phase.

Chapter 2

References

1. D.J. Thouless, M. Kohmoto, M.P. Nightingale, and M. den Nijs, Phys. Rev. Lett. **49**, 405 (1982).
2. J. Zak, Phys. Rev. Lett. **62**, 2747 (1989). [Reprint (2.2)]
3. M. Atala *et al*, Nat. Phys. **9**, 795 (2013).
4. M.C. Chang, and Q. Niu, Phys. Rev. B **53**, 7010 (1996). [Reprint (2.3)] also see M.C. Chang, and Q. Niu, Phys. Rev. Lett. **75**, 1348 (1995).
5. G. Sundaram and Q. Niu, Phys. Rev. B **59**, 14915 (1999). [Reprint (2.4)]
6. J. Bellissard, cond-mat/9504030, unpublished; M.Y. Lee, M.C. Chang, and T.M. Hong, Phys. Rev. B **57**, 11895 (1998).
7. G.P. Mikitik and Yu. V. Sharlai, Phys. Rev. Lett. **82**, 2147 (1999). [Reprint (2.5)]
8. G.P. Mikitik and Yu. V. Sharlai, Phys. Rev. Lett. **93**, 106403-1 (2004). [Reprint (2.6)]
9. D. Xiao, Y. Yao, Z. Fang, and Q. Niu, Phys. Rev. Lett. **97**, 026603 (2006).
10. D. Culcer, Y. Yao, and Q. Niu, Phys. Rev. B **72**, 85110 (2005). [Reprint (2.7)]
11. R. Shindou and K.-I. Imura, Nucl. Phys. B **720** 339 (2005).
12. S. Murakami, N. Nagaosa, and S. C. Zhang, Science **301**, 1348 (2003).
13. R.G. Littlejohn, Phys. Rev. A **38**, 6034 (1988). [Reprint (2.1)]

3

Pumping, polarization, and orbital magnetization

In the previous chapter, it was shown that the Berry phase of the Bloch wave function can modify the semiclassical dynamics of electrons through an anomalous velocity term, first derived by Karplus and Luttinger [1]. The most obvious consequence of the Berry phase is to give rise to a Hall current. A less obvious but equally important consequence is that the Berry phase also plays a role in equilibrium properties of solids, most notably electric polarization and orbital magnetization, two of the most important material properties. In hindsight, this should not be surprising because both quantities can be related to the adiabatic current, which is expressed in terms of the Berry curvature.

3.1 Adiabatic pumping

In 1983, Thouless [2] first proposed the concept of adiabatic pumping, which refers to the process in which a DC current is generated by the cyclic adiabatic evolution of the system. Before we delve into detailed discussions, it is useful to first consider a simple case for such a pumping process. Let us consider a one-dimensional insulating crystal with periodic boundary condition, i.e., we put the system on a ring. Suppose the periodic potential has a time-dependence

$$V(x,t) = V(x - vt) . \tag{3.1}$$

We further assume that v is sufficiently small such that inter-band transition can be neglected, i.e., we are in the adiabatic limit. Since the system is insulating, it can be described by a set of Wannier functions and any current must result from the movement of the Wannier center. As $V(x)$ changes with time t, the Wannier center must be moving along with the potential (Fig. 3.1). On the other hand, at $t = a/v$, where a is the lattice constant, the system returns to itself, which means that the Wannier center must move by a lattice constant. This example shows that the pumped charge must be quantized upon the finish of a complete cycle.

Once we make the connection of the adiabatic pumping to Wannier function, the derivation becomes straightforward. Consider a crystal described by a time-dependent Hamiltonian $H(t)$. We denote by $\psi_{n\boldsymbol{k}}(t)$ the instantaneous Bloch functions of the system. By definition, the Wannier function is given by

$$|W_n(\boldsymbol{R};t)\rangle = \frac{V}{(2\pi)^3}\int_{\mathrm{BZ}} d\boldsymbol{k}\, e^{-i\boldsymbol{k}\cdot\boldsymbol{R}}|\psi_{n\boldsymbol{k}}(t)\rangle \; . \tag{3.2}$$

For simplicity, we are going to trace the motion of $|W_n(0;t)\rangle$. The Wannier center of $|W_n(0;t)\rangle$ is

$$\begin{aligned}\langle \boldsymbol{r}_n\rangle &= \langle W_n(0;t)|\boldsymbol{r}|W_n(0,t)\rangle \\ &= i\frac{V}{(2\pi)^3}\int d\boldsymbol{k}\,\langle u_{n\boldsymbol{k}}(t)|\partial_{\boldsymbol{k}}|u_{n\boldsymbol{k}}(t)\rangle \; .\end{aligned} \tag{3.3}$$

The adiabatic current density is obtained by taking the time-derivative of $\langle \boldsymbol{r}_n\rangle$, which is given by

$$\begin{aligned}\boldsymbol{j}_n &= -\frac{e}{V}\frac{d\langle \boldsymbol{r}_n\rangle}{dt} \\ &= -\frac{e}{(2\pi)^3}\int d\boldsymbol{k}\, i[\langle \partial_t u_{n\boldsymbol{k}}|\partial_{\boldsymbol{k}} u_{n\boldsymbol{k}}\rangle - \langle \partial_{\boldsymbol{k}} u_{n\boldsymbol{k}}|\partial_t u_{n\boldsymbol{k}}\rangle] \\ &= -\frac{e}{(2\pi)^3}\int d\boldsymbol{k}\, \Omega^n_{t\boldsymbol{k}} \; .\end{aligned} \tag{3.4}$$

This is the same expression first obtained by Thouless [2], albeit with a different method. If there are multiple bands, a sum over all occupied bands is required to obtain the total current. For simplicity, in the following we consider only one occupied band and drop the band index n.

Quantization of the pumped charge.—An immediate conclusion from Eq. (3.4) is that for a 1D system the pumped charge per cycle must be quantized. The pumped charge is given by

$$Q = \int_0^T dt\, j = -e\int_0^T dt \int_{-\pi}^{\pi}\frac{dk}{2\pi}\Omega_{tk} \; . \tag{3.5}$$

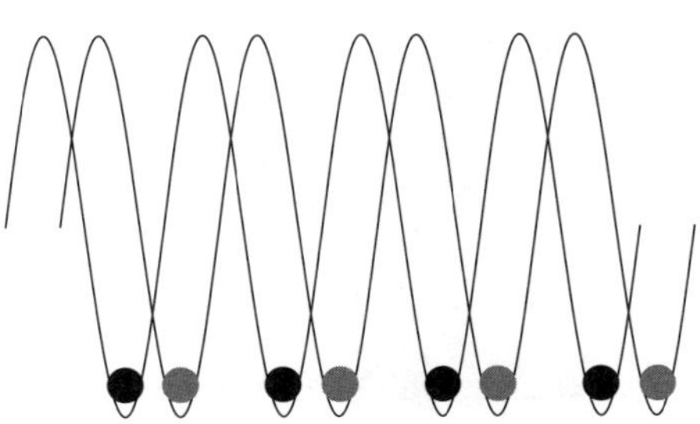

Fig. 3.1. An adiabatic shift of the periodic potential generates a current.

The integration of the Berry curvature over a closed manifold, in the above case spanned by k and t, is always quantized in the units of 2π. A brief proof is sketched below. The parameter space k, t is a torus. Introduce $x = t/T$ and $y = k/2\pi$. We begin by cutting the torus open and writing Eq. (3.5) in terms of the Berry connection using Stokes' theorem, i.e.,

$$\begin{aligned} c = \frac{1}{2\pi}\Big\{ \int_A^B dx\, \mathcal{A}_x(x,0) + \int_B^C dy\, \mathcal{A}_y(1,y) \\ + \int_C^D dx\, \mathcal{A}_x(x,1) + \int_D^A dy\, \mathcal{A}_y(0,y) \Big\} \\ = \frac{1}{2\pi}\Big\{ \int_0^1 dx\, [\mathcal{A}_x(x,0) - \mathcal{A}_x(x,1)] \\ - \int_0^1 dy\, [\mathcal{A}_y(0,y) - \mathcal{A}_y(1,y)] \Big\} \,, \end{aligned} \tag{3.6}$$

Let us consider the integration over x. By definition, $\mathcal{A}_x(x,y) = \langle u(x,y)|i\nabla_x|u($ Recall that $|u(x,0)\rangle$ and $|u(x,1)\rangle$ describe physically equivalent states, therefore they can only differ by a phase factor, i.e., $e^{i\theta_x(x)}|u(x,1)\rangle = |u(x,0)\rangle$. We thus have

$$\int_0^1 dx\, [\mathcal{A}_x(x,0) - \mathcal{A}_x(x,1)] = \theta_x(1) - \theta_x(0) \,. \tag{3.7}$$

Similarly,

$$\int_0^1 dy\, [\mathcal{A}_y(0,y) - \mathcal{A}_y(1,y)] = \theta_y(1) - \theta_y(0) \,, \tag{3.8}$$

where $e^{i\theta_y(y)}|u(y,1)\rangle = |u(y,0)\rangle$. The total integral is

$$c = \frac{1}{2\pi}[\theta_x(1) - \theta_x(0) + \theta_y(0) - \theta_y(1)] \,. \tag{3.9}$$

On the other hand, using the phase matching relations at the four corners A, B, C, and D,

$$\begin{aligned} e^{i\theta_x(0)}|u(0,1)\rangle &= |u(0,0)\rangle \,, \\ e^{i\theta_x(1)}|u(1,1)\rangle &= |u(1,0)\rangle \,, \\ e^{i\theta_y(0)}|u(1,0)\rangle &= |u(0,0)\rangle \,, \\ e^{i\theta_y(1)}|u(1,1)\rangle &= |u(0,1)\rangle \,, \end{aligned}$$

we obtain

$$|u(0,0)\rangle = e^{i[\theta_x(1)-\theta_x(0)+\theta_y(0)-\theta_y(1)]}|u(0,0)\rangle \,. \tag{3.10}$$

The single-valuedness of $|u\rangle$ requires that the exponent must be an integer multiple of 2π. Therefore the transported particle number c, given in Eq. (3.9), must be quantized. This integer is called the first Chern number.

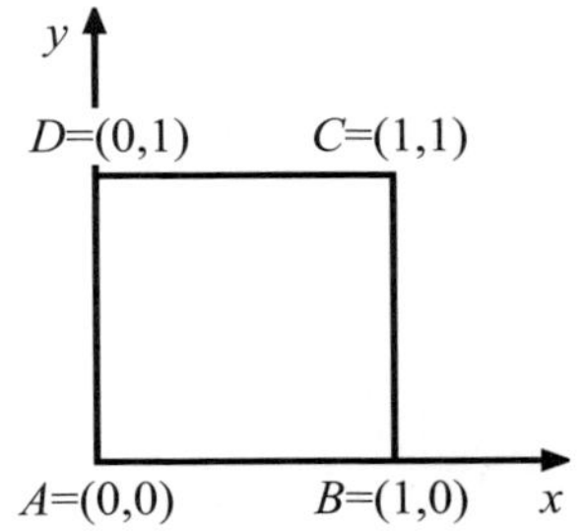

Fig. 3.2. The counter integral of Eq. (3.6).

Niu and Thouless [3] further showed that the quantization of the pumped charge is a robust phenomena, and is unaffected by many-body interactions and disorder as long as the system remains an insulator in the pumping cycle. Later Niu [4] suggested that the exact quantization of the adiabatic transport can be used as a standard for charge current and proposed an experimental realization in nanodevices, which could serve as a charge pump.

Existence of the edge state.—So far our result is derived under the assumption that the evolution is adiabatic; that is, the system remains insulating throughout the process. On the other hand, to actually pump an electron out of the system, there must be some electronic states crossing the Fermi energy. How to reconcile this obvious paradox? The answer lies at the edge. In our derivation, we have used the periodic boundary condition. The system is closed. For actual pumping to happen, the system must be open, therefore it has an edge. Even though during the pumping process, the bulk of the system remains insulating, there will be edge states that cross the Fermi energy.

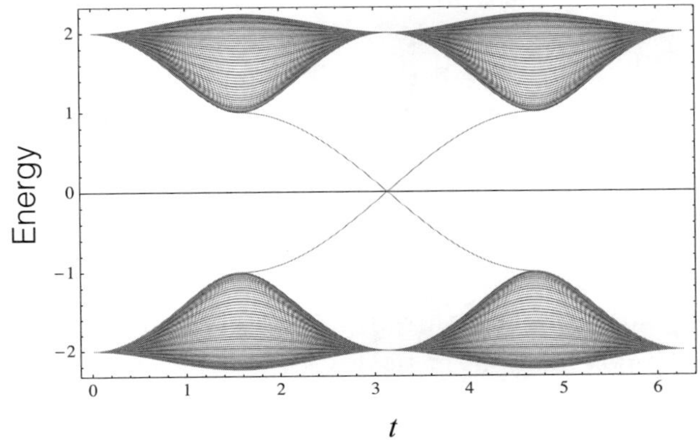

Fig. 3.3. Energy spectrum of the Hamiltonian (3.11) with its time-dependence given by Eq. (3.12) on a chain with 100 atomic sites. $t_0 = \Delta_0 = \delta_0 = 1$.

To give a concrete example of the edge state, we consider a one-dimensional dimerized lattice model described by the following Hamiltonian [5, 6]

$$H = \sum_j (\frac{t_0}{2} + (-1)^j \frac{\delta}{2})(c_j^\dagger c_{j+1} + \text{h.c.}) + \Delta(-1)^j c_j^\dagger c_j \,, \tag{3.11}$$

where t_0 is the uniform hopping amplitude, δ is the dimerization order, and Δ is a staggered sublattice potential. It is the prototype Hamiltonian for a class of one-dimensional ferroelectrics. At half-filling, the system is a metal if $\Delta = \delta = 0$, and an insulator otherwise. Now consider a pumping process described by

$$\Delta(t) = \Delta_0 \sin(\omega t) \,, \quad \delta(t) = \delta_0 \cos(\omega t) \,. \tag{3.12}$$

Obviously, the system remains an insulator throughout the pumping cycle. Using Eq. (3.5), one can show that after one cycle, precisely one electron will be pumped out of the system.

Next we place the system on a finte chain. The energy spectrum as a function of t is shown in Fig. 3.3. It is clear that two edge states, located at separate ends of the chain, appear as t varies. At one end, the edge state raises from the valence band to the conduction band, while at the other end, the edge state drops from the conduction band to the valence band. Hence, the overall effect is that the system will take one electron from one end and release an electron from the other end. Therefore even though the system returns to itself after one cycle, an electron has been transported.

3.2 Electric polarization

An important application of adiabatic charge pumping is the theory of electric polarization, which is essential to any proper description of dielectric phenomena of matter. Early attempts to develop a microscopic theory of electric polarization in crystalline solids has met with the difficulty that the dipole moment is not well defined in a periodic system [7, 8]. Of course, for any finite system such a difficulty does not exist; in principle one can simply integrate $\rho(\boldsymbol{r})\boldsymbol{r}$ over the system size, where $\rho(\boldsymbol{r})$ is the charge density. However, when the system size becomes macroscopically large, such a calculation becomes impractical, it is thus desirable to develop a theory using periodic boundary conditions. It is with this boundary condition the position operator $\boldsymbol{r}$ becomes ill-defined.

The breakthrough came when Resta [9] realized that in experiments, it is the change of the polarization, instead of its absolute value, being measured. The simplest setup of such measurements is made by putting the sample between two metal plates, which are connected by an ammeter. As the polarization being switched, the ammeter records how much current has passed, which is then converted to the change of polarization. To describe this process, we begin with the relation

$$\nabla \cdot \boldsymbol{P}(\boldsymbol{r}) = -\rho(\boldsymbol{r}) , \tag{3.13}$$

where $\boldsymbol{P}(\boldsymbol{r})$ is the polarization density and $\rho(\boldsymbol{r})$ is the charge density. Coupled with the continuity equation

$$\frac{\partial \rho(\boldsymbol{r})}{\partial t} + \nabla \cdot \boldsymbol{j} = 0 , \tag{3.14}$$

Eq. (3.13) leads to

$$\nabla \cdot (\frac{\partial \boldsymbol{P}}{\partial t} - \boldsymbol{j}) = 0 . \tag{3.15}$$

Therefore up to a divergence-free part, the change in the polarization density is given by

$$\Delta P_\alpha = \int_0^T dt\, j_\alpha . \tag{3.16}$$

The change of the polarization can be expressed as the integrated adiabatic charge current. This result was obtained by King-Smith and Vanderbilt [10].

How to apply Eq. (3.16) to realistic systems? Let us consider a perovskite structure in which many ferroelectric crystalize. A perovskite has the chemical formula ABO_3. In its ideal structure, the A site atom occupies the cubic center, B site atom at the corners of the cubic, and the oyxgen atom sits at side center. In some materials the A site atom can have a small shift off its center position, giving rise to a nonzero $\boldsymbol{P}$. One can thus imagine a process in which the A site atom slowly moves from the center, which has inversion symmetry, to its actual position. The path should be chosen such that the band gap remains open along the path. Then one can calculate the adiabatic current and the change in polarization.

In first-principles band structure calculations of $\boldsymbol{P}$, the above method is not desirable because it requires a series calculation of the band structure for a set of points along the adiabatic path. King-Smith and Vanderbilt showed that, using the periodic gauge, the polarization can be written as [10]

$$\boldsymbol{P} = \frac{e}{V} \sum_n \int_{\mathrm{BZ}} d\boldsymbol{k}\, \langle u_{n\boldsymbol{k}} | i\nabla_{\boldsymbol{k}} | u_{n\boldsymbol{k}} \rangle . \tag{3.17}$$

This expression is much simpler since it only involves the integration of $\boldsymbol{k}$. A comparison with Eq. (3.3) shows that the above expression has a very intuitive interpretation: the electric polarization is given by the dipole moment of the Wannier charge density. However, such a simplication has a price to pay. Equation (3.17) only determines the polarization up to integer multiple of the polarization quantum $e\boldsymbol{a}/V_0$, where $\boldsymbol{a}$ is the Bravais lattice vector and V_0 is the volume of the unit cell. This is also reflected by the fact that the Wannier center position is defined only up to a lattice vector due to the translation symmetry.

3.3 Orbital magnetization

Next we discuss the theory of orbital magnetization. The development of a microscopic theory of the orbital magnetization has the same problem as electric polarization. By definition, the orbital magnetization is the expectation value of $-e\boldsymbol{r} \times \boldsymbol{v}$. For a periodic system, we again faces the problem of an ill-defined position operator $\boldsymbol{r}$.

To overcome this difficulty, we consider an equivalent definition of $\boldsymbol{M}$ as the response of the free energy with respect to an applied magnetic field $\boldsymbol{B}$. To carry out the linear response calculation, we employ the semiclassical theory discussed in previous chapter. In the semiclassical formalism, each electron is described by a wave packet, which will acquire an energy of $-\boldsymbol{B} \cdot \boldsymbol{m}(\boldsymbol{k})$ in the presence of a magnetic field. However, that's not the complete story; Xiao, Shi, and Niu [11] showed that there is a modification to the density of states due to the Berry curvature. To see this, we begin with the semiclassical equations of motion:

$$\dot{\boldsymbol{r}} = \frac{\partial \varepsilon_M(\boldsymbol{k})}{\hbar \partial \boldsymbol{k}} - \dot{\boldsymbol{k}} \times \boldsymbol{\Omega}(\boldsymbol{k}) , \tag{3.18a}$$

$$\hbar \dot{\boldsymbol{k}} = -e\boldsymbol{E} - e\dot{\boldsymbol{r}} \times \boldsymbol{B} , \tag{3.18b}$$

where $\varepsilon_M(\boldsymbol{k}) = \varepsilon(\boldsymbol{k}) - \boldsymbol{B} \cdot \boldsymbol{m}(\boldsymbol{k})$. Let us follow the time evolution of a phase-space volume element $\Delta V = \Delta \boldsymbol{r} \Delta \boldsymbol{k}$, which is given by

$$\frac{1}{\Delta V} \frac{\partial \Delta V}{\partial t} = \boldsymbol{\nabla}_{\boldsymbol{r}} \cdot \dot{\boldsymbol{r}} + \boldsymbol{\nabla}_{\boldsymbol{k}} \cdot \dot{\boldsymbol{k}} . \tag{3.19}$$

Inserting the equations of motion Eq. (3.18) into the above equation yields

$$\Delta V = \frac{V_0}{1 + (e/\hbar) \boldsymbol{B} \cdot \boldsymbol{\Omega}} . \tag{3.20}$$

Although the phase-space volume element is no longer a constant, the number of quantum states must be a conserved quantity. Thus to compensate the change of ΔV, we must introduce a modified density of states

$$D(\boldsymbol{r}, \boldsymbol{k}) = \frac{1}{(2\pi)^d} \left(1 + \frac{e}{\hbar} \boldsymbol{B} \cdot \boldsymbol{\Omega}\right) . \tag{3.21}$$

With this in mind, we can write down the grand canonical potential to the first order of $\boldsymbol{B}$:

$$\begin{aligned} F &= -\frac{1}{\beta} \sum_{\boldsymbol{k}} \log(1 + e^{-\beta(\varepsilon_M - \mu)}) \\ &= -\frac{1}{\beta} \int \frac{d\boldsymbol{k}}{(2\pi)^d} (1 + \frac{e}{\hbar} \boldsymbol{B} \cdot \boldsymbol{\Omega}) \log(1 + e^{-\beta(\varepsilon_M - \mu)}) , \end{aligned} \tag{3.22}$$

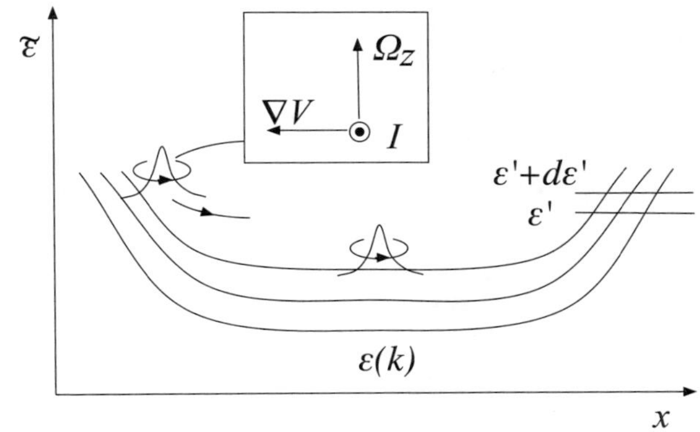

Fig. 3.4. Electron energy in a slowly varying confining potential $V(\boldsymbol{r})$. In addition to the self-rotation, wave packets near the boundary will also move along the boundary due to the potential V. Level spacings between different bulk $\boldsymbol{k}$ states are exaggerated; they are continuous in the semiclassical limit. The inset shows directions of the Berry curvature, the effective force, and the current carried by a wave packet on the left boundary.

where the electron energy ε_M includes a correction due to the orbital magnetic moment $\boldsymbol{m}(\boldsymbol{k})$. The magnetization is then the field derivative of F at fixed temperature and chemical potential, $\boldsymbol{M} = -(\partial F/\partial \boldsymbol{B})_{\mu,T}$, with the result

$$\begin{aligned}\boldsymbol{M}(\boldsymbol{r}) = &\int \frac{d\boldsymbol{k}}{(2\pi)^d}\, f(\boldsymbol{k})\boldsymbol{m}(\boldsymbol{k}) \\ &+ \frac{1}{\beta}\int \frac{d\boldsymbol{k}}{(2\pi)^d}\,\frac{e}{\hbar}\boldsymbol{\Omega}(\boldsymbol{k})\log(1+e^{-\beta(\varepsilon-\mu)})\,.\end{aligned} \tag{3.23}$$

Integration by parts of the second term will give us,

$$M_z = \int \frac{d\boldsymbol{k}}{(2\pi)^d}\, f(\boldsymbol{k})m_z(\boldsymbol{k}) + \frac{1}{e}\int d\varepsilon\, f(\varepsilon)\sigma_{xy}(\varepsilon)\,, \tag{3.24}$$

where $\sigma_{xy}(\varepsilon)$ is the zero-temperature Hall conducitvity for a system with Fermi energy ε:

$$\sigma_{xy}(\varepsilon) = \frac{e^2}{\hbar}\int \frac{d\boldsymbol{k}}{(2\pi)^d}\Theta(\varepsilon-\varepsilon(\boldsymbol{k}))\Omega_z(\boldsymbol{k})\,. \tag{3.25}$$

The linear response calculation can be carried out quantum mechanically with the same result [12]. For insulators with zero Chern number, Thonhauser et al [13] has shown that one can construct a set of Wannier functions and calculate the expectation value of $\boldsymbol{r}\times\boldsymbol{v}$ directly using the Wannier basis. Gat and Avron [14] obtained an equivalent result for the special case of the Hofstadter model.

Equation (3.23) has an intuitive interpretation. The first term is a contribution due to the self-rotation of the wave packet, whereas the second one is from the center-of-mass motion. To see this, we place the system in a confining potential $V(\boldsymbol{r})$. We assume $V(\boldsymbol{r})$ varies slowly at atomic length scale such that to a good approximation, we can write the wave-packet energy as $\varepsilon(\boldsymbol{k}) + V(\boldsymbol{r})$. At the boundary, the wave packet feels the confining potential and acquires a transverse velocity

$$\frac{1}{\hbar}\nabla V(\boldsymbol{r}) \times \boldsymbol{\Omega}(\boldsymbol{k}) , \tag{3.26}$$

which in turn leads to a boundary charge current given by

$$I = \frac{e}{\hbar}\int dx \int \frac{d\boldsymbol{k}}{(2\pi)^2}\frac{dV}{dx} f[\varepsilon(\boldsymbol{k}) + V(\boldsymbol{r})]\Omega_z(\boldsymbol{k}) . \tag{3.27}$$

where x is in the direction perpendicular to the boundary, and the integration range is taken from deep into the bulk to outside the sample. Recall that for a current I flowing in a closed circuit enclosing a sufficiently small area A, the circuit carries a magnetic moment given by IA. Therefore the magnetization (magnetic moment per unit area) has the magnitude of the current I. Integration by parts yield the second term of Eq. (3.23).

The above interpretation has an important consequence on thermoelectric transport. Since $\boldsymbol{m}(\boldsymbol{k})$ behaves like electron spin, the magnetization current due to the self-rotation of the wave packet should not contribute to any transport phenomena. On the other hand, the center-of-mass motion should. In fact, in our derivation above, we can assume the chemical potential on the left and right is different, which will then result in a net current, given by

$$\boldsymbol{j} = \frac{\partial M_{\mathrm{COM}}}{\partial \mu}\nabla\mu . \tag{3.28}$$

Insert Eq. (3.27) into the above equation and invoke the Einstein relation that $-e\boldsymbol{E}$ is equivalent to $\nabla\mu$; we then obtain the Hall conductivity given by

$$\sigma_H = \frac{e^2}{\hbar}\int \frac{d\boldsymbol{k}}{(2\pi)^2} f(\varepsilon_{\boldsymbol{k}})\Omega_z(\boldsymbol{k}) . \tag{3.29}$$

Following this observation, Xiao et al [15] showed that the orbital magnetization is essential to understand Berry-phase effect in transport driven by statistical forces.

Chapter 3

References

1. R. Karplus and J. M. Luttinger, Phys. Rev. **95**, 1154 (1954).
2. D. J. Thouless, Phys. Rev. B **27**, 6083 (1983). [Reprint (3.1)]
3. Q. Niu and D. J. Thouless, J. Phys. A **17**, 2453 (1984).
4. Q. Niu, Phys. Rev. Lett. **64**, 1812 (1990). [Reprint (3.2)]
5. W. P. Su, J. R. Schrieffer, and A. J. Heeger, Phys. Rev. Lett. **42**, 1698 (1979).
6. M. J. Rice and E. J. Mele, Phys. Rev. Lett. **49**, 1455 (1982).
7. R. M. Martin, Phys. Rev. B **5**, 1607 (1972).
8. R. M. Martin, Phys. Rev. B **9**, 1998 (1974).
9. R. Resta, Ferroelectrics **136** 51 (1992).
10. R. D. King-Smith and D. Vanderbilt, Phys. Rev. B **47**, 1651(R) (1993). [Reprint (3.3)]
11. D. Xiao, J. Shi, and Q. Niu, Phys. Rev. Lett. **95**, 137204 (2005). [Reprint (3.6)]
12. J. Shi, G. Vignale, D. Xiao, and Q. Niu, Phys. Rev. Lett. **99**, 197202 (2007). [Reprint (3.7)]
13. T. Thonhauser, D. Ceresoli, D. Vanderbilt, and R. Resta, Phys. Rev. Lett. **95**, 137205 (2005). [Reprint (3.5)]
14. O. Gat and J.E. Avron, Phys. Rev. Lett. **91**, 186801 (2003). [Reprint (3.4)]
15. D. Xiao., Y. Yao, Z. Fang, and Q. Niu, Phys. Rev. Lett. **97**, 026603 (2006).

4

The Hall effects

The Hall effect is a vast topic in condensed matter physics with a long history. Since its initial discovery more than a century ago, the study of the Hall effect has led to a string of major discoveries, accompanied by the development of many elegant yet powerful concepts. In the following we provide a chronicled account of how the Berry phase came to the fore in the world of Hall physics.

4.1 The Predawn Era

The Berry phase was discovered in 1983 [1] while the Hall effect has a much longer history than that. Nonetheless it is still useful to briefly review the early activities, which not only gives a historical context of our discussion, but in hindsight, already exposed some of the connections between the Berry phase and the Hall effect.

In 1978 Edwin Hall [2], then a graduate student at John Hopkins University, discovered that a transverse voltage would appear when a current-carrying conductor is placed in a perpendicular magnetic field. This effect, now known as the ordinary Hall effect (OHE), is due to the reflection of conduction electrons by the Lorentz force. Since the Hall coefficient, defined by $R_H = E_y/(j_x B_z)$, is inversely proportional to the carrier density, the OHE is widely used as an accurate characterization tool in laboratories. It also has applications in magnetic field sensors.

One year later, Hall [3] found that in irons the transverse voltage is an order of magnitude larger than that in normal metals. It turns out that this anomalous Hall effect (AHE) is a general feature of ferromagnetic metals and persists even in the absence of the magnetic field. Despite the similarities between the OHE and the AHE, the microscopic origin of the latter has remained a controversial subject for decades. In 1954, Karplus and Luttinger (KL) [4] carried out the first serious calculation of the AHE. By including the spin-orbit coupling (SOC) in the energy bands, they showed that the Bloch electrons can acquire an anomalous velocity that is transverse to the applied electric field, giving rise to a nonzero Hall current when summed over both spin components in a ferromagnet. However, their result was met with great skepticism because the anomalous velocity is entirely a band structure effect and the resulting Hall current is independent of impurity scattering. Soon after Karplus and Luttinger's work, Smit [5, 6] argued that the AHE can be explained by the asymmetric scattering amplitudes for spin-up and -down electrons in the presence of disorder and the SOC (skew scattering). Later, Berger [7] showed that the SOC also causes a sudden transverse shift (side-jump) of the electron wave packet during the scattering event, which contributes to the Hall current. Both mechanisms are of extrinsic nature, and soon gained great traction in the AHE community. Since the seventies most of the debate on the origin of the AHE have been centered around the skew scattering and side-jump mechanisms, and the KL work was largely forgotten. It was until much later that the KL mechanism was interpreted as a Berry phase effect.

4.2 The rise of the quantum Hall effect

Almost a century after Hall's original discovery, and independent of the progress made in the AHE study, the integer and fractional quantum Hall effect (QHE) were discovered in the early eighties [8, 9]. These effects appear in a two-dimensional electron gas in the presence of a strong magnetic field. The defining feature of the QHE is the exact quantization of the Hall conductance, i.e,

$$\sigma_{xy} = \nu \frac{e^2}{h}\,, \quad (\nu \text{ is an integer or a fraction}) \tag{4.1}$$

accompanied by the vanishing longitudinal conductance. In the following we shall confine ourselves to the integer QHE (ν is an integer). The quantization of σ_{xy} is astonishing; it is accurate to nearly one part in a billion. Thouless and collaborators (TKNN) [10] were the first to point out that σ_{xy} can be expressed as a topological number, the Chern number, therefore it must be quantized. Their argument is based on the Kubo-formula expression of the Hall conductance for a two-dimensional band insulator

$$\sigma_{xy} = i\frac{e^2}{\hbar}\sum_{n,n'}\int_{\mathrm{BZ}} \frac{d^2k}{(2\pi)^2} \frac{\langle u_n|v_x|u_{n'}\rangle\langle u_{n'}|v_y|u_n\rangle - \text{c.c.}}{(\epsilon_{nk}-\epsilon_{n'k})^2}\,, \tag{4.2}$$

where n and n' are the band indices of occupied and unoccupied bands, respectively, $u_n(\boldsymbol{k})$ is the periodic part of the Bloch function, ϵ_{nk} is the band energy, and the integration over $\boldsymbol{k}$ is performed over the entire Brillouin zone. In the $u_n(\boldsymbol{k})$ basis, the velocity operator is given by $v_\alpha = \partial H(\boldsymbol{k})/\partial k_\alpha$. To see why σ_{xy} is quantized, we use the identity $\langle u_n|\partial H/\partial k_\alpha|u'_n\rangle = (\epsilon_{n'k}-\epsilon_{nk})\langle\partial_{k_\alpha}u_n|u_{n'}\rangle$

to recast Eq. (4.2) into

$$\sigma_{xy} = \frac{e^2}{\hbar} \sum_n \int_{\mathrm{BZ}} \frac{d^2k}{(2\pi)^2} \boldsymbol{\nabla}_{\boldsymbol{k}} \times \boldsymbol{\mathcal{A}}_n(\boldsymbol{k}) \;, \tag{4.3}$$

where $\boldsymbol{\mathcal{A}}_n(\boldsymbol{k}) = i\langle u_n(\boldsymbol{k})|\boldsymbol{\nabla}_{\boldsymbol{k}(\boldsymbol{k})} u_n\rangle$ is the Berry connection. Keeping in mind that the Brillouin zone has the topology of a torus, we can interpret Eq. (4.3) as the integration of the curl of a vector field over a closed surface. According to the Gauss-Bonnet theorem, it must be quantized. This is similar to the quantization of the magnetic monopole charge. The integer ν is called the Chern number, also known as the TKNN number in the physics community.

The result by TKNN shows that the Hall conductance of any two-dimensional band insulator must be quantized. However, their theory is based on the band picture (or the magnetic Bloch band in the presence of a magnetic field), which does not take into account the effect of disorder and, to some extend, the electron-electron interaction, both of which cannot be ignored when describing a real physical system. To answer this question, Niu, Wu, and Thouless [11] showed that the Hall conductance is quantized for *any* two-dimensional insulators. In their derivation, they used the so-called twisted boundary condition, which is equivalent to the following process. Consider a two-dimensional system and assume periodic boundary condition in both directions, i.e., the system is placed on a torus. Now thread the torus with two fluxes through both holes, then the Hamiltonian becomes dependent on the flux ϕ_1 and ϕ_2. For a given ϕ_1 and ϕ_2, the Hall conductance is

$$\sigma_{xy} = i\frac{e^2}{\hbar} \sum_{n>0} \frac{\langle \Phi_0|v_1|\Phi_n\rangle\langle\Phi_n|v_2|\Phi_0\rangle - \text{c.c.}}{(\epsilon_0 - \epsilon_n)^2} \;, \tag{4.4}$$

where Φ_n is the many-body wave function with Φ_0 the ground state. In the presence of the flux, the velocity operator is given by $v_\alpha = \partial H(\kappa_1, \kappa_2)/\partial\kappa_\alpha$, where $\kappa_\alpha = (e/\hbar)\phi_\alpha/L_\alpha$ with L_α being the dimensions of the system. The existence of a bulk energy gap guarantees that the Hall conductivity remains unchanged after thermodynamic averaging, which is written as

$$\sigma_{xy} = i\frac{e^2}{\hbar} \sum_n \int_0^{2\pi/L_1} d\kappa_1 \int_0^{2\pi/L_2} d\kappa_2 \frac{\langle \Phi_0|v_1|\Phi_n\rangle\langle\Phi_n|v_2|\Phi_0\rangle - \text{c.c.}}{(\epsilon_0 - \epsilon_n)^2} \;. \tag{4.5}$$

Note that the Hamiltonian $H(\kappa_1, \kappa_2)$ is periodic in both κ_α with period $2\pi/L_\alpha$ because the system returns to its original state after the flux is changed by a flux quantum h/e. So we again have the situation of integrating the Berry curvature over a closed surface, hence the quantization of the Hall conductance, even in the presence of disorder and many-body interactions. In Ref. [11], the authors also showed that if the ground state is degenerate, then the Hall conductance becomes fractal as in the fractional quantum Hall effect.

An remarkable aspect of the above derivation is that the quantization of the Hall conductance actually does not depend on the existence of a magnetic

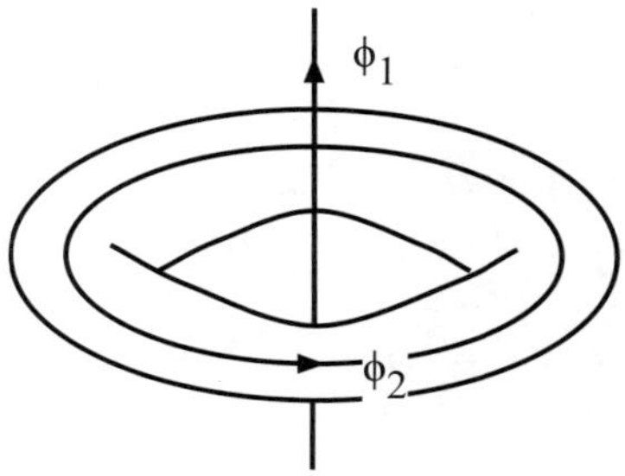

Fig. 4.1. Magnetic flux going through the holes of the torus.

field at all, even though it was the only known way to have a nonzero QHE back then. This poses an interesting question: can we have the QHE in the absence of Landau levels? In 1988, Haldane [12] wrote down a tight-binding model on a honeycomb lattice in which the average magnetic flux per unit cell is zero, yet the system displays a nonzero quantized Hall conductance. Haldane's result can be understood as follows. It is well known that the band structure of an electron hopping on a honeycomb lattice consists of two Dirac points located at the corners of the hexagonal Brillouin zone. The low-energy effective Hamiltonian is given by

$$H = \tau k_x \sigma_x + k_y \sigma_y \ , \tag{4.6}$$

where $\tau = \pm 1$ labels the K and $-K$ points, and the Pauli matrices act in the pseudo-spin space of the two sublattice. The Berry phase around each Dirac point is π and the Berry curvature is zero everywhere except at the Dirac points where it becomes singular. To get rid of the singularity, one can open a band gap by introducing a field Δ that couples to σ_z. If $\Delta\sigma_z$ has the same sign at the two Dirac points, then the time-reversal symmetry is preserved. In this case the Berry curvature has opposite sign around the two Dirac points, and the Hall conductance vanishes as required by time-reversal symmetry. However, if $\Delta\sigma_z$ has opposite sign at the Dirac points, the Berry curvature will have the same sign. Consequently the Hall conductance must be nonzero and, according to the TKNN theorem, quantized. Haldane showed that such a gap-opening term can be obtained by introducing a complex hopping integral between next-nearest neighbors.

Even though Haldane's model is only a toy model, it has far reaching consequences. Traditionally, the QHE is always associated with strong magnetic field and low temperatures; Haldane's result shows that it is possible to realize the quantum Hall physics even in "ordinary" band insulators. It also inspired the discovery of topological insulators, which will be discussed in Chapter 5.

4.3 Modern interpretation of the anomalous Hall effect and beyond

With the insight gained in the study of the QHE, the AHE attracted renewed attention in 21st century. In the QHE, the Hall conductance is expressed as the sum over the Berry curvature of the occupied states. If the system is a metal, then the integration is no longer performed on a closed manifold, and the Hall conductance is not quantized anymore. But it still can be nonzero. This is precisely the AHE. Using the semiclassical wave-packet formalism, Sundaram and Niu showed that the KL anomalous velocity can be written in terms of the Berry curvature, which puts the theory on a firm ground. The intrinsic contribution to the AHE is

$$\sigma_{xy} = \frac{e^2}{\hbar}\sum_n \int \frac{d\boldsymbol{k}}{(2\pi)^d} f(\epsilon_{n\boldsymbol{k}})\Omega^n_{k_x k_y} \,, \tag{4.7}$$

where $f(\epsilon)$ is the Fermi-Dirac distribution and $\Omega^n_{k_x k_y}$ the Berry curvature of the nth band. The KL mechanism is further confirmed by first-principles calculations of the anomalous Hall conductivity in a number of different material classes, including semiconductors [13], metals [14], and oxides [15]. It is important to emphasize that while our focus here is on the Berry phase contribution to the AHE, the other two mechanisms, i.e., skew scattering and side jump, are equally important. The competition between these different mechanisms, and the dominant source in different material systems are still topics of ongoing research [20]. Readers who are interested in the AHE are encouraged to check out recent reviews on this topic [21].

While it is satisfying to finally have a complete understanding of the AHE, there is still one puzzling piece. The transport phenomena in metals is usually understood as a Fermi surface phenomena because it should only involve states close to the Fermi surface due to the Pauli exclusive principle. However, so far the Hall conductivity is given as a summation of all occupied states, including these buried deep below the Fermi energy. This issue was first raised by Haldane [16], and he showed that the Hall conductivity can be written as the Berry phase of quasiparticles on the Fermi surface, up to an integer multiple of e^2/h. Therefore the intrinsic Hall conductivity is also a Fermi surface property. This observation suggests that the Berry phase on the Fermi surface should be added as a topological ingredient to the Landau Fermi-liquid theory.

In the following we briefly sketch Haldane's derivation for a two-dimensional system. We assume there is only one partially filled band. If we write the Berry curvature in terms of the Berry vector potential and integrate Eq. (4.7) by parts, we have

$$\sigma_{xy} = \frac{e^2}{\hbar}\int \frac{d^2k}{(2\pi)^2}\Big(\frac{\partial f}{\partial k_y}\mathcal{A}_{k_x} - \frac{\partial f}{\partial k_x}\mathcal{A}_{k_y}\Big) \tag{4.8}$$

Note that the Fermi distribution function f is a step function at the Fermi energy. If we assume the Fermi surface is a closed loop in the Brillouin zone, then

$$\sigma_{xy} = \frac{e^2}{2\pi\hbar} \oint d\boldsymbol{k} \cdot \boldsymbol{\mathcal{A}}_{\boldsymbol{k}} \,. \tag{4.9}$$

The integral is nothing but the Berry phase along the Fermi circle in the Brillouin zone. The three-dimensional case is more complicated, which has been worked out by Haldane.

The success of the Berry phase theory for the AHE not only solves a long-standing problem in condensed matter physics, but also points out the importance the Berry phase effect in general and provides plenty of ideas to study new phenomena. For instance, it is possible that the Berry curvature of opposite spins have opposite sign, which in a paramagnetic material will yield a vanishing charge Hall current, but finite spin Hall current [17, 18]. To properly address this question requires the generalization of the Berry phase to non-Abelian cases. Furthermore, the essential ingredient to our discussion, the anomalous velocity, is simply a band structure effect. Similar phenomena should arise whenever a band structure can be defined. This has led to the photon [19, 22, 23], phonon [24, 25], and magnon Hall [26, 27, 28] effect. In the future, we expect that the Hall effect will continue to bring out exiting and surprising physics, with the Berry phase playing an important role.

Chapter 4

References

1. M. V. Berry, Proc. R. Soc. London, Ser. A **392**, 45 (1984).
2. E. H. Hall, Am. J. Math. **2**, 287 (1879).
3. E. H. Hall, Philos. Mag. **12**, 157 (1881).
4. R. Karplus and J. M. Luttinger, Phys. Rev. **95**, 1154 (1954).
5. J. Smit, Physica (Amsterdam) **21**, 877 (1955).
6. J. Smit, Physica (Amsterdam)**24**, 39 (1958).
7. L. Berger, Phys. Rev. B **2**, 4559 (1970).
8. K. v. Klitzing, G. Dorda, and M. Pepper, Phys. Rev. Lett. **45**, 494 (1980).
9. D. C. Tsui, H. L. Stormer, and A. C. Gossard, Phys. Rev. Lett. **48**, 1559 (1982).
10. D. J. Thouless, M. Kohmoto, M. P. Nightingale, and M. den Nijs, Phys. Rev. Lett. **49**, 405 (1982). [Reprint (4.1)]
11. Q. Niu, D. J. Thouless, and Y.-S. Wu, Phys. Rev. B **31**, 3372 (1985). [Reprint (4.2)]
12. F.D.M. Haldane, Phys. Rev. Lett. **61**, 2015 (1988).
13. T. Jungwirth, Q. Niu, and A.H. MacDonald, Phys. Rev. Lett. **88**, 207208 (2002). [Reprint (4.3)]
14. Y. Yao, L. Kleinman, A. H. MacDonald, J. Sinova, T. Jungwirth, D.-S. Wang, E. Wang, and Q. Niu, Phys. Rev. Lett. **92**, 037204 (2004). [Reprint (4.7)]
15. Z. Fang et al., Science **302**, 92 (2003).
16. F. D. M. Haldane, Phys. Rev. Lett. **93**, 206602 (2004). [Reprint (4.6)]
17. S. Murakami, N. Nagaosa, and S.-C. Zhang, Science **301**, 1348 (2003).
18. J. Sinova, D. Culcer, Q. Niu, N. A. Sinitsyn, T. Jungwirth, and A. H. MacDonald, Phys. Rev. Lett. **92**, 126603 (2004). [Reprint (4.4)]
19. M. Onoda, S. Murakami, and N. Nagaosa, Phys. Rev. Lett. **93**, 083901 (2004). [Reprint (4.5)]
20. D. Hou, G. Su, Y. Tian, X. Jin, S. A. Yang, and Q. Niu, Phys. Rev. Lett. **114**, 217203 (2015).
21. N. Nagaosa, J. Sinova, S. Onoda, A. H. MacDonald, and N. P. Ong, Rev. Mod. Phys. **82**, 1539 (2010).
22. F. D. M. Haldane and S. Raghu, Phys. Rev. Lett. **100**, 013904 (2008). [Reprint (4.8)]
23. K. Y. Bliokh, A. Niv, V. Kleiner, and E. Hasman, Nature Photonics **2**, 748 (2008).
24. L. Zhang, J. Ren, J.-S. Wang, B. Li, Phys. Rev. Lett. **105**, 225901 (2010).
25. T. Qin, J. Zhou, J. Shi, Phys. Rev. B **86**, 104305 (2012).
26. H. Katsura, N. Nagaosa, and P. A. Lee, Phys. Rev. Lett. **104**, 066403 (2010).
27. R. Matsumoto and S. Murakami, Phys. Rev. Lett. **106**, 197202 (2011).
28. R. Matsumoto and S. Murakami, Phys. Rev. B **84**, 184406 (2011)

5

Topological insulators

The discovery of quantum Hall effect has great impacts on related theories and experiments. This naturally motivates people to search for *quantized* version of spin Hall effect. Efforts along this direction leads to the discovery of topological insulators, which is one of the most important condensed-matter finding in the past decade. Subsequent development leads to other discoveries, such as the Weyl semimetal, the Majorana fermion, and Kitaev's periodic table. In this chapter, we collect essential papers on this subject during its early stage of development.

The topological invariant of a geometric object refers to the properties that are invariant under continuous deformation. For example, a spherical balloon with radius R has a constant Gaussian curvature $G = 1/R^2$ at every point on the surface. When you squeeze it (without popping), the curvature distribution changes. However, the total curvature integrated over the whole surface M remains a constant,

$$\int_M da\ G = 4\pi. \tag{5.1}$$

That is, the total curvature of a closed, sphere-like surface is a topological invariant. This is the simplest example of the Gauss-Bonnett theorem.

Similar topological invariants exist for other two-dimensional surfaces. For example, the total curvature of a donut-like surface is 0, independent of continuous deformation. In general, if a closed surface has a number of g holes (e.g., $g = 1$ for a torus), then the total curvature is $4\pi(1 - g)$.

If a surface is not closed, but has an edge, then in addition to the surface curvature, one also has to consider the geodesic curvature κ along the edge ∂M, and the Gauss-Bonnett theorem can be generalized as,

$$\int_M da\ G + \int_{\partial M} ds\ \kappa = 2\pi\chi(M, \partial M), \tag{5.2}$$

where χ is an integer (the Euler characteristic) that depends on the topology of the surface (M) and its edge (∂M). For example, a disk-like surface has $\chi = 1$, no matter how the disk is deformed.

5.1 Quantum Hall system

Similar mathematical structure exists in Bloch state, in which Berry curvature plays the role of the Gaussian curvature mentioned above (see Chapters 2 and 4). In 2D, the integral of the Berry curvature Ω_z over the whole Brillouin zone (BZ) is an integer multiple of 2π,

$$\int_{BZ} d^2k\ \Omega_z = 2\pi C_1. \tag{5.3}$$

The integer C_1 is called the first Chern number, which is invariant under the deformation of an energy band (as long as the band gaps remain open). The most celebrated application of this would be the quantum Hall effect: the Hall conductivity σ_H of a filled band can only be the quantized value $C_1 e^2/h$ [1].

In mathematical terms, the $U(1)$ phase of the Bloch state, together with its domain (the BZ), constitute a *fiber bundle*. The BZ is the base of the fiber bundle, while the $U(1)$ phases are the fibers grown on the base. Locally, a fiber bundle is a product of the base space and the fiber space. However, globally it cannot be decomposed as a direct product because of the nontrivial topology. The Chern number C_1 characterizes the topology of a fiber bundle, just like the Euler characteristic for a two-dimensional surface.

In experiments, the plateaus of Hall conductivity are located at integer multiples of e^2/h to extremely high precision for different sample in different laboratories. The formulation above is built upon Bloch states with lattice translation symmetry. In the presence of disorders and electron-electron interactions, the Bloch momentum and the Brillouin zone are no longer suitable for topological characterization. To fix this, Niu, Thouless, and Wu [2] reformulated the theory so that a manybody wave function could also have a Chern number, and this again guarantees the quantization of the Hall conductivity (see Chapter 4).

5.2 Searching for topological invariants

It is told that W. Pauli once said that solid state physics is 'physics of dirt'. The outcome of a measurement would often differ from sample to sample, due to different fabrication processes, ambient conditions... etc. However, if the quantity being measured has a topological nature (such as the quantum Hall conductivity), then the result would be almost universal, independent of samples, laboratories, and dates of measurement. That is why physicists like to search for such topological quantities.

One of such a topological quantity is found by Avron *et al* [3], who considers a half-odd-integer spin $\mathbf{J}$ coupled with a quadrupole field Q_{ij},

$$H = \sum_{i,j=1}^{3} Q_{ij} J_i J_j. \tag{5.4}$$

The quadrupole field is a symmetric, traceless 3×3 matrix with 5 real parameters. Under time reversal, $\mathbf{J} \to -\mathbf{J}$, and H remains invariant. Because of the Kramer degeneracy, each energy level is two-fold degenerate. Therefore, the non-Abelian Berry connection has a $SU(2)$ gauge structure (see chap 2). Even though its first Chern number vanishes, the second Chern number C_2 is nonzero, indicating nontrivial topology. In particular, if $J = 3/2$, then (Ω is the Berry curvature) [3]

$$C_2 \equiv \frac{1}{8\pi^2} \int_{S^4} \mathrm{tr}\ \Omega^2 = \pm 1. \tag{5.5}$$

It is an integral over S^4, a sphere in the five-dimensional parameter space of Q_{ij}.

Haldane's work in Ref. [4] is another attempt to find non-trivial topology in a system without magnetic field. This goal is achieved only partially, since a periodic magnetic field with zero average flux is still required. He considers a graphene model with three tunable parameter: next-nearest hopping t_2 that breaks particle-hole symmetry, on-site energies $\pm M$ for different sublattices that break inversion symmetry, and periodic flux ϕ that breaks time reversal symmetry. Non-zero values of t_2 and ϕ shift the zero-mode away from zero. As a result, the 'zero mode' is either filled or empty (in the absence of bias), instead of half-filled. Therefore, different from the half-integer quantum Hall effect in the usual graphene, the Hall conductivity becomes $\sigma_H = \pm 1(e^2/h)$. A generalization of this graphene model (including spin-orbit interactions) leads to the discovery of quanum spin Hall effect [5], which is the 2D version of topological insulator (TI).

5.3 Topological insulator

As far as conductivity goes, insulators are boring and all the same. However, the physical mechanisms that hinder the electron motion in different materials could be vastly different. In textbooks, one can find the Wilson insulator: the electrons in a filled band cannot move, unless hopping to a higher band. Other than this, there is the Anderson insulator, in which the electrons are trapped by sufficient amount of disorders. There is also the Mott insulator, in which the electrons are locked by electron-electron interactions. The TI is a new and important addition to the zoo of insulators.

The essential ingredients of a TI are lattice fermions with spin-orbit couplings (and time-reversal symmetry). The particles being fermions is crucial since Kramer degeneracy plays a crucial role in TI. Under proper conditions (explained below), such a system could have non-trivial Z_2 topology. The most important consequence of it is the emergence of metallic surface state with peculiar property: the surface state has an odd number of pairs, in each pair opposite spins move along opposite directions. This is called the *helical* surface state. Protected by topology, the surface state is robust against perturbations. The topology can be changed, and the surface state destroyed, only if the insulating gap is closed (or the time-reversal symmetry is broken).

To understand the Z_2 topology, we follow Moore and Balent's argument and start from a 2D TI [6]. Because of time-reversal symmetry, the Bloch states for $\mathbf{k}$ and $-\mathbf{k}$ in a Brillouin zone are time-reversal conjugate (see Fig. 1(a)). Their Berry curvatures cancel with each other, and the first Chern number for a filled band vanishes. Since the domain of independent states covers only half of the BZ (called effective BZ, or EBZ), one may wonder if the integral of Berry curvature over the EBZ could be quantized.

Unfortunately, since the EBZ does not form a closed surface (see Fig. 1(b)), no quantization is guaranteed. To fix this, one can put two fictitious caps to close the EBZ. This closed surface should have an integer C_1, but its value depends on the caps added. Nevertheless, Moore and Balents proved that C_1 (mod 2) is independent of the caps of choice. Therefore, it is an intrinsic property of the EBZ itself. We thus have two distinctive topological classes: 0 being the usual insulator, and 1 being the topological insulator.

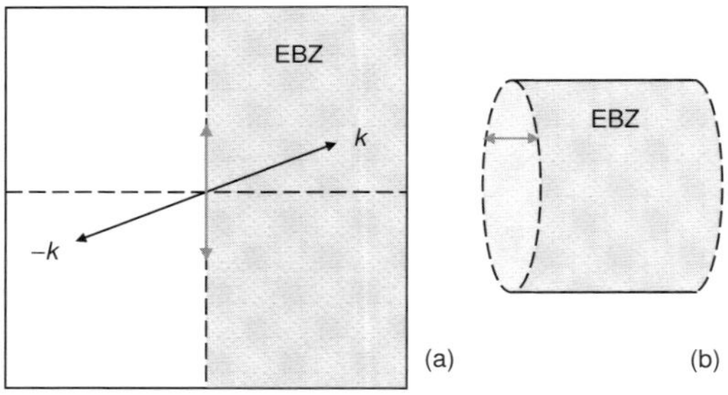

Fig. 5.1. (a) Time reversal conjugate pairs and effective Brillouin zone. (b) Folding the EBZ into a cylinder, with open edges.

To put it in concrete terms, the Z_2 integer is given by [7]

$$\nu = \frac{1}{2\pi}\left(\int_{EBZ} d^2k\ \Omega_z - \oint_{\partial(EBZ)} d\mathbf{k}\cdot\mathbf{A}\right) \bmod 2. \tag{5.6}$$

The second term is the Berry connection integrated around the boundary of the EBZ. This formula is analogues to the Euler characteristic for a manifold with edge (see Eq. 5.2).

What about three dimensions? For simplicity, consider a cubical BZ (see Fig. 2). Similar time-reversal conjugate exists between $\mathbf{k}$ and $-\mathbf{k}$ states. In particular, since the set of points on xy plane map to themselves under time reversal, according to the 2D argument above, this time-reversal-invariant plane should have a corresponding Z_2 index z_0. By symmetry, the yz plane and the zx plane also have their own Z_2 indices, x_0 and y_0. In addition, the planes at front, right, and top side of the cube also map to themselves under time reversal. So there are three more Z_2 indices, x_+, y_+, and z_+. Moore and Balents showed that these numbers are constrained by two relations, $x_+ - x_0 = y_+ - y_0 = z_+ - z_0$. Therefore, there are only 4 independent Z_2 integers. The same conclusion is reached by Kane and Mele [8, 9], and Roy [10] at about the same time.

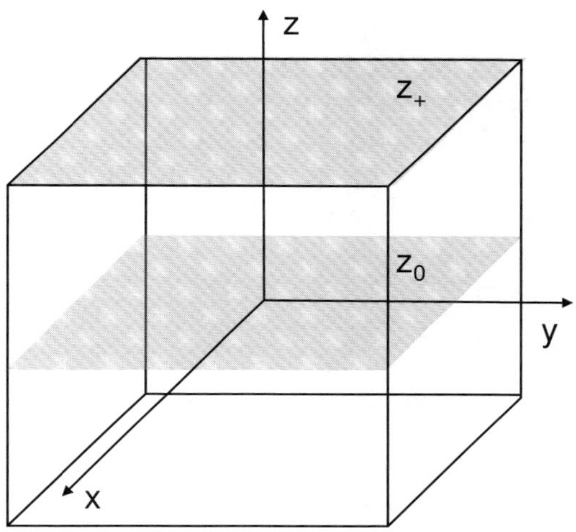

Fig. 5.2. Two of the time-reversal-invariant planes in a Brillouin zone.

Among these 4 indices, $\nu_0 = z_+ - z_0$ is called the strong TI index; x_0, y_0, z_0 are called weak TI indices. The later are similar to the indices obtained from stacking 2D TIs along x, y, and z directions, while the former is intrinsic to the 3D structure.

If the TI has inversion symmetry, then these indices can be determined easily from the parities of the Bloch states at the **k**-points that are invariant under time reversal. These points are called time-reversal invariant momenta, or TRIM. For a 2D square BZ, the TRIM are at the center (0,0), and $(0, \pi), (\pi, 0), (\pi, \pi)$ (and their equivalent points). Because of the Kramer degeneracy, the energy level must be two-fold degenerate at these points. Let's designate these points as $\Gamma_i, i = 1, \cdots, 4$, and the parity of the n-th band at Γ_i as $\xi_n(\Gamma_i)$. Each Γ_i has a cumultative parity from half of the filled Kramer pairs,

$$\delta_i = \prod_{\text{filled } n} \xi_{2n}(\Gamma_i). \tag{5.7}$$

From them one can determine the Z_2 index ν,

$$\delta_1\delta_2\delta_3\delta_4 = (-1)^\nu. \tag{5.8}$$

The strong and weak indices for 3D TI can be determined similarly. For example, the strong index is related to the parities of the 8 TRIM at the corners of the cube in the first octant. All we need to do is to multiply the δ_i's from these 8 points to get ν_0,

$$\prod_{i=1}^{8} \delta_i = (-1)^{\nu_0}. \tag{5.9}$$

What type of materials could be TIs? They are usually narrow gap insulators with strong spin-orbit interaction. The latter, when strong enough, could switch the characters of conduction electrons and valence electrons. For example, in an usual semiconductor, the valence electrons are p-like, and the conduction electrons s-like. Band inversion would make the valence electrons s-like, while the conduction electrons p-like near the energy gap. A different way to put this is that, the *mass* of the energy gap changes from positive to negative, accompanied by a change of parity at TRIM. As a result, the insulator becomes topological.

5.4 Helical surface state

To change the topology, the energy gap needs to be closed (the mass needs to go through zero in order to change sign). That is, if two solids occupying the spaces of $z > 0$ and $z < 0$ have different topologies, then near the interface, the conduction band and valence band must cross each other, producing a gapless region with low-lying electronic states. These states appear near the open surface of a TI (since the vacuum is an insulator with trivial topology). Similar states protected by topology can be found at the edges of quantum Hall systems.

As we have emphasized, these surface (or edge) states are robust against perturbations that preserve the topology. In TI, the protection is a result of the surface states being helical. If there is time-reversal symmetry, then the energy levels of surface states must cross each other at TRIM, producing a two-fold point degeneracy. In a topological insulator, away from the TRIM, the degeneracy is lifted by spin-orbit coupling of the type $\lambda\mathbf{k}\cdot\boldsymbol{\sigma}$ or $\lambda\mathbf{k}\times\boldsymbol{\sigma}\cdot\hat{z}$, for example. Such a coupling would force the spin for $\mathbf{k}$ state to be opposite to the spin for $-\mathbf{k}$ state. As a result, opposite spins move along opposite directions. Such a traffic rule guarantees that a spin-up electron moving along the surface cannot be back-scattered, unless its spin is flipped, e.g., by magnetic impurities. The crossing of the surface-state energy levels at TRIM and their helical structure have been confirmed in spin-resolved ARPES [11, 12].

Bernevig *et al* first propose a 2D topological insulator built upon quantum well structure [13], which is soon confirmed experimentally [14]. The measurement is performed on a HgTe/(Hg,Cd)Te quantum well (QW). Bulk HgTe has an intrinsic band inversion. A thin layer of HgTe is sandwitched between CdTe, which has no band inversion and has a larger band gap. By changing the width of the quantum well, one can tune the QW in and out of the TI phase. When the width of the QW is below a critical value, the band inversion in HgTe is frustrated by the neighboring CdTe's. On the other hand, there is band inversion (TI phase) when the QW is above a critical width. What is being observed in the TI phase is the saturation of four-terminal resistance in the ballistic transport regime, which corresponds to a conductance $G \simeq 2e^2/h$, and indicates the emergence of helical edge channels.

5.5 Electromagnetic response and axion angle

The TI materials discovered so far are far from ideal. In an ideal TI, the Fermi level intersect only with the surface state, but not with the bulk states. If this can be done, then electrons would only conduct through the surface states, while the bulk inside remains an insulator. Better yet, if the Fermi energy happens to be exactly at the Dirac point, then we have a two-dimensional metallic state similar to the graphene. Unlike graphene, which has an even number of Dirac points, such a TI surface state has an odd number of Dirac points (on each surface). This offers a chance to observe genuine half-integer quantum Hall effect and other topological magneto-electric phenomena [15].

The electromagnetic Lagrangian density for such an ideal 3D TI is given as,

$$L = \frac{1}{8\pi^2}\left(E^2 - B^2\right) + \alpha\frac{\Theta}{4\pi^2}\mathbf{E}\cdot\mathbf{B}, \tag{5.10}$$

where α is the fine structure constant. The coefficient Θ of the magneto-electric coupling is called the axion angle, which is equal to π for a TI. This value is closely linked with the value of the half-integer Hall conductivity.

The axion angle can be given as the Chern-Simons integral of the non-Abelian Berry connection,

$$\Theta = \frac{1}{4\pi}\int_{BZ} d^3k\epsilon_{\alpha\beta\gamma}\mathrm{tr}\left(A_\alpha\partial_\beta A_\gamma - \frac{2}{3}iA_\alpha A_\beta A_\gamma\right), \tag{5.11}$$

where $(A_\alpha)_{mn} = i\langle u_{mk}|\partial/\partial k_\alpha|u_{nk}\rangle$, and u_{nk} is the cell-periodic Bloch function for the n-th band. This result could be reached from different routes of investigation: 1. The electric polarization of an inhomogeneous crystal [16] (see Chap. 2). 2. Dimensional reduction from four-dimensional quantum Hall effect [15]. 3. Nonlinear response of the insulator in an electromagnetic field [17]. Furthermore, Wang *et al.* have shown that, with time-reversal symmetry, the value of this integral can only be 0 or π [18]. An insulator with trivial topology has $\Theta = 0$, while a TI has $\Theta = \pi$.

Chapter 5

References

1. D.J. Thouless, M. Kohmoto, M.P. Nightingale, and M. den Nijs, Phys. Rev. Lett. **49**, 405 (1982).
2. Qian Niu, D.J. Thouless, and Yong-Shi Wu, Phys. Rev. B **31**, 3372 (1985).
3. J.E. Avron, L. Sadun, J. Segert, and B. Simon, Phys. Rev. Lett. **61**, 1329 (1988). [Reprint (5.2)]
4. F.D.M. Haldane, Phys. Rev. Lett. **61**, 2015 (1988). [Reprint (5.1)]
5. C. L. Kane and E.J. Mele, Phys. Rev. Lett. **95**, 146802 (2005). [Reprint (5.3)]; C. L. Kane and E. J. Mele, Phys. Rev. Lett. **95**, 226801 (2005).
6. J. E. Moore and L. Balents, Phys. Rev. B **75**, 121306(R) (2007). [Reprint (5.4)]
7. L. Fu and C.L. Kane, Phys. Rev. B **74**, 195312 (2006).
8. L. Fu, C.L. Kane, and E.J. Mele, Phys. Rev. Lett. **98**, 106803 (2007). [Reprint (5.5)]
9. L. Fu and C.L. Kane, Phys. Rev. B **76**, 045302 (2007).
10. R. Roy, Phys. Rev. B **79**, 195321 (2009)
11. D. Hsieh, D. Qian, L. Wray, Y. Xia, Y. Hor, R. J. Cava, and M. Z. Hasan, Nature **452**, 970 (2008).
12. D. Hsieh, Y. Xia, L. Wray, D. Qian, A. Pal, J. H. Dil, F. Meier, J. Osterwalder, G. Bihlmayer, C.L. Kane, Y.S. Hor, R. J. Cava, and M.Z. Hasan, Science **323**, 919 (2009).
13. B. A. Bernevig, T.L. Hughes, and S.-C. Zhang, Science **314**, 1757 (2006).
14. M. König, S. Wiedmann, C. Brüne, A. Roth, H. Buhmann, L.W. Molenkamp, X.L. Qi, and S.C. Zhang, Science **318**, 766 (2007).
15. X.-L. Qi, R. Li, J. Zang, and S.-C. Zhang, Science **323**, 1184 (2009).
16. Di Xiao, Junren Shi, Dennis P. Clougherty, and Qian Niu, Phys. Rev. Lett. **102**, 087602 (2009).
17. Andrew M. Essin, Ari M. Turner, Joel E. Moore, and David Vanderbilt Phys. Rev. B **81**, 205104 (2010); A. Malashevich, I. Souza, S. Coh, and D. Vanderbilt, New J. Phys. **12**, 053032 (2010).
18. Z. Wang, X.L. Qi, S.C. Zhang, New J. Phys. **12**, 065007 (2010).

6

Artificial gauge fields in real space

Preceding chapters focus on the topological nature of Berry's phase, which refers to the *global* property of the entire system. On the other hand, the *local* effect that depends on the Berry curvature, namely, the role it plays as an artificial gauge field in the configuration space of slow variables, also gives rise to many intriguing physical phenomena. One of the recurring interests on this route is the Berry curvature in real spacetime, where it mimics a genuine electromagnetic field and affects the dynamic of slow variables through an effective Lorentz force.

In this chapter, we introduce in a pure generic sense the basic formalism of artificial gauge fields in real spacetime, focusing on the underlying physics and its implications. We discuss several salient features of artificial gauge fields exhibited in magnetic materials and cold atoms. We will follow conventions and terminology used in the field of spintronics. An exhausted review flavored in cold atoms can be found in Ref. [1].

6.1 Geometric Phase and Geometric Force

We have been obsessed by the adiabatic approximation and the associated geometric phase in preceding chapters. To capture the geometric force that virtually refers to the back-action exerted on the adiabatic parameter, we now shift gear to a different context and reformulate the whole problem. Consider a so-called quantum-classical hybrid system [2], where the quantum subsystem bears an intrinsic dynamics much faster than the classical subsystem. Strictly speaking, the quantum part does not need to be really quantum mechanical, nor does the classical part to be purely classical. As long as the time scales of the two coupled subsystems are quite off each other, one can adopt the adiabatic approximation [3] to simplify the calculation. That is where geometric effects come into play.

An emblematic example in this scenario is the Born-Oppenheimer problem [1, 3], say, a slowly-moving nucleus couples to fast orbiting electrons via

Coulomb's interaction. The electron energy levels can be solved to a good approximation by freezing the position of the nucleus. Then, by treating thus-obtained electron wave functions as functions of the nucleus position, one can solve the reaction of electrons on the nucleus in the form of geometric force.

Generalizing the above idea, we consider a quantum Hamiltonian $H = H(\boldsymbol{r}(t))$ that depends on time through a slow parameter $\boldsymbol{r}(t)$. The criterion for slowness will become clear later. At each instant of time, the Hamiltonian fosters a complete set of eigenstates $\{|\psi_j(\boldsymbol{r}(t))\rangle\}$ that is also parametrically depends on $\boldsymbol{r}(t)$. The wave vector of the system can be expanded by its eigenstates as

$$|\Psi\rangle = \sum_j c_j |\psi_j(\boldsymbol{r}(t))\rangle, \tag{6.1}$$

where $c_j = |c_j|e^{i\theta_j}$ is complex and respects the normalization condition $\sum_j |c_j|^2 = 1$. The Schrödinger equation that governs the system dynamics is equivalent to the variational derivative of the action $S = \int dtL$ ($\hbar = 1$ for simplicity) with respect to $|c_j|$ and θ_j, where the Lagrangian reads

$$L = \langle\Psi|(i\partial_t - H)|\Psi\rangle. \tag{6.2}$$

What distinguishes the variational approach from the usual Schrödinger equation is that one is able to take functional derivative of S with respect to the slow parameter $\boldsymbol{r}$. In other words, the variational approach puts the dynamics of the slow parameter and the quantum evolution on an equal footing.

Inserting Eq. (6.1) into the Lagrangian Eq. (6.2), and considering the orthogonal-normalization condition of eigenstates $\langle\psi_i(\boldsymbol{r}(t))|\psi_j(\boldsymbol{r}(t))\rangle = \delta_{ij}$ (for $\forall\ t$), we obtain

$$\begin{aligned} L =& \frac{1}{2} i\partial_t \left(\sum_j |c_j|^2\right) - \sum_i |c_j|^2 \dot{\theta}_j \\ &+ i\sum_{ij} |c_i||c_j| e^{i(\theta_j - \theta_i)} \dot{\boldsymbol{r}} \cdot \langle\psi_i(\boldsymbol{r})|\boldsymbol{\nabla}\psi_j(\boldsymbol{r})\rangle - \sum_j |c_j|^2 E_j(\boldsymbol{r}), \end{aligned} \tag{6.3}$$

where $\boldsymbol{\nabla}$ is the gradient over $\boldsymbol{r}$, and $E_i(\boldsymbol{r})$ is defined by

$$H(\boldsymbol{r})|\psi_j(\boldsymbol{r})\rangle = E_j(\boldsymbol{r})|\psi_j(\boldsymbol{r})\rangle. \tag{6.4}$$

The first term of Eq. (6.3) vanishes due to the normalization condition. The third term is a little bit tricky. If the eigenvalues $\{E_i\}$ are non-degenerate and $\boldsymbol{r}(t)$ changes with time sufficiently slowly, the phase factor $e^{i(\theta_j - \theta_i)}$ for $i \neq j$ would have undergone multiple rotations on the complex plane while $\langle\psi_i(\boldsymbol{r})|\nabla\psi_j(\boldsymbol{r})\rangle$ has only developed a slight change. If we take a time average over one period of the phase oscillation, all terms are nearly zero except the diagonal terms with $i = j$. Therefore, to the lowest order in the deviation from a frozen $\boldsymbol{r}(t)$, we keep only $i = j$ terms in the summation. This is known as the adiabatic approximation. Strictly speaking, the slowness of the adiabatic parameter $\boldsymbol{r}(t)$ makes sense only when it is compared to the time scale of the phase oscillation $\theta_i - \theta_j$ (for $i \neq j$), *i.e.*, the gap of the quantum subsystem.

Let $I_j = |c_j|^2$, and define the Berry connection on the j-th level as

$$\boldsymbol{\mathcal{A}}_j(\boldsymbol{r}) \equiv i\langle\psi_j(\boldsymbol{r})|\boldsymbol{\nabla}\psi_j(\boldsymbol{r})\rangle, \tag{6.5}$$

the adiabatic Lagrangian (ignoring all $i \neq j$ terms) can then be written as

$$L \approx \sum_j I_j[-\dot{\theta}_j + \dot{\boldsymbol{r}} \cdot \boldsymbol{\mathcal{A}}_j(\boldsymbol{r}) - E_j(\boldsymbol{r})]. \tag{6.6}$$

As mentioned earlier, the system is labeled by three sets of parameters $\{I_j\}$, $\{\theta_j\}$, and $\boldsymbol{r} = \{r_1, r_2, \cdots\}$, where the first two characterize the fast quantum part and the last one labels the slow classical part. Taking variational derivatives over these parameters amounts to describing the dynamics of both fast and slow degrees of freedom in a unified fashion. Specifically, the variational derivatives give three sets of Euler-Lagrange equations:

(1) $\frac{\partial L}{\partial \theta_j} - \frac{\partial}{\partial t}\left(\frac{\partial L}{\partial \dot{\theta}_j}\right) = 0$ gives $\dot{I}_j = 0$ or I_j =constant, which means the probability of finding the system in any eigenstate remains invariant. This is consistent with the adiabatic assumption.

(2) $\frac{\partial L}{\partial I_j} - \frac{\partial}{\partial t}\left(\frac{\partial L}{\partial \dot{I}_j}\right) = 0$ gives the dynamics of the phase angle

$$\dot{\theta}_j = \dot{\boldsymbol{r}} \cdot \boldsymbol{\mathcal{A}}_j(\boldsymbol{r}) - E_j(\boldsymbol{r}). \tag{6.7}$$

After an integration over time, Eq. (6.7) becomes

$$\theta_j(t) = \int d\boldsymbol{r} \cdot \boldsymbol{\mathcal{A}}_j(\boldsymbol{r}) - \int dt E_j(\boldsymbol{r}), \tag{6.8}$$

where dt drops out in the first term. As a result, the first term only depends on the geometry of the path of $\boldsymbol{r}$ in the parameter space; it does not depend on the rate of change of $\boldsymbol{r}(t)$ at any instant of time. For a closed path C in the parameter space, this geometric term can be expressed as

$$\oint_C d\boldsymbol{r} \cdot \boldsymbol{\mathcal{A}}_j(\boldsymbol{r}) = \iint_{\text{enclosure } C} d\boldsymbol{s} \cdot \boldsymbol{\nabla} \times \boldsymbol{\mathcal{A}}_j(\boldsymbol{r}) \tag{6.9}$$

by Stokes' theorem. The curl term $\boldsymbol{\mathcal{B}}_j(\boldsymbol{r}) = \boldsymbol{\nabla} \times \boldsymbol{\mathcal{A}}_j(\boldsymbol{r})$ is known as the Berry curvature, which amounts to a fictitious magnetic field in the parameter space. As discussed in preceding chapters, it can be expressed as

$$\boldsymbol{\mathcal{B}}_j = -\text{Im} \sum_{i \neq j} \frac{\langle\psi_j|\boldsymbol{\nabla} H|\psi_i\rangle \times \langle\psi_i|\boldsymbol{\nabla} H|\psi_j\rangle}{(E_i - E_j)^2}. \tag{6.10}$$

The flux of this artificial magnetic field determines the geometric phase (or Berry's phase) of the system. While the above derivation requires that the eigenstates are non-degenerate, a non-zero geometric phase, however, usually originates from the magnetic flux of a fictitious monopole located at the degenerate point. What should be bear in mind is that the actual evolution of the system can never touch the degenerate point (unless a phase transition takes place) in order to respect the adiabatic approximation. But the degenerate point must exist somewhere in the parameter space.

(3) Before taking derivatives with respect to $\boldsymbol{r}$, we first clarify the physical origin of $\boldsymbol{r}$. A typical realization is that $\boldsymbol{r}$ plays the role of a classical position. As mentioned earlier, in the prototype Born-Oppenheimer problem, the mass of the nucleus far exceeds that of the electrons, thus it is a good approximation to describe the nucleus' motion classically while keeping the electron dynamics quantum mechanical. Therefore, $\boldsymbol{r}$ represents the position of the nucleus. In this case, we need to add the kinetic term of the classical part to the Lagrangian: $L_{tot} = L + \frac{1}{2}M\dot{\boldsymbol{r}}^2$, where M is the mass of the classical particle (the nucleus). Then the Euler-Lagrange equation $\frac{\partial L_{tot}}{\partial \boldsymbol{r}} - \frac{\partial}{\partial t}\left(\frac{\partial L_{tot}}{\partial \dot{\boldsymbol{r}}}\right) = 0$ gives

$$M\ddot{\boldsymbol{r}} = \dot{\boldsymbol{r}} \times \boldsymbol{\mathcal{B}}(\boldsymbol{r}) - \boldsymbol{\nabla} E(\boldsymbol{r}), \tag{6.11}$$

where $\boldsymbol{\mathcal{B}}(\boldsymbol{r}) = \sum_j I_j \boldsymbol{\mathcal{B}}_j(\boldsymbol{r})$ is the total (fictitious) magnetic field and $E(\boldsymbol{r}) = \sum_j I_j E_j(\boldsymbol{r})$ is the average energy. The first term of Eq. (6.11) is a Lorentz-like force ascribed to the magnetic field—this is what we call the geometric force [2, 4, 5]. The second term of Eq. (6.11) is known as the Born-Oppenheimer force. Those two terms reflect different back-actions that the quantum subsystem exerts on the classical subsystem. The former is typically much smaller than the latter and is often ignored. However, in various magnetic systems, the Born-Oppenheimer force vanishes whereas the geometric force plays a crucial role.

What has been demonstrated above is the simplest case of the adiabatic dynamics. If the Hamiltonian depends on time both through $\boldsymbol{r}(t)$ and explicitly, $H = H(t, \boldsymbol{r}(t))$, the eigenstates and eigenvalues are $|\psi_j(t, \boldsymbol{r}(t))\rangle$ and $E_j(t, \boldsymbol{r}(t))$, respectively. In this case, besides the vector potential defined in Eq. (6.5), we can also define a scalar potential

$$\phi_j(t, \boldsymbol{r}) \equiv i\langle \psi_j(t, \boldsymbol{r}) | \partial_t \psi_j(t, \boldsymbol{r}) \rangle, \tag{6.12}$$

which gives rise to an electric component to the Berry curvature along with the magnetic component,

$$\boldsymbol{\mathcal{E}}(t, \boldsymbol{r}) = \partial_t \boldsymbol{\mathcal{A}}(t, \boldsymbol{r}) - \boldsymbol{\nabla}\phi(t, \boldsymbol{r}), \tag{6.13}$$

$$\boldsymbol{\mathcal{B}}(t, \boldsymbol{r}) = \boldsymbol{\nabla} \times \boldsymbol{\mathcal{A}}(t, \boldsymbol{r}), \tag{6.14}$$

where $\phi(t,\boldsymbol{r}) = \sum_j I_j \phi_j(t,\boldsymbol{r})$ and $\boldsymbol{\mathcal{A}}(t,\boldsymbol{r}) = \sum_j I_j \boldsymbol{\mathcal{A}}_j(t,\boldsymbol{r})$. Then Eq. (6.11) becomes

$$M\ddot{\boldsymbol{r}} = \boldsymbol{\mathcal{E}}(t,\boldsymbol{r}) + \dot{\boldsymbol{r}} \times \boldsymbol{\mathcal{B}}(t,\boldsymbol{r}) - \boldsymbol{\nabla} E(t,\boldsymbol{r}). \tag{6.15}$$

As a matter of fact, the electric and magnetic fields can be recast in a unified description by virtue of the joint spacetime coordinate $r_\mu \equiv (t,\boldsymbol{r})$ with $\mu = 0,1,2,3,\cdots$. Regarding the effective speed of light as 1, we can group the scalar and vector potentials in a single gauge potential as $\mathcal{A}_\mu \equiv (\phi, \boldsymbol{\mathcal{A}})$. This allows us to define the electromagnetic field

$$\mathcal{F}_{\mu\nu} = \partial_\mu \mathcal{A}_\nu - \partial_\nu \mathcal{A}_\mu, \tag{6.16}$$

which is an antisymmetric tensor $\mathcal{F}_{\mu\nu} = -\mathcal{F}_{\nu\mu}$.

The appearance of artificial electromagnetic fields in the parameter space can be easily understood by gauge invariance. In the adiabatic approximation, the system does not make transitions between different energy levels, thus the dynamics of the quantum subsystem refers only to the phase evolution of each individual eigenstates. As a result, an arbitrary *local* phase change $|\psi_j\rangle \to e^{i\chi_j(r_\mu)}|\psi_j\rangle$ yields the Lagrangian

$$L = -I_j[\dot{\theta}_j + E_j] + \dot{r}_\mu \mathcal{A}_\mu \tag{6.17}$$

invariant, where repeated indices are summed. This is known as the induced $U(1)$ symmetry by the adiabatic approximation. It guarantees the appearance of the artificial electromagnetic field $\mathcal{F}_{\mu\nu}$ in the dynamics of slow variables, in exactly the same sense as a charged particle that couples to a real electromagnetic field. The Bianchi identity of the antisymmetric tensor $\mathcal{F}_{\mu\nu}$ requires $\varepsilon_{\alpha\lambda\mu\nu}\partial_\lambda \mathcal{F}_{\mu\nu} = 0$ for $\forall\alpha$. This translates directly into Faraday's law

$$\boldsymbol{\nabla} \times \boldsymbol{\mathcal{E}} + \partial_t \boldsymbol{\mathcal{B}} = 0, \tag{6.18}$$

which has important consequences.

A famous example is an electron moving in a magnetic texture that varies slowly in space and time [6, 7, 8, 9, 10, 11, 12, 13]. With a sufficiently strong exchange coupling, the electron spin is locked to the direction of the local magnetization $\boldsymbol{m}(t,\boldsymbol{r})$. The precession of the electron spin then generates a geometric phase in spacetime, producing an effective electromagnetic field that depends on the gradient and the time-derivative of the magnetization

$$\boldsymbol{\mathcal{E}} = \frac{1}{2}\varepsilon_{ijk} m_i \partial_t m_j \boldsymbol{\nabla} m_k = \frac{1}{2}\sin\alpha(\partial_t\alpha\boldsymbol{\nabla}\beta - \boldsymbol{\nabla}\alpha\partial_t\beta), \tag{6.19}$$

$$\boldsymbol{\mathcal{B}} = -\frac{1}{4}\varepsilon_{ijk} m_i \boldsymbol{\nabla} m_j \times \boldsymbol{\nabla} m_k = -\frac{1}{2}\sin\alpha\boldsymbol{\nabla}\alpha \times \boldsymbol{\nabla}\beta, \tag{6.20}$$

where $\alpha(t, \boldsymbol{r})$ and $\beta(t, \boldsymbol{r})$ are the two spherical angles specifying the direction of $\boldsymbol{m}$:

$$\boldsymbol{m} = \{\sin\alpha\cos\beta, \sin\alpha\sin\beta, \cos\alpha\}. \tag{6.21}$$

In ferromagnets with slowly-varying texture, Eq. (6.19) and (6.20) are responsible for the spin-motive force [7, 8, 9] and the topological Hall effect [10, 11, 12, 13], respectively. For example, when a magnetic domain wall precesses under an external field, the effective $\boldsymbol{\mathcal{E}}$ field generates a voltage across the wall that can be utilized as a battery by external circuits. This process converts the magnetic energy stored in the domain wall into electrical power. Moreover, in a two dimensional magnetic skyrmion lattice, stationary swirls lead to an out-of-plane $\boldsymbol{\mathcal{B}}$ field that curves the electron orbits, by which an additional Hall voltage is detectable on top of the anomalous Hall signal. Besides magnetic systems, recent investigations on quantum plasma have also found similar terms in a generalized magnetohydrodynamics where electron spins adiabatically adjust to the vectorial mean field [14, 15].

6.2 Non-Abelian generalization

It often happens that energy levels are degenerate. When a group of degenerate eigenstates is far from other states in the spectrum, the system will remain in the subspace of the degenerate states if it is initially prepared so. A cyclic motion of the slow variables, then, causes a unitary rotation within the degenerate subspace [16, 17], which marks a generalization of the geometric Berry phase.

Suppose the Hamiltonian is parameterized by the general spacetime coordinate $H = H(r_\mu)$. We decompose the wave vector in terms of the degenerate eigenstates as

$$|\psi(r_\mu)\rangle = \sum_{i=1}^{D} c_i |\psi_i(r_\mu)\rangle, \tag{6.22}$$

where D is the degree of degeneracy. Defining $\eta = [c_1, c_2, \cdots, c_D]^T$ as the column vector collecting the decomposition coefficients within the degenerate subspace, we obtain the effective Lagrangian as

$$L = i\eta^\dagger\dot{\eta} + \dot{r}_\mu[\eta^\dagger\mathcal{A}_\mu\eta] - E(r_\mu), \tag{6.23}$$

where $\dot{r}_\mu$ is understood as $(1, \dot{\boldsymbol{r}})$, and the Berry connection $\mathcal{A}_\mu$ is now a $D \times D$ matrix with the element

$$[\mathcal{A}_\mu]_{\alpha\beta} = i\langle\psi_\alpha|\partial_\mu|\psi_\beta\rangle, \quad \alpha, \beta \in 1, 2, \cdots, D. \tag{6.24}$$

The system dynamics involves a set of coupled equations of motion for both the internal variable η and the external variable r_μ, which correspond to the

Euler-Lagrange equations:

$$i\dot{\eta} = \dot{r}_\mu \eta^\dagger \mathcal{A}_\mu, \tag{6.25}$$

$$M\ddot{r}_\mu = \dot{r}_\nu[\eta^\dagger \mathcal{F}_{\mu\nu}\eta] - \partial_\mu E, \tag{6.26}$$

where the gauge field $\mathcal{F}_{\mu\nu}$ is also a $D \times D$ matrix. It is defined as

$$\mathcal{F}_{\mu\nu} = \partial_\mu \mathcal{A}_\nu - \partial_\nu \mathcal{A}_\mu - i[\mathcal{A}_\mu, \mathcal{A}_\nu], \tag{6.27}$$

where the commutation term $[\mathcal{A}_\mu, \mathcal{A}_\nu]$ is in general non-zero. It is this term that exhibits the non-Abelian feature, which literally refers to the inequality $\mathcal{A}_\mu \mathcal{A}_\nu \neq \mathcal{A}_\nu \mathcal{A}_\mu$.

It worths emphasizing that two important properties can distinguish the non-Abelian gauge field from its Abelian counterpart. First, even when the (matrix) gauge potential $\mathcal{A}_\mu$ is homogeneous in spacetime, we could get a non-zero gauge field according to Eq. (6.27). Second, although the effective Lagrangian Eq. (6.23) is invariant under an arbitrary $U(D)$ transformation in the degenerate subspace: $|\psi_j\rangle \to |\psi'_j\rangle = U_{jk}|\psi_k\rangle$ with U_{jk} the element of a unitary matrix, the gauge field itself is not invariant under the same gauge transformation. Instead, it is gauge covariant:

$$\mathcal{F}_{\mu\nu} \to \mathcal{F}'_{\mu\nu} = U^\dagger \mathcal{F}_{\mu\nu} U. \tag{6.28}$$

At the same time, the column vector η undergoes a unitary transformation that exactly compensates the change of $\mathcal{F}_{\mu\nu}$ so that the combination $[\eta^\dagger \mathcal{F}_{\mu\nu}\eta]$ appearing in the equation of motion (6.26) is gauge invariant. The first term on the right hand side of Eq. (6.26) is known as Wong's force [18], which is the non-Abelian version of the Lorentz-like geometric force. What makes it unique is the internal dynamics determined by Eq. (6.25). It cannot be separated from the external dynamics—they always have to be solved together.

The non-Abelian gauge field has aroused renewed interests in both magnetic materials and cold atoms. In antiferromagnetic metals with slowly-varying texture, the conduction band is doubly degenerate, in which the two sub-bands have opposite spins with different spatial Bloch waves [19]. The internal dynamics between the two sub-bands then leads to spin misalignment with the background Néel order parameter even in the adiabatic limit. A direct consequence is that a known spacetime dependence of the background Néel order is insufficient to fix the artificial Lorentz force, one has to tell its instantaneous projections on the two sub-bands, which tracks back to the entire history of the electron orbit.

6.3 Quantum theory

In the above, the slow variables have been treated as purely classical, but a full quantum description of the slow dynamics is available in the Born-Oppenheimer formalism [1, 3]. Consider a system with separable slow and fast degrees of freedom. Denote $\boldsymbol{r}$ and $\boldsymbol{p}$ as the position and momentum operators for the fast variables; $\boldsymbol{R}$ and $\boldsymbol{P}$ as operators for the slow variables, respectively. The quantum Hamiltonian of the entire hybrid system is

$$H = \frac{\boldsymbol{P}^2}{2M} + \frac{\boldsymbol{p}^2}{2m} + V(\boldsymbol{R}, \boldsymbol{r}) = \frac{\boldsymbol{P}^2}{2M} + H_f(\boldsymbol{R}), \tag{6.29}$$

where $H_f(\boldsymbol{R})$ is the Hamiltonian of the fast degrees of freedom that depend parametrically on the position of the slow variable $\boldsymbol{R}$. Assume that $H_f(\boldsymbol{R})$ determines N eigenstates $|\phi_j(\boldsymbol{R})\rangle$ $(j = 1, 2, \cdots, N)$ with corresponding eigenvalues $E_j(\boldsymbol{R})$, then in the position representation we can expand the system wave function as

$$\Psi(\boldsymbol{R}, \boldsymbol{r}) = \sum_{j=1}^{N} \chi_j(\boldsymbol{R}) \phi_j(\boldsymbol{r}, \boldsymbol{R}), \tag{6.30}$$

where $\phi_j(\boldsymbol{r}, \boldsymbol{R}) = \langle \boldsymbol{r} | \phi_j(\boldsymbol{R}) \rangle$, and $\chi_j(\boldsymbol{R})$ is the wave function of the slow subsystem when the fast one is in state $|\phi_j(\boldsymbol{R})\rangle$. Eq. (6.30) is known as the Born-Oppenheimer decomposition [20].

If some of those eigenstates are degenerate, we can formulate an effective description of the dynamics within the degenerate subspace. Without loss of generality, let us assume that the first D states are degenerate with energy $E(\boldsymbol{R})$, the stationary Schrödinger equation reads

$$-\frac{1}{2M} \frac{\partial^2}{\partial \boldsymbol{R}^2} \Psi + E(\boldsymbol{R}) \Psi = \varepsilon \Psi. \tag{6.31}$$

In the adiabatic limit, we ignore transitions to any state outside the degenerate group, which allows us to project Eq. (6.31) onto the degenerate subspace. Defining the wave vector

$$\tilde{\psi} \equiv (\chi_1, \chi_2, \cdots, \chi_D)^T, \tag{6.32}$$

and regarding the orthogonal-normal condition $\langle \phi_i(\boldsymbol{R}) | \phi_j(\boldsymbol{R}) \rangle = \delta_{ij}$, we obtain the projected Schrödinger equation as

$$i \frac{\partial \tilde{\psi}}{\partial t} = \left[\frac{1}{2M} (\boldsymbol{P}\hat{1} - \boldsymbol{A}(\boldsymbol{R}))^2 + E(\boldsymbol{R})\hat{1} + W(\boldsymbol{R}) \right] \tilde{\psi}, \tag{6.33}$$

where $\boldsymbol{P} = -i\frac{\partial}{\partial \boldsymbol{R}}$, and $\hat{1}$ is the D dimensional identity matrix. The induced vector potential $\boldsymbol{A}(\boldsymbol{R})$ is a $D \times D$ matrix with its element

$$[\boldsymbol{A}(\boldsymbol{R})]_{mn} = i\langle \phi_m(\boldsymbol{R}) | \frac{\partial}{\partial \boldsymbol{R}} \phi_n(\boldsymbol{R}) \rangle = i \int d^3r \; \phi_m^*(\boldsymbol{r}, \boldsymbol{R}) \frac{\partial}{\partial \boldsymbol{R}} \phi_n(\boldsymbol{r}, \boldsymbol{R}), \tag{6.34}$$

which is the same thing that Eq. (6.24) has defined. What is new to the quantum theory lies in the induced scalar potential $W(\boldsymbol{R})$, which is a $D \times D$

matrix of scalar elements

$$[W(\boldsymbol{R})]_{mn} = \frac{1}{2M} \sum_{l=D+1}^{N} [\boldsymbol{A}(\boldsymbol{R})]_{ml} \cdot [\boldsymbol{A}(\boldsymbol{R})]_{ln}, \quad m, n \in 1, 2, \cdots, D. \tag{6.35}$$

This scalar potential captures the correction to the adiabatic result from *virtue* quantum transitions that connect the degenerate states and states outside the degenerate subspace. It seems quite surprising that transitions we have neglected at the very beginning should appear in such a subtle way in the final result. In particular, when $D = 1$ (i.e., the state we care is non-degenerate), the scalar potential $W(\boldsymbol{R})$ is just the trace of the quantum geometric tensor [21].

6.4 Realization in cold atoms

The artificial gauge fields are not restricted to magnetic systems, they are applicable to general cases whenever the internal state of a particle follows the background order parameter adiabatically. For example, in cold atomic gases where the spin (hyperfine structure) of neutral particles adjusts to a prescribed background spacetime, (Abelian) electromagnetic fields are induced in a similar manner as the adiabatic electrons in ferromagnetic textures. This method provides a plausible way to synthesize electromagnetic fields for charge neutral atoms that would otherwise decouple from real electromagnetic fields [22, 23, 24]. In Ref.[25], an artificial magnetic field is generated for a ^{87}Rb Bose-Einstein condensate (BEC). Two perpendicular Raman lasers are used to couple the three spin states of Rb's electronic state $^5S_{1/2}$, F $=$ 1, which are Zeeman shifted by a real magnetic field with a constant spatial gradient. The strength of the artificial field can be tuned by changing the gradient of the applied field. Vortices were generated and observed in the BEC when the artificial magnetic field was tuned strong enough.

The non-Abelian property of the artificial gauge field, namely, the coupling between internal and external degrees of freedom, can be harnessed to invent controllable spin-orbit coupling in atoms free of real spin-orbit interactions. The key issue of generating non-Abelian gauge structure in ultracold atoms is how to prepare reliable degenerate states [26, 27, 28]. One proposal is to use atoms with a multi-pod electronic energy levels where there is one excited energy state and a group of $N(\geq 3)$ ground states [29]. The excited state is coupled to the N ground states with N laser beams. This scheme can create N degenerate dark states and, therefore, an artificial non-Abelian gauge field with $U(3)$ symmetry. However, the dark states are not the lowest dressed states and are thus unstable. Therefore, it is quite challenging to realize this proposal in experiments.

There have been many other proposals on how to generate effective spin-orbit coupling for charge-neutral atoms (see Ref. [1] and references therein). What worths mentioning is that there is no need to have absolute degeneracy.

It is sufficient in practice to have quasi-degenerate states, where two or more energy levels are close to each other while far away from other states. Liu *et al.* noticed this fact and proposed a much simpler scheme to generate non-Abelian gauge field (or spin-orbit coupling) [27]. The idea was adopted and realized experimentally by the Maryland group [30] (with a similar set-up as used in Ref. [25]). In this experiment, a quasi-degeneracy for the two lowest energy states was created by tuning the laser frequency and the Zeeman shift. The artificial spin-orbit coupling realized in this experiment is one-dimensional. But a two dimensional spin-orbit coupling has also been realized recently [31]. All these experimental and theoretical efforts have not only put geometric phase in a broader perspective, and more importantly have given us a brand new system to play–bosons with spin-orbit coupling. Recalling how much wonderful physics we have learned by playing liquid helium, a boson system with no spin, we are confident that much more exciting physics awaits to be discovered by exploring bosons with spin-orbit coupling. In fact, many interesting phenomena have already been predicted [32, 33].

Chapter 6

References

1. J. Dalibard and F. Gerbier, G. Juzeliūnas, and P. Öhberg, Rev. Mod. Phys. **83**, 1523 (2011).
2. Q. Zhang and B. Wu, Phys. Rev. Lett. **97**, 190401 (2006). [Reprint (6.3)]
3. A. Messiah, *Quantum Mechanics Vol. I and II*, 1958.
4. Y. Aharonov and A. Stern, Phys. Rev. Lett. **69**, 3593 (1992). [Reprint (6.1)]
5. M. V. Berry and J. M. Robbins, Proc. R. Soc. Lond. A **442**, 641 (1993).
6. G. E. Volovik, J. Phys. C **20**, L83 (1987).
7. S. E. Barnes and S. Maekawa, Phys. Rev. Lett. **98**, 246601 (2007). [Reprint (6.4)]
8. S. A. Yang *et al.*, Phys. Rev. Lett. **102**, 067201 (2009). [Reprint (6.5)]; S. A. Yang *et al.*, Phys. Rev. B **82**, 054410 (2010).
9. Y. Yamane, Phys. Rev. Lett. **107**, 236602 (2011).
10. M. Lee, W. Kang, Y. Onose, Y. Tokura, and N. P. Ong, Phys. Rev. Lett. **102**, 186601 (2009).
11. A. Neubauer *et al.*, Phys. Rev. Lett. **102**, 186602 (2009).
12. P. Bruno, V. K. Dugaev, and M. Taillefumier, Phys. Rev. Lett. **93**, 096806 (2004). [Reprint (6.2)]
13. J. Ye, Y. B. Kim, A. J. Millis, B. I. Shraiman, P. Majumdar, and Z. Tešanović, Phys. Rev. Lett. **83**, 3737 (1999).
14. S. M. Mahajan and F. A. Asenjo, Phys. Rev. Lett. **107**, 195003 (2011).
15. S. Braun, F. A. Asenjo, and S. M. Mahajan, Phys. Rev. Lett. **109**, 175003 (2012).
16. F. Wilczek and A. Zee, Phys. Rev. Lett. **52**, 2111 (1984).
17. J. Moody, A. Shapere, and F. Wilczek, Phys. Rev. Lett. **56**, 893 (1986).
18. S. K. Wong, IL Nuovo Cimento **A 65**, 689 (1970); Int. J. Theor. Phys. **5**, 221 (1972).
19. R. Cheng and Q. Niu, Phys. Rev. B **86**, 245118 (2012). [Reprint (6.6)]; *ibid*, **89**, 081105(R) (2014).
20. C. -P. Sun and M. -L. Ge, Phys. Rev. D **41**, 1349 (1990).
21. J. P. Provost and G. Vallee, Commun. Math. Phys. **76**, 289 (1980).
22. A. Lamacraft, Phys. Rev. A **77**, 063622 (2008).
23. R. A. Duine and H. T. C. Stoof, Phys. Rev. Lett. **103**, 170401 (2009).
24. G. Juzeliūnas, J. Ruseckas, P. Öhberg, and M. Fleischhauer, Phys. Rev. A **73**, 025602 (2006). [Reprint (6.8)]

25. Y. -J. Lin, R. L. Compton, A. R. Perry, W. D. Phillips, J. V. Porto, and I. B. Spielman, Phys. Rev. Lett. **102**, 130401 (2009). [Reprint (6.10)]
26. J. Ruseckas, G. Juzeliūnas, P. Öhberg, and M. Fleischhauer, Phys. Rev. Lett. **95**, 010404 (2005).
27. X. -J. Liu, M. F. Borunda, X. Liu, and J. Sinova, Phys. Rev. Lett. **102**, 046402 (2009). [Reprint (6.9)]
28. V. Pietilä and M. Möttönen, Phys. Rev. Lett. **102**, 080403 (2009).
29. G. Juzeliuńas, J. Ruseckas, and J. Dalibard, Phys. Rev. A **81**, 053403 (2010).
30. Y.-J. Lin, K. Jiménez-García, and I. B. Spielman, Nature **471**, 83 (2011).
31. Z. Wu, L. Zhang, W. Sun, X.-T. Xu, B.-Z. Wang, S.-C. Ji, Y. Deng, S. Chen, X.-J. Liu, and J.-W. Pan, Science **354**, 83 (2016).
32. C. Wu, I. Mondragon-Shem, and X. Zhou, Chin. Phys. Lett. **28**, 097102 (2011).
33. Q. Zhu, C. Zhang, and B. Wu, Europhys. Lett. **100**, 50003 (2012).

7

Non-canonical phase space and quantization

One of the main themes threading through previous chapters is the theory of semiclassical dynamics. In standard condensed matter textbooks, it is used, for example, to study the trajectories of cyclotron orbits on a Fermi surface. However, in order to explain the de Haas-van Alphen oscillation, the cyclotron orbit needs be quantized. Such a re-quantization (of semiclassical dynamics) can be achieved with the Bohr-Sommerfeld quantization (see Chapter 2). In this chapter, we investigate this issue of re-quantization further using an alternative tool: the canonical quantization.

In the beginning, we give a brief review of the fundamentals of classical mechanics. A dynamical system can be described with at least three layers of formulation: the Newtonian theory, the Lagrangian theory, and the Hamiltonian theory [1]. In the first layer, we have the Newtonian equation of motion (EOM) based on experimental observation. For example, for a charged particle in an electromagnetic field, one has

$$m\frac{d^2\mathbf{r}}{dt^2} = -\frac{\partial V}{\partial \mathbf{r}} + \frac{d}{dt}\frac{\partial V}{\partial \dot{\mathbf{r}}}, \tag{7.1}$$

in which $V = q\phi(\mathbf{r}) - q\dot{\mathbf{r}}\cdot\mathbf{A}(\mathbf{r})$, and ϕ and $\mathbf{A}$ are the electromagnetic potentials.

In the second layer, the EOM can be derived from the Euler-Lagrange equation. The Lagrangian needs be chosen in such a way that the Euler-Lagrange equation produces the correct Newtonian EOM. For the example of a charged particle in an electromagnetic field, the Lagrangian is,

$$L(\mathbf{r}, \dot{\mathbf{r}}) = \frac{m}{2}\dot{\mathbf{r}}^2 - q\phi + q\dot{\mathbf{r}}\cdot\mathbf{A}. \tag{7.2}$$

In the third layer, instead of the Lagrangian, a dynamical system is described by the Hamiltonian,

$$H(\mathbf{r}, \mathbf{p}) = \frac{1}{2m}(\mathbf{p} - q\mathbf{A}(\mathbf{r}))^2 + q\phi(\mathbf{r}), \tag{7.3}$$

which is derived from the Lagrangian using the Legendre transformation. The dynamical variables are now $\mathbf{r}$ and momentum $\mathbf{p}$, instead of $\mathbf{r}$ and velocity $\dot{\mathbf{r}}$. The EOMs are given by Hamilton's equations of motion,

$$\dot{\mathbf{r}} = \frac{\partial H}{\partial \mathbf{p}}; \quad \dot{\mathbf{p}} = -\frac{\partial H}{\partial \mathbf{r}}. \tag{7.4}$$

It determines the trajectory of a particle $(\mathbf{r}(t), \mathbf{p}(t))$ in phase phase.

In the symplectic formulation of Hamilton's dynamics, the Poisson bracket is introduced,

$$\{f, g\} = \sum_i \left(\frac{\partial f}{\partial r_i}\frac{\partial g}{\partial p_i} - \frac{\partial f}{\partial p_i}\frac{\partial g}{\partial r_i} \right). \tag{7.5}$$

For example, the Poisson brackets of the phase-space variables are,

$$\begin{aligned} \{r_i, r_j\} &= 0, \\ \{p_i, p_j\} &= 0, \\ \{r_i, p_j\} &= \delta_{ij}. \end{aligned} \tag{7.6}$$

This is similar to the commutation relations in quantum mechanics. The conjugate pair $(\mathbf{r}, \mathbf{p})$ of variables with such a canonical form of Poisson brackets are called the canonical variables.

Furthermore, if the canonical variables $\mathbf{r}$ and $\mathbf{p}$ are considered as parts of a generalized variable $\boldsymbol{\xi} = (\mathbf{r}, \mathbf{p})^T$. Then the Hamilton's EOM can be written as a single equation,

$$\dot{\boldsymbol{\xi}} = \{\boldsymbol{\xi}, H\}. \tag{7.7}$$

This is similar in form to the Heisenberg's equation of motion in quantum mechanics.

7.1 Physical variables in semiclassical theory

The classical theory described by canonical variables can be upgraded to its corresponding quantum theory directly. The only step one needs to take is to replace the Poisson bracket by the commutator,

$$\{f, g\} \to \frac{[f, g]}{i\hbar}, \tag{7.8}$$

so that

$$\begin{aligned} [r_i, r_j] &= 0, \\ [p_i, p_j] &= 0, \\ [r_i, p_j] &= i\hbar\delta_{ij}. \end{aligned} \tag{7.9}$$

In Ref. [2], it was found that the Poisson brackets of the dynamical variables in the semiclassical theory of wavepacket are,

$$\begin{aligned}\{r_i, r_j\} &= \epsilon_{ijk}\Omega_k/\kappa,\\ \{k_i, k_j\} &= -\epsilon_{ijk}eB_k/\kappa,\\ \{r_i, k_j\} &= (\delta_{ij} + eB_i\Omega_j)/\kappa,\end{aligned} \tag{7.10}$$

in which $\mathbf{B}$ is the magnetic field, $\boldsymbol{\Omega}$ is the Berry curvature, and $\kappa \equiv 1+e\mathbf{B}(\mathbf{r})\cdot\boldsymbol{\Omega}(\mathbf{k})$. This obviously is not of the standard form in Eq. (7.6). Therefore, these variables are not canonical variables.

Having non-canonical variables poses no particular problem, since we still have the correct equations of motion. Still, the structure of the phase space now becomes non-standard. For example, the Liouville theorem — the conservation of a volume element $dV = d^Drd^Dk$ in phase space during dynamical evolution — is no longer valid. Nevertheless, it is possible to introduce a modified density of states in the phase space [2],

$$D(\mathbf{r}, \mathbf{k}) = (2\pi)^{-D}(1 + e\mathbf{B}\cdot\boldsymbol{\Omega}), \tag{7.11}$$

such that $D(\mathbf{r}, \mathbf{k})dV$ remains conserved during dynamical evolution.

Because the volume with modified density of states can still be conserved, it appears that, with proper re-scaling of the variables, the original Liouville theorem can still be valid [3]. If so, then the re-scaled variables would be canonical variables. Indeed, the Darboux theorem guarantees the existence of such canonical variables [4]. Once found, then the dynamical system can be quantized with the substitution mentioned in Eq. (7.8). However, exact canonical variables may not be easy to find, and approximation is often required.

7.2 Canonical variables in semiclassical theory

We use a simple model to illustrate the relation between canonical and non-canonical variables in semiclassical theory. The Rice-Mele model describes a one-dimensional dimerized lattice with staggered on-site energies $\pm\Delta$ and alternating bond strengths $t \pm \delta$. Even though it has no Berry curvature Ω_{kk} in momentum space, its mixed Berry curvature $\Omega_{kr} = \partial_k A_r - \partial_r A_k$ can be non-zero if the dimerization parameter $\delta(r)$ is allowed to have a non-uniform spatial profile (see Ref. [5]).

Because of the non-zero Ω_{kr}, the EOM of a wavepacket with energy E is no longer of the canonical form. Instead,

$$\begin{aligned}\dot{r} &= \frac{1}{1+\Omega_{kr}}\frac{\partial E}{\partial k},\\ \dot{k} &= -\frac{1}{1+\Omega_{kr}}\frac{\partial E}{\partial r},\end{aligned} \tag{7.12}$$

and r and k are not canonical variables.

The Lagrangian for a wavepacket in the Rice-Mele model is,

$$L = k\dot{r} + \dot{k}A_k + \dot{r}A_r - E(r,k). \tag{7.13}$$

The Lagrangian and the EOM are closely related. Therefore, if one can find new variables x and q, such that the Lagrangian appears in the standard form,

$$L_0 = q\dot{x} - E(x,q), \tag{7.14}$$

then the EOM would automatically be those in Eq. (7.4), and the variables x and q would naturally be canonical variables.

It is not difficult to see that, if

$$\begin{aligned} q &= k + A_r, \\ x &= r - A_k, \end{aligned} \tag{7.15}$$

then

$$L = L_0 + \frac{d}{dt}(kA_k) + O(A^2). \tag{7.16}$$

The total time derivative can be ignored since it would not change the EOM. If the higher order term $O(A^2)$ can be neglected, then the Lagrangian L is of the standard form L_0.

To re-quantize the semiclassical version of the Rice-Mele model, we need to impose the commutation relation on the canonical variables,

$$[x,q] = i\hbar. \tag{7.17}$$

Then, accurate to linear order of $A_{x,q}$, we substitute the r and k in E with q and x using

$$\begin{aligned} k &= q - A_x, \\ r &= x + A_q. \end{aligned} \tag{7.18}$$

The effective quantum Hamiltonian would then be

$$H(x,q) = E(x + A_q, q - A_x)\,. \tag{7.19}$$

7.3 Generalized Peierls substitution

We now consider realistic solid state systems. For a wavepacket moving under the influence of momentum-space Berry curvature and electromagnetic field, the relations between non-canonical ($\mathbf{r}_c$ and $\mathbf{k}_c$) and canonical ($\mathbf{r}$ and $\mathbf{p}$) variables are found to be [6],

$$\begin{aligned} \mathbf{r}_c &= \mathbf{r} + \mathbf{R}(\boldsymbol{\pi}), \\ \hbar\mathbf{k}_c &= \boldsymbol{\pi} + e\mathbf{B} \times \mathbf{R}(\boldsymbol{\pi}), \end{aligned} \tag{7.20}$$

in which $\mathbf{R}$ is the Berry connection, $\mathbf{A}$ is the electromegnetic vector potential, and $\boldsymbol{\pi} = \mathbf{p} + e\mathbf{A}(\mathbf{r})$ is the covariant momentum. Such a formulation is accurate to linear order of the electromagnetic field.

From Eq. (7.11), we see that a phase space has non-canonical structure only if $\mathbf{B} \cdot \boldsymbol{\Omega}$ is nonzero. For a system with trivial gauge structure, $\mathbf{R}$ is a pure gauge and can be set as zero. Consequently, Eqs. (7.20) reduces to the Peierls substitution, $\hbar \mathbf{k}_c = \mathbf{p} + e\mathbf{A}(\mathbf{r})$: Once the Bloch energy $E(\mathbf{k})$ is known, the system can be re-quantized by replacing $\hbar\mathbf{k}$ with the operator $-i\hbar\partial/\partial\mathbf{r} + e\mathbf{A}$ [7]. This is the well-known practice in studying, e.g., the Landau levels in a Bloch band. But one has to bear in mind that, if the Bloch states have non-zero Berry curvature, then the usual Peierls substitution is incomplete for the quantization.

7.4 Semiclassical Dirac theory

We use the Dirac theory of relativistic electrons to illustrate the process of re-quantization. A Dirac electron has positive and negative branches with energies $E_0^{\pm}(q) = \pm\sqrt{c^2\hbar^2 q^2 + m^2c^4}$. A Dirac wavepacket constructed from positive energy states has non-trivial gauge structure. Its Berry connection and self-spinning angular momentum are found to be [6]

$$\mathbf{R} = \frac{\lambda_c^2}{2\gamma(\gamma+1)}\mathbf{q} \times \langle\boldsymbol{\sigma}\rangle, \tag{7.21}$$

$$\mathbf{L} = \frac{\hbar}{\gamma^2}\left(\langle\boldsymbol{\sigma}\rangle + \lambda_c^2 \frac{\mathbf{q}\cdot\langle\boldsymbol{\sigma}\rangle}{\gamma+1}\mathbf{q}\right), \tag{7.22}$$

where $\lambda_c = \hbar/mc$ is the Compton wavelength, and $\gamma(q) = \sqrt{1 + (\hbar q/mc)^2}$ is the dilation factor.

The energy of a wavepacket moving under the influence of external fields is,

$$E(\mathbf{r}_c, \mathbf{k}_c) = E_0(\mathbf{r}_c) - e\phi(\mathbf{r}_c) + \frac{e}{2m}\mathbf{L}(\mathbf{k}_c)\cdot\mathbf{B}. \tag{7.23}$$

To get the quantum Hamiltonian H, one first replaces the variables using Eq. (7.20). After a lengthy calculation, it is possible to show that [6],

$$E(\mathbf{r}, \mathbf{p}) = \gamma(\boldsymbol{\pi})mc^2 - e\phi(\mathbf{r}) \tag{7.24}$$

$$+ \frac{\mu_B}{\gamma}\langle\boldsymbol{\sigma}\rangle\cdot\mathbf{B} + \frac{\mu_B}{\gamma(\gamma+1)}\frac{\boldsymbol{\pi}}{mc^2} \times \langle\boldsymbol{\sigma}\rangle\cdot\mathbf{E}. \tag{7.25}$$

Afterwards, the classical canonical variables $\mathbf{r}$ and $\mathbf{p}$ are promoted to quantum canonical variables, and $\langle\boldsymbol{\sigma}\rangle$ to Pauli matrices. The result is the effective quantum Hamiltonian for an electron in the positive-energy branch [8]. At low velocity, it reduces to the familiar Pauli Hamiltonian.

7.5 Identifying physical variables

Both in the semiclassical theory and in the quantum theory, we have encountered two sets of variables describing the same physical entity. The first is non-canonical and cannot be quantized directly, but has clear physical interpretations (e.g., the center of mass of a wavepacket). The second set is canonical and can be quantized directly, but seems to have no clear interpretations. Indeed, canonical variables such as $\mathbf{r} = i\hbar\partial/\partial\mathbf{k}$ in Eq. (7.20) is gauge dependent in the quantum theory. Such a gauge dependence is required to cancel with the gauge dependence in $\mathbf{R}$ so that $\mathbf{r}_c$ can be gauge invariant. Because of its gauge dependence, canonical variables are generally non-physical. For example, the variable $\mathbf{r}$ in the Pauli Hamiltonian is not the physical position. It is a physical position only if one neglects the underlying gauge structure.

In semiclassical theory, one can have two sets of variables living in domains with different extensions. For example, in Eq. (7.20), one can treat $\mathbf{r}$ as a fundamental observable in the re-quantized Pauli theory. In the semi-classical version of the Pauli theory, $\mathbf{r}_a \equiv \langle\mathbf{r}\rangle$ would be the center of mass of a Pauli wavepacket. However, when viewing from a higher standing point, we know that $\mathbf{r}$ differs from the center of a Dirac wavepacket by a shift $\mathbf{R}$. Furthermore, since the shift itself is time-dependent, the two semi-classical theories would predict different particle trajectories. In Ref. [6], we have shown that the transverse velocities in the two theories differ by a factor of two [9]. Naturally, one expects the trajectory of the Dirac wavepacket to be more accurate.

7.6 Alternative theories of effective quantum Hamiltonian

We have derived the effective quantum Hamiltonian via the route of the wavepacket theory and canonical re-quantization. Essentially the same effective Hamiltonian can also be obtained by different approaches. A common feature of these methods is that the variables are manipulated at the operator level. Therefore, no re-quantization is required.

For example, the Pauli Hamiltonian can be obtained from the Dirac Hamiltonian via the Foldy-Wouthuysen transformation [10]. One first performs an unitary transformation to separate the free-particle Dirac Hamiltonian to positive-energy and negative-energy blocks. An external electromagnetic field would induce couplings between these two blocks, and further unitary transformation is performed to eliminate the off-diagonal matrix elements. The resulting positive-energy (negative-energy) block then describes the effective dynamics of an electron (a hole), which is the Pauli Hamiltonian.

W. Kohn applies similar transformation to solid state systems [7]. The unitary transformation is quite complicated since one has to deal with (essentially) infinite number of energy bands, and finding the unitary transformation to block-diagonalize the Hamiltonian is not an easy task. Such an approach had been simplified by various researchers, such as Blount and Roth, during the 60's [11, 12]. In recent years, there is renewed interest on this subject. Notable works have been done by Gosselin and Mohrbach, among others [13]. Even though higher order terms could be systematically extracted, the procedure remains complicated.

7.7 Higher order effect

So far our discussion of the Berry phase effect on electron dynamics has been limited to first orders. However, response functions such as electric polarizability, magnetic susceptibility, and magnetoelectric polarizability (which relates the change of electric polarization to an applied magnetic field, or the change of magnetization to an applied electric field) would require a theory that is accurate up to second order in external fields.

Let us first highlight the difficulties and challenges of developing a second order theory by considering the problem of inhomogeneity-induced polarization [14]. Suppose we have an insulating crystal with an order parameter $\boldsymbol{h}(\boldsymbol{r})$ that varies slowly in space, and we are interested in the polarization induced by $\boldsymbol{\nabla}\cdot\boldsymbol{h}(\boldsymbol{r})$. The strategy in Ref. [14] is to make use of the relation between electric polarization and adiabatic current, while keep terms up to first order of gradient of $\boldsymbol{h}(\boldsymbol{r})$. The Hamiltonian "felt" by a wavepacket located at $\boldsymbol{r}_c$, to first order, is

$$H = H_c + \nabla^r_\alpha h_\beta (r_\alpha - r_{c,\alpha})\frac{\partial H}{\partial h_\beta}\,, \tag{7.26}$$

where H_c is obtained from the full Hamiltonian by taking the value of $\boldsymbol{h}(\boldsymbol{r})$ at $\boldsymbol{r}_c$, and the second term is the gradient correction with $\nabla^r_\alpha \equiv \partial/\partial r_\alpha$ (summation over repeated indices is implied throughout our derivation). The eigenstates of H is denoted by $|\tilde{u}_{nk}\rangle = |u_{nk}\rangle + |\delta u_{nk}\rangle$, where $|\delta u_{nk}\rangle$ is proportional to the gradient.

In principle, one can then construct a wavepacket using $|\tilde{u}_{nk}\rangle$, derive the equation of motion, and calculate the induced polarization. The result be divided into two parts: a perturbative contribution stemming from a correction to the basis functions

$$P^{(1)}_\alpha = 2e\int \frac{d\boldsymbol{k}}{(2\pi)^d}\mathrm{Im}\langle \delta u_{nk}|\nabla^k_\alpha|u_{nk}\rangle\,, \tag{7.27}$$

and a topological contribution described in terms of the Chern-Simons form of the Berry gauge fields

$$P^{(2)}_\alpha = e\int \frac{d\boldsymbol{k}}{(2\pi)^d}(A^k_\alpha \nabla^r_\beta A^k_\beta + A^k_\beta \nabla^k_\alpha A^r_\beta + A^r_\beta \nabla^k_\beta A^k_\alpha)\,, \tag{7.28}$$

where $A^k_a = \langle u|i\nabla^k_\alpha|u\rangle$ is the Berry connection. Since Eq. (7.28) is already linear in ∇^r_α, the Berry connection can be evaluated using the zeroth order eigenfunction $|u_{nk}\rangle$. We recognize that the integrand in the above equation is of the Chern-Simons form. An important application of the above result is the magnetic-field-induced polarization. Essin *et al* [15] consider the case where the inhomogeneity is caused by the vector potential, and showed that in the presence of time-reversal symmetry, the magnetoelectric polarizability, defined by $\theta = (2\pi h/e^2)P/B$, is either 0 or π. When $\theta = \pi$, the corresponding insulator is known as a 3D Z_2 topological insulator (see Chapter 5).

However, even though we can symbolically write the eigenstates of H as $|\tilde{u}_{nk}\rangle = |u_{nk}\rangle + |\delta u_{nk}\rangle$, it is not clear how to actually obtain $|\delta u_{nk}\rangle$ because the gradient term is unbounded. Gao *et al* [16] consider the special case of uniform electromagnetic field. They found that, to first order, there is a positional shift of the wavepacket center due to the interband mixing induced by external electromagnetic fields. They also derived the semiclassical equations of motion to second order. Surprisingly, the form of the equations of motion is the same as before, provided that the wavepacket energy is derived up to the same order. This result, while justified the 'naive' approach of Ref. [14], also leads to many interesting predictions such as a nonlinear anomalous Hall effect. Presently, the higher order theory of semiclassical dynamics still remains an open problem waiting to be further explored.

Chapter 7

References

1. H. Goldstein, C.P. Poole Jr., and J.L. Safko, Classical Mechanics (3rd ed.), Addison-Wesley (2001).
2. D. Xiao, J. Shi, and Q. Niu, Phys. Rev. Lett. **95**, 137204 (2005).
3. C. Duval, Z. Horvath, P. A. Horvathy, L. Martina, P. Stichel, Mod.Phys.Lett. B20, 373 (2006).
4. V. Arnold, Mathematical Methods of Classical Mechanics, Springer, Berlin (1989).
5. D. Xiao, M.C. Chang, and Q. Niu, Rev. Mod. Phys. **82**, 1959 (2010).
6. M.C. Chang, and Q. Niu, J. Phys.: Condens. Matter **20**, 193202 (2008). [Reprint (7.5)]
7. Walter Kohn, Phys. Rev. **115**, 1460 (1959). [Reprint (7.2)]
8. Also see Eq.(33) in A.J. Silenko, J. Math. Phys. **44**, 2952 (2003).
9. Also see K.Y. Bliokh, Europhys. Lett. **72** 7 (2005).
10. L. Foldy and S.A. Wouthuysen, Phys. Rev. **78**, 29 (1950). [Reprint (7.1)]
11. E. I. Blount, Phys. Rev. **126**, 1636 (1962). [Reprint (7.3)]; E. I. Blount, *Solid State Physics*, edited by Seitz F and Turnbull D (Academic Press Inc., New York, 1962), Vol. 13
12. L.M. Roth, J. Phys. Chem. Solids, **23**, 433 (1962); L.M. Roth, Phys. Rev. **145**, 434 (1966).
13. P. Gosselin, F. Menas, A. Berard and H. Mohrbach, Europhys. Lett **76**, 651 (2006); P. Gosselin, A. Berard, and H. Mohrbach, Eur. Phys. J. **B 58**, 137 (2007). [Reprint (7.4)]; P. Gosselin and H. Mohrbach, arXiv:0806.1509.
14. D. Xiao, J. Shi, D. P. Clougherty, and Q. Niu, Phys. Rev. Lett. **102**, 087602 (2009). [Reprint (7.6)]
15. A. M Essin, J. E. Moore, and D. Vanderbilt, Phys. Rev. Lett. **102**, 146805 (2009).
16. Y. Gao, S. A. Yang, and Q. Niu, Phys. Rev. Lett. **112**, 166601 (2014). [Reprint (7.7)]

8

Nonlinear quantum systems

Shortly after Berry's seminal work in 1984 [1], there were various attempts to generalize his geometric phase. Wilczek and Zee generalized it to degenerate eigenstates [2] as already mentioned in early chapters. Hannay found its classical counterpart, which is now known as Hannay's angle [3]. The Aharonov-Anandan phase is a generalization of Berry phase to non-adiabatic evolutions [4]. In this chapter we discuss another generalization, geometric phases in nonlinear systems, where phase is an inherent part of their dynamics. We call them nonlinear quantum systems, which can arise in many situations, for example, the mean-field description of many-body quantum systems [5] and nonlinear optical waveguides [6]. For the former, the nonlinear geometric phase can be regarded as a mean-field approximation of its quantum many-body counterpart.

8.1 Mean-field theory of Bose-Einstein condensates and its classical Hamiltonian structure

Although our results are general, it is more instructive to focus on Bose-Einstein condensate (BEC) and use it to introduce nonlinear geometric phase. The mean-field Hamiltonian of a BEC is given by

$$H = \int d\mathbf{r}\Big\{\Psi^*(\mathbf{r},t)\big(-\frac{\hbar^2}{2m}\nabla^2\big)\Psi(\mathbf{r},t) + V(\mathbf{r})|\Psi(\mathbf{r},t)|^2 + \frac{c}{2}|\Psi(\mathbf{r},t)|^4\Big\}, \quad (8.1)$$

where $\Psi(\mathbf{r},t)$ is often called macroscopic wave function, m is the atomic mass, and $V(\mathbf{r})$ is the external potential. The phase of $\Psi(\mathbf{r},t)$ is an inherent part of the BEC's dynamics, which is governed by the nonlinear Gross-Pitaevskii (GP) equation,

$$i\hbar\frac{\partial}{\partial t}\Psi(\mathbf{r},t) = -\frac{\hbar^2}{2m}\nabla^2\Psi(\mathbf{r},t) + V(\mathbf{r})\Psi(\mathbf{r},t) + c|\Psi(\mathbf{r},t)|^2\Psi(\mathbf{r},t)\,. \quad (8.2)$$

As the GP equation reduces to the Schrödinger equation at $c = 0$, we shall compare our nonlinear geometric phase with Berry phase whenever possible.

Berry phase is defined only for the adiabatic evolution of an energy eigenstate. The validity of this definition lies in that there is no transition between instantaneous energy eigenstates during the adiabatic evolution. If the system starts in the nth eigenstate, it stays in this eigenstate. Mathematically, it is to say that the occupation probability at each eigenstate is an adiabatic invariant. This definition is also sufficient. For a general quantum state, with the superposition principle, it is clear how Berry phase enters into its adiabatic evolution. In contrast, in nonlinear quantum systems, there is no superposition principle and it is not clear what are the adiabatic invariants. These difficulties can be overcome by noticing that nonlinear quantum systems possess a classical Hamiltonian structure. This allows us to borrow the familiar concepts in classical adiabatic dynamics, for example, actions and Hannay's angle.

It was noticed long time ago that linear quantum systems have a classical Hamiltonian structure: one can construct a classical Hamiltonian and define a set of Poisson brackets, and finally derive the Schrödinger equation as a set of Hamiltonian equations of motion [7, 8]. Nonlinear quantum systems also possess classical Hamiltonian structure [8, 9]. We expand the state Ψ in a complete orthonormal basis $\phi_n(\mathbf{r})$

$$\Psi(\mathbf{r}, t) = \sum_n a_n(t)\phi_n(\mathbf{r}) \,. \tag{8.3}$$

The Hamiltonian in Eq.(8.1) becomes $H(a_n^*, a_n)$. After defining a set of Poisson brackets

$$\{a_n^*, a_m\} = \frac{i}{\hbar}\delta_{nm}, \qquad \{a_n, a_m\} = \{a_n^*, a_m^*\} = 0 \,, \tag{8.4}$$

we obtain the Hamiltonian equations of motion

$$\frac{da_n}{dt} = \{a_n, H\} = -\frac{i}{\hbar}\frac{\delta H}{\delta a_n^*} \,, \tag{8.5}$$

$$\frac{da_n^*}{dt} = \{a_n^*, H\} = \frac{i}{\hbar}\frac{\delta H}{\delta a_n} \,. \tag{8.6}$$

One can readily check that these equations of motion are identical to the GP equation in Eq.(8.2). This shows that the GP equation possesses a classical Hamiltonian structure; all the tools and knowledge in classical mechanics can now be borrowed and applied to the GP equation. Note that this quantum-classical relation is purely mathematical and is not the usual quantum-classical relation at the limit of $\hbar \to 0$.

As an example, we consider the non-interacting case $c = 0$, where the system becomes linear. We write $a_n = \sqrt{I_n}e^{i\theta_n}$. It is easy to show that

$$\{I_n, \theta_m\} = \frac{i}{\hbar}\delta_{nm} \,, \tag{8.7}$$

and other Poisson brackets are zero. So, $\{I_n, \theta_n\}$ are pairs of conjugate canonical valuables. If $\phi_n(\mathbf{r})$'s are chosen to be the energy eigenstates, we have $a_n(t) = e^{-iE_n t/\hbar}|a_n|$ with E_n being the corresponding eigen-energies. This means that $I_n = |a_n|^2$ is a constant of motion and an action while $\theta_n = -iE_n t/\hbar$ is its conjugate angle. Suppose that this system is changing slowly with an external parameter. According to the classical adiabatic theorem, the actions I_n are adiabatic invariants. Physically, it implies that the occupation probability at each eigenstate $\phi_n(\mathbf{r})$ does not change during the adiabatic driving. This is just the quantum adiabatic theorem. In this way, quantum adiabatic theorem mathematically can be regarded as a special case of the classical adiabatic theorem [9].

8.2 Adiabatic Invariants and Nonlinear geometric phase

The adiabatic evolution in classical mechanics is well understood for integrable systems or a limited phase space in a non-integrable system, where the dynamics is regular and action and angle variables can be found. In these cases the adiabatic invariants are actions and, besides the usual dynamical angles, there are geometric Hannay's angles. As discussed above, nonlinear quantum systems can mathematically be regarded as classical Hamiltonian systems and, therefore, their adiabatic evolutions can be completely understood with the language of classical adiabatic evolutions. In a mathematical sense, we need to say nothing further about the adiabatic evolution of nonlinear quantum systems. Nevertheless, we find it very rewarding to "translate" the classical language into quantum language with wave functions. As we will see, the classical adiabatic invariant, action, and the Hannay's angle are in fact linked to the Aharonov-Anandan (AA) phase and Berry phase, respectively.

Without loss of generality, we again consider a BEC system and describe it with the mean-field theory. Its adiabatic driving, characterized by the parameter R, can be achieved by either changing the external potential or the interaction strength slowly. In general, a BEC system is not integrable and its dynamics is chaotic. For example, while its two-state model is integrable, its three-state model is already chaotic [10, 11]. However, even in a chaotic system, there can be a region in the Hilbert space, where the dynamics is regular, e.g., the region near the ground state. We focus on the integrable systems or regular regions in chaotic systems.

In a regular dynamics, there are plenty of cyclic motions, where the BEC state $\Psi(\mathbf{r},t)$ returns to its original state after a period of time T. The AA phase arises in this kind of cyclic motion. We write $\Psi(\mathbf{r},t) = e^{i\lambda(t)}\Phi(\mathbf{r},t)$ such that $\Phi(\mathbf{r},t) = \Phi(\mathbf{r},t+T)$. We then have from Eq.(8.2)

$$\frac{d\lambda}{dt} = i\int d\mathbf{r}\Phi^*\frac{\partial\Phi}{\partial t} - \int\frac{d\mathbf{r}}{\hbar}\Big\{-\Phi^*\frac{\hbar^2}{2m}\nabla^2\Phi + V(\mathbf{r})|\Phi|^2 + c|\Phi|^4\Big\}. \tag{8.8}$$

When the first term on the right hand is integrated over the period of time, we have the AA phase

$$\gamma_{\mathrm{AA}} = i \oint dt \int d\mathbf{r} \Phi^* \frac{\partial \Phi}{\partial t} \,. \tag{8.9}$$

This phase is geometric as it is independent of how the overall phase λ is chosen. It is easy to check that the AA phase reduces to Berry phase when the cyclic motion is adiabatic. The second term on the right hand side of Eq.(8.8) is usually called dynamical phase.

We come back to the adiabatic evolution and see how the AA phase arises as adiabatic invariant. We assume that R is fixed. We expand the wave function $\Psi(\mathbf{r}, t)$ as in Eq.(8.3) and get a set of canonical conjugate pairs a_n^*, a_n. We choose the overall phase λ such that $a_1 e^{-i\lambda}$ is always real (this choice of λ is for convenience, not essential). Writing $a_n = |a_n| e^{i\theta_n}$, we have

$$i \int d\mathbf{r}\, \Psi^* \frac{d\Psi}{dt} = i \sum_n a_n^* \frac{da_n}{dt} = -\chi \frac{d\lambda}{dt} - \sum_{n>1} |a_n|^2 \frac{d\tilde{\theta}_n}{dt} \,, \tag{8.10}$$

where $\chi = \sum_n |a_n|^2$ and $\tilde{\theta}_n = \theta_n - \lambda$. This means that $\{|a_n|^2, \tilde{\theta}_n\}$ $(n > 1)$ are pairs of conjugate variables as well as χ and λ. However, as χ does not change with time and λ is the overall phase, this pair is trivial and can be removed from the dynamics. Noticing $\Psi(\mathbf{r}, t) = e^{i\lambda(t)} \Phi(\mathbf{r}, t)$, we further have

$$i \int d\mathbf{r}\, \Phi^* \frac{d\Phi}{dt} = -\sum_{n>1} |a_n|^2 \frac{d\tilde{\theta}_n}{dt} \,. \tag{8.11}$$

When both sides are integrated over the period of time T, the left hand side gives the AA phase while the right hand side yields the action as $\{|a_n|^2, \tilde{\theta}_n\}$ are pairs of conjugate variables. This shows that the AA phase is the quantum "translation" of the action in the classical dynamics. It is just like "Xiangwei" is the Chinese translation of phase.

For a regular motion of a BEC, there always exist a set of action and angle variables $\boldsymbol{I}, \boldsymbol{\theta}$. According to the theory of classical mechanics, when R traverses slowly over a closed loop, the actions $\boldsymbol{I}$ remains constant while the angles $\boldsymbol{\theta}$ evolves rapidly with time. The time evolution of the angles is composed of two parts, one is dynamical and the other is geometric; the latter is called Hannay's angle. To express the Hannay's angle in terms of wave functions, we notice that there is a canonical transformation between a_n^*, a_n and $\boldsymbol{I}, \boldsymbol{\theta}$ so that we can write formally $\Psi = \Psi(\boldsymbol{I}, \boldsymbol{\theta}; R)$. We define the geometric phase for a general state as [13]

$$\gamma(\mathcal{O}) = \int_{\mathcal{O}} \overline{dR \langle \Psi(\boldsymbol{I}, \boldsymbol{\theta}; R) | i \frac{\partial}{\partial R} | \Psi(\boldsymbol{I}, \boldsymbol{\theta}; R) \rangle} \,, \tag{8.12}$$

where $\mathcal{O}$ is the loop traversed in the parameter space R and the overbar indicates an average over all initial states with the same actions. The averaging is necessary since the state Ψ in general does not come back to its original state after traversing the loop $\mathcal{O}$. As a result of this averaging , $\gamma(\mathcal{O})$ does not depend on $\boldsymbol{\theta}$ and becomes geometric. It can be shown that the Hannay's angle $\alpha_n(\mathcal{O})$ can be extracted from $\gamma_n(\mathcal{O})$ with $\alpha_n(\mathcal{O}) = -\partial\gamma/\partial I_n$.

We illustrate this geometric phase with two examples. In the first example, Ψ is the ground state Ψ_0 of the system. In this case, we have only one pair of action and angle: the norm of the wave function and the overall phase. As a result, we have

$$\gamma_0(\mathcal{O}) = \int_{\mathcal{O}} dR\langle\Psi_0(R)|i\frac{\partial}{\partial R}|\Psi_0(R)\rangle\,, \tag{8.13}$$

which is just the Berry phase.

In the second example, we come back to linear quantum systems ($c = 0$), where as already discussed above the action and angle variables can be explicitly found: the action is the probability at each energy-eigenstate ϕ_n while the angle is the corresponding phase. With straightforward computation, we find that

$$\gamma(\mathcal{O}) = \sum_n I_n\gamma_n(\mathcal{O})\,, \tag{8.14}$$

where $\gamma_n(\mathcal{O}) = \int_{\mathcal{O}} d\mathbf{r}\langle\phi_n(R)|i\frac{\partial}{\partial R}|\phi(R)\rangle$ is just the usual Berry phase. For the Hannay's angle, we have $\alpha_n(\mathcal{O}) = -\gamma_n(\mathcal{O})$.

8.3 Relation to many-body quantum geometric phase

A BEC is a many-body quantum system which in principle can be described by a many-body wave function. For a given many-body energy eigenstate $\Phi_n(\mathbf{r}_1, \mathbf{r}_2, \cdots, \mathbf{r}_N; R)$, the Berry phase is given by

$$\gamma_{N,n}(\mathcal{O}) = \int_{\mathcal{O}} dR\langle\Phi_n(\mathbf{r}_1, \mathbf{r}_2, \cdots, \mathbf{r}_N; R)|i\frac{\partial}{\partial R}|\Phi_n(\mathbf{r}_1, \mathbf{r}_2, \cdots, \mathbf{r}_N; R)\rangle\,. \tag{8.15}$$

In the mean-field theory of a BEC, we have

$$\Phi_n(\mathbf{r}_1, \mathbf{r}_2, \cdots, \mathbf{r}_N; R) \approx \phi_n(\mathbf{r}_1; R)\phi_n(\mathbf{r}_2; R)\cdots\phi_n(\mathbf{r}_N; R)\,, \tag{8.16}$$

which leads to

$$\gamma_{N,n}(\mathcal{O}) \approx N\int_{\mathcal{O}} dR\langle\phi_n(\mathbf{r}; R)|i\frac{\partial}{\partial R}|\phi_n(\mathbf{r}; R)\rangle\,. \tag{8.17}$$

This shows that the nonlinear geometric phase of a mean-field eigenstate $\phi_n(\mathbf{r})$ is an approximation of its many-body counterpart. So far it is not clear how the geometric phase of the general state in Eq.(8.12) and its associated Hannay's angles are related to the many-body quantum geometric phases.

We illustrate the relation between the nonlinear geometric phase of a mean-field eigenstate and its many-body counterpart with a nonlinear two-state model (the simplest model of this kind). The mean-field Hamiltonian of this two-state model is given by [12]

$$H = \frac{X}{2}(a^*b + ab^*) + \frac{iY}{2}(ab^* - a^*b) + \frac{Z}{2}(|a|^2 - |b|^2) - \frac{c}{4}(|a|^2 - |b|^2)^2 . \quad (8.18)$$

We focus on the ground state of this two-state system. The ground state is given by

$$|\phi\rangle \equiv \begin{pmatrix} a \\ b \end{pmatrix} = \begin{pmatrix} \sqrt{\frac{1-p}{2}} \\ -\sqrt{\frac{1+p}{2}}\frac{X+iY}{\sqrt{X^2+Y^2}} \end{pmatrix}, \quad (8.19)$$

where p is the solution of the following equation,

$$p\sqrt{X^2 + Y^2} = (Z + cp)\sqrt{1 - p^2}. \quad (8.20)$$

This equation has one real root when $\sqrt{X^2 + Y^2} \geq c$. When $X^2 + Y^2 < c^2$, it can have three real roots. In particular, when $Z = 0$, two of the three real roots given by $p = \pm\sqrt{1 - (X^2 + Y^2)/c^2}$ have the same energy and are for the ground states. For convenience, we compute the corresponding curvature instead of the geometric phase. The curvature $\boldsymbol{\mathcal{B}}$ is found to be

$$\boldsymbol{\mathcal{B}} = \frac{p^3}{2(cp + Z)^2(cp^3 + Z)}(\mathbf{R} + cp\hat{z}), \quad (8.21)$$

where $\mathbf{R} = \{X, Y, Z\}$ and $\hat{z}$ is the unit vector for the Z direction.

This two-state model can also be described by a second-quantized Hamiltonian

$$H = \frac{X}{2}(\hat{a}^\dagger\hat{b} + \hat{a}\hat{b}^\dagger) + \frac{iY}{2}(\hat{a}\hat{b}^\dagger - \hat{a}^\dagger\hat{b}) + \frac{Z}{2}(\hat{a}^\dagger\hat{a} - \hat{b}^\dagger\hat{b}) - \frac{c}{4N}(\hat{a}^\dagger\hat{a} - \hat{b}^\dagger\hat{b})^2 . \quad (8.22)$$

As a result, $\boldsymbol{\mathcal{B}}_q$, the many-body quantum version of $\boldsymbol{\mathcal{B}}$, can be computed. It was found that $\boldsymbol{\mathcal{B}}_q$ and $\boldsymbol{\mathcal{B}}$ agree with each other very well when the number of particle is large [12]. This agreement shows that the definition of nonlinear geometric phase involving Eq.(8.13) is legitimate and can be regarded as an approximation of the many-body quantum geometric phase.

There was an interesting development of nonlinear geometric phase. It was found in Ref. [14] that, for a given nonlinear eigenstate, there is an additional term besides the usual Berry phase as in Eq.(8.13). This additional term is a direct result of nonlinearity and has no quantum many-body correspondence. When a dynamical system starts at its eigenstate (or fixed point in the classical language), under an adiabatic driving, it will always be driven slightly away from the eigenstate. As a result, it oscillates around this eigenstate. In a linear system, the oscillation frequencies are independent of how far the system is away from the eigenstate, which determines the actions of the oscillations. However, in a nonlinear system, the oscillation frequencies depend on actions. Although the actions are invariant during an adiabatic evolution, there are always slight deviations. These deviations will in turn slightly modify the oscillation frequencies, and cause a small change in phase. The accumulation of the phase change gives rise to the additional geometric term found in Ref. [14]. Therefore, this additional geometric part comes in fact from the dynamical phase and it arises solely due to nonlinearity. The detail of this additional geometric phase can be found in Ref. [14], which has been included as a reprint here, and in Ref. [15, 18].

It is possible to observe the nonlinear geometric phase experimentally. Berry phase was first demonstrated experimentally with optical fiber about three decades ago [16]. Berry gave a detailed theoretical account how his phase arises in optical fiber [17]. If one repeats such an experiment with nonlinear optical fiber, one should be able to observe the nonlinear geometric phase. The nonlinear Berry phase as in Eq.(8.13) must be observable while the additional term discussed above is not clear. Unfortunately, we are not aware of any experiment of this kind.

8.4 Nonlinear artificial gauge field

In Chapter 6, we have discussed how the geometric phase can be used to generate artificial gauge field for neutral atoms in non-uniform external fields. The nonlinear geometric phase can in fact be used to achieve the same feat for neutral atoms even with uniform external fields. As a result, without any external rotation, the ground state of such a system may have a vortex. Consider a spinor BEC where the interaction between particles depends on the atomic hyperfine states [19]. When such an atom moves in space, its magnetic moments feels two fields: one is the external field and the other is an internal field generated by surrounding atoms. As a result, the atom feels a nonlinear gauge field at the mean-field level. Note that as the atomic density varies spatially this nonlinear gauge field can be non-zero even when the external field is spatially uniform.

We put this into mathematical form by considering an ultracold boson gas where the atom has more than one relevant hyperfine states and has different s-wave scattering lengths for different hyperfine state channels. With the mean-field approach, we have

$$\begin{aligned} i\hbar\frac{\partial}{\partial t}|\Psi(\mathbf{r},t)\rangle &= -\frac{\hbar^2}{2m}\nabla^2|\Psi(\mathbf{r},t)\rangle + \hat{U}(\mathbf{r})|\Psi(\mathbf{r},t)\rangle \\ &+ c_0\langle\Psi(\mathbf{r},t)|\Psi(\mathbf{r},t)\rangle|\Psi(\mathbf{r},t)\rangle + \langle\Psi(\mathbf{r},t)|\hat{\mathbf{F}}|\Psi(\mathbf{r},t)\rangle\cdot\hat{\mathbf{F}}|\Psi(\mathbf{r},t)\rangle\,, \end{aligned} \tag{8.23}$$

where c_0 represents the interaction that is independent of hyperfine states and $\hat{\mathbf{F}}$ is of matrix form describing the internal-state-dependent interaction. Each atom feels an effective total field

$$\hat{U}_{\text{eff}}(\mathbf{r},t) = \hat{U}(\mathbf{r}) + \langle\Psi(\mathbf{r},t)|\hat{\mathbf{F}}|\Psi(\mathbf{r},t)\rangle\cdot\hat{\mathbf{F}}\,. \tag{8.24}$$

Note that this effective field varies with time as the wave function $|\Psi(\mathbf{r},t)\rangle$ changes with time. We assume that each atom locally stays in the ground state of the effective field $\hat{U}_{\text{eff}}$ and write $|\Psi(\mathbf{r},t)\rangle = \phi(\mathbf{r},t)|E_0(\mathbf{r},t)\rangle$ where

$$\hat{U}_{\text{eff}}|E_0\rangle = E_0|E_0\rangle \tag{8.25}$$

with $\langle E_0|E_0\rangle = 1$. By substituting this to Eq.(8.23), we obtain

$$i\hbar\frac{\partial}{\partial t}\phi = \frac{(\mathbf{p}-\mathbf{A})^2}{2m}\phi + \widetilde{W}\phi + E_0\phi\,, \tag{8.26}$$

where

$$\mathbf{A}(\mathbf{r}) = i\hbar\langle E_0|\nabla|E_0\rangle\,, \tag{8.27}$$

$$\widetilde{W}(\mathbf{r}) = \frac{\hbar^2}{2m}(\nabla\langle E_0|)\cdot\nabla|E_0\rangle - i\hbar\langle E_0|\frac{\partial}{\partial t}|E_0\rangle\,. \tag{8.28}$$

Here emerges an artificial gauge field $\mathbf{A}(\mathbf{r})$, which is nonlinear as it depends on the quantum state of the boson gas. There are a number of points to note. (*i*) The equation of motion Eq.(8.26) is not self-contained and it has be solved with Eq.(8.25) self-consistently. (*ii*) It is less challenging to solve numerically Eq.(8.26) with Eq.(8.25) than to solve the original equation (8.23). This is because Eq.(8.26) involves only one wave function while Eq.(8.23) has multiple wave functions for different hyperfine states. (*iv*) Eq.(8.26) offers a better intuitive physical picture. The emergence of the vector potential $\mathbf{A}$ indicates that the ground state can have vortices without any external fields. A similar scheme has been discussed in Ref.[20].

8.5 Non-Hermitian systems

So far we have considered only Hermitian systems; generalization to non-Hermitian systems is possible. One typical non-Hermitian system is the Bogoliubov quasi-particle for boson systems. These quasi-particles satisfy the following dynamical equation

$$i\frac{d}{dt}|\Psi\rangle = \sigma_z H|\Psi\rangle\,, \tag{8.29}$$

where $|\Psi\rangle$ is a vector of $2n$ components and H is a $(2n)\times(2n)$ Hermitian matrix while

$$\sigma_z = \begin{pmatrix} I_{n\times n} & 0 \\ 0 & -I_{n\times n}\end{pmatrix}\,. \tag{8.30}$$

In this Bogoliubov dynamics, the conserved norm is $\langle\Psi|\sigma_z|\Psi\rangle = 1$ and two states $|\Psi\rangle$ and $|\Psi\rangle'$ are said to orthogonal to each other when $\langle\Psi|\sigma_z|\Psi'\rangle = 0$.

We can similarly define eigenstates as

$$\omega_j|\Psi_j\rangle = \sigma_z H|\Psi_j\rangle \,. \tag{8.31}$$

When the system is driven by external parameters R, we can introduce a connection

$$\mathbf{A}_j = i\langle\Psi_j(R)|\sigma_z \frac{\partial}{\partial R}|\Psi_j(R)\rangle \,, \tag{8.32}$$

and the geometric phase $\beta_j = \oint dR\mathbf{A}_j$. This geometric phase and associated curvature were used to compute the transverse force acting on a vortex in a BEC [21].

We may call the dynamics in Eq.(8.30) Lorentz quantum mechanics [22]. In contrast to the unitary transformation in Hilbert space in the usual quantum mechanics, the dynamical evolution in Lorentz quantum mechanics is complex Lorentz transformation in complex Minkowski space. Geometric phase in this kind of Lorentz dynamics has been analyzed in detail in Ref. [22].

Chapter 8

References

1. M. V. Berry, Proc. R. Soc. London, Ser. A **392**, 45 (1984).
2. F. Wilczek and A. Zee, Phys. Rev. Lett. **52**, 2111 (1984).
3. J. H. Hannay, J. Phys. A **18**, 221 (1985).
4. Y. Aharonov and J. Anandan, Phys. Rev. Lett. **58**, 1593 (1987).
5. F. Dalfovo, S. Giorgini, L.P. Pitaevskii, and S. Stringari, Rev. Mod. Phys. **71**, 463 (1999).
6. X. Luo, Q. Xie, and B. Wu, Phys. Rev. A **76**, 051802 (2007).
7. A. Heslot, Phys. Rev. D **31**, 1341 (1985).
8. S. Weinberg, Ann. Phys. (N.Y.) **194**, 336 (1989).
9. J. Liu, B. Wu, and Q. Niu, Phys. Rev. Lett. **90**, 170404 (2003). [Reprint (8.1)]
10. B. Liu, L.B.Fu, S.P.Yang, and J.Liu, Phys. Rev. A. **75**, 033601 (2007).
11. Q. Guo, X. Chen, and B. Wu, Optics Express **22**, 19237 (2014).
12. B. Wu, Q. Zhang, and J. Liu, Physics Letters A **375**, 545 (2011).
13. B. Wu, J. Liu, and Q. Niu, Phys. Rev. Lett. **94**, 140402 (2005). [Reprint (8.2)]
14. J. Liu and L. B. Fu, Physical Review A **81**, 052112 (2010). [Reprint (8.3)]
15. L. B. Fu and J. Liu, Annals of Physics **325**, 2425 (2010).
16. A. Tomita and R. Y. Chiao, Phys. Rev. Lett. **57**, 937 (1986).
17. M. V. Berry, Nature **326**, 277 (1987); Journal of Mod. Optics, **34**, 1401 (1987).
18. Q. Zhang, J. B. Gong, and C. H. Oh, Phys. Rev. Lett. **110** (2013).
19. Tin-Lun Ho, Phys. Rev. Lett. **81**, 742 (1998).
20. Jun-hui Zheng, B. Xiong, Gediminas Juzeliūnas, and Daw-Wei Wang, Phys. Rev. A **92**, 013604 (2015).
21. C. Zhang, A. M. Dudarev,and Q. Niu, Phys. Rev. Lett **97**, 040401 (2006). [Reprint (8.4)]
22. Q. Zhang and B. Wu, arXiv:1701.00057 (2017).

9

Geometric phase in collective coordinates

The formulation of geometric phase, Berry curvature, and Chern number applies to any quantum system. For many-body quantum systems, exact results can sometimes be obtained from such considerations without detailed solutions. One example is the quantum Hall effect of interacting systems, with the possibility of disorder, as discussed in Chapter 4. The Hall conductance can be written as a Berry curvature in the space of boundary condition phases. If the system is insulating in the bulk, the Berry curvature can be regarded as a constant in the boundary phases if one ignores errors exponentially small in the system size. The quantization of Hall conductance then follows from the periodicity of the ground state with respect to the boundary phases.

In this chapter, we introduce another application which yields an exact result on the transverse force on a quantized vortex in a superfluid [1]. If the vortex is made to move by trapping it into a pinning potential that slowly translates in space, the force from the pinning potential can be calculated by adiabatic perturbation expansion with respect to the velocity. To first order, the transverse part of the force is just a Berry curvature in the vortex coordinates. At zero temperature, this is the Berry curvature of the ground state of the superfluid in the topological sector of a single vortex. At finite temperatures, a thermal average over excitations also needs to be included. Amazingly, this Berry curvature is given by the momentum circulation around the vortex on a path far away from the vortex center. If the circulation is carried entirely by the superfluid component of the many-body system, the transverse force must then be proportional to the superfluid density.

In this example, the vortex center position plays the role of collective coordinates of the superfluid system, which are controlled by external means (the pinning potential in this case). Sometimes, usually within certain approximations, one can write down the explicit many-body wave function with some parameters, which then play the role of collective coordinates. For example, the discussions on quasiparticles and quasiholes in the fractional quantum Hall states have been carried out using the Laughlin wave functions. Also, earlier formulations of transverse force on vortices were done similarly using Feynman wave functions.

Most often, one discusses the dynamics of collective coordinates by first deriving a nonlinear equation of motion for the order parameter field under some approximation. The nonlinear Schrödinger equation discussed in the last chapter is such an example. Within this approach, one can still study the motion of topological defects by considering some topological structures of the order parameter field. The order parameter dynamics also allows one to study the long wavelength excitations corresponding to the Goldstone modes.

A formally exact approach for deriving the order parameter dynamics is illustrated in the work for itinerant spin systems when the magnetization field evolves at a rate much slower than the other degrees of freedom [2]. Here the magnetization is defined as the expectation value of the spin-density at each point in space. In the adiabatic limit and at zero temperature, one can make the assumption that the many-body system stays in the manifold of constrained ground states. (A constrained ground state is the lowest energy state for a given configuration of the magnetization.) The time-evolution of the magnetization field should be such that the constrained ground state satisfies the time-dependent variational principle for the Schrödinger equation.

By definition, the constrained ground state is a functional of the magnetization field. In principle, it can be found from the time-independent spin-density functional theory. In practice, one needs to introduce constraining fields in order to achieve various configurations of the magnetization. The equations of motion turn out to depend only on two properties of the constrained ground state, its energy and Berry curvature in the space of magnetization configurations.

The Landau-Lifshitz equation for spin dynamics is born out of this formulation in the tight-binding limit, where the many-body wave function is just a product of single particle spin-orbital wave functions for each atom. In this limit, the magnetization configuration is specified by the spin orientations for each atom, and there are no additional degrees of freedom within each atom. Due to the tight-binding nature, there is no component of the Berry curvature involving different atoms. However, interatomic Berry curvatures may become important for itinerant electronic spins, leading to modifications of the Landau-Lifshitz equations. This subject deserves further investigation.

One can also specify the magnetization configurations by their Fourier components [3]. This is particularly suitable for the study of spin waves, which are the normal modes of small amplitude spin excitations about the true ground state. Spin dynamics is then represented by the equations of motion for the Fourier components, which can be similarly formulated with the knowledge of energy and Berry curvature in the space of the Fourier components. In the special case of collinear magnetization, either ferromagnetic or antiferromagnetic, one can deduce a particularly simple formula for the spinwave spectrum in terms of a ratio of the frozen spinwave energy and the magnetization reduction. *Ab initio* calculations based on this formula have yielded excellent agreement with experimental measurement of the ferromagnetic Fe spinwave spectrum [4].

Chapter 9

References

1. D. J. Thouless, Ping Ao, and Qian Niu. Phys. Rev. Lett. **76**, 3758 (1996). [Reprint (9.1)]
2. Q. Niu, Xindong Wang, L. Kleinman, Wu-Ming Liu, D. M. C. Nicholson, and G. M. Stocks. Phys. Rev. Lett. **83**, 207 (1999). [Reprint (9.3)]
3. Qian Niu and Leonard Kleinman, Phys. Rev. Lett. **80**, 2205 (1998). [Reprint (9.2)]
4. D. M. Bylander, Qian Niu, and Leonard Kleinman, Phys. Rev. B **61**, R11875 (2000). [Reprint (9.4)]

Reprint Pages

2

Geometric phases in momentum space

Phase anholonomy in the classical adiabatic motion of charged particles

Robert G. Littlejohn*
Center for Nonlinear Studies, Los Alamos National Laboratory, Los Alamos, New Mexico 87545
(Received 10 December 1987; revised manuscript received 1 August 1988)

The classical adiabatic motion of charged particles in strong magnetic fields, otherwise known as guiding center motion, exhibits an anholonomic phase similar to Berry's phase and the phase discovered in a general context by Hannay. Analysis of this effect reveals that there often is no best way to define the gyrophase when magnetic field lines are curved. Instead, a change in the definition of the phase is a kind of gauge transformation, the study of which leads to new insight into guiding center theory. The path-dependent phase that occurs in this problem is coupled with the metrical structure of physical space, giving rise to a transport process for vectors and frames which is similar to parallel transport in non-Euclidean geometry. Strong analogies with Fermi-Walker transport and Thomas precession in special relativity are pointed out.

I. INTRODUCTION

Anholonomic effects are common in physics, and have come under under special scrutiny lately. Generally speaking, they occur when a physical system is transported through some sequence of states, and when the result of the transport process depends not only on the final state, but also on the history or sequence of intermediate states. A familiar example from thermodynamics is the heat absorbed by a system during a transport through a sequence of equilibrium states. An anholonomic effect is generally nonvanishing when the process is cyclic, on account of the path dependence. Mathematically, anholonomies are often described by inexact differentials (such as the differential of the heat $đQ$ in thermodynamics), whose line integrals are path dependent. This will be the case with the phase anholonomies discussed below, which give rise to inexact differentials of phase.

Recently there has been considerable interest in various phase anholonomies that occur in adiabatic processes. Probably the most familiar of these is Berry's phase,[1] otherwise known as the quantal adiabatic phase, which is an anholonomic shift in the overall phase of an energy eigenfunction in an adiabatic process in quantum mechanics. The classical analog of Berry's phase is found in Hannay's angle,[2] which is an anholonomic increment in the angle variable (i.e., the variable conjugate to the action) in adiabatic processes in integrable classical systems. (The assumption of integrability is necessary when the system has more than one degree of freedom.) Berry[3] has shown that Hannay's angle is the semiclassical limit of his adiabatic phase, and Gozzi and Thacker[4] have examined the canonical and geometrical structure associated with Hannay's angle. Experimental prediction and confirmation of these phase shifts has recently been achieved by Chiao and co-workers[5,6] using photons whose wave vectors are adiabatically transported as the photon follows a helical optical fiber.

In this paper I shall describe a phase anholonomy which occurs in the classical adiabatic motion of a charged particle in a strong magnetic field, which is also known as guiding center motion. Since this is a classical problem, one might expect the phase anholonomy which occurs here to be a special case of Hannay's angle, and at one level this is certainly true. Guiding center motion, however, does not precisely fit the model examined by Hannay, so the comparison is somewhat rough. More importantly, guiding center motion gives rise to anholonomic features not envisioned in Hannay's analysis, including a transport process for frames of unit vectors in three-dimensional physical space. This transport process involves the metrical structure of physical space, and bears a strong resemblance to Fermi-Walker transport as it occurs in Thomas precession in special relativity.

There are two main points of interest in this paper. The first is that guiding center motion provides a novel example of a phase anholonomy, and one with several unique features relating to gauge theories and other topics of general interest in physics. The second is that guiding center motion is a basic theoretical tool of plasma physics, and therefore any new phenomenon which concerns it is of interest in this field. Some of the issues surrounding this phase anholonomy have caused puzzlement or difficulties of various kinds in certain theoretical endeavors in plasma physics. For example, a number of plasma physicists over the years have wondered how best to define the gyrophase in geometries involving curved magnetic field lines; this is a question that should largely be put to rest by this paper. (It is put to rest by showing that there is no best way, at least in the absence of symmetry.) In addition, there is the fact that any perturbation analysis of a problem involving curved magnetic field lines will inevitably lead to terms which do not have an immediate physical or geometrical interpretation, namely, those terms which depend on the perpendicular unit vectors. This paper will deal directly with such terms, and will provide the missing interpretation. These questions have arisen, for example, in the recent work on gyrokinetic equations by Hagan and Frieman.[7]

Finally, one should note that the value of this analysis goes beyond the classical motion of charged particles, as shown by Berry's recent quantal and semiclassical treatment[8] of spin precession in the Stern-Gerlach experiment. Berry studies in detail a simplified version of this problem, in which the spin precession is produced by the magnetic field at the position of the particle. (This is of course reasonable from a physical point of view, although since the spin is treated quantally and the position of the particle classically, it is something of a hybrid approach. It should be noted that there apparently does not exist a "quantal guiding center theory," in which the coupling between the spin and spatial degrees of freedom in an inhomogeneous magnetic field is systematically taken into account. The classical theory is well developed, however, and provides corrections to the picture in which the gyration of the particle is viewed as if due to a purely time-dependent magnetic field. It would be desirable to generalize the classical theory into a quantal version.) Berry examines the expectation value of the spin operator, which of course obeys classical equations of motion, and shows that its evolution involves an adiabatic phase which is a special case of Hannay's angle. He also shows that the phase two-form is identical with the vector **N** of Eq. (3.17) below (reported earlier in Ref. 9).

Because of these different points of interest, this paper spans several fields, and I have therefore written it for a general audience. For example, in Secs. II and III I have provided enough background on Hannay's angle and guiding center motion to bring the relevant issues into focus. Nor have I employed much of the abstract mathematics which is common in gauge theories; a slightly more mathematical treatment of some of the matters discussed here may be found in my earlier work on this subject.[9]

The outline of this paper is as follows. Section II contains a review of Hannay's angle, as well as some commentary to connect it with the analysis to follow. Section III presents an introduction to guiding center motion and discusses in detail the phase anholonomy which occurs there. The latter subject is the main novel element in this paper. In Sec. IV various generalizations are considered, including extensions to time-dependent fields and relativistic motion. Section IV also demonstrates the close analogy between the phase anholonomy in guiding center motion and the rotations which occur in Thomas precession. Finally, the Appendix clarifies certain points in the derivation of Hannay's angle.

II. BACKGROUND ON HANNAY'S ANGLE

Hannay's angle emerges from asking new questions about the old adiabatic theorem. Following Hannay, we consider a classical Hamiltonian $H(q,p,\mathbf{X})$, where $\mathbf{X}=(X_1,\ldots,X_n)$ represents a collection of parameters, considered to be slow functions of time. For simplicity, we will work with a system of one degree of freedom; the analysis is easily transcribed to an arbitrary number of degrees of freedom, so long as the crucial requirement of integrability is met. We consider the canonical transformation taking us from (q,p) to action-angle variables, parametrized by $\mathbf{X}$,

$$\theta=\theta(q,p,\mathbf{X}), \quad I=I(q,p,\mathbf{X}) . \tag{2.1}$$

To generalize this to more than one degree of freedom, one must assume that the system is integrable for all values of $\mathbf{X}$, in order that this transformation exist.

Transforming $H(q,p,\mathbf{X})$ to action-angle variables, we obtain the new Hamiltonian,

$$K(\theta,I,\mathbf{X})=K_0(I,\mathbf{X})+K_1(\theta,I,\mathbf{X}) , \tag{2.2}$$

where K_0 is simply H, reexpressed in terms of the action-angle variables, and K_1 is the time derivative of the generating function $S=S(q,I,\mathbf{X})$ of the canonical transformation, i.e.,

$$K_1(\theta,I,\mathbf{X})=\frac{\partial}{\partial t}S(q,I,\mathbf{X}) . \tag{2.3}$$

The time dependence occurs only through $\mathbf{X}$, so if the parameters $\mathbf{X}$ are frozen, then K_1 vanishes.

In that case, the action I is constant in time, and the angle θ evolves with the constant frequency $\omega_0=\omega_0(I,\mathbf{X})=\partial K_0/\partial I$. On the other hand, if the parameters $\mathbf{X}$ are allowed to be functions of time (but slow ones), then the classical adiabatic theorem states that the action will still be constant, at least in an averaged sense. Hannay's contribution is to ask what happens to the angle in this case.

The answer can be extracted from Hamilton's equation for the time evolution of the angle,

$$\dot\theta=\omega_0(I,\mathbf{X}(t))+\Delta\dot\theta , \tag{2.4}$$

where the term

$$\Delta\dot\theta=\frac{\partial K_1(\theta,I,\mathbf{X}(t))}{\partial I} \tag{2.5}$$

is the one coming from the time dependence. Integrating, we obtain

$$\theta(t)=\int_0^t \omega_0(I,\mathbf{X}(t'))dt'+\Delta\theta , \tag{2.6}$$

where the first term is called the "dynamical phase." The second term $\Delta\theta$ has a time evolution consisting of small amplitude, rapid oscillations, superimposed on a slow or secular evolution, which we may denote by $\langle\Delta\theta\rangle$. Thus, after a long time, the dominant contributions to the phase θ are the dynamical phase and the quantity $\langle\Delta\theta\rangle$, which is known as "Hannay's angle."

Hannay computes $\langle\Delta\theta\rangle$ in the following way. First, he notes that $\Delta\dot\theta$ can be written in the form,

$$\Delta\dot\theta=\frac{\partial K_1}{\partial I}=\frac{\partial\theta(q,p,\mathbf{X}(t))}{\partial t}=\sum_i\frac{\partial\theta(q,p,\mathbf{X})\dot X_i}{\partial X_i} . \tag{2.7}$$

(The second equality here is not immediately obvious, nor does Hannay elaborate on it. Therefore I have provided a short proof of it in the Appendix.) Next, to get $\langle\Delta\theta\rangle$, Hannay averages over θ. This requires that $\partial\theta(q,p,\mathbf{X})/\partial X_i$ be expressed as a function of $(\theta,I,\mathbf{X})$, whereupon the averaging gives an n-dimensional vector **A** with components

$$A_i(I,\mathbf{X})=\frac{1}{2\pi}\int_0^{2\pi}d\theta\frac{\partial\theta(\theta,I,\mathbf{X})}{\partial X_i} . \tag{2.8}$$

Therefore

$$\langle \Delta\theta \rangle = \int_0^t \sum_i A_i(I,\mathbf{X})\dot{X}_i dt = \int_{X_0}^{X_1} \mathbf{A}(I,\mathbf{X})\cdot d\mathbf{X} , \qquad (2.9)$$

where $\mathbf{X}_0$ and $\mathbf{X}_1$ are the initial and final values of the parameters.

Thus Hannay's angle is given by a line integral in parameter space, one which depends on the path taken, but not on the temporal rate at which the path is traversed (so long as it is slow). Except in special cases, this integral will be path dependent. Therefore Hannay's angle will in general be nonvanishing when the path $\mathbf{X}(t)$ through parameter space is closed, taking the functional form of the Hamiltonian (but not the actual dynamical state) back to its starting point. For such closed paths, the line integral can be transformed into a surface integral by Stokes's theorem, which, in the case that $\mathbf{X}$ and $\mathbf{A}$ are three-dimensional, can be expressed in terms of a new vector field,

$$\mathbf{B} = \mathbf{B}(I,\mathbf{X}) = \nabla \times \mathbf{A} , \qquad (2.10)$$

which represents a kind of "angle flux" in parameter space. (More generally, $\mathbf{B}$ is a two-form in this space.) Hannay[2] has given some nice examples of the vector fields $\mathbf{A}$ and $\mathbf{B}$ and the angles associated with them in simple physical problems.

The vector fields $\mathbf{A}$ and $\mathbf{B}$ have obvious analogs in electromagnetic theory, encouraged by the notation, and, in the same analogy, they possess a kind of gauge invariance. One of the features of gauge degrees of freedom in physical systems is that they provide a mathematically redundant description of a single physical reality, as seen in the possibility of many vector potentials $\mathbf{A}$ for a given magnetic field $\mathbf{B}$. The choice of gauge in Hannay's problem is not in the action I, which is given by a definite loop integral in phase space and which is not subject to redefinition, but rather in the angle θ. Unlike the action, the angle is subject to redefinition, since there is no privileged choice for the point on the phase-space loop (or invariant torus) which should correspond to the origin of phase, i.e., the point $\theta = 0$. One is always free to shift or redefine this point of origin, possibly in a different way from one loop (or torus) to the next. If the system is time-dependent, as here, one can even make the redefinition depend on time.

In the present case, such a redefinition corresponds to the canonical transformation

$$I' = I, \quad \theta' = \theta + f(I,\mathbf{X}(t)) , \qquad (2.11)$$

where (in one degree of freedom) the function f is arbitrary. This is a kind of gauge transformation, showing that I is gauge invariant, but θ is not. Further, from Eq. (2.8) we see that its effect on the vector $\mathbf{A}$ is given by

$$\mathbf{A}' = \mathbf{A} + \nabla f , \qquad (2.12)$$

and therefore

$$\mathbf{B}' = \mathbf{B} , \qquad (2.13)$$

just as in magnetostatics. Not surprisingly, we find that Hannay's angle for a closed adiabatic transport process in $\mathbf{X}$ space is gauge invariant, although for an open one it is not.

III. PHASE ANHOLONOMY IN GUIDING CENTER MOTION

In this section we will examine the phase anholonomy which occurs in nonrelativistic guiding center motion. The relativistic case is quite similar in most respects, although there are some interesting differences. We will comment briefly on these in Sec. IV. We begin by reviewing some of the elementary features of guiding center motion.

The motion of a charged particle in a nonuniform magnetic field $\mathbf{B} = \mathbf{B}(x)$ is a nontrivial problem. The equations of motion are nonlinear and generally nonintegrable, they exhibit chaotic behavior, etc. In the event, however, that the magnetic field is "strong" (in a sense which can be made precise), there is a separation of time scales in the three degrees of freedom, making the equations amenable to a perturbation analysis. The most rapid time scale is that of the gyrating of the particle about a magnetic field line; this degree of freedom acts somewhat like an oscillator which is adiabatically transported from place to place by the other, slower degrees of freedom. In this manner, adiabatic processes occur even in an autonomous (i.e., time-independent) system. The perturbation expansion, essentially in inverse powers of $B = |\mathbf{B}|$, is called the "guiding center expansion." The best general reference on guiding center motion is the book by Northrop,[10] and the Hamiltonian and canonical structure of this problem have been examined by myself and others.[11–14] In this paper we will not pay much attention to the details of this expansion, except to note how it leads to phase anholonomies.

To establish terminology, let us consider first the elementary problem of a charged particle in a uniform magnetic field, say $\mathbf{B} = B_0\hat{\mathbf{z}}$ (see Fig. 1). Let $\mathbf{x}$ be the position of the charged particle, which follows a helical orbit around a field line (dotted in the figure). The projection of the motion onto the plane perpendicular to $\mathbf{B}$ is circular, with frequency $\Omega = eB_0/mc$. This frequency is called the "gyrofrequency," and is usually considered large. The projection of $\mathbf{x}$ onto the field line about which it spirals is denoted $\mathbf{X}$ and is called the "guiding center"; it is in a sense the average of $\mathbf{x}$ over the rapid oscillations. It should not be confused with Hannay's parameter vector $\mathbf{X}$, although as we shall see it plays a somewhat similar role. Of course, the guiding center moves along the field line with constant velocity. The vector running from $\mathbf{X}$ to $\mathbf{x}$ is called the "gyroradius" vector; its magnitude is $v_\perp/\Omega$, where $v_\perp$ is the magnitude of the component of the particle's velocity perpendicular to the magnetic field. Finally, we define the "gyrophase" θ as the angle in the perpendicular plane between some reference direction, denoted $\hat{\mathbf{e}}$ in Fig. 1, and the gyroradius vector. For the case of the uniform field, $\hat{\mathbf{e}}$ may conveniently be taken to be any constant perpendicular direction, such as $\hat{\mathbf{x}}$.

Let us turn now to the case of a nonuniform (but time-independent) magnetic field, in which the particle motion

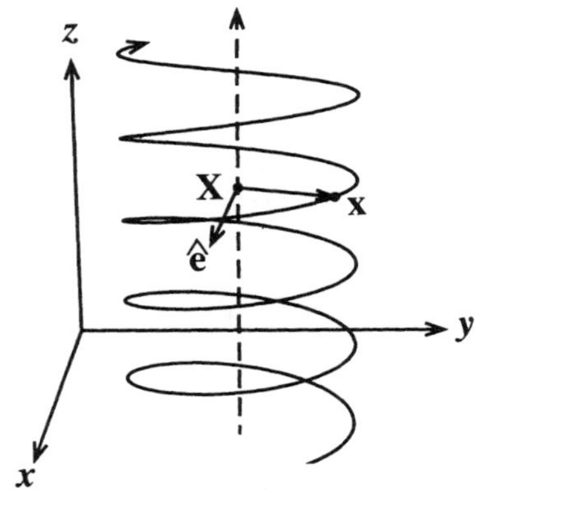

FIG. 1. Motion of a charged particle in a uniform magnetic field, $\mathbf{B}=B\hat{\mathbf{z}}$. $\mathbf{x}$ is the particle position, $\mathbf{X}$ is the guiding center position, and $\hat{\mathbf{e}}$ is a reference direction to which the definition of the gyrophase is tied.

acquires new features. (See Fig. 2. The case of time-dependent fields will be considered briefly in Sec. IV.) Now the guiding center position $\mathbf{X}(t)$ no longer follows a field line, but slowly "drifts" away from it. This is important to keep in mind, because when we come to certain transport processes below, we will want to consider arbitrary paths of transport, not just those along field lines. In addition, the gyrofrequency is now a function of position, i.e., $\Omega=\Omega(\mathbf{x})=eB(\mathbf{x})/mc$. Therefore it is not surprising that now the dominant contribution in the time evolution of the gyrophase is given by

$$\theta(t)=\int_0^t \Omega(\mathbf{X}(t'))dt' , \tag{3.1}$$

which we may call the "dynamical phase" in analogy with Eq. (2.6). Note that in this equation we have evaluated the gyrofrequency at the guiding center position rather than the particle position, in order to represent an average or secular contribution to the gyrophase.

But, just as with Hannay's angles, the dynamical phase is not the whole story, as one can easily see by reflecting on how the gyrophase is to be defined in nonuniform magnetic fields. Unlike the case of uniform fields, it is no longer possible to choose a constant reference direction in the perpendicular plane to represent the origin of the gyrophase. Instead, it is reasonable to introduce an orthonormal triad of vectors $(\hat{\mathbf{e}}_1,\hat{\mathbf{e}}_2,\hat{\mathbf{b}})$ as shown in Fig. 3 , where $\hat{\mathbf{b}}=\mathbf{B}/B$, and $\hat{\mathbf{e}}_1$ and $\hat{\mathbf{e}}_2$ are unit vectors spanning the perpendicular plane. Note that since $\mathbf{B}$ itself is a function of position, all three unit vectors will be also. Filling up space with such triads, we obtain a field of orthonormal frames. Then, using these frames, we might, for example, define the gyrophase θ to be the angle between $\hat{\mathbf{e}}_1$ and the gyroradius vector. The vectors $\hat{\mathbf{e}}_1$ and $\hat{\mathbf{e}}_2$ are not arbitrary, because they are constrained to be perpendicular to $\hat{\mathbf{b}}$, but the requirement that they form an orthonormal frame with $\hat{\mathbf{b}}$ still leaves them free to be rotated in the perpendicular plane about $\hat{\mathbf{b}}$ by an arbitrary angle. Therefore the dynamical phase in Eq. (3.1) cannot be the only contribution to the gyrophase, because the definition of the gyrophase depends on the choice of $(\hat{\mathbf{e}}_1,\hat{\mathbf{e}}_2)$, and we do not see any such dependence in Eq. (3.1).

We will elaborate on this in a moment, but for now let us ask the following question, which arises naturally at this point: Is there some "privileged" choice of $(\hat{\mathbf{e}}_1,\hat{\mathbf{e}}_2)$ that is somehow better than the rest? A suggestion that has often been made in the past is that $(\hat{\mathbf{e}}_1,\hat{\mathbf{e}}_2)$ should be chosen to be the principal normal and binormal vectors of the field line, i.e.,

$$\hat{\mathbf{e}}_1=\frac{\hat{\mathbf{b}}\cdot\nabla\hat{\mathbf{b}}}{|\hat{\mathbf{b}}\cdot\nabla\hat{\mathbf{b}}|}, \quad \hat{\mathbf{e}}_2=\hat{\mathbf{b}}\times\hat{\mathbf{e}}_1 . \tag{3.2}$$

This has the advantage that it ties $(\hat{\mathbf{e}}_1,\hat{\mathbf{e}}_2)$ to $\hat{\mathbf{b}}$, which is determined by the physics and is therefore unambiguous. Unfortunately, it has few other advantages. For example, Eq. (3.2) leaves $(\hat{\mathbf{e}}_1,\hat{\mathbf{e}}_2)$ undefined for straight field lines, and undefined and discontinuous at an inflection point of a curved field line. Further, there is no formula or ex-

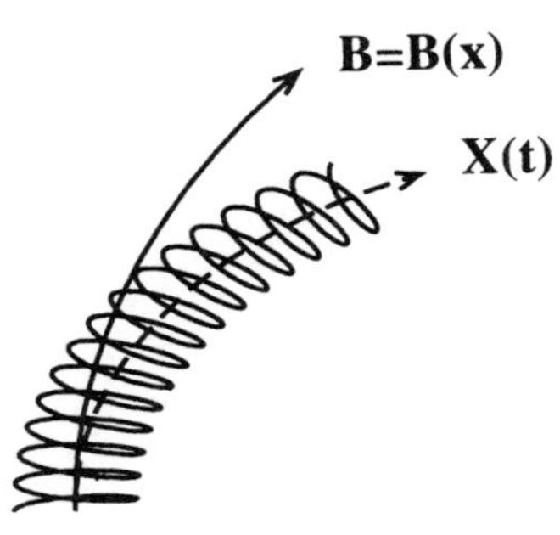

FIG. 2. In a nonuniform magnetic field, the guiding center $\mathbf{X}$ gradually drifts away from a field line.

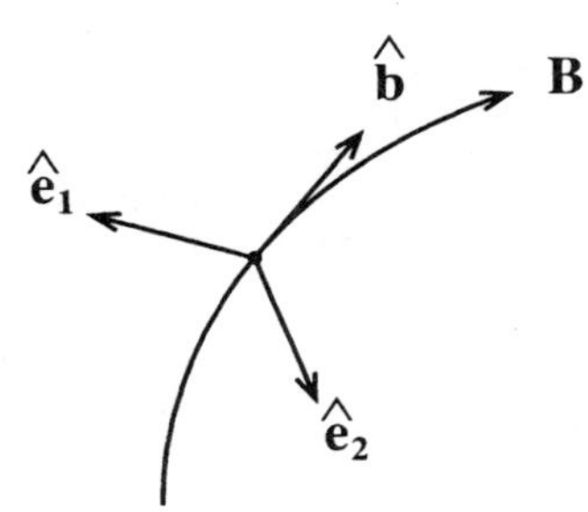

FIG. 3. To coordinatize the perpendicular plane, it is convenient to introduce a field of orthonormal triads, $(\hat{\mathbf{e}}_1,\hat{\mathbf{e}}_2,\hat{\mathbf{b}})$.

pression in the subsequent analysis of the guiding center perturbation expansion or any version of the gyrokinetic expansion for which this choice of $(\hat{\mathbf{e}}_1,\hat{\mathbf{e}}_2)$ causes any simplification or clarification. Those who have worked on guiding center or gyrokinetic problems will know that the principal effect of making the substitution (3.2) is to complicate all the subsequent algebra. Furthermore, the choice (3.2) is not "rotationless," in a sense which will be defined presently, even along magnetic field lines.

Instead, we shall regard Eq. (3.2) as only one choice out of many, none of which is to be preferred over another. In effect, Eq. (3.2) is a choice of gauge, and every other choice of gauge is related to this one by a rotation in the perpendicular plane. Thus we are led to define what I shall call a "gyrogauge transformation,"

$$\begin{aligned} \hat{\mathbf{e}}_1' &= \hat{\mathbf{e}}_1 \cos\psi + \hat{\mathbf{e}}_2 \sin\psi \,, \\ \hat{\mathbf{e}}_2' &= -\hat{\mathbf{e}}_1 \sin\psi + \hat{\mathbf{e}}_2 \cos\psi \,, \end{aligned} \tag{3.3}$$

which maps any choice of gauge $(\hat{\mathbf{e}}_1,\hat{\mathbf{e}}_2)$ onto another choice $(\hat{\mathbf{e}}_1',\hat{\mathbf{e}}_2')$. Note that the angle $\psi=\psi(\mathbf{x})$ is allowed to be a function of position; there is no reason why the angle of rotation cannot differ from point to point.

[In some circumstances there are, in fact, reasons for considering certain choices of $(\hat{\mathbf{e}}_1,\hat{\mathbf{e}}_2)$ to be privileged. For example, in a system with azimuthal symmetry (rotational symmetry about some axis), there are advantages in choosing $(\hat{\mathbf{e}}_1,\hat{\mathbf{e}}_2)$ to have the same symmetry. This causes the entire guiding center Lagrangian to have azimuthal symmetry, so that Noether's theorem can be applied. The proper law of conservation of angular momentum for such systems can be derived in this way. I have elaborated on this point in Ref. 13. Similarly, in electromagnetic theory, if the electromagnetic fields have azimuthal symmetry, then one is always free to choose the potentials to have the same symmetry. Of course, one is not obliged to do so. Translational symmetry is more subtle, in that it involves issues of "coadjoint equivariance;" see Ref. 15. In other circumstances the physics may dictate the choice of gauge, as in the work of Cohen, Rowlands, and Foote.[16] In still other cases, such as in computer integrations of the drift equations, matters of convenience may dictate a choice. Finally, one can certainly use Eq. (3.2) if one wants to; although it may not be any better than other choices, it may be no worse, either. In any case, the main point of this paper is to study the gauge freedom in this problem, rather than to try to work with a particular gauge.]

Having postulated our gyrogauge transformation, we can begin to investigate which quantities are gyrogauge invariant, and how those which are not transform. We certainly expect all truly physical quantities, i.e., those independent of the convention involved in $(\hat{\mathbf{e}}_1,\hat{\mathbf{e}}_2)$, to be gyrogauge invariant, and to be expressible purely in terms of $\hat{\mathbf{b}}$, $\mathbf{B}$, etc. The most obvious quantity which is not gyrogauge invariant is the gyrophase itself; it transforms according to

$$\theta' = \theta + \psi(\mathbf{x}) \,. \tag{3.4}$$

(It is convenient to reckon the gyrophase in a clockwise direction, the same as the direction of rotation of a positive particle. This convention gives a plus sign in this equation.)

Returning now to Eq. (3.1), we know that since θ is not gyrogauge invariant, $\dot{\theta}$ cannot be either, so there must be terms in addition to the dynamical phase reflecting this fact. Indeed, the guiding center expansion produces an equation of evolution for the gyrophase of the form[13]

$$\dot{\theta} = \Omega(\mathbf{X}) + \mathbf{R}\cdot\dot{\mathbf{X}} + (\;\cdots\;) \,, \tag{3.5}$$

where the ellipsis indicates terms which are gyrogauge invariant, and where the vector $\mathbf{R}$ is defined by

$$\mathbf{R} = \nabla\hat{\mathbf{e}}_1\cdot\hat{\mathbf{e}}_2 \,. \tag{3.6}$$

(This means $R_i = e_{1j,i}e_{2j}$. The ∇ acts only on the operand immediately following, and dot products or contractions act only on adjacent symbols.) Equation (3.5) is actually an averaged equation; all rapid oscillations have been averaged out.

The term $\mathbf{R}\cdot\dot{\mathbf{X}}$ will be of special interest to us. Note that it is independent of the charge and mass of the particle, as well as other physical constants. This is a sign that it represents a purely geometrical effect. This term does depend on $(\hat{\mathbf{e}}_1,\hat{\mathbf{e}}_2)$, through $\mathbf{R}$ itself, but this in itself does not necessarily indicate that $\mathbf{R}$ is gyrogauge dependent. For example, $\hat{\mathbf{e}}_1\times\hat{\mathbf{e}}_2=\hat{\mathbf{b}}$ gyrogauge invariant, and in a moment we will see more complicated expressions involving $(\hat{\mathbf{e}}_1,\hat{\mathbf{e}}_2)$ which are also gyrogauge invariant, and which can also be reexpressed in terms of $\hat{\mathbf{b}}$. Nevertheless, a short calculation, combining Eqs. (3.3) and (3.6), shows that

$$\mathbf{R}' = \nabla\hat{\mathbf{e}}_1'\cdot\hat{\mathbf{e}}_2' = \mathbf{R} + \nabla\psi \,. \tag{3.7}$$

Therefore $\mathbf{R}$ is not gyrogauge invariant, and it cannot be expressed in terms of $\hat{\mathbf{b}}$. Note, however, that $\nabla\times\mathbf{R}$ is gyrogauge invariant; this suggests that it should be possible to express the vector $\nabla\times(\nabla\hat{\mathbf{e}}_1\cdot\hat{\mathbf{e}}_2)$ purely in terms of $\hat{\mathbf{b}}$. This is not an obvious conclusion, but it is correct, as we shall presently show.

Now when we combine Eqs. (3.5) and (3.7) to get the transformation properties of $\dot{\theta}$, we find

$$\dot{\theta}' = \dot{\theta} + \nabla\psi\cdot\dot{\mathbf{X}} \,, \tag{3.8}$$

which is completely consistent with Eq. (3.4). In other words, the term $\mathbf{R}\cdot\dot{\mathbf{X}}$ in Eq. (3.5) is necessary in order for both sides of the equation to have the same transformation properties under a gyrogauge transformation.

It is now clear that $\mathbf{R}$ is the analog of Hannay's vector $\mathbf{A}$, and that the time integral of the second term of Eq. (3.5), which can be written as a line integral along the guiding center trajectory,

$$\Delta\theta = \int_{\mathbf{X}_0}^{\mathbf{X}_1} \mathbf{R}\cdot d\mathbf{X} \,, \tag{3.9}$$

is the analog of Hannay's angle. (We omit the averaging brackets, since it is understood at this point that we are dealing with averaged quantities.) So long as $\nabla\times\mathbf{R}\neq\mathbf{0}$, this angle will be anholonomic, i.e., dependent on the path $\mathbf{X}(t)$ of the guiding center. We also see that the guiding center position $\mathbf{X}$, in determining the slowly changing environment for the degree of freedom corresponding to the rapid gyrations (i.e., the direction and magnitude of the magnetic field), is like Hannay's parameter vector $\mathbf{X}$. We have not derived our anholonomic term $\mathbf{R}\cdot\dot{\mathbf{X}}$ from the canonical structure of the original equations of motion, as Hannay did, because the canonical structure of the guiding center problem is complicated by the separation of time scales and by the coupling to the magnetic field. (See Refs. 12–14 for a discussion of this canonical structure.) Nevertheless, there is little doubt that a formula for $\mathbf{R}$ such as Eq. (2.8) could be extracted from such an approach.

Now we shall explore properties of the gyrophase which go beyond Hannay's analysis. These arise from the fact that the gyrophase is not only an angle in phase space, but also one in real physical space. Unlike phase space, physical space possesses a metrical structure, which enriches the picture we are exploring. For the time being, we will imagine that some definite, smooth choice has been made for the fields of vectors $\hat{\mathbf{e}}_1(\mathbf{x}),\hat{\mathbf{e}}_2(\mathbf{x})$ over some region of space.

We begin with the statement, which we shall gradually make more precise, that the changes in $\hat{\mathbf{e}}_1(\mathbf{X}),\hat{\mathbf{e}}_2(\mathbf{X})$, as the guiding center $\mathbf{X}$ moves from place to place, are due to two causes. The first is that $(\hat{\mathbf{e}}_1,\hat{\mathbf{e}}_2)$ are forced to be perpendicular to $\hat{\mathbf{b}}$, and since $\hat{\mathbf{b}}$ is changing, there is some kind of "irreducible minimum" by which $(\hat{\mathbf{e}}_1,\hat{\mathbf{e}}_2)$ must change. And the second is that $(\hat{\mathbf{e}}_1,\hat{\mathbf{e}}_2)$ can also change by rotating in their plane of definition. This second part, it turns out, corresponds precisely to the term $\mathbf{R}\cdot\dot{\mathbf{X}}$ in the equation for $\dot{\theta}$.

But what sense does it make to talk about a rotation in a plane which itself is changing from point to point? The following answer may be given. First we compare $(\hat{\mathbf{e}}_1,\hat{\mathbf{e}}_2)$ at two neighboring points $\mathbf{X}$ and $\mathbf{X}+d\mathbf{X}$, in order to determine the infinitesimal "angle of rotation" $d\alpha$ between the two pairs of vectors. Since the planes spanned by the two pairs are not identical, the answer is not immediate. We proceed by first moving $\hat{\mathbf{e}}_1(\mathbf{X})$ and $\hat{\mathbf{e}}_2(\mathbf{X})$ parallel to themselves over to the new point $\mathbf{X}+d\mathbf{X}$. In general, these transported vectors will not coincide with $\hat{\mathbf{e}}_1(\mathbf{X}+d\mathbf{X})$ and $\hat{\mathbf{e}}_2(\mathbf{X}+d\mathbf{X})$, because they still span different planes. Next, we project the transported vectors down onto the plane spanned by $\hat{\mathbf{e}}_1(\mathbf{X}+d\mathbf{X})$ and $\hat{\mathbf{e}}_2(\mathbf{X}+d\mathbf{X})$, and call the projected vectors $\hat{\mathbf{f}}_1,\hat{\mathbf{f}}_2$. Because the projection is perpendicular, the norm of the vectors is preserved through first order in $d\mathbf{X}$, as is their orthogonality. Therefore $\hat{\mathbf{f}}_1$ and $\hat{\mathbf{f}}_2$ are still orthonormal. However, $\hat{\mathbf{f}}_1$ and $\hat{\mathbf{f}}_2$ still do not in general coincide with $\hat{\mathbf{e}}_1(\mathbf{X}+d\mathbf{X})$ and $\hat{\mathbf{e}}_2(\mathbf{X}+d\mathbf{X})$, even though now they all lie in the same plane, simply because there is no reason for them to do so. Given an arbitrary choice for the fields $(\hat{\mathbf{e}}_1,\hat{\mathbf{e}}_2)$, such as Eq. (3.2), we should not expect the vectors which have been transported and projected to coincide with the vectors that are already at the new location. Therefore there is some angle between $(\hat{\mathbf{f}}_1,\hat{\mathbf{f}}_2)$ and $(\hat{\mathbf{e}}_1(\mathbf{X}+d\mathbf{X}),\hat{\mathbf{e}}_2(\mathbf{X}+d\mathbf{X}))$, which we define to be $d\alpha$. This is our definition of the angle of rotation of $(\hat{\mathbf{e}}_1,\hat{\mathbf{e}}_2)$ between points of infinitesimal separation. For points of finite separation, we simply use infinitesimal process step by step along some curve joining the points, to get an accumulated angle of rotation. As we shall see, for points of finite separation, the result depends on the path.

The definition of $d\alpha$ can be stated in another way. The increment in a vector such as $\hat{\mathbf{e}}_1$ or $\hat{\mathbf{e}}_2$ on moving from $\mathbf{X}$ to $\mathbf{X}+d\mathbf{X}$ consists of a part parallel to $\hat{\mathbf{b}}$, representing both the "irreducible minimum" mentioned above and the projection we have just described, and a part perpendicular to $\hat{\mathbf{b}}$, representing the angle of rotation $d\alpha$ within the moving perpendicular plane. For example, let us take $\hat{\mathbf{e}}_1$ and decompose its increment $d\hat{\mathbf{e}}_1$ with respect to the three unit vectors. Of course, $d\hat{\mathbf{e}}_1$ has no component along $\hat{\mathbf{e}}_1$ itself, since $\hat{\mathbf{e}}_1$ is a unit vector. The other two components are given by

$$\begin{aligned} d\hat{\mathbf{e}}_1 &= d\mathbf{X}\cdot\nabla\hat{\mathbf{e}}_1 \\ &= (d\mathbf{X}\cdot\nabla\hat{\mathbf{e}}_1\cdot\hat{\mathbf{e}}_2)\hat{\mathbf{e}}_2+(d\mathbf{X}\cdot\nabla\hat{\mathbf{e}}_1\cdot\hat{\mathbf{b}})\hat{\mathbf{b}} \\ &= (\mathbf{R}\cdot d\mathbf{X})\hat{\mathbf{e}}_2-(d\mathbf{X}\cdot\nabla\hat{\mathbf{b}}\cdot\hat{\mathbf{e}}_1)\hat{\mathbf{b}} , \end{aligned} \tag{3.10}$$

where we have used the orthonormality of the unit vectors to write, for example, $\nabla\hat{\mathbf{e}}_1\cdot\hat{\mathbf{b}}=-\nabla\hat{\mathbf{b}}\cdot\hat{\mathbf{e}}_1$.

The first term on the right shows that the angle $d\alpha$ can be written

$$d\alpha=\mathbf{R}\cdot d\mathbf{X} , \tag{3.11}$$

which is identical with the increment in the gyrophase resulting from the second term in Eq. (3.5) when the guiding center moves through the displacement $d\mathbf{X}$. In other words, we have not only succeeded in defining an angle of rotation in a moving plane, but we have also shown that the analog of Hannay's angle for guiding center motion, given by Eq. (3.9), is precisely the accumulation of this angle of rotation as the guiding center moves from some initial to some final point.

There has been a tendency in previous studies of the guiding center or gyrokinetic expansion to regard a term such as $\mathbf{R}\cdot d\mathbf{X}$ in Eq. (3.5) as undesirable, since it is not gyrogauge invariant and therefore not completely "physical." Some people have attempted to eliminate such terms by making some judicious choice for $(\hat{\mathbf{e}}_1,\hat{\mathbf{e}}_2)$. This is a reasonable approach to take at this point, because we have just determined that the infinitesimal increment in $\hat{\mathbf{e}}_1$ or $\hat{\mathbf{e}}_2$ consists of an irreducible minimum, i.e., a component parallel to $\hat{\mathbf{b}}$, and another part representing a rotation in the perpendicular plane. [These first and second parts correspond, respectively, to the second and first terms of Eq. (3.10).] Therefore we may ask whether it is possible to redefine $\hat{\mathbf{e}}_1$ and $\hat{\mathbf{e}}_2$ so that the second part vanishes. The answer turns out to be yes, so long as we confine ourselves to a single curve.

Therefore instead of filling up space arbitrarily with a field of triads, let us attempt to construct a field that is as "rotationless" as possible. We begin by choosing a point $\mathbf{x}_0$ of space and an arbitrary orientation for $(\hat{\mathbf{e}}_1,\hat{\mathbf{e}}_2)$ at $\mathbf{x}_0$. (See Fig. 4. We shall henceforth write $\mathbf{x}$ for both the particle and guiding center position, since the distinction will no longer be important.) This gives us a single triad, not a field. Next, we draw a curve $\mathbf{x}=\mathbf{x}(s)$, parameterized by the distance s, passing through $\mathbf{x}_0$. This curve need not be a field line, as the figure indicates ($\hat{\mathbf{b}}$ is not

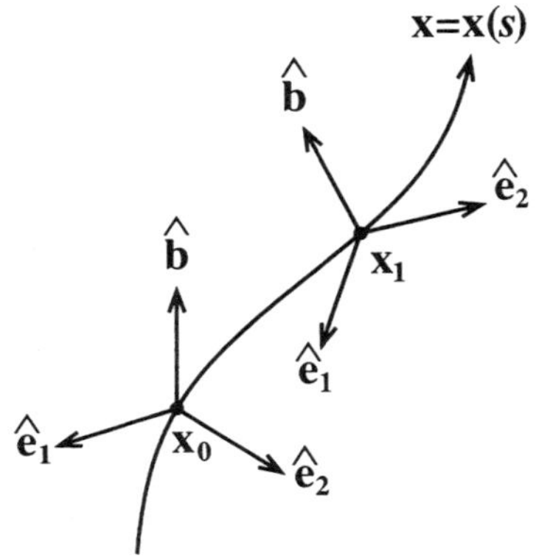

FIG. 4. A rotationless frame can be established along a given curve $\mathbf{x}=\mathbf{x}(s)$ by transporting $\hat{\mathbf{e}}_1$ and $\hat{\mathbf{e}}_2$ from some initial point $\mathbf{x}_0$.

tangent to it). In order to extend the definition of $(\hat{\mathbf{e}}_1,\hat{\mathbf{e}}_2)$ to remaining points on the curve in such a way that the triad is rotationless along the curve, we may proceed in the following way. First we move $\hat{\mathbf{e}}_1$ and $\hat{\mathbf{e}}_2$ parallel to themselves a small distance ds down the curve. Then we project the transported $\hat{\mathbf{e}}_1$ and $\hat{\mathbf{e}}_2$ onto the perpendicular plane at the new point, and call the resulting vectors the new $\hat{\mathbf{e}}_1$ and $\hat{\mathbf{e}}_2$ at that point. Repeating these steps for successive increments of distance, we fill up the curve with triads.

We have just described a transport process, which amounts to demanding that the increments in $\hat{\mathbf{e}}_1$ or $\hat{\mathbf{e}}_2$ be parallel to $\hat{\mathbf{b}}$ [effectively, the first term of Eq. (3.10) is forced to vanish]. To write this as a differential equation, let $\mathbf{u}$ be any vector which lies in the perpendicular plane at $\mathbf{x}_0$ and which we wish to transport. ($\mathbf{u}$ might be $\hat{\mathbf{e}}_1$ or $\hat{\mathbf{e}}_2$ or something else.) To transport $\mathbf{u}$ and extend it to a function $\mathbf{u}(s)$, we solve the transport differential equation,

$$\frac{d\mathbf{u}}{ds}=-\hat{\mathbf{b}}\left[\frac{d\hat{\mathbf{b}}}{ds}\cdot\mathbf{u}\right], \tag{3.12}$$

which comes from an adaptation of the last term in Eq. (3.10).

This transport equation has several notable properties. First, one can easily show that the quantity $\mathbf{u}\cdot\hat{\mathbf{b}}$ is constant along the curve. Since we require $\mathbf{u}$ to be perpendicular to $\hat{\mathbf{b}}$ at the initial point, our equation guarantees that it will remain perpendicular to $\hat{\mathbf{b}}$ at all successive points. Next, let $\mathbf{u}_1$ and $\mathbf{u}_2$ be any two perpendicular vectors, both of which are transported by Eq. (3.12). Then again it is easy to show that $\mathbf{u}_1\cdot\mathbf{u}_2$ is constant along the curve. Letting $\mathbf{u}_1=\mathbf{u}_2=\hat{\mathbf{e}}_1$ or $\hat{\mathbf{e}}_2$, we see that our transported unit vectors remain unit vectors; and letting $\mathbf{u}_1=\hat{\mathbf{e}}_1$, $\mathbf{u}_2=\hat{\mathbf{e}}_2$, we see that they remain orthogonal. This shows that our transport equation does indeed produce an orthonormal frame along the given curve, one which is rotationless.

(Although in the guiding center problem we only need to transport perpendicular vectors, it is natural to ask whether our transport process can be generalized to accommodate vectors $\mathbf{u}$ which are not purely perpendicular. The answer is yes; the most reasonable way to do this is to demand that the component of $\mathbf{u}$ along $\hat{\mathbf{b}}$ should remain constant throughout the transport. In this way, $\hat{\mathbf{b}}$ itself could be transported, and it would always be taken into itself. For arbitrary vectors $\mathbf{u}$, the generalized transport equation would be

$$\frac{d\mathbf{u}}{ds}=\left[\frac{d\hat{\mathbf{b}}}{ds}\hat{\mathbf{b}}-\hat{\mathbf{b}}\frac{d\hat{\mathbf{b}}}{ds}\right]\cdot\mathbf{u}\,. \tag{3.13}$$

This equation is the direct analog of Fermi-Walker transport in special relativity,[17] with $\hat{\mathbf{b}}$ being replaced by a unit timelike vector. It has the property that it preserves all scalar products of transported vectors. The importance of this kind of transport for the phase anholonomy in guiding center motion was pointed out in Ref. 9, and more recently Haldane,[18] Berry,[19] and Segert[20] have discussed the role of Fermi-Walker transport in connection with the experiments of Chiao and co-workers[5,6] on light propagation in optical fibers.)

We now have triads along one curve, but not a field of triads, so $\mathbf{R}$ is not defined. If, however, the definition of $(\hat{\mathbf{e}}_1,\hat{\mathbf{e}}_2)$ is extended to a finite region around the curve, without changing their values on the curve itself, then we would find that $\mathbf{R}\cdot d\mathbf{x}/ds=0$ on the curve. If the curve were an actual guiding center trajectory, then the second term in Eq. (3.5) would vanish. More generally, one can show that a single component of $\mathbf{R}$ can always be made to vanish over a region by an appropriate choice of $(\hat{\mathbf{e}}_1,\hat{\mathbf{e}}_2)$, where the component chosen can be a function of position. (One simply transports perpendicular vectors along the integral curves of the direction field which represents the chosen component, starting from some initial value surface transverse to the integral curves.) For example, it is possible to choose a gauge such that $\hat{\mathbf{b}}\cdot\mathbf{R}=0$; this possibility has been explored by Hagan and Frieman.[7] However, such choices may not be practical; for example, if field lines are closed, then the demand $\hat{\mathbf{b}}\cdot\mathbf{R}=0$ will usually lead to a discontinuity in $(\hat{\mathbf{e}}_1,\hat{\mathbf{e}}_2)$ at some point.

Although we can always make one component of $\mathbf{R}$ vanish by an appropriate choice of gauge, we cannot make all three components vanish (at least over a finite region). For if $\mathbf{R}=0$, then $\nabla\times\mathbf{R}=\mathbf{0}$. But $\nabla\times\mathbf{R}$ is gyrogauge invariant, so if it vanishes for one choice of gauge, it vanishes for all. Nor it is very useful in plasma problems even to make one component of $\mathbf{R}$ vanish. Although we can make our analog of Hannay's angle vanish along the guiding center trajectory of a single particle, a different particle will have a different trajectory, so we do not gain much. Instead, it is better to acknowledge the gauge dependence and even exploit it.

An interesting question is the relation which Eq. (3.12) bears to parallel transport as it occurs in Riemannian geometry or general relativity. The two cannot be identical, because we have a nontrivial transport process in Euclidean space, where the usual parallel transport is trivial. If we define a connection (or Christoffel symbols) by

$$\Gamma^i_{jk} = b^i \frac{\partial b_k}{\partial x^j} , \tag{3.14}$$

then our transport equation becomes

$$\frac{du^i}{ds} + \Gamma^i_{jk} \frac{dx^j}{ds} u^k = 0 . \tag{3.15}$$

Although this is in the usual form of parallel transport, the connection has torsion, i.e., $\Gamma^i_{jk} \neq \Gamma^i_{kj}$, unless $\nabla \times \hat{\mathbf{b}} = 0$. But that condition is more than sufficient for the existence of a family of two-dimensional surfaces which are everywhere perpendicular to $\hat{\mathbf{b}}$. Assuming $\nabla \times \hat{\mathbf{b}} = 0$, we can imagine restricting our transport process to curves which lie in one of these surfaces. In that case, our transport is identical to parallel transport in that surface. (Actually, the condition $\hat{\mathbf{b}} \cdot \nabla \times \hat{\mathbf{b}} = 0$ is sufficient for the existence of such surfaces, by the Frobenius theorem. Although this does not guarantee that $\Gamma^i_{jk} = \Gamma^i_{kj}$ for all j,k, it does guarantee symmetry in the lower two indices for a Γ which is appropriately restricted to the surface. Indeed, this restricted Γ is the usual connection of parallel transport on the surface.) The fact that our transport coincides with parallel transport when restricted to one of these surfaces can be seen geometrically, since parallel transport of a vector tangent to a two-dimensional surface embedded in three-dimensional Euclidean space can be thought of as a sequence of small parallel displacements in the larger space, each followed by a projection onto the plane tangent to the surface. Altogether, we can say that our transport equation is a generalization of parallel transport, in that it allows arbitrary paths in three-dimensional space and it does not require the existence of such surfaces.

Let us now consider transporting a vector $\mathbf{u}$ perpendicular to $\mathbf{B}$ around a closed loop. We do not do this because we expect guiding center trajectories to follow a closed path—usually they do not—but because, in the spirit of Stokes's theorem, we wish to explore further the issue of the path dependence of Hannay's angle for guiding center motion, Eq. (3.9). In particular, let us take for our closed loop an infinitesimal parallelogram, spanned by vectors $\Delta\mathbf{x}_1$ and $\Delta\mathbf{x}_2$. If we take a perpendicular vector $\mathbf{u}$ at one corner of the parallelogram and transport it around, it must still lie in the perpendicular plane when it returns to the original corner. But the final vector need not be identical to the original vector; in general, it will differ from it by some angle $\Delta\alpha$. To calculate this angle, we simply expand the solution of the transport equation, Eq. (3.12), out to second order in s for each of the four legs of the parallelogram. It is necessary to go to second order, because the first-order terms cancel one another. Then one finds that the net angle of rotation after one circuit can be written in the form,

$$\Delta\alpha = \mathbf{N} \cdot \Delta\mathbf{S} , \tag{3.16}$$

where $\Delta\mathbf{S} = \Delta\mathbf{x}_1 \times \Delta\mathbf{x}_2$ is the infinitesimal area element, and where

$$\mathbf{N} = \tfrac{1}{2}\hat{\mathbf{b}}[(b_{i,j}b_{j,i}) - (\nabla\cdot\hat{\mathbf{b}})^2] + (\nabla\cdot\hat{\mathbf{b}})\hat{\mathbf{b}}\cdot\nabla\hat{\mathbf{b}} - \hat{\mathbf{b}}\cdot\nabla\hat{\mathbf{b}}\cdot\nabla\hat{\mathbf{b}} . \tag{3.17}$$

It is interesting that this vector field $\mathbf{N}$ only involves the first derivatives of $\hat{\mathbf{b}}$, even though we expanded to second order in s. The reason is that all the terms involving second derivatives cancel.

On the other hand, if our parallelogram is embedded in a region in which $(\hat{\mathbf{e}}_1, \hat{\mathbf{e}}_2)$ have been defined, then we can integrate $\mathbf{R}\cdot d\mathbf{x}$ around the loop, and, independent of the choice of gauge, we should get the same answer. Therefore by Stokes's theorem we conclude that $\mathbf{N} = \nabla \times \mathbf{R}$, and our earlier guess, that $\nabla \times (\nabla\hat{\mathbf{e}}_1 \cdot \hat{\mathbf{e}}_2)$ could be expressed purely in terms of $\hat{\mathbf{b}}$, is shown to be true. One can also verify $\mathbf{N} = \nabla \times \mathbf{R}$ directly, using various vector identities. We see that the divergence-free vector $\mathbf{N}$ is the analog of Hannay's vector $\mathbf{B}$, and it represents a kind of angle flux in ordinary space. It is also something of a generalization of the Riemann tensor for two-dimensional surfaces embedded in Euclidean three-dimensional space.

There is another interesting geometrical fact connected with the vectors $\mathbf{R}$ and $\mathbf{N}$. Let us choose a fixed set of spherical coordinates in space, to which directions can be referred, and let us denote the spherical angles of $\hat{\mathbf{b}}$ at some point $\mathbf{x}$ by (θ,ϕ). (One can think of θ and ϕ as being functions of $\mathbf{x}$). Then we can let $\hat{\mathbf{e}}_1 = \hat{\boldsymbol{\theta}}, \hat{\mathbf{e}}_2 = \hat{\boldsymbol{\phi}}$, i.e.,

$$\begin{aligned} \hat{\mathbf{e}}_1 &= \hat{\mathbf{x}} \cos\theta \cos\phi + \hat{\mathbf{y}} \cos\theta \sin\phi - \hat{\mathbf{z}} \sin\theta , \\ \hat{\mathbf{e}}_2 &= -\hat{\mathbf{x}} \sin\phi + \hat{\mathbf{y}} \cos\phi . \end{aligned} \tag{3.18}$$

This is essentially a choice of gauge, as before, except that now we may think of $(\hat{\mathbf{e}}_1, \hat{\mathbf{e}}_2)$ as being defined over the unit sphere instead of over three-dimensional space, i.e., as functions of (θ,ϕ). These vectors are discontinuous at the poles, where ϕ is undefined, but nevertheless they are suitable for subsequent calculations. (Discontinuities like these are inevitable, in view of the "hair on the coconut" theorem, which denies the existence of smooth, nonvanishing vector fields tangent to a sphere.) In the gauge (3.18) it is a short calculation to show that

$$\mathbf{R}\cdot d\mathbf{x} = \cos\theta \, d\phi . \tag{3.19}$$

Therefore the integral of $\mathbf{R}\cdot d\mathbf{x}$ around a closed loop in real space must be equal to the integral of $\cos\theta \, d\phi$ around the corresponding closed loop on the unit sphere traced out by $\hat{\mathbf{b}}$. But this latter integral is just the solid angle contained in the cone traced out by $\hat{\mathbf{b}}$, which therefore is also equal to the net angle of rotation of the triad on being transported around the given loop in real space. This is a strikingly simple geometrical fact, for which there may be a shorter proof.

There are several instances in the literature on Berry's phase in which the solid angle on a sphere (sometimes multiplied by one half) has turned out to be equal to an anholonomic phase in a physical problem, and now we see that guiding center motion provides yet another example. The reason for this repetition has been pointed out by Segert,[20] who has shown that such a result is the only one consistent with rotational symmetry.

Just as Eq. (3.18) effectively transfers the meaning of the vector $\mathbf{R}$ to the unit sphere, we can do the same for the vector $\mathbf{N}$. The result must be expressed in terms of differential two-forms, in the following way:

$$\mathbf{N}\cdot d\mathbf{S}=\tfrac{1}{2}\epsilon_{ijk}N_i dx_j \wedge dx_k = \sin\theta\, d\phi \wedge d\theta \ , \tag{3.20}$$

which is just the area element on the surface of the sphere.

It is also interesting to convert our transport equation, Eq. (3.13), into its analog on the unit sphere, for use in carrying out rotationless transport along curves on the sphere. We will not write this new transport equation down, but it is clear geometrically, in view of the infinitesimal description we have given for the transport process, that the new equation will represent ordinary parallel transport on the unit sphere. Altogether, we have shown that the phase anholonomy which occurs in guiding center motion can be reformulated completely in terms of the non-Euclidean geometry of the unit sphere, a two-dimensional surface of constant positive curvature.

In Sec. IV we will discuss Thomas precession in some detail, but it is interesting at this point to note the analog of our most recent observation. It is that the net three-dimensional rotation experienced by an accelerated particle when its velocity vector is taken around a closed loop is identical with the rotation suffered by a frame which has been parallel transported around a closed loop on a three-dimensional surface of constant negative curvature. This surface is simply the unit mass shell of special relativity, i.e., the space of all unit timelike vectors. This observation allows an interpretation of Thomas precession in terms of Lobachevskian non-Euclidean geometry. For example, if the orbit of the particle in ordinary space lies in a plane, it turns out that the loop traced out by its unit timelike vector on the unit mass shell also lies in a Lobachevskian plane. In this case, the transport is essentially two-dimensional, as on the surface of the sphere, so the rotations are described by a single angle. (That is, the gauge theory becomes Abelian.) Not surprisingly, the angle of rotation is just the Lobachevskian area in this plane, in direct analogy with the solid angle which occurs in the guiding center frame transport, and the axis of rotation is perpendicular to the plane. Transport around orbits which are not planar is more complicated, and involves noncommuting rotations. These are relatively new insights into Thomas precession, which have also been recently made by Segert[20] and Jordan.[21]

To complete this discussion, we will now derive some identities satisfied by **N**, which are useful in practical calculations involving the guiding center or gyrokinetic expansions.

Of course, we must have $\nabla\cdot\mathbf{N}=0$, but if we directly take the divergence of Eq. (3.17), we find

$$\nabla\cdot\mathbf{N}=\tfrac{3}{2}(\nabla\cdot\hat{\mathbf{b}})(b_{i,j}b_{j,i})-\tfrac{1}{2}(\nabla\cdot\hat{\mathbf{b}})^3-(b_{j,i}b_{k,j}b_{i,k}) \ . \tag{3.21}$$

It is not obvious that this vanishes, but to prove that it does we may call on the following fact. Let $\underline{\mathbf{M}}$ be any 3×3 matrix with components M_{ij}. Then

$$\det\underline{\mathbf{M}}=\frac{1}{3!}[(\mathrm{Tr}\mathbf{M})^3-3(\mathrm{Tr}\mathbf{M})(\mathrm{Tr}\mathbf{M}^2)+(\mathrm{Tr}\mathbf{M}^3)] \ . \tag{3.22}$$

Therefore by setting $M_{ij}=b_{j,i}$, we have

$$\nabla\cdot\mathbf{N}=-\tfrac{1}{3}\det(b_{j,i}) \ . \tag{3.23}$$

This in turn vanishes because the matrix $b_{j,i}$ has an eigenvector with eigenvalue zero. This is simply $\hat{\mathbf{b}}$ itself, since $\nabla\hat{\mathbf{b}}\cdot\hat{\mathbf{b}}=0$.

The matrix $b_{j,i}$ must also have a left eigenvector of eigenvalue zero, which turns out to be **N**. One can show this by considering the characteristic polynomial of the matrix $b_{j,i}$. Let the eigenvalues of this matrix be λ_0, λ_1, and λ_2. Since at least one of these must vanish, let us assume that $\lambda_0=0$. Then Eq. (3.17) can be written

$$\mathbf{N}=-\lambda_1\lambda_2\hat{\mathbf{b}}+(\lambda_1+\lambda_2)\hat{\mathbf{b}}\cdot\nabla\hat{\mathbf{b}}-\hat{\mathbf{b}}\cdot\nabla\hat{\mathbf{b}}\cdot\nabla\hat{\mathbf{b}} \ . \tag{3.24}$$

If we multiply $\nabla\hat{\mathbf{b}}$ on the left by this vector, the result must vanish, because any matrix satisfies its own secular equation. Therefore

$$\mathbf{N}\cdot\nabla\hat{\mathbf{b}}=0 \ . \tag{3.25}$$

The fact that such a left null eigenvector exists means that there is a direction in space along which $\hat{\mathbf{b}}$ is stationary; and the fact that **N** is this eigenvector means that this direction is also the one in which a normal surface intercepts the maximum amount of "angle flux."

The most important message of this analysis for guiding center and gyrokinetic problems is that there will inevitably occur terms depending on the perpendicular unit vectors which cannot be transformed away, and that it is best to live with these terms and to understand their gauge dependence. However, we have dealt only with geometrical issues, and ignored dynamical ones. For example, one would also like to know if it is possible to define, say, the guiding center position or a transformed, gyrokinetic version of the Vlasov equation which is gyrogauge invariant to all orders. The answer is yes, as is shown in Refs. 9 and 13. Another question is whether the only gyrogauge-dependent quantities we will ever encounter in the guiding center or gyrokinetic expansions are the ones considered here, or whether there may arise new ones, independent of these. The answer to this is that there exist perturbation expansions, based on gyrogauge invariant Lie generators, which guarantee that all gyrogauge-dependent quantities that one will ever encounter can be expressed in terms of those already considered here. This assertion is also proved in Refs. 9 and 13.

IV. GENERALIZATIONS, ANALOGIES, AND DISCUSSION

We obtain a rather simple generalization of these results by considering guiding center motion in time-dependent magnetic fields, for which $\hat{\mathbf{b}}=\hat{\mathbf{b}}(\mathbf{x},t)$. Now the parameters which determine the environment of the guiding center are $(\mathbf{x},t)$, allowing us to identify Hannay's parameter space with space-time. But we may still wish to treat the problem nonrelativistically; in that case, we need only augment the "vector potential" **R** with a "scalar potential" σ, defined by

$$\sigma=\frac{\partial\hat{\mathbf{e}}_2}{\partial t}\cdot\hat{\mathbf{e}}_1 \ . \tag{4.1}$$

Under a time-dependent gyrogauge transformation, σ transforms according to

$$\sigma' = \sigma - \frac{\partial \psi}{\partial t} , \tag{4.2}$$

and $\mathbf{R}$ and σ together produce a new gyrogauge-invariant field,

$$\mathbf{F} = -\nabla\sigma - \frac{\partial \mathbf{R}}{\partial t} . \tag{4.3}$$

Just like $\mathbf{N}$, $\mathbf{F}$ can also be expressed purely in terms of $\hat{\mathbf{b}}$; the result is

$$\mathbf{F} = \nabla\hat{\mathbf{b}} \cdot \left[\hat{\mathbf{b}} \times \frac{\partial \hat{\mathbf{b}}}{\partial t} \right] . \tag{4.4}$$

Finally, $\mathbf{N}$ and $\mathbf{F}$ together satisfy the analog of the Faraday equation,

$$\nabla \times \mathbf{F} = -\frac{\partial \mathbf{N}}{\partial t} . \tag{4.5}$$

Relativistic guiding center motion leads to another interesting generalization. Just as the nonrelativistic magnetic field $\mathbf{B}$ causes a local decomposition of three-dimensional space (more properly, the tangent space) into the direct sum of a one-dimensional and a two-dimensional subspace, namely, the direction parallel to $\hat{\mathbf{b}}$ and the plane perpendicular to it, so also the relativistic field tensor $F_{\mu\nu}$ causes a decomposition of four-dimensional space-time into the direct sum of two two-dimensional subspaces. One of these is the relativistic analog of the nonrelativistic plane perpendicular to $\hat{\mathbf{b}}$, and the other is a two-dimensional subspace spanned by one spacelike and one timelike vector. These two subspaces can be thought of as the degenerate eigenspaces of the electromagnetic stress-energy tensor; their importance for relativistic guiding center theory was first recognized by Fradkin.[22] Furthermore, just as we have an arbitrary choice for the orientation of unit vectors in the perpendicular plane in the nonrelativistic problem, we now have an arbitrary choice for two orientations, one in each of the two two-dimensional subspaces. Changing one of these amounts to a two-dimensional rotation, as in Eq. (3.3), and changing the other is equivalent to a Lorentz transformation in one space and one time dimension. This gauge structure and the fields associated with it have recently been examined in detail by Boghosian.[23]

Equation (3.3) shows that nonrelativistic guiding center motion is an SO(2)=U(1) gauge theory, representing two-dimensional rotations in the perpendicular plane. Hannay's analysis leads to a $U(1)^N$ gauge theory, where N is the number of degrees of freedom, so it looks like nonrelativistic guiding center motion is just the special case $N=1$. [The $U(1)^N$ gauge transformation represents a redefinition of the point on the N-dimensional invariant torus where all the angles vanish.] But the relativistic guiding center problem is a surprise; it is an $SO(2)\times SO(1,1)$ gauge theory. All these examples are Abelian, but another example, which provides a geometrical picture remarkably like that found in guiding center motion, contains a non-Abelian SO(3) gauge theory. This is Thomas precession, which I shall now discuss for the purpose of drawing the analogy with the phase anholonomy in guiding center motion.

Thomas precession is a special relativistic effect which involves a nontrivial transport process, namely, Fermi-Walker transport, in spite of the fact that the four-dimensional space-time of special relativity is flat. This is analogous to the fact that the guiding center problem gives rise to a nontrivial transport process in flat three-dimensional space. In the case of guiding center motion, it is the magnetic field which provides the extra structure needed for the transport process, and in the case of Thomas precession, it is the world line of a particle (see Fig. 5). This world line is analogous to a single magnetic field line in the guiding center problem, and Fermi-Walker transport is necessarily confined to this line. In this sense, Thomas precession is simpler than the guiding center problem, where transport can occur along any curve in space.

The Lorentz frame (e_0, e_1, e_2, e_3) of the figure is an instantaneous rest frame of the particle, and is analogous to the frame $(\hat{\mathbf{e}}_1, \hat{\mathbf{e}}_2, \hat{\mathbf{b}})$ of the guiding center problem. (The subscripts on the unit vectors of this Lorentz frame are not covariant indices, but merely labels to distinguish the vectors.) In order to qualify as a rest frame, the timelike unit vector e_0 must be given by

$$e_0^\mu = \frac{dx^\mu}{d\tau} , \tag{4.6}$$

where τ is proper time, but the other three spacelike unit vectors are arbitrary, apart from the fact that they must form an orthonormal tetrad with e_0. Therefore any frames which qualify as rest frames at a given point must have identical time axes, and can differ from one another only by a purely spatial rotation. We may regard an assignment of rest frames along the world line of the particle as a choice of gauge, which differs from other choices by a spatial rotation at each point. Of course, the rotation can differ from point to point, i.e., it can be a function of τ, just as the two-dimensional angle of rotation ψ in the guiding center problem is allowed to be a function

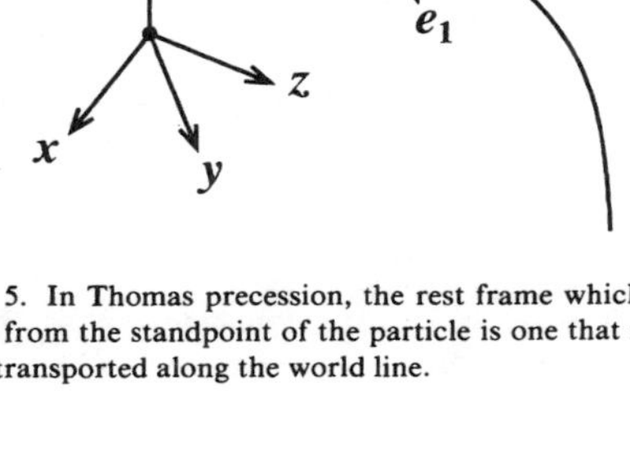

FIG. 5. In Thomas precession, the rest frame which is rotationless from the standpoint of the particle is one that is Fermi-walker transported along the world line.

of $\mathbf{x}$. A discussion of these rotations from the standpoint of Wigner's little group has recently been given by Han, Kim, and Son.[24]

Accounts of Thomas precession often refer to "the" rest frame of the particle, which in the present context must be understood as a particular choice of gauge. This is usually specified by demanding that the Lorentz transformation connecting some fixed lab frame, indicated by (t,x,y,z) in Fig. 5, and the rest frame of the particle be a pure boost. If $A^{\mu}=(A^0, \mathbf{A})$ represent the components of some contravariant vector in the particle rest frame, and $(A_0^0, \mathbf{A}_0)$ the components in the lab frame, then this Lorentz transformation is given by

$$\begin{aligned} A^0 &= \gamma(A_0^0 - \mathbf{v}\cdot\mathbf{A}_0) , \\ \mathbf{A} &= \mathbf{A}_0 + \frac{\gamma-1}{v^2}(\mathbf{v}\cdot\mathbf{A}_0)\mathbf{v} - \gamma\mathbf{v}A_0^0 , \end{aligned} \tag{4.7}$$

where $\mathbf{v}$ is the velocity of the particle and where $c=1$. This Lorentz transformation uniquely specifies an assignment of rest frames along the particle's world line, which we shall identify with (e_0,e_1,e_2,e_3) and call the "conventional" rest frames.

This assignment of rest frames is not rotationless, however, if we define the rotationless transport of a perpendicular four-vector u (i.e., one for which $e_0\cdot u=0$) by the transport equation

$$\frac{du}{d\tau} = -e_0\left[\frac{de_0}{d\tau}\cdot u\right] \tag{4.8}$$

(essentially Fermi-Walker transport), in analogy with Eq. (3.12). (Here I have used a metric with signature $+---$.) The fact that the infinitesimal change in u given by this equation is parallel to e_0 means that the change is governed by a pure boost. This transport equation can be used to define a rotationless frame in exactly the same manner as in guiding center motion; we simply choose an arbitrary orthonormal tetrad $(e_0,\tilde{e}_1,\tilde{e}_2,\tilde{e}_3)$ at some initial point, and then solve the transport equation with u identified with the $\tilde{e}_i$. The tildes distinguish the spatial unit vectors of the rotationless frame from those of the conventional frame.

The conventional and rotationless frames, as seen by the particle, must be related by a spatial rotation, whose time derivative is represented by an angular velocity vector $\boldsymbol{\omega}$. This vector $\boldsymbol{\omega}$ is the analog of $\mathbf{R}$ in the guiding center problem, or, more exactly, $\boldsymbol{\omega}\, d\tau$ is the analog of $\mathbf{R}\cdot d\mathbf{x}$. ($\boldsymbol{\omega}\, d\tau$ is a vector, representing an infinitesimal rotation in three dimensions, and $\mathbf{R}\cdot d\mathbf{x}=d\theta$ is a scalar, representing a two-dimensional rotation. Note that they are both gauge dependent.) Let us now compute $\boldsymbol{\omega}$, in order to bring out this analogy.

Let u be a rotationless vector satisfying Eq. (4.8), with components $(0,u^i)$ in the conventional frame. The components u^i are, of course, time dependent, since the conventional frame is not rotationless. This vector is, however, stationary as seen in the rotationless frame, and plays somewhat the same role as a vector fixed in a rigid body. Therefore the angular velocity $\boldsymbol{\omega}$ of the rotationless frame with respect to the conventional frame satisfies

$$\frac{du^i}{d\tau} = \epsilon_{ijk}\omega_j u^k , \tag{4.9}$$

where ω_j are the components of $\boldsymbol{\omega}$ also with respect to the conventional frame. (It would seem more logical to compute the rate of rotation of the conventional frame with respect to the rotationless frame, but since we have a definite convention for the former and not for the latter, we will do it backwards.)

Since the four-vector u has a vanishing component along e_0, we have $u=u^i e_i$ (repeated Latin indices are summed over 1,2,3). Combining this with Eqs. (4.8) and (4.9), we have

$$\begin{aligned} \frac{du}{d\tau} = -e_0\left[\frac{de_0}{d\tau}\cdot u\right] &= \frac{d}{d\tau}(u^i e_i) \\ &= \epsilon_{ijk}e_i\omega_j u^k + u^k\frac{de_k}{d\tau} . \end{aligned} \tag{4.10}$$

Forming the dot product of this with e_m and noting that u^k represents an arbitrary spatial vector, we can solve for ω_i to get

$$\omega_i = \tfrac{1}{2}\epsilon_{ijk}\frac{de_j}{d\tau}\cdot e_k . \tag{4.11}$$

This should be compared with the equation, $\mathbf{R}=\nabla\hat{\mathbf{e}}_1\cdot\hat{\mathbf{e}}_2$.

It is now straightforward to obtain the components of the unit vectors e_i of the conventional frame with respect to the lab frame, and to use them in Eq. (4.11) to compute ω. The calculation is simplified by disregarding all terms symmetric in (j,k), since these will cancel anyway. The result is

$$\boldsymbol{\omega} = \frac{\gamma^3}{1+\gamma}\mathbf{a}\times\mathbf{v} , \tag{4.12}$$

where $\mathbf{a}$ is the acceleration of the particle. This differs from the usual formula for the angular velocity of Thomas precession[25] by a factor of γ, since we have computed the rate of rotation with respect to proper time instead of lab time. (It is more logical to do it this way, since the rotation occurs in the particle's rest frame.) This is something of a hybrid expression, since $\mathbf{a}$ and $\mathbf{v}$ are measured in the lab frame, whereas $\boldsymbol{\omega}$ is measured in the conventional rest frame. Most elementary discussions of Thomas precession do not make it clear that $\boldsymbol{\omega}$ does not represent an angular velocity in the lab frame, but rather in the particle rest frame.

Several further analogies could be drawn between the guiding center phase anholonomy and similar phenomena. For example, Berry's phase in quantum mechanics can be thought of as an angle of rotation in a moving complex plane (a one-dimensional complex linear space) in Hilbert space, with a transport equation that is virtually identical in structure to those presented here. But since these similarities do not add much to the understanding of guiding center motion, they will not be considered here.

ACKNOWLEDGMENTS

I am grateful for useful discussions with Ted Frankel, Alan Weinstein, and Darryl Holm. I would also like to thank Allan Kaufman for a careful reading of the manuscript. This work was supported by the U.S. Department of Energy under Contract Nos. DE-AC03-76F00098 and W7405/ENG-36, and by the National Science Foundation under Grant No. NSF-PYI-84-51276.

APPENDIX

In this appendix we will prove that

$$\frac{\partial K_1}{\partial I} = \frac{\partial \theta}{\partial t}(q,p,\mathbf{X}(t)) , \tag{A1}$$

where we follow the notation of Sec. II. The easiest way to do this is to note that

$$\dot{\theta} = \frac{\partial \theta}{\partial q}\dot{q} + \frac{\partial \theta}{\partial p}\dot{p} + \frac{\partial \theta}{\partial t} = \{\theta, H\} + \frac{\partial \theta}{\partial t} , \tag{A2}$$

where the curly bracket is the Poisson bracket. But since the Poisson bracket can be evaluated in any set of canonical coordinates, and since H and K_0 are the same functions, expressed in terms of two different sets of independent variables, we have

$$\{\theta, H\}_{qp} = \{\theta, K_0\}_{\theta I} = \frac{\partial K_0}{\partial I} , \tag{A3}$$

where the subscripts indicate the canonical variables with respect to which the Poisson bracket is evaluated. Therefore we have

$$\dot{\theta} = \frac{\partial K_0}{\partial I} + \frac{\partial \theta}{\partial t} = \frac{\partial K_0}{\partial I} + \frac{\partial K_1}{\partial I} , \tag{A4}$$

since the total Hamiltonian in the (θ, I) coordinates is $K = K_0 + K_1$. This proves Eq. (A1).

---

*Permanent address: Department of Physics, University of California, Berkeley, California 94720.

[1]M. V. Berry, Proc. Roy. Soc. London, Ser. A **392**, 45 (1984).
[2]J. H. Hannay, J. Phys. A **18**, 221 (1985).
[3]M. V. Berry, J. Phys. A **18**, 15 (1985).
[4]E. Gozzi and W. D. Thacker, Phys. Rev. D **35**, 2398 (1987).
[5]R. Y. Chiao and Y. S. Wu, Phys. Rev. Lett. **57**, 933 (1986).
[6]A. Tomita and R. Y. Chiao, Phys. Rev. Lett. **57**, 937 (1986).
[7]W. K. Hagan and E. A. Frieman, Phys. Fluids **28**, 2641 (1985).
[8]M. Berry, in *Fundamental Aspects of Quantum Theory,* edited by V. Gorini and A. Frigerio (Plenum, New York, 1986).
[9]R. G. Littlejohn, in *Contemporary Mathematics,* edited by J. E. Marsden (American Mathematical Society, Providence, RI, 1984), Vol. 28, p. 151.
[10]T. G. Northrop, *The Adiabatic Motion of Charged Particles* (Interscience, New York, 1963).
[11]C. S. Gardner, Phys. Rev. **115**, 791 (1959).
[12]R. G. Littlejohn, J. Math. Phys. **20**, 2445 (1979).
[13]R. G. Littlejohn, J. Plasma Phys. **29**, 111 (1983).
[14]J. R. Cary and R. G. Littlejohn, Ann. Phys. **151**, 1 (1983).
[15]R. Abraham and J. E. Marsden, *Foundations of Mechanics* (Benjamin-Cummings, Reading, MA, 1978), p. 279.
[16]R. H. Cohen, G. Rowlands, and J. H. Foote, Phys. Fluids **21**, 627 (1978).
[17]C. W. Misner, K. S. Thorne, and J. A. Wheeler, *Gravitation* (Freeman, New York, 1973), p. 176.
[18]F. D. M. Haldane, Opt. Lett. **11**, 730 (1986).
[19]M. V. Berry, Nature **326**, 277 (1987).
[20]Jan Segert, Phys. Rev. A **36**, 10 (1987).
[21]Thomas F. Jordan, J. Math. Phys. **29**, 2042 (1988).
[22]D. M. Fradkin, J. Phys. A **11**, 1069 (1978).
[23]Bruce M. Boghosian, Ph.D. dissertation, University of California, Berkeley, 1987. Available as Lawrence Berkeley Laboratory Report No. LBL-23241.
[24]D. Han, Y. S. Kim, and D. Son, Class. Quantum Grav. **4**, 1777 (1987).
[25]J. D. Jackson, *Classical Electrodynamics,* 2nd ed. (Wiley, New York, 1975), p. 546.

Berry's Phase for Energy Bands in Solids

J. Zak
Department of Physics, Technion–Israel Institute of Technology, Haifa 32000, Israel
(Received 28 December 1988)

Berry's phase is defined for the dynamics of electrons in periodic solids and an explicit formula is derived for it. Because of the special torus topology of the Brillouin zone a nonzero Berry phase is shown to exist in a one-dimensional parameter space. Symmetry of the Bloch functions in the Brillouin zone leads to the quantization of Berry's phase. A connection is established between the latter and the Wyckoff positions in the crystal in the framework of band representations of space groups. Berry's phase can therefore be used for labeling energy bands in solids.

PACS numbers: 71.25.−s, 03.65.Bz, 61.50.Em

The concept of Berry's phase[1] has recently caused a great deal of interest in a variety of fields in physics. Thus, in the last two years experiments have been carried out for measuring this phase in neutron spin rotation,[2] in nuclear quadrupole resonance,[3] in rotation of polarized light,[4] in fractional quantization of orbital quasirotation of molecules,[5] and in electronic states of Jahn-Teller systems.[6] In addition, a considerable number of theoretical papers have been published dealing with both quantum[7] and classical[8] aspects of the problem. It seems, however, that one important and natural system for the appearance of Berry's phase was left out. We have in mind the motion of an electron in a periodic solid. That such a system should be of interest can be seen in the following way.

In solids the energy spectrum has a band structure, e.g., it is piecewise continuous. The energy in each continuous piece depends on the Bloch quasimomentum $\mathbf{k}$, which varies in the Brillouin zone. What is most characteristic for the concept of Berry's phase is the existence of a continuous parameter space in which the state of the system can travel on a closed path. Such a parameter space exists naturally in the band structure of solids. This parameter space is the Brillouin zone which is a torus, and in which the Bloch $\mathbf{k}$ vector is used for describing an energy band.[9] In a periodic solid $\mathbf{k}$ is a conserved quantity and the Bloch function $\psi_{nk}(\mathbf{r})$ is specified by a band index n and $\mathbf{k}$. If by applying a perturbation to the solid one can make $\mathbf{k}$ vary on a closed path in the Brillouin zone, then $\psi_{nk}(\mathbf{r})$ should, in general, pick up a Berry phase. Since the Brillouin zone is a torus, one can vary $\mathbf{k}$ in a given direction and when the edge of the zone is reached the path closes automatically. This is best seen in a one-dimensional solid where the Brillouin zone is the interval $[-\pi/a,\pi/a]$ with the end points $-\pi/a$ and π/a identified (a is the lattice constant). In the particular case of a one-dimensional crystal, when k is made to vary in the whole Brillouin zone, the Bloch functions $\psi_{nk}(x)$ correspond to the whole energy band. One should therefore expect that when k sweeps the interval $[-\pi/a,\pi/a]$, the Bloch function $\psi_{nk}(x)$ will pick up a Berry phase. This is a very amazing situation because in a one-dimensional crystal $\psi_{nk}(x)$ can be chosen to be periodic in k with the period $2\pi/a$ of the reciprocal lattice. Nevertheless, because of the torus topology of the Brillouin zone a Berry phase is picked up when k is forced to vary by an external perturbation through the entire zone. In solids the continuous parameter space is therefore already built in as a part of the unperturbed problem. What the perturbation does is that it makes k vary in this parameter space (the Brillouin zone). It should be pointed out that in all the above-mentioned physical systems for the observation of the Berry phase the parameter space is multidimensional.[1-8] Because of this very unique situation in the Bloch dynamics of solids, we shall concentrate in what follows on one-dimensional crystals. Keeping in mind that an entire energy band corresponds to the variation of k in the Brillouin zone $[-\pi/a,\pi/a]$, one is to expect that the Berry phase which is picked up by $\psi_{nk}(x)$ in such a variation process should have to do with some characteristic feature of the energy band as a whole entity. Such a characteristic feature is known to exist in the framework of band representations of space groups[10] and it is described by the band-center operator.[11] The eigenvalues of the latter correspond to the Wyckoff positions in the solid. A connection should therefore be anticipated between the Berry phase and the band center in the dynamics of electrons in solids. This is shown in what follows.

In this Letter I derive a formula for the Berry phase for an electron in a periodic potential (one dimension) in the presence of an externally applied time-dependent vector potential $A(t)$. The formula turns out to coincide with the expression for the band center in the framework of the band representations of space groups.[10,11] When the linear chain has no symmetry, the Berry phase can assume any value. However, when inversion symmetry is present, Berry's phase becomes quantized, and it can assume only the values 0 and π (modulo 2π). For three-dimensional solids, the Brillouin zone is a three-dimensional torus and one expects to obtain a nonvanishing

Berry phase for any closed path in the Brillouin zone. When symmetry is present one can make **k** vary through the entire Brillouin zone in a given symmetry direction. As a consequence of symmetry in the Brillouin zone (parameter space) Berry's phase turns out to be quantized and it assumes well defined discrete values that correspond to the Wyckoff positions in the crystal.[12] Our conclusion is that Berry's phase can therefore be utilized for specifying entire bands in solids, in very much the same way as the band center is used in the framework of band representations of space groups.

Consider a Bloch electron in one dimension under the influence of a time-dependent vector potential $A(t)$. The Schrödinger equation for this problem is

$$i\hbar\frac{\partial\psi}{\partial t}=\left[\frac{1}{2m}\left[p-\frac{e}{c}A(t)\right]^2+V(x)\right]\psi\,, \tag{1}$$

where $V(x+a)=V(x)$. At this stage there is no need to specify the source of the potential $A(t)$. However, the assumption will be made that $A(t)$ is time dependent only and that the variation of $A(t)$ with time is adiabatic [the frequencies corresponding to the relevant gaps in the energy spectrum of the solid are much larger than the frequencies in the Fourier expansion of $A(t)$]. Having defined the problem, one could directly use Berry's closed formula[1] for the dynamics of electrons in solids. However, it is instructive to rederive it again because, unlike in Ref. 1, the spectrum of Eq. (1) is bandlike (quasicontinuum), and the parameter space is one dimensional. Following the adiabatic approximation one is to solve the equation

$$\left[\frac{1}{2m}\left[p-\frac{e}{c}A(t)\right]^2+V(x)\right]\psi_{nt}(x)=\epsilon_n(t)\psi_{nt}(x)\,, \tag{2}$$

where n labels the energy band and $\epsilon_n(t)$ is the energy function (its meaning is given below). Equations (1) and (2) are well known in the literature of a Bloch electron in an electric field,[13,14] and one can seek a solution of Eq. (2) in the form

$$\psi_{nt}(x)=\exp(ikx)u_{nk(t)}(x)\,, \tag{3}$$

where k is the quasimomentum [in $\exp(ikx)$ it is time independent, while in the function u, $k(t)=k-(e/c\hbar)\times A(t)$] and $u_{nk}(x)$ is the periodic part of the Bloch function. It satisfies the equation

$$\left[\frac{1}{2m}\left[p+\hbar k-\frac{e}{c}A(t)\right]^2+V(x)\right]u_{nk(t)}=\epsilon_n(k(t))u_{nk(t)}\,, \tag{4}$$

where k in the Hamiltonian is time independent, and where $\epsilon_n(t)$ of Eq. (2) is now written as $\epsilon_n(k(t))$ [$\epsilon_n(k)$ is the energy spectrum for the band n]. With these notations, it now follows that the solution of Eq. (1) can be given the adiabatic form[1,13]

$$\psi(x,t)=\exp\left[i\gamma_n(t)-\frac{i}{\hbar}\int_0^t\epsilon_n(k(t'))dt'\right]\psi_{nt}\,, \tag{5}$$

where $\gamma_n(t)$ is a time-dependent phase for the energy band n. By substituting this solution into Schrödinger's equation [Eq. (1)] and by using Eqs. (2) and (4), one finds [$u_{nk}(x)$ is normalized as $(2\pi/a)\int_0^a|u_{nk}(x)|^2dx=1$] for Berry's phase γ_n the expression

$$\left[\dot{\gamma}_n(t)=i\int_0^a u^*_{nk(t)}(x)\frac{\partial}{\partial t}u_{nk(t)}(x)dx\right]\,,$$

$$\gamma_n=\int_{-\pi/a}^{\pi/a}X_{nn}(k)dk\,, \tag{6}$$

where the quantity $X_{nn}(k)$ is widely used in the band theory of solids and is given by[11,14]

$$X_{nn}(k)=\frac{2\pi}{a}\int_0^a u^*_{nk}(x)i\frac{\partial u_{nk}(x)}{\partial k}dx\,. \tag{7}$$

It is clear that the result for γ_n [Eq. (6)] is obtained when k is made to vary over the whole Brillouin zone $[-\pi/a,\pi/a]$ by the external perturbation $A(t)$. Here the following remark is appropriate. Originally, Berry's phase was defined for a system with a cyclic Hamiltonian $H(t+T)=H(t)$, where T is the period of the cycle. Our Hamiltonian does not satisfy this cyclic condition. However, for k to vary over the Brillouin zone, $A(t)$ has to change by $2\pi/a$ [see Eq. (4)]. In the Bloch problem such a change of k in Eq. (4) should be considered cyclic because this is a gauge change which is compensated by the fact that $u_{nk+2\pi/a}(x)=\exp(-i2\pi x/a)u_{nk}(x)$. The Bloch problem is therefore a generalization of the Berry phase to a noncyclic change of the Hamiltonian, up to a gauge transformation. We would like to point out that the result for Berry's phase [relation (6)] is gauge invariant. As an example, consider the constant electric field E, when $A=-cE$. Then by a gauge transformation $\exp[-(1/\hbar)eExt]$, the Hamiltonian in Eq. (1) will become $(1/2m)p^2+V(x)-eEx$. A corresponding change will appear in Eq. (4). We can still seek the solution in the form of Eqs. (3) and (5) but now k is time dependent in $\exp(ikx)$ of Eq. (3). By using the acceleration theorem $\hbar\dot{k}=-eE$ one recovers the result (6) for Berry's phase. Formula (6) coincides in form with the original expression of Berry.[1] The novel feature of this formula is in the fact that we have here, for the first time, the appearance of Berry's phase for a one-dimensional parameter space in a purely quantum-mechanical system (the Bloch problem has no analog for a particle in classical mechanics). It will soon be shown that γ_n is, in general, nonzero.

By using the periodicity properties of $u_{nk}(x)$ it follows that $X_{nn}(k)$ [Eq. (7)] is periodic in k with the period $2\pi/a$ (the constant of the reciprocal lattice). It also follows from Eq. (7) that $X_{nn}(k)$ is phase dependent;

e.g., when $u_{nk} \rightarrow \bar{u}_{nk} = \exp(i\phi)u_{nk}$, $X_{nn}(k) \rightarrow \bar{X}_{nn}(k) = X_{nn}(k) - \partial\phi/\partial k$. Since the general form for ϕ is $mka + \alpha(k)$, where m is an integer and $\alpha(k)$ is a periodic function, the phase γ_n in Eq. (6) changes by $-2\pi m$ under such a general phase transformation. This means that $\exp(i\gamma_n)$ [in Eq. (5) for the wave function] is invariant under any phase transformation and it is therefore a well defined quantity for each energy band.

By using the Wannier function $a_n(x)$ for the band n, an alternative expression can be derived for γ_n. Thus, with the help of the relation between $u_{nk}(x)$ and $a_n(x)$, which is[15]

$$\exp(ikx)u_{nk}(x) = \left(\frac{a}{2\pi}\right)^{1/2} \sum_m \exp(ikma)a_n(x-ma) \,, \quad (8)$$

γ_n assumes the form [Eqs. (6)–(8)]

$$\gamma_n = \int_{-\pi/a}^{\pi/a} X_{nn}(k)dk = \left(\frac{a}{2\pi}\right)^{-1} \int_{-\infty}^{\infty} x\,|a_n(x)|^2 dx \,. \quad (9)$$

Formula (9) gives Berry's phase for the nth energy band expressed either via the Bloch function $u_{nk}(x)$ or the Wannier function $a_n(x)$ of this particular band.

Let us now connect the result in Eq. (9) to the band center[11] and the Wyckoff position[12] of the energy band. In Ref. 11 the band center for the energy band n was defined by the following formula [Eq. (5) of Ref. 11]:

$$q_n = \frac{a}{2\pi} \int_{-\pi/a}^{\pi/a} X_{nn}(k)dk = \int_{-\infty}^{\infty} x\,|a_n(x)|^2 dx \,. \quad (10)$$

By comparing q_n with γ_n we see that the following connection holds:

$$\gamma_n = (2\pi/a)q_n \,. \quad (11)$$

When there is no symmetry in the one-dimensional chain, q_n as well as γ_n can assume any value as a function of the band index n. However, when inversion symmetry is present, it follows from the symmetry of the Wannier functions[11] that q_n can assume two values only, $q_n = 0$ or $a/2$. This follows from the following argument. When the crystal possesses inversion symmetry, the phase of the Bloch function can always be chosen in such a way as to make the Wannier function even or odd around either of the two centers[16] $x = 0$ or $a/2$. Thus, assume that $a_n(-x) = \pm a_n(x)$, it then follows that $q_n = 0$ [see Eq. (10)]. In the other case, when $a_n(-x+a) = \pm a_n(x)$ (inversion around $a/2$), then it follows from Eq. (10) that $q_n = a/2$. 0 and $a/2$ are also the Wyckoff positions of the one-dimensional crystal. From Eq. (11) it follows that Berry's phase for a one-dimensional crystal with inversion symmetry can be 0 or π only. Equation (11) establishes therefore a connection between Berry's phase, the band center, and the Wyckoff position.

Having established the connection between Berry's phase and the label of an energy band as a whole entity, let us now come back to the question of the specification of the vector potential $A(t)$ in Schrödinger's equation [Eq. (1)]. It was required that $A(t)$ changes adiabatically and that it makes k vary over the entire Brillouin zone. This requirement is not very restrictive and is practically met by any homogeneous external electric field E whose frequency is much smaller than the relevant band gaps.[17] Such an electric field is derived from the following vector potential: $A(t) = -c\int_0^t E(t') \times dt'$. The variation of k with time is then given by the relation[9] $\hbar\dot{k} = -eE(t)$ which holds also for a time-dependent adiabatically varying electric field. A particular case is the constant electric field which has been widely investigated in connection with the Wannier-Stark ladder problem.[18] It is of interest to point out that the Wannier-Stark spectrum contains the band center q_n [Eq. (10)] explicitly.[16] Since γ_n is connected to q_n, this establishes a connection of Berry's phase to the well known Wannier-Stark ladder problem in solids.[19] Bearing in mind that different energy bands, say n and n', have, in general, different γ's (when symmetry is present γ assumes the values 0 and π only), electrons in these two bands will pick up different Berry phases in the presence of an electric field. In principle, any interference experiment with electrons from two such energy bands should enable one to measure the phase difference between γ_n and $\gamma_{n'}$. This is so very much in principle, because one has to make sure that the phase is fully controlled by the electric field and that scattering by imperfections can be neglected. As is well known this is not a simple problem.[18]

The generalization of the results to three-dimensional crystals is straightforward conceptually, but technically the problem becomes much more involved because of the energy-band topology in three dimensions.[10,20] No attempt will be made here to go into any details, but one should point out that the Berry's phase in the framework of energy bands in three-dimensional crystals is a very exciting and rich subject. In particular, the whole band-labeling problem will assume the geometrical form of topological invariants.

In conclusion, we have shown how to define Berry's phase in the dynamics of electrons in solids. There are a number of unique features that appear in this problem. First, the parameter space is furnished by the system itself, namely, the Brillouin zone of the solid. This is unlike the other physical examples of the appearance of Berry's phase[1-8] where the parameter space is introduced externally, e.g., the parameters of the magnetic field,[2,3] the optical fiber parameters,[4] etc. In the case of the solid the parameter space (the Brillouin zone) exists naturally and what the external field does is to change the Bloch quasimomentum **k** inside this space. Second, the Brillouin zone is a torus and this is why even in a one-dimensional parameter space a nonvanishing Berry phase can appear. Third, the Bloch states possess symmetry properties in the Brillouin zone (parameter space), and, as a consequence of this symmetry, Berry's phase becomes quantized. Finally, we would like to comment on the possible significance of Berry's phase in solids. We have already mentioned the Stark ladder where Berry's phase γ_n appears explicitly in the energy spectrum. Thus, when $\gamma_n = \pi$, $q_n = a/2$ [see Eq. (11)] and the Wannier-Stark ladders is half integer. This is reminiscent of the influence of Berry's phase on the molecular spectrum, where it leads to fractional quantization of the angular momentum.[5] In three-dimensional crystals, because of the torus topology of the Brillouin zone, there are, in general, nonvanishing Berry phases along any path that covers a vector of the reciprocal lattice. This by itself contains a very rich structure of Berry phases which should lead to a better understanding of the topology of energy bands. In addition, one can also define Berry's phase on closed circuits in the Brillouin zone. As will be shown in a separate publication[21] if a perturbation in the solid makes **k** vary on a closed orbit (the impurity problem or the dynamics in a magnetic field), then the Berry phase assumes the form

$$\gamma_n(C) = \oint_C \mathbf{X}_{nn}(\mathbf{k}) \cdot d\mathbf{k}\,, \tag{12}$$

where C is the path of integration and $\mathbf{X}_{nn}(\mathbf{k})$ is a three-dimensional generalization of Eq. (7). As was proven by Berry,[1] $\gamma_n(C)$ is nonvanishing when the integration path lies in the vicinity of a point of degeneracy. This is very often the case in the valence and conduction bands of semiconductors.[22] We expect therefore Berry's topological phase to play an important role in the whole dynamics of Bloch electrons in solids.

The author would like to acknowledge useful discussions on the subject with Professor H. Bacry and Professor L. Michel. This research was supported in part by the Fund for the Promotion of Research at Technion.

[1]M. V. Berry, Proc. Roy. Soc. London A **392**, 451 (1984).
[2]T. Bitter and D. Dubbers, Phys. Rev. Lett. **59**, 251 (1987).
[3]R. Tycko, Phys. Rev. Lett. **58**, 2281 (1987).
[4]A. Tomita and R. Y. Chiao, Phys. Rev. Lett. **57**, 937 (1986); R. Y. Chiao, A. Antaramian, K. M. Gaugge, H. Jiao, S. R. Wilkinson, and H. Nathel, Phys. Rev. Lett. **60**, 1214 (1988); R. Simon, H. J. Kimble, and E. C. G. Sundarshan, Phys. Rev. Lett. **61**, 19 (1988).
[5]G. Delacretaz, E. R. Grant, R. L. Whetten, L. Wöste, and J. W. Zwanziger, Phys. Rev. Lett. **56**, 259 (1986).
[6]F. S. Ham, Phys. Rev. Lett. **58**, 725 (1987).
[7]I. J. R. Aitchison, Phys. Scr. **T23**, 12 (1988), and references therein.
[8]M. Kugler and S. Shtrikman, Phys. Rev. D **37**, 934 (1988); A. Bhattacharjee and T. Sen, Phys. Rev. A **38**, 4389 (1988); M. V. Berry and J. H. Hannay, J. Phys. A **21**, 325 (1988).
[9]C. Kittel, in *Quantum Theory of Solids* (Wiley, New York, 1963).
[10]J. Zak, Phys. Rev. Lett. **45**, 1025 (1980); H. Bacry, L. Michel, and J. Zak, Phys. Rev. Lett. **61**, 1005 (1988).
[11]J. Zak, Phys. Rev. Lett. **48**, 359 (1982).
[12]*International Tables for Crystallography*, edited by T. Hahn (Reidel, Dordrecht, 1983).
[13]W. V. Houston, Phys. Rev. **57**, 184 (1940).
[14]J. Zak, in *Solid State Physics*, edited by H. Ehrenreich, F. Seitz, and D. Turnbull (Academic, New York, 1972), Vol. 27.
[15]G. H. Wannier, Phys. Rev. **52**, 191 (1937).
[16]W. Kohn, Phys. Rev. **115**, 809 (1959).
[17]R. Simon and N. Kumar, J. Phys. A **21**, 1725 (1988).
[18]J. E. Avron, Ann. Phys. (N.Y.) **143**, 33 (1982).
[19]G. H. Wannier, Rev. Mod. Phys. **34**, 645 (1962).
[20]J. des Cloizeaux, Phys. Rev. **129**, 554 (1963); **135**, A685 (1964).
[21]J. Zak (to be published).
[22]G. Dresselhaus, A. F. Kip, and C. Kittel, Phys. Rev. **98**, 3681 (1955).

ARTICLES

Berry phase, hyperorbits, and the Hofstadter spectrum: Semiclassical dynamics in magnetic Bloch bands

Ming-Che Chang and Qian Niu
Department of Physics, University of Texas at Austin, Austin, Texas 78712
(Received 16 October 1995)

We have derived a set of semiclassical equations for electrons in magnetic Bloch bands. The velocity and energy of magnetic Bloch electrons are found to be modified by the Berry phase and magnetization. This semiclassical approach is used to study general electron transport in a dc or ac electric field. We also find a close connection between the cyclotron orbits in magnetic Bloch bands and the energy subbands in the Hofstadter spectrum. Based on this formalism, the pattern of band splitting, the distribution of Hall conductivities, and the positions of energy subbands in the Hofstadter spectrum can be understood in a simple and unified picture.

I. INTRODUCTION

The semiclassical method has played a very important role in studying electron dynamics in periodic systems.[1] In this approach, the effect of a periodic potential is treated by quantum-mechanical methods and yields usual band structure for energy spectrum, while an extra electromagnetic field is treated as a classical perturbation. The velocity of an electron in the one-band approximation is given by

$$\dot{\mathbf{r}}=\frac{\partial \mathscr{E}_n(\mathbf{k})}{\hbar\,\partial \mathbf{k}}, \tag{1.1}$$

where $\mathscr{E}_n$ is the energy spectrum for the nth band. The dynamics of quasimomentum $\mathbf{k}$ is governed by the Lorentz-force formula

$$\hbar\dot{\mathbf{k}}=-e\mathbf{E}-e\dot{\mathbf{r}}\times\mathbf{B}, \tag{1.2}$$

where $\mathbf{E}$ and $\mathbf{B}$ are the external electric and magnetic fields. These equations may be regarded as the equations of motion for the center of mass of a wave packet in the $\mathbf{r}$ and $\mathbf{k}$ spaces. A tremendous amount of work has been done to justify these simple looking formulas and their quantization.[2]

These formulas, however, become invalid if the magnetic field is so strong that it is no longer appropriate to be treated as a perturbation. In this case, we need to solve the Schrödinger equation with the following Hamiltonian:

$$H_0=\frac{1}{2m}\left(-i\hbar\frac{\partial}{\partial \mathbf{r}}+e\mathbf{A}_0(\mathbf{r})\right)^2+V(\mathbf{r}), \tag{1.3}$$

where $\mathbf{A}_0(\mathbf{r})$ is the vector potential of a homogeneous magnetic field,[3] and $V(\mathbf{r})$ is a periodic potential. The eigenenergies of Eq. (1.3) will be called magnetic Bloch bands, and its energy eigenstates, magnetic Bloch states. A crucial difference between a Bloch state and a magnetic Bloch state lies in their translational properties. The Hamiltonian H_0 is not invariant under lattice translation because $\mathbf{A}_0(\mathbf{r})$ cannot be a periodic function if the mean value of $\mathbf{B}$ is not zero. However, H_0 can be made invariant under "magnetic" translation operators, which are the usual translation operators multiplied by a position-dependent phase factor.[4]

We first give a brief review of the magnetic translation symmetry. In order to simplify the discussion, we assume the motion of electrons is confined in a plane $[\mathbf{r}=(x,y)]$ and the magnetic field is along the z direction. A magnetic Bloch state is the state that satisfies

$$H_0\Psi_{n\mathbf{k}}(\mathbf{r})=\mathscr{E}_n(\mathbf{k})\Psi_{n\mathbf{k}}(\mathbf{r}), \tag{1.4}$$

as well as

$$\tilde{T}_1(R_1)\Psi_{n\mathbf{k}}(\mathbf{r})=e^{ik_1R_1}\Psi_{n\mathbf{k}}(\mathbf{r}),$$

$$\tilde{T}_2(R_2)\Psi_{n\mathbf{k}}(\mathbf{r})=e^{ik_2R_2}\Psi_{n\mathbf{k}}(\mathbf{r}), \tag{1.5}$$

where $\tilde{T}_1$ and $\tilde{T}_2$ are magnetic translation operators. Although $\tilde{T}_1$ and $\tilde{T}_2$ commute with the Hamiltonian by construction, they do not commute with each other unless there is an integer number of flux quantum ϕ_0 enclosed by $|\mathbf{R}_1\times\mathbf{R}_2|$. Therefore, when the magnetic flux is a rational multiple p/q of the flux quantum ϕ_0 per unit cell of the lattice (plaquette), we must choose a "magnetic" unit cell containing q plaquettes in order that both k_1 and k_2 be good quantum numbers. $\Psi_{n\mathbf{k}}$ thus defined forms a complete set and satisfies the orthogonality condition

$$\langle\Psi_{n'\mathbf{k}'}|\Psi_{n\mathbf{k}}\rangle=\delta_{n'n}\delta_{\mathbf{k}'\mathbf{k}}. \tag{1.6}$$

The domain of $\mathbf{k}$ is a magnetic Brillouin zone (MBZ), which is q times smaller than a usual Brillouin zone. Furthermore, because of the magnetic translation symmetry, the MBZ has exactly a q-fold degeneracy. We will call each repetition unit a "reduced" MBZ.

One example of magnetic Bloch bands is the subbands split from Bloch bands due to a magnetic field. The number of subbands into which a band splits depends on the mag-

TABLE I. Comparison between properties of the usual and magnetic Bloch bands.

	Bloch band	Magnetic Bloch band $[(p/q)\phi_0$ per plaquette$]$
Unperturbed Hamiltonian	$H_0=\frac{1}{2m}\left(-i\hbar\frac{\partial}{\partial\mathbf{r}}\right)^2+V(\mathbf{r})$	$H_0=\frac{1}{2m}\left[-i\hbar\frac{\partial}{\partial\mathbf{r}}+e\mathbf{A}_0(\mathbf{r})\right]^2+V(\mathbf{r})$
Translation operators	$T(\mathbf{R})=e^{\mathbf{R}\cdot\partial/\partial\mathbf{r}}$	$\tilde{T}(\mathbf{R})=e^{ie/\hbar\int_0^{\mathbf{R}}d\mathbf{r}'\cdot\mathbf{A}_0(\mathbf{r}+\mathbf{r}')}e^{\mathbf{R}\cdot\partial/\partial\mathbf{r}}$
Number of plaquettes per unit cell	1 plaquette	q plaquettes
Range of $\mathbf{k}$ vector	One Brillouin zone	One magnetic Brillouin zone (one Brillouin zone divided by q)
Perturbing fields	**E**,**B**	$\mathbf{E},\delta\mathbf{B}$
Velocity of electron	$\dot{\mathbf{r}}=\partial\mathscr{E}_n(\mathbf{k})/\hbar\partial\mathbf{k}$	$\dot{\mathbf{r}}=\partial E_n(\mathbf{k})/\hbar\partial\mathbf{k}-\dot{\mathbf{k}}\times\mathbf{\Omega}_n(\mathbf{k})$, $E_n(\mathbf{k})=\mathscr{E}_n^{\mathrm{mag}}(\mathbf{k})+(e/2m)\delta\mathbf{B}\cdot\mathbf{L}_n(\mathbf{k})$
Dynamics for $\mathbf{k}$	$\hbar\dot{\mathbf{k}}=-e\mathbf{E}-e\dot{\mathbf{r}}\times\mathbf{B}$	$\hbar\dot{\mathbf{k}}=-e\mathbf{E}-e\dot{\mathbf{r}}\times\delta\mathbf{B}$
Quantization condition for cyclotron orbits	$\mathrm{Area}(C_m)=2\pi(m+\frac{1}{2})eB/\hbar$	$\mathrm{Area}(C_m)=2\pi\left(m+\frac{1}{2}-\frac{\Gamma(C_m)}{2\pi}\right)e\delta B/\hbar$

netic flux per plaquette in an intricate way.[5] If $\phi=p/q$ (in units of ϕ_0), a Bloch band will split into q magnetic subbands. On the other hand, if the magnetic field is very strong, it is more appropriate to treat the lattice potential as a perturbation, then a Landau level will be broadened and split into p subbands.

For a usual solid with a lattice constant $a=5$ Å, the magnetic field has to be as large as 10^4 T in order for p/q to be of order unity. This is the reason the splitting was once considered impossible to observe. However, the field strength can be greatly reduced to a few tesla if we use an artificial lattice with a much larger (say, 500 Å) lattice constant. Evidence for such splitting has appeared in recent transport measurements.[6] It is expected that more evidence will emerge in the future by using a very pure sample in a very low temperature environment. Under such circumstances, what is the dynamics for electrons in such magnetic Bloch bands?

Using the magnetic Bloch states as an unperturbed basis, we found the following semiclassical dynamics in magnetic Bloch bands:[7]

$$\dot{\mathbf{r}}=\frac{\partial E_n(\mathbf{k})}{\hbar\partial\mathbf{k}}-\dot{\mathbf{k}}\times\mathbf{\Omega}_n(\mathbf{k}) \tag{1.7}$$

and

$$\hbar\dot{\mathbf{k}}=-e\mathbf{E}-e\dot{\mathbf{r}}\times\delta\mathbf{B}, \tag{1.8}$$

where $E_n(\mathbf{k})$ consists of a band energy $\mathscr{E}_n(\mathbf{k})$ and a correction from the magnetic moment of the wave packet (this correction did not appear in Ref. 7). $\mathbf{\Omega}_n(\mathbf{k})$ is the "Berry curvature," whose integral over an area bounded by a path C in $\mathbf{k}$ space is the Berry phase $\Gamma_n(C)$.[8] $\mathbf{E}$ and $\delta\mathbf{B}$ are external fields added to the already present B_0 field. These equations will be derived and explained in detail in Sec. II.

Despite the similarities between Eqs. (1.1), (1.2) and Eqs. (1.7), (1.8), there are several essential differences. See Table I for a comparison between this semiclassical dynamics and the conventional one. The last item in Table I, about the quantization of orbits, will be explained in Sec. IV. We have to emphasize that the δB in Eq. (1.8) is the field applied to the magnetic Bloch states; it is not the total magnetic field applied to the sample. This separation is particularly useful when $B(x,y)$ is composed of a large constant part B_0 and a small nonuniform part $\delta B(x,y)$. In this case, we can calculate the effect of B_0 exactly and treat $\delta B(x,y)$ as a classical perturbation.

This paper is organized as follows. Section II is devoted to the derivation of Eqs. (1.7) and (1.8). Their use in calculating transport properties in a dc or ac electric field is demonstrated in Sec. III. The presence of δB will lead to formation of cyclotron orbits in magnetic Bloch bands, similar to the formation of cyclotron orbits in usual Bloch bands. This is explained in Sec. IV. In Sec. V, we explore the connection between these cyclotron orbits and the Hofstadter spectrum. In Sec. VI, we estimate the energy levels in the Hofstadter spectrum by calculating the cyclotron energies in magnetic Bloch bands. Finally, this paper is summarized in Sec. VII.

II. DERIVATION OF THE SEMICLASSICAL DYNAMICS

The method we use is to construct a wave packet out of $\Psi_{n\mathbf{k}}$ (hence it has already included the effect of B_0), and study its motion governed by the following Hamiltonian:

$$H=\frac{1}{2m}\left[-i\hbar\frac{\partial}{\partial\mathbf{r}}+e\mathbf{A}_0(\mathbf{r})+e\,\delta\mathbf{A}(\mathbf{r},t)\right]^2+V(\mathbf{r}), \quad (2.1)$$

where $-\partial\delta\mathbf{A}/\partial t=\mathbf{E}$, and $\nabla\times\delta\mathbf{A}=\delta\mathbf{B}$. For simplicity we assume both $\mathbf{E}$ and $\delta\mathbf{B}$ are uniform; the derivation is still valid if they are slowly varying in space and/or time.

A. Wave packet in a magnetic Bloch band

Our derivation will be confined to one energy band by neglecting interband transitions; therefore, the band index n is henceforth dropped. Consider the following wave packet centered at $\mathbf{r}_c$ which is formed from the superposition of magnetic Bloch states,

$$|W_0\rangle=\int_{\mathrm{MBZ}}d^2\mathbf{k}\;w(\mathbf{k})|\Psi(\mathbf{k})\rangle, \quad (2.2)$$

where $w(\mathbf{k})$ is a function localized around $\mathbf{k}_c$ (see Fig. 1 for an illustration). It has to be chosen such that

$$\int d^2\mathbf{k}\;\mathbf{k}|w(\mathbf{k})|^2=\mathbf{k}_c, \quad (2.3)$$

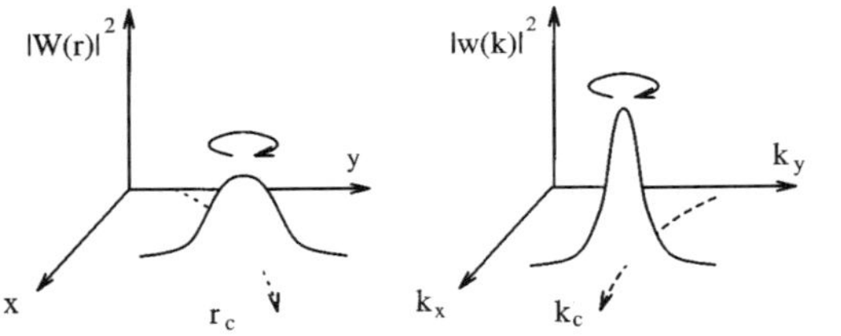

FIG. 1. Schematic plots for the motion of wave packets in **r** and **k** space.

and

$$\langle W_0|\mathbf{r}|W_0\rangle=\mathbf{r}_c. \quad (2.4)$$

By defining $u_{\mathbf{k}}(\mathbf{r})=e^{-i\mathbf{k}\cdot\mathbf{r}}\Psi_{\mathbf{k}}(\mathbf{r})$, the mean position of W_0 can be written as

$$\begin{aligned}\langle W_0|\mathbf{r}|W_0\rangle&=\int d^2\mathbf{k}'\int d^2\mathbf{k}\;w^*(\mathbf{k}')w(\mathbf{k})\left\langle \Psi(\mathbf{k}')\middle|\left(-i\frac{\partial}{\partial\mathbf{k}}e^{i\mathbf{k}\cdot\mathbf{r}}\right)\middle|u(\mathbf{k})\right\rangle\\&=\int d^2\mathbf{k}'\int d^2\mathbf{k}\;w^*(\mathbf{k}')w(\mathbf{k})\left[\left(-i\frac{\partial}{\partial\mathbf{k}}\right)\delta(\mathbf{k}-\mathbf{k}')+\delta(\mathbf{k}-\mathbf{k}')\left\langle u(\mathbf{k})\middle|i\frac{\partial}{\partial\mathbf{k}}\middle|u(\mathbf{k})\right\rangle_{\mathrm{cell}}\right]\\&=\int d^2\mathbf{k}\left[w^*(\mathbf{k})i\frac{\partial}{\partial\mathbf{k}}w(\mathbf{k})+|w(\mathbf{k})|^2\left\langle u(\mathbf{k})\middle|i\frac{\partial}{\partial\mathbf{k}}\middle|u(\mathbf{k})\right\rangle_{\mathrm{cell}}\right],\end{aligned} \quad (2.5)$$

where we have used the identity[9]

$$\left\langle u(\mathbf{k}')\middle|e^{i(\mathbf{k}-\mathbf{k}')\cdot\mathbf{r}}i\frac{\partial}{\partial\mathbf{k}}\middle|u(\mathbf{k})\right\rangle=\delta(\mathbf{k}-\mathbf{k}')\left\langle u(\mathbf{k})\middle|i\frac{\partial}{\partial\mathbf{k}}\middle|u(\mathbf{k})\right\rangle_{\mathrm{cell}}. \quad (2.6)$$

The subscript means that the spatial integration is restricted to a magnetic unit cell. By defining

$$\mathscr{A}(\mathbf{k})=i\left\langle u(\mathbf{k})\middle|\frac{\partial}{\partial\mathbf{k}}\middle|u(\mathbf{k})\right\rangle_{\mathrm{cell}}, \quad (2.7)$$

Eq. (2.4) can be written as

$$\int d^2\mathbf{k}\left[w^*(\mathbf{k})i\frac{\partial}{\partial\mathbf{k}}w(\mathbf{k})+|w(\mathbf{k})|^2\mathscr{A}(\mathbf{k})\right]=\mathbf{r}_c. \quad (2.8)$$

B. Effective Lagrangian for a moving wave packet

The dynamics of a moving wave packet is governed by the following effective Lagrangian

$$L(\mathbf{r}_c,\mathbf{k}_c,\dot{\mathbf{r}}_c,\dot{\mathbf{k}}_c)=\left\langle W\middle|i\hbar\frac{\partial}{\partial t}\middle|W\right\rangle-\langle W|H|W\rangle, \quad (2.9)$$

where W is a wave packet centered at $\mathbf{r}_c$ and $\mathbf{k}_c$ in the presence of external electromagnetic fields ($\mathbf{k}_c$ is treated as a generalized coordinate here). We can always choose a gauge such that the vector potential $\delta\mathbf{A}$ is locally gauged away at a chosen point $\mathbf{r}=\mathbf{r}_c$. At this particular point, the moving wave packet W is the same as the W_0 in Eq. (2.2).[10] The value of W near $\mathbf{r}_c$ can be approximated as

$$W(\mathbf{r})=e^{-ie/\hbar\,\delta\mathbf{A}(\mathbf{r}_c,t)\cdot\mathbf{r}}W_0(\mathbf{r}). \quad (2.10)$$

First, we evaluate the energy of this wave packet, which is $\langle W|H|W\rangle=\langle W_0|H'|W_0\rangle$, with

$$\begin{aligned}H'&=\frac{1}{2m}\left\{-i\hbar\frac{\partial}{\partial\mathbf{r}}+e\mathbf{A}_0(\mathbf{r})+e[\delta\mathbf{A}(\mathbf{r},t)-\delta\mathbf{A}(\mathbf{r}_c,t)]\right\}^2\\&\quad+V(\mathbf{r})\\&\simeq H_0+\frac{e}{2m}\{[\delta\mathbf{A}(\mathbf{r},t)-\delta\mathbf{A}(\mathbf{r}_c,t)]\cdot\mathbf{P}+\mathrm{H.c.}\}.\end{aligned} \quad (2.11)$$

$\mathbf{P}$ is the mechanical momentum operator corresponding to H_0. For simplicity, we choose the circular gauge for δB, which gives $\delta\mathbf{A}(\mathbf{r},t)=-\mathbf{E}t+\frac{1}{2}\delta\mathbf{B}\times\mathbf{r}$, and leads to

$$H'\simeq H_0+\frac{e}{2m}\delta\mathbf{B}\cdot\mathbf{L}, \tag{2.12}$$

where $\mathbf{L}=(\mathbf{r}-\mathbf{r}_c)\times\mathbf{P}$ is the mechanical angular momentum of the wave packet about its center of mass. Therefore,

$$\langle W|H|W\rangle\simeq\mathscr{E}(\mathbf{k}_c)+\frac{e}{2m}\delta\mathbf{B}\cdot\langle W_0|\mathbf{L}|W_0\rangle. \tag{2.13}$$

The second term represents the energy correction due to magnetic moment of the wave packet. Notice that while a wave packet in an ordinary Bloch band does not rotate, a wave packet in a magnetic Bloch band usually does.

For the first term on the right hand side of Eq. (2.9), we have

$$\begin{aligned}\left\langle W_0\left|i\hbar\frac{\partial}{\partial t}\right|W_0\right\rangle&=\langle W_0|e\,\delta\dot{\mathbf{A}}(\mathbf{r}_c,t)\cdot\mathbf{r}|W_0\rangle\\&\quad+\int d^2\mathbf{k}\;w^*(\mathbf{k})i\hbar\frac{\partial}{\partial t}w(\mathbf{k})\\&=e\,\delta\dot{\mathbf{A}}(\mathbf{r}_c,t)\cdot\mathbf{r}_c+\int d^2\mathbf{k}|w(\mathbf{k})|^2\hbar\frac{\partial}{\partial t}\gamma(\mathbf{k},t)\\&\simeq e\,\delta\dot{\mathbf{A}}(\mathbf{r}_c,t)\cdot\mathbf{r}_c+\hbar\frac{\partial}{\partial t}\gamma(\mathbf{k}_c,t),\end{aligned} \tag{2.14}$$

where $w(\mathbf{k})\equiv|w(\mathbf{k})|e^{-i\gamma(\mathbf{k},t)}$. Up to terms of total time derivative, which have no effect on the dynamics, the last line can be written as

$$-e\,\delta\mathbf{A}\cdot\dot{\mathbf{r}}_c-\hbar\dot{\mathbf{k}}_c\cdot\frac{\partial}{\partial\mathbf{k}^c}\gamma(\mathbf{k}_c,t). \tag{2.15}$$

Using the condition in Eq. (2.8) and neglecting another term of total time derivative, we can write Eq. (2.15) as

$$-e\,\delta\mathbf{A}\cdot\dot{\mathbf{r}}_c+\hbar\mathbf{k}_c\cdot\dot{\mathbf{r}}_c+\hbar\dot{\mathbf{k}}_c\cdot\mathscr{A}(\mathbf{k}_c). \tag{2.16}$$

Combining Eqs. (2.13) and (2.16), we have a final form for the effective Lagrangian (omitting subscript c)

$$L(\mathbf{r},\mathbf{k},\dot{\mathbf{r}},\dot{\mathbf{k}})=-e\,\delta\mathbf{A}(\mathbf{r},t)\cdot\dot{\mathbf{r}}+\hbar\mathbf{k}\cdot\dot{\mathbf{r}}+\hbar\mathscr{A}(\mathbf{k})\cdot\dot{\mathbf{k}}-E(\mathbf{k}), \tag{2.17}$$

where $E(\mathbf{k})\equiv\mathscr{E}(\mathbf{k})+(e/2m)\delta\mathbf{B}\cdot\mathbf{L}(\mathbf{k})$. Under a gauge transformation for $\mathbf{A}_0(\mathbf{r})$, the Berry potential $\mathscr{A}(\mathbf{k})$ will be changed by a term like $\partial_{\mathbf{k}}\chi(\mathbf{k})$, and this will only change L by a total time derivative. This is also true if a different gauge is chosen for $\delta\mathbf{B}$. Therefore, the dynamics is invariant under gauge transformation.

The dynamical equations in Eqs. (1.7) and (1.8) can be obtained straightforwardly from this Lagrangian by using the Euler-Lagrange equation. The relation between $\mathbf{\Omega}(\mathbf{k})$ and $\mathscr{A}(\mathbf{k})$ is

$$\mathbf{\Omega}(\mathbf{k})=\nabla\times\mathscr{A}(\mathbf{k}), \tag{2.18}$$

which can also be written as (z-component)

$$\Omega(\mathbf{k})=i\left(\left\langle\frac{\partial u_n}{\partial k_1}\middle|\frac{\partial u_n}{\partial k_2}\right\rangle-\left\langle\frac{\partial u_n}{\partial k_2}\middle|\frac{\partial u_n}{\partial k_1}\right\rangle\right). \tag{2.19}$$

This is the familiar Berry curvature in the study of the quantum Hall effect.[11]

III. TRANSPORT IN MAGNETIC BLOCH BANDS

A. Transport by an electric field

The next step is to combine the semiclassical equations with the Boltzmann equation to study the transport properties of magnetic Bloch electrons. The Boltzmann equation is

$$\dot{\mathbf{r}}\cdot\frac{\partial f}{\partial\mathbf{r}}+\dot{\mathbf{k}}\cdot\frac{\partial f}{\partial\mathbf{k}}=\left(\frac{\partial f}{\partial t}\right)_{\text{coll}}, \tag{3.1}$$

where $f=f(\mathbf{r},\mathbf{k})$ is a distribution function. The effect of impurities is included in the collision term $(\partial f/\partial t)_{\text{coll}}$. We use the relaxation time approximation to replace it by $-(f-f_0)/\tau(\mathbf{k})$. For a random distribution of δ impurities $v_0\Sigma_i\delta(\mathbf{r}-\mathbf{r}_i)$, the impurity scattering rate is

$$\tau(\mathscr{E}_F)^{-1}=\frac{\pi}{\hbar}\rho(\mathscr{E}_F)n_iv_0^2, \tag{3.2}$$

where n_i is the area density of impurities. This rate is proportional to the density of states $\rho(\mathscr{E}_F)$ at Fermi energy, which varies wildly with $\mathscr{E}_F$ because the energy spectrum is discrete. However, since the following calculation is confined to only one band, τ will be approximated by a constant.

The equations of motion of an electron subject to a uniform electric field are

$$\dot{\mathbf{r}}=\frac{\partial\mathscr{E}(\mathbf{k})}{\hbar\partial\mathbf{k}}+\frac{e}{\hbar}\mathbf{E}\times\mathbf{\Omega}(\mathbf{k}),\quad\dot{\mathbf{k}}=-e\mathbf{E}, \tag{3.3}$$

where $\mathscr{E}(\mathbf{k})$ is the reduced form of $E(\mathbf{k})$ in the absence of δB. Substituting the expressions for $\dot{\mathbf{r}}$ and $\dot{\mathbf{k}}$ in Eq. (3.3) into Eq. (3.1), and setting $f=f_0$ on the left hand side of Eq. (3.1), we obtain

$$f=f_0-\tau\left(\frac{\partial\mathscr{E}}{\hbar\partial\mathbf{k}}\cdot\frac{\partial f_0}{\partial\mathbf{r}}+\frac{e}{\hbar}(\mathbf{E}\times\mathbf{\Omega})\cdot\frac{\partial f_0}{\partial\mathbf{r}}-\frac{e}{\hbar}\mathbf{E}\cdot\frac{\partial f_0}{\partial\mathbf{k}}\right), \tag{3.4}$$

where f_0 is the distribution function in equilibrium. Electric current is given by

$$\mathbf{J}=-e\int\frac{d^2\mathbf{k}}{(2\pi)^2}f\dot{\mathbf{r}}. \tag{3.5}$$

It can be decomposed into three parts, $\mathbf{J}^\Omega+\mathbf{J}^\tau+\mathbf{J}^\mu$, with the following definitions:

$$\begin{aligned}\mathbf{J}^\Omega&=-\mathbf{E}\times\frac{e^2}{\hbar}\int\frac{d^2\mathbf{k}}{(2\pi)^2}f_0\mathbf{\Omega}(\mathbf{k}),\\\mathbf{J}^\tau&=e^2\tau\int\frac{d^2\mathbf{k}}{(2\pi)^2}\left(-\frac{\partial f_0}{\partial\mathscr{E}}\right)\mathbf{v_b}(\mathbf{v_b}\cdot\mathbf{E}),\\\mathbf{J}^\mu&=e\tau\int\frac{d^2\mathbf{k}}{(2\pi)^2}\left(-\frac{\partial f_0}{\partial\mathscr{E}}\right)\mathbf{v}\left(\mathbf{v}\cdot\frac{\partial\mu}{\partial\mathbf{r}}\right).\end{aligned} \tag{3.6}$$

In these expressions, $\mathbf{v_b}$ is the velocity $\partial\mathscr{E}/\hbar\partial\mathbf{k}$ due to energy dispersion, and $\mathbf{v}$ is the total velocity in Eq. (3.3). The meaning of these currents is explained below.

First, $\mathbf{J}^\Omega$ is the Hall current. This is most evident considering a filled band with $f_0=1$. In this case, both $\mathbf{J}^\tau$ and $\mathbf{J}^\mu$ vanish, and only $\mathbf{J}^\Omega$ is nonzero. The integral of $\Omega(\mathbf{k})$ over one magnetic Brillouin zone divided by 2π is always an integer, which is the topological Chern number discovered by Thouless *et al.*[11] Therefore we have

$$\mathbf{J}^\Omega=-C\frac{e^2}{h}\mathbf{E}\times\hat{z} \quad (C\in Z). \tag{3.7}$$

This formula represents the quantization of Hall current for a magnetic Bloch band.

Second, $\mathbf{J}^\tau$ is the diffusion current due to disorder scatterings. It can be put in the following form:

$$\mathbf{J}^\tau=e^2\int\frac{d^2\mathbf{k}}{(2\pi)^2}\left(-\frac{\partial f_0}{\partial\mathscr{E}}\right)\int_0^\infty dt\ \mathbf{v_b}(t)\mathbf{v_b}(0)\cdot\mathbf{E}, \tag{3.8}$$

where $\mathbf{v_b}(t)\equiv e^{-t/\tau}\mathbf{v_b}(0)$ is the current relaxed by scatterings after time t. At very low temperature, it can be simplified to

$$\mathbf{J}^\tau=e^2g(\mathscr{E}_F)\mathsf{D}\cdot\mathbf{E}, \tag{3.9}$$

where $g(\mathscr{E}_F)$ is the density of states at Fermi energy. $\mathsf{D}\equiv\int_0^\infty dt\langle\mathbf{v_b}(t)\mathbf{v_b}(0)\rangle$ is the diffusion tensor, and the angular bracket $\langle\rangle$ means averaging over the Fermi surface.[12]

Third, $\mathbf{J}^\mu$ is the current due to density gradient. It can be put in a form similar to $\mathbf{J}^\tau$ in Eq. (3.9) at low temperature, but with two changes. (1) The velocity $\mathbf{v_b}$ in D is replaced by the total velocity $\mathbf{v}$ that includes the curvature term. (2) The driving force $e\mathbf{E}$ is replaced by $\partial\mu/\partial\mathbf{r}$. To evaluate this current, we need to know the explicit form of $\Omega(\mathbf{k})$. This in general requires numerical calculation (see Sec. VI).

We remark that, even though the derivation of Eq. (3.8) is based on a uniform electric field, its validity goes beyond that. It is actually a kinetic formulation, first proposed by Chambers, that is also valid in the presence of a magnetic field (for magnetic Bloch bands, the magnetic field is δB).[13]

B. Perturbation by an ac electric field

The semiclassical method is much simpler to use than full quantum-mechanical approaches. This is most evident when the perturbation is changing in time. We illustrate this by considering a magnetic Bloch electron in an ac electric field. To simplify the discussion, we will neglect the effect of disorder and focus on the dynamics itself.

Assuming that a uniform electric field along the x direction oscillates with a low frequency ω, we then have

$$\dot{\mathbf{r}}=\frac{\partial\mathscr{E}(\mathbf{k})}{\hbar\partial\mathbf{k}}-\frac{e}{\hbar}E_0e^{i\omega t}\Omega(\mathbf{k})\hat{y},\quad \hbar\dot{\mathbf{k}}=-eE_0e^{i\omega t}\hat{x}. \tag{3.10}$$

Considering a square lattice with the following energy spectrum:

$$\mathscr{E}(\mathbf{k})=2[\cos(k_1a)+\cos(k_2a)], \tag{3.11}$$

and substituting the solution $\mathbf{k}(t)$ into the $\dot{\mathbf{r}}$ equation in (3.10), we have

$$\hbar\dot{x}=2a\sin\left(\frac{eE_0a}{\hbar\omega}\sin(\omega t)-k_{0x}a\right)$$

$$\hbar\dot{y}=-2a\sin(k_{0y}a)-\frac{eE_0}{\hbar}\Omega(\mathbf{k}(t))\cos(\omega t), \tag{3.12}$$

where (k_{0x},k_{0y}) is the initial value of $\mathbf{k}$. It is not difficult to see that after many cycles of oscillation, there is a net drift along the x direction with average velocity

$$\langle\dot{x}\rangle=-\frac{2a}{\hbar}\sin(k_{0x}a)J_0(z), \tag{3.13}$$

where J_0 is the zeroth order Bessel function, and $z\equiv eE_0a/\hbar\omega$ is a ratio between two energy scales.[15] It can be seen that the original band transport velocity $-2a/\hbar\ \sin(k_{0x}a)$ is modified by $J_0(z)$ because of the ac field. The electron is immobile along the x direction when z is a zero of the Bessel function. This resembles the collapse of the usual Bloch band in the ac Wannier-Stark ladder problem.[16]

C. How to divide B into B_0 and δB

Finally, we comment that the use of semiclassical equations is based on the assumption that impurities do not alter the band structure. Therefore, the electron dynamics between collisions can be nicely described by Eqs. (1.7) and (1.8). This assumption is no longer valid when q is large. In that case, disorder broadening tends to merge the subbands and wash out the fine structure. (This will be clearer after the discussion of hierarchical structure of the energy bands in Sec. V.) However, it was found that despite the energy spectrum having singular B dependence, the density of states appears as a continuous function of the magnetic field.[14] Therefore, we divide total magnetic field B into B_0 and δB, where B_0 is related to the band structure undestroyed by disorder, and δB is a small perturbation. In this case, the semiclassical dynamics in the magnetic Bloch band of B_0, driven by E and δB, will be employed in the Boltzmann equation.

IV. MAGNETIC PERTURBATION AND HYPERORBITS

The usual Bloch electron will circulate around the Fermi surface along a constant energy contour in the presence of a magnetic field. It is well known that the quantization of cyclotron orbits leads to the famous de Haas–van Alphen effect. In this section, we study a similar type of cyclotron motion in magnetic Bloch bands. It will be seen that this investigation yields very fruitful results. In particular, it offers a very simple explanation for the complex Hofstadter spectrum, which will be shown in Sec. V and Sec. VI.

A. General properties of hyperorbits

Combining the two equations in (3.3), we can eliminate $\dot{\mathbf{r}}$ to obtain

$$\hbar\dot{\mathbf{k}}=-eZ_{\delta B}(\mathbf{k})\frac{\partial E(\mathbf{k})}{\hbar\partial\mathbf{k}}\times\delta\mathbf{B}, \tag{4.1}$$

where $Z_{\delta B}(\mathbf{k}) \equiv [1+\Omega(\mathbf{k})\delta Be/\hbar]^{-1}$ is a curvature correction factor. This equation determines the trajectories of magnetic Bloch electrons in **k** space. It is not difficult to see that **k** moves along a constant energy contour of $E(\mathbf{k})$ [which is slightly different from $\mathscr{E}(\mathbf{k})$]. In a classical picture, it is the drifting-center trajectory of the tighter cyclotron orbit formed from B_0. However, we have to emphasize that the existence of hyperorbits is of quantum origin and cannot be explained classically. To differentiate them from the usual orbits of Bloch electrons, we will call them "hyperorbits."[17] The hyperorbit in real space is derived from $\dot{\mathbf{r}} = \dot{\mathbf{k}} \times \hat{z}(\hbar/e\delta B)$, which is the **k** orbit rotated by $\pi/2$ and scaled by the factor $\hbar/e\delta B$. It is also possible to define an effective cyclotron mass according to its frequency. However, this frequency will be very sensitive to its energy if the magnetic Bloch band is narrow, which is usually the case.

There are several ways to verify the existence of hyperorbits. One way is through the measurement of magnetoresistance oscillation that originates from the quantization of hyperorbits. This oscillation has a much shorter period than the usual de Haas–van Alphen oscillation because the effective magnetic field δB is much smaller. The other way of verifying it is by using an electron focusing device to detect its real-space orbit.[18] This method has been used to map out the shape of a Fermi surface by measuring the shape of cyclotron orbits.[19] Another possible approach is to observe the ultrasonic absorption spectrum of the sample. The energy of an ultrasonic wave will be absorbed when it is in resonance with the hyperorbits. Similar method has been used to detect the existence of composite fermions in half-filled quantum Hall systems.[20]

B. Quantization of hyperorbits

In a previous paper, we have derived the quantization condition using Lagrangian formulation combined with the path-integral method.[7] Here, it will be rederived using a slightly different approach. Substituting $\dot{\mathbf{r}} = \dot{\mathbf{k}} \times \hat{z}(\hbar/e\delta B)$ into the Lagrangian in Eq. (2.17), we will obtain an effective Lagrangian for the quasimomentum **k**,

$$L(\mathbf{k},\dot{\mathbf{k}}) = \frac{\hbar^2}{2e\delta B}(k_1\dot{k}_2 - k_2\dot{k}_1) + \hbar\,\mathscr{A}\cdot\dot{\mathbf{k}} - E(\mathbf{k}). \quad (4.2)$$

It can be easily shown that Eq. (4.1) does follow from this L. The generalized momentum for coordinate **k** is equal to

$$\boldsymbol{\pi} = \frac{\partial L}{\partial \dot{\mathbf{k}}} = -\frac{\hbar^2}{2e\delta B}\mathbf{k}\times\hat{z} + \hbar\mathscr{A}(\mathbf{k}). \quad (4.3)$$

(This leads to the following effective Hamiltonian:

$$H(\mathbf{k},\boldsymbol{\pi}) \equiv \boldsymbol{\pi}\cdot\dot{\mathbf{k}} - L(\mathbf{k},\dot{\mathbf{k}}) = E(\mathbf{k}). \quad (4.4)$$

Notice that because H does not depend on $\boldsymbol{\pi}$, the coordinate **k** will be a constant of motion and the dynamics is trivial. A Hamiltonian that gives correct dynamics will be given in Appendix A. Since it is not central to our derivation, we will not discuss it here.)

The quantization of hyperorbits is given by $\oint \boldsymbol{\pi}\cdot d\mathbf{k} = (m+\gamma)h$, where m is a non-negative integer and γ will be taken to be 1/2. It leads to area quantization in **k** space,

$$\tfrac{1}{2}\oint_{C_m}(\mathbf{k}\times d\mathbf{k})\cdot\hat{z} = 2\pi\left(m+\frac{1}{2}-\frac{\Gamma(C_m)}{2\pi}\right)\frac{e\delta B}{\hbar}, \quad (4.5)$$

where

$$\Gamma(C_m) = \oint_{C_m}\mathscr{A}\cdot d\mathbf{k} \quad (4.6)$$

is the Berry phase for orbit C_m. The orientation of C_m is chosen such that the sign of the area on the left hand side of Eq. (4.5) equals the sign of δB.

The total number of hyperorbits in a MBZ is determined by requiring the area of the outermost orbit be smaller than the area of a MBZ. Assume the flux before perturbation is $B_0 a^2 = p/q$, then the number of hyperorbits in a MBZ is equal to $|1/(q\delta\phi)+\sigma|$,[7] where $\delta\phi \equiv \delta B a^2 e/h$, and σ is the Hall conductivity of the parent band. These hyperorbits are the lowest order approximation to the split energy subbands. They will be broadened by tunnelings between degenerate orbits. Since the MBZ is q-fold degenerate, the above number has to be divided by q to get the actual number of daughter bands,

$$D = \frac{|1/(q\delta\phi)+\sigma|}{q}. \quad (4.7)$$

This formula is essential for understanding the splitting pattern of the Hofstadter spectrum.

The Hall conductivity for a subband can be calculated in the following way: In the presence of both **E** and $\delta\mathbf{B}$, the velocity of a magnetic Bloch electron consists of two parts,[1]

$$\dot{\mathbf{r}} = \frac{\hbar}{e\delta B}\dot{\mathbf{k}}\times\hat{z} + \frac{\mathbf{E}\times\hat{z}}{\delta B}. \quad (4.8)$$

The first term is the velocity of revolution, and the second term is the velocity of drifting along $\mathbf{E}\times\mathbf{B}$ direction. The current density for a filled subband is

$$\mathbf{J} = -\int\frac{d^2\mathbf{k}}{(2\pi)^2}\frac{\hbar}{\delta B}\dot{\mathbf{k}}\times\hat{z} - e\int\frac{d^2\mathbf{k}}{(2\pi)^2}\frac{\mathbf{E}\times\hat{z}}{\delta B}. \quad (4.9)$$

The first integral is zero for a closed orbit. Therefore, the Hall conductivity is obtained from the drifting term, which leads to σ_{yx}(or simplyσ)$=\rho e/\delta B$, where ρ is the electron density per unit area. This will be used in the next section to determine the Hall conductivities for subbands in the Hofstadter spectrum.

V. THE HOFSTADTER SPECTRUM

A. Hierarchical structure of the spectrum

The discussion in the preceding section presumes that we know δB. But this is not apparent if the magnetic field B is homogeneous. In this case, we can still divide it into two parts, but where is the dividing point between B_0 and δB? A natural way of dividing B (or ϕ) is to write it as a continued fraction,

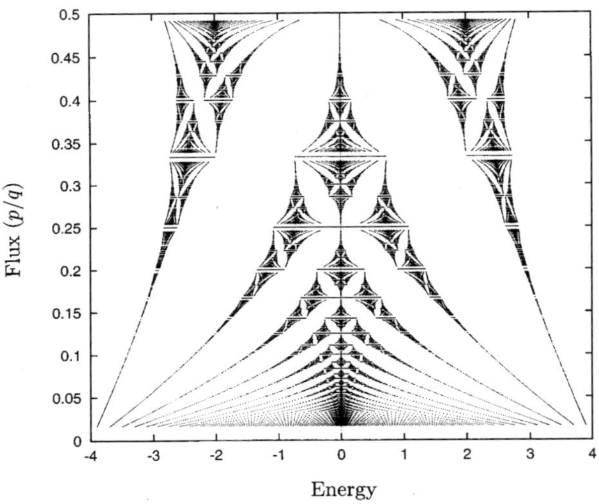

FIG. 2. Hofstadter spectrum for a square lattice generated from a tight-binding model ($q \leq 60$). Only the $\phi \leq 1/2$ part is shown since the spectrum is symmetric with respect to $\phi = 1/2$.

$$\phi = \cfrac{1}{f_1 + \cfrac{1}{f_2 + \cfrac{1}{f_3 + \cdots}}}, \tag{5.1}$$

and truncate it according to the accuracy we need. The rth order approximation of ϕ will be written as p_r/q_r.[21] For example, if $\phi = 1/(2+\sqrt{2})$, we will have $p_1/q_1 = 1/3$, $p_2/q_2 = 2/7, p_3/q_3 = 5/17, \ldots$, etc., which are truncations of

$$\frac{1}{2+\sqrt{2}} = \cfrac{1}{3 + \cfrac{1}{2 + \cfrac{1}{2 + \cdots}}}. \tag{5.2}$$

The q_r's satisfy the following recursion relation:

$$q_{r+1} = f_{r+1} q_r + q_{r-1}, \tag{5.3}$$

which relates the number of subbands at neighboring orders. There is also a relation between p's and q's,

$$p_{r+1} q_r - p_r q_{r+1} = (-1)^r. \tag{5.4}$$

It follows that the extra magnetic flux between the rth order and the $(r+1)$th order truncation is

$$\delta\phi_r = \frac{p_{r+1}}{q_{r+1}} - \frac{p_r}{q_r} = \frac{(-1)^r}{q_{r+1} q_r}. \tag{5.5}$$

Since the sign of $\delta\phi_r$ alternates from one order to the next, the direction of δB also alternates.[22]

Notice that for a chosen fraction p_r/q_r, the size of a MBZ is fixed. Without perturbation from the part that is truncated away, a wave packet will move on a straight line. The trajectory is curved because of $\delta\phi_r$. The higher the order of approximation we use, the smaller the $\delta\phi$ gets and the larger the radius of the hyperorbit becomes. In the ideal case, without any complications due to disorder, thermal broadening, etc., we expect there will be a hierarchical structure of hyperorbits due to different orders of approximation. This structure finds its correspondence in the hierarchical structure of the Hofstadter spectrum[23] (Fig. 2).

B. Distribution of Hall conductivities and splitting of energy bands

A magnetic Bloch band carries quantized Hall current, and this current will redistribute among daughter bands in such a way that the total Hall current for subbands equals the original current.[24] We will call this the "sum rule." The current distribution among subbands, which is also quantized in each subband, was obtained by Thouless *et al.*[11] In their famous paper, they found that the subband Hall conductivities are the integer-valued solutions of the Diophantine equation. Here we show that semiclassical dynamics offers an alternative and very heuristic solution to this problem. The Hall conductivities calculated will be used in Eq. (4.7) to calculate the number of magnetic subbands after splitting.

The general expression for the Hall conductivity of a "closed" subband is $\sigma = \rho e/\delta B$ [see Eq. (4.9) and below]. Therefore, σ can be determined if ρ and δB are known. Consider a subband at the rth order of splitting. Since all subbands at this level share the same number of states, each subband will have $\rho_r = \rho_0/q_r$, where ρ_0 is the density of states for the original Bloch band. The perturbation field for a subband at this level is $\delta B_r = h\,\delta\phi_{r-1}/(ea^2)$. Therefore, we have

$$\sigma_r^{\text{close}} = \frac{e\rho_r}{\delta B_r} = (-1)^{r-1} q_{r-1} \quad (\text{in units of } e^2/h). \tag{5.6}$$

(Since $\sigma_1 = 1$ for a closed subband at the first level, q_0 will be set to 1.) We have to emphasize that Eq. (5.6) is valid for *every closed subbands* at the rth order. Combining Eq. (4.7) with (5.6), we can determine the number of daughter bands being split from an rth order parent band, which is

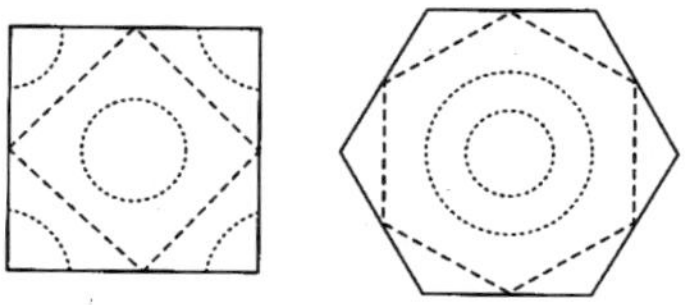

FIG. 3. Schematic plots of the constant energy contours in the reduced MBZ of a square lattice and a triangular lattice. Dashed lines are the open orbits.

$$\mathscr{D}_r^{\rm close} = \frac{|(-1)^r q_{r+1} + \sigma_r^{\rm close}|}{q_r} = f_{r+1}. \tag{5.7}$$

The Hall conductivity for an open subband is more difficult to obtain. It requires the knowledge of the exact **k** trajectory to figure out the first integral in Eq. (4.9). However, for a square or a triangular lattice, there is an easier way of calculating it. This is so because there is only one open orbit in every MBZ for either lattice (see Fig. 3). Therefore, there is only one open daughter band for every parent band. Its Hall conductivity can be figured out by using the "sum rule":

$$\sigma_{\rm parent} = \sum \sigma_{\rm daughter}. \tag{5.8}$$

For example, at the first order, we have $\sigma_1^{\rm open} = \sigma_0 - (f_1 - 1)\sigma_1^{\rm close} = -(f_1 - 1)$, where $\sigma_0 = 0$ since it is a Bloch band at the very beginning. For $r \geqslant 2$, the Hall conductivity of an open daughter band at the rth order is

$$\begin{aligned}\sigma_r^{\rm open} &= \sigma_{r-1}^{\rm close} - (f_r - 1)\sigma_r^{\rm close} \\ &= (-1)^{r-2} q_{r-2} - (f_r - 1)(-1)^{r-1} q_{r-1} \\ &= (-1)^{r-1} q_{r-1} + (-1)^r q_r,\end{aligned} \tag{5.9}$$

where we have assumed that its parent is a "closed" band. Using Eq. (4.7), we see that this open band (now as a parent) will split into

$$\mathscr{D}_r^{\rm open} = f_{r+1} + 1 \tag{5.10}$$

subbands under perturbation. It can be shown that the same result as Eq. (5.9) is obtained if its parent is an "open" band (with $f_r + 1$ daughters). This checks the consistency of this calculation. It is clear that the extra splitting of one subband from each open parent band (there are q_{r-1} of them) accounts for the extra q_{r-1} in the recursion relation Eq. (5.3)

We give one example to demonstrate the use of these rules. Consider a square lattice with $\phi = 1/(2+\sqrt{2})$. Because $\sigma_1^{\rm close} = 1$, and $\sigma_1^{\rm open} = -2$, the distribution of σ's for the three subbands at the first order is

$$\boldsymbol{\sigma}_1 = (1, -2, 1), \tag{5.11}$$

where we have put $\sigma_1^{\rm open}$ in the middle since for a square lattice the open subband is located at the center of a parent band (Fig. 3).

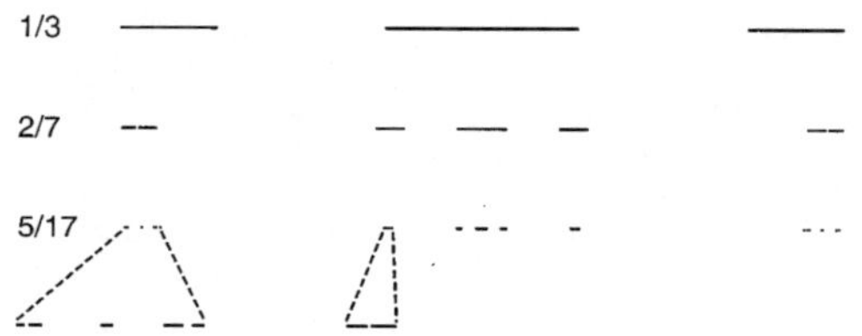

FIG. 4. Pattern of energy splitting for ϕ in Eq. (5.2). Different orders of approximation give 3, 7, and 17 subbands, respectively.

These three bands will be split into seven subbands due to the extra flux $\delta\phi_1 = p_2/q_2 - p_1/q_1 = 2/7 - 1/3$. Since $f_2 = 2$, the pattern of splitting will be, according to Eqs. (5.7) and (5.10),

$$\mathscr{D}_1 = (2, 3, 2). \tag{5.12}$$

Furthermore, since $\sigma_2^{\rm close} = -3$ and $\sigma_2^{\rm open} = 4$ according to Eqs. (5.6) and (5.9), the Hall conductivity distribution is[25]

$$\boldsymbol{\sigma}_2 = (-3, 4, -3, 4, -3, 4, -3). \tag{5.13}$$

Consequently, we have

$$\mathscr{D}_2 = (2, 3, 2, 3, 2, 3, 2). \tag{5.14}$$

The Hall conductivities we just obtained are the same as those derived from the Diophantine equation. Actual pattern of splitting is shown in Fig. 4 for comparison. The distribution in Fig. 4 for the left (or right) five subbands in Eq. (5.14) appears to be (2,1,2), instead of (2,3) [or (3,2)]. However, closer examination reveals that the left (right) three subbands actually come from the same parent. In fact, slight asymmetry in the distribution is inevitable because when ϕ is changed by a small amount, an electron state cannot suddenly jump out of the band edge to the middle of a gap.

Equations (5.6)–(5.10) also apply to a triangular lattice. The only difference is that $\sigma_{\rm open}$ no longer locates at the center of a parent band. Given the same $\phi = 1/(2+\sqrt{2})$, we now have

$$\begin{aligned}&\boldsymbol{\sigma}_1 = (1, 1, -2), \\ &\mathscr{D}_1 = (2, 2, 3), \\ &\boldsymbol{\sigma}_2 = (4, -3, -3, 4, 4, -3, -3), \\ &\mathscr{D}_2 = (3, 2, 2, 3, 3, 2, 2).\end{aligned} \tag{5.15}$$

This again conforms with the actual spectrum and the solutions of the Diophantine equation with subsidiary constraints suitable for a triangular lattice.[26]

VI. CALCULATIONS OF ENERGY SPECTRUM, CURVATURE, MAGNETIC MOMENT, AND CYCLOTRON ENERGY

In this section, we give detailed calculations of $\mathscr{E}_n(\mathbf{k})$, $\Omega_n(\mathbf{k})$, and $L_n(\mathbf{k})$. They are used in calculating the cyclotron energies using the quantization formula Eq. (4.5). The cyclotron energies will be used to estimate the subband energies in the Hofstadter spectrum. In doing so, we not only presume the one-band approximation (on which Fig. 2 is based), but also neglect the interorbit transitions that broaden (and may slightly shift) the energy levels. The latter approximation leads to negligible error if the bandwidths under consideration are very small. Similar calculations have been done by Wilkinson, and his results have been very successful.[27] However, we believe that the approach proposed here is conceptually simpler, and is easier to generalize to other types of lattices.[28]

A. Calculation of energy spectrum for parent bands

The following calculation is based on the tight-binding model. We will only give a very brief explanation of this approach. For more details, we request readers to refer to Ref. 29. In the tight-binding approximation, a Bloch state is expanded as (for $\phi=p/q$)

$$\Psi(\mathbf{k})=\sum_{l=1}^{q} a_l(\mathbf{k})\psi_l(\mathbf{k}), \tag{6.1}$$

where $\mathbf{k}$ is restricted to a reduced MBZ. The basis $\psi_l(\mathbf{k})$ is defined to be $\psi(k_1+2\pi\phi l,k_2)$, where $\psi(\mathbf{k})$ is a Bloch state before the magnetic perturbation. A tight-binding Hamiltonian in the absence of δB, when being expressed on the basis of ψ_l, is a $q\times q$ matrix (the lattice constant a is set to 1),

$$H_{l',l}=-\begin{pmatrix} \cos(k_1+2\pi/q) & e^{ik_2} & 0 & \cdot & 0 & e^{-ik_2} \\ e^{ik_2} & \cos(k_1+4\pi/q) & e^{ik_2} & \ddots & \ddots & 0 \\ 0 & e^{ik_2} & \ddots & \ddots & \ddots & \cdot \\ \cdot & \ddots & \ddots & \ddots & e^{ik_2} & 0 \\ 0 & \ddots & \ddots & e^{ik_2} & \cos\left(k_1+(q-1)\dfrac{2\pi}{q}\right) & e^{ik_2} \\ e^{-ik_2} & 0 & \cdot & 0 & e^{ik_2} & \cos(k_1) \end{pmatrix}. \tag{6.2}$$

$\mathscr{E}_n(\mathbf{k})$ and $a_l^n(\mathbf{k})$ are nothing but the eigenvalues and eigenvectors of this matrix. For example, if $p/q=1/3$, then a straightforward calculation shows that the $\mathscr{E}_n(\mathbf{k})$'s are solutions of the following characteristic equation:

$$-\mathscr{E}^3+6\mathscr{E}=2[\cos(3k_1)+\cos(3k_2)]. \tag{6.3}$$

There are three roots for each $\mathbf{k}$, and variation of $\mathbf{k}$ over the MBZ leads to the three energy bands for $\phi=1/3$ (see Fig. 2). It is not difficult to see that the band edges are located at (from high to low) $\mathscr{E}_1(\mathbf{g}),\mathscr{E}_1(\mathbf{0}),\mathscr{E}_2(\mathbf{0}),\mathscr{E}_2(\mathbf{g}),\mathscr{E}_3(\mathbf{g})$, and $\mathscr{E}_3(\mathbf{0})$, where $\mathbf{g}=(\pi/3,\pi/3)$.

B. Calculation of Berry curvature

To calculate the Berry curvature in Eq. (2.19), we need to know the eigenvectors of $H_{l'l}$. Before doing that, we will try to rewrite Eq. (2.19) in a form suitable for the tight-binding calculation. First, we insert a complete state $\Sigma_{n'}|u_{n'}\rangle\langle u_{n'}|$ inside the dot products that appear in Eq. (2.19). Since we are using the one-band approximation, n' only runs through subbands in the same parent band, and

$$\Omega_n(\mathbf{k})=i\sum_{n'=1}^{q}{}' \left[\left\langle\frac{\partial u_n}{\partial k_1}\middle|u_{n'}\right\rangle\left\langle u_{n'}\middle|\frac{\partial u_n}{\partial k_2}\right\rangle-\text{c.c.}\right], \tag{6.4}$$

where we have dropped a term with $n'=n$ since $\langle u_n|\partial/\partial\mathbf{k}|u_n\rangle$ is purely imaginary and does not contribute to the curvature. With the help of the identity

$$\left\langle u_{n'}\middle|\frac{\partial}{\partial\mathbf{k}}\middle|u_n\right\rangle=\frac{\left\langle u_{n'}\middle|\dfrac{\partial\tilde{H}}{\partial\mathbf{k}}\middle|u_n\right\rangle}{\mathscr{E}_{n'}-\mathscr{E}_n}, \tag{6.5}$$

where $\tilde{H}\equiv e^{-i\mathbf{k}\cdot\mathbf{r}}He^{i\mathbf{k}\cdot\mathbf{r}}$ [this and the following H's are the unperturbed Hamiltonian in Eq. (1.3); the subscript 0 is dropped for brevity], we can rewrite Eq. (6.4) in the form

$$\Omega_n(\mathbf{k})=i\sum_{n'}{}'\left[\frac{\left\langle u_n\middle|\dfrac{\partial\tilde{H}}{\partial k_1}\middle|u_{n'}\right\rangle\left\langle u_{n'}\middle|\dfrac{\partial\tilde{H}}{\partial k_2}\middle|u_n\right\rangle}{(\mathscr{E}_{n'}-\mathscr{E}_n)^2}-\text{c.c.}\right]. \tag{6.6}$$

Expanding u_n by $\tilde{\psi}_{l\mathbf{k}}(\mathbf{r})$, which is defined to be $e^{-i\mathbf{k}\cdot\mathbf{r}}\psi_{l\mathbf{k}}(\mathbf{r})$, we have

$$\begin{aligned}\left\langle u_{n'}\middle|\frac{\partial\tilde{H}}{\partial\mathbf{k}}\middle|u_n\right\rangle&=\sum_{l',l}a_{l'}^{n'}{}^*a_l^n\left\langle\tilde{\psi}_{l'}\middle|\frac{\partial\tilde{H}}{\partial\mathbf{k}}\middle|\tilde{\psi}_l\right\rangle\\&=\sum_{l',l}a_{l'}^{n'}{}^*a_l^n\frac{\partial}{\partial\mathbf{k}}\langle\tilde{\psi}_{l'}|\tilde{H}|\tilde{\psi}_l\rangle\\&\quad+(\epsilon_n-\epsilon_{n'})\sum_{l',l}a_{l'}^{n'*}a_l^n\left\langle\tilde{\psi}_{l'}\middle|\frac{\partial\tilde{\psi}_l}{\partial\mathbf{k}}\right\rangle.\end{aligned} \tag{6.7}$$

For a Bloch state in the absence of a magnetic field, we can choose the phase such that the inner product $\langle\tilde{\psi}_{l'}|\partial\tilde{\psi}_l|\partial\mathbf{k}\rangle$ is zero. Therefore,

$$\left\langle u_{n'}\middle|\frac{\partial\tilde{H}}{\partial\mathbf{k}}\middle|u_n\right\rangle=\sum_{ll'}a_{l'}^{n'}{}^*a_l^n\frac{\partial H_{l',l}}{\partial\mathbf{k}}. \tag{6.8}$$

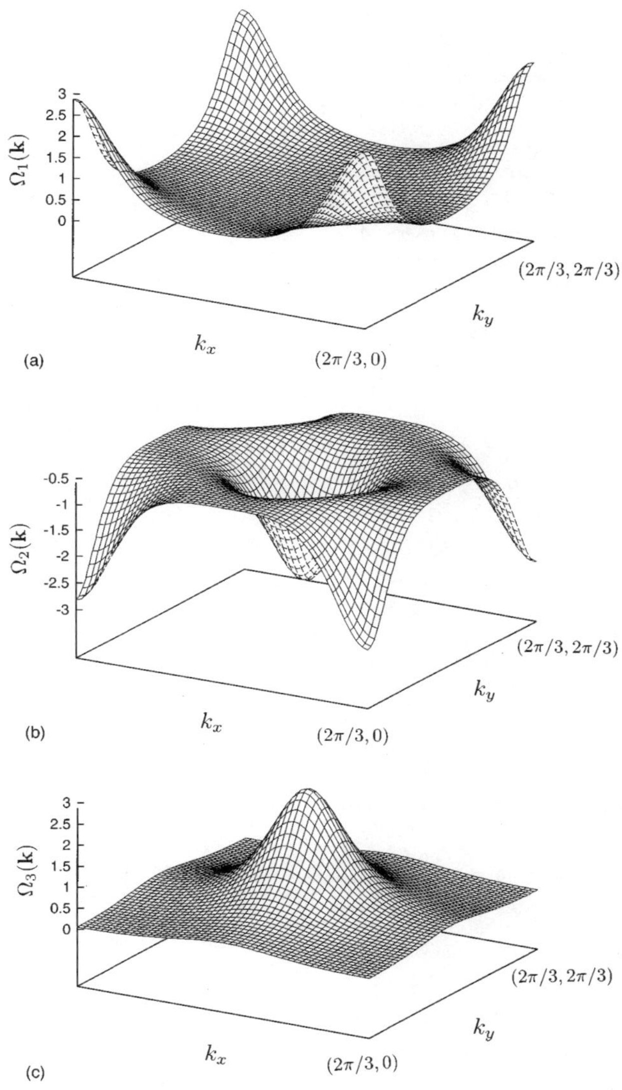

FIG. 5. Distributions of Berry curvature $\Omega_n(\mathbf{k})$ (in units of e^2/h). $\Omega_1(\mathbf{k})$ is equal to $\Omega_3(\mathbf{k})$ shifted by $(\pi/3, \pi/3)$.

Beyond this stage, the calculation is straightforward since we only need to calculate a_l^n from Eq. (6.2) and combine Eqs. (6.6) and (6.8) to obtain $\Omega_n(\mathbf{k})$.

Again we choose a simple fraction $p/q = 1/3$ and calculate the Berry curvature distributions for the three magnetic subbands. The result is shown in Fig. 5, in which the range of $\mathbf{k}$ vector is one reduced MBZ (the basic unit of repetition). Note that the curvature tends to concentrate on four inner band edges because the electron states near inner gaps are changed the most from the original Bloch states that have zero Berry curvature. The curvatures from the three bands cancel locally, that is,

$$\sum_{n=1}^{3} \Omega_n(\mathbf{k}) = 0, \quad \forall \mathbf{k}. \tag{6.9}$$

This is in general true for any q and can be easily proved from Eq. (6.6). This is a stronger condition than the (global)

TABLE II. Hofstadter spectrum (for $\phi=22/67$) and the cyclotron energies calculated from Eq. (4.5). Only the top ten subbands for the middle parent band are shown. The last column is the subband closest to the parent band edge $\mathscr{E}_2(\mathbf{0})=0.7321$.

$E_{\rm Hofst}$	0.0618	0.1067	0.1566	0.2124	0.2747	0.3443	0.4221	0.5098	0.6098	0.7266
	0.0678	0.1085	0.1570	0.2125						
$E_{\rm cyclo}$	0.0632	0.1063	0.1558	0.2115	0.2738	0.3435	0.4212	0.5086	0.6082	0.7240

sum rule $\Sigma_n \sigma_n=0$. Finally, integration of $\Omega_n(\mathbf{k})$ over a MBZ divided by 2π gives us integers $(1,-2,1)$. These are indeed the Chern numbers we expected [see Eq. (5.11)].

C. Calculation of magnetic moment

The self-rotating angular momentum of the wave packet $\langle \mathbf{L}\rangle_{n\mathbf{k}}$ in Eq. (2.13) can be written in a from that is more tractable for calculation:

$$L_n(\mathbf{k})=i\frac{m}{\hbar}\left[\left\langle\frac{\partial u_n}{\partial k_1}\middle|\tilde{H}-\mathscr{E}_n\middle|\frac{\partial u_n}{\partial k_2}\right\rangle-\text{c.c.}\right]. \tag{6.10}$$

Derivation of this formula is given in Appendix B. Equation (6.10) can also be rewritten as

$$L_n(\mathbf{k})=i\frac{m}{\hbar}\sum_{n'}{}'\left[\frac{\left\langle u_n\middle|\frac{\partial \tilde{H}}{\partial k_1}\middle|u_{n'}\right\rangle\left\langle u_{n'}\middle|\frac{\partial \tilde{H}}{\partial k_2}\middle|u_n\right\rangle}{\mathscr{E}_{n'}-\mathscr{E}_n}-\text{c.c.}\right]. \tag{6.11}$$

Derivation of Eq. (6.11) is very similar to the derivation of Eq. (6.6). The only change is that the extra factor of $\tilde{H}-\mathscr{E}_n$ in the numerator cancels a $\mathscr{E}_{n'}-\mathscr{E}_n$ in the denominator.

$L_n(\mathbf{k})$ can be readily calculated by combining Eq. (6.8) with Eq. (6.11). The result is shown in Fig. 6 (again for $\phi=1/3$). Similar to the distribution of $\Omega_n(\mathbf{k})$, $L_n(\mathbf{k})$ also has peaks near inner band edges. The total magnetizations from all three bands cancel each other. However, unlike the curvature, they do not cancel locally at each **k** point.

We remark that this magnetization energy first appeared in a paper by Kohn.[2] His objective was to study the effective Hamiltonian for Bloch electrons in a weak electromagnetic field. This term is in general zero for Bloch bands, but can be nonzero in the presence of spin-orbit interaction. In the latter case, it contributes an extra g factor to Bloch electrons.[29] An expression that is the same as the right hand side of Eq. (6.10) has also been obtained by Rammal and Bellissard.[30] Without the $m/\hbar$ factor (and apart from a factor of 2), it is called the Rammal-Wilkinson form.

D. Calculation of cyclotron energy

After obtaining $\mathscr{E}_n(\mathbf{k})$, $\Omega_n(\mathbf{k})$, and $L_n(\mathbf{k})$, we can determine the cyclotron energies according to the quantization formula Eq. (4.5). In the following example, we add $\delta\phi=-1/201$ to $\phi=1/3$. This gives $\phi'=22/67=1/(3+1/22)$. According to the simple rules derived in the preceding section, the original magnetic bands are expected to split into 22, 23, and 22 subbands, respectively. In Table II, we compare the exact spectrum with the quantized cyclotron energies. Only part of the subbands from the parent band in the middle is shown. We show two numbers (for two band edges) in the first row when the bandwidth for $E_{\rm Hofst}$ is larger than 10^{-4}. The $E_{\rm cyclo}$'s in the second row are obtained by fine-tuning the path C_m in Eq. (4.5), with uncertainty on the order of 10^{-4}. It can be seen that the match between $E_{\rm cyclo}$ and $E_{\rm Hofst}$ is quite satisfying. We have done calculations for subbands from other parent bands, and they also show similar accuracy.

Notice that the energy levels from fractions like $\phi=1/f$ are broadened cyclotron levels in an ordinary Bloch band. They do not split from a magnetic parent band. In this case $\Omega_n(\mathbf{k})$ and $L_n(\mathbf{k})$ are zero, and Eq. (4.5) reduces to the usual Onsager quantization formula. Numerical result based on this simplified formula for the cyclotron energies also agrees very well with the positions of subbands in Fig. 2 as expected.

VII. SUMMARY

Electron states in a lattice subject to a homogeneous magnetic field satisfy magnetic translation symmetry and have bandlike energy spectrum similar to the usual Bloch band. However, to our knowledge, the semiclassical dynamics of magnetic Bloch electrons has never been studied explicitly. One reason is that the observation of band-splitting remains an experimental challenge to date; the other reason might be due to the fact that magnetic Bloch bands, unlike Bloch bands, can be changed easily by varying an external magnetic field. However, our study has shown that the inquiry of magnetic Bloch bands can be very rewarding in itself. Major findings in this paper are summarized below:

The Berry curvature of magnetic bands plays a crucial role in the dynamics. It gives electrons an extra velocity in the direction of $\mathbf{E}\times\mathbf{B}$, and this term directly relates to the quantization of Hall conductivity. This semiclassical dynamics, combined with the Boltzmann equation, is used to study electron transport in a dc or ac electric field.

In the presence of δB, the energy dispersion $E(\mathbf{k})$ is shifted from the usual band energy because of the nonzero magnetic moment. Similar to usual Bloch electrons, magnetic Bloch electrons execute cyclotron motion on the constant energy surface of $E(\mathbf{k})$. However, the quantization condition for cyclotron orbits has to be modified from the usual Onsager condition because of the Berry phase. Based on this modified formula, we obtain a simple rule that calculates the number of daughter bands for every parent band in the Hofstadter spectrum. Furthermore, a fairly heuristic explanation for the distribution of Hall conductivities is given using the picture of cyclotron orbit drifting.

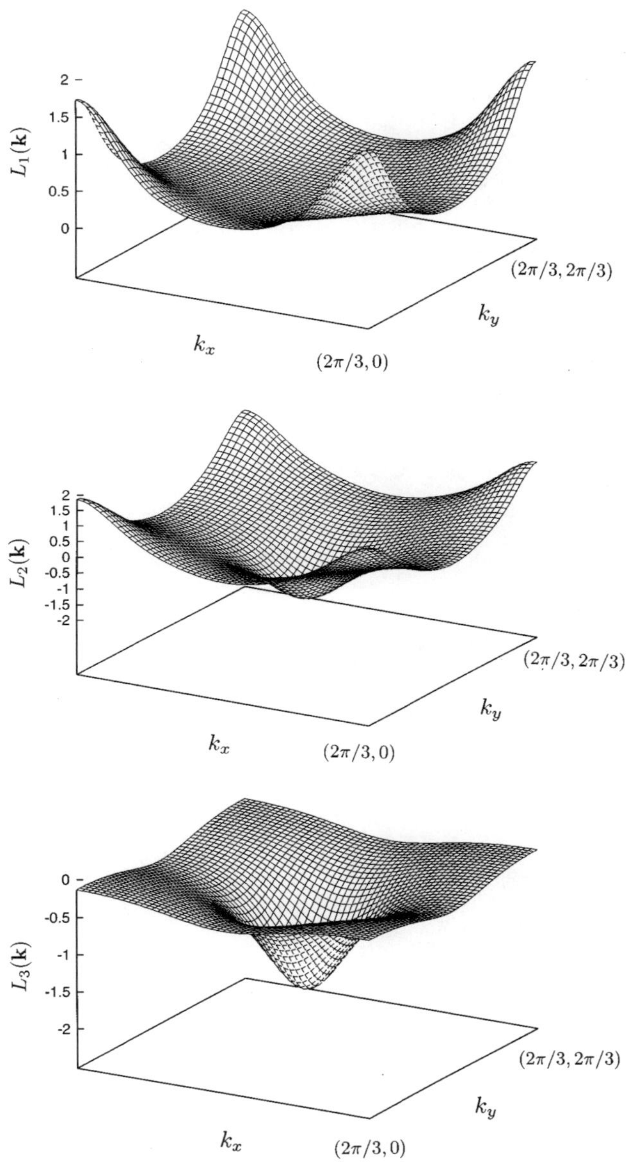

FIG. 6. Distributions of angular momentum $L_n(\mathbf{k})$ (in units of $2m/\hbar$).

These quantized orbits are closely related to the energy bands in the Hofstadter spectrum. We give detailed numerical calculations for $\Omega_n(\mathbf{k})$ and $L_n(\mathbf{k})$ based on the tight-binding model for the case of $\phi = 1/3$. They are used in the calculation of cyclotron energies using the quantization condition, and the result is in very good agreement with the actual spectrum. This shows that the complex pattern of the Hofstadter spectrum is nothing more than the broadened cyclotron energy spectrum in magnetic Bloch bands.

ACKNOWLEDGMENTS

The authors wish to thank A. Barr, J. Bellissard, F. Claro, E. Demircan, G. Georgakis, M. Kohmoto, W. Kohn, M. Marder, G. Sundaram, and M. Wilkinson for many helpful discussions. This work is supported by the R. A. Welch Foundation.

APPENDIX A

Note that there is no $\dot{\mathbf{k}}$ dependence in $\boldsymbol{\pi}$ [because there is no kinetic energy term $\dot{\mathbf{k}}^2$ in $L(\mathbf{k},\dot{\mathbf{k}})$]. In this case, it is more rtinent to treat Eq. (4.3) as a constraint on the variables $\mathbf{k}$ and $\boldsymbol{\pi}$,

$$\boldsymbol{\theta}(\mathbf{k},\boldsymbol{\pi})\equiv\boldsymbol{\pi}+\frac{\hbar^2}{2e\,\delta B}\mathbf{k}\times\hat{z}-\hbar\mathscr{A}(\mathbf{k})\simeq 0. \tag{A1}$$

Strictly speaking, this constraint cannot be used before obtaining the equations of motion, therefore we use $\simeq$ to distinguish it from a real identity.[31] A general Hamiltonian for a system with constraints is given by

$$H^*(\mathbf{k},\boldsymbol{\pi})=H(\mathbf{k},\boldsymbol{\pi})+\boldsymbol{\lambda}\cdot\boldsymbol{\theta}=E(\mathbf{k})+\boldsymbol{\lambda}\cdot\boldsymbol{\theta}, \tag{A2}$$

where $\boldsymbol{\lambda}=(\lambda_1,\lambda_2)$ are arbitrary functions of $\mathbf{k}$ and $\boldsymbol{\pi}$. The dynamical equations for this Hamiltonian are

$$\dot{\mathbf{k}}=\frac{\partial H^*}{\partial\boldsymbol{\pi}}=\boldsymbol{\lambda},$$

$$\dot{\boldsymbol{\pi}}=-\frac{\partial H^*}{\partial\mathbf{k}}=-\frac{\partial E}{\partial\mathbf{k}}+\frac{\hbar^2}{2e\,\delta B}\boldsymbol{\lambda}\times\hat{z}+\hbar\sum_i\lambda_i\frac{\partial\mathscr{A}_i}{\partial\mathbf{k}}, \tag{A3}$$

where we have discarded a term $\partial\boldsymbol{\lambda}/\partial\mathbf{k}\cdot\boldsymbol{\theta}=0$. According to Eq. (4.3), we should also have

$$\dot{\boldsymbol{\pi}}=-\frac{\hbar^2}{2e\,\delta B}\dot{\mathbf{k}}\times\hat{z}+\hbar\sum_i\dot{k}_i\frac{\partial\mathscr{A}}{\partial k_i}. \tag{A4}$$

Equating (A3) to (A4), and replacing $\boldsymbol{\lambda}$ by $\dot{\mathbf{k}}$, we will get Eq. (4.1).

APPENDIX B

We will rewrite the angular momentum of a wave packet in Eq. (2.13) in terms of magnetic Bloch functions. By defining $\tilde{w}(\mathbf{k})=e^{i\mathbf{k}\cdot\mathbf{r}_c}w(\mathbf{k})$, we have

$$\begin{aligned}\mathbf{L}_n(\mathbf{k}_c)&=\int d^2\mathbf{k}'\int d^2\mathbf{k}\;w^*(\mathbf{k}')w(\mathbf{k})\langle\Psi_n(\mathbf{k}')|(\mathbf{r}-\mathbf{r}_c)\times\mathbf{P}|\Psi_n(\mathbf{k})\rangle\\&=\int d^2\mathbf{k}'\int d^2\mathbf{k}\;\tilde{w}^*(\mathbf{k}')\tilde{w}(\mathbf{k})\langle u_n(\mathbf{k}')|e^{i(\mathbf{k}-\mathbf{k}')\cdot(\mathbf{r}-\mathbf{r}_c)}(\mathbf{r}-\mathbf{r}_c)\times\tilde{\mathbf{P}}(\mathbf{k})|u_n(\mathbf{k})\rangle,\end{aligned} \tag{B1}$$

where $\tilde{\mathbf{P}}$ is the momentum operator on the $|u_n\rangle$ basis. Since

$$\tilde{\mathbf{P}}(\mathbf{k})|u_n(\mathbf{k})\rangle=\sum_{n'}|u_{n'}(\mathbf{k})\rangle\langle u_{n'}(\mathbf{k})|\tilde{\mathbf{P}}(\mathbf{k})|u_n(\mathbf{k})\rangle, \tag{B2}$$

and

$$\begin{aligned}\langle u_n(\mathbf{k}')|e^{i(\mathbf{k}-\mathbf{k}')\cdot(\mathbf{r}-\mathbf{r}_c)}(\mathbf{r}-\mathbf{r}_c)|u_{n'}(\mathbf{k})\rangle&=i\,\delta_{n,n'}\frac{\partial}{\partial\mathbf{k}'}\delta(\mathbf{k}-\mathbf{k}')-i\left\langle\frac{\partial u_n}{\partial\mathbf{k}'}\right|e^{i(\mathbf{k}-\mathbf{k}')\cdot(\mathbf{r}-\mathbf{r}_c)}\left|u_{n'}\right\rangle\\&=i\,\delta_{n,n'}\frac{\partial}{\partial\mathbf{k}'}\delta(\mathbf{k}-\mathbf{k}')-i\,\delta(\mathbf{k}-\mathbf{k}')\left\langle\frac{\partial u_n}{\partial\mathbf{k}'}\right|u_{n'}\right\rangle,\end{aligned} \tag{B3}$$

we have

$$\mathbf{L}_n(\mathbf{k}_c)=-i\int d^2\mathbf{k}\left[\frac{\partial}{\partial\mathbf{k}}\tilde{w}^*(\mathbf{k})\right]\tilde{w}(\mathbf{k})\times\langle\mathbf{P}\rangle_n-i\int d^2\mathbf{k}|\tilde{w}(\mathbf{k})|^2\left\langle\frac{\partial u_n}{\partial\mathbf{k}}\right|\times\tilde{\mathbf{P}}(\mathbf{k})\left|u_n\right\rangle. \tag{B4}$$

Because $\tilde{\mathbf{P}}(\mathbf{k})=(m/\hbar)\partial\tilde{H}/\partial\mathbf{k}$, the integrand of the second term can be written as

$$\begin{aligned}\frac{m}{\hbar}\left\langle\frac{\partial u_n}{\partial k_1}\right|\frac{\partial\tilde{H}}{\partial k_2}\left|u_n\right\rangle-(k_1\leftrightarrow k_2)&=\frac{m}{\hbar}\frac{\partial}{\partial k_2}\left\langle\frac{\partial u_n}{\partial k_1}\right|\tilde{H}\left|u_n\right\rangle-\frac{m}{\hbar}\left\langle\frac{\partial u_n}{\partial k_1}\right|\tilde{H}\left|\frac{\partial u_n}{\partial k_2}\right\rangle-(k_1\leftrightarrow k_2)\\&=\frac{m}{\hbar}\left\langle\frac{\partial u_n}{\partial k_1}\middle|\frac{\partial u_n}{\partial k_2}\right\rangle\mathscr{E}_n+\frac{m}{\hbar}\left\langle\frac{\partial u_n}{\partial k_1}\middle|u_n\right\rangle\frac{\partial\mathscr{E}_n}{\partial k_2}-\frac{m}{\hbar}\left\langle\frac{\partial u_n}{\partial k_1}\right|\tilde{H}\left|\frac{\partial u_n}{\partial k_2}\right\rangle\\-(k_1\leftrightarrow k_2)&=\frac{m}{\hbar}\left\langle\frac{\partial u_n}{\partial\mathbf{k}}\right|\times(\mathscr{E}_n-\tilde{H})\left|\frac{\partial u_n}{\partial\mathbf{k}}\right\rangle+\left\langle\frac{\partial u_n}{\partial\mathbf{k}}\middle|u_n\right\rangle\times\langle\mathbf{P}\rangle_n.\end{aligned} \tag{B5}$$

Therefore we have

$$\mathbf{L}_n(\mathbf{k}_c)=i\frac{m}{\hbar}\left\langle\frac{\partial u_n}{\partial\mathbf{k}}\right|\times(\tilde{H}-\mathscr{E}_n)\left|\frac{\partial u_n}{\partial\mathbf{k}}\right\rangle\Bigg|_{\mathbf{k}=\mathbf{k}_c}-i\int d^2\mathbf{k}|\tilde{w}(\mathbf{k})|^2\left\langle\frac{\partial u_n}{\partial\mathbf{k}}\right|u_n\right\rangle\times\mathbf{P}\langle\mathbf{P}_n\rangle-i\int d^2\mathbf{k}\left[\frac{\partial}{\partial\mathbf{k}}\tilde{w}^*(\mathbf{k})\right]\tilde{w}(\mathbf{k})\times\mathbf{P}\langle\mathbf{P}\rangle_{\mathbf{n}}. \quad \text{(B6)}$$

The last two terms cancel because of Eq. (2.8), and this leads to Eq. (6.10) Notice that this result is independent of the way a wave packet is constructed since there is no $w(\mathbf{k})$ dependence in the new expression.

---

[1] N. W. Ashcroft and N. D. Mermin, *Solid State Physics* (W. B. Saunders Co., Philadelphia, 1976), Chap. 12.

[2] J. M. Luttinger and W. Kohn, Phys. Rev. **97**, 869 (1955); W. Kohn, *ibid.* **115**, 1460 (1959); L. M. Roth, J. Phys. Chem. Solids **23**, 433 (1962); Phys. Rev. **145**, 434 (1969); R. G. Chambers, Proc. Phys. Soc. **89**, 695 (1966); J. Zak, Phys. Rev. **168**, 686 (1967). A recent attempt to justify the semiclassical dynamics in a short derivation is given by A. Manohar, Phys. Rev. B **34**, 1287 (1986).

[3] In general, the magnetic field can be periodic. This would only change the band structure, but not the dynamics itself. See M. C. Chang and Q. Niu, Phys. Rev. B **50**, 10 843 (1994).

[4] J. Zak, Phys. Rev. **136**, A776 (1964); **134**, A1602 (1964). Also see M. Y. Azbel, Sov. Phys. JETP **17**, 665 (1963).

[5] P. G. Harper, Proc. Phys. Soc. London Ser. A **68**, 874 (1955); A. Rau, Phys. Status Solidi B **65**, K131 (1974); **69**, K9 (1975).

[6] R. R. Gerhardts, D. Weiss, and U. Wulf, Phys. Rev. B **43**, 5192 (1991); I. N. Harris *et al.*, Europhys. Lett. **29**, 333 (1995).

[7] M. C. Chang and Q. Niu, Phys. Rev. Lett. **75**, 1348 (1995).

[8] M. V. Berry, Proc. R. Soc. London Ser. A **392**, 45 (1984); B. Simon, Phys. Rev. Lett. **51**, 2167 (1983). Also see A. Bohm, *Quantum Mechanics,* 3rd ed. (Springer-Verlag, New York, 1993), Chap. 22 for an extensive discussion.

[9] See, for example, E. I. Blount, in *Solid State Physics*, edited by F. Seitz and D. Turnbull (Academic, New York, 1962), Vol. 13. The derivation there is for usual Bloch states, but it also applies to magnetic Bloch states.

[10] Notice that since $\mathbf{r}_c$ is slowly varying with time (in the adiabatic sense), the gauge condition is also slowly changing with time.

[11] D. J. Thouless, M. Kohmoto, M. P. Nightingale, and M. den Nijs, Phys. Rev. Lett. **49**, 405 (1982); M. Kohmoto, Ann. Phys. **160**, 343 (1985); J. Phys. Soc. Jpn. **62**, 659 (1993).

[12] C. W. J. Beenakker and H. van Houten in *Solid State Physics*, edited by H. Ehrenreich and D. Turnbull (Academic Press, New York, 1991), Vol. 44.

[13] R. G. Chambers, Proc. Phys. Soc. London Ser. A **65**, 458 (1952); H. Budd, Phys. Rev. **127**, 4 (1962). A clear presentation is given by J. Callaway, *Quantum Theory of the Solid State* 2nd ed. (Academic Press, New York, 1991), p. 651.

[14] D. Pfannkuche and R. R. Gerhardts, Phys. Rev. B **46**, 12 606 (1992); U. Wulf and A. H. MacDonald, *ibid.* **47**, 6566 (1993).

[15] Similar analysis of the oscillating part in $\dot{y}$ shows that the correction from Ω does not lead to a net drift in the y direction.

[16] D. H. Dunlap and V. M. Kenkre, Phys. Rev. B **34**, 3625 (1986); X. G. Zhao, Phys. Lett. **155**, 299 (1991); J. Phys. Condens. Matter **6**, 2751 (1994); M. Holthaus, Phys. Rev. Lett. **69**, 351 (1992).

[17] A. B. Pippard, in *Physics of Solids in Intense Magnetic Fields*, edited by E. D. Haidemenakis (Plenum, New York, 1969), p. 370. Also in *Magnetoresistance in Metals* (Cambridge University Press, Cambridge, 1989), p. 155.

[18] H. van Houten *et al.*, Phys. Rev. B **39**, 8556 (1989).

[19] J. J. Heremans, M. B. Santos, and M. Shayegan, Surf. Sci. **305**, 348 (1994).

[20] R. L. Willett *et al.*, Phys. Rev. B **47**, 7344 (1993).

[21] Mathematically it can be proved that p_r/q_r is the closest approximation to ϕ among all fractions with $q\leqslant q_r$. See A. Y. Khinchin, *Continued Fractions* (The University of Chicago Press, Chicago, 1964).

[22] Another consequence of Eq. (5.5) is that q_r and q_{r+1} cannot both be even, for otherwise there will be a common factor of 2 for both numerator and denominator of $p_{r+1}/q_{r+1}-p_r/q_r$. This implies that the total number of daughter bands cannot be even if the total number of parent bands is even.

[23] D. R. Hofstadter, Phys. Rev. **14**, 2239 (1976); F. H. Claro and G. H. Wannier, Phys. Rev. B **19**, 6068 (1979). The connection between continued fraction and the hierarchical spectrum was first suggested in M. Y. Azbel, Zh. Eskp. Teor. Fiz. **46**, 939 (1964) [Sov. Phys. JETP **19**, 634 (1964)].

[24] J. E. Avron, R. Seiler, and B. Simon, Phys. Rev. Lett. **51**, 51 (1983); J. E. Avron and B. Simon, J. Phys. A **18**, 2199 (1985).

[25] It is not $(4,-3,-3,4,-3,-3,4)$ because the subbands near the edges always have σ_{close}.

[26] D. J. Thouless, Surf. Sci. **142**, 147 (1984). In Eq. (5.15) we have estimated the location of σ_{open} by the position of an open orbit in a MBB.

[27] M. Wilkinson, Proc. R. Soc. London Ser. A **391**, 305 (1984); J. Phys. A **17**, 3459 (1984); **20**, 4337 (1987); **23**, 2529 (1990).

[28] M. Kohmoto, Phys. Rev. B **39**, 11 943 (1989); Y. Hatsugai and M. Kohmoto, *ibid.* **42**, 8282 (1990).

[29] E. H. Cohen and E. I. Blount, Philos Mag. **5**, 115 (1960); L. M. Roth, Phys. Rev. **118**, 1534 (1960).

[30] R. Rammal and J. Bellissard, J. Phys. France **51**, 1803 (1990); Y. Tan, J. Phys. A **28**, 4163 (1995); M. Wilkinson and R. Kay (unpublished).

[31] P. A. M. Dirac, *Lectures on Quantum Mechanics* (Belfer Graduate School of Science, 1964).

Errata

1. In Eq. (3.3), $\dot{\mathbf{k}} = -e\mathbf{E}$ should be $\hbar\dot{\mathbf{k}} = -e\mathbf{E}$.

2. In Eq. (6.2), the diagonal matrix elements should be multiplied by 2, and the elements in the lower left triangle should be complex conjugated. Eq. (6.3) and the following numerical calculations are from the correct matrix and need not be changed.

3. On the last line of Eq. (6.7), $\langle\dot{\bar{\psi}}_{l'}|$ should be $\langle\tilde{\psi}_{l'}|$.

4. In Eq. (B6), the $\mathbf{P}\langle\mathbf{P}_n\rangle$ and $\mathbf{P}\langle\mathbf{P}\rangle_\mathbf{n}$ in the last two terms should both be $\langle\mathbf{P}\rangle_n$.

The arXiv version cond-mat/9511014 fixed points 3 and 4, but not 1 and 2.

ARTICLES

Wave-packet dynamics in slowly perturbed crystals: Gradient corrections and Berry-phase effects

Ganesh Sundaram and Qian Niu
Department of Physics, University of Texas at Austin, Austin, Texas 78712-1081
(Received 8 June 1998)

We present a unified theory for wave-packet dynamics of electrons in crystals subject to perturbations varying slowly in space and time. We derive the wave-packet energy up to the first-order gradient correction and obtain all kinds of Berry phase terms for the semiclassical dynamics and the quantization rule. For electromagnetic perturbations, we recover the orbital magnetization energy and the anomalous velocity purely within a single-band picture without invoking interband couplings. For deformations in crystals, besides a deformation potential, we obtain a Berry-phase term in the Lagrangian due to lattice tracking, which gives rise to new terms in the expressions for the wave-packet velocity and the semiclassical force. For multiple-valued displacement fields surrounding dislocations, this term manifests as a Berry phase, which we show to be proportional to the Burgers vector around each dislocation. [S0163-1829(99)07023-X]

I. INTRODUCTION

Our understanding of electronic properties of crystalline solids is primarily based on the Bloch theory for periodic systems.[1] It has been of great interest to extend this theory to situations where crystals are perturbed in various ways. So far, the most useful description has been the semiclassical theory for electron dynamics within a band supplemented by the semiclassical quantization rule or the Boltzmann transport equations. For example, the equations of motion of Bloch electrons in electromagnetic fields are given by[2]

$$\dot{\mathbf{x}}=\frac{1}{\hbar}\frac{\partial\mathcal{E}_{0,n}(\mathbf{k})}{\partial\mathbf{k}},$$

$$\hbar\dot{\mathbf{k}}=-e\mathbf{E}-e\dot{\mathbf{x}}\times\mathbf{B}, \tag{1.1}$$

where $\mathcal{E}_{0,n}(\mathbf{k})$ is the energy of the nth band of an unperturbed crystal. These equations have played a fundamental role in the physics of metals and semiconductors.

The derivation of Eq. (1.1) dates back to Bloch, Peierls, Jones and Zener in the early 1930s.[3] By assuming that the transition probabilities to other bands are negligible, they showed that Eqs. (1.1) describe the motion of a narrow wave packet obtained by superposing the Bloch states of a band. Various extensions of the theory have been made to deal with perturbations of more general nature and to obtain corrections to Eqs. (1.1) in high fields.

Peierls[4] pioneered the effort of constructing an effective one-band Hamiltonian to describe the quantum dynamics of a Bloch electron. By using the tight-binding model, he was able to show that the effective Hamiltonian in the presence of a magnetic field may be obtained by replacing the crystal momentum $\hbar\mathbf{k}$ by the gauge invariant momentum operator $[-i\hbar\nabla+e\mathbf{A}(\hat{\mathbf{x}})]$ in the unperturbed band energy

$$\hat{H}_{\text{eff}}=\mathcal{E}_{0,n}\left[-i\nabla+\frac{e}{\hbar}\mathbf{A}(\hat{\mathbf{x}})\right], \tag{1.2}$$

which later came to be known as the Peierls substitution. Two decades later, Slater[5] and Luttinger[6] gave a more rigorous derivation of the effective Hamiltonian for electromagnetic perturbations, by expanding the wave function in the basis of Wannier functions

$$\Psi(\mathbf{x},t)=\sum_{l} f_l(t)W(\mathbf{x}-\mathbf{R}_l), \tag{1.3}$$

where $\{\mathbf{R}_l\}$ are the lattice positions. They showed that the envelope function $f(\mathbf{x},t)$, defined by $f(\mathbf{R}_l,t)=f_l(t)$ and a smooth interpolation between the atomic positions, satisfies the effective Schrödinger equation

$$i\hbar\frac{\partial}{\partial t}f=\left\{\mathcal{E}_{0,n}\left[-i\nabla+\frac{e}{\hbar}\mathbf{A}(\mathbf{x})\right]-e\phi(\mathbf{x})\right\}f, \tag{1.4}$$

where $\phi(\mathbf{x})$ is a slowly varying scalar potential. The equations of motion (1.1) then follow from Eq. (1.4) and the correspondence principle.

Further development of the theory was made by taking into account the effects of interband coupling. Adams[7] extended the works of Slater and Luttinger to many-band operator formalism. Karplus, Luttinger, and Kohn derived a correction to the velocity, known as the anomalous velocity, and predicted a spontaneous Hall effect in ferromagnetic materials.[8] Later, Adams and Blount[9,10] showed that this term arises from the noncommutability between the Cartesian components of the intraband position operator. Recently, Chang and Niu[11,12] related the anomalous velocity correction to the Berry phase associated with the electron motion in an energy band.[13–15] Corrections to the effective Hamiltonian as an asymptotic series in the field strength were obtained by eliminating the interband matrix elements with unitary transformations by Kohn, Blount, and Roth in the early 1960s.[16] Later, Brown[17] extended the Wannier function method to crystals under high magnetic fields using magnetic-translation symmetry. A decade later, a variant of the Wannier function method that treats position and momentum in a symmetric way, known as kq-representation, was developed by Zak.[18] Recently, Rammal and Bellissard used an algebraic approach[19] and Wilkinson an operator approach[20] to derive the first-order field correction for the special case of the Harper's equation.[21,22]

Another approach to this problem involves the use of the WKB expansion to derive a Hamilton-Jacobi equation, and then making the correspondence from the classical variables to the quantum operators. This method was applied by many researchers to understand the nature of the spectrum and the wave function of electrons described by the Harper's equation, particularly in the semiclassical limit.[22] A more general treatment of the problem is based on a two-scale expansion in which the electron coordinate and the slowly varying vector potential are regarded as independent variables.[23]

In this paper, we come back to the original consideration of a wave packet in a band and use a time-dependent variational principle[24,25] to derive the wave-packet energy up to first order in the gradient of the perturbations and Berry-phase corrections to the semiclassical dynamics and the quantization rule. We are able to obtain the magnetization energy and the anomalous velocity entirely from the single-band point of view. Also, our method can be directly extended to the case of slowly perturbed magnetic bands, where methods based on the usual Wannier function approach break down because of the nonexistence of localized Wannier functions for such bands in general.[26]

This program was started with Chang and Niu[11,12] for the special case of a two-dimensional periodic system in a strong magnetic field. Here we establish a unified framework for slowly perturbed crystals whose Hamiltonian can be expressed in the form

$$H[\hat{\mathbf{x}},\hat{\mathbf{p}};\beta_1(\hat{\mathbf{x}},t),\ldots,\beta_r(\hat{\mathbf{x}},t)], \tag{1.5}$$

where $\{\beta_i(\mathbf{x},t)\}$ are the modulation functions characterizing the perturbations. They may represent either deformation strain fields, gauge potentials of electromagnetic fields, or slowly varying impurity potentials. They also appear in model potentials for modulated and incommensurate crystals,[27] and for graded semiconductors.[28]

We shall illustrate our formalism with two special cases of perturbations: electromagnetic fields and deformations in crystals. In the first case, in addition to the corrections of orbital magnetization energy and anomalous velocity mentioned above, we discuss the Peierls substitution, Berry-phase-modified Landau levels, and Zak-phase-modified Wannier-Stark levels. For deformational perturbations, we show that the deformation correction to the wave-packet energy can be obtained from the differential shift in the band energy under uniform strain. Then we obtain for the Lagrangian a Berry phase term due to lattice tracking, which gives rise to new terms in the expressions for the wave packet velocity and the semiclassical force; for multiple-valued displacement fields in the presence of dislocations, this term manifests as a Berry phase, which we show to be proportional to the Burgers vector around each dislocation, and thus, in a sense keeps track of the lattice position. We also discuss the consequences of the Berry phase term on electron transport, and the Aharonov-Bohm-type effects in dislocated crystals.

The paper is organized as follows. We present our formalism in Sec. II, treat electromagnetic and deformational perturbations in Secs. III and IV, respectively, and conclude with a summary in Sec. V.

II. FORMALISM

We shall begin by constructing a basis local to the wave packet and describe the wave packet in detail. Then we derive the Lagrangian, the semiclassical equations of motion, and Berry-phase correction to the semiclassical quantization, and discuss some aspects of formal quantization through the Hamiltonian formalism.

A. The local basis and the wave packet

Consider a wave packet centered at $\mathbf{x}_c$ at a given time, with its spread small compared to the length scale of the perturbations. Then the approximate Hamiltonian that the wave packet "feels" may be obtained by linearizing the perturbations about the wave-packet center as

$$\hat{H}\approx\hat{H}_c+\left\{\begin{array}{c}\text{sum of terms}\propto\\ (\hat{\mathbf{x}}-\mathbf{x}_c)\cdot\mathrm{grad}_{\mathbf{x}_c}\beta_i(\mathbf{x}_c,t)\end{array}\right\}, \tag{2.1}$$

where $\hat{H}_c\equiv H(\hat{\mathbf{x}},\hat{\mathbf{p}};\{\beta_i(\mathbf{x}_c,t)\})$ will be called the local Hamiltonian. The terms within the braces are small in the neighborhood of the wave packet and may be treated perturbatively. The local Hamiltonian $\hat{H}_c$ has the required periodicity of the unperturbed crystal, and has an energy spectrum of bands (Bloch bands) with Bloch wave eigenstates satisfying

$$\hat{H}_c(\mathbf{x}_c,t)|\psi_{\mathbf{q}}(\mathbf{x}_c,t)\rangle=\mathcal{E}_c(\mathbf{x}_c,\mathbf{q},t)|\psi_{\mathbf{q}}(\mathbf{x}_c,t)\rangle, \tag{2.2}$$

where $\mathbf{q}$ is the Bloch wave vector and $\mathcal{E}_c(\mathbf{x}_c,\mathbf{q},t)$ is the band energy. Since we will be concerned with only a single band, we have omitted the band index for simplicity of notation. We note that both the wave-packet center $\mathbf{x}_c$ and time t enters in the Bloch states and the band energy parametrically. We shall see that the dependency on the center position of the wave packet will manifest as new types of Berry-phase terms in the equations of motion.

These eigenstates form a convenient basis to expand the wave packet. Specifically, we write

$$|\Psi\rangle=\int d^3q\,a(\mathbf{q},t)|\psi_{\mathbf{q}}(\mathbf{x}_c,t)\rangle, \tag{2.3}$$

where $a(\mathbf{q},t)$ is the amplitude with the normalization

$$\int d^3q|a(\mathbf{q},t)|^2=\langle\Psi|\Psi\rangle=1. \tag{2.4}$$

Here, we have taken the convention that $\langle\Psi_{\mathbf{q}'}|\Psi_{\mathbf{q}}\rangle=\delta(\mathbf{q}'-\mathbf{q})$. It is assumed that the distribution $|a(\mathbf{q},t)|^2$ is narrow compared to the size of the Brillouin zone and has the mean wave vector

$$\mathbf{q}_c = \int d^3q\mathbf{q}|a(\mathbf{q},t)|^2. \tag{2.5}$$

To be consistent, the wave packet must yield the preassigned center position

$$\mathbf{x}_c = \langle\Psi|\hat{\mathbf{x}}|\Psi\rangle. \tag{2.6}$$

This condition can be expressed in terms of other wave-packet parameters as follows. Writing the amplitude in the form $a(\mathbf{q},t) = |a(\mathbf{q},t)|\exp[-i\gamma(\mathbf{q},t)]$, and using the matrix elements of position operator $\hat{\mathbf{x}}$ between the Bloch states [Eq. (A1) of Appendix A], we find that

$$\langle\Psi|\hat{\mathbf{x}}|\Psi\rangle = \int d^3q|a|^2\left[\frac{\partial\gamma}{\partial\mathbf{q}} + \left\langle u\left|i\frac{\partial u}{\partial\mathbf{q}}\right.\right\rangle\right], \tag{2.7}$$

where $|u\rangle \equiv |u(\mathbf{x}_c,\mathbf{q},t)\rangle = e^{-i\mathbf{q}\cdot\hat{\mathbf{x}}}|\psi_{\mathbf{q}}(\mathbf{x}_c,t)\rangle$ is the periodic part of the Bloch wave, and inner products involving the periodic part $|u\rangle$ mean an integration over the unit-cell volume v_c with a factor of $(2\pi)^3/v_c$, which implies the normalization $\langle u|u\rangle = 1$. According to our assumption of a narrow wave packet in the $\mathbf{q}$ space, Eq. (2.6) becomes

$$\mathbf{x}_c = \frac{\partial\gamma_c}{\partial\mathbf{q}_c} + \left\langle u\left|i\frac{\partial u}{\partial\mathbf{q}_c}\right.\right\rangle, \tag{2.8}$$

where $|u\rangle$ now stands for $|u(\mathbf{x}_c,\mathbf{q}_c,t)\rangle$, and γ_c for $\gamma(\mathbf{q}_c,t)$.

In writing down the expansion (2.3), we have assumed that the wave packet that is initially in a band always lies in the same band. This is justified if the band is separated from other bands by finite gaps, and if the time and the length scales of the perturbations are long compared to those associated with these gaps.[29]

B. The Lagrangian and dynamics

The dynamics of the mean position $\mathbf{x}_c$ and the crystal momentum $\hbar\mathbf{q}_c$ can in principle be derived from the Schrödinger equation for the wave packet. It is more conveniently obtained using a time-dependent variational principle[24,25] with the Lagrangian given by

$$L = \left\langle\Psi\left|i\frac{d}{dt} - \hat{H}\right|\Psi\right\rangle, \tag{2.9}$$

where and hereafter we use the convention $\hbar = 1$. We use d/dt to mean the derivative with respect to the time dependence of the wave function explicitly or implicitly through $\mathbf{x}_c$ and $\mathbf{q}_c$. The partial derivative $\partial/\partial t$, is reserved for those with $\mathbf{x}_c$ and $\mathbf{q}_c$ held fixed.

Under the previously discussed conditions on the widths of the wave packet, we can evaluate the Lagrangian as a function of $\mathbf{x}_c$ and $\mathbf{q}_c$, their time derivatives, and the time t

$$L \approx L(\mathbf{x}_c,\dot{\mathbf{x}}_c,\mathbf{q}_c,\dot{\mathbf{q}}_c,t). \tag{2.10}$$

The terms involving higher moments of the wave packet, which specify its width and shape, are of higher order in the gradient of the perturbations and hence are neglected.[30]

Accordingly, we obtain for the first term in Eq. (2.9),

$$\left\langle\Psi\left|i\frac{d\Psi}{dt}\right.\right\rangle = \frac{\partial\gamma_c}{\partial t} + \left\langle u\left|i\frac{\partial u}{\partial t}\right.\right\rangle + \dot{\mathbf{x}}_c\cdot\left\langle u\left|i\frac{\partial u}{\partial\mathbf{x}_c}\right.\right\rangle. \tag{2.11}$$

The first term comes from the explicit time dependence of γ_c. The contribution from $|a(\mathbf{q},t)|$ is zero because of the normalization condition (2.4) on the amplitude. The second and third terms come about because of the dependence of the basis functions on time explicitly and implicitly through $\mathbf{x}_c$, respectively. Using the relation

$$\frac{\partial\gamma_c}{\partial t} = \frac{d\gamma_c}{dt} - \dot{\mathbf{q}}_c\cdot\frac{\partial\gamma_c}{\partial\mathbf{q}_c} \tag{2.12}$$

and Eq. (2.8), Eq. (2.11) can be recast into the form

$$\left\langle\Psi\left|i\frac{d\Psi}{dt}\right.\right\rangle = \frac{d\gamma_c}{dt} - \dot{\mathbf{q}}_c\cdot\mathbf{x}_c + \dot{\mathbf{q}}_c\cdot\left\langle u\left|i\frac{\partial u}{\partial\mathbf{q}_c}\right.\right\rangle + \dot{\mathbf{x}}_c\cdot\left\langle u\left|i\frac{\partial u}{\partial\mathbf{x}_c}\right.\right\rangle + \left\langle u\left|i\frac{\partial u}{\partial t}\right.\right\rangle, \tag{2.13}$$

where γ_c appears only in a total time derivative.

The expectation value of the Hamiltonian, which gives the wave-packet energy $\mathcal{E}$, may be evaluated up to first order in the perturbation gradients using the linearized Hamiltonian (2.1)

$$\mathcal{E} = \langle\Psi|\hat{H}|\Psi\rangle \approx \langle\Psi|\hat{H}_c|\Psi\rangle + \langle\Psi|\Delta\hat{H}|\Psi\rangle, \tag{2.14}$$

where the gradient correction $\Delta\hat{H}$ may be written as[31]

$$\Delta\hat{H} = \frac{1}{2}\left[(\hat{\mathbf{x}}-\mathbf{x}_c)\cdot\frac{\partial\hat{H}_c}{\partial\mathbf{x}_c} + \frac{\partial\hat{H}_c}{\partial\mathbf{x}_c}\cdot(\hat{\mathbf{x}}-\mathbf{x}_c)\right]. \tag{2.15}$$

The expectation value of the local Hamiltonian is just the band energy at the mean wave vector,

$$\langle\Psi|\hat{H}_c|\Psi\rangle = \mathcal{E}_c(\mathbf{x}_c,\mathbf{q}_c,t), \tag{2.16}$$

while the gradient correction requires some calculations [Appendix A], but the result has the simple form

$$\Delta\mathcal{E} = -\mathrm{Im}\left[\left\langle\frac{\partial u}{\partial\mathbf{x}_c}\right|\bullet(\mathcal{E}_c - \hat{H}_c)\left|\frac{\partial u}{\partial\mathbf{q}}\right\rangle\right]\Bigg|_{\mathbf{q}=\mathbf{q}_c}, \tag{2.17}$$

where "•" denotes a scalar product between the vectors formed by gradients with respect to $\mathbf{x}_c$ and $\mathbf{q}$.

The Lagrangian thus takes the form

$$L = -\mathcal{E} + \mathbf{q}_c\cdot\dot{\mathbf{x}}_c + \dot{\mathbf{q}}_c\cdot\left\langle u\left|i\frac{\partial u}{\partial\mathbf{q}_c}\right.\right\rangle + \dot{\mathbf{x}}_c\cdot\left\langle u\left|i\frac{\partial u}{\partial\mathbf{x}_c}\right.\right\rangle + \left\langle u\left|i\frac{\partial u}{\partial t}\right.\right\rangle. \tag{2.18}$$

where we have neglected a term of total time derivative $d(\gamma - \mathbf{x}_c\cdot\mathbf{q}_c)/dt$ in the Lagrangian, as it does not affect the equations of motion and the quantization rule. The last three terms may be grouped into a single term $\langle u|idu/dt\rangle$, which turns out to be the net rate of change of Berry phase for wave-packet motion within the band. We note that under the transformations $|u\rangle \to \exp[i\varphi(\mathbf{x}_c,\mathbf{q},t)]|u\rangle$ or $\mathbf{q}\to\mathbf{q}+\mathbf{K}$, $\mathbf{K}$ being a reciprocal lattice vector, the Lagrangian remains invariant up to a total time derivative of some function of $\mathbf{x}_c$,

$\mathbf{q}_c$, and t. The former corresponds to gauge invariance while the latter to periodicity in the reciprocal space.

From the Lagrangian (2.18) we obtain the following equations of semiclassical motion:

$$\dot{\mathbf{x}}_c = \frac{\partial \mathcal{E}}{\partial \mathbf{q}_c} - (\overleftrightarrow{\Omega}_{\mathbf{qx}} \cdot \dot{\mathbf{x}}_c + \overleftrightarrow{\Omega}_{\mathbf{qq}} \cdot \dot{\mathbf{q}}_c) + \mathbf{\Omega}_{t\mathbf{q}},$$

$$\dot{\mathbf{q}}_c = -\frac{\partial \mathcal{E}}{\partial \mathbf{x}_c} + (\overleftrightarrow{\Omega}_{\mathbf{xx}} \cdot \dot{\mathbf{x}}_c + \overleftrightarrow{\Omega}_{\mathbf{xq}} \cdot \dot{\mathbf{q}}_c) - \mathbf{\Omega}_{t\mathbf{x}}. \tag{2.19}$$

The components of the tensor $\overleftrightarrow{\Omega}_{\mathbf{qq}}$ are defined by

$$(\overleftrightarrow{\Omega}_{\mathbf{qq}})_{\alpha\beta} \equiv \Omega_{q_\alpha q_\beta} = i\left[\left\langle \frac{\partial u}{\partial q_{c\alpha}} \middle| \frac{\partial u}{\partial q_{c\beta}} \right\rangle - \left\langle \frac{\partial u}{\partial q_{c\beta}} \middle| \frac{\partial u}{\partial q_{c\alpha}} \right\rangle\right], \tag{2.20}$$

and those of the vector $\mathbf{\Omega}_{t\mathbf{x}}$ by

$$(\mathbf{\Omega}_{t\mathbf{x}})_\alpha \equiv \Omega_{tx_\alpha} = i\left[\left\langle \frac{\partial u}{\partial t} \middle| \frac{\partial u}{\partial x_{c\alpha}} \right\rangle - \left\langle \frac{\partial u}{\partial x_{c\alpha}} \middle| \frac{\partial u}{\partial t} \right\rangle\right], \tag{2.21}$$

where α and β are Cartesian indices. The other tensors $\overleftrightarrow{\Omega}_{\mathbf{xx}}, \overleftrightarrow{\Omega}_{\mathbf{xq}}$, and $\overleftrightarrow{\Omega}_{\mathbf{qx}}$ and the vector $\mathbf{\Omega}_{t\mathbf{q}}$ are defined similarly. These quantities are known as Berry curvatures.[13,14] We note that these equations involve Berry curvatures between every pair of parameters and that they have symplectic symmetry in the absence of time dependence.

C. Formal and semiclassical quantization

We mentioned in the Introduction that the equations of motion were usually derived from the effective Hamiltonian upon using the correspondence principle. Here, we consider the reverse process to obtain the effective quantum Hamiltonian from the semiclassical dynamics. This requires a knowledge of canonical structure of the wave-packet dynamics. Following the standard procedure of analytical mechanics, we introduce the canonical momenta conjugate to the generalized coordinates

$$\mathbf{P}_1 = \frac{\partial L}{\partial \dot{\mathbf{x}}_c} = \mathbf{q}_c + \left\langle u \middle| i\frac{\partial u}{\partial \mathbf{x}_c} \right\rangle, \tag{2.22}$$

$$\mathbf{P}_2 = \frac{\partial L}{\partial \dot{\mathbf{q}}_c} = \left\langle u \middle| i\frac{\partial u}{\partial \mathbf{q}_c} \right\rangle, \tag{2.23}$$

and the semiclassical Hamiltonian $\mathcal{H}$ by the Legendre transformation

$$\mathcal{H} = \dot{\mathbf{x}}_c \cdot \mathbf{P}_1 + \dot{\mathbf{q}}_c \cdot \mathbf{P}_2 - L = \mathcal{E}(\mathbf{x}_c, \mathbf{q}_c, t) - \left\langle u \middle| i\frac{\partial u}{\partial t} \right\rangle. \tag{2.24}$$

The semiclassical Hamiltonian is independent of $\mathbf{P}_1$ and $\mathbf{P}_2$, because the Lagrangian is linear in the generalized velocities.

Starting with $\mathcal{H}$, regarded formally as a function of $\mathbf{x}_c$, $\mathbf{q}_c$, $\mathbf{P}_1$, and $\mathbf{P}_2$, one cannot obtain the equations of motion (2.19) from the Hamilton equations. This is because the Eqs. (2.22) and (2.23) defining $\mathbf{P}_1$ and $\mathbf{P}_2$ do not depend on the generalized velocities, and hence they should be treated as constraints between the canonical variables.

In the simple case where the Berry-phase terms are zero, these constraints become $\mathbf{P}_1 = \mathbf{q}_c$, and $\mathbf{P}_2 = 0$, and $\mathcal{H} = \mathcal{E}$. These suggest that we treat $(\mathbf{x}_c, \mathbf{q}_c)$ as a canonical pair and forget about the other degrees of freedom. By doing so, we can indeed obtain the equations of motion (2.19) from the Hamilton equations. Having identified the canonical pair, one can proceed with a formal procedure of quantization (''requantization''), $\mathbf{q}_c \to -i\partial/\partial \mathbf{x}_c$, to obtain an effective quantum Hamiltonian. A slightly more general case will be encountered in the case of electromagnetic perturbations in the next section of this article.

When Berry-phase terms are present, constraints (2.22) and (2.23) still imply some hidden canonical relations between $\mathbf{x}_c$ and $\mathbf{q}_c$, but these are entangled in a complicated manner that cannot be expressed explicitly in general. This clearly shows the difficulty of the Hamiltonian formalism in the presence of Berry-phase terms. If one insists on using the Hamiltonian approach, one can employ the method of Lagrange multipliers, which allows the spurious degrees of freedom to be formally treated as independent,[32] and obtain the equations of motion (2.19). The ''requantization'' procedure for this case is quite complicated and needs further investigation.

The semiclassical quantization, on the other hand, is quite straight forward. In order that stationary states and energy levels can be talked about, we shall restrict ourselves to static perturbations. For a wave-packet motion that is regular and is described by closed orbits in the phase space $(\mathbf{x}_c, \mathbf{q}_c)$, semiclassical energy levels are obtained using the quantization procedure[33] due to Einstein, Brillouin, and Keller,

$$\oint_C \mathbf{P}_1 \cdot d\mathbf{x}_c + \oint_C \mathbf{P}_2 \cdot d\mathbf{q}_c = 2\pi\left[m + \frac{\nu}{4}\right], \tag{2.25}$$

where C denotes an orbit of constant energy $\mathcal{E}$, m an integer that labels the eigenvalue, and ν the number of caustics traversed. With Eqs. (2.22) and (2.23) for $\mathbf{P}_1$ and $\mathbf{P}_2$, the above condition reduces to

$$\oint_C \mathbf{q}_c \cdot d\mathbf{x}_c = 2\pi\left[m + \frac{\nu}{4} - \frac{\Gamma(C)}{2\pi}\right], \tag{2.26}$$

where

$$\Gamma(C) = \oint_C d\mathbf{x}_c \cdot \left\langle u \middle| i\frac{\partial u}{\partial \mathbf{x}_c} \right\rangle + \oint_C d\mathbf{q}_c \cdot \left\langle u \middle| i\frac{\partial u}{\partial \mathbf{q}_c} \right\rangle \tag{2.27}$$

is the Berry phase acquired by the wave packet upon going round the closed orbit once. The Berry phase correction to the quantization condition made its first appearance in Wilkinson's work[34] on Harper's equation, in Kuratsuji and Iida's work[35] on adiabatic nuclear motion, and also recently in the work of Chang and Niu[12] on wave-packet dynamics in magnetic Bloch bands.

III. ELECTROMAGNETIC FIELDS

So far, our treatment of perturbations has been in general terms, and our results are in an abstract form. Their physical meaning will become clear through the consideration of two special cases in this and the next section. For a class of perturbations for which the Hamiltonian is of the special form

$$H_0[\hat{\mathbf{x}}+\boldsymbol{\beta}_1(\hat{\mathbf{x}},t),\hat{\mathbf{p}}+\boldsymbol{\beta}_2(\hat{\mathbf{x}},t)]+\beta_3(\hat{\mathbf{x}},t), \tag{3.1}$$

all the results can be expressed in terms of the unperturbed Bloch wave basis. In this section, we shall consider electromagnetic perturbations for which $\boldsymbol{\beta}_1(\hat{\mathbf{x}},t)=0$.

A. The gauge invariant crystal momentum

Let $\hat{H}_0(\mathbf{q})$ denote the Hamiltonian for the bare crystal, with the eigenstate $|u(\mathbf{q})\rangle$ (the periodic part of the Bloch wave) and the band energy $\mathcal{E}_0(\mathbf{q})$ for a particular band. The Hamiltonian gets modified by the gauge potentials $[\mathbf{A}(\mathbf{x},t),\phi(\mathbf{x},t)]$ of an electromagnetic field to

$$\hat{H}=H_0[\mathbf{q}+e\mathbf{A}(\hat{\mathbf{x}},t)]-e\phi(\hat{\mathbf{x}},t). \tag{3.2}$$

This has the form (1.5) and (3.1), with the gauge potentials playing the role of the modulation functions, and hence the local Hamiltonian must have the form

$$\hat{H}_c=\hat{H}_0[\mathbf{q}+e\mathbf{A}(\mathbf{x}_c,t)]-e\phi(\mathbf{x}_c,t). \tag{3.3}$$

As $e\mathbf{A}(\mathbf{x}_c,t)$ is only an additive constant to the crystal momentum $\mathbf{q}$, the basis states have the form $|u(\mathbf{x}_c,\mathbf{q},t)\rangle=|u(\mathbf{k})\rangle$, where $\mathbf{k}=\mathbf{q}+e\mathbf{A}(\mathbf{x}_c,t)$ is the gauge invariant or mechanical crystal momentum. In terms of the gauge invariant crystal momentum $\mathbf{k}$, a number of simplifications can be obtained. First, the eigenenergy can be written in the form

$$\mathcal{E}_c(\mathbf{x}_c,\mathbf{k},t)=\mathcal{E}_0(\mathbf{k})-e\phi(\mathbf{x}_c,t). \tag{3.4}$$

Second, the gradient correction (2.17) becomes the orbital magnetization energy of the wave packet,

$$-\mathbf{M}\cdot\mathbf{B}, \tag{3.5}$$

where $\mathbf{B}\equiv\mathrm{curl}_{\mathbf{x}_c}\mathbf{A}(\mathbf{x}_c,t)$ is the magnetic field, and

$$\mathbf{M}=e\,\mathrm{Im}\left[\left\langle\frac{\partial u}{\partial\mathbf{k}}\right|\times(\mathcal{E}_0-\hat{H}_0(\mathbf{k}))\left|\frac{\partial u}{\partial\mathbf{k}}\right\rangle\right]\Bigg|_{\mathbf{k}=\mathbf{k}_c} \tag{3.6}$$

is the orbital magnetic moment of Bloch electrons. Third, the last three terms of the Lagrangian (2.18) simply become the single term $\dot{\mathbf{k}}_c\cdot\langle u|i\partial u/\partial\mathbf{k}_c\rangle$. Finally, the Lagrangian takes the form

$$L=-\mathcal{E}_M+e\phi(\mathbf{x}_c,t)+\dot{\mathbf{x}}_c\cdot\mathbf{k}_c-e\dot{\mathbf{x}}_c\cdot\mathbf{A}(\mathbf{x}_c,t)+\dot{\mathbf{k}}_c\cdot\left\langle u\left|i\frac{\partial u}{\partial\mathbf{k}_c}\right.\right\rangle, \tag{3.7}$$

where $\mathcal{E}_M\equiv\mathcal{E}_0(\mathbf{k}_c)-\mathbf{M}\cdot\mathbf{B}$.

B. The reciprocal magnetic field and orbital magnetization energy

The equations of motion can either be derived variationally from the above Lagrangian or directly from Eq. (2.19) derived for the general case in the previous section. They have the form

$$\dot{\mathbf{x}}_c=\frac{\partial\mathcal{E}_M}{\partial\mathbf{k}_c}-\dot{\mathbf{k}}_c\times\boldsymbol{\Omega},$$

$$\dot{\mathbf{k}}_c=-e\mathbf{E}-e\dot{\mathbf{x}}_c\times\mathbf{B}, \tag{3.8}$$

where $\mathbf{E}\equiv-\mathrm{grad}_{\mathbf{x}_c}\phi(\mathbf{x}_c,t)-\partial\mathbf{A}(\mathbf{x}_c,t)/\partial t$ is the electric field, and

$$(\boldsymbol{\Omega})_\alpha\equiv\tfrac{1}{2}\epsilon_{\alpha\beta\gamma}(\overleftrightarrow{\Omega}_{\mathbf{kk}})_{\beta\gamma}, \tag{3.9}$$

are the components of the vector form of the antisymmetric tensor $\overleftrightarrow{\Omega}_{\mathbf{kk}}$ given by Eq. (2.20). In Eq. (3.9) and henceforth, repeated Cartesian indices are taken to be summed. Because $\boldsymbol{\Omega}$ occupies a similar position as the magnetic field in the equations of motion, it will be called the reciprocal magnetic field.

The above equations differ from Eqs. (1.1) in two respects. Firstly, the energy $\mathcal{E}$ contains a correction term from the magnetic moment (3.6) of Bloch electron. This term has been derived earlier as a first order correction in the theory of Bloch electrons subject to magnetic fields.[36] A similar term has also been found in the theory of electrons in incommensurate lattices,[19,20] and in the theory of wave-packet dynamics in magnetic Bloch bands.[12] Secondly, the correction term to the velocity, $-\dot{\mathbf{k}}_c\times\boldsymbol{\Omega}$, is the anomalous velocity that was predicted to give rise to a spontaneous Hall conductivity in ferromagnetic materials.[8,9] In the context of the quantum Hall effect, the integral of the Berry curvature $\Omega_{k_\alpha k_\beta}$ over the Brillouin zone was shown to be proportional to the Hall conductivity for a full band and to be quantized (Chern's topological invariant).[37] Recently, Chang and Niu[12] proved this result semiclassically. It seems more appropriate to call this term Hall velocity than anomalous velocity.

The semiclassical Eq. (3.8) should be invariant under time reversal, spatial inversion, or certain rotations if these are symmetries of the unperturbed crystal. Such symmetries impose severe restrictions on the behavior of the reciprocal magnetic field $\boldsymbol{\Omega}$ and the magnetic moment $\mathbf{M}$ as functions of $\mathbf{k}$. Under time reversal, $\dot{\mathbf{x}}_c$, $\mathbf{k}_c$, and $\mathbf{B}$ change sign while $\mathbf{x}_c$, $\dot{\mathbf{k}}_c$, and $\mathbf{E}$ are invariant. If the bare crystal is invariant under time reversal, we must have $\boldsymbol{\Omega}(-\mathbf{k})=-\boldsymbol{\Omega}(\mathbf{k})$, and $\mathbf{M}(-\mathbf{k})=-\mathbf{M}(\mathbf{k})$, which implies in particular that they must vanish at $\mathbf{k}=0$. Under spatial inversion, $\mathbf{E}$, $\mathbf{x}_c$, $\mathbf{k}_c$, and the time derivatives of the last two change sign while $\mathbf{B}$ remains unchanged. If the bare crystal has inversion symmetry, we must have $\boldsymbol{\Omega}(-\mathbf{k})=\boldsymbol{\Omega}(\mathbf{k})$, and $\mathbf{M}(-\mathbf{k})=\mathbf{M}(\mathbf{k})$. Finally, if the system is invariant under certain proper rotations, the reciprocal magnetic field and the magnetic moment should transform like vectors under these rotations.

For monatomic nonmagnetic crystals, both time reversal and spatial inversion symmetries are present, rendering $\boldsymbol{\Omega}$ and $\mathbf{M}$ null everywhere in the Brillouin zone. However, it is not entirely justified to ignore these quantities for magnetic crystals or nonmagnetic crystals without inversion symmetry (such as GaAs). Investigations have been undertaken to see whether the presence of the reciprocal magnetic field and the orbital magnetization lead to observable effects.

The Lagrangian (3.7) and the equations of motion (3.8) were derived earlier by Chang and Niu[12] for perturbed magnetic Bloch electrons in two dimensions in the gauge where

$\phi(\mathbf{x},t)=0$. They used a wave packet that gauges away the vector potential locally so that their magnetic Bloch wave vector is the same as the corresponding gauge invariant crystal momentum $\mathbf{q}+e\mathbf{A}(\mathbf{x}_c,t)=\mathbf{k}$ here. Their derivation was less general in that it is only for two dimensions and more general in that it provides a description of electrons in rational magnetic fields $\mathbf{B}_0$, for which the flux through a unit cell equals a rational fraction $[\sim\mathcal{O}(1)]$ of the flux quantum (h/e). However, our formalism can easily be generalized to this situation, for a more general three-dimensional case, by assuming a background of constant rational magnetic field $\mathbf{B}_0$. All we need to do is to interpret $\mathbf{q}$ as the wave vector of magnetic Bloch states defined within a reduced Brillouin zone (the magnetic Brillouin zone),[12] and to replace $\mathbf{B}$ in (3.8) by $\mathbf{B}-\mathbf{B}_0$.

C. Peierls substitution and Landau levels

It follows from the Lagrangian (3.7) that the canonical momenta are given by

$$\mathbf{P}_1=\mathbf{k}_c-e\mathbf{A}(\mathbf{x}_c,t), \tag{3.10}$$

$$\mathbf{P}_2=\left\langle u\left| i\frac{\partial u}{\partial \mathbf{k}_c}\right.\right\rangle, \tag{3.11}$$

and the Hamiltonian by $\mathcal{H}=\mathcal{E}_M(\mathbf{k}_c)-e\phi(\mathbf{x}_c,t)$. In the absence of the Berry-phase term (3.11), we may obtain the Hamiltonian as a function of the canonical pair $(\mathbf{x}_c,\mathbf{P}_1)$ as

$$\mathcal{H}=\mathcal{E}_M[\mathbf{P}_1+e\mathbf{A}(\mathbf{x}_c,t)]-e\phi(\mathbf{x}_c,t). \tag{3.12}$$

The quantization of this Hamiltonian by setting $\mathbf{P}_1=-i\partial/\partial\mathbf{x}_c$ amounts to the Peierls substitution. However, it is not clear how to deal with the case with a Berry-phase term using the Hamiltonian approach.

When only a uniform magnetic field is present, the two equations of motion (3.8) can be combined into a single one for the $\mathbf{k}$-space motion. It is evident that a $\mathbf{k}$-space orbit must lie in a plane normal to $\mathbf{B}$ and must be on a constant energy surface of $\mathcal{E}(\mathbf{k})$. If such an orbit is closed, known as a cyclotron orbit, the EBK formula yields

$$\frac{1}{2}\hat{\mathbf{B}}\cdot\oint_C \mathbf{k}_c\times d\mathbf{k}_c=\frac{e|\mathbf{B}|}{\hbar}\left[m+\frac{1}{2}-\frac{\Gamma(C)}{2\pi}\right], \tag{3.13}$$

where

$$\Gamma(C)=\oint_C d\mathbf{k}\cdot\left\langle u\left| i\frac{\partial u}{\partial \mathbf{k}_c}\right.\right\rangle \tag{3.14}$$

is the Berry phase accumulated by the wave packet upon completing a circuit along the loop C, and we have restored the Planck constant. The left-hand side of Eq. (3.13) is just the $\mathbf{k}$-space area enclosed by the orbit C. As this phase influences energy levels, it affects the density of states. It is shown in Ref. 12 that Γ plays an important role in determining the spectral splitting pattern of magnetic bands.

D. Zak phase and Wannier-Stark ladder

In this subsection, we restrict our discussion to one dimension for simplicity. The semiclassical motion under a uniform electric field is described by the Hamiltonian $\mathcal{H}=\mathcal{E}_0(k_c)+eEx_c$. It follows from the boundedness and the periodicity of the band energy that the motion in real space is also bounded and periodic. Such a closed motion in the real space is known as Bloch oscillations. In the reduced zone scheme, the motion is closed also in the phase space (x_c,k_c), which is a cylinder. Quantizing this motion according to Eq. (2.25) gives the condition

$$-\int_{-\pi/a}^{\pi/a} dk_c x_c(k_c)=2\pi\left(m+\frac{\nu}{4}-\frac{\Gamma}{2\pi}\right), \tag{3.15}$$

where a stands for the lattice constant, and

$$\Gamma=\int_{-\pi/a}^{\pi/a} dk\left\langle u\left| i\frac{\partial u}{\partial k}\right.\right\rangle \tag{3.16}$$

is known as the Zak phase,[38] and $x_c(k_c)$ is the constant energy curve for the mth energy level defined by $W_m=\mathcal{E}_0(k_c)+eEx_c$, m being an integer. Averaging this expression over the orbit, we obtain from Eq. (3.15)

$$W_m=\bar{\mathcal{E}}_0+eEa\left(-m-\frac{\nu}{4}+\frac{\Gamma}{2\pi}\right), \tag{3.17}$$

where $\bar{\mathcal{E}}_0$ is the average of the band energy over the Brillouin zone, and m is any integer between $-\infty$ to ∞, since the mean value of x_c can be anywhere on the cylinder. This spectrum, known as the Wannier-Stark ladder, was first derived by Wannier[39] without the Berry-phase term. The correction was due to Zak, who later interpreted it as a Berry phase.[40]

IV. DEFORMATIONS IN CRYSTALS

We shall now come to deformational perturbations. It turns out that the model Hamiltonian for a deformed crystal also has the special form (3.1) with $\boldsymbol{\beta}_2(\hat{\mathbf{x}},t)=0$, which is $H_0[\hat{\mathbf{x}}+\boldsymbol{\beta}_1(\hat{\mathbf{x}},t),\hat{\mathbf{p}}]+\beta_3(\hat{\mathbf{x}},t)$, and hence all the corrections are expressible in terms of the undeformed basis for this case too.

A. The translated crystal basis

A deformed crystal with atomic displacements $\{\mathbf{u}_l\}$ may be described by the Hamiltonian[41,42]

$$H=\frac{\hat{\mathbf{p}}^2}{2m}+V_0[\hat{\mathbf{x}}-\mathbf{u}(\hat{\mathbf{x}})]+s_{\alpha\beta}(\hat{\mathbf{x}})\mathcal{V}_{\alpha\beta}[\hat{\mathbf{x}}-\mathbf{u}(\hat{\mathbf{x}})], \tag{4.1}$$

where $\mathbf{u}(\mathbf{x})$ is a smooth displacement field[43] satisfying $\mathbf{u}(\mathbf{R}_l+\mathbf{u}_l)=\mathbf{u}_l$, and $s_{\alpha\beta}=\partial u_\alpha/\partial x_\beta$ is the unsymmetrized strain. The justification of the above Hamiltonian and an explicit expression for the last term are given in Appendix B. While the last term of the Hamiltonian, being proportional to the strain, can be treated perturbatively, the other terms are of the form Eq. (1.5), with the displacement field playing the role of the modulation function. The local Hamiltonian is then given by

$$\hat{H}_c=\frac{\hat{\mathbf{p}}^2}{2m}+V_0[\hat{\mathbf{x}}-\mathbf{u}(\mathbf{x}_c)]. \tag{4.2}$$

This is nothing but the Hamiltonian of an undeformed crystal shifted in position by the displacement field at the center of the wave packet, $\mathbf{u}(\mathbf{x}_c)$. The band energy is therefore the same as that of the undeformed crystal, $\mathcal{E}_0(\mathbf{k})$, and the eigenstates are the translated Bloch waves $\{\psi_{\mathbf{k}}[\mathbf{x}-\mathbf{u}(\mathbf{x}_c)]\}$.

Our wave packet will thus be formed out of these translated Bloch states of the undeformed but translated crystal. This procedure is valid so long as the strain is weak, so that the variation of the displacement within the spatial width of the wave packet is small. When the first-order corrections to the Hamiltonian are taken into account, our method should give the same physical results (to the same order) as obtained using a strained basis. However, our formulation should be simpler and easier to interpret, because it avoids the necessity of transformation between the lab and lattice frames of reference.

Although the small strain regime covers the vast majority of practical situations, it is some times necessary to consider the effect of large strains. In the case where a large uniform and static strain is superposed on top of a small varying strain, our formulation can still be applied; one only needs to interpret the basis as that of the uniformly strained crystal. There can be a third possibility in which the strain variation is large over large distances but is small over the size of the wave packet. In this case, it is more appropriate to use a strained local basis, that is, a basis of a homogeneously strained crystal with the strain value given by the actual strain at the center of the wave packet.

B. The crystal deformation potential

The wave-packet energy is obtained by summing the expectation values of the local Hamiltonian (4.2), the gradient correction (2.15), and the last term of Eq. (4.1). Because of the functional form of the basis states, we have

$$\langle \Psi|\hat{H}_c|\Psi\rangle=\mathcal{E}_0(\mathbf{k}_c), \tag{4.3}$$

while the gradient correction becomes

$$-m s_{\alpha\beta}(\mathbf{x}_c)[\langle \hat{v}_\alpha \hat{v}_\beta\rangle-\langle \hat{v}_\alpha\rangle\langle \hat{v}_\beta\rangle]. \tag{4.4}$$

The angular brackets in the above expression represent the expectation value of the enclosed operators in the Bloch state at $\mathbf{k}=\mathbf{k}_c$, and $\hat{v}_\alpha=\partial\hat{H}(\mathbf{k})/\partial k_\alpha$ is the velocity operator. As is well known, $\langle \hat{v}_\alpha\rangle=\partial\mathcal{E}_0/\partial k_\alpha\equiv v_\alpha$ is the group velocity of Bloch electrons. As for the last term of the Hamiltonian (4.1), we may write, in accordance with our approximation [Eq. (2.10)],

$$\langle \Psi|s_{\alpha\beta}(\hat{\mathbf{x}})\hat{\mathcal{V}}_{\alpha\beta}|\Psi\rangle\approx s_{\alpha\beta}(\mathbf{x}_c)\langle \hat{\mathcal{V}}^c_{\alpha\beta}\rangle, \tag{4.5}$$

where $\mathcal{V}^c_{\alpha\beta}=\mathcal{V}_{\alpha\beta}[(\mathbf{x}-\mathbf{u}(\mathbf{x}_c)]$. We again write the energy of the wave packet in the form

$$\mathcal{E}=\mathcal{E}_0(\mathbf{k}_c)+\Delta\mathcal{E}(\mathbf{x}_c,\mathbf{k}_c), \tag{4.6}$$

where, this time, the correction $\Delta\mathcal{E}$, which is to be called the deformation potential for the wave packet,[44] has two contributions [Eqs. (4.4) and (4.5)]. As the latter is proportional to the local deformation, it may be written in the form

$$\Delta\mathcal{E}=s_{\alpha\beta}(\mathbf{x}_c)D^w_{\alpha\beta}(\mathbf{k}_c), \tag{4.7}$$

with

$$D^w_{\alpha\beta}=m[v_\alpha v_\beta-\langle \hat{v}_\alpha\hat{v}_\beta\rangle]+\langle \hat{\mathcal{V}}^c_{\alpha\beta}\rangle. \tag{4.8}$$

We note that this quantity vanishes in the free-electron limit $(V_0\to 0)$, which is quite satisfying from a physical point of view: a wave packet should not feel the effect of a deformation of the lattice in the absence of electron-lattice coupling.

Further, from the differential band structure[45] under a uniform strain $\overleftrightarrow{\epsilon}$,

$$\mathcal{E}'(\mathbf{k}')-\mathcal{E}_0(\mathbf{k})=\epsilon_{\alpha\beta}D^b_{\alpha\beta}(\mathbf{k})=\epsilon_{\alpha\beta}[-m\langle \hat{v}_\alpha\hat{v}_\beta\rangle+\langle \hat{\mathcal{V}}^c_{\alpha\beta}\rangle], \tag{4.9}$$

where $\mathbf{k}'=(\overleftrightarrow{1}+\overleftrightarrow{s})^{-1}\cdot\mathbf{k}$ $\Delta\mathcal{E}$ can be determined by the relation

$$D^w_{\alpha\beta}=D^b_{\alpha\beta}+m v_\alpha v_\beta, \tag{4.10}$$

according to Eqs. (4.8) and (4.9). The same expression has been obtained for the deformation potential for electron-phonon interaction at long wavelengths, showing the equivalence of the latter with our wave-packet deformation potential (4.7).[46]

C. Equations of motion and lattice tracking

Before we proceed further, let us first make the displacements time dependent, $\{\mathbf{u}_l(t)\}$, as we are discussing the dynamical aspect. Accordingly, we extend the results of the previous subsections with the replacements $\mathbf{u}(\mathbf{x})\to\mathbf{u}(\mathbf{x},t)$ and $s_{\alpha\beta}(\mathbf{x})\to s_{\alpha\beta}(\mathbf{x},t)$, and with the Bloch wave basis given by the states $\{\psi_{\mathbf{k}}[\mathbf{x}-\mathbf{u}(\mathbf{x}_c,t)]\}$.

As for the Lagrangian, the functional form of the basis states imply that the last two terms of its general expression (2.18) become

$$\left\langle u\left|i\frac{\partial u}{\partial \mathbf{x}_c}\right.\right\rangle=f_\alpha\frac{\partial u_\alpha}{\partial \mathbf{x}_c},\quad \left\langle u\left|i\frac{\partial u}{\partial t}\right.\right\rangle=f_\alpha\frac{\partial u_\alpha}{\partial t}, \tag{4.11}$$

where

$$\mathbf{f}(\mathbf{k})=\frac{m}{\hbar}\frac{\partial\mathcal{E}_0}{\partial\mathbf{k}}-\hbar\mathbf{k}, \tag{4.12}$$

in which $\hbar$ has been restored. The reader is warned that the displacement field ($\mathbf{u}$ or u_α) should not be confused with the periodic part of the Bloch state, $|u\rangle$. The quantity $\mathbf{f}(\mathbf{k})$, being the difference between the group momentum of a Bloch electron and the momentum of a free electron, denotes that part of the momentum arising from the lattice interaction alone. It has the desired property of vanishing in the free electron limit, where the lattice deformation should not be felt. By substituting Eq. (4.11) into Eq. (2.18), we obtain

$$L=-\mathcal{E}+\dot{\mathbf{x}}_c\cdot\mathbf{k}_c+\dot{\mathbf{k}}_c\cdot\left\langle u\left|i\frac{\partial u}{\partial\mathbf{k}_c}\right.\right\rangle+\dot{\mathbf{u}}\cdot\mathbf{f}(\mathbf{k}_c), \tag{4.13}$$

where $\dot{\mathbf{u}}$ denotes $d\mathbf{u}/dt$. The last term is new and represents the rate of change of the Berry phase due to lattice tracking. We shall see below that there is a tendency for the lattice to drag the electron with its displacement motion, hence the word "tracking." This term also gives rise to the Burgers vector when integrated around a dislocation, so in a sense it keeps track of the lattice position.

Similarly, the Berry curvatures in Eq. (2.19) take values

$$\Omega_{k_\alpha x_\beta} = -\Omega_{x_\beta k_\alpha} = \frac{\partial u_\gamma}{\partial x_{c\beta}} \frac{\partial f_\gamma}{\partial k_{c\alpha}},$$

$$\Omega_{tk_\alpha} = -\frac{\partial f_\gamma}{\partial k_{c\alpha}} \frac{\partial u_\gamma}{\partial t}, \quad \Omega_{x_\alpha x_\beta} = \Omega_{tx_\beta} = 0, \tag{4.14}$$

and by expressions (4.6) and (4.7) for $\mathcal{E}$, the equations of motion become

$$\dot{x}_{c\alpha} = \frac{\partial \mathcal{E}_0}{\partial k_{c\alpha}} + s_{\beta\gamma} \frac{\partial D^w_{\beta\gamma}}{\partial k_{c\alpha}} - (\dot{\mathbf{k}}_c \times \mathbf{\Omega})_\alpha - \dot{u}_\beta \frac{\partial f_\beta}{\partial k_{c\alpha}}, \tag{4.15}$$

$$\dot{k}_{c\alpha} = -D^w_{\beta\gamma} \frac{\partial s_{\beta\gamma}}{\partial x_{c\alpha}} - s_{\beta\alpha} \frac{\partial f_\beta}{\partial k_{c\gamma}} \dot{k}_{c\gamma}. \tag{4.16}$$

In the above equations, the semiclassical force has two contributions: the first term arises from the deformation potential, and the second term, which arises from the lattice tracking term, will be called the tracking force. For uniform strains the semiclassical force vanishes, so that $\mathbf{k}_c$ is a good quantum number in such a case.

The velocity has three contributions in addition to the usual term given by the gradient of the band energy. The second term corresponds to the group velocity arising from the deformation potential, and the third term, as was seen earlier in Eq. (3.8) of Sec. III, is the Hall velocity. The last term arises from the lattice tracking term of the Lagrangian, and will be similarly called the tracking velocity. To understand this term better, let us rewrite the term using Eq. (4.12) as

$$\dot{u}_\alpha - m\dot{u}_\beta \frac{\partial^2 \mathcal{E}_0}{\partial k_{c\beta} \partial k_{c\alpha}}. \tag{4.17}$$

In the free-electron limit, where electrons and lattice decouple, the adiabatic velocity vanishes identically, for

$$\frac{\partial^2 \mathcal{E}_0}{\partial k_{c\alpha} \partial k_{c\beta}} \to \frac{\delta_{\alpha\beta}}{m}. \tag{4.18}$$

On the other hand, when the band under consideration is full, the second term of Eq. (4.17) averaged over the band is zero, implying a complete adiabatic following of the lattice motion, which confirms an earlier result on adiabatic particle transport.[47]

D. Dislocations and Berry phase

Our formalism is also applicable to dislocation strain fields, which are well defined except in a region of a few atomic spacings around the line of dislocation. Outside this region, the displacement field $\mathbf{u}(\mathbf{x})$ is a smooth but multiple valued function. The change in the displacement field along a closed loop around the line of dislocation, $\Delta \mathbf{u} = \int_C dx_\alpha \partial \mathbf{u} / \partial x_\alpha = \mathbf{b}$, is known as the Burgers vector; it equals one of the Bravais lattice vectors. On account of this multiple-valuedness, a wave packet of incident wave vector $\mathbf{k}$ taken around the line of dislocation acquires a Berry phase

$$\Gamma = \oint_C d\mathbf{x}_c \cdot \left\langle u \middle| i \frac{\partial u}{\partial \mathbf{x}_c} \right\rangle = \oint_C d\mathbf{u} \cdot \mathbf{f}(\mathbf{k}_c) \approx \mathbf{b} \cdot \mathbf{f}(\mathbf{k}), \tag{4.19}$$

where we have assumed $\langle u | i \partial u / \partial \mathbf{k} \rangle$ to be zero[48] and the corrections arising from the changes in the wave vector to be negligible. In other words, the action integral of the last term of Eq. (4.13) around a dislocation gives us a Berry phase that is proportional to the Burgers vector.

Note that this Berry phase is independent of the path as long as it encloses the dislocation line. What we have is a situation similar to the Aharonov-Bohm effect,[49] with the dislocation playing the role of the solenoid, and the Berry curvature $\mathbf{\Omega}_{\mathbf{xx}}$ the role of the magnetic field. Just as in the case of the solenoidal field, $\mathbf{\Omega}_{\mathbf{xx}} = 0$ everywhere except for the core region where it should be taken to be singular.

The Berry phase (4.19) affects the scattering of electrons by a dislocation,[50] and our result can be used to compute the shift in the scattering fringe pattern due the Berry phase. The intensity distribution has interference terms of the form $\cos(\theta + \Gamma)$ due to each pair of paths along the two sides of the dislocation, where θ is given by the path length difference per wave length of the incident beam. The Berry phase and hence the shift are maximal when the incident wave vector $\mathbf{k}$ is such that $\mathbf{f}(\mathbf{k}) = m\mathbf{v} - \hbar\mathbf{k}$ is parallel to the Burgers vector $\mathbf{b}$. To fix ideas, let us make the assumption that $\mathbf{f} \| \mathbf{k}$. For an edge dislocation, where $\mathbf{b}$ lies in the normal plane of the dislocation axis, we expect that maximal effect of the Berry phase is seen when the direction of the incident beam is perpendicular to the axis and coincide with the direction of $\mathbf{b}$. For a screw dislocation, the maximal Berry phase occurs when $\mathbf{k}$ is parallel to the axis along which $\mathbf{b}$ lies. However, since the beam must pass the dislocation in order to produce the interference pattern, the maximal Berry-phase effect should actually occur when $\mathbf{k}$ is along some finite angle away from the axis.

The Berry phase can also affect the electron diffraction pattern of a deformed crystal with or without a dislocation. When an electron beam of wave vector $\mathbf{k}$ is sent through a crystal, it propagates as Bloch waves in different bands of the crystal, all with the same Bloch wave vector, which is equal to $\mathbf{k}$ modulo a reciprocal lattice vector. Because of the energy differences of the bands, the Bloch waves grow out of phase from one another as they propagate. Therefore, when they exit the crystal, they recombine to produce not only the incident beam but also the diffracted beams. For a deformed crystal, the phase change of a Bloch wave is given by the time integral of the Lagrangian (4.13). The contribution from the $\dot{\mathbf{x}}_c \cdot \mathbf{k}_c$ term may be dropped, because it is independent of the band index. For weak strains, we may also neglect the deformation potential and the third term in Eq. (4.13). The dominant correction to the dynamical phase arises from the last term: $\Delta \mathbf{u} \cdot \mathbf{f}(\mathbf{k})$, where $\Delta \mathbf{u}$ is the change in the displacement field over the path of propagation of the Bloch waves. To show the soundness of these general ideas, we have calculated the diffraction pattern from a thin slab of crystal containing a screw dislocation, correctly reproducing earlier experimental and theoretical results.[51]

V. SUMMARY

We provide a unified framework for wave-packet dynamics of electrons in slowly perturbed crystals, which consists of constructing a wave packet using the Bloch states belonging to a single band of the local periodic Hamiltonian, obtained by replacing the perturbations by their value at the center of the wave packet, and deriving its dynamics in a general form, based on the time-dependent variational principle. We derive the wave-packet energy up to first order in the gradient of the perturbations and all kinds of Berry-phase corrections to the semiclassical dynamics and the quantization rule. Also, we give a discussion of a formal quantization procedure through the Hamiltonian formalism with the semiclassical dynamics as the starting point.

We illustrate our framework with two cases of perturbations. For electromagnetic fields, previous results of orbital magnetization and anomalous velocity are obtained purely from a single-band point of view. For deformations in crystals, we obtain a Berry-phase term in the Lagrangian due to lattice tracking, which gives to the new terms of tracking velocity and force in the equations of motion of the wave packet. For multiple-valued displacement fields in the presence of dislocations, this term manifests as a Berry phase, which we show to be proportional to the Burgers vector around each dislocation. Also, we relate the deformation correction to the wave-packet energy to the shift in band energy under uniform strain, which turns out to be the same as the deformation potential for electron-phonon interaction at long wavelengths.

The combined effects of electromagnetic and deformational perturbations are yet to be studied. Given that the perturbed Hamiltonian is of the form Eq. (3.1), the equations of motion for the wave-packet dynamics can be completely determined in terms of the properties of the unperturbed crystal and the fields of perturbations. Such a theory will provide a basis for electron transport in deformed crystals and should be pursued in the near future.

ACKNOWLEDGMENTS

The authors wish to thank M. C. Chang, R. Diener, E. Demircan, S. Fishman, D. Gavenda, G. A. Georgakis, L. Kleinman, M. Marder, F. Nori, R. Resta, E. C. G. Sudarshan, Bala Sundaram, and J. Zak for many insightful discussions. This work was supported by the NSF (Grants Nos. DMR 9705406 and PHY 9722610), and by the R. A. Welch Foundation.

APPENDIX A: THE GRADIENT CORRECTION $\Delta\mathcal{E}$

In the derivation below, the band indices for the Bloch states and the energy bands have been restored.

We begin with the matrix elements of the position operator,[10]

$$\langle \psi_{n\mathbf{q}}|\hat{\mathbf{x}}|\psi_{n'\mathbf{q}'}\rangle = \left[i\frac{\partial}{\partial\mathbf{q}}\delta_{nn'} + \left\langle u_n \middle| i\frac{\partial u_{n'}}{\partial\mathbf{q}} \right\rangle \right]\delta(\mathbf{q}'-\mathbf{q}), \tag{A1}$$

and the identity

$$\begin{aligned}\left\langle \psi_{n\mathbf{q}} \middle| \frac{\partial \hat{H}_c}{\partial \mathbf{x}_c} \middle| \psi_{n'\mathbf{q}'} \right\rangle &= \delta(\mathbf{q}-\mathbf{q}')\left\langle u_n \middle| \frac{\partial \hat{H}_c}{\partial \mathbf{x}_c} \middle| u_{n'} \right\rangle \\ &= \delta(\mathbf{q}-\mathbf{q}')\left[(\mathcal{E}_{cn}-\mathcal{E}_{cn'})\left\langle \frac{\partial u_{n\mathbf{q}}}{\partial \mathbf{x}_c} \middle| u_{n'\mathbf{q}} \right\rangle \right. \\ &\quad \left. + \frac{\partial \mathcal{E}_{cn}}{\partial \mathbf{x}_c}\delta_{nn'} \right],\end{aligned} \tag{A2}$$

which are easily verified.

From Eqs. (A1), (A2), and the completeness relation

$$\sum_n \int d^3q\,|\psi_{n\mathbf{q}}\rangle\langle\psi_{n\mathbf{q}}| = \hat{1}, \tag{A3}$$

we can show that

$$\begin{aligned}&\left\langle \psi_{n\mathbf{q}} \middle| \frac{\partial \hat{H}_c}{\partial \mathbf{x}_c}\cdot\hat{\mathbf{x}} \middle| \psi_{n\mathbf{q}'} \right\rangle \\ &= \frac{\partial \mathcal{E}_{cn}}{\partial \mathbf{x}_c}\cdot\langle\psi_{n\mathbf{q}}|\hat{\mathbf{x}}|\psi_{n\mathbf{q}'}\rangle \\ &\quad + i\left\langle \frac{\partial u_{n\mathbf{q}}}{\partial \mathbf{x}_c} \middle| \cdot(\mathcal{E}_{cn}-\hat{H}_c) \middle| \frac{\partial u_{n\mathbf{q}}}{\partial \mathbf{q}} \right\rangle \delta(\mathbf{q}-\mathbf{q}').\end{aligned} \tag{A4}$$

If we assume that the wave packet is constructed from the Bloch states of the nth band, we have

$$\left\langle \Psi \middle| \frac{\partial \hat{H}_c}{\partial \mathbf{x}_c} \middle| \Psi \right\rangle = \int d^3q|a|^2\frac{\partial \mathcal{E}_{cn}}{\partial \mathbf{x}_c} = \left.\frac{\partial \mathcal{E}_{cn}}{\partial \mathbf{x}_c}\right|_{\mathbf{q}=\mathbf{q}_c}, \tag{A5}$$

as only the second term of Eq. (A2) contribute to the expectation value. By the same token, from Eqs. (A5) and (A4), we obtain

$$\left\langle \Psi \middle| \frac{\partial \hat{H}_c}{\partial \mathbf{x}_c}\cdot(\hat{\mathbf{x}}-\mathbf{x}_c) \middle| \Psi \right\rangle = i\left.\left\langle \frac{\partial u_{n\mathbf{q}}}{\partial \mathbf{x}_c} \middle| \bullet(\mathcal{E}_{cn}-\hat{H}_c) \middle| \frac{\partial u_{n\mathbf{q}}}{\partial \mathbf{q}} \right\rangle\right|_{\mathbf{q}=\mathbf{q}_c}. \tag{A6}$$

Hence, the expectation value (2.17) of Eq. (2.15) is just half of the sum of (A6) and its complex conjugate.

APPENDIX B: THE DEFORMED CRYSTAL POTENTIAL

Given the atomic displacements $\{\mathbf{u}_l\}$, which could be due to either strain or rotations, let us denote the deformed crystal potential regarded as a function of the coordinates of the electron and the equilibrium atomic positions by $V(\mathbf{x};\{\mathbf{R}_l+\mathbf{u}_l\})$. If the displacements are small, we can expand the potential in powers of $\{\mathbf{u}_l\}$ and do away with the perturbation theory. When they are large, we have to adopt a different technique. For this purpose, we introduce a smooth displacement field $\mathbf{u}(\mathbf{x})$ such that $\mathbf{u}(\mathbf{R}_l+\mathbf{u}_l)=\mathbf{u}_l$, and write the potential as

$$V[\mathbf{x}-\mathbf{u}(\mathbf{x});\{\mathbf{R}_l+\mathbf{u}_l-\mathbf{u}(\mathbf{x})\}], \tag{B1}$$

where we have used the invariance property of a potential under a simultaneous translation of electronic and atomic positions. We may now expand the potential in powers of $\mathbf{u}_l-\mathbf{u}(\mathbf{x})$ as

$$V_0[\mathbf{x}-\mathbf{u}(\mathbf{x})]+\sum_l [\mathbf{u}_l-\mathbf{u}(\mathbf{x})]_\alpha V_{l\alpha}+\cdots. \quad (B2)$$

where $V_0(\mathbf{x}-\mathbf{u})\equiv V(\mathbf{x}-\mathbf{u};\{\mathbf{R}_l\})$ is the potential used in the deformable ion model in which the approximated potential at $\mathbf{x}$ equals the undeformed crystal potential at the undeformed electron coordinate $\mathbf{x}-\mathbf{u}(\mathbf{x})$, and

$$V_{l\alpha}\equiv\frac{\partial V[\mathbf{x}-\mathbf{u}(\mathbf{x});\{\mathbf{R}_l\}]}{\partial R_{l\alpha}}=V_{l\alpha}[\mathbf{x}-\mathbf{u}(\mathbf{x})-\mathbf{R}_l]. \quad (B3)$$

The equality in the last equation is meant to suggest that the center of fall off of $V_{l\alpha}$ is at the zero of the expression $\mathbf{x}-\mathbf{u}(\mathbf{x})-\mathbf{R}_l$, viz., $\mathbf{x}=\mathbf{u}_l+\mathbf{R}_l$.

The expansion (B2) is meaningful only if $V_{l\alpha}$ decreases sufficiently rapidly with increasing $|\mathbf{x}-\mathbf{u}(\mathbf{x})-\mathbf{R}_l|$. This condition holds for metals because of Coulomb screening; it has been argued that $V_{l\alpha}$ is short ranged[52] also for nonpolar semiconductors and insulators.

Under such a condition, we may write

$$\begin{aligned} u_{l\alpha}&=u_\alpha(\mathbf{x}+\mathbf{R}_l+\mathbf{u}_l-\mathbf{x})\\ &\approx u_\alpha(\mathbf{x})+(\mathbf{R}_l+\mathbf{u}_l-\mathbf{x})_\beta s_{\alpha\beta}(\mathbf{x})\approx u_\alpha(\mathbf{x})\\ &\quad+[\mathbf{R}_l+\mathbf{u}(\mathbf{x})-\mathbf{x}]_\beta s_{\alpha\beta}(\mathbf{x}), \end{aligned} \quad (B4)$$

whence the summation in Eq. (B2) can be put into the form $s_{\alpha\beta}\mathcal{V}_{\alpha\beta}$ with

$$\mathcal{V}_{\alpha\beta}[\mathbf{x}-\mathbf{u}(\mathbf{x})]=\sum_l [\mathbf{R}_l+\mathbf{u}(\mathbf{x})-\mathbf{x}]_\beta V_{l\alpha}. \quad (B5)$$

---

[1] N. W. Ashcroft and N. D. Mermin, *Solid State Physics* (Saunders, Philadelphia, 1976), Chaps. 8–11.
[2] See *Solid State Physics*, (Ref. 1), Chap. 12.
[3] H. Jones and C. Zener, Proc. R. Soc. London **144** , 101 (1934). References to works of Bloch and Peierls, and their approximations have been given here.
[4] R. Peierls, Z. Phys. **80**, 763 (1933).
[5] J. C. Slater, Phys. Rev. **76**, 1592 (1949).
[6] J. M. Luttinger, Phys. Rev. **84**, 814 (1951).
[7] E. N. Adams, Phys. Rev. **85**, 41 (1952).
[8] R. Karplus and J. M. Luttinger, Phys. Rev. **95**, 1154 (1954); W. Kohn and J. M. Luttinger, *ibid.* **108**, 590 (1957); J. M. Luttinger, *ibid.* **112**, 739 (1958).
[9] E. N. Adams and E. I. Blount, J. Chem. Phys. **10**, 286 (1959).
[10] E. I. Blount, in *Solid State Physics*, edited by F. Seitz and D. Turnbull (Academic Press, New York, 1962), Vol. 13, p. 305.
[11] M. C. Chang and Q. Niu, Phys. Rev. Lett. **75**, 1348 (1995).
[12] M. C. Chang and Q. Niu, Phys. Rev. B **53**, 7010 (1996).
[13] M. V. Berry, Proc. R. Soc. London, Ser. A **392**, 45 (1984).
[14] B. Simon, Phys. Rev. Lett. **51**, 2167 (1983).
[15] J. Zak, Phys. Rev. B **40**, 3156 (1989).
[16] W. Kohn, Phys. Rev. **115**, 1460 (1959); E. I. Blount, *ibid.* **126**, 1636 (1962); L. M. Roth, J. Chem. Phys. **23**, 433 (1963). Wannier and Fredkin also developed an effective Hamiltonian formalism; however, their method does not give corrections explicitly as a power series in field strengths; G. H. Wannier, Phys. Rev. **117**, 432 (1960); G. H. Wannier and D. R. Fredkin, *ibid.* **125**, 1910 (1962); G. H. Wannier, Rev. Mod. Phys. **34**, 645 (1962).
[17] E. Brown, Phys. Rev. **166**, 626 (1967).
[18] J. Zak, Phys. Rev. B **15**, 771 (1977); **16**, 4154 (1977).
[19] R. Rammal and J. Bellissard, J. Phys. (France) (France) **51**, 1803 (1990).
[20] M. Wilkinson and R. J. Kay, Phys. Rev. Lett. **76**, 1896 (1996).
[21] P. G. Harper, Proc. R. Soc. London, Ser. A **86**, 874 (1955).
[22] J. B. Sokoloff, Phys. Rep. **126**, 189 (1985).
[23] G. Nenciu, Rev. Mod. Phys. **63**, 91 (1991).
[24] P. Kramer and M. Saraceno, *Geometry of the Time-Dependent Variational Principle in Quantum Mechanics*, Lecture Notes in Physics No. 140 (Springer-Verlag, 1981).
[25] A. K. Pattanayak, Ph.D. dissertation, The University of Texas, Austin, 1994.
[26] See Ref. 23. However, the existence of a generalized type of Wannier functions with the required localization properties has been proved by M. Wilkinson, J. Phys.: Condens. Matter **10**, 7407 (1998).
[27] G. Venkataraman, D. Sahoo and V. Balakrishnan, *Beyond the Crystalline State—An Emerging Perspective* (Springer-Verlag, Berlin, 1989).
[28] M. R. Geller and W. Kohn, Phys. Rev. Lett. **70**, 3103 (1993).
[29] The time scale associated with an energy gap is given by the uncertainty time h/E_g, where E_g is the gap size. The length scale of an gap is given by $1/\kappa_m$, where κ_m is the maximum size of the imaginary part of the wave vector in the gap.
[30] Our results are valid only in the time scale in which the effects of spreading may be ignored. For studies on the dynamics of higher order moments of a wave packet, see Ref. 25. A systematic study of wave-packet dynamics was initiated by E. J. Heller, J. Chem. Phys. **62**, 1544 (1975).
[31] This result can be verified for Hamiltonians of the form $\hat{H}=(1/2m)[\hat{\mathbf{p}}+e\mathbf{A}(\hat{\mathbf{x}},t)]^2+V(\hat{\mathbf{x}},\{\beta_i(\hat{\mathbf{x}},t)\})$, the only types considered in this paper.
[32] P. A. M. Dirac, *Lectures on Quantum Mechanics*, Belfer Graduate School of Science (Yeshiva University, New York 1964); Can. J. Math. **2**, 129 (1950).
[33] Michael Tabor, *Chaos and Integrability in Nonlinear Dynamical Systems: An Introduction* (Wiley Interscience, New York, 1989).
[34] M. Wilkinson, J. Phys. C **17**, 3459 (1984).
[35] H. Kuratsuji and S. Iida, Prog. Theor. Phys. **74**, 439 (1985).
[36] In addition to the works cited in Ref. 16 see L. M. Roth, Phys. Rev. **118**, 1534 (1960); E. H. Cohen and E. I. Blount, Philos. Mag. **5**, 115 (1960), derive the orbital magnetic moment for an effective mass Hamiltonian.
[37] D. J. Thouless, M. Kohmoto, M. P. Nightingale, and M. den Nijs, Phys. Rev. Lett. **49**, 405 (1982); Q. Niu, D. J. Thouless, and Y. Wu, Phys. Rev. B **31**, 3372 (1985).
[38] J. Zak, Phys. Rev. Lett. **62**, 2747 (1989), and Ref. 15. For centrosymmetric crystals, the Berry phase is quantized and equals 0 or π; this corresponds to centers of Wannier functions being either at $x=la$ or $(l+1/2)a$, where l is an integer.
[39] G. H. Wannier, Phys. Rev. **117**, 432 (1960). Though Wannier-

Stark ladder was predicted in the early 1960s, only in the late 1980s that it was confirmed by experiments. For details, see E. E. Mendes and G. Bastard, Phys. Today **46** (6), 34 (1993).

[40] J. Zak, Phys. Rev. Lett. **20**, 1477 (1969).

[41] M. Lax (unpublished).

[42] L. Kleinman, Phys. Rev. B **24**, 7412 (1981), has pointed out the difficulty in defining absolute potentials and deformation potentials for infinite crystals. We defer a more detailed consideration to a future publication.

[43] This definition corresponds to the Eulerian convention. In the Lagrangian convention one defines another displacement field $\mathbf{u}^L(\mathbf{x})$ such that $\mathbf{u}^L(\mathbf{R}_l)=\mathbf{u}_l$. In either convention, strains are of the same order of magnitude, which makes comparison with other works reasonable.

[44] J. Bardeen and W. Shockley, Phys. Rev. **80**, 72 (1950), introduced the concept of deformation potential as the strain-induced shift in electron energy to simplify the computation of the matrix elements of the electron-phonon interaction at long wavelengths.

[45] F. S. Khan and P. B. Allen, Phys. Rev. B **29**, 3341 (1984), use Fermi energy as the reference point rather than the band bottom in their definition of the deformation potential, and include a term for the shift in the Fermi energy due the applied strain. However, this reference can be incorporated at the level of Hamiltonian itself by a suitable mapping, which result in band energies modified accordingly.

[46] Referred to as the deformation potential theorem, the result was first derived by Bardeen and Shockley for Bloch states near band edges where the term $m\mathbf{v}\mathbf{v}$ is zero (Ref. 44). The exact result (4.10) was obtained by Khan and Allen (Ref. 45) and later confirmed by E. Kartheuser and S. Rodriguez, Phys. Rev. B **33**, 772 (1986).

[47] Q. Niu, Phys. Rev. Lett. **64**, 1812 (1990); Mod. Phys. Lett. A **5**, 923 (1991).

[48] This can always be done as long as the Berry curvature $\overleftrightarrow{\Omega}_{\mathbf{kk}}$ is zero everywhere. This condition is satisfied for centrosymmetric crystals in the absence of magnetic fields.

[49] Y. Aharonov and D. Bohm, Phys. Rev. **115**, 485 (1959).

[50] K. Kawamura, Z. Phys. B **29**, 101 (1978).

[51] D. M. Bird and A. R. Preston, Phys. Rev. Lett. **61**, 2863 (1988). See Eqs. (4a) and (4b) for the calculated Berry phase.

[52] R. Resta and L. Colombo, Phys. Rev. B **41**, 12 358 (1990).

Manifestation of Berry's Phase in Metal Physics

G. P. Mikitik and Yu. V. Sharlai
B.Verkin Institute For Low Temperature Physics & Engineering, National Ukrainian Academy of Sciences, Kharkov 310164, Ukraine
(Received 24 November 1997; revised manuscript received 3 November 1998)

It is shown that in crystals the semiclassical quantization condition for energy levels of electrons in the magnetic field depends on Berry's phase. When the electron orbit links to the band-contact line of the metal (i.e., surrounds it), Berry's phase is nonzero and the quantization condition differs from that commonly used. This result is closely analogous to the Aharonov-Bohm's effect provided the band-contact line plays the role of the infinitely thin "solenoid" with the fixed "magnetic flux." The predicted effect must manifest itself in oscillation phenomena for a number of metals. [S0031-9007(99)08623-8]

PACS numbers: 71.70.Di, 03.65.Bz

In 1984 Berry [1] discovered the important result which aroused considerable interest in many fields of physics (see, e.g., [2,3]). According to Berry, if a Hamiltonian of a quantum system depends on parameters and the parameters undergo adiabatic changes so that they eventually return to the original values, then the wave function of the system can acquire the so-called geometrical phase in addition to the familiar dynamical one. This additional phase is completely determined by a closed trajectory of the system in the parameter space and does not depend on details of the temporal evolution. The manifestations of Berry's phase were found in molecular physics [4], optics [5], resonance phenomena in nuclear quadrupole systems [6], etc. As for solid state physics, Zak [7] argued that Berry's results are applicable to an electron moving in a crystal, with the Brillouin zone playing the role of the parameter space. In this paper, we consider an electron of a metal in the external magnetic field $\mathbf{H}$ and show that Berry's phase appears in the semiclassical quantization condition for its energy levels. Interestingly, when the phase is nonzero, the condition turns out to be different from that commonly used [8,9]. Since the semiclassical energy levels of electrons provide the basis for the analysis of oscillation phenomena in metals [8,10], the Berry's phase must manifest itself in a number of the well-known physical effects (e.g., the de Haas–van Alphen effect, the Shubnikov–de Haas effect, etc.) and, hence, can be detected through their experimental investigation. In our subsequent analysis, we shall assume that the crystal under study has a center of inversion (most metals fall into this class of solids). Besides, to elucidate the heart of the matter, we completely neglect the spin-orbit interaction and spin of the electron [11].

It is well known (see, e.g., Refs. [9,12]) that the correspondence between a semiclassical electron trajectory in a crystal (i.e., in the coordinate space) and the orbit in the space of wave vectors $\mathbf{k}$ (i.e., in the Brillouin zone) exists. The latter orbit is the intersection of the constant-energy surface, $\varepsilon(\mathbf{k}) = \text{const}$, with the plane, $k_z = \text{const}$, where z is the direction of the magnetic field $\mathbf{H}$. In the case of the closed orbit, the quantization condition for energy levels of the electron looks like [13,14]

$$S(\varepsilon, k_z) = \left(\frac{2\pi|e|H}{\hbar c}\right)(n + \gamma), \tag{1}$$

where S is the cross-sectional area of the orbit in $\mathbf{k}$ space, n is the large integer ($n > 0$), γ is the constant ($0 \le \gamma < 1$), and e is the electron charge. In what follows we shall consider only those orbits for which probabilities of intraband and interband magnetic breakdowns are negligible. In other words, the orbit under study does not come close to any other trajectory with the same k_z, and its shape differs noticeably from an intersecting one. In this case, according to Zilberman [15] (see also Ref. [9]), γ always has the value

$$\gamma = \frac{1}{2}. \tag{2}$$

It is this value that is commonly used in describing oscillation phenomena in metals [8]. If a magnetic breakdown occurs, γ essentially depends on ε and k_z [16] but, as noted above, we shall not consider this situation.

Now we express the quantity γ in terms of the Berry's phase. Denote the Bloch wave function of the electron in the band with index l by $\psi_{\mathbf{k}l}(\mathbf{r})$:

$$\psi_{\mathbf{k}l}(\mathbf{r}) = \exp(\imath\mathbf{k}\mathbf{r})u_{\mathbf{k}l}(\mathbf{r}),$$

where $u_{\mathbf{k}l}(\mathbf{r})$ is the periodic function of $\mathbf{r}$. Let the closed semiclassical orbit in $\mathbf{k}$ space, Γ, correspond to some band with $l = 0$, i.e., it is intersection of the surface $\varepsilon_0(\mathbf{k}) = \text{const}$ with the plane $k_z = \text{const}$. Then γ is determined by the formula that follows from Eqs. (42) and (43) of Ref. [17]:

$$\gamma - \frac{1}{2} = -\frac{1}{2\pi}\oint_\Gamma \frac{M_0(\mathbf{k})}{v_\perp(\mathbf{k})}\, d\kappa, \tag{3}$$

where $v_\perp$ is the absolute value of the projection of the electron velocity $\mathbf{v} = \hbar^{-1}\nabla_{\mathbf{k}}\varepsilon_0$ on the plane of the orbit; $d\kappa$ is the length of an infinitesimal element of Γ; M_0 is the matrix element,

$$M_0 = \frac{1}{2}\langle \mathbf{k}0| \left[\left(\frac{\hat{\mathbf{p}}}{m} + \mathbf{v} \right) \times \hat{\mathbf{r}} \right]_z |\mathbf{k}0\rangle ,$$

calculated with the use of the Bloch function $\psi_{\mathbf{k}0}(\mathbf{r})$. Here, m is the electron mass, $\hat{\mathbf{p}}$ is the momentum operator whose diagonal part is $m\mathbf{v}$, and $\hat{\mathbf{r}}$ is the periodic (in $\mathbf{k}$) part of the coordinate operator $\hat{\mathbf{R}}$ in the Crystal Momentum Representation [9,18], $\hat{\mathbf{R}} = \imath\nabla_{\mathbf{k}} + \hat{\mathbf{r}}$. It is well known [9,18] that the contribution of interband matrix elements of $\hat{\mathbf{p}}$ and $\hat{\mathbf{r}}$ in M_0 is equal to zero in crystals with the inversion symmetry. This follows from the fact that electron states $|\mathbf{k}l\rangle$ are invariant under the transformation $U = KI$, where K and I are the operators of complex conjugation and inversion, respectively. Hence,

$$M_0 = [\mathbf{v} \times \mathbf{\Omega}]_z ,$$

where $\mathbf{\Omega}$ is the intraband matrix element of $\hat{\mathbf{r}}$ which has the form [see Eq. (44) in Ref. [17]]

$$\mathbf{\Omega}(\mathbf{k}) = \imath \int d\mathbf{r}\, u^*_{\mathbf{k}0}(\mathbf{r}) \nabla_k u_{\mathbf{k}0}(\mathbf{r}) .$$

Here the integration is carried out over a unit cell of the crystal. Substitution of M_0 in Eq. (3) gives

$$\gamma - \frac{1}{2} = -\frac{1}{2\pi} \oint_\Gamma \mathbf{\Omega}\, d\mathbf{k} , \tag{4}$$

where $d\mathbf{k} \equiv d\kappa[\mathbf{i}_z \times \mathbf{v}]/v_\perp$, and $\mathbf{i}_z$ is the unit vector parallel to $\mathbf{H}$ ($d\mathbf{k}$ is aligned with the tangent to Γ and $|d\mathbf{k}| = d\kappa$). The integral in the right-hand side of Eq. (4) is just the Berry's phase for the orbit Γ [7] . When this phase is equal to zero, we arrive at Eq. (2).

The difference of Berry's phase from zero is usually due to degeneracy of electron states [1,19]. In this connection, let us list the various possible types of degeneracy of the electron energy bands in crystals with a center of inversion. It is common knowledge that the contact of the bands in a metal can occur *at symmetry points* and *along symmetry axis* of its Brillouin zone. Besides, as was shown by Herring [20], there are *lines of an accidental contact* between the bands in crystals. The term "accidental" means that the degeneracy of electron states is not caused by their symmetry. Such band-contact lines must exist in many metals. This statement is easily understood when one takes into account Herring's result: If there is a point of an intersection of two energy bands in an axis of symmetry of the Brillouin zone, and the interband matrix element of the velocity operator is nonzero at this point, then a band-contact line has to pass through the point. Intersection of bands at points in axes are known to occur in many metals [21]. As for the matrix element of the velocity operator, the necessary information on it follows from the irreducible representations of the appropriate bands. The simple analysis of literature data shows that the lines of the accidental contact must exist, for example, in Be, Mg, Zn, Cd, Al, and other metals (see Figs. 1a and 1b).

As early as 1962, Blount [22] established properties of the integral in the right-hand side of Eq. (4). He found that, if the contour Γ surrounds a line of the contact between the band under study and some other one, and the energies of the bands separate linearly in $\mathbf{k}$ in the vicinity of the line, then

$$\int_\Gamma \mathbf{\Omega}\, d\mathbf{k} = \pm\pi , \tag{5}$$

where the sign in the right-hand side of Eq. (5) is determined by a direction of the integration [23]. If in the vicinity of the band-contact line the energy splitting of electron states is quadratic in the distance of the point $\mathbf{k}$ from the line, then

$$\int_\Gamma \mathbf{\Omega}\, d\mathbf{k} = 0 . \tag{6}$$

It should be emphasized that the integrals in Eqs. (5) and (6) do not depend on the shape and the size of the contour Γ. This is not surprising, since the equation

$$\nabla_k \times \mathbf{\Omega}(\mathbf{k}) = 0 \tag{7}$$

holds everywhere outside the band-contact line [22] (but in the line $u_{\mathbf{k}0}$ is the nonanalytical function of $\mathbf{k}$, and $\mathbf{\Omega}$ is undefined).

Now we are able to find γ for any semiclassical electron orbit Γ. The above-mentioned condition of the linear separation of the energy bands is fulfilled in the vicinity of any line of the accidental contact [20] and also near a threefold symmetry axis when the degeneracy of

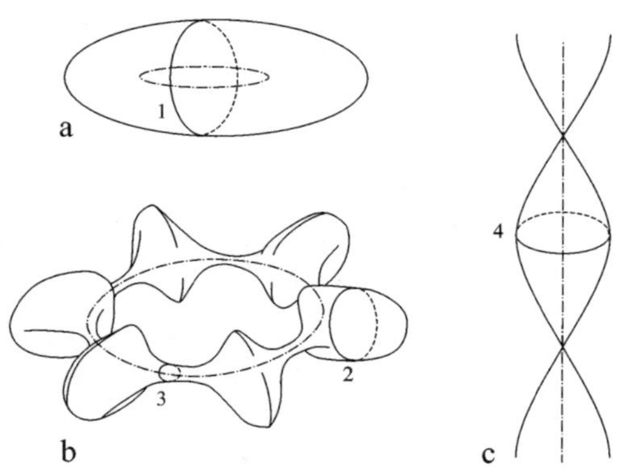

FIG. 1. The schematic sketch of Fermi surfaces for several metals with band-contact lines: the third-band electron "lens" of Zn and Cd (a); the second-band hole "coronet" ("monster") of Be and Mg (b); the self-intersecting Fermi surface of graphite (c). The band-contact lines are shown as the dash-dotted lines. The semiclassical orbits 3 and 4 link to the band-contact lines while the orbits 1 and 2 do not.

states occurs in it. If Γ links to such a line or an axis (see, e.g., the orbits 3 and 4 in Fig. 1), it follows from Eqs. (4) and (5) that

$$\gamma = 0 \tag{8}$$

($\gamma = 1$ and $\gamma = 0$ are equivalent). If the linking is absent (the orbits 1 and 2 in Fig. 1), a surface with boundary Γ necessary exists which does not intersect the band-contact line [for the surface in the case of the orbit 1 (or 2) we can take a part of the constant-energy one shown in Fig. 1]. Transforming the right-hand side of Eq. (4) into the integral over this surface with the use of Stokes' theorem and taking into account Eq. (7), we arrive at formula (2). Interestingly, Eq. (2) is also obtained when Γ links to an even number of band-contact lines (such a situation takes place, e.g., for any central cross section of the second-band Fermi surface of Al). In the case of a symmetry axis different from a threefold one, the separation of the bands is quadratic in a distance of $\mathbf{k}$ from the axis and one always has Eq. (2). The same result is true for symmetry points in the Brillouin zone. This follows from the same considerations that have been used in the case of the absence of the linking.

The obtained result for γ is closely analogous to the Aharonov-Bohm's effect [24]. As pointed out by Blount [18], the quantity $\mathbf{\Omega}$ is similar to a vector potential for a magnetic field. Indeed, the phase transformation,

$$u_{\mathbf{k}0} \to u'_{\mathbf{k}0} = u_{\mathbf{k}0} \exp[\imath\varphi(\mathbf{k})], \tag{9}$$

where $\varphi(\mathbf{k})$ is some regular function of $\mathbf{k}$, results in the change of $\mathbf{\Omega}$:

$$\mathbf{\Omega} \to \mathbf{\Omega}' = \mathbf{\Omega} - \nabla_{\mathbf{k}}\varphi\,, \tag{10}$$

which is characteristic of a vector potential. Taking into account Eqs. (5) and (7), we can treat a band-contact line as an infinitely thin "solenoid" which carries the fixed flux of the "field" $[\nabla_k \times \mathbf{\Omega}]$. With this in mind, the above-mentioned analogy becomes apparent. Although the semiclassical electron moving round the band-contact line does not reach the region in which the "field" is nonzero, it experiences the vector potential $\mathbf{\Omega}$ that cannot be made to vanish along the whole length of the orbit linking to the solenoid. The semiclassical electron state with the energy determined by Eq. (1) is the standing wave. If the electron orbit surrounds the band-contact line (i.e., the solenoid), the interference picture corresponding to this wave is shifted as compared to the case when the line is absent. This Aharonov-Bohm's shift manifests itself as the change in γ.

Berry [1] analyzed the geometrical phase of a quantum system when the orbit of this system in the parameter space is located near the point of degeneracy of its states. The Hamiltonian of the system was assumed to be a Hermitian matrix which is linear in deviations of the parameters from the point. Berry presented his results in the pictorial form. He found that such a point can be considered as the "monopole" in the parameter space when the geometrical phase is calculated. In other words, the point "generates" the field which coincides in the form with that of the monopole, and the flux of this field through the contour Γ gives the geometrical phase of the system. In crystals with the inversion symmetry, the Berry's phase of electrons has the specific features which are due to the fact that electron states are invariant under the transformation U (for the definition of U, see above). First, in any point of the Brillouin zone the Hermitian Hamiltonian can be transformed into the real one. As a consequence, the character of the energy-band splitting near the point of the degeneracy is changed as compared with the general case considered by Berry (now the splitting does not occur along some direction), and the monopole in $\mathbf{k}$ space disappears. Second, the field of Berry $[\nabla \times \mathbf{\Omega}]$ satisfies Eqs. (5) and (7), and Berry's phase does not depend on the shape and the *size* of the electron orbit but is specified by its topological characteristics (there is a linking or it is absent). Thus, it is valid to say that, instead of the monopoles, the solenoids associated with band-contact lines appear in metals.

The value γ can be experimentally determined through the investigation of oscillation effects in metals [8]. Since the measurement of γ is easiest to make for semiclassical orbits corresponding to small extremal cross sections of a Fermi surface, we point out that such orbits exist, e.g., in beryllium, magnesium, and graphite, and in these metals they link to the band-contact lines (see Fig. 1). In Be and Mg, the accidental contact between the second and third bands occurs in the basal plane of the crystals. If $\mathbf{H}$ lies, e.g., in this plane also, Eq. (8) must be valid for the orbits on the "necks" of the second-band hole coronet (or monster). It should be noted that in Zn and Cd which are akin, in many respects, to Be and Mg the same band-contact line is located in the third-band electron lens and does not link to the semiclassical orbits (therefore, in this case $\gamma = 1/2$). In graphite the degeneracy of two bands takes place along the vertical edge *HKH* of the Brillouin zone (i.e., along the threefold symmetry axis). Thus Eq. (8) is expected to be true for the extremal orbit surrounding the point *K* (see orbit 4 in Fig. 1).

In summary, we have shown that in quantization condition (1) γ is equal to zero when in $\mathbf{k}$ space the closed electron orbit associated with a certain energy band $\varepsilon_0(\mathbf{k})$ surrounds the line of degeneracy of this band with some other one. This result depends neither on the form of $\varepsilon_0(\mathbf{k})$ in the neighborhood of the orbit nor on the shape and the size of the electron trajectory and is topological in nature. It is due to the fact that the electron orbit *links* to the band-contact line which is the line of singularities for the Bloch wave functions. If the linking is absent, the conventional result, $\gamma = 1/2$, holds. Measurements of γ can provide a possibility of detecting band-contact lines in metals (beryllium, magnesium, and graphite appear to have considerable promise on this point).

[1] M. V. Berry, Proc. R. Soc. London A **392**, 45 (1984).
[2] I. J. R. Aitchison, Phys. Scr. **T23**, 12 (1988).
[3] A. Shapere and F. Wilczek, *Geometrical Phase in Physics* (World Scientific, Singapore, 1989).
[4] G. Delacretaz *et al.,* Phys. Rev. Lett. **56**, 2598 (1986).
[5] A. Tomita and R. Y. Chiao, Phys. Rev. Lett. **57**, 937 (1986).
[6] R. Tycko, Phys. Rev. Lett. **58**, 2281 (1987).
[7] J. Zak, Phys. Rev. Lett. **62**, 2747 (1989).
[8] D. Shoenberg, *Magnetic Oscillations in Metals* (Cambridge University Press, Cambridge, England, 1984).
[9] E. M. Lifshitz and L. P. Pitaevskii, *Statistical Physics,* Course in Theoretical Physics Vol. 9 (Permagon, Oxford, 1986), Pt. 2.
[10] I. M. Lifshitz, M. Ya. Azbel', and M. I. Kaganov, *Electron Theory of Metals* (Nauka, Moscow, 1971).
[11] Inclusion of the weak spin-orbit coupling leaves the results for the magnetic energy levels of electrons practically unchanged {G. P. Mikitik and Yu. V. Sharlai, Zh. Eksp. Teor. Fiz. **114**, 1375 (1998) [Sov. Phys. JETP **87**, 747 (1998)]}.
[12] J. M. Ziman, *Principles of the Theory of Solids* (University Press, Cambridge, England, 1972).
[13] L. Onsager, Philos. Mag. **43**, 1006 (1952).
[14] I. M. Lifshitz and A. M. Kosevich, Zh. Eksp. Teor. Fiz. **29**, 730 (1955) [Sov. Phys. JETP **2**, 636 (1956)].
[15] G. E. Zilberman, Zh. Eksp. Teor. Fiz. **32**, 296 (1957) [Sov. Phys. JETP **5**, 208 (1957)].
[16] G. E. Zilberman, Zh. Eksp. Teor. Fiz. **34**, 748 (1958) [Sov. Phys. JETP **7**, 513 (1958)]; M. Ya. Azbel', Zh. Eksp. Teor. Fiz. **39**, 1276 (1960) [Sov. Phys. JETP **12**, 891 (1961)]; A. A. Slutskin, Zh. Eksp. Teor. Fiz. **53**, 767 (1967) [Sov. Phys. JETP **26**, 474 (1968)].
[17] L. M. Roth, Phys. Rev. **145**, 434 (1966).
[18] E. I. Blount, *Solid State Physics* (Academic Press, New York, 1962), Vol. 13, pp. 305–373.
[19] Berry's phase in crystals can be also nonzero for open electron trajectories [7]. However, energies for such orbits are not quantized and we do not consider these trajectories.
[20] C. Herring, Phys. Rev. **52**, 365 (1937).
[21] D. A. Papaconstantopoulos, *Handbook of the Band Structure of Elemental Solids* (Plenum, New York, 1986).
[22] See pages 327, 328, and 361 in Ref. [18].
[23] In general, the term $2\pi q$ must be added to π in this equation [22] where q is some integer. However, $q = 0$ when spin-orbit interaction is taken into account [11]. (Besides, nonzero q would modify n and not affect γ.)
[24] Y. Aharonov and D. Bohm, Phys. Rev. **115**, 485 (1959).

Berry Phase and de Haas–van Alphen Effect in $LaRhIn_5$

G. P. Mikitik and Yu. V. Sharlai

B. Verkin Institute for Low Temperature Physics & Engineering, Ukrainian Academy of Sciences, Kharkov 61103, Ukraine

(Received 19 April 2004; published 2 September 2004)

We explain the experimental data on the magnetization of $LaRhIn_5$ recently published by Goodrich *et al.* [Phys. Rev. Lett. **89**, 026401 (2002)]. We show that the magnetization of a small electron group associated with a band-contact line was detected in that Letter. These data provide the first observation of the Berry phase of electrons in metals via the de Haas–van Alphen effect.

DOI: 10.1103/PhysRevLett.93.106403 PACS numbers: 71.18.+y, 03.65.Vf

In recent years the concept of the so-called Berry phase [1] has attracted considerable attention thanks to its fundamental origin; see, e.g., Refs. [2,3], and references therein. According to Berry, if a Hamiltonian of a quantum system depends on parameters, and if the parameters undergo adiabatic changes so that they eventually return to their original values, the wave function of the system can acquire the so-called geometrical phase in addition to the familiar dynamical one. This additional phase (the Berry phase) differs from zero when the trajectory Γ of the system in the parameter space is located near a *point* at which the states of the system are degenerate [1]. In analyzing this situation, Berry assumed that the Hamiltonian of the system is a Hermitian matrix which is linear in deviations of the parameters from the point, and he presented his final result in the pictorial form. He found that such a point can be considered as a "monopole" in the parameter space when the geometrical phase is calculated. In other words, the point "generates" a field which coincides in the form with that of the monopole, and the flux of this Berry field through the contour Γ gives the geometrical phase of the system. Evidence for this phase was obtained in experiments with various physical systems [2,3]. However, an experimental observation of the Berry phase for electrons in crystals has proved a challenging problem (some progress in this direction was achieved only recently [4–6]).

It is well known (see, e.g., Ref. [7]) that the semiclassical motion of an electron in a crystal in the magnetic field H can be represented as the motion of the wave vector $\mathbf{k}$ in an orbit in the Brillouin zone. This orbit is the intersection of the constant-energy surface, $\varepsilon(\mathbf{k}) =$ const, with the plane, $k_z =$ const, where z is the direction of the magnetic field $\mathbf{H}$ and $\varepsilon(\mathbf{k})$ is the electron dispersion relation in the crystal. Berry's result is applicable to such an electron, with the Brillouin zone playing the role of the parameter space [8]. However, in crystals with the inversion symmetry and a weak spin-orbit interaction, the Berry phase of the electrons has specific features [9] which are due to the fact that the electron states are invariant under the simultaneous inversion of time and spatial coordinates. This invariance permits one to transform the Hermitian Hamiltonian of the electron into the real form for any point of the Brillouin zone. As a consequence, the character of the energy-band degeneracy differs from that considered by Berry. Now the electron energy bands $\varepsilon_l(\mathbf{k})$ contact along *lines* in the Brillouin zone, and the lines need not be symmetry axes [10]. In other words, the degeneracy is not lifted along these lines, and the monopole in the $\mathbf{k}$ space disappears. As it was shown in our Letter [9], in such a situation the band-contact lines play the role of infinitely thin "solenoids" which generate the Berry field with the flux $\pm\pi$. Although this field is zero outside the solenoids, if the electron orbit surrounds the line, the flux threads the orbit, and the electron acquires the Berry phase $\phi_B = \pm\pi$ when it moves around this line. It is clear that in this case the Berry phase does not depend on the shape and the size of the electron orbit but is specified only by its topological characteristics (there is either a linking of the orbit with the band-contact line or there is not).

The Berry phase of the electron modifies [9] the well-known semiclassical quantization rule [11] for the electron energy in the magnetic field, ε,

$$S(\varepsilon, k_z) = \frac{2\pi eH}{\hbar c}(n + \gamma), \tag{1}$$

where S is the cross-sectional area of the closed electron orbit in the $\mathbf{k}$ space, n is a large integer ($n > 0$), e is the absolute value of the electron charge, and the constant γ is given by

$$\gamma = \frac{1}{2} - \frac{\phi_B}{2\pi}. \tag{2}$$

The meaning of formula (2) is the following: When the electron makes a complete circuit in its orbit, the change of the phase of its wave function consists of the usual semiclassical part $\hbar cS/eH$, the shift $-\pi$ associated with the so-called turning points of the orbit where the semiclassical approximation fails, and the Berry phase. Equating this change to $2\pi n$, one arrives at Eqs. (1) and (2). Thus, when the electron orbit links to the band-contact line, one obtains $\gamma = 0$ (the values $\gamma = 0$ and $\gamma = 1$ are equivalent) instead of the usual value [11]

$\gamma = 1/2$. This change of γ has to manifest itself in the de Haas–van Alphen effect [9].

In a recent experimental investigation [12] of the de Haas–van Alphen effect in $LaRhIn_5$, the oscillations of magnetization associated with a small cross section of the Fermi surface of this metal were detected. The electron cyclotron mass m^* corresponding to this cross section was also small compared with electron mass m, $|m^*| \approx 0.067m$. Authors of that paper attributed these oscillations to a small electron pocket of the Fermi surface. Besides, in that paper, the magnetization of the electrons of the pocket was studied in the ultraquantum limit, and the following intriguing contradiction between the obtained experimental data and the existing theory was discovered: Since in the ultraquantum limit the electrons occupy only the lowest Landau level, they have to migrate into large sheets of the Fermi surface when this level is raised above the Fermi energy. Hence, the magnetization M of the electrons of the pocket has to vanish with increasing magnetic field H. However, the experimental data [12] reveal a finite contribution to the magnetization even in such magnetic fields.

In this Letter, we resolve this contradiction. We show that in fact a small electron group associated with a band-contact line was detected in Ref. [12], and the results of Ref. [12] provide the first observation of the Berry phase of the electrons via the de Haas–van Alphen effect.

A small cross section of a Fermi surface appears near that point $\mathbf{k}_0$ of the Brillouin zone for which the two conditions are fulfilled: Topology of this surface changes at the $\mathbf{k}_0$ if the Fermi energy ε_F is shifted past some critical energy ε_0, and this ε_0 is close to the initial Fermi level of the crystal. In the case of the degeneracy of two electron energy bands of the crystal [say, $\varepsilon_+(\mathbf{k})$ and $\varepsilon_-(\mathbf{k})$] along a line in the Brillouin zone, these bands near the $\mathbf{k}_0$ always can be represented in the form [13]

$$\varepsilon_\pm(\mathbf{k}) = \varepsilon_0 + \frac{\hbar^2 k_3^2}{2m_3} + \hbar(\mathbf{v}_\perp \cdot \mathbf{k}) \pm \sqrt{(\hbar V_1 k_1)^2 + (\hbar V_2 k_2)^2}, \quad (3)$$

where the wave vector $\mathbf{k}$ is measured from the $\mathbf{k}_0$; the constants m_3, V_1, V_2, and $\mathbf{v}_\perp = (v_1, v_2, 0)$ are some parameters of the spectrum; the k_3 axis coincides with the tangent to the band-contact line at the point $\mathbf{k}_0$; and the $\mathbf{k}_0$ is defined by the condition that the band energies in the line [$\varepsilon_+(\mathbf{k}) = \varepsilon_-(\mathbf{k})$ there] reach the extremal value ε_0 at this point. A small extremal cross section can exist only under the condition $a_\perp^2 \equiv (v_1/V_1)^2 + (v_2/V_2)^2 < 1$, which we imply to hold below.

If $\mathrm{sgn}(m_3)(\varepsilon_F - \varepsilon_0) < 0$, the Fermi surface has the shape of a neck, with the band-contact line being inside the neck [Fig. 1(b)]. Here $\mathrm{sgn}(z) = 1$ for $z > 0$ and $\mathrm{sgn}(z) = -1$ if $z < 0$. As the Fermi energy passes the critical energy, $\mathrm{sgn}(m_3)(\varepsilon_F - \varepsilon_0) > 0$, the neck is broken,

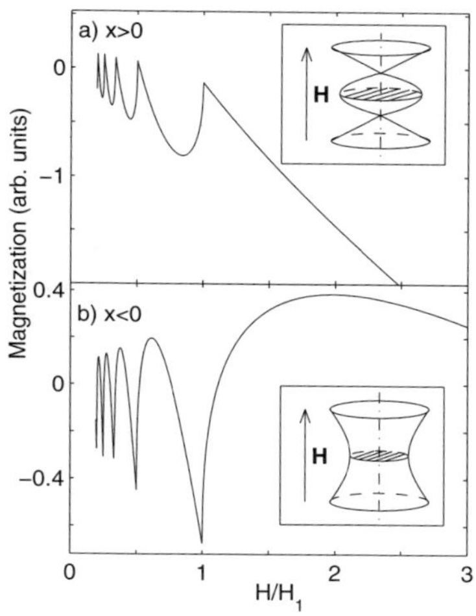

FIG. 1. The electron magnetization associated with the band-contact line, Eq. (4), for $x > 0$ (a) and for $x < 0$ (b). A sign of x coincides with the sign of $(\varepsilon_F - \varepsilon_0)m_3$. The electron spectrum is described by Eq. (3) with $a_\perp^2 \equiv (v_1/V_1)^2 + (v_2/V_2)^2 < 1$. The field H_1 is given by Eq. (6). The insets show the appropriate Fermi surfaces and the extremal cross sections; the dash-dotted lines depict the band-contact line.

and immediately a new pocket appears, i.e., the Fermi surface takes the self-intersecting shape, with the band-contact line still lying inside it [Fig. 1(a)]. Thus, in the case of the degeneracy of the bands, the $\mathbf{k}_0$ is the point where a self-intersecting Fermi surface appears (or disappears) at $\varepsilon_F = \varepsilon_0$.

Let the magnetic field H be in the k_3 direction. In this case, the exact Landau levels were found in a vicinity of the critical energy ε_0 [14]. Interestingly, these exact levels obtained from the appropriate Schrödinger equation coincide with the semiclassical levels, given by Eq. (1) and $\gamma = 0$, at *all* n (even at $n \sim 1$), and not just for $n \gg 1$. On the basis of this spectrum, the magnetization M of the electrons with spectrum (3) was calculated for an *arbitrary* strength of H [14]:

$$M = -\left(\frac{e}{c}\right)^2 \frac{2^{3/2}|m_3(\varepsilon_F - \varepsilon_0)|^{1/2}}{\pi^2 \hbar |m^*|} H_1^{1/4} H^{3/4} f(x), \quad (4)$$

where

$$x = \mathrm{sgn}[m_3(\varepsilon_F - \varepsilon_0)](H_1/H)^{1/2},$$

and the universal function $f(x)$ is completely independent of the spectrum,

$$f(x) = \frac{1}{4}\int_{-x}^{\infty} dt \left(\frac{1}{2} - \{t^2\}\right) \mathrm{sgn}(t) \frac{7t + 6x}{\sqrt{x + t}}. \quad (5)$$

Here $\{z\}$ means the fractional part of the number z. The

H_1 is one of the fields given by the formula

$$\frac{e}{c\hbar}H_n = \frac{S_{ex}}{2\pi}\frac{1}{n} = \frac{m^*}{2\hbar^2}\frac{(\varepsilon_F - \varepsilon_0)}{n}, \tag{6}$$

where $n = 1, 2, \ldots$, S_{ex} is the area of the extremal cross section of the Fermi surface (see Fig. 1), and the cyclotron mass m^* is proportional to $\varepsilon_F - \varepsilon_0$,

$$m^* = \frac{\varepsilon_F - \varepsilon_0}{V_1V_2(1 - a_\perp^2)^{3/2}}, \tag{7}$$

and is small compared to the electron mass m at $|\varepsilon_F - \varepsilon_0| \ll mV_1V_2 \sim 1$–$10$ eV. The meaning of the fields H_n will become clear from the subsequent analysis. Note that we consider $M(H)$ at a fixed ε_F since, in a metal, large sheets of its Fermi surface provide such large density of states that ε_F is practically independent of the magnetic field.

If $|x| \gg 1$, the magnetization (4) splits into the oscillation part, which completely agrees with the well-known Lifshits-Kosevich formula [11,15], and the smooth contribution χH, with the magnetic susceptibility χ coinciding with that of Ref. [13]. At low temperatures,

$$\delta\varepsilon_H \equiv \frac{e\hbar H}{|m^*|c} \gg 2\pi^2 T, \tag{8}$$

the oscillations of the magnetization M prevail over the smooth part, many harmonics in the Lifshits-Kosevich formula are relevant, and sharp peaks of M occur when the Landau levels cross the Fermi energy. It is the field H_n defined by Eq. (6) that gives the position of the peak at crossing ε_F by the nth Landau level. Note that at $x > 0$ the oscillations of M result from the *maximum* cross section of the pocket with $k_3 = 0$, and the peaks of the magnetization are directed *upward* [Fig. 1(a)], while at $x < 0$ the oscillations result from the *minimum* cross section of the neck with $k_3 = 0$, and the peaks are directed *downward* [Fig. 1(b)].

Formula (6) provides the possibility to find the band-contact lines in metals, using the Shoenberg procedure [11]. Plotting experimental values of $1/H_n$ versus n, one can state that a band-contact line has been detected if this dependence is extrapolated to the origin of the coordinate. If the γ were different from zero, the dependence would be extrapolated to $-\gamma$. We emphasize that since for the electrons near the point $\mathbf{k}_0$ the exact spectrum in the magnetic field coincides with the semiclassical spectrum, formula (6) defines the positions of the peaks in $M(H)$ not only for large n but also for $n \sim 1$. Thus, it is sufficient to use several last oscillations of $M(H)$ in this detection. This enables one to find γ with maximal accuracy. Note also that the observation of the sharp peaks for the last oscillations ($n \sim 1$) is the most favorable since $\delta\varepsilon_H \sim |\varepsilon_F - \varepsilon_0|/n$ in Eq. (8).

The ultraquantum limit occurs when $|x| \leq 1$, i.e., when $H \geq H_1$. In this limit the magnetization $M(H)$ cannot be decomposed into the oscillation parts and the term χH [e.g., $f(x) \approx 0.156$ at $|x| \ll 1$]. In this case the magnetization is determined by the Landau levels of the lower band $\varepsilon_-(\mathbf{k})$, which are *all* occupied by the electrons. It is important that at $H \sim H_1$ the magnetization far exceeds that of the usual small pockets and necks which contain no band-contact line. Indeed, for such the pockets and necks the appropriate magnitude of M is of the order of the prefactor before the function $f(x)$ in Eq. (4), but m^* in such situations is not proportional to $\varepsilon_F - \varepsilon_0$ and generally not small, $m^* \sim m$. Moreover, at $H \sim H_1$ the M in Eq. (4) generally *is not small* compared with the smooth part of the magnetization, $\chi_0 H_1$, caused by the *large electron groups* in metals. Estimating the smooth part of the magnetic susceptibility of the large electron groups, χ_0, by the Landau formula for the electron gas [11], $\chi_0 \sim (e/c)^2(\varepsilon_F)^{1/2}/\hbar\sqrt{m}$, we find that, at $H \sim H_1$,

$$\frac{M(H)}{\chi_0 H} \sim \frac{mV_1V_2}{(\varepsilon_F|\varepsilon_F - \varepsilon_0|)^{1/2}} \gg 1.$$

In the last inequality, we have assumed that mV_1V_2 and ε_F are of the order of the characteristic energies in metals, 1–10 eV.

Above we have neglected the spin-orbit interaction in the crystal. With this interaction, the semiclassical quantization rule (1) is modifies as follows [11]:

$$S(\varepsilon, k_z) = \frac{2\pi eH}{\hbar c}\left(n + \gamma \pm \frac{gm^*}{4m}\right), \tag{9}$$

where $\gamma = 1/2$, and g is the so-called g factor of the electron orbit. Besides, the spin-orbit interaction generally lifts the degeneracy of the bands $\varepsilon_+(\mathbf{k})$ and $\varepsilon_-(\mathbf{k})$. But if this interaction is not too strong, so that the gap between these bands is essentially smaller than the energy gaps between $\varepsilon_\pm(\mathbf{k})$ and other bands of the crystal, the concept of the band-contact line is still valid approximately. As it was shown in our paper [16], if the semiclassical electron orbit in the magnetic field surrounds such a band-contact line, one has $g \approx 2m/m^*$, and formula (9) is equivalent to Eq. (1) with $\gamma = 0$ for all n. In other words, Eq. (6) for the peak positions is robust to "switching on" the spin-orbit interaction. Note that the g factor is *large even for a very weak spin-orbit interaction*, and this result is the other manifestation of the nonzero Berry phase (instead of $\gamma = 0$). The spin-orbit interaction also modifies formula (4) [14]. However, if the splitting of the bands $\varepsilon_\pm(\mathbf{k})$ is small near the point $\mathbf{k}_0$, the modification is negligible, and it increases rather *slowly* with the strength of the spin-orbit interaction.

We now apply the above results to the experimental data of Ref. [12]. These data obtained at a low temperature (1.5 K) reveal the sharp peaks in the magnetization of $LaRhIn_5$ when the magnetic field H is parallel to the [001] direction of this tetragonal compound (Fig. 2). The analysis of the peak positions (see the inset of Fig. 2)

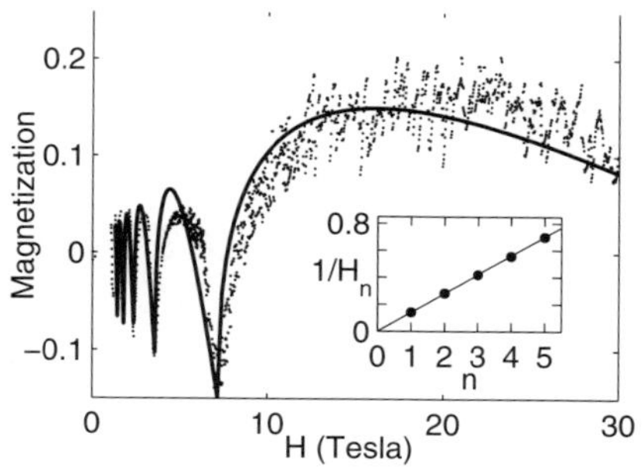

FIG. 2. The experimental data [12] on the magnetization of $LaRhIn_5$ (dots) and the magnetization calculated from Eqs. (4) and (10) (solid line) for $\mathrm{sgn}(m_3)(\varepsilon_F - \varepsilon_0) = -25$ meV, $|m^*| = 0.067m$ ($H_1 = 7$ T), and χ_0 given in the text. The inset shows the dependence of the experimental values of $1/H_n$ [12] on n. This dependence gives $\gamma = 0$.

gives $\gamma = 0$ [17]; i.e., we conclude that the oscillations in the magnetization result from some small electron group near the band-contact line [18]. Using the experimental value [12] of the cyclotron mass, $|m^*| = 0.067m$, and the position of the last peak $H_1 \approx 7$ T, we find from Eq. (6) that $|\varepsilon_F - \varepsilon_0| \approx 25$ meV. The downward peaks mean that we deal with the situation shown in Fig. 1(b).

To verify this conclusion, we also compare the theoretical $M(H)$ with the experimental data (Fig. 2). The experimental magnetization M_{exp} has been approximated by

$$M_{\mathrm{exp}}(H) = \chi_0 H + M(H), \tag{10}$$

where $M(H)$ is given by Eq. (4), while χ_0 is the smooth part of the magnetic susceptibility of the large electron groups in $LaRhIn_5$. Thus, we have only the two *constants* to fit the experimental data: the prefactor in Eq. (4) and χ_0. In Fig. 2 we show the theoretical curve calculated under the condition $\chi_0 H_1/M(H_1) = 0.14$. Note that $M(H_1)$ is noticeably larger than $\chi_0 H_1$. It is also evident that the curve reproduces the experimental data sufficiently well even without any corrections to M due to the spin-orbit interaction.

Although the band structure of $LaRhIn_5$ was calculated in Ref. [19], the data presented in that paper do not permit one to find the band-contact lines in this crystal. To locate these lines, it would be well to calculate the bands lying near ε_F over the Brillouin zone and to trace the evolution of these bands with the strength of the spin-orbit interaction. Such an analysis could also clarify one more point: It turns out that two small and almost equal cross sections determine the oscillations of M in $LaRhIn_5$ [12]. This can occur if the direction of the magnetic field slightly differs from the [001] axis, and if the directions of the band-contact lines at the equivalent points $\mathbf{k}_0$ do not coincide with this axis. Note that, when the magnetic field is tilted away from the k_3 axis, the component H_3 has to be inserted in the above formulas [14]. Thus, an experimental investigation of the angular dependences of the two cross sections could also assist in clarifying this result of Ref. [12].

In summary, we resolve the contradiction discovered in Ref. [12]. It turns out that a small neck of the Fermi surface with *the band-contact line inside the neck* [see the inset of Fig. 1(b)] was discovered in Ref. [12]. In this case the magnetization in the ultraquantum limit does not vanish, while the positions of the peaks in the oscillation part of $M(H)$ depend on the nonzero Berry phase for the electron orbits in the magnetic field. In other words, the results of Ref. [12] provide essentially the first observation of the Berry phase via the de Haas–van Alphen effect.

[1] M.V. Berry, Proc. R. Soc. London A **392**, 45 (1984).
[2] A. Shapere and F. Wilczek, *Geometrical Phase in Physics* (World Scientific, Singapore, 1989).
[3] R. Resta, J. Phys. Condens. Matter **12**, R107 (2000).
[4] Jeng-Bang Yau, E. P. De Poortere, and M. Shayegan, Phys. Rev. Lett. **88**, 146801 (2002).
[5] T. Jungwirth, Q. Niu, and A. H. MacDonald, Phys. Rev. Lett. **88**, 207208 (2002).
[6] Y. Yao, L. Kleinman, A. H. MacDonald, J. Sinova, T. Jungwirth, Ding-sheng Wang, E. Wang, and Q. Niu, Phys. Rev. Lett. **92**, 037204 (2004).
[7] J. M. Ziman, *Principles of the Theory of Solids* (Cambridge University Press, Cambridge, 1972).
[8] J. Zak, Phys. Rev. Lett. **62**, 2747 (1989).
[9] G. P. Mikitik and Yu. V. Sharlai, Phys. Rev. Lett. **82**, 2147 (1999).
[10] C. Herring, Phys. Rev. **52**, 365 (1937).
[11] D. Shoenberg, *Magnetic Oscillations in Metals* (Cambridge University Press, Cambridge, England, 1984).
[12] R. G. Goodrich, D. L. Maslov, A. F. Hebard, J. L. Sarrao, D. Hall, Z. Fisk, Phys. Rev. Lett. **89**, 026401 (2002).
[13] G. P. Mikitik and I. V. Svechkarev, Fiz. Nizk. Temp. **15**, 295 (1989) [Sov. J. Low Temp. Phys. **15**, 165 (1989)].
[14] G. P. Mikitik and Yu. V. Sharlai, Fiz. Nizk. Temp. **22**, 762 (1996) [Low Temp. Phys. **22**, 585 (1996)].
[15] I. M. Lifshits and A. M. Kosevich, Zh. Eksp. Teor. Fiz. **29**, 730 (1955) [Sov. Phys. JETP **2**, 636 (1956)].
[16] G. P. Mikitik and Yu. V. Sharlai, Zh. Eksp. Teor. Fiz. **114**, 1375 (1998) [Sov. Phys. JETP **87**, 747 (1998)].
[17] A close inspection of the inset of Fig. 2 of Ref. [12] leads to the same conclusion.
[18] It is possible that this line is almost straight and parallel to the [001] direction, with these properties of the line being due to the quasi-two-dimensional electron energy spectrum of the crystal.
[19] H. Shishido *et al.*, J. Phys. Soc. Jpn. **71**, 162 (2002).

Coherent wave-packet evolution in coupled bands

Dimitrie Culcer,[1,2] Yugui Yao,[1,2,3] and Qian Niu[1,2]
[1]*Department of Physics, The University of Texas at Austin, Austin, Texas 78712-1081, USA*
[2]*International Center for Quantum Structures, Chinese Academy of Sciences, Beijing 100080, China*
[3]*Institute of Physics, Chinese Academy of Sciences, Beijing 100080, China*
(Received 10 November 2004; revised manuscript received 16 March 2005; published 5 August 2005)

We develop a formalism for treating *coherent* wave-packet dynamics of charge and spin carriers in degenerate and nearly degenerate bands. We consider the two-band case carefully in view of spintronics applications, where transitions between spin-split bands often occur even for relatively weak electromagnetic fields. We demonstrate that much of the semiclassical formalism developed for the single-band case can be generalized to multiple bands, and examine the nontrivial non-Abelian corrections arising from the additional degree of freedom. Along with the center of mass motion in crystal momentum and real space, one must also take into account the probability amplitudes to characterize the dynamics between the bands. We derive the wave packet energy up to the first order gradient correction and obtain the equations of motion for the real- and k-space center of the wave packet, as well as for the probability amplitudes. These equations include the non-Abelian Berry curvature terms and a non-Abelian correction to the group velocity. As an example, we apply our formalism to describe coherent wave packet evolution under the action of an electric field, demonstrating that it leads to electrical separation of spins. A sizable separation will be observed, with a large degree of tunability, making this mechanism a practical method of generating a spin polarization. We then turn our attention to a magnetic field, where we recover Larmor precession, which cannot be obtained from a single-band point of view. In this case, the gradient energy correction can be regarded as due to a magnetic moment from the self-rotation of the wave packet, and we calculate its value for the light holes in the spherical four-band Luttinger model.

DOI: 10.1103/PhysRevB.72.085110 PACS number(s): 72.10.Bg, 72.20.My, 72.25.Dc, 73.23.−b

I. INTRODUCTION

It often happens, in transport phenomena, that one has to consider carrier dynamics in bands which are coupled together. This coupling arises either through strong interband scattering or as a result of the bands being degenerate, or both. The nearly degenerate case is particularly relevant in transport theory as transitions often occur between bands even at relatively weak electromagnetic fields. Such situations include two-dimensional systems described by the Rashba Hamiltonian[1] with strong scattering, the doubly degenerate heavy and light hole bands in the Luttinger model,[2] which is frequently used to model the valence bands of bulk zinc blende semiconductors, and the conduction bands of wurtzite structures.[3] The case of nearly degenerate bands has not, to date, received the attention it deserves,[4–6] despite the important role played by such bands in semiconductor spintronics systems,[7,8] whether in dealing with spin currents,[9] spin generation,[10] and relaxation,[11] or spin injection across a semiconductor interface.[12,13]

Spintronics systems lend themselves to a semiclassical treatment, as the external electromagnetic fields vary on scales that are considerably larger than atomic size. The semiclassical formalism has had much success in describing carrier dynamics and transport phenomena in condensed matter physics. In the nondegenerate case, the carrier dynamics can be obtained semiclassically then combined with the Boltzmann equation to produce accurate descriptions of the transport properties of many materials. This approximation is used in the descriptions of cyclotron orbits, conduction in solids, the Hall effect and magnetoresistance.[14] An essential application of the semiclassical model, which is specifically relevant to our discussion, is in treating external fields that are not represented by bounded operators, so that a perturbative expansion will not converge.[15] The most common example is provided by uniform electric and magnetic fields, where the potential is linear in position.

We therefore develop, in this paper, a semiclassical description of transport in degenerate and nearly degenerate bands. One of our main purposes is to extend the semiclassical approach, as developed by Sundaram and Niu,[16] to the case of coupled Bloch bands, in order to take into account the spin degree of freedom. We illustrate the underlying physics by treating two bands, without loss of generality. Two-band models are frequently an adequate description of the conduction bands of many semiconductors.[17] In experiments on spin transport in semiconductors the carriers have traditionally been electrons,[18] as the strong spin-orbit coupling in the valence band causes holes to lose spin information much faster.[8] However, in recent years research has also focused on spin currents in the valence bands of semiconductors,[9] with a degeneracy which is usually greater than 2, and the formalism we outline is straightforwardly extended to multiple bands.

To formulate a description of coherent transport in coupled bands we may no longer work with each band individually but must instead treat the coupled-band manifold as a whole. The condition for our theory to be valid, which in the one-band case states that there must be no transitions out of that band,[14] translates into the requirement that there be no transitions out of the manifold under consideration. We will consider a wave packet made up of two bands, which is a suitable description of *coherent* transport, when the density matrix has off-diagonal terms and the relative phase of the two wave functions plays a crucial role. This approach allows us to retain the notion of the real-space center of the wave packet, $\mathbf{r}_c$, which remains well defined. Moreover, in extending the formalism to two bands we are able, in the presence of a magnetic field, to recover Larmor precession, which is not possible from a one band picture. The additional degree of freedom of the two-band system can be taken into account by defining a wave function with the Bloch periodicity in such a way as to incorporate both bands, which allows us to derive the dynamics from a single-band point of view. The coefficients of the bands can then be grouped into a vector, the structure and dynamics of which makes clear the gauge structure of the problem. An interesting fact which will emerge from our analysis is that the effect of the external perturbations can be incorporated entirely into the Berry curvatures,[16] which in turn are generated by a set of connections in real and reciprocal space as well as in time. The Berry curvatures acquire additional terms needed to ensure gauge covariance, and in the framework we present they take the form of field strength tensors associated with the connections.

The organization of this paper is as follows: In Sec. II we develop the semiclassical formalism for coherent transport in the presence of electromagnetic fields, deriving the Lagrangian, based on a time-dependent variational principle, and the equations of motion. In Sec. III we use our formalism to show how coherent wave-packet evolution under the action of an electric field leads to the separation of up and down spins. This idea is similar in principle to the spin transistor proposed by Datta and Das.[19] We demonstrate that a large degree of tunability can be achieved by varying the gate field and number density. Finally, in Sec. IV we examine the case of a magnetic field. We show that the gradient correction to the energy can be interpreted as an intrinsic angular momentum of the wave packet,[16] and we calculate this angular momentum correction for the light holes in the spherical four-band model of the Luttinger Hamiltonian.

II. DEVELOPMENT OF THE FORMALISM

The semiclassical model describes the dynamics of wave packets. The wave packet we consider is well localized in reciprocal space, and it is assumed it sees only a small part of the lattice at any one time. It is chosen in such a way that its spread in wave vector is much smaller than the size of the Brillouin zone, so that its motion at any moment is dependent only on the local properties of the band structure. In order for this to happen, the uncertainty principle dictates that the spread in real space must be greater than the size of the lattice constant.

We consider a systems whose Hamiltonian is a function of slowly varying parameters, such as the potentials of weak external electromagnetic fields, which vary on larger length scales than that of the wave packet, and are treated classically. The periodic potential of the ions on the other hand, changing over dimensions small compared to the wave-packet spread, must be treated quantum mechanically.[14] Given these conditions, we define the *local* Hamiltonian $\hat{H}_c(\mathbf{r}_c, t)$ as the Hamiltonian with the slowly varying potentials evaluated at the center of the wave packet, which we denote by $\mathbf{r}_c$, and time t. The Hamiltonian may be expanded[16] about $\mathbf{r}_c$ and if the external fields vary on spatial scales much larger than that of the wave packet we may truncate the expansion after the gradient term, which we define by $\Delta\hat{H}$,

$$\Delta\hat{H} = \frac{1}{2}\left[(\hat{\mathbf{r}} - \mathbf{r}_c)\cdot\frac{\partial\hat{H}_c}{\partial\mathbf{r}_c} + \text{c.c.}\right]. \tag{1}$$

The gradient term gives rise to a correction to the energy, which will play an important role in our discussion below.

The energy spectrum of the local Hamiltonian $\hat{H}_c$ consists, as usual, of a series of bands, of which several are close together in energy and are separated from the others by larger gaps. It is the subset spanned by these bands that constitutes the focus of our attention. We regard the fields in this problem as small enough that Zener tunneling to the remote bands is negligible, but they may still be strong enough to induce transitions within the subset. For an energy spectrum with such a structure we may further decompose the local Hamiltonian into a degenerate part, $\hat{H}_d$, which, when restricted to the subset of bands closely spaced in energy, is proportional to the identity matrix, and a nondegenerate part, $\hat{H}_n$, which is assumed small and treated perturbatively. The local Hamiltonian of (1) is then

$$\hat{H}_c = \hat{H}_d + \hat{H}_n. \tag{2}$$

The gradient correction to $\hat{H}_c$ can also be expressed in terms of the degenerate and nondegenerate contributions,

$$\Delta\hat{H} = \Delta\hat{H}_d + \Delta\hat{H}_n. \tag{3}$$

Since $\hat{H}_n$ is treated as a perturbation, the gradient correction to it, $\Delta\hat{H}_n$ will be second order in smallness. We will therefore neglect this correction henceforth.

When the external fields are smoothly varying the states move within the subset of bands which are close in energy and which henceforth, for simplicity and without loss of generality, we take to be two-dimensional. The subset is spanned by two basis functions, which are eigenstates of $\hat{H}_d$, the degenerate part of the local Hamiltonian, evaluated at $\mathbf{r}_c$, which has the periodicity of the unperturbed crystal,

$$\hat{H}_d|\Psi_i(\mathbf{r}_c,\mathbf{q},t)\rangle = \varepsilon|\Psi_i(\mathbf{r}_c,\mathbf{q},t)\rangle, \tag{4}$$

where ε is independent of the band index i within the degenerate subset. For a given $\mathbf{r}_c$, therefore, these eigenstates have the Bloch form, with the functions $|u_i\rangle$ representing the lattice periodic parts of the wave functions,

$$|\Psi_1(\mathbf{r}_c,\mathbf{q},t)\rangle = e^{i\mathbf{q}\cdot\hat{\mathbf{r}}}|u_1(\mathbf{r}_c,\mathbf{q},t)\rangle, \tag{5}$$

$$|\Psi_2(\mathbf{r}_c,\mathbf{q},t)\rangle = e^{i\mathbf{q}\cdot\hat{\mathbf{r}}}|u_2(\mathbf{r}_c,\mathbf{q},t)\rangle. \tag{6}$$

The wave functions $|u_i(\mathbf{r}_c,\mathbf{q},t)\rangle$ are spinors with the full periodicity of the lattice. Despite the fact that the two bands are spin split, it cannot be assumed that their local spin quantization axes are antiparallel, as the interactions with neighboring bands may affect the direction of quantization. Therefore, in principle, a finite overlap exists between the spinors corresponding to the two bands and it is not revealing to make a further decomposition of the eigenfunctions into an orbital and a spin part. Additionally, the Hamiltonian contains terms describing the spin-orbit interaction, which may depend on wave vector and position.

Employing the crystal momentum representation, the wave packet is therefore expanded in the basis of Bloch eigenstates,

$$|w\rangle = \int d^3q\{a(\mathbf{q},t)[\eta_1(\mathbf{q},t)|\Psi_1\rangle + \eta_2(\mathbf{q},t)|\Psi_2\rangle]\}. \quad (7)$$

As the wave packet depends only on the local properties of the band structure, the basis functions $|\Psi_1\rangle$, $|\Psi_2\rangle$ are functions of the position of the wave-packet center, $\mathbf{r}_c$, wave vector and time, although implicit in the ket notation is dependence on position. The function $a(\mathbf{q},t) = |a(\mathbf{q},t)|e^{-i\Gamma(\mathbf{q},t)/2}$, which incorporates the overall phase term, is a narrow distribution function describing the extent of the wave packet in reciprocal space and is sharply peaked at the center of the wave packet, denoted by $\mathbf{q}_c$, as discussed by Sundaram and Niu.[16] The functions η_1 and η_2 describe the composition of the wave packet in terms of the two bands. The wave packet satisfies the normalization conditions,

$$\int d^3q|a|^2 = 1, \quad |\eta_1|^2 + |\eta_2^2| = 1. \quad (8)$$

The wave packet can be rewritten by grouping together the coefficients in an overall wave function $|u\rangle$, which retains the Bloch periodicity,

$$|w\rangle = \int d^3q|a|e^{-i\Gamma/2}e^{i\mathbf{q}\cdot\hat{\mathbf{r}}}|u\rangle. \quad (9)$$

Note that $|u\rangle$ is not an eigenstate of the local Hamiltonian $\hat{H}_c = \hat{H}_d + \hat{H}_n$, but an expansion in eigenstates of $\hat{H}_d$, a crucial difference from the one-band situation. In addition, the wave vector and time dependence of $|u\rangle$ come both from the wave vector and time dependence of the Bloch states and that of the coefficients.

We require the real-space center of the wave packet to be given by

$$\mathbf{r}_c = \langle w|\hat{\mathbf{r}}|w\rangle = \frac{1}{2}\frac{\partial\Gamma_c}{\partial\mathbf{q}_c} + \mathbf{R}_c. \quad (10)$$

The subscript c signifies that the quantity is evaluated at the center of the wave packet in reciprocal space, that is $\mathbf{q}=\mathbf{q}_c$. The vector $\mathbf{R}$, representing a connection in reciprocal space, is defined as follows:

$$\mathbf{R} = \langle u|i\frac{\partial}{\partial\mathbf{q}}|u\rangle. \quad (11)$$

The energy of the wave packet is given by the expectation value

$$\begin{aligned}\langle w|\hat{H}|w\rangle &= \langle w|\hat{H}_d|w\rangle + \langle w|\hat{H}_n|w\rangle + \langle w|\Delta\hat{H}_d|w\rangle \\ &\equiv \varepsilon + \Delta_n + \Delta_d \equiv \mathcal{E}.\end{aligned}$$

Both Δ_n and Δ_d are expressible entirely in terms of the Bloch wave function $|u\rangle$. Δ_n is given by

$$\Delta_n = \langle u|\widetilde{H}_n|u\rangle = \eta_i^*\Delta_{ij}^n\eta_j, \quad \Delta_{ij}^n = \langle u_i|\widetilde{H}_n|u_j\rangle,$$

while Δ_d is

$$\begin{aligned}\Delta_d &= \frac{i}{2}\left(\left\langle u\left|\frac{\partial\widetilde{H}_d}{\partial\mathbf{r}_c}\cdot\right|\frac{\partial u}{\partial\mathbf{q}_c}\right\rangle - \text{c.c.}\right) - \frac{\partial\varepsilon}{\partial\mathbf{r}_c}\cdot\mathbf{R} \\ &= \left\langle\frac{\partial u}{\partial\mathbf{r}_c}\right|\cdot(\widetilde{H}_d - \varepsilon)\left|\frac{\partial u}{\partial\mathbf{q}_c}\right\rangle = \eta_i^*\Delta_{ij}^d\eta_j,\end{aligned}$$

$$\Delta_{ij}^d = \left\langle\frac{\partial u_i}{\partial\mathbf{r}_c}\right|\cdot(\widetilde{H}_d - \varepsilon)\left|\frac{\partial u_j}{\partial\mathbf{q}_c}\right\rangle.$$

In the above, the operator $\widetilde{H}_n = e^{-i\mathbf{q}\cdot\hat{\mathbf{r}}}\hat{H}_n e^{i\mathbf{q}\cdot\hat{\mathbf{r}}}$, while $\widetilde{H}_d = e^{-i\mathbf{q}\cdot\hat{\mathbf{r}}}\hat{H}_d e^{i\mathbf{q}\cdot\hat{\mathbf{r}}}$. The energy correction Δ_d is identical to the expression obtained by Sundaram and Niu.[16] It takes on an additional significance when a magnetic field is present, as will be seen in the last section.

The Lagrangian $\mathcal{L}$ is obtained semiclassically by means of a variational principle

$$\mathcal{L} = \langle w|\left(i\hbar\frac{d}{dt} - \hat{H}\right)|w\rangle. \quad (12)$$

Its use is justified by the fact that the Euler-Lagrange equation of motion for $|w\rangle$ derived from it is the time-dependent Schrödinger equation. Following the method used by Sundaram and Niu,[16] the following expression is found for the Lagrangian:

$$\begin{aligned}\mathcal{L} &= \left\langle u\left|i\hbar\frac{du}{dt}\right.\right\rangle + \hbar\mathbf{q}_c\cdot\dot{\mathbf{r}}_c - \mathcal{E} \\ &= i\hbar\eta_i^*\frac{d\eta_i}{dt} + \hbar\eta_i^*\left\langle u_i\left|i\frac{du_j}{dt}\right.\right\rangle\eta_j + \hbar\mathbf{q}_c\cdot\dot{\mathbf{r}}_c - \varepsilon \\ &\quad - \eta_i^*(\Delta_{ij}^n + \Delta_{ij}^d)\eta_j.\end{aligned} \quad (13)$$

In the above, d/dt represents the total time derivative, including both the explicit time dependence and the implicit, which is due to dependence on $\mathbf{q}_c$ and $\mathbf{r}_c$. The Lagrangian depends only on the values of η_i and $d\eta_i/dt$ along the trajectory $\mathbf{q}=\mathbf{q}_c(t)$. Since $\mathbf{q}_c$ is a function of time only, we may regard η_i in the Lagrangian as an independent variable, $\eta_i(t)$. The equations of motion derived from the Lagrangian are

$$\hbar\dot{\mathbf{q}}_c = -\frac{\partial\mathcal{E}}{\partial\mathbf{r}_c} + (\mathbf{\Omega}_{\mathbf{rr}}\dot{\mathbf{r}}_c + \mathbf{\Omega}_{\mathbf{rq}}\dot{\mathbf{q}}_c) - \mathbf{\Omega}_{t\mathbf{r}},$$

$$\hbar\dot{\mathbf{r}}_c = \frac{\partial\mathcal{E}}{\partial\mathbf{q}_c} - (\mathbf{\Omega}_{\mathbf{qr}}\dot{\mathbf{r}}_c + \mathbf{\Omega}_{\mathbf{qq}}\dot{\mathbf{q}}_c) + \mathbf{\Omega}_{t\mathbf{q}},$$

$$i\hbar\frac{d\eta_i}{dt} = \left(\mathcal{H}_{ij} - \hbar\left\langle u_i \middle| i\frac{du_j}{dt}\right\rangle\right)\eta_j.$$

The curvature tensor $\Omega_{\mathbf{rr}}^{\alpha\beta}$ is defined by

$$\Omega_{\mathbf{rr}}^{\alpha\beta} = i\left(\left\langle \frac{\partial u}{\partial r_\alpha} \middle| \frac{\partial u}{\partial r_\beta}\right\rangle - \left\langle \frac{\partial u}{\partial r_\beta} \middle| \frac{\partial u}{\partial r_\alpha}\right\rangle\right) \quad (14)$$

and the vector $\mathbf{\Omega}_{t\mathbf{q}}$ by

$$\Omega_{t\mathbf{q}}^{\alpha} = i\left(\left\langle \frac{\partial u}{\partial t} \middle| \frac{\partial u}{\partial q_\alpha}\right\rangle - \left\langle \frac{\partial u}{\partial q_\alpha} \middle| \frac{\partial u}{\partial t}\right\rangle\right). \quad (15)$$

The others can be deduced analogously. These quantities have exactly the same form as the curvatures defined in the paper by Sundaram and Niu.[16]

We specialize in the case of an external electromagnetic field. The effect of such an external field is discussed thoroughly by Sundaram and Niu.[16] The wave vector $\mathbf{q}$ must be replaced by $\mathbf{k}=\mathbf{q}+(e/\hbar)\mathbf{A}(\mathbf{r},t)$, which is the gauge invariant crystal momentum (for electrons with charge $-e$), and therefore the Hamiltonian will have the form $\hat{H}(\mathbf{k})+eV(\mathbf{r},t)$. Provided the magnetic or exchange field is constant and uniform, so that the Zeeman term has no time or space dependence, the basis states $\{|u_i\rangle\}$ will depend only on $\mathbf{k}$. The reason for this is that all the spatial and time dependence of the wave functions will only come from the spatial and time dependence of the vector potential $\mathbf{A}(\mathbf{r},t)$. We will therefore restrict our attention to constant uniform magnetic fields, while the electric fields may be space- and time-dependent. As the electromagnetic fields vary on a spatial scale which is large compared to that of the wave packet, the local Hamiltonian will have the form $\hat{H}[\mathbf{q}+(e/\hbar)\mathbf{A}(\mathbf{r}_c,t)]+eV(\mathbf{r}_c,t)$. The band eigenstates $\{|\Psi_{n\mathbf{k}}\rangle\}$ take the form $|\psi_{n\mathbf{k}}\rangle=e^{i\mathbf{q}\cdot\mathbf{r}}|u_{n\mathbf{k}}\rangle =e^{i(\mathbf{k}-(e\mathbf{A}/\hbar))\cdot\mathbf{r}}|u_{n\mathbf{k}}\rangle$. The time dependence of $|u\rangle$ comes both from the Bloch wave functions $\{|u_i\rangle\}$, which depend only on $\mathbf{k}$, and from the coefficients, which depend only on time. Therefore, the Lagrangian in the presence of electromagnetic fields can be written as

$$\mathcal{L} = \hbar\left\langle u \middle| i\frac{du}{dt}\right\rangle + [\hbar\mathbf{k}_c - e\mathbf{A}(\mathbf{r}_c,t)]\cdot\dot{\mathbf{r}}_c - \varepsilon - \Delta_n - \Delta_d - eV(\mathbf{r}_c,t). \quad (16)$$

The equations of motion now take the following form:

$$\hbar\dot{\mathbf{k}}_c = -e(\mathbf{E} + \dot{\mathbf{r}}_c \times \mathbf{B}),$$

$$\hbar\dot{\mathbf{r}}_c = \frac{\partial\mathcal{E}}{\partial\mathbf{k}_c} - \hbar\dot{\mathbf{k}}_c \times \mathbf{\Omega} + \mathbf{\Omega}_{t\mathbf{k}},$$

$$i\hbar\frac{d\eta_i}{dt} = \left(\mathcal{H}_{ij} - \hbar\dot{\mathbf{k}}_c\cdot\left\langle u_i \middle| i\frac{\partial u_j}{\partial\mathbf{k}_c}\right\rangle\right)\eta_j, \quad (17)$$

where $\mathbf{\Omega}=i\langle\partial u/\partial\mathbf{k}|\times|\partial u/\partial\mathbf{k}\rangle$. Note that the position-vector equation of motion is very similar to the one band case[16] excepting the presence of the vector $\mathbf{\Omega}_{t\mathbf{k}}$, which is nonzero due to the time dependence of $|u\rangle$ through the coefficients. The equation of motion for $|u\rangle$, if a magnetic field is present, leads to the formula for Larmor precession. The equations may be solved to any desired order in the external fields and are not limited to the linear response regime (the fields are weak enough that they do not induce transitions to remote bands).

III. THE PROBABILITY AMPLITUDES

The treatment we have presented so far is an exact analogy with the single-band dynamics. The equations of motion (17) are complete. Nevertheless, the equations of motion can be made more explicit in terms of the coefficients η_i, and the non-Abelian quantities emerging in the process illustrate the gauge structure of the Hilbert space.

The coefficients η_1, η_2 give the composition of the wave packet in terms of the two bands, and it is natural to think of them as a vector, $\binom{\eta_1}{\eta_2}$, which will be called η. The connection $\mathbf{R}$ can be expanded in terms of η:

$$R^\alpha = \eta^\dagger\mathcal{R}^\alpha\eta + i\eta^\dagger\frac{\partial\eta}{\partial q^\alpha}, \quad \text{where} \quad \mathcal{R}_{ij}^\alpha = \left\langle u_i \middle| i\frac{\partial u_j}{\partial q_\alpha}\right\rangle, \quad (18)$$

and we will also introduce the time connection $\mathcal{T}_{ij} =\langle u_i|i(\partial u_j/\partial t)\rangle$. The Lagrangian in this picture takes the form,

$$\mathcal{L} = i\hbar\eta^\dagger\frac{D\eta}{Dt} + \hbar\mathbf{q}_c\cdot\dot{\mathbf{r}}_c - \eta^\dagger\mathcal{H}\eta, \quad (19)$$

where $\mathcal{H}_{ij}=\langle u_i|\hat{H}|u_j\rangle$ and the covariant derivative with respect to time, defined as $D/Dt=(d/dt)-i(\mathcal{T}+\dot{\mathbf{q}}_c\cdot\mathcal{R})$, has been introduced. Specializing in electromagnetic fields, we end up with the following Lagrangian:

$$\mathcal{L} = \eta^\dagger\left(i\hbar\frac{D}{Dt}\right)\eta + [\hbar\mathbf{k}_c - e\mathbf{A}(\mathbf{r}_c,t)]\cdot\dot{\mathbf{r}}_c - \eta^\dagger\mathcal{H}\eta - eV(\mathbf{r}_c,t). \quad (20)$$

The equations of motion derived from the electromagnetic Lagrangian are as follows:

$$\hbar\dot{\mathbf{k}}_c = -e(\mathbf{E} + \dot{\mathbf{r}}_c \times \mathbf{B}),$$

$$\hbar\dot{\mathbf{r}}_c = \eta^\dagger\left[\frac{D}{D\mathbf{k}},\mathcal{H}\right]\eta - \hbar\dot{\mathbf{k}}_c \times \eta^\dagger\mathcal{F}\eta,$$

$$i\hbar\frac{D\eta}{Dt} = \mathcal{H}\eta. \quad (21)$$

The covariant derivative with respect to the wave vector, which has the form $(D/Dk_\alpha)=(\partial/\partial k_\alpha)-i\mathcal{R}^\alpha$, has been introduced. The non-Abelian Berry curvature matrix, $\mathcal{F}_{ij}^\gamma$, is expressed in terms of the field strength tensor corresponding to the covariant wave vector derivatives:

$$\mathcal{F}_{ij}^\gamma = \tfrac{1}{2}\epsilon^{\alpha\beta\gamma}\mathcal{F}_{ij}^{\alpha\beta}, \quad (22)$$

where the tensor $\mathcal{F}_{ij}^{\alpha\beta}$ is

$$\mathcal{F}_{ij}^{\alpha\beta} = i\left[\frac{D}{Dk_\alpha},\frac{D}{Dk_\beta}\right]_{ij} = \frac{\partial \mathcal{R}_{ij}^{\beta}}{\partial k_\alpha} - \frac{\partial \mathcal{R}_{ij}^{\alpha}}{\partial k_\beta} - i[\mathcal{R}^\alpha,\mathcal{R}^\beta]_{ij}. \tag{23}$$

This form, which includes the non-Abelian correction from the commutator of the connection matrices, makes evident its gauge covariance with respect to unitary transformations of η. The curvature tensor is antisymmetric under interchange of α and β, while the indices i and j satisfy $\mathcal{F}_{ij}^{\alpha\beta}=(\mathcal{F}_{ji}^{\alpha\beta})^*$.

It is seen from the equations of motion that working in the coupled-band manifold entails the presence of non-Abelian quantities such as the modified Berry curvature and gauge covariant group velocity $(1/\hbar)[D/D\mathbf{k},\mathcal{H}]$, which are corrections to the one band equations of motion needed to ensure gauge covariance. The matrix $\mathcal{H}$ is not necessarily diagonal, as it may include energy gradient corrections.

We note that equivalent results can be derived using an argument based on the Ehrenfest theorem, as was done in the extensive work of Shindou and Imura.[20]

IV. CONSTANT ELECTRIC FIELD

We will examine first the case of a constant uniform electric field acting on two degenerate bands. We choose a gauge such that the scalar electric potential need not be included in the Hamiltonian, and the electric field is represented purely by the vector potential **A**. With experiment in mind, we take $\mathbf{E}=(0,0,E)$, modeling a gate field, and study its effect on transport in the xy-plane.

A. Electrical spin separation

We choose as an example the spherical four-band model:

$$\hat{H}_{Lutt} = \frac{\hbar^2}{2m}\left[\left(\gamma_1+\frac{5}{2}\gamma_2\right)k^2 - 2\gamma_2(\mathbf{k}\cdot\hat{\mathbf{J}})^2\right], \tag{24}$$

where $\hat{\mathbf{J}}$ is the total angular momentum operator, m is the bare electron mass and γ_1 and γ_2 are material-specific parameters. The wave-functions are eigenstates of the helicity operator $\mathbf{k}\cdot\hat{\mathbf{J}}$ and have the form $|u_m\rangle=e^{-i\phi J_z}e^{-i\theta J_y}|m\rangle$ where $|m\rangle$ are eigenstates of the orbital angular momentum operator J_z while θ and ϕ are the polar and azimuthal angles of the wave-vector, respectively. We shall treat the twofold degenerate heavy and light hole manifolds separately and we shall denote the probability amplitudes in the heavy hole subspace by η^H and those in the light hole subspace by η^L.

In these subspaces, the equations of motion for the probability amplitudes take the form

$$i\hbar\frac{d\eta^H}{dt} = (\mathcal{H}^H - eE\mathcal{R}^{zH})\eta^H,$$

$$i\hbar\frac{d\eta^L}{dt} = (\mathcal{H}^L - eE\mathcal{R}^{zL})\eta^L, \tag{25}$$

where the superscripts H and L represent restrictions to the heavy and light hole subspaces, respectively. The reciprocal-space connection matrix $\mathcal{R}$ is given by the following expression:

$$\mathcal{R} = \frac{\partial\theta}{\partial\mathbf{k}}J^y + \frac{\partial\phi}{\partial\mathbf{k}}(J^z\cos\theta - J^x\sin\theta). \tag{26}$$

In the heavy hole sector $\mathcal{R}^z=0$ and the bands decouple, therefore no spin separation can be achieved electrically in the heavy hole manifold. Henceforth we shall concentrate only on the light hole manifold, where the connection matrix $\mathcal{R}^z=-(k_\perp/k^2)\sigma^y$ has off-diagonal elements only, with $\mathbf{k}_\perp=(k_x,k_y)$ and σ^y a Pauli spin matrix. We shall suppress the index L in what follows.

The equations of motion for the position and wave vector are (suppressing the c index):

$$\hbar\dot{\mathbf{k}} = e\mathbf{E}, \quad \hbar\dot{\mathbf{r}} = \frac{\partial\varepsilon_l}{\partial\mathbf{k}} - e\mathbf{E}\times\eta^\dagger\mathcal{F}\eta, \tag{27}$$

in which $\mathbf{k}_0$ is the initial value of $\mathbf{k}$, $\varepsilon_l=\hbar^2k^2/2m_l$ is the light hole energy, m_l is the light hole effective mass, and the curvature $\mathcal{F}=\frac{3}{2}(\mathbf{k}/k^3)\sigma^z$. The wave vector equation of motion is readily integrated to give $\mathbf{k}=\mathbf{k}_0+(e\mathbf{E}t/\hbar)$. Since the Berry curvature is parallel to $\mathbf{k}$, there are two limiting cases to consider: the case $\mathbf{k}_0//\mathbf{E}$ is trivial because the curvature correction vanishes and the bands decouple, so we will focus on the more interesting case $\mathbf{k}_0\perp\mathbf{E}$.

The equations of motion can be solved exactly. η is given by

$$\eta = \begin{pmatrix}\eta_1^{(0)}\cos\alpha + \eta_2^{(0)}\sin\alpha\\ \eta_2^{(0)}\cos\alpha - \eta_1^{(0)}\sin\alpha\end{pmatrix}, \tag{28}$$

with the angle $\alpha(\tau)=\arctan((\tau+\cos\theta_0)/\sin\theta_0)-((\pi/2)-\theta_0)$, where we have introduced the dimensionless time $\tau=eEt/\hbar k_0$ and θ_0 is the polar angle of $\mathbf{k}_0$, and where $\eta_i^{(0)}$ are the values of η at $\tau=0$.

In this system, the contraction $\langle\hat{\sigma}^i\rangle\equiv\eta^\dagger\hat{\sigma}^i\eta$ (with $i=1,2,3$) is the expectation value of the components of the pseudo-spin. Its components evolve in time as

$$\langle\hat{\sigma}^1\rangle = \langle\hat{\sigma}^1\rangle_{\tau=0}\cos 2\alpha - \langle\hat{\sigma}^3\rangle_{\tau=0}\sin 2\alpha,$$

$$\langle\hat{\sigma}^2\rangle = \langle\hat{\sigma}^2\rangle_{\tau=0},$$

$$\langle\hat{\sigma}^3\rangle = \langle\hat{\sigma}^3\rangle_{\tau=0}\cos 2\alpha + \langle\hat{\sigma}^1\rangle_{\tau=0}\sin 2\alpha. \tag{29}$$

The electric field therefore only rotates the 1 and 3 components of the pseudo-spin into combinations of each other, while the 2 component remains unaffected. To understand the significance of these results we will examine a concrete example, taking initially a positive helicity eigenstate so that $\eta_1^{(0)}=1$, $\eta_2^{(0)}=0$, and fixing the initial wave vector along the x-axis such that $\mathbf{k}_0=k_0\hat{\mathbf{x}}$, which means that $\theta_0=\pi/2$. The full time evolution of the pseudo-spin components are

$$\langle\hat{\sigma}^1\rangle = \langle\hat{\sigma}^1\rangle_{\tau=0}\frac{1-\tau^2}{1+\tau^2} - \langle\hat{\sigma}^3\rangle_{\tau=0}\frac{2\tau}{1+\tau^2},$$

$$\langle\hat{\sigma}^2\rangle = \langle\hat{\sigma}^2\rangle_{\tau=0},$$

$$\langle\hat{\sigma}^3\rangle = \langle\hat{\sigma}^3\rangle_{\tau=0}\frac{1-\tau^2}{1+\tau^2} + \langle\hat{\sigma}^1\rangle_{\tau=0}\frac{2\tau}{1+\tau^2}.$$

As $\tau\to\infty$, α reaches the limiting value of $\pi/2$ and the components of the pseudo-spin become

$$\langle\hat{\sigma}^1\rangle = -\langle\hat{\sigma}^1\rangle_{\tau=0}, \quad \langle\hat{\sigma}^2\rangle = \langle\hat{\sigma}^2\rangle_{\tau=0},$$

$$\langle\hat{\sigma}^3\rangle = -\langle\hat{\sigma}^3\rangle_{\tau=0}. \tag{30}$$

Thus the 1 and 3 components of the pseudo-spin are reversed while the 2 component is conserved.

The time evolution of the wave vector is described entirely by means of the time evolution of the angle θ, which is most conveniently expressed as

$$\cos\theta = \frac{k_z}{k} = \frac{\frac{eEt}{\hbar}}{\sqrt{k_0^2+\left(\frac{eEt}{\hbar}\right)^2}} = \frac{\tau}{\sqrt{1+\tau^2}},$$

$$\sin\theta = \frac{k_\perp}{k} = \frac{k_0}{\sqrt{k_0^2+\left(\frac{eEt}{\hbar}\right)^2}} = \frac{1}{\sqrt{1+\tau^2}}.$$

Therefore, initially we have $\cos\theta=0$ and $\sin\theta=1$ while as $\tau\to\infty$, $\cos\theta\to 1$ and $\sin\theta\to 0$.

The expectation value of a spin component operator $\hat{s}^\alpha$ in the wave packet $|w\rangle$ is given by $\langle w|\hat{s}^\alpha|w\rangle = \eta^\dagger s^\alpha\eta$, where $s_{ij}^\alpha = \langle u_i|\hat{s}^\alpha|u_j\rangle$. The time evolution of the spin of one electron can thus be found by knowing the time evolution of its pseudo-spin. Since our goal is to separate spins of opposite orientations, it is sufficient to know only the value of the pseudo-spin. The bands being spin-split, holes with pseudo-spin up also have spin up and holes with pseudo-spin down have spin down. However, it is instructive to follow the motion of the spin as time progresses, as well as the time evolution of the helicity. The expectation values of $\hat{s}^x$, $\hat{s}^y$, and $\hat{s}^z$ are

$$\langle\hat{s}^x\rangle = \frac{\hbar}{3}\left(\frac{1}{2}\sin\theta\cos\phi\langle\hat{\sigma}^3\rangle + \cos\theta\cos\phi\langle\hat{\sigma}^1\rangle - \sin\phi\langle\hat{\sigma}^2\rangle\right),$$

$$\langle\hat{s}^y\rangle = \frac{\hbar}{3}\left(\frac{1}{2}\sin\theta\sin\phi\langle\hat{\sigma}^3\rangle + \cos\theta\sin\phi\langle\hat{\sigma}^1\rangle + \cos\phi\langle\hat{\sigma}^2\rangle\right),$$

$$\langle\hat{s}^z\rangle = \frac{\hbar}{3}\left(\frac{1}{2}\cos\theta\langle\hat{\sigma}^3\rangle - \sin\theta\langle\hat{\sigma}^1\rangle\right). \tag{31}$$

We assume the carriers have been polarized (by optical means, for example as done in the experiments of Malajovich *et al.*,[12,21] although those utilized electrons) so that $\eta^{(0)}$ is either $\binom{1}{0}$ or $\binom{0}{1}$. Therefore, the initial expectation values of $\{\hat{\sigma}^i\}$ are

$$\langle\hat{\sigma}^1\rangle_\uparrow = \langle\hat{\sigma}^1\rangle_\downarrow = 0, \quad \langle\hat{\sigma}^2\rangle_\uparrow = \langle\hat{\sigma}^2\rangle_\downarrow = 0,$$

$$\langle\hat{\sigma}^3\rangle_\uparrow = -\langle\hat{\sigma}^3\rangle_\downarrow = 1,$$

and the initial spin expectation values are given by

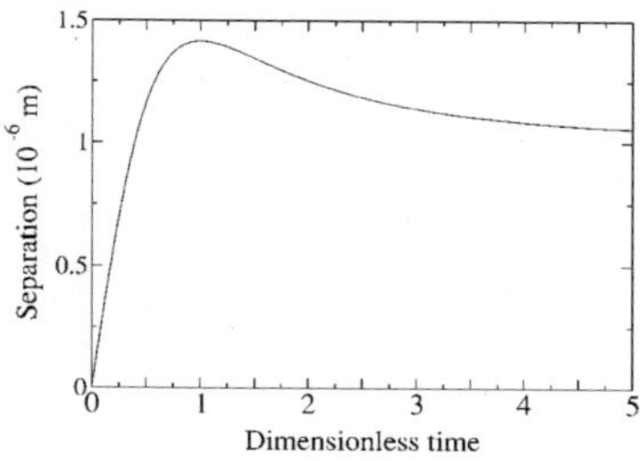

FIG. 1. Separation along the y-direction between light holes of opposite helicities as a function of τ, the dimensionless time. The maximum separation occurs for $\tau=1$.

$$\langle\hat{s}^x\rangle_0 = \pm\frac{\hbar}{6}, \quad \langle\hat{s}^y\rangle_0 = \langle\hat{s}^z\rangle_0 = 0.$$

It can then be easily seen that the y component of the spin is zero at all times. Substituting for θ and $\{\hat{\sigma}^i\}$ in Eq. (30), we obtain the time evolution of the other two spin components

$$\langle\hat{s}^x\rangle = \frac{\hbar}{3}\frac{1-5\tau^2}{2(1+\tau^2)^{3/2}}, \quad \langle\hat{s}^z\rangle = \frac{\hbar}{3}\frac{\tau(5-\tau^2)}{2(1+\tau^2)^{3/2}}.$$

As $\tau\to\infty$, the expectation values of the spin components are

$$\langle\hat{s}^x\rangle = \langle\hat{s}^y\rangle = 0, \quad \langle\hat{s}^z\rangle = \mp\frac{\hbar}{6}.$$

The spin in this case is not conserved. However, a closer look at (30) reveals that $\langle\hat{s}^x\rangle$, $\langle\hat{s}^y\rangle$, and $\langle\hat{s}^z\rangle$ cannot be obtained from $\langle\hat{\sigma}^1\rangle$, $\langle\hat{\sigma}^2\rangle$, and $\langle\hat{\sigma}^3\rangle$ by a rotation, as the matrix describing the transformation is not unitary. The reason for this is evident from (30), where it is seen that the spin cannot be obtained by a rotation of the pseudo-spin. Therefore, one should not think of the projection of the spin onto the light-hole subspace as a vector.

Finally, the helicity is given by

$$\chi = \frac{\mathbf{k}\cdot\langle\hat{\mathbf{s}}\rangle}{k} = \frac{k_x\langle\hat{s}^x\rangle + k_z\langle\hat{s}^z\rangle}{k} = \frac{\hbar}{6}\langle\hat{\sigma}^3\rangle = \pm\frac{\hbar}{6}\frac{1-\tau^2}{1+\tau^2}. \tag{32}$$

The helicity is proportional to the expectation value of the third component of the pseudo-spin. It is therefore not conserved for the light holes in an electric field. This conclusion has also been reached by Jiang *et al.*[22]

The $\mathbf{r}$ equation of motion can be integrated to give the trajectories of the carriers:

$$\mathbf{r} = \frac{\hbar^2 k_0^2}{eEm_l}\left(\tau\hat{\mathbf{x}} + \frac{\tau^2}{2}\hat{\mathbf{z}}\right) - \frac{\tau(3+\tau^2)(|\eta_1^{(0)}|^2 - |\eta_2^{(0)}|^2)}{2k_0(1+\tau^2)^{3/2}}\hat{\mathbf{y}}. \tag{33}$$

We have omitted a term proportional to $\eta_1^{(0)}\eta_2^{(0)}$ since in our setup either one of them will be zero. The second term in Eq. (32) will have opposite signs for the carriers with η initially up and those with η initially down. Therefore, these carriers will be separated in the y-direction. From the above and Fig. 1 it can be seen that the maximum separation in the y-direction occurs at $\tau=1$ while as $\tau\to\infty$ this separation tends to $1/k_0$.

B. Experimental observation

We discuss an experimental setup in which the effect we have described can be measured. We propose using a three-dimensional semiconductor slab containing a nondegenerate hole gas. The sample must be clean in order for the hole spin relaxation time to be long, specifically of the order of picoseconds. Carriers are excited optically from the conduction band into the valence bands by using a laser beam. Provided the laser beam is sharp, only a narrow range of k-space will be excited around $\mathbf{k}=0$. The optically excited holes will have wave vectors lying in a narrow spot about the origin. We assume they have been excited into a state of definite spin. Both light and heavy holes are excited but, as shown in the previous section, the heavy holes do not separate according to spin under the action of an electric field. A source and a drain will be positioned along the x-direction on the two faces of the sample while a gate terminal will be present on top. After the optical excitation, the magnitude of the holes' wave vector can be increased by applying a source-drain field E_x in the form of a picosecond pulse, which will accelerate the carriers along the x-axis, its magnitude tuned to ensure k_0 has the desired value. This source-drain field provides an additional advantage. In the process of optical excitation electrons as well as holes will be excited in the sample and the field which drives the holes one way will drive the electrons the other way, ensuring that the effect observed is indeed due to holes. By adjusting the magnitude of the source-drain electric field pulse the initial wave vector k_0 of the holes incident upon the interface is tunable over several orders of magnitude. We will choose a source-drain electric field in such a way that the wave-vector $\mathbf{k}_0$ will have an x-component which overwhelms the y- and z-components. We will also choose the magnitude of k_0 to be approximately $1/b$, where b is the real-space thickness of the laser beam. The reason for this is that in the limit of large τ the spins are separated by a distance of approximately $1/k_0$, therefore the separation of the spins will be approximately the same as the width of the laser beam. Once excited the carriers will be subjected to the action of the gate field E along z, which will lead to the separation of spins as described above. The spin accumulation at the other end of the sample can be measured by Faraday or Kerr rotation. It will be position dependent along the y-direction, that is, as one moves along y the spin-z polarization will change sign.

We take the dimensions of the slab to be 50 nm$\times$5 μm $\times$5 μm and the width of the laser beam is taken as 1 μm. The optically excited holes will be accelerated until their wave vector reaches the value of $k_0=10^6$ m^{-1}. For a source-drain field E_x of 500 V m^{-1} and a light hole mass of $0.1m_0$, where m_0 is the bare electron mass, the distance traveled by the light holes along the x-axis will be $\hbar^2k_0^2/2m_leE_x$ $=7.2$ nm. This will happen after a time of 1.25 ps, which can be achieved in samples in which the holes have longer spin lifetimes. Therefore, the source-drain field must be a 1.25 ps pulse of amplitude 500 V m^{-1}.

We will take the gate electric field E=25 000 V m^{-1}. If one waits for the value of τ to reach 50, then the magnitude of the spin polarization along the z-direction will be approximately $\hbar/6$ while along the x-direction it will be negligible. The separation between the carriers with spin-z up and spin-z down will be approximately 1 μm, which is the same as the real-space width of the laser beam and thus observable. The waiting time will be approximately 1.3 ps. Finally, the distances traveled in the x- and z-directions under the action of the gate electric field are 35 nm and 850 nm, respectively.

This phenomenon is similar to effects such as the spin Hall effect since the carriers with different helicities are separated in the xy-plane by the electric field normal to the plane.

V. CONSTANT MAGNETIC FIELD

When a constant uniform magnetic field is present, the gradient correction to the degenerate part of the Hamiltonian gives rise to an energy correction which takes the form

$$\Delta_d = -\mathbf{M}\cdot\mathbf{B}. \tag{34}$$

$\mathbf{M}$, which is identified with the intrinsic magnetic moment of the wave packet,[16,23,24] is given by the expression

$$\begin{aligned}\mathbf{M} &= \pm\frac{e}{2}\Re\langle u|\hat{\mathbf{v}}\times\left(i\frac{\partial}{\partial\mathbf{k}}-\mathbf{R}\right)|u\rangle \\ &= \pm\frac{e}{2}\Re\left(\eta_i^*\langle u_i|\hat{\mathbf{v}}\times\left(i\frac{\partial}{\partial\mathbf{k}}-\mathbf{R}\right)|u_j\rangle\eta_j\right),\end{aligned} \tag{35}$$

where the sign is negative for electrons and positive for holes and $\Re$ stands for the real part. The operator $\hat{\mathbf{v}}=(1/\hbar)$ $\times(\partial\hat{H}_d/\partial\mathbf{k})$ is the velocity operator corresponding to the degenerate part of the Hamiltonian. The form of Δ_d shows that it can be regarded as a correction to the Zeeman term. Written explicitly in component form and restricting our attention to holes, the magnetic moment is

$$M^\alpha = -\frac{e}{4}\epsilon^{\alpha\beta\gamma}\eta_i^*\langle u_i|\left\{\hat{v}_\beta,\left(i\frac{\partial}{\partial k_\gamma}-R^\gamma\right)\right\}|u_j\rangle\eta_j. \tag{36}$$

$\epsilon^{\alpha\beta\gamma}$ represents the antisymmetric tensor. It is straightforward to prove that

$$\eta_i^*\langle u_i|\hat{\mathbf{v}}|u_j\rangle\eta_j \equiv \eta^\dagger\mathbf{v}\,\eta = \frac{1}{\hbar}\frac{\partial\varepsilon}{\partial\mathbf{k}}, \tag{37}$$

in which $\mathbf{v}_{ij}=\langle u_i|\hat{\mathbf{v}}|u_j\rangle$. Therefore the second term in $\mathbf{M}$ is

$$\mathbf{M}_2 = \frac{e}{2}\eta^\dagger\mathbf{v}\,\eta\times\mathbf{R} = -\frac{e}{2\hbar}\frac{\partial\varepsilon}{\partial\mathbf{k}}\times\mathbf{R}. \tag{38}$$

The first term is

$$\mathbf{M}_1 = -\frac{e}{2\hbar}\frac{\partial\varepsilon}{\partial\mathbf{k}}\times\mathbf{R} - \frac{e}{2}\Re\sum_{i,j}^{in}\sum_{l}^{out}\eta_i^*\mathbf{v}_{il}\times\mathcal{R}_{lj}\eta_j, \tag{39}$$

where "out" means the sum runs over all bands outside the degenerate subspace, that is $l\neq i,j$. The first term exactly cancels $\mathbf{M}_2$, so the final result is

$$\mathbf{M} = -\frac{e}{2}\Re\sum_{i,j}^{in}\sum_{l}^{out}\eta_i^*\mathbf{v}_{il}\times\mathcal{R}_{lj}\eta_j. \qquad (40)$$

Thus the magnetic moment can be expressed purely in terms of matrix elements connecting the degenerate subspace to bands outside the subspace.

We take as an example once again the light-hole manifold of the four-band Luttinger model in the spherical approximation in the presence of a constant uniform magnetic field. The Hamiltonian in this case is

$$\hat{H} = \hat{H}_{Lutt} - \frac{ge}{m}\hat{\mathbf{S}}\cdot B, \qquad (41)$$

where $\hat{H}_{Lutt}$ has been defined in (24). The first part of the Hamiltonian is $\hat{H}_d$ while the Zeeman term is $\hat{H}_n$. The Zeeman interaction between the spin and the magnetic field does not contribute to the velocity operator and therefore the magnetic moment. The light-hole intrinsic magnetic moment in the spherical four-band model is given by the following expression:

$$\mathbf{M} = \frac{3e\hbar\gamma_2\hat{\mathbf{k}}}{2m}\langle\hat{\sigma}^3\rangle. \qquad (42)$$

The magnetic moment is proportional to the expectation value of the third component of the pseudo-spin and therefore to the helicity, as shown in (31). Depending on the weight of each band in the wave packet the intrinsic magnetic moment can be positive or negative and if the bands are equally represented it will be zero.

ACKNOWLEDGMENTS

Thanks are due to Z. F. Jiang and C. P. Chuu for enlightening discussions. D.C. was supported by the NSF under Grant No. DMR-0404252. Y.Y. acknowledges the support of the National Science Foundation of China under Grant No. 10404035 and that of NSF Grant No. DMR-0306239. Q.N. was supported by the DOE under Grant No. DE-FG03-02ER45958 and the Welch Foundation of Texas.

[1] Yu. A. Bychkov and E. I. Rashba, JETP Lett. **39**, 66 (1984).

[2] See, for example, *Semiconductor Laser Fundamentals: Physics of the Gain Materials*, edited by W. W. Chow and S. W. Koch (Springer, Berlin, 1999), Chap. 5.

[3] L. C. Lew Yan Voon, M. Willatzen, M. Cardona, and N. E. Christensen, Phys. Rev. B **53**, 10703 (1996).

[4] Y. Qi and S. Zhang, Phys. Rev. B **65**, 214407 (2002); Y. Qi and S. Zhang, *ibid.* **67**, 052407 (2003).

[5] M. Q. Weng and M. W. Wu, J. Appl. Phys. **93**, 410 (2003).

[6] M. Q. Weng, M. W. Wu, and Q. W. Shi, Phys. Rev. B **69**, 125310 (2004).

[7] For a review of spin injection in metals and semiconductors, see M. Johnson, Semicond. Sci. Technol. **17**, 298 (2002), and references therein.

[8] For a review of spin injection into semiconductors, see G. Schmidt and L. W. Molenkamp, Semicond. Sci. Technol. **17**, 310 (2002), and references therein.

[9] There is already a vast theoretical literature on spin currents from the past few years. The following is a selection: D. Culcer, J. Sinova, N. A. Sinitsyn, T. Jungwirth, A. H. MacDonald, and Q. Niu, Phys. Rev. Lett. **93**, 046602 (2004); S. Murakami, N. Nagaosa, and S.-C. Zhang, Science **301**, 1348 (2003); S. Murakami, N. Nagaosa, and S.-C. Zhang, Phys. Rev. B **69**, 235206 (2004); S. Zhang and Z. Yang, cond-mat/0407704 (2004); T. Damker, H. Bottger, and V. V. Bryksin, Phys. Rev. B **69**, 205327 (2004); L. Hu, J. Gao, and S.-Q. Shen, *ibid.* **70**, 235323 (2004); E. G. Mishchenko, A. V. Shytov, and B. I. Halperin, Phys. Rev. Lett. **93**, 226602 (2004); A. G. Aronov, Yu. B. Lyanda-Geller, and G. E. Pikus, Sov. Phys. JETP **73**, 537 (1991); J. I. Inoue, G. E. W. Bauer, and L. W. Molenkamp, Phys. Rev. B **67**, 033104 (2003).

[10] Recent experiments on spin generation include: Y. Kato, R. C. Myers, A. C. Gossard, and D. D. Awschalom, Phys. Rev. Lett. **93**, 176601 (2004); Y. Kato, R. C. Myers, A. C. Gossard, and D. D. Awschalom, Nature (London) **427**, 50 (2004); J. Stephens, J. Berezovsky, J. P. McGuire, L. J. Sham, A. C. Gossard, and D. D. Awschalom, cond-mat/0401197 (2004). Theoretical efforts include: D. Culcer, Y. Yao, A. H. MacDonald, and Q. Niu, Phys. Rev. B **72**, 045215 (2005); L. I. Magarill, A. V. Chaplik, and M. V. Entin, Semiconductors **35** (9), 1081 (2001); V. M. Edelstein, Solid State Commun. **73**, 233 (1990).

[11] O. Bleibaum, Phys. Rev. B **69**, 205202 (2004); M. Q. Weng, M. W. Wu, and L. Jiang, *ibid.* **69**, 245320 (2004); S. Saikin *et al.*, J. Appl. Phys. **94**, 1769 (2003); Y. Pershin, Physica E (Amsterdam) **23**, 226 (2004); S. Saikin *et al.*, J. Phys.: Condens. Matter **16**, 5071 (2004).

[12] I. Malajovich, J. M. Kikkawa, D. D. Awschalom, J. J. Berry, and N. Samarth, Phys. Rev. Lett. **84**, 1015 (2000).

[13] R. Fiederling, P. Grabs, W. Ossau, G. Schmidt, and L. W. Molenkamp, Appl. Phys. Lett. **82**, 2160 (2003).

[14] N. W. Ashcroft and N. D. Mermin, *Solid State Physics*, 2nd ed. (Saunders College, New York, 1976).

[15] J. Callaway, *Quantum Theory of the Solid State* (Academic, New York, 1976).

[16] G. Sundaram and Q. Niu, Phys. Rev. B **59**, 14915 (1999).

[17] E. G. Mishchenko and B. I. Halperin, Phys. Rev. B **68**, 045317 (2003); D. Culcer, A. H. MacDonald, and Q. Niu, *ibid.* **68**, 045327 (2003); Y. S. Gui, C. R. Becker, J. Liu, V. Daumer, V. Hock, H. Buhmann, and L. W. Molenkamp, Europhys. Lett. **65** (3), 393 (2004); see also Refs. 2 and 3.

[18] Y. Ohno, R. Terauchi, T. Adachi, F. Matsukura, and H. Ohno, Phys. Rev. Lett. **83**, 4196 (1999); Y. Ohno, D. K. Young, B. Beschoten, F. Matsukura, H. Ohno, and D. D. Awschalom, Nature (London) **402**, 790 (1999); V. Daumer, I. Golombek, M. Gbordzoe, E. G. Novik, V. Hock, C. R. Becker, H. Buhmann, and L. W. Molenkamp, Appl. Phys. Lett. **83**, 1376 (2003).

[19] S. Datta and B. Das, Appl. Phys. Lett. **56**, 665 (1990).
[20] R. Shindou and K. Imura, Nucl. Phys. B **720**, 399 (2005).
[21] I. Malajovich, J. J. Berry, N. Samarth, and D. D. Awschalom, Nature (London) **411**, 770 (2001).
[22] Z. F. Jiang, R. D. Li, S.-C. Zhang, and W. M. Liu, Phys. Rev. B **72**, 045201 (2005).
[23] R. B. Diener, A. M. Dudarev, G. Sundaram, and Q. Niu, cond-mat/0306184 (2003).
[24] M.-C. Chang and Q. Niu, Phys. Rev. B **53**, 7010 (1996).

3

Pumping, polarization, and orbital magnetization

Quantization of particle transport

D. J. Thouless
Department of Physics, FM-15, University of Washington, Seattle, Washington 98195
(Received 4 February 1983)

The integrated particle current produced by a slow periodic variation of the potential of a Schrödinger equation is evaluated. It is shown that in a finite torus the integral of the current over a period can vary continuously, but in an infinite periodic system with full bands it must have an integer value. This quantization of particle transport is used to classify the energy gaps in a one-dimensional system with competing or incommensurate periods. It is also used to rederive Prange's results for the fractional charge of a soliton.

I. INTRODUCTION

In this paper the effect of an adiabatic change of the potential on an electron or a system of electrons is considered. The question that is posed is "If the potential is changed slowly in such a way that it returns to its starting value in time T, is the integrated current of electrons across a boundary quantized?" This question, which is formulated in Sec. II, is answered negatively for electrons in a finite torus in Sec. III. In Sec. IV it is shown that for electrons in filled bands in an infinite periodic system the particle transport is quantized. This result is closely related to our earlier work on the quantized Hall effect in a periodic potential.[1] The result may be applicable to problems of sliding charge-density waves in a solid, but that application is not considered here. In Sec. V the result is used to classify the gaps in a system with competing or incommensurate periodicities. In Sec. VI the problem of one-dimensional solitons is considered.[2,3] It is shown how the quantization of charge transport leads to Prange's theorem for soliton charge.[4]

II. ADIABATIC TRANSPORT BY A POTENTIAL

In this paper the motion of particles which satisfy the Schrödinger equation with a slowly varying time-dependent potential $V(\vec{r},t)$ is considered. The potential is taken to be periodic both in time t with period T and in one of the space variables x with period L. For example, we could consider a potential of the form

$$V(\vec{r},t)=V_0(\vec{r})+V_1(x-vt,y,z)\ , \tag{2.1}$$

where V_0 and V_1 have the common period L and v is small. In this case L/v must be a multiple of T. The particle current integrated over the period T gives the total number of particles transported in a period.

Two versions of the problem are examined. In the first case the system is supposed to have periodic boundary conditions in x, so that the solutions satisfy

$$\psi(x+L,y,z)=\psi(x,y,z) \tag{2.2}$$

for all times, while in the second case the system is taken to be infinite in x. In both cases the potential is taken to confine the particles close to the x axis, so the first case is equivalent to a torus of circumference L, while the second case gives an infinite one-dimensional periodic system whose unit cell is of length L.

Since the potential is slowly varying the adiabatic approximation is used for the wave functions—in fact the validity of the adiabatic approximation is the criterion for how slow the potential variation must be. Expansion of the wave functions in terms of the instantaneous normalized eigenfunctions $\psi_j(t)$, with eigenvalues $\epsilon_j(t)$, gives the solution close to $\psi_0(t)$ as[5]

$$|\psi(t)\rangle\approx\exp\left[-(i/\hbar)\int^t\epsilon_0(t')dt'\right]\left[|\psi_0(t)\rangle+i\hbar\sum_{j\neq0}|\psi_j(t)\rangle(\epsilon_j-\epsilon_0)^{-1}\langle\psi_j(t)|\dot\psi_0(t)\rangle\right] \tag{2.3}$$

to first order in the time derivatives, where the phase of $\psi_0(t)$ is chosen in such a way that its time derivative $\dot\psi_0$ is orthogonal to ψ_0. To the same order the particle current density produced by the moving potential is

$$\frac{\hbar^2}{2m}\sum_{j\neq 0}\frac{1}{(\epsilon_j-\epsilon_0)}\left[\langle\psi_j\,|\,\dot{\psi}_0\rangle\left[\psi_0^*\frac{\partial\psi_j}{\partial x}-\frac{\partial\psi_0^*}{\partial x}\psi_j\right]+\langle\dot{\psi}_0\,|\,\psi_j\rangle\left[-\psi_j^*\frac{\partial\psi_0}{\partial x}+\frac{\partial\psi_j^*}{\partial x}\psi_0\right]\right]. \tag{2.4}$$

For the problem of a torus (periodic boundary conditions) this can be integrated over all space to get the current as

$$J_0=\frac{\hbar^2}{mL}\sum_{j\neq 0}\frac{1}{(\epsilon_j-\epsilon_0)}\left[-\langle\psi_j\,|\,\dot{\psi}_0\rangle\left\langle\frac{\partial\psi_0}{\partial x}\,\middle|\,\psi_j\right\rangle-\langle\dot{\psi}_0\,|\,\psi_j\rangle\left\langle\psi_j\,\middle|\,\frac{\partial\psi_0}{\partial x}\right\rangle\right]$$
$$=\frac{\hbar^2}{mL}\left[\left\langle\frac{\partial\psi_0}{\partial x}\,\middle|(\epsilon_0-H)^{-1}\,|\,\dot{\psi}_0\right\rangle+\left\langle\dot{\psi}_0\,|\,(\epsilon_0-H)^{-1}\middle|\,\frac{\partial\psi_0}{\partial x}\right\rangle\right]. \tag{2.5}$$

After some manipulation this can be written in terms of the Green function as

$$J=-\frac{\hbar^2}{2\pi imL}\oint dE\int dx\int dx'\frac{\partial}{\partial x}G(x,x';E)\frac{\partial}{\partial t}G(x',x;E)\,, \tag{2.6}$$

where the contour goes round the eigenvalue ϵ_0.

For an infinite periodic system with filled bands Eq. (2.6) remains valid if the integration over x is restricted to one period and if the integration over E surrounds the energies of the filled bands only. Equation (2.5) can be replaced by

$$J=\frac{\hbar^2}{2\pi m}\sum_{\lambda\mu}\int dk\,\frac{f_\lambda(1-f_\mu)}{\epsilon_\lambda(k)-\epsilon_\mu(k)}\left[\left\langle\frac{\partial\psi_{\lambda k}}{\partial x}\,\middle|\,\psi_{\mu k}\right\rangle\langle\psi_{\mu k}\,|\,\dot{\psi}_{\lambda k}\rangle+\langle\dot{\psi}_{\lambda k}\,|\,\psi_{\mu k}\rangle\left\langle\psi_{\mu k}\,\middle|\,\frac{\partial\psi_{\lambda k}}{\partial x}\right\rangle\right], \tag{2.7}$$

where k is the Bloch wave number, λ and μ are band indices, f_λ is unity for full bands and zero for empty bands, and the Bloch wave functions $\psi_{\lambda k}$ are normalized in a unit cell.

III. TRANSFER AROUND A TORUS

If the wave functions in Eq. (2.5) satisfy the periodic boundary conditions (2.2) the integral of the current is a continuously variable quantity. This can be shown by evaluation of the integral for special cases. For example, in a pure one-dimensional problem with a potential of the form

$$V(x,t)=-A\delta(x-vt)\quad\text{for } 0<t<L/v \tag{3.1}$$

an explicit expression for the Green function can be obtained. The qualitative nature of the result can be studied by taking A to be sufficiently small that perturbation theory is applicable. The eigenfunctions can be written as $\phi_n(x-vt)$, where, for the ground state,

$$\phi_0(x)\approx L^{-1/2}[1+mA(x-\tfrac{1}{2}L)^2/\hbar^2L] \tag{3.2}$$

up to first order in A. The overlap of $\partial\phi_0/\partial x$ with ϕ_n is given by

$$(\sqrt{2}/L)(2mA/\hbar^2L)\int_0^L(x-\tfrac{1}{2}L)\sin(2\pi nx/L)dx$$
$$=-mA\sqrt{2}/\pi\hbar^2n\,. \tag{3.3}$$

The integral of the current J_0 over the period L/v gives the particle transfer as

$$C_0=(\hbar^2/m)\sum_{n=1}^{\infty}4(mA/\pi\hbar^2n)^2 2m(L/2\pi n\hbar)^2$$
$$=(mAL/\hbar^2)^2/45\,. \tag{3.4}$$

For the other states the degeneracy is broken by the potential, so that they come in pairs separated by an energy $2A/L$. The integral of the gradient of one function multiplied by the other is

$$-(4\pi n/L^2)\int_0^1\sin^2(2\pi nx/L)dx=-2\pi n/L\,, \tag{3.5}$$

so the leading term in the integral of the current is

$$C_n=\pm 4\pi^2\hbar^2 n/mAL\,, \tag{3.6}$$

with the plus sign for the lower state and the minus sign for the upper state. This also depends continuously on the strength A of the potential, but diverges as A goes to zero whereas (3.4) tends to zero.

IV. FILLED BANDS IN A PERIODIC SYSTEM

In Eq. (2.7) it is possible to replace

$$[\epsilon_\lambda(k)-\epsilon_\mu(k)]^{-1}\partial/\partial x$$

by $(im/\hbar^2)\partial/\partial k$, as might be expected from the formal relations between x and $i\partial/\partial k$, and between the commutator of x with the Hamiltonian and $\partial/\partial x$. To show this explicitly one can write

$$\psi_k = u_k e^{ikx} , \tag{4.1}$$

where u_k satisfies periodic boundary conditions and the equation

$$\left[\frac{1}{2m} \left[-i\hbar\frac{\partial}{\partial x} + \hbar k \right]^2 - \frac{\hbar^2}{2m} \left[\frac{\partial^2}{\partial y^2} + \frac{\partial^2}{\partial z^2} \right] + V - \epsilon(k) \right] u_k = 0 . \tag{4.2}$$

From this it follows that

$$\frac{\partial u_{\lambda k}}{\partial k} = ig(k)u_{\lambda k} + \sum_{\mu \neq \lambda} u_{\mu k} \frac{1}{\epsilon_\lambda(k) - \epsilon_\mu(k)} \times \frac{\hbar^2}{m} \left\langle u_{\mu k} \left| \left[-i\frac{\partial}{\partial x} + k \right] \right| u_{\lambda k} \right\rangle , \tag{4.3}$$

where $g(k)$ is an arbitrary real function. With Eq. (4.1) this gives

$$\frac{\partial \psi_{\lambda k}}{\partial k} = i[g(k) + x]\psi_{\lambda k} + \sum_{\mu \neq \lambda} \psi_{\mu k} \frac{1}{\epsilon_\lambda(k) - \epsilon_\mu(k)} \times -\frac{i\hbar^2}{m} \left\langle \psi_{\mu k} \left| \frac{\partial \psi_{\lambda k}}{\partial x} \right. \right\rangle . \tag{4.4}$$

Substitution into Eq. (2.7) and integration over t (which eliminates the term involving the matrix element of x) give the particle transfer as

$$C = \frac{i}{2\pi} \sum_\lambda f_\lambda \int_0^T dt \int_0^{2\pi/L} dk \left[\left\langle \frac{\partial \psi_{\lambda k}}{\partial t} \middle| \frac{\partial \psi_{\lambda k}}{\partial k} \right\rangle - \left\langle \frac{\partial \psi_{\lambda k}}{\partial k} \middle| \frac{\partial \psi_{\lambda k}}{\partial t} \right\rangle \right] . \tag{4.5}$$

In the paper of Thouless *et al.*[1] it was argued that such an integral over two parameters in which the Hamiltonian is periodic should give $2\pi i$ times an integer, since the integral is equal to i times the change of phase of the wave function round the perimeter of the region of integration. The integral defines the first Chern class of the mapping of the torus in tk space on the complex projective space of normalized wave functions.[6]

There is a slightly different way of understanding the integral. Rather than allowing ψ_k to be analytic, but possibly multiple-valued functions of t and k on the torus, one can make them single-valued by demanding that the functions at one particular reference point in space be real and positive. This definition introduces singularities at those values of t,k for which ψ is zero at the chosen point. The integral then reduces to a sum of contributions from these singularities, and it can be shown that the contribution of a singularity is $\pm 2\pi i$, with a sign that depends on the sign of $\mathrm{Im}(\psi^* \partial\psi/\partial k)$ at the reference point.

It is also possible to change from the variable k to the variable $E = \epsilon_\lambda(k)$. The integral over a period of k becomes a contour integral over E which surrounds the energy band. This integral can be continued into the complex plane by restricting $|\psi_E\rangle$ to those solutions that satisfy the Floquet condition with a negative real part of the exponent; $\langle \psi_E |$ is then the adjoint solution with a positive real part of the exponent. The particle transfer is

$$C = \frac{i}{2\pi} \int_0^T dt \oint dE \left[\left\langle \frac{\partial \psi}{\partial t} \middle| \frac{\partial \psi}{\partial E} \right\rangle - \left\langle \frac{\partial \psi}{\partial E} \middle| \frac{\partial \psi}{\partial t} \right\rangle \right] . \tag{4.6}$$

For the one-dimensional Schrödinger problem this topological invariant can be expressed in a form that is more transparent. Firstly, ψ_E can only have zeros for E on the real axis, since, from the Floquet condition, a zero at one point implies zeros at any multiple of L from that point, and so implies a solution of the homogeneous eigenvalue problem in a system of length L. Secondly, since $|\psi_E\rangle$ can be constructed by integrating the Schrödinger equation from an arbitrarily large value X of x to small values of x and taking the limit $X \to \infty$, the nodes of $|\psi_E\rangle$ move monotonically toward larger values of x as E is increased. This implies that $\partial\psi/\partial x$ and $\partial\psi/\partial E$ have opposite signs at the nodes, so that if $\partial\psi/\partial t$ and $\partial\psi/\partial E$ have the same sign the node is moving to the right, while if they have opposite signs at a node it is moving to the left. If the contour in Eq. (4.6) crosses the real axis at E_F and at some point below all the bands, then the integral is just the sum of contributions from all those values of t for which there is a node at the reference point of the wave function at E_F; the contribution is $-2\pi i$ for a node moving right and $+2\pi i$ for a node moving left. The particle transfer C is just the net flux of nodes moving through a given point at energy E_F during a period T.

V. COMPETING AND INCOMMENSURATE PERIODS

This method can be used to classify the energy gaps of a one-dimensional Schrödinger equation with competing periods. The potential is taken to have the form

$$V(x)=\sum_i V_i(x/\xi_i-\eta_i) \quad \text{with } V_i(x+1)=V_i(x) \tag{5.1}$$

so that ξ_i is the period of the component i and η_i gives its phase. In the general case, according to the Wigner–von Neumann theorem,[7] an energy gap will not close when a single parameter η_i is varied continuously, so, if the contour of integration crosses the real axis at E_F in an energy gap, then the gap is characterized by an integer t_i which gives the particle transfer when η_i is increased adiabatically by unity; this integer is given by

$$t_i=\frac{1}{2\pi i}\oint dE \int_0^1 d\eta_i \left[\left\langle \frac{\partial\psi}{\partial\eta_i}\middle|\frac{\partial\psi}{\partial E}\right\rangle - \left\langle \frac{\partial\psi}{\partial E}\middle|\frac{\partial\psi}{\partial\eta_i}\right\rangle\right] . \tag{5.2}$$

The path of integration may have to be shifted as η_i is varied to keep it within an energy gap.

If the gap remains open for all values of the phases and if there is a lowest common multiple $\xi=p_i\xi_i$ of the periods, then a uniform translation of the whole potential by a distance ξ is equivalent to successive translations of the η_i by p_i. This gives

$$\sum_i p_i t_i = m , \tag{5.3}$$

where m is the total number of particles in the length ξ; all the numbers in this equation are integers. This is a generalization of Eq. (5) of Ref. 1.

If the gap closes for some values of the η_j, the values of t_i may depend on the other η_j. A closed path that goes through a single-level crossing leads to a half-integer value of C, since the wave function changes sign round such a path.[8]

For incommensurate potentials continuity suggests that these relations should still hold. If the gap remains open for all values of the phases, then the generalization of Eq. (5.3) is

$$\sum_i t_i/\xi_i = \rho , \tag{5.4}$$

where ρ is the average particle density.

VI. SOLITON CHARGES

There are two types of discrete solitons in one-dimensional systems:

(a) solitons such that each soliton must be succeeded by some sort of antisoliton. An example of this is a domain wall in a ferroelectric polarized along the chain. In this example the soliton charge is just twice the component of the dipole moment along the chain divided by the spacing between dipoles. There is no constraint on the value of such a soliton charge.

(b) solitons which can occur in succession without any intervening antisoliton. The solitons in polyacetylene [$(CH)_x$] discussed by Su, Schrieffer, and Heeger[2] are of this sort—they are breaks in the alternating sequence of double and single bonds along the chain of carbon atoms.

Polyacetylene does not have fractionally charged solitons because the molecular orbital states are occupied by two electrons. A model in which the magnetic field is so strong that only one spin state is occupied for the valence electron can be considered. For example, a regular chain of alternating B and Be atoms joined by alternating double and single bonds has solitons of type (a), as is shown in Fig. 1(a), since the soliton illustrated, with two double bonds on a B atom, must be succeeded either by two single bonds on a B atom or by two double bonds on a Be atom. The B-Be pairs carry a dipole moment, so movement of the soliton one unit cell (two atoms) to the right, as shown in Fig. 1(b), involves a charge transfer which is not just one electron going half a unit cell to the right, as the figure suggests; but also there is an additional amount which depends on the polarization of the bonds.

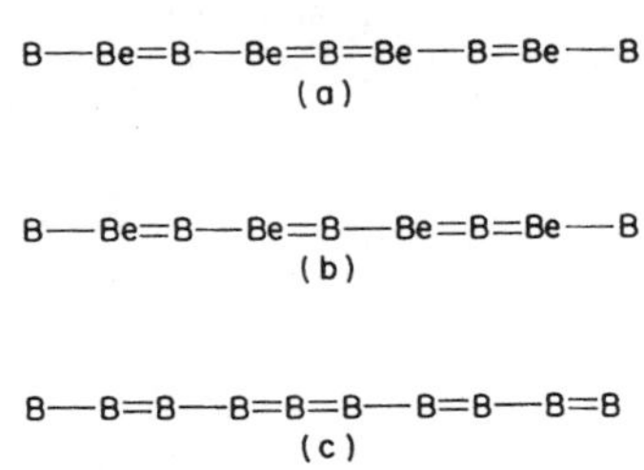

FIG. 1. Possible configurations for polymers with alternating single and double bonds when the magnetic field is so strong that only one spin state of the valence electron is allowed. In (a) and (b) the bonds have an electric dipole moment that is reversed on the two sides of the soliton. In (c) there is no dipole moment, but the units carry a net charge of one unit for each pair of atoms.

In the polyboron shown in Fig. 1(c) there is no polarization of the bonds, but there is a net charge of one proton on each pair of boron atoms. This soliton can be succeeded either by a similar soliton or by the antisoliton with two single bonds. The results of Sec. IV can be applied to such a soliton.

The potentials in which the electrons move are, of course, associated with ionic charges, so movement of the potentials involves also movement of these ionic charges, which are multiples of the proton charge and treated classically. A periodic system is constructed which has a large number N unit cells, each of length a (except near the soliton) and charge m units, with a soliton at the end of each group of N unit cells. The period of the system is

$$L = Na + d , \tag{6.1}$$

so that d is the length taken up by the soliton, and the net charge per period is $Nm + m'$ units, so that m' is the extra charge associated with the soliton. From the theorem of Sec. IV translation of the whole system by a distance L produces transport of an integer number of electrons (and certainly involves transport of an integer number of ionic charges), provided the Fermi energy lies in a band gap, so that m and m' are integers.

A second type of periodic motion of the system that can be considered consists of adiabatic displacement of the soliton one unit cell to the right, followed by uniform displacement of the whole system one unit cell to the left. If q_s is the soliton charge, this process involves the displacement of a charge $q_s - mN - m'$ in each length L a distance a, so the total charge transport, which again must be an integer, is

$$(q_s - mN - m')a/(Na + d) = -m . \tag{6.2}$$

The value of this integer must, of course, be $-m$ since N is large. This leads to the result

$$q_s = m' - md/a . \tag{6.3}$$

For the type (b) solitons fractional charge is only possible if there is a net charge on the chain, and if the fractional part of the charge is equal to minus the charge per unit cell multiplied by the ratio of the length taken by the soliton to the length of a unit cell. These results are identical to those originally derived by Prange.[4]

ACKNOWLEDGMENT

This work was supported in part by the National Science Foundation under Grant No. DMR-79-20785.

[1]D. J. Thouless, M. Kohmoto, P. Nightingale, and M. den Nijs, Phys. Rev. Lett. 49, 405 (1982).

[2]W. P. Su, J. R. Schrieffer, and A. J. Heeger, Phys. Rev. Lett. 42, 1698 (1979); Phys. Rev. B 22, 2099 (1980).

[3]W. P. Su and J. R. Schrieffer, Phys. Rev. Lett. 46, 741 (1981).

[4]R. E. Prange, Phys. Rev. B 26, 991 (1982).

[5]A. Messiah, *Quantum Mechanics* (North-Holland, Amsterdam, 1962), Vol. II, p. 752.

[6]I wish to thank Professor T. E. Duchamp (private communication) for pointing this out to me. See Y. Choquet-Bruhat, C. DeWitt-Morette, and M. Dillard-Bleick, *Analysis, Manifolds, and Physics,* revised ed. (North-Holland, Amsterdam, 1982), p. 393.

[7]J. von Neumann and E. P. Wigner, Z. Phys. 30, 467 (1929).

[8]G. H. Herzberg and H. C. Longuet-Higgins, Discuss. Faraday Soc. 35, 77 (1963); H. C. Longuet-Higgins, Proc. R. Soc. London, Ser. A 344, 147 (1975).

Towards a Quantum Pump of Electric Charges

Q. Niu
Department of Physics, University of California, Santa Barbara, California 93106
(Received 4 December 1989)

Here we give a critical examination of the possibility of realizing a quantum pump of electric charges. The physics is based on the theory of quantum adiabatic particle transport initially due to Thouless. We present theoretical guidelines on the experimental conditions for observing this phenomenon. We argue that potentially very accurate quantization of charge transport or electron current can be achieved, once these conditions are approximately satisfied. An example of the experimental setup is outlined to demonstrate the practical possibilities. Some comments are also made on the scientific significance of such a quantum device.

PACS numbers: 73.20.Dx, 06.20.Hq, 73.40.−c, 85.30.Tv

We have a quantum voltage standard from the Josephson effect, and we also have a quantum resistance standard from the quantum Hall effect.[1] Can we also have an independent quantum standard for electric charges or currents? Here we propose a possible quantum charge pump that has precise control on the amount of pumped charge or electric current. As will be explained below, this idea may be realized through the effect of quantized adiabatic particle transport initially due to Thouless.[2]

Ideally, we consider a one-dimensional noninteracting electron gas in a potential, $U(x,t)$, with spatial and temporal periodicities of a and τ, respectively. We assume zero-temperature, adiabatic time variation of the potential, and an integer number of filled Bloch bands of the instantaneous Hamiltonian,

$$H(t) = -\frac{\hbar^2}{2m}\frac{\partial^2}{\partial x^2} + U(x,t) . \tag{1}$$

Then, according to Thouless, the number of electrons transported through any cross section during time τ is an integer, provided that the Fermi gap[3] remains open at all times. The corresponding dc electric current is then quantized as

$$I = nev , \tag{2}$$

where $v = 1/\tau$ is the operating frequency of the potential.

We will first give a brief exposition of the physics behind this effect, and answer some important questions in the above model. We will then present more realistic considerations, and outline a possible experimental setup. Finally, we will discuss the scientific significance of such a device.

The quantization of particle transport can be easily understood in the simplest case of a rigidly sliding potential $U(x-vt)$, where the temporal period is related to the spatial period as $\tau = a/v$. The adiabatic theorem and the Galilean principle ensure that the electron number density of the filled bands shifts all together with the potential. The integral of the number density over the spatial period l is an integer n for n filled bands,[4] which is also the number of electrons transported in time τ.

In the more general case where the Galilean principle does not hold, the result is not trivial at all. The original proof of quantization was based on topological arguments regarding the phase of a wave function in multiply connected parameter space. Here we present a new and physically more transparent proof.[5] Consider, for simplicity, the case that only the lowest band is filled. The corresponding many-body state is a determinant of Bloch waves, but it can also be written as a determinant of single-particle Wannier states.[6] The particle density can then be written as

$$\rho(x,t) = \sum_{-\infty<j<\infty} |W(x-ja,t)|^2 , \tag{3}$$

where $W(x,t)$ is a Wannier function for the filled band of the instantaneous Hamiltonian $H(t)$. It is convenient to define $W(x,t)$ as an eigenstate of the position operator projected onto the subspace of the Bloch band,

$$\hat{P}(t)\hat{x}\hat{P}(t)W(x,t) = l(t)W(x,t) , \tag{4}$$

where $\hat{P}(t)$ is the projection operator. The eigenposition, $l(t)$, is also the center of mass of the density profile $|W(x,t)|^2$.[7] Such a Wannier function is exponentially localized in space. The other eigenstates of $\hat{P}(t)\hat{x}\hat{P}(t)$ are of the form $W(x-ja,t)$, with their eigenpositions given by $l_j(t) = l(t) + ja$. It can be shown that, if the Fermi gap does not close, $l(t)$ varies continuously with $U(x,t)$.[5] We can therefore follow the positions of the Wannier functions. Since the set, $\{l_j(t)\}$, of eigenpositions must return to its initial configuration when $U(x,t)$ returns its initial form, each Wannier state must have shifted its center of mass by $l(t+\tau) - l(t) = na$. The particle transport, namely, the time integral of the current due to the motion of these localized densities, is just equal to n, thus quantized.

In an article by the author,[8] the integer of particle transport was calculated for a number of nontrivial cases. One important question asked there was the following: Suppose the potential $U(x,t)$ consists of an array of mutually disjoint local potentials, can bounded movements of the local potentials induce global transport of the particles? In other words, can the particle transport be a nonzero integer? One type of local movements considered was coherent positional oscillations as in a traveling sound wave. The answer is yes in general. The other type was phase-coherent amplitude oscillations with a traveling envelope. The answer is also positive in the generic case.

Here we examine an example of the latter type, partly to illustrate ideas, and partly for its relevance to later discussions. Suppose the total potential is given by

$$U(x,t) = \sum_{-\infty < j < \infty} A_j(t) u(x - ja/3) , \tag{5}$$

where $u(x)$ is a local potential (a square barrier, say), with the amplitudes varying coherently in time as

$$A_j(t) = C + C_1 \cos[2\pi(j/3 - t/\tau)] . \tag{6}$$

The spatial period of $U(x,t)$ is a and contains three local potentials. Figure 1 illustrates how the potential progresses in time. It is seen that, after a third of the temporal period $\tau/3$, the potential configuration has been shifted to the right by $a/3$. Each of the Wannier functions in a given Bloch band must have shifted a distance of $ma + a/3$, where m is an integer. The total shift during a whole period τ is $3ma + a$. The corresponding particle transport is then $3m + 1$, a nonzero integer. In fact, if the Fermi gap is mainly determined by the corresponding Fourier component of the potential, then the integer of particle transport is 1, -1, or 0, for the lowest one, two, or three filled bands, respectively.[8] It is easy to understand, by following the positions of the minima of the potential shown in Fig. 1, how the particle density in the lowest band is squeezed to the right.

Having briefly described the physics of quantized charge transport in the ideal situation, we now proceed to consider a number of major issues that one must face in reality. First of all, the formula (2) of the quantized current involves an operational frequency ν. However, as for the voltage standard using the ac Josephson effect, the accuracy in the frequency control can be very high. Thus, the main question is the correction to the integer. From experience in the theoretical development for the integral quantum Hall effect, we expect that a finite Fermi gap will make the electrons inert to all kinds of perturbations: thermal activation, nonadiabatic excitation, static disorder, many-body interaction, finite size, etc. The correction due to thermal activation will be of the order of $e^{-\Delta/kT}$, which can be very small if the thermal energy gets smaller than the Fermi gap Δ. The correction due to nonadiabatic excitations is an exponential of $-\Delta/h\nu$,[9] and can therefore be easily controlled. It has also been shown by Niu and Thouless[10] that quantization of charge transport is not affected by disorder and many-body interaction, so long as the adiabatic ground state of the system is separated from the excitations by a finite-energy gap. Finally, current leads at the ends of the system will certainly close the Fermi gap, but the states in the gap are localized in the leads. In order for these states to have an effect, they must tunnel all the way from one end to the other of the system. The correction to quantization is therefore of the order of $e^{-L/l}$, where L is the system size, and l is the localization length of the edge states. In summary, once the system lies near the perfect situation, every conceivable correction to the quantization is either identically zero, or exponentially small in the perturbation. Thus, the effect of quantized charge transport can, in principle, lead to a potentially very accurate standard for electric charge and current measurements.

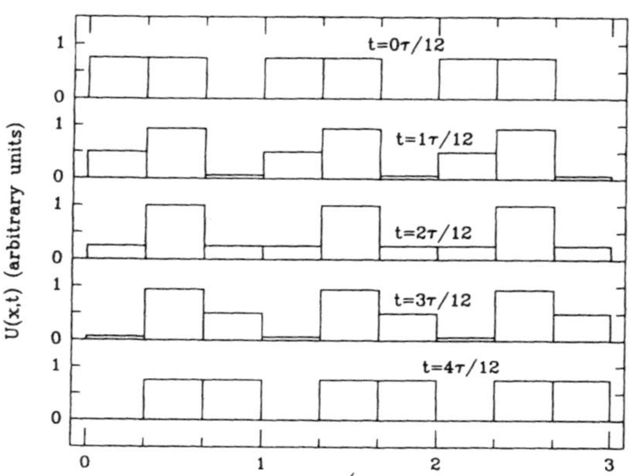

FIG. 1. The time evolution of the potential $U(x,t)$ in Eq. (5) during the first third temporal period.

The analogy with the quantum Hall effect goes further. It was shown by Laughlin[11] that the integer of the quantized Hall conductance is equal to the number of electrons transported from one edge of a cylinder to the other when a magnetic flux through the center of the cylinder is adiabatically increased by a flux quantum. The number of transported electrons is an integer, because, in the bulk, the extended states below the Fermi energy map to one another, and the localized states about the Fermi energy map to themselves. Consider in our case a situation that the Fermi energy lies in a range of localized states, which map to themselves when the potential evolves adiabatically back to its initial configuration. We will call such states impurity-locked states, and their energy range the impurity-locked mobility gap.[12] The particle transport should still be quantized by Laughlin's argument. Since the impurity-locked states contribute nothing to the particle transport, thermal or nonadiabatic smearing of the Fermi energy within the impurity-locked mobility gap should introduce no correction to the integer of particle transport. Corrections due to excitations beyond the impurity-

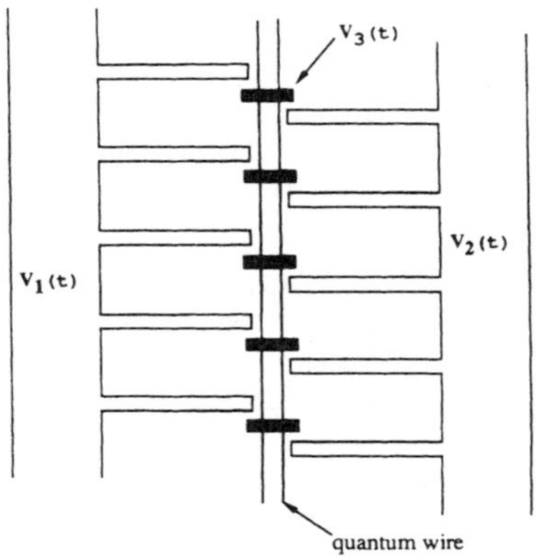

FIG. 2. The quantum wire and the voltage leads.

locked mobility gap Δ_i should be exponentially small in Δ_i/kT or $\Delta_i/h\nu$. Errors due to Mott-type inelastic hops over the localized states are also exponentially small.

The lesson of the quantum Hall effect also tells us that the one dimensionality of the system is not necessary. We may consider a quantum wire with several lateral modes below the Fermi energy, so long as we have a big enough Fermi gap to beat down all types of perturbations.

How can one create a potential that can induce a nonzero and quantized particle transport? Here we suggest an example to illustrate the practical possibility. The details will be presented elsewhere.[13] The setup is illustrated in Fig. 2. We first make a quantum wire on a substrate. We then connect to the wire two sets of voltage leads located at $\{x=na\}$ and $\{x=na+a/3\}$, respectively, where x is the coordinate along the wire, and a would be the spatial period. The leads within a set should be identical. Let the first set lie on the left of the wire, with the individual leads connected together internally, and with just one common lead L_1 coming out of the device. The second set will be put on the right of the wire, connected internally and coming out the device with a common lead L_2. We then ground the substrate, and apply to L_1 and L_2 the voltages $V_1(t)$ and $V_2(t)$, respectively. The whole quantum wire is also going to be biased by an additional potential $V_b(t)$ with respect to the substrate. We choose these voltages to be such that

$$V_j(t)=A(t)+B\cos[2\pi(j/3-t/\tau)],\quad j=1,2,3\,, \tag{7}$$

where $V_3(t)$ is the voltage at the positions $\{x=na+2a/3\}$ (marked gray in Fig. 2) along the quantum wire. The common term $A(t)$ should be chosen to regulate the zero point of the electronic energy spectrum in the quantum wire, such that a given energy gap always contains in its middle the Fermi energy (or the chemical potential maintained by the substrate).

The potential energy created along the quantum wire is then similar to the one given in (5) and (6). The actual potential does not need to be perfect, as has been emphasized before. It is, however, important to always lock the Fermi energy in a gap. How? Suppose we want to find the appropriate value of $A(t)$ at $t=t_0$. Take a static voltage configuration:

$$V_j=A'+B\cos[2\pi(j/3-t_0/\tau)],\quad j=1,2,3\,. \tag{8}$$

Measure the conductance of the wire at a low enough temperature. By looking at the pattern that the conductance changes with A', we should know for which value of A' the Fermi energy is in the desired gap. We then assign the appropriate value of A' to $A(t_0)$.

Suppose the wire has only one lateral mode below the Fermi energy. The energy gap above the first Bloch band should be of the order of $\Delta=|eB|$, if the thickness of each of the leads is about $a/3$. The condition for $\Delta>k_BT$ can be fufilled for $T<10$ K if $|B|$ is greater than a millivolt. For a frequency as high as $\nu<10^{12}$ Hz, the adiabatic condition, $h\nu\ll\Delta$, is well satisfied.

We may take the period a of the order of 1000 Å, and we need about twenty or so such periods with good periodicity. More periods and better periodicity are preferred. Finally, to observe a nanoampere current, we need $\nu\sim10^{10}$ Hz, a frequency corresponding to a wavelength of 3 cm of an electromagnetic wave. A lower frequency can be used to produce the same amount of current, if more quantum channels (additional lateral or longitudinal modes, or more quantum wires) are used.

If the requirement of high frequency cannot be satisfied at the present stage of technology, we can nevertheless stick to a lower frequency. We may not observe a steady current, but we can store the pumped charges to a capacitor in series with the quantum wire. We then discharge the capacitor at time intervals τ_i bigger than τ, and measure the current pulses. The relaxation time τ_c of the capacitor should be smaller than τ_i. The current at the pulses is enhanced by a factor of τ_i/τ_c. The integrated charge of a pulse should be quantized in units of $Q=e\tau_i/\tau$.

From the above considerations, the observation of the quantized charge transport should not be too difficult, and is certainly possible. Better designs from experienced expermentalists are expected, once the possibility has been demonstrated. It is emphasized again that great potential of refining the accuracy of measurements exists due primarily to the exponential dependence of errors on small perturbations.

How can we make use of such a quantum device? Here we just list a few foreseeable applications. First of all, the device is a tunable, high-precision dc current source, if it is driven at a frequency constant in time. The internal resistance is exponentially large in the length of the quantum wire of the pump, until the voltage becomes comparable to Δ/e. The device can, therefore, also be used as a direct-current standard, assuming the electronic charge is accurately known. Second, the device can serve as a precise pump of electric charges. The amount of pumped charges can be conveniently and accurately controlled by the number of operation cycles. Consequently, we can use it to measure capacitances with high precision.[14] Third and alternatively, the device can be used to measure the fundamental electron charge e.[15]

If the device as a current standard is realized, then we have completed the triangle of voltage, resistance, and current involved in Ohm's law, with three direct quantum standards derived from three physically different quantum effects: the Josephson effect, the quantum Hall effect, and the effect of quantum adiabatic particle transport. Such a triangle of quantum metrology has been discussed by Likharev and Zorin, in connection to a possible quantum current standard through the effect of Bloch oscillations in Josephson-effect devices.[16]

On a more fundamental level, the quantum charge pump generates the charge quantum e, a Josephson device generates the magnetic-flux quantum $h/2e$, while a quantum Hall system generates the resistance quantum h/e^2. These satisfy another Ohm's law: $e(h/e^2)=2h/2e$. Therefore, these devices provide three different ways to measure two independent fundamental constants. We can check whether the fundamental constants really appear the same in different systems, although they should be so according to the present theoretical understanding of these systems.

The author is grateful to G. M. Mozurkewich for many discussions over the past years. He thanks W. Kohn, J. Rehr, E. Fradkin, D. J. Thouless, A. Gossard, K. Ensslin, and P. Petroff for their enthusiastic encouragement. This work is supported at the University of California, Santa Barbara, by NSF Grant No. DMR87-03434 and ONR Grant No. N00014-84-K-0548. He also acknowledges the valuable stimulation provided by the regular meetings devoted to theoretical issues relevant to the NSF Center for Quantized Electronic Structures (QUEST).

---

[1]B. N. Taylor, Phys. Today **42**, No. 8, 23 (1989).

[2]D. J. Thouless, Phys. Rev. B **27**, 6083 (1983); A. Zee, Phys. Lett. **135B**, 307 (1984); J. E. Avron and M. C. Cross, Phys. Rev. B **39**, 756 (1989); J. E. Avron and L. Sudun, Phys. Rev. Lett. **62**, 3082 (1989); J. E. Avron, A. Raveh, and B. Zur, Rev. Mod. Phys. **60**, 873 (1988).

[3]Here and hereafter, Fermi gap refers to the energy gap between the filled and empty bands of $H(t)$.

[4]Here and hereafter, a factor of 2 due to the spin degeneracy is omitted, for the sake of concentrating on the essentials.

[5]Details will be given elsewhere.

[6]The Wannier functions are essentially the Fourier transforms of the Bloch waves with respect to the wave number k. They can be made exponentially localized by suitable choice of the phase of the Bloch functions. See, for example, W. Kohn, Phys. Rev. **115**, 809 (1959).

[7]S. Kivelson, Phys. Rev. B **26**, 4269 (1982).

[8]Q. Niu, Phys. Rev. B **34**, 5093 (1986).

[9]L. D. Landau and E. M. Lifshitz, *Quantum Mechanics —Nonrelativistic Theory*, english translation by J. B. Sykes and J. S. Bell (Pergamon, New York, 1977), pp. 194–196.

[10]Q. Niu and D. J. Thouless, J. Phys. A **17**, 2453 (1984).

[11]R. B. Laughlin, Phys. Rev. B **23**, 5632 (1981).

[12]In the analogy, the extended states in the Hall problem corresponds here to the states that are not locked by the impurities, although all the states may be localized at a given time.

[13]Q. Niu and K. Ensslin (to be published).

[14]Connect a capacitor in series with the quantum charge pump. Operate the charge pump for N cycles. The total charge on the capacitor is then neN, where n is an integer. Measure the voltage, V, across the capacitor. The capacitance is then given by neN/V.

[15]Pump a given number of electrons, then measure their total charge by other methods.

[16]K. K. Likharev and A. B. Zorin, J. Low Temp. Phys. **59**, 347 (1985).

Theory of polarization of crystalline solids

R. D. King-Smith and David Vanderbilt
Department of Physics and Astronomy, Rutgers University, P. O. Box 849, Piscataway, New Jersey 08855-0849
(Received 10 June 1992)

We consider the change in polarization $\Delta\mathbf{P}$ which occurs upon making an adiabatic change in the Kohn-Sham Hamiltonian of the solid. A simple expression for $\Delta\mathbf{P}$ is derived in terms of the valence-band wave functions of the initial and final Hamiltonians. We show that physically $\Delta\mathbf{P}$ can be interpreted as a displacement of the center of charge of the Wannier functions. The formulation is successfully applied to compute the piezoelectric tensor of GaAs in a first-principles pseudopotential calculation.

Experimentally changes in the electrical polarization of solids can be induced by various means including application of a strain (piezoelectricity) or changes in temperature (pyroelectricity). Ferroelectrics are a technologically important class of materials whose polarization can be switched by application of an electric field.[1] To date there have been relatively few theoretical attempts to calculate these quantities from a quantum-mechanical starting point. In this paper we derive simple formulas for calculating finite changes in the polarization of a crystalline solid. The method is ideally suited to first-principles density-functional investigations of polarization effects.

We begin by considering the change in the electronic polarization per unit volume of a crystal which is induced upon making an adiabatic change in the self-consistent Kohn-Sham potential. We parametrize the change in the potential with a variable λ which is arranged to have values 0 and 1 at initial and final values of the potential, respectively. In the following we shall specialize to the case where the change in potential preserves the translational symmetry of the solid. The formalism developed below will therefore be applicable for computing $\Delta\mathbf{P}$ with macroscopic electric field $\mathbf{E}$ held to be zero. If the material is an insulator for all values of λ in the range 0–1 then we have

$$\partial P_\alpha/\partial\lambda = -\frac{ifq_e\hbar}{N\Omega m_e}\sum_{\mathbf{k}}\sum_{n=1}^{M}\sum_{m=M+1}^{\infty}\frac{\langle\psi_{\mathbf{k}n}^{(\lambda)}|\hat{p}_\alpha|\psi_{\mathbf{k}m}^{(\lambda)}\rangle\langle\psi_{\mathbf{k}m}^{(\lambda)}|\partial V_{\mathrm{KS}}^{(\lambda)}/\partial\lambda|\psi_{\mathbf{k}n}^{(\lambda)}\rangle}{(\epsilon_{\mathbf{k}n}^{(\lambda)}-\epsilon_{\mathbf{k}m}^{(\lambda)})^2}+\mathrm{c.c.}\ , \quad (1)$$

where α is a Cartesian direction, m_e and q_e are the electron mass and charge, N is the number of unit cells in the crystal, Ω is the volume of a unit cell, f is the occupation number of states in the valence band (in spin-degenerate systems $f=2$), M is the number of occupied bands, $\hat{p}$ is the momentum operator, and $V_{\mathrm{KS}}^{(\lambda)}$ is the Kohn-Sham potential. A heuristic derivation of Eq. (1) has recently been given by Resta.[2] Resta proposes that we compute the total change in polarization per unit volume, $\Delta\mathbf{P}$, using

$$\Delta\mathbf{P}=\int_0^1(\partial\mathbf{P}/\partial\lambda)d\lambda\ . \quad (2)$$

Physically, $\Delta\mathbf{P}$ arises from the flow of polarization currents in the solid and Eq. (1) may also be regarded as the adiabatic limit of a Kubo formula for the current.[3] It is somewhat surprising that the change in polarization can be computed without explicitly stating how the crystal is terminated. The fundamental justification for taking the thermodynamic limit in Eq. (1) rests with the fact that the current response of an insulator depends only on the local environment.[4,5]

The expression for $\partial\mathbf{P}/\partial\lambda$ can be recast in a form in which conduction-band states do not explicitly appear using an argument developed by Thouless *et al.*[6] in their analysis of the quantum Hall effect. We introduce a set of cell-periodic functions, $u_{\mathbf{k}n}^{(\lambda)}$, with a choice of phases such that $u_{\mathbf{k}n}^{(\lambda)}$ are analytic in both $\mathbf{k}$ and λ. The matrix elements in Eq. (1) can be rewritten as

$$\langle\psi_{\mathbf{k}n}^{(\lambda)}|\hat{p}_\alpha|\psi_{\mathbf{k}m}^{(\lambda)}\rangle=\frac{m_e}{\hbar}\langle u_{\mathbf{k}n}^{(\lambda)}|[\partial/\partial k_\alpha,\hat{H}_{\mathbf{k}}^{(\lambda)}]|u_{\mathbf{k}m}^{(\lambda)}\rangle \quad (3a)$$

and

$$\langle\psi_{\mathbf{k}n}^{(\lambda)}|\partial V_{\mathrm{KS}}^{(\lambda)}/\partial\lambda|\psi_{\mathbf{k}m}^{(\lambda)}\rangle=\langle u_{\mathbf{k}n}^{(\lambda)}|[\partial/\partial\lambda,\hat{H}_{\mathbf{k}}^{(\lambda)}]|u_{\mathbf{k}m}^{(\lambda)}\rangle\ , \quad (3b)$$

where the $\hat{H}_{\mathbf{k}}^{(\lambda)}$ is the cell-periodic Hamiltonian

$$\hat{H}_{\mathbf{k}}^{(\lambda)}=(1/2m_e)(-i\hbar\nabla+\hbar\mathbf{k})^2+V_{\mathrm{KS}}^{(\lambda)}(r)\ . \quad (3c)$$

Substituting Eqs. (3a) and (3b) into Eq. (1) it is straightforward to show by analogy with Ref. 6 that

$$\Delta P_\alpha=-(ifq_e/8\pi^3)\sum_{n=1}^{M}\int_{\mathrm{BZ}}d\mathbf{k}\int_0^1 d\lambda[\langle\partial u_{\mathbf{k}n}^{(\lambda)}/\partial k_\alpha|\partial u_{\mathbf{k}n}^{(\lambda)}/\partial\lambda\rangle-\langle\partial u_{\mathbf{k}n}^{(\lambda)}/\partial\lambda|\partial u_{\mathbf{k}n}^{(\lambda)}/\partial k_\alpha\rangle]\ , \quad (4)$$

where the integral over $\mathbf{k}$ extends over any primitive cell in reciprocal space. The one-dimensional (1D) analogue of Eq. (4) has been derived previously by Thouless[7] in a slightly different context for the case of noninteracting electrons where the formal similarities with the quantum Hall effect are particularly striking. For example, in a

1D system with period a, Stokes's theorem allows us to write the change in polarization per unit length as

$$\Delta P = -\frac{fq_e}{2\pi}\sum_{n=1}^{M}\left\{ i\oint_C \sum_{j=1}^{2} d\tau_j \langle u_{kn}^{(\lambda)}|\partial/\partial\tau_j|u_{kn}^{(\lambda)}\rangle \right\}, \tag{5}$$

where τ is a two-component vector with elements (λ,k) and the contour of integration C is around the loop in τ space from $(0,\pi/a)\rightarrow(1,\pi/a)\rightarrow(1,-\pi/a)\rightarrow(0,-\pi/a)\rightarrow(0,\pi/a)$. The quantity in curly brackets can be recognized as the change in Berry phase for fictitious adiabatic evolution of the cell-periodic wave function around the loop C.[8,9] Thouless has observed that the contour integral in Eq. (5) is quantized in the special case where the potential is the same at $\lambda=1$ and 0. In these circumstances the quantity in curly brackets measures the change in the phase of the wave function at any given real-space point as (λ,k) is taken around C. Given that the cell-periodic parts of the wave function can be chosen to be analytic in k and λ this change in phase must be an integer multiple of 2π. We therefore conclude that the polarization per unit length of a 1D system can only change by an integer multiple of fe for adiabatic changes in the Hamiltonian for which $V_{\rm KS}^{(0)}=V_{\rm KS}^{(1)}$. An analogous result for 3D systems will be derived below.

The physical content of Eq. (4) can be made more readily apparent by working in a gauge where the wave functions are periodic in reciprocal space, i.e., $\psi_{\mathbf{k},n}^{(\lambda)}(\mathbf{r})=\psi_{\mathbf{k}+\mathbf{G},n}^{(\lambda)}(\mathbf{r})$ for all reciprocal lattice vectors $\mathbf{G}$. In terms of the cell-periodic functions in such a gauge we have

$$u_{\mathbf{k},n}^{(\lambda)}(\mathbf{r})=e^{i\mathbf{G}\cdot\mathbf{r}}u_{\mathbf{k}+\mathbf{G},n}^{(\lambda)}(\mathbf{r}) . \tag{6}$$

We remark that the gauge condition of Eq. (6) does not uniquely define the phase of the wave functions. Integrating Eq. (4) by parts we find that

$$\Delta P_\alpha = \frac{ifq_e}{8\pi^3}\sum_{n=1}^{M}\int_{\rm BZ} d\mathbf{k}\left\{ [\langle u_{\mathbf{k}n}^{(\lambda)}|\partial/\partial k_\alpha|u_{\mathbf{k}n}^{(\lambda)}\rangle]_0^1 - \int_0^1 d\lambda \frac{\partial}{\partial k_\alpha}\langle u_{\mathbf{k}n}^{(\lambda)}|\partial/\partial\lambda|u_{\mathbf{k}n}^{(\lambda)}\rangle \right\} . \tag{7}$$

With our choice of gauge, $\langle u_{\mathbf{k}n}^{(\lambda)}|\partial/\partial\lambda|u_{\mathbf{k}n}^{(\lambda)}\rangle$ is periodic in $\mathbf{k}$. The gradient of this quantity integrated over the Brillouin zone (BZ) is therefore zero, so the second term in Eq. (7) makes no contribution. In the periodic gauge we therefore arrive at the conclusion that

$$\Delta\mathbf{P}=\mathbf{P}^{(1)}-\mathbf{P}^{(0)} , \tag{8a}$$

where

$$P_\alpha^{(\lambda)}=(ifq_e/8\pi^3)\sum_{n=1}^{M}\int_{\rm BZ} d\mathbf{k}\langle u_{\mathbf{k}n}^{(\lambda)}|\partial/\partial k_\alpha|u_{\mathbf{k}n}^{(\lambda)}\rangle . \tag{8b}$$

The integral on the right-hand side of Eq. (8b) is closely related to the Berry phase of band n, a quantity which has been recently introduced by Zak and co-workers.[10] The form of Eq. (8) is particularly simple when written in terms of the Wannier functions $W_n^{(\lambda)}(\mathbf{r})$ of the occupied bands.[10] The Wannier functions depend on the particular choice of phases used in the periodic gauge. We define the Wannier function using

$$W_n^{(\lambda)}(\mathbf{r}-\mathbf{R})=(\sqrt{N}\,\Omega/8\pi^3)\int_{\rm BZ} d\mathbf{k}\, e^{i\mathbf{k}\cdot(\mathbf{r}-\mathbf{R})}u_{\mathbf{k}n}^{(\lambda)}(\mathbf{r}) \tag{9a}$$

which implies that

$$u_{\mathbf{k}n}^{(\lambda)}(\mathbf{r})=(1/\sqrt{N})\sum_{\mathbf{R}} e^{-i\mathbf{k}\cdot(\mathbf{r}-\mathbf{R})}W_n^{(\lambda)}(\mathbf{r}-\mathbf{R}) , \tag{9b}$$

where the sum over $\mathbf{R}$ runs over all real-space lattice vectors. Substituting Eq. (9b) into (8b) we find the simple result that

$$\mathbf{P}^{(\lambda)}=(fq_e/\Omega)\sum_{n=1}^{M}\int \mathbf{r}|W_n^{(\lambda)}(\mathbf{r})|^2 d\mathbf{r} . \tag{10}$$

Physically, Eqs. (8a) and (10) state that the change in polarization of the solid is proportional to the displacement of the center of charge of the Wannier functions induced by the adiabatic change in the Hamiltonian.

Returning to the case where the Hamiltonians at $\lambda=0$ and 1 are identical, $u_{\mathbf{k}n}^{(0)}(\mathbf{r})$ and $u_{\mathbf{k}n}^{(1)}(\mathbf{r})$ can at most differ by a phase factor so that

$$u_{\mathbf{k}n}^{(1)}(\mathbf{r})=e^{i\theta_{\mathbf{k}n}}u_{\mathbf{k}n}^{(0)}(\mathbf{r}) . \tag{11}$$

In this limit Eq. (8) reduces to

$$\Delta P_\alpha = -(fq_e/8\pi^3)\sum_{n=1}^{M}\int_{\rm BZ} d\mathbf{k}\,\partial\theta_{\mathbf{k}n}/\partial k_\alpha . \tag{12}$$

With our periodic choice of gauge $e^{i\theta_{\mathbf{k}n}}$ must be periodic in $\mathbf{k}$. The most general form for the phase angle under these circumstances is $\theta_{\mathbf{k}n}=\beta_{\mathbf{k}n}+\mathbf{k}\cdot\mathbf{R}_n$, where $\beta_{\mathbf{k}n}$ is periodic in $\mathbf{k}$. We thus conclude that

$$\Delta\mathbf{P}=(fq_e/\Omega)\sum_{n=1}^{M}\mathbf{R}_n . \tag{13}$$

The change in polarization per unit volume for paths where the Hamiltonian returns to itself is therefore quantized in units of $(fq_e/\Omega)\mathbf{R}$. A particularly simple case to consider is the magnitude of $\Delta\mathbf{P}$ for paths of the form $V_{\rm KS}^{(\lambda)}(\mathbf{r})=V_{\rm KS}^{(0)}(\mathbf{r}-\lambda\mathbf{R})$, which physically corresponds to a translation of the crystal. In this case it is straightforward to verify by explicit calculation that Eq. (8) yields $\Delta\mathbf{P}=(fq_e/\Omega)M\mathbf{R}$, as one would expect on physical grounds.

We have in Eqs. (8) and (13) the rather remarkable result that $\Delta\mathbf{P}$ for a crystal can in principle be determined, to within a factor of $(fe/\Omega)\mathbf{R}$, from a knowledge of the valence-band Kohn-Sham wave functions at $\lambda=0$ and 1. In practice the arbitrary factor of $(fe/\Omega)\mathbf{R}$ can often be eliminated by inspection because one is usually interested in polarization changes where $|\Delta\mathbf{P}| \ll |(fe/\Omega)\mathbf{R}_1|$ where $\mathbf{R}_1$ is the shortest nonzero real-space lattice vector. In other cases any uncertainties introduced by this factor can always be removed by dividing the change in the Hamiltonian into a number of subintervals.

Direct evaluation of $\Delta\mathbf{P}$ via Eq. (8) is cumbersome in numerical calculations, because in practice we only compute the wave functions at a finite number of points in the Brillouin zone, and in general there will be no particular phase relationship between the eigenvectors generated by the diagonalization routine. In actual calculations we circumvent this difficulty using the following strategy. First we pick a direction parallel to a short reciprocal-lattice vector of the solid, $\mathbf{G}_{\parallel}$. We choose the primitive cell for the $\mathbf{k}$-space integration to be a prism with its axis aligned along $\mathbf{G}_{\parallel}$. The component of $\Delta\mathbf{P}$ directed along $\mathbf{G}_{\parallel}$ can be written

$$\Delta P_{\parallel} = P_{\parallel}^{(1)} - P_{\parallel}^{(0)} , \tag{14a}$$

where, in an obvious notation

$$P_{\parallel}^{(\lambda)} = \frac{ifq_e}{8\pi^3} \int_A d\mathbf{k}_{\perp} \sum_{n=1}^{M} \int_0^{|\mathbf{G}_{\parallel}|} dk_{\parallel} \left\langle u_{\mathbf{k}n}^{(\lambda)} \left| \frac{\partial}{\partial k_{\parallel}} \right| u_{\mathbf{k}n}^{(\lambda)} \right\rangle . \tag{14b}$$

The integration in the perpendicular direction poses no special problems and can be performed by sampling over a 2D mesh of k points generated, for example, using the Monkhorst-Pack method.[12] To perform the integral over $k_{\parallel}$ at each point in the $k_{\perp}$ mesh we compute the cell-periodic parts of the wave functions at the string of J k points at $\mathbf{k}_j = \mathbf{k}_{\perp} + j\mathbf{G}_{\parallel}/J$ where j runs from 0 to $J-1$. We then compute the variable $\phi_J^{(\lambda)}(\mathbf{k}_{\perp})$ defined through

$$\phi_J^{(\lambda)}(\mathbf{k}_{\perp}) = \mathrm{Im}\left\{ \ln \prod_{j=0}^{J-1} \det\left(\langle u_{\mathbf{k}_j,m}^{(\lambda)} | u_{\mathbf{k}_{j+1},n}^{(\lambda)} \rangle\right) \right\} , \tag{15}$$

where it is understood that $u_{\mathbf{k}_J,n}^{(\lambda)} = e^{-i\mathbf{G}_{\parallel}\cdot\mathbf{r}} u_{\mathbf{k}_0,n}^{(\lambda)}$. The determinant in Eq. (15) is that of the $M \times M$ matrix formed by allowing n and m to run over all valence bands. With an analytic choice of cell-periodic wave functions it can be verified that

$$\begin{aligned} \phi^{(\lambda)}(\mathbf{k}_{\perp}) &\equiv \lim_{J\to\infty} \phi_J^{(\lambda)}(\mathbf{k}_{\perp}) \\ &= -i \sum_{n=1}^{M} \int_0^{|\mathbf{G}_{\parallel}|} dk_{\parallel} \langle u_{\mathbf{k}n}^{(\lambda)} | \partial/\partial k_{\parallel} | u_{\mathbf{k}n}^{(\lambda)} \rangle \end{aligned} \tag{16a}$$

so our expression for $P_{\parallel}^{(\lambda)}$ becomes

$$P_{\parallel}^{(\lambda)} = -(fq_e/8\pi^3) \int_A d\mathbf{k}_{\perp} \phi^{(\lambda)}(\mathbf{k}_{\perp}) . \tag{16b}$$

It is straightforward to confirm that the product over j in Eq. (15) is independent of how the phases of the wave functions are chosen. Changes of the phase of $u_{\mathbf{k}n}^{(\lambda)}$ can change the value of the integral in Eq. (16) by an integer multiple of 2π. Correspondingly the arbitrary constant in the definition of $\phi_J^{(\lambda)}(\mathbf{k}_{\perp})$ given in Eq. (15) arises from the fact that the imaginary part of the log of a complex number is only defined up to a constant multiple of 2π. In practice the arbitrary constant is removed by comparing $\phi_J^{(1)}(\mathbf{k}_{\perp})$ with $\phi_J^{(0)}(\mathbf{k}_{\perp})$ using the argument outlined in the previous paragraph.

We have in Eqs. (14)–(16) all the ingredients necessary for computing polarization changes in a practical calculation. Within this approach the need for supercells or linear-response techniques is completely avoided.[11] The method is ideally suited to modern electronic structure methods based on iterative diagonalization techniques, which concentrate on computing the valence-band wave functions only.[13]

We illustrate the approach by computing the transverse effective charge tensor and piezoelectric constant of GaAs in a first-principles pseudopotential calculation. The effective charge of GaAs may be determined by computing the change in polarization which is induced on making a small displacement of one sublattice with the boundary condition $\mathbf{E}=\mathbf{0}$. For example, if we move the Ga sublattice by a vector $\mathbf{u}$, then the electronic contribution to the polarization difference between the distorted and undistorted structures is

$$\Delta\mathbf{P} = (e/\Omega) Z_{\mathrm{Ga}}^{*(\mathrm{el})} \mathbf{u} , \tag{17}$$

where $Z_{\mathrm{Ga}}^{*(\mathrm{el})}$ is the electronic contribution to the effective charge. The piezoelectric tensor γ is the strain derivative of the polarization under boundary conditions of $\mathbf{E}=\mathbf{0}$.[4] In the zinc-blende structure there is only one independent component of the piezoelectric tensor, γ_{14}. The piezoelectric tensor can be thought of as the sum of two independent terms. The first term, which we denote by $\gamma_{14}^{(0)}$ following Ref. 11, arises from the change in polarization when the ions are subjected to a homogeneous strain. The second contribution owes its origin to the relative displacement of the sublattices, and can be expressed in terms of the effective charges and internal strain parameters.[4,11] It is shown in Ref. 11 that

$$(a^2/e)\gamma_{14} \equiv \overline{\gamma}_{14} = (a^2/e)\gamma_{14}^{(0)} + Z_{\mathrm{Ga}}^{*}\xi , \tag{18}$$

where ξ is the internal strain parameter.

Our first-principles calculations used norm-conserving nonlocal pseudopotentials.[14] We note in passing that, strictly speaking, a nonlocal potential causes a modification to the momentum operator in Eq. (1).[2] However, in this situation there is a precisely compensating change to the Hamiltonian for the cell-periodic part of the wave function, and equations from Eq. (4) on remain correct at they stand. Our calculation treated exchange and correlation in the local-density approximation using the Wigner form. The wave functions were expanded using a 20-Ry plane-wave cutoff. All calculations of the self-consistent Kohn-Sham potential were performed with a (4,4,4) Monkhorst-Pack mesh.[12] Calculations in the cubic structure were performed at the theoretical lattice constant a, which came out to be 5.576 Å with the above parameters. For the calculations of the effective charge we displaced the Ga atom a distance of $0.01a$ in the (001) direction and computed the polarization change in the z direction. The integration mesh for computing $\Delta\mathbf{P}$ used 16 k points in the $\mathbf{k}_{\perp}$ mesh and a string of 10 k points in the parallel direction. We obtained $\gamma_{14}^{(0)}$ by computing the change in polarization in the z directions induced by applying a 1% xy shear strain to the crystal.

The results of our calculation are summarized in Table

I. The total value of $Z^*_{\rm Ga}$ (electronic plus ionic contributions) came out to be 1.984, in excellent agreement with the value of 1.994 obtained from pseudopotential linear-response calculations.[11] Both sets of theoretical values of $Z^*_{\rm Ga}$ are about 8% smaller than the experimental value. Our calculations on the strained crystal yielded a $(a^2/e)\gamma^{(0)}_{14}$ of -1.352. The value agrees to better than 5% with the result obtained from linear-response methods.[11] Our overall value for the piezoelectric constant $\bar{\gamma}_{14}$ was -0.28, compared with an experimental value of -0.32. The agreement between our calculation and experiment is reasonable, given that the two terms in Eq. (18) show a strong tendency to cancel. We have checked that our calculation is converged with respect to k-point set and plane-wave cutoff, and that the polarization response is linear in the applied perturbation. We attribute the small differences between our results and those of Gironcoli, Baroni, and Resta[11] to the use of different pseudopotentials and parametrizations of the exchange and correlation potential.

Before closing we note that it is tempting to physically identify the quantity $\mathbf{P}^{(\lambda)}$ defined in Eq. (8b) as the absolute polarization of the perturbed crystal. Of course it would have to be understood that the polarization, defined in this way, would only be well defined modulo $ef\mathbf{R}/\Omega$. The conditions under which such an identification is useful will be the subject of a future communication.

TABLE I. Theoretical and experimental piezoelectric response of GaAs.

	This work	Linear response[a]	Experiment
a (Å)	5.576	5.496	5.642
ξ	0.542	0.528	0.55
$Z^*_{\rm Ga}$	1.984	1.994	2.16
$(a^2/e)\gamma^{(0)}_{14}$	-1.352	-1.405	
$\bar{\gamma}_{14}$	-0.28	-0.35	-0.32

[a]Reference 11.

In conclusion, we have shown that adiabatic changes in the Kohn-Sham Hamiltonian lead to polarization changes in the solid which can be computed in terms of the initial and final valence-band wave functions of the system. This result forms the basis for a scheme for computing polarization changes of solids within the context of first-principles total-energy calculations.

We would like to thank Karin Rabe and Raffaele Resta for stimulating discussions at the inception of this work. This work was supported by the Office of Naval Research under Contract No. N00014-91-J-1184.

[1]M. E. Lines and A. M. Glass, *Principles and Applications of Ferroelectrics and Related Materials* (Clarendon, Oxford, 1977).
[2]R. Resta, Ferroelectrics (to be published).
[3]E. Fradkin, *Field Theories of Condensed Matter Systems* (Addison Wesley, Redwood City, CA, 1991).
[4]R. M. Martin, Phys. Rev. B **5**, 1607 (1972); **6**, 4874 (1972).
[5]C. Kallin and B. I. Halperin, Phys. Rev. B **29**, 2175 (1984).
[6]D. J. Thouless, M. Kohmoto, M. P. Nightingale, and M. den Nijs, Phys. Rev. Lett. **49**, 405 (1982).
[7]D. J. Thouless, Phys. Rev. B **27**, 6083 (1983).
[8]M. V. Berry, Proc. R. Soc. London Ser. A **392**, 45 (1984).
[9]B. Simon, Phys. Rev. Lett. **51**, 2167 (1983).
[10]J. Zak, Phys. Rev. Lett. **62**, 2747 (1989); L. Michel and J. Zak, Europhys. Lett. **18**, 239 (1992).
[11]S. de Gironcoli, S. Baroni, and R. Resta, Phys. Rev. Lett. **62**, 2853 (1989).
[12]H. J. Monkhorst and J. D. Pack, Phys. Rev. B **13**, 5188 (1976).
[13]R. Natarajan and D. Vanderbilt, J. Comput. Phys. **82**, 218 (1989).
[14]D. R. Hamann, M. Schlüter, and C. Chiang, Phys. Rev. Lett. **43**, 1494 (1979).

Semiclassical Analysis and the Magnetization of the Hofstadter Model

O. Gat and J. E. Avron
Department of Physics, Technion 32000, Haifa, Israel
(Received 12 June 2003; published 29 October 2003)

The magnetization and the de Haas–van Alphen oscillations of Bloch electrons are calculated near commensurate magnetic fluxes. Two phases that appear in the quantization of mixed systems—the Berry phase and a phase first discovered by Wilkinson—play a key role in the theory.

DOI: 10.1103/PhysRevLett.91.186801 PACS numbers: 73.43.–f, 75.20.–g, 75.70.–i

The magnetization of a free electron gas was calculated by Landau in 1930 in the early days of quantum mechanics [1]. Considerable efforts have since been devoted to extending these results to Bloch electrons. Most of the progress made was in the region of weak magnetic fields [2,3] where the flux Φ through a unit cell is small. This is adequate for most solid state applications. There is, however, also considerable interest in a better understanding of phenomena that have to do with commensuration in condensed matter physics [4–8]. This is the case when Φ is close to a rational number. The magnetization of Bloch electrons near rational fluxes, other than $\Phi = 0$, remained an open challenge which we solve here. The difficulty lies in the delicate spectral properties resulting from commensuration [5].

The problem of magnetization near fractional flux becomes tractable by an idea that goes back to Wilkinson [8]. Namely, that near a rational flux the Hamiltonian can be understood as the semiclassical quantization of a mixed system: In mixed systems some, but not all, degrees of freedom may be treated semiclassically. As a consequence the "classical Hamiltonian" is matrix (or operator) valued. Pauli and Dirac equations for a spinning electron in a slowly varying potential, the Born-Oppenheimer theory of molecules, and the Hofstadter model near rational flux [8] are examples of mixed systems. In the Hofstadter model the role of the Planck constant is played by the deviation from a nearby rational

$$h = \Phi - p/q. \qquad (1)$$

Littlejohn and Flynn [9] developed an elegant geometric formalism for the quantization of mixed systems. They show that in order $\hbar$ the quantization of mixed systems gives rise to two phases: One is the Berry phase [10] and the other is a phase that is sometimes known as the "no-name phase" [11] and sometimes as the Wilkinson-Rammal (WR) phase [12]. For both phases to appear, the classical Hamiltonian must have nontrivial commutation properties in both coordinates and momenta. These phases play a central role in determining the magnetization; see, e.g., Eq. (25) below.

Our results are closely related to recent progress made in the semiclassical dynamics of Bloch electrons under slowly varying electric and magnetic fields [12]. When one goes beyond the leading order expressed by Peierls substitution [2,12] one finds that the Berry phase and the WR phase play a role in the dynamics. This led [12] to identify the WR term with the magnetization of a wave packet. Although related, the notions of wave packet vs thermodynamic magnetization, expressed in Eq. (25), are distinct; for example, wave packet magnetization is not defined in the gaps, while the thermodynamic magnetization and, of course, the de Haas–van Alphen oscillations have a nontrivial dependence on the chemical potential also in the gaps.

Figure 1 shows the zero temperature magnetization at $\Phi = 1/[3 + (1/40)]$. The complexity of the magnetization is due to the multiplicity of scales: The big scale is determined by the denominator $q = 3$, and the small scale by h. On the small scale one sees the rapid de Haas–van Alphen oscillations. On the big scales one sees continuous features: the linear pieces in the (big) gaps, the envelopes of the de Haas–van Alphen oscillations, and their mean. Our theory of magnetization accounts for all these features.

While the amplitude of the oscillation is determined solely by the leading terms in the semiclassical expansion, the mean magnetization requires knowledge of the terms beyond leading order, thus depending on the fine

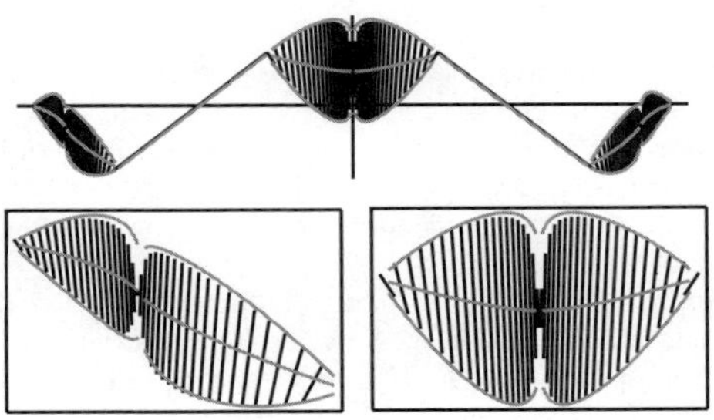

FIG. 1 (color online). The de Haas–van Alphen oscillations of the magnetization as a function of the chemical potential in the Hofstadter model for flux $\Phi = 40/121$, compared with the limiting envelopes and their averages, and the gap magnetization for $\Phi \to 1/3$. The two insets are details of the left and center envelopes.

details of the spectrum. Intriguingly, it is the *latter* quantity which is stable against perturbations. Finite temperatures larger than the typical eigenvalue spacing wash out the de Haas–van Alphen oscillations of the small scale but leave intact the mean magnetization. Semiclassical approximations that retain only the leading order yield no magnetization at all at finite temperatures.

Let us start by describing the semiclassical quantization of mixed systems [9]. The classical Hamiltonian, $\mathcal{H}(x, k)$, is a Hermitian *matrix* which depends on x and k. We denote by $\varepsilon_j(x, k)$ its jth band of eigenvalues and by $|u_j(x, k)\rangle$ the corresponding eigenvectors. We also assume that bands do not cross. (In the Hofstadter model, this is guaranteed by Chambers relation.) The corresponding quantum Hamiltonian is $\mathcal{H}(\hat{x}, \hat{k})$ with $[\hat{x}, \hat{k}] = i\hbar$.

Let $S(E; H, \rho)$ denote the classical action associated with a closed orbit of energy E of a classical Hamiltonian function H, with phase space area form ρ. Since phase space is two dimensional the action is the area enclosed by the orbit. Note that H, unlike $\mathcal{H}$, is a scalar valued function. The Bohr-Sommerfeld quantization rule in mixed systems says that the semiclassical approximation of the eigenvalue E_n is given by

$$S(E_n; H_\hbar, \rho_\hbar) = h(n + \gamma_m), \qquad n \in \mathbb{Z}, \tag{2}$$

where γ_m is the Maslov index of the orbit [13]. $H_\hbar$ has an expansion in powers of $\hbar$, $H_\hbar = H^{(0)} + \hbar H^{(1)} + \cdots$. Peierls substitution sets $H_j^{(0)}(x, k) = \varepsilon_j(x, k)$, and the next order is [9]

$$H_j^{(1)}(x, k) = \mathrm{Im}\langle \partial_x u_j | (\mathcal{H} - \varepsilon_j) | \partial_k u_j \rangle (x, k). \tag{3}$$

The expansion $\rho_\hbar = \rho^{(0)} + \hbar \rho^{(1)} + \cdots$ begins with the canonical form $\rho_j^{(0)} = dk \wedge dx$ and the subleading term is the Berry curvature form

$$\begin{aligned} \rho_j^{(1)} &= 2\omega(x, k) dk \wedge dx, \\ \omega(x, k) &= \mathrm{Im}\langle \partial_x u_j | \partial_k u_j \rangle (x, k). \end{aligned} \tag{4}$$

This formulation is manifestly gauge invariant (independent of the choice of phases for $|u_j\rangle$) and preserves the symmetry properties of $\mathcal{H}$, which is useful when one wants to correctly count the dimension of the Hilbert space of the quantized operator, as we now proceed to explain.

Suppose that $\mathcal{H}$ is periodic in both x and k up to gauge transformations and hence describes (classical) motion on a phase space torus $\mathcal{T}$. $H_j^{(0)}$, $H_j^{(1)}$, and $\rho_j^{(1)}$ are all well-defined functions on $\mathcal{T}$. The Chern number of the jth band is the integer

$$C_j = \frac{1}{\pi} \int_{\mathcal{T}} \omega_j dk \wedge dx. \tag{5}$$

It follows from Eqs. (2) and (4) that the dimension of the Hilbert space associated with the jth band is

$$\frac{|\mathcal{T}|}{h} + C_j, \tag{6}$$

where $|\mathcal{T}|$ denotes the area of $\mathcal{T}$. Since the dimension of the Hilbert space is necessarily an integer, quantization on the torus is possible only for certain values of h. Equation (6) goes beyond the classical Weyl law which determines only the leading, $|\mathcal{T}|/h$, behavior. The Chern numbers shift states between the spaces of different bands since $\sum C_j = 0$ [14].

The $\hbar$ corrections to the action can be moved from the left-hand side of the Bohr-Sommerfeld relation to the right-hand side, where they acquire an interpretation as two additional phases:

$$\begin{aligned} S_j(E) &= h[n + \gamma_m + \Gamma_j(E)], \\ -2\pi\Gamma_j(E) &= \gamma_j^{\mathrm{B}}(E) + \gamma_j^{\mathrm{WR}}(E), \end{aligned} \tag{7}$$

where $S_j(E) = S(E; \varepsilon_j, dk \wedge dx)$. $\gamma_j^{\mathrm{B}}(E)$ is the Berry phase [10],

$$\gamma_j^{\mathrm{B}}(E) = 2 \int \omega_j \theta(E - \varepsilon_j) dk \wedge dx, \tag{8}$$

and $\gamma_j^{\mathrm{WR}}(E)$ is the Wilkinson-Rammal phase

$$\gamma_j^{\mathrm{WR}}(E) = -\int H_j^{(1)} \delta(E - \varepsilon_j) dk \wedge dx. \tag{9}$$

It is noteworthy that the WR phase *need not* vanish at band edges.

Let us now recall some basic facts about the Hofstadter model [5,8]. When the magnetic flux Φ is a rational number p/q, the model is represented by [8]

$$\mathcal{H}(x, k) = e^{2\pi i x} U + e^{2\pi i k} V + \mathrm{H.c.}, \tag{10}$$

where U and V are the $q \times q$ matrices

$$U = \begin{pmatrix} 0 & 1 & \cdots & 0 \\ & \ddots & \ddots & \\ & & \ddots & 1 \\ 1 & & & 0 \end{pmatrix}, \quad V = \begin{pmatrix} 1 & & & \\ & e^{2\pi i p/q} & & \\ & & \ddots & \\ & & & e^{2\pi i (q-1)p/q} \end{pmatrix}. \tag{11}$$

The magnetic bands of the Hofstadter Hamiltonian at $\Phi = p/q$ are given by $\varepsilon_j(x, k)$ on the Brillouin zone

$$\mathrm{BZ} = \{x, k | 0 \le x \le 1, 0 \le k \le 1/q\}. \tag{12}$$

Evidently, $\mathcal{H}(x, k)$ is periodic with period 1 in both variables. Moreover, $\mathcal{H}$ is periodic with smaller periods up to unitary transformations:

$$\mathcal{H}(x,k) = G\mathcal{H}(x+1/q,k)G^{\dagger} = T\mathcal{H}(x,k+1/q)T^{\dagger}. \tag{13}$$

G is a gauge transformation (a diagonal unitary) and T is a shift. This makes the band dispersion functions $\varepsilon_j(x,k)$ periodic with periods $1/q$ in each variable and with q periods in $\mathcal{T} = \text{BZ}$.

The spectrum of the Hofstadter model for other values of Φ is obtained by setting $[\hat{x}, \hat{k}] = i\hbar$ in $\mathcal{H}$ with $\hbar$ given by Eq. (1). BZ is the minimal torus on which $\mathcal{H}$ may be quantized. The dimension of the Hilbert space associated with it is then $\frac{1}{qh} - C_j$. Since the band functions are q periodic on BZ, the number of *distinct* eigenvalues is of order $1/(q^2h)$. The semiclassical approximation is valid provided this number is large; i.e., $q^2\hbar \ll 1$.

We now turn to the magnetization of the model. Recall that the Hofstadter model approximates the Schrödinger equation in two dual limits: When the magnetic field is weak relative to the periodic potential and also in the opposite limit where the magnetic field dominates all other interactions. The two limits have related but different thermodynamics. For the sake of concreteness we consider the tight-binding interpretation. The magnetization of the "split Landau level" follows from the duality transformation of [15].

The thermodynamic potential per lattice site of the Hofstadter model for rational flux and zero temperature is [15]

$$\Omega(\mu,\Phi) = \int_{\text{BZ}} dk \wedge dx \text{Tr}(\mu - \mathcal{H})_+, \tag{14}$$

where $x_+ = x\theta(x)$. When μ is in a spectral gap, the thermodynamic potential can be written as a sum of the potential of the occupied bands, $\Omega = \sum_< \Omega_<$.

The magnetization per unit area m is

$$m(\mu,\Phi) = -\left(\frac{\partial\Omega}{\partial\Phi}\right)_{\mu}. \tag{15}$$

(To translate the magnetization to ordinary units one needs to divide our dimensionless magnetization by the unit of quantum flux.) The magnetization in the gaps can likewise be expressed as a sum of the magnetization of the occupied bands

$$m(\mu,\Phi) = \sum_{<} m_<(\Phi), \tag{16}$$

where the magnetization of a full band is, as we see below (see also [15]),

$$m_j(\mu,\Phi) = -\frac{1}{2\pi}\int_{\text{BZ}} [2(\mu - \varepsilon_j)\omega_j - H^{(1)}] dk \wedge dx. \tag{17}$$

The term proportional to μ is the Chern number of the band. It follows that the magnetization as a function of μ has quantized slopes in the gaps.

The envelope of the de Haas–van Alphen oscillations of the jth band, as we show below, is given by

$$L_j(\mu,\Phi) = \langle m\rangle_j(\mu,\Phi) \pm \delta m_j(\mu,\Phi). \tag{18}$$

$\langle m\rangle_j(\mu,\Phi)$ is the natural restriction of Eq. (17) to a partially filled band; i.e.,

$$\langle m\rangle_j(\mu,\Phi) = -\frac{1}{2\pi}\int_{\text{BZ}} \theta(\mu-\varepsilon_j) \times [2(\mu-\varepsilon_j)\omega_j - H_j^{(1)}] dk \wedge dx. \tag{19}$$

It describes the mean value of the magnetization, averaged over the de Haas–van Alphen oscillations.

The width of the envelope is given in terms of the classical action associated with ε_j:

$$\delta m_j(\mu) = \frac{q}{2}\frac{S_j(\mu)}{S_j'(\mu)}. \tag{20}$$

$\frac{1}{q}S'(\mu)$ is proportional to the density of states. Since the density of states in two dimensions diverges logarithmically near the separatrix, the width δm shrinks to zero logarithmically there. Near the bottom of the band S_j vanishes linearly while the density of states approaches a positive value. This shows that δm vanishes linearly at band edges. These properties characterize the universal liplike shape of the envelopes.

We conclude with an outline of the derivation of Eqs. (18)–(20). Consider the zero temperature thermodynamic potential associated with one fixed band j. It follows from the Chambers relation [16] and the square symmetry of the Hofstadter Hamiltonian that for all energies except the separatrix, the level sets of $\varepsilon_j(x,k)$ are deformed circles, and therefore $\gamma_m = 1/2$. All spectral quantities below refer to the same band, and we may therefore suppress the index j without risk of confusion.

Suppose that μ is such that n spectral points of the split jth band are occupied. Recall that each spectral point is q-fold degenerate and that, by Eq. (6), $nq \le \frac{1}{qh} + C$. Suppose for definiteness that μ is below the separatrix. By the Bohr-Sommerfeld rule the thermodynamic potential is (to order $\hbar^2$)

$$\Omega(\Phi,\mu) = qh\left(\mu n - \sum_{\ell=0}^{n-1} S^{-1}\{[\ell + 1/2 + \Gamma(E_\ell)]h\}\right). \tag{21}$$

The overall factor qh comes from the degeneracy per unit area of each eigenvalue. Approximating the sum with the second Euler-Maclaurin sum formula gives (again to order $\hbar^2$)

$$\Omega(\Phi,\mu) = q\mu nh - q\int_0^{nh} dx S^{-1}\{x + h\Gamma[S^{-1}(x)]\}. \tag{22}$$

Taking a derivative with respect to Φ is the same as

taking a derivative with respect to h. The magnetization is therefore given (to order $\hbar$) by

$$m(\Phi, \mu) = -qn\{\mu - S^{-1}[(n+\Gamma)h]\} + q\int^{\mu} \Gamma(E)dE$$
$$= qn\left(\frac{E_n + E_{n-1}}{2} - \mu\right) + q\int^{\mu} \Gamma(E)dE, \qquad (23)$$

where the E integration is over the jth band energies below μ.

The first term describes the de Haas–van Alphen oscillations: It vanishes in the middle of each spectral gap $[E_{n-1}, E_n]$ and reaches its maximum magnitude at the band edges. $\delta m(\mu)$ is half the variation of m across a spectral gap. Therefore

$$\delta m(\Phi, \mu) = qn\frac{E_n - E_{n-1}}{2} \underset{h\to 0}{\longrightarrow} \frac{qS(\mu)}{2S'(\mu)}. \qquad (24)$$

The first term in Eq. (23) has zero average over the gap, while the second term is nearly constant. The mean magnetization is therefore

$$\langle m\rangle(\Phi, \mu) = -\frac{q}{2\pi}\int^{\mu}[\gamma^{\mathrm{B}}(E) + \gamma^{\mathrm{WR}}(E)]dE. \qquad (25)$$

The Berry phase contributes

$$q\int^{\mu}\gamma^{\mathrm{B}}(E)dE = 2\int_{\mathrm{BZ}} \omega dk \wedge dx \int^{\mu} \theta(E-\varepsilon)dE$$
$$= 2\int_{\mathrm{BZ}} (\mu - \varepsilon)\omega\theta(\mu - \varepsilon)dk \wedge dx.$$

The WR phase contributes

$$q\int^{\mu}\gamma^{\mathrm{WR}}(E)dE = -\int_{\mathrm{BZ}} dk \wedge dx H^{(1)} \int^{\mu} dE\delta(E-\varepsilon)$$
$$= -\int_{\mathrm{BZ}} H^{(1)}\theta(E-\varepsilon)dk \wedge dx.$$

Together, they add up to give Eq. (19).

Finally, let us present a streamlined derivation of the rules for band splitting [8]. Consider, for example,

$$\Phi = \frac{1}{q - \frac{1}{n}} = \frac{n}{qn-1} = \frac{1}{q} + h, \qquad h = \frac{1}{q(qn-1)},$$

with q odd and n even. We demonstrate the following splitting rule: Of the q bands associated with the flux $1/q$, the center band splits into $n-1$ subbands and the rest into n subbands, together accounting for the $qn-1$ band associated with the flux Φ. The Diophantine equation of [17] at flux $1/q$ bands implies that the Chern number of the center band is $1-q$, and all other bands have Chern number 1. Recalling that the area of the Brillouin zone is $1/q$, Eq. (6) gives that the center band splits into $q(n-1)$ levels and the other bands split into qn levels each. We recall also that the band dispersion functions have q periods in the Brillouin zone, and therefore there are only $qn-1$ distinct levels, each q-fold degenerate. (The levels are broadened into bands by tunneling, which we do not discuss. This does not modify the counting of dimensions.) This example illustrates the algorithm which generates the hierarchical structure of the Hofstadter butterfly.

This work is supported by the Technion fund for promotion of research and by EU Grant No. HPRN-CT-2002-00277.

[1] L. D. Landau and E. M. Lifshitz, *Statistical Physics*, edited by E. M. Lifshitz and L. P. Pitaevskii (Pergamon, Oxford, 1980), Pt. 1, 3rd ed.

[2] R. Peierls, Z. Phys. **3**, 1055 (1933); P. G. Harper, Proc. Phys. Soc. London, Sect. A **68**, 874–892 (1955).

[3] W. Kohn, Phys. Rev. **115**, 1460 (1959); B. I. Blount, Phys. Rev. **126**, 1636 (1962); J. Zak, Phys. Rev. **168**, 686 (1968); G. Nenciu, Rev. Mod. Phys. **63**, 91 (1991).

[4] F. Axel, *Beyond Quasicarystals* (Springer, New York, 1995); R. Saito, G. Dresselhaus, and M. Dresselhaus, *Physical Properties of Carbon Nanotubes* (Imperial College Press, London, 1998).

[5] M. Ya. Azbel, Sov. Phys. JETP **19**, 634–645 (1964); D. Hofstadter, Phys. Rev. B **14**, 2239 (1976).

[6] B. Helffer and J. Sjöstrand, Mem. Soc. Math. France **39**, 1–139 (1989); Y. Last, Commun. Math. Phys. **164**, 421–432 (1994); A. Gordon, S. Jitomirskaya, Y. Last, and B. Simon, Acta Math. **178**, 169–183 (1997).

[7] J. Bellissard, I. Guarneri, H. Schulz-Baldes, Commun. Math. Phys. **227**, 515–539 (2002).

[8] M. Wilkinson, J. Phys. A **20**, 4337–4354 (1987); M. Wilkinson, J. Phys. A **17**, 3459–3476 (1984); R. Ramal and J. Bellissard, J. Phys. (Paris) **51**, 1803–1830 (1990); J. Bellissard, C. Kreft, and R. Seiler, J. Phys. A **24**, 2329–2353 (1991); B. Helffer and J. Sjöstrand, Mem. Soc. Math. France **40**, 1 (1990).

[9] R. G. Littlejohn and W. G. Flynn, Phys. Rev. A **44**, 5239 (1991).

[10] M. V. Berry, Proc. R. Soc. London, Ser. A **392**, 45 (1984).

[11] C. Emmrich and A. Weinstein, Commun. Math. Phys. **176**, 701–712 (1996); J. Bolte and R. Glaser, math-ph/0204018.

[12] G. Panati, H. Spohn, and S. Teufel, Phys. Rev. Lett. **88**, 250405 (2002); math-ph/0212041; G. Sundaram and Q. Niu, Phys. Rev. B **59**, 14915 (1999); T. Jungwirth, Q. Niu, and A. H. MacDonald, Phys. Rev. Lett. **88**, 207208 (2002).

[13] V. P. Maslov and M. V. Fedoryuk, *Semiclassical Approximation in Quantum Mechanics* (D. Reidel, Dordrecht, 1981).

[14] J. Avron, R. Seiler, and B. Simon, Phys. Rev. Lett. **51**, 51 (1983).

[15] O. Gat and J. Avron, New J. Phys. **5**, 44.1–44.8 (2003).

[16] W. Chambers, Phys. Rev. **140**, A135 (1965).

[17] D. J. Thouless, M. Kohmoto, M. P. Nightingale, and M. den Nijs, Phys. Rev. Lett. **49**, 405 (1982).

Orbital Magnetization in Periodic Insulators

T. Thonhauser,[1] Davide Ceresoli,[2] David Vanderbilt,[1] and R. Resta[3]

[1] *Department of Physics and Astronomy, Rutgers University, Piscataway, New Jersey 08854, USA*
[2] *International School for Advanced Studies (SISSA/ISAS) and INFM-DEMOCRITOS, via Beirut 4, 34014, Trieste, Italy*
[3] *Dipartimento di Fisica Teorica Università di Trieste and INFM-DEMOCRITOS, strada Costiera 11, 34014, Trieste, Italy*
(Received 20 May 2005; published 22 September 2005)

Working in the Wannier representation, we derive an expression for the orbital magnetization of a periodic insulator. The magnetization is shown to be comprised of two contributions, an obvious one associated with the internal circulation of bulklike Wannier functions in the interior, and an unexpected one arising from net currents carried by Wannier functions near the surface. Each contribution can be expressed as a bulk property in terms of Bloch functions in a gauge-invariant way. Our expression is verified by comparing numerical tight-binding calculations for finite and periodic samples.

DOI: 10.1103/PhysRevLett.95.137205 PACS numbers: 75.10.Lp, 73.20.At, 73.43.−f

Recent years have seen a surge of interest in issues of charge and spin transport in magnetic materials and nanostructures, notably the development of a theory of the intrinsic anomalous Hall conductivity and some controversies surrounding the spin-Hall effect [1]. In this context it is quite surprising that the theory of orbital magnetization, essential for any proper description of magnetism, has remained in a primitive state. Linear-response methods allow calculations of magnetization *changes* [2–5], but not of the magnetization itself.

Hirst [6] has emphasized that a knowledge of the bulk local current density $\mathbf{J}(\mathbf{r})$ is insufficient, even in principle, to determine the macroscopic orbital magnetization $\mathbf{M}$, just as the density $\rho(\mathbf{r})$ cannot be used to determine the electric polarization $\mathbf{P}$. Thus, the theory of $\mathbf{M}$ today is in a condition very similar to that of $\mathbf{P}$ in the early 1990s, when the problem of computing *finite* polarization changes was solved by the introduction of the Berry-phase theory [7,8]. The essential difficulty, that the matrix elements of the position operator $\mathbf{r}$ are not well defined in the Bloch representation, could be overcome by reformulating the problem in the Wannier representation. Because Wannier functions (WFs) are exponentially localized in an insulator, matrix elements of $\mathbf{r}$ between WFs are indeed well defined.

Here we show that it is possible to formulate a corresponding theory of the orbital magnetization for an insulator with broken time-reversal symmetry. The problem is analogous, with the circulation operator $\mathbf{r} \times \mathbf{v}$ now being ill defined in the Bloch representation. Working instead in the Wannier representation, we write the orbital magnetization as a gauge-invariant Brillouin-zone integral over occupied Bloch functions. It contains two terms, the first of which describes the internal circulation of bulklike WFs [9]. The second is much more subtle, arising only from surface WFs and reflecting the fact that the information about surface currents needed to define the macroscopic magnetization is actually contained in the bulk band structure. The resulting formula is consistent with a recent semiclassical argument [10] based on an expression for the magnetization of a wave packet [11] and can easily be implemented in first-principles codes.

For our derivation, we restrict ourselves to the case of an insulator described by a one-particle Hamiltonian with broken time-reversal symmetry. While the restriction to insulators is essential for the theory of polarization, we suspect that it is less so here, so that future generalizations to metals are not ruled out. We also require a vanishing macroscopic magnetic field (or, more generally, an integer number of flux quanta per unit cell) so that the Bloch wave vector $\mathbf{k}$ remains a good quantum number. We have in mind cases in which a staggered magnetic field averages to zero over the unit cell, or in which the time-reversal breaking comes about through spin-orbit coupling to a background of ordered local moments [12–16]. For simplicity we work with spinless electrons (the generalization to the spin-unrestricted case being straightforward) and furthermore restrict ourselves to zero-Chern-number insulators [12,13].

Let us consider a finite sample representing a fragment of a larger crystalline system. We assume that the occupied states can be represented in terms of well-localized orthonormal orbitals $|\phi_i\rangle$, which we will refer to as Wannier functions. If we introduce the velocity operator as

$$\mathbf{v} = -\frac{i}{\hbar}[\mathbf{r}, H] \tag{1}$$

then the total magnetic moment of the finite system involves the matrix elements $\langle\psi_i|\mathbf{r} \times \mathbf{v}|\psi_i\rangle$, where the $|\psi_i\rangle$ are the occupied eigenstates of H. By invariance of the trace, this can be written in terms of WFs as

$$\mathbf{m} = -\frac{e}{2c}\sum_i \langle\phi_i|\mathbf{r} \times \mathbf{v}|\phi_i\rangle, \tag{2}$$

where $-e$ is the electron charge. The magnetization $\mathbf{M}$ can then be defined as the magnetic moment $\mathbf{m}$ per unit volume. For large but finite samples, all $|\phi_i\rangle$ that are sufficiently far from the surface become exponentially similar to bulk WFs, which we will denote as $|w_i\rangle$. For the electric polarization, the transition $\langle\phi_i|\mathbf{r}|\phi_i\rangle \to \langle w_i|\mathbf{r}|w_i\rangle$ describes the polarization of periodic systems correctly [7]. Thus, it is tempting to assume that the magnetization should be expressible in a similar way in terms of the circulation $\langle w_i|\mathbf{r}\times\mathbf{v}|w_i\rangle$ of bulk WFs.

We set out to verify this hypothesis in the context of numerical tight-binding calculations. For simplicity, we chose the Haldane model [12], which is comprised of a honeycomb lattice with two tight-binding sites per cell with site energies $\pm E_0$, real first-neighbor hoppings t_1, and complex second-neighbor hoppings $t_2e^{\pm i\varphi}$, as shown in Fig. 1(a). For our tests, we have chosen a lattice constant equal to unity, $E_0 = \pm 2$, $t_1 = 1$, and $t_2 = 1/3$ and allowed φ to vary [17]. We treat the upper band as empty and the lower band as occupied. The corresponding WFs were obtained by acting with the band projector on a set of δ functions located on the sites with $E_0 = -2$ and applying a subsequent symmetric orthonormalization.

The results of our numerical calculations for a finite 30×30 sample are depicted as the symbols in Fig. 2. First, we calculated the total magnetic moment according to Eq. (2) and divided by the total sample area A to obtain the magnetization M indicated as circles in Fig. 2. Next, we evaluated the contribution to Eq. (2) from a single WF deep in the bulk of the finite sample and divided by the unit-cell area A_0 to obtain a "local circulation" (LC) magnetization $M_{\rm LC}$ plotted as triangles in Fig. 2. We expected these two quantities to agree with each other within some numerical tolerance. On the contrary, the results indicate no agreement whatsoever.

This surprising result forced us to reconsider our entire line of argument, revealing a profound oversight. It is easily shown that each bulk band of an insulating crystal must carry no net current, even in the absence of time-reversal symmetry. This means that each bulklike WF, such as the one that we chose from the deep interior of the sample, must carry no net current, as is easily confirmed in the numerical calculation. We had assumed that the WFs at the boundary of the finite sample would likewise carry no net current, but *this assumption is incorrect*. In fact, the WFs near the boundary do carry a net current, and the total circulation associated with these net currents provides just the needed contribution to resolve the discrepancy.

This is illustrated in Fig. 1(b), where we have plotted the net current $-e\mathbf{v}_i = -e\langle\phi_i|\mathbf{v}|\phi_i\rangle$ located at the Wannier center $\bar{\mathbf{r}}_i = \langle\phi_i|\mathbf{r}|\phi_i\rangle$ for each WF in the sample. While confirming that the net currents in the deep interior are exponentially small, the results reveal that the WFs near the surface do carry a substantial current that contributes extensively to the total magnetic moment of the sample. Dividing by the sample area, we obtain an "itinerant circulation" (IC) contribution $M_{\rm IC}$ to the magnetization that is plotted as squares in Fig. 2. A glance at the figure suggests, and numerical tests confirm, that $M = M_{\rm LC} + M_{\rm IC}$ within numerical precision [18].

It is not surprising that $M_{\rm LC}$, corresponding to the local circulation of a bulk WF, is a bulk property of the insulator [9]. It is far less clear whether $M_{\rm IC}$ is also a bulk property. To show that it is, we go back to Eq. (2). For simplicity, we restrict ourselves henceforth to the case of a two-dimensional system with a single occupied band (the generalization to three dimensions is straightforward, while the multiband treatment is more subtle). Then Eq. (2) can be rewritten as

$$M = -\frac{e}{2Ac}\sum_i\Big[\underbrace{\langle\phi_i|(\mathbf{r}-\bar{\mathbf{r}}_i)\times\mathbf{v}|\phi_i\rangle}_{\rm LC} + \underbrace{\bar{\mathbf{r}}_i\times\langle\phi_i|\mathbf{v}|\phi_i\rangle}_{\rm IC}\Big], \tag{3}$$

where again LC and IC correspond to local and itinerant circulation contributions, respectively. We divide the finite sample into an "interior" and a "surface" region in such a way that the latter occupies a nonextensive fraction of the total sample area in the thermodynamic limit. We label WFs from the surface and interior regions as $|\phi_s\rangle$ and $|\mathbf{R}\rangle$, respectively, where $\mathbf{R}$ is a lattice vector.

Next, we note that the WFs in the surface region make a negligible contribution to the LC term of Eq. (3) since they occupy a nonextensive fraction of the area in the thermodynamic limit. The LC term then becomes

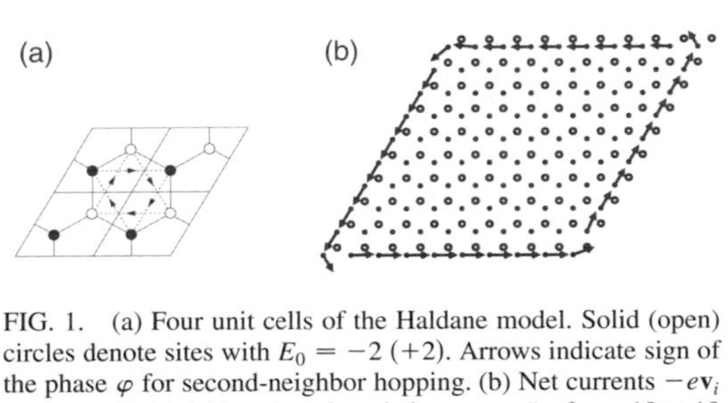

FIG. 1. (a) Four unit cells of the Haldane model. Solid (open) circles denote sites with $E_0 = -2$ $(+2)$. Arrows indicate sign of the phase φ for second-neighbor hopping. (b) Net currents $-e\mathbf{v}_i$ associated with WFs, plotted at their centers $\bar{\mathbf{r}}_i$, for a 10×10 sample with $\varphi = \pi/4$. Currents decrease rapidly into the bulk, so that only surface currents are visible.

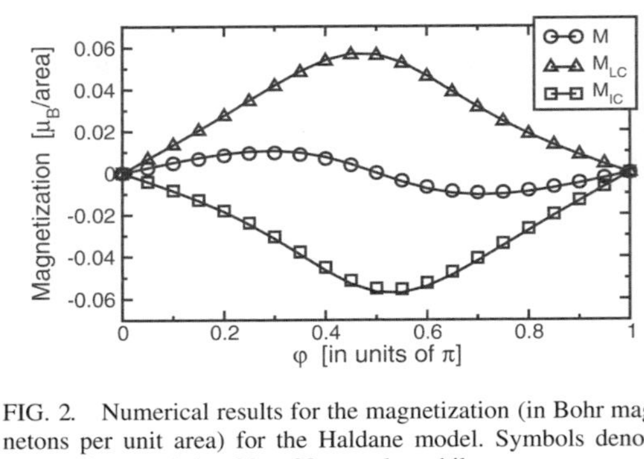

FIG. 2. Numerical results for the magnetization (in Bohr magnetons per unit area) for the Haldane model. Symbols denote results from a finite 30×30 sample, while curves represent results of a k-space calculation on a periodic system.

$$M_{\rm LC} = -\frac{e}{2Ac}\sum_{\mathbf{R}}\langle \mathbf{R}|(\mathbf{r}-\bar{\mathbf{r}}_{\mathbf{R}})\times \mathbf{v}|\mathbf{R}\rangle = -\frac{e}{2A_0 c}\langle \mathbf{0}|\mathbf{r}\times\mathbf{v}|\mathbf{0}\rangle. \quad (4)$$

Here we have used translational symmetry and the fact that $\langle \mathbf{R}|\mathbf{v}|\mathbf{R}\rangle = 0$ (since bulk bands carry no net current). Equation (4) shows that $M_{\rm LC}$ can be expressed simply in terms of the bulk WF $|\mathbf{0}\rangle$ in the home unit cell. Turning to the IC term, the interior WFs now make no contribution (again because $\langle \mathbf{R}|\mathbf{v}|\mathbf{R}\rangle = 0$) so that

$$M_{\rm IC} = -\frac{e}{2Ac}\sum_{s}\bar{\mathbf{r}}_s \times \mathbf{v}_s, \quad (5)$$

where the sum runs over surface WFs only.

We now concentrate on this IC contribution and consider a vertical strip of which one horizontal section is sketched in Fig. 3, and choose vertical boundaries (dashed lines) to discriminate between interior and surface regions. Focusing on the right edge, we use labels s and s' to label WFs in the interior ($x < x_0$) and surface ($x > x_0$) regions, respectively. The vertical macroscopic current flowing in the right surface is

$$I_y = -\frac{e}{\Delta l}\sum_{s'}{}' v_{s',y}, \quad (6)$$

where the primed sum is further limited to WFs whose centers are inside a vertical segment of length Δl, equal to the vertical repeat unit. The current carried by the ith WF can be written in terms of contributions from its neighbors as

$$\mathbf{v}_i = -\frac{i}{\hbar}\langle \phi_i|[\mathbf{r}, H]|\phi_i\rangle = \sum_j \mathbf{v}_{\langle ji\rangle}, \quad (7)$$

where $\mathbf{v}_{\langle ji\rangle} = (2/\hbar)\,{\rm Im}\,\mathbf{r}_{ij}H_{ji}$ has the interpretation of a current "donated from WF j to WF i." Since $\mathbf{v}_{\langle ji\rangle} = -\mathbf{v}_{\langle ij\rangle}$, the total current carried by any subset of WFs can be computed as the sum of all $\mathbf{v}_{\langle ji\rangle}$ for which i is inside and j is outside the subset. Applying this to the piece of surface region considered above, Eq. (6) becomes

$$I_y = -\frac{e}{\Delta l}\sum_{s}\sum_{s'}{}' v_{\langle s,s'\rangle,y}. \quad (8)$$

Setting the boundary deep enough below the surface to be

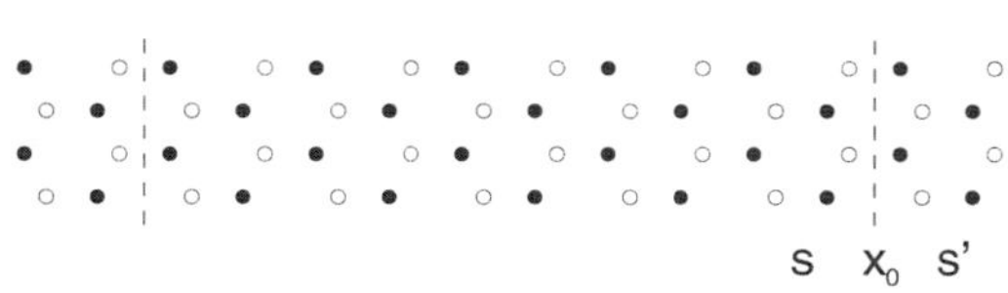

FIG. 3. Horizontal slice from a sample that extends indefinitely in the vertical direction but is otherwise similar to the one in Fig. 1. Vertical dashed lines delimit bulk and surface regions in which WFs are labeled by s and s', respectively.

in a bulklike region and invoking the exponential localization of the WFs and of derived quantities like $\mathbf{v}_{\langle ji\rangle}$, we can identify $|\phi_s\rangle$ and $|\phi_{s'}\rangle$ with the corresponding bulk WFs. Exploiting translational symmetry, $\mathbf{v}_{\langle \mathbf{R},\mathbf{R}'\rangle} = \mathbf{v}_{\langle \mathbf{0},\mathbf{R}'-\mathbf{R}\rangle}$, Eq. (8) becomes

$$I_y = -\frac{e}{\Delta l}\sum_{R_x<x_0}\sum_{R'_x>x_0}{}' v_{\langle \mathbf{0},\mathbf{R}'-\mathbf{R}\rangle,y}, \quad (9)$$

where the sum is still restricted to a segment of height Δl. The number of terms in Eq. (9) having a given value of $\mathbf{R}'-\mathbf{R}$ is just $(R'_x - R_x)\Delta l/A_0$ if $(R'_x - R_x) > 0$ and zero otherwise. With a change of summation index, Eq. (9) becomes

$$I_y = -\frac{e}{2A_0}\sum_{\mathbf{R}} R_x v_{\langle \mathbf{0},\mathbf{R}\rangle,y}, \quad (10)$$

where the factor of 2 enters because the sum has been extended to all $\mathbf{R}$. For a boundary fragment of arbitrary orientation, Eq. (10) generalizes to $I_\alpha = \sum_\beta G_{\alpha\beta}\hat{n}_\beta$, where $\hat{\mathbf{n}}$ is the unit normal to the boundary, and

$$G_{\alpha\beta} = -\frac{e}{A_0\hbar}\sum_{\mathbf{R}}{\rm Im}\langle \mathbf{R}|r_\alpha|\mathbf{0}\rangle\langle \mathbf{0}|H|\mathbf{R}\rangle R_\beta. \quad (11)$$

The contribution of this itinerant current to the magnetic moment $(1/2c)\oint(\mathbf{r}\times\mathbf{I})dl$ is easily seen to be related to the antisymmetric part [19] of G, so that

$$M_{\rm IC} = \frac{1}{c}G^A_{xy} = \frac{1}{2c}(G_{yx} - G_{xy}). \quad (12)$$

Eqs. (11) and (12) constitute our first major result, showing that the itinerant circulation contribution to the orbital magnetization can indeed be expressed as a bulk property in terms of the bulk WFs alone.

In the remainder of this Letter, we show that the two contributions $M_{\rm LC}$ and $M_{\rm IC}$ can both be converted into k-space expressions that can be evaluated directly in the Bloch representation. The WFs are defined via

$$|\mathbf{R}\rangle = \frac{A_0}{(2\pi)^2}\int d^2k\, e^{i\mathbf{k}\cdot(\mathbf{r}-\mathbf{R})}|u_{\mathbf{k}}\rangle, \quad (13)$$

where $|u_{\mathbf{k}}\rangle = e^{-i\mathbf{k}\cdot\mathbf{r}}|\psi_{\mathbf{k}}\rangle$ is the cell-periodic part of the Bloch function $|\psi_{\mathbf{k}}\rangle$. Inserting Eq. (1) into Eq. (4) and using $\mathbf{r}\times\mathbf{r} = 0$, it follows that

$$M_{\rm LC} = \frac{e}{2A_0\hbar c}{\rm Im}\langle \mathbf{0}|\mathbf{r}\times H\mathbf{r}|\mathbf{0}\rangle. \quad (14)$$

Defining $H_{\mathbf{k}} = e^{-i\mathbf{k}\cdot\mathbf{r}}He^{i\mathbf{k}\cdot\mathbf{r}}$ and using that

$$\mathbf{r}|\mathbf{R}\rangle = i\frac{A_0}{(2\pi)^2}\int d^2k\, e^{i\mathbf{k}\cdot(\mathbf{r}-\mathbf{R})}|\partial_{\mathbf{k}}u_{\mathbf{k}}\rangle, \quad (15)$$

Eq. (14) can be written [9] as a cross product

$$M_{\rm LC} = \frac{e}{2\hbar c}{\rm Im}\int\frac{d^2k}{(2\pi)^2}\langle \partial_{\mathbf{k}}u_{\mathbf{k}}|\times H_{\mathbf{k}}|\partial_{\mathbf{k}}u_{\mathbf{k}}\rangle, \quad (16)$$

between three-component bra and ket states.

In order to convert $M_{\rm IC}$ to k space, we note that the matrix elements appearing in Eq. (11) are given by

$$\langle \mathbf{0}|\mathbf{r}|\mathbf{R}\rangle = \frac{A_0}{(2\pi)^2}\int d^2k \mathbf{A}_\mathbf{k} e^{-i\mathbf{k}\cdot\mathbf{R}}, \qquad (17)$$

$$\langle \mathbf{0}|H|\mathbf{R}\rangle = \frac{A_0}{(2\pi)^2}\int d^2k E_\mathbf{k} e^{-i\mathbf{k}\cdot\mathbf{R}}, \qquad (18)$$

where the Berry connection $\mathbf{A}_\mathbf{k} = i\langle u_\mathbf{k}|\partial_\mathbf{k}|u_\mathbf{k}\rangle$ is real and $E_\mathbf{k}$ is the band energy. After some algebra including an integration by parts, we find

$$G_{\alpha\beta} = -\frac{e}{\hbar}\int \frac{d^2k}{(2\pi)^2} E_\mathbf{k}\partial_{k_\beta}A_{\mathbf{k}\alpha}. \qquad (19)$$

Inserting in Eq. (12) gives

$$M_{\rm IC} = -\frac{e}{2\hbar c}\int \frac{d^2k}{(2\pi)^2} E_\mathbf{k}\Omega_\mathbf{k}, \qquad (20)$$

where $\Omega_\mathbf{k} = \partial_\mathbf{k} \times \mathbf{A}_\mathbf{k}$ is the Berry curvature.

Interestingly, both magnetization contributions (16) and (20) are individually gauge invariant, i.e., insensitive to the choice of phases of the Bloch functions used to construct the WFs. This was shown for Eq. (16) in Ref. [9], and it follows immediately for Eq. (20) because $\Omega(\mathbf{k})$ is a gauge-invariant quantity. Each contribution is also invariant with respect to a shift of the zero of the Hamiltonian; the effect of such a shift is proportional to $\int d^2k\Omega_\mathbf{k} = 2\pi C$, where C is the Chern number which has been assumed to vanish.

Adding $M_{\rm LC}$ and $M_{\rm IC}$, placing both in a common form, and returning to three dimensions, we find that the total magnetization of the crystalline solid can be expressed as

$$\mathbf{M} = \frac{e}{2\hbar c}\,{\rm Im}\int \frac{d^3k}{(2\pi)^3}\langle\partial_\mathbf{k}u_\mathbf{k}| \times (H_\mathbf{k} + E_\mathbf{k})|\partial_\mathbf{k}u_\mathbf{k}\rangle. \quad (21)$$

This is our principal result. Equation (21) is consistent with Eq. (11) of Ref. [10], thereby providing a direct, fully quantum derivation of a result inferred there on the basis of semiclassical arguments alone. Note that Eq. (21) differs from the magnetization of an individual wave packet discussed in Ref. [11] in that $H_\mathbf{k}$ and $E_\mathbf{k}$ enter with the same relative sign.

To confirm the correctness of our formulation, we used discretized [9,21] versions of Eqs. (16) and (20) to calculate $M_{\rm LC}$, $M_{\rm IC}$, and M for the Haldane model using a 300×300 k-point mesh. The results, drawn as the lines in Fig. 2, are entirely consistent with the results for finite samples. We also carried out numerical tests which confirmed that the extrapolation of the results from $N \times N$ finite samples for $N = 10$, 20, and 30 to $N \to \infty$ shows very precise agreement with the k-space calculation on a dense mesh. We can thus be confident that the formal derivations are correct and that there is no longer any possibility that terms in the magnetization are being overlooked. We have carried out similar tests for other tight-binding models with reduced symmetry [18], with similar results.

In conclusion, we have derived a formula for the orbital magnetization of a crystalline system by working in the Wannier representation, and we have demonstrated its correctness via numerical tests. While limited to the case of a noninteracting zero-Chern-number insulator in a vanishing (or commensurate) magnetic field, our result nonetheless represents significant progress towards a more general theory of orbital magnetization. The resulting formula is easily evaluated in the context of a k-space electronic-structure code. The generalization to the multiband case will be discussed in a forthcoming publication. It remains tantalizingly uncertain whether such a Wannier-based approach can also be generalized to handle insulators with nonzero Chern numbers, metals, or arbitrary magnetic fields.

This work was supported by ONR Grant No. N00014-03-1-0570 and NSF Grant No. DMR-0233925.

[1] Charles Day, Phys. Today **58**, No. 2, 17 (2005).
[2] F. Mauri and S. G. Louie, Phys. Rev. Lett. **76**, 4246 (1996).
[3] D. Sebastiani and M. Parrinello, J. Phys. Chem. A **105**, 1951 (2001).
[4] F. Mauri, B.G. Pfrommer, and S.G. Louie, Phys. Rev. Lett. **77**, 5300 (1996); C.J. Pickard and F. Mauri, Phys. Rev. Lett. **88**, 086403 (2002).
[5] D. Sebastiani, G. Goward, I. Schnell, and M. Parrinello, Comput. Phys. Commun. **147**, 707 (2002).
[6] L. L. Hirst, Rev. Mod. Phys. **69**, 607 (1997).
[7] R.D. King-Smith and D. Vanderbilt, Phys. Rev. B **47**, R1651 (1993); D. Vanderbilt and R. D. King-Smith, Phys. Rev. B **48**, 4442 (1993).
[8] R. Resta, Rev. Mod. Phys. **66**, 899 (1994).
[9] R. Resta, D. Ceresoli, T. Thonhauser, and D. Vanderbilt, Chem. Phys. Chem. 6, 1815 (2005).
[10] D. Xiao, J. Shi, and Q. Niu, Phys. Rev. Lett. **95**, 137204 (2005).
[11] G. Sundaram and Q. Niu, Phys. Rev. B **59**, 14 915 (1999).
[12] F. D. M. Haldane, Phys. Rev. Lett. **61**, 2015 (1988).
[13] K. Ohgushi, S. Murakami, and N. Nagaosa, Phys. Rev. B **62**, R6065 (2000).
[14] T. Jungwirth, Q. Niu, and A. H. MacDonald, Phys. Rev. Lett. **88**, 207208 (2002).
[15] S. Murakami, N. Nagaosa, and S.-C. Zhang, Science **301**, 1348 (2003).
[16] Y. Yao, L. Kleinman, A. H. MacDonald, J. Sinova, T. Jungwirth, D.-S. Wang, E. Wang, and Q. Niu, Phys. Rev. Lett. **92**, 037204 (2004).
[17] This choice of parameters results in a zero Chern number.
[18] The antisymmetry of the total magnetization M about $\varphi = \pi/2$ in Fig. 2 is related to the presence of a subtle particle-hole symmetry in the Haldane model [9,12].
[19] The corresponding symmetric part G^S_{xy} is related to the time derivative of the spread of Wannier functions [20].
[20] N. Marzari and D. Vanderbilt, Phys. Rev. B **56**, 12 847 (1997).
[21] N. Sai, K. M. Rabe, and D. Vanderbilt, Phys. Rev. B **66**, 104108 (2002).

Berry Phase Correction to Electron Density of States in Solids

Di Xiao, Junren Shi, and Qian Niu

Department of Physics, The University of Texas, Austin, Texas 78712-0264, USA
(Received 14 February 2005; published 22 September 2005; corrected 5 October 2005)

Liouville's theorem on the conservation of phase-space volume is violated by Berry phase in the semiclassical dynamics of Bloch electrons. This leads to a modification of the phase-space density of states, whose significance is discussed in a number of examples: field modification of the Fermi-sea volume, connection to the anomalous Hall effect, and a general formula for orbital magnetization. The effective quantum mechanics of Bloch electrons is also sketched, where the modified density of states plays an essential role.

DOI: 10.1103/PhysRevLett.95.137204 PACS numbers: 73.43.−f, 72.15.−v, 75.20.−g

Semiclassical dynamics of Bloch electrons in external fields has provided a powerful theoretical framework to account for various properties of metals, semiconductors, and insulators [1]. In recent years, it has become increasingly clear that essential modification of the semiclassical dynamics is necessary for a proper understanding of a number of phenomena. It was known earlier that global geometric phase effects [2,3] on Bloch states are very important for insulators in our understanding of the quantum Hall effect [4], quantized adiabatic pumps [5], and electric polarization [6,7]. It was shown [8,9] later that geometric phase also modifies the local dynamics of Bloch electrons and thus affects the transport properties of metals and semiconductors. Recently these ideas have been successfully applied to the anomalous Hall effect in ferromagnetic semiconductors and metals [10–13], as well as spin transport [14,15].

In this Letter, we reveal a general property of the Berry phase modified semiclassical dynamics which has been overlooked so far: the violation of Liouville's theorem for the conservation of phase-space volume. Liouville's theorem was originally established for standard classical Hamiltonian dynamics, and its importance cannot be overemphasized as it serves as a foundation for classical statistical physics. The Berry phase makes, in general, the equations of motion noncanonical [8,9,16–18], rendering the violation of Liouville's theorem. Nevertheless, we are able to remedy the situation by modifying the density of states in the phase space.

This modified phase-space density of states enters naturally in the semiclassical expression for the expectation value of physical quantities, and has profound effects on equilibrium as well as transport properties. We demonstrate this with several examples. First, we consider a Fermi sea of electrons in a weak magnetic field, and show that the Fermi-sea volume can be changed linearly by the field. Second, we show how the Berry phase formula for the intrinsic anomalous Hall conductivity may be derived from equilibrium thermodynamics using the Středa formula [19]. Third, we provide a general derivation of an orbital-magnetization formula which is convenient for first-principles calculations.

In addition, we present an effective quantum mechanics for Bloch electrons in solids by quantizing the semiclassical dynamics with the geometric phase. The density of states enters in a nontrivial manner into the commutators of the phase-space coordinates, and relates directly to the minimal uncertainty volume in the phase space.

To begin with, we write down the semiclassical equations of motion for a Bloch electron in weak electric and magnetic fields [9]

$$\dot{\boldsymbol{r}} = \frac{1}{\hbar}\frac{\partial \varepsilon_n(\boldsymbol{k})}{\partial \boldsymbol{k}} - \dot{\boldsymbol{k}} \times \boldsymbol{\Omega}_n(\boldsymbol{k}), \tag{1a}$$

$$\hbar\dot{\boldsymbol{k}} = -e\boldsymbol{E}(\boldsymbol{r}) - e\dot{\boldsymbol{r}} \times \boldsymbol{B}(\boldsymbol{r}), \tag{1b}$$

where $\boldsymbol{\Omega}_n(\boldsymbol{k})$ is the Berry curvature of electronic Bloch states defined by $\boldsymbol{\Omega}_n(\boldsymbol{k}) = i\langle \boldsymbol{\nabla}_k u_n(\boldsymbol{k})| \times |\boldsymbol{\nabla}_k u_n(\boldsymbol{k})\rangle$ with $|u_n(\boldsymbol{k})\rangle$ being the periodic part of Bloch waves in the nth band; $\varepsilon_n(\boldsymbol{k})$ is the band energy with a correction due to the orbital magnetic moment [see Eq. (10) and above]. For crystals with broken time-reversal symmetry (such as ferromagnetic materials) or spatial inversion symmetry (such as GaAs), the Berry curvature $\boldsymbol{\Omega}_n(\boldsymbol{k})$ is nonzero.

To show the violation of Liouville's theorem, we consider the time evolution of a volume element $\Delta V = \Delta\boldsymbol{r}\Delta\boldsymbol{k}$ in the phase space. The equation of motion for ΔV is given by $(1/\Delta V)d\Delta V/dt = \boldsymbol{\nabla}_r \cdot \dot{\boldsymbol{r}} + \boldsymbol{\nabla}_k \cdot \dot{\boldsymbol{k}}$ [20]. A straightforward but somewhat tedious calculation shows that the right-hand side is equal to $-d\ln(1 + e\boldsymbol{B}\cdot\boldsymbol{\Omega}/\hbar)/dt$, which is a total time derivative. Therefore we can solve for the time evolution of the volume element and obtain

$$\Delta V = \Delta V_0/(1 + e\boldsymbol{B}\cdot\boldsymbol{\Omega}_n/\hbar). \tag{2}$$

The fact that the Berry curvature is generally $\boldsymbol{k}$ dependent (and the magnetic field can also depend on $\boldsymbol{r}$) implies that the phase-space volume element changes during time evolution of the state variables $(\boldsymbol{r}, \boldsymbol{k})$.

Nevertheless, we have a remedy to this breakdown of Liouville's theorem. Equation (2) shows that the volume element is a local function of the state variables (through the magnetic field and the Berry curvature) and has nothing to do with the history of time evolution. We can thus introduce a modified density of states

$$D_n(\boldsymbol{r}, \boldsymbol{k}) = (2\pi)^{-d}(1 + e\boldsymbol{B} \cdot \boldsymbol{\Omega}_n/\hbar), \tag{3}$$

such that the number of states in the volume element, $D_n(\boldsymbol{r}, \boldsymbol{k})\Delta V$, remains constant in time, where d is the spatial dimensionality of the system. The prefactor $(2\pi)^{-d}$ is obtained by demanding that the density of states $D_n(\boldsymbol{r}, \boldsymbol{k})$ reduces to the conventional form when the Berry curvature vanishes. As will be shown later, this density of states corresponds to the minimal quantum uncertainty volume of the state variables. Therefore, it does serve as the semiclassical measure for the number of quantum states per unit volume in the phase space. Based on this understanding, we write the classical phase-space probability density as

$$\rho_n(\boldsymbol{r}, \boldsymbol{k}, t) = D_n(\boldsymbol{r}, \boldsymbol{k}) f_n(\boldsymbol{r}, \boldsymbol{k}, t), \tag{4}$$

with $f_n(\boldsymbol{r}, \boldsymbol{k}, t)$ being the occupation number of the state labeled by $(\boldsymbol{r}, \boldsymbol{k})$. Probability conservation demands that $\rho_n(\boldsymbol{r}, \boldsymbol{k}, t)$ satisfies the continuity equation in phase space. On the other hand, our density of states satisfies $dD_n/dt = -(\nabla_{\boldsymbol{r}} \cdot \dot{\boldsymbol{r}} + \nabla_{\boldsymbol{k}} \cdot \dot{\boldsymbol{k}})D_n$. It then follows that the occupation number introduced above has the desired property of being invariant along the trajectory, i.e., $df_n/dt = 0$ [21].

We can thus write the real space density of a physical observable $\hat{O}$ in the form [22]

$$\bar{O}(\boldsymbol{R}) = \sum_n \int d\boldsymbol{k} D_n(\boldsymbol{r}, \boldsymbol{k}) f_n(\boldsymbol{r}, \boldsymbol{k}, t) \langle \hat{O}\delta(\hat{\boldsymbol{r}} - \boldsymbol{R})\rangle_{\boldsymbol{r}\boldsymbol{k}n} \tag{5}$$

where $\langle\cdots\rangle_{\boldsymbol{r}\boldsymbol{k}n}$ denotes the expectation value in the wave-packet state centered at $(\boldsymbol{r}, \boldsymbol{k})$ with the band index n. In the spatially homogeneous case, it reduces to

$$\bar{O} = \sum_n \int d\boldsymbol{k} D_n(\boldsymbol{k}) f_n(\boldsymbol{k}) O_n(\boldsymbol{k}), \tag{6}$$

where $O_n(\boldsymbol{k})$ is the expectation value of $\hat{O}$ in a Bloch state. For simpler notation, we will drop the band index n and assume that the integral over $\boldsymbol{k}$ includes the sum over n.

We now discuss the magnitude of the correction term $e\boldsymbol{B} \cdot \boldsymbol{\Omega}/\hbar$ to the density of states in Eq. (3). The Berry curvature for several materials has been calculated before using the first-principles method [11,12]. Over large regions of the Brillouin zone, its magnitude is on the order of a^2 with a being the lattice constant. Thus, $e\boldsymbol{B} \cdot \boldsymbol{\Omega}/\hbar \sim eBa^2/\hbar$ is the ratio of the magnetic flux through a unit cell to the magnetic flux quantum, and can be 10^{-2} to 10^{-3} for a magnetic field of 1 T. In the vicinity of some isolated points, the Berry curvature can be several orders of magnitude higher, leading to bigger effects for measurement. In the following, we will present a number of applications of our formula Eq. (6).

In our first example, we consider the quantity of electron density and show that the Fermi-sea volume can be changed linearly by a magnetic field when the Berry curvature is nonzero. Assuming zero temperature and using Eq. (3), we have the electron density as

$$n_e = \int^{\mu} \frac{d\boldsymbol{k}}{(2\pi)^d}\left(1 + \frac{e\boldsymbol{B} \cdot \boldsymbol{\Omega}}{\hbar}\right), \tag{7}$$

where the upper limit means that the integral is over states with energies below the chemical potential μ. Noting that the electron density is fixed by the background charge density, we conclude that the Fermi volume must change with the magnetic field. To first order, this change is given by

$$\delta V_F = -\int^{\mu_0} d\boldsymbol{k} \frac{e\boldsymbol{B} \cdot \boldsymbol{\Omega}}{\hbar}. \tag{8}$$

We note that while Landau levels make the Fermi-sea volume oscillate with the field, the effect described above gives an overall shift on average. Such a shift has important implications to those Fermi-surface related behaviors such as transport properties. For instance, in metals, it can induce a magnetoresistance linearly depending on the magnetic field. On the other hand, in band insulators, the $\boldsymbol{k}$ space is limited to the Brillouin zone. Electrons must populate a higher band if $(e/\hbar)\int_{\text{BZ}} d\boldsymbol{k}\boldsymbol{B} \cdot \boldsymbol{\Omega}$ is negative. When this quantity is positive, holes must appear at the top of the valence bands. Discontinuous behavior of physical properties in a magnetic field is therefore expected for band insulators with a nonzero integral of the Berry curvatures (Chern numbers).

In our second example, we show a connection between our phase-space density of states to the intrinsic anomalous Hall effect, which is due to spin-orbit coupling in the band structure of a ferromagnetic crystal. In the context of the quantum Hall effect, Středa derived a formula relating the Hall conductivity to the field derivative of the electron density at a fixed chemical potential [19], $\sigma_{xy} = -e(\partial n_e/\partial B_z)_\mu$. There is a simple justification of this relation by a thermodynamic argument by considering the following adiabatic process in two dimensions. A time dependent magnetic flux generates an electric field with an emf around the boundary of some region; and the Hall current leads to a net flow of electrons across the boundary and thus a change of electron density inside. This argument can be straightforwardly applied to the case of anomalous Hall effect and to three dimensions. By taking the derivative of the electron density (7) with respect to $\boldsymbol{B} = B\hat{z}$ at fixed chemical potential, we find that

$$\sigma_{xy} = -\frac{e^2}{\hbar}\int^{\mu} \frac{d\boldsymbol{k}}{(2\pi)^d}\Omega_z. \tag{9}$$

This is an intrinsic effect because it is independent of scattering, and thus differs from conventional skew scattering and side jump mechanisms [10–13].

As a third example of application, we now derive a semiclassical formula for orbital magnetization. In the semiclassical picture, a Bloch electron is modeled by a wave packet in a Bloch band, which is found to rotate about its center of mass in general, yielding an intrinsic magnetic moment given by $\boldsymbol{m}(\boldsymbol{k}) = -\mathrm{i}(e/2\hbar)\langle\boldsymbol{\nabla}_{\boldsymbol{k}}u| \times [\hat{H}_0(\boldsymbol{k}) - \varepsilon_0(\boldsymbol{k})]|\boldsymbol{\nabla}_{\boldsymbol{k}}u\rangle$, where $\hat{H}_0$ is the Hamiltonian [23]. In the presence of a weak magnetic field $\boldsymbol{B}$, the electron band structure energy $\varepsilon_0(\boldsymbol{k})$ (which may already include Zeeman energy from spin magnetization) acquires a correction term from this intrinsic orbital moment [8,9], $\varepsilon(\boldsymbol{k}) = \varepsilon_0(\boldsymbol{k}) - \boldsymbol{m}(\boldsymbol{k}) \cdot \boldsymbol{B}$. For an equilibrium ensemble of electrons, the total orbital magnetization can be found from the total energy, which is given by Eq. (6) as,

$$E = \int^{\mu} \frac{d\boldsymbol{k}}{(2\pi)^d}\left(1 + \frac{e\boldsymbol{B}\cdot\boldsymbol{\Omega}}{\hbar}\right)\left(\varepsilon_0(\boldsymbol{k}) - \boldsymbol{m}(\boldsymbol{k})\cdot\boldsymbol{B}\right). \quad (10)$$

Taking the differential of E with respect to $\boldsymbol{B}$, we obtain the magnetization at zero magnetic field to be

$$\begin{aligned}\boldsymbol{M} &= \int^{\mu_0} \frac{d\boldsymbol{k}}{(2\pi)^d}\left(\boldsymbol{m}(\boldsymbol{k}) + \frac{e\boldsymbol{\Omega}}{\hbar}[\mu_0 - \varepsilon_0(\boldsymbol{k})]\right) \\ &= \frac{e}{2\hbar}\int^{\mu_0} \frac{d\boldsymbol{k}}{(2\pi)^d}\mathrm{i}\left\langle\frac{\partial u}{\partial \boldsymbol{k}}\right| \times [2\mu_0 - \varepsilon_0(\boldsymbol{k}) - \hat{H}_0]\left|\frac{\partial u}{\partial \boldsymbol{k}}\right\rangle.\end{aligned} \quad (11)$$

In the upper line of the above expression, the first term is the contribution from the intrinsic orbital moment of each Bloch electron, and the second term comes from the explicit field dependence of the density of states and the resulting change in the Fermi volume in Eq. (8). We expect this effect to be important in ferromagnetic materials with strong spin-orbit coupling.

Gat and Avron obtained an equivalent result for the special case of the Hofstadter model [24]. Our derivation provides a more general formula that is applicable to other systems. Following the discussions on band insulators in our first example, there will be a discontinuity of the orbital magnetization if the integral of the Berry curvature over the Brillouin zone, or the anomalous Hall conductivity, is nonzero and quantized. Depending on the direction of the field, the chemical potential μ_0 in the above formula should be taken at the top of the valence bands or the bottom of the conduction bands. The size of the discontinuity is given by the quantized anomalous Hall conductivity times E_g/e, where E_g is the energy gap. For insulators with zero Chern numbers, the orbital magnetization can be directly evaluated from Wannier functions, with results consistent with our general formula [25]. Our general formula can also be derived from a full quantum mechanical linear response analysis [26].

The central result of this Letter, Eq. (3), can be extended to the more general case when Berry curvature includes the components of $\overleftrightarrow{\Omega}^{kr}$ as well as $\overleftrightarrow{\Omega}^{kk}$ and $\overleftrightarrow{\Omega}^{rr}$ [9]. In this case, we introduce the Berry curvature in phase space,

$$\overleftrightarrow{\Omega} = \begin{pmatrix} \overleftrightarrow{\Omega}^{rr} & \overleftrightarrow{\Omega}^{rk} \\ \overleftrightarrow{\Omega}^{kr} & \overleftrightarrow{\Omega}^{kk} \end{pmatrix}, \quad (12)$$

where each block is a 3×3 matrix; $\overleftrightarrow{\Omega}^{rk} = -(\overleftrightarrow{\Omega}^{kr})^T$. The phase-space density of states then reads,

$$D = (2\pi)^{-d}\sqrt{\det(\overleftrightarrow{\boldsymbol{\Omega}} - \overleftrightarrow{\mathbf{J}})}. \quad (13)$$

with

$$\overleftrightarrow{\mathbf{J}} = \begin{pmatrix} 0 & \overleftrightarrow{I} \\ -\overleftrightarrow{I} & 0 \end{pmatrix}.$$

In the special case of electromagnetic perturbations with $\overleftrightarrow{\Omega}^{kk}_{ab} = \epsilon_{abc}\Omega_c$, $\overleftrightarrow{\Omega}^{rr}_{ab} = -(e/\hbar)\epsilon_{abc}B_c$ and $\overleftrightarrow{\Omega}^{kr} = 0$, it reduces to (3). On the other hand, when either $\overleftrightarrow{\Omega}^{kk}$ or $\overleftrightarrow{\Omega}^{rr}$ vanishes, it has a simpler form

$$D = (2\pi)^{-d}\det(\overleftrightarrow{I} - \overleftrightarrow{\Omega}^{rk}). \quad (14)$$

This result has found application in the study of spin-force induced charge-Hall effect [27].

Finally, we show how the density of states emerges naturally in the effective quantum mechanics of Bloch electrons. Although our system is not canonical, it can nevertheless be quantized following a standard procedure developed for nonholonomic systems with second class constraints [28,29]. First, one redefines the Poisson bracket $\{f, g\}^* = (\partial f/\partial\xi^a)M_{ab}(\partial g/\partial\xi^b)$, where ξ^a are the components of phase-space coordinates $\boldsymbol{\xi} \equiv (\boldsymbol{r}, \boldsymbol{k})$ and $\overleftrightarrow{\mathbf{M}} = (\overleftrightarrow{\boldsymbol{\Omega}} - \overleftrightarrow{\mathbf{J}})^{-1}$. Our equations of motion (1) can then be written as $\dot{\xi}^a = \{\xi^a, \varepsilon\}^*$, where the energy $\varepsilon(\boldsymbol{\xi})$ plays the role as the Hamiltonian function. Then, one promotes the Poisson brackets into quantum commutators:

$$[\hat{\xi}^a, \hat{\xi}^b] = \mathrm{i}M_{ab}, \quad (15)$$

where $\hat{\xi}^a$ is the quantum operator corresponding to the phase-space coordinates. It then follows that a phase-space point acquires a minimal uncertainty volume given by [30]

$$\min\left(\prod_a \Delta\xi^a\right) = 2^{-d}[\det(\overleftrightarrow{\boldsymbol{\Omega}} - \overleftrightarrow{\mathbf{J}})]^{-1/2}. \quad (16)$$

This can be understood as the phase-space volume occupied by a single quantum state, therefore Eq. (13), which is proportional to the reciprocal of this volume, can naturally be regarded as the semiclassical expression for the number of quantum states per unit volume in the phase space.

Equation (15) presents the effective quantum mechanics of Bloch electrons. As a demonstration for the validity of the quantization scheme as well as the quantum effect of the phase-space density of states, we consider a simple toy model of a two-dimensional electron system with a constant Berry curvature, subjected to a uniform magnetic field. The commutators read,

$$\begin{aligned}&[\hat{x}, \hat{y}] = \mathrm{i}\frac{\Omega}{1 + (e/\hbar)B\Omega}, \qquad [\hat{k}_x, \hat{k}_y] = -\mathrm{i}\frac{(e/\hbar)B}{1 + (e/\hbar)B\Omega}, \\ &[\hat{x}, \hat{k}_x] = [\hat{y}, \hat{k}_y] = \mathrm{i}\frac{1}{1 + (e/\hbar)B\Omega}.\end{aligned} \quad (17)$$

In the absence of the Berry curvature, we reduce the problem to a known case with the familiar nontrivial commutator $[\hat{k}_x, \hat{k}_y] = -\mathrm{i}(e/\hbar)B$. In the absence of the B field, we have the nontrivial commutator $[\hat{x}, \hat{y}] = \mathrm{i}\Omega$ discussed extensively in the literature on noncommutative geometry. It is interesting to see that in the presence of both fields, we do not just have a combination of these nontrivial commutators. Instead, we have a nontrivial density of states which enters into all of the commutators.

Assuming $\varepsilon(\boldsymbol{k}) = \hbar^2\boldsymbol{k}^2/2m$, the system can be solved algebraically to yield the energy spectrum and degeneracy. We found that the spectrum consists of a set of Landau levels with the renormalized cyclotron frequency $\omega_c = \omega_c^0/[1 + (e/\hbar)B\Omega]$, where $\omega_c^0 = eB/m$ is the usual cyclotron frequency [31]. At the same time, it is more important to note that each Landau level still has the same degeneracy of eB/h as in the absence of the Berry curvature. It is known that this degeneracy is directly related to the quantized Hall conductance e^2/h for a filled Landau level [32]. Had the density of states not entered in the commutators, the Landau level degeneracy would be modified, violating the topological requirement that the Hall conductance for a filled Landau level is quantized.

Before closing, we note that the phase-space density of states also enters naturally in the alternative quantization scheme with the Feynman path integral. The S matrix is calculated by [28]

$$\langle \mathrm{out}|S|\mathrm{in}\rangle = \int \prod_t [D(\boldsymbol{\xi})d\boldsymbol{\xi}] \exp\left[\frac{\mathrm{i}}{\hbar}\int L dt\right]. \tag{18}$$

where L is the Lagrangian for our system [9],

$$L = \frac{1}{2}\dot{\xi}^a J_{ab}\xi^b - \varepsilon(\boldsymbol{\xi}) + \dot{\xi}^a \mathcal{A}_a(\boldsymbol{\xi}) \tag{19}$$

with $\mathcal{A}_a(\boldsymbol{\xi}) \equiv \mathrm{i}\langle u(\boldsymbol{\xi})|\nabla_\alpha u(\boldsymbol{\xi})\rangle$ being the phase-space gauge potentials associated with the Berry curvature field $\vec{\boldsymbol{\Omega}}$.

In summary, we have found a Berry phase correction to the phase-space density of states for Bloch electrons. This correction emerges naturally in both semiclassical and quantum mechanics of Bloch electrons, and has profound effects on the equilibrium and transport properties. Because of the fundamental change introduced by this correction, it could have important implications on other aspects of condensed matter physics, such as the Fermi liquid theory. For instance, in the presence of a magnetic field, interaction between electrons can change the Fermi-sea volume by modifying the Berry curvature and thus the phase-space density of states.

We acknowledge useful discussions with M. C. Chang, R. Resta, G. Y. Guo, Y. G. Yao, and D. Vanderbilt. This work is supported by DOE Grant No. DE-FG03-02ER45958.

[1] N. W. Ashcroft and N. D. Mermin, *Solid State Physics* (Saunders, Philadelphia, 1976).
[2] M. V. Berry, Proc. R. Soc. A **392**, 45 (1984).
[3] A. Bohm, A. Mostafazadeh, H. Koizumi, Q. Niu, and J. Zwanziger, in *The Geometric Phase in Quantum Systems: Foundations, Mathematical Concepts, and Applications in Molecular and Condensed Matter Physics* (Springer-Verlag, Berlin, 2003).
[4] D. J. Thouless, M. Kohmoto, M. P. Nightingale, and M. den Nijs, Phys. Rev. Lett. **49**, 405 (1982).
[5] D. J. Thouless, Phys. Rev. B **27**, 6083 (1983).
[6] R. D. King-Smith and D. Vanderbilt, Phys. Rev. B **47**, R1651 (1993).
[7] R. Resta, Rev. Mod. Phys. **66**, 899 (1994).
[8] M.-C. Chang and Q. Niu, Phys. Rev. B **53**, 7010 (1996).
[9] G. Sundaram and Q. Niu, Phys. Rev. B **59**, 14 915 (1999).
[10] T. Jungwirth, Q. Niu, and A. H. MacDonald, Phys. Rev. Lett. **88**, 207208 (2002).
[11] Z. Fang *et al.*, Science **302**, 92 (2003).
[12] Y. Yao *et al.*, Phys. Rev. Lett. **92**, 37 204 (2004).
[13] F. D. M. Haldane, Phys. Rev. Lett. **93**, 206602 (2004).
[14] S. Murakami, N. Nagaosa, and S.-C. Zhang, Science **301**, 1348 (2003).
[15] J. Sinova *et al.*, Phys. Rev. Lett. **92**, 126603 (2004).
[16] H. Kuratsuji and S. Iida, Phys. Rev. D **37**, 441 (1988).
[17] R. G. Littlejohn and W. G. Flynn, Phys. Rev. A **44**, 5239 (1991).
[18] G. Panati, H. Spohn, and S. Teufel, Commun. Math. Phys. **242**, 547 (2003).
[19] P. Středa, J. Phys. C **15**, L717 (1982).
[20] L. E. Reichl, *A Modern Course in Statistical Physics* (University of Texas, Austin, 1980).
[21] In the presence of collision, the function satisfies the usual Boltzmann equation:

$$\frac{df_n}{dt} \equiv \left(\frac{\partial}{\partial t} + \dot{\boldsymbol{r}}\cdot\nabla_{\boldsymbol{r}} + \dot{\boldsymbol{k}}\cdot\nabla_{\boldsymbol{k}}\right)f_n(\boldsymbol{k},\boldsymbol{r},t) = \left(\frac{\partial f_n}{\partial t}\right)_{\mathrm{coll}}.$$

The right-hand side denotes the collision contribution.
[22] D. Culcer *et al.*, Phys. Rev. Lett. **93**, 046602 (2004).
[23] There is a typo in Ref. [9]. The expression for orbital magnetic moment in Eq. (3.6) misses a factor of $-1/2$. This term is equivalent to the Rammal-Wilinson term as discussed in Ref. [18].
[24] O. Gat and J. E. Avron, Phys. Rev. Lett. **91**, 186801 (2003).
[25] R. Resta, D. Ceresoli, T. Thonhauser, and D. Vanderbilt, Chem. Phys. Chem. **6**, 1815 (2005); T. Thonhauser, D. Ceresoli, D. Vanderbilt, and R. Resta, Phys. Rev. Lett. **95**, 137205 (2005).
[26] J. Shi, D. Xiao, and Q. Niu (to be published).
[27] P. Zhang, J. Shi, D. Xiao, and Q. Niu (to be published).
[28] P. Senjanovic, Ann. Phys. (N.Y.) **100**, 227 (1976).
[29] D. M. Gitman and I. V. Tyutin, *Quantization of Fields with Constraints* (Springer-Verlag, New York, 1990).
[30] H. P. Robertson, Phys. Rev. **46**, 794 (1934).
[31] P. A. Horváthy, L. Martina, and P. C. Stichel, Phys. Lett. B **615**, 87 (2005).
[32] The Hall conductance is given by $\sigma = ne/B$, where n is the electron density. The quantized value is obtained if we substitute the degeneracy density for n.

Quantum Theory of Orbital Magnetization and Its Generalization to Interacting Systems

Junren Shi,[1] G. Vignale,[2] Di Xiao,[3] and Qian Niu[3]

[1]*Institute of Physics and ICQS, Chinese Academy of Sciences, Beijing 100080, China*
[2]*Department of Physics, University of Missouri, Columbia, Missouri 65211, USA*
[3]*Department of Physics, The University of Texas at Austin, Austin, Texas 78712, USA*

(Received 28 April 2007; published 9 November 2007)

Based on standard perturbation theory, we present a full quantum derivation of the formula for the orbital magnetization in periodic systems. The derivation is generally valid for insulators with or without a Chern number, for metals at zero or finite temperatures, and at weak as well as strong magnetic fields. The formula is shown to be valid in the presence of electron-electron interaction, provided the one-electron energies and wave functions are calculated self-consistently within the framework of the exact current and spin-density functional theory.

DOI: 10.1103/PhysRevLett.99.197202 PACS numbers: 75.10.Lp, 73.20.At

Magnetism is one of the most important properties of materials. Both spin and orbital motion of electrons can contribute to the total magnetization. While the spin magnetization can already be calculated from first principles with high accuracy by state-of-art methods such as the spin-density functional theory (SDFT), the study of orbital magnetization is still in a comparatively primitive stage.

A first difficulty arises from the fact that there is still no theoretically well-established formula for calculating the orbital magnetization of a crystalline solid. The nonlocality of the orbital magnetization operator $\hat{\boldsymbol{M}} = -(e/2)\hat{\boldsymbol{r}} \times \hat{\boldsymbol{v}}$ is the major obstacle to obtaining a closed formula for an extended periodic system. Recently, Xiao *et al.* [1] and, independently, Thonhauser *et al.* [2,3], obtained an orbital magnetization formula which avoids the nonlocality problem and looks very promising for applications. However, up to date, there exists no general quantum mechanical derivation of this formula. The derivation presented in Ref. [1] relies on the semiclassical wave-packet dynamics of Bloch electrons [1,4,5], and its validity in the quantum context is not completely clear. On the other hand, the derivation presented in Ref. [2] is quantum mechanical but relies on the existence of localized Wannier functions, and it cannot be easily generalized to metals or insulators with nonzero Chern number. In addition, both derivations are limited to noninteracting systems. The shortcomings of these approaches call for a full quantum mechanical and many-body theory of the orbital magnetization.

A second difficulty is that a first principle calculation of the orbital magnetization (taking into account many-body effects) should be based on the spin current density functional theory (SCDFT) [6] rather than the conventional SDFT. Unfortunately, SCDFT has been hindered so far by the lack of reliable expressions for the magnetization-dependent effective potentials. This may partly explain why the orbital moments of ferromagnetic transition metals such as Fe, Co, and Ni calculated in SCDFT were found to be significantly smaller than the experimentally determined values [7]. How problematic these calculations are is well explained in the review article by Richter [8]. The situation, however, has been rapidly changing in recent years. The advent of optimized effective potentials [9,10] which treat exchange exactly and may systematically be improved for correlations opens new avenues to the study of magnetic materials. Applications to atoms and molecules have already appeared in the literature, and applications to periodic systems are the obvious next step.

Against this background, the present Letter serves a dual purpose. First, we present a general derivation of orbital magnetization in periodic systems based on the standard perturbation theory of quantum mechanics. The derivation clarifies the origin of the novel aspects of the semiclassical derivation, such as the Berry phase correction to the density of states. It is generally valid for metals and insulators with or without a Chern number, at zero or finite temperatures, in weak or strong magnetic field, and, of course, in the presence of spin-orbit coupling. Second, and most important, we combine this derivation with the exact current and spin-density functional theory [6,11], proving the validity of the magnetization formula for interacting systems. We believe that the magnetization formula, in combination with the recent advances the construction of optimized effective potential for SCDFT, will turn out to be a powerful practical tool for the study of systems that have long defied traditional *ab initio* methods [8].

We start from the standard thermodynamic definition of the orbital magnetization density:

$$\boldsymbol{M} = -\frac{1}{V}\left(\frac{\partial \Omega}{\partial \boldsymbol{B}}\right)_{T,\mu}, \tag{1}$$

where $\Omega = E - TS - \mu N$ is the grand thermodynamic potential, V is the total volume of the system, and $\boldsymbol{B}$ is a magnetic field that only couples to the orbital motion of electrons (but does not contribute to the Zeeman energy) [12]. For convenience of derivation, we will first calculate the auxiliary quantity

$$\tilde{\boldsymbol{M}} \equiv -\frac{1}{V}\left(\frac{\partial K}{\partial \boldsymbol{B}}\right)_{T,\mu}, \tag{2}$$

where $K = E - \mu N$. We have $\boldsymbol{M} = \tilde{\boldsymbol{M}} + T(\partial S/\partial \boldsymbol{B})_{T,\mu}$ and, making use of the Maxwell relation $(\partial S/\partial \boldsymbol{B})_{T,\mu} = (\partial \boldsymbol{M}/\partial T)_{\mu,\boldsymbol{B}}$, we have the simple relation back to the orbital magnetization:

$$\frac{\partial(\beta \boldsymbol{M})}{\partial \beta} = \tilde{\boldsymbol{M}}, \tag{3}$$

where $\beta \equiv 1/kT$.

In principle, one can evaluate $\tilde{\boldsymbol{M}}$ by employing the standard perturbation theory of quantum mechanics to calculate the energy correction due to a uniform magnetic field. However, such an approach will again hit the difficulty of the nonlocality of the orbital magnetization operator. To go around the difficulty, we apply an external magnetic field that has an infinitely slow spatial variation [13,14]:

$$\boldsymbol{B}(\boldsymbol{r}) = B\cos(qy)\hat{\boldsymbol{z}}. \tag{4}$$

The slow spatial variation of the field is controlled by the wave vector q, which will tend to zero at the end of the calculation. The correction to the energy density in this situation can be written, up to linear order in B, as

$$\delta K(\boldsymbol{r}) = -\tilde{\boldsymbol{M}} \cdot \boldsymbol{B}(\boldsymbol{r}). \tag{5}$$

We can then readout $\tilde{\boldsymbol{M}}$ and take the limit of $q \to 0$ for a uniform magnetic field.

Noninteracting periodic systems.—For clarity, we first carry out the perturbation calculation for noninteracting periodic systems. The single-particle Hamiltonian can be expanded as $\hat{H} = \hat{H}_0 + \hat{V}_{\boldsymbol{B}}$, where $\hat{H}_0$ is the unperturbed Hamiltonian, which yields the band dispersion ϵ_{nk} and the corresponding Bloch wave function $\psi_{nk}(\boldsymbol{r}) = \exp(i\boldsymbol{k}\cdot\boldsymbol{r})u_{nk}(\boldsymbol{r})$, and $\hat{V}_{\boldsymbol{B}}$ denotes the coupling to the external magnetic field:

$$\hat{V}_{\boldsymbol{B}} = \frac{e}{2}[\hat{\boldsymbol{v}}\cdot \boldsymbol{A}(\boldsymbol{r}) + \boldsymbol{A}(\boldsymbol{r})\cdot\hat{\boldsymbol{v}}], \tag{6}$$

where $\hat{\boldsymbol{v}}$ is the velocity operator, and $\boldsymbol{A}(\boldsymbol{r})$ is the vector potential

$$\boldsymbol{A}(\boldsymbol{r}) = -B\frac{\sin qy}{q}\hat{\boldsymbol{x}}, \tag{7}$$

which corresponds to the magnetic field discussed earlier. It is natural to define the grand-canonical ensemble energy density as $K(\boldsymbol{r}) = \sum_{nk} f_{nk}\mathrm{Re}\{\psi^*_{nk}(\boldsymbol{r})\hat{K}\psi_{nk}(\boldsymbol{r})\}$, where $\hat{K} = \hat{H} - \mu\hat{N}$, and f_{nk} is the occupation number of the single-electron states of band index n and crystal-momentum $\boldsymbol{k}$ [15]. To first order in the perturbation, three kinds of terms arise from changes in the occupation number, the operator $\hat{K}$, and the wave function:

$$\delta K(\boldsymbol{r}) = \mathrm{Re}\sum_{nk}\{\delta f_{nk}\psi^*_{nk}\hat{K}_0\psi_{nk} + f_{nk}\psi^*_{nk}\hat{V}_{\boldsymbol{B}}\psi_{nk} + f_{nk}(\psi^*_{nk}\hat{K}_0\delta\psi_{nk} + \delta\psi^*_{nk}\hat{K}_0\psi_{nk})\}. \tag{8}$$

The orbital magnetization can be determined from the appropriate Fourier component of the energy density, i.e.,

$$\tilde{M}_z = -\frac{2}{VB}\int d\boldsymbol{r}\,\delta K(\boldsymbol{r})\cos qy. \tag{9}$$

It is easy to verify that the first two terms of Eq. (8) does not contribute to $\tilde{M}_z$, and only the contribution from the change of the wave functions remains. The first order perturbation to the wave function reads:

$$\delta\psi_{nk}(\boldsymbol{r}) = -\frac{eB}{4iq}\sum_{n'}\left\{\frac{e^{i(\boldsymbol{k}+\boldsymbol{q})\cdot\boldsymbol{r}}|u_{n'k+q}\rangle\langle u_{n'k+q}|\hat{v}_x(\boldsymbol{k}) + \hat{v}_x(\boldsymbol{k}+\boldsymbol{q})|u_{nk}\rangle}{\epsilon_{nk} - \epsilon_{n'k+q}} - (\boldsymbol{q}\to -\boldsymbol{q})\right\}, \tag{10}$$

where $\hat{\boldsymbol{v}}(\boldsymbol{k}) \equiv \partial\hat{H}_0(\boldsymbol{k})/\partial(\hbar\boldsymbol{k})$ is the velocity operator and $\hat{H}_0(\boldsymbol{k})$ is defined from the unperturbed Hamiltonian by shifting the momentum operator with $\hbar\boldsymbol{k}$. The transformed Hamiltonian acts on the periodic functions, with u_{nk} being its eigenfunctions and the band energy ϵ_{nk} its eigenvalues. Making use of Eqs. (8)–(10), we have:

$$\tilde{\boldsymbol{M}}_z = \frac{e}{4q}\mathrm{Im}\sum_{nn'k}\frac{(\epsilon_{nk} + \epsilon_{n'k+q} - 2\mu)\langle u_{nk}|u_{n'k+q}\rangle\langle u_{n'k+q}|\hat{v}_x(\boldsymbol{k}) + \hat{v}_x(\boldsymbol{k}+\boldsymbol{q})|u_{nk}\rangle}{\epsilon_{nk} - \epsilon_{n'k+q}}(f_{nk} - f_{n'k+q}). \tag{11}$$

Taking the long-wavelength limit $q \to 0$, we obtain:

$$\begin{aligned}\tilde{M}_z = {} & \frac{e}{2}\mathrm{Im}\sum_{n\neq n'k}\frac{(\epsilon_{nk} + \epsilon_{n'k} - 2\mu)\langle u_{nk}|\partial u_{n'k}/\partial k_y\rangle\langle u_{n'k}|\hat{v}_x(\boldsymbol{k})|u_{nk}\rangle}{\epsilon_{nk} - \epsilon_{n'k}}(f_{nk} - f_{n'k}) \\ & + e\sum_{nk}(\epsilon_{nk} - \mu)\mathrm{Im}\left[v^n_x(\boldsymbol{k})\left\langle u_{nk}\,\middle|\,\frac{\partial u_{nk}}{\partial k_y}\right\rangle + \left\langle\frac{\partial u_{nk}}{\partial k_y}\,\middle|\,\hat{v}_x(\boldsymbol{k})\,\middle|\,u_{nk}\right\rangle + \frac{1}{2}\left\langle u_{nk}\,\middle|\,\frac{\partial\hat{v}_x(\boldsymbol{k})}{\partial k_y}\,\middle|\,u_{nk}\right\rangle\right]f'_{nk},\end{aligned} \tag{12}$$

where $f'_{nk} \equiv \partial f(\epsilon_{nk})/\partial\epsilon_{nk}$. The second term comes from the intraband contribution with $n = n'$. Equation (12) can be further simplified with the help of the relations $\langle u_{n'k}|\hat{v}_x(\boldsymbol{k})|u_{nk}\rangle = (1/\hbar)(\epsilon_{nk} - \epsilon_{n'k})\langle u_{n'k}|\partial u_{nk}/\partial k_x\rangle$ for $n \neq n'$, and $\mathrm{Im}[\ldots] = (1/\hbar)\mathrm{Im}\langle\partial u_{nk}/\partial k_y|\epsilon_n(\boldsymbol{k}) - \hat{H}_0(\boldsymbol{k})|\partial u_{nk}/\partial k_x\rangle$, where $[\ldots]$ denotes the expression inside the square bracket in Eq. (12). Combining these relations, and generalizing the result to the other components of $\tilde{\boldsymbol{M}}$, we obtain finally:

$$\tilde{M} = -\frac{e}{2\hbar} i \sum_{nk} \left\{ \left\langle \frac{\partial u_{nk}}{\partial k} \middle| [\epsilon_n(k) + \hat{H}_0(k) - 2\mu] \middle| \frac{\partial u_{nk}}{\partial k} \right\rangle f_{nk} - (\epsilon_{nk} - \mu) \left\langle \frac{\partial u_{nk}}{\partial k} \middle| [\epsilon_n(k) - \hat{H}_0(k)] \middle| \frac{\partial u_{nk}}{\partial k} \right\rangle f'_{nk} \right\}. \tag{13}$$

The auxiliary and proper orbital magnetization become the same at zero temperature. In this case, the second term in the above expression Eq. (13) vanishes because f' becomes a δ function of $(\epsilon_{nk} - \mu)$. The result is in perfect agreement with the semiclassical formula of zero temperature orbital magnetization of Xiao *et al.* [1]. For finite temperatures, we integrate Eq. (3) and obtain:

$$M = \sum_{nk} \left\{ m_n(k) f_{nk} + \frac{e}{\hbar} \Omega_n(k) \frac{1}{\beta} \ln(1 + e^{-\beta(\epsilon_{nk} - \mu)}) \right\}, \tag{14}$$

where $m_n(k) \equiv (e/2\hbar) i \langle \nabla_k u_{nk} | [\epsilon_n(k) - \hat{H}_0(k)] \times | \nabla_k u_{nk} \rangle$ is the orbital moment of state n, k and $\Omega_n(k) \equiv i \langle \nabla_k u_{nk} | \times | \nabla_k u_{nk} \rangle$ is the Berry curvature. The same expression was also obtained in Ref. [16].

Thus, all the previously known results are recovered by our fully quantum mechanical formalism. These results are valid not only for insulators with or without a Chern number, but also for metals at zero or finite temperatures. This implies that the semiclassical results are in general valid to linear order in the external fields. In hindsight, this should have been expected, because the semiclassical theory is designed to be exact in the limit of long length scales in the perturbation to the Hamiltonian. In our case, this length scale (through the vector potential) does diverge in the limit of vanishing magnetic field.

Generalization to interacting systems.—It is very desirable to generalize the above results to an interacting system. This can be done exactly within the framework of the current and spin-density functional theory (CSDFT) [6,11]. CSDFT is a generalization of the spin-density functional theory which includes the current density as an independent variable for the energy functional and thus provides direct access, via a variational principle of the Hohenberg-Kohn type, to the current density and the orbital magnetization of the thermodynamic equilibrium ensemble [7]. Following the formalism of Ref. [6], the many-body problem can be reduced to solving an effective one-body Schrödinger equation:

$$\left\{ \frac{1}{2m} [-i\hbar\nabla + eA'_\sigma(r)]^2 + V'_\sigma(r) \right\} \psi_{i\sigma}(r) = \epsilon_{i\sigma} \psi(r), \tag{15}$$

with

$$V'_\sigma = V_\sigma + V_H + V_{xc\sigma} + \frac{e^2}{2m}(A_\sigma^2 - A_\sigma'^2), \tag{16}$$

$$A'_\sigma = A_\sigma + A_{xc\sigma}. \tag{17}$$

Here V_σ and A_σ are the external scalar and vector potential, respectively, acting on the σ component of the spin ($\sigma = \uparrow$ or $\downarrow$); $V_H = e^2 \int dr' n(r')/|r - r'|$ is the Hartree potential, and $V_{xc\sigma}$ and $A_{xc\sigma}$ are the exchange-correlation (xc) scalar and vector potentials derived from the xc energy functional $\Omega_{xc}[n_\sigma, j_{p\sigma}]$ according to the formulas $V_{xc\sigma} = \delta\Omega_{xc}/\delta n_\sigma$, $eA_{xc\sigma} = \delta\Omega_{xc}/\delta j_{p\sigma}$. The density, $n_\sigma(r)$, and the paramagnetic current density, $j_{p\sigma}(r)$, are to be determined self-consistently from the solutions of the above equation according to the formulas $n_\sigma(r) = \sum_i |\psi_{i\sigma}(r)|^2 f(\epsilon_{i\sigma})$ and $j_{p\sigma}(r) = (-i\hbar/2m) \sum_i [\psi^*_{i\sigma}(r) \nabla \psi_{i\sigma}(r) - \nabla \psi^*_{i\sigma}(r) \psi_{i\sigma}(r)] f(\epsilon_{i\sigma})$. At finite temperature, the thermodynamic potential functional can be written as [6]:

$$\Omega = -\frac{1}{\beta} \sum_{i\sigma} \ln[1 + e^{-\beta(\epsilon_{i\sigma} - \mu)}] - \frac{1}{2} e^2 \iint dr dr' \frac{n(r) n(r')}{|r - r'|}$$
$$- \sum_\sigma \int dr\, n_\sigma(r) V_{xc\sigma}(r) - e \sum_\sigma \int dr\, j_{p\sigma}(r) \cdot A_{xc\sigma}(r)$$
$$+ \Omega_{xc}[n_\sigma, j_{p\sigma}], \tag{18}$$

which is a functional of four fields: the densities n_σ and $j_{p\sigma}$, and the external potentials V_σ and A_σ—the last two enter the expression through the eigenvalues $\epsilon_{i\sigma}$.

To calculate the orbital magnetization one needs to evaluate the variation $\delta\Omega$ of the thermodynamic potential in response to a variation of the external magnetic field, which, in turn, is generated by a variation in the external vector potentials δA_σ. In general, $\delta\Omega$ can be separated into two contributions: the primary one ($\delta\Omega|_{n_\sigma, j_{p\sigma}}$) arises directly from the variation of the vector potentials, keeping n_σ and $j_{p\sigma}$ constant at their unperturbed values; the secondary one ($\delta\Omega|_{V_\sigma, A_\sigma}$) might arise from the changes of n_σ and $j_{p\sigma}$ at constant external potentials (these changes would affect Ω via the modification of the effective potentials V_H, $V_{xc\sigma}$, and $A_{xc\sigma}$):

$$\delta\Omega = \delta\Omega|_{n_\sigma, j_{p\sigma}} + \delta\Omega|_{V_\sigma, A_\sigma}. \tag{19}$$

It is easy to see that $\delta\Omega|_{n_\sigma, j_{p\sigma}}$ contributes an orbital magnetization that is exactly given by Eq. (14), as if the system were a noninteracting system with eigenfunctions and eigenvalues determined by Eq. (15). This is because the variation of the external vector potential affects only the eigenvalues $\epsilon_{i\sigma}$ in the one-body term of Eq. (18). To evaluate $\delta\Omega|_{V_\sigma, A_\sigma}$ it is sufficient to observe that for given external potentials the thermodynamic potential Ω is stationary against small changes of the density and the current about their equilibrium values: $\delta\Omega/\delta n_\sigma = \delta\Omega/\delta j_{p\sigma} = 0$.

Thus we have

$$\delta\Omega|_{V_\sigma, A_\sigma} = 0. \tag{20}$$

We then conclude that in the context of the CSDFT, we can

treat the system as an effective one-body system and use Eq. (14) to calculate the orbital magnetization, albeit using the dispersion and wave functions derived from Eq. (15).

We stress that this conclusion would not hold true if the one-electron orbitals and their energies were calculated within the framework of the *ordinary* density functional theory [17] (as opposed to CSDFT). In such a formulation the xc energy functional would be a functional of density *and* magnetic field: $\Omega_{xc}[n_\sigma, \boldsymbol{B}]$. Then the formula for the orbital magnetization would include a functional derivative of Ω_{xc} with respect to $\boldsymbol{B}$, at variance to the one-electron formula of Eq. (14).

Finite fields.—Our formula can be applied rigorously to finite magnetic fields if these fields are rational in the sense that fluxes through the faces of a unit cell are fractional multiples of the flux quantum h/e. In this case, one can define Bloch-like eigenstates with respect to magnetic translations, which are ordinary translations on the crystal combined with gauge transformations. Orbital magnetization at such a field can then be calculated perturbatively by adding a small change δB to this field. Both the semiclassical theory and quantum perturbation with respect to δB give the same expression for the orbital magnetization, provided we use the magnetic Bloch wave functions as a basis. Likewise, the CSDFT can also be formulated for systems with periodic boundary conditions with respect to magnetic translations, so the justification of our results for interacting systems is straightforward [6]. Indeed, current density functional theory has been used to study the formation of an electron crystal (Wigner crystal) at very high magnetic field [18].

The situation for irrational fields is a bit tricky if one insists on rigorous results [19]. One may consider an irrational field as a limit of sequence of rational fields. This is possible if the magnetization depends on the field continuously, which is expected to be the case when the temperature is finite. Indeed, when the fast de Haas-van Alphen oscillations are smeared out, the average magnetization changes continuously with the Fermi energy [19]. For a fixed Fermi energy, the average magnetization also changes continuously with the magnetic field. Therefore, we expect that the average magnetization is a continuous function of magnetic field at a fixed density of electrons.

In summary, we have presented a full quantum derivation of the orbital magnetization formula. The derivation is generally valid for insulators with or without a Chern number, for metals at zero or finite temperatures, and at weak as well as strong magnetic fields. We also find that the resulting formula is directly applicable to interacting systems provided one uses one-electron energies and wave functions obtained from the self-consistent solution of the Kohn-Sham equation of current and spin-density functional theory.

This work was supported by National Science Foundation of China No. 10604063 and No. 10610335. G. V. was supported by NSF (No. DMR-031368), D. X. was supported by NSF (No. DMR-0404252/0606485), and Q. N. by the Welch Foundation and DOE (No. DE-FG03-02ER45958).

[1] D. Xiao, J. Shi, and Q. Niu, Phys. Rev. Lett. **95**, 137204 (2005).
[2] T. Thonhauser, D. Ceresoli, D. Vanderbilt, and R. Resta, Phys. Rev. Lett. **95**, 137205 (2005).
[3] D. Ceresoli, T. Thonhauser, D. Vanderbilt, and R. Resta, Phys. Rev. B **74**, 024408 (2006).
[4] M.-C. Chang and Q. Niu, Phys. Rev. Lett. **75**, 1348 (1995).
[5] G. Sundaram and Q. Niu, Phys. Rev. B **59**, 14 915 (1999).
[6] G. Vignale and M. Rasolt, Phys. Rev. B **37**, 10 685 (1988).
[7] H. Ebert, M. Battocletti, and E. K. U. Gross, Europhys. Lett. **40**, 545 (1997).
[8] M. Richter, *Handbook of Magnetic Materials*, edited by H. H. J. Buschow (Elsevier, New York, 2001), Vol. 13, p. 87.
[9] S. Rohra and A. Görling, Phys. Rev. Lett. **97**, 013005 (2006).
[10] S. Pittalis, S. Kurth, N. Helbig, and E. K. U. Gross, Phys. Rev. A **74**, 062511 (2006).
[11] G. Vignale and M. Rasolt, Phys. Rev. Lett. **59**, 2360 (1987).
[12] The separation of spin and orbital degrees of freedom is only possible in a semirelativistic framework, such as the one provided by the Pauli equation. In a fully relativistic treatment orbital and spin magnetizations merge together, and our formula gives the total magnetization.
[13] G. Vignale, Phys. Rev. Lett. **67**, 358 (1991).
[14] F. Mauri and S. G. Louie, Phys. Rev. Lett. **76**, 4246 (1996).
[15] The definition here differs slightly in the kinetic energy density from the conventional one by a total divergence. Its perturbation by the magnetic field vanishes in the limit of $q \to 0$ and therefore does not affect our final result.
[16] D. Xiao, Y. Yao, Z. Fang, and Q. Niu, Phys. Rev. Lett. **97**, 026603 (2006).
[17] C. J. Grayce and R. A. Harris, Phys. Rev. A **50**, 3089 (1994).
[18] G. Vignale, Phys. Rev. B **47**, 10 105 (1993).
[19] O. Gat and J. E. Avron, Phys. Rev. Lett. **91**, 186801 (2003).

4

The Hall effects

Quantized Hall Conductance in a Two-Dimensional Periodic Potential

D. J. Thouless, M. Kohmoto,[a] M. P. Nightingale, and M. den Nijs
Department of Physics, University of Washington, Seattle, Washington 98195
(Received 30 April 1982)

The Hall conductance of a two-dimensional electron gas has been studied in a uniform magnetic field and a periodic substrate potential U. The Kubo formula is written in a form that makes apparent the quantization when the Fermi energy lies in a gap. Explicit expressions have been obtained for the Hall conductance for both large and small $U/\hbar\omega_c$.

PACS numbers: 72.15.Gd, 72.20.Mg, 73.90.+b

The experimental discovery by von Klitzing, Dorda, and Pepper[1] of the quantization of the Hall conductance of a two-dimensional electron gas in a strong magnetic field has led to a number of theoretical studies of the problem.[2-6] It has been concluded that a noninteracting electron gas has a Hall conductance which is a multiple of e^2/h if the Fermi energy lies in a gap between Landau levels, or even if there are tails of localized states from the adjacent Landau levels at the Fermi energy. However, it can be concluded from Laughlin's[2] argument that the Hall conductance is quantized whenever the Fermi energy lies in an energy gap, even if the gap lies within a Landau level. For example, it is known that if the electrons are subject to a weak sinusoidal perturbation as well as to the uniform magnetic field, with $\varphi=p/q$ magnetic-flux quanta per unit cell of the perturbing potential, each Landau level is split into p subbands of equal weight.[7] One might expect each of these subbands to give a Hall conductance equal to e^2/ph, and that is what the classical theory of the Hall current suggests, but according to Laughlin each subband must carry an integer multiple of the Hall current carried by the entire Landau level. This result appears even more paradoxical when it is realized that p, the number of subbands, can become arbitrarily large by an arbitrarily small change of the flux density. This paper contains a calculation of the Hall conductance for such a system, both in the limit of a weak periodic potential and in the tight-binding limit of a strong periodic potential. We have derived explicit expressions for the Hall currents carried by the various subbands, and show how the paradox is resolved.

We consider electrons in a potential $U(x,y)$ which is periodic in x,y with periods a, b, and in a uniform magnetic induction B perpendicular to the plane of the electrons. The band structure of such a system depends critically on $\varphi = abeB/h$, which is the number of flux quanta per unit cell. We take φ to be a rational number p/q; the behavior for irrational values of φ can be deduced by taking an appropriate limit. We use the Landau gauge in which the vector potential has components $(0, eBx)$. In this gauge the eigenfunctions of the Schrödinger equation can be chosen to satisfy the generalized Bloch condition

$$\psi_{k_1k_2}(x+qa,y)\exp(-2\pi ipy/b - ik_1qa) = \psi_{k_1k_2}(x,y+b)\exp(-ik_2b) = \psi_{k_1k_2}(x,y), \tag{1}$$

where k_1 (modulo $2\pi/aq$) and k_2 (modulo $2\pi/b$) are good quantum numbers.[8] We can now define functions $u_{k_1k_2} = \psi_{k_1k_2}\exp(-ik_1x - ik_2y)$ which satisfy the generalized periodic boundary conditions

$$u_{k_1k_2}(x+qa,y)e^{-2\pi ipy/b} = u_{k_1k_2}(x,y+b) = u_{k_1k_2}(x,y), \tag{2}$$

and are eigenfunctions of a Hamiltonian

$$\hat{H}(k_1,k_2) = \frac{1}{2m}\left(-i\hbar\frac{\partial}{\partial x} + \hbar k_1\right)^2 + \frac{1}{2m}\left(-i\hbar\frac{\partial}{\partial y} + \hbar k_2 - eBx\right)^2 + U(x,y). \tag{3}$$

The components of the velocity operator are then given by $\hbar^{-1}$ times the partial derivatives of $\hat{H}$ with respect to k_1, k_2.

There are two quite different approaches to the problem of calculating the Hall conductance σ_{H}. Laughlin[2] and Halperin[6] have studied the effects produced by changes in the vector potential on the states at the edges of a finite system. By this technique the quantization of the conductance is made explicit, but it is not obvious that the result is insensitive to boundary conditions. An alternative approach is to use the Kubo formula for a bulk two-dimensional conductor. In previous work using this method[3-5] it has not been made obvious that an integer value for the conductance must be obtained.

Because of the relation between the velocity operator and the derivatives of $\hat{H}$, the Kubo formula can be written as

$$\sigma_{\mathrm{H}} = \frac{ie^2}{A_0\hbar}\sum_{\epsilon_\alpha<E_F}\sum_{\epsilon_\beta>E_F}\frac{(\partial\hat{H}/\partial k_1)_{\alpha\beta}(\partial\hat{H}/\partial k_2)_{\beta\alpha} - (\partial\hat{H}/\partial k_2)_{\alpha\beta}(\partial\hat{H}/\partial k_1)_{\beta\alpha}}{(\epsilon_\alpha - \epsilon_\beta)^2}, \tag{4}$$

where A_0 is the area of the system and $\epsilon_\alpha, \epsilon_\beta$ are eigenvalues of the Hamiltonian. This can be related to the partial derivatives of the wave functions u, and gives

$$\begin{aligned}\sigma_{\mathrm{H}} &= \frac{ie^2}{2\pi h}\sum\int d^2k\int d^2r\left(\frac{\partial u^*}{\partial k_1}\frac{\partial u}{\partial k_2} - \frac{\partial u^*}{\partial k_2}\frac{\partial u}{\partial k_1}\right)\\ &= \frac{ie^2}{4\pi h}\sum\oint dk_j\int d^2r\left(u^*\frac{\partial u}{\partial k_j} - \frac{\partial u^*}{\partial k_j}u\right),\end{aligned} \tag{5}$$

where the sum is over the occupied electron subbands and the integrations are over the unit cells in r and k space. The integral over the k-space unit cell has been converted to an integral around the unit cell by Stokes's theorem. For nonoverlapping subbands ψ is a single-valued analytic function everywhere in the unit cell, which can only change by an r-independent phase factor θ when k_1 is changed by $2\pi/aq$ or k_2 by $2\pi/b$. The integrand reduces to $\partial\theta/\partial k_j$. The integral is $2i$ times the change in phase around the unit cell and must be an integer multiple of $4\pi i$.

The problem of evaluating this quantum number remains. We have considered the potential

$$U(x,y) = U_1\cos(2\pi x/a) + U_2\cos(2\pi y/b), \tag{6}$$

both in the limit of a weak periodic potential ($|U| \ll \hbar\omega_c$) and in the tight-binding limit of a strong periodic potential. In the weak-potential limit the wave function can be written as a superposition of the nearly degenerate Landau functions in the same Landau level N:

$$u_{k_1k_2} = \sum_{n=1}^{p} d_n \sum_{l=-\infty}^{\infty} \chi_N\left(x - \frac{\hbar k_2}{eB} - lqa - \frac{nqa}{p}\right) \exp\left[-ik_1\left(x - lqa - \frac{nqa}{p}\right) + 2\pi i y \frac{(lp+n)}{b}\right], \tag{7}$$

where χ_N is the appropriate oscillator wave function. Since the term $U_1 \cos(2\pi x/a)$ is diagonal in the Landau functions and $U_2 \cos(2\pi y/b)$ changes the value of n by unity, the amplitudes d_n satisfy the secular equation of the form

$$V \exp(-iqak_1/p)d_{n-1} + 2V' \cos(qbk_2/p + 2\pi nq/p)d_n + V \exp(iqak_1/p)d_{n+1} = Ed_n, \tag{8}$$

with V and V' proportional to U_2 and U_1,[7] and with $d_{n+p} = d_n$. This is known as Harper's equation.[9] Its spectrum has been studied in detail by Hofstadter[10] for the iostropic case $V = V'$. We have made considerable use of his results.

Numerical solution of the eigenvalue problem (8) and direct substitution in Eqs. (7) and (5) are possible but laborious. We have done this for a number of cases. The results obtained in this way are entirely concordant with the general results we have obtained by examining the limit $V \ll V'$. The quantum number of a subband cannot change unless the gaps close up; we believe that none of the gaps closes when V/V' is varied, but we have no proof of this. For small V/V' only one component d_n is appreciably different from zero, except for qbk_2 close to a multiple of π, where there is a changeover in the dominant component. These values of k_2 are the locations of the energy gaps which have opened for small V/V'. In the interval $0 \leq k_2 < 2\pi/b$, the rth gap repeats itself q times, at values of qbk_2 equal to an odd (even) multiple of π, if $p - r$ is odd (even). Here the rth and $(r+1)$th band change their dominant component d_n according to $n_r \to n_r + s_r$, $n_{r+1} \to n_{r+1} - s_r$. The value of s_r is independent of k_2 and determined by the Diophantine equation

$$r = s_r q + t_r p, \tag{9}$$

where $|s| \leq p/2$. The rth gap is of order $(V/V')^{|s_r|}$. Each time, the wave function of band r picks up a phase from the off-diagonal terms in Eq. (8): q times qak_1s_r/p and q times $-qak_1s_{r-1}/p$. The total phase change in d_n is

$$d_n(k_1, k_2) = d_n(k_1 + 2\pi/aq, k_2)$$
$$= d_n(k_1, k_2 + 2\pi/b)\exp[-i\theta_d(k_1)] \tag{10a}$$
$$\theta_d(k_1) = qak_1(t_{r-1} - t_r) + qak_1/p. \tag{10b}$$

The term qak_1/p in θ_d cancels against the phase change which is already explicit in Eq. (7). It represents the classical Hall current. The total Hall current carried by the rth band is quantized according to $\sigma_H = (e^2/h)(t_r - t_{r-1})$. If the Fermi surface is located in the rth gap of the Nth Landau level, the total Hall conductance is equal to

$$\sigma_H = (e^2/h)(t_r + N - 1), \tag{11}$$

with t_r the solution of Eq. (9). This has an unambiguous solution, expect for the case of p even, $r = \frac{1}{2}p$, where there is no gap.

For $q = 1$ this gives $t = 0$ in the first half of the Landau level and $t = 1$ in the second half, so that only the central subband of each Landau level carries the Hall current. For $q = 2$, the values of t are alternately 1 and 0, so that alternate subbands carry ± 1 times the Hall current of the Landau level. In general each subband carries one of two possible Hall currents which differ from one another by q units. For example, in the case $q/p = \frac{11}{7}$, the first 11 values of t are -3, 5, 2, -1, -4, 4, 1, -2, 6, 3, 0, so that the Hall current is proportional to -3 or 8 in each subband.

Figure 1 gives an intuitive explanation of the results. The abscissa represents the variable k_2, or equivalently the position of the center coordinate of the state in the x direction, while the ordinate gives the energy as a function of this variable. Application of an electric field in the y direc-

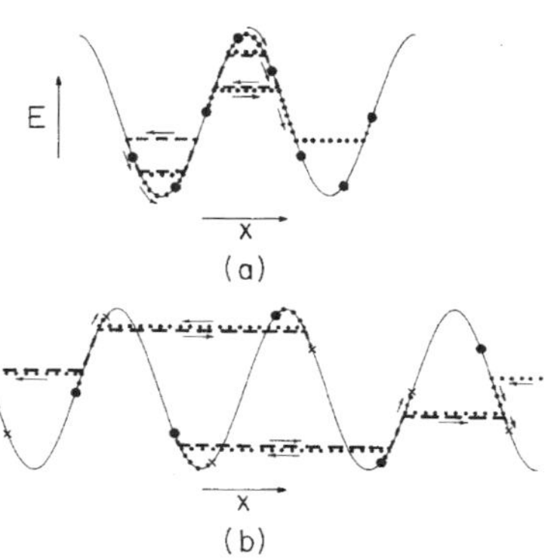

FIG. 1. Motion of electrons in the x direction under the influence of an electric field in the y direction for $V << V'$, for the two cases (a) $\varphi = 5$ and (b) $\varphi = \frac{5}{3}$.

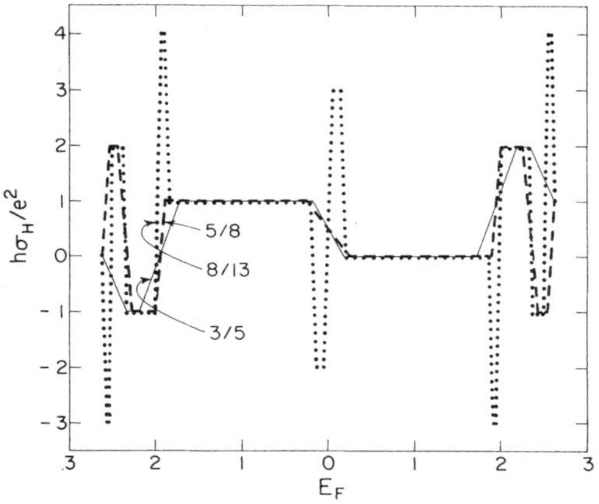

FIG. 2. Hall current as a function of Fermi energy for a weak isotropic substrate. Truncating the continued-fraction expansion of $\frac{1}{2}\sqrt{5}-\frac{1}{2}$, we obtained $q/p=\frac{3}{5}$, $\frac{5}{8}$, $\frac{8}{13}$, ... Within the energy bands, the Hall current was not calculated; the interpolation merely serves as a guide to the eye.

tion gives a steady change of k_2, and therefore a steady motion of a representative point along the $E(k_2)$ curve, until the energy gets very close to the energy of a state whose k_2 value differs by a multiple of $2\pi/b$. Then the two states exchange places rather than crossing in energy. Figure 1(a) shows the case $p=5$, $q=1$. Four of the orbits are closed and give no motion in the x direction, while a Hall current of +1 is given by the middle orbit. Figure 1(b) shows $p=5$, $q=3$, where three orbits give a negative Hall current equal to -1, and the other two give a positive current +2.

In the opposite limit of a strong potential U the same Eq. (8) can be obtained, where V and V' are the tight-binding matrix elements that take an electron from a site to its neighbors in the x and y directions, and φ is replaced by $1/\varphi$, so that p and q are interchanged.[9,10] The result is that the Hall conductance σ_H is equal to te^2/h. Again, t is given by Eq. (9), but now s is unconstrained and t must lie between $-\frac{1}{2}q$ and $+\frac{1}{2}q$. In the case $p/q=\frac{7}{11}$, the values of t are -3, 5, 2, -1, -4, 4, 1, -2, -5, 3, 0, which are the same as the weak-potential limit gives except for the ninth gap, where + 6 has been replaced by -5. This pattern is easy to understand, as p subbands constitute one Landau level, so we get a Hall current t if we count up tp (modulo q) levels from the bottom of the band, and $-t$ if we count down tp (modulo q) levels from the top of the band.

It is generally true that the larger the integer part of tp/q, the smaller will be the corresponding energy gap. The complexity of the structure is associated with the smaller energy gaps as illustrated in Fig. 2; for these smaller energy gaps, the electric field has to be very small for linear response theory to be valid. Stronger fields will cause tunneling across the gap. It is this restriction, and corresponding restrictions on disorder and deviations from the simple sinusoidal potential, that resolves the paradox of the sensitivity of the Hall conductance to the precise value of φ. In fact, as Hofstadter[10] and Wannier[11] have shown, each energy gap persists over a continuous range of φ. The Hall conductance is constant in a particular energy gap, even for irrational values of φ.

Streda[12] has recently obtained a result for the Hall conductance that is in agreement with our results.

This work was supported by the National Science Foundation under Grant No. DMR7920785.

(a) Present address: Department of Physics, University of Illinois, Urbana, Ill. 61801.

1K. von Klitzing, G. Dorda, and M. Pepper, Phys. Rev. Lett. 45, 494 (1980).

2R. B. Laughlin, Phys. Rev. B 23, 5632 (1981).

3H. Aoki and T. Ando, Solid State Commun. 38, 1079 (1981).

4R. E. Prange, Phys. Rev. B 23, 4802 (1981).

5D. J. Thouless, J. Phys. C 14, 3475 (1981).

6B. I. Halperin, to be published.

7A. Rauh, G. H. Wannier, and G. Obermair, Phys. Status Solidi (b) 63, 215 (1974).

8J. Zak, Phys. Rev. 134, A1607 (1964).

9P. G. Harper, Proc. Phys. Soc., London, Sect. A 68, 874 (1955).

10D. Hofstadter, Phys. Rev. B 14, 2239 (1976).

11G. H. Wannier, Phys. Status Solidi (b) 88, 757 (1978).

12P. Streda, to be published.

Quantized Hall conductance as a topological invariant

Qian Niu, D. J. Thouless,* and Yong-Shi Wu†
Department of Physics FM-15, University of Washington, Seattle, Washington 98195
(Received 21 September 1984)

Whenever the Fermi level lies in a gap (or mobility gap) the bulk Hall conductance can be expressed in a topologically invariant form showing the quantization explicitly. The new formulation generalizes the earlier result by Thouless, Kohmoto, Nightingale, and den Nijs to the situation where many-body interaction and substrate disorder are also present. When applying to the fractional quantized Hall effect, we draw the conclusion that there must be a symmetry breaking in the many-body ground state. The possibility of writing the fractionally quantized Hall conductance as a topological invariant is also discussed.

I. INTRODUCTION

In the experiments on both integral[1] and fractional[2] quantized Hall effect it is found that the appearance of a plateau in the Hall conductance is always accompanied by a dip in the longitudinal conductance. This well-observed fact suggests that the existence of the Fermi gap (the energy gap or mobility gap in which the Fermi energy of the system lies) is a necessary condition for the quantization of the Hall conductors. On the other hand, since the phenomenon is quite independent of the details of the devices used in the experiments, this condition must also be sufficient (of course, at zero temperature and in weak electric field).

By now, in the integral case, this relationship has been quite established by perturbation theory[3,4] or by gauge-invariance argument.[5] The latter is more profound for it only uses global properties of the electron system in the external fields. But the solenoid device typically employed in this theory seem to be artificial to most of the known experiments.

Another nonperturbative approach was proposed by Thouless *et al.*[6] (henceforth referred to as TKNdN), who considered an infinite two-dimensional electron gas in a periodic substrate potential commensurate to the perpendicular magnetic field. The Hall conductance calculated from the Kubo formula was rewritten into an integral which shows quantization explicitly. This expression has the advantage that it is independent of the detailed structure of the periodic potential. Later this integral was recognized as the first Chern class of a U(1) principal fiber bundle on a torus.[7,8] The fibers are the magnetic Bloch waves and the torus corresponds to the magnetic Brillouin zone.

Unfortunately, this theory cannot allow either impurity disorder or many-body interactions, because the use of Bloch waves is quite essential to their derivations. In this paper we generalize TKNdN's idea so that an invariant expression can still be constructed in the general case.

The method we are going to use is quite parallel to the generalized formulation of Niu and Thouless[9] for the quantization of particle transport induced by a potential varying slowly and periodically. We use the same geometry as used by TKNdN, consequently we share with them the same deficiency of ignoring the edge effect.[10] The many-body wave functions are required to satisfy a particular boundary condition described by two fixed phase parameters. We then prove that the Hall conductance becomes independent of the phase parameters in the thermodynamic limit, so that it can be averaged over all the phases that prescribe different boundary conditions. The averaged quantity which equals the Hall conductance possesses an expression whose value is quantized explicitly. This expression is of the same form as TKNdN's integral, except that the roles played by the Bloch wave numbers are now played by the phase parameters in the boundary conditions. Thus the same topological identification can be made in the new theory.

In our generalized formulation it is found that the Hall conductance is quantized in an integer times e^2/h as long as the Fermi gap is finite and the many-body ground-state energy is nondegenerate. This quantized value is topological in the sense that it is unchanged under a variation of the potentials so long as the Fermi gap is kept open. Also, the result does not depend on how this gap is generated. It could be generated by the action of the magnetic field alone (Landau gap), together with a periodic substrate potential (gaps between the subbands), or with the many-body interactions.

To obtain a fractional quantization, we have to require, in addition to a finite Fermi gap, that the ground-state energy is degenerate and the ground states have a discrete symmetry breaking. In this case, the Fermi gap must be generated by the many-body interactions, since otherwise the degeneracy cannot be obtained, nor can the symmetry breaking. Recently, Tao and Wu[11] generalized Laughlin's gauge-invariance argument; our result agrees with theirs.

The degeneracy in the ground-state energy at fractional fillings has been clearly demonstrated by the numerical calculation of Su[12] for a few small systems with torus geometry. On the other hand, such degeneracy was not found in Haldane's[13] numerical calculations with spherical geometry. At the present time we cannot conclude (although we suspect) whether the Haldane system will eventually present a degeneracy in the thermodynamic limit, because it is not clear how the spherical geometry could be fitted into our formulation.

Recently, the topological nature of the quantized Hall effect has been revealed in a different approach by Levine, Libby, and Pruisken.[14] They showed that the effective Lagrangian in a replica treatment of a two-dimensional (2D) disordered electron system in a transverse magnetic field contains a nonperturbative topological term which gives rise to the quantization of the Hall conductance. Unfortunately, this theory seems unable to deal with the many-body interaction and therefore the fractionally quantized Hall effect.

In Sec. II we present the formulation generalizing TKNdN's theory in the integral quantized Hall effect. In Sec. III we illustrate the idea of how the fractional quantized Hall conductance could be expressed as a topological invariant. Finally, in Sec. IV we give a brief discussion of the relation between our formulation and Laughlin's gauge-invariance argument.

II. THE INTEGER CASE

The method we will use relies on the fact that for an energy in the Fermi gap the Green function falls off exponentially with distance, and so the current induced by an electric field is a local function of the field. Therefore the response to a field in the interior (away from the edge) of a two-dimensional system can be calculated with only exponentially small errors by replacing the realistic boundary conditions with convenient artificial ones.

Edge currents are a separate problem and cannot be calculated in this way, since they depend on a delicate balance between the diamagnetic currents at the two edges. However, it is possible to devise geometrical conditions under which there are no edge currents. For example, we could consider a Corbino disc or a cylinder with the emf applied in the azimuthal direction.

We now consider a two-dimensional interacting electron system in both a magnetic field $B\hat{\mathbf{z}}$ perpendicular to the plane and an electric field $E\hat{\mathbf{x}}$ in the plane. The substrate potential may or may not be periodic in space. The Hall current which flows in the **y** direction can be calculated by the Kubo formula[15] derived from a linear-response theory as

$$\sigma=\frac{ie^2\hbar}{A}\sum_{n\,(>0)}\frac{(v_1)_{0n}(v_2)_{n0}-(v_2)_{0n}(v_1)_{n0}}{(E_0-E_n)^2}\,, \tag{2.1}$$

where $A=L_1L_2$ is the area of the system; the subscripts 0 and n label the ground state and the excited states of the N-body Hamiltonian in the absence of the external electric field:

$$H=\sum_{i=1}^{N}\left[\frac{1}{2m_i}\left[-i\hbar\frac{\partial}{\partial x_i}\right]^2+\frac{1}{2m_i}\left[-i\hbar\frac{\partial}{\partial y_i}-eBx_i\right]^2\right]$$
$$+\sum_{i=1}^{N}U(x_i,y_i)+\sum_{j=1}^{N}\sum_{\substack{i=1\\ i<j}}^{N}V(|\mathbf{r}_i-\mathbf{r}_j|)\,. \tag{2.2}$$

Correspondingly, E_0 and E_n are the eigenenergies. The velocity operators appearing in the Kubo formula are given by

$$v_1=\sum_{i=1}^{N}\frac{1}{m_i}\left[-i\hbar\frac{\partial}{\partial x_i}\right],$$
$$v_2=\sum_{i=1}^{N}\frac{1}{m_i}\left[-i\hbar\frac{\partial}{\partial y_i}-eBx_i\right]. \tag{2.3}$$

A realistic boundary condition is

$$\psi(x_i=L_1)=\psi(x_i=0)=0\,, \tag{2.4a}$$
$$\psi(y_i+L_2)=e^{i\beta L_2}\psi(y_i)\,, \tag{2.4b}$$
$$i=1,2,\ldots,N$$

where the phase parameter β is independent of the particle indices, as is required by the total antisymmetry. But we are only interested in the bulk contribution to the Hall conductance; the condition (2.4a) can then be relaxed to the following form:

$$\psi(x_i+L_1)=e^{i\alpha L_1}e^{i(eB/\hbar)y_iL_1}\psi(x_i)\,, \tag{2.4c}$$

where the y-dependent phase factor is necessary for the Hamiltonian to be Hermitian. At this moment one should note that the boundary conditions (2.4b) and (2.4c) are appropriate only when the particular Landau gauge (2.2) is chosen, but in general we can use instead the gauge-covariant boundary conditions of the form

$$\mathcal{T}_i(L_1,\hat{\mathbf{x}})\psi(x_i)=e^{i\alpha L_1}\psi(x_i)\,, \tag{2.4d}$$
$$\mathcal{T}_i(L_2,\hat{\mathbf{y}})\psi(y_i)=e^{i\beta L_2}\psi(y_i)\,, \tag{2.4e}$$
$$i=1,2,\ldots,N$$

where $\mathcal{T}_i(L_1,\hat{\mathbf{x}})$ and $\mathcal{T}_i(L_2,\hat{\mathbf{y}})$ are the single-particle magnetic translation operators in the **x** and **y** directions.[16] With this generalization the argument will follow the same line, so we would rather stay with the special case (2.4b) and (2.4c).

Now we make the unitary transformation

$$\phi_n=\exp[-i\alpha(x_1+\cdots x_N)]\times\exp[-i\beta(y_1+\cdots+y_N)]\psi_n\,. \tag{2.5}$$

Then (2.1) becomes

$$\sigma=\frac{ie^2}{A\hbar}\sum_{n\,(>0)}\frac{\left\langle\phi_0\left|\frac{\partial\tilde{H}}{\partial\alpha}\right|\phi_n\right\rangle\left\langle\phi_n\left|\frac{\partial\tilde{H}}{\partial\beta}\right|\phi_0\right\rangle-\left\langle\phi_0\left|\frac{\partial\tilde{H}}{\partial\beta}\right|\phi_n\right\rangle\left\langle\phi_n\left|\frac{\partial\tilde{H}}{\partial\alpha}\right|\phi_0\right\rangle}{(E_0-E_n)^2}\,, \tag{2.6}$$

where $\widetilde{H}$ is the transformed Hamiltonian. Equivalently, $\widetilde{H}$ can be obtained from H by the following replacement:

$$-i\frac{\partial}{\partial x_i}\to -i\frac{\partial}{\partial x_i}+\alpha,\quad -i\frac{\partial}{\partial y_i}\to -i\frac{\partial}{\partial y_i}+\beta\,. \tag{2.7}$$

It is clear that $(1/\hbar)\partial\widetilde{H}/\partial\alpha$ and $(1/\hbar)\partial\widetilde{H}/\partial\beta$ are just the transformed velocity operators. By a simple manipulation we can express (2.6) in terms of the partial derivatives of the transformed wave function for the ground state of the many-body system:

$$\sigma=\frac{ie^2}{\hbar A}\left[\left\langle\frac{\partial\phi_0}{\partial\alpha}\middle|\frac{\partial\phi_0}{\partial\beta}\right\rangle-\left\langle\frac{\partial\phi_0}{\partial\beta}\middle|\frac{\partial\phi_0}{\partial\alpha}\right\rangle\right]$$

$$=\frac{ie^2}{\hbar}\left[\left\langle\frac{\partial\phi_0}{\partial\theta}\middle|\frac{\partial\phi_0}{\partial\varphi}\right\rangle-\left\langle\frac{\partial\phi_0}{\partial\varphi}\middle|\frac{\partial\phi_0}{\partial\alpha}\right\rangle\right], \tag{2.8}$$

where $\theta=\alpha L_1$ and $\varphi=\beta L_2$.

So far the derivatives are formal and we still cannot see why the Hall conductance should be quantized. To proceed further, we need to assume that there is always a finite energy gap between the ground state and the excitations under any given boundary conditions of the form in (2.4b) and (2.4c). Also, it is plausible to say that the bulk conductance as given by the Kubo formula should be insensitive to the boundary conditions if the particles do not have long-range correlations in the ground state. We leave the justification of this point to the Appendix. In fact, in the special case of zero interaction and flat substrate potential, one can explicitly show that the above expression is indeed independent of the parameters, even without taking the thermodynamic limit.

Consequently we can equate σ with its average over all the phases $(0\le\theta<2\pi,\ 0<\varphi\le2\pi)$ that specify different boundary conditions, i.e.,

$$\sigma=\bar{\sigma}=\frac{e^2}{h}\int_0^{2\pi}\int_0^{2\pi}d\theta\,d\varphi\frac{1}{2\pi i}\left[\left\langle\frac{\partial\phi_0}{\partial\varphi}\middle|\frac{\partial\phi_0}{\partial\theta}\right\rangle-\left\langle\frac{\partial\phi_0}{\partial\theta}\middle|\frac{\partial\phi_0}{\partial\varphi}\right\rangle\right]. \tag{2.9}$$

This is of the same form as the integral that appeared in TKNdN's original theory, except that positions of the Bloch wave numbers are now taken by the phase parameters θ and φ. Because of the energy gap, the ground state must go back to itself (up to an overall phase factor) as θ or φ changes by 2π, unless the ground state is not uniquely determined by the boundary condition. Thus the Hall conductance is quantized into an integer times e^2/h whenever the ground state is nondegenerate and is separated from the excited states by a finite energy gap.

The integral in (2.9) is actually a topological invariant. It is the first Chern class of a U(1) principal fiber bundle of the ground-state wave functions on the base manifold of a torus $\mathbf{T}^2$ parametrized by the phases θ and φ. Originally such a recognition was made by Avron, Seiler, and Simon[7] in the context of TKNdN's original theory. The base manifold was the magnetic Brillouin zone, and the fibers were the single-particle Bloch waves. Recently, Kohmoto[8] finished a detailed analysis showing how the abstract topological idea is applied to their expression for the quantized Hall conductance. But since the use of single-particle Bloch waves is essential in TKNdN's theory, this topological idea loses sense as soon as the many-body interaction and the substrate disorder are taken into account.

Fortunately, all these can be recovered by manipulating the phases describing the boundary conditions. Apparently, the Hall conductance should be calculated under a fixed boundary condition as given in (2.8). But the insensitivity of the physical quantity to the boundary conditions allows us to make an average over the phases. In this way an expression for the Hall conductance similar to that of TKNdN's is obtained, so the same topological words can apply to the problem in the rather general situation.

In fact, the recognition of the Hall conductance as a topological invariant is not only of mathematical formality, but also of physical content, because we can explain the stability of the quantization of the Hall conductance against various kinds of perturbations. The reason lies in the fact that the existence of a finite Fermi gap above the ground state is a discrete property which does not depend upon the potentials continuously. Also, in the presence of slight disorder in the substrate, the Fermi level can be locked into the impurity spectra and allows the mobility gap to open in a finite range of the magnetic field. This explains the plateaus of the Hall conductance at the quantized values.

III. THE FRACTIONAL CASE

As mentioned in the introduction, the observation of fractional quantized Hall effect[2] (FQHE) is also accompanied with a vanishing of the longitudinal conductance in the zero-temperature limit. This suggests that a Fermi gap must also exist at the fractional fillings near which FQHE is observed. Several theoretical calculations[17–19] have already justified this point. Furthermore, the ground state is shown to be liquidlike, so correlation between electrons decays rapidly as their separation becomes large. Thus we can continue to use the method employed in the preceding section to equate the Hall conductance to its average over all different boundary conditions as in (2.9). Since a nondegenerate ground state always leads to an integral quantization, we must require a degeneracy in order to explain the fractional quantization. In this case, Eq. (2.9) should be written as

$$\sigma=\bar{\sigma}$$

$$=\frac{e^2}{hd}\sum_{K=1}^{d}\int_0^{2\pi}\int_0^{2\pi}d\theta\,d\varphi\frac{1}{2\pi i}\left[\left\langle\frac{\partial\phi_K}{\partial\varphi}\middle|\frac{\partial\phi_K}{\partial\theta}\right\rangle-\left\langle\frac{\partial\phi_K}{\partial\theta}\middle|\frac{\partial\phi_K}{\partial\varphi}\right\rangle\right], \tag{3.1}$$

where d is the degree of the degeneracy, and $\{\phi_K\}$ is an orthogonal basis spanning the ground-state Hilbert space. In the above expression we have also used the fact that there is no coupling between different ground states, because they are macroscopically separated in the sense that they cannot be obtained from one another by a few number of single-particle excitations.

Unlike the nondegenerate case, the integral over θ and φ is no longer a topological invariant, since the variation of θ or φ by 2π does not necessarily lead each ground state back to itself. But the summation over the integrals may still be a topological invariant. We consider this possibility in the following.

Consider the $1/p$ fillings first. We start from the parent states similar to those proposed by Tao and Thouless.[17] The single-particle states in the ground Landau level are

$$e^{i\lambda y}u(x-\lambda b^2),\quad b^2=\frac{\hbar}{eB}\ . \tag{3.2}$$

In order to satisfy the boundary conditions (2.4b) and (2.4c), we make the following linear combinations:

$$W_m(\alpha,\beta)=\sum_{n=-\infty}^{\infty} e^{i\delta_m\lambda_n b^2}e^{i\lambda_n y}u(x-\lambda_n b^2)\ , \tag{3.3}$$

where

$$\lambda_n=\beta+\frac{2\pi}{L_2}pn,\quad \delta_n=\alpha+\frac{2\pi}{L_1}m \tag{3.4}$$

and the linear dimensions of the system is chosen to give $1/p$ filling of the ground Landau level:

$$\frac{N}{L_1L_2}=\frac{1}{2\pi pb^2}\ . \tag{3.5}$$

Then one can show that these satisfy the following boundary conditions:

$$W_m(x+L_1)=e^{i\alpha L_1}e^{iL_1y/b^2}W_m(x)\ , \tag{3.6}$$

$$W_m(y+L_2)=e^{i\beta L_2}W_m(y)\ . \tag{3.7}$$

Also one should notice that there are N different states of the form (3.3) (for fixed α and β), since W_{m+N} differs from W_m by only an overall phase factor. This is not surprising because we only used those states in (3.2) whose centers (in the x direction) are separated by a multiple of $(2\pi/L_2)pb^2$ to construct the linear combinations. The parent state of the many-body system is then made by taking the determinant of these single-particle states. We claim that this parent state is equivalent to that proposed by Tao and Thouless,[17] because different linear combinations of the single-particle states give the same determinantal wave function. One can check explicitly that the same boundary conditions as in (3.6) and (3.7) are satisfied by the parent state just constructed, i.e.,

$$\psi(x_l+L_1)=e^{i\alpha L_1}e^{iL_1y_l/b^2}\psi(x_l)\ , \tag{3.8}$$

$$\psi(y_l+L_2)=e^{i\beta L_2}\psi(y_l)\ , \tag{3.9}$$

$$l=1,2,\ldots,N\ .$$

Now we come to the interesting point. Clearly the above boundary condition is specified by $\theta=\alpha L_1$ (mod 2π) and $\varphi=\beta L_2$ (mod 2π). But the parent state cannot be uniquely determined in the same way, since a variation of αL_1 by 2π does not lead the parent state back to itself. In other words, under a given boundary condition we can construct many different parent states from which the true ground states would be generated by the adiabatic turning on of the many-body interactions. Since a variation of βL_2 by 2π and/or of αL_1 by $p2\pi$ do lead the state back to itself, p different parent states are obtained:

$$\psi(\theta,\varphi),\psi(\theta+2\pi,\varphi),\ldots,\psi(\theta+(p-1)2\pi,\varphi)\ . \tag{3.10}$$

These states are orthogonal since they are constructed from different sets of single-particle states. In fact, the true ground states generated from them are also orthogonal to one another. The reason is that these parent states belong to different eigenstates (with different eigenvalues) of the magnetic translation operator $\mathcal{T}((2\pi/L_1)b^2,\hat{\mathbf{y}})$ which translates all the particles by $(2\pi/L_1)b^2$ along the $\mathbf{y}$ direction, while the operator commutes with the total Hamiltonian and its interaction part from which the S matrix is constructed.[18] In other words the true ground states obtained by acting the S matrix on the parent states must also belong to the eigenstates of $\mathcal{T}((2\pi/L_1)b^2,\hat{\mathbf{y}})$ with different eigenvalues, therefore they must be orthogonal to one another.

On the other hand, different parent states are connected by the total magnetic translation $\mathcal{T}((2\pi/L_2)b^2,\hat{\mathbf{x}})$ and its powers. Since this operator also commutes with the S matrix, the states generated from different parent states must have the same energy.

Let us now look back to formula (3.1), where ϕ_K are just the true ground states transformed by relation (2.5). If there are no other ground states other than those generated from our parent states, then (3.1) can be written as

$$\sigma=\bar{\sigma}=\frac{e^2}{hp}\int_0^{2\pi p}d\theta\int_0^{2\pi}d\varphi\frac{1}{2\pi i}\left[\left\langle\frac{\partial\phi_1}{\partial\varphi}\middle|\frac{\partial\phi_1}{\partial\theta}\right\rangle-\left\langle\frac{\partial\phi_1}{\partial\theta}\middle|\frac{\partial\phi_1}{\partial\varphi}\right\rangle\right]\ , \tag{3.11}$$

where we have absorbed the summation into the integration over the extended range $[0\le\theta\le p2\pi]$, because different ground states can be obtained from ϕ_1 by continuous variation of θ by 2π, 4π, etc. We can regard the extended zone $[0\le\theta<p2\pi,\ 0\le\varphi<2\pi]$ as a torus, because ϕ_1 is led back to itself (up to an overall factor) as θ change by $2\pi p$ or φ by 2π. Thus

$$\sigma=\frac{e^2}{h}\frac{c}{p}\ , \tag{3.12}$$

where c is the integer given by the integral. As in the nondegenerate case we can continue to attach a topological meaning to this integer if the extended zone is regarded as the base manifold.

To determine the integer c we may turn off the varying part of the substrate potential; the integer is unchanged assuming the Fermi gap is unclosed by this process. We then transform to the moving frame in which the external electric field becomes zero, we will find no electron current in this frame. Consequently the Hall conductance is directly related to the filling factor $1/p$, so c is unity.

In the above argument we used the concept of parent states to illustrate the idea how the fractional quantization can be obtained within our frame. We do not mean to give a rigorous proof, but we suggest that the following scenario might correspond to the physical reality:

(1) At a $1/p$ filling with p odd and small, the ground-state energy of the system has a p-fold degeneracy. As the phase parameter θ changes by 2π, 4π, etc., the system goes from one ground state to another, and it comes back to the original ground state after θ varies by $2\pi p$.

(2) At a q/p filling, again with p odd and small, the degeneracy is jp-fold with $j \leq q$. But these states fall into j groups, and the states in each group transform in the same way as in (1).

(3) As the density of electrons deviates a small but finite amount from one of the fillings considered above, the ground states may seek a similar structure as in the neighboring case in order to gain the commensurate energy (which is negative) achieved at the neighboring small denominator fillings. If this is right, then the finite plateaus of the Hall conductance observed near the small and odd denominator fillings can be understood.

(4) Since p is even, conjectures (1) and (2) might still be true, but (3) may break down by the fractional statistics recently suggested by a number of people.[20]

Before closing this section we would like to add one more comment. The argument about the degeneracies of the ground state has been presented in terms of the wave function proposed by Tao and Thouless,[17] but it can also be presented in terms of Laughlin's wave function.[19] With the boundary condition (2.4b) (periodicity in the **y** direction) a wave function confined to the region $0 < x_1 < hNp/eBL_2$ can be written in the form

$$\phi_1 = \exp\left[-\frac{1}{2b^2}\sum_{i=1}^{N} x_i^2 + \beta \sum_{i=1}^{N} z_i\right]$$

$$\times \prod_{j=1}^{N} \prod_{\substack{i=1 \\ i<j}}^{N} (e^{2\pi z_i/L_2} e^{2\pi z_j/L_2})^p \tag{3.13}$$

where $z_j = x_j + iy_j$. This has the same local properties as the wave function Laughlin writes in cylindrical geometry or Haldane[21] in spherical geometry. This state can be modified to satisfy the condition (2.4c) by expanding it in terms of the single-particle Landau states (3.2), and replacing each of these by its periodic continuation in the **x** direction (3.3). This state is also degenerate with the $p-1$ orthogonal states which can be constructed by the action of the magnetic translation operator on it.

IV. RELATION TO LAUGHLIN'S GAUGE ARGUMENT

According to Laughlin's theory,[5] the Hall conductance can be expressed in terms of the number of electrons transported from one edge of the sample to the other after an adiabatic change of the gauge flux by one quantum. We now try to relate this charge transport to the Kubo formula we have been using. The Hamiltonian with a time-dependent gauge parameter is given by (2.2) with the replacement

$$-i\frac{\partial}{\partial y_i} \to -i\frac{\partial}{\partial y_i} + \beta(t) \ . \tag{4.1}$$

The current in the **x** direction induced by the adiabatic variation of $\beta(t)$ can be calculated from the formula[9,22]

$$J_x = \frac{i\hbar^2\dot{\beta}(t)}{L_1} \sum_{n\,(>0)} \frac{(v_2)_{0n}(v_1)_{n0} - (v_1)_{0n}(v_2)_{n0}}{(E_0 - E_n)^2} , \tag{4.2}$$

where the indices 0 and n indicate the instantaneous eigenstates of the Hamiltonian. The charge transport in a period T during which $\beta(t)$ changes by $2\pi/L_2$ (corresponding to a flux quantum) is thus

$$C = \int_0^T dt\, J_x$$

$$= \int_0^{2\pi/L_2} d\beta \frac{i\hbar^2}{L_1} \sum_{n\,(>0)} \frac{(v_2)_{0n}(v_1)_{n0} - (v_1)_{0n}(v_2)_{n0}}{(E_0 - E_n)^2}$$

$$= \frac{h}{e^2} \int_0^{2\pi} \frac{d\varphi}{2\pi} \sigma(\varphi) , \tag{4.3}$$

or

$$\bar{\sigma} = \frac{e^2}{h} C , \tag{4.4}$$

where $\sigma(\varphi)$ is the Hall conductance calculated from the Kubo formula under a fixed gauge $\beta \equiv \varphi L_2$.

Note that it is the averaged Hall conductance that corresponds to the charge transport, so Laughlin's argument really involved an approximation. In fact, as he pointed out in his paper, when he made use of the Faraday's theorem, he actually replaced the adiabatic derivative $dU/d\Phi$ by the fraction $\Delta U/\Phi_0 \propto C$. Here U is the total energy of the electron system, Φ the flux, and Φ_0 the flux quantum. In other words, his approximation is of the same nature as ours.

One final comment. When Laughlin[5] tried to establish the quantization of the Hall conductance, he actually based his argument upon the belief that as the electronic states in the bulk go back to themselves, the particle transport from one edge to the other must be an integral number. Although this idea is physically intuitive, it is still not obvious because of the wave nature of the electrons. Our formulation presents a rigorous proof of this idea.

ACKNOWLEDGMENTS

This work was supported by the National Science Foundation under Grant No. DMR-83-19301, by the Royal Society of London, and by the U.S. Department of Energy under Contract No. DE-AC06-81ER40048.

APPENDIX

The insensitivity of the Hall conductance to the boundary conditions is most easily understood in the noninteracting case. Then the Kubo formula (2.6) can be written in terms of the single-particle quantities as

$$\sigma=\frac{ie^2}{A\hbar}\sum_{\epsilon_m\,(<\epsilon_f)}\sum_{\epsilon_n\,(>\epsilon_f)}\frac{\left\langle\varphi_m\left|\frac{\partial\tilde{h}}{\partial\alpha}\right|\varphi_n\right\rangle\left\langle\varphi_n\left|\frac{\partial\tilde{h}}{\partial\beta}\right|\varphi_m\right\rangle-\left\langle\varphi_m\left|\frac{\partial\tilde{h}}{\partial\beta}\right|\varphi_n\right\rangle\left\langle\varphi_n\left|\frac{\partial\tilde{h}}{\partial\alpha}\right|\varphi_m\right\rangle}{(\epsilon_m-\epsilon_n)^2} \tag{A1}$$

where $\tilde{h}$ is the transformed single-particle Hamiltonian and φ_n are the single-particle wave functions. With some manipulations, this can in turn be written in terms of the Green functions as

$$\sigma=\frac{ie^2}{A\hbar}\oint\frac{dz}{2\pi i}\mathrm{Tr}\left[g\frac{\partial\tilde{h}}{\partial\alpha}g\frac{\partial\tilde{h}}{\partial\beta}g\right], \tag{A2}$$

where the integral contour surrounds the filled state energies. The derivative of the Hall conductance σ with respect to θ is

$$\frac{\partial\sigma}{\partial\theta}=\frac{ie^2}{A\hbar L_1}\oint\frac{dz}{2\pi i}\mathrm{Tr}\left[2g\frac{\partial\tilde{h}}{\partial\alpha}g\frac{\partial\tilde{h}}{\partial\alpha}g\frac{\partial\tilde{h}}{\partial\beta}g+g\frac{\partial\tilde{h}}{\partial\alpha}g\frac{\partial\tilde{h}}{\partial\beta}g\frac{\partial\tilde{h}}{\partial\alpha}g+g\frac{\partial^2\tilde{h}}{\partial\alpha^2}g\frac{\partial\tilde{h}}{\partial\beta}g\right]. \tag{A3}$$

Now, because of the existence of the Fermi gap the energy parameter z can be chosen away from the spectrum of the extended states, hence the Green function $g(\mathbf{r},\mathbf{r}')$ is exponentially bounded as $|\mathbf{r}-\mathbf{r}'|$ becomes large.[4] Thus $\partial\sigma/\partial\theta$ is of the order of $(l/L_1)e^2/h$, with l being the localization length of the Green function. Similarly we have

$$\frac{\partial\sigma}{\partial\varphi}\sim\frac{l}{L_2}\frac{e^2}{h}. \tag{A4}$$

When the many-body interactions are taken into account, we can use the method used in Ref. 9 to estimate $\partial\sigma/\partial\theta$ and $\partial\sigma/\partial\varphi$. Although the manipulations are complicated, the same result can be obtained under the assumption that the electrons do not have long-range correlations in the ground state.

---

*Also at Cavendish Laboratory, University of Cambridge, Madingley Road, Cambridge CB3 0HE, United Kingdom.

†On leave from Department of Physics, University of Utah, Salt Lake City, UT 84112 (address after April 1, 1985).

[1]K. V. Klitzing, G. Dorda, and M. Pepper, Phys. Rev. Lett. **45**, 494 (1980).

[2]D. C. Tsui, H. L. Störmer, and A. C. Gossard, Phys. Rev. Lett. **48**, 1559 (1982); H. L. Störmer, D. C. Tsui, A. C. Gossard, and J. C. M. Hwang, in Proceedings of the 16th International Conference on Physics of Semiconductors (unpublished).

[3]H. Aoki and T. Ando, Solid State Commun. **38**, 1079 (1981); R. E. Prange, Phys. Rev. B **23**, 4802 (1981); R. E. Prange and R. Joynt, *ibid.* **25**, 2943 (1982).

[4]D. J. Thouless, J. Phys. C **14**, 3475 (1981).

[5]R. B. Laughlin, Phys. Rev. B **23**, 5632 (1981).

[6]D. J. Thouless, M. Kohmoto, M. P. Nightingale, and M. den Nijs, Phys. Rev. Lett. **49**, 405 (1982).

[7]J. Avron, R. Seiler, and B. Simon, Phys. Rev. Lett. **51**, 51 (1983); B. Simon, Phys. Rev. Lett. **51**, 2167 (1983).

[8]M. Kohmoto (unpublished).

[9]Q. Niu and D. J. Thouless, J. Phys. A **17**, 2453 (1984).

[10]B. I. Halperin, Phys. Rev. B **25**, 2185 (1982).

[11]R. Tao and Y. S. Wu, Phys. Rev. B **30**, 1097 (1984).

[12]W. P. Su, Phys. Rev. B **30**, 1069 (1984).

[13]F. D. M. Haldane (private communication).

[14]H. Levine, S. B. Libby, and A. M. M. Pruisken, Nucl. Phys. B **240**, 30 (1984); **240**, 49 (1984); **240**, 71 (1984).

[15]At zero temperature this form of the Kubo formula is most easily derived by first-order static perturbation expansion.

[16]J. Zak, Phys. Rev. **134**, A1607 (1964).

[17]R. Tao and D. J. Thouless, Phys. Rev. B **28**, 1142 (1983).

[18]R. Tao, Phys. Rev. B **29**, 636 (1984).

[19]R. B. Laughlin, Phys. Rev. Lett. **50**, 1395 (1983); D. Yoshioka, B. I. Halperin, and P. A. Lee, Surf. Sci. **142**, 155 (1984).

[20]B. I. Halperin, Phys. Rev. Lett. **52**, 1583 (1984); D. Arovas, J. R. Schrieffer, and F. Wilczek, Phys. Rev. Lett. **53**, 722 (1984); R. Tao and Y. S. Wu (unpublished).

[21]F. D. M. Haldane, Phys. Rev. Lett. **51**, 605 (1983).

[22]D. J. Thouless, Phys. Rev. B **27**, 6083 (1983).

Anomalous Hall Effect in Ferromagnetic Semiconductors

T. Jungwirth,[1,2] Qian Niu,[1] and A. H. MacDonald[1]
[1]*Department of Physics, The University of Texas, Austin, Texas 78712*
[2]*Institute of Physics ASCR, Cukrovarnická 10, 162 53 Praha 6, Czech Republic*
(Received 3 October 2001; published 6 May 2002)

We present a theory of the anomalous Hall effect in ferromagnetic (III, Mn)V semiconductors. Our theory relates the anomalous Hall conductance of a homogeneous ferromagnet to the Berry phase acquired by a quasiparticle wave function upon traversing closed paths on the spin-split Fermi surface. The quantitative agreement between our theory and experimental data in both (In, Mn)As and (Ga, Mn)As systems suggests that this disorder independent contribution to the anomalous Hall conductivity dominates in diluted magnetic semiconductors. The success of this model for (III, Mn)V materials is unprecedented in the longstanding effort to understand origins of the anomalous Hall effect in itinerant ferromagnets.

DOI: 10.1103/PhysRevLett.88.207208 PACS numbers: 75.50.Mm, 73.50.Jt

In recent years the semiconductor research community has enjoyed a remarkable achievement, making III-V compounds ferromagnetic by doping them with magnetic elements. The 1992 discovery [1] of hole-mediated ferromagnetic order in (In, Mn)As has motivated research [2] on GaAs and other III-V host materials. Ferromagnetic transition temperatures in excess of 100 K [3,4] and long spin-coherence times in GaAs [5,6] have fueled hopes that a new magnetic medium is emerging that could open radically new pathways for information processing and storage technologies. Recent reports [7] of the room temperature ferromagnetism predicted [8] for (Ga, Mn)N have added to interest in this class of materials. In both (In, Mn)As and (Ga, Mn)As systems, measurements of the anomalous Hall effect have played a key role in establishing ferromagnetism in the studied samples, and in providing evidence for the essential role of hole-mediated coupling between Mn local moments in establishing long-range order [1,2,9]. Despite the importance of the anomalous Hall effect (AHE) [10–15] for sample characterization, a theory which allows these experiments to be interpreted quantitatively has not been available. In this article we present a theory of the AHE in ferromagnetic III-V semiconductors that appears to account for existing observations.

The Hall resistivity of ferromagnets has an ordinary contribution, proportional to the external magnetic field strength, and an anomalous contribution often assumed to be proportional to the sample magnetization. Theories of the anomalous Hall effect (AHE) in a metal [13] usually start from a mean-field theory description of its ferromagnetic state, in which current is carried by quasiparticles in spontaneously spin-split bands. Recent theoretical studies of the AHE in colossal magnetoresistance manganites [16], pyrochlore ferromagnets [17], or noncoplanar antiferromagnets [18] relate the AHE to Berry phase effects caused by carrier hopping in a *noncollinear* spin-splitting effective field produced by a spin-lattice background. Noncollinear magnetic order is, however, not required to produce an anomalous Hall effect and indeed it is not present in most materials in which the effect has been measured, including those of interest to us here. In this paper we relate the AHE in collinear ferromagnets to a Berry phase in momentum space, rather than real space. The AHE we evaluate is a ground state property which depends on the way in which spin-orbit coupled Bloch band wave functions evolve with wave vector [19]. Our theory of the Hall effect is not without precedent. It is related to the Kubo formula for the Hall conductance derived by Thouless and co-workers [20] in their analysis of the quantum Hall effect, and to the master-equation quantum-transport analysis of Luttinger [11]. What is without precedent in this work, is the demonstration that it quantitatively explains the large anomalous Hall effect observed in this new class of magnetic materials.

Our theory of the anomalous Hall effect in (III, Mn)V ferromagnets is built on a mean-field description that has recently been developed [21–25] and used successfully to interpret many magnetic and transport properties. In this mean-field theory the host semiconductor valence bands are split by an effective field that results from exchange interactions with polarized Mn moments. The field makes a wave vector independent contribution,

$$H_{\text{split}} = h\hat{m} \cdot \vec{s} \tag{1}$$

to the band Hamiltonian. Here $\hat{m}$ is the polarization direction of the local moments and $\vec{s}$ is the electron spin operator. The effective field h is proportional to the average local moment magnetization and is nonzero only in the ferromagnetic state. The antiferromagnetic interaction [26,27] between localized and itinerant spins implies that $h > 0$. When Mn spins are fully polarized, $\hat{m}$ is *uniform* and $h = N_{\text{Mn}} S J_{pd}$, where N_{Mn} is the density of Mn ions with spin $S = 5/2$, and $J_{pd} = 50 \pm 5$ meV nm^3 is the strength of the exchange coupling between the local moments and the valence band electrons [9]. In the (In, Mn)As and (Ga, Mn)As AHE measurements, [1,2] $\hat{m}$ is in the $\langle 00\bar{1}\rangle$ direction for positive external magnetic fields.

From a symmetry point of view, the AHE is made possible by this effective magnetic field, and by the spin-orbit coupling present in the host semiconductor valence band.

In the standard model of the AHE in metals, skew-scattering [10] and side-jump [12] scattering give rise to contributions to the Hall *resistivity* proportional to the diagonal resistivity ρ and ρ^2, respectively, with the latter process tending to dominate in alloys because ρ is larger. We present our anomalous Hall effect theory using the language of semiclassical transport theory, namely the expressions for wave-packet dynamics developed previously by one of us [28,29]. These imply a contribution to the Hall *conductivity* independent of the kinetic equation scattering term. Identical expressions can be derived using theoretical approaches of Thouless and co-workers [20] or Luttinger [11]. In the early stages of the development of AHE theory disagreements arose between Smit [10] and Luttinger [11], that do not appear to have ever been fully resolved, on whether AHE can occur in a perfectly periodic lattice. While Smit [10] argued that the explanation has to be based on the anisotropic scattering we follow Luttinger [11] in taking the view that there is a contribution to the AHE due to the change in wave packet group velocity that occurs when an electric field is applied to a ferromagnet.

Our focus on this anomalous wave-pocket velocity contribution is motivated in part by practical considerations, since our current understanding of (III, Mn)V ferromagnets is not sufficient to permit confident modeling of quasiparticle scattering, and in part by estimates [30] of a typical disorder spectral broadening which is smaller than the spin-orbit coupling strength. Since the Hall resistivity is invariably smaller than the diagonal resistivity, a temperature independent value of the Hall conductivity corresponds to a Hall resistivity proportional to ρ^2, usually interpreted as evidence for dominant side-jump scattering. As we explain below, we find quantitative agreement between our Hall conductance values and experiment.

The Bloch electron group velocity correction is conveniently evaluated using expressions derived by Sundaram and Niu [28,29]:

$$\dot{x}_c = \frac{\partial \epsilon}{\hbar \partial \vec{k}} + (e/\hbar)\vec{E} \times \vec{\Omega}\,. \tag{2}$$

The first term on the right-hand side of Eq. (2) is the standard Bloch band group velocity. Our anomalous Hall conductivity is due to the second term, proportional to the $\vec{k}$-space Berry curvature $\vec{\Omega}$. It follows from symmetry considerations that for a cubic semiconductor under lattice-matching strains and with $\hat{m}$ aligned by external fields along the $\langle 001\rangle$ growth direction, only $\Omega_z \neq 0$:

$$\Omega_z(n,\vec{k}) = 2\,\mathrm{Im}\left[\left\langle \frac{\partial u_n}{\partial k_y}\middle|\frac{\partial u_n}{\partial k_x}\right\rangle\right]. \tag{3}$$

Here $|u_n\rangle$ is the periodic part of the nth Bloch band wave function with the mean-field spin-splitting term included in the Hamiltonian. The anomalous Hall conductivity that results from this velocity correction is

$$\sigma_{AH} = -\frac{e^2}{\hbar}\sum_n \int \frac{d\vec{k}}{(2\pi)^3} f_{n,\vec{k}}\Omega_z(n,\vec{k}), \tag{4}$$

where $f_{n,\vec{k}}$ is the equilibrium Fermi occupation factor for the band quasiparticles. We have taken the convention that a positive σ_{AH} means that the anomalous Hall current is in the same direction as the normal Hall current.

This Berry phase contribution occurs for any itinerant electron ferromagnet. To assess its importance for (III, Mn)V compounds, we first explore a simplified model that yields parabolic dispersion for both heavy-hole and light-hole bands and a spin-orbit coupling [23,31] strength that is much larger than the hole Fermi energy. Detailed numerical calculations that account for mixing of the spin-orbit split-off bands, and warping of the occupied heavy-hole and light-hole bands [23,31] in (In, Mn)As and (Ga, Mn)As samples [1,3] will follow this general and qualitative discussion. Within a 4-band model, the spin operator $\vec{s} = \vec{j}/3$ in Eq. (1), and the spherical model Hamiltonian for holes in III-V host semiconductors can be written as

$$H_0 = \frac{\hbar^2}{2m}\left[\left(\gamma_1 + \frac{5}{2}\gamma_2\right)k^2 - 2\gamma_2(\vec{k}\cdot\vec{j})^2\right], \tag{5}$$

where $\vec{j}$ is the total angular momentum operator, γ_1 and γ_2 are Luttinger parameters [31,32]. In the unpolarized case ($h = 0$), the total Hamiltonian, $H = H_0 + H_{\mathrm{split}}$, is diagonalized by spinors $|\, j_{\hat{k}}\rangle$ where, e.g., $j_{\hat{k}} \equiv \vec{j}\cdot\hat{k} = \pm 3/2$ for the two degenerate heavy-hole bands with effective mass $m_{hh} = m/(\gamma_1 - 2\gamma_2)$. The Berry curvature (3), which can be rewritten as $\vec{\nabla}_{\vec{k}} \cdot [\hat{z} \times \langle u_n|\partial u_n/\partial\vec{k}\rangle]$, is familiar in this case since the Bloch eigenstates are $j = 3/2$ spin coherent states [33]. Integrating over planes of occupied states at fixed k_z we find that $\int d^2k f_{n,\vec{k}}\Omega_z(n,\vec{k}) = \pm 3/2(\cos\theta_{\vec{k}} - 1)$ where $\cos(\theta_{\vec{k}}) \equiv k_z/k_{hh}$ and k_{hh} is the Fermi wave vector. The anomalous Hall conductivity (4) vanishes in the $h = 0$ limit because the contributions from the two heavy hole bands, and also from the two light hole bands, cancel. In the ferromagnetic state, on the other hand, majority and minority spin heavy and light hole Fermi surfaces differ and also the Berry phases are modified when $h \neq 0$. Both effects contribute to σ_{AH}. Up to linear order in h we obtain that $k_{hh}^{\pm} = k_{hh} \pm \cos\theta_{\vec{k}} h m_{hh}/(2\hbar^2 k_{hh})$ and the Berry phase is altered by the factor $1 \mp 2mh/(9\gamma_2\hbar^2 k_{hh}^2)$. A similar analysis for the light-hole bands leads to a total net contribution to the AHE from the four bands whose lower and upper bounds are the following:

$$\frac{e^2}{2\pi\hbar}\frac{h}{2\pi\hbar^2}(3\pi^2 p)^{-1/3} m_{hh} < \sigma_{AH} < \frac{e^2}{2\pi\hbar}\frac{h}{2\pi\hbar^2}(3\pi^2 p)^{-1/3} 2^{2/3} m_{hh}\,. \tag{6}$$

Here $p = k_{hh}^3/3\pi^2[1 + (m_{lh}/m_{hh})^{3/2}]$ is the total hole density and $m_{lh} = m/(\gamma_1 + 2\gamma_2)$ is the light-hole effective mass. The lower bound in Eq. (6) is obtained assuming $m_{lh} \ll m_{hh}$ while the upper bound is reached when $m_{lh} \approx m_{hh}$.

Based on the above analysis we conclude that the Berry phase anomalous velocity can yield a sizable AHE in (III, Mn)V ferromagnets. The dot-dashed line in Fig. 1 shows our analytic results for GaAs effective masses $m_{hh} = 0.5m$ and $m_{lh} = 0.08m$. Note that in experiment, anomalous Hall conductances are of order 1–10 Ω^{-1} cm^{-1} and the effective exchange field $h \sim 10 - 100$ meV. According to Eq. (6) *larger* σ_{AH} *values should be expected in systems with larger heavy-hole effective masses and in systems with the ratio* m_{lh}/m_{hh} *close to unity.*

So far we have discussed the limit of infinitely strong spin-orbit coupling with an exchange field that is small relative to the hole Fermi energy. In the opposite limits of zero spin-orbit coupling or large h, σ_{AH} vanishes. This implies that the anomalous Hall conductivity is generally nonlinear in the exchange field and the magnetization. To explore the intermediate regime we numerically diagonalized the 6-band Luttinger Hamiltonian [23,31] with the spin-orbit gap $\Delta_{so} = 1$ eV, and for the GaAs value $\Delta_{so} = 341$ meV. The results shown in Fig. 1 confirm that *a smaller* σ_{AH} *is expected in systems with smaller* Δ_{so} *and suggest that both positive and negative signs of* σ_{AH} *can occur in general.*

The curves in Fig. 1 are obtained by neglecting band warping in III-V semiconductor compounds. The property that the valence bands in these materials are strongly nonparabolic, even in the absence of the field h and even in the large Δ_{so} limit, is accurately captured by introducing the third phenomenological Luttinger parameter γ_3 [23,31]. Our numerical results indicate that *warping* $(\gamma_2 \neq \gamma_3)$ *leads to an increase of* σ_{AH}, as seen when comparing the solid curves in Fig. 1 and in the top panel of Fig. 2. The hole-density dependence of σ_{AH}, illustrated in Fig. 2, is qualitatively consistent with the spherical model prediction (6). The numerical data in Fig. 2 are also consistent with the trends for dependence on host parameters, highlighted in italics in the preceeding paragraphs, suggesting a large positive AHE coefficient for (Al, Mn)As ($m_{hh} = 0.66m, m_{hh}/m_{lh} = 3.96, \gamma_2/\gamma_3 = 1.73$), an intermediate positive σ_{AH} in (Ga, Mn)As ($m_{hh} = 0.5m, m_{hh}/m_{lh} = 6.05, \gamma_2/\gamma_3 = 1.42$), and a relatively weak AHE in (In, Mn)As ($m_{hh} = 0.43m, m_{hh}/m_{lh} = 21.5, \gamma_2/\gamma_3 = 1.08$) with a sign that may be sensitive to strain and other details of a particular sample.

We now compare our σ_{AH} theory with the experimental data available in (In, Mn)As and (Ga, Mn)As

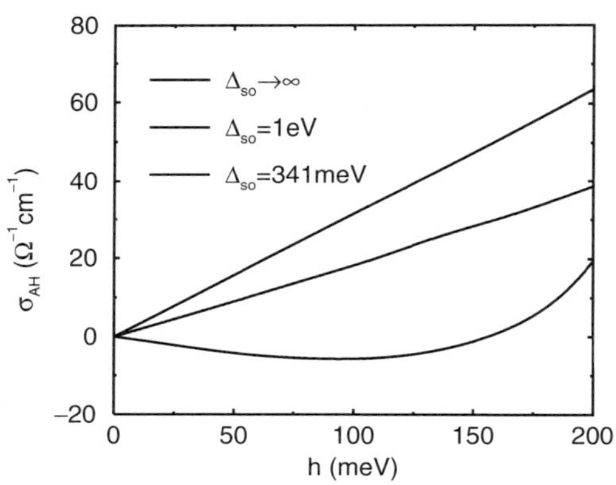

FIG. 1. Illustrative calculations of the anomalous Hall conductance as a function of polarized Mn ions field for hole density $p = 0.35$ nm^{-1}. The dot-dashed curve was obtained assuming infinitely large spin-orbit coupling. The decrease of theoretical σ_{AH} with decreasing spin-orbit coupling strength is demonstrated for $\Delta_{so} = 1$ eV (dashed line) and $\Delta_{so} = 341$ meV (solid line).

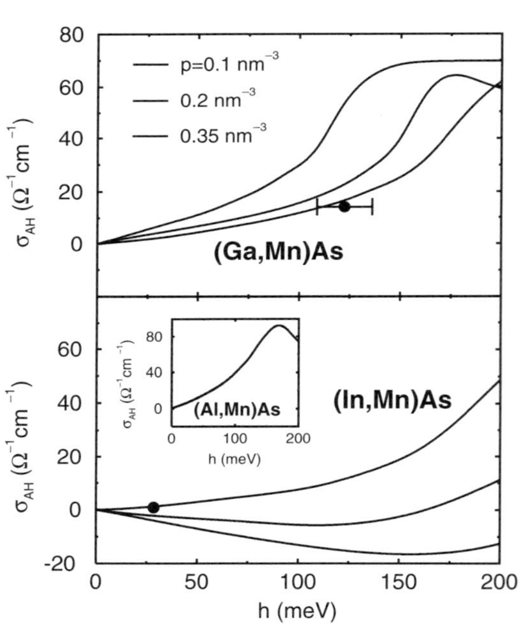

FIG. 2. Full numerical simulations of σ_{AH} for GaAs host (top panel), InAs host (bottom panel), and AlAs host (inset) with hole densities $p = 0.1$ nm^{-1} (dotted lines), $p = 0.2$ nm^{-1} (dashed lines), and $p = 0.35$ nm^{-1} (solid lines). Luttinger parameters of the valence bands were obtained from Ref. [32]. Filled circles in the top and bottom panels represent measured AHE [1,3,9]. The saturation mean-field h values for the two points were estimated from nominal sample parameters [1,3,9]. Horizontal error bars correspond to the experimental uncertainty of the J_{pd} coupling constant. Experimental hole density in the (Ga, Mn)As sample is $p = 0.35$ nm^{-1}; for (In, Mn)As, $p = 0.1$ nm^{-1} was determined indirectly from sample's transition temperature.

samples, studied extensively by Ohno and co-workers [1,3,4,9]. The nominal Mn densities in these two systems are $N_{\rm Mn} = 0.23$ nm^{-3} for the InAs host and $N_{\rm Mn} = 1.1$ nm^{-3} for the GaAs host, yielding saturation values of the effective field $h = 25 \pm 3$ meV and $h = 122 \pm 14$ meV, respectively. The low-temperature hole density of the (Ga, Mn)As sample, $p = 0.35$ nm^{-3}, was unambiguously determined [9] from the ordinary Hall coefficient measured at high magnetic fields. Since similar experiments have not been reported for the (In, Mn)As sample, we estimated the hole density, $p = 0.1$ nm^{-3}, by fitting the density-dependent mean-field theory T_c to the measured value $T_c = 7.5$ K. The use of a mean-field theory description of the ferromagnetic state in both samples is justified [34,35] by the homogeneity of the samples and by the relatively small Fermi energy density of states. Indeed, the measured ferromagnetic transition temperature for the (Ga, Mn)As sample, $T_c = 110$ K, is in an excellent agreement with the calculated transition temperature [22,25,36], and mean-field theory also successfully explains the magnetic anisotropy of both systems [22,23]. Luttinger parameters for the two host semiconductors are well known [32] and are listed in the caption of Fig. 2. As demonstrated in Fig. 2, our theory explains the order of magnitude difference between AHE's in the two materials [$\sigma_{AH} \approx 1\ \Omega^{-1}\mathrm{cm}^{-1}$ in (In, Mn)As and $\sigma_{AH} \approx 14\ \Omega^{-1}\mathrm{cm}^{-1}$ in (Ga, Mn)As]. The calculations are also consistent with the positive sign and monotonic dependences of σ_{AH} on sample magnetizations [9].

We take the agreement in both magnitude and sign of the AHE as a strong indication that the anomalous velocity contribution dominates AHE in homogeneous (III, Mn)V ferromagnets. This Berry phase conductivity, which is independent of quasiparticle scatterers, is relatively easily evaluated with high accuracy. According to our theory, comparison of theoretical and experimental Hall conductivity values provides information not only on the magnetization but also on the character of the itinerant electron wave functions that participate in the magnetism. For example, we predict that size quantization effects in quantum wells that inhibit heavy-light hole mixing will reduce the $\vec{k}$-space Berry curvatures and hence anomalous Hall conductivities. The extreme strength of the spin-orbit interaction compared to other characteristic energy scales in DMS ferromagnets, may be responsible for the relatively simple physics that evidently determines their AHE. Nevertheless, the success reported here motivates a reexamination of this effect in all itinerant electron ferromagnets.

We are grateful for helpful discussions with T. Dietl, J. Furdyna, F. Matsukura, and H. Ohno. Our work was supported by the Department of Energy under Grant No. DE-FG03-02ER45958, the Welch Foundation, the Ministry of Education of the Czech Republic under Grant No. OC P5.10, and the Grant Agency of the Czech Republic under Grant No. 202/02/0912.

[1] H. Ohno *et al.*, Phys. Rev. Lett. **68**, 2664 (1992).
[2] H. Ohno *et al.*, Appl. Phys. Lett. **69**, 363 (1996).
[3] F. Matsukura *et al.*, Phys. Rev. B **57**, R2037 (1998).
[4] H. Ohno, Science **281**, 951 (1998).
[5] J. M. Kikkawa and D. D. Awschalom, Nature (London) **397**, 139 (1999).
[6] I. Malajovich *et al.*, Nature (London) **411**, 770 (2001).
[7] M. L. Reed *et al.*, Appl. Phys. Lett. **79**, 3473 (2001); S. Sonoda *et al.*, e-print cond-mat/0108159.
[8] T. Dietl *et al.*, Science **287**, 1019 (2000).
[9] H. Ohno, J. Magn. Magn. Mater. **200**, 110 (1999).
[10] J. Smit, Physica (Amsterdam) **21**, 877 (1955); *ibid.* **21**, 39 (1958).
[11] J. M. Luttinger, Phys. Rev. **112**, 739 (1958).
[12] L. Berger, Phys. Rev. B **2**, 4559 (1970).
[13] *The Hall Effect and Its Applications,* edited by C. L. Chien and C. R. Westgate (Plenum, New York, 1980).
[14] S. H. Chun *et al.*, Phys. Rev. Lett. **84**, 757 (2000).
[15] A. Crépieux and P. Bruno, Phys. Rev. B **64**, 014416 (2001).
[16] J. Ye *et al.*, Phys. Rev. Lett. **83**, 3737 (1999).
[17] Y. Taguchi *et al.*, Science **291**, 2573 (2001).
[18] R. Shindou *et al.*, Phys. Rev. Lett. **87**, 116 801 (2001).
[19] It vanishes in the limit relevant to colossal magnetoresistance manganites [16], that of a single band without spin-orbit coupling, consistent with the vanishing $T = 0$ AHE of these materials.
[20] D. J. Thouless *et al.*, Phys. Rev. Lett. **49**, 405 (1982).
[21] T. Jungwirth *et al.*, Phys. Rev. B **59**, 9818 (1999).
[22] T. Dietl *et al.*, Science **287**, 1019 (2000).
[23] M. Abolfath *et al.*, Phys. Rev. B **63**, 054418 (2001).
[24] T. Dietl, H. Ohno, and F. Matsukura, Phys. Rev. B **63**, 195205 (2001).
[25] J. König, J. Schliemann, T. Jungwirth, and A. H. MacDonald, in *Electronic Structure and Magnetism of Complex Materials,* edited by D. J. Singh and D. A. Papaconstantopoulos (Springer-Verlag, Berlin, 2002).
[26] T. Dietl, in *Handbook of Semiconductors* **3B**, 1264–1267 (North-Holland, Amsterdam, 1994).
[27] J. Okabayashi *et al.*, Phys. Rev. B **58**, R4211 (1998).
[28] G. Sundaram and Q. Niu, Phys. Rev. B **59**, 14 915 (1999).
[29] M. P. Marder, *Condensed Matter Physics* (Wiley, New York, 1999), pp. 425–429.
[30] M. Abolfath, T. Jungwirth, J. Sinova, J. Kučera, and A. H. MacDonald (to be published).
[31] W. W. Chow, S. W. Koch, and M. Sargent III, *Semiconductor Laser Physics* (Springer-Verlag, Berlin, 1999), pp. 179–192.
[32] I. Vurgaftman, J. R. Meyer, and L. R. Ram-Mohan, J. Appl. Phys. **89**, 5815 (2001).
[33] A. Auerbach, *Interacting Electrons and Quantum Magnetism* (Springer-Verlag, New York, 1994).
[34] J. König, H. H. Lin, and A. H. MacDonald, Phys. Rev. Lett. **84**, 5628 (2000).
[35] J. Schliemann *et al.*, Appl. Phys. Lett. **78**, 1550 (2001).
[36] T. Jungwirth, J. König, J. Sinova, J. Kučera, and A. H. MacDonald, cond-mat/0201157 [Phys. Rev. B (to be published)].

Universal Intrinsic Spin Hall Effect

Jairo Sinova,[1,2] Dimitrie Culcer,[2] Q. Niu,[2] N. A. Sinitsyn,[1] T. Jungwirth,[2,3] and A. H. MacDonald[2]
[1]*Department of Physics, Texas A&M University, College Station, Texas 77843-4242, USA*
[2]*Department of Physics, University of Texas at Austin, Austin, Texas 78712-1081, USA*
[3]*Institute of Physics ASCR, Cukrovarnická 10, 162 53 Praha 6, Czech Republic*
(Received 27 July 2003; published 25 March 2004)

We describe a new effect in semiconductor spintronics that leads to dissipationless spin currents in paramagnetic spin-orbit coupled systems. We argue that in a high-mobility two-dimensional electron system with substantial Rashba spin-orbit coupling, a spin current that flows perpendicular to the charge current is intrinsic. In the usual case where both spin-orbit split bands are occupied, the intrinsic spin-Hall conductivity has a universal value for zero quasiparticle spectral broadening.

DOI: 10.1103/PhysRevLett.92.126603 PACS numbers: 72.10.–d, 72.15.Gd, 73.50.Jt

The science of devices whose operation is based in part on manipulation of the electronic spin degree of freedom, spintronics, has emerged [1] as an active subfield of condensed matter physics because of its potential impact on information technology and because of the challenging basic questions that it poses. Many spintronic concepts involve ferromagnets, in which spins are easier to manipulate because they behave collectively. Spintronic magnetoresistive [2–5] sensors based on the properties of ferromagnetic metals, for example, have reinvented the hard-disk industry over the past several years. Spintronics in semiconductors is richer scientifically than spintronics in metals because doping, gating, and heterojunction formation can be used to engineer key material properties and because of the intimate relationship in semiconductors between optical and transport properties. Practical spintronics in semiconductors has appeared, however, to be contingent on either injection of spin-polarized carriers [6–12] from ferromagnetic metals combined with long spin lifetimes [13] or on room-temperature semiconductor ferromagnetism [14]. In this Letter we explain a new effect [15] that might suggest a new direction for semiconductor spintronics research.

In the following paragraphs we argue that in high-mobility two-dimensional electron systems (2DES) that have substantial Rashba [16] spin-orbit coupling, spin currents always accompany charge currents. The Hamiltonian of a 2DES with Rashba spin-orbit coupling is given by [16]

$$H = \frac{p^2}{2m} - \frac{\lambda}{\hbar}\vec{\sigma}\cdot(\hat{z}\times\vec{p}), \tag{1}$$

where λ is the Rashba coupling constant, $\vec{\sigma}$ is the Pauli matrices, m is the electron effective mass, and $\hat{z}$ is the unit vector perpendicular to the 2DES plane. The Rashba coupling strength in a 2DES can be modified by as much as 50% by a gate field [17]. Recent observations of a spin-galvanic effect [18,19] and a spin-orbit coupling induced metal-insulator transition in these systems [20] illustrate the potential importance of this tunable interaction in semiconductor spintronics [21]. The spin current we discuss is polarized in the direction perpendicular to the two-dimensional plane and flows in the planar direction that is perpendicular to the charge current direction. It is therefore a spin Hall effect, but unlike the effect conceived by Hirsch [22], it is purely intrinsic and does not rely on anisotropic scattering by impurities. Remarkably, in the usual case when both spin-orbit split Rashba bands are occupied, the spin Hall conductivity has a universal value independent of both the 2DES density and the Rashba coupling strength.

The basic physics of this effect is illustrated schematically in Fig. 1. In a translationally invariant 2DES, electronic eigenstates have definite momentum and, because of spin-orbit coupling, a momentum-dependent effective magnetic field that causes the spins (red or dark gray arrows) to align perpendicularly to the momenta (green or light gray arrows), as illustrated in Fig. 1(a). In the presence of an electric field, which we take to be in the $\hat{x}$ direction as shown in Fig. 1(b), electrons are accelerated and drift through momentum space at the rate $\dot{\vec{p}} = -eE\hat{x}$. Our spin Hall effect arises from the time dependence of the effective magnetic field experienced by the spin because of its motion in momentum space. For the Rashba Hamiltonian case of interest here, the effect can be understood most simply by considering the Bloch equation of a spin-1/2 particle, as we explain in the following paragraph. More generally the effect arises from nonresonant interband contributions to the Kubo-formula expression for the spin Hall conductivity that survive in the static limit.

The dynamics of an electron spin in the presence of time-dependent Zeeman coupling is described by the Bloch equation:

$$\frac{\hbar d\hat{n}}{dt} = \hat{n}\times\vec{\Delta}(t) + \alpha\frac{\hbar d\hat{n}}{dt}\times\hat{n}, \tag{2}$$

where $\hat{n}$ is direction of the spin and α is a damping parameter that we assume is small. For the application we have in mind the $\vec{p}$ dependent Zeeman coupling term in the spin Hamiltonian is $-\vec{s}\cdot\vec{\Delta}/\hbar$, where $\vec{\Delta} = 2\lambda/\hbar(\hat{z}\times\vec{p})$. For a Rashba effective magnetic field with

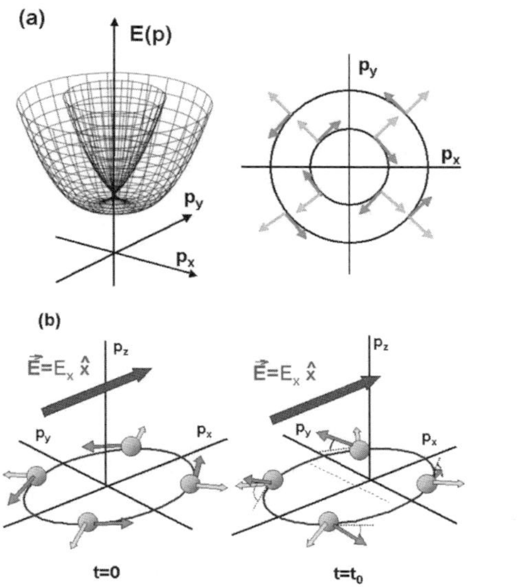

FIG. 1 (color online). (a) The 2D electronic eigenstates in a Rashba spin-orbit coupled system are labeled by momentum (green or light gray arrows). For each momentum the two eigenspinors point in the azimuthal direction (red or dark gray arrows). (b) In the presence of an electric field the Fermi surface (circle) is displaced an amount $|eE_x t_0/\hbar|$ at time t_0 (shorter than typical scattering times). While moving in momentum space, electrons experience an effective torque which tilts the spins up for $p_y > 0$ and down for $p_y < 0$, creating a spin current in the y direction.

magnitude Δ_1 that initially points in the $\hat{x}_1$ direction, then tilts (arbitrarily slowly) slightly toward $\hat{x}_2$, where $\hat{x}_1$ and $\hat{x}_2$ are orthogonal in-plane directions, it follows from the linear response limit of Eq. (2) that

$$\frac{\hbar dn_2}{dt} = n_z\Delta_1 + \alpha dn_z/dt,$$
$$\frac{\hbar dn_z}{dt} = -\Delta_1 n_2 - \alpha dn_2/dt + \Delta_2, \tag{3}$$

where $\Delta_2 = \vec{\Delta}\cdot\hat{x}_2$. Solving these inhomogeneous coupled equations using a Green function technique, it follows that to leading order in the slow-time dependences $n_2(t) = \Delta_2(t)/\Delta_1$, i.e., the $\hat{x}_2$ component of the spin rotates to follow the direction of the field, and that

$$n_z(t) = \frac{1}{\Delta_1^2}\frac{\hbar d\Delta_2}{dt}. \tag{4}$$

Our intrinsic spin Hall effect follows from Eq. (4). When a Bloch electron moves through momentum space, its spin orientation changes to follow the momentum-dependent effective field and also acquires a momentum-dependent $\hat{z}$ component. We now show that in the case of Rashba spin-orbit coupling, this effect leads to an intrinsic spin-Hall conductivity that has a universal value in the limit of zero quasiparticle spectral broadening.

For a given momentum $\vec{p}$, the spinor originally points in the azimuthal direction. An electric field in the $\hat{x}$ direction ($\dot{p}_x = -eE_x$) changes the y component of the $\vec{p}$-dependent effective field. Applying the adiabatic spin dynamics expressions explained above, identifying the azimuthal direction in momentum space with $\hat{x}_1$ and the radial direction with $\hat{x}_2$ we find that the z component of the spin direction for an electron in a state with momentum $\vec{p}$ is

$$n_{z,\vec{p}} = \frac{-e\hbar^2 p_y E_x}{2\lambda p^3}. \tag{5}$$

(Linear response theory applies for $eE_x r_s \ll \Delta_1$, where r_s is the interparticle spacing in the 2DES.) Summing over all occupied states the linear response of the $\hat{z}$ spin-polarization component vanishes because of the odd dependence of n_z on p_y, as illustrated in Fig. 1, but the spin current in the $\hat{y}$ direction is finite.

The Rashba Hamiltonian has two eigenstates for each momentum with eigenvalues $E_\pm = p^2/2m \mp \Delta_1/2$; the discussion above applies for the lower energy (labeled + for majority spin Rashba band) eigenstate while the higher energy (labeled −) eigenstate has the opposite value of $n_{z,\vec{p}}$. Since Δ_1 is normally much smaller than the Fermi energy [17], only the annulus of momentum space that is occupied by just the lower energy band contributes to the spin current. In this case we find that the spin current in the $\hat{y}$ direction is [23]

$$j_{s,y} = \int_{\text{annulus}} \frac{d^2\vec{p}}{(2\pi\hbar)^2}\frac{\hbar n_{z,\vec{p}}}{2}\frac{p_y}{m} = \frac{-eE_x}{16\pi\lambda m}(p_{F+} - p_{F-}), \tag{6}$$

where p_{F+} and p_{F-} are the Fermi momenta of the majority and minority spin Rashba bands. We find that when both bands are occupied, i.e., when $n_{2D} > m^2\lambda^2/\pi\hbar^4 \equiv n_{2D}^*$, $p_{F+} - p_{F-} = 2m\lambda/\hbar$ and then the spin Hall (sH) conductivity is

$$\sigma_{sH} \equiv -\frac{j_{s,y}}{E_x} = \frac{e}{8\pi}, \tag{7}$$

independent of both the Rashba coupling strength and of the 2DES density. For $n_{2D} < n_{2D}^*$ the upper Rashba band is depopulated. In this limit p_{F-} and p_{F+} are the interior and exterior Fermi radii of the lowest Rashba split band, and σ_{sH} vanishes linearly with the 2DES density:

$$\sigma_{sH} = \frac{e}{8\pi}\frac{n_{2D}}{n_{2D}^*}. \tag{8}$$

The intrinsic spin Hall conductivity of the Rashba model can also be evaluated by using transport theories that are valid for systems with multiple spin-orbit spit bands, either the linear-response-theory Kubo-formula approach [24] or a generalized Boltzmann equation approach that accounts for anomalous contributions to wave-packet dynamics [25]. In the Kubo formalism approach, our universal intrinsic spin Hall effect comes from the static $\omega = 0$ limit of the nondissipative reactive term in the expression for the spin-current response to an electric field [26]:

$$\sigma_{xy}^{\mathrm{sH}}(\omega) = \frac{e\hbar}{V} \sum_{\mathbf{k},n\neq n'} (f_{n',k} - f_{n,k}) \times \frac{\mathrm{Im}[\langle n'k|\hat{j}^{z}_{\mathrm{spin}\,x}|nk\rangle\langle nk|v_y|n'k\rangle]}{(E_{nk} - E_{n'k})(E_{nk} - E_{n'k} - \hbar\omega - i\eta)}, \tag{9}$$

where n, n' are band indices, $\hat{j}^{z}_{\mathrm{spin}} = \frac{\hbar}{4}\{\sigma_z, \vec{v}\}$ is the spin-current operator, ω and η are set to zero in the dc clean limit, and the velocity operators at each $\vec{p}$ are given [7] by

$$\begin{aligned} \hbar v_x &= \hbar \partial H(\vec{p})/\partial p_x = \hbar p_x/m - \lambda\sigma_y, \\ \hbar v_y &= \hbar \partial H(\vec{p})/\partial p_y = \hbar p_y/m + \lambda\sigma_x. \end{aligned} \tag{10}$$

For the Rashba model $n, n' = \pm$ and the eigenspinors are spin coherent states given explicitly by

$$|\mp, p\rangle = \frac{1}{\sqrt{2}}\begin{pmatrix} \pm i e^{-i\phi} \\ 1 \end{pmatrix}, \tag{11}$$

where $\phi = \arctan p_x/p_y$. The energy denominators in Eq. (9) in this case are given by the Rashba splitting, and the velocity matrix elements can be evaluated using Eqs. (10) and (11). The integral over momentum space can be evaluated easily because the energy denominators are independent of orientation and the intrinsic spin Hall conductivity expressions in Eqs. (7) and (8) are recovered. This principal result of the Letter is summarized in Fig. 2 where σ_{sH} is plotted as a function of carrier density and Rashba coupling strength in the zero quasiparticle spectral broadening limit. The Kubo-formula analysis also makes it clear that, unlike the universal Hall conductivity value on a 2DES quantum Hall plateau, the universality of the intrinsic spin Hall effect is not robust against disorder and will be reduced whenever the disorder broadening is larger than the spin-orbit coupling splitting [27–29].

The intrinsic character of our spin Hall effect, compared to the extrinsic character of the effect discussed originally by Hirsch [22], is analogous to the intrinsic character that we have recently proposed for the anomalous Hall effect in some ferromagnets and strongly polarized paramagnets [27,30–36]. In both cases the skew-scattering contributions to the Hall conductivities can become important [37,38], when the overall electron scattering rate is small and the steady state distribution function of the current-carrying state is strongly disturbed compared to the equilibrium one. In the Kubo-formula approach these contributions to the spin-Hall conductivity appear as dissipative contributions from disorder scattering of Fermi-energy quasiparticles. A general and quantitative analysis of the disorder-potential-dependent interplay between intrinsic and skew-scattering contributions to anomalous Hall and spin-Hall effects is a subject of current research that is beyond the scope of this Letter.

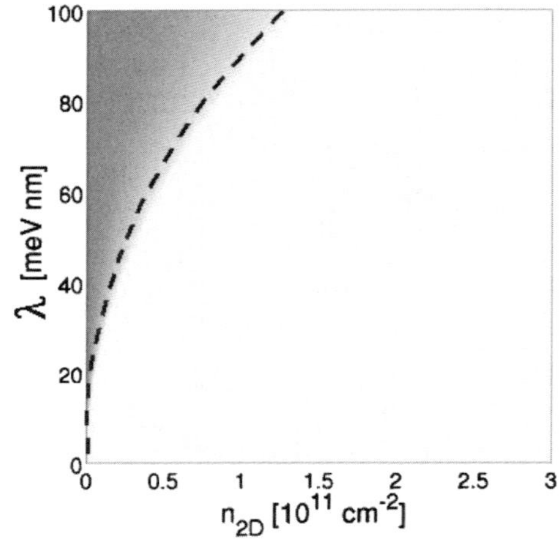

FIG. 2 (color online). Plot of σ_{sH} vs the two-dimensional electron density n_{2D} and the Rashba coupling constant λ. The scale is yellow/light gray $= e/8\pi$ and dark red/dark gray $= 0$. The dashed line indicates the boundary between the universal region (constant σ_{sH}) and the linear density dependence region. Most of the current experimental samples are deep in the universal value region. We expect the intrinsic spin Hall contribution to be present only when the Rashba splitting is larger than the disorder broadening of the quasiparticle energy levels, i.e., when $\lambda p_F \tau/\hbar^2 > 1$.

Several schemes have been proposed for measuring the spin Hall effect in metals [22,39,40], and these can be generalized to the semiconductor case. In semiconductors the close relationship between optical properties [6,7] and spin polarizations opens up new possibilities for detecting nonequilibrium spins accumulated near contacts or near the sample perimeter by spin currents. Spatially resolved Faraday or Kerr effects should be able to detect spin accumulations induced by the spin currents we have evaluated. As in the case of the ferromagnetic semiconductor anomalous Hall effect, the origin of the intrinsic spin Hall-type effect is strong spin-orbit coupling. A sizable intrinsic spin Hall effect will occur in any paramagnetic material with strong spin-orbit coupling, including hole-doped bulk semiconductors [15], although the universal value we obtain here is a unique property of Rashba systems.

The authors thank S. Murakami, N. Nagaosa, and S.-C. Zhang for sharing their results with us prior to publication and acknowledge stimulating interactions with G. Bauer, J. Inoue, J. Hu, D. Loss, L. Molenkamp, S. Murakami, N. Nagaosa, E. Rashba, J. Schliemann, W. Teizer, B. van Wess, R. Winkler, S.-C. Zhang, and X. Zhang. J. S. and A. H. M. gratefully acknowledge the invaluable help and support of M. M. Trudys and G. T. B. Sapphire throughout this research. This work was supported by the Welch Foundation, by the DOE under Grant No. DE-FG03-02ER45958, and by the Grant Agency of the Czech Republic under Grant No. 202/02/0912, by NSF under Grant No. DMR0072115, by DOE No. DE-FG03-96ER45598, and by the Telecommunication and Information Task Force at TAMU.

[1] S. A. Wolf *et al.*, Science **294**, 1488 (2001).
[2] M. N. Baibich *et al.*, Phys. Rev. Lett. **61**, 2472 (1988).
[3] G. Binasch *et al.*, Phys. Rev. B **39**, 4828 (1989).
[4] J. S. Moodera *et al.*, Phys. Rev. Lett. **74**, 3273 (1995).
[5] W. J. Gallagher *et al.*, J. Appl. Phys. **81**, 3741 (1997).
[6] B. T. Jonker, Proc. IEEE **91**, 727 (2003), and references therein.
[7] R. Fiederling *et al.*, Nature (London) **402**, 787 (1999).
[8] G. Schmidt and L.W. Molenkamp, Semicond. Sci. Technol. **17**, 310 (2002), and references therein.
[9] R. Fiederling *et al.*, Nature (London) **402**, 787 (1999).
[10] Y. Ohno *et al.*, Nature (London) **402**, 790 (1999).
[11] H. J. Zhu *et al.* Phys. Rev. Lett. **87**, 016601 (2001).
[12] G. Schmidt *et al.*, Phys. Rev. Lett. **87**, 227203 (2001).
[13] J. M. Kikkawa and D. D. Awschalom, Phys. Rev. Lett. **80**, 4313 (1998).
[14] H. Ohno, Science **281**, 951 (1998).
[15] A related discussion of an intrinsic spin Hall effect for hole-doped bulk semiconductors came to our attention as this work was nearing completion. S. Murakami, A. Nagaosa, and S.-C. Zhang (private communication). Their argument is based on the quantum holonomy of hole-doped bulk semiconductors.
[16] Y. A. Bychkov and E. I. Rashba, J. Phys. C **17**, 6039 (1984).
[17] J. Nitta, T. Akazaki, H. Takayanagi, and T. Enoki, Phys. Rev. Lett. **78**, 1335 (1997).
[18] S. D. Ganichev *et al.*, Nature (London) **417**, 153 (2002).
[19] S. D. Ganichev *et al.*, Phys. Rev. B **68**, 035319 (2003).
[20] T. Koga, J. Nitta, T. Akazaki, and H. Takayanagi, Phys. Rev. Lett. **89**, 046801 (2002).
[21] J. Inoue, G. E.W. Bauer, and L.W. Molenkamp, Phys. Rev. B **67**, 033104 (2003).
[22] J. E. Hirsch, Phys. Rev. Lett. **83**, 1834 (1999).
[23] The electron group velocity includes a term proportional to spin [see Eq. (10)] which is dropped because in Eq. (6) it is the spin current which is included. This spin current is formed by the anticommutator of the spin and the group velocity, and the second term hence vanishes.
[24] J. Sinova, A. H. MacDonald, and Q. Niu, *et al.* (unpublished).
[25] D. Culcer *et al.* (to be published).
[26] The derivation of this expression is identical to the derivation of the frequency-dependent Hall conductivity expression [see, for example, M. Marder, *Condensed Matter Physics* (John Wiley & Sons, Inc., New York, 2000), pp. 573–574] except that the x component of the charge current is replaced by the x component of the spin current.
[27] Jairo Sinova, T. Jungwirth, J. Kucera, and A. H. MacDonald, Phys. Rev. B **67**, 235203 (2003).
[28] J. Schliemann and D. Loss, cond-mat/0310108 [Phys. Rev. B (to be published)].
[29] N. A. Sinitsyn, E. H. Hankiewicz, W. Teizer, and J. Sinova, cond-mat/0310315.
[30] T. Jungwirth, Q. Niu, and A. H. MacDonald, Phys. Rev. Lett. **88**, 207208 (2002).
[31] T. Jungwirth *et al.*, Appl. Phys. Lett. **83**, 320 (2003).
[32] G. Sundaram and Q. Niu, Phys. Rev. B **59**, 14 915 (1999).
[33] K. Ohgushi, S. Murakami, and N. Nagaosa, Phys. Rev. B **62**, R6065 (2000).
[34] M. Onoda and N. Nagaosa, J. Phys. Soc. Jpn. **71**, 19 (2002).
[35] D. Culcer, A. H. MacDonald, and Q. Niu, Phys. Rev. B **68**, 045327 (2003)
[36] R. Karplus and J. M. Luttinger, Phys. Rev. **95**, 1154 (1954).
[37] A. Crepieux, J. Wunderlich, V. K. Dugaev, and P. Bruno, cond-mat/0105614.
[38] T. Dietl, F. Matsukura, H. Ohno, J. Cibert, D. Ferrand, cond-mat/0306484.
[39] S. Zhang, Phys. Rev. Lett. **85**, 393 (2000).
[40] T. P. Pareek, Phys. Rev. Lett. **92**, 076601 (2004).

Hall Effect of Light

Masaru Onoda,[1,*] Shuichi Murakami,[2,†] and Naoto Nagaosa[1,2,‡]

[1]*Correlated Electron Research Center (CERC), National Institute of Advanced Industrial Science and Technology (AIST), Tsukuba Central 4, Tsukuba 305-8562, Japan*

[2]*CREST, Department of Applied Physics, University of Tokyo, Bunkyo-ku, Tokyo 113-8656, Japan*

(Received 6 March 2004; published 18 August 2004)

We derive the semiclassical equation of motion for the wave packet of light taking into account the Berry curvature in momentum-space. This equation naturally describes the interplay between orbital and spin angular momenta, i.e., the conservation of the total angular momentum of light. This leads to the shift of wave-packet motion perpendicular to the gradient of the dielectric constant, i.e., the polarization-dependent Hall effect of light. An enhancement of this effect in photonic crystals is also proposed.

DOI: 10.1103/PhysRevLett.93.083901 PACS numbers: 42.25.Bs, 03.65.Sq, 03.65.Vf, 42.15.–i

The similarity between geometrical optics and particle dynamics has been the guiding principle to develop the quantum mechanics in its early stage. One can consider a trajectory or ray of light at the length-scale much larger than the wavelength λ. By setting $\hbar = c = 1$, the equation for the eikonal ψ in the geometrical optics reads $[\partial_\mu \psi(x)]^2 = [\nabla\phi(\boldsymbol{r})]^2 - n(\boldsymbol{r})^2 = 0$, where $x_\mu = (\boldsymbol{r}, t)$ is the four-dimensional coordinates, $\psi(x) = \phi(\boldsymbol{r}) - n(\boldsymbol{r})t$, and $n(\boldsymbol{r})$ is the refractive index slowly varying within the wavelength [1]. This equation is identical to the Hamilton-Jacobi equation for the particle motion by replacing ψ by the action S. The equation for the ray of light can be derived from the eikonal equation in parallel to the Hamiltonian equation of motion. Deviation from this geometrical optics is usually treated in terms of the diffraction theory. However, most of the analyses have been done in terms of the scalar diffraction theory, neglecting its vector nature. However, the light has the degree of freedom of the polarization, which is represented by the spin $S = 1$ parallel or antiparallel to the wave vector $\boldsymbol{k}$. It is known that this spin produces the Berry-phase when the light is guided by the optical fiber with torsion [2]. Also the adiabatic change of the polarization even without the change of $\boldsymbol{k}$ produces the phase change called the Pancharatnam phase [3]. In this Letter, we will show that the trajectory or ray of light itself is affected by the Berry phase [4], which leads to various nontrivial effects including the Hall effect.

For simplicity, we focus on an isotropic, nonmagnetic medium; the refractive index $n(\mathbf{r})$ is real and scalar, but is not spatially uniform. Let us consider the wave function for the wave packet with the wave vector centered at $\boldsymbol{k}_c$ and the position centered at $\boldsymbol{r}_c$. Because of the conjugate relation of the position and wave vector, both of them inevitably have a finite width of distribution. Therefore, the relative phase of the wave functions at $\boldsymbol{k}$ and $\boldsymbol{k} + d\boldsymbol{k}$ matters, which is the Berry connection or gauge field $\boldsymbol{\Lambda}_{\boldsymbol{k}}$ defined by $[\boldsymbol{\Lambda}_{\boldsymbol{k}}]_{\lambda\lambda'} = -i\boldsymbol{e}^\dagger_{\lambda\boldsymbol{k}} \nabla_{\boldsymbol{k}} \boldsymbol{e}_{\lambda'\boldsymbol{k}}$ where $\boldsymbol{e}_{\lambda\boldsymbol{k}}$ is the polarization vector of λ-polarized photon, and $\lambda = \pm$ correspond to the right- and left-circular polarization. Here, it is noted that the SU(2) gauge field $\boldsymbol{\Lambda}_{\boldsymbol{k}}$ is 2×2 matrix. The corresponding Berry curvature (BC) or field strength is given by $\boldsymbol{\Omega}_{\boldsymbol{k}} = \nabla_{\boldsymbol{k}} \times \boldsymbol{\Lambda}_{\boldsymbol{k}} + i\boldsymbol{\Lambda}_{\boldsymbol{k}} \times \boldsymbol{\Lambda}_{\boldsymbol{k}}$. Because of the masslessness of a photon, it is diagonal in the basis of the right and left-circular polarization as given by $\boldsymbol{\Omega}_{\boldsymbol{k}} = \sigma_3 \boldsymbol{k}/k^3$ (a massive case will be discussed later). It corresponds to the field radiated from the monopole with strength ± 1 located at $\boldsymbol{k} = 0$. We pick up a correction to the geometric optics up to the linear order in $|\nabla n(\boldsymbol{r}_c)|/k_c$. An equation of motion (EOM) is derived including an effect of $\boldsymbol{\Lambda}_{\boldsymbol{k}}$ in the similar way as the Bloch waves of electrons [5]. By considering the effective Lagrangian for the center coordinates $\boldsymbol{r}_c$ and $\boldsymbol{k}_c$, the variational principle leads to the following EOM [5,6]:

$$\dot{\boldsymbol{r}}_c = v(\boldsymbol{r}_c)\frac{\boldsymbol{k}_c}{k_c} + \dot{\boldsymbol{k}}_c \times (z_c|\boldsymbol{\Omega}_{\boldsymbol{k}_c}|z_c), \tag{1}$$

$$\dot{\boldsymbol{k}}_c = -[\nabla v(\boldsymbol{r}_c)]k_c, \tag{2}$$

$$|\dot{z}_c) = -i\dot{\boldsymbol{k}}_c \cdot \boldsymbol{\Lambda}_{\boldsymbol{k}_c}|z_c), \tag{3}$$

where $v(\boldsymbol{r}) = 1/n(\boldsymbol{r})$ is the velocity of light, and $|z) = {}^t[z_+, z_-]$ represents the polarization state. This is the central result of this Letter, from which the Hall effect of light is derived below in a straightforward way.

The limit of geometrical optics is obtained by neglecting the terms containing $\boldsymbol{\Lambda}_{\boldsymbol{k}}$ or $\boldsymbol{\Omega}_{\boldsymbol{k}}$, reproducing Fermat's principle. Let us study phenomena caused by $\boldsymbol{\Lambda}_{\boldsymbol{k}}$ and $\boldsymbol{\Omega}_{\boldsymbol{k}}$. First, the $\boldsymbol{\Lambda}_{\boldsymbol{k}}$ term in (4) causes the phase shift by the directional change of the propagation discussed in ref. [2], as seen in the following way. The equation for $|\dot{z}_c)$ gives the solution $|z_c^{\rm out}) = {}^t[e^{-i\Theta}z_+^{\rm in}, e^{i\Theta}z_-^{\rm in}]$, where $|z_c^{\rm in}) = {}^t[z_+^{\rm in}, z_-^{\rm in}]$ is the initial state of polarization. Θ is the solid angle made by the trajectory of momentum: $\Theta = \oint d\boldsymbol{k} \cdot [\boldsymbol{\Lambda}_{\boldsymbol{k}}]_{++} = \int_S d\boldsymbol{S}_{\boldsymbol{k}} \cdot [\boldsymbol{\Omega}_{\boldsymbol{k}}]_{++}$ where $d\boldsymbol{S}_{\boldsymbol{k}}$ is the surface element in $\boldsymbol{k}$ space and S is a surface surrounded by the trajectory. This is the phase shift by the directional change of propagation [2].

On the other hand, the second term in Eq. (1) induces a new effect found in this Letter; the trajectory of light is affected by $\boldsymbol{\Omega}_k$. We note that this term guarantees the conservation of total angular momentum of photon $j_z = [\boldsymbol{r}_c \times \boldsymbol{k}_c + (z_c|\sigma_3|z_c)(\boldsymbol{k}_c/k_c)]_z$, assuming the rotational symmetry around the z axis. From Eqs. (2)–(4), one can prove $dj_z/dt = 0$ when $\epsilon(\boldsymbol{r}) = \epsilon(\sqrt{x^2+y^2}, z)$. Note that the orbital angular momentum $\boldsymbol{r}_c \times \boldsymbol{k}_c$ is defined only when the position $\boldsymbol{r}_c$ is well defined, which necessarily leads to the finite distribution of the wave vector and hence to the Berry phase. This argument applies to a light beam, whereas it does not for the plane wave state.

This Berry-phase term induces the Hall effect of light. Let us first consider reflection and refraction of light at a flat interface between two media with different dielectric constants. Let $z = 0$ be the boundary between the two dielectric media with $n = v_0/v_1$ where v_0 and v_1 are the velocities of light in the media below and above the boundary, and $\boldsymbol{r}_c$ and $\boldsymbol{k}_c$ of the incident light are within the $y = 0$ plane (Fig. 1).

The Snell's law tells us $\sin\theta_T/v_1 = \sin\theta_I/v_0$ for the transmitted light and $\theta_R = \pi - \theta_I$ for the reflected light, where $\theta_{I,T,R}$ are the angle between the z axis and $\boldsymbol{k}_c$ of the incident, transmitted and reflected lights. This relation can be derived from Fermat's principle in geometrical optics. The question is the following: What is the deviation from this Snell's law due to the Berry-phase? The answer is the transverse shifts of the centers of the reflected and the transmitted wave packets. This phenomenon could be regarded as the "Hall effect of light," wherein the "force" to the light, i.e., the change of the dielectric constant, is perpendicular to the shift. The shift of the transmitted light can be easily calculated from the EOM. Strictly speaking, the EOM can be applied only when the refractive index n varies much slower than the wavelength; thus the EOM is not reliable when the interface is sharp. Nonetheless, it turns out to give the correct answer for the shift of the transmitted light even for the sharp interface, because the shift is governed by the conservation law of the z component j_z of total angular momentum. We assume that j_z is conserved in reflection and in refraction separately: $j_z^I = j_z^T$, $j_z^I = j_z^R$, where the superscripts I, T, and R stand for incident, transmitted, and reflected lights, respectively. This is required when the light is regarded as a quantized object, photon; each photon is reflected or refracted with some probabilities. From this conservation of j_z, the transverse shifts of transmitted and reflected wave packets are estimated as

$$\delta y_c^A(\theta_I) = \frac{[(z_c^A|\sigma_3|z_c^A)\cos\theta_A - (z_c^I|\sigma_3|z_c^I)\cos\theta_I]}{k_c^I \sin\theta_I}, \quad (4)$$

where $A = T$ or R, k_c^I is the momentum of the incident light, $|z_c^{I,T,R})$ stand for the polarizations of the incident, transmitted and reflected photons. We note that $|z_c^{T(R)})$ is obtained from $|z_c^I)$ and the transmittances (reflectances) of two eigenmodes with linear polarizations.

It has been already known that the totally reflected light beam undergoes shifts of position in the plane of incidence (longitudinal) or normal to the plane of incidence (transverse). The longitudinal shift was first studied by Goos and Hänchen [7]. The transverse shift was predicted by Fedorov [8], and have been observed experimentally by Imbert [9], followed by a number of theoretical approaches [10,11]. The transverse shift occurs not only in total reflection, but also in partial reflection and refraction [12]. While there are several papers on theoretical predictions [12–14], they contain rather complicated calculations and the issue remains still controversial. The shift of the transmitted light has not been

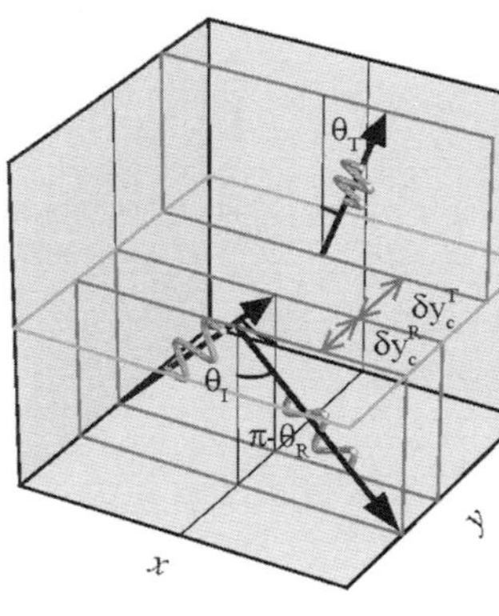

FIG. 1 (color). Transverse shift of light beams in the refraction and reflection at an interface.

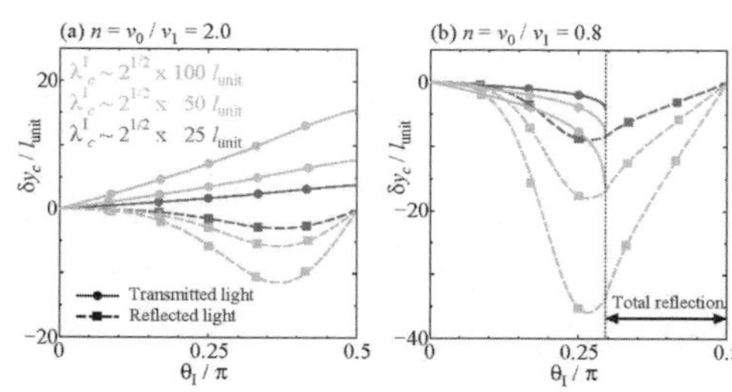

FIG. 2 (color). Transverse shift in the center of the transmitted and reflected wave packets of light for (a) $n = 2.0$ and (b) $n = 0.8$ with the incident wave packet being right-circularly polarized. The solid and dashed lines for the analytic results Eq. (4). The solid circles and squares are the results of simulation. $\lambda_c^I = 2\pi/k_c^I$ is the wavelength of incident wave packet. l_{unit} is an arbitrary length scale. The sign of the shifts is reversed for the left-circularly polarized light.

experimentally measured to the authors' knowledge. Our derivation based on the EOM is much clearer and simpler. Also the result is supported by the numerics described below, showing a wide applicability of our EOM.

In order to test our prediction Eq. (4), we have numerically solved the Maxwell equation. It is found that both the reflected and transmitted wave packets are shifted along the y or $-y$ direction depending on the circular polarization, which perfectly agrees with Eq. (4) as shown in Fig. 2. This phenomenon is analogous to the spin Hall effect [15,16], where the Berry-phase from band structure gives rise to spin-dependent motion of electrons.

It is noted that the similar effect is expected also for massive particle waves with intrinsic angular momentum. However, the effect is suppressed by the mass factor, m. For example, for the spin-1/2 massive field, the BC in helicity basis is given by $\boldsymbol{\Omega}_k = \frac{1}{2E_k^2}[\frac{m}{E_k}(\sigma_1 \boldsymbol{e}_\theta + \sigma_2 \boldsymbol{e}_\phi) + \sigma_3 \boldsymbol{e}_k]$, where $E_k = \sqrt{k^2 + m^2}$ and $\boldsymbol{e}_{k,\theta,\phi}$ are the unit vectors of the spherical coordinate system in $\boldsymbol{k}$ space.

The BC of light is proportional to $1/k^2$. Consequently, the transverse shift is a fraction of the wavelength, and is small for the visible light. In the following we propose a method to enhance the Hall effect of light by use of photonic crystals [17]. As is known, the periodic modulation of the dielectric constant leads to formation of Bloch waves of light analogous to those of electron in solids. The degeneracy of right- and left-circularly polarized light is generally lifted. Hence, the SU(2) gauge invariance breaks down to U(1) (Abelian). This is similar to the anomalous Hall effect in ferromagnetic materials [18]. In photonic crystals, monopoles in $\boldsymbol{k}$-space can exist away from $\boldsymbol{k} = \boldsymbol{0}$. The anomalous velocity is enhanced when $\boldsymbol{k}$ is near the monopole.

For an explicit example, we consider a two-dimensional (2D) photonic crystal with the dielectric function $\epsilon(\mathbf{r})$, $\frac{1}{\epsilon(r)} = \{4/[3(5 + 12|\xi| + 8\xi^2)]\}\sum_{i=1}^{3}\{[\xi - \cos(\boldsymbol{b}_i \cdot \boldsymbol{r} + \frac{2\pi}{3})]^2 + [\xi + \cos(\boldsymbol{b}_i \cdot \boldsymbol{r} - \frac{2\pi}{3})]^2\}$, where $\boldsymbol{b}_1 = (\frac{2\pi\sqrt{3}}{3a}, -\frac{2\pi}{3a})$, $\boldsymbol{b}_2 = (0, \frac{4\pi}{3a})$, and $\boldsymbol{b}_3 = -\boldsymbol{b}_1 - \boldsymbol{b}_2$. The spatial distribution of $\epsilon(\mathbf{r})$ is shown in Fig. 3(a). For simplicity, we shall focus on the case with $k_z = 0$. In other words, we shall consider wave packets extended in z direction, i.e., wave ribbons. The eigenmodes are classified into transverse magnetic (TM) modes and transverse electric (TE) modes. For the TM modes, the magnetic field is in the xy-plane, while for the TE modes it is in the z direction. The photonic bands are shown in Fig. 3(b). The BC is along the z direction in 2D, and its z component is shown in Fig. 4. It is seen that $\boldsymbol{\Omega}_k$ is strongly enhanced near the corners of the Brillouin zone, which can be interpreted as a 2D cut of the monopole structure in the extended space including the parameters, e.g., ξ in the present case (ξ represents the degree of inversion-symmetry breaking).

To induce the Hall effect of light in photonic crystals, we have to introduce a gradient of the refractive index, as seen in Eq. (2). Therefore, we shall introduce a slow variation of the envelope of the refractive index as $n(\boldsymbol{r}) \to n(\boldsymbol{r})/\gamma(x)$ where $1/\gamma(x)$ changes from 1.0 to 1.2 within the range of $w = 10a$. We have solved the EOM for the wave packets in the photonic crystal, $\dot{\boldsymbol{r}}_c = \gamma(x_c)\boldsymbol{\nabla}_{\boldsymbol{k}_c}E_{nk_c} + \dot{\boldsymbol{k}}_c \times \boldsymbol{\Omega}_{nk_c}$, $\dot{\boldsymbol{k}}_c = -[\boldsymbol{\nabla}\gamma(x_c)]E_{nk_c}$, where E_{nk} and $\boldsymbol{\Omega}_{nk}$ are the energy and the BC of the nth band in the unperturbed crystal. The EOM for $|z\rangle$ brings about the phase shift for each mode. We set the initial $\boldsymbol{k}_c$ near the corners of the Brillouin zone. This is because the BC is large at the corners and the effect by the anomalous velocity is expected to be prominent. The obtained trajectories are shown in Fig. 5. It is found that the shift of $\boldsymbol{r}_c$ reaches to dozens of times the lattice constant especially for the TE second band.

We note that the BC is generally present in photonic crystals if the inversion-symmetry is broken. Our EOM offers an easy method to calculate the trajectory of wave packets for generic photonic crystals [19]. By contriving crystal structures, this effect can be enhanced consider-

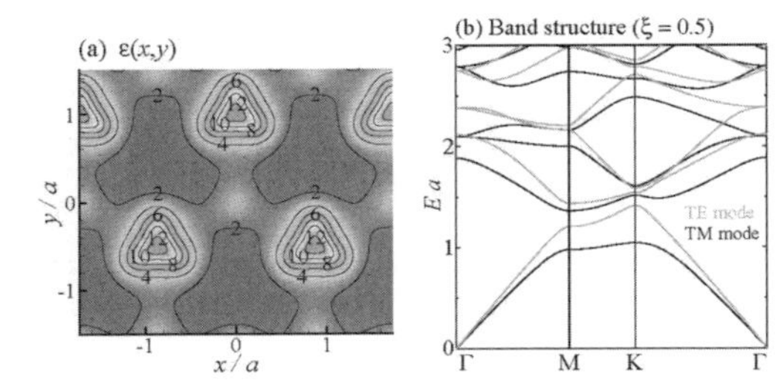

FIG. 3 (color). (a) Dielectric function and (b) band structure of the 2D photonic crystal. The Brillouin zone is shown in Figs. 4 and 5(b).

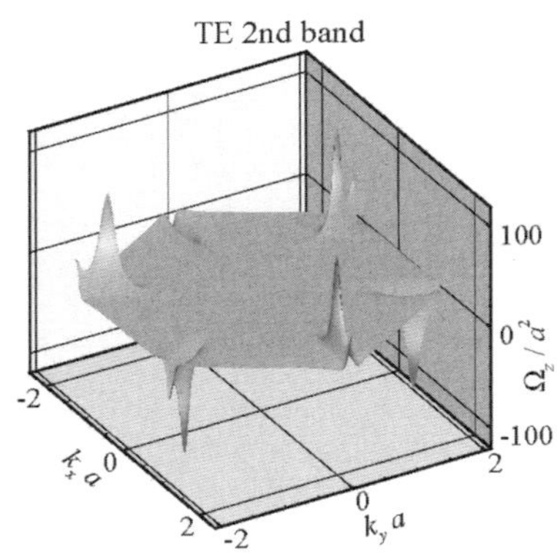

FIG. 4 (color). Example of the BC of the 2D photonic crystal. It is prominent at the corners of the Brillouin zone.

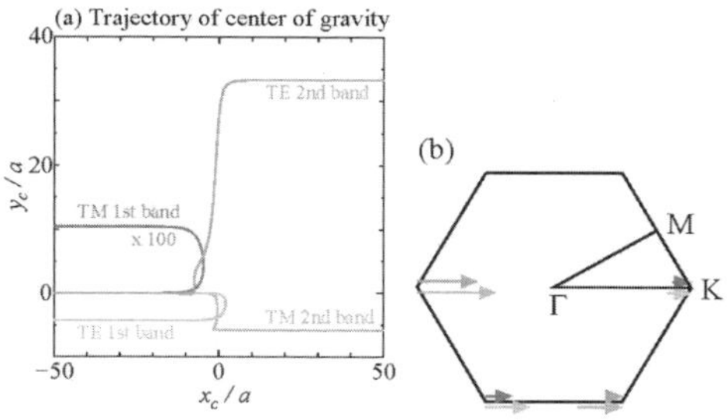

FIG. 5 (color). (a) Real-space trajectory of the center of the wave packet. (b) Momentum-space trajectory of the center of the wave packet. The arrows in the same colors in (a) and (b) refer to the same photonic modes.

ably as follows. The BC around the zone corners is determined mostly by the splitting, $2|\Delta|$, between the neighboring bands. Suppose, for example, at $\boldsymbol{k} = \boldsymbol{k}_c$ another band comes very close in energy to the one considered. The BC is approximately given by $\Omega_z \sim [v^2\Delta/(\Delta^2 + v^2|\boldsymbol{k} - \boldsymbol{k}_0|^2)^{3/2}]$, where v is a nominal velocity of light. Thus when the light traverses this $\boldsymbol{k}_0$ point, the shift is given by $\delta y_c \sim \mathrm{sgn}[\nabla_{x_c}\gamma(x_c)]v/\Delta$. Therefore $|\delta y_c|$ is larger for smaller $|\Delta|$ as shown in Fig. 5; the second gap is smaller and hence the wave packet in the second band shows larger $|\delta y_c/a|$.

Let us discuss an effect of disorder in real photonic crystals. It has been discussed that the photonic band with lower index is more robust against disorder [20]. Hence, in order to see the Hall effect of light, it is better to use the first band than the second band. While in Fig. 5 the shift of the first band is smaller than the second, it can be enhanced by reducing the gap between the two bands. The first band is not affected appreciably when fluctuations of system parameters are less than a few percent. For visible or near infrared light, this requires fabrication precision within ~50 nm, which is comparable to the current state of the art. Experimentally, the band splitting Δ of the order of 10^{-5} nm^{-1} can be controlled in the photonic crystal for the visible light [21]. This corresponds to $\delta y_c \sim 1/\Delta \sim 100a$. Thus, we can expect the shift as large as a hundred times the lattice constant a.

To conclude, we have derived the semiclassical equation of motion for the wave packet of light including the Berry curvature and resultant anomalous velocity. This gives a natural generalization of geometrical optics including the wave nature of light, and leads to the Hall effect of light. This also offers the natural framework to describe the ray of light in the photonic crystal without the inversion-symmetry where the Hall effect can be magnified tremendously.

The authors thank S.C. Zhang, A.V. Balatsky, K. Miyano, and A. Furusawa for fruitful discussions. M.O. is supported by the Japan Society for the Promotion of Science. This work is financially supported by NAREGI Grant, Grant-in-Aids from the Ministry of Education, Culture, Sports, Science and Technology of Japan, and by NEDO Grant.

Note added.—A related argument on a spin transport of photons came to our attention after the submission of this work. K.Yu. Bliokh and Yu.P. Bliokh, physics/0402110 (unpublished).

*Electronic address: m.onoda@aist.go.jp
†Electronic address: murakami@appi.t.u-tokyo.ac.jp
‡Electronic address: nagaosa@appi.t.u-tokyo.ac.jp

[1] M. Born and E. Wolf, *Principles of Optics* (Cambridge University Press, Cambridge, 1999), 7th ed.
[2] R.Y. Chiao and Y.S. Wu, Phys. Rev. Lett. **57**, 933 (1986); A. Tomita and R.Y. Chiao, *ibid.* **57**, 937 (1986); M.V. Berry, Nature (London) **326**, 277 (1987).
[3] S. Pancharatnam, The Proceedings of the Indian Academy of Sciences, 1956, Vol. XLIV, No. 5, Sec. A, p. 247.
[4] M.V. Berry, Proc. R. Soc. London A **392**, 45 (1984); *Geometrical Phases in Physics*, edited by A. Shapere and F. Wilczek (World Scientific, Singapore, 1989).
[5] M.-C. Chang and Q. Niu, Phys. Rev. B **53**, 7010 (1996); G. Sundaram and Q. Niu, *ibid.* **59**, 14915 (1999).
[6] M. Onoda, S. Murakami, and N. Nagaosa (unpublished).
[7] F. Goos and M. Hänchen, Ann. Phys. (Leipzig) **1**, 333 (1947).
[8] F.I. Fedorov, Dokl. Akad. Nauk SSSR **105**, 465 (1955).
[9] C. Imbert, Phys. Rev. D **5**, 787 (1972).
[10] D.G. Boulware, Phys. Rev. D **7**, 2375 (1973); N. Ashby and S.C. Miller, Jr., Phys. Rev. D **7**, 2383 (1973).
[11] O. Costa de Beauregard, Phys. Rev. **139**, B1443 (1965); J. Ricard, Nouv. Rev. Opt. **1**, 275 (1970).
[12] H. Schilling, Ann. Phys. (Leipzig) **16**, 122 (1965).
[13] V.G. Fedoseev, Opt. Spektrosk. **71**, 829 (1991) [Opt. Spectrosc. (USSR) **71**, 483 (1991)]; Opt. Spektrosk. **71**, 992 (1991) [Opt. Spectrosc. (USSR) **71**, 570 (1991)].
[14] M.A. Player, J. Phys. A **20**, 3667 (1987); V.G. Fedoseev, *ibid.* **21**, 2045 (1988); W.N. Hugrass, J. Mod. Opt. **37**, 339 (1990).
[15] S. Murakami, N. Nagaosa, and S.-C. Zhang, Science **301**, 1348 (2003).
[16] J. Sinova *et al.*, Phys. Rev. Lett. **92**, 126603 (2004).
[17] J.D. Joannopoulos, R.D. Meade, and J.N. Winn, *Photonic Crystals* (Princeton University Press, Princeton, 1995).
[18] Z. Fang *et al.*, Science **302**, 92 (2003).
[19] The EOM holds with good accuracy when the spatial variation of the envelope $\gamma(x)$ of the refractive index is slower than the wavelength. On the other hand, the periodic variation of the refractive index in the photonic crystal need not to be slow.
[20] A.A. Asatryan *et al.*, Phys. Rev. E **62**, 5711 (2000).
[21] A. Rosenberg *et al.*, J. Appl. Phys. **82**, 6354 (1997).

Berry Curvature on the Fermi Surface: Anomalous Hall Effect as a Topological Fermi-Liquid Property

F. D. M. Haldane

Department of Physics, Princeton University, Princeton New Jersey 08544-0708, USA
(Received 28 June 2004; revised manuscript received 20 October 2004; published 11 November 2004)

The intrinsic anomalous Hall effect in metallic ferromagnets is shown to be controlled by Berry phases accumulated by adiabatic motion of quasiparticles on the Fermi surface, and is purely a Fermi-liquid property, not a bulk Fermi sea property like Landau diamagnetism, as has been previously supposed. Berry phases are a new topological ingredient that must be added to Landau Fermi-liquid theory in the presence of broken inversion or time-reversal symmetry.

DOI: 10.1103/PhysRevLett.93.206602 PACS numbers: 72.15.–v, 73.43.–f

Renewed interest in the anomalous Hall effect (AHE) in metallic ferromagnets has lead to a reinterpretation of the classic Karplus-Luttinger formula [1] for the "anomalous velocity" in terms of the Berry curvature of occupied electronic Bloch states [2,3]. This gives an intrinsic contribution to the Hall conductivity in the low-temperature clean limit of metallic ferromagnets when the quasiparticle lifetimes become long, and now appears to be the dominant contribution to the AHE [4,5].

The expression [1–3] for the intrinsic Hall conductivity of metals with broken time-reversal symmetry has the all the appearance of a "bulk" band structure property that depends on all the filled electronic states, not just the ones at the Fermi level. However, this seems at odds with the spirit of Landau's Fermi-liquid theory, which holds that charge transport in metals involves only quasiparticles with energies within k_BT of the Fermi level.

In this Letter, I show that, despite appearances, the *nonquantized part* of the intrinsic Hall conductivity is completely expressible in terms of Berry phases [6] of quasiparticles moving on the Fermi surface, and thus fully consistent with Fermi-liquid theory. This exposes a new topological ingredient that must be added to Fermi-liquid theory unless both inversion and time-reversal symmetry are present: quasiparticle Berry phases.

The "anomalous velocity" is an extra term [7,8] in the semiclassical equations of motion of a Bloch electron in weak electric and magnetic fields; ignoring Zeeman couplings, these are

$$\hbar\frac{dk_a}{dt} = eE_a(\mathbf{x}) + eF_{ab}(\mathbf{x})\frac{dx^b}{dt}, \tag{1}$$

$$\frac{dx^a}{dt} = \frac{1}{\hbar}\nabla_k^a\varepsilon_n(\mathbf{k}) + \mathcal{F}_n^{ab}(\mathbf{k})\frac{dk_b}{dt}, \tag{2}$$

where $F_{ab} = \nabla_a A_b - \nabla_b A_a = \epsilon_{abc}B^c$ is the magnetic flux density written as an antisymmetric tensor, $\nabla_a \equiv \partial/\partial x^a$, $\varepsilon_n(\mathbf{k})$ is the energy of a Bloch electron in band n, $\nabla_k^a \equiv \partial/\partial k_a$ and $\mathcal{F}_n^{ab}$ is the antisymmetric "Berry curvature" tensor in $\mathbf{k}$ space, described below. The anomalous velocity in (2) is the $\mathbf{k}$-space dual of the Lorentz force.

If one writes the electron occupations $\langle n_{kn}\rangle$ as $n_n^0(\mathbf{k},\mu) + \delta n_{kn}$, where $n_n^0(\mathbf{k},\mu)$ is the ground-state occupation function at chemical potential μ, the linear current response to a uniform electric field $\mathbf{E}$ (with $\mathbf{B} = 0$) is

$$J_e^a = \frac{e}{\hbar}\frac{1}{\Omega N}\sum_{kn}\nabla_k^a\varepsilon_n(\mathbf{k})\delta n_{kn} + \sigma_0^{ab}(\mu)E_b, \tag{3}$$

where N is the number of primitive unit cells, which have volume Ω. Here $\sigma_0^{ab}(\mu)$ is an intrinsic ground-state property describing a dissipationless Hall conductivity

$$\sigma_0^{ab}(\mu) = \frac{e^2}{\hbar}\frac{1}{\Omega N}\sum_{kn}\mathcal{F}_n^{ab}n_n^0(\mathbf{k},\mu). \tag{4}$$

This also controls the low-temperature limit of the thermal Hall conductivity κ^{ab} (Righi-Leduc effect) and the Peltier coefficient α^{ab}; their "intrinsic" parts are

$$\kappa_0^{ab}(\mu) = \frac{\pi^2}{3}\frac{k_B^2T}{e^2}\sigma_0^{ab}(\mu), \qquad \alpha_0^{ab}(\mu) = e\frac{\partial\kappa_0^{ab}(\mu)}{\partial\mu}. \tag{5}$$

A heat current $J_Q^a = T\alpha_0^{ab}E_b$ flows with the Hall current.

If time-reversal symmetry is present, the electronic bands have the property

$$\varepsilon_n(-\boldsymbol{k}) = \varepsilon_n(\boldsymbol{k}), \qquad \mathcal{F}_n^{ab}(-\boldsymbol{k}) = -\mathcal{F}_n^{ab}(-\boldsymbol{k}), \tag{6}$$

the sum (4) cancels, and the intrinsic Hall conductivity vanishes. If inversion symmetry is also unbroken, $\mathcal{F}_n^{ab}(-\boldsymbol{k}) = \mathcal{F}_n^{ab}(\boldsymbol{k})$, and the Berry curvature vanishes.

The Berry curvature is obtained from a "vector potential" derived from the 1-particle Bloch states $|\psi_n(\boldsymbol{k})\rangle$:

$$\mathcal{A}_n^a(\boldsymbol{k}) = -i\langle\psi_n(\boldsymbol{k})|\nabla_k^a\psi_n(\boldsymbol{k})\rangle, \tag{7}$$

$$\mathcal{F}_n^{ab}(\boldsymbol{k}) = \nabla_k^a\mathcal{A}_n^b(\boldsymbol{k}) - \nabla_k^b\mathcal{A}_n^a(\boldsymbol{k}), \tag{8}$$

$$\epsilon_{abc}\nabla_k^a\mathcal{F}_n^{bc}(\boldsymbol{k}) = \sum_i q_{ni}\delta^3(\boldsymbol{k}-\boldsymbol{k}_{ni}), \qquad q_{ni} = \pm 2\pi. \tag{9}$$

The last equation is the divergence of the $\boldsymbol{k}$-space Berry curvature field $\mathcal{F}_n^{ab}(\boldsymbol{k})$: it is divergence-free except for quantized "monopole" sources with a "charge quantum" 2π, which are associated with band degeneracies. These occur at isolated $\boldsymbol{k}$ points: in complex Hermitian eigenproblems, it is sufficient to vary three parameters (here, the components of $\boldsymbol{k}$) to encounter degeneracies.

Assuming that the one-electron energy $\varepsilon_n(\boldsymbol{k})$ is nondegenerate, the wave functions are completely defined, except for an arbitrary phase factor. If both time-reversal and inversion symmetry are present, this can consistently be chosen real, but otherwise, is an arbitrary complex factor that can vary continuously with $\boldsymbol{k}$. The Berry vector potential $\mathcal{A}_n^a(\boldsymbol{k})$ depends on this "gauge choice" but the curvature $\mathcal{F}_n^{ab}(\boldsymbol{k})$ is a well-defined gauge-invariant quantity with physical significance. The Berry phase [6] for a closed path Γ is also gauge invariant:

$$\exp i\phi_n(\Gamma) = \exp i \oint_\Gamma \mathcal{A}_n^a(\boldsymbol{k}) dk_a. \qquad (10)$$

The integral is the curvature flux linked through Γ: the 2π ambiguity of the Berry phase quantizes the "charge" of the monopole sources of the curvature field.

In a 3D band structure, the integral over the Brillouin zone (BZ) of the Berry curvature of a nondegenerate band is a topological invariant [9] that is a generalization of the better-known 2D Chern number [10,11]

$$\frac{1}{2\pi}\int d^3\boldsymbol{k}\mathcal{F}_n^{ab}(\boldsymbol{k})P_{\text{BZ}}(\boldsymbol{k}) = C_n\epsilon^{abc}G_{cn}^C, \qquad (11)$$

where C_n is an integer Chern number, $1/2\pi$ times the integral of the Berry curvature over a compact 2D surface, e.g., a 2D BZ (a torus), and $\mathbf{G}_n^C$ is a primitive reciprocal lattice vector; $P_{\text{BZ}}(\boldsymbol{k}) = 1$ inside the BZ, 0 outside it.

The intrinsic Hall conductivity may be parametrized as $\sigma_0^{ab} = (e/\Phi_0)\epsilon^{abc}(K_c/2\pi)$, $\Phi_0 = h/e$, where $\boldsymbol{k}$ is dimensionally a wave vector: in one-electron band theory,

$$\frac{1}{2\pi}\sum_n\int d^3\boldsymbol{k}\mathcal{F}_n^{ab}P_{\text{BZ}}(\boldsymbol{k})n_n^0(\boldsymbol{k},\mu) = \epsilon^{abc}K_c. \qquad (12)$$

If band n is completely below the Fermi level, it contributes a quantized amount $C_n\boldsymbol{G}_n^C$ to $\boldsymbol{k}$. This produces an integer quantum Hall effect (QHE) with "filling factor" $\nu = C_n$ on the lattice planes indexed by $\boldsymbol{G}_n^C$.

The QHE is usually discussed in the context of strong magnetic fields where the electronic states are split up into Landau levels, so it might be wondered how "simple" Bloch electrons could exhibit a QHE without Landau levels. In fact, only broken time-reversal symmetry is required: the possibility of a $\mathbf{B} = 0$ "zero-field QHE" was first demonstrated in Ref. [12], using a model that, in retrospect, exhibits both a nonmetallic QHE phase and a metallic AHE phase.

It will be useful to understand the process by which the Chern invariants of a 3D band structure can change. While a band remains nondegenerate, its Chern invariant is "quantized" to be a reciprocal lattice vector. As a control parameter is varied, two bands may come into contact at some point in the BZ, and this initial degeneracy point then subsequently splits into two "Dirac point" singularities (near which the energy dispersion is linear). The bands are now tightly coupled by a "Berry flux loop" where Berry curvature flux 2π passes from one band to the other through one Dirac point, then returns through the other one. There is a striking analogy to the idea of "wormholes" connecting different universes, where here the two "universes" are Bloch bands, and "space" is $\boldsymbol{k}$ space. Each band has one positive and one negative monopole source of Berry curvature; at each Dirac point the two bands have opposite-sign sources. Eventually, after relative displacement by a reciprocal lattice vector $\boldsymbol{G}$, the monopoles may recombine, allowing the bands to split apart. This process conserves the sum of their invariants, but individually they change by $\pm\boldsymbol{G}$.

It is useful to first examine (4) in the simpler 2D case. Then $\sigma_0^{xy} = \nu e/\Phi_0$, where $\nu = \sum_n \nu_n$, and

$$\nu_n(\mu) = \frac{1}{2\pi}\int d^2\boldsymbol{k}\mathcal{F}_n^{xy}(\boldsymbol{k})P_{\text{BZ}}(\boldsymbol{k})n_n(\boldsymbol{k},\mu). \qquad (13)$$

For simplicity, assume that the occupied region does not touch the Brillouin zone boundary (BZB), and drop $P_{\text{BZ}}(\boldsymbol{k})$. Using the Berry vector potential representation and integrating by parts gives

$$\nu_n = \frac{1}{2\pi}\int d^2\boldsymbol{k}(\mathcal{A}_n^x\nabla_k^y n_n^0(\boldsymbol{k}) - A_n^y\nabla_k^x n_n^0(\boldsymbol{k})), \qquad (14)$$

which is clearly a Fermi-surface integral if the band is partially filled, since $n_n^0(\boldsymbol{k})$ has a step discontinuity at the Fermi surface, and is constant everywhere else. If the Fermi surface is a simple closed loop, this can be recognized as the integral giving the Berry phase ϕ_F for an adiabatic path around the Fermi surface (the $\boldsymbol{k}$-space version of the Bohm-Aharonov effect)

$$\nu_n = \frac{1}{2\pi}\oint \mathcal{A}_n^a(\boldsymbol{k}_F)d\boldsymbol{k}_{Fa} = \frac{\phi_F}{2\pi}. \qquad (15)$$

Since the Berry phase is ambiguous by a multiple of 2π, only the nonquantized part of the intrinsic Hall conductivity is determined at the Fermi surface. If the system has evolved adiabatically along some path in a parameter space from one with time-reversal symmetry, the QHE can be determined from the history of ϕ_F during that process.

Armed with the insight that the AHE is a Fermi-surface property, I now examine the 3D problem. In general, there may be multiple sheets S_α of the Fermi surface, with both simple and multiply connected topology, with pieces "glued together" at degeneracy points where a line of high-symmetry intersects the surface, or along lines where a plane of high-symmetry intersects it; if time-reversal symmetry is unbroken, there may also be

Kramers degeneracy [13]. Complexes of intrinsically connected sheets $\boldsymbol{k}_F^{(\alpha)}(s)$, where $s = \{s^1, s^2\}$ is a surface parametrization, will be referenced by a single label α, specified implicitly if $\mathbf{s} \in S_\alpha$. The outward normal unit vector $\hat{\boldsymbol{n}}(s)$ is also the direction of the Fermi velocity.

The key Fermi-surface property is the Luttinger sum rule relating particle number to Fermi-surface volume. The Fermi-surface geometry fixes the particle density modulo integer multiples of the "density quantum" $\rho_0 = 1/\Omega$ associated with filled bands. The change in particle density $\delta\rho_\alpha(\boldsymbol{r})$ associated with a local fluctuation $\delta\boldsymbol{k}_F(\boldsymbol{r}, s) = \boldsymbol{\nabla}\varphi(\boldsymbol{r}, s)$ [where $\varphi(\boldsymbol{r}, s)$ is the quasiparticle phase] is just proportional to the $\boldsymbol{k}$-space volume swept out by the changing Fermi surface:

$$\partial_\mu \boldsymbol{k}_F(s) \times \partial_\nu \boldsymbol{k}_F(s) = \ell_{\mu\nu}(s)\hat{\boldsymbol{n}}(s), \tag{16}$$

$$\delta\rho_\alpha(\boldsymbol{r}) = \int_{S_\alpha} \frac{ds^\mu \wedge ds^\nu}{(2\pi)^3} \ell_{\mu\nu}(s)\hat{\boldsymbol{n}}(s) \cdot \boldsymbol{\nabla}\varphi(\boldsymbol{r}, s), \tag{17}$$

$$\int_{S_\alpha} \frac{ds^\mu \wedge ds^\nu}{(2\pi)^3} \ell_{\mu\nu}(s)\hat{\boldsymbol{n}}(s) = \rho_0 \sigma_\alpha \boldsymbol{R}_\chi. \tag{18}$$

Here $\partial_\mu \equiv \partial/\partial s^\mu$; a nonzero integer σ_α signals a "chiral anomaly": the system then has a quasi-1D character where the primitive real-space lattice vector $\boldsymbol{R}_\chi$ defines a special direction of lattice lines along which quasi-1D electrons predominantly move (such a band structure may also have nonchiral Fermi-surface sheets.)

The change in density $\delta\rho_\alpha$ if the Fermi-surface sheet α is rigidly displaced by a constant shift $\delta\boldsymbol{k}_F$ is $\sigma_\alpha \rho_0 \boldsymbol{R}_\chi \cdot \delta\boldsymbol{k}_F$. Note that the absolute value of $\boldsymbol{k}_F(s)$ is not invariant under position-space gauge transformations $\boldsymbol{k}_F \rightarrow \boldsymbol{k}_F - (e/\hbar)\boldsymbol{A}(\boldsymbol{r})$ and only the relative displacements (modulo reciprocal lattice vectors) between Fermi vectors are physically meaningful. Gauge invariance requires that the total Fermi-surface chiral anomaly must vanish. Typical Fermi-surface sheets are nonchiral but the possibility of chiral sheets needs to be kept in mind; the integer σ_α is a measure of how many distinct chiral sheets are glued together.

It is straightforward to repeat the integration by parts to expose the nonquantized part of the 3D intrinsic Hall conductivity as a Fermi-surface property. The Berry vector potential and associated gauge-invariant Berry curvature for paths restricted to lie in a surface $\boldsymbol{k}(s)$ are

$$\mathcal{A}_\mu(s) = \mathcal{A}^a(\boldsymbol{k}(s))\partial_\mu k_a(s), \quad \mathcal{F}_{\mu\nu} = \partial_\mu \mathcal{A}_\nu - \partial_\nu \mathcal{A}_\mu. \tag{19}$$

The surface integral that results from integration by parts of the band-n contribution is over a surface S_n that is divided into a set of one or more outward-oriented compact surfaces enclosing occupied regions within the BZ:

$$\boldsymbol{K}_n = \frac{1}{2\pi} \int_{S_n} ds^\mu \wedge ds^\nu \mathcal{F}_{\mu\nu}(s)\boldsymbol{k}(s). \tag{20}$$

Some parts of S_n are on the Fermi surface, but there may also be contributions from states below the Fermi level which are on the BZ boundary (BZB). The Fermi surface of band n will be divided into one or more disjoint oriented surfaces $S_{n\alpha}$, and the intersections of these with the BZB define a set of conjugate pairs of closed directed paths $C_{n\alpha i\pm}$ which have Berry phases $\pm\phi_{n\alpha i}$; the "+" path is displaced relative to its partner on the opposite side of the BZB by a primitive reciprocal lattice vector $\boldsymbol{G}_{n\alpha i}$. The contribution to the integral (20) from conjugate BZB intersections is $\phi_{n\alpha i}\boldsymbol{G}_{n\alpha i}/2\pi$, which can be apportioned equally to the "+" and "−" paths.

Eliminating the integrals over the BZB allows band indices n to be dropped, and $\sum_n \boldsymbol{K}_n$ can be written (modulo $\boldsymbol{G}$) as a sum $\sum_\alpha \boldsymbol{K}_\alpha$ of Fermi-surface integrals

$$\boldsymbol{K}_\alpha = \frac{1}{2\pi} \int_{S_\alpha} d^2\mathcal{F}\, \boldsymbol{k}_F + \frac{1}{4\pi} \sum_i \boldsymbol{G}_{\alpha i} \int_{\partial S_\alpha^i} d\mathcal{A}, \tag{21}$$

where $d^2\mathcal{F} \equiv \mathcal{F}_{\mu\nu}(s) ds^\mu \wedge ds^\nu$ is the Berry curvature 2-form and $d\mathcal{A}$ is the connection 1-form $\mathcal{A}_\mu(s)ds^\mu$; ∂S_α^i are the 1-manifolds where S_α intersects the BZB, across which $\boldsymbol{k}_F(s)$ jumps by $\boldsymbol{G}_{\alpha i}$. These boundary terms are Berry-gauge dependent, but if S_α is nonchiral, the BZ can be chosen so all ∂S_α^i are closed paths, and the gauge ambiguity is an (unknown) quantized QHE contribution. [If S_α is chiral, (21) is valid in a Berry gauge where $\mathcal{A}_\mu(s)$ is periodic, and the boundary terms cancel.]

The integral of $1/2\pi$ times the Berry curvature over Fermi-surface sheet α defines an integer Chern number C_α. If a gauge transformation shifts $\boldsymbol{k}_F(s)$ by a constant $\delta\boldsymbol{k}_F$, $\boldsymbol{k}_\alpha$ changes by $C_\alpha \delta\boldsymbol{k}_F$. Gauge invariance requires that the total sum of Fermi-surface Chern numbers must vanish.

If a nonchiral compact piece of Fermi surface has nonzero Chern number, it must enclose a source of Berry curvature, and has a hidden "wormhole" connection through which "spectral flow" can occur. Consider two bands linked by a "Berry flux loop" through a pair of Dirac points, one below the Fermi level, the other above it (both time-reversal and inversion symmetries must be broken). A particlelike Fermi surface with Chern number $C = \pm 1$ surrounds the lower Dirac point, and a holelike one with opposite-sign Chern number surrounds the other. A "spectral flow" process driven by spatial inhomogeneities can "pump" states into a band through one Dirac point and out through the other, conserving the total number of states per band. States that flow carry their occupations with them: the volumes of the holelike and particlelike Fermi surfaces shrink or expand by the same amount, conserving total charge.

The additional Berry phase terms in (21) are there for a very concrete reason: the choice of the Brillouin zone defined by $P_{\mathrm{BZ}}(\boldsymbol{k})$ is yet another kind of arbitrary gauge choice: since $\boldsymbol{k}_F(s)$ is defined to be in the BZ, however it is chosen, there must be BZB lines on a multiply connected surface across which $\boldsymbol{k}_F(s)$ jumps discontinuously back into the BZ. The Berry phase counterterms merely guarantee that the value of $\boldsymbol{K}_\alpha$ is unchanged by any continuous deformation of the standard BZ into any other primitive cell.

The Hall conductivity also controls the charge density $\rho_e = e\delta\rho$ induced when a uniform magnetic flux density is applied, keeping the Fermi energy μ fixed:

$$\lim_{B\to 0}\frac{\partial \rho_e}{\partial F_{ab}}\bigg|_{\mu} = \sigma_0^{ab}(\mu), \qquad F_{ab} \equiv \epsilon_{abc}B^c. \tag{22}$$

Such a Streda-type formula [14] should hold separately for each Fermi-surface sheet. This can easily be found from semiclassical quantization with a Fermi-surface Berry phase. For $\boldsymbol{B} = B\hat{\boldsymbol{n}}$ (technically parallel to a lattice translation), let the set of Fermi-surface orbits in the $\boldsymbol{k}$-space plane $\hat{\boldsymbol{n}}\cdot\boldsymbol{k} = k$ have cross-sectional areas $A_\alpha(k)$ and Berry phases $\phi_\alpha(k)$.

Recall that σ_0^{ab} is parametrized by $\boldsymbol{k}\alpha$: here

$$\hat{\boldsymbol{n}}\cdot\boldsymbol{K}_\alpha = \frac{1}{2\pi}\int_{-\infty}^{\infty} dk P_{\mathrm{BZ}}(k\hat{\boldsymbol{n}})\phi_\alpha(k). \tag{23}$$

The semiclassical quantization condition is

$$A_\alpha(k^i_{F\alpha})\ell_B^2 - \phi_\alpha(k^i_{F\alpha}) = 2\pi\left(n^i_\alpha + \frac{1}{2}\right), \tag{24}$$

where $\ell_B^2 = \hbar/|eB|$, and n^i_α are integers. The $k^i_{F\alpha}$ (ordered so $k^i_{F\alpha} > k^j_{F\alpha}$ for $i > j$) are the Fermi momenta of sets of 1D Fermi gases of particles on sheet α, moving along field lines with a 2D density $(2\pi\ell_B^2)^{-1}$ (one line per flux quantum). The induced charge density is completely determined by Fermi-surface geometry:

$$\rho_e = \frac{e}{(2\pi)^3}\sum_{\alpha i}\frac{1}{2}(A^i_\alpha + A^{i+1}_\alpha)\Delta k^{i,i+1}_\alpha - v^{i,i+1}_\alpha, \tag{25}$$

where $A^i_\alpha = A_\alpha(k^i_{F\alpha}) - (eB/\hbar)\phi_\alpha(k^i_{F\alpha})$, $v^{i,i+1}_\alpha$ is the integral of $A_\alpha(k)$ from $k^i_{F\alpha}$ to $k^{i+1}_{F\alpha}$, and $\Delta k^{i,i+1}_\alpha = k^{i+1}_{F\alpha} - k^i_{F\alpha}$. The Streda formula (23) is now easily obtained when $B \to 0$. Note the differences between density and energy shifts due to Landau quantization: the latter derive from changes to states deep below the Fermi level, and Landau diamagnetism is not a Fermi-surface effect.

In summary, the intrinsic Hall conductivity of a metal with broken time-reversal symmetry can be written as $\sigma_0^{ab} = (e/\Phi_0)\epsilon^{abc}(K_c/2\pi)$, $\Phi_0 = h/e$. I have shown that the nonquantized part of the wave vector $\boldsymbol{K}$ (i.e., the part modulo a reciprocal vector $\boldsymbol{G}$) is a topological Fermi surface property given by a sum of terms $\boldsymbol{K}_\alpha$ (21) associated with each distinct Fermi-surface sheet. Separate batically conserved currents are associated with each such sheet (or group of sheets mutually coupled by "wormholes"): this generalizes the "extra" conservation laws at each distinct chiral Fermi point of a 1D Luttinger liquid. In the absence of BCS pairing processes, breakdown of these extra conservation laws can occur only through nonadiabatic impurity or surface scattering. A "topological Fermi liquid theory" that includes Berry phases of quasiparticles adiabatically moving on Fermi surface manifolds must now be developed.

This work was supported in part by the U.S. National Science Foundation (under MRSEC Grant No. DMR02-13706) at the Princeton Center for Complex Materials, and under Grant No. PHY99-07949 at the Kavli Institute for Theoretical Physics, UC Santa Barbara.

Note added.—The simplicity of the Fermi-surface formula (21) strongly suggests that it is a fundamental Fermi liquid property, also valid in an interacting system, like the relation between electron density and Fermi surface volume [used above to derive (22)]. As I will describe elsewhere, this is indeed the case: the key point is that (only) at the Fermi surface, the Bloch states $|\psi_n(\boldsymbol{k}_F(\boldsymbol{s}))\rangle$ retain their meaning as the zero-mode eigenfunctions of the inverse (exact) one-electron propagator $G^{-1}(\boldsymbol{k}_F(s), \omega = 0)$, which in interacting Fermi-liquid theory remains Hermitian at $T = 0$. The formal proof uses a 3D version of the Ward-Takahashi identity used previously to relate the 2D integer QHE Hall conductance to the exact (interacting) one-electron propagator [15]. I thank P. B. Wiegmann for bringing Ref. [15] to my attention.

[1] R. Karplus and J. M. Luttinger, Phys. Rev. **95**, 1154 (1954); J. M. Luttinger, Phys. Rev. **112**, 739 (1958).

[2] T. Jungwirth, Q. Niu, and A. H. MacDonald, Phys. Rev. Lett. **88**, 207208 (2002).

[3] M. Onoda and N. Nagaosa, J. Phys. Soc. Jpn. **71**, 19 (2002).

[4] Z. Fang *et al.*, Science **302**, 92 (2003).

[5] Wei-Li Lee, Satoshi Watauchi, R. J. Cava, and N. P. Ong, Science **303**, 1647 (2004).

[6] M. V. Berry, Proc. R. Soc. London A **392**, 45 (1984).

[7] G. Sundaram and Q. Niu, Phys. Rev. B **59**, 14915 (1999).

[8] M. P. Marder, *Condensed Matter Physics* (Wiley, New York, 1999), pp. 425–429.

[9] M. Kohmoto, B. I. Halperin, and Y.-S. Wu, Phys. Rev. B **45**, 13488 (1992).

[10] D. J. Thouless, M. Kohmoto, M. P. Nightingale, and M. den Nijs, Phys. Rev. Lett. **49**, 405 (1982).

[11] J. E. Avron, R. Seiler, and B. Simon, Phys. Rev. Lett. **51**, 51 (1983).

[12] F. D. M. Haldane, Phys. Rev. Lett. **61**, 2015 (1988).

[13] Kramers degeneracy with spin-orbit coupling and inversion symmetry leads to non-Abelian SU(2) Berry curvature on the Fermi surface, to be described elsewhere.

[14] P. Streda, J. Phys. C **15**, L717 (1982).

[15] K. Ishikawa and T. Matsuyama, Z. Phys. C **33**, 41 (1986); Nucl. Phys. **B280**, 23 (1987).

First Principles Calculation of Anomalous Hall Conductivity in Ferromagnetic bcc Fe

Yugui Yao,[1,2,3] Leonard Kleinman,[1] A. H. MacDonald,[1] Jairo Sinova,[4,1] T. Jungwirth,[5,1] Ding-sheng Wang,[3] Enge Wang,[2,3] and Qian Niu[1]

[1]*Department of Physics, University of Texas, Austin, Texas 78712, USA*
[2]*International Center for Quantum Structure, Chinese Academy of Sciences, Beijing 100080, China*
[3]*Institute of Physics, Chinese Academy of Sciences, Beijing 100080, China*
[4]*Department of Physics, Texas A&M University, College Station, Texas 77843-4242, USA*
[5]*Institute of Physics ASCR, Cukrovarnická 10, 162 53 Praha 6, Czech Republic*
(Received 14 July 2003; published 22 January 2004)

We perform a first principles calculation of the anomalous Hall effect in ferromagnetic bcc Fe. Our theory identifies an intrinsic contribution to the anomalous Hall conductivity and relates it to the k-space Berry phase of occupied Bloch states. This dc conductivity has the same origin as the well-known magneto-optical effect, and our result accounts for experimental measurement on Fe crystals with no adjustable parameters.

DOI: 10.1103/PhysRevLett.92.037204 PACS numbers: 75.47.–m, 71.15.–m, 72.15.Eb, 78.20.Ls

In ferromagnets the Hall resistivity, ρ_H, exhibits an anomalous contribution proportional to the magnetization of the material, in addition to the ordinary contribution proportional to the applied magnetic field, $\rho_H = R_0 B + R_s 4\pi M$ [1–3]. The anomalous Hall effect (AHE) has played an important role in the investigation and characterization of itinerant electron ferromagnets because R_s is usually at least one order of magnitude larger than the ordinary constant R_0. Although the effect has been recognized for more than a century [2], it is still somewhat poorly understood, a circumstance reflected by the controversial and sometimes confusing literature on the subject. Previous theoretical work has failed to explain the magnitude of the observed effect even in well understood materials like Fe [4].

Karplus and Luttinger [5] pioneered the theoretical investigation of this effect, by showing how spin-orbit coupling in Bloch bands can give rise to an anomalous Hall conductivity (AHC) in a perfect ferromagnetic crystal. This conclusion was questioned by Smit [6], who argued that R_s must vanish in a periodic lattice. Smit proposed an alternative mechanism, skew scattering, in which spin-orbit coupling causes spin polarized electrons to be scattered preferentially to one side by impurities. The skew-scattering mechanism predicts an anomalous Hall resistivity linearly proportional to the longitudinal resistivity; this is in accord with experiment in some cases, but an approximately quadratic proportionality is more common. Later, Berger [7] proposed yet another mechanism, the side jump, in which the trajectories of scattered electrons shift to one side at impurity sites because of spin-orbit coupling. The side jump mechanism does predict a quadratic dependence of the AHC on the longitudinal resistivity. However, because of inevitable difficulties in modeling impurity scattering in real materials, it has not been possible to compare quantitatively with experiment. It appears to us that the AHE has generally been regarded as an extrinsic effect due solely to impurity scattering, even though this notion has never been critically tested, and that the intrinsic contribution initially proposed by Karplus and Luttinger has been discounted.

Several years ago, the scattering free contribution of Karplus and Luttinger was rederived in a semiclassical framework of wave packet motion in Bloch bands by taking into account Berry phase effects [8,9]. According to this work, there is an AHC purely from the equilibrium distribution given by the sum of Berry curvatures [see Eqs. (2) and (6) below] of the occupied Bloch states [10]. Recently, Jungwirth *et al.* [11,12] applied this picture of the AHE to (III, Mn)V ferromagnetic semiconductors and found very good agreement with experiment. (III, Mn)V ferromagnets are unusual, however, because they are strongly disordered and have extremely strong spin-orbit interactions. In this Letter, we report on an evaluation of the intrinsic AHC in a classic transition metal ferromagnet, bcc Fe. Our calculation is based on spin-density functional theory and the linearized augmented plane wave method. The close agreement between theory and experiment that we find leads us to conclude that the AHC in transition metal ferromagnets is intrinsic in origin, except possibly at low temperature in highly conductive samples.

We begin our discussion by briefly reviewing the semiclassical transport theory. By including the Berry phase correction to the group velocity [8,9], one can derive the following equations of motion:

$$\dot{\mathbf{x}} = \frac{1}{\hbar}\frac{\partial \varepsilon_n(\mathbf{k})}{\partial \mathbf{k}} - \dot{\mathbf{k}} \times \mathbf{\Omega}_n(\mathbf{k}), \qquad (1)$$

$$\hbar\dot{\mathbf{k}} = -e\mathbf{E} - e\dot{\mathbf{x}} \times \mathbf{B},$$

where $\mathbf{\Omega}_n$ is the Berry curvature of the Bloch state defined by

$$\mathbf{\Omega}_n(\mathbf{k}) = -\text{Im}\langle\nabla_{\mathbf{k}} u_{n\mathbf{k}}| \times |\nabla_{\mathbf{k}} u_{n\mathbf{k}}\rangle, \quad (2)$$

with $u_{n\mathbf{k}}$ being the periodic part of the Bloch wave in the nth band. We are interested in the case of $\mathbf{B} = 0$, for which $\varepsilon_n(\mathbf{k})$ is just the band energy. The distribution function satisfies the Boltzmann equation with the usual drift and scattering terms, and can be written as $f_n(\mathbf{k}) + \delta f_n(\mathbf{k})$, where f_n is the equilibrium Fermi-Dirac distribution function and δf_n is a shift proportional to the electric field and relaxation time. The electric current is given by the average of the velocity over the distribution function, yielding to first order in the electric field [13]

$$-\frac{e^2}{\hbar}\mathbf{E} \times \int d^3k \sum_n f_n \mathbf{\Omega}_n(\mathbf{k}) - \frac{e}{\hbar}\int d^3k \sum_n \delta f_n(\mathbf{k}) \frac{\partial \varepsilon_n}{\partial \mathbf{k}}. \quad (3)$$

The first term is the anomalous Hall current originally derived by Karplus and Luttinger [5], and can lead to quantized Hall conductivity for full magnetic bands [9,14]. In the second term, apart from the longitudinal current, there can also be a Hall current in the presence of skew scattering because the distribution function can acquire a shift in the transverse direction. This shift will be small when normal scattering processes are strong which keep the distribution very close to equilibrium. This is the case for pure crystals at temperature ranges where phonon scattering dominates, because phonon scattering is known to have very little skew part [1]. At low temperatures, where impurity scattering dominates, there can be a significant skew part that shifts the distribution in the transverse direction. In general, because the shift is proportional to the relaxation time, the skew-scattering contribution to the Hall conductivity goes up linearly with the longitudinal conductivity. This contribution can thus be identified, when it is dominant, by the traditional test, $\rho_{xy} \propto \rho_{xx}$. On the other hand, the Berry curvature contribution to the Hall conductivity is independent of scattering, and should lead to a quadratic dependence, $\rho_{xy} \propto \rho_{xx}^2$.

We now discuss our scheme for calculating the Berry curvature and the AHC. For a cubic material with magnetization aligned along [001], only the z component $\Omega^z(\mathbf{k}) \neq 0$. In our calculation, we find it convenient to use a different but equivalent expression for the Berry curvature that arises naturally from the Kubo-formula derivation [14],

$$\Omega_n^z(\mathbf{k}) = -\sum_{n' \neq n} \frac{2\text{Im}\langle\psi_{n\mathbf{k}}|v_x|\psi_{n'\mathbf{k}}\rangle\langle\psi_{n'\mathbf{k}}|v_y|\psi_{n\mathbf{k}}\rangle}{(\omega_{n'} - \omega_n)^2}, \quad (4)$$

where $E_n = \hbar\omega_n$, and v's are velocity operators. It is also instructive to introduce the sum (for each $\mathbf{k}$) of Berry curvatures over the occupied bands:

$$\Omega^z(\mathbf{k}) = \sum_n f_n \Omega_n^z(\mathbf{k}). \quad (5)$$

Then the intrinsic AHC is an integration over the Brillouin zone (BZ):

$$\sigma_{xy} = -\frac{e^2}{\hbar}\int_{\text{BZ}} \frac{d^3k}{(2\pi)^3} \Omega^z(\mathbf{k}). \quad (6)$$

The recent development of highly accurate *ab initio* electronic structure calculation methods enables us to complete the work of Karplus and Luttinger by evaluating their intrinsic Hall conductivity and comparing it with experiment. We employ the full-potential linearized augmented plane-wave method [15] with the generalized gradient approximation (GGA) for the exchange-correlation potential [16]. Fully relativistic band calculations were performed using the program package WIEN2K [17]. A converged ground state with magnetization in the [001] direction was obtained using 20 000 $\mathbf{k}$ points in the first Brillouin zone and $K_{\max}R_{MT} = 10$, where R_{MT} represents the muffin-tin radius and $K_{\max}$ the maximum size of the reciprocal-lattice vectors. Wave functions and potentials inside the atomic sphere are expanded in spherical harmonics up to $l = 10$ and 4, respectively, and $3s$ and $3p$ semicore local orbitals are included in the basis set. The calculations were performed using the experimental lattice constant of 2.87 Å. The spin magneton number was found to be 2.226, compared to the experimental value of 2.12 as deduced from measurements of the magnetization [18] and of the g (=2.09) factor. The calculated energy bands are very similar to those obtained in Ref. [19]. If the spin-orbit interaction is parametrized as $\xi\,\mathbf{l}\cdot\mathbf{s}$, its strength ξ is found to be approximately 5.1 mRy from the band splitting near the H point and the Fermi energy.

After obtaining the self-consistent potential with 20 000 $\mathbf{k}$ points, we calculated the Berry curvature with several larger sets of $\mathbf{k}$ points in order to achieve the convergence for σ_{xy} shown in Fig. 1. The Monkhorst-Park special-point method [20] was used for the integration in Eq. (6). To go beyond 2×10^6 points, we adopted a method of adaptive mesh refinement; i.e., when $\Omega^z(\mathbf{k})$ is large at a certain $\mathbf{k}$ point, we construct a finer mesh by adding 26 additional points around it. This procedure yields a converged value of $\sigma_{xy} = 751\ (\Omega\,\text{cm})^{-1}$ at zero temperature (using a step function for the Fermi-Dirac distribution) and a slightly smaller value of $\sigma_{xy} = 734\ (\Omega\,\text{cm})^{-1}$ at room temperature (300 K). Our result is in fair agreement with the value 1032 $(\Omega\,\text{cm})^{-1}$ extracted from Dheer's data on iron whiskers [21] at room temperature.

The slow convergence is caused by the appearance of large contributions of both signs to Ω^z which occur in very small regions of $\mathbf{k}$ space. Spin-orbit effects are small except when they mix states that are degenerate or nearly degenerate. A pair of mixed states contributes negligibly if both are occupied or are unoccupied. Therefore, only when the Fermi surface lies in a spin-orbit induced gap is

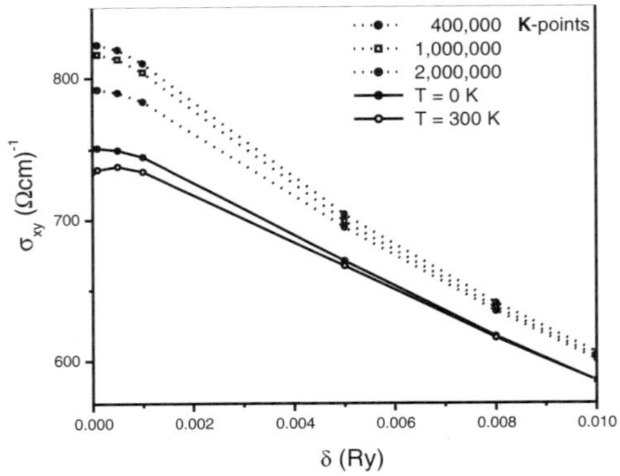

FIG. 1. Anomalous Hall effect vs δ with different numbers of **k** points in full Brillouin zone. Here δ is introduced by adding δ^2 to the denominator in Eq. (4). The dotted lines are obtained (for zero temperature) using a different number of **k** points. The solid lines are obtained by an adaptive mesh refinement method.

there a large contribution. For example, as shown in Fig. 2, the large spike near $H(1, 0, 0)$ in the direction of $P(\frac{1}{2}, \frac{1}{2}, \frac{1}{2})$ is due to a pair of spin-orbit coupled bands, one occupied and one unoccupied in a small **k** interval. The small energy gap gives rise to a small energy denominator, making the contribution to the Berry curvature very large in this small interval. The largest peaks and valleys in the distribution of the total Berry curvature are, however, located off the **k**-space symmetry lines. For example, as can be seen in Fig. 3, the Berry curvature shows sharp peaks and valleys of several orders of magnitude in height and depth at general **k** points of the (010) plane.

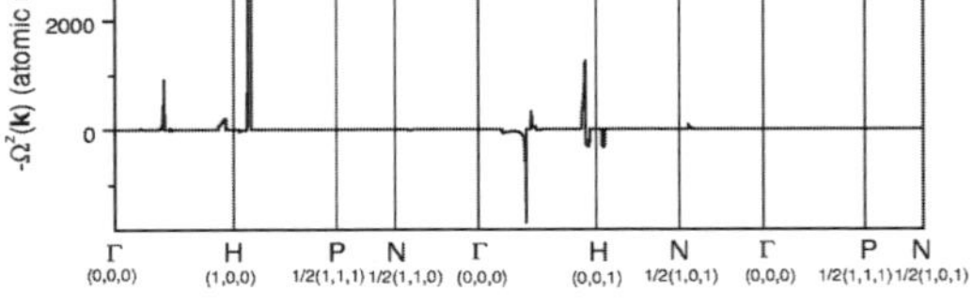

FIG. 2. Band structure near Fermi energy (upper panel) and Berry curvature $\Omega^z(\mathbf{k})$ (lower panel) along symmetry lines.

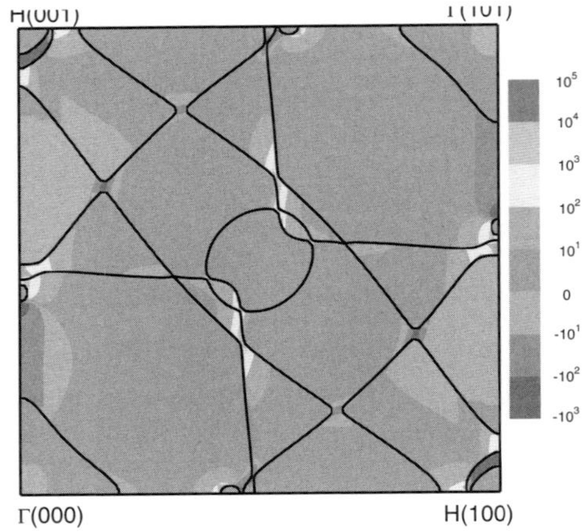

FIG. 3 (color). Fermi surface in (010) plane (solid lines) and Berry curvature $-\Omega^z(\mathbf{k})$ in atomic units (color map).

In order to further understand the role of spin-orbit coupling in the AHE, we artificially varied the speed of light, thereby changing the spin-orbit coupling strength $\xi \propto c^{-2}$. As shown in Fig. 4, σ_{xy} is linear in ξ for small coupling, but not for large coupling. For iron, nonlinearity becomes significant for $\xi/\xi_0 > 1/2$, which means that the spin-orbit interaction in iron cannot be accurately treated in a perturbative manner.

It is straightforward to extend our calculation to the ac Hall case by using the Kubo-formula [22] approach:

$$\sigma(\omega)_{xy} = \frac{e^2}{\hbar}\int_{V_G}\frac{d^3k}{(2\pi)^3}\sum_{n\neq n'}(f_{n,\mathbf{k}} - f_{n',\mathbf{k}}) \times \frac{\mathrm{Im}\langle\psi_{n\mathbf{k}}|v_x|\psi_{n'\mathbf{k}}\rangle\langle\psi_{n'\mathbf{k}}|v_y|\psi_{n\mathbf{k}}\rangle}{(\omega_{n'} - \omega_n)^2 - (\omega + i\delta)^2}, \quad (7)$$

where δ is a positive infinitesimal. In the upper panel in

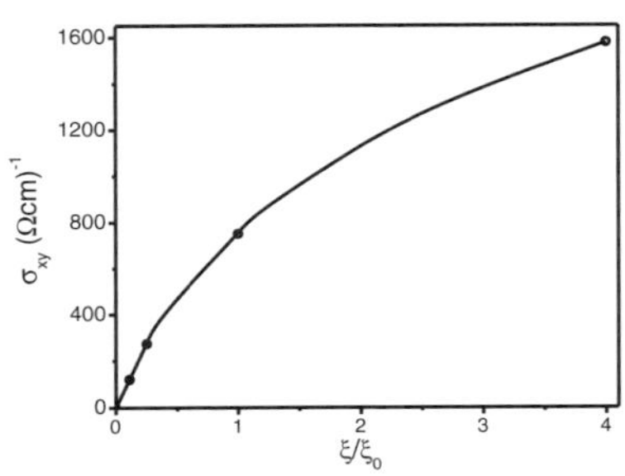

FIG. 4. Calculated anomalous Hall conductivity (open circles) vs the effective spin-orbit coupling strength relative to its value for iron. The line is a guide to the eye.

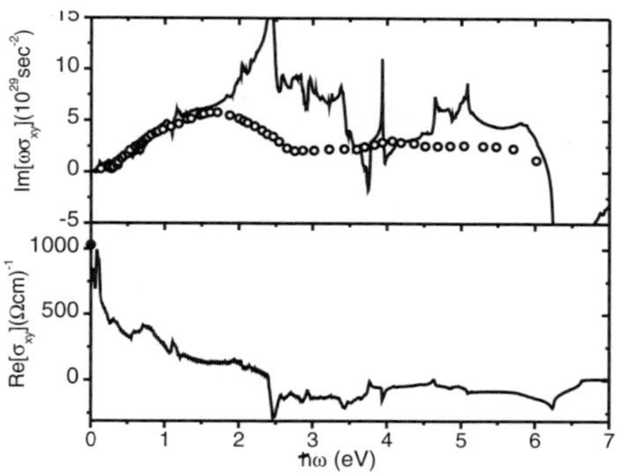

FIG. 5. Frequency dependence of the Hall conductivity at zero temperature. In the upper panel, the calculated imaginary part of $\omega\sigma_{xy}$ (solid curve) is compared with experimental results ○; see Ref. [23]. In the lower panel, the real part of σ_{xy} is shown together with the dc experiment value ● extracted from Ref. [21].

Fig. 5, we show results for the imaginary part of $\omega\sigma_{xy}$ as a function of frequency that are in agreement with earlier calculations [24]. Experimental results [23] are in excellent agreement below 1.7 eV but become smaller at higher energies. In the lower panel of the figure, the real part of the Hall conductivity, obtained from the imaginary part by a Kramers-Kronig relation, is shown as a function of frequency. The dc limit result, $\sigma(\omega = 0)_{xy} = 750.8\ (\Omega\,\mathrm{cm})^{-1}$, is essentially identical to that obtained from Eq. (6). Despite the small discrepancy with theory in the dc limit, the experimental point ● [21] seems to agree rather well with the overall trend of the frequency dependence of the calculated AHC.

In conclusion, we have shown that the AHC of bcc Fe, and presumably all other transition metal ferromagnets, is primarily intrinsic. [The only previous evaluation of the AHC of which we are aware [4] found that $\sigma_{xy} = 20.9\ (\Omega\,\mathrm{cm})^{-1}$ for Fe.] The remaining discrepancy between theory and experiment is likely due to shortcomings of the GGA, neglect of scattering effects, and experimental uncertainties.

This work was supported by DoE/DE-FG03-02ER45958, the Welch Foundation, NNSF of China (10134030, 60021403), and the Grant Agency of the Czech Republic (202/02/0912). L. K. was also supported by his NSF Grant No. DMR-0073546. A. H. M. and J. S. acknowledge helpful discussions with P. Bruno.

Note added.—After submission we became aware of two recent works on Kubo-formula evaluation of the anomalous Hall conductivity in ferromagnetic crystals with similar conclusions [25].

[1] *The Hall Effect and Its Application*, edited by C. L. Chien and C. R. Westgate (Plenum, New York, 1980).
[2] E. H. Hall, Philos. Mag. **10**, 301 (1880); **12**, 157 (1881).
[3] A.W. Smith and R.W. Sears, Phys. Rev. **34**, 1466 (1929).
[4] Henri Leribaux, Phys. Rev. **150**, 384 (1966).
[5] R. Karplus and J. M. Luttinger, Phys. Rev. **95**, 1154 (1954). It appears to us that there is an error in Eq. (2.18) of this paper, which depends on an arbitrary phase choice of the wave functions.
[6] J. Smit, Physica (Utrecht) **21**, 877 (1955); **24**, 39 (1958). In his criticism of Karplus and Luttinger, Smit attributed the anomalous velocity to a field induced polarization of the band by identifying $\langle u_{nk}|i\nabla_k|u_{nk}\rangle$ as the polarization. This physical interpretation is questionable in our opinion because of its gauge dependence.
[7] L. Berger, Phys. Rev. B **2**, 4559 (1970).
[8] G. Sundaram and Q. Niu, Phys. Rev. B **59**, 14 915 (1999).
[9] The semiclassical formalism can also be applied to magnetic Bloch bands which incorporate a strong magnetic field; see Ming-Che Chang and Q. Niu, Phys. Rev. B **53**, 7010 (1996).
[10] This is distinct from the Berry phase effect in real space discussed in J. Ye *et al.*, Phys. Rev. Lett. **83**, 3737 (1999); Y. Taguchi *et al.*, Science **291**, 2573 (2001); R. Shindou *et al.*, Phys. Rev. Lett. **87**, 116801 (2001).
[11] T. Jungwirth, Q. Niu, and A. H. MacDonald, Phys. Rev. Lett. **88**, 207208 (2002).
[12] T. Jungwirth *et al.*, Appl. Phys. Lett. **83**, 320 (2003).
[13] J. M. Luttinger, Phys. Rev. **112**, 739 (1958).
[14] D. J. Thouless, M. Kohmoto, M. P. Nightingale, and M. den Nijs, Phys. Rev. Lett. **49**, 405 (1982).
[15] D. J. Singh, *Planewaves, Pseudopotentials and the LAPW Method* (Kluwer Academic, Boston, 1994).
[16] J. P. Perdew, K. Burke, and M. Ernzerhof, Phys. Rev. Lett. **77**, 3865 (1996); **78**, 1396(E) (1997).
[17] P. Blaha *et al.*, in WIEN2K, Karlheinz Schwarz, Technische Universität Wien, Austria, 2001.
[18] H. Danan, A. Herr, and A. J. P. Meyer, J. Appl. Phys. **39**, 669 (1968).
[19] M. Singh, C. S. Wang, and J. Callaway, Phys. Rev. B **11**, 287 (1975).
[20] H. J. Monkhorst and J. D. Park, Phys. Rev. B **13**, 5188 (1976).
[21] P. N. Dheer, Phys. Rev. **156**, 637 (1967). It was found that $R_{yx} = 43.1 \times 10^{-13}\ \Omega\,\mathrm{cm/G}$ at room temperature, $4\pi M_s = 21.6$ kG, and $\rho_{xx} = 9.5 \times 10^{-6}\ \Omega\,\mathrm{cm}$, yielding $\sigma_{xy} = 4\pi M_s R_{yx}/\rho_{xx}^2 = 1032\ (\Omega\,\mathrm{cm})^{-1}$.
[22] J. Sinova *et al.*, Phys. Rev. B **67**, 235203 (2003).
[23] G. S. Krinchik and V. A. Artem'ev, Zh. Eksp. Teor. Fiz. **53**, 1901 (1967) [Sov. Phys. JETP **26**, 1080 (1967)].
[24] N. Mainkar, D. A. Browne, and J. Callaway, Phys. Rev. B **53**, 3692 (1996); G. Y. Guo and H. Ebert, Phys. Rev. B **51**, 12 633 (1995).
[25] I. V. Solovyev, Phys. Rev. B **67**, 174406 (2003); Z. Fang *et al.*, Science **302**, 92 (2003).

Possible Realization of Directional Optical Waveguides in Photonic Crystals with Broken Time-Reversal Symmetry

F. D. M. Haldane and S. Raghu*

Department of Physics, Princeton University, Princeton, New Jersey 08544-0708, USA

(Received 23 March 2005; revised manuscript received 30 May 2007; published 10 January 2008)

We show how, in principle, to construct analogs of quantum Hall edge states in "photonic crystals" made with nonreciprocal (Faraday-effect) media. These form "one-way waveguides" that allow electromagnetic energy to flow in one direction only.

DOI: 10.1103/PhysRevLett.100.013904 PACS numbers: 42.70.Qs, 03.65.Vf

In this Letter, we describe a novel effect involving an interface between two magneto-optic photonic crystals (periodic "metamaterials" that transmit electromagnetic waves) which can theoretically act as a "one-way waveguide", i.e., a channel along which electromagnetic energy can propagate in only a single direction, with no possibility of being backscattered at bends or imperfections. The unidirectional photonic modes confined to such interfaces are the direct analogs of the "chiral edge states" of electrons in the quantum Hall effect (QHE) [1,2]. The key enabling ingredient is the presence of "nonreciprocal" (Faraday-effect) media that breaks time-reversal symmetry in the metamaterial.

Just as in the electronic case, every two-dimensional photonic band is characterized by a topological invariant known as the Chern number [3], an integer that vanishes identically unless time-reversal symmetry is broken. If the material contains a photonic band gap (PBG), the Chern number, summed over all bands below the gap, plays a role similar to that of the same quantity summed over all *occupied* bands in the electronic case. In particular, if the total Chern number *changes* across an interface separating two PBG media, there necessarily will occur states localized to the interface having a nonzero net current along the interface [1,2]. In the photonic case, such states would comprise our "one-way waveguide".

Such an interface between two PBG media can be realized as a domain wall in a 2D periodic photonic metamaterial, across which the direction of the Faraday axis reverses. Unidirectional edge states are guaranteed in this system provided that the Faraday effect generates photonic bands with nonzero Chern numbers. Here, we construct photonic bands with nonzero Chern invariants in a hexagonal array of dielctric rods with a Faraday effect. We then show that as a consequence of topology of the single-particle photon bands in the Brillouin zone, the edge states of light occur along domain walls (where the Faraday effect vanishes).

It may seem surprising that the physics of the QHE can have analogs in photonic systems. The QHE is exhibited by incompressible quantum fluid states of electrons—conserved strongly interacting charged fermions—in high magnetic fields, while photons are nonconserved neutral bosons which do not interact in linear media; furthermore, photonic bands can be described classically, in terms of Maxwell's equations. However, the integer QHE can, in principle, occur without any uniform magnetic flux density (just with broken time-reversal symmetry) as has explicitly shown by one of us in a graphenelike model of noninteracting Bloch electrons [4]; thus Landau-level quantization is not an essential requirement for the quantum Hall effect.

We have transcribed the key features of the electronic model of Ref. [4] to the photonic context. The edge states are a property of a one-particle eigenstate problem similar to the Maxwell normal-mode problem, so they are replicated in the photonics problem. (The QHE itself has no photonic analog, as it follows from the Pauli principle of filling all one-particle states below the Fermi level.)

The Maxwell normal-mode problem in loss-free linear media with spatially periodic local frequency-dependent constitutive relations is a generalized self-consistent Hermitian eigenproblem, somewhat different from the standard Hermitian eigenproblem. The nonreciprocal parts of the local Hermitian permittivity and permeability tensors $\boldsymbol{\epsilon}(\boldsymbol{r}, \omega)$ and $\boldsymbol{\mu}(\boldsymbol{r}, \omega)$ are odd imaginary functions of frequency, so frequency dependence is unavoidable. The generalized eigenproblem has the structure

$$\boldsymbol{U}^{\dagger}(\boldsymbol{k})\boldsymbol{A}\boldsymbol{U}(\boldsymbol{k})|u_n(\boldsymbol{k})\rangle = \omega_n(\boldsymbol{k})\boldsymbol{B}(\omega_n(\boldsymbol{k}))|u_n(\boldsymbol{k})\rangle, \quad (1)$$

where $\boldsymbol{U}(\boldsymbol{k})$ is a unitary operator that defines the Bloch vector $\boldsymbol{k}$; $\boldsymbol{A}$ and $\boldsymbol{B}(\omega)$ are Hermitian operators, with the real-eigenvalue stability condition that the Hermitian operator $\boldsymbol{B}_0(\omega) \equiv (\partial/\partial\omega)[\omega\boldsymbol{B}(\omega)]$ is positive definite [this assumes that the periodic medium coupled to the electromagnetic fields has a linear response described by harmonic-oscillator modes, none of which have natural frequency $\omega_n(\boldsymbol{k})$—a detailed derivation has been presented in Ref. [5]]. The eigenfunctions $\langle\boldsymbol{r}|u_n(\boldsymbol{k})\rangle$ are the spatially periodic factors of the Bloch states. The electronic band-structure problem is a simplification of Eq. (1), with $\boldsymbol{A}$ replaced by the one-electron Hamiltonian, $\boldsymbol{B}$ by the identity operator $\mathbb{1}$, and ω_n by the energy eigenvalue.

In this formulation of Maxwell's equations, the eigenfunction $\boldsymbol{u}_n(\boldsymbol{k},\boldsymbol{r}) \equiv \langle \boldsymbol{r}|u_n(\boldsymbol{k})\rangle$ is the 6-component vector of complex electromagnetic fields

$$\boldsymbol{u}_n(\boldsymbol{k},\boldsymbol{r}) = \begin{pmatrix} \boldsymbol{E}_n(\boldsymbol{k},\boldsymbol{r}) \\ \boldsymbol{H}_n(\boldsymbol{k},\boldsymbol{r}) \end{pmatrix}. \quad (2)$$

In this basis, $\boldsymbol{U}(\boldsymbol{k},\boldsymbol{r}) = \exp i\boldsymbol{k}\cdot\boldsymbol{r}$ and $\boldsymbol{A} = -i\boldsymbol{J}^a\nabla_a$ (with $\nabla_a \equiv \partial/\partial r^a$, and repeated indices summed), where

$$\boldsymbol{J}^a = \begin{pmatrix} 0 & i\boldsymbol{L}^a \\ -i\boldsymbol{L}^a & 0 \end{pmatrix}, \qquad \boldsymbol{B} = \begin{pmatrix} \boldsymbol{\epsilon}(\boldsymbol{r},\omega) & 0 \\ 0 & \boldsymbol{\mu}(\boldsymbol{r},\omega) \end{pmatrix}; \quad (3)$$

here $\boldsymbol{L}^a$ are the 3×3 spin-1 matrices, $(L^b)^{ac} = i\epsilon^{abc}$. If the physical electromagnetic fields are given by the real parts of $\boldsymbol{u}_n(\boldsymbol{k},\boldsymbol{r})\exp i[\boldsymbol{k}\cdot\boldsymbol{r} - \omega_n(\boldsymbol{k})t]$, the spatially periodic time-averaged energy density and energy current are the quadratic forms $[\boldsymbol{u}_n^*, \boldsymbol{B}_0(\omega_n)\boldsymbol{u}_n]$, $(\boldsymbol{u}_n^*, \boldsymbol{J}^a\boldsymbol{u}_n)$. For our purposes, the key photonic band-structure quantity is the *Berry connection* $\mathcal{A}_n^a(\boldsymbol{k})$, a real function of $\boldsymbol{k}$ given by

$$\mathcal{A}_n^a = \frac{\langle u_n|\boldsymbol{B}_0(\omega_n)|\nabla_k^a u_n\rangle - \langle \nabla_k^a u_n|\boldsymbol{B}_0(\omega_n)|u_n\rangle}{2i\langle u_n|\boldsymbol{B}_0(\omega_n)|u_n\rangle}, \quad (4)$$

where $\nabla_k^a \equiv \partial/\partial k_a$ is the k-space derivative. We obtained (4) as a generalization of the $\boldsymbol{B} = \mathbb{1}$ expression [3] by deriving (1) from a standard Hermitian eigenproblem where electromagnetic fields are explicitly coupled to harmonic-oscillator degrees of freedom of the medium [5].

The solution of the normal-mode eigenproblem only determines $\boldsymbol{u}_n(\boldsymbol{k},\boldsymbol{r})$ up to an arbitrary $\boldsymbol{k}$-dependent phase factor; if the replacement $\boldsymbol{u}_n(\boldsymbol{k},\boldsymbol{r}) \to \boldsymbol{u}_n(\boldsymbol{k},\boldsymbol{r})\exp i\chi_n(\boldsymbol{k})$ is made, $\mathcal{A}_n^a(\boldsymbol{k}) \to \mathcal{A}_n^a(\boldsymbol{k}) + \nabla_k^a\chi_n(\boldsymbol{k})$. The Berry connection is a "gauge-dependent" analog of the electromagnetic vector potential; the associated gauge-invariant function (analogous to the magnetic flux density) is the k-space *Berry curvature* $\mathcal{F}_n^{ab}(\boldsymbol{k}) = \nabla_k^a\mathcal{A}_n^b - \nabla_k^b\mathcal{A}_n^a$. The *Berry phase* $\exp i\phi_n(\Gamma) = \exp i\oint \mathcal{A}_n^a dk_a$ associated [6] with adiabatic evolution around a closed path Γ (here in k space) is the gauge-invariant analog of the Bohm-Aharonov phase factor and can be expressed in terms of the integral of $\mathcal{F}_n^{ab}$ over a surface bounded by Γ [3].

The Berry curvature satisfies a k-space analog of the Gauss law, except that "monopole" singularities emitting total "Berry flux" $\pm 2\pi$ can occur at k-space points where there are "accidental degeneracies" between bands (this quantization of the monopole charge ensures that the expression for the Berry phase in terms of Berry curvature on a surface bounded by Γ is independent of how that surface is chosen [3]). The integer Chern invariant associated with any compact surface (2-manifold) Σ in k space is

$$C_n^{(1)}(\Sigma) = \frac{1}{2\pi}\iint_\Sigma dk_a \wedge dk_b \mathcal{F}_n^{ab}. \quad (5)$$

In the case of a 2D band structure, Σ may be taken to be the 2D Brillouin zone (BZ) itself, and $C_n^{(1)}$ is a property of the 2D band [7]. If time-reversal symmetry is unbroken, $\mathcal{F}_n^{ab}(-\boldsymbol{k}) = -\mathcal{F}_n^{ab}(\boldsymbol{k})$, and Chern numbers vanish.

We now wish to construct a 2D photonic band structure where some bands have a nonzero Chern number. The key idea is to start with a band structure that has both time-reversal symmetry *and* inversion symmetry, which allows the existence of pairs of "Dirac points" in the 2D BZ. These are isolated points where two bands become degenerate, but split apart with a linear dispersion (resembling that of the massless Dirac equation) for nearby Bloch vectors. Dirac points are generically allowed because if inversion symmetry is present, $\mathcal{F}_n^{ab}(-\boldsymbol{k}) = \mathcal{F}_n^{ab}(\boldsymbol{k})$; in combination with time-reversal symmetry, this means that $\mathcal{F}_n^{ab}(\boldsymbol{k}) = 0$. It is then possible to choose a phase convention such that the eigenfunctions $\boldsymbol{u}_n(\boldsymbol{k},\boldsymbol{r})$ of (1) are real for all $\boldsymbol{k}$. The eigenproblem is then of the real-symmetric type, where it is possible to find an "accidental" degeneracy between two bands by varying just two parameters (in this case, the 2D Bloch vector $\boldsymbol{k}$); in contrast, in the general complex-Hermitian case, *three* parameters must be varied to find a degeneracy, which cannot be done by merely "fine-tuning" a 2D $\boldsymbol{k}$.

Dirac points can exist in a 2D band structure with both spatial-inversion and time-reversal symmetry, but a gap opens if either symmetry is broken. While breaking of inversion symmetry leads to nonzero Berry curvature (and hence corrections to the "semiclassical" equations for the trajectories of light rays in adiabatically varying media [8]), it does not lead to nontrivial topology of the bands. In contrast, when a gap opens at Dirac points due to time-reversal breaking, the two bands that split apart inevitably acquire nonzero Chern numbers.

We can now give an in-principle demonstration that "one-way waveguides" can be constructed using nonreciprocal photonic crystals. Consider a system with a uniform isotropic permeability tensor $\mu_0\delta^{ab}$, and an isotropic but spatially varying permittivity tensor $\epsilon(\boldsymbol{r})\epsilon_0\delta^{ab}$, with

$$\epsilon(\boldsymbol{r}) = \epsilon[1 + \lambda V_G(\boldsymbol{r})], \qquad V_G(\boldsymbol{r}) = 2\sum_{n=1}^{3}\cos(\boldsymbol{G}_n\cdot\boldsymbol{r}), \quad (6)$$

where $\boldsymbol{G}_n$, $n = 1, 2, 3$ are three equal-length reciprocal vectors in the xy plane, rotated 120° relative to each other. For small λ, this problem can be solved analytically in a "nearly-free-photon" approximation. This system has continuous translational invariance in the z direction, and we will restrict attention to wave numbers $k_z = 0$, with Bloch vector $\boldsymbol{k}$ in the xy plane. The electromagnetic fields then separate into decoupled "TE" and "TM" sets, $\{E_x, E_y, H_z\}$ and $\{H_x, H_y, E_z\}$; we specialize to the TE set. The six corners of the (first) BZ are at $\pm\boldsymbol{K}_n$, where $\boldsymbol{K}_1 = (\boldsymbol{G}_2 - \boldsymbol{G}_3)/3$, etc., and $|K| = |G|/\sqrt{3}$; since $\boldsymbol{K}_2 - \boldsymbol{K}_1 = \boldsymbol{G}_3$, the three wave vectors $\boldsymbol{K}_i$ are equivalent as Bloch vectors. To leading order in λ and the deviation $\delta\boldsymbol{k} = \boldsymbol{k} - \boldsymbol{K}_i$ of the 2D Bloch vector from the BZ corner, the three "free photon" TE plane waves with speed c_0 split into a "Dirac-point" doublet with $\omega = \omega_D \pm v_D|\delta \boldsymbol{k}| + O(|\delta \boldsymbol{k}|^2)$, where $\omega_D = c_0|K|[1 - \lambda/4 + O(\lambda)^2]$, $v_D = c_0/2 + O(\lambda)$, and a singlet $\omega = \omega_0 + O(|\delta \boldsymbol{k}|^2)$, $\omega_0 = c_0|K|[1 + \lambda/2 + O(\lambda^2)]$.

We now perturb the Dirac points by a Faraday term (which explicitly breaks time-reversal symmetry), with an axis normal to the xy plane, added to the permittivity tensor: $\epsilon^{xy} = -\epsilon^{yx} = i\epsilon_0\epsilon\eta(\boldsymbol{r}, \omega)$, where

$$\eta(\boldsymbol{r}, \omega) = \eta_0(\omega) + \eta_1(\omega)V_G(\boldsymbol{r}); \tag{7}$$

$\eta_0(\omega)$, $\eta_1(\omega)$ are real odd functions of ω. We assume that, for $\omega \approx \omega_D$, $|\eta_0(\omega)|, |\eta_1(\omega)| \ll |\lambda| \ll 1$, with negligible frequency dependence. The Dirac points now split, with dispersion $\omega = \omega_D \pm v_D(|\delta \boldsymbol{k}|^2 + \kappa^2)^{1/2}$, where, to leading order in η, $\kappa = |K|[\frac{3}{2}\eta_1(\omega_D) - 3\lambda\eta_0(\omega_D)]$.

For small κ, the Berry curvatures of the upper and lower $k_z = 0$ bands near the split Dirac points are

$$F^{xy}_{\pm}(\delta \boldsymbol{k}) = \pm\tfrac{1}{2}\kappa(|\delta \boldsymbol{k}|^2 + \kappa^2)^{-3/2}. \tag{8}$$

There is a total integrated Berry curvature of $\pm\pi$ near each Dirac point, giving total Chern numbers ± 1 for the split bands. By inversion symmetry, the Berry curvatures at the two Dirac points have the same sign; if the gap was opened by broken inversion symmetry, with unbroken time-reversal invariance, they would have opposite sign, and the Chern number would vanish.

We now consider an adiabatically spatially varying Faraday term parametrized by a $\kappa(\boldsymbol{r})$ that is positive in some regions and negative in other regions. The splitting of the Dirac points vanishes locally on the line where $\kappa(\boldsymbol{r}) = 0$. It is necessary that, in the perfectly periodic structure with $\kappa = 0$, there are no photonic modes at other Bloch vectors that are degenerate with the modes at the Dirac points.

Such frequency isolation of the Dirac points cannot occur in the weak-coupling "nearly-free photon" limit, but can be achieved, at least for $k_z = 0$ modes, in hexagonal arrays of infinitely long dielectric rods parallel to the z axis. An example can be seen in Fig. 1(a) of Ref. [9]. That figure was exhibited to demonstrate a frequency gap between the first and second TE bands, but incidentally also shows that the second and third TE bands are separated by a substantial gap except in the vicinity of the BZ corners, where they touch at Dirac points. The corresponding TM bands were not given in Ref. [9], but we found that the Dirac-point frequency ω_D is also inside a large gap of the TM spectrum (see Fig. 1). When a Faraday term is added, the bands forming the Dirac point in Fig. 1 split apart, and each now nondegenerate band will have associated with it a nonzero Chern number (see Ref. [5]).

The Faraday effect incorporated to the hexagonal array of rods explicitly breaks time-reversal symmetry on the scale of the unit cell of the metamaterial: the permittivity tensor acquires an imaginary off-diagonal component having the periodicity of the unit cell, as described above. A hexagonal array consisting of a material having a large Verdet coefficient, such as a rare-earth garnet with ferromagnetically ordered domains, would give rise to such an effect.

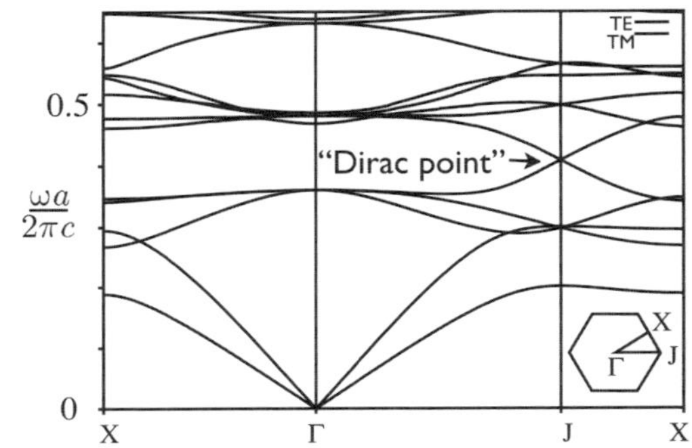

FIG. 1. Photon bands for $k_z = 0$ electromagnetic waves propagating normal to the axis of a hexagonal 2D array of cylindrical dielectric rods; a is the lattice constant. As in Fig. 1(a) of Ref. [9], the rods fill a fraction $f = 0.431$ of the volume, with dielectric constant $\epsilon = 14$, and they are embedded in an $\epsilon = 1$ background. The lowest five 2D bands are well separated from higher bands, except near a pair of "Dirac points" at the two distinct Brillouin zone corners (J).

While these $k_z = 0$ Dirac-point modes are not degenerate with any other $k_z = 0$ modes, they *are* degenerate with $k_z \neq 0$ modes. To fully achieve a "one-way" edge-mode structure, the light must also be confined in the z direction, with Dirac points at a frequency that is nondegenerate with *any* other modes. To design such structures, it will be necessary to vary the filling factor of the rods along the z direction so that light remains confined to regions of relatively larger filling factors. The technical challenge would be to vary the filling factors without introducing any modes into the bulk TE gaps surrounding the Dirac points.

Let $|u_\sigma(\pm \boldsymbol{k}_D)\rangle$, $\sigma = \pm$ be the degenerate solutions of Eq. (1) at a pair of isolated Dirac points, normalized so $\langle u_\sigma(\pm \boldsymbol{k}_D)|\boldsymbol{B}_0(\omega_D)|u_{\sigma'}(\pm \boldsymbol{k}_D)\rangle = B_0\delta_{\sigma\sigma'}$. Now add a Faraday perturbation $\delta \boldsymbol{B}(\boldsymbol{r}, \omega)$: in degenerate perturbation theory, normal modes with small $\delta\omega = \omega - \omega_D$ have the form $\sum_{\sigma,\pm}\psi^{\pm}_{\sigma}(\boldsymbol{r})\boldsymbol{U}(\pm \boldsymbol{k}_D, \boldsymbol{r})\boldsymbol{u}_\sigma(\pm \boldsymbol{k}_D, \boldsymbol{r})$. For slow spatial variation, there is negligible mixing between modes at different Dirac points, and $\psi^{\pm}_{\sigma}(\boldsymbol{r})$ is the solution of

$$\sum_{\sigma'}(-iJ^a_{\perp}\nabla_a - \omega_D\delta B(\boldsymbol{r}))^{\pm}_{\sigma\sigma'}\psi^{\pm}_{\sigma'}(\boldsymbol{r}) = \delta\omega B_0\psi^{\pm}_{\sigma}(\boldsymbol{r}), \tag{9}$$

where $J^a_{\perp}$ and $\delta B(\boldsymbol{r})$ are 2×2 matrices given by

$$\begin{aligned}(J^a_{\perp})^{\pm}_{\sigma\sigma'} &= \langle u_\sigma(\pm \boldsymbol{k}_D)|\boldsymbol{J}^a|u_{\sigma'}(\pm \boldsymbol{k}_D)\rangle,\\ a &= x, y, (\delta B(\boldsymbol{r}))\\ &= \langle u_\sigma(\pm \boldsymbol{k}_D)|\delta \boldsymbol{B}(\boldsymbol{r}, \omega_D)|u_{\sigma'}(\pm \boldsymbol{k}_D)\rangle.\end{aligned} \tag{10}$$

For a straight line interface, this equation has the form $v_D\hat{K}|\psi\rangle = \delta\omega|\psi\rangle$, with $v_D > 0$, and

$$\hat{K} = -i\boldsymbol{\sigma}^x\nabla_x + \delta k_{\parallel}\boldsymbol{\sigma}^y + \kappa(x)\boldsymbol{\sigma}^z, \qquad (11)$$

where $\boldsymbol{\sigma}^a$ are Pauli matrices. Here $k_{Dy} + \delta k_{\parallel}$ is the conserved Bloch vector parallel to the interface; we take $\kappa(x)$ to be monotonic, with $\kappa(x) \to \pm\kappa^{\infty}$ as $x \to \pm\infty$.

It is instructive to first consider the exactly-solvable case $\kappa(x) = \kappa^{\infty}\tanh(x/\xi)$, $\xi > 0$, where $\hat{K}^2$ is essentially the integrable Pöschl-Teller Hamiltonian [10]. The spectrum of modes bound to the interface is

$$\omega_0(\delta k_{\parallel}) = \omega_D + s_{\kappa}v_D\delta k_{\parallel}, \qquad s_{\kappa} \equiv \mathrm{sgn}\,(\kappa^{\infty}), \qquad (12a)$$

$$\omega_{n\pm}(\delta k_{\parallel}) = \omega_D \pm v_D(\delta k_{\parallel}^2 + \kappa_n^2)^{1/2}, \qquad n > 0, \qquad (12b)$$

with $|\kappa_n| < |\kappa^{\infty}|$; for the integrable model, κ_n^2 is given by $2n|\kappa^{\infty}|/\xi$, $n < |\kappa^{\infty}|\xi/2$. There is always a unidirectional $n = 0$ mode with speed v_D and a direction determined by the sign of κ^{∞}; in the small-ξ (or sharp-wall) limit $|\kappa^{\infty}|\xi < 2$, this is the only interface mode.

Let $\phi(\kappa^2)$ be the dimensionless area in the $x - k_x$ phase plane enclosed by a closed constant-frequency orbit $(k_x)^2 + [\kappa(x)]^2 = \kappa^2 < |\kappa^{\infty}|^2$, corresponding to a bound state. For the integrable model, this has the simple form $\phi(\kappa^2) = \pi\kappa^2\xi/|\kappa^{\infty}|$; the $n > 0$ bidirectional modes thus satisfy a constructive-interference condition

$$\phi(\kappa_n^2) = 2\pi n. \qquad (13)$$

This contrasts with the usual semiclassical condition $\phi = 2\pi(n + \frac{1}{2})$; the change is needed for the $n = 0$ "zero mode" (12a) to exist, and it can be interpreted as deriving from an extra Berry phase factor of -1 because the orbit encloses a Dirac degeneracy point at $(x, k_x) = (0, 0)$. For general $\kappa(x)$, the $n = 0$ eigenfunction is

$$\psi_{\sigma}^0(\boldsymbol{r}) \propto \varphi_{\sigma}(s_{\kappa})\exp\left[i\delta k_{\parallel}y - s_{\kappa}\int^x \kappa(x')dx'\right], \qquad (14)$$

$\boldsymbol{\sigma}^y\boldsymbol{\varphi}(s) = s\boldsymbol{\varphi}(s)$. For slowly varying $\kappa(x)$, the condition (13) will determine κ_n^2 for any $n > 0$ interface modes.

Since there are two Dirac points, there are two such unidirectional edge modes at a boundary across which the Faraday axis reverses. The crucial feature is that both modes propagate in the *same* direction and cannot disappear, even if the interface becomes sharp, bent, or disordered. As in the QHE, the difference between the number of modes moving in the two directions along the interface is topologically determined by the difference of the total Chern number of bands at frequencies below the bulk photonic band gap in the regions on either side of the interface; in this case $|\Delta C^{(1)}| = 2$.

For $|\delta\omega| < v_D|\kappa^{\infty}|$ a Faraday interface has no counter-propagating modes into which elastic backscattering can take place, so the "one-way waveguide" that it forms is immune to localization effects, just like electronic transport in the QHE. In the QHE, the number of electrons is strictly conserved; in photonics, the photons only propagate ballistically if absorption and nonlinear effects are absent. These effects do allow degradation of the electromagnetic energy current flowing along the interface, so the analogy with the QHE is not perfect.

Even if a 2D metamaterial with isolated Dirac points can be designed, the problem of finding a suitable magneto-optic material to provide the Faraday effect must be addressed. The effect must be large enough to induce a gap that overcomes the effect of any inversion-symmetry breaking. The parameter $|\kappa^{\infty}|$ is the inverse length that controls the width of the unidirectionally propagating channel (and the unidirectional frequency range); in order to keep the wave confined to the interface and prevent leakage, the Faraday coupling must be strong enough so that this width is significantly smaller than the physical dimensions of the sample of metamaterial.

In summary, we have shown that analogs of quantum Hall effect edge modes can, in principle, occur in two-dimensional photonic crystals with broken time-reversal symmetry. The electromagnetic energy in these modes travel in a single direction. Explicit theoretical examples of such modes have been constructed in Ref. [5]. Such quasilossless unidirectional channels are a novel possibility that might one day be physically realized in "photonic metamaterials" with nonreciprocal constituents.

This work was supported in part by the US National Science Foundation (under MRSEC Grant No. DMR02-13706) at the Princeton Center for Complex Materials. Part of this work was carried out at the Kavli Institute for Theoretical Physics, UC Santa Barbara, with support from KITP's NSF Grant No. PHY99-07949.

---

*Current address: Department of Physics, Stanford University, Stanford, CA 94305.

[1] B. I. Halperin, Phys. Rev. B **25**, 2185 (1982).
[2] X.-G. Wen, Phys. Rev. B **43**, 11 025 (1991).
[3] B. Simon, Phys. Rev. Lett. **51**, 2167 (1983).
[4] F. D. M. Haldane, Phys. Rev. Lett. **61**, 2015 (1988).
[5] S. Raghu and F. D. M. Haldane, arXiv:cond-mat/0602501.
[6] M. V. Berry, Proc. R. Soc. A **392**, 45 (1984).
[7] D. J. Thouless, M. Kohmoto, M. P. Nightingale, and M. den Nijs, Phys. Rev. Lett. **49**, 405 (1982).
[8] M. Onoda, S. Murakami, and N. Nagaosa, Phys. Rev. Lett. **93**, 083901 (2004).
[9] M. Plihal and A. A. Maradudin, Phys. Rev. B **44**, 8565 (1991).
[10] See, e.g., L. D. Landau and E. M. Lifschitz, *Quantum Mechanics: Non-Relativistic Theory* (Pergamon, New York, 1977), p. 73, Problem 5.

5

Topological insulators

Model for a Quantum Hall Effect without Landau Levels: Condensed-Matter Realization of the "Parity Anomaly"

F. D. M. Haldane

Department of Physics, University of California, San Diego, La Jolla, California 92093

(Received 16 September 1987)

A two-dimensional condensed-matter lattice model is presented which exhibits a nonzero quantization of the Hall conductance σ^{xy} in the *absence* of an external magnetic field. Massless fermions *without spectral doubling* occur at critical values of the model parameters, and exhibit the so-called "parity anomaly" of (2+1)-dimensional field theories.

PACS numbers: 05.30.Fk, 11.30.Rd

The quantum Hall effect[1] (QHE) in two-dimensional (2D) electron systems is usually associated with the presence of a uniform externally generated magnetic field, which splits the spectrum of electron energy levels into Landau levels. In this Letter I show how, in principle, a QHE may also result from breaking of time-reversal symmetry (i.e., magnetic ordering) *without* any net magnetic flux through the unit cell of a periodic 2D system. In this case, the electron states retain their usual Bloch state character.

The model presented here is also interesting in that if its parameters are on a critical line at which its ground state changes from the normal semiconductor state to this new type of QHE state, its low-energy states simulate a "(2+1)-dimensional" relativistic quantum field theory exhibiting the so-called "parity anomaly"[2] and a (2+1)-D analog of "chiral" fermions *without* the opposite-chirality anomaly-canceling partners[3] that usually accompany them in lattice realizations of field theories ("fermion doubling").

In the zero-temperature limit, the transverse conductivity σ^{xy} of a periodic 2D electron system with a gap in the single-particle density of states at the Fermi level takes quantized values $\nu e^2/h$, where ν is generally rational, but can only take *integer* values in the absence of electron interactions.[4] This property of a pure system is stable against sufficiently weak disorder effects. Since σ^{xy} is odd under time reversal, a nonzero value can only occur if time-reversal invariance is broken.

In the usual QHE, the gap at the Fermi level results from the splitting of the spectrum into Landau levels by an external magnetic field. The scenario considered here is different, and involves a 2D semimetal where there is a degeneracy at isolated points in the Brillouin zone between the top of the valence band and the bottom of the conduction band, that is associated with the presence of both inversion symmetry and time-reversal invariance. If inversion symmetry is broken, a gap opens and the system becomes a normal semiconductor ($\nu=0$), but if the gap opens because time-reversal invariance is broken the system becomes a $\nu=\pm 1$ integer QHE state. If both perturbations are present, their relative strengths determine which type of state is realized.

To model a 2D semimetal, I use the "2D graphite" model investigated previously by Semenoff[5] as a possible lattice realization of a (2+1)-D field theory with the anomaly. 2D graphite has the honeycomb net structure, consisting of two interpenetrating triangular lattices ("A" and "B" sublattices) with one lattice point of each type per unit cell (Fig. 1). A 2D inversion (i.e., a rotation in the plane by π) interchanges the two sublattices. Since spin-orbit coupling effects will not be included, the electron spin will (for the moment) be suppressed.

Semenoff[5] investigated the tight-binding model with one orbital per site and a real hopping matrix element t_1 between nearest neighbors on different sublattices, and also considered the effect of an inversion-symmetry-breaking on-site energy $+M$ on A sites and $-M$ on B sites. The model has point group C_{6v} ($M=0$) or C_{3v} ($M\neq 0$). In this original version of the model, time-reversal invariance is present, and Semenoff[5] found complete cancellation of the anomaly in the $M=0$ model due to fermion doubling, and normal semiconductor behavior for $M\neq 0$.

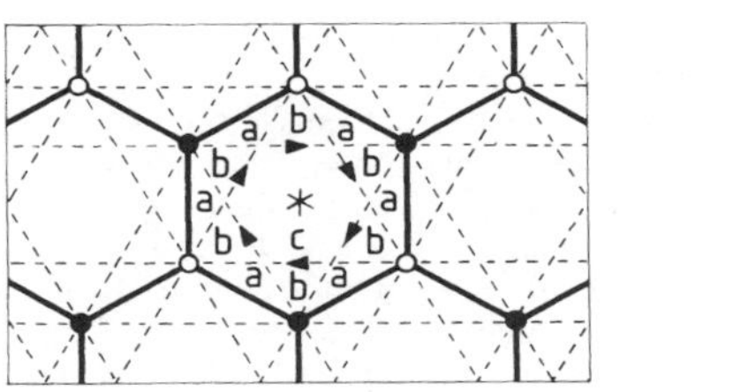

FIG. 1. The honeycomb-net model ("2D graphite") showing nearest-neighbor bonds (solid lines) and second-neighbor bonds (dashed lines). Open and solid points, respectively, mark the A and B sublattice sites. The Wigner-Seitz unit cell is conveniently centered on the point of sixfold rotation symmetry (marked "*") and is then bounded by the hexagon of nearest-neighbor bonds. Arrows on second-neighbor bonds mark the directions of positive phase hopping in the state with broken time-reversal invariance.

I now include a second real hopping term t_2 between second-neighbor sites (i.e., between nearest-neighbor sites on the *same* sublattice). This does not change the space group, though it does eliminate a particle-hole symmetry of the energy bands of the original model. To break time-reversal invariance, I also add a periodic local magnetic-flux density $B(\mathbf{r})$ in the $\hat{\mathbf{z}}$ direction normal to the 2D plane, with the full symmetry of the lattice, and with *zero total flux through the unit cell.*

Since the net flux per unit cell vanishes, the vector potential $\mathbf{A}(\mathbf{r})$ can be chosen to be periodic. The effect of this local field is to multiply the matrix element for hopping between sites by the unimodular phase factor $\exp[i(e/\hbar)\int \mathbf{A}\cdot d\mathbf{r}]$ where the integral is along the hopping path, which I take to be rectilinear. The phases can be chosen with any consistent convention such that the total phase accumulated around a closed path adds up to the flux enclosed in units of the flux quantum $\Phi_0 = |h/e|$.

Since closed paths of first-neighbor hops enclose complete unit cells (and hence no net flux) the t_1 matrix elements are unaffected. The t_2 matrix elements acquire a phase $\phi = 2\pi(2\Phi_a + \Phi_b)/\Phi_0$, where Φ_a and Φ_b are the fluxes through the regions of the unit cell marked a and b in Fig. 1. The hopping directions for which the amplitudes are $t_2\exp(+i\phi)$ are shown in Fig. 1: It can be seen that the Hamiltonian has acquired a chirality if the local field is present.

It is useful to consider a possible model for the origin of such an internal magnetic field. Magnetic dipole moments μ, ordered ferromagnetically normal to the plane, are placed at the center of each hexagonal cell of the honeycomb net, and $B(\mathbf{r})$ is the sum of their dipole fields. Note that ferromagnetic ordering in 2D does *not* generate a uniform component of the magnetic-flux density. The absolute value of ϕ is $C\alpha^2\mu/a$, where α is the fine-structure constant, μ is the dipole moment in Bohr magneton units, a is the lattice spacing Bohr radii, and C is a dimensionless constant (of order unity) which depends on the lattice structure. If μ and a are of order unity in their natural units, the phase ϕ in this model will be a small quantity controlled by the fine-structure constant.

To diagonalize the Hamiltonian, I use a basis of two-component "spinors" $(\psi_{\mathbf{k}A},\psi_{\mathbf{k}B})$ of Bloch states constructed on the two sublattices. Let $\mathbf{a}_1,\mathbf{a}_2,\mathbf{a}_3$ be the displacements from a B site to its three nearest-neighbor A sites, defined so that $\hat{\mathbf{z}}\cdot\mathbf{a}_1\times\mathbf{a}_2$ is positive. I also define $\mathbf{b}_1 = \mathbf{a}_2 - \mathbf{a}_3$, $\mathbf{b}_2 = \mathbf{a}_3 - \mathbf{a}_1$, etc.; the set of displacements to the six nearest neighbors on the same sublattice is $\{\pm\mathbf{b}_i\}$. In this representation, the Hamiltonian becomes

$$\mathbf{H}(\mathbf{k}) = 2t_2\cos\phi\left(\textstyle\sum_i \cos(\mathbf{k}\cdot\mathbf{b}_i)\right)\mathbf{I} + t_1\left(\textstyle\sum_i [\cos(\mathbf{k}\cdot\mathbf{a}_i)\sigma^1 + \sin(\mathbf{k}\cdot\mathbf{a}_i)\sigma^2]\right) + \left[M - 2t_2\sin\phi\left(\textstyle\sum_i \sin(\mathbf{k}\cdot\mathbf{b}_i)\right)\right]\sigma^3, \qquad (1)$$

where σ^i are Pauli matrices. The Brillouin zone is a hexagon rotated 90° with respect to the Wigner-Seitz unit cell: At its six corners $(\mathbf{k}\cdot\mathbf{a}_1, \mathbf{k}\cdot\mathbf{a}_2, \mathbf{k}\cdot\mathbf{a}_3)$ is a permutation of $(0, 2\pi/3, -2\pi/3)$. The two distinct corners $\mathbf{k}_\alpha^0$ are defined so that $\mathbf{k}_\alpha^0\cdot\mathbf{b}_i = (2\pi/3)\alpha$, $\alpha = \pm 1$.

The energy bands are easily obtained. There are two bands which only touch if all three Pauli matrix terms in (1) have vanishing coefficients. This can only occur at zone corners $\mathbf{k}_\alpha^0$, and then only if $M = 3\sqrt{3}\alpha t_2\sin\phi$. I will assume $|t_2/t_1| < \frac{1}{3}$, which guarantees that the two bands never overlap, and are separated by a finite gap unless they touch.

If both M and $t_2\sin\phi$ vanish, the bands touch at both zone corners, where the group of the wave vector[6] has the unitary subgroup C_{3v}, which contains a reflection that interchanges the A and B sublattices. Apart from the zone center, these are the only points in the Brillouin zone where this group has irreducible representations with dimensions greater than unity, and the degenerate states at these points belong to the two-dimensional representation. The touching of the bands at *two* distinct points in the Brillouin zone is a manifestation of fermion doubling.[3,5] The degeneracy of the bands at these points is lifted either by nonzero M or nonzero $t_2\sin\phi$, either of which reduce the unitary subgroup to C_3, which has only one-dimensional irreducible representations.

When the Fermi level lies in a gap between two bands, σ_{xy} is quantized at $T=0$, and its value can be obtained through the thermodynamic relation[7] $\sigma^{xy} = \partial\sigma/\partial B_0|_{\mu,T}$ evaluated at $B_0 = 0$, where σ is the 2D electric-charge density, and B_0 is the flux density of a uniform external magnetic field in the $\hat{\mathbf{z}}$ direction. To calculate the induced charge density σ to a weak external magnetic field, it is convenient to expand the Hamiltonian in the neighborhood of the band extrema at the zone corners $\mathbf{k}_\alpha^0$ to linear order in $\delta\mathbf{k} = \mathbf{k} - \mathbf{k}_\alpha^0$, and make the Landau-Peierls substitution $\hbar\delta\mathbf{k} \to \mathbf{\Pi}$, where $\mathbf{\Pi} = (\Pi^x, \Pi^y)$ is the dynamical momentum with components satisfying the commutation relation $[\Pi^x, \Pi^y] = i\hbar e B_0$.

For weak B_0, coupling between the two distinct zone corners can be neglected, and two independent effective Hamiltonians $\mathbf{H}_\alpha$ are obtained, where

$$\mathbf{H}_\alpha = c(\Pi_\alpha^1\sigma^2 - \Pi_\alpha^2\sigma^1) + m_\alpha c^2\sigma^3. \qquad (2)$$

Here $c = \frac{3}{2}t_1|\mathbf{a}_i|/\hbar$ and $m_\alpha c^2 = M - 3\sqrt{3}\alpha t_2\sin\phi$; Π_α^1 and Π_α^2 are Hermitian operators with the commutation relation $[\Pi_\alpha^1, \Pi_\alpha^2] = i\alpha e B_0\hbar$, defined by

$$(\Pi_\alpha^1 + i\Pi_\alpha^2) = \tfrac{2}{3}\textstyle\sum_i e^{-i\mathbf{k}_\alpha^0\cdot\mathbf{a}_i}(\mathbf{a}_i\cdot\mathbf{\Pi})/|a_i|. \qquad (3)$$

After second-quantization, (2) is precisely the Hamiltonian of a free-fermion-field theory studied by Jackiw[2] as an (2+1)-D analog of the Dirac Hamiltonian.

The spectrum of (2) is relativistic; for $B_0 = 0$,

$$\epsilon_{\alpha\pm}(\mathbf{k}) = \pm[(\hbar ck)^2 + (m_\alpha c^2)^2]^{1/2},$$

while for $B_0 \neq 0$, relativistic Landau levels are obtained[2] as follows:

$$\epsilon_{an\pm} = \pm[(m_a c^2)^2 + n\hbar|eB_0|c^2]^{1/2} \quad (n \geq 1), \tag{4a}$$

$$\epsilon_{a0} = am_a c^2 \operatorname{sgn}(eB_0). \tag{4b}$$

Every $n \geq 1$ level that evolves out of the upper band as B_0 is turned on is balanced by a level that evolves from the lower band. However, the $n=0$ "zero-mode" energy is not symmetric under $B_0 \to -B_0$: It evolves from the *upper* band if $am_a eB_0$ is *positive*, and from the *lower* band if it is *negative*.

In the time-reversal symmetric case $t_2 \sin\phi = 0$, the two masses m_+ and m_- are equal, and the *sum* of the Landau-level spectra derived from the two distinct zone corners is particle-hole symmetric, *and* invariant under $B_0 \to -B_0$. In this case, $\sigma^{xy} = 0$ by time-reversal invariance. As the Hamiltonian is changed, σ^{xy} remains invariant, provided the Fermi level remains in a gap.[4] When $B_0 = 0$, models where the Fermi level is in the gap and m_+ and m_- have the same sign can evolve continuously from the time-reversal invariant case, and hence have $\sigma^{xy} = 0$.

To calculate σ^{xy} for models where m_+ and m_- have *opposite* signs, I continuously turn on the external field, then vary m_+ and m_- until they become equal, at the same time varying the Fermi level so at all times it lies in a gap. Comparison of the occupation numbers of the Landau levels obtained this way with those obtained by continuously applying the field to the time-reversal invariant system shows that they differ by the complete filling of one Landau level. Thus at $T=0$ and with a fixed chemical potential, the application of a weak external magnetic field to a system where m_+ and m_- have opposite signs induces an extra *field-dependent* ground-state charge density $\Delta\sigma = \pm e^2 B_0/h$ relative to the field-independent charge density when these parameters have

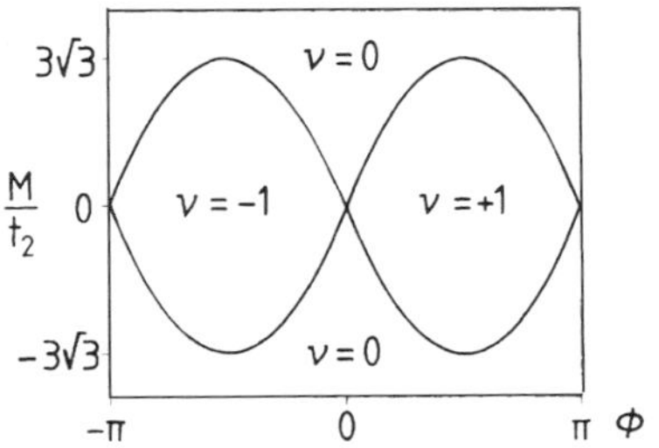

FIG. 2. Phase diagram of the spinless electron model with $|t_2/t_1| < \frac{1}{3}$. Zero-field quantum Hall effect phases ($\nu = \pm 1$, where $\sigma^{xy} = \nu e^2/h$) occur if $|M/t_2| < 3\sqrt{3}|\sin\phi|$. This figure assumes that t_2 is positive; if it is negative, ν changes sign. At the phase boundaries separating the anomalous and normal ($\nu = 0$) semiconductor phases, the low-energy excitations of the model simulate undoubled massless chiral relativistic fermions.

the same sign. This allows σ^{xy} in the limit $B_0 = 0$ to be evaluated as $\nu e^2/h$, where $\nu = \frac{1}{2}[\operatorname{sgn}(m_-) - \operatorname{sgn}(m_+)] = \pm 1$ or 0. The phase diagram of ν for the spinless electron model as a function of M/t_2 and ϕ is shown in Fig. 2.

I note that when the model has *neither* an inversion center *nor* time-reversal invariance (i.e., when both M and $t_2 \sin\phi$ are nonzero), so $|m_+| \neq |m_-|$, the spectrum is no longer invariant under $\mathbf{k} \to -\mathbf{k}$, and the fermion-doubling principle is defeated. In particular, along the critical lines in the phase diagram where one of m_+ or m_- vanishes, the model has a low-lying massless spectrum simulating nondegenerate relativistic *chiral* fermions.

When $m_a = 0$, the fermion field theory derived from the expansion (2) about the Fermi point with vanishing gap has a charge-conjugation symmetry (particle-hole symmetry) which is *not* present in the lattice model with $t_2 \neq 0$ from which it is derived. In the continuum field theory, there is no lower bound to the Dirac sea of filled electron states, and the establishment of *absolute* as opposed to *relative* values of σ^{xy} is ambiguous. Jackiw[2] invokes the charge-conjugation symmetry of (2) with $m_a = 0$ to assign the value $\sigma^{xy} = 0$ in the case of a particle-hole symmetric Fermi level, where the "zero-mode" Landau level (4b) is half filled. This would imply a quantum Hall effect with $\nu = \frac{1}{2}\alpha$ if the zero mode is filled, and $\nu = -\frac{1}{2}\alpha$ if it is empty. This suggests "charge fractionalization," and violates the principle that a noninteracting electron system can only exhibit an *integral* QHE. The model studied here shows how the high-energy cutoff structure of a model with undoubled fermions described by the relativistic Hamiltonian (2) at low energies *must* break the charge-conjugation symmetry, and give an extra contribution of $\pm\frac{1}{2}$ to ν, restoring an integral QHE. Thus even if the low-energy spectrum consists of undoubled chiral fermions, their partners must be present at high energies to restore a properly integral QHE.

When electron spin is included without any other change, there is an equal contribution from both spin components, and σ^{xy} is doubled. However, a periodic local magnetic field with the full symmetry of the lattice will also couple to electrons with a Zeeman term $H' = \gamma\phi S^z$, where S^z is the azimuthal electron spin. This term will relatively displace the up-spin and down-spin bands by an energy $\gamma\hbar\phi$, and if this exceeds the gap at the Fermi level, the system will become a partially spin-polarized metal. If $\frac{1}{2}|\gamma|\hbar$ exceeds $3\sqrt{3}|t_2|$, the QHE phases are completely eliminated, but if it is smaller, they survive for small enough M and $t_2 \sin\phi$. (The direct transition from the normal to the anomalous semiconductor phase as M is varied is then replaced by an intermediate spin-polarized metallic phase.) For the realization of the internal field proposed earlier, $\gamma\hbar$ (in units of the rydberg) is given by $C'g/a^2$, where C' is another geometrical constant of order unity, and g is the Landé g factor for the electrons.

While the particular model presented here is unlikely to be directly physically realizable, it indicates that, at least in principle, the QHE can be placed in the wider context of phenomena associated with broken time-reversal invariance, and does not necessarily require external magnetic fields, but could occur as a consequence of magnetic ordering in a quasi-two-dimensional system.

This requirement is *not* fulfilled by the physical system (a domain wall in a PbTe-type semiconductor) in which Fradkin, Dagotto, and Boyanovsky[8] (FDB) have recently proposed related effects may be realized. In this model, spin-orbit coupling is supposed to give rise to the effect, but this does *not* break time-reversal symmetry. In fact, in "simplifying" the p bands of the Hamiltonian that describes PbTe, FDB introduce an unphysical effective spin-dependent hopping term that is *odd* under time reversal, and thus break the time-reversal invariance of the original physically motivated model. This, rather than any topological character of the domain wall, is the reason that FDB find the "parity anomaly" at the end of their calculation.

I thank E. Fradkin and T. A. L. Ziman for very useful discussions. The author would like to thank the Alfred P. Sloan Foundation for financial support.

[1]K. Von Klitzing, G. Dorda, and M. Pepper, Phys. Rev. Lett. **45**, 494 (1980).

[2]R. Jackiw, Phys. Rev. D **27**, 2375 (1984).

[3]H. B. Nielsen and M. Ninomiya, Nucl. Phys. **B185**, 20 (1981), and **B193**, 173 (1981).

[4]D. J. Thouless, M. Kohmoto, M. P. Nightingale, and M. den Nijs, Phys. Rev. Lett. **49**, 405 (1982).

[5]G. Semenoff, Phys. Rev. Lett. **53**, 2449 (1984).

[6]The full group of the wave vector at zone corners contains additional *antiunitary* operations, but these do not give rise to extra degeneracies.

[7]P. Strěda, J. Phys. C **15**, L717 (1982); A. Widom, Phys. Lett. **90A**, 474 (1982).

[8]E. Fradkin, E. Dagotto, and D. Boyanovsky, Phys. Rev. Lett. **57**, 2967 (1986), and **58**, 961(E) (1987); D. Boyanovsky, E. Dagotto, and E. Fradkin, Nucl. Phys. **B285**, 340 (1987).

Topological Invariants in Fermi Systems with Time-Reversal Invariance

J. E. Avron,[a] L. Sadun, J. Segert, and B. Simon

Division of Physics, Mathematics, and Astronomy, California Institute of Technology, Pasadena, California 91125

(Received 6 July 1988)

We discuss topological invariants for Fermi systems that have time-reversal invariance. The TKN^2 integers (first Chern numbers) are replaced by second Chern numbers, and Berry's phase becomes a unit quaternion, or equivalently an element of SU(2). The canonical example playing much the same role as spin $\frac{1}{2}$ in a magnetic field is spin $\frac{3}{2}$ in a quadrupole electric field. In particular, the associated bundles are nontrivial and have ± 1 second Chern number. The connection that governs the adiabatic evolution coincides with the symmetric SU(2) Yang-Mills instanton.

PACS numbers: 03.65.Bz, 02.40.+m

Much interest has focused recently on topological and geometric invariants in eigenvalue perturbation theory, namely on TKN^2 integers[1,2] and on Berry's phase.[3,4] The folk wisdom is that nontrivial values of these invariants require broken time-reversal invariance such as occurs in a magnetic field. This belief is correct in the sense that if all Hamiltonians can be made simultaneously real, the invariants vanish. But such simultaneous reality is a feature of time-reversal invariance for integer spins, i.e., Bose systems only. Our goal here is to discuss the rich structure present for time-reversal–invariant half-odd-integer spins, i.e., Fermi systems. Additional details will appear in Ref. 5. Non-Abelian phases in Kramers degenerate systems have previously been considered in Ref. 6.

Let **J** be the angular momentum operator, which is, of course, odd under time reversal. In the usual representation (with J_x and J_z real and J_y imaginary) time reversal, which we call T, may be implemented by the product of complex conjugation and rotation about the y axis by π. This is in fact the only antiunitary operator, up to phases, that reverses **J**. In representations where J is half-odd integral, rotation by 2π is -1, and so $T^2 = -1$.

This holds in general. That is, in Fermi systems with time-reversal invariance, there exists an antiunitary operator T which commutes with the Hamiltonian (or family of Hamiltonians if some external parameters are varied) and obeys $T^2 = -1$. Since T anticommutes with multiplication by i, it gives the Hilbert space a quaternionic structure,[7,8] if we think of T as multiplication by j and iT as multiplication by k.

H is time-reversal invariant if it commutes with T.[9] This says that H is Hermitian in the quaternionic sense. The eigenspaces are quaternionic vector spaces and so are even dimensional as complex vector spaces. Kramers' degeneracy[7] follows immediately from this structure. An eigenvalue is called *quaternionically simple* if it has the minimum degeneracy consistent with Kramers, namely double, degeneracy in the sense of a complex Hilbert space.

As for complex Hermitian matrices, where generically all eigenvalues are simple in the complex sense, the eigenvalues of quaternionic Hermitian matrices are generically simple in the quaternionic sense. (The usual definition but with complex conjugate replaced by quaternionic conjugate.) Whereas in the complex Hermitian case eigenvalue crossings have codimension 3, in the quaternionic Hermitian case they have codimension 5. Simple parametric counting via Wigner and von Neumann[10] yields these numbers. Codimension 5 means that, generically, there are no level crossings on four-dimensional surfaces.

What we have said so far is not new, and is presented for background. In the complex case, a basic role is played by the homotopy groups of $M_n(\mathbb{C})$, the space of nondegenerate, complex Hermitian $n\times n$ matrices. This space is homotopic to the quotient $U(n)/[U(1)\times\cdots\times U(1)]$ of the unitary matrices. As shown in Ref. 2, $\pi_2[M_n(\mathbb{C})]=\mathbb{Z}^{n-1}$, which reflects the fact that there is a TKN^2 integer associated with each eigenvalue, and the integers satisfy a zero sum rule.

The quaternionic analog is $M_n(\mathbb{H})$, the nondegenerate quaternionic Hermitian $n\times n$ matrices, which is homotopic to the quotient $Sp(n)/[Sp(1)\times\cdots\times Sp(1)]$ of the unitary symplectic matrices by the diagonal unitary symplectic matrices.[9] Now $\pi_j[M_n(\mathbb{H})]=0$ for $j=1,2,3$ and $\pi_4[M_n(\mathbb{H})]=\mathbb{Z}^{n-1}$. These results follow from the homotopy exact sequences. $M_n(\mathbb{H})$ turns out to have a richer homotopy than $M_n(\mathbb{C})$ for the simple reason that $Sp(1)\sim S^3$ has a richer homotopy than $U(1)\sim S^1$, so unlike the complex case, there are higher homotopy invariants in the quaternionic case.[5]

The TKN^2 integers associated to a map of S^2 to $M_n(\mathbb{C})$ can be computed as the integral of the first Chern two-form ω_1. In terms of the projection $P(Q)$ onto the one-dimensional eigenspace for $H(Q)$ depending on parameter Q, the curvature Ω is given by $\Omega=iP(dP\wedge dP)$, and

$$\omega_1=\frac{1}{2\pi}\mathrm{Tr}(\Omega).$$

The TKN^2 numbers are the first Chern numbers, $C_1\equiv\int_{S^2}\omega_1$, where S^2 is a closed two-surface in parameter space. For the canonical example $H(B)=\mathbf{J}\cdot\mathbf{B}$, $C_1=2m$, where m is the azimuthal quantum number. Clearly, if $H(Q)$ commutes with T, then $\omega_1\equiv0$.

In the quaternionic case, the basic object is the second Chern four-form (in this context more appropriately called the first symplectic Pontryagin form[11])

$$\omega_2=\frac{1}{8\pi^2}[\mathrm{Tr}(\Omega^2)-\mathrm{Tr}(\Omega)^2]=\frac{1}{8\pi^2}\mathrm{Tr}(\Omega^2),$$

so that, for example, if Q parametrizes points in S^4, then the second Chern integer is given by

$$C_2=\int_{S^4}\omega_2. \tag{1}$$

As in the complex case, solving the Schrödinger equation adiabatically (or in any other smooth way that is guaranteed to return to the initial state[12]) produces a holonomy identical to the natural one induced by the metric on the Hilbert space on the bundle of eigenspaces.[4] In the complex case, this holonomy is given by a complex number of magnitude one—the Berry's phase. In the quaternionic case it is given by a unit quaternion, or equivalently by an element of SU(2). In the general twofold-degenerate case considered by Wilczek and Zee,[3] the holonomy is an element of U(2).

These rather abstract considerations are made concrete in a set of elementary examples. Since $\mathbf{J}$ is odd under T, the simplest time-reversal-invariant Hamiltonian is

$$H(Q)\equiv\sum_{mn}Q_{mn}J_mJ_n, \tag{2}$$

with Q a real symmetric second-rank tensor, which may be taken traceless. The space of such tensors is a five-dimensional real vector space, with an invariant inner product given by $\langle Q,Q'\rangle\equiv\frac{3}{2}\mathrm{Tr}(QQ')$, where the trace is taken as the trace of linear operators on $\mathbb{R}^3$. A standard orthonormal basis Q_a, $a=0,\ldots,4$, can be defined such that, using $e_a=H(Q_a)$, $e_0\equiv J_z^2-J^2/3$, $e_1\equiv3^{-1/2}(J_xJ_z+J_zJ_x)$, $e_2\equiv3^{-1/2}(J_yJ_z+J_zJ_y)$, $e_3\equiv3^{-1/2}(J_x^2-J_y^2)$, and $e_4\equiv3^{-1/2}(J_xJ_y+J_yJ_x)$.

While this example looks specialized, by the Wigner-Eckart theorem it will apply to any spin-J multiplet perturbed by a second-rank tensor. For example, our Berry's phase calculation below will be relevant to any spin-atomic $\frac{3}{2}$ level under perturbation by an electric quadrupole.

At $Q=0$, $H(Q)=0$ is obviously highly degenerate. One can show that for $Q\neq0$ all levels are simple in the quaternionic sense, i.e., only doubly degenerate in the complex sense.[5] If we take for S^4 in Eq. (1) the unit sphere in the space of tensor operators, i.e., $\frac{3}{2}\mathrm{Tr}(Q^2)=1$, all the $J+\frac{1}{2}$ energy levels have well defined second Chern integers, satisfying a zero sum rule. Moreover, a simple argument shows that C_2 of the nth eigenvalue is $-C_2$ of the $(J+\frac{3}{2}-n)$th eigenvalue. This is in some ways like the situation discussed by Berry,[3] except that for $J>\frac{3}{2}$, there is not enough symmetry to facilitate a simple calculation of the C_2's.

The case $J=\frac{3}{2}$ is basic and has some remarkable symmetries, as we will explain. The sense that it is basic is that, as in Ref. 4, the generic situation for quaternion degeneracies as one varies five parameters is twofold degeneracy removed to first order. Just as the prototypical complex situation is spin-$\frac{1}{2}$ $\mathbf{J}\cdot\mathbf{B}$, the prototypical quaternionic situation is this spin-$\frac{3}{2}$ case.

Here are the main features of the spin-$\frac{3}{2}$ case.[5]

(a) The time-reversal-invariant traceless operators on $\mathbb{C}^4$ (equivalently traceless quaternionic Hermitian operators on $\mathbb{H}^2$) are exactly the family of $H(Q)$. This can be seen by just counting dimensions. Each is exactly five-dimensional and each $H(Q)$ is quaternionic, Hermitian, and traceless.

(b) Each nonzero $H(Q)$, being traceless, Hermitian, and 2×2 (quaternionically), can be written uniquely as

$$H=c(P_+-P_-)=c(2P_+-1)=c(1-2P_-),$$

where c is a positive real number and $P_\pm$ is the spectral projection onto the positive (or negative) eigenspace. Conversely, to each projection P and positive number c we can associate the Hamiltonian $c(2P-1)$. As a result, there is a one-to-one correspondence between the unit sphere of $H(Q)$'s and the set of spectral projections P_+ (or P_-), yielding the first proof that that the two

Chern integers are ± 1.

(c) $\mathrm{Tr}[H(Q)^2] = 6\mathrm{Tr}(Q^2)$. By (a), the unit sphere of $H(Q)$'s is left invariant by the full group of 2×2 symplectic matrices, $\mathrm{Sp}(2) \cong \mathrm{Spin}(5)$, the twofold universal cover of SO(5). Spin(5) acts on the unit sphere of Hamiltonians by SO(5) rotations. Thus, this family of operators, which *a priori* only has SO(3) symmetry, has SO(5) symmetry. Any Hamiltonian can be mapped to any other Hamiltonian by an SO(5) rotation, even though there are pairs of Hamiltonians that are not related by SO(3).

(d) Because all unit Hamiltonians are equivalent by the SO(5) symmetry, the second Chern form is a multiple of the area form. The constant of proportionality can be computed at any point [say $(J_z^2 - \frac{1}{3}J^2)$] to yield another proof that the two Chern integers are ± 1.

(e) Using the explicit matrices of J_x, J_y, and J_z in the spin-$\frac{3}{2}$ representation, one can easily check that the matrices e_α defined above form a Clifford algebra, i.e., $e_\alpha e_\beta + e_\beta e_\alpha = 2\delta_{\alpha\beta}$. The commutators $[e_\alpha, e_\beta]$ generate the Spin(5) action. By a unitary change of basis, these matrices can be transformed to

$$\hat{e}_0 = \begin{bmatrix} 1 & 0 \\ 0 & -1 \end{bmatrix}, \quad \hat{e}_1 = \begin{bmatrix} 0 & 1 \\ 1 & 0 \end{bmatrix}, \quad \hat{e}_2 = \begin{bmatrix} 0 & -i \\ i & 0 \end{bmatrix},$$

$$\hat{e}_3 = \begin{bmatrix} 0 & j \\ -j & 0 \end{bmatrix}, \quad \hat{e}_4 = \begin{bmatrix} 0 & -k \\ k & 0 \end{bmatrix},$$

which evidently span the traceless quaternionic Hermitian 2×2 matrices. The Clifford property makes the fact $P_\pm(Q) = \frac{1}{2}[1 \pm H(Q)]$ of (b) evident.

(f) For many paths on the four-sphere of Hamiltonians, the SO(5) symmetry enables one to compute the holonomy explicitly.[5] Specifically, consider paths of the form $H(\tau) = U(\tau)e_0U(-\tau)$, $0 \le \tau \le 2\pi$, where $U(\tau)$ is a one-parameter group of symplectic matrices with $U(2\pi) = -1$. Paths of this form are closed, as $H(0) = H(2\pi) = e_0$. By (e), such one-parameter groups are of the form $U(\tau) = \exp(\tau C_{\alpha\beta}[e_\alpha, e_\beta])$ for appropriate constants $C_{\alpha\beta}$. In particular, the physical SO(3) rotations are of this type. Suppose the system is initially in an eigenstate of $J_z = m$. The probability of the final state being $J_z = -m$ is easily recovered from the holonomy, and is a physical observable amenable to experimental measurement. Consider a path generated by the physical rotation $U(\tau) = \exp\{-i\tau[\cos(\phi)J_z + \sin(\phi)J_x]\}$. For any path generated by physical rotation the probability of an initial state $J_z = \frac{3}{2}$ going to a final state $J_z = -\frac{3}{2}$ is zero. This result is based on an analysis of the geometry of SO(3) representations. The SO(3) symmetry further allows one to compute the probability of the $J_z = \frac{1}{2}$ state going to the $J_z = -\frac{1}{2}$, which is given by[6,13]

$$\frac{(4\sin^2\phi)\{\sin^2[\pi(1+3\sin^2\phi)^{1/2}]\}}{1+3\sin^2\phi}.$$

For paths which are generated by SO(5) rotations not corresponding to the physical rotations, the probability of $J_z = \frac{3}{2}$ going to $J_z = -\frac{3}{2}$ need not be zero. In fact, there exist paths for which this probability is unity, for example, the path generated by $U(\tau) = \exp[(\tau/8)e_1(\sqrt{3}e_0 + 2e_3)]$.

(g) This example is closely related to the symmetric SU(2) instanton.[14] The connection governing the adiabatic time evolution for the spin $\frac{3}{2}$ has self-dual curvature, and thus is a classical solution of the SU(2) Yang-Mills equations.

(h) Connections with self-dual curvature are classified via twistor theory,[14] and the spin-$\frac{3}{2}$ representation arises naturally in this framework. The rotation group SO(3) acts on the four-sphere of unit quadrupoles. The twistor space of S^4 is $\mathbb{C}P^3$, which now has an induced SO(3) action. This action extends to a projective representation of SO(3) on $\mathbb{C}^4$, which is precisely the spin-$\frac{3}{2}$ representation of SU(2).

The second Chern numbers may be thought of as quantum numbers that arise from topology rather than symmetry. Besides labeling the spectrum, quantum numbers that come from symmetry give various selection rules forbidding certain transitions. Similarly, the topological quantum numbers, besides labeling the spectrum give selection rules for adiabatic processes that can force transitions whenever the eigenstate bundle is nontrivial, at least for some paths. This holds on general grounds and is also illustrated by (f) above.

We thank Peter Weichman for useful discussions. The research of one of us (J.E.A.) was partially supported by National Science Foundation Grants No. DMS-8416049 and No. BSF-84-00376. The research of another one of us (B.S.) was partially supported by National Science Foundation Grant No. DMS-8416049.

---

[a] On leave from Physics Department, Technion, Haifa, 32000 Israel.

[1] D. Thouless *et al.*, Phys. Rev. Lett. **49**, 405 (1982); B. A. Dubrovin *et al.*, Zh. Eksp. Teor. Fiz. **79**, 1006 (1980) [Sov. Phys. JETP **52**, 511 (1980)]; M. Kohmoto, Ann. Phys. **160**, 343 (1985).

[2] J. Avron, R. Seiler, and B. Simon, Phys. Rev. Lett. **51**, 51 (1983).

[3] M. Berry, Proc. Roy. Soc. London A **392**, 45 (1984); F. Wilczek and A. Zee, Phys. Rev. Lett. **52**, 2111 (1984); "Geometric Phases in Physics," edited by F. Wilczek and A. Shapere (to be published).

[4] B. Simon, Phys. Rev. Lett. **51**, 2167 (1983).

[5] J. Avron, L. Sadun, J. Segert, and B. Simon, to be published; some results will appear in J. Avron, R. Seiler, and B. Simon, to be published.

[6] C. Alden Mead, Phys. Rev. Lett. **59**, 161 (1987).

[7] F. Dyson, J. Math. Phys. **3**, 140 (1964); we are indebted to M. Berry for raising the issue of fermions with time-reversal invariance and pointing out the connection with quaternions.

[8]E. P. Wigner, Göttinger Nachr. **31**, 546 (1932). See also E. P. Wigner, *Group Theory* (Academic, New York, 1959). The connection between quaternions and antilinear operators with $T^2 = -1$ and the fact that time-reversal invariance without reality has such a structure both have a long history in the mathematical literature on group representations. If U is an irreducible representation unitarily equivalent to its complex conjugate, then either U can be made simultaneously real or it is a quarternionic representation. Moreover, if the group is compact, which it is can be found by averaging the square of the group character over the group (this average is 1 in the real case and -1 in the quaternionic case). For finite groups, this result is called the Frobenius-Schur theorem, and goes back to the first decade of this century. In the terminology of modern representation theory, T is called a structure map. Applied to SU(2), it shows that the integral spin representations are real, while the half-odd-integral spin case is quaternionic.

[9]The unitary symplectic group Sp(n) (often called simply the symplectic group) consists of the $n \times n$ quaternion-valued matrices which preserve the usual norm on $\mathbb{H}^n$. The generators of Sp(n) are the Hermitian quaternionic matrices. If $\mathbb{H}^n$ is identified with $\mathbb{C}^{2n}$, Sp(n) is that subset of the complex unitary matrices U($2n$) which commute with T, and the quaternionic Hermitian matrices are those complex Hermitian matrices that commute with T. That $\mathrm{Sp}(1) \cong \mathrm{SU}(2)$, and $\mathrm{Sp}(2) \cong \mathrm{Spin}(5)$ is well known.

[10]J. von Neumann and E. Wigner, Phys. Z **30**, 467 (1929); to see the answer, look at 2×2 traceless matrices over the base field, F. Their dimension is the sum of 1 for the diagonal and dim(F) for the off-diagonal, yielding dimension 2,3,5 for the case $F = \mathbb{R}, \mathbb{C}, \mathbb{H}$.

[11]A. Borel and F. Hirzebruch, Am. J. Math. **80**, 458 (1958).

[12]Y. Aharonov and J. Anandan, Phys. Rev. Lett. **58**, 1593 (1987).

[13]J. Segert, J. Math. Phys. **28**, 2102 (1987).

[14]M. Atiyah, N. Hitchin, V. Drinfeld, and Y. Manin, Phys. Lett. **65A**, 185 (1978); M. Atiyah, N. Hitchin, and I. Singer, Proc. Roy. Soc. London Ser. A **362**, 425 (1978).

Z_2 Topological Order and the Quantum Spin Hall Effect

C. L. Kane and E. J. Mele
Department of Physics and Astronomy, University of Pennsylvania, Philadelphia, Pennsylvania 19104, USA
(Received 22 June 2005; published 28 September 2005)

The quantum spin Hall (QSH) phase is a time reversal invariant electronic state with a bulk electronic band gap that supports the transport of charge and spin in gapless edge states. We show that this phase is associated with a novel Z_2 topological invariant, which distinguishes it from an ordinary insulator. The Z_2 classification, which is defined for time reversal invariant Hamiltonians, is analogous to the Chern number classification of the quantum Hall effect. We establish the Z_2 order of the QSH phase in the two band model of graphene and propose a generalization of the formalism applicable to multiband and interacting systems.

DOI: 10.1103/PhysRevLett.95.146802 PACS numbers: 73.43.−f, 72.25.Hg, 73.61.Wp, 85.75.−d

The classification of electronic states according to topological invariants is a powerful tool for understanding many body phases which have bulk energy gaps. This approach was pioneered by Thouless, Kohmoto, Nightingale, and den Nijs [1] (TKNN), who identified the topological invariant for the noninteracting integer quantum Hall effect. The TKNN integer, n, which gives the quantized Hall conductivity for each band $\sigma_{xy} = ne^2/h$, is given by an integral of the Bloch wave functions over the magnetic Brillouin zone, and corresponds to the first Chern class of a $U(1)$ principal fiber bundle on a torus [2,3]. An equivalent formulation, generalizable to interacting systems, is to consider the sensitivity of the many body ground state to phase twisted periodic boundary conditions [4,5]. This topological classification distinguishes a simple insulator from a quantum Hall state, and explains the insensitivity of the Hall conductivity to weak disorder and interactions. Nonzero TKNN integers are also intimately related to the presence of gapless edge states on the sample boundaries [6].

Since the Hall conductivity violates time reversal ($\mathcal{T}$) symmetry, the TKNN integer must vanish in a $\mathcal{T}$-invariant system. Nonetheless, we have recently shown that the spin-orbit interaction in a single plane of graphene leads to a $\mathcal{T}$-invariant quantum spin Hall (QSH) state which has a bulk energy gap, and a pair of gapless spin filtered edge states on the boundary [7]. In the simplest version of our model (a π-electron tight-binding model with mirror-symmetry about the plane) the perpendicular component of the spin, S_z, is conserved. Our model then reduces to independent copies for each spin of a model introduced by Haldane [8], which exhibits an integer quantum Hall effect even though the average magnetic field is zero. When S_z is conserved the distinction between graphene and a simple insulator is thus easily understood. Each spin has an independent TKNN integer $n_\uparrow$, $n_\downarrow$. $\mathcal{T}$ symmetry requires $n_\uparrow + n_\downarrow = 0$, but the difference $n_\uparrow - n_\downarrow$ is nonzero and defines a quantized spin Hall conductivity.

This characterization breaks down when S_z nonconserving terms are present. Such terms will inevitably arise due to coupling to other bands, mirror-symmetry breaking terms, interactions, or disorder. Though these perturbations destroy the quantization of the spin Hall conductance, we argued that they do not destroy the topological order of the QSH state because Kramers' theorem prevents $\mathcal{T}$-invariant perturbations from opening a gap at the edge [7]. Thus, even though the single defined TKNN number (the total Hall conductance) is zero, the QSH ground state is distinguishable from that of a simple insulator. This suggests that there must be an additional topological classification for $\mathcal{T}$-invariant systems.

In this Letter we clarify the topological order of the QSH phase and introduce a Z_2 topological index that characterizes $\mathcal{T}$-invariant systems. This classification is similar to the TKNN classification, and gives a simple test which can be applied to Bloch energy bands to distinguish the insulator from the QSH phase. It may also be formulated as a sensitivity to phase twisted boundary conditions. We will begin by describing our model of graphene and demonstrate that the QSH phase is robust even when S_z is not conserved. We will then analyze the constraints of $\mathcal{T}$ invariance and derive the Z_2 index.

Consider the tight-binding Hamiltonian of graphene introduced in Ref. [7], which generalizes Haldane's model [8] to include spin with $\mathcal{T}$-invariant spin-orbit interactions.

$$H = t\sum_{\langle ij\rangle} c_i^\dagger c_j + i\lambda_{SO}\sum_{\langle\langle ij\rangle\rangle} \nu_{ij} c_i^\dagger s^z c_j + i\lambda_R\sum_{\langle ij\rangle} c_i^\dagger (\mathbf{s}\times\hat{\mathbf{d}}_{ij})_z c_j + \lambda_v \sum_i \xi_i c_i^\dagger c_i. \tag{1}$$

The first term is a nearest neighbor hopping term on the honeycomb lattice, where we have suppressed the spin index on the electron operators. The second term is the mirror symmetric spin-orbit interaction which involves spin dependent second neighbor hopping. Here $\nu_{ij} = (2/\sqrt{3})(\hat{\mathbf{d}}_1 \times \hat{\mathbf{d}}_2)_z = \pm 1$, where $\hat{\mathbf{d}}_1$ and $\hat{\mathbf{d}}_2$ are unit vectors along the two bonds the electron traverses going from site j to i. s^z is a Pauli matrix describing the electron's spin. The third term is a nearest neighbor Rashba term, which explicitly violates the $z \to -z$ mirror symmetry, and will arise due to a perpendicular electric field or interaction with a substrate. The fourth term is a staggered sublattice potential ($\xi_i = \pm 1$), which we include to describe the transition between the QSH phase and the simple insulator. This term violates the symmetry under twofold rotations in the plane.

H is diagonalized by writing $\phi_s(\mathbf{R} + \alpha\mathbf{d}) = u_{\alpha s}(\mathbf{k})e^{i\mathbf{k}\cdot\mathbf{R}}$. Here s is spin and $\mathbf{R}$ is a bravais lattice vector built from primitive vectors $\mathbf{a}_{1,2} = (a/2)(\sqrt{3}\hat{\mathbf{y}} \pm \hat{\mathbf{x}})$. $\alpha = 0, 1$ is the sublattice index with $\mathbf{d} = a\hat{\mathbf{y}}/\sqrt{3}$. For each $\mathbf{k}$ the Bloch wave function is a four component eigenvector $|u(\mathbf{k})\rangle$ of the Bloch Hamiltonian matrix $\mathcal{H}(\mathbf{k})$. The 16 components of $\mathcal{H}(\mathbf{k})$ may be written in terms of the identity matrix, 5 Dirac matrices Γ^a and their 10 commutators $\Gamma^{ab} = [\Gamma^a, \Gamma^b]/(2i)$ [9]. We choose the following representation of the Dirac matrices: $\Gamma^{(1,2,3,4,5)} = (\sigma^x \otimes I, \sigma^z \otimes I, \sigma^y \otimes s^x, \sigma^y \otimes s^y, \sigma^y \otimes s^z)$, where the Pauli matrices σ^k and s^k represent the sublattice and spin indices. This choice organizes the matrices according to $\mathcal{T}$. The $\mathcal{T}$ operator is given by $\Theta|u\rangle \equiv i(I \otimes s^y)|u\rangle^*$. The five Dirac matrices are even under $\mathcal{T}$, $\Theta\Gamma^a\Theta^{-1} = \Gamma^a$ while the 10 commutators are odd, $\Theta\Gamma^{ab}\Theta^{-1} = -\Gamma^{ab}$. The Hamiltonian is thus

$$\mathcal{H}(\mathbf{k}) = \sum_{a=1}^{5} d_a(\mathbf{k})\Gamma^a + \sum_{a<b=1}^{5} d_{ab}(\mathbf{k})\Gamma^{ab}, \qquad (2)$$

where the $d(\mathbf{k})$'s are given in Table I. Note that $\mathcal{H}(\mathbf{k} + \mathbf{G}) = \mathcal{H}(\mathbf{k})$ for reciprocal lattice vectors $\mathbf{G}$, so $\mathcal{H}(\mathbf{k})$ is defined on a torus. The $\mathcal{T}$ invariance of $\mathcal{H}$ is reflected in the symmetry (antisymmetry) of d_a (d_{ab}) under $\mathbf{k} \to -\mathbf{k}$.

Equation (2) gives four energy bands, of which two are occupied. For $\lambda_R = 0$ there is an energy gap with magnitude $|6\sqrt{3}\lambda_{SO} - 2\lambda_v|$. For $\lambda_v > 3\sqrt{3}\lambda_{SO}$ the gap is dominated by λ_v, and the system is an insulator. $3\sqrt{3}\lambda_{SO} > \lambda_v$ describes the QSH phase. Though the Rashba term violates S_z conservation, for $\lambda_R < 2\sqrt{3}\lambda_{SO}$ there is a finite region of the phase diagram in Fig. 1 that is adiabatically connected to the QSH phase at $\lambda_R = 0$. Figure 1 shows the energy bands obtained by solving the lattice model in a zigzag strip geometry [7] for representative points in the insulating and QSH phases. Both phases have a bulk energy gap and edge states, but in the QSH phase the edge states traverse the energy gap in pairs. At the transition between the two phases, the energy gap closes, allowing the edge states to "switch partners."

The behavior of the edge states signals a clear difference between the two phases. In the QSH phase for each energy in the bulk gap there is a single time reversed pair of eigenstates on each edge. Since $\mathcal{T}$ symmetry prevents the mixing of Kramers' doublets these edge states are robust against small perturbations. The gapless states thus persist even if the spatial symmetry is further reduced [for instance, by removing the C_3 rotational symmetry in (1)]. Moreover, weak disorder will not lead to localization of the edge states because single particle elastic backscattering is forbidden [7].

In the insulating state the edge states do not traverse the gap. It is possible that for certain edge potentials the edge states in Fig. 1(b) could dip below the band edge, reducing—or even eliminating—the edge gap. However, this is still distinct from the QSH phase because there will necessarily be an *even* number of Kramers' pairs at each energy. This allows elastic backscattering, so that these edge states will in general be localized by weak disorder. The QSH phase is thus distinguished from the simple insulator by the number of edge state pairs modulo 2. Recently two-dimensional versions [10] of the spin Hall insulator models [11] have been introduced, which under conditions of high spatial symmetry exhibit gapless edge states. These models, however, have an even number of edge state pairs. We shall see below that they are topologically equivalent to simple insulators.

The QSH phase is *not* generally characterized by a quantized spin Hall conductivity. Consider the rate of spin accumulation at the opposite edges of a cylinder of circumference L, which can be computed using Laughlin's argument [12]. A weak circumferential electric field E can be induced by adiabatically threading magnetic flux through the cylinder. When the flux increases by h/e each momentum eigenstate shifts by one unit: $k \to k + 2\pi/L$. In the insulating state [Fig. 1(b)] this has no effect, since the valence band is completely full. However, in the QSH state a particle-hole excitation is produced at the Fermi energy E_F. Since the particle and hole states do not have the same spin, spin accumulates at the edge. The rate of spin accumulation defines a spin Hall conductance $d\langle S_z\rangle/dt = G_{xy}^s E$, where

TABLE I. The nonzero coefficients in Eq. (2) with $x = k_x a/2$ and $y = \sqrt{3}k_y a/2$.

d_1	$t(1 + 2\cos x \cos y)$	d_{12}	$-2t \cos x \sin y$
d_2	λ_v	d_{15}	$\lambda_{SO}(2\sin 2x - 4\sin x \cos y)$
d_3	$\lambda_R(1 - \cos x \cos y)$	d_{23}	$-\lambda_R \cos x \sin y$
d_4	$-\sqrt{3}\lambda_R \sin x \sin y$	d_{24}	$\sqrt{3}\lambda_R \sin x \cos y$

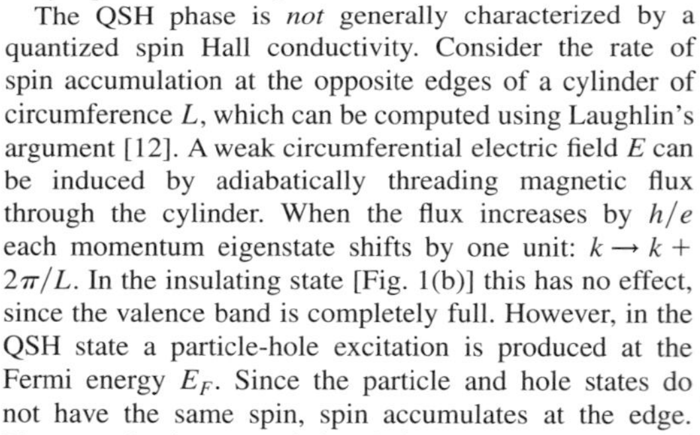

FIG. 1 (color online). Energy bands for a one-dimensional "zigzag" strip in the (a) QSH phase $\lambda_v = 0.1t$ and (b) the insulating phase $\lambda_v = 0.4t$. In both cases $\lambda_{SO} = .06t$ and $\lambda_R = .05t$. The edge states on a given edge cross at $ka = \pi$. The inset shows the phase diagram as a function of λ_v and λ_R for $0 < \lambda_{SO} \ll t$.

$$G_{xy}^s = \frac{e}{h}(\langle S_z\rangle_L - \langle S_z\rangle_R)|_{E_F}. \quad (3)$$

Here the expectation value of S_z is evaluated for the left and right moving states at E_F. Since the edge states are not necessarily S_z eigenstates this spin Hall conductance is not quantized. G_{xy}^s is zero in the insulating phase, though, provided E_F is in the gap at the edge. If in the insulator the edge states cross E_F, then in a clean system there could be spin accumulation at the edge (resulting from the acceleration of the edge electrons in response to E). However, if the edge states are localized then there will be no spin accumulation. Thus the nonzero spin accumulation persists only for the QSH phase, justifying the term quantum (but not quantized) spin Hall effect.

In the quantum Hall effect, the states with zero and one flux quantum threading the cylinder are distinguished by the charge polarization. The two states cannot be connected by any operator that locally conserves charge. In the QSH effect there is no simple conserved quantity distinguishing the two states. However, the states *are* distinguishable, because the state with an edge particle-hole excitation at E_F cannot be connected to the ground state with a local $\mathcal{T}$ symmetric operator. Note, however, that if a second flux is added, then there will be $\mathcal{T}$ invariant interactions which do connect the state with the zero flux state. This suggests that the state with one flux added is distinguished by a Z_2 "$\mathcal{T}$ polarization."

The classification of quantum Hall states on the cylinder according to Laughlin's argument is intimately related to the TKNN classification of the Bloch wave functions [4]. To establish the corresponding topological classification for $\mathcal{T}$-invariant systems we consider $\mathcal{T}$ constraints on the Bloch wave functions for the two occupied bands $|u_{i=1,2}(\mathbf{k})\rangle$. $|u_i(\mathbf{k})\rangle$ form a rank 2 vector bundle over Brillouin zone torus. $\mathcal{T}$ introduces an involution on the torus which identifies pairs of points $\mathbf{k}$ and $-\mathbf{k}$. Wave functions at the identified points are related by $|u_i(-\mathbf{k})\rangle = \Theta|u_i(\mathbf{k})\rangle$, implying that the bundle is "real." Since $\Theta^2 = -1$, Θ has period 4, so that the real bundle is "twisted." These bundles are classified within the mathematical framework of twisted Real K theory [13]. It is found that such bundles have a $Z \times Z_2$ classification on a torus [14]. The first integer gives the rank of the bundle (i.e., the number of occupied bands). The Z_2 index is related to the mod 2 index of the real Dirac operator [15]. In the following we will explicitly construct this Z_2 index from the Bloch wave functions and show that it distinguishes the QSH phase from the simple insulator.

$\mathcal{T}$ symmetry identifies two important subspaces of the space of Bloch Hamiltonians $\mathcal{H}(\mathbf{k})$ and the corresponding occupied band wave functions $|u_i(\mathbf{k})\rangle$. The "even" subspace, for which $\Theta H(\mathbf{k})\Theta^{-1} = H(\mathbf{k})$, have the property that $\Theta|u_i(\mathbf{k})\rangle$ is equivalent to $|u_i(\mathbf{k})\rangle$ up to a $U(2)$ rotation. From Eq. (2) it is clear that in this subspace $d_{ab}(\mathbf{k}) = 0$. $\mathcal{T}$ symmetry requires that $H(\mathbf{k})$ belong to the even subspace at the Γ point $\mathbf{k} = \mathbf{0}$ as well as the three M points shown in Figs. 2(a) and 2(b). The odd subspace has wave functions with the property that the space spanned by $\Theta|u_i(\mathbf{k})\rangle$ is *orthogonal* to the space spanned by $|u_i(\mathbf{k})\rangle$. We will establish the Z_2 classification by studying the set of $\mathbf{k}$ which belong to the odd subspace.

The special subspaces can be identified by considering the matrix of overlaps, $\langle u_i(\mathbf{k})|\Theta|u_j(\mathbf{k})\rangle$. From the properties of Θ it is clear that this matrix is antisymmetric, and may be expressed in terms of a single complex number as $\epsilon_{ij}P(\mathbf{k})$. $P(\mathbf{k})$ is in fact equal to the *Pfaffian*

$$P(\mathbf{k}) = \mathrm{Pf}\,[\langle u_i(\mathbf{k})|\Theta|u_j(\mathbf{k})\rangle], \quad (4)$$

which for a 2×2 antisymmetric matrix A_{ij} simply picks out A_{12}. We shall see below that the Pfaffian is the natural generalization when there are more than two occupied bands. $P(\mathbf{k})$ is not gauge invariant. Under a $U(2)$ transformation $|u_i'\rangle = U_{ij}|u_j\rangle$, $P' = P\,\mathrm{det}U$. Thus P is unchanged by a $SU(2)$ rotation, but under a $U(1)$ transformation $U = e^{i\theta}$, $P' = Pe^{2i\theta}$. In the even subspace $\Theta|u_i\rangle$ is equivalent to $|u_i\rangle$ up to a $U(2)$ rotation, and we have $|P(\mathbf{k})| = 1$. In the odd subspace $P(\mathbf{k}) = 0$.

If no spatial symmetries constrain its form, the zeros of $P(\mathbf{k})$ are found by tuning two parameters, and generically occur at points in the Brillouin zone. First order zeros occur at time reversed pairs of points $\pm\mathbf{k}^*$ with opposite "vorticity," where the phase of $P(\mathbf{k})$ advances in opposite directions around $\pm\mathbf{k}^*$. For $\lambda_v \neq 0$ the QSH phase is distinguished from the simple insulator by the presence of a single pair of first order zeros of $P(\mathbf{k})$. The C_3 rotational symmetry of our model constrains $\mathbf{k}^*$ to be at the corner of the Brillouin zone as shown in Fig. 2(a). If the C_3 symmetry is relaxed, $\mathbf{k}^*$ can occur anywhere *except* the

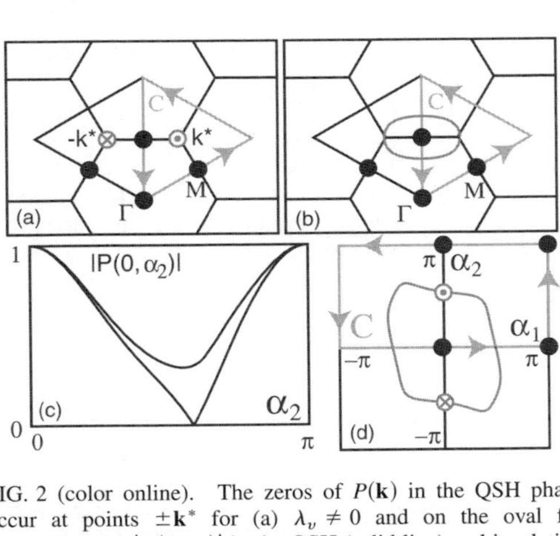

FIG. 2 (color online). The zeros of $P(\mathbf{k})$ in the QSH phase occur at points $\pm\mathbf{k}^*$ for (a) $\lambda_v \neq 0$ and on the oval for (b) $\lambda_v = 0$. (c) $|P(0, \alpha_2)|$ in the QSH (solid line) and insulating (dashed line) phases for a 2×2 supercell using parameters in Fig. 1. (d) Point ($\lambda_v \neq 0$) and line ($\lambda_v = 0$) zeros of $P(\vec{\alpha})$ for the 2×2 supercell. In (a), (b), and (d) the solid dots are $\mathcal{T}$-symmetric points, which cannot be zeros of P, and C is the contour of integration for Eq. (5).

four symmetric points where $|P(\mathbf{k})| = 1$. The number of pairs of zeros is a Z_2 topological invariant. This can be seen by noting that two pairs $\pm\mathbf{k}_{1,2}^*$ can come together to annihilate each other when $\mathbf{k}_1^* = -\mathbf{k}_2^*$. However a single pair of zeros at $\pm\mathbf{k}^*$ cannot annihilate because they would have to meet at either Γ or M, where $|P(\mathbf{k})| = 1$. If $\mathcal{T}$ symmetry is broken then the zeros are no longer prevented from annihilating, and the topological distinction of the QSH phase is lost.

The Z_2 index can thus be determined by counting the number of pairs of complex zeros of P. This can be accomplished by evaluating the winding of the phase of $P(\mathbf{k})$ around a loop enclosing *half* the Brillouin zone (defined so that $\mathbf{k}$ and $-\mathbf{k}$ are never both included).

$$I = \frac{1}{2\pi i}\oint_C d\mathbf{k} \cdot \nabla_{\mathbf{k}} \log[P(\mathbf{k}) + i\delta], \qquad (5)$$

where C is the path shown in Figs. 2(a) and 2(b).

When $\lambda_v = 0$ (as it is in graphene) H has a C_2 rotational symmetry, which when combined with $\mathcal{T}$ constrains the form of $\mathcal{H}(\mathbf{k})$, and allows $P(\mathbf{k})$ to be chosen to be real. The zeros of $P(\mathbf{k})$ then occur along lines, rather than at points. We find that the zeros are absent in the insulating phase, but enclose the M point in the QSH phase as shown in Fig. 2(b). In this case we find that Eq. (4) also determines the Z_2 index (given by $1/2$ the number of sign changes along the path C), provided we include the convergence factor δ. Note that though the sign of I depends on the sign of δ, I mod 2 does not. We thus conclude that the QSH phase and the insulator are distinguished by the Z_2 index I.

The spin Hall insulator models studied in Refs. [10,11] are simple insulators with $I = 0$. Their Hamiltonian, when expressed in the form of Eq. (2), has $d_{ab}(\mathbf{k}) = 0$, so that $|u_i(\mathbf{k})\rangle$ is in the even subspace and $|P(\mathbf{k})| = 1$ for all $\mathbf{k}$. Reference [16] introduces a model which does appear to exhibit a QSH effect.

Having established the topological classification of the Bloch wave functions we now ask whether, in analogy with Ref. [4], the classification can be formulated in terms of the sensitivity of the ground state wave function to phase twisted periodic boundary conditions. Such a formulation will address the topological stability of the many body ground state with respect to weak disorder and electron interactions. It also provides the appropriate generalization of (4) and (5) for multiband Hamiltonians. Consider a $\mathbf{L}_1 \times \mathbf{L}_2$ sample with boundary condition $\Psi(\ldots, \mathbf{r}_i + \mathbf{L}_k, \ldots) = e^{i\alpha_k}\Psi(\ldots, \mathbf{r}_i, \ldots)$. For concreteness we consider a rectangular geometry, with $\mathbf{L}_1 = N_1(\mathbf{a}_1 + \mathbf{a}_2)$ and $\mathbf{L}_2 = N_2(\mathbf{a}_1 - \mathbf{a}_2)$. For noninteracting electrons, we may view the entire sample as a large unit cell with $N_a = 4N_1N_2$ atoms imbedded in an even larger crystal. Then $\vec{\alpha}$ plays the role of $\mathbf{k}$, and the occupied single particle eigenstates $\phi_i(\vec{\alpha})$ play the role of $u_i(\mathbf{k})$. $\phi_i(\vec{\alpha})$ form a rank N_a bundle on the torus defined by $\alpha_{1,2}$. The Z_2 classification can be obtained by studying the zeros of

$$P(\vec{\alpha}) = \mathrm{Pf}\,[\langle\phi_i(\vec{\alpha})|\Theta|\phi_j(\vec{\alpha})\rangle]. \qquad (6)$$

Figure 2(c) compares $|P(\vec{\alpha})|$ in the QSH and insulating phases. For a 16-site sample with $N_1 = N_2 = 2$. In the insulating phase there are no zeros. In the QSH phase the structure of the zeros in Fig. 2(d) is similar to Figs. 2(a) and 2(b). For $\lambda_v \neq 0$ the first order zeros are at points, while for $\lambda_v = 0$ they are on a loop. The zeros cannot be at the four $\mathcal{T}$ symmetric points. This structure persists in the QSH phase for any cell size. The Z_2 index I can be computed by performing the integral analogous to (5) along the contour C in Fig. 2(d).

A many body formulation requires the index to be expressed in terms of the many particle ground state $|\Phi(\vec{\alpha})\rangle$. It is interesting to note that for noninteracting electrons $\langle\Phi(\vec{\alpha})|\Theta|\Phi(\vec{\alpha})\rangle = \det[\langle\phi_i(\vec{\alpha})|\Theta|\phi_j(\vec{\alpha})\rangle] = P(\vec{\alpha})^2$. This suggests a many body generalization

$$P(\vec{\alpha}) = \sqrt{\langle\Phi(\vec{\alpha})|\Theta|\Phi(\vec{\alpha})\rangle}. \qquad (7)$$

We suspect that with this definition the topological structure $P(\vec{\alpha})$ in Figs. 2(c) and 2(d) will remain in the presence of weak electron interactions.

To conclude, we have introduced a Z_2 topological classification of $\mathcal{T}$ invariant systems, analogous to the TKNN classification of quantum Hall states. This shows that the QSH phase of graphene has a topological stability that is insensitive to weak disorder and interactions.

We thank Tony Pantev for many helpful discussions. This work was supported by the NSF under MRSEC Grant No. DMR-00-79909 and the DOE under Grant No. DE-FG02-ER-0145118.

[1] D.J. Thouless, M. Kohmoto, M.P. Nightingale, and M. den Nijs, Phys. Rev. Lett. **49**, 405 (1982).
[2] J.E. Avron, R. Seiler, and B. Simon, Phys. Rev. Lett. **51**, 51 (1983).
[3] M. Kohmoto, Ann. Phys. (Berlin) **160**, 343 (1985).
[4] Q. Niu, D.J. Thouless, and Y.S. Wu, Phys. Rev. B **31**, 3372 (1985).
[5] D.P. Arovas *et al.*, Phys. Rev. Lett. **60**, 619 (1988).
[6] Y. Hatsugai, Phys. Rev. Lett. **71**, 3697 (1993).
[7] C.L. Kane and E.J. Mele, cond-mat/0411737.
[8] F.D.M. Haldane, Phys. Rev. Lett. **61**, 2015 (1988).
[9] S. Murakami, N. Nagaosa, and S.C. Zhang, Science **301**, 1348 (2003); Phys. Rev. B **69**, 235206 (2004).
[10] X.L. Qi, Y.S. Wu, and S.C. Zhang, cond-mat/0505308; M. Onoda and N. Nagaosa, Phys. Rev. Lett. **95**, 106601 (2005).
[11] S. Murakami, N. Nagaosa, and S.C. Zhang, Phys. Rev. Lett. **93**, 156804 (2004).
[12] R.B. Laughlin, Phys. Rev. B **23**, R5632 (1981).
[13] M. Atiyah, Quart. J. Math. Oxford **17**, 367 (1966); M. Atiyah and G. Segal, in *Michael Atiyah Collected Works* (Clarendon, Oxford, 2004), Vol. 6, p. 983.
[14] We are indebted to Tony Pantev for explaining these connections to us.
[15] M. Atiyah and I. Singer, Ann. Math. **93**, 139 (1971).
[16] B.A. Bernevig and S.C. Zhang, cond-mat/0504147.

Topological invariants of time-reversal-invariant band structures

J. E. Moore[1,2] and L. Balents[3]
[1]Department of Physics, University of California, Berkeley, California 94720, USA
[2]Materials Sciences Division, Lawrence Berkeley National Laboratory, Berkeley, California 94720, USA
[3]Department of Physics, University of California, Santa Barbara, California 93106, USA
(Received 28 July 2006; revised manuscript received 1 February 2007; published 26 March 2007)

The topological invariants of a time-reversal-invariant band structure in two dimensions are multiple copies of the $\mathbb{Z}_2$ invariant found by Kane and Mele. Such invariants protect the "topological insulator" phase and give rise to a spin Hall effect carried by edge states. Each pair of bands related by time reversal is described by one $\mathbb{Z}_2$ invariant, up to one less than half the dimension of the Bloch Hamiltonians. In three dimensions, there are four such invariants per band pair. The $\mathbb{Z}_2$ invariants of a crystal determine the transitions between ordinary and topological insulators as its bands are occupied by electrons. We derive these invariants using maps from the Brillouin zone to the space of Bloch Hamiltonians and clarify the connections between $\mathbb{Z}_2$ invariants, the integer invariants that underlie the integer quantum Hall effect, and previous invariants of $\mathcal{T}$-invariant Fermi systems.

DOI: 10.1103/PhysRevB.75.121306 PACS number(s): 73.43.−f, 73.20.At, 85.75.−d

In a remarkable pair of papers, Kane and Mele[1,2] proposed a $\mathbb{Z}_2$ topological invariant of time-reversal-invariant insulators in two dimensions, showed that the nontrivial "topological insulator" phase created by spin-orbit coupling has an intrinsic spin Hall effect distinct from earlier proposals,[3,4] and discussed graphene as a candidate system for this effect. Here $\mathbb{Z}_2 \equiv \mathbb{Z}/2\mathbb{Z}$ is the cyclic group of two elements ("even" and "odd"); it was argued that every band insulator in two dimensions belongs to either the even class or the odd class, where the even class describes ordinary insulators and the odd class describes topological insulators.

Both spin-orbit coupling and breaking of inversion symmetry are required to generate a topological insulator. A direct and experimentally relevant characterization of the invariant was given in terms of edge states at the boundary of a two-dimensional (2D) insulator: the topological insulator has an odd number of Kramers pairs of edge modes, while the ordinary insulator has an even number. Two explanations for the invariant as a property of the bulk band structure with spin-orbit coupling were also given.

The goals of this paper are to explain how $\mathbb{Z}_2$ invariants arise in terms of the Thouless–Kohmoto–Nightingale–den Nijs (TKNN) integers[5,6] or "Chern numbers" familiar from the ordinary quantum Hall effect, and then to generalize the results to an arbitrary multiband band structure in two or three spatial dimensions. In the derivation below, we stress analogies to previous work[6] on understanding the integer quantum Hall effect (IQHE) via topological invariants of band structures in a commensurate magnetic field. Time-reversal-invariant ($\mathcal{T}$-invariant) 2D insulators are shown to have multiple $\mathbb{Z}_2$ invariants that are analogous to the band TKNN integers or Chern numbers in IQHE and can be understood using similar methods. This derivation differs from the two original descriptions of a single bulk $\mathbb{Z}_2$ invariant, which do not follow the standard homotopy paradigm of most topological invariants in condensed-matter physics, and whose connection to the IQHE is opaque. Intuitively, our results suggest that the topological insulator can be thought of as an IQHE that exists without a magnetic field or other $\mathcal{T}$ breaking.

Bands in a time-reversal-invariant fermion band structure come in pairs related by time reversal, with twofold Kramers degeneracies at certain special points of the Brillouin zone. We find that there are multiple $\mathbb{Z}_2$ invariants in such systems, associated with band pairs rather than individual bands, and perhaps most interestingly, four invariants per band pair in three dimensions. In two dimensions, the physical significance of these invariants is straightforward: just as two IQHE states with the same *sum* of Chern numbers for occupied bands are adiabatically connected, two $\mathcal{T}$-invariant band insulators are adiabatically connected if and only if they have the same $\mathbb{Z}_2$ sum of individual $\mathbb{Z}_2$ invariants for occupied band pairs.

The approach in this paper can be summarized as follows. The basic objects of homotopy theory are the homotopy groups $\pi_n(M)$ that describe equivalence classes under smooth deformations of mappings from the sphere S^n to a manifold M. A band structure can be thought of as a map from the Brillouin zone (a torus rather than a sphere) to the space of Bloch Hamiltonians. The importance of time-reversal symmetry is as follows: $\mathcal{T}$ symmetry means that only "half of the Brillouin zone" (which we define formally below as an "effective Brillouin zone," or EBZ) needs to be assigned Bloch Hamiltonians, as then the other half of the Brillouin zone is determined by symmetry. Then we study the topological classes of maps from the EBZ to the space of Bloch Hamiltonians.

The EBZ in two dimensions is almost but not quite a sphere: a key object in our analysis is a "contraction," defined below, that extends a mapping from the EBZ to one from the sphere. Once we have such a mapping from the sphere, then the ordinary Chern number is defined. Surprisingly, a given mapping from the EBZ has infinitely many inequivalent contractions in this sense; there are only two connected classes of band structures, for the case of two occupied bands connected by $\mathcal{T}$. These classes correspond to the ordinary insulator and the topological insulator. For one band pair, an ordinary (topological) insulator becomes the equivalence class of all mappings from the sphere with even (odd) Chern numbers. This construction requires only standard homotopy results, makes no assumptions about details of band structure or the existence of additional commuting operators such as spin, and generalizes to multiple bands and higher dimensions.

We begin the derivation by recalling some facts about topological invariants of 2D band structures on the "magnetic Brillouin zone" (an effective unit cell in a lattice system with a magnetic field).[5,6] For 2D $\mathcal{T}$-breaking systems, in a nondegenerate band structure each band is associated with a TKNN integer that is invariant under smooth perturbations of the Bloch Hamiltonians. More precisely, if the Hilbert space of the nondegenerate Bloch Hamiltonians has dimension n, there are $n-1$ independent integer-valued invariants[6] because the invariants sum to zero. The Chern integer for a band is a Brillouin-zone integral involving the projection operator,[6]

$$n_i = \frac{i}{2\pi}\int_{BZ} \mathrm{Tr}(dP_i P_i dP_i), \quad P_i = |\psi_i\rangle\langle\psi_i|. \tag{1}$$

Here $dP_i = dx\partial_x P_i + dy\partial_y P_i$ and $dxdy = -dydx$. A powerful way to prove that these integer invariants are exhaustive[6] is by considering mappings of the torus to Bloch Hamiltonians, assumed nondegenerate. Denoting the space of such nondegenerate Hermitian matrices as $\mathcal{M}$, the TKNN integers follow from the homotopy groups

$$\pi_1(M) = 0, \quad \pi_2(M) = \mathbb{Z}^{n-1}, \tag{2}$$

where the second formula indicates $n-1$ copies of the infinite cyclic group $\mathbb{Z}$. The second result follows from the same exact sequence as used in the theory of topological defects.[7] For a pair of bands i,j that are degenerate with each other but with no other bands, there is a single integer-valued invariant[6] (total Chern number) obtained by replacing P_i in Eq. (1) with $P_{ij} = P_i + P_j$. In a time-reversal-invariant system, bands come in pairs, and the bands within a pair are required to be degenerate at some points by Kramers degeneracy. We define $\mathcal{C}$ as the set of Hermitian matrices allowing possible pair degeneracies (i.e., eigenvalues 1 and 2 may be degenerate, eigenvalues 3 and 4 may be degenerate, and so forth).

Now consider consequences of invariance under the time-reversal operator $\mathcal{T}$. For fermions, $\mathcal{T}^2 = -1$ and $\mathcal{T}$ is represented by an antiunitary operator Θ in the Hilbert space of Bloch Hamiltonians. Time-reversal connects both pairs of points in the Brillouin zone $(k,-k)$ and the associated Bloch Hamiltonians:

$$H(-k) = \Theta H(k)\Theta^{-1}. \tag{3}$$

Time-reversal invariance forces degeneracies at the four points where k goes to $-k$, and as a result the Bloch Hamiltonians in the band structure cannot be assumed nondegenerate (as in the IQHE case); instead, bands come in pairs related by time-reversal, and the only generic degeneracies are those within a pair.[23]

Figure 1 shows the original toroidal Brillouin zone and one possible EBZ (the right half of the original zone). Specifying the Hamiltonians for the points in the EBZ determines them everywhere. Note that *any* 2D Brillouin zone is topologically a torus: this follows from the requirement that multiple copies of the Brillouin zone tile the 2D reciprocal-lattice plane. There are four points of the 2D Brillouin zone that are invariant under $k \to -k$, and we assume below that the EBZ boundaries are chosen to connect pairs of these points. The Bloch Hamiltonians can be specified independently on the EBZ except at the boundaries, where points are conjugate under $\mathcal{T}$ as shown. Clearly points at which $k = -k$, such as **Γ**, **A**, **B**, **C** in Fig. 1, are special: at these points the Bloch Hamiltonian commutes with Θ. We denote the set of such Hamiltonians that commute with Θ, with the additional assumption of no degeneracies other than the twofold degeneracies required by $\mathcal{T}$, as $\mathcal{Q}$ (Ref. 8). The twofold Kramers degeneracies exist because, for any energy eigenstate $|\psi\rangle$, $T|\psi\rangle$ is an orthogonal state with the same energy (since $T^2 = -1$). $\mathcal{T}$ invariance requires an even number of bands $2n$, so $\mathcal{Q}$ consists of $2n \times 2n$ Hermitian matrices for which H commutes with Θ. $\mathcal{Q}$ is the even subspace in the language of Ref. 1.

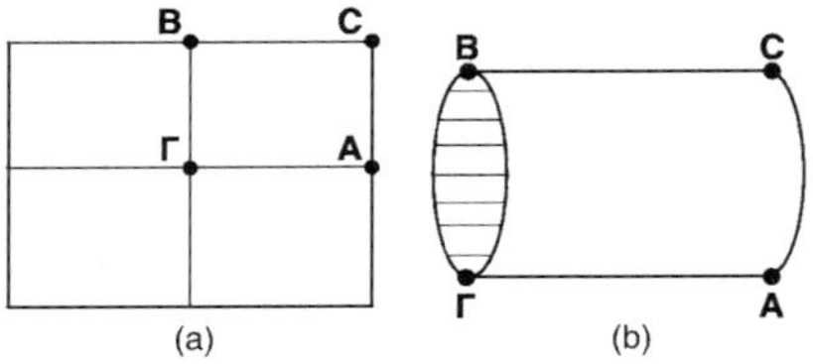

FIG. 1. The topology of the effective Brillouin zone (EBZ): if the original Brillouin zone is the torus in (a), then $\mathcal{T}$ invariance reduces the independent degrees of freedom to live on half the Brillouin zone; one choice of "half the Brillouin zone" is the manifold in (b) (here we have chosen the right half of the original Brillouin zone to form an EBZ). Points on the boundary circles that are connected by horizontal lines are conjugate under $\mathcal{T}$; the points Γ, *A*, *B*, *C* are self-conjugate, and their Bloch Hamiltonians are therefore in the even subspace $\mathcal{Q}$. Note that the EBZ has been chosen so that each boundary passes through two of these special points.

In general, a $\mathcal{T}$-invariant system need not have Bloch Hamiltonians in $\mathcal{Q}$ except at these special points as long as inversion symmetry is broken, which allows $H(k) \neq H(-k)$. There is no obvious topological invariant for a degenerate band with time-reversal symmetry, because the Chern number for the whole Brillouin zone vanishes. It is simplest to see this for the projection operator P_i for a single nondegenerate band: writing $P_i = |\psi_i\rangle\langle\psi_i|$,

$$\begin{aligned} n_i &= \frac{i}{2\pi}\,\mathrm{Im}\int [\langle\partial_x\psi_i|\partial_y\psi_i\rangle + \langle\partial_x\Theta\psi_i|\partial_y\Theta\psi_i\rangle] \\ &= \frac{i}{2\pi}\,\mathrm{Im}\int [\langle\partial_x\psi_i|\partial_y\psi_i\rangle + \langle\partial_y\psi_i|\partial_x\psi_i\rangle] = 0. \end{aligned} \tag{4}$$

The total Chern number similarly vanishes for two degenerate bands connected by $\mathcal{T}$ in a $\mathcal{T}$-invariant system, but there is still a topological invariant,[1] as we explain.

As a quick example of homotopy arguments, suppose that inversion symmetry is unbroken, which implies that the Bloch Hamiltonians are everywhere in $\mathcal{Q}$. Any mapping of

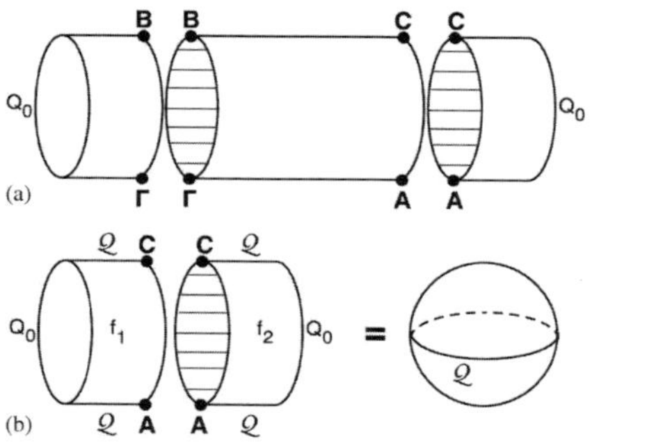

FIG. 2. (a) Contracting the extended Brillouin zone to a sphere. (b) Two contractions can be combined according to Eq. (5) to give a mapping from the sphere, but this sphere has a special property: points in the northern hemisphere are conjugate under $\mathcal{T}$ to those in the southern hemisphere, in such a way that overall every band pair's Chern number must be even.

the torus T^2 to $\mathcal{Q}$ with the condition that points k and $-k$ go to the same point is determined by its behavior on one point from each $(k,-k)$ pair, and the EBZ under this condition is topologically identical to a sphere [stitching $\mathcal{T}$-conjugate points together in Fig. 1(b)]: the classes of such mappings are given by[8] $\pi_2(\mathcal{Q})=0$. Hence there is no homotopy invariant for 2D band structures with both $\mathcal{T}$ and inversion symmetry, although $\pi_4(\mathcal{Q})\neq 0$ and higher-dimensional invariants can exist.

Now consider possible invariants without inversion symmetry. We seek to classify mappings from T^2 to Bloch Hamiltonians that are consistent with $\mathcal{T}$ invariance and have no accidental degeneracies. Such a mapping is determined by a mapping from the EBZ, i.e., a mapping from the cylinder C to $\mathcal{C}$ with certain conditions on the two circular boundaries reflecting time reversal. The image of a boundary point must be the Θ conjugate of the image of the point related by $k \leftrightarrow -k$.

We show that any mapping from C to $\mathcal{C}$, even if the elements at a boundary are not all the same, can be deformed ("contracted") to one in which the boundary elements are identical to an arbitrary reference element $Q_0 \in \mathcal{Q}$ [Fig. 2(a)]; the resulting map from the *sphere* has well-defined Chern numbers. It is required that at each stage of the contraction, the boundary has the same conjugacy under $\mathcal{T}$ as the original EBZ boundary, to guarantee that two maps from the EBZ that can be contracted to maps from the sphere with the same Chern numbers are homotopic (deformable to each other).

Then it is shown that different contractions differ by arbitrary even Chern numbers, so that there are only two invariant classes per band pair according to whether the Chern number with any contraction is an odd or even integer. A contraction is formally defined as a mapping from the cylinder with angular coordinate $\theta \in [0,2\pi)$ and length coordinate $\lambda \in [0,1]$ to $\mathcal{C}$, $f(\theta,\lambda)$, such that $f(\theta,\lambda)$ and $f(2\pi-\theta,\lambda)$ are $\mathcal{T}$ conjugates [which implies that $f(0,\lambda)$ and $f(\pi,\lambda)$ are in $\mathcal{Q}$]. Now $f(\theta,0)$ should agree with the initial specification of Bloch Hamiltonians on the EBZ boundary, while $f(\theta,1)=Q_0$ is constant. If both boundaries are contracted, the resulting sphere has a well-defined Chern number n_B^i for each pair i of bands. The existence of one contraction follows from $\pi_1(\mathcal{C})=0$: contract one side of the boundary circle to the point Q_0, then determine the other side by $\mathcal{T}$ conjugacy.

Many topologically inequivalent contractions exist, and this reduces the integer-valued Chern numbers on the sphere to $\mathbb{Z}_2$ invariants on the EBZ. Let f_1 and f_2 be two different contractions. Then define a mapping $g(\theta,\lambda)$, again with coordinates $\theta \in [0,2\pi)$, $\lambda \in [0,1]$, which combines contractions f_1 and f_2 [Fig. 2(b)]:

$$g(\theta,\lambda) = \begin{cases} f_1(\theta,1-2\lambda) & \text{if } 0 \leq \lambda < 1/2 \\ f_2(\theta,2\lambda-1) & \text{if } 1/2 \leq \lambda \leq 1 \end{cases}. \tag{5}$$

Although the domain of the mapping g is topologically a sphere because the circles at both ends go to the same point, g differs in its $\mathcal{T}$ symmetry from the contracted half of the Brillouin zone, which in its interior has no $\mathcal{T}$ symmetry relating different values of θ: the values of g at the points (θ,λ) and $(2\pi-\theta,\lambda)$ are $\mathcal{T}$ conjugate.

Now we show that g has *even* Chern numbers. This can be verified in two steps: map the equator of the sphere S to the reference element Q_0, which is possible since $\pi_1(\mathcal{Q})=0$ and topologically unique since $\pi_2(\mathcal{Q})=0$, then note that each hemisphere has a well-defined Chern number for each band pair and that the Chern numbers of the two hemispheres are *equal*, rather than *opposite* as in the case of the original Brillouin zone. The reason for this equality can be understood easily in the cylindrical coordinates above, where the equator is at $\theta=\pi$ and $\theta=0$. The identification under $\mathcal{T}$ of θ and $2\pi-\theta$ means that $d\theta$ changes sign between a point and its time-reversal conjugate, but $d\lambda$ does not, giving an additional change of sign in Eq. (4). Another minus sign comes from $\mathcal{T}$ conjugacy, as in Eq. (4), so overall the contributions from the two hemispheres add. Finally, the change Δn_B^i in the band pair Chern numbers n_B^i that results by changing from contraction f_1 to contraction f_2 is given by the Chern numbers n_g^i of the mapping g (note that f_1 appears in g with one of its coordinates reversed, but not the other, so that the sign of its Chern number is flipped). Since the n_g^i are twice those of one hemisphere n_S^i,

$$\Delta n_B^i = n_g^i = 2n_S^i. \tag{6}$$

Finally, in a band structure with n pairs of bands, there is one integer for each pair with a zero sum rule. Hence there is one $\mathbb{Z}_2$ invariant for each band pair, with a $\mathbb{Z}_2$ zero sum rule; there must be an even number of "odd" pairs in a complete band structure.

We now sketch the generalization to three-dimensional Brillouin zones, where there are significant differences between $\mathbb{Z}_2$ invariants and the 3D integer-valued TKNN invariants.[5,6] There are four independent $\mathbb{Z}_2$ invariants per pair of bands, even though there are only three Chern numbers for a pair of degenerate bands. The set of mappings from the Brillouin zone T^3 to $\mathcal{C}$ is determined by three ordinary Chern numbers since $\pi_3(\mathcal{C})=0$: the three integers correspond to the xy, yz, and xz planes.[6]

Now consider possible $\mathbb{Z}_2$ invariants for a time-reversal-invariant system in three dimensions. Suppose that the Brillouin zone is $x,y,z\in[-1,1]$, and construct an EBZ by taking the part of this three-torus with $z\geqslant 0$. Then slices at constant z for $0<z<1$ have the topology of the torus T^2, while at $z=0$ and $z=1$ there are additional $\mathcal{T}$ constraints that reduce the degrees of freedom to the 2D BZ. The $\mathcal{T}$ constraint means that the xy Chern number is zero.

The EBZ boundaries at $z=0$ and $z=1$ are characterized by one $\mathbb{Z}_2$ invariant each: one boundary may be even while the other is odd because the boundary slices have the same Chern number (zero) and thus are homotopic as maps to $\mathcal{C}$ in the EBZ interior. Once the two $\mathbb{Z}_2$ invariants at the boundaries are fixed, two contractions of an original 3D EBZ to the torus T^3 differ by two even Chern numbers, one for the xz slices and one for the yz slices. These are even because a slice of the contraction has the same symmetries that force even Chern numbers in the 2D case. There are four additive $\mathbb{Z}_2$ invariants per band pair in three dimensions and 16 insulating phases.

To understand these four invariants using the original Brillouin zone, note that the six planes $x=0$, $x=\pm 1$, $y=0$, $y=\pm 1$, $z=0$, $z=\pm 1$ have the symmetries of the 2D BZ and hence have a $\mathbb{Z}_2$ invariant. If the six invariants with values ± 1 are x_0, x_1, y_0, y_1, z_0, z_1, then there are two relations $x_0x_1 = y_0y_1 = z_0z_1$. As a 3D example, a model can be designed on the 3D NaCl lattice to reduce to a previously introduced[9] 2D square lattice topological insulator in the $k_y=0$ or $k_z=0$ planes: it has a phase with $y_0=y_1=z_0=z_1=-1$, $x_0=x_1=1$. In two dimensions, both insulating phases can occur in models with a conserved quantity (e.g., S_z) that allows a definition of ordinary Chern integers. In three dimensions, the eight phases with $x_0x_1=y_0y_1=z_0z_1=-1$ cannot be realized in this way.

Since the results here are for Hilbert spaces of arbitrary dimension, they apply to many-body problems with an odd number of fermions[8] if there are two periodic parameters in the Hamiltonian that are connected by time reversal in the same way as the momentum components (k_x,k_y). In two dimensions, just as the Chern number predicts the number of edge states in the IQHE,[10] the class of even (odd) Chern numbers in the bulk corresponds to edges with an even (odd) number of Kramers pairs of modes.

A 2D $\mathbb{Z}_2$ invariant was also obtained by Haldane.[11] The bulk-edge connection has been derived when ordinary Chern integers are defined[12–14] (see also Ref. 15). Recent papers define the 2D $\mathbb{Z}_2$ invariant as an obstruction[14] or as a noninvariant Chern integral plus a formal integral on the EBZ boundary that is defined up to addition of an even integer.[15] Defining this integral is equivalent to our prescription of choosing any contraction to define a Chern integer. The full counting of $\mathbb{Z}_2$ invariants in two or three dimensions has not been obtained previously.

These bulk invariants and the stability of total $\mathbb{Z}_2$ to edge interactions and scattering[16–18] confirm that the topological insulator is a robust phase with a deep connection to the quantum Hall effect.

Recently, consistent results in three dimensions were found by others.[19,20] Other important results include 3D examples with nonzero fourth invariant z_0z_1 (Ref. 21) and an algorithm to obtain the invariants.[22]

The authors acknowledge valuable correspondence with C. L. Kane and financial support from NSF Contract No. DMR-0238760 (J.E.M.), NSF Contract No. DMR04-57440 (L.B.), and the Packard Foundation (L.B.).

---

[1] C. L. Kane and E. J. Mele, Phys. Rev. Lett. **95**, 146802 (2005).
[2] C. L. Kane and E. J. Mele, Phys. Rev. Lett. **95**, 226801 (2005).
[3] S. Murakami, N. Nagaosa, and S.-C. Zhang, Science **301**, 1348 (2003).
[4] J. Sinova, D. Culcer, Q. Niu, N. A. Sinitsyn, T. Jungwirth, and A. H. MacDonald, Phys. Rev. Lett. **92**, 126603 (2004).
[5] D. J. Thouless, M. Kohmoto, M. P. Nightingale, and M. den Nijs, Phys. Rev. Lett. **49**, 405 (1982).
[6] J. E. Avron, R. Seiler, and B. Simon, Phys. Rev. Lett. **51**, 51 (1983).
[7] N. D. Mermin, Rev. Mod. Phys. **51**, 591 (1979).
[8] J. E. Avron, L. Sadun, J. Segert, and B. Simon, Phys. Rev. Lett. **61**, 1329 (1988).
[9] R. Roy, cond-mat/0603271 (unpublished).
[10] Y. Hatsugai, Phys. Rev. Lett. **71**, 3697 (1993).
[11] F. D. M. Haldane (unpublished).
[12] D. N. Sheng, Z. Y. Weng, L. Sheng, and F. D. M. Haldane, Phys. Rev. Lett. **97**, 036808 (2006).
[13] X.-L. Qi, Y.-S. Wu, and S.-C. Zhang, Phys. Rev. B **74**, 045125 (2006).
[14] R. Roy, cond-mat/0604211 (unpublished).
[15] L. Fu and C. L. Kane, Phys. Rev. B **74**, 195312 (2006).
[16] I. A. Gruzberg, N. Read, and S. Vishveshwara, Phys. Rev. B **71**, 245124 (2005).
[17] C. Xu and J. E. Moore, Phys. Rev. B **73**, 045322 (2006).
[18] C. Wu, B. A. Bernevig, and S.-C. Zhang, Phys. Rev. Lett. **96**, 106401 (2006).
[19] R. Roy, cond-mat/0607531 (unpublished).
[20] B. A. Bernevig and H.-D. Chen, cond-mat/0611766 (unpublished).
[21] L. Fu, C. L. Kane, and E. J. Mele, cond-mat/0607699 (unpublished).
[22] T. Fukui and Y. Hatsugai, cond-mat/0611423 (unpublished).
[23] Note that just as touching bands can transfer the Chern number in the IQHE, interpair degeneracies allow $\mathbb{Z}_2$ to be transferred between pairs with total $\mathbb{Z}_2$ conserved.

Topological Insulators in Three Dimensions

Liang Fu, C. L. Kane, and E. J. Mele
Department of Physics and Astronomy, University of Pennsylvania, Philadelphia, Pennsylvania 19104, USA
(Received 26 July 2006; published 7 March 2007)

We study three-dimensional generalizations of the quantum spin Hall (QSH) effect. Unlike two dimensions, where a single Z_2 topological invariant governs the effect, in three dimensions there are 4 invariants distinguishing 16 phases with two general classes: weak (WTI) and strong (STI) topological insulators. The WTI are like layered 2D QSH states, but are destroyed by disorder. The STI are robust and lead to novel "topological metal" surface states. We introduce a tight binding model which realizes the WTI and STI phases, and we discuss its relevance to real materials, including bismuth.

DOI: 10.1103/PhysRevLett.98.106803 PACS numbers: 73.43.−f, 72.25.Hg, 73.20.−r, 85.75.−d

The advent of spintronics has motivated the study of the effects of spin orbit interactions (SOI) on the electronic structure of solids. SOI leads to the spin Hall effect [1,2], which has been observed in GaAs [3,4]. We proposed [5] that in graphene the SOI leads to the quantum spin Hall (QSH) effect. The QSH phase has a bulk excitation gap and gapless spin-filtered edge states. It is distinguished from a band insulator by a Z_2 topological invariant [6], analogous to the TKNN invariant of the integer quantum Hall effect [7]. Because of the weak SOI in carbon, the SOI induced energy gap in graphene is likely to be small [8]. However, Murakami has recently suggested that bismuth bilayers may provide an alternative venue for the QSH effect [9], providing a new direction for the experimental observation of this phase.

In this Letter we consider the generalization of the QSH effect to three dimensions. Recent work has shown that time reversal invariant energy bands in 3D are characterized by four Z_2 invariants, leading to 16 classes of "topological insulators"[10,11]. Here, we will explain the physical meaning of these invariants and characterize the phases they distinguish. One of the invariants distinguishes what we will refer to as weak (WTI) and strong (STI) topological insulators. With disorder, the WTI is equivalent to a band insulator, while the STI remains robust. We show WTIs and STIs have surface states with an even and odd number of Dirac points, respectively. The latter case leads to a new "topological metal" surface phase, which we characterize. We introduce a tight binding model on a distorted diamond lattice, which realizes both the WTI and STI phases, allowing the surface states to be studied explicitly.

In Ref. [12] we established the connection between the Z_2 invariant for the bulk QSH phase and the spin-filtered edge states. We begin by reviewing that argument in a way which makes the generalization to three dimensions transparent. The 2D invariant can be understood using a Laughlin type construction [13] on a cylinder threaded by magnetic flux $\Phi = 0$ or π (in units of $\hbar/e$). The invariant characterizes the change in the time reversal polarization (TRP), which signals the presence of a Kramers degeneracy at the ends, when Φ is changed from 0 to π. If the cylinder consists of a single unit cell in the circumferential (x) direction, then the magnetic flux threading the cylinder plays the role of the crystal momentum k_x in band theory. The spectrum of the discrete end states of the cylinder as a function of flux then reflects the edge state spectrum as a function of momentum. The change in the TRP as a function of flux determines the way the Kramers degenerate end states at the edge time reversal invariant momenta (TRIM) $k_x = \Lambda_1 = 0$ and $k_x = \Lambda_2 = \pi$ are connected to each other. In the QSH phase the Kramers pairs "switch partners" [Fig. 1(a)], reflecting the change in the TRP, while in the conventional insulator [Fig. 1(b)] they do not. It follows that in the QSH phase edge states traverse the bulk energy gap, crossing the Fermi energy E_F an odd number of times between Λ_a and Λ_b. In the insulating phase, the edge states are not topologically protected and cross E_F an even number of times if at all.

The change in the TRP between Λ_1 and Λ_2 is related to the bulk band structure, defined for a 2D system with periodic boundary conditions in both directions. The 2D Brillouin zone has four TRIM, Γ_i, which are related to $-\Gamma_i$ by a reciprocal lattice vector. For an edge perpendicular to $\mathbf{G}$ the 1D edge TRIM $\Lambda_{a=1,2}$ are projections of pairs $\Gamma_{i=a1}$, $\Gamma_{i=a2}$, which satisfy $\Gamma_{a1} - \Gamma_{a2} = \mathbf{G}/2$, onto the line perpendicular to $\mathbf{G}$.

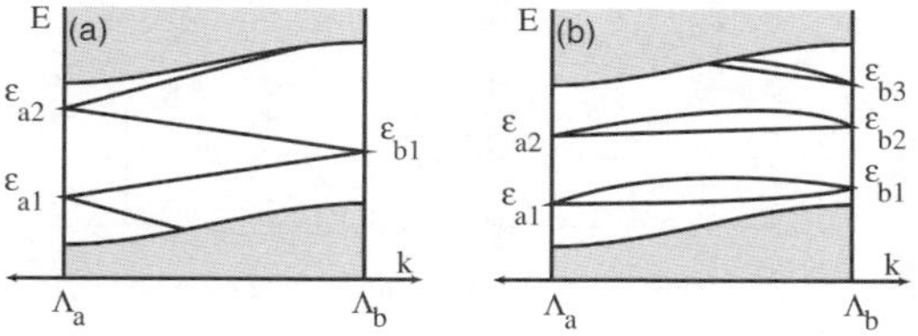

FIG. 1. Schematic surface (or edge) state spectra as a function of momentum along a line connecting Λ_a to Λ_b for (a) $\pi_a\pi_b = -1$ and (b) $\pi_a\pi_b = +1$. The shaded region shows the bulk states. In (a) the TRP changes between Λ_a and Λ_b, while in (b) it does not.

The TRP associated with Λ_a can be expressed as $\pi_a = \delta_{i=a1}\delta_{i=a2}$, where [12]

$$\delta_i = \sqrt{\det[w(\Gamma_i)]}/\text{Pf}[w(\Gamma_i)] = \pm 1. \tag{1}$$

Here the unitary matrix $w_{ij}(\mathbf{k}) = \langle u_i(-\mathbf{k})|\Theta|u_j(\mathbf{k})\rangle$. At $\mathbf{k} = \Gamma_i$, $w_{ij} = -w_{ji}$, so the Pfaffian Pf$[w]$ is defined. π_a is free of the ambiguity of the square root in (1), provided the square root is chosen continuously as a function of $\mathbf{k}$. However, π_a is not gauge invariant. A $\mathbf{k}$ dependent gauge transformation can change the sign of any pair of δ_i's. This reflects the physical fact that the end Kramers degeneracy depends on how the crystal is terminated. It is similar to the ambiguity of the charge polarization [12]. The product, $\pi_1\pi_2 = \delta_1\delta_2\delta_3\delta_4$, is gauge invariant, and characterizes the change in TRP due to changing the flux from $\Lambda_1 = 0$ to $\Lambda_2 = \pi$. This defines the single Z_2 invariant in 2D, and using the above argument, determines the connectivity of the edge state spectrum.

In three dimensions there are 8 distinct TRIM, which are expressed in terms of primitive reciprocal lattice vectors as $\Gamma_{i=(n_1n_2n_3)} = (n_1\mathbf{b}_1 + n_2\mathbf{b}_2 + n_3\mathbf{b}_3)/2$, with $n_j = 0$, 1. They can be visualized as the vertices of a cube as in Fig. 2. A gauge transformation can change the signs of δ_i associated with any four Γ_i that lie in the same plane. Modulo these gauge transformations, there are 16 invariant configurations of δ_i. These can be distinguished by 4 Z_2 indices ν_0; $(\nu_1\nu_2\nu_3)$, which we define as

$$(-1)^{\nu_0} = \prod_{n_j=0,1} \delta_{n_1n_2n_3}, \tag{2}$$

$$(-1)^{\nu_{i=1,2,3}} = \prod_{n_{j\neq i}=0,1;n_i=1} \delta_{n_1n_2n_3}. \tag{3}$$

ν_0 is independent of the choice of $\mathbf{b}_k$. $(\nu_1\nu_2\nu_3)$ are not, but they can be identified with $\mathbf{G}_\nu \equiv \sum_i \nu_i\mathbf{b}_i$, which belongs to the 8 element mod 2 reciprocal lattice, in which vectors that differ by $2\mathbf{G}$ are identified. $(\nu_1\nu_2\nu_3)$ can be interpreted as Miller indices for $\mathbf{G}_\nu$.

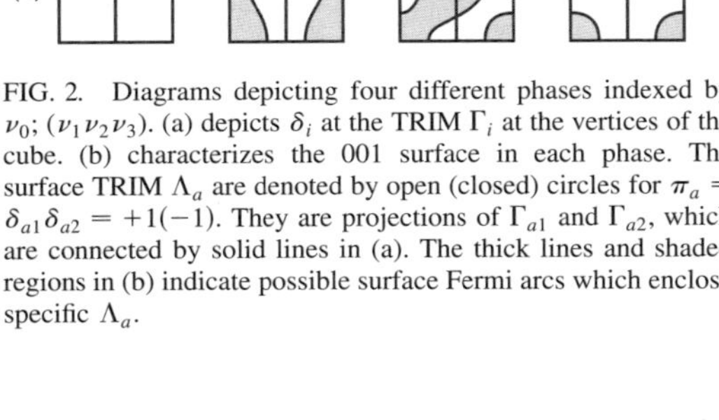

FIG. 2. Diagrams depicting four different phases indexed by ν_0; $(\nu_1\nu_2\nu_3)$. (a) depicts δ_i at the TRIM Γ_i at the vertices of the cube. (b) characterizes the 001 surface in each phase. The surface TRIM Λ_a are denoted by open (closed) circles for $\pi_a = \delta_{a1}\delta_{a2} = +1(-1)$. They are projections of Γ_{a1} and Γ_{a2}, which are connected by solid lines in (a). The thick lines and shaded regions in (b) indicate possible surface Fermi arcs which enclose specific Λ_a.

$\nu_{0\text{–}4}$ are equivalent to the four invariants introduced by Moore and Balents [10] using general homotopy arguments. The power of the present approach is that it allows us to characterize the surface states on an arbitrary crystal face. Generalizing the Laughlin argument to three dimensions, consider a system with open ends in one direction and periodic boundary conditions in the other two directions. This can be visualized as a torus with a finite thickness (a "Corbino donut"), which has an inside and an outside surface. Viewed as a 1D system, we then seek to classify the changes in the Kramers degeneracy associated with the surfaces as a function of two fluxes threading the torus (or equivalently as a function of the two components of the surface crystal momentum).

For a surface perpendicular to $\mathbf{G}$, the surface Brillouin zone has four TRIM Λ_a which are the projections of pairs Γ_{a1}, Γ_{a2}, that differ by $\mathbf{G}/2$, into the plane perpendicular to $\mathbf{G}$. Because of Kramers' degeneracy, the surface spectrum has two dimensional Dirac points at Λ_a. The relative values of $\pi_a = \delta_{a1}\delta_{a2}$ determine how these Dirac points are connected to one another, as illustrated in Fig. 1. For any path connecting Λ_a to Λ_b, the surface band structure will resemble Fig. 1(a) and 1(b) for $\pi_a\pi_b = -1(+1)$, and the surface bands will intersect E_F an odd (even) number of times. It follows that the surface Fermi arc divides the surface Brillouin zone into two regions. The Dirac points at the TRIM Λ_a with $\pi_a = +1$ are on one side, while those with $\pi_a = -1$ are on the other side.

In Fig. 2 we depict δ_i for four different topological classes, along with the predictions for the edge state spectrum for a 001 face. The surface Fermi arc encloses either 0(4), 1(3), or 2 Dirac points. When the number of Dirac points is not 0(4), there must be surface states which connect the bulk conduction and valence bands.

There are two classes of phases depending on the parity of ν_0. For $\nu_0 = 0$ each face has either 0(4) or 2 enclosed Dirac points. For a face $\mathbf{G} = \sum_i m_i\mathbf{b}_i$ there are 0(4) Dirac points for $m_i = \nu_i \bmod 2 (i = 1, 2, 3)$ and 2 Dirac points otherwise. These phases can be interpreted as layers of 2D QSH states stacked in the $\mathbf{G}_\nu$ direction. They resemble 3D quantum Hall phases [14], which are indexed by a triad of Chern integers that define a reciprocal lattice vector $\mathbf{G}$ perpendicular to the layers and give the conductivity $\sigma_{ij} = (e^2/h)\varepsilon_{ijk}G_k/(2\pi)$. In the present case, $\mathbf{G}_\nu$ is defined modulo $2\mathbf{G}$, so that layered QSH phases stacked along $\mathbf{G}_\nu$ and $\mathbf{G}_\nu + 2\mathbf{G}$ are equivalent.

The presence or absence of surface states in the $\nu_0 = 0$ phases is delicate. For the 0;(001) phase in Fig. 2, the 100 face has two Dirac points, while the 801 face has 0(4). This sensitivity is a symptom of the fact that the topological distinction of these phases relies on the translational symmetry of the lattice. Indeed, if the unit cell is doubled, the two Dirac points fold back on one another. A weak periodic potential then opens a gap. It is thus likely that disorder will eliminate the topological distinction between these phases and simple insulators. Surface states will generically be localized. For this reason, we refer to the $\nu_0 = 0$ phases as "weak" topological insulators. Nonetheless, the weak invariants have important implications for clean surfaces.

The $\nu_0 = 1$ phases are more robust, and we refer to them as "strong" topological insulators. In this case the surface Fermi arc encloses 1(3) Dirac points on all faces. If the Fermi energy is exactly at the Dirac point this provides a time reversal invariant realization of the 2 + 1-dimensional parity anomaly [15–18] without fermion doubling. This can occur because the Dirac point partners reside on opposite surfaces. For a generic Fermi energy inside the bulk gap the surface Fermi arc will enclose a single Dirac point. This defines a two-dimensional "topological metal" that is topologically protected because a quantized Berry's phase of π is acquired by an electron circling the Fermi arc. This Berry's phase implies that with disorder the surface is in the symplectic universality class [19,20], which is not localized by weak disorder.

To develop an explicit model of these phases we consider a 4 band tight binding model of s states on a diamond lattice with SOI, that generalizes the 2D honeycomb lattice model [5,16,18].

$$H = t\sum_{\langle ij\rangle} c_i^\dagger c_j + i(8\lambda_{SO}/a^2)\sum_{\langle\langle ij\rangle\rangle} c_i^\dagger \mathbf{s}\cdot(\mathbf{d}_{ij}^1 \times \mathbf{d}_{ij}^2)c_j. \quad (4)$$

The first term is a nearest neighbor hopping term. The second term connects second neighbors with a spin dependent amplitude. $\mathbf{d}_{ij}^{1,2}$ are the two nearest neighbor bond vectors traversed between sites i and j. a is the cubic cell size. The energy bands are shown in Fig. 3(a). Because of inversion symmetry, each band is doubly degenerate. The

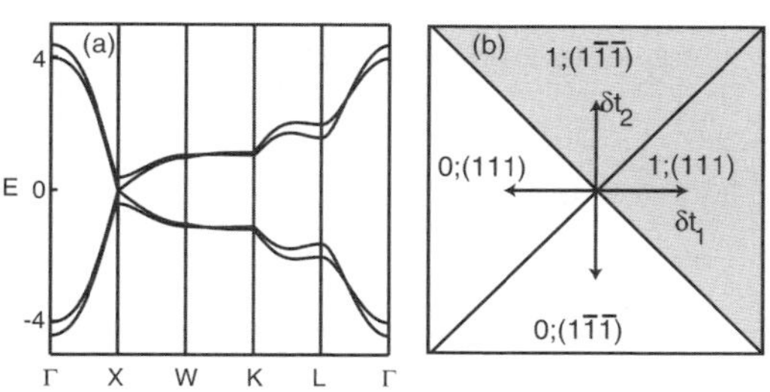

FIG. 3. Energy bands for (a) the model (4) with $t = 1$, $\lambda_{SO} = 0.125$. The symmetry points are $\Gamma = (0, 0, 0)$, $X = (1, 0, 0)$, $W = (1, 1/2, 0)$, $K = (3/4, 3/4, 0)$, and $L = (1/2, 1/2, 1/2)$ in units of $2\pi/a$. The dashed line shows the energy gap due to $\delta t_1 = 0.4$. (b) shows the phase diagram as a function of δt_1 and δt_2 (for bonds in the 111 and $1\bar{1}\bar{1}$ directions) with phases indexed according to cubic Miller indices for $\mathbf{G}_\nu$. The shaded region is the STI phase.

conduction and valence bands meet at 3D Dirac points at the three inequivalent X points $X^r = 2\pi\hat{r}/a$, where $r = x$, y, z. This degeneracy is lifted by symmetry lowering modulations of the four nearest neighbor bonds $t \to t + \delta t_p$, with $p = 1, \ldots, 4$.

Near X^z the low energy effective mass model has the form of a 3 + 1-dimensional Dirac equation,

$$\mathcal{H}_{\rm eff}^z = ta\sigma^y q_z + 4\lambda_{SO}a\sigma^z(s^x q_x - s^y q_y) + m^z\sigma^x. \quad (6)$$

Here $\mathbf{q} = \mathbf{k} - X^z$ and $m^z = \sum_p \delta t_p {\rm sgn}[\mathbf{d}_p \cdot \hat{z}]$. $\mathbf{d}_p$ is the bond vector associated with the pth nearest neighbor bond. The Pauli matrices σ^i are associated with the sublattice degree of freedom, while s^i describe the spin. $\mathcal{H}_{\rm eff}^{x,y}$ are the same with x, y, and z permuted in q_i and s^i, but not σ^i. Transitions between different phases occur when the masses at any of the X^r vanish. $\delta t_p = 0$ is thus a multicritical point separating 8 different phases.

Determining ν_i for these phases using (1–3) requires eigenvectors that are defined continuously throughout the Brillouin zone [12]. Since the Chern integers vanish, this is always possible. Determining the appropriate phases numerically, however, is nontrivial. An alternative numerical approach would be to characterize the Pfaffian function introduced in Ref. [6]. Generalizing the results of Ref. [12] it can be shown that the product of 4 δ_i in any plane is related to the zeros of the Pfaffian in that plane, which can be identified without choosing phases. For the present problem, however, we are fortunate because δ_i can be determined analytically. We find

$$\delta_i = {\rm sgn}\left[\sum_p (t + \delta t_p)\cos\Gamma_i \cdot (\mathbf{d}_p - \mathbf{d}_1)\right]. \quad (7)$$

For small δt_p, $\delta = 1$ at $\mathbf{k} = 0$ and at 3 of the L points. $\delta = -1$ at the 4th L point. At X^r, $\delta = {\rm sgn}[m^r]$. When one of the four bonds is weaker than the others ($\delta t_p < 0$, $\delta t_{(p'\neq p)} = 0$ for instance) the system is in a WTI phase, which may be interpreted as a QSH state layered in the $\mathbf{d}_p$ direction. There are 4 such states, depending on p, of which two are shown in Fig. 3(b). We labeled the phases with the conventional cubic Miller indices for $\mathbf{G}_\nu$. 111 and $1\bar{1}\bar{1}$ are distinct elements of the fcc mod 2 reciprocal lattice. When one of the bonds is stronger than the others the system is in one of four STI phases. The band insulator 0;(000) is not perturbatively accessible from this critical point. However, in the tight binding model it occurs when one bond is turned up so that $\delta t_1 > 2t$. A staggered sublattice potential also leads to a band insulator, but the strength must exceed a finite value (set by λ_{SO}) before that transition occurs.

To study the surface states, we solve (4) in a slab geometry. Figure 4 shows the 2D band structures of the four phases in Fig. 3 for a slab parallel to the 111 surface along lines that visit each of the four TRIM. The plots with the same ν_0 can be viewed as different faces of the same

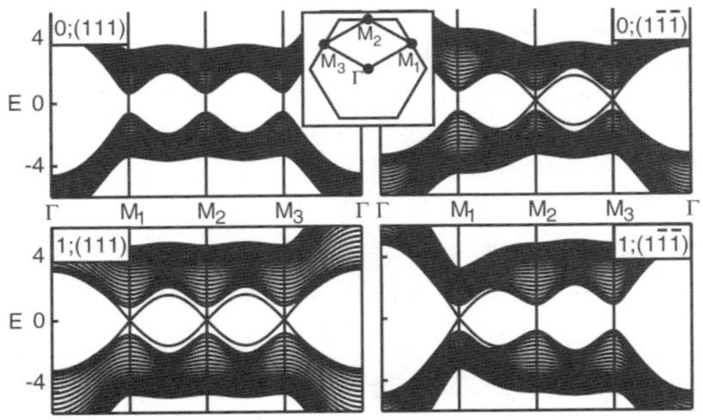

FIG. 4. 2D band structures for a slab with a 111 face for the four phases in Fig. 3. The states crossing the bulk energy gap are localized at the surface. In the WTI (STI) phases there are an even (odd) number of Dirac points in the surface spectrum. The inset shows the surface Brillouin zone.

state. The bulk states above the bandgap are clearly seen. In addition, there are surface states which traverse the gap. In the WTI phases 0;(111) and 0; $(1\bar{1}\,\bar{1})$ there are 0 and 2 2D Dirac points, on both the top and bottom surfaces, as expected from the general arguments given above. In the STI phases 1;(111) and 1; $(\bar{1}\,\bar{1}\,1)$ there is 1(3) Dirac point on each surface. In each case, the nondegenerate surface states near the Dirac points are spin filtered, such that $\langle\vec{s}(-\mathbf{k})\rangle = -\langle\vec{s}(\mathbf{k})\rangle$. Spin density and charge current are thus coupled.

Though the 4 band diamond lattice model is simple, it is probably not directly relevant to any specific material. However, it may give insight into the behavior of real crystals. Consider a sequence of crystal structures obtained from diamond by continuously displacing the fcc sublattices in the (111) direction:

$$\text{diamond} \longrightarrow \text{graphite}(ABC) \longrightarrow \text{cubic}.$$

Starting with diamond, the 111 nearest neighbor bond is stretched, leading to the 0;(111) WTI phase. As the sublattice is displaced further both sublattices eventually reside in the same plane with a structure similar to *ABC* stacked graphite. Displacing further, the lattice eventually becomes cubic. At this point, the gap closes, and the system is metallic. The *s* state model remains in the WTI phase up to the cubic point.

Bismuth has the rhombohedral A7 structure, which can be viewed as a cubic lattice distorted "toward diamond", along with a trigonal distortion of the fcc Bravais lattice. Murakami showed that a bilayer of bismuth, whose structure is similar to a single plane of graphene, is in the QSH phase. This suggests that for weak coupling between bilayers bismuth is in the 0;(111) WTI phase. While this agrees with the simple model presented above, a realistic description of bismuth requires a theory which incorporates bismuth's five valence bands [21].

It will be interesting to search for materials in the STI phase, which occur on the "other side of diamond" in our sequence. We hope that the exotic surface properties predicted for this phase will stimulate further experimental and theoretical efforts.

It is a pleasure to thank Joel Moore and Leon Balents for helpful discussions. This work was supported by NSF Grants No. DMR-0079909 and No. DMR-0605066 and DOE Grant No. DE-FG02-ER-0145118.

Note added.—In subsequent work we have predicted that a number of specific materials are STI's [22]. These include the semiconducting alloy $Bi_{1-x}Sb_x$ as well as α-Sn and HgTe under uniaxial strain.

[1] S. Murakami, N. Nagaosa, and S. C. Zhang, Science **301**, 1348 (2003).
[2] J. Sinova *et al.*, Phys. Rev. Lett. **92**, 126603 (2004).
[3] Y. K. Kato *et al.*, Science **306**, 1910 (2004).
[4] J. Wunderlich, B. Kaestner, J. Sinova, and T. Jungwirth, Phys. Rev. Lett. **94**, 047204 (2005).
[5] C. L. Kane and E. J. Mele Phys. Rev. Lett. **95**, 226801 (2005).
[6] C. L. Kane and E. J. Mele, Phys. Rev. Lett. **95**, 146802 (2005).
[7] D. J. Thouless, M. Kohmoto, M. P. Nightingale, and M. den Nijs, Phys. Rev. Lett. **49**, 405 (1982).
[8] Y. Yao *et al.*, Phys. Rev. B **75**, 041401(R) (2007); H. Min *et al.*, Phys. Rev. B **74**, 165310 (2006).
[9] S. Murakami, Phys. Rev. Lett. **97**, 236805 (2006).
[10] J. E. Moore and L. Balents, cond-mat/0607314 [Phys. Rev. B. (to be published)].
[11] R. Roy, cond-mat/0607531.
[12] L. Fu and C. L. Kane, Phys. Rev. B **74**, 195312 (2006).
[13] R. B. Laughlin, Phys. Rev. B **23**, R5632 (1981).
[14] M. Kohmoto, B. I. Halperin, and Y.-S. Wu, Phys. Rev. B **45**, 13 488 (1992).
[15] R. Jackiw, Phys. Rev. D **29**, 2375 (1984).
[16] G. W. Semenoff, Phys. Rev. Lett. **53**, 2449 (1984).
[17] E. Fradkin, E. Dagotto, and D. Boyanovsky, Phys. Rev. Lett. **57**, 2967 (1986).
[18] F. D. M. Haldane, Phys. Rev. Lett. **61**, 2015 (1988).
[19] H. Suzuura and T. Ando, Phys. Rev. Lett. **89**, 266603 (2002).
[20] S. Hikami, A. I. Larkin, and Y. Nagaoka, Prog. Theor. Phys. **63**, 707 (1980).
[21] Y. Liu and R. E. Allen, Phys. Rev. B **52**, 1566 (1995).
[22] L. Fu and C. L. Kane, cond-mat/0611341.

6

Artificial gauge fields in real space

Origin of the Geometric Forces Accompanying Berry's Geometric Potentials

Yakir Aharonov and Ady Stern
Raymond and Beverly Sackler Faculty of Exact Sciences, School of Physics and Astronomy, Tel Aviv University, Tel Aviv, Israel 69978
(Received 12 May 1992)

We consider the dynamics of a particle carrying a magnetic moment in a strong magnetic field whose direction varies slowly in space. In particular, we discuss the geometric Lorentz-type and electric-type forces that were discovered in studies of Berry's phase. We show that the Lorentz-type force felt by the particle is caused by a small misalignment of the magnetic moment with respect to the magnetic field while the electric-type force is the time average of a strong oscillatory force induced by the precession of the magnetic moment.

PACS numbers: 03.65.−w, 41.20.Gz

The investigation of the geometric phase [1], initiated by Berry in 1984 [2], has led to a series of studies of the appearance of gauge fields in adiabatic dynamics. The canonical example discussed in these studies is the problem of a neutral particle carrying a magnetic moment, moving in a magnetic field whose direction varies in space. These studies, to be reviewed below, have revealed surprising results regarding the dynamics of the particle [3]. In this work we discuss this problem through its equations of motion, with the goal of exploring the physical origins of these results.

The problem we analyze is that of a magnetic moment in a static, space-dependent, magnetic field. The Hamiltonian describing the system is

$$H=\mathbf{P}^2/2m-\gamma|\mathbf{B}(\mathbf{x})|\mathbf{l}\cdot\hat{\mathbf{b}}(\mathbf{x}), \tag{1}$$

where $\mathbf{l}$ is the internal angular momentum carried by the magnetic moment, $\boldsymbol{\mu}\equiv\gamma\mathbf{l}$ is the magnetic moment of the particle, $|\mathbf{B}(\mathbf{x})|$ is the magnitude of the magnetic field, and $\hat{\mathbf{b}}(\mathbf{x})$ is a unit vector in the direction of the magnetic field. For brevity, from now on we absorb the factor γ into the magnitude of the field $|\mathbf{B}(\mathbf{x})|$. We also omit the $\mathbf{x}$ dependence of the magnetic field wherever this does not lead to confusion. Particular examples of the Hamiltonian (1) were considered in various contexts by several authors [3–10], with the following similar results. If the magnetic field varies slowly enough in space, the adiabatic approximation can be employed. Employing that approximation, one finds that the dynamics of the internal angular momentum $\mathbf{l}$ (the fast degree of freedom of the problem) is characterized by the action $l_\parallel\equiv\mathbf{l}\cdot\hat{\mathbf{b}}$. This action, which is the component of $\mathbf{l}$ parallel to the magnetic field, is an adiabatic invariant. The effective Hamiltonian governing the dynamics of the particle (the slow degree of freedom of the problem) is found to include geometric vector and scalar potentials, induced by the spatial variation of $\hat{\mathbf{b}}$ [3]:

$$H_{\rm eff}=\frac{[\mathbf{P}-\mathbf{A}(\hat{\mathbf{b}}(\mathbf{x}),l_\parallel)]^2}{2m}+\Phi(\hat{\mathbf{b}}(\mathbf{x}),l^2-l_\parallel^2) - l_\parallel|\mathbf{B}(\mathbf{x})|, \tag{2}$$

where $l\equiv|\mathbf{l}|$. The vector and scalar potentials depend on the direction of the magnetic field, but do not depend on the magnitude of the field $|\mathbf{B}|$ [3]. The dynamics of the particle is affected by the vector potential $\mathbf{A}(\hat{\mathbf{b}},l_\parallel)$ in a way similar to the effect of an electromagnetic vector potential on a charged particle. If the problem is solved quantum mechanically, one finds effects analogous to the Aharonov-Bohm effect, namely, geometric phases induced by the vector potential in interference phenomena. If the problem is treated classically, the vector potential affects the dynamics only through its curl, i.e., it induces a Lorentz-type force. The term "Lorentz type" refers to a force whose direction is perpendicular to that of the particle's velocity, and whose magnitude is proportional to that velocity. The scalar potential $\Phi(\hat{\mathbf{b}},l^2-l_\parallel^2)$ affects the classical dynamics of the particle through its gradient, that induces an "electric-type" force. The scalar potential and the electric-type force are proportional to $l^2 - l_\parallel^2$ and are *inversely proportional to the mass of the particle.* We conclude this review of previous works by emphasizing that the induced Lorentz-type and electric-type forces are independent of the magnitude of the magnetic field $|\mathbf{B}|$ once that magnitude is strong enough to justify an adiabatic approximation.

The Lorentz-type force acting on a neutral particle due to the interaction of its magnetic moment with a space-dependent magnetic field is a surprising consequence of the spatial variation of $\hat{\mathbf{b}}$. It is surprising because the equations of motion derived from the full Hamiltonian (1) [see Eqs. (3) and (4) below] do not seem to include any velocity-dependent forces, and do seem to indicate that any force exerted on the particle should scale with the magnitude of the magnetic field $|\mathbf{B}|$. In this work we study these equations of motion. We first show that an approximation in which the magnetic moment is assumed to be aligned with the magnetic field at any point along the particle's path necessarily misses the two geometric forces. To reveal the origin of the Lorentz-type force, we then study the equation of motion for $\mathbf{l}$, expressed in a rotating frame of reference in which the magnetic field always aligns with the z axis. In that frame of reference the magnetic moment is subject to the influence of an effective magnetic field $\mathbf{B}'=\mathbf{B}+\dot{\hat{\mathbf{b}}}\times\hat{\mathbf{b}}$, where the time dependence of $\hat{\mathbf{b}}$ originates from the motion of the particle in a region of space where $\hat{\mathbf{b}}$ varies. The Lorentz-type force then results from approximating the magnetic moment as following the direction of $\mathbf{B}'$, rather than that of $\mathbf{B}$; i.e., it results from a slight misalignment of the magnetic moment relative to the direction of the magnetic field. This misalignment is of the order of $\dot{\mathbf{x}}/|\mathbf{B}|$, where $\dot{\mathbf{x}}$ is the velocity of the particle. We show that the force acting on the particle is proportional to the product of the magnitude of the magnetic field $|\mathbf{B}|$, with the misalignment of the magnetic moment, and that this product does indeed yield a $|\mathbf{B}|$-independent, velocity-dependent, Lorentz-type force. After that discussion, we turn to analyze the electric-type force, and show that this force is a time average of a strong oscillatory force, induced by the precession of the magnetic moment. We also explore the origin of the unique mass dependence of that force.

Our analysis of the problem starts from the equations of motion governing the dynamics of the particle and its internal angular momentum, as they are derived from the full Hamiltonian (1):

$$F_i = m\frac{d^2x_i}{dt^2} = l_\parallel \frac{\partial|\mathbf{B}(\mathbf{x})|}{\partial x_i} + |\mathbf{B}(\mathbf{x})|\,\mathbf{l}\cdot\frac{\partial\hat{\mathbf{b}}(\mathbf{x})}{\partial x_i}, \qquad (3)$$

$$\frac{d\mathbf{l}}{dt} = \mathbf{l}\times\mathbf{B}(\mathbf{x}(t)), \qquad (4)$$

where F_i is the ith component of the force exerted on the particle. The force F_i is composed of two components. The first component originates in the variation of the magnitude of the magnetic field $|\mathbf{B}|$. It is proportional to $l_\parallel$, and to the derivative of the magnitude of the magnetic field. The second component originates from spatial variations of the direction of the field, and is therefore the relevant term for the understanding of the Lorentz-type and electric-type forces. However, since $\partial\hat{\mathbf{b}}(\mathbf{x})/\partial x_i$ is perpendicular to $\hat{\mathbf{b}}(\mathbf{x})$, this second term is proportional to the part of $\mathbf{l}$ perpendicular to the magnetic field. *Therefore, the roots of the Lorentz-type and electric-type forces lie in the projection of the magnetic moment* $\mathbf{l}$ *on the plane perpendicular to the magnetic field.*

Following these observations, we now turn to an approximation scheme for the solutions to Eqs. (3) and (4). We start with Eq. (4), trying to obtain a solution for the angular momentum $\mathbf{l}(t)$ in terms of the path of the particle. Then, we substitute that solution in the equation for the position (3). As mentioned above, in the adiabatic limit, i.e., when $|\mathbf{B}|\to\infty$, the angle between $\mathbf{l}$ and the magnetic field $\mathbf{B}(\mathbf{x}(t))$ is an adiabatic invariant. We first focus on the case in which this angle is 0, namely, the case in which the magnetic moment approaches a direction parallel to the magnetic field. As shown below, for any finite value of the magnetic field, there is a component of the magnetic moment $\mathbf{l}$ perpendicular to the magnetic field $\mathbf{B}$, and this component is the origin of the Lorentz-type force.

We turn to Eq. (4), assume that the time-dependent magnetic field $\mathbf{B}(\mathbf{x}(t))$ is known, and try to solve for $\mathbf{l}$. The main difficulty in solving Eq. (4) is, of course, the time dependence of $\hat{\mathbf{b}}$. Attempting to "freeze" this time dependence, we express the problem in terms of a time-dependent coordinate frame, in which $\hat{\mathbf{b}}$ aligns with the z axis. We denote by $\mathbf{l}'$ the vector $\mathbf{l}$ described in terms of this time-dependent reference frame, and by $\boldsymbol{\omega}(t)$ the generator of the infinitesimal rotation that makes the z axis align with $\hat{\mathbf{b}}$. In other words, $\mathbf{r}'=\mathbf{r}+\mathbf{r}\times\boldsymbol{\omega}(t)dt$ is the infinitesimal rotation transforming a vector $\mathbf{r}$ in the coordinate frame where the z axis aligns with $\hat{\mathbf{b}}(t)$ to the vector $\mathbf{r}'$ in a coordinate frame where the z axis aligns with $\hat{\mathbf{b}}(t+dt)=\hat{\mathbf{b}}(t)+\dot{\hat{\mathbf{b}}}dt$. The alignment of $\hat{\mathbf{b}}$ with the z axis does not uniquely define the time-dependent frame, since it does not specify the directions of the time dependent x,y axes. Out of the infinitely many possible choices for these directions we choose the one that yields the smallest $\boldsymbol{\omega}(t)$, i.e., the one in which $\boldsymbol{\omega}(t)$ is perpendicular both to $\hat{\mathbf{b}}(t)$ and to $\hat{\mathbf{b}}(t+dt)$. Consequently,

$$\boldsymbol{\omega}(t) = \dot{\hat{\mathbf{z}}}\times\hat{\mathbf{z}}, \qquad (5)$$

where $\hat{\mathbf{z}}$ is the unit vector $\hat{\mathbf{b}}$ as seen in the time-dependent frame, and $\dot{\hat{\mathbf{z}}}$ is the vector $\dot{\hat{\mathbf{b}}}$ as seen in the time-dependent frame. This choice of $\boldsymbol{\omega}(t)$ makes the coordinate frame follow a parallel transport trajectory [3,11]. The equation of motion for the vector $\mathbf{l}'$ is given by [12,13]

$$\frac{d\mathbf{l}'}{dt} = \mathbf{l}'\times[|\mathbf{B}(\mathbf{x}(t))|\hat{\mathbf{z}} + \dot{\hat{\mathbf{z}}}\times\hat{\mathbf{z}}]. \qquad (6)$$

The equation of motion for $\mathbf{l}'$ is similar to that of $\mathbf{l}$ [Eq.

(4)], but with the magnetic field **B** replaced by $\mathbf{B}' = |\mathbf{B}(\mathbf{x}(t))|\hat{\mathbf{z}}+\dot{\hat{\mathbf{z}}}\times\hat{\mathbf{z}}$. Hence, the effective magnetic field felt by the vector $\mathbf{l}'$ is composed of two components: One is the adiabatic component, which is the original magnetic field, transformed to align with the z axis. The other is the nonadiabatic component, which arises from the time dependence of the reference frame, i.e., from the time dependence of $\hat{\mathbf{b}}$. The nonadiabatic component of $\mathbf{B}'$ is perpendicular to the adiabatic one. Approximating the solution to Eq. (6) by $\mathbf{l}' = l\hat{\mathbf{z}}$ amounts then to neglecting a component of magnitude $\dot{\hat{\mathbf{b}}}$ relative to a component of magnitude $|\mathbf{B}|$. When attempting to improve this approximation, it is important to note that although the direction of $\mathbf{B}'$ is generally time dependent, its time dependence is, in the limit $|\mathbf{B}|\to\infty$, slower than that of $\hat{\mathbf{b}}$. While the latter is independent of the magnitude of the magnetic field $|\mathbf{B}|$, the former is at least of the order of $|\mathbf{B}|^{-1}$. Thus, a refined approximation to the solution of Eq. (6) is

$$\mathbf{l}' = (l/|\mathbf{B}'|)(|\mathbf{B}|\hat{\mathbf{z}}+\dot{\hat{\mathbf{z}}}\times\hat{\mathbf{z}})\,, \tag{7}$$

i.e., the magnetic moment is approximated to follow the direction of $\mathbf{B}'$, rather than that of $\mathbf{B}$. As can be seen by direct substitution of Eq. (7) in Eq. (6), our refined approximation is indeed closer to the exact solution. The expression for the vector $\mathbf{l}$ in the "laboratory," static, frame is obtained from Eq. (7) by the replacements $\hat{\mathbf{z}}\to\hat{\mathbf{b}}$ and $\dot{\hat{\mathbf{z}}}\to\dot{\hat{\mathbf{b}}}$. Note that $\hat{\mathbf{b}}$ depends on time only through the time dependence of the particle's path, $\mathbf{x}(t)$. Thus, $\dot{\hat{\mathbf{b}}} = \dot{x}_j(t)\,\partial\hat{\mathbf{b}}/\partial x_j$ (a summation over repeated indices is understood). *Therefore, the magnetic moment has a component perpendicular both to the direction of the magnetic field* $\hat{\mathbf{b}}$, *and to the time derivative of this direction* $\dot{\hat{\mathbf{b}}}$. *The magnitude of this component is proportional to the velocity of the particle, and inversely proportional to the strength of the magnetic field.*

Although this conclusion appears, at first glance, to contradict the adiabatic invariance of the action, it actually does not. The adiabatic invariance states that $d(\mathbf{l}\cdot\hat{\mathbf{b}})/dt = 0$. However, using Eq. (4) we see that

$$\frac{d}{dt}(\mathbf{l}\cdot\hat{\mathbf{b}}) = \mathbf{l}\cdot\dot{\hat{\mathbf{b}}}\,; \tag{8}$$

i.e., adiabatic invariance requires that the magnetic moment $\mathbf{l}$ should not have a component along $\dot{\hat{\mathbf{b}}}$, but does not force its component along the direction $\dot{\hat{\mathbf{b}}}\times\hat{\mathbf{b}}$ to vanish. It is indeed this component that is responsible for the Lorentz-type force.

To calculate F_i, the ith component of the force acting on the particle [Eq. (3)], we have to take the scalar product of the magnetic moment $\mathbf{l}$ with the vector $\partial B/\partial x_i$. This scalar product can be taken in either the static or the time-dependent frame. We choose the static frame, in which

$$\frac{\partial B}{\partial x_i} = \frac{\partial|\mathbf{B}|}{\partial x_i}\hat{\mathbf{b}} + |\mathbf{B}|\frac{\partial\hat{\mathbf{b}}}{\partial x_i}\,; \tag{9}$$

i.e., it is composed of a component in the direction of the magnetic field, originating from variations in the magnitude of the magnetic field, and a component perpendicular to the field, originating from variations in the direction of the magnetic field. While the first component is proportional to the derivative of $|\mathbf{B}|$, the second is proportional to $|\mathbf{B}|$ itself. It is the product of that second component with the components of $\mathbf{l}$ perpendicular to the magnetic field which yields a force independent of $|\mathbf{B}|$. Calculating the force we get

$$F_i = \frac{l|\mathbf{B}|}{|\mathbf{B}'|}\left[\frac{\partial|\mathbf{B}|}{\partial x_i} + \frac{\partial\hat{\mathbf{b}}}{\partial x_i}\cdot(\dot{\hat{\mathbf{b}}}\times\hat{\mathbf{b}})\right]. \tag{10}$$

In the adiabatic limit, when $|\mathbf{B}|\to\infty$, the magnitude of $\mathbf{B}'$ approaches that of $\mathbf{B}$. Taking that limit, and expressing $\dot{\hat{\mathbf{b}}}$ in terms of the particle's velocity, we finally arrive at

$$F_i = l\frac{\partial|\mathbf{B}|}{\partial x_i} + l\dot{x}_j\left[\frac{\partial\hat{\mathbf{b}}}{\partial x_i}\times\frac{\partial\hat{\mathbf{b}}}{\partial x_j}\right]\cdot\hat{\mathbf{b}}\,. \tag{11}$$

The first term of Eq. (11) is the force utilized in Stern-Gerlach-type experiments. The second term is the Lorentz-type force. It is indeed proportional and perpendicular to the velocity, and independent of $|\mathbf{B}|$. It is instructive, at this point, to express the unit vector $\hat{\mathbf{b}}$ as $\hat{\mathbf{b}} = (\sin\alpha\cos\beta, \sin\alpha\sin\beta, \cos\beta)$ where α,β are functions of $\mathbf{x}$. In terms of α and β, Eqs. (10),(11) are

$$F_i = \frac{l|\mathbf{B}|}{|\mathbf{B}'|}\left[\frac{\partial|\mathbf{B}|}{\partial x_i} + (\dot{\alpha}\nabla\beta - \dot{\beta}\nabla\alpha)\sin\alpha\right]$$

$$= l\frac{\partial|\mathbf{B}|}{\partial x_i} + l\dot{\mathbf{x}}\times(\nabla\alpha\times\nabla\beta)\sin\alpha\,. \tag{12}$$

Note that the refinement of the adiabatic approximation to include a component of $\mathbf{l}$ perpendicular to $\hat{\mathbf{b}}$ does not affect the Stern-Gerlach-type force [the first term in Eqs. (10)–(12)]. The component of $\mathbf{l}$ perpendicular to $\hat{\mathbf{b}}$ is of the order of $|\mathbf{B}|^{-1}$, so that it yields a correction of the order of $|\mathbf{B}|^{-2}$ to the projection of $\mathbf{l}$ on $\hat{\mathbf{b}}$, and this correction does not affect the force in the limit $|\mathbf{B}|\to\infty$.

Up to this stage we have limited our attention to the case in which the magnetic moment approaches a complete alignment with the magnetic field. We now discuss the case in which $l_\parallel \neq l$, i.e., the case in which the Lorentz-type force is accompanied by an electric-type one. The important new point that should be discussed in this case is the precession of the magnetic moment around the effective magnetic field $\mathbf{B}'$. When a magnetic moment precesses around a magnetic field, it has a time-independent component parallel to the field, and an oscillatory time-dependent (precessing) component in the two directions perpendicular to the field. Now, as Eq. (3) shows, when the vector $\mathbf{l}$ has a component parallel to $\partial\hat{\mathbf{b}}/\partial x_i$, this component yields a force. Since $\partial\hat{\mathbf{b}}/\partial x_i$ is perpendicular to the original magnetic field $\mathbf{B}$, and hence nearly perpendicular to the effective field $\mathbf{B}'$, the component of $\boldsymbol{l}$ along $\partial\hat{\mathbf{b}}/\partial x_i$ is expected to oscillate in time too, and thus to yield an oscillatory force on the particle. This force can be expected to be large [proportional to $|\mathbf{B}|$—see Eq. (3)], but its rapid oscillations, of frequency $|\mathbf{B}|$, can be expected to result in a zero average. A careful analysis of that force, to be described below, shows that it does not average to zero, but rather yields the $|\mathbf{B}|$-independent electric-type force.

We start our analysis by writing an expression for $\boldsymbol{l}$ as a function of time. Following our refined approximation, $\boldsymbol{l}$ is expected to precess around the effective magnetic field $\mathbf{B}'$. Then, the internal angular momentum $\boldsymbol{l}$ is

$$\boldsymbol{l}(t)=l_{\parallel}\hat{\mathbf{b}}'+l_{\perp}[\cos(|\mathbf{B}|t)\hat{\mathbf{c}}+\sin(|\mathbf{B}|t)(\hat{\mathbf{b}}'\times\hat{\mathbf{c}})] , \quad (13)$$

where $l_{\parallel}^2+l_{\perp}^2=l^2$, $\hat{\mathbf{b}}'$ is a unit vector in the direction of $\mathbf{B}'$, and $\hat{\mathbf{c}}$ is an arbitrarily chosen unit vector perpendicular to $\hat{\mathbf{b}}'$. In the framework of our refined approximation $l_{\parallel}$ and $l_{\perp}$ are constants of motion. The $l_{\parallel}$ term of Eq. (13) yields the Lorentz-type force given by Eqs. (10)–(12), with l replaced by $l_{\parallel}$. Since this force was discussed above, we disregard it now and focus on the contribution of the $l_{\perp}$ term. For that term we neglect the difference between $\hat{\mathbf{b}}$ and $\hat{\mathbf{b}}'$. This difference induces a force that vanishes in the $|\mathbf{B}|\rightarrow\infty$ limit. We choose $\hat{\mathbf{c}}$ to be in the direction of $\partial\hat{\mathbf{b}}/\partial x_i$, and substitute Eq. (13) in the expression for the force, Eq. (3). We get

$$F_i^{\perp}=|\mathbf{B}|l_{\perp}\frac{\partial\hat{\mathbf{b}}}{\partial x_i}\cos|\mathbf{B}|t , \quad (14)$$

where $F_i^{\perp}$ denotes the $l_{\perp}$-dependent part of the force. As expected, the precession of the magnetic moment induces a big, though rapidly oscillating, force on the particle. As a result of that force, the smooth path of the particle, dominated by the Stern-Gerlach and Lorentz-type forces, is modulated onto rapid oscillations, dictated by the force (14). Altogether, the position of the particle is given by $\mathbf{x}(t)=\mathbf{x}_0(t)+\delta\mathbf{x}(t)$, where $\mathbf{x}_0(t)$ is a slowly varying smooth function of the time determined by the forces in Eq. (10) and the rapidly oscillating part $\delta\mathbf{x}(t)$ satisfies the following equation of motion:

$$m\frac{d^2\delta\mathbf{x}}{dt^2}=|\mathbf{B}|l_{\perp}\frac{\partial\hat{\mathbf{b}}}{\partial x_i}\cos|\mathbf{B}|t . \quad (15)$$

The solution of Eq. (15) is given by

$$\delta\mathbf{x}(t)=-\frac{l_{\perp}}{m|\mathbf{B}|}\frac{\partial\hat{\mathbf{b}}}{\partial x_i}\cos|\mathbf{B}|t .$$

Two points should be stressed regarding this expression for $\delta\mathbf{x}$. First, $\delta\mathbf{x}$ is inversely proportional to the mass. Second, although the force is proportional to $|\mathbf{B}|$, its rapid oscillations make the amplitude of $\delta\mathbf{x}$ *inversely* proportional to $|\mathbf{B}|$. Thus, in the adiabatic limit this amplitude becomes very small. However, as we now show, these oscillations are the source for the nonvanishing value of the time average of $F_i^{\perp}$. The period of precession is $2\pi/|\mathbf{B}|$, and thus the average of the force over one period is given by

$$\langle F_i\rangle(t)=\left[\frac{2\pi}{|\mathbf{B}|}\right]^{-1}\int_t^{t+2\pi/|\mathbf{B}|}l_{\perp}|\mathbf{B}|\frac{\partial\hat{\mathbf{b}}(\mathbf{x}(t'))}{\partial x_i}\cos|\mathbf{B}|t'dt' . \quad (16)$$

Substituting $\mathbf{x}(t)=\mathbf{x}_0(t)+\delta\mathbf{x}(t)$ and taking the adiabatic limit, the amplitude of the oscillations becomes small, and we can expand

$$\frac{\partial\hat{\mathbf{b}}(\mathbf{x}(t))}{\partial x_i}=\frac{\partial\hat{\mathbf{b}}_0}{\partial x_i}-\frac{\partial^2\hat{\mathbf{b}}_0}{\partial x_i\partial x_j}\frac{l_{\perp}}{m|\mathbf{B}|}\frac{\partial\hat{\mathbf{b}}_0}{\partial x_j}\cos|\mathbf{B}|t \quad (17)$$

[where $\hat{\mathbf{b}}_0\equiv\hat{\mathbf{b}}(\mathbf{x}_0(t))$] and substitute in Eq. (16). The first term in the expansion averages to zero but the second term yields

$$\langle F_i\rangle=-\frac{l_{\perp}^2}{2m}\frac{\partial\hat{\mathbf{b}}}{\partial x_j}\frac{\partial^2\hat{\mathbf{b}}}{\partial x_i\partial x_j} , \quad (18)$$

which is, of course, just the gradient of the scalar potential

$$\Phi=\frac{l_{\perp}^2}{4m}\sum_j\left[\frac{\partial\hat{\mathbf{b}}}{\partial x_j}\right]^2 ,$$

the scalar potential appearing in the effective Hamiltonian (2). The scalar potential is always positive, and it repels the particle from regions in which $\hat{\mathbf{b}}$ strongly varies.

The source of the electric-type force is then in the rapidly oscillating force resulting from the precession of the magnetic moment. If the mass of the particle is infinite, it does not respond to the oscillatory force, and the time average of the latter is zero. However, if the mass is finite, this force induces a small-amplitude oscillatory motion of the particle, and the combination of that motion with the precession makes the average of the oscillatory force nonzero.

Finally, we note that while the Lorentz-type force does not do work on the particle, the electric-type force does. It then transforms kinetic energy, $P^2/2m$, into potential energy, $-|\mathbf{B}|\boldsymbol{l}\cdot\hat{\mathbf{b}}(\mathbf{x})$, and vice versa. Consequently, it also changes the value of the action $\boldsymbol{l}\cdot\hat{\mathbf{b}}$. However, the work done by the electric-type force is independent of $|\mathbf{B}|$. Thus, the change in the action is of the order of $|\mathbf{B}|^{-1}$, i.e., it vanishes in the adiabatic limit.

We thank M. V. Berry, M. Kugler, Y. Imry, O. Entin-Wohlman, J. Hannay, J. Robbins, and R. Littlejohn for instructive discussions. This work was supported by the Wolfson foundation, by the German–Israel foundation for scientific research and development (GIF) and by the Minerva foundation, Munich, Germany. One of us (A.S.) acknowledges the hospitality of the Weizmann Institute of Sciences, where part of this work was done.

[1] *Geometric Phases in Physics*, edited by A. Shapere and F. Wilczek (World Scientific, Singapore, 1989).

[2] M. V. Berry, Proc. R. Soc. London A **392**, 45 (1984).
[3] A particularly useful reference for our purpose is M. V. Berry's paper in *Geometric Phases in Physics* (Ref. [1]).
[4] M. V. Berry and R. Lim, J. Phys. A **23**, L655 (1990).
[5] C. A. Mead, Rev. Mod. Phys. **64**, 51 (1992).
[6] B. Zygelman, Phys. Lett. A **125**, 476 (1987).
[7] R. Littlejohn, J. Plasma Phys. **29**, 111 (1983).
[8] Y. Aharonov, E. Ben-Reuven, S. Popescu, and D. Rohrlich, Phys. Rev. Lett. **65**, 3065 (1990).
[9] D. Loss, P. Goldbart, and A. V. Balatsky, Phys. Rev. Lett. **65**, 1655 (1990).
[10] A. Stern, Phys. Rev. Lett. **68**, 1022 (1992).
[11] J. Anandan and L. Stodolsky, Phys. Rev. D **35**, 2597 (1987).
[12] H. Goldstein, *Classical Mechanics* (Addison-Wesley, Reading, MA, 1980).
[13] V. I. Arnold, *Mathematical Methods of Classical Mechanics* (Springer-Verlag, New York, 1978), Sec. 26.

Topological Hall Effect and Berry Phase in Magnetic Nanostructures

P. Bruno,[1] V. K. Dugaev,[1,2] and M. Taillefumier[1,3]

[1]*Max-Planck-Institut für Mikrostrukturphysik, Weinberg 2, 06120 Halle, Germany*
[2]*Institute for Problems of Materials Science, NASU, Vilde 5, 58001 Chernovtsy, Ukraine*
[3]*Laboratoire Louis Néel, CNRS, Boite Postale 166, 38042 Grenoble CEDEX 09, France*

(Received 21 October 2003; published 27 August 2004)

We discuss the anomalous Hall effect in a two-dimensional electron gas subject to a spatially varying magnetization. This topological Hall effect does not require any spin-orbit coupling and arises solely from Berry phase acquired by an electron moving in a smoothly varying magnetization. We propose an experiment with a structure containing 2D electrons or holes of diluted magnetic semiconductor subject to the stray field of a lattice of magnetic nanocylinders. The striking behavior predicted for such a system (of which all relevant parameters are well known) allows one to observe unambiguously the topological Hall effect and to distinguish it from other mechanisms.

DOI: 10.1103/PhysRevLett.93.096806 PACS numbers: 73.20.Fz, 72.10.Fk, 72.15.Rn

After half a century of theoretical efforts, the Hall effect of ferromagnets (usually called *anomalous* or *extraordinary* Hall effect) remains a puzzling and controversial topic. Until recently, it was considered that it originates from the combined effect of exchange and spin-orbit (SO) interactions. Two mechanisms of anomalous Hall effect (AHE) have been identified (the *skew scattering* [1,2] and *side jump* [3]) and studied thoroughly [4]. Recently, a new point of view has been proposed [5], in which the AHE is expressed in terms of a Berry curvature in momentum space. However, one should note that a generally accepted theory treating on an equal footing all the above mentioned contributions to the SO-induced AHE is still missing.

Recently, it has been suggested that (in addition to the above mentioned SO-induced mechanism) a new mechanism may give rise to a nonvanishing Hall effect in ferromagnets having a topologically nontrivial (chiral) spin texture, such as manganites or pyrochlore-type compounds [6]. To distinguish this mechanism from the SO-based mechanism, we shall refer to it hereafter as the *topological Hall effect* (THE).

Several theoretical papers have been devoted to the THE. In order to explain the AHE observed in manganites, a model of 3D ferromagnet with thermally excited Skyrmion strings (topological dipoles) has been proposed [7], showing that a THE can be induced by the Berry phase [8] related to the spatial variation of magnetization in the vicinity of the string. The case of disordered ferromagnets in the limit of small exchange splitting has been addressed in Ref. [9]. In both cases, in order to get a net topological field (or chirality), the SO coupling must be invoked. It should be noted that in cases such as the 2D Kagomé lattice or a 3D pyrochlore lattice the net topological field vanishes; a nonvanishing THE may nevertheless be obtained [10], but the sign and strength of the effect depends in a complicated manner on fine details of the magnetic and electronic structures, which makes a quantitative experimental check of the theory questionable. In this respect it would be desirable to have systems having a net topological field that can be controlled by means of some external parameter.

In addition, all the above mentioned discussions of the THE concern systems with spin chirality at the microscopic scale (e.g., pyrochlore lattice) or due to Skyrmion strings. In both cases, quantitative experimental information on the chirality is not easily available. Furthermore, the SO mechanism is usually also present, which complicates the quantitative interpretation of the observations.

In the present Letter, we propose to investigate the THE in nanostructures, namely, in a 2D electron (or hole) gas, in which a well-controlled artificial chirality can be induced from the stray field of a lattice of magnetic nanocylinders. The great advantage of such a model system is that, in contrast to the above mentioned cases, all relevant parameters are well known and (to some extent) adjustable. We show that this yields a significant net topological field (equivalent to a magnetic field of about 5 kG), and that the latter has a nontrivial characteristic variation with respect to a (uniform) external magnetic field, providing an unambiguous signature of the THE. Finally, we suggest a system appropriate for an experimental check of our theory.

We start from a model of two-dimensional electron gas (2DEG) in a smoothly varying magnetization $\mathbf{M}(\mathbf{r})$. The Hamiltonian has the following form

$$H = -\frac{\hbar^2}{2m}\frac{\partial^2}{\partial \mathbf{r}^2} - g\boldsymbol{\sigma}\cdot\mathbf{M}(\mathbf{r}), \tag{1}$$

where g is the coupling constant and $\boldsymbol{\sigma}$ is the vector of Pauli matrices. We assume that the amplitude of magnetization is constant, $\mathbf{M}(\mathbf{r}) = M\mathbf{n}(\mathbf{r})$, and that the 3D unit vector $\mathbf{n}(\mathbf{r})$ is a slowly varying function of coordinates.

We use a gauge transformation $T(\mathbf{r})$, which makes the quantization axis oriented along vector $\mathbf{n}(\mathbf{r})$ at each point [11]. It transforms the last term in (1) as $T^\dagger(\mathbf{r})\times[\boldsymbol{\sigma}\cdot\mathbf{n}(\mathbf{r})]T(\mathbf{r})=\sigma_z$, corresponding to a local rotation of the quantization axis from z axis to the axis along $\mathbf{n}(\mathbf{r})$.

The transformed Hamiltonian describes the electrons moving in a (spinor) gauge potential $\mathbf{A}(\mathbf{r})$:

$$H' = -\frac{\hbar^2}{2m}\left[\frac{\partial}{\partial\mathbf{r}} - \frac{ie}{\hbar c}\mathbf{A}(\mathbf{r})\right]^2 - gM\sigma_z, \qquad (2)$$

where $A_i(\mathbf{r}) = -2\pi i\phi_0 T^\dagger(\mathbf{r})\partial_i T(\mathbf{r})$, $\phi_0 = hc/|e|$ is the flux quantum, and $i = x, y$. For convenience, we define the gauge potential $\mathbf{A}(\mathbf{r})$ to have the same dimension as the electromagnetic vector potential. The components of $\mathbf{A}(\mathbf{r})$ can be found easily using an explicit form of $T(\mathbf{r})$.

Hamiltonian (2) with the matrix $\mathbf{A}(\mathbf{r})$ contains terms inducing transitions between the spin-polarized states. We consider the case when spin-flip transitions can be neglected, i.e., where the spin adiabatically follows $\mathbf{n}(\mathbf{r})$. It corresponds to the condition $\lambda \equiv (\varepsilon_F/\varepsilon_0)(k_F\xi)^{-1} \ll 1$, where ξ is a characteristic length of the variation of $\mathbf{n}(\mathbf{r})$, ε_F is the Fermi energy, and $\varepsilon_0 = 2gM$ is the spin splitting.

We also assume that the 2DEG is half metallic with the Fermi level located in the spin-up subband. Then we can neglect the spin-down electrons, and we obtain the following effective Hamiltonian of spinless electrons:

$$\tilde{H} = -\frac{\hbar^2}{2m}\left[\frac{\partial}{\partial\mathbf{r}} - \frac{ie}{\hbar c}\mathbf{a}(\mathbf{r})\right]^2 + V(\mathbf{r}), \qquad (3)$$

where

$$a_i(\mathbf{r}) = \frac{\pi\phi_0(n_x\partial_i n_y - n_y\partial_i n_x)}{1+n_z}, \qquad (4)$$

the potential $V(\mathbf{r}) = (\hbar^2/8m)(\partial_i n_\mu)^2$ results from the second order in $\mathbf{A}(\mathbf{r})$ terms, and $\mu = x, y, z$. For spin-down electrons the sign of the gauge field $\mathbf{a}(\mathbf{r})$ in Eq. (3) is reversed.

The topological field is defined as $B_t = \partial_x a_y - \partial_y a_x$. It acts on the electrons within the spin-polarized subband like the ordinary magnetic field, and, in particular, gives rise to a Lorentz-type force [12]. Using (4) we find

$$B_t = \frac{\phi_0}{4\pi}\epsilon_{\mu\nu\lambda}n_\mu(\partial_x n_\nu)(\partial_y n_\lambda), \qquad (5)$$

where $\epsilon_{\mu\nu\lambda}$ is the unit antisymmetric tensor. The integral of (5) over an area S_0 enclosed by an arbitrary contour L_0 is proportional to the Berry phase calculated as the spherical angle spanned by area S inside the contour L in $\mathbf{n}$ space. This results from the mapping of L_0 onto the contour L in the mapping space S_2. In the 2D case with a constant magnetization at infinity, we can compactify 2D plane to a sphere S_2, and the integral of (5) over S_2 is proportional to topological invariant corresponding to the number of covering the mapping space.

We consider a mesoscopic system consisting of a 2DEG subject to the stray field of a periodic 2D lattice of magnetic nanocylinders [13] (uniformly magnetized along the z axis), as sketched on Fig. 1. The stray field induced in the 2DEG equals zero on average, but its spacial fluctuations induce in the 2DEG a topological field with a nonzero average. The topological field distribution is solely determined by a few geometric parameters such as the lattice type, cylinder diameter, intercylinder spacing, and the distance between the 2DEG and the bottom of the cylinders. In any practical realization, all these parameters (and hence the topological field) are well known and (to some extent) easily adjustable. From the discussion following Eq. (5), one concludes immediately that the net topological flux per unit cell is equal to an integer multiple of ϕ_0, i.e., the average topological field is $\overline{B}_t = I\phi_0/s_0$, where I is an integer and s_0 the unit cell area. Using the semiclassical Drude theory of magnetotransport (which is valid in the weak field limit) and using the mean field approximation (i.e., neglecting spacial fluctuations of the topological and magnetic stray fields), one obtains $\sigma_{xx} = \sigma_{xx}^\uparrow + \sigma_{xx}^\downarrow$ and $\sigma_{xy} = \sigma_{xy}^\uparrow + \sigma_{xy}^\downarrow$ (since the two spin channels are decoupled in the rotated spin-frame) with:

$$\sigma_{xx}^s = \frac{n_s e^2\tau_s}{m}, \qquad (6)$$

$$\sigma_{xy}^s = s\sigma_{xx}^s\frac{e\overline{B}_t\tau_s}{m}, \qquad (7)$$

where $s = +1$ (respectively, $s = -1$) for $\uparrow$ spin (resp. $\downarrow$ spin).

We have calculated the stray field and topological field for a triangular lattice of nanocylinders (of magnetization M_s along the z axis), using realistic values of parameters for the nanocylinder lattice [13]. Namely, we take a lattice constant of 100 nm, a cylinder radius of 37 nm (i.e., a filling ratio of 50%), and the gap D between the lattice and semiconductor equal to 20 nm. The distribu-

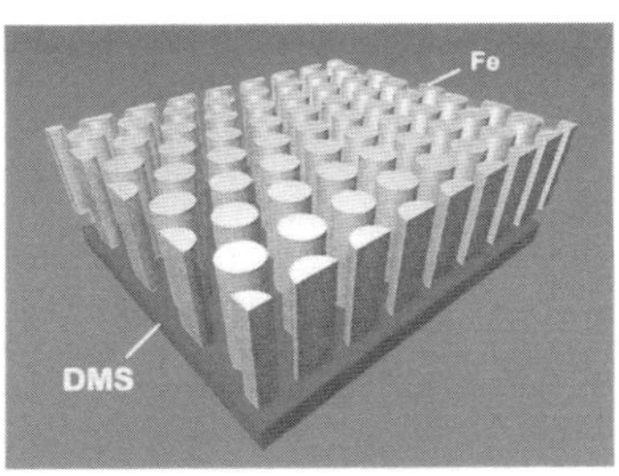

FIG. 1 (color). The proposed structure consisting of a triangular lattice of magnetic nanocylinders on top of 2D diluted magnetic semiconductor.

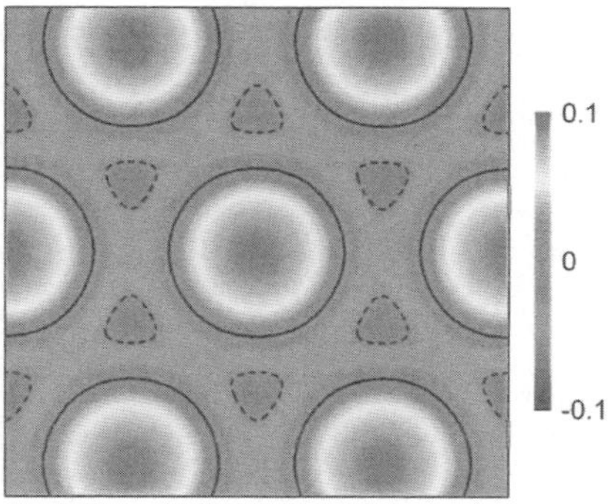

FIG. 2 (color). Distribution of the z-component of dipolar field $B/4\pi M_s$ inside the semiconductor film for the triangular lattice of magnetic nanocylinders, for a zero external field. The black solid circles correspond to the lines with $B_z = 0$. Dashed lines correspond to the lines with $B_z = 0$ under a uniform external magnetic field $B_{\text{ext}}/4\pi M_s = +0.058$.

tion of z component of the dipolar field (normal to the semiconductor surface) is shown in Fig. 2. The corresponding topological field calculated from Eq. (5) is presented in Fig. 3. For Fe cylinders, the dipolar field B inside the semiconductor is about 2 kG in absolute value. The characteristic length for the variations of $\mathbf{n}(\mathbf{r})$ is of the order of the gap D, i.e., $\xi \simeq 20$ nm.

One can see quite easily (for example by considering the Berry phase from the mapping of the plane onto the sphere S_2) that the net topological flux per unit cell ϕ_t has to be an integer multiple of ϕ_0, and that ϕ_t/ϕ_0 is given by the number (per unit cell) of lines $B_z = 0$ enclosing a region with $B_z < 0$ minus the number of lines $B_z = 0$ enclosing a region with $B_z > 0$. From Fig. 2, one thus sees that we have a triangular lattice of lines $B_z = 0$ enclosing regions with $B_z < 0$, yielding $\phi_t = \phi_0$ for a vanishing external field. For the considered geometry this corresponds to an average topological field of about 5 kG, with local values ranging between -5 kG and $+15$ kG.

If a negative external field is applied (assuming that the magnetization of the nanocylinders remains unchanged, due to some large coercivity), the lines with $B_z = 0$ shrink without changing their topology until they collapse to a single point and eventually disappear at a critical field B_1, beyond which $\phi_t = 0$. If a positive field is applied, the lines with $B_z = 0$ expand, until they connect each other at a critical field B_2 and change their topology to a honeycomb lattice (the dual of the triangular lattice) of lines $B_z = 0$ enclosing regions with $B_z > 0$ (dashed lines in Fig. 2), giving $\phi_t = -2\phi_0$. Increasing further the external field leads to a collapse of the lines with $B_z = 0$ at a critical field B_3, beyond which $\phi_t = 0$. For Fe nanocylinders in the above geometry, the critical fields are, respectively, $B_1 \simeq -2$ kG, $B_2 \simeq +0.9$ kG, and $B_3 \simeq +1.3$ kG. This change of topology under application of an external magnetic field results in a striking field dependence of the Hall conductivity, as sketched in Fig. 4, and constitutes an unambiguous signature of the THE. Note that, in any realistic case, the adiabaticity condition would not be satisfied in the vicinity of the critical fields $B_{1,2,3}$, leading to a "rounding" of the jumps in the Hall conductivity.

For the practical realization, we propose to use II–VI dilute magnetic semiconductors (DMS) which exhibit giant Zeeman splitting; p-type DMS are best suited since the exchange constants for holes are much larger than for electrons [14]. For the estimation of the topological field acting on holes, we use the Luttinger Hamiltonian [15]

$$H_h = \frac{\hbar^2}{2m_0}\left[\left(\gamma_1 + \frac{5}{2}\gamma_2\right)\nabla^2 - 2\gamma_2(\mathbf{J}\cdot\nabla)^2\right] + E_{ex}\mathbf{J}\cdot\mathbf{n}(\mathbf{r}), \tag{8}$$

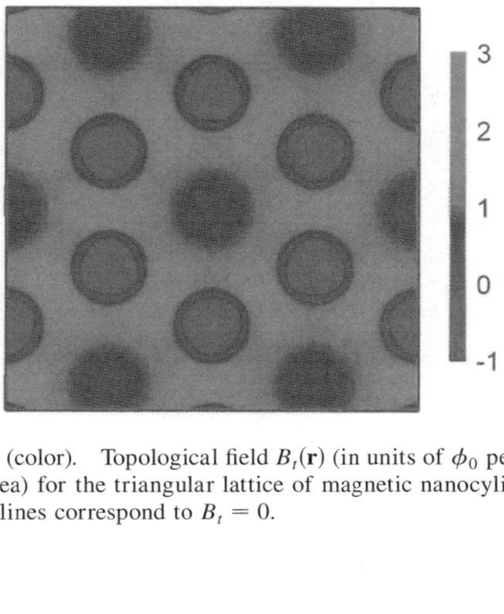

FIG. 3 (color). Topological field $B_t(\mathbf{r})$ (in units of ϕ_0 per unit cell area) for the triangular lattice of magnetic nanocylinders. Black lines correspond to $B_t = 0$.

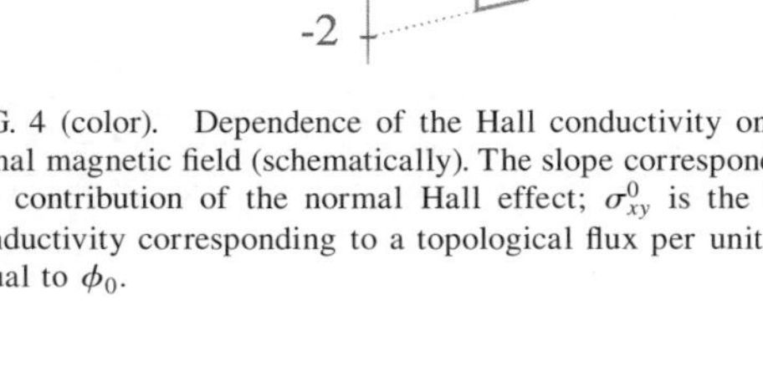

FIG. 4 (color). Dependence of the Hall conductivity on external magnetic field (schematically). The slope corresponds to the contribution of the normal Hall effect; σ^0_{xy} is the Hall conductivity corresponding to a topological flux per unit cell equal to ϕ_0.

where J_μ are the matrices of momentum $J = 3/2$, γ_1 and γ_2 are the Luttinger parameters, m_0 is the free electron mass, $\mathbf{n}(\mathbf{r})$ is the unit vector along the direction of exchange field acting on 2D gas of holes, E_{ex} is the average exchange field created by magnetic impurities,

$$E_{ex} = -JSxN_0\beta\mathcal{B}_S(g_L\mu_B B/k_B T), \tag{9}$$

$N_0\beta$ is the $p-d$ exchange constant, S is the magnetic moment of impurity, xN_0 is the concentration of magnetic atoms, $\mathcal{B}_S(z)$ is the modified Brillouin function, and g_L is the Landé factor of magnetic atom.

We calculate the topological field for holes described by Hamiltonian (8), using the transformation $T(\mathbf{r}) = \exp[i\mathbf{J}\cdot\mathbf{\Omega}(\mathbf{r})/2]$, where $\mathbf{\Omega}(\mathbf{r})$ is the vector of rotation. We find the topological field B_t considering small deviations of the vector $\mathbf{n}(\mathbf{r})$ in a vicinity of some point $\mathbf{r}$. Assuming that the magnetic splitting under exchange field is strong, we find that the gauge potential and the topological field for holes in the valence band, being proportional to J, are given by the same expressions, (4) and (5), as for electrons, multiplied by a factor 3.

After transforming (8) and restricting ourselves by the subband $J_z = -3/2$, we obtain for spin-polarized holes, the same Hamiltonian as (3), with $-1/m$ replaced by $1/m^\star = (\gamma_1 + \gamma_2)/m_0$.

The spin splitting in II–VI semiconductor can be large enough to provide 100% polarization of holes under the dipolar field of 2 kG. To show this we make the following estimations, taking the exchange coupling $N_0\beta = -1.2$ eV, which is the typical magnitude for different compounds ($Cd_{1-x}Mn_xSe$, $Zn_{1-x}Mn_xSe$, etc.), and the atomic density of magnetic atoms $x = 0.05$. For $B = 2$ kG, $g_L = 2$, and $T = 4.2$ K the Brillouin function $\mathcal{B}_{5/2}(g\mu_B B/k_B T) \simeq 0.075$. Then using (8) we obtain $E_{ex} \simeq 11$ meV . The Fermi energy of 2D holes in the spin-splitted subband is determined by the density of holes n_p, $E_F = 2\pi\hbar^2 n_p/m^*$, where m^* is the effective mass of holes. By taking $m^* = 0.5m_0$, and $n_p = 10^{11}$ cm^{-2}, we obtain $E_F \simeq 1$ meV. Thus, the condition of 100% polarization of holes, $E_F < E_{ex}$, is quite realistic for II–VI diluted magnetic semiconductors. The adiabaticity parameter for this choice of parameters is $\lambda \simeq 0.04 \ll 1$. It should be noted that much stronger splitting can be reached for $Zn_{1-x}Cr_xTe$ compounds, for which $N_0\beta = 3.6$ eV [16].

For electrons, on the other hand, in order to fulfill the conditions of full polarization and adiabaticity, lower temperatures and/or very low electron densities would be needed. Taking the exchange constant of $s-d$ interaction $N_0\alpha = 0.22$ eV [14], the electron effective mass $m_e = 0.22m_0$, $T = 4.2$ K, and the concentration of electrons $n_e = 10^{11}$ cm^{-2}, we obtain $E_F \simeq 4.4$ meV, $E_{ex} \simeq 2$ meV, and $\lambda \simeq 1$; the condition of adiabaticity is poorly fulfilled, resulting in a reduction of the effect. Also the gas is not fully polarized, reducing further the THE due to the partial compensation of contributions from spin-up and -down subbands. Although not fully developed, the THE should nevertheless be observable with electrons.

In conclusion, we have proposed that the THE can be observed in suitably chosen nanostructures, and that its striking behavior under an external magnetic field provides an unambiguous experimental signature of the THE.

We thank J. Cibert for discussions. V. D. thanks Polish State Committee for Scientific Research for support under Grants PBZ/KBN/044/P03/2001 and 2P03B05325. P. B. acknowledges financial support from BMBF (Grant No. 01BM924).

[1] J. Smit, Physica (Amsterdam) **24**, 39 (1958).
[2] R. Karplus and J. M. Luttinger, Phys. Rev. **95**, 1154 (1954); J. M. Luttinger, Phys. Rev. **112**, 739 (1958).
[3] L. Berger, Phys. Rev. B **2**, 4559 (1970); **5**, 1862 (1972).
[4] C. M. Hurd, *The Hall Effect in Metals and Alloys* (Plenum, New York, 1972); *The Hall Effect and Its Applications*, edited byC. L. Chien and C. R. Westgate (Plenum, New York, 1979); A. Crépieux and P. Bruno, Phys. Rev. B **64**, 014416 (2001); V. K. Dugaev *et al.*, Phys. Rev. B **64**, 104411 (2001).
[5] M. Onoda and N. Nagaosa, J. Phys. Soc. Jpn. **71**, 19 (2002); Phys. Rev. Lett. **90**, 206601 (2003); T. Jungwirth *et al.*, Phys. Rev. Lett. **88**, 207208 (2002); Z. Fang *et al.*, Science **302**, 92 (2003); D. Culcer *et al.*, Phys. Rev. B **68**, 045327 (2003); Y. Yao *et al.*, Phys. Rev. Lett. **92**, 037204 (2004).
[6] P. Matl *et al.*, Phys. Rev. B **57**, 10248 (1998); Y. Taguchi *et al.*, Science **291**, 2573 (2001); Y. Taguchi and Y. Tokura, Europhys. Lett. **54**, 401 (2001); S. H. Chun *et al.*, Phys. Rev. Lett. **84**, 757 (2000).
[7] J. Ye *et al.*, Phys. Rev. Lett. **83**, 3737 (1999).
[8] M.V. Berry, Proc. R. Soc. London A **392**, 45 (1984).
[9] G. Tatara and H. Kawamura, J. Phys. Soc. Jpn. **71**, 2613 (2002).
[10] K. Ohgushi *et al.*, Phys. Rev. B **62**, R6065 (2000); S. Onoda and N. Nagaosa, Phys. Rev. Lett. **90**, 196602 (2003).
[11] G. Tatara and H. Fukuyama, Phys. Rev. Lett. **78**, 3773 (1997).
[12] Y. Aharonov and A. Stern, Phys. Rev. Lett. **69**, 3593 (1992).
[13] K. Nielsch *et al.*, Appl. Phys. Lett. **79**, 1360 (2001); J. Magn. Magn. Mater. **249**, 234 (2002).
[14] J. K. Furdyna, J. Appl. Phys. **64**, R29 (1988); J. Kossut and W. Dobrowolski, in *Handbook of Magnetic Materials*, edited by K. H. J. Buschow (Elsevier, Holland, 1993), Vol. 7, p. 231.
[15] J. M. Luttinger, Phys. Rev. **102**, 1030 (1956).
[16] W. Mac *et al.*, Phys. Rev. B **54**, 5528 (1996).

General Approach to Quantum-Classical Hybrid Systems and Geometric Forces

Qi Zhang and Biao Wu
Institute of Physics, Chinese Academy of Sciences, Beijing 100080, China
(Received 8 March 2006; published 7 November 2006)

We present a general theoretical framework for a hybrid system that is composed of a quantum subsystem and a classical subsystem. We approach such a system with a simple canonical transformation which is particularly effective when the quantum subsystem is dynamically much faster than the classical counterpart, which is commonly the case in hybrid systems. Moreover, this canonical transformation generates a vector potential which, on one hand, gives rise to the familiar Berry phase in the fast quantum dynamics and, on the other hand, yields a Lorentz-like geometric force in the slow classical dynamics.

DOI: 10.1103/PhysRevLett.97.190401 PACS numbers: 03.65.Vf, 03.65.Ca

Single spin detection was recently achieved with magnetic-resonance-force microscopy [1]. Technically, this remarkable experiment marks a major step in man's ability to control objects at the atomic scale and has great potential for application in future technology [2]. On the scientific side, this experiment also invokes many interesting questions of fundamental interest.

We first notice that the system used in the experiment is a hybrid system composed of a quantum subsystem (single spin) and a classical subsystem (cantilever). In Refs. [3,4], both subsystems are treated classically. Although this is adequate to address specific issues and systems related to the current experiments [1,5], this method is inadequate to treat a general hybrid system, for example, when the quantum subsystem is not a spin or when one tries to discuss phase-related issues. This kind of hybrid system was also studied by Berry and Robbins [6]. However, their formalism is not satisfying, either, as the classical dynamics never appears explicitly in the formalism and it cannot deal with the situation in which the quantum system is in a coherent noneigenstate. Then the question is how to treat adequately a general hybrid system. This question will grow more important since various techniques are being developed or explored to control objects at the atomic scale. These techniques are certainly all based on hybrid systems where a classical sensor interacts with a quantum object.

We also notice that, in the experiment of single spin detection [1], the quantum subsystem (spin) is dynamically much faster than the classical counterpart (cantilever). In other words, it is a Born-Oppenheimer-type system (one subsystem is fast and the other slow). This is, of course, typical of a hybrid system. Born-Oppenheimer-type systems have been studied by many in different settings [6–11]. One general scene in these Born-Oppenheimer-type systems is that vector potentials related to geometric phases are found to arise and generate geometric forces in the slow subsystem. It is then interesting to ask whether the geometric force can be detected with current experimental techniques in light of this successful single spin detection [1]. It is also interesting to know how to formulate the vector potential and the geometric force when the quantum subsystem is in a noneigenstate, since there is no reason to assume that the quantum subsystem is only in an eigenstate. In previous studies, the quantum subsystem is always assumed in an eigenstate [6–9].

In this Letter, we present a general theoretical framework for hybrid systems. For this kind of system, we use the well-known fact [12–14] that a quantum system possesses mathematically a canonical classical Hamiltonian structure. In this way, we can describe the hybrid system with a unified classical Hamiltonian. We emphasize that in our Hamiltonian the quantum subsystem is reduced only mathematically to a classical system and no physics is lost. This is in contrast to the Hamiltonian of Ref. [4], where the quantum subsystem is reduced physically to be classical and some physics may be lost.

We try to decipher the complicated dynamics of the hybrid system with a special canonical transformation. This transformation is particularly effective when the quantum subsystem is fast and the classical subsystem is slow. After the canonical transformation, it becomes clear that the two subsystems influence each other not only via interaction but also through a vector potential. This vector potential generates the familiar Berry phase [15] in the fast quantum subsystem while it produces a Lorentz-like geometric force in the slow classical subsystem.

We shall use a simple example, the coupling of a heavy magnetic particle with a single spin, to illustrate our theory. We also use this example to show how big the geometric force can be and whether it is detectable with current experimental techniques.

In addition, we point out that our method can also be regarded as an alternative way of deriving the Berry phase [15]. Our method has the advantage that the fast quantum subsystem need not be in an eigenstate. In other words, the Berry phase can be defined for a general quantum state. Our method can also be generalized to derive Hannay's angles [16] or the geometric phase proposed in Ref. [17] for nonlinear quantum systems. In this sense, our approach shows that these different geometric phases (Berry phase and Hannay's angles) are just different manifestations of the same mathematical concept in different dynamical systems.

We describe the aforementioned hybrid coupled system with the following Hamiltonian:

$$H = \langle \boldsymbol{\Psi} | \hat{H}_1(\mathbf{q}_2) | \boldsymbol{\Psi} \rangle + H_2(\mathbf{p}_2, \mathbf{q}_2), \tag{1}$$

where $\hat{H}_1$ is the Hamiltonian operator of the linear N-level fast quantum subsystem and $|\boldsymbol{\Psi}\rangle = (\Psi_1, \Psi_2, \ldots, \Psi_N)^T$ is its quantum state. The Hamiltonian H_2 governs a heavy classical subsystem that moves slowly and $\mathbf{p}_2$, $\mathbf{q}_2$ are its momenta and coordinates, respectively. The dependence of $\hat{H}_1$ on $\mathbf{q}_2$ indicates the coupling between the two subsystems.

We first focus on the quantum subsystem, assuming temporarily that $\mathbf{q}_2$ are just some fixed parameters. The Schrödinger equation of the quantum subsystem is $i\hbar d|\boldsymbol{\Psi}\rangle/dt = \hat{H}_1|\boldsymbol{\Psi}\rangle$, which can be rewritten as

$$i\hbar \frac{d\Psi_j}{dt} = \frac{\partial}{\partial \Psi_j^*} H_1(\boldsymbol{\Psi}, \boldsymbol{\Psi}^*, \mathbf{q}_2), \tag{2}$$

where $H_1 = \langle \boldsymbol{\Psi} | \hat{H}(\mathbf{q}_2) | \boldsymbol{\Psi} \rangle$. This shows that the quantum system has a classical Hamiltonian structure. This fact is known to many people and was discussed in detail in Refs. [12,13]. To have a more "classical" look for this quantum system, we define $p_{1j} = \sqrt{i\hbar}\Psi_j^*$, $q_{1j} = \sqrt{i\hbar}\Psi_j$ and write Eq. (2) in an apparent canonical Hamiltonian formalism

$$\frac{dq_{1j}}{dt} = \frac{\partial \tilde{H}_1}{\partial p_{1j}}, \qquad \frac{dp_{1j}}{dt} = -\frac{\partial \tilde{H}_1}{\partial q_{1j}}, \tag{3}$$

where $\tilde{H}_1 = \tilde{H}_1(\mathbf{p}_1, \mathbf{q}_1, \mathbf{q}_2) = H_1(\boldsymbol{\Psi}, \boldsymbol{\Psi}^*, \mathbf{q}_2)$. In this way, we have classically reformulated quantum systems.

The quantum state $|\boldsymbol{\Psi}\rangle$ can be expanded in terms of instantaneous eigenstates

$$|\boldsymbol{\Psi}\rangle = \sum_{n=1}^{N} a_n |\varphi_n(\mathbf{q}_2)\rangle, \tag{4}$$

where $\hat{H}_1(\mathbf{q}_2)|\varphi_n(\mathbf{q}_2)\rangle = E_n(\mathbf{q}_2)|\varphi_n(\mathbf{q}_2)\rangle$. The quantum system can be described alternatively by these expansion coefficiencies a_n's. Define $I_{1n} = \hbar |a_n|^2$ and $\Theta_{1n} = -\arg(a_n)$; one can prove readily that $\mathbf{I}_1$ and $\boldsymbol{\Theta}_1$ are another set of canonical variables for the Hamiltonian H_1. According to the standard classical theory [18], there is a canonical transformation between $\mathbf{p}_1$, $\mathbf{q}_1$ and $\mathbf{I}_1$, $\boldsymbol{\Theta}_1$, and this transformation is given by a generating function $F_1(\mathbf{q}_1, \mathbf{I}_1, \mathbf{q}_2)$ that satisfies

$$\mathbf{p}_1 = \frac{\partial F_1}{\partial \mathbf{q}_1}, \qquad \boldsymbol{\Theta}_1 = \frac{\partial F_1}{\partial \mathbf{I}_1}. \tag{5}$$

With this transformation, the Hamiltonian H_1 becomes $\mathcal{H}_1 = \mathcal{H}_1(\mathbf{I}_1, \mathbf{q}_2) = \sum_n E_n(\mathbf{q}_2) I_{1n}/\hbar$, independent of the angles $\boldsymbol{\Theta}_1$.

We go back to the hybrid coupled system, where $\mathbf{q}_2$ are dynamical variables instead of some fixed parameters. The Hamiltonian for the hybrid system can now be expressed in a pure classical formalism

$$\tilde{H} = \tilde{H}_1(\mathbf{p}_1, \mathbf{q}_1, \mathbf{q}_2) + H_2(\mathbf{p}_2, \mathbf{q}_2). \tag{6}$$

We introduce a canonical transformation from $\mathbf{p}_1$, $\mathbf{q}_1$, $\mathbf{p}_2$, $\mathbf{q}_2$ to $\mathbf{I}_1$, $\boldsymbol{\Theta}_1$, $\mathbf{P}_2$, $\mathbf{Q}_2$ with the following generating function:

$$F = F_1(\mathbf{q}_1, \mathbf{I}_1, \mathbf{q}_2) + \mathbf{q}_2 \mathbf{P}_2. \tag{7}$$

The canonical transformation is then given by

$$\mathbf{p}_1 = \frac{\partial F}{\partial \mathbf{q}_1} = \frac{\partial F_1}{\partial \mathbf{q}_1}, \qquad \boldsymbol{\Theta}_1 = \frac{\partial F}{\partial \mathbf{I}_1} = \frac{\partial F_1}{\partial \mathbf{I}_1}, \tag{8}$$

$$\mathbf{p}_2 = \frac{\partial F}{\partial \mathbf{q}_2} = \frac{\partial F_1}{\partial \mathbf{q}_2} + \mathbf{P}_2, \qquad \mathbf{Q}_2 = \frac{\partial F}{\partial \mathbf{P}_2} = \mathbf{q}_2. \tag{9}$$

The transformation does two things: (i) Since Eq. (8) is identical to Eq. (5), the transformation changes $\mathbf{p}_1$, $\mathbf{q}_1$ to $\mathbf{I}_1$, $\boldsymbol{\Theta}_1$ as if it is generated by F_1; (ii) it puts an additional vector function $\mathbf{A} = -\partial F_1/\partial \mathbf{q}_2$ in the momenta $\mathbf{p}_2$ while keeping the coordinates $\mathbf{q}_2$ unchanged. After the transformation, the total Hamiltonian in Eq. (6) becomes

$$\mathcal{H} = \mathcal{H}_1(\mathbf{I}_1, \mathbf{Q}_2) + H_2(\mathbf{P}_2 - \mathbf{A}, \mathbf{Q}_2). \tag{10}$$

In the second subsystem, $\mathbf{A}$ appears very much like a vector potential. However, it is not a true vector potential since it also depends on variables other than $\mathbf{q}_2$. For convenience, we shall call it a pseudovector potential. As we shall see, this pseudovector potential can lead to a true vector potential.

By assuming that the classical subsystem has the usual Hamiltonian $H_2 = \mathbf{p}_2^2/2M + V_2(\mathbf{q}_2)$, we write down the equations of motion for the whole system. For the quantum part, we have

$$\dot{I}_{1j} = \dot{\mathbf{q}}_2 \cdot \frac{\partial \mathbf{A}}{\partial \Theta_{1j}}, \tag{11}$$

$$\dot{\Theta}_{1j} = \frac{\partial \mathcal{H}_1}{\partial I_{1j}} - \dot{\mathbf{q}}_2 \cdot \frac{\partial \mathbf{A}}{\partial I_{1j}}. \tag{12}$$

For the classical part, we obtain

$$\dot{P}_{2j} = -\frac{\partial \mathcal{H}_1}{\partial q_{2j}} - \frac{\partial V_2}{\partial q_{2j}} + \dot{\mathbf{q}}_2 \cdot \frac{\partial \mathbf{A}}{\partial q_{2j}}, \tag{13}$$

$$\dot{Q}_{2j} = \dot{q}_{2j} = (P_{2j} - A_j)/M. \tag{14}$$

Most of the hybrid systems that we encounter in concrete problems are of Born-Oppenheimer type; that is, the quantum subsystem is much faster than the classical subsystem. We now apply the general formalism and approach to this typical type of hybrid systems. In this case, the classical variables $\mathbf{q}_2$ can be considered as the adiabatic parameters for the quantum subsystem.

We first analyze the pseudovector potential **A**. To do this, we make the potential an explicit function of the new dynamical variables $\mathbf{I}_1$, $\mathbf{\Theta}_1$, $\mathbf{P}_2$, and $\mathbf{Q}_2$ $(=\mathbf{q}_2)$ by defining a new function

$$\tilde{F}_1(\mathbf{I}_1, \mathbf{\Theta}_1, \mathbf{q}_2) = F_1(\mathbf{q}_1(\mathbf{I}_1, \mathbf{\Theta}_1, \mathbf{q}_2), \mathbf{I}_1, \mathbf{q}_2). \tag{15}$$

This leads to

$$A_j = \mathbf{p}_1 \frac{\partial \mathbf{q}_1}{\partial q_{2j}} - \frac{\partial \tilde{F}_1}{\partial q_{2j}} = i\hbar\langle\mathbf{\Psi}|\frac{\partial}{\partial q_{2j}}|\mathbf{\Psi}\rangle - \frac{\partial \tilde{F}_1}{\partial q_{2j}}. \tag{16}$$

Since the variables $\mathbf{q}_2$ change very slowly compared to the dynamics of the quantum subsystem, we are allowed to apply the standard averaging technique in the study of adiabatic evolution [16,18]. It also implies that the probabilities $\mathbf{I}_1$ are conserved during the evolution according to the quantum adiabatic theorem [19]. After averaging and using the quantum adiabatic theorem, the pseudovector potential becomes

$$\begin{aligned}\bar{A}_j &= \oint \frac{d\mathbf{\Theta}_1}{(2\pi)^N}\left[i\hbar\langle\mathbf{\Psi}|\frac{\partial}{\partial q_{2j}}|\mathbf{\Psi}\rangle - \frac{\partial \tilde{F}_1}{\partial q_{2j}}\right] \\ &= i\sum_{n=1}^{N} I_{1n}\langle\varphi_n|\frac{\partial}{\partial q_{2j}}|\varphi_n\rangle - \overline{\frac{\partial \tilde{F}_1}{\partial q_{2j}}},\end{aligned} \tag{17}$$

where the overline indicates that the average has been done for the variable. The function $\bar{\mathbf{A}}$ is now a true vector potential as it no longer depends on $\mathbf{\Theta}_1$ and at the same time $\mathbf{I}_1$ are constant. After ignoring the trivial gradient term in Eq. (17), we obtain a true vector potential

$$\bar{\mathbf{A}} = \sum_{n=1}^{N} I_{1n}\mathcal{A}_n, \qquad \mathcal{A}_n = i\langle\varphi_n|\frac{\partial}{\partial \mathbf{q}_2}|\varphi_n\rangle. \tag{18}$$

Substituting it into Eq. (12), we arrive at

$$\dot{\Theta}_{1j} = \frac{\partial \mathcal{H}_1}{\partial I_{1j}} - \mathcal{A}_j \cdot \dot{\mathbf{q}}_2, \tag{19}$$

where the integration of the last term produces exactly the Berry phase of the jth eigenstate. One can regard this as a new way to derive the Berry phase; in this new way, the quantum system does not need to be in an eigenstate.

For the slow classical subsystem, we take the averaging over Eq. (13) and rewrite it in a physically more transparent form

$$M\ddot{\bar{\mathbf{q}}}_2 = -\frac{\partial \mathcal{H}_1}{\partial \mathbf{q}_2} - \frac{\partial V_2}{\partial \mathbf{q}_2} + \dot{\bar{\mathbf{q}}}_2 \times \mathcal{B}, \tag{20}$$

where $\mathcal{B} = \nabla \times \bar{\mathbf{A}} = \sum_n I_{1n}\nabla \times \mathcal{A}_n$ is a magneticlike gauge field. Similar to the usual magnetic field, the gauge field influences the dynamics in terms of a Lorentz-like force. So the Berry phase is shown to be linked to a physical force. We shall have more discussion with this force later via an example.

We have applied our method to the cantilever-spin system in the single spin detection experiment [1,5] and recovered the theoretical result in Ref. [4]. The pity is that, since the cantilever vibrates only in one dimension, the force associated with $\bar{\mathbf{A}}$ is always zero. To have a nonzero $\mathcal{B}$, we consider an example where the classical subsystem is a magnetic particle and the quantum subsystem is a spin of 1/2 as shown in Fig. 1. The magnetic particle has a magnetic moment $\mathbf{m}_F$ and mass m; it moves freely in the xy plane. We assume that the magnetic moment $\mathbf{m}_F$ always points in the negative z direction. A single spin with magnetic moment $\boldsymbol{\mu}$ is placed below the plane at the distance d. For simplicity, we place the origin of our coordinate system in the particle plane and directly above the spin.

Because of the magnetic dipolar interaction, the spin feels a magnetic field from the classical magnetic particle. The field is given by

$$\{B_x, B_y, B_z\} = -\frac{\mu_0 m_F\{3xd, 3yd, 2d^2 - r^2\}}{4\pi(d^2 + r^2)^{5/2}}, \tag{21}$$

where $r^2 = x^2 + y^2$. So the Hamiltonian operator for the spin is

$$\hat{H}_1 = -\mu\begin{pmatrix} B_z & B_x - iB_y \\ B_x + iB_y & -B_z \end{pmatrix}. \tag{22}$$

This Hamiltonian has two eigenstates $|\pm\rangle$, whose eigenenergies are, respectively, $\mp\mu B$, with $B = \sqrt{B_x^2 + B_y^2 + B_z^2}$. The total Hamiltonian is

$$H = \langle\mathbf{\Psi}|\hat{H}_1|\mathbf{\Psi}\rangle + \mathbf{p}^2/2m, \tag{23}$$

where $|\mathbf{\Psi}\rangle = (\Psi_1, \Psi_2)^T$ is the spin wave function and p is the momentum of the magnetic particle.

Following the general procedure described above, we can transform the above Hamiltonian to

$$\mathcal{H} = (|a_-|^2 - |a_+|^2)\mu B + \frac{(\mathbf{P} - \bar{\mathbf{A}})^2}{2m}, \tag{24}$$

where $|a_\pm|^2$ are the probabilities on the two spin eigenstates $|\pm\rangle$, respectively. The vector potential is given by

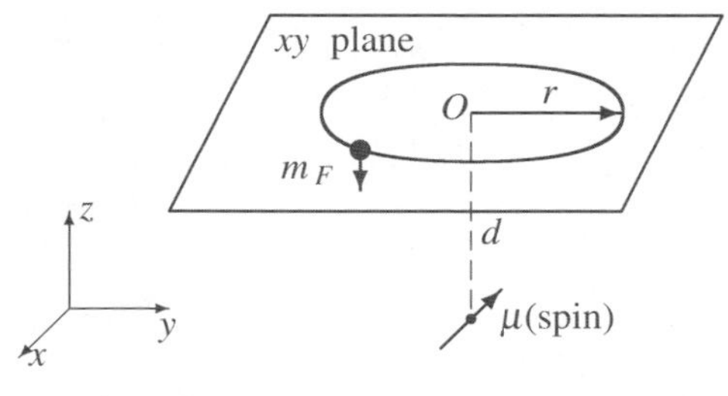

FIG. 1. A schematic setup of a magnetic particle interacting with a spin. The particle moves freely in the xy plane with a magnetic moment of $\mathbf{m}_F$ always pointing in the negative z direction. A single spin with magnetic moment $\boldsymbol{\mu}$ is placed beneath the plane with a distance of d.

$$\bar{\mathbf{A}} = i\hbar|a_+|^2\langle+|\frac{\partial}{\partial\mathbf{r}}|+\rangle + i\hbar|a_-|^2\langle-|\frac{\partial}{\partial\mathbf{r}}|-\rangle. \quad (25)$$

The equations of motion for the particle are

$$m\ddot{x} = \frac{3\mu\mu_0 m_F(5d^2 + r^2)(|a_-|^2 - |a_+|^2)}{4\pi\sqrt{4d^2 + r^2}(d^2 + r^2)^3}x + \mathcal{B}\dot{y}, \quad (26)$$

$$m\ddot{y} = \frac{3\mu\mu_0 m_F(5d^2 + r^2)(|a_-|^2 - |a_+|^2)}{4\pi\sqrt{4d^2 + r^2}(d^2 + r^2)^3}y - \mathcal{B}\dot{x}. \quad (27)$$

The magneticlike $\mathcal{B}$ field always points in the z direction. The field strength is

$$\mathcal{B} = \frac{9\hbar d^2(r^2 + 2d^2)}{2[(r^2 + d^2)(r^2 + 4d^2)]^{3/2}}(|a_+|^2 - |a_-|^2). \quad (28)$$

We make two observations. First, the field is geometric, depending only on the position of the magnetic particle besides $\hbar$ and independent of the strength of the dipolar interaction. Second, it curves motion in an unexpected way. For instance, if the initial conditions of the slow particle are $x(0) = 0$, $y(0) = 0$, $\dot{x}(0) > 0$, and $\dot{y}(0) = 0$, then everything in the real space is symmetric with respect to the inverse of y, including the dipolar force and the spin direction. One would then expect intuitively that the particle moves in a straight line along the x direction. However, the slow particle will curve owing to $\mathcal{B}$, breaking the left-right symmetry in the real space. What is interesting is that the force breaking of this real-space symmetry comes from the dynamics of the spin wave function's phase, a non-real-space variable.

To detect such a force experimentally, the best way may be to measure the frequency associated with this force, such as the frequency change in the single spin detection [1]. For the simple system in Fig. 1, if the spin is only an electron and $\mu < 0$, we notice that the slow particle can move in a circle if $|a_+|^2 < |a_-|^2$. If somehow one can fix the radius r of the circle, the frequency that the particle circulates clockwise is different from that when it circulates anticlockwise. Simple calculations show that the frequency difference is $\Delta\nu = \mathcal{B}/2\pi m$. For estimation of the value of the frequency difference, the following parameters are used: $\mu_0 m_F \sim 2.0 \times 10^{-21}$ T m^3, $m \sim 2.5 \times 10^{-15}$ kg, $r \sim 1$ nm, and $d \sim 1000$ nm. For these values of parameters, we obtain, for the case $|a_+|^2 = 0$ and $|a_-|^2 = 1$, $\mathcal{B} \sim -1.20 \times 10^{-22}$ kg/s (if it were for an electron, it would be equivalent to a magnetic field of 7.5×10^{-4} T) and $Bz \sim 3.2 \times 10^{-4}$ T. The frequency difference is $\Delta\nu \sim 0.7 \times 10^{-8}$ Hz, which is a challenging task for the current technique [1]. Moreover, in a real experiment, the magnetic particle needs to be attached to something such as a string so that it can oscillate in two dimensions.

A similar vector potential can also arise in connection to Hannay's angle [16]. The theoretical framework starting at Eq. (6) can be readily generalized to the cases where the fast subsystem is a classical integrable system. In this generalization, one needs only to regard $\mathbf{I}_1$ and $\mathbf{\Theta}_1$ as a set of action-angle variables. Such a coupled system was also discussed in Ref. [20]. The same extension can be done for the geometric phase proposed for nonlinear quantum systems [17].

In the general case where the two subsystems are not necessarily one fast and one slow, one may use the pseudovector potential $\mathbf{A}$ to define a "geometric" phase for nonadiabatic processes. However, this general phase may be of little use since it does not provide any insight into the dynamics. We finally note that the quantum-classical hybrid system has been studied extensively in the context of quantum measurement [21].

We thank C. P. Sun for drawing our attention to the single-spin-detection experiment. We are also grateful to Junren Shi, Di Xiao, and Qian Niu for helpful discussion. This work is supported by the "BaiRen" program of the CAS, NSF of China (No. 10504040), and the 973 project (No. 2005CB724508).

[1] D. Rugar, R. Budakian, H. J. Mamin, and B. W. Chui, Nature (London) **430**, 329 (2004).
[2] P. C. Hammel, Nature (London) **430**, 300 (2004).
[3] C. P. Slichter, *Principles of Magnetic Resonance* (Springer, Berlin, 1990).
[4] G. P. Berman, D. I. Kamenev, and V. I. Tsifrinovich, Phys. Rev. A **66**, 023405 (2002).
[5] H. J. Mamin, R. Budakian, B. W. Chui, and D. Rugar, Phys. Rev. Lett. **91**, 207604 (2003).
[6] M. V. Berry and J. M. Robbins, Proc. R. Soc. A **442**, 659 (1993).
[7] C. A. Mead and D. G. Truhlar, J. Chem. Phys. **70**, 2284 (1979).
[8] Y. Aharonov, E. Ben-Reuven, S. Popescu, and D. Rohrlich, Phys. Rev. Lett. **65**, 3065 (1990).
[9] H. Kuratsuji and S. Iida, Prog. Theor. Phys. **74**, 439 (1985).
[10] M. Stone, Phys. Rev. D **33**, 1191 (1986).
[11] R. G. Littlejohn and S. Weigert, Phys. Rev. A **48**, 924 (1993).
[12] A. Heslot, Phys. Rev. D **31**, 1341 (1985).
[13] S. Weinberg, Ann. Phys. (N.Y.) **194**, 336 (1989).
[14] J. Liu, B. Wu, and Q. Niu, Phys. Rev. Lett. **90**, 170404 (2003).
[15] M. V. Berry, Proc. R. Soc. A **392**, 45 (1984).
[16] J. H. Hannay, J. Phys. A **18**, 221 (1985).
[17] B. Wu, J. Liu, and Q. Niu, Phys. Rev. Lett. **94**, 140402 (2005).
[18] V. I. Arnold, *Mathematical Methods of Classical Mechanics* (Springer-Verlag, Berlin, 1978).
[19] A. Messiah, *Quantum Mechanics* (Dover, New York, 1958).
[20] E. Gozzi and W. D. Thacker, Phys. Rev. D **35**, 2398 (1987).
[21] M. J. W. Hall and M. Reginatto, Phys. Rev. A **72**, 062109 (2005).

Generalization of Faraday's Law to Include Nonconservative Spin Forces

S. E. Barnes[1,2] and S. Maekawa[1,3]

[1]*Institute for Materials Research, Tohoku University, Sendai 980-8577, Japan*
[2]*Physics Department, University of Miami, Coral Gables, Florida 33124, USA*
[3]*CREST, Japan Science and Technology Agency (JST), Kawaguchi 332-0012, Japan*
(Received 22 November 2006; published 15 June 2007)

The usual Faraday's Law $\mathcal{E} = -d\Phi/dt$ determines an electromotive force $\mathcal{E}$ which accounts only for forces resulting from the charge of electrons. In ferromagnetic materials, in general, there exist nonconservative spin forces which also contribute to $\mathcal{E}$. These might be included in Faraday's Law if the magnetic flux Φ is replaced by $[\hbar/(-e)]\gamma$, where γ is a Berry phase suitably averaged over the electron spin direction. These contributions to $\mathcal{E}$ represent the requirements of energy conservation in itinerant ferromagnets with time dependent order parameters.

DOI: 10.1103/PhysRevLett.98.246601 PACS numbers: 72.25.Ba, 03.65.Vf, 75.60.Ch, 85.75.−d

Faraday's $\mathcal{E} = -d\Phi/dt$ is a fundamental law which relates the electromotive force (emf) $\mathcal{E}$ to the time derivative of the total magnetic flux Φ. By definition [1], $q\mathcal{E} \equiv \oint_C \vec{f} \cdot d\vec{r}$, i.e., is the work done on a test charge q, at $\vec{r}$, as it is carried around a contour C, which may or may not coincide with a physical electrical circuit. Usually [1] the force $\vec{f}$ is taken to be $q(\vec{E} + \vec{v} \times \vec{B})$, where $\vec{v}$ is the velocity of C. In principle, this definition of $\mathcal{E}$ includes all contributions to the emf $\mathcal{E}$ which arise from the charge q. However, in nature, the charge carriers are elementary particles, usually electrons, which carry a magnetic moment μ, in addition to charge. Omitted in this definition of the emf $\mathcal{E}$ are the forces of spin origin.

In this Letter, in order to include nonconservative spin forces, it is proposed to generalize Faraday's Law as

$$\mathcal{E} = -\frac{\hbar}{(-e)}\frac{d\gamma}{dt}, \tag{1}$$

where γ is a spin average of the geometric or Berry phase [2] of both charge and spin origin. Of electromagnetic origin, for an electron, when it is transported around a contour C, [2,3] $\gamma_e = \frac{(-e)}{\hbar}\Phi$, is just the phase of Aharonov-Bohm and, with $\gamma = \gamma_e$, Eq. (1) reproduces the usual $\mathcal{E} = -d\Phi/dt$. The calculation of the spin contribution to γ is here illustrated by the simple example of a ferromagnetic wire containing a free domain wall. Domain walls are not essential, as will be shown elsewhere; spin contributions to $\mathcal{E}$ occur in all situations where the order parameter is time dependent.

The usual formulation of an itinerant ferromagnet needs to be extended to include the Berry phase [2,3]. It is usual to define $\varphi_s^\pm$, a potential energy which is different for majority and minority electrons. In the superscript the upper (lower) sign corresponds majority (minority) electrons, a convention used throughout this Letter. In a simple situation, $\varphi_s^\pm = \mp\mu B_i(\vec{r})$, where $B_i(\vec{r})$ is the magnitude of the position, $\vec{r}$, dependent internal field. However, the evident force $-\vec{\nabla}_{\vec{r}}\varphi_s^\pm$, operative in the Stern-Gerlach experiment, is conservative and cannot contribute to $\mathcal{E}$. As will be seen explicitly below, the spin Berry phase $\gamma_s^\pm = (1/2)\int^{\vec{r}} \vec{A}_s^\pm \cdot d\vec{r}$ can be written in terms of a spin vector potential [2,3] $\vec{A}_s^\pm$, which makes a contribution to the total spin force

$$\vec{f}_s^{\,\pm} = -\frac{\hbar}{2}\frac{\partial \vec{A}_s^\pm}{\partial t} - \vec{\nabla}_{\vec{r}}\varphi_s^\pm. \tag{2}$$

The nonconservative spin forces which contribute to $\mathcal{E}$ are reflected in $\gamma_s^\pm$, $\vec{A}_s^\pm$, and thereby $\vec{f}_s^{\,\pm}$. In Eq. (2), $\hbar/2$ is to be recognized as the "spin charge" of an electron.

The very simple example studied here, see Fig. 1, is a ferromagnetic wire, of cross-sectional area $\mathcal{A}$, which lies along the z direction and which contains a single free domain wall. The easy axis is along the wire and perpendicular anisotropy is zero. The ferromagnetic order parameter is indicated by the arrows and the unit vector $\hat{\mathbf{n}}(\vec{r}, t)$. This defines the direction of the internal exchange field $JM\hat{\mathbf{n}}(\vec{r}, t)$. For such a wall centered at $\langle z_0\rangle$, $\hat{\mathbf{n}}$ makes an angle $\theta(z, t) = 2\cot^{-1} e^{-([z-\langle z_0\rangle(t)]/w)}$ to the z axis, where w is the wall width. For a density n of spin $\frac{1}{2}$ electrons with a polarization p, for a centrally located

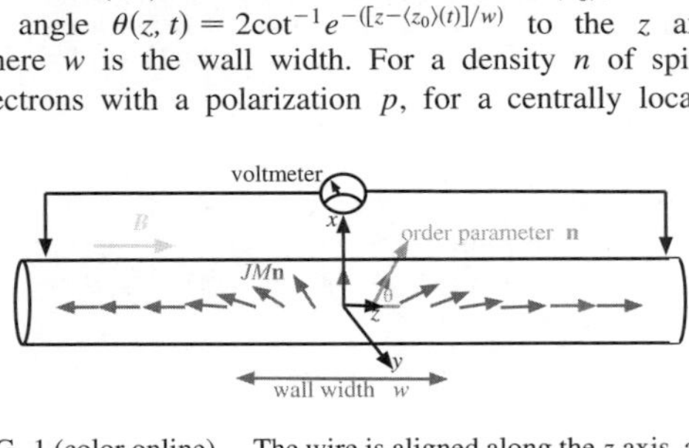

FIG. 1 (color online). The wire is aligned along the z axis, as is the longitudinal field $\vec{B} = B_0\hat{\mathbf{z}}$. The wire contains a single domain wall near its center. The unit vector $\hat{\mathbf{n}}$ defines the direction of the ferromagnetic order parameter, and the internal exchange field $JM\hat{\mathbf{n}}$, within this wall. This vector makes an angle $\theta(z) = 2\cot^{-1} e^{-[(z-\langle z_0\rangle)/w]}$ to the z axis. The emf developed along the wire is measured by an ideal voltmeter connected as shown. In this illustration the domain wall lies $x - z$ plane. In fact, the wall processes so that it makes a time dependent angle ϕ with this plane, with $d\phi/dt = 2\mu B_0/\hbar$.

wall, the total z component of the magnetization $\langle S_z \rangle = \frac{1}{2}\hbar np\mathcal{A} \int dz \cos\theta = \hbar np\mathcal{A}\langle z_0 \rangle$, following an integration by parts, i.e., the position operator $z_0 = S_z/\hbar np\mathcal{A}$, is to an excellent approximation proportional to that for the total spin, but see below. The wall makes a uniform angle $\phi(t)$ to $x - z$ plane and undergoes a Larmor precession, i.e., $d\phi/dt = 2\mu B_0/\hbar$, due to the pressure exerted on the wall by a static external magnetic field $\vec{B} = B_0\hat{\mathbf{z}}$ which is assumed small compared to the internal field, i.e., $2\mu B_0 \ll \hbar JM$.

The Berry phase is usually defined in terms of a Hamiltonian $\mathcal{H}(\vec{R})$ which is dependent on the time through the parameters comprising $\vec{R}(t)$. The time dependent $\psi_{m\pm}(\vec{R})$, where $+(-)$ corresponds to majority (minority) spin, are solutions to the eigenenergy problem: $\mathcal{H}(\vec{R})\psi_{m\pm}(\vec{R}) = E_m(\vec{R})\psi_{m\pm}(\vec{R})$, and the Berry phase is [3]

$$\gamma^{\pm} = \frac{1}{2}\int^{\vec{R}} \vec{a}_s^{\mp} \cdot d\vec{R}; \qquad \vec{a}_s^{\pm}(\vec{R}, t) = 2i\langle \psi_{m\pm}|\vec{\nabla}_{\vec{R}}\psi_{m\pm}\rangle. \quad (3)$$

For an electron subject to a time dependent magnetic field along the direction $\hat{\mathbf{n}}(t)$, the spin Berry phase $\gamma_s^{\mp} = \mp(1/2)\Omega$, where Ω, Fig. 2(a), is the solid angle subtended by the path of $\hat{\mathbf{n}}(t)$ on a unit sphere [2,3]. In the frame of an electron, as it passes through a domain wall, the effective magnetic field follows the direction of the order parameter $\hat{\mathbf{n}}(\vec{r}, t)$. This rotates, Fig. 2(b), from the South to North Pole. By a change of variables, this motion in spin space can be mapped to that in real space, as illustrated by Figs. 2(b) and 2(c). Using Eq. (3), but with $\vec{R}$ identified simply as $\vec{r}$, is defined a real space vector potential $\vec{A}_s^{\mp} \equiv 2i\langle \psi_{\vec{k}\pm}|\vec{\nabla}_{\vec{r}}\psi_{\vec{k}\pm}\rangle$, such that the Berry phase $\gamma_s^{\mp} = (1/2) \times \int^{\vec{r}} \vec{A}_s^{\mp} \cdot d\vec{r}$. While $\vec{A}_s^{\mp}$ and $\gamma_s^{\pm}$ for the path of Fig. 2(b) and 2(c) are gauge dependent, the force $-\partial\vec{A}_s^{\mp}/\partial t$ and hence the spin emf are gauge independent, as will be discussed below.

In the present context the usual result $\gamma_s^{\mp} = \mp(1/2)\Omega$ is established within the widely accepted ferromagnetic Stoner model [4]. There is an internal field proportional to the order parameter $M\hat{\mathbf{n}}(\vec{r}, t)$, and the appropriate Schrödinger's equation is $i\hbar\frac{\partial\psi(\vec{r},t)}{\partial t} = \mathcal{H}\psi(\vec{r}, t)$, with

$$\mathcal{H} = \left[\frac{p^2}{2m} + V(\vec{r}) - JM\vec{s}\cdot\hat{\mathbf{n}} - \frac{2\mu B_0}{\hbar}s_z\right], \quad (4)$$

where $V(\vec{r})$ is a potential including any disorder and $\vec{s}$ the electron spin vector. First a z-axis rotation $u_\phi(t) = e^{-is_z\phi/\hbar}$, with $d\phi/dt = 2\mu B_0/\hbar$ eliminates $(2\mu B_0/\hbar)s_z$ and then [5] the axis of quantization is made parallel to $\hat{\mathbf{n}}$ by the y-axis rotation $u_\theta = e^{-is'_y\theta/\hbar}$. The result for $\psi'(\vec{r}, t) = u_\theta u_\phi \psi(\vec{r}, t)$ is $i\hbar\frac{\partial\psi'(\vec{r},t)}{\partial t} = \mathcal{H}'\psi'(\vec{r}, t)$

$$\mathcal{H}' = \left[\frac{(\vec{p} - \frac{\hbar}{2}\vec{A}_s^{\prime t})^2}{2m} + V(\vec{r}) - JMs'_z\right], \quad (5)$$

where a prime indicated a quantity in this new frame. The

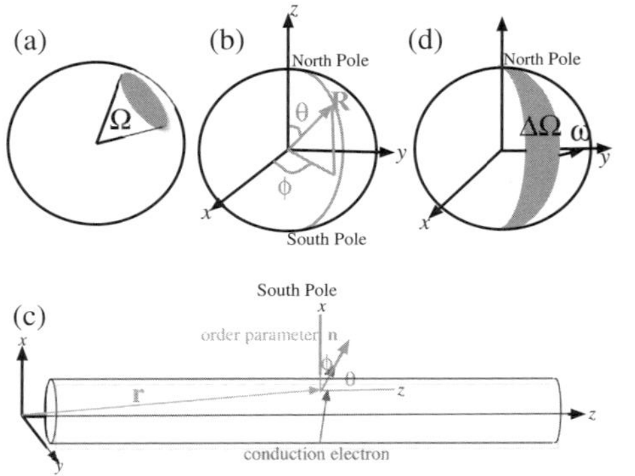

FIG. 2 (color online). The spin Berry phase $\gamma_s^{\mp}$ is $\mp\Omega/2$, where Ω is the solid angle subtended by the path in spin space, as illustrated in (a). The electron in (c) is at position $\vec{r}$ and sees a field with a direction defined by the Euler angles θ, ϕ, while, as shown in (b), in spin space, the vector $\vec{R}$, defined by the same angles, passes from the South to North pole. The solid angle Ω for such a path is gauge dependent; however, the solid angle $\Delta\Omega$ sweep out in a time Δt is well defined, as illustrated in red in (d). Reflecting the precession of the wall, the path has an angular speed $\omega = \dot{\phi} = 2\mu B_0/\hbar$ and the derivative $\dot{\gamma}_s^{\mp} = \mp(1/2)\dot{\Omega} = \mp\dot{\phi}$. While this is a "motional motive force" in spin space it is not so in real space since in (c) the electron has no transverse velocity.

spin transverse vector potential $\vec{A}_s^{\prime t} = -\frac{2}{\hbar}s'_y\frac{\partial\theta}{\partial z}\hat{\mathbf{z}}$. In the absence of this potential, Eq. (5) is the Stoner model without a wall. The stationary solutions with label m, $\psi'_{m\pm}(\vec{r}, t)$ have energies $E_m^{\pm} = E_m^0 \mp (\hbar/2)JM$, as the spin eigenvalue $s'_z = \pm\hbar/2$. These time reversal conjugate pairs correspond to majority or minority electrons as $s'_z = \hbar/2$ or $-\hbar/2$, respectively. Their energies differ by $\hbar JM$. If $V(\vec{r}) = 0$ (or is periodic), these are also eigenstates of the (crystal) momentum $\vec{p} = \hbar\vec{k}$; i.e., the total force is zero and $\vec{p}$ is a constant of motion.

The results for the current example reflect the adiabatic approximation and follow when $\vec{A}_s^{\prime t}$ is neglected, in Eq. (5). There $(\hbar/2m)\vec{p}\cdot\vec{A}_s^{\prime t}$ represents an effective transverse field $\sim\hbar v_F\pi/w$, v_F the Fermi velocity, and it is necessary that this field be small compared to the internal field, i.e., $\hbar JM > \hbar v_F\pi/w$, and implying the matrix elements of $(\hbar/2m)\vec{p}\cdot\vec{A}_s^{\prime t}$ between the $\pm$ eigenstates $\psi'_{m\pm}(\vec{r})$ are small compared to their energy difference $\hbar JM$. That the energy spread of these coupled states $\Delta E \propto v_F/w$ is also small compared to $\hbar JM$ is implicit. It is still the case that $\vec{A}_s^{\prime t}$ is important since it leads to a small but highly important correction to the wall position which will be calculated below.

The initial $\mathcal{H}$, Eq. (4), is time dependent only through the single parameter $\phi(t)$, i.e., the Berry phase

$$\gamma_s^{\mp}(\vec{r}) = i\int_0^{\phi}\left\langle\psi_{m\pm}|\frac{\partial}{\partial\phi}\psi_{m\pm}\right\rangle d\phi, \quad (6)$$

where the expectation value in Eq. (6) is with respect to the laboratory frame spinor wave function, $\psi_{m\pm}(\vec{r},t) = u_\phi^{-1}u_\theta^{-1}\psi'_{m\pm}(\vec{r},t)$, and $\gamma_s^\pm(\vec{r})$ is defined at each real space point, explicitly,

$$\gamma_s^\mp = \pm(1/2)\int_0^\phi \cos\theta d\phi = \pm(1/2)\cos\theta\phi, \quad (7)$$

which is just $\mp(1/2)\Omega$ but with an unusual specific gauge in which half of the Dirac string [1] leaves each of the South and North Poles; i.e., a path which encircles either Pole picks up a π phase. The real space $\vec{A}_s^\pm(\vec{r},t)$ is then determined by writing $\gamma_s^\pm = (1/2)\int^{\vec{r}} \vec{A}_s^\pm(\vec{r},t)\cdot d\vec{r}$, whence using the result $\gamma_s^\mp = \pm(1/2)\cos\theta\phi$, it follows $\vec{A}_s^\mp = 2(\partial/\partial z)\gamma_s^\mp\hat{\mathbf{z}} = \pm\phi(\partial/\partial z)\cos\theta\hat{\mathbf{z}}$.

The mathematical equivalence, to within the definition of charge, of the Berry phases of electromagnetic and spin origin [3] is a sufficient proof that there is a spin force $\vec{f}_{\rm nc} = -(\hbar/2)(\partial\vec{A}_s^\pm/\partial t)$, as in Eq. (2), acting on an electron. Explicitly the $\vec{f}_{\rm nc} = \mp(\hbar/2)(d\phi/dt)(\partial/\partial z)\cos\theta\hat{\mathbf{z}}$, using $\vec{A}_s^\pm$ from above. The spin contribution to the motive force is therefore $\mathcal{E}_s^\mp = (\hbar/2e)\oint(\partial\vec{A}_s^\mp/\partial t)\cdot d\vec{r}$; however, $\vec{A}_s^\mp$ is only finite in the vicinity of the wall, so this contour integral simplifies to $\int_{-\infty\hat{z}}^{\infty\hat{z}}(\partial\vec{A}_s^\mp/\partial t)\cdot d\vec{r}$. This is a gauge invariant quantity since it is given by the limit $\Delta t\to 0$ of $\Delta I = \int_{-\infty\hat{z}}^{\infty\hat{z}}[\vec{A}_s^\mp(\vec{r},t+\Delta t) - \vec{A}_s^\mp(\vec{r},t)]\cdot d\vec{r}$. Mapped to spin space, $I(t) = \int_{-\infty\hat{z}}^{\infty\hat{z}}\vec{A}_s^\mp(\vec{r},t)\cdot d\vec{r} = \int_{C_\ell}\vec{a}_s^\mp(\vec{R},t)\cdot d\vec{R}$, where C_ℓ is the longitude with $\phi = 0$, while for $I(t+\Delta t)$, $\phi = (d\phi/dt)\Delta t$. It follows $\Delta I = I(t+\Delta t) - I(t) = \oint_{\Delta C}\vec{a}_s^\mp\cdot d\vec{R}$, where the contour ΔC corresponds to the edge of the shaded region in Fig. 2(d). Since ΔC is *closed*, from the definition Eq. (3) this indeed corresponds to a *gauge invariant* infinitesimal Berry phase $d\gamma_s^\mp = dI(t)/2$ and the result,

$$\mathcal{E}_s^\mp = -\frac{\hbar}{(-e)}\frac{d\gamma_s^\pm}{dt}. \quad (8)$$

Alternatively, using the relationship $\gamma_s^\mp = \mp(1/2)\Omega$, this can be written as

$$\mathcal{E}_s^\mp = \pm\frac{\hbar}{2(-e)}\frac{d\Omega}{dt}, \quad (9)$$

i.e., in terms of the rate at which solid angle is swept out in spin space. Since $(d\phi/dt) = 2\mu B_0/\hbar$ this leads to $\mathcal{E}_s^\mp = \pm 2\mu B_0/(-e)$.

Given $\mathcal{E}_e = -d\Phi/dt$ is the motive force due to charge, the total $\mathcal{E}_e + \mathcal{E}_s^\mp$ is different for majority and minority electrons. When the transport in an electric field $\vec{E}$ alone can be described by conductivities, $\sigma^\pm$, such that the current densities $\vec{J}^\pm = \sigma^\pm\vec{E}$, then trivially $J = J^+ + J^- = \sigma E$ with $\sigma = \sigma^+ + \sigma^-$ while the spin equivalent $J_s = p\sigma E = pJ$, where $p = (\sigma^+ - \sigma^-)/\sigma$ is the polarization of the spin current. It follows that $\mathcal{E}_e + p\mathcal{E}_s^+$ is an emf which drives the charge current while $\mathcal{E}_s^+$ drives a pure spin current. Faraday's Law for the emf is therefore Eq. (1) with

$$\gamma = \gamma_e + p\gamma_s^+ \quad (10)$$

as the average Berry phase. This is the principal result.

The nature of this spin contribution to Faraday's Law is subtle; it is necessary to account for the very small wall displacements $\Delta z_0^\mp$ induced by those of the individual electrons, an effect which is contained in the vector potential $\vec{A}_s^{\prime t}$. Since this is irrotational within the wall, the formal solution of Eq. (5) is $u_A\psi'_{m\pm}$, where $u_A = e^{i(1/2)\int_{\vec{r}_0}^{\vec{r}}\vec{A}_s^t\cdot d\vec{r}'}$ and $\vec{r}_0$ is arbitrary. Consider the full Slater determinant Ψ' constructed with the $\psi'_{m\pm}$ and corresponding to a typical state. Define U_A, U_θ, and U_ϕ as the product over all electrons of the u_A, u_θ, and u_ϕ, respectively. These are the rotations, etc. appropriate to the full Hilbert space. Because the $\psi'_{m\pm}$ are spin eigenstates, with the approximation $U_A = 1$, the $\Delta z_0^\mp = 0$. It follows that these corrections are given by the expectation value of $\Delta \equiv U_A^{-1}\tilde{z}_0U_A - \tilde{z}_0 = U_A^{-1}[\tilde{z}_0, U_A]$. Here z_0 is the wall position operator and $\tilde{z}_0 \equiv U_\theta U_\phi z_0 U_\phi^{-1}U_\theta^{-1}$. Reflecting the smallness of the $\Delta z_0^\mp$, Δ can be separated into a contributions $\Delta z_0 \equiv i(1/2)[\int_{\vec{r}_0}^{\vec{r}}\vec{A}_s^t\cdot d\vec{r}', \tilde{z}_0]$ for each electron. Then using the zeroth approximation, $z_0 = S_z/\hbar\mathcal{A}pn$, where S_z is the z component of the total spin, the σ_y in $\vec{A}_s^t$ commutes with all the other spin operators contained in S_z, and so $\Delta z_0 = -i(1/2)(1/\hbar\mathcal{A}pn)[\int_{\vec{r}_0}^{\vec{r}}\vec{A}_s^t\cdot d\vec{r}', \tilde{s}_z]$, where $\tilde{s}_z = U_\theta U_\phi s_z U_\phi^{-1}U_\theta^{-1} = u_\theta u_\phi s_z u_\phi^{-1}u_\theta^{-1} = \cos\theta s_z - \sin\theta s_x$. Taking the expectation value of the expression for Δz_0 with $\psi'_{m\pm}$ gives

$$\Delta z_0^\mp = \pm\frac{1}{2\mathcal{A}pn}[\cos\theta(\vec{r}) - \cos\theta(\vec{r}_0)]. \quad (11)$$

While the derivation is difficult, the result is easily understood. As an electron moves from $\vec{r}_0$ to $\vec{r}$, its orientation changes, causing the z component of the angular momentum to change from $\pm(\hbar/2)\cos\theta(\vec{r}_0)$ to $\pm(\hbar/2)\cos\theta(\vec{r})$. In order to conserve angular momentum, this change must be transferred to the wall which thereby undergoes the displacement $\Delta z_0^\mp$.

The existence of an nonconservative force $\vec{f}_{\rm nc} = f_{\rm nc}\hat{z}$ then follows from the conservation of energy. There is a pressure $P_z = n\mu B_0$ exerted on the wall by the field B_0 and the displacements $\Delta z_0^\mp$ imply an amount of work $P_z\mathcal{A}\Delta z_0^\mp$ is done on the wall. In order to conserve energy, there must be a corresponding nonconservative force $\vec{f}_{\rm nc} = f_{\rm nc}\hat{z}$ acting on the electron such that $f_{\rm nc} = -\partial/\partial z P_z\mathcal{A}\Delta z_0^\mp = \mp\mu B_0(\partial/\partial z)\cos\theta$. This is the force $-(\hbar/2)(\partial\vec{A}_s^\mp/\partial t)$ which leads to the $\mathcal{E}_s^\mp$.

Corresponding to the second term in Eq. (2) there is a distinct conservative force $\vec{f}_c = -\vec{\nabla}\varphi_s^\mp(\vec{r})$, which is contained in the dynamic phase [3] and reflects the position dependent Zeeman term in the effective spin potential energy $\varphi_s^\mp(\vec{r}) = \mp(\hbar JM + \mu B_0\cos\theta)$. Directly $\vec{f}_c = \pm\mu B_0(\partial/\partial z)\cos\theta\hat{\mathbf{z}}$ and which is the exact negative of $\vec{f}_{\rm nc}$.

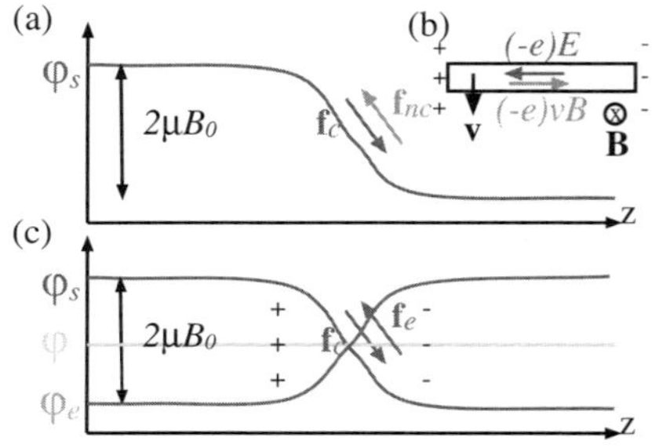

FIG. 3 (color online). The balance of forces. In (a), shown in blue, is φ_s the spin potential energy for the majority spins. The difference in potential energy at the ends is $2\mu B_0$. The net force $\vec{f}_c + \vec{f}_{\rm nc} = 0$ corresponding to equilibrium. (b) Compare this with the forces which occur when there is the usual motional emf. Charge accumulates at the ends of the bar producing an electrical potential energy φ_e the equivalent of the φ_s. The resulting conservative force $(-e)E$ is equal and opposite to the nonconservative force $(-e)vB$, to give a null net force. (c) If the wall is fixed so that $\Delta z_0^{\mp} = 0$, the initial potential is again that due to the Zeeman energy shown in blue. However, since now $\vec{f}_{\rm nc} = 0$ there is a force $\vec{f}_c$ on the electrons which leads to an accumulation of charge such that the total force $\vec{f}_c + \vec{f}_e = 0$, corresponding again to equilibrium. In purple is the electrical potential, the mirror image of the Zeeman energy, and the flat sum, shown in green.

This is no accident. Considered is an open circuit wire with no currents. This is an equilibrium situation for which it must be the case that the total spin force $\vec{f}_c + \vec{f}_{\rm nc} = 0$, as is illustrated in Fig. 3(a). This situation might be compared with that for the usual motional emf, Fig. 3(b). In this case, for an open circuit wire, the nonconservative force $(-e)vB_0$ is equal and opposite to the conservative force $(-e)E$ generated by charge accumulation at the ends of the wire.

Consider, for contrast, the case when the wall is pinned so that $\Delta z_0^{\mp} = 0$ and imagine for simplicity a half-metal (majority spins only). Now, Fig. 3(c), $\vec{f}_{\rm nc} = 0$, so the total force is just $\vec{f}_c \neq 0$. This is not an equilibrium situation and currents will flow. There will again be an accumulation of charge and the resulting internal electric field produces a force $\vec{f}_e = (-e)\vec{E}$. When equilibrium is reached the new total force $\vec{f}_c + \vec{f}_e = 0$. Both $\vec{f}_c$ and $\vec{f}_e$ derive from potential functions and are conservative. In equilibrium the total potential energy $\varphi(\vec{r})$ is therefore a constant. Since there are no nonconservative forces there is no emf. A time dependent order parameter is essential to the existence of a spin derived emf.

Stern [6] has already shown a radio frequency (rf) motive force and an associated rf current are induced in mesoscopic ring pumped by an rf field. This situation has been considered in more detail, e.g., by Ryu [7]. Berger [8] and others [9] have discussed the generation of direct current (dc) spin currents by rf fields and the associated motive forces. In each of these cases the source of energy driving the motive force is external. In contrast, the motive forces discussed here reflect internal degrees of freedom. In none of these earlier works were the implications for Faraday's Law clearly recognized. It is emphasized, in the absence of the motive forces discussed here, that the "classical" theory of current transport in ferromagnetic materials with time dependent order parameters does not conserve energy.

The equivalent of the vector potential, $\vec{A}_s^t$, in the context of domain walls, was first introduced by Bazaliy *et al.* [10]. This gauge field reflects the spatial dependence of the axis of quantization. In contrast, the Berry phase vector potential $\vec{A}_s^{\pm}$, of importance here, arises because these axes are time dependent. This vector potential and associated forces $f_s^{\mp}$ did not occur in this earlier time independent gauge theory.

The divergence of the differential form of Ampère's Law must reduce to the continuity equation for charge and that of Faraday's Law to the corresponding equation for magnetic charge [1]. In the absence of magnetic monopoles the latter becomes $(\partial/\partial t)\vec{\nabla}\cdot\vec{B} = 0$. Here the equivalent of the differential Faraday's Law results from taking the curl of the total force $\vec{f} + \vec{f}_s^{\pm}$ and taking the divergence of this results in $(\partial/\partial t)\vec{\nabla}\cdot(\vec{B} + \vec{B}_s^{\mp}) = 0$, where $\vec{B}_s^{\mp} = \vec{\nabla}\times\vec{A}_s^{\pm}$ and using $\vec{\nabla}\cdot(\vec{\nabla}\times\vec{A}) = 0$ for any $\vec{A}$. This reflects the absence of sources for $\vec{B}_s^{\pm}$.

In conclusion, in order to account for nonconservative spin derived forces reflecting the conversion of stored magnetic into electrical energy, Faraday's Law should involve the derivative of the spin average Berry phase and not just the magnetic flux. An additional contribution to the motive force occurs, with *static* external fields, in circumstances where there is no usual electromagnetic induction. The magnitude of $\mathcal{E}_s \sim 10^{-4}$ V/T and very recently emfs of this order have been observed [11]. This emf is important in spintronic applications [12].

[1] See, e.g., J. D. Jackson, *Classical Electrodynamics* (Wiley, New York, 1998).
[2] M. V. Berry, Proc. R. Soc. A **392**, 45 (1984).
[3] For details see, e.g., D. J. Griffiths, *Introduction to Quantum Mechanics* (Prentice-Hall, New Jersey, 1995).
[4] E. C. Stoner, Proc. R. Soc. A **165**, 372 (1938).
[5] S. E. Barnes and S. Maekawa, Phys. Rev. Lett. **95**, 107204 (2005).
[6] A. Stern, Phys. Rev. Lett. **68**, 1022 (1992).
[7] C. M. Ryu, Phys. Rev. Lett. **76**, 968 (1996).
[8] L. Berger, Phys. Rev. B **59**, 11 465 (1999).
[9] A. Brataas *et al.*, Phys. Rev. B **66**, 060404(R) (2002).
[10] Y. B. Bazaliy *et al.*, Phys. Rev. B **57**, R3213 (1998).
[11] E. Saitoh (private communication).
[12] S. E. Barnes, J. Ieda, and S. Maekawa, Appl. Phys. Lett. **89**, 122507 (2006).

Universal Electromotive Force Induced by Domain Wall Motion

Shengyuan A. Yang, Geoffrey S. D. Beach, Carl Knutson, Di Xiao, Qian Niu, Maxim Tsoi, and James L. Erskine
Department of Physics, The University of Texas, Austin, Texas, 78712-0264, USA
(Received 29 July 2008; published 9 February 2009)

The electromotive force induced by a moving magnetic domain wall in a nanostrip has been calculated theoretically and detected experimentally. It is found that the emf depends only on the domain wall transformation frequency through a universal Josephson type relation, which is closely related to the topological nature of the domain wall. Our experimental measurements confirm the theoretical prediction.

DOI: 10.1103/PhysRevLett.102.067201 PACS numbers: 75.75.+a, 72.25.Ba, 75.47.−m, 75.60.Ch

The interplay between charge transport and spin dynamics is a central theme of spintronics research. It is now well established that an electric current can drive magnetic domain wall (DW) motion via coupling between conduction electrons and local magnetic moments [1–5]. The reverse of this effect, i.e., an emf induced by a DW moving through a stationary electron gas, has also been predicted [6]. Recently there has been renewed interest in this effect [7–9], and its connection with Berry phase [10] has highlighted important topological aspects. However, prior theoretical work on DW-induced emf has only treated simple 1D walls, rather than more realistic 2D wall geometries [11,12], and no experimental detection has yet been reported. In this Letter, we formalize Berger's original result, generalize it to 2D DWs, and report the first experimental evidence for DW-driven emf.

To describe electron dynamics in ferromagnetic metals we adopt the *s-d* model, in which conduction electrons interact with local magnetic moments through an exchange coupling $H_{sd} = -J\hat{\boldsymbol{n}}(\boldsymbol{r}, t) \cdot \boldsymbol{\sigma}$. The unit vector $\hat{\boldsymbol{n}}$ denotes the local spin direction, $\boldsymbol{\sigma}$ is the itinerant electron spin, and J is the coupling strength. For typical ferromagnetic metals (Co, Fe, Ni and their alloys), *s-d* coupling is strong ($J \sim 1$ eV), and the conduction electron spin follows the local spin direction during its motion. Then the spatial variation of the local spin texture gives rise to the well-known Berry curvature field $\boldsymbol{C}(\boldsymbol{r}, t) = (1/2)\sin\theta(\nabla\theta \times \nabla\phi)$, where ($\theta$, ϕ) are the spherical angles specifying the direction of $\hat{\boldsymbol{n}}$. The $\boldsymbol{C}$-field acts on the electrons as an effective magnetic field whose flux enclosed by a loop gives the Berry phase of the spin-1/2 particle that moves around the loop. This is analogous to the Aharonov-Bohm phase acquired by a charged particle moving around a magnetic flux. Furthermore, in analogy with the electric field that arises from motion of a magnetic field, the $\boldsymbol{C}$-field of a moving spin texture also generates an effective electric field $\boldsymbol{D}(\boldsymbol{r}, t) = (1/2)\sin\theta(\partial_t\phi\nabla\theta - \partial_t\theta\nabla\phi)$ [13]. These fields can be derived directly from the semiclassical equation of motion for an electron wave packet [14], where the force on the conduction electrons from a moving spin texture can be shown to be $\boldsymbol{F} = -\hbar(\dot{\boldsymbol{r}} \times \boldsymbol{C} + \boldsymbol{D})$. This $\boldsymbol{D}$-field is the origin of the emf generated by a moving DW.

In this work, we consider a DW confined in a ferromagnetic nanostrip. Depending on the strip size, the static configuration of the DW can be either a (1D) transverse DW or a vortex DW [11]. Berger's phenomenological approach [6] predicted a voltage $V_x = (\hbar/e)d\psi/dt$ generated across a transverse DW-driven by a field beyond the Walker precessional threshold [15], which was termed ferro-Josephson effect due to its resemblance to the Josephson relation in superconductivity. Here subscript x labels the longitudinal direction of the nanostrip, $\hbar$ is the reduced Planck constant, $-e$ is the electron charge, and ψ is the tilt angle of the rigidly precessing DW plane. Berger's prediction follows rigorously from our Berry phase approach. The $\boldsymbol{D}$-field of a moving transverse wall, integrated along the direction of its motion, exactly reproduces Berger's result and is independent of the detailed form of the wall.

The large magnetostatic energy associated with rigid wall plane canting causes 1D DW models to fail at relatively low fields [16,17]. In nanostrips (width w, thickness t, where $w \gg t$), well before the 1D Walker field is reached [16], a vortex core periodically nucleates in a DW at the wire edge and slowly travels across the wire [18]. Therefore the dynamic wall-structure evolution, via gyrotropic vortex motion, requires a fully 2D treatment of the vortex structure. A vortex DW is characterized by a small core region with radius a ($\sim$ several nm), in which $\hat{\boldsymbol{n}}$ points out of the 2D plane. The Berry curvature field $\boldsymbol{C}$ is nonzero only inside the vortex core and its integral over the 2D plane gives the topologically invariant Skyrmion charge associated with the vortex [19], i.e., $\int \boldsymbol{C} d^2r = pq\pi\hat{\boldsymbol{e}}_z$, which is independent of the detailed profile of the vortex. Here p, $q = \pm 1$ are the polarization and vorticity of the vortex, respectively.

We calculate the emf induced by the motion of a 2D vortex DW by solving the Poisson equation $\nabla^2 V = (\hbar/e)\nabla \cdot \boldsymbol{D}$ with Neumann boundary conditions (no current leaving the sample). We find that the $\boldsymbol{D}$-field is that of an electric dipole $\mathbf{P}$ located within the vortex core,

$$\mathbf{P} = \varepsilon_0 \frac{\hbar}{e} \int \boldsymbol{D} d^2r = \varepsilon_0 \frac{\hbar}{e} pq\pi\hat{\boldsymbol{e}}_z \times \boldsymbol{v}, \quad (1)$$

where we use $\boldsymbol{D} = \boldsymbol{C} \times \boldsymbol{v}$ for rigid DW motion with $\boldsymbol{v}$ being the velocity of the vortex core. Since $a \ll w$, **P** can be treated as a point dipole. By the method of image charge, we find that the longitudinal voltage drop along the direction of the DW propagation is

$$V_x = \pi \frac{\hbar}{e} \frac{|v_y|}{w}. \quad (2)$$

Transverse vortex motion results from the Magnus force [20]. At low fields, this force is balanced by the confining potential of the wire, and $v_y = 0$. Above the breakdown transition, the Magnus force overcomes the confining potential and the vortex core within a DW is driven to the wire edge. There, it annihilates and a new vortex with opposite polarization is emitted, and is driven to the other edge [21,22]. Since both p and v_y change sign at the edge while q is kept, the sign of the product pqv_y is conserved during the periodic process, leading to a nonzero dc voltage across the wall whose polarity is unchanged for a unidirectional DW propagation (along the x direction). Its time average depends only on the frequency f of transverse motion, i.e. $\bar{V}_x = 2\pi(\hbar/e)f$. This result, together with the result for 1D DW, reveal a universal Josephson type relation with the Josephson frequency being the frequency of DW transformation. Moreover, our derivation demonstrates clearly that this remarkable result is intimately related to the topological nature of DW.

If the DW is driven by an applied magnetic field H, it has been shown that this frequency is the Larmor frequency, $2\pi f = \gamma H$ [22,23], where γ is the gyromagnetic ratio. Then the average voltage is linear in the applied field, with $\bar{V}_x/H = (\hbar/e)\gamma \approx 11.6$ nV/Oe. Accounting for spin minority carriers amounts to multiplying the predicted voltage by the spin-polarization factor P. Additional voltages from nonadiabatic or dissipative effects, estimated through force balance considerations [8,24], are found to be at least 10 times smaller than the adiabatic contribution above breakdown, though they are the only contribution below breakdown.

We have detected the longitudinal voltage generated by the motion of a single DW using the technique outlined in Fig. 1. The DW was confined within a $Ni_{80}Fe_{20}$ (Permalloy, Py) nanowire, 500 nm wide, 20 nm thick, and 35 μm long, that was joined to a large Py "injection pad" at one end and tapered to a point at the other. A field applied antiparallel to the wire magnetization (which lies along its axis) reverses the magnetically-soft pad and generates a DW at the pad-wire junction which is of vortex type according to the phase diagram [11]. If the field exceeds the DW injection field H_{inj}, the DW enters and traverses the wire from left to right. A nonmagnetic Pt bridge connected the pointed end of the wire to a second large Py pad. Electrical contacts made to the pads allowed the voltage across the wire to be measured as the DW traveled along it.

A high-frequency ($\sim$ 320 kHz) square-wave drive-field of amplitude H_d was used to repeatedly inject and drive DWs along the wire. The drive-field half-period $T_{1/2}$ exceeded the DW transit time τ_{DW} [see Fig. 1(e)], allowing each DW to reach the end of the wire before a new wall was injected. The high slew-rate fields used to inject and drive DWs generated large inductive voltages (tens of mV) in the measurement circuit. To extract the comparatively small voltage induced by DW motion ($\sim$100 nV), a modulation scheme was employed. In this approach, a secondary magnet was used to apply two additional fields: a 317 Hz square-wave modulation field (amplitude $H_m = 11$ Oe)

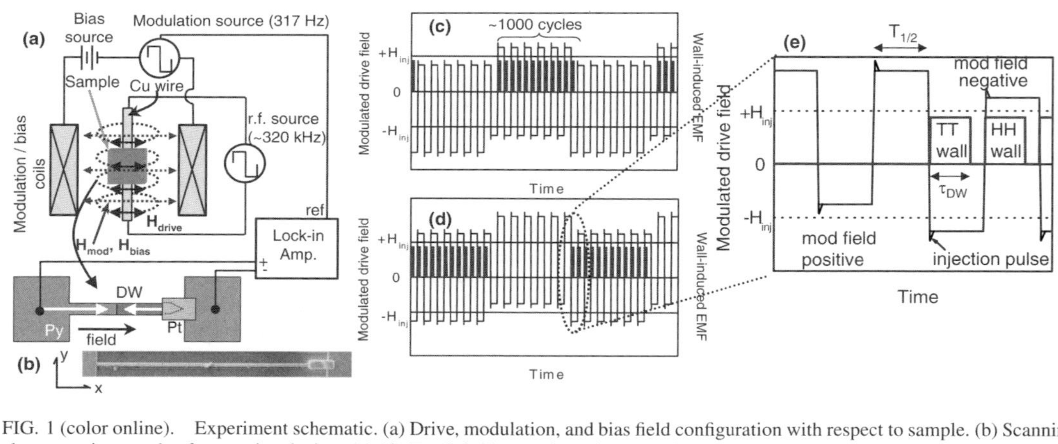

FIG. 1 (color online). Experiment schematic. (a) Drive, modulation, and bias field configuration with respect to sample. (b) Scanning electron micrograph of nanowire device. (c),(d) Total field waveform for two bias field levels. Injection field H_{inj} indicated by horizontal lines. Black solid lines indicate field waveform; blue indicates voltage signal expected from DW moving along nanowire. (e) Detail of several drive-field cycles immediately before and after a step in the modulation field. Before the step, the nanowire is saturated and no DWs travel along it. After the step, each drive-field half-cycle (duration $T_{1/2}$) injects a DW that moves along the wire, generating a voltage until it reaches the end of the wire after a time τ_{DW}.

plus a dc bias field H_{dc} which can be tuned. These two fields were used to modulate the probability that DWs were injected by H_d. Their combined effect is like a "switch" which can periodically turn the DW voltage signal "on" and "off."

The sum of these fields as a function of time is shown schematically in Fig. 1 for two values of $H_{\text{dc}}(= \pm H_m)$. For the negatively biased ($H_{\text{dc}} < 0$) waveform of Fig. 1(c), a DW is injected and driven across the wire at the beginning of the negative phase of the modulation cycle. Because the positive-going drive-field steps are biased below the DW injection threshold, no further DW motion occurs within the wire until the positive phase of the modulation cycle. During that phase, H_{total} oscillates symmetrically about zero and DWs, alternately head-to-head and tail-to-tail, are repeatedly injected and driven to the end of the nanowire. Each DW travels in the same direction, leading to a rapid sequence of voltage pulses with the same polarity and nonzero dc level. The voltage across the nanowire was measured using a lock-in amplifier phase-locked to the 317 Hz modulation field source [Fig. 1(a)], thereby rejecting rf inductive pickup at the drive frequency. The in-phase (real) lock-in voltage component, V_L, is proportional to the difference in the average wall-induced voltage between the positive and negative phases of the modulation field cycle.

The system has a rather high injection threshold $H_{\text{inj}} \approx$ 37 Oe. In order to study the emf for a driving field with $H_d < H_{\text{inj}}$, we add to the high-frequency drive-field waveform a short "injection pulse" (45 Oe peak, ~60 ns FWHM) [16] preceding each half-cycle [see Fig. 1(e)]. Each pulse briefly exceeded H_{inj}, injecting a DW which then propagated under a lower field H_d. This permitted the study of walls driven by a range of fields H_d, while maintaining identical injection characteristics. DW injection and propagation were probed using high-bandwidth scanning Kerr polarimetry [16]. The probability of wall injection, P_{inj}, as a function of H_{dc} was detected (with H_m set to zero here). At $H_{\text{dc}} = 0$, each injection pulse exceeds H_{inj}, and $P_{\text{inj}} = 1$. When a bias $|H_{\text{dc}}| > 8$ Oe is applied, P_{inj} drops as either the positive or negative-going injection pulse is shifted below the (full) injection threshold. When the injection pulse is omitted, at least one side of the drive waveform is always below the DW injection threshold for any H_{dc}, so $P_{\text{inj}} = 0$ and no DW-induced voltage is expected.

The modulation field $H_m = 11$ Oe was used to toggle P_{inj}, thereby modulating the DW-induced voltage. The detected lock-in voltage V_L should be proportional to the difference in P_{inj} between the positive and negative phases of the modulation cycle, weighted by the average DW-induced voltage in each of those phases. The lock-in voltage V_L at $H_d = 20$ Oe is shown in Fig. 2(a) as a function of the dc bias field H_{dc}. We observe that it is more complex than that expected from motion of a single DW along the nanowire. The large offset voltage is due primarily to inductive pickup of the modulation field by the

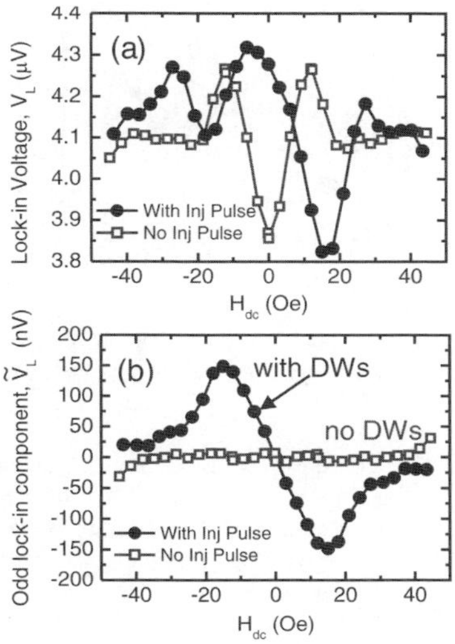

FIG. 2 (color online). (a) Lock-in voltage versus H_{dc} for the drive-field waveforms with $H_d = 20$ Oe with/and without injection pulse. (b) Odd components of the curves in (a).

measurement loop, and it scales linearly with the modulation field frequency and amplitude. The dependence of V_L on H_{dc}, particularly when injection pulses are omitted from the drive-field waveform, points to additional sources of modulated voltage. The contact or injection pads are magnetic, and likely contribute in nontrivial ways to the measured signal. For example, DW motion within the pads could contribute a net emf, and stray magnetostatic fields could lead to a bias field-dependant flux through the measurement loop.

To discriminate between voltage generated by DW motion along the nanowire, and other parasitic contributions, we exploit the fact that unidirectional DW motion generates a voltage with definite polarity that does not depend on the sign of the total driving field [cf. discussion following Eq. (2)]. Hence any DW-induced contribution to V_L must

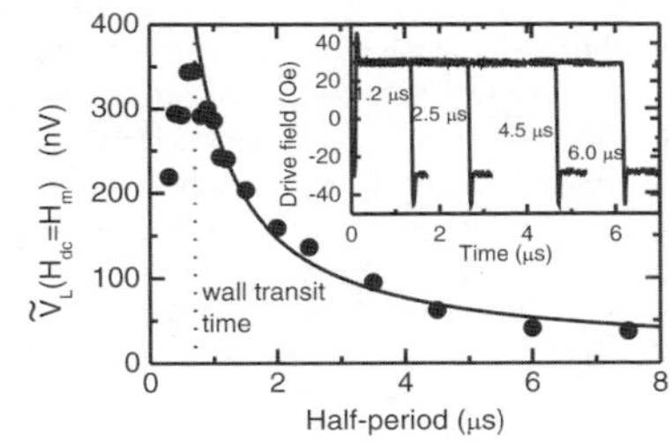

FIG. 3. Odd-symmetry component of lock-in voltage at $H_{\text{dc}} = H_m$ for drive-field waveforms of varying half-period $T_{1/2}$ (half-cycles shown in inset). Solid line is fit to $1/T_{1/2}$.

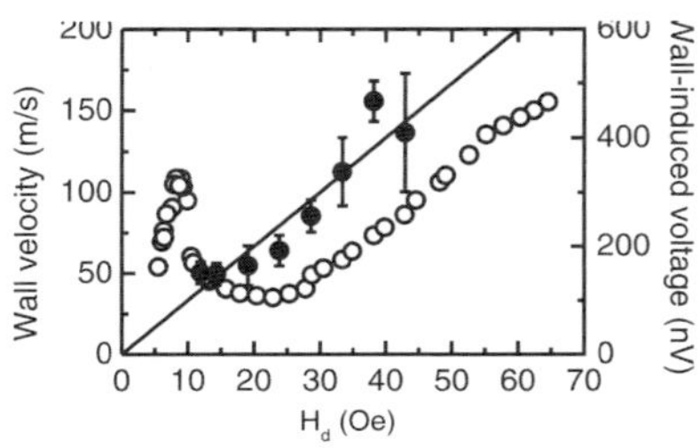

FIG. 4. Measured wall velocity (open symbols) and wall-induced voltage (solid symbols) versus drive-field. Solid line is fit with slope 10 nV/Oe.

be odd in H_{dc}: comparing Figs. 1(c) and 1(d), the modulated DW voltage has the same amplitude for $\pm H_{dc}$, but it is in-phase with the modulation field for negative bias, and out of phase for positive bias. This phase difference manifests as a sign change in the DW contribution to V_L.

Figure 2(b) shows the odd component $\tilde{V}_L$ of the data in Fig. 2(a). When the drive-field includes injection pulses, V_L contains an odd component that closely follows the modulated DW injection probability measured experimentally. When injection pulses are omitted from the drive waveform, and hence DW motion does not occur within the nanowire, there is no voltage contribution with this symmetry. Based on its symmetry and its correlation with P_{inj}, we interpret $\tilde{V}_L$ as arising from DW motion along the nanowire.

This identification is supported by additional systematic tests that correlate the magnitude of $\tilde{V}_L$ with the motion of the DW. The contribution to $\tilde{V}_L$ by DWs traversing the nanowire is proportional to the ratio $\tau_{DW}/T_{1/2}$ [cf. Fig. 1(e)]. τ_{DW} will be the same for each drive-field half-cycle if we set $H_{dc} = H_m$ such that the drive-field waveform has zero offset during the on phase of the modulation cycle, and each injected wall is driven by a field of amplitude $\pm H_d$ [cf. Figs. 1(c) and 1(d)]. Figure 3 shows $\tilde{V}_L$ measured at $H_{dc} = H_m$ for a series of drive-field waveforms with varying period. $\tilde{V}_L$ varies inversely with $T_{1/2}$, except for $T_{1/2} < 0.7$ μs, where $\tilde{V}_L$ drops precipitously. This crossover time corresponds to the independently measured wall transit time τ_{DW}. At $T_{1/2} = \tau_{DW}$, DWs generate voltage over the full duration of each drive-field cycle. For $T_{1/2} < \tau_{DW}$, a DW injected into a uniform wire generates voltage during the first drive-field half-period, but before it reaches the end of the wire, H_d changes polarity. A new DW is injected and moves forward while the initial wall moves back, giving zero net voltage. Averaged over a full drive-field cycle, $\tilde{V}_L$ should drop, consistent with Fig. 3.

$\tilde{V}_L(H_{dc})$ was also measured for a series of drive-field waveforms with the same injection pulse, but with varying H_d. The curves followed Fig. 2(b), varying only in amplitude. At $H_{dc} = H_m$, $\tilde{V}_L$ is proportional to the voltage V_{DW} generated by a DW driven by a field $|H_d|$, i.e., $V_{DW}(H_d) = C(T_{1/2}/\tau_{DW})\tilde{V}_L[H_{dc} = H_m]$, where $C^{-1} = 4/\pi$ is the weight of the fundamental Fourier component of a square-wave. The average DW velocity, shown in Fig. 4, was independently-measured, yielding τ_{DW}. The dependence of V_{DW} on H_d is shown in Fig. 4 [25]. V_{DW} varies in proportion to field, with a slope of $\sim$10 nV/Oe. This is close to the predicted slope $\gamma\hbar/e = 11.6$ nV/Oe, and suggests a spin-polarization $P \sim 0.85$, somewhat higher than expected but within reasonable experimental bounds.

In summary, we have developed a general theory to calculate the emf generated by the motion of an arbitrary spin texture and obtained a universal ferro-Josephson relation for a DW moving in a nanostrip. The theory yields a universal result $V_x/H = 11.6$ nV/Oe for a field-driven DW, independent of the details of its internal structure. We have detected experimentally the voltage generated during the motion of a single domain wall over a range of driving field and find a result that agrees well with this prediction.

The authors thank C. Xu, W. Li, C.-P. Chuu, W. Yao, D. Clougherty, and S. Zhang for valuable discussions. S. Y., G. B., C. K., M. T., and J. E. were supported by NSF DMR-0404252, DX was supported by NSF DMR-0606485, G. B. and J. E. by the Welch Foundation, and Q. N. by DOE (DE-FG03-02ER45958).

[1] P. P. Freitas and L. Berger, J. Appl. Phys. **57**, 1266 (1985).
[2] A. Yamaguchi *et al.*, Phys. Rev. Lett. **92**, 077205 (2004).
[3] M. Kläui *et al.*, Phys. Rev. Lett. **95**, 026601 (2005).
[4] G. S. D. Beach *et al.*, Phys. Rev. Lett. **97**, 057203 (2006).
[5] M. Hayashi *et al.*, Phys. Rev. Lett. **96**, 197207 (2006).
[6] L. Berger, Phys. Rev. B **33**, 1572 (1986).
[7] S. E. Barnes *et al.*, Appl. Phys. Lett. **89**, 122507 (2006); S. E. Barnes and S. Maekawa, Phys. Rev. Lett. **98**, 246601 (2007).
[8] R. A. Duine, Phys. Rev. B **77**, 014409 (2008).
[9] W. M. Saslow, Phys. Rev. B **76**, 184434 (2007).
[10] M. V. Berry, Proc. R. Soc. A **392**, 45 (1984).
[11] R. D. McMichael and M. J. Donahue, IEEE Trans. Magn. **33**, 4167 (1997).
[12] M. Kläui *et al.*, Appl. Phys. Lett. **85**, 5637 (2004).
[13] G. E. Volovik, J. Phys. C **20**, L83 (1987).
[14] G. Sundaram and Q. Niu, Phys. Rev. B **59**, 14 915 (1999).
[15] N. L. Schryer and L. R. Walker, J. Appl. Phys. **45**, 5406 (1974).
[16] G. S. D. Beach *et al.*, Nature Mater. **4**, 741 (2005).
[17] G. S. D. Beach, M. Tsoi, and J. L. Erskine, J. Magn. Magn. Mater. **320**, 1272 (2008).
[18] Y. Nakatani *et al.*, Nature Mater. **2**, 521 (2003).
[19] A. A. Belavin and A. M. Polyakov, Pis'ma Zh. Eksp. Teor. Fiz. **22**, 503 (1975) [JETP Lett. **22**, 245 (1975)].
[20] J. Shibata *et al.*, Phys. Rev. B **73**, 020403(R) (2006).
[21] J. He *et al.*, Phys. Rev. B **73**, 184408 (2006).
[22] J.-Y. Lee *et al.*, Phys. Rev. B **76**, 184408 (2007).
[23] M. Hayashi *et al.*, Nature Phys. **3**, 21 (2007).
[24] L. Berger, J. Appl. Phys. **55**, 1954 (1984).
[25] Voltage generated during the injection pulse adds to the calculated $V_{DW}(H_d)$, and has not been corrected for.

Electron dynamics in slowly varying antiferromagnetic texture

Ran Cheng[1,*] and Qian Niu[1,2]
[1]Department of Physics, University of Texas at Austin, Austin, Texas 78712, USA
[2]International Center for Quantum Materials, Peking University, Beijing 100871, China
(Received 23 March 2012; revised manuscript received 8 November 2012; published 18 December 2012)

Adiabatic dynamics of conduction electrons in antiferromagnetic (AFM) materials with slowly varying spin texture is developed. Quite different from the ferromagnetic (FM) case, adiabaticity in AFM texture does not imply perfect alignment of conduction electron spins with background profile, instead, it introduces an internal dynamics between degenerate bands. As a result, the orbital motion of conduction electrons becomes spin dependent and is affected by two emergent gauge fields: one of them is the non-Abelian version of what has been discovered in FM systems; the other leads to an anomalous velocity that has no FM counterpart. Two examples with experimental predictions are provided.

DOI: 10.1103/PhysRevB.86.245118 PACS number(s): 75.50.Ee, 03.65.Vf, 72.10.Bg, 72.25.−b

I. INTRODUCTION

Interplay between current and magnetization is an essential issue underpinning the field of spintronics,[1] which consists of two reciprocal problems: control of current through magnetization with a known configuration, and its converse, i.e., control of magnetization dynamics via applied current. In ferromagnetic (FM) materials with slowly varying spin texture $\boldsymbol{m}(\boldsymbol{r},t)$ over space and time, these issues can be solved by assuming that conduction electron spins always follow the background texture profile, known as the adiabatic approximation.[2,3] The microscopic basis underlying adiabaticity is the strong exchange coupling $H = -J\sigma \cdot \boldsymbol{m}(\boldsymbol{r},t)$ between conduction electron spins and local magnetic moments, through which spin mistracking with the background causes large energy penalty and becomes highly unfavorable.[4]

Under adiabatic approximation, the current-magnetization interaction is recast into an emergent electrodynamics, in which its reciprocal influence boils down to a simple electromagnetic problem. Specifically, by diagonalizing the local exchange Hamiltonian via local unitary transformation, fictitious electric and magnetic fields are generated into the orbital dynamics[4–7]

$$E_i = \tfrac{1}{2}\boldsymbol{m} \cdot (\partial_t \boldsymbol{m} \times \partial_i \boldsymbol{m}) = \tfrac{1}{2}\sin\theta(\partial_t\theta\partial_i\phi - \partial_i\theta\partial_t\phi),$$
$$B_i = -\tfrac{1}{4}\varepsilon_{ijk}\boldsymbol{m} \cdot (\partial_j \boldsymbol{m} \times \partial_k \boldsymbol{m}) = -\tfrac{1}{2}\varepsilon_{ijk}\sin\theta\partial_j\theta\partial_k\phi,$$

where $\theta(\boldsymbol{r},t)$ and $\phi(\boldsymbol{r},t)$ are spherical angles specifying the direction of $\boldsymbol{m}$. As a consequence, the influence of background texture is represented by an effective Lorentz force $\boldsymbol{F} = s\hbar(\boldsymbol{E} + \dot{\boldsymbol{r}} \times \boldsymbol{B})$ exerted on conduction electrons, where $s = +1(-1)$ denotes spin-up (-down) bands. The electric and magnetic components of the Lorentz force are responsible for the spin motive force[4,6] and the topological Hall effect,[8] respectively. In turn, back-reaction of the Lorentz force provides an interpretation to the current-induced spin torque exerted on magnetic texture.[9–11] In a formal language, the adiabaticity induces an effective gauge interaction $\mathcal{L}_{\text{int}} = j_\mu \mathcal{A}_\mu$, where current j_μ acquires gauge charge according to $s = \pm 1$, and $\mathcal{A}_\mu = \mathcal{A}_\mu(\boldsymbol{m},\partial\boldsymbol{m})$ is the effective electromagnetic potential representing space-time dependence of the texture. Variation over the current $\delta\mathcal{L}_{\text{int}}/\delta j_\mu = 0$ yields the effective Lorentz force; and varying over the magnetization $\delta\mathcal{L}_{\text{int}}/\delta\boldsymbol{m} = 0$ produces the spin-transfer torque. Thus the current-magnetization interaction is reciprocal.

However, the above picture apparently fails in antiferromagnetic (AFM) materials where neighboring magnetic moments are antiparallel. Conduction electrons are not able to adjust spins with local moments that change orientation on atomic scale. Nevertheless, the staggered order parameter $\boldsymbol{n} = (\boldsymbol{M}_A - \boldsymbol{M}_B)/2M_s$ can be slowly varying over space-time, where $\boldsymbol{M}_A$ and $\boldsymbol{M}_B$ are the alternating local moments and M_s denotes their magnitudes. A natural question is whether a slowly varying staggered order still renders adiabatic dynamics of conduction electrons in some other sense. This is desired knowledge for studying spin transport in AFM materials, especially the quest for current-magnetization interaction as that for FM materials.

In spite of recent theoretical[12] and experimental[13] progress, this problem has never been addressed. But, at the same time, AFM materials are believed to be promising candidates for new thrusts of spintronics,[14] partly due to their tiny anisotropy, robustness against external magnetic perturbations, and absence of demagnetization, which bring prevailing advantages for experimental control. In this paper, we develop the effective electron dynamics in a smooth AFM texture by applying the non-Abelian Berry phase theory[15–17] on energy bands that are doubly degenerate. The physics of adiabaticity in AFM materials is found to be an internal dynamics between degenerate bands which can be attributed to a SU(2) Berry curvature. When translating into spin dynamics, the adiabaticity no more indicates spin alignment with the background, but a totally new evolution principle [Eq. (8a)]. Aside from spin dynamics, the orbital motion of conduction electrons is coupled to two different gauge fields: one leads to the non-Abelian generalization of the effective Lorentz force; the other results in an anomalous velocity that is truly new and unique to AFM systems. With comparisons to FM materials, this paper provides a general framework on how a given AFM texture affects the dynamics of conduction electrons. The other side of the story, i.e., back-reaction of current on the AFM texture, will appear in a forthcoming publication.

The paper is organized as follows. In Sec. II, the general formalism is presented where isospin is introduced. In Sec. III, our central results (8) are derived, followed by discussions on non-Abelian Berry phase and monopole, spin and orbital dynamics of conduction electrons, and comparisons of AFM electron dynamics with its FM counterparts. In Sec. IV, two examples are provided, and the paper is summarized in Sec. V. Mathematical derivations are included in the Appendixes.

II. FORMALISM

Consider an AFM system on a bipartite lattice with local magnetic moments labeled by alternating $\boldsymbol{M}_A$ and $\boldsymbol{M}_B$. The spin of a conduction electron couples to the local moments by the exchange interaction $J(\boldsymbol{M}/M_s)\cdot\boldsymbol{\sigma}$, where $\boldsymbol{\sigma}$ denotes the spin operator of the conduction electron, and $\boldsymbol{M}$ flips sign on neighboring A and B sublattice sites. In spite of antiparallel of neighboring moments, the staggered order parameter $\boldsymbol{n} = (\boldsymbol{M}_A - \boldsymbol{M}_B)/2M_s$ usually varies slowly over space and time, and we can treat it as a continuous function $\boldsymbol{n}(\boldsymbol{r},t)$. Accordingly, the conduction electron is described by a nearest-neighbor tight-binding Hamiltonian locally defined around $\boldsymbol{n}(\boldsymbol{r},t)$:

$$\mathcal{H}(\boldsymbol{n}(\boldsymbol{r},t)) = \begin{bmatrix} -J\boldsymbol{n}\cdot\boldsymbol{\sigma} & \gamma(\boldsymbol{k}) \\ \gamma^*(\boldsymbol{k}) & J\boldsymbol{n}\cdot\boldsymbol{\sigma} \end{bmatrix}, \tag{1}$$

where $\gamma(\boldsymbol{k}) = -t\sum_{\boldsymbol{\delta}} e^{i\boldsymbol{k}\cdot\boldsymbol{\delta}}$ is the hopping term with $\boldsymbol{\delta}$ connecting nearest-neighboring A-B sites (we set $\hbar = 1$). In general, J can be negative, but we assume a positive J throughout this paper.

The local band structure can be easily solved as $\pm\varepsilon(\boldsymbol{k})$ with $\varepsilon(\boldsymbol{k}) = \sqrt{J^2 + |\gamma(\boldsymbol{k})|^2}$, and in the adiabatic limit we neglect transitions between ε and $-\varepsilon$. Each of the two bands is doubly degenerate, and without loss of generality we will focus on the lower band $-\varepsilon$ with the two subbands labeled by A and B, the wave functions of which are $|\psi_a\rangle = e^{i\boldsymbol{k}\cdot\boldsymbol{r}}|u_a\rangle$ and $|\psi_b\rangle = e^{i\boldsymbol{k}\cdot\boldsymbol{r}}|u_b\rangle$. The Bloch waves $|u_a\rangle = |A(\boldsymbol{k})\rangle|\uparrow(\boldsymbol{r},t)\rangle$ and $|u_b\rangle = |B(\boldsymbol{k})\rangle|\downarrow(\boldsymbol{r},t)\rangle$ maintain local periodicity around $(\boldsymbol{r},t)$, where

$$|\uparrow(\boldsymbol{r},t)\rangle = \begin{bmatrix} e^{-i\frac{\phi}{2}}\cos\frac{\theta}{2} \\ e^{i\frac{\phi}{2}}\sin\frac{\theta}{2} \end{bmatrix}, \quad |\downarrow(\boldsymbol{r},t)\rangle = \begin{bmatrix} -e^{-i\frac{\phi}{2}}\sin\frac{\theta}{2} \\ e^{i\frac{\phi}{2}}\cos\frac{\theta}{2} \end{bmatrix} \tag{2}$$

are local spin wave functions with $\theta = \theta(\boldsymbol{r},t)$ and $\phi = \phi(\boldsymbol{r},t)$ being spherical angles specifying the orientation of $\boldsymbol{n}(\boldsymbol{r},t)$. The periodic parts are spinors in the pseudospin space furnished by the A-B sublattices

$$|A(\boldsymbol{k})\rangle = \frac{[\varepsilon(\boldsymbol{k}) + J,\ |\gamma(\boldsymbol{k})|]^{\mathrm{T}}}{\sqrt{[\varepsilon(\boldsymbol{k}) + J]^2 + |\gamma(\boldsymbol{k})|^2}}, \tag{3a}$$

$$|B(\boldsymbol{k})\rangle = \frac{[\varepsilon(\boldsymbol{k}) - J,\ |\gamma(\boldsymbol{k})|]^{\mathrm{T}}}{\sqrt{[\varepsilon(\boldsymbol{k}) - J]^2 + |\gamma(\boldsymbol{k})|^2}}, \tag{3b}$$

which exhibit opposite spatial patterns schematically illustrated in Fig. 1. While $\langle\psi_a|\psi_b\rangle = 0$ due to the orthogonality of local spin eigenstates, $\langle A(\boldsymbol{k})|B(\boldsymbol{k})\rangle$ does not vanish, and we define this overlap as

$$\xi(\boldsymbol{k}) = \langle A(\boldsymbol{k})|B(\boldsymbol{k})\rangle = \frac{|\gamma(\boldsymbol{k})|}{\sqrt{J^2 + |\gamma(\boldsymbol{k})|^2}} = \frac{\sqrt{\varepsilon^2 - J^2}}{\varepsilon}, \tag{4}$$

which is a key parameter in our theory and $\xi < 1$. It reaches maximum at the Brillouin zone (BZ) center and vanishes at the BZ boundary. From Eq. (4), we know $\xi(\boldsymbol{k})$ is a system parameter determined by the band structure, and it is constant since the energy conservation $\dot{\varepsilon} = 0$ requires $\dot{\xi} = 0$. If J tends to infinity, the overlap $\xi(\boldsymbol{k})$ will vanish and the two subbands will be effectively decoupled, by which the system will become a simple combination of two independent FM subsystems.

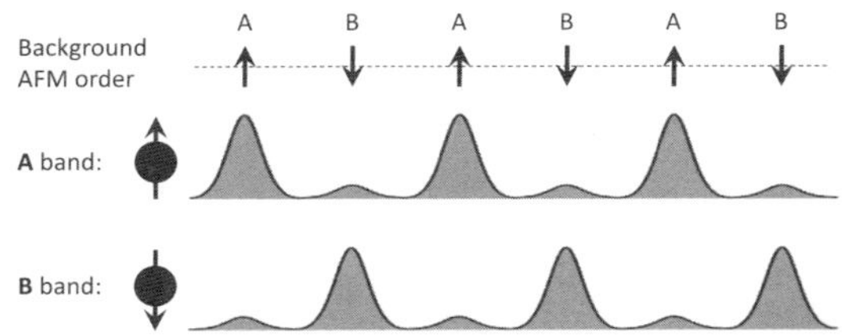

FIG. 1. A schematic view of Bloch waves in the lower band. Subband A means a local spin-up electron has a larger probability on the A sites and a smaller probability on the B sites; subband B means the opposite case. They are degenerate in energy, and their wave functions have a finite overlap depending on the ratio of J/ε.

We adopt the semiclassical approach to construct effective electron dynamics,[15] where an individual electron is described by a wave packet $|W\rangle = \int d\boldsymbol{k}\, w(\boldsymbol{k} - \boldsymbol{k}_c)[c_a|\psi_a\rangle + c_b|\psi_b\rangle]$, where $\int d\boldsymbol{k}\,\boldsymbol{k}|w(\boldsymbol{k} - \boldsymbol{k}_c)|^2 = \boldsymbol{k}_c$ gives the center-of-mass momentum, and $\langle W|\boldsymbol{r}|W\rangle = \boldsymbol{r}_c$ is the center-of-mass position. The coefficients c_a and c_b reflect relative contributions from the two subbands and $|c_a|^2 + |c_b|^2 = 1$. Since now the band is degenerate, non-Abelian formalism[3,15–17] must be invoked (see Appendix A), where dynamics between the A and B subbands introduces an internal degree of freedom represented by the isospin vector

$$\begin{aligned} \boldsymbol{\mathcal{C}} &= \{c_1, c_2, c_3\} \\ &= \{2\,\mathrm{Re}(c_a c_b^*),\ -2\,\mathrm{Im}(c_a c_b^*), |c_a|^2 - |c_b|^2\}. \end{aligned} \tag{5}$$

The electron dynamics is characterized by the equations of motion of the three parameters $\boldsymbol{k}_c$, $\boldsymbol{r}_c$, and $\boldsymbol{\mathcal{C}}$, which can be obtained from variational principles with the effective Lagrangian $\mathcal{L} = \langle W|(i\frac{\partial}{\partial t} - \mathcal{H})|W\rangle$.[3,15] The detailed derivations are presented in Appendix A; here we only write the results

$$\dot{\boldsymbol{\mathcal{C}}} = 2\boldsymbol{\mathcal{C}} \times \left(\boldsymbol{\mathcal{A}}^r_\mu \dot{r}_\mu + \boldsymbol{\mathcal{A}}^k_\mu \dot{k}_\mu\right), \tag{6a}$$

$$\dot{k}_\mu = \partial^r_\mu \varepsilon + \boldsymbol{\mathcal{C}}\cdot\left[\boldsymbol{\Omega}^{rr}_{\mu\nu}\dot{r}_\nu + \boldsymbol{\Omega}^{rk}_{\mu\nu}\dot{k}_\nu\right], \tag{6b}$$

$$\dot{r}_\mu = -\partial^k_\mu \varepsilon - \boldsymbol{\mathcal{C}}\cdot\left[\boldsymbol{\Omega}^{kr}_{\mu\nu}\dot{r}_\nu + \boldsymbol{\Omega}^{kk}_{\mu\nu}\dot{k}_\nu\right], \tag{6c}$$

where $r_\mu = (t, \boldsymbol{r}_c)$, but $k_\mu = (0, \boldsymbol{k}_c)$ has no temporal component. In Eqs. (6), the centerdot and multi symbols denote scalar and cross products in the isospin vector space. The Berry curvatures $\boldsymbol{\Omega}$ are obtained from the gauge potentials $\boldsymbol{\mathcal{A}}$ defined on the A and B subbands, for instance,

$$\left[\boldsymbol{\mathcal{A}}^r_\mu \cdot \boldsymbol{\tau}\right]_{ij} = i\langle u_i|\partial^r_\mu|u_j\rangle, \tag{7a}$$

$$\boldsymbol{\Omega}^{rr}_{\mu\nu} = \partial^r_\mu \boldsymbol{\mathcal{A}}^r_\nu - \partial^r_\nu \boldsymbol{\mathcal{A}}^r_\mu + 2\boldsymbol{\mathcal{A}}^r_\mu \times \boldsymbol{\mathcal{A}}^r_\nu, \tag{7b}$$

where $\boldsymbol{\tau}$ is a vector of Pauli matrices representing the isospin and i,j run between a,b. Other components of the Berry curvatures are explained in Appendix A.

III. ELECTRON DYNAMICS

Equipped with Eqs. (6), we are now ready to derive the dynamics of an individual electron. Note that the isospin vector $\boldsymbol{C}$ itself is not gauge invariant in the sense that different choice of spin wave functions results in different $\boldsymbol{C}$. Therefore, in deriving electron dynamics, we need to relate $\boldsymbol{C}$ to the real spin defined by $\boldsymbol{s} = \langle W|\boldsymbol{\sigma}|W\rangle$ (in units of $\frac{1}{2}$) which is a physical variable and fully gauge invariant. While detailed derivations are lengthy and complicated, which is left for Appendix B, the final results are quite simple and elegant:

$$\dot{\boldsymbol{s}} = (1-\xi^2)(\boldsymbol{s}\cdot\boldsymbol{n})\dot{\boldsymbol{n}}, \tag{8a}$$

$$\dot{\boldsymbol{k}} = -\tfrac{1}{2}\boldsymbol{n}\cdot(\nabla\boldsymbol{n}\times\dot{\boldsymbol{s}}), \tag{8b}$$

$$\dot{\boldsymbol{r}} = -\partial_{\boldsymbol{k}}\varepsilon - \tfrac{1}{2}(\boldsymbol{s}\times\boldsymbol{n})\cdot\dot{\boldsymbol{n}}\,\partial_{\boldsymbol{k}}\ln\xi, \tag{8c}$$

where $\dot{\boldsymbol{n}} = \partial_t\boldsymbol{n} + (\dot{\boldsymbol{r}}\cdot\nabla)\boldsymbol{n}$, and we have omitted subscript c of $\boldsymbol{r}_c$ and $\boldsymbol{k}_c$ for convenience of following discussions. Equations (8) are the fundamental equations of motion of a conduction electron in an AFM material with slowly varying texture, which are represented by joint evolutions of three variables $(\boldsymbol{s},\boldsymbol{k},\boldsymbol{r})$. An essential feature distinguishing the AFM electron dynamics from its FM counterpart lies in Eq. (8a), from which we know that the real spin $\boldsymbol{s}$ does not follow the background order parameter $\boldsymbol{n}$ in the adiabatic limit.

Spin dynamics. The motion of $\boldsymbol{s}$ can be decomposed into a superposition of two motions: one strictly follows $\boldsymbol{n}$ (for stationary $\boldsymbol{C}$) and the other represents mistracking with $\boldsymbol{n}$ (for dynamical $\boldsymbol{C}$), where the latter originates from dynamics between the A and B subbands and is unique to AFM materials (see examples in Sec. IV). It is worth emphasizing that the mistracking between $\boldsymbol{s}$ and $\boldsymbol{n}$ has nothing to do with any nonadiabatic process, but is entirely due to the non-Abelian nature of the problem. The overall spin evolution can be attributed to the accumulation of a SU(2) non-Abelian Berry phase $\mathcal{P}\exp[-i\int\boldsymbol{\mathcal{A}}^r_\mu\cdot\boldsymbol{\tau}\,dr_\mu]$ along the electron trajectory.[17] As compared with its U(1) counterpart in FM materials,[3–11] which can be regarded as the magnetic flux of a Dirac monopole located at the center of the sphere spanned by $\boldsymbol{m}$, the SU(2) Berry phase here can be related to a 't Hooft-Polyakov monopole in the parameter space. Detailed discussions on the monopole can be found in Appendix C, and the key point here is that a recent proposal of artificial 't Hooft-Polyakov monopole[18] can be realized in our AFM texture systems.

Since $|\uparrow(\boldsymbol{r},t)\rangle$ and $|\downarrow(\boldsymbol{r},t)\rangle$ form local spin bases with quantization axis being $\boldsymbol{n}(\boldsymbol{r},t)$, the isospin $\boldsymbol{C}$ can be pictured as a vector in the local frame moving with $\boldsymbol{n}(\boldsymbol{r},t)$. In Appendix B, we have shown that in the particular gauge marked by $\chi = 0$ [see Eq. (B18)],

$$\boldsymbol{C} = c_1\boldsymbol{\theta} + c_2\boldsymbol{\phi} + c_3\boldsymbol{n} = \frac{1}{\xi}(s_1\boldsymbol{\theta} + s_2\boldsymbol{\phi}) + s_3\boldsymbol{n}, \tag{9}$$

where $\{\boldsymbol{\theta},\boldsymbol{\phi},\boldsymbol{n}\}$ form local bases associated with the local order parameter $\boldsymbol{n}(\boldsymbol{r},t)$ (Fig. 2, left). Equation (9) indicates two important properties: (1) $\boldsymbol{C}$ is coplanar with $\boldsymbol{n}$ and $\boldsymbol{s}$, which is specific to the particular gauge $\chi = 0$; (2) while the isospin vector is constrained on the unit sphere $c_1^2 + c_2^2 + c_3^2 = 1$, the physical spin satisfies $\frac{s_1^2+s_2^2}{\xi^2} + s_3^2 = 1$, which constrains the tip of $\boldsymbol{s}$ on an prolate spheroid with semimajor axis being $\boldsymbol{n}(\boldsymbol{r},t)$ and semiminor axis on its equator having length ξ (Fig. 2, right). The latter property is gauge independent and can be justified directly from Eq. (8a) without using Eq. (9) (see Appendix D), which, when written in gauge-invariant from, becomes

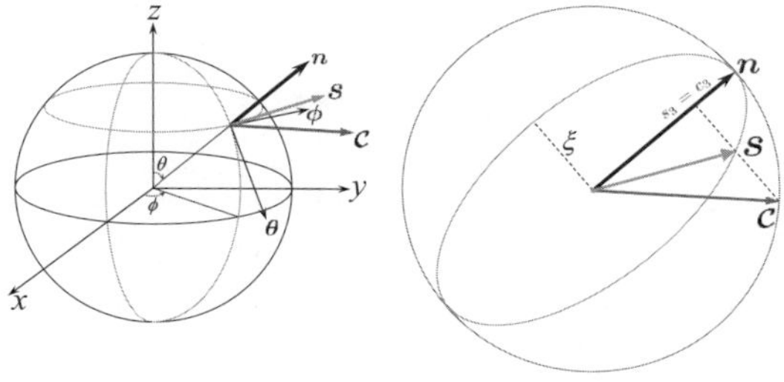

FIG. 2. (Color online) Left panel: The isospin vector $\boldsymbol{C}$ (blue arrow) in the local frame: $\boldsymbol{C} = c_1\boldsymbol{\theta} + c_2\boldsymbol{\phi} + c_3\boldsymbol{n}$, where $\boldsymbol{\theta}$ and $\boldsymbol{\phi}$ are spherical unit vectors. Right panel: In our particular gauge, $\boldsymbol{C}$ (blue) is coplanar with $\boldsymbol{n}$ and $\boldsymbol{s}$ (red). The tip of $\boldsymbol{C}$ moves on a unit sphere, whereas the tip of $\boldsymbol{s}$ is constrained on the ellipsoid whose semimajor axis is $\boldsymbol{n}$ and semiminor axis has length ξ.

$$(\boldsymbol{s}\cdot\boldsymbol{n})^2 + \frac{(\boldsymbol{s}\times\boldsymbol{n})^2}{\xi^2} = s_3^2 + \frac{s_1^2+s_2^2}{\xi^2} = 1. \tag{10}$$

For arbitrary gauges with $\chi \neq 0$, it is easy to show that we always have $s_3 = c_3$ and $s_1^2 + s_2^2 = \xi^2(c_1^2 + c_2^2)$, but the angles between $s_{1,2}$ and $c_{1,2}$ will be different, i.e., $\boldsymbol{C}$ will not be coplanar with $\boldsymbol{n}$ and $\boldsymbol{s}$.

Now, the physical picture of adiabatic spin evolution is clear: as the background order parameter $\boldsymbol{n}(\boldsymbol{r},t)$ moves slowly in space-time, the prolate spheroid moves with it. The motion of physical spin $\boldsymbol{s}$ is a superposition of the relative motion on the spheroid and the motion of the spheroid itself. The overall motion of $\boldsymbol{s}$ described by Eq. (8a) is purely geometrical as dt can be eliminated on both sides, as a result, a given path of $\boldsymbol{n}$ uniquely determines a path of $\boldsymbol{s}$ on the spheroid which is independent of the Hamiltonian. Associated with this geometric motion, a SU(2) Berry phase is accumulated by the electron wave function, which can be regarded as the (non-Abelian) gauge flux of a 't Hooft-Polyakov monopole at the center of the unit sphere spanned by $\boldsymbol{n}$ (Table I).

A further remark: It seems to be a surprise that the magnitude of $\boldsymbol{s}$ varies on the spheroid since $\xi \leqslant 1$, but how can the physical spin have a nonconstant magnitude? We answer this question by studying the *reduced* density matrix for the spin degree of freedom. It is a 2×2 matrix and can be written as $\rho_s = \frac{1}{2}(1 + \boldsymbol{a}\cdot\boldsymbol{\sigma})$, thus the expectation value of physical spin is $\boldsymbol{s} = \mathrm{Tr}[\rho_s\boldsymbol{\sigma}] = \boldsymbol{a}$. Now, since $s^2 \leqslant 1$, thus $a^2 \leqslant 1$, and what follows is $\mathrm{Tr}\rho_s^2 \leqslant \mathrm{Tr}\rho_s$, which suggests that the electron is *effectively* in a mixed spin state. This can be attributed to the entanglement of spin and sublattice degrees of freedom, specifically, because $s_3 = c_3$, we are able to infer the spin projection along $\boldsymbol{n}$ by measuring the probability difference on neighboring A-B sites (vice versa). The entanglement provides us with partial information of spin orientation from the knowledge of sublattice; this destroys full coherence of the spin states.

TABLE I. Comparison of effective electron dynamics in FM and AFM textures. In the FM case, spin dynamics is trivial; along closed path Γ, the electron acquires a U(1) Berry phase which is the magnetic flux of a Dirac monopole; A Lorentz force is resulted in the orbital motion. In the AFM case, spin dynamics is nontrivial due to the mixture of degenerate sub-bands through a SU(2) non-Abelian Berry phase, which is the gauge flux generated by a 't Hooft-Polyakov monopole. The orbital dynamics is subject to a spin-dependent Lorentz force and an anomalous velocity.

FM spin texture	AFM spin texture
$s = n$	$\dot{s} = (1-\xi^2)(s\cdot n)\dot{n}$
U(1) Abelian Berry phase:	SU(2) non-Abelian Berry phase:
$\gamma(\Gamma) = \oint_\Gamma \mathcal{A}_\mu dr_\mu$	$U(\Gamma) = \mathcal{P}\exp[-i\oint_\Gamma \mathcal{A}^r_\mu \cdot \tau dr_\mu]$
Dirac monopole	't Hooft-Polyakov monopole
$\dot{k} = E + B\times\dot{r}$	$\dot{k} = (1-\xi^2)(s\cdot n)(E + B\times\dot{r})$
$\dot{r} = -\partial_k\varepsilon$	$\dot{r} = -\partial_k\varepsilon - \frac{1}{2}(s\times n)\cdot\dot{n}\,\partial_k\ln\xi$

Orbital dynamics. In correspondence with the novel spin dynamics, the orbital dynamics of an individual electron also becomes nontrivial. By substituting Eq. (8a) into (8b), we get (see also Appendix B)

$$\dot{k} = (1-\xi^2)(s\cdot n)(E + \dot{r}\times B), \tag{11}$$

$$E = \tfrac{1}{2}\sin\theta(\partial_t\theta\nabla\phi - \nabla\theta\partial_t\phi), \tag{12}$$

$$B = -\tfrac{1}{2}\sin\theta(\nabla\theta\times\nabla\phi), \tag{13}$$

the E and B fields here are the same as their FM counterparts, where they are responsible for the spin motive force[6] and the topological Hall effect,[8] respectively. Also, as in FM systems, it is easy to check that Eqs. (12) and (13) satisfy the Faraday's relation $\nabla\times E + \frac{\partial B}{\partial t} = 0$.

However, quite different from the FM case, the gauge charge $s\cdot n$ in Eq. (11) is not just a constant, but involves the internal dynamics. In other words, the orbital motion is accompanied by a time-dependent gauge charge which should be determined by solving the coupled equations (8) all together. Moreover, the factor ξ^2 results from the noncommutative term $2\mathcal{A}^r_\mu\times\mathcal{A}^r_\nu$ in Eq. (7b); it also reflects the coupling between spin and orbital dynamics. The parameter $\xi\in(0,1)$ plays a key role here: in the $\xi\to 1$ limit, $1-\xi^2$ vanishes thus from Eqs. (8a) and (11), we get null results $\dot{s}=0$ and $\dot{k}=0$. In the other limit where $\xi\to 0$, the solution of Eq. (8a) reduces to $s=\pm n$ if initial condition is $s(0)=\pm n(0)$, and Eq. (11) reduces to the FM Lorentz force equation, by which the system loses the manifest non-Abelian feature and behaves as two decoupled FM subsystems. It deserves attention that in real AFM materials, both A and B subbands host majority carriers, but they are subject to effective Lorentz forces of opposite directions, which may lead to nontrivial spin transport.

Furthermore, the real-space dynamics governed by Eq. (8c) exhibits spin-orbit coupling through the anomalous velocity term $\frac{1}{2}(s\times n)\cdot\dot{n}\,\partial_k\ln\xi$. It is along the same direction as $\partial_k\varepsilon$, so Eq. (8c) amounts to give a modified group velocity. We mention that this term is unique to AFM textures and has nothing to do with the anomalous velocity studied in FM or quantum Hall systems.[3] It originates from the $\Omega^{kr}_{\mu\nu}$ curvature that joints real space with BZ, the importance of which has

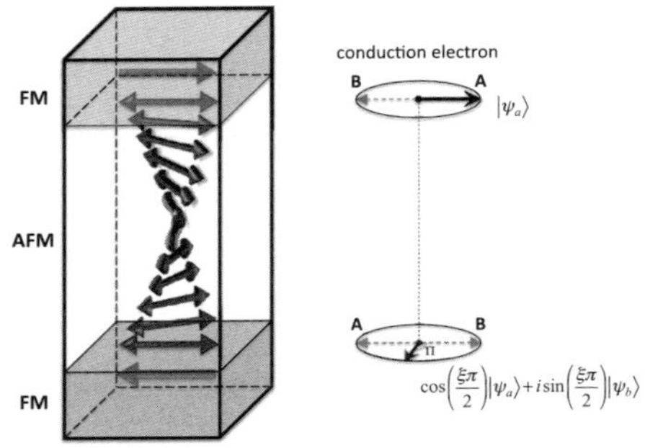

FIG. 3. (Color online) Left: FM/AFM/FM trilayer with opposite FM orientations on two sides. The black double arrows represent the A-B sublattices of the AFM layer, which is dragged into a spiraling texture due to exchange bias on the interfaces. Right: incoming electrons only enter the A subband due to the upper FM polarizer, the outgoing electrons partially occupy the B subband depending on the value of ξ.

been overlooked before. For better comparison, we summarize the fundamental electron dynamics of FM and AFM textures in Table I.

A final remark on the theory part: In real materials with impurities, our fundamental equations (8) are valid so long as spin coherence length is as large as, if not more than, the typical width of the texture. While this is quite true in FM materials, its validity in AFM materials awaits experimental verification. At extremely low temperatures, spin-flip scattering is dominated by magnetic impurities which can be made negligibly small in clean samples. Besides, spin-independent scattering processes (e.g., electron-phonon scattering) do not destroy our essential conclusions if $\dot{r}$ is understood as the drift velocity of carriers. We mention that AFM spintronics is an emerging field where very little is known. While it shares some similarities with the established FM spintronics, it is not always correct to copy ideas from FM systems. The adiabatic electron dynamics studied in this paper is one example.

IV. EXAMPLES

First, consider a spiraling AFM texture sandwiched by two ferromagnetic layers (see Fig. 3). This magnetic structure has been realized in Co/FeMn/Py trilayers in a recent experiment[19] where the FM order of Co layer is nearly fixed, but that of Py can be rotated by external magnetic field. The AFM order is dragged into a spiral due to the exchange bias effect on the AFM/FM interfaces. The layer thickness of FeMn is roughly $10\sim 20$ nm and can be made even larger, which far exceeds the lattice constant, thus adiabatic approximation is valid. Meanwhile, typical spin coherence length is larger than the layer thickness at low temperatures so that spin evolution is governed by Eq. (8a).

When an electron flows from top to bottom with applied current, the top FM layer polarizes its spin so that it enters the A subband across the interface. According to Eq. (8a), the physical spin orientation of the electron after passing through the AFM layer is rotated by $\Pi = \pi - \arctan[\xi\tan\xi\pi]$ if

$\xi < \frac{1}{2}$, and $\Pi = -\arctan[\xi \tan \xi\pi]$ is $\xi > \frac{1}{2}$. This is a topological result that only depends on the initial and final directions of $\boldsymbol{n}$, but is *independent* of the texture's profile detail. When $\xi \to 0$, Π reduces to π, which means the electron spin follows $\boldsymbol{n}$ and remains in the A subband, thus it flows into the bottom FM layer with a lower resistance; in the $\xi \to 1$ limit, Π vanishes and the electron completely evolves into the B subband, thus experiencing a higher resistance. For an arbitrary ξ and an arbitrary total rotation of the spiral denoted by Φ, the electron will partially evolve into the B subband with the wave function $\cos(\xi\Phi/2)|\psi_a\rangle + i\sin(\xi\Phi/2)|\psi_b\rangle$, thus the total resistance is

$$\rho = \rho_0 + \tfrac{1}{2}\Delta\rho[1 - \cos(\xi\Phi)], \tag{14}$$

where ρ_0 is the intrinsic resistance of the AFM texture itself, which depends monotonously but not too much on Φ. $\Delta\rho$ represents the magnetoresistance of the spin valve which is determined by material details of the two FM layers and is independent of Φ. If Φ is increased beyond π, ρ will reach a maximum at $\Phi_m = \pi/\xi$ and then reduces. The resistance maximum, if observed, serves as an experimental verification of Eq. (8a). Moreover, measuring Φ_m also enables us to find ξ without calculating the band structure.

We remark that the above results survive in the presence of diffusive processes so long as spin-flip scattering is ignored. The reason is that spin-independent scattering only deflects $\boldsymbol{k}$-space orbit, whereas the $\boldsymbol{s}$ dynamics is determined by the variation of $\boldsymbol{n}$ that is blind to $\boldsymbol{k}$ in one dimension. In addition, FeMn is a noncollinear antiferromagnet that has more than two sublattices. To test our theory unambiguously, we can replace FeMn by the collinear IrMn which is feasible for the current technique.[20] Moreover, we are aware of the experiment[21] where the spiraling AFM texture exhibits spatial *periodic* patterns: it provides a better way of realizing large Φ's.

Consider a second example where $\boldsymbol{n}(t)$ is varying round a cone of constant semiangle θ in the laboratory frame, which can be realized in a spin wave (see Fig. 4). According to Eq. (8a), we know that $ds_z = 0$ due to $dn_z = 0$, thus, the tip of $\boldsymbol{s}$ should stay in the bottom plane of the cone. On the other hand, we learn from Eq. (10) that the tip is constrained on the spheroid that moves with the instantaneous $\boldsymbol{n}(t)$. Therefore, the actual orbit traversed by the tip is contained in the intersection of the two constraints. Through some straightforward geometric analysis, we know that $\boldsymbol{s}$ is bounded between the $\boldsymbol{n}$ cone and an inner cone whose semiangle depends on ξ. Figure 4 depicts the actual orbits of $\boldsymbol{s}$ for three different ξ's: they all exhibit precession and nutation, which can be easily read out from the bird's eye view. Remarkably, the motion of $\boldsymbol{s}$ falls into two topologically distinct classes separated by the critical condition

$$\xi_c^2 = \frac{\cos^2\theta}{(1 + \cos^2\theta)}. \tag{15}$$

In a real spin wave, θ is nearly zero, thus $\xi_c \approx 1/\sqrt{2}$, and for real materials we expect $t \leqslant J$; thus from Eq. (4) we know that for a partially filled band, ξ is always smaller than $1/\sqrt{2}$. Therefore, the $\xi < \xi_c$ phase is more realistic.

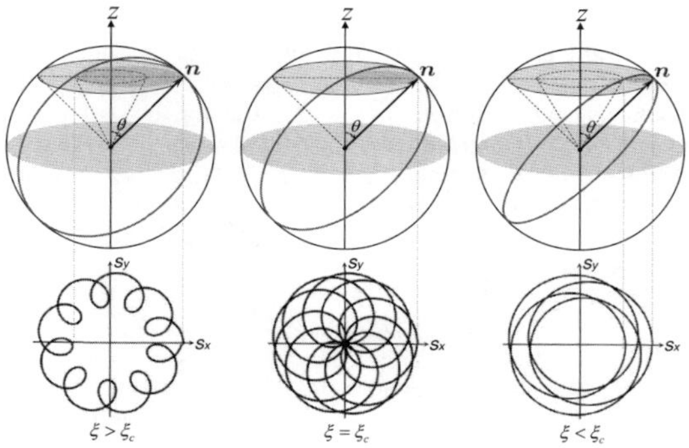

FIG. 4. (Color online) Spin evolutions for three different ξ's when $\boldsymbol{n}(t)$ is moving round a cone with constant angle θ from the z axis. Upper panels: the tip of $\boldsymbol{s}$ respects two constraints: it stays both on the cone's bottom (small gray slab) and on the spheroid described by Eq. (10) (blue ellipsoid), thus the vector $\boldsymbol{s}$ is confined in between two cones with different semiangles. Lower panels: orbits of the tip from bird's eye view. The topology of the orbits is separated into two classes (left and right) by the critical case (middle) where the inner cone's semiangle shrinks to zero. The orbits are not necessarily commensurate with $\boldsymbol{n}$.

V. CONCLUSIONS

In this paper, we find that a slowly varying AFM texture renders adiabatic dynamics of conduction electrons, which are described by three coupled equations of motion [Eqs. (8)]. Quite different from the FM case, the adiabaticity in AFM materials does not imply strict alignment between conduction electron spins and the profile of background texture. Instead, the adiabatic spin evolution is a superposition of a motion following the background order plus a motion on a prolate spheroid attached to the local order, where the latter originates from internal dynamics between degenerate bands. The overall motion of the spin is still geometric; it can be attributed to the accumulation of a SU(2) non-Abelian Berry phase originating from the gauge flux of an effective 't Hooft-Polyakov monopole in the parameter space.

The corresponding orbital dynamics shares some similarities with FM materials in that the $\boldsymbol{k}$-space dynamics can be described by an effective Lorentz force equation. However, two prominent differences in the orbital dynamics distinguish an AFM system from its FM counterpart: first, the gauge charge is dynamical rather than constant, by which spin and orbital motions no longer separate; second, the group velocity is renormalized by a spin-dependent anomalous velocity, which is quite different from what has been studied before.

Our theory lays the foundation for charge and spin transports in AFM texture systems, which will be applied to real materials in the future. The validity of our theory needs to be tested experimentally since available data on AFM spintronic materials are very rare. Theoretically, this paper solves only the first half of the whole story; the other half, i.e., the converse effect regarding the back-reaction of current on background AFM order, will appear in a forthcoming paper.

ACKNOWLEDGMENTS

We are grateful for Y. You and B. Wu for numerous help on detailed calculations, and K. Everschor for insightful comments. We thank M. Tsoi and J. Cai for discussions on possible experimental observations. We also acknowledge help from J. Zhou, S. A. Yang, A. H. MacDonald, X. Li, Y. Gao, and D. Xiao for comments from diverse aspects. This work is supported by DOE-DMSE, NBRPC, NSFC, and the Welch Foundation.

APPENDIX A

For simplicity, we denote the joint space-time coordinate as $r_\mu \equiv (t, \boldsymbol{r}_c)$, so that the local Hamiltonian $\mathcal{H}(r_\mu)$ is parametrized by r_μ and the wave packet of the electron can be expressed as

$$\begin{aligned}|W(r_\mu,\boldsymbol{k}_c)\rangle &= \int d\boldsymbol{k}\, w(\boldsymbol{k}-\boldsymbol{k}_c)[c_a|\psi_a\rangle + c_b|\psi_b\rangle] \\ &= \int d\boldsymbol{k}\, w(\boldsymbol{k}-\boldsymbol{k}_c)[c_a|A(\boldsymbol{k})\rangle|\uparrow(r_\mu)\rangle \\ &\quad + c_b|B(\boldsymbol{k})\rangle|\downarrow(r_\mu)\rangle], \qquad \text{(A1)}\end{aligned}$$

where the profile function $w(\boldsymbol{k}-\boldsymbol{k}_c)$ satisfies $\int d\boldsymbol{k}|w(\boldsymbol{k}-\boldsymbol{k}_c)|^2 = 1$ and $\int d\boldsymbol{k}\,\boldsymbol{k}|w(\boldsymbol{k}-\boldsymbol{k}_c)|^2 = \boldsymbol{k}_c$. The coefficients c_a and c_b reflect relative contributions from the two subbands. Here, we define the isospinor $\tilde{c} = [c_a, c_b]^T$, then the normalization condition $|c_a|^2 + |c_b|^2 = 1$ becomes $\tilde{c}^\dagger\tilde{c} = 1$.[22] The effective Lagrangian for the wave packet is then written as[15]

$$\begin{aligned}\mathcal{L} &= \langle W|\left(i\frac{\partial}{\partial t} - \mathcal{H}\right)|W\rangle = \mathcal{L}(r_\mu,k_\mu,\tilde{c};\, \dot{r}_\mu,\dot{k}_\mu,\dot{\tilde{c}}) \\ &= \varepsilon + \boldsymbol{k}_c\cdot\dot{\boldsymbol{r}}_c + i\tilde{c}^\dagger\dot{\tilde{c}} + \tilde{c}^\dagger\left(A^r_\mu\dot{r}_\mu + A^k_\mu\dot{k}_\mu\right)\tilde{c}, \qquad \text{(A2)}\end{aligned}$$

where $k_\mu = (0, \boldsymbol{k}_c)$ has no temporal component in contrast to $r_\mu = (t, \boldsymbol{r}_c)$, but it is still written this way just to simplify symbols. Here, the Berry connections are 2×2 matrices and are functions of both r_μ and k_μ. The real space components are defined by[15–17]

$$\begin{aligned}A^r_\mu &= i\begin{bmatrix}\langle u_a|\partial^r_\mu|u_a\rangle, & \langle u_a|\partial^r_\mu|u_b\rangle \\ \langle u_b|\partial^r_\mu|u_a\rangle, & \langle u_b|\partial^r_\mu|u_b\rangle\end{bmatrix} \\ &= \frac{1}{2}\begin{bmatrix}\cos\theta\partial_\mu\phi & \xi(-i\partial_\mu\theta - \sin\theta\partial_\mu\phi) \\ \xi(i\partial_\mu\theta - \sin\theta\partial_\mu\phi) & -\cos\theta\partial_\mu\phi\end{bmatrix} \\ &= \frac{1}{2}[-\tau_1\xi\sin\theta\partial_\mu\phi + \tau_2\xi\partial_\mu\theta + \tau_3\cos\theta\partial_\mu\phi], \qquad \text{(A3)}\end{aligned}$$

where $\{\tau_1, \tau_2, \tau_3\} \equiv \boldsymbol{\tau}$ is a vector of Pauli matrices representing isospin; it should not be confused with the real spin operator $\boldsymbol{\sigma}$. Similarly, the momentum-space components are

$$A^k_\mu = i\begin{bmatrix}\langle u_a|\partial^k_\mu|u_a\rangle, & \langle u_a|\partial^k_\mu|u_b\rangle \\ \langle u_b|\partial^k_\mu|u_a\rangle, & \langle u_b|\partial^k_\mu|u_b\rangle\end{bmatrix} = \begin{bmatrix}0 & 0\\ 0 & 0\end{bmatrix}, \qquad \text{(A4)}$$

which in general should not vanish if spin-orbit coupling terms are added to the original Hamiltonian. To avoid complicated manipulations on matrix products in the final results, we define the isospin vector

$$\begin{aligned}\boldsymbol{C} &= \{c_1, c_2, c_3\} = \tilde{c}^\dagger\boldsymbol{\tau}\tilde{c} \\ &= \{2\,\mathrm{Re}(c_a c_b^*),\, -2\,\mathrm{Im}(c_a c_b^*), |c_a|^2 - |c_b|^2\}, \qquad \text{(A5)}\end{aligned}$$

with which the Berry connection A^r_μ can be expressed as a vector in the isospin vector space (the adjoint representation)

$$\begin{aligned}\boldsymbol{\mathcal{A}}^r_\mu &= \tfrac{1}{2}\mathrm{Tr}[\boldsymbol{\tau}A^r_\mu] = \left\{\left(\mathcal{A}^r_\mu\right)^1, \left(\mathcal{A}^r_\mu\right)^2, \left(\mathcal{A}^r_\mu\right)^3\right\} \\ &= \tfrac{1}{2}\{-\xi\sin\theta\partial_\mu\phi,\, \xi\partial_\mu\theta,\, \cos\theta\partial_\mu\phi\}, \qquad \text{(A6)}\end{aligned}$$

where we have used $\frac{1}{2}\mathrm{Tr}[\tau_i\tau_j] = \delta_{ij}$. In a similar way, $\boldsymbol{\mathcal{A}}^k_\mu = \frac{1}{2}\mathrm{Tr}[\boldsymbol{\tau}A^k_\mu] = 0$, but we still keep it at this step for the completeness of the entire formalism. The effective Lagrangian now becomes

$$\mathcal{L} = \varepsilon + k_\mu\dot{r}_\mu + i\tilde{c}^\dagger\dot{\tilde{c}} + \boldsymbol{C}\cdot\left[\boldsymbol{\mathcal{A}}^r_\mu\dot{r}_\mu + \boldsymbol{\mathcal{A}}^k_\mu\dot{k}_\mu\right], \qquad \text{(A7)}$$

where the $i\tilde{c}^\dagger\dot{\tilde{c}}$ term can not be expressed in terms of $\boldsymbol{C}$ and $\dot{\boldsymbol{C}}$, but this poses no problem in the following. Both Eqs. (A2) and (A7) are useful.

To obtain the equations of motion for the three parameters $(\boldsymbol{C}, r_\mu, k_\mu)$, we resort to the following variational principles.

(i) $\delta\mathcal{L}/\delta\tilde{c} = 0$ on Eq. (A2) gives

$$\dot{\tilde{c}} = i\left[A^r_\mu\dot{r}_\mu + A^k_\mu\dot{k}_\mu\right]\tilde{c}, \qquad \text{(A8a)}$$

$$\dot{\tilde{c}}^\dagger = -i\tilde{c}^\dagger\left[A^r_\mu\dot{r}_\mu + A^k_\mu\dot{k}_\mu\right]; \qquad \text{(A8b)}$$

then from $\boldsymbol{C} = \tilde{c}^\dagger\boldsymbol{\tau}\tilde{c}$ we have

$$\begin{aligned}\dot{\boldsymbol{C}} &= \dot{\tilde{c}}^\dagger\boldsymbol{\tau}\tilde{c} + \tilde{c}^\dagger\boldsymbol{\tau}\dot{\tilde{c}} \\ &= i\dot{r}_\mu\tilde{c}^\dagger\left[\boldsymbol{\tau}A^r_\mu - A^r_\mu\boldsymbol{\tau}\right]\tilde{c} + i\dot{k}_\mu\tilde{c}^\dagger\left[\boldsymbol{\tau}A^k_\mu - A^k_\mu\boldsymbol{\tau}\right]\tilde{c}, \qquad \text{(A9)}\end{aligned}$$

where $\boldsymbol{\tau}A_\mu$ terms are matrix products. In view of the decomposition Eq. (A6), we take a specified component of Eq. (A9):

$$\begin{aligned}\dot{C}_\alpha &= i\dot{r}_\mu\left(\mathcal{A}^r_\mu\right)^\beta\tilde{c}^\dagger(\tau_\alpha\tau_\beta - \tau_\beta\tau_\alpha)\tilde{c} \\ &\quad + i\dot{k}_\mu\left(\mathcal{A}^k_\mu\right)^\beta\tilde{c}^\dagger(\tau_\alpha\tau_\beta - \tau_\beta\tau_\alpha)\tilde{c} \\ &= -2\varepsilon_{\alpha\beta\gamma}\left[\left(\mathcal{A}^r_\mu\right)^\beta\dot{r}_\mu + \left(\mathcal{A}^k_\mu\right)^\beta\dot{k}_\mu\right](\tilde{c}^\dagger\tau_\gamma\tilde{c}). \qquad \text{(A10)}\end{aligned}$$

When written in the isospin vector form, the above equation becomes

$$\dot{\boldsymbol{C}} = 2\boldsymbol{C}\times\left(\boldsymbol{\mathcal{A}}^r_\mu\dot{r}_\mu + \boldsymbol{\mathcal{A}}^k_\mu\dot{k}_\mu\right), \qquad \text{(A11)}$$

which proves Eq. (6a).

(ii) $\delta\mathcal{L}/\delta r_\mu = 0$ on Eq. (A7) requires some care:

$$\frac{\delta\mathcal{L}}{\delta r_\mu} = \frac{\partial\mathcal{L}}{\partial r_\mu} - \frac{d}{dt}\frac{\partial\mathcal{L}}{\partial\dot{r}_\mu} = 0 \text{ with}$$

$$\frac{\partial\mathcal{L}}{\partial r_\mu} = \partial^r_\mu\varepsilon + \boldsymbol{C}\cdot\left[(\partial^r_\mu\boldsymbol{\mathcal{A}}^r_\nu)\dot{r}_\nu + (\partial^r_\mu\boldsymbol{\mathcal{A}}^k_\nu)\dot{k}_\nu\right], \qquad \text{(A12a)}$$

$$\begin{aligned}\frac{d}{dt}\frac{\partial\mathcal{L}}{\partial\dot{r}_\mu} &= \dot{k}_\mu + \left(\dot{\boldsymbol{C}}\cdot\boldsymbol{\mathcal{A}}^r_\mu + \boldsymbol{C}\cdot\frac{d}{dt}\boldsymbol{\mathcal{A}}^r_\mu\right) \\ &= \dot{k}_\mu + 2\left[\dot{r}_\nu\boldsymbol{C}\cdot\left(\boldsymbol{\mathcal{A}}^r_\nu\times\boldsymbol{\mathcal{A}}^r_\mu\right) + \dot{k}_\nu\boldsymbol{C}\cdot\left(\boldsymbol{\mathcal{A}}^k_\nu\times\boldsymbol{\mathcal{A}}^r_\mu\right)\right] \\ &\quad + \boldsymbol{C}\cdot\left[(\partial^r_\nu\boldsymbol{\mathcal{A}}^r_\mu)\dot{r}_\nu + (\partial^k_\nu\boldsymbol{\mathcal{A}}^r_\mu)\dot{k}_\nu\right], \qquad \text{(A12b)}\end{aligned}$$

where in the last line Eq. (A11) has been used. Hence, we obtain the equation of motion

$$\dot{k}_\mu = \partial^r_\mu\varepsilon + \boldsymbol{C}\cdot\left[\boldsymbol{\Omega}^{rr}_{\mu\nu}\dot{r}_\nu + \boldsymbol{\Omega}^{rk}_{\mu\nu}\dot{k}_\nu\right], \qquad \text{(A13)}$$

where the Berry curvatures are defined as

$$\boldsymbol{\Omega}^{rr}_{\mu\nu} \equiv \partial^r_\mu \boldsymbol{\mathcal{A}}^r_\nu - \partial^r_\nu \boldsymbol{\mathcal{A}}^r_\mu + 2\boldsymbol{\mathcal{A}}^r_\mu \times \boldsymbol{\mathcal{A}}^r_\nu, \tag{A14a}$$

$$\boldsymbol{\Omega}^{rk}_{\mu\nu} \equiv \partial^r_\mu \boldsymbol{\mathcal{A}}^k_\nu - \partial^k_\nu \boldsymbol{\mathcal{A}}^r_\mu + 2\boldsymbol{\mathcal{A}}^r_\mu \times \boldsymbol{\mathcal{A}}^k_\nu. \tag{A14b}$$

They are antisymmetric tensors with permutations of r_μ and k_μ, and at the same time they are vectors in the isospin vector space, the internal space unique to non-Abelian gauge theory.

(iii) $\delta\mathcal{L}/\delta k_\mu = 0$ on Eq. (A7) follows quite similar procedures as above, and the equation of motion is

$$\dot{r}_\mu = -\partial^k_\mu \varepsilon - \boldsymbol{C} \cdot \left[\boldsymbol{\Omega}^{kr}_{\mu\nu}\dot{r}_\nu + \boldsymbol{\Omega}^{kk}_{\mu\nu}\dot{k}_\nu\right], \tag{A15}$$

where the Berry curvatures are

$$\boldsymbol{\Omega}^{kk}_{\mu\nu} \equiv \partial^k_\mu \boldsymbol{\mathcal{A}}^k_\nu - \partial^k_\nu \boldsymbol{\mathcal{A}}^k_\mu + 2\boldsymbol{\mathcal{A}}^r_\mu \times \boldsymbol{\mathcal{A}}^r_\nu, \tag{A16a}$$

$$\boldsymbol{\Omega}^{kr}_{\mu\nu} \equiv \partial^k_\mu \boldsymbol{\mathcal{A}}^r_\nu - \partial^r_\nu \boldsymbol{\mathcal{A}}^k_\mu + 2\boldsymbol{\mathcal{A}}^k_\mu \times \boldsymbol{\mathcal{A}}^r_\nu, \tag{A16b}$$

which, together with Eqs. (A14), form a generalized matrix of Berry curvature jointing real space and BZ into a unified parameter space

$$\tilde{\boldsymbol{\Omega}}_{\mu\nu} = \begin{bmatrix} \boldsymbol{\Omega}^{rr}_{\mu\nu} & \boldsymbol{\Omega}^{rk}_{\mu\nu} \\ \boldsymbol{\Omega}^{kr}_{\mu\nu} & \boldsymbol{\Omega}^{kk}_{\mu\nu} \end{bmatrix}. \tag{A17}$$

Equations (A13), (A15), and (A11) justify the main conclusions in Sec. II. It is worth mentioning that the Abelian version of the above formalism is a well-established field, whose great success has been proved by compelling evidences in the last decade.[3] However, the non-Abelian version is still underdeveloped in previous work.[15]

APPENDIX B

Before deriving Eqs. (8), special attention should be paid on the fact that gauge fields (Berry curvatures) in non-Abelian theory are not gauge invariant, but gauge covariant. It is the isospin *scalars* $\boldsymbol{C} \cdot \boldsymbol{\Omega}_{\mu\nu}$ appearing in Eqs. (8) that respect gauge invariance. Specifically, as we make a gauge transformation on the wave functions $|\psi_a\rangle$ and $|\psi_b\rangle$, change of $\boldsymbol{\Omega}$ just compensates that of $\boldsymbol{C}$. However, to perform real calculations, we have to work in a specified gauge. In this paper, the gauge is fixed in the choice of local spin eigenstates, which are obtained by acting $U(\boldsymbol{r},t) = e^{-i\sigma_z\phi/2}e^{-i\sigma_y\theta/2}e^{-i\sigma_z\chi/2}$ on the eigenstates of σ_z. While $\theta(\boldsymbol{r},t)$ and $\phi(\boldsymbol{r},t)$ are physical, $\chi(\boldsymbol{r},t)$ is not and can be chosen arbitrarily; the gauge is fixed by setting $\chi = 0$.

From Eq. (A6) the real-space curvature is obtained:

$$\begin{aligned}\boldsymbol{\Omega}^{rr}_{\mu\nu} &\equiv \partial^r_\mu \boldsymbol{\mathcal{A}}^r_\nu - \partial^r_\nu \boldsymbol{\mathcal{A}}^r_\mu + 2\boldsymbol{\mathcal{A}}^r_\mu \times \boldsymbol{\mathcal{A}}^r_\nu \\ &= \left\{0,0,(\xi^2-1)\tfrac{1}{2}\sin\theta\left(\partial^r_\mu\theta\partial^r_\nu\phi - \partial^r_\nu\theta\partial^r_\mu\phi\right)\right\} \\ &= \left\{0,\,0,\,(\xi^2-1)\tfrac{1}{2}\boldsymbol{n}\cdot\left(\partial^r_\mu\boldsymbol{n}\times\partial^r_\nu\boldsymbol{n}\right)\right\},\end{aligned} \tag{B1}$$

where $\boldsymbol{n} = \{\sin\theta\cos\phi, \sin\theta\sin\phi, \cos\theta\}$ is the local order parameter. We see that only the third component is nonzero in our particular gauge marked by $\chi = 0$. But, one can check that in any gauge with $\chi \neq 0$, the first two components do not vanish. However, the third component is actually gauge invariant and it has the form of skyrmion density. The cross components of Berry curvature are obtained in a similar way:

$$\begin{aligned}\boldsymbol{\Omega}^{rk}_{\mu\nu} = -\boldsymbol{\Omega}^{kr}_{\nu\mu} &= \partial^r_\mu \boldsymbol{\mathcal{A}}^k_\nu - \partial^k_\nu \boldsymbol{\mathcal{A}}^r_\mu + 2\boldsymbol{\mathcal{A}}^r_\mu \times \boldsymbol{\mathcal{A}}^k_\nu \\ &= \tfrac{1}{2}\left\{\partial^k_\nu\xi\sin\theta\partial^r_\mu\phi\ -\partial^k_\nu\xi\partial^r_\mu\theta,\,0\right\},\end{aligned} \tag{B2}$$

where again the first two components are changeable subject to gauge transformations, whereas the third is gauge invariant. Moreover, due to $\boldsymbol{\mathcal{A}}^k_\mu = 0$, the BZ space Berry curvature $\boldsymbol{\Omega}^{kk}_{\mu\nu}$ vanishes and we will not mention it in the following. Now let us turn to the equations of motion.

(i) Substitute Eq. (A6) into Eq. (A11), noting that $\dot{\theta} = \dot{r}_\mu\partial_\mu\theta = \partial_t\theta + \dot{\boldsymbol{r}}_c\cdot\nabla\theta$ and the same for $\dot{\phi}$:

$$\frac{d}{dt}\begin{bmatrix} c_1 \\ c_2 \\ c_3 \end{bmatrix} = \begin{bmatrix} 0 & \cos\theta\dot{\phi} & -\xi\dot{\theta} \\ -\cos\theta\dot{\phi} & 0 & -\xi\sin\theta\dot{\phi} \\ \xi\dot{\theta} & \xi\sin\theta\dot{\phi} & 0 \end{bmatrix}\begin{bmatrix} c_1 \\ c_2 \\ c_3 \end{bmatrix}. \tag{B3}$$

Our target is to transform the dynamics of $\boldsymbol{C}$ to the dynamics of the physical spin defined by

$$\boldsymbol{s} = \langle W(r_\mu)|\boldsymbol{\sigma}|W(r_\mu)\rangle, \tag{B4}$$

which respects gauge invariance. From Eqs. (2), (A1), (A5), and (B4), we know the components of $\boldsymbol{s}$ in the *laboratory* frame after some tedious algebra,

$$s_x = c_3\sin\theta\cos\phi + \xi[c_1\cos\theta\cos\phi - c_2\sin\phi], \tag{B5a}$$

$$s_y = c_3\sin\theta\sin\phi + \xi[c_1\cos\theta\sin\phi + c_2\cos\phi], \tag{B5b}$$

$$s_z = c_3\cos\theta - \xi c_1\sin\theta, \tag{B5c}$$

then we take the total time derivative over each component of $\boldsymbol{s}$, for example,

$$\begin{aligned}\dot{s}_x &= \dot{c}_3\sin\theta\cos\phi + c_3(\cos\theta\cos\phi\dot{\theta} - \sin\theta\sin\phi\dot{\phi}) \\ &\quad +\xi[\dot{c}_1\cos\theta\cos\phi - c_1(\sin\theta\cos\phi\dot{\theta} + \cos\theta\sin\phi\dot{\phi})] \\ &\quad -\xi[\dot{c}_2\sin\phi + c_2\cos\phi\dot{\phi}] \\ &= c_3(1-\xi^2)(\cos\theta\cos\phi\dot{\theta} - \sin\theta\sin\phi\dot{\phi}) \\ &= c_3(1-\xi^2)\dot{n}_x,\end{aligned} \tag{B6}$$

where in deriving the second equality above Eq. (B3) has been used. Similarly,

$$\dot{s}_y = c_3(1-\xi^2)\dot{n}_y, \quad \dot{s}_z = c_3(1-\xi^2)\dot{n}_z. \tag{B7}$$

To eliminate c_3 in the above equations, we reverse Eqs. (B5) and obtain

$$c_3 = s_x\sin\theta\cos\phi + s_y\sin\theta\sin\phi + s_z\cos\theta = \boldsymbol{s}\cdot\boldsymbol{n}; \tag{B8}$$

then, from Eqs. (B6) and (B7), we obtain a simple and elegant equation of motion for the physical spin

$$\dot{\boldsymbol{s}} = (1-\xi^2)(\boldsymbol{s}\cdot\boldsymbol{n})\dot{\boldsymbol{n}}, \tag{B9}$$

which justifies Eq. (8a).

(ii) To justify Eq. (8b), we substitute Eqs. (B1) and (B2) into Eq. (A13), regarding that k_μ has only spatial but no temporal components, and we arrive at

$$\begin{aligned}\dot{\boldsymbol{k}}_c &= \tfrac{1}{2}c_3(\xi^2-1)\boldsymbol{n}\cdot(\nabla\boldsymbol{n}\times\dot{\boldsymbol{n}}) + \tfrac{1}{2}\xi[c_1\sin\theta\nabla\phi - c_2\nabla\theta] \\ &= \tfrac{1}{2}\boldsymbol{n}\cdot\{\nabla\boldsymbol{n}\times[(\xi^2-1)(\boldsymbol{s}\cdot\boldsymbol{n})\dot{\boldsymbol{n}}]\} \\ &= -\tfrac{1}{2}\boldsymbol{n}\cdot(\nabla\boldsymbol{n}\times\dot{\boldsymbol{s}}),\end{aligned} \tag{B10}$$

where $\dot{\xi} = 0$ and Eq. (B9) has been used. We also have ignored the $\partial^r_\mu\varepsilon$ term in Eq. (B10) since the band structure is only a function of $\boldsymbol{k}$ and independent of space-time in the adiabatic approximation. As in Sec. III, we will omit the superscript c in the following for simplicity. To make better comparisons with the spin motive force[4,6] and the topological Hall effect[8] discovered in FM materials, we also derive another suggestive form of Eq. (8b). Take an arbitrary component i of Eq. (B10),

$$\begin{aligned}\dot{k}_i &= \tfrac{1}{2}c_3(\xi^2-1)\sin\theta(\partial_i\theta\dot{\phi}-\dot{\theta}\partial_i\phi)\\ &= \tfrac{1}{2}c_3(\xi^2-1)\sin\theta\{[\partial_i\theta\partial_t\phi-\partial_t\theta\partial_i\phi]\\ &\quad+[\partial_i\theta(\dot{r}_j\partial_j\phi)-(\dot{r}_j\partial_j\theta)\partial_i\phi]\}\\ &= \tfrac{1}{2}c_3(\xi^2-1)\{\sin\theta[\partial_i\theta\partial_t\phi-\partial_t\theta\partial_i\phi]\\ &\quad+\sin\theta\varepsilon_{ijk}\varepsilon_{klm}\dot{r}_j\partial_l\theta\partial_m\phi\},\end{aligned} \tag{B11}$$

where $\dot{\theta}=\partial_t\theta+\dot{r}_i\partial_i\theta$ (the same for $\dot{\phi}$) and the identity $\varepsilon_{ijk}\varepsilon_{klm}=\delta_{il}\delta_{jm}-\delta_{im}\delta_{jl}$ has been used. Equation (B11) can be written in a concise way as

$$\dot{\boldsymbol{k}}=(1-\xi^2)(\boldsymbol{s}\cdot\boldsymbol{n})(\boldsymbol{E}+\dot{\boldsymbol{r}}\times\boldsymbol{B}), \tag{B12}$$

$$\begin{aligned}\boldsymbol{E}&=\tfrac{1}{2}\sin\theta(\partial_t\theta\nabla\phi-\nabla\theta\partial_t\phi),\\ \boldsymbol{B}&=-\tfrac{1}{2}\sin\theta(\nabla\theta\times\nabla\phi),\end{aligned} \tag{B13}$$

which proves Eqs. (11)–(13) in Sec. III.

(iii) From Eqs. (A15) and (B2), we have

$$\dot{\boldsymbol{r}}=-\partial_{\boldsymbol{k}}\varepsilon+\tfrac{1}{2}\partial_{\boldsymbol{k}}\xi(c_1\sin\theta\dot{\phi}-c_2\dot{\theta}). \tag{B14}$$

There is a smart way to eliminate $c_{1,2}$ in terms of physical spin. Notice that in the special gauge marked by $\chi=0$, the isospin vector $\boldsymbol{C}$ can be pictured as a vector in the local frame extended by $\boldsymbol{\theta}$, $\boldsymbol{\phi}$, and $\boldsymbol{n}$ (see Fig. 2): $\boldsymbol{C}=c_1\boldsymbol{\theta}+c_2\boldsymbol{\phi}+c_3\boldsymbol{n}$, and components of the real spin $\boldsymbol{s}$ in this local frame are

$$s_1=s_x\cos\theta\cos\phi+s_y\cos\theta\sin\phi-s_z\sin\theta=\xi c_1, \tag{B15}$$

$$s_2=-s_x\sin\phi+s_y\cos\phi=\xi c_2, \tag{B16}$$

$$s_3=s_x\sin\theta\cos\phi+s_y\sin\theta\sin\phi+s_z\cos\theta=c_3, \tag{B17}$$

where Eqs. (B5) have been used, and we obtain the important relation

$$\boldsymbol{s}=\xi(c_1\boldsymbol{\theta}+c_2\boldsymbol{\phi})+c_3\boldsymbol{n}. \tag{B18}$$

Since $\dot{\boldsymbol{n}}=\dot{\theta}\boldsymbol{\theta}+\sin\theta\dot{\phi}\boldsymbol{\phi}$, we immediately have

$$\boldsymbol{n}\cdot(\boldsymbol{s}\times\dot{\boldsymbol{n}})=\xi(c_1\sin\theta\dot{\phi}-c_2\dot{\theta}), \tag{B19}$$

thus Eq. (B14) becomes

$$\dot{\boldsymbol{r}}=-\partial_{\boldsymbol{k}}\varepsilon-\tfrac{1}{2}(\boldsymbol{s}\times\boldsymbol{n})\cdot\dot{\boldsymbol{n}}\partial_{\boldsymbol{k}}\ln\xi, \tag{B20}$$

which justifies Eq. (8c). Moreover,

$$\partial_{\boldsymbol{k}}\ln\xi=\frac{1-\xi^2}{\xi^2}\frac{\partial_{\boldsymbol{k}}\varepsilon}{\varepsilon}, \tag{B21}$$

thus the term $\frac{1}{2}(\boldsymbol{s}\times\boldsymbol{n})\cdot\dot{\boldsymbol{n}}\partial_{\boldsymbol{k}}\ln\xi$ has the same direction as the group velocity: it represents the modification of group velocity due to AFM texture.

A further point that should be added is that in the most general case, the effective Lagrangian (A2) should also contain a term representing self-rotation of the wave packet $-\mathrm{Im}[\tilde{c}_i\langle\partial^r_\mu u_i|(\varepsilon-\mathcal{H})|\partial^r_\mu u_j\rangle\tilde{c}_j]$, but after some boring manipulations, one can show that this term vanishes for similar reasons as the vanishing $\boldsymbol{\Omega}^{kk}_{\mu\nu}$.

APPENDIX C

To study the monopole, we should turn to a different coordinate system. By assigning a variable magnitude to $\boldsymbol{n}$, we define the dimensionless order parameter $\boldsymbol{R}\equiv R\boldsymbol{n}=\frac{J}{t}\boldsymbol{n}$. Then, the Berry connection can be equivalently defined in the $\boldsymbol{R}$ space, which relates to the original one by $A_\mu dr_\mu=A_i dR_i$, where

$$A_i=i\begin{bmatrix}\langle\uparrow|\partial_i|\uparrow\rangle & \xi\langle\uparrow|\partial_i|\downarrow\rangle\\ \xi\langle\downarrow|\partial_i|\uparrow\rangle & \langle\downarrow|\partial_i|\downarrow\rangle\end{bmatrix}, \tag{C1}$$

where $\xi=|\tilde{\gamma}|/\sqrt{R^2+|\tilde{\gamma}|^2}$ is also a function of R, and $|\tilde{\gamma}|=\sum_{\boldsymbol{\delta}}e^{i\boldsymbol{k}\cdot\boldsymbol{\delta}}$ depends on the position in BZ. Written in spherical coordinates, components of Eq. (C1) are

$$\begin{aligned}A_{\mathcal{R}}&=0,\quad A_\theta=\frac{\xi}{2R}\begin{bmatrix}0&-i\\ i&0\end{bmatrix},\\ A_\phi&=\frac{1}{2R}\begin{bmatrix}\cot\theta&-\xi\\ -\xi&-\cot\theta\end{bmatrix}.\end{aligned} \tag{C2}$$

To see the monopole, we should further make a singular gauge transformation on the potential[23]

$$\begin{aligned}A'_\theta&=SA_\theta S^\dagger+i\frac{1}{R}S\partial_\theta S^\dagger\\ &=i\frac{[1-\xi(R)]}{2R}\begin{bmatrix}0&e^{-i\phi}\\ -e^{i\phi}&0\end{bmatrix},\end{aligned} \tag{C3a}$$

$$\begin{aligned}A'_\phi&=SA_\phi S^\dagger+i\frac{1}{R\sin\theta}S\partial_\phi S^\dagger\\ &=\frac{[1-\xi(R)]}{2R}\begin{bmatrix}-\sin\theta&e^{-i\phi}\cos\theta\\ e^{i\phi}\cos\theta&\sin\theta\end{bmatrix}\end{aligned} \tag{C3b}$$

with the unitary matrix

$$S=\begin{bmatrix}e^{-i\phi/2}\cos\frac{\theta}{2}&-e^{-i\phi/2}\sin\frac{\theta}{2}\\ e^{i\phi/2}\sin\frac{\theta}{2}&e^{i\phi/2}\cos\frac{\theta}{2}\end{bmatrix}. \tag{C4}$$

Finally, expressing the gauge potential in Cartesian coordinates, we obtain

$$\begin{aligned}A_x&=A'_\theta\cos\theta\cos\phi-A'_\phi\sin\phi=\frac{(1-\xi)}{2R^2}\begin{bmatrix}y&iz\\ -iz&-y\end{bmatrix},\\ A_y&=A'_\theta\sin\phi+A'_\phi\cos\phi=\frac{(1-\xi)}{2R^2}\begin{bmatrix}-x&z\\ z&x\end{bmatrix},\\ A_z&=A'_\theta\sin\theta=\frac{(1-\xi)}{2R^2}\begin{bmatrix}0&-y-ix\\ -y+ix&0\end{bmatrix}.\end{aligned}$$

They are nothing but the "hedgehog" gauge potential of a 't Hooft-Polyakov monopole at $\boldsymbol{R}=0$ (Ref. 23):

$$A'_i=\frac{1-\xi(\mathcal{R})}{2R^2}\varepsilon_{ijk}R_j\sigma_k, \tag{C5}$$

where $R_i=\{x,y,z\}$ and σ_i's are Pauli matrices. The radial profile of Eq. (C5) is determined by the factor $1-\xi(R)$. For fixed nonzero $\tilde{\gamma}$, $1-\xi(R)$ tends to 1 as $R\to\infty$, and $1-\xi(R)\sim R^2$ as $R\to 0$, which cancels the R^2 in the denominator, thus the gauge potential is *regular* at origin. The form and behaviors of Eq. (C5) are all the same as the gauge potential proposed by Sonner and Tong for realizing the artificial 't Hooft-Polyakov monopole.[18] In fact, our parameter ξ can be understood as the $f(B)$ factor in Ref. 18; they both reflect the overlap of (partial) wave functions from doubly degenerate bands. However, at BZ boundary $\tilde{\gamma}$ is zero, thus $\xi = 0$ regardless of R. In this case, Eq. (C5) becomes the gauge potential of a Wu-Yang monopole and the origin $\boldsymbol{R} = 0$ becomes singular.

Along with the gauge potential, we can define the associated Higgs field

$$\phi_H \equiv \begin{bmatrix} \langle A|\sigma_3|A\rangle & \langle A|\sigma_3|B\rangle \\ \langle B|\sigma_3|A\rangle & \langle B|\sigma_3|B\rangle \end{bmatrix} = \sqrt{1-\xi^2}\,\sigma_3, \quad \text{(C6)}$$

which is the pseudospin polarization of AFM systems. It describes the extent to which the conduction electrons with opposite spins are spatially separated on alternating A and B sites. In other words, it represents how much those electrons respect the staggered order. Upon the same gauge transformation with matrix (C4),

$$\phi_H \to \phi'_H = S\phi_H S^\dagger = \frac{\boldsymbol{R}\cdot\boldsymbol{\sigma}}{\sqrt{|\tilde{\gamma}|^2 + r^2}} = \boldsymbol{\Phi}_H \cdot \boldsymbol{\sigma}. \quad \text{(C7)}$$

The SU(2) gauge field associated with Eq. (C1) is broken into an Abelian magnetic field due to *effective* Higgs mechanism

$$F_{ij} = \partial_i(\boldsymbol{\Phi}\cdot\boldsymbol{A}_j) - \partial_j(\boldsymbol{\Phi}\cdot\boldsymbol{A}_i) + 2\boldsymbol{\Phi}\cdot(\partial_i\boldsymbol{\Phi}\times\partial_j\boldsymbol{\Phi}), \quad \text{(C8)}$$

$$B_i = \frac{1}{2}\varepsilon_{ijk}F_{jk} = \frac{R_i}{2R^3}, \quad \text{(C9)}$$

where $\boldsymbol{\Phi} = \boldsymbol{\Phi}_H/|\boldsymbol{\Phi}_H|$, and Eq. (C9) is the magnetic field of a Dirac monopole.

An unsolved issue is the Bogomol'nyi relation.[18,23] For the non-Abelian gauge field $\Omega_{ij} = \partial_i A_j - \partial_j A_i - i[A_i, A_j]$, and the covariant derivative $\mathcal{D}_i = \partial_i - i[A_i, \,]$, it is straightforward to derive

$$\Omega_{ij} = \frac{1}{2}\varepsilon_{ijk}\left[\mathcal{D}_k\phi_H - \frac{|\tilde{\gamma}| + \sqrt{|\tilde{\gamma}|^2 + R^2}}{(|\tilde{\gamma}| + R^2)^{3/2}}\sigma_k\right]. \quad \text{(C10)}$$

If not for the last term, Eq. (C10) reproduces the Bogomol'nyi relation. One can show that only for a profile function $\xi(R) = 2R/\sinh(R)$ (the case of true 't Hooft-Polyakov monopole) that the last term vanishes. While our $\xi(R)$ asymptotically resembles $2R/\sinh(R)$, it gives a different profile for finite R. As a result, the last term in Eq. (C10) vanishes only when $R \to \infty$.

APPENDIX D

We eliminate dt on both sides of Eq. (8a):

$$d\boldsymbol{s} = (1-\xi^2)(\boldsymbol{s}\cdot\boldsymbol{n})d\boldsymbol{n}. \quad \text{(D1)}$$

First, dot product both sides of Eq. (D1) with $\boldsymbol{n}$, regarding the constraint $\boldsymbol{n}^2 = 1$,

$$\boldsymbol{n}\cdot d\boldsymbol{s} = (1-\xi^2)(\boldsymbol{s}\cdot\boldsymbol{n})\tfrac{1}{2}d(\boldsymbol{n}^2) = 0, \quad \text{(D2)}$$

which gives us the relation $d(\boldsymbol{n}\cdot\boldsymbol{s}) = \boldsymbol{s}\cdot d\boldsymbol{n}$. Second, dot product with $\boldsymbol{s}$ on both sides of Eq. (D1) and we get

$$d(\boldsymbol{s}^2) = 2(1-\xi^2)(\boldsymbol{s}\cdot\boldsymbol{n})(\boldsymbol{s}\cdot d\boldsymbol{n}) = (1-\xi^2)d(\boldsymbol{s}\cdot\boldsymbol{n})^2. \quad \text{(D3)}$$

From Pythagorean theorem, we know that $\boldsymbol{s}^2 = (\boldsymbol{s}\cdot\boldsymbol{n})^2 + (\boldsymbol{s}\times\boldsymbol{n})^2$; then, taking derivative on both sides, regarding Eq. (D3), we arrive at

$$\xi^2 d(\boldsymbol{s}\cdot\boldsymbol{n})^2 = -d(\boldsymbol{s}\times\boldsymbol{n})^2. \quad \text{(D4)}$$

Assume the initial condition $(\boldsymbol{s}\cdot\boldsymbol{n})|_0 = 1$, and the above equation can be integrated into

$$(\boldsymbol{s}\cdot\boldsymbol{n})^2 + \frac{(\boldsymbol{s}\times\boldsymbol{n})^2}{\xi^2} = s_3^2 + \frac{s_1^2 + s_2^2}{\xi^2} = 1, \quad \text{(D5)}$$

which justifies Eq. (10).

*rancheng@physics.utexas.edu

[1] I. Žutić, J. Fabian, and S. D. Sarma, Rev. Mod. Phys. **76**, 323 (2004), and the references therein.

[2] M. V. Berry, Proc. R. Soc. London, Ser. A **392**, 45 (1984).

[3] D. Xiao, M.-C. Zhang, and Q. Niu, Rev. Mod. Phys. **82**, 1959 (2010), and the references therein.

[4] S. A. Yang *et al.*, Phys. Rev. Lett. **102**, 067201 (2009); Phys. Rev. B **82**, 054410 (2010).

[5] G. E. Volovik, J. Phys. C: Solid State Phys. **20**, L83 (1987).

[6] S. E. Barnes and S. Maekawa, Phys. Rev. Lett. **98**, 246601 (2007).

[7] T. Schulz *et al.*, Nat. Phys. **8**, 301 (2012); K. Everschor, M. Garst, R. A. Duine, and A. Rosch, Phys. Rev. B **84**, 064401 (2011).

[8] M. Lee, W. Kang, Y. Onose, Y. Tokura, and N. P. Ong, Phys. Rev. Lett. **102**, 186601 (2009); A. Neubauer *et al.*, *ibid.* **102**, 186602 (2009); P. Bruno, V. K. Dugaev, and M. Taillefumier, *ibid.* **93**, 096806 (2004); J. Ye *et al.*, *ibid.* **83**, 3737 (1999).

[9] Y. B. Bazaliy, B. A. Jones, and S.-C. Zhang, Phys. Rev. B **57**, R3213 (1998).

[10] C. H. Wong and Y. Tserkovnyak, Phys. Rev. B **80**, 184411 (2009); Y. Tserkovnyak and C. H. Wong, *ibid.* **79**, 014402 (2009).

[11] J. Zang, M. Mostovoy, J. H. Han, and N. Nagaosa, Phys. Rev. Lett. **107**, 136804 (2011); F. Jonietz *et al.*, Science **330**, 1648 (2010).

[12] A. C. Swaving and R. A. Duine, J. Phys.: Condens. Matter **24**, 024223 (2012); K. M. D. Hals, Y. Tserkovnyak, and A. Brataas, Phys. Rev. Lett. **106**, 107206 (2011); R. Wieser, E. Y. Vedmedenko, and R. Wiesendanger, *ibid.* **106**, 067204 (2011); A. C. Swaving and R. A. Duine, Phys. Rev. B **83**, 054428 (2011); Y. Xu, S. Wang, and K. Xia, Phys. Rev. Lett. **100**, 226602 (2008); P. M. Haney and A. H. MacDonald, *ibid.* **100**, 196801 (2008).

[13] S. Urazhdin and N. Anthony, Phys. Rev. Lett. **99**, 046602 (2007); R. Jaramillo *et al.*, *ibid.* **98**, 117206 (2007); Z. Wei *et al.*, *ibid.* **98**, 116603 (2007).

[14] A. H. MacDonald and M. Tsoi, Philos. Trans. R. Soc., A **369**, 3098 (2011).

[15] D. Culcer, Y. Yao, and Q. Niu, Phys. Rev. B **72**, 085110 (2005).

[16] J. Dalibard *et al.*, Rev. Mod. Phys. **83**, 1523 (2011).

[17]F. Wilczek and A. Zee, Phys. Rev. Lett. **52**, 2111 (1984); J. Moody, A. Shapere, and F. Wilczek, *ibid.* **56**, 893 (1986); C. A. Mead, *ibid.* **59**, 161 (1987); B. Zygelman, *ibid.* **64**, 256 (1990).

[18]J. Sonner and D. Tong, Phys. Rev. Lett. **102**, 191801 (2009).

[19]F. Y. Yang and C. L. Chien, Phys. Rev. Lett. **85**, 2597 (2000).

[20]J.-W. Cai (private communication).

[21]M. Bode *et al.*, Nature (London) **447**, 190 (2007).

[22]Different from the peudospin furnished by the A-B sublattices, isospin refers to the superposition coefficients c_a and c_b. Only in the limit $\xi \rightarrow 0$ where $A\rangle$ and $B\rangle$ are orthogonal do isospin and peudospin become equivalent.

[23]Ya. Shnir, *Magnetic Monopoles* (Springer, Berlin, 2005).

Phase-space Berry phases in chiral magnets: Dzyaloshinskii-Moriya interaction and the charge of skyrmions

Frank Freimuth,[1,*] Robert Bamler,[2] Yuriy Mokrousov,[1] and Achim Rosch[2]
[1]*Peter Grünberg Institut and Institute for Advanced Simulation, Forschungszentrum Jülich and JARA, 52425 Jülich, Germany*
[2]*Institute for Theoretical Physics, Universität zu Köln, D-50937 Köln, Germany*
(Received 5 September 2013; revised manuscript received 22 November 2013; published 11 December 2013)

The semiclassical motion of electrons in phase space $\mathbf{x} = (\mathbf{R},\mathbf{k})$ is influenced by Berry phases described by a six-component vector potential $\mathbf{A} = (\mathbf{A}^{\mathrm{R}},\mathbf{A}^{\mathrm{k}})$. In chiral magnets, Dzyaloshinskii-Moriya (DM) interactions induce slowly varying magnetic textures (helices and skyrmion lattices) for which all components of $\mathbf{A}$ are important, inducing effectively a curvature in mixed position and momentum space. We show that for smooth textures and weak spin-orbit coupling, phase-space Berry curvatures determine the DM interactions and give important contributions to the charge. Using *ab initio* methods, we calculate the strength of DM interactions in MnSi in good agreement with experiment and estimate the charge of skyrmions.

DOI: 10.1103/PhysRevB.88.214409 PACS number(s): 75.10.Lp, 03.65.Vf, 71.15.Mb, 71.20.Lp

I. INTRODUCTION

In chiral magnets without inversion symmetry, spin-orbit interaction (SOI) effects described by Dzyaloshinskii-Moriya (DM) interactions[1,2] induce the formation of magnetic textures. For smooth textures and cubic systems such as MnSi, the leading DM contribution to the free-energy density is given by the term $D\hat{\mathbf{n}} \cdot (\nabla \times \hat{\mathbf{n}})$, where $\hat{\mathbf{n}}$ is the direction of the magnetization. This term describes that energy can be gained when the magnetic structure twists. In small magnetic fields, these interactions (in combination with thermal fluctuations) can stabilize lattices of topologically quantized magnetic whirls, so called skyrmions.[3,4] Skyrmions couple due to their topological winding extremely efficiently to electric currents resulting in ultralow critical currents for the motion of skyrmions.[5–7]

In this paper, we argue that Berry phases in phase space provide not only a natural framework to understand the physical properties of skyrmions and other magnetic textures, but also generate DM interactions and act therefore as the main driving force inducing magnetic textures in chiral magnets. We focus on Berry curvatures in mixed position and momentum space which lead both to DM interactions and also to an electric charge of skyrmions. Thereby, we naturally link skyrmions in chiral magnets to skyrmions in quantum Hall systems with filling close to $\nu = 1$, which are characterized by a quantized electric charge.[8–12]

Berry phases are quantum mechanical phases picked up by a quantum system when the wave function changes adiabatically.[13–15] They can strongly affect the semiclassical motion of electrons. For each electronic band n, the effects of smoothly varying magnetic textures can efficiently be described by a six-component vector potential $\mathbf{A}_n = (\mathbf{A}_n^{\mathrm{R}},\mathbf{A}_n^{\mathrm{k}})$, with

$$A_{n,j}(\mathbf{x}) = \langle \mathbf{x},n|i\frac{\partial}{\partial x_j}|\mathbf{x},n\rangle, \quad j = 1,\ldots,6 \qquad (1)$$

where $\mathbf{x} = (\mathbf{R},\mathbf{k})$ is the position in phase space and $|\mathbf{x},n\rangle = |\hat{\mathbf{n}}(\mathbf{R}),\mathbf{k},n\rangle$ is the Bloch function, which depends not only on lattice momentum $\hbar\mathbf{k}$, but also on the orientation $\hat{\mathbf{n}}(\mathbf{R})$ of the magnetization. Here, we use the letter $\mathbf{R}$ to denote smooth variations on length scales much larger than the lattice spacing.

Two aspects of Berry phase physics have been well studied in the context of chiral magnets. First, Berry phases in momentum space, described by the $\mathbf{k}$ dependence of $\mathbf{A}_n^{\mathrm{k}}(\mathbf{x})$, give rise to the anomalous Hall effect,[16] which dominates the Hall response for a wide range of temperatures and fields in materials such as MnSi.[17] Powerful *ab initio* methods have been developed to calculate the anomalous Hall effect quantitatively.[18–22] Second, real-space Berry phases give rise to the so-called topological Hall effect. For weak SOI, each skyrmion contributes due to their topology one flux quantum of an emergent magnetic flux,[23] arising from the effective magnetic field $B_{n,i}^{\mathrm{R}} = \frac{\hbar}{e}\epsilon_{ijk}\partial_{R_j}A_{n,k}^{\mathrm{R}} \approx \pm\frac{\hbar}{4e}\epsilon_{ijk}\hat{\mathbf{n}} \cdot (\partial_{R_j}\hat{\mathbf{n}} \times \partial_{R_k}\hat{\mathbf{n}})$ with positive (negative) sign for majority (minority) band n, respectively. This real-space emergent magnetic field acts similar to the "real" magnetic field and has been observed in MnSi (Refs. 3 and 23) and other materials (see, e.g., Refs. 24 and 25) as an extra contribution to the Hall signal. The same effect is also responsible for the efficient coupling of skyrmions to electric currents.[5–7]

Much less studied are systems with Berry phases in phase space, where the $\mathbf{R}$ dependence of $\mathbf{A}_n^{\mathrm{k}}(\mathbf{x})$ and the $\mathbf{k}$ dependence of $\mathbf{A}_n^{\mathrm{R}}(\mathbf{x})$ become important. It has been argued that such a situation arises in smoothly deformed crystals[14] or in the presence of spatially varying external magnetic fields. Also, in antiferromagnets with slowly varying spin texture the mixed Berry phase was suggested to crucially influence the adiabatic dynamics of electrons.[26] Some of us[27] have recently pointed out that DM interactions arise from certain Berry phases. Here, we will provide a purely semiclassical derivation of the Berry phase contribution to the DM interaction showing that DM interactions can be viewed as a phase-space Berry phase effect. By the same mechanism, magnetic skyrmions also obtain a charge. In general, chiral magnets and their magnetic phases turn out to be ideal model systems to study phase-space Berry phases due to their smoothly varying magnetic textures driven by DM interactions.

II. PHASE-SPACE BERRY PHASES

As has been shown by Niu *et al.*[14,28] phase-space Berry phases effectively lead to a curvature of phase space described by the antisymmetric 6×6 Berry-curvature tensor[28]

$$\Omega_{n,ij} = \frac{\partial A_{n,j}}{\partial x_i} - \frac{\partial A_{n,i}}{\partial x_j} = \begin{pmatrix} \Omega_n^{\mathrm{RR}} & \Omega_n^{\mathrm{Rk}} \\ \Omega_n^{\mathrm{kR}} & \Omega_n^{\mathrm{kk}} \end{pmatrix}_{ij}. \tag{2}$$

Here, $\Omega^{\mathrm{RR}}_{n,ij} = \frac{e}{\hbar}\epsilon_{ijk}B^{\mathrm{R}}_{n,k}$ describes real-space Berry phases while $\Omega^{\mathrm{kk}}_{n,ij}$ encodes the momentum-space Berry phases also discussed above. The 3×3 matrix $\Omega^{\mathrm{Rk}}_{n,ij} dR_i dk_j$ is the Berry phase which is picked up when an electron moves along a loop in the phase-space plane spanned by the coordinates R_i and k_j.

The Berry phases influence the semiclassical description of the system in three points. First, the combination of smooth variations in both position and momentum space leads to a shift of the semiclassical energy levels[14] $\epsilon_n(\mathbf{x}) = \epsilon_n^{(0)}(\mathbf{x}) + \delta\epsilon_n(\mathbf{x})$ where $\epsilon_n^{(0)}(\mathbf{x}) = \langle \mathbf{x},n|H(\mathbf{x})|\mathbf{x},n\rangle$ and

$$\delta\epsilon_n(\mathbf{x}) = -\mathrm{Im}\left[\frac{\partial\langle\mathbf{x},n|}{\partial R_i}\left[\epsilon_n^{(0)}(\mathbf{x}) - H(\mathbf{x})\right]\frac{\partial|\mathbf{x},n\rangle}{\partial k_i}\right]. \tag{3}$$

Second, the Berry phases modify the semiclassical equations of motion,[14] which read as $(\Omega_n - J)\dot{\mathbf{x}} = \frac{\partial\epsilon_n}{\partial\mathbf{x}}$ where $J = \begin{pmatrix} 0 & \mathbb{1} \\ -\mathbb{1} & 0 \end{pmatrix}$. Third, the curvature of phase space leads in semiclassical approximation to a modified density of states in phase space[28,29]

$$\begin{aligned} W_n(\mathbf{x}) &= \sqrt{\det(\Omega_n - J)} \\ &= \frac{\epsilon_{ijklrs}}{48}(\Omega_n - J)_{ij}(\Omega_n - J)_{kl}(\Omega_n - J)_{rs}. \end{aligned} \tag{4}$$

Only for this modified density of states the above semiclassical equations of motion satisfy the Liouville theorem. A derivation of Eq. (4) is given in Appendix B.

III. SEMICLASSICS FOR CHIRAL MAGNETS

While the Berry phase description has originally been developed for noninteracting systems, it can also be used to calculate ground-state properties in the presence of interactions using that, within density functional theory (DFT), the interacting system is mapped to a noninteracting one. In the following, we describe an *ab initio* method for the calculation of the DM interaction strength and the charge density in chiral magnets with weak SOI. The method is based on DFT calculations in the homogeneous (i.e., ferromagnetic) state. Berry phase effects coming from the inhomogeneous magnetic texture are then taken into account by means of a gradient expansion.

Within DFT, one can describe the ground state of a ferromagnetic many-particle system with magnetization $\mathbf{M}$ parallel to the unit vector $\hat{\mathbf{n}}$ by an effective single-particle Kohn-Sham Hamiltonian

$$H_{\hat{\mathbf{n}}} = \frac{\mathbf{p}^2}{2m} + V(\mathbf{r}) - \mathbf{M}\cdot\mathbf{B}(\mathbf{r}) - \frac{1}{2mc^2}\mathbf{M}\cdot[\mathbf{E}(\mathbf{r})\times\mathbf{p}], \tag{5}$$

parametrized by an effective potential $V(\mathbf{r})$, an exchange field $\mathbf{B}(\mathbf{r})$, and an electric field $\mathbf{E}(\mathbf{r})$. We use the letter $\mathbf{r}$ for variations on the atomic length scale. To obtain such a ferromagnetic state in a chiral magnet, one has to apply a small external field $\mathbf{B}^{\mathrm{ext}}$ in the direction of $\hat{\mathbf{n}}$ (implicitly included in $\mathbf{B}$) (see following). In the absence of SOI, V and $\mathbf{E}$ are independent of $\hat{\mathbf{n}}$, while $\mathbf{B} \parallel \hat{\mathbf{n}}$.

A. DM interaction

Starting from the eigenstates $|\hat{\mathbf{n}},\mathbf{k},n\rangle$ of the *uniform* Hamiltonian (5) for fixed $\hat{\mathbf{n}}$, we can obtain the change of the free-energy density $\delta F^{(1)}(\mathbf{R})$ to leading order in an adiabatic approximation. We assume that $\hat{\mathbf{n}}(\mathbf{R})$ slowly varies in space and use Eqs. (3) and (4) with $|\mathbf{x},n\rangle = |\hat{\mathbf{n}}(\mathbf{R}),\mathbf{k},n\rangle$ to obtain

$$\begin{aligned} \delta F^{(1)}(\mathbf{R}) = \sum_n \int \frac{d^3k}{(2\pi)^3}\Big[& f_{\mathbf{k}n}\delta\epsilon_n(\mathbf{x}) \\ & + \frac{1}{\beta}\ln(1 + e^{-\beta(\epsilon_{\mathbf{k}n}-\mu)})\Omega^{\mathrm{Rk}}_{n,ii}(\mathbf{x})\Big]. \end{aligned} \tag{6}$$

Note that to leading order all contributions arise from mixed position- and momentum-space Berry curvatures (see following). They contribute only when both inversion symmetry is broken and SOI is present.

To calculate $\delta\epsilon_n(\mathbf{x})$ and the Berry curvature $\Omega^{\mathrm{Rk}}_{n,ii}(\mathbf{x})$ directly, we use that the change of an eigenstate $|n\rangle$ upon changing a parameter λ of H is given by $\partial_\lambda|n\rangle = \sum_{m\neq n}\frac{|m\rangle}{E_n - E_m}\langle m|\frac{\partial H}{\partial\lambda}|n\rangle$ and therefore the Berry curvature and the energy shift read as

$$\Omega_{n,ij} = -2\sum_{m\neq n}\mathrm{Im}\frac{\langle\mathbf{x},n|\frac{\partial H}{\partial x_i}|\mathbf{x},m\rangle\langle\mathbf{x},m|\frac{\partial H}{\partial x_j}|\mathbf{x},n\rangle}{\left[\epsilon_n^{(0)}(\mathbf{x}) - \epsilon_m^{(0)}(\mathbf{x})\right]^2}, \tag{7}$$

$$\delta\epsilon_n = -\sum_{m\neq n}\mathrm{Im}\frac{\langle\mathbf{x},n|\frac{\partial H}{\partial R_i}|\mathbf{x},m\rangle\langle\mathbf{x},m|\frac{\partial H}{\partial k_i}|\mathbf{x},n\rangle}{\epsilon_n^{(0)}(\mathbf{x}) - \epsilon_m^{(0)}(\mathbf{x})}. \tag{8}$$

The derivative with respect to crystal momentum is identified with the velocity $\mathbf{v}$, while the derivative with respect to position arises from the $\mathbf{R}$ dependence of $\hat{\mathbf{n}}$:

$$\frac{1}{\hbar}\frac{\partial H}{\partial k_i} = v_i, \quad \frac{\partial H}{\partial R_i} = \frac{\partial H}{\partial\hat{\mathbf{n}}}\cdot\frac{\partial\hat{\mathbf{n}}}{\partial R_i} = \mathbf{T}(\mathbf{r})\cdot\left(\hat{\mathbf{n}}\times\frac{\partial\hat{\mathbf{n}}}{\partial R_i}\right), \tag{9}$$

where $\mathbf{T}(\mathbf{r}) = \hat{\mathbf{n}}\times\frac{\partial H}{\partial\hat{\mathbf{n}}}$ is the torque operator. Thus, we arrive at

$$\delta F^{(1)}(\mathbf{R}) = D_{ij}\hat{\mathbf{e}}_i\cdot\left(\hat{\mathbf{n}}\times\frac{\partial\hat{\mathbf{n}}}{\partial R_j}\right), \tag{10}$$

with

$$\begin{aligned} D_{ij} &= \sum_n\int\frac{d^3k}{(2\pi)^3}\left[f_{\mathbf{k}n}A_{\mathbf{k}nij} + \frac{\ln[1+e^{-\beta(\epsilon_{\mathbf{k}n}-\mu)}]B_{\mathbf{k}nij}}{\beta}\right], \\ A_{\mathbf{k}nij} &= \hbar\sum_{m\neq n}\mathrm{Im}\left[\frac{\langle\mathbf{k}n|T_i|\mathbf{k}m\rangle\langle\mathbf{k}m|v_j(\mathbf{k})|\mathbf{k}n\rangle}{\epsilon_{\mathbf{k}m} - \epsilon_{\mathbf{k}n}}\right], \\ B_{\mathbf{k}nij} &= -2\hbar\sum_{m\neq n}\mathrm{Im}\left[\frac{\langle\mathbf{k}n|T_i|\mathbf{k}m\rangle\langle\mathbf{k}m|v_j(\mathbf{k})|\mathbf{k}n\rangle}{(\epsilon_{\mathbf{k}m} - \epsilon_{\mathbf{k}n})^2}\right], \end{aligned} \tag{11}$$

where $A_{\mathbf{k}nij}$ describes the Berry energy (3), $\delta\epsilon_n = A_{\mathbf{k}ni'i}(\hat{\mathbf{n}}\times\frac{\partial\hat{\mathbf{n}}}{\partial R_i})_{i'}$, and $B_{\mathbf{k}nij}$ the Berry curvature $\Omega^{\mathrm{Rk}}_{n,ij} = B_{\mathbf{k}ni'j}(\hat{\mathbf{n}}\times\frac{\partial\hat{\mathbf{n}}}{\partial R_i})_{i'}$, respectively.

Equation (10) can directly be identified with the DM interaction[1,2] in the continuum limit. We have therefore shown that for smooth textures and weak SOI, the DM interaction arises from mixed momentum- and position-space Berry curvatures and the corresponding energy shifts obtained from the Kohn-Sham Hamiltonian. Our semiclassical derivation yields the same expression for D_{ij} as obtained from quantum mechanical perturbation theory.[27]

At $T = 0$, to linear order in $\nabla\hat{\mathbf{n}}$ and for weak SOI, the formula (10) is exact even for a fully interacting quantum system [provided the exact Kohn-Sham Hamiltonian is used in Eq. (5)]. To calculate changes of the ground-state energy to linear order, changes of H due to $\nabla\hat{\mathbf{n}}$ can be neglected (a manifestation of the magnetic force theorem[30]). Therefore, the only remaining source of errors is the external magnetic field $\boldsymbol{B}^{\text{ext}}$ needed to stabilize the ferromagnetic solution underlying Eq. (5). As this field can be chosen to be weak (second order in SOI strength), it does not affect the value of the DM interaction to leading order in SOI.

Aside from DM interactions, also the current-induced spin-orbit torque relies on broken inversion symmetry. Recently, it has been shown that the intrinsic contribution to this torque is related to the Berry curvature $B_{\mathbf{k}nij}$.[31]

B. Charge density

The change of charge density to first order in the gradients of the magnetization also arises from both the change of the density of states and the energy levels from phase-space curvatures. It is given by

$$\delta\rho^{(1)}(\mathbf{R}) = e\sum_n \int \frac{d^3k}{(2\pi)^3}\left[\frac{\partial f_{\mathbf{k}n}}{\partial \epsilon}\delta\epsilon_n(\mathbf{x}) - f_{\mathbf{k}n}\,\Omega^{\text{Rk}}_{nii}(\mathbf{x})\right], \tag{12}$$

where $e = -|e|$ is the electron charge. We obtain

$$\delta\rho^{(1)}(\mathbf{R}) = e\,G_{ij}\hat{\mathbf{e}}_i \cdot \left(\hat{\mathbf{n}} \times \frac{\partial\hat{\mathbf{n}}}{\partial R_j}\right),$$
$$G_{ij} = \int \frac{d^3k}{(2\pi)^3}\sum_n \left[\frac{\partial f_{\mathbf{k}n}}{\partial \epsilon_{\mathbf{k}n}} A_{\mathbf{k}nij} - f_{\mathbf{k}n} B_{\mathbf{k}nij}\right]. \tag{13}$$

In metals, extra charges are screened on the length scale set by the Thomas-Fermi screening length λ_{TF}, resulting in a strongly suppressed charge density $\rho_{\text{tot}} \approx -\lambda_{\text{TF}}^2\nabla^2(\delta\rho^{(1)})$. Since the metal screens extra charge by changing the occupation of the states on the Fermi surface, while the Fermi sea participates in the formation of $\delta\rho^{(1)}$, the calculation of the unscreened $\delta\rho^{(1)}$ is interesting to understand how many electrons are energetically redistributed between Fermi surface and Fermi sea due to the phase-space Berry phases.

The results of our semiclassical derivation can be reproduced by a gradient expansion of the Green's function similar to the technique used by Yang *et al.* in Ref. 32 for insulators (see Appendix C). While to leading order in $\nabla\hat{\mathbf{n}}$ the semiclassical formulas (3) and (4) are reproduced, higher orders give for metals rise to additional contributions to the density of states that are not captured by the higher-order terms of Eq. (4) as we have checked explicitly.

In insulators, the situation is different. Neither energy shifts [Eq. (3)] nor the term linear in Ω_n in Eq. (4) contribute to the total charge of a single skyrmion. The integral $\int \frac{dR_x dk_x}{2\pi}\Omega^{\text{Rk}}_{n,xx}$, for example, has to be quantized (first Chern number). As it evaluates to 0 for $y \to \infty$, it vanishes everywhere. All contributions to the charge arise from higher-order terms in the gradient expansion. As the charge has to be a topological invariant in an insulator, it can be calculated from an adiabatically deformed band structure where all occupied bands are completely flat and degenerate. In this limit, one can use standard arguments[33] to show that the total accumulated charge due to smooth variations in phase space, e.g., in two space dimensions, is given by the second Chern number

$$\delta Q^{(2)} \equiv \int d^2R\,\delta\rho^{(2)}$$
$$= -e\int \frac{d^2R\,d^2k}{(2\pi)^2}\frac{\epsilon_{ijkl}}{8}\text{Tr}[\Omega_{ij}\Omega_{kl}], \tag{14}$$

where Ω is a matrix in the space of occupied bands. We refer to Appendix C 3 for the derivation of Eq. (14).

Therefore, either Abelian or non-Abelian winding numbers can occur. In cases when all non-Abelian winding numbers vanish, the right-hand side of Eq. (14) coincides with the integral over the term quadratic in Ω_n on the right-hand side of Eq. (4). For Abelian situations, $\delta Q^{(2)}$ can be expressed as a product of two simple real-space and momentum-space winding numbers (i.e., two first Chern numbers) $\delta Q^{(2)} = \sigma_{xy}\Phi_0$, where $\sigma_{xy} = \frac{e^2}{\hbar}\int \frac{d^2k}{(2\pi)^2}\Omega^{\text{kk}}_{xy}$ is the quantized Hall conductivity and $\Phi_0 = \frac{\hbar}{|e|}\int d^2R\,\Omega^{\text{RR}}_{xy}$ the quantized total flux arising from the real-space Berry phases. This can be shown by rewriting Eq. (14) as a surface integral and using, for example, that for $\mathbf{R} \to \infty$ both Ω^{RR} and Ω^{Rk} vanish (see Appendix C 4 for details).

IV. RESULTS

A. Qualitative dependency on the SO strength

In order to investigate first qualitatively how the accumulated charge in metals depends on the strength of SOI, we consider the simple two-dimensional two-band toy model

$$H = \epsilon_{\mathbf{k}} + [\mathbf{b}^{\text{ex}}(\mathbf{R}) + \mathbf{g}^{\text{so}}(\mathbf{k})]\cdot\boldsymbol{\sigma} = \epsilon_{\mathbf{k}} + \mathbf{n}(\mathbf{R},\mathbf{k})\cdot\boldsymbol{\sigma}, \tag{15}$$

where $\boldsymbol{\sigma}$ is the vector of Pauli matrices, $\mathbf{b}^{\text{ex}}(\mathbf{R})$ the exchange field arising from the magnetic texture, $\mathbf{g}^{\text{so}}(\mathbf{k})$ the SOI field, and $\mathbf{n} = \mathbf{b}^{\text{ex}} + \mathbf{g}^{\text{so}}$. From Eqs. (1)–(3), one finds

$$\Omega_{\pm,ij} = \mp\frac{1}{2}\hat{\mathbf{n}}\cdot\left(\frac{\partial}{\partial x_i}\hat{\mathbf{n}} \times \frac{\partial}{\partial x_j}\hat{\mathbf{n}}\right), \tag{16}$$

$$\delta\epsilon_+ = \delta\epsilon_- = |\mathbf{n}|\sum_{i=1}^{3}\Omega^{\text{Rk}}_{+,ii}, \tag{17}$$

where $\pm$ labels the minority and majority bands, respectively. To analyze the model (15) analytically, we consider the limit of weak SOI parameter λ_{so} with $|\mathbf{g}^{\text{so}}|/|\mathbf{b}^{\text{ex}}| \sim \lambda_{\text{so}}$. SOI also controls the size of skyrmions as their formation is driven by DM interactions. For skyrmion lattices in chiral magnets, the diameter of the skyrmions is proportional to $1/\lambda_{\text{so}}$ (Ref. 3). Expanding Eqs. (16) and (17) in λ_{so} shows that both $\delta\epsilon_\pm$ and *all* components of Ω are of order λ_{so}^2. This argument shows that the expansion in powers of Ω, which lead to Eq. (6), is valid for chiral magnets with weak SOI. Remarkably, all factors of λ_{so} cancel, when the total charge of a single skyrmion $\delta Q^{(1)} = \int \delta\rho^{(1)}\,d^2R$ is calculated using Eq. (12). Assuming that the exchange field $\mathbf{b}^{\text{ex}}$ is small compared to the Fermi

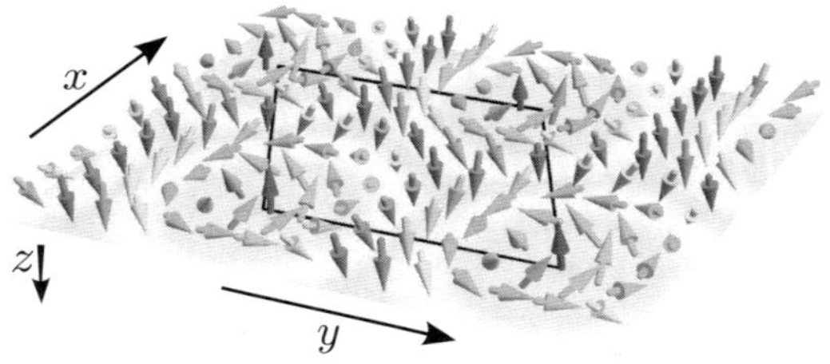

FIG. 1. (Color online) Illustration of the exchange field $\mathbf{b}^{\text{ex}}(\mathbf{R})$ in a skyrmion lattice [Eq. (19)]. Arrows are colored according to their z component and spaced in an arbitrary distance much larger than the atomic lattice constant in MnSi. The black polygon encloses the magnetic unit cell. In the three-dimensional system, the magnetic texture is translationally invariant in the z direction.

energy, we find

$$
\begin{aligned}
D_{ij} &\approx \frac{|\mathbf{b}^{\text{ex}}|^2}{3} \int \frac{d^2k}{(2\pi)^2} \frac{\partial g_i^{\text{so}}}{\partial k_j} f''(\epsilon_{\mathbf{k}}), \\
\delta Q^{(1)} &\approx \frac{2}{3} e \int \frac{d^2R\, d^2k}{(2\pi)^2} |\mathbf{b}^{\text{ex}}|^3\, \Omega^{\text{Rk}}_{+,ii}\, f'''(\epsilon_{\mathbf{k}}).
\end{aligned}
\tag{18}
$$

To obtain a qualitative estimate, we assume $|\partial g^{\text{so}}/\partial k| \sim \lambda_{\text{so}} E_F a$ where E_F is the Fermi energy and a the lattice constant. According to neutron scattering experiments,[3,34] the skyrmion lattice in MnSi is well described by (see Fig. 1 for an illustration)

$$
\begin{aligned}
\mathbf{b}^{\text{ex}}(\mathbf{R}) = B_0 \hat{\mathbf{z}} + B_1 \sum_{n=0}^{2} [(\hat{\mathbf{z}} \times \hat{\boldsymbol{\xi}}_n) \sin(q_0\, \hat{\boldsymbol{\xi}}_n \cdot \mathbf{R}) \\
+ \hat{\mathbf{z}} \cos(q_0\, \hat{\boldsymbol{\xi}}_n \cdot \mathbf{R})],
\end{aligned}
\tag{19}
$$

where $q_0 \approx 2\pi/190$ Å, $\hat{\mathbf{z}} = (0,0,1)$ is the unit vector parallel to a small magnetic field stabilizing the skyrmion lattice, and $\hat{\boldsymbol{\xi}}_n = [\cos(2\pi n/3), \sin(2\pi n/3), 0]$. From mean-field calculations,[34] $B_1/B_0 \approx -1.5$ is obtained. As q_0 is linear in λ_{so}, we set $q_0 = \lambda_{\text{so}} 2\pi/a$. In this model, we obtain

$$
D_{ij} \sim \lambda_{\text{so}}\, \delta_{ij}\, \frac{E_F}{a} \frac{B_0^2}{E_F^2}, \quad \delta Q^{(1)} \sim e \frac{B_0^2}{E_F^2}. \tag{20}
$$

As expected, the DM interaction is linear in SOI and quadratic in the magnetization. Interestingly, the skyrmion charge is *independent* of the SOI strength (when screening is ignored) but proportional to the square of the local magnetization. These main conclusions remain valid when we calculate the charge with *ab initio* methods using the real band structure of a complex material (see following).

B. Quantitative results for the DM interaction strength and the skyrmion charge in MnSi

Based on the electronic structure of MnSi obtained within the local density approximation (LDA), we compute D_{ij} and G_{ij} at $T = 0$ using Wannier functions[35,36] to reduce the computational burden (see Appendix A for computational details). Furthermore, we approximate V, $\mathbf{E}$, and $\mathbf{B}$ in Eq. (5) by their value for vanishing spin-orbit coupling. This allows us to perform the calculation at $\mathbf{B}^{\text{ext}} = 0$ using that $\mathbf{B} \parallel \hat{\mathbf{n}}$. The

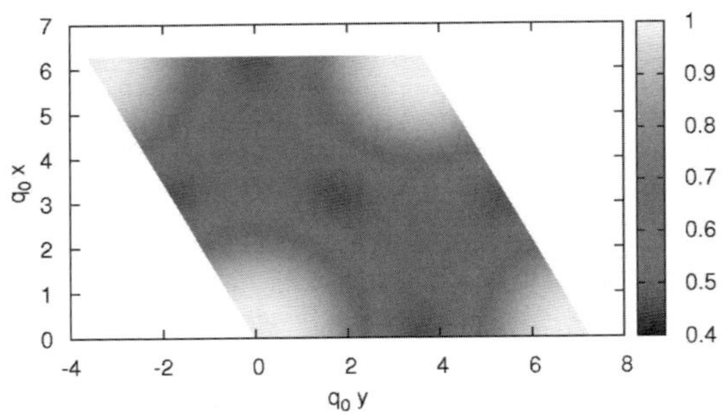

FIG. 2. (Color online) Normalized free-energy density $\delta F^{(1)}(\mathbf{R})/\delta F^{(1)}(0)$ and normalized charge density $\delta\rho^{(1)}(\mathbf{R})/\delta\rho^{(1)}(0)$ within the magnetic unit cell. The minimal free-energy density is given by $\delta F^{(1)}(0) = -0.0018$ meV/Å^3, the total free energy is reduced by 231 meV per skyrmion and layer. The maximal charge density amounts to $\delta\rho^{(1)}(0) = 1.95 \times 10^{-6} e/\text{Å}^3$, the total charge per skyrmion and layer is $0.246e$. The skyrmion center is located at the origin, as in Eq. (19) (cf. Fig. 1).

torque is then simply given by $\mathbf{T} = \mathbf{M} \times \mathbf{B}$. For a left-handed crystal structure, we obtain $D_{ij} = -D\delta_{ij}$ with

$$
D = -4.1 \text{ meV Å per eight-atom cell.} \tag{21}
$$

An experimental value for D can be obtained from neutron scattering in the helical phase of MnSi because a finite D shifts the minimum of $E(q) = Dq + Jq^2$ from $q = 0$ to $-D/(2J)$ for a left-handed spiral. Using $J = 52$ meV Å^2 per eight-atom cell[37] and $q = 2\pi/190$ Å leads to an experimental value of $D = -3.43$ meV Å in good agreement with our result.

Next, we discuss the manifestations of phase-space Berry phases on the skyrmions in MnSi. As $G_{ij} \propto D_{ij} \propto \delta_{ij}$ by symmetry, $\delta F^{(1)}(\mathbf{R})$ and $\delta\rho^{(1)}(\mathbf{R})$ are proportional to each other and can therefore be shown in a single plot (see Fig. 2), where we used $\hat{\mathbf{n}}(\mathbf{R}) = \mathbf{b}^{\text{ex}}/|\mathbf{b}^{\text{ex}}|$ [with $\mathbf{b}^{\text{ex}}$ from Eq. (19)]. Integrating $\delta F^{(1)}(\mathbf{R})$ over the magnetic unit cell, we obtain a free-energy reduction of 231 meV. Both charge density and free-energy density are maximal in the center of the skyrmion located at (0,0). Integrating $\delta\rho^{(1)}(\mathbf{R})$ over the magnetic unit cell, we obtain the charge of $0.246e$. However, $\delta\rho^{(1)}(\mathbf{R})$ is strongly

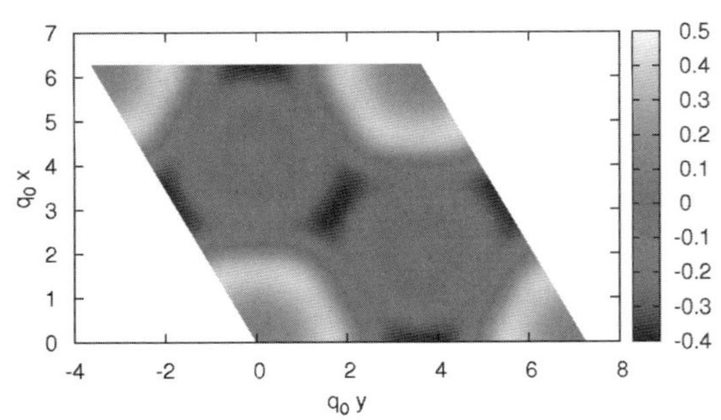

FIG. 3. (Color online) Screened charge density $\rho_{\text{tot}}/[\delta\rho^{(1)}(0)\lambda_{\text{TF}}^2 q_0^2]$ within the magnetic unit cell, with $\delta\rho^{(1)}(0)\lambda_{\text{TF}}^2 q_0^2 = 1.07 \times 10^{-10} e/\text{Å}^3$.

screened due to the short $\lambda_{\rm TF} = \sqrt{\epsilon_0/(e^2 N_F)} \approx 0.224$ Å, where $N_F \approx 0.11/(\text{eV Å}^3)$ is the density of states at the Fermi level obtained in our LDA calculations. The resulting screened charge density varies between $\rho_{\rm tot}^{\rm max} \approx 4.5 \times 10^{-11} e/\text{Å}^3$ close to the core and $\rho_{\rm tot}^{\rm min} \approx -4.1 \times 10^{-11} e/\text{Å}^3$ between two skyrmions (see Fig. 3 for illustration).

V. CONCLUSIONS

Our analysis has shown that mixed real-space/momentum-space Berry phases are quantitatively important in materials such as MnSi. Energetically, they are the driving force for the formation of magnetic textures and lead to a redistribution of charge in the skyrmion phase which we calculated using *ab initio* methods. For the future, it will be interesting to investigate how the phase-space Berry curvature $\Omega^{\rm Rk}$ affects the Hall effect. As in MnSi, the contributions arising from the topological Hall effect, i.e., from $\Omega^{\rm RR}$, and the anomalous Hall effect due to $\Omega^{\rm kk}$, are of similar magnitude, we also expect substantial contributions from $\Omega^{\rm Rk}$.

ACKNOWLEDGMENTS

We thank S. Blügel, H. Geiges, C. Pfleiderer, M. Zirnbauer, and, especially, A. Altland for illuminating discussions. Financial support of the DFG (SFB TR 12, FOR 960), funding under the HGF-YIG programme VH-NG-513 and from Deutsche Telekom Stiftung (R.B.), and computing time on the supercomputers JUQUEEN and JUROPA at Jülich Supercomputing Center are gratefully acknowledged.

APPENDIX A: COMPUTATIONAL DETAILS OF THE *AB INITIO* CALCULATIONS

From the full-potential linearized augmented-plane-wave code FLEUR (Ref. 38), the electronic structure of MnSi was obtained within the local density approximation[39] to density functional theory. The atomic coordinates and lattice parameter ($a = 4.558$ Å) of the eight-atom unit cell of MnSi as given in Ref. 40, muffin-tin radii of $2.12a_0$ for both Mn and Si, and a plane-wave cutoff of $3.7a_0^{-1}$ were used in the calculations ($a_0 = 0.529\,177$ Å is Bohr's radius). The basis set was supplemented with local orbitals for the Mn 3s and 3p states. The unconstrained spin moment per formula unit is $0.94\mu_{\rm B}$ and thus larger than the measured spin moment by more than a factor of 2. We constrained the spin moment per formula unit to the value of $0.4\mu_{\rm B}$. From the relativistic first-principles Bloch functions of 100 bands given on an $8\times8\times8$ **k** mesh, we constructed 64 relativistic maximally localized Wannier functions using disentanglement within the WANNIER90 code.[41] The lowest 40 bands in the valence window are the 32 local orbitals plus 8 Mn 4s bands. These were skipped, i.e., the 100 bands from which the Wannier functions were disentangled are bands 41 to 140.

Based on Wannier interpolation,[42,43] we evaluated $D_{ij}(\hat{\mathbf{n}})$ for the 001, 111, and 110 directions $\hat{\mathbf{n}}$ of magnetization using a $512\times512\times512$ interpolation mesh. We find that to very good approximation

$$\mathbf{D}_j(\hat{\mathbf{n}}) = D_{ij}(\hat{\mathbf{n}})\hat{\mathbf{e}}_i = D\hat{\mathbf{n}} \times (\hat{\mathbf{n}} \times \hat{\mathbf{e}}_j), \tag{A1}$$

where $\hat{\mathbf{e}}_1 = \hat{\mathbf{x}}$, $\hat{\mathbf{e}}_2 = \hat{\mathbf{y}}$, and $\hat{\mathbf{e}}_3 = \hat{\mathbf{z}}$ are unit vectors of the Cartesian coordinate system and a single parameter $D = -7.69a_0/V$ meV describes the amplitude of DMI, with $V = a^3$ the volume of the unit cell. Equation (A1) neglects the anisotropy of $\mathbf{D}_j(\hat{\mathbf{n}})$, which is small according to our calculations.

Using Eq. (A1), we can express $\delta F^{(1)}(\mathbf{R})$ as follows:

$$\begin{aligned}\delta F^{(1)}(\mathbf{R}) &= \mathbf{D}_i(\hat{\mathbf{n}}) \cdot \left(\hat{\mathbf{n}} \times \frac{\partial \hat{\mathbf{n}}}{\partial R_i}\right)\\ &= D\hat{\mathbf{n}} \cdot [\nabla \times \hat{\mathbf{n}}].\end{aligned} \tag{A2}$$

In the skyrmion lattice in MnSi, only the derivatives $\partial_x \hat{\mathbf{n}}$ and $\partial_y \hat{\mathbf{n}}$ contribute and $\partial_z \hat{\mathbf{n}} = 0$. Thus, we have

$$\nabla \times \hat{\mathbf{n}} = \begin{pmatrix} -\sin\theta\,\frac{\partial\theta}{\partial y} \\ \sin\theta\,\frac{\partial\theta}{\partial x} \\ \cos\theta\sin\phi\frac{\partial\theta}{\partial x} + \sin\theta\cos\phi\frac{\partial\phi}{\partial x} - \\ \cos\theta\cos\phi\frac{\partial\theta}{\partial y} + \sin\theta\sin\phi\frac{\partial\phi}{\partial y} \end{pmatrix}, \tag{A3}$$

yielding an alternative expression for $\delta F^{(1)}(\mathbf{R})$ in terms of the azimuthal and polar angles of the exchange field and their derivatives:

$$\begin{aligned}\delta F^{(1)}(\mathbf{R}) &= D\bigg[\sin\phi\frac{\partial\theta}{\partial x} - \cos\phi\frac{\partial\theta}{\partial y}\\ &\quad + \sin\theta\cos\theta\left(\cos\phi\frac{\partial\phi}{\partial x} + \sin\phi\frac{\partial\phi}{\partial y}\right)\bigg]\\ &= Dq_0\mathcal{Q}(q_0x, q_0y),\end{aligned} \tag{A4}$$

where

$$\begin{aligned}\mathcal{Q}(q_0x, q_0y) &= \bigg[\sin\phi\frac{\partial\theta}{\partial(q_0x)} - \cos\phi\frac{\partial\theta}{\partial(q_0y)}\\ &\quad + \sin\theta\cos\theta\left(\cos\phi\frac{\partial\phi}{\partial(q_0x)} + \sin\phi\frac{\partial\phi}{\partial(q_0y)}\right)\bigg].\end{aligned} \tag{A5}$$

We define the free energy per skyrmion $\delta E^{(1)}$ as integral of $\delta F^{(1)}(\mathbf{R})$ over the magnetic unit cell, where we set the extension of the magnetic cell in the z direction equal to the lattice parameter a of the eight-atom unit cell of MnSi. We obtain

$$\begin{aligned}\delta E^{(1)} &= a\int \delta F^{(1)}(\mathbf{R})dx\,dy\\ &= \frac{aD}{q_0}\int d(q_0x)d(q_0y)\mathcal{Q}(q_0x, q_0y)\\ &= 39\frac{aD}{q_0} = -231 \text{ meV}.\end{aligned} \tag{A6}$$

Determining the tensor

$$t_{ij} = e\int \frac{d^3k}{(2\pi)^3}\sum_n f_{\mathbf{k}n} B_{\mathbf{k}nij} \tag{A7}$$

from Wannier interpolation, we get

$$t_{ij}(\hat{\mathbf{n}}) = t\hat{\mathbf{e}}_i \cdot [\hat{\mathbf{n}} \times (\hat{\mathbf{n}} \times \hat{\mathbf{e}}_j)], \tag{A8}$$

with $t = -0.091ea_0/V$, where small anisotropies of t_{ij} have been neglected. t_{ij} describes the intrinsic component of the SOI-mediated spin torque per volume to an applied electric field in the homogeneous system.[31] One contribution to $\delta\rho^{(1)}(\mathbf{R})$ is given by

$$-t_{ij}\hat{\mathbf{e}}_i \cdot \left(\hat{\mathbf{n}} \times \frac{\partial \hat{\mathbf{n}}}{\partial R_j}\right). \tag{A9}$$

However, due to the additional Fermi surface term, the complete expression for $\delta\rho^{(1)}(\mathbf{R})$ is given by

$$\delta\rho^{(1)}(\mathbf{R}) = eG_{ij}\hat{\mathbf{e}}_i \cdot \left(\hat{\mathbf{n}} \times \frac{\partial \hat{\mathbf{n}}}{\partial R_j}\right), \tag{A10}$$

where according to our calculations

$$eG_{ij} = g\hat{\mathbf{e}}_i \cdot [\hat{\mathbf{n}} \times (\hat{\mathbf{n}} \times \hat{\mathbf{e}}_j)], \tag{A11}$$

with $g = 0.0082\frac{ea_0}{V}$, neglecting again the small anisotropies. Similar to rewriting the free-energy density above, we obtain

$$\delta\rho^{(1)}(\mathbf{R}) = g\hat{\mathbf{n}} \cdot [\nabla \times \hat{\mathbf{n}}] = gq_0\mathcal{Q}(q_0x, q_0y). \tag{A12}$$

We define the charge per skyrmion $\delta Q^{(1)}$ as integral of $\delta\rho^{(1)}(\mathbf{R})$ over the magnetic unit cell, where we set the extension of the magnetic cell in the z direction equal to the lattice parameter a of the eight-atom unit cell of MnSi. This yields

$$\begin{aligned}\delta Q^{(1)} &= a \int \delta\rho^{(1)}(\mathbf{R})dx\,dy \\ &= \frac{ag}{q_0}\int d\,(q_0x)d\,(q_0y)\mathcal{Q}(q_0x, q_0y) \\ &= 39\frac{ag}{q_0} = 0.246e.\end{aligned} \tag{A13}$$

APPENDIX B: COORDINATE INDEPENDENT FORMULATION OF THE PHASE-SPACE VOLUME

In this section, we review the derivation of the volume element [Eq. (4)] of the main text following mostly the review by Morrison[29] and rewrite some of our formulas using differential forms. This makes the derivation transparent and is manifestly independent of the chosen coordinate system.[44,45]

Consider a transformation from canonical coordinates X_i with standard Poisson brackets $\{f,g\} = \frac{\partial f}{\partial X_i}J_{ij}\frac{\partial g}{\partial X_j}$ to a new set of coordinates $x_i = x_i(\mathbf{X})$. The Poisson brackets of the (noncanonical) coordinates x_i are given by

$$\{x_i, x_j\} = \frac{\partial x_i}{\partial X_{i'}}J_{i'j'}\frac{\partial x_j}{\partial X_{j'}} = (\omega^{-1})_{ij}. \tag{B1}$$

The natural volume element of the $2d$-dimensional phase space is obtained from the Jacobi determinant $|\partial X_i/\partial x_j|$:

$$dV = \frac{d^{2d}X}{(2\pi)^d} = \left|\frac{\partial X_i}{\partial x_j}\right|\frac{d^{2d}x}{(2\pi)^d} = \sqrt{\det\omega}\frac{d^{2d}x}{(2\pi)^d}, \tag{B2}$$

where we used that $\det\omega = (|\partial x_i/\partial X_j|^2 \det J)^{-1} = |\partial X_i/\partial x_j|^2$ as $\det J = 1$ and $\det M^{-1} = 1/\det M$.

It is useful to rewrite Eq. (B2) using that the phase-space volume dV is independent of the coordinate system. In canonical coordinates, we define the 2-form $\hat{\omega}$ from the inverse of $J^{-1} = -J$ using

$$\begin{aligned}\hat{\omega} &= \frac{1}{2}(J^{-1})_{ij}\,dX^i \wedge dX^j \qquad (B3)\\ &= \frac{1}{2}\frac{\partial X_{i'}}{\partial x_i}(J^{-1})_{i'j'}\frac{\partial X_{j'}}{\partial x_j}\,dx^i \wedge dx^j \\ &= \frac{1}{2}\omega_{ij}\,dx^i \wedge dx^j, \qquad (B4)\end{aligned}$$

where we used the definition of ω from Eq. (B1). As in Eq. (B3), $\hat{\omega}$ is expressed in canonical coordinates, the phase-space volume is directly obtained from the d-fold wedge product $\hat{\omega}^d = \hat{\omega} \wedge \ldots \wedge \hat{\omega}$:

$$dV = \frac{\hat{\omega}^d}{d!(2\pi)^d}. \tag{B5}$$

While Eqs. (B2) and (B5) are equivalent, Eq. (B5) is much easier to handle due to the missing square root.

A remarkable aspect is the close relation of Poisson brackets, phase-space volume, Berry connections, and Chern classes. The semiclassical equations of motion for an electron in band n in the presence of phase-space Berry phases are given in the main text as

$$\dot{x}_i = [(\Omega_n - J)^{-1}]_{ij}\frac{\partial\epsilon_n}{\partial x_j} \equiv \{x_i, \epsilon_n\}, \tag{B6}$$

where the Poisson brackets are defined by Eq. (B1) with $\omega = \Omega_n - J$. From Eq. (B4) follows

$$\hat{\omega} = \tfrac{1}{2}[(J^{-1})_{ij} + \Omega_{ij}]\,dx^i \wedge dx^j = \hat{\omega}_0 + \hat{\Omega}, \tag{B7}$$

where $\hat{\omega}_0 = \frac{1}{2}(J^{-1})_{ij}\,dx^i \wedge dx^j$ is the "canonical" 2-form which obtains a correction from the Abelian Berry curvature

$$\hat{\Omega} = d\hat{A} = \tfrac{1}{2}\Omega_{ij}\,dx^i \wedge dx^j, \tag{B8}$$

where $\hat{A} = A_i\,dx_i$ and we have omitted all band indices. From Eqs. (B5) and (B7), one obtains Eq. (4) of the main text.

The Berry curvature directly gives the first Chern form

$$\hat{c}_1 = \frac{\hat{\Omega}}{2\pi}. \tag{B9}$$

Integrals of the wedge product of m such Chern forms, $\int \hat{c}_1^m = \int \hat{c}_1 \wedge \ldots \wedge \hat{c}_1$, over compact $2m$-dimensional manifolds without boundary define Chern numbers which are quantized to integers.[46] Such wedge products directly show up when expanding dV in powers of $\hat{\Omega}$ using Eqs. (B5) and (B7):

$$dV = \sum_{m=0}^{d}\frac{1}{m!(d-m)!}\hat{c}_1^m \wedge \frac{\hat{\omega}_0^{d-m}}{(2\pi)^{d-m}}. \tag{B10}$$

APPENDIX C: CHARGE DENSITY FROM A GRADIENT EXPANSION IN THE QUANTUM MECHANICAL SYSTEM

In this Appendix, we demonstrate that the results for the charge density obtained from the semiclassical arguments in the main text can be confirmed by using a gradient expansion in the quantum mechanical system. We show that in metals the semiclassical formulas (3) and (4) are correct to first order in the spatial gradients. In insulators, the charge density vanishes to first order in the spatial gradients, but the second-order term of Eq. (4) leads to the correct (quantized) total charge per skyrmion.

1. Gradient expansion of the Green's function

We follow the notation used in Ref. 47. For a spatially inhomogeneous system with Green's function $G(\omega;\mathbf{r}_1,\mathbf{r}_2)$, we introduce the Wigner transform of the Green's function

$$\tilde{G}(\omega;\mathbf{x}) \equiv \tilde{G}(\omega;\mathbf{R},\mathbf{k}) = \int d^3r\, e^{-i\mathbf{k}\cdot\mathbf{r}} G\left(\omega;\mathbf{R}+\frac{\mathbf{r}}{2},\mathbf{R}-\frac{\mathbf{r}}{2}\right), \qquad \text{(C1)}$$

where $\mathbf{R} = \frac{1}{2}(\mathbf{r}_1 + \mathbf{r}_2)$ and $\mathbf{r} = \mathbf{r}_1 - \mathbf{r}_2$ are the center-of-mass and relative coordinates, respectively. The inverse K of the Green's function is defined by $\int d^3r' G(\omega;\mathbf{r}_1,\mathbf{r}')K(\omega;\mathbf{r}',\mathbf{r}_2) = \delta(\mathbf{r}_1 - \mathbf{r}_2)$. In the Wigner representation, this relation becomes[47]

$$e^{\frac{i}{2}J_{ij}\partial_i^K\partial_j^G}\tilde{K}(\omega;\mathbf{x})\tilde{G}(\omega;\mathbf{x}) = \mathbb{1}, \qquad \text{(C2)}$$

where $J = \begin{pmatrix} 0 & \mathbb{1} \\ -\mathbb{1} & 0 \end{pmatrix}$, the derivative ∂_i^K (∂_j^G) acts on $\tilde{K}$ $(\tilde{G})$ only, and the symbol $\mathbb{1}$ on the right-hand side of Eq. (C2) is the unit matrix in band space. For a smooth spatial variation, one can expand the exponential in Eq. (C2) in powers of spatial derivatives and solve order by order for $\tilde{G}$. We write

$$\tilde{G} \approx \tilde{G}_0 + \tilde{G}_1 + \tilde{G}_2 + \mathcal{O}\left(\partial_{R_i}^3\right), \qquad \text{(C3)}$$

where the subscript indicates the order of spatial derivatives. Here, we truncate the gradient expansion after the second order to capture only the leading-order contribution to the charge density, which turns out to be linear in spatial gradients for metals and quadratic for insulators (see following).

The zero-order contribution $\tilde{G}_0$ is given by the pointwise inverse of $\tilde{K}$, i.e., $\tilde{G}_0(\omega,\mathbf{x}) = [\tilde{K}(\omega,\mathbf{x})]^{-1}$. In the limit of a homogeneous system, spatial gradients vanish and $\tilde{G} = \tilde{G}_0$. For a noninteracting homogeneous system $\tilde{K}(\omega,\mathbf{x}) \equiv \tilde{K}(\omega,\mathbf{k}) = \hbar\omega - H(\mathbf{k})$ where $H(\mathbf{k})$ is the band Hamiltonian.

If the system varies smoothly in space gradient corrections to the Green's function arise. The part of Eq. (C2) linear in spatial gradients reads as

$$\tilde{K}\tilde{G}_1 + \frac{i}{2}J_{ij}(\partial_i\tilde{K})(\partial_j\tilde{G}_0) = 0, \qquad \text{(C4)}$$

where we omitted the dependency on ω and $\mathbf{x}$ to improve readability. With $\tilde{K} = \tilde{G}_0^{-1}$ follows

$$\tilde{G}_1 = \frac{i}{2}J_{ij}\tilde{G}_0\left(-\partial_i\tilde{G}_0^{-1}\right)\tilde{G}_0\left(-\partial_j\tilde{G}_0^{-1}\right)\tilde{G}_0$$

$\equiv$ 

(C5)

The last equality of Eq. (C5) introduces a graphical shorthand notation where a solid line represents $\tilde{G}_0$, a vertex labeled with i represents $(-\partial_i\tilde{G}_0^{-1})$ (where a minus sign is included in anticipation of the structure of higher-order terms) and a dashed arrow from a vertex i to a vertex j represents a factor of $\frac{i}{2}J_{ij}$.

The second-order part of Eq. (C2) reads as

$$\tilde{K}\tilde{G}_2 + \frac{i}{2}J_{ij}(\partial_i\tilde{K})(\partial_j\tilde{G}_1) + \frac{1}{2}\left(\frac{i}{2}\right)^2 J_{ij}J_{kl}(\partial_i\partial_k\tilde{K})(\partial_j\partial_l\tilde{G}_0) = 0. \qquad \text{(C6)}$$

Solving Eq. (C6) for $\tilde{G}_2$ we arrive at [using the graphical notation introduced in Eq. (C5)]

$\tilde{G}_2 =$ 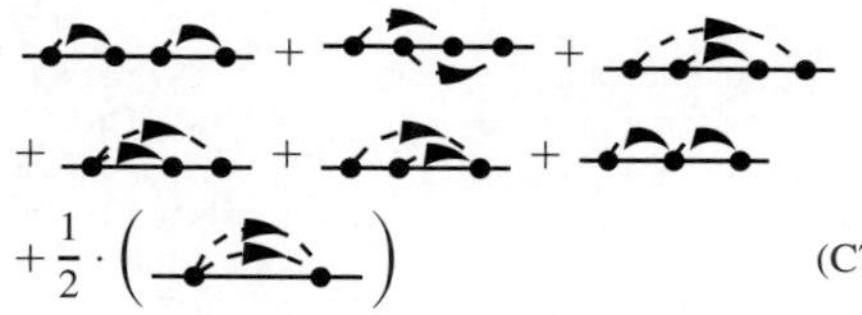
(C7)

where, e.g., the fourth diagram represents the term $(\frac{i}{2})^2 J_{ij}J_{kl}\tilde{G}_0(-\partial_i\partial_k\tilde{G}_0^{-1})\tilde{G}_0(-\partial_j\tilde{G}_0^{-1})\tilde{G}_0(-\partial_l\tilde{G}_0^{-1})\tilde{G}_0$.

2. Charge density in metals

The charge density is obtained from

$$\rho(\mathbf{R}) = e\frac{1}{\beta}\sum_{\omega_n}\int\frac{d^3k}{(2\pi)^3}\,\mathrm{Tr}[\tilde{G}(i\omega_n;\mathbf{R},\mathbf{k})], \qquad \text{(C8)}$$

where $\tilde{G}$ is the Green's function in Wigner representation as defined in Eq. (C1), e is the electron charge, $\beta = 1/(k_BT)$ the inverse temperature, and $\hbar\omega_n = \frac{2\pi}{\beta}(n+\frac{1}{2})$. We use the gradient expansion Eq. (C3) for $\tilde{G}$. As $\tilde{G}_0(i\omega_n;\mathbf{R},\mathbf{k}) = 1/[i\hbar\omega_n - H(\mathbf{R},\mathbf{k})]$, the zero-order charge density $\rho^{(0)}(\mathbf{R})$ depends only on the local properties of the system at the point $\mathbf{R}$ and it is the same as the charge density of a homogeneous system described by the band Hamiltonian $H(\mathbf{R},\mathbf{k})$ at fixed $\mathbf{R}$.

In metals, the leading-order correction $\delta\rho^{(1)}(\mathbf{R})$ to the charge density is obtained by inserting $\tilde{G}_1$ from Eq. (C5) for $\tilde{G}$ into Eq. (C8). Standard techniques for the evaluation of Matsubara summations lead to

$$\delta\rho^{(1)}(\mathbf{R}) = -e\frac{i}{2}J_{ij}\int\frac{d^3k}{(2\pi)^3}\int_{-\infty}^{\infty}\frac{d\epsilon}{2\pi i}f(\epsilon) \times \mathrm{Tr}\left[\tilde{G}_0^R\left(-\partial_i\tilde{G}_0^{-1}\right)\tilde{G}_0^R\left(-\partial_j\tilde{G}_0^{-1}\right)\tilde{G}_0^R - \tilde{G}_0^A\left(-\partial_i\tilde{G}_0^{-1}\right)\tilde{G}_0^A\left(-\partial_j\tilde{G}_0^{-1}\right)\tilde{G}_0^A\right], \qquad \text{(C9)}$$

where $f(\epsilon)$ is the Fermi function and $G^{R/A}$ is the retarded (advanced) Green's function, i.e., evaluated at frequencies $\epsilon/\hbar \pm i0^+$ just above (below) the real axis. We use the cyclicity of the trace and the relations $2i\,\mathrm{Im}\tilde{G} = \tilde{G}^R - \tilde{G}^A$ and $\partial\tilde{G}^{R/A}/\partial\epsilon = -(\tilde{G}^{R/A})^2$ to rewrite Eq. (C9):

$$\delta\rho^{(1)}(\mathbf{R}) = -ieJ_{ij}\int\frac{d^3k}{(2\pi)^3}\int_{-\infty}^{\infty}\frac{d\epsilon}{2\pi}f(\epsilon) \times \mathrm{Tr}\Big[(\tilde{G}^A)^2\left(-\partial_i\tilde{G}_0^{-1}\right)\mathrm{Im}\tilde{G}\left(-\partial_j\tilde{G}_0^{-1}\right) - \frac{\partial\,\mathrm{Im}\tilde{G}}{\partial\epsilon}\left(-\partial_i\tilde{G}_0^{-1}\right)\tilde{G}_0^R\left(-\partial_j\tilde{G}_0^{-1}\right)\Big]. \qquad \text{(C10)}$$

Integrating by parts over ϵ in the second term in the trace and using the antisymmetry of J_{ij} leads to

$$\begin{aligned}\delta\rho^{(1)}(\mathbf{R}) &= -ieJ_{ij}\int\frac{d^3k}{(2\pi)^3}\int_{-\infty}^{\infty}\frac{d\epsilon}{2\pi}\Big(-f(\epsilon)\mathrm{Tr}\big[\mathrm{Im}\tilde{G}\ (-\partial_i\tilde{G}_0^{-1})\\ &\times\big((\tilde{G}_0^R)^2+(\tilde{G}_0^A)^2\big)(-\partial_j\tilde{G}_0^{-1})\big]+\frac{\partial f(\epsilon)}{\partial\epsilon}\\ &\times\mathrm{Tr}\big[\mathrm{Im}\tilde{G}\ (-\partial_i\tilde{G}_0^{-1})\tilde{G}_0^R(-\partial_j\tilde{G}_0^{-1})\big]\Big).\end{aligned}\tag{C11}$$

In a noninteracting system $\tilde{G}_0^{R/A}(\omega;\mathbf{x}) = [\hbar\omega - H(\mathbf{x}) \pm i0^+]^{-1}$ and $\mathrm{Im}\tilde{G}(\omega;\mathbf{x}) = -\pi\delta[\hbar\omega - H(\mathbf{x})]$. Direct evaluation of the two traces in Eq. (C11) in the local eigenbasis of $H(\mathbf{x})$ confirms the semiclassical result (12) where $\Omega_{n,ij}$ and $\delta\epsilon_n$ are given in Eqs. (7) and (8).

For the noninteracting system, the semiclassical approach is therefore exact to leading order in the spatial gradients. This is not the case when higher orders are considered. Already, the next higher-order contribution to the semiclassical charge density [coming from terms of order $\mathcal{O}(\Omega^2)$ in Eq. (4)] does not coincide with the higher-order terms of the gradient expansion [obtained by inserting $\tilde{G}_2$ into Eq. (C8)]. This is obvious from the appearance of second-order derivatives (i.e., vertices with two attached dashed arrows) on the right-hand side of Eq. (C7), which have no semiclassical counterpart.

The free energy can be calculated in a similar way by inserting the gradient expansion (C3) into $F = -T\sum_{\omega_n}\mathrm{Tr}\ \ln[-T\tilde{G}]$. The result confirms Eq. (6) to first order.

3. Quantized skyrmion charge in insulators

In insulators, as argued in the main text, the first-order correction to the charge density $\delta\rho^{(1)}(\mathbf{R})$ does not contribute to the total charge of a skyrmion. The leading-order contribution to the skyrmion charge in insulators is second order in spatial derivatives. In this section, we present the derivation of Eq. (14) using the gradient expansion method and give an explicit definition of the non-Abelian Berry curvature tensor. The derivation is similar to the calculation in Ref. 32, Appendix 2.

In a metal, $\tilde{G}(\omega)$ is analytical at $\omega = 0$ and therefore Eq. (C8) simplifies (at $T = 0$) to

$$\rho(\mathbf{R}) = e\hbar\int\frac{d^3k}{(2\pi)^3}\int_{-\infty}^{\infty}\frac{d\omega}{2\pi}\ \mathrm{Tr}[\tilde{G}(i\omega;\mathbf{R},\mathbf{k})].\tag{C12}$$

The second-order correction to the charge density is given by inserting $\tilde{G}_2$ from Eq. (C7) into (C8). The contribution from the last term in Eq. (C7) is proportional to

$$\begin{aligned}&\hbar J_{ij}J_{kl}\int\frac{d\omega}{2\pi}\ \mathrm{Tr}[\tilde{G}_0(-\partial_i\partial_k\tilde{G}_0^{-1})\tilde{G}_0(-\partial_j\partial_l\tilde{G}_0^{-1})\tilde{G}_0]\\ &= -J_{ij}J_{kl}\int\frac{d\omega}{2\pi}\ \mathrm{Tr}\Big[(-\partial_i\partial_k\tilde{G}_0^{-1})\tilde{G}_0(-\partial_j\partial_l\tilde{G}_0^{-1})\frac{\partial\tilde{G}_0}{\partial\omega}\Big]\\ &= J_{ij}J_{kl}\int\frac{d\omega}{2\pi}\ \mathrm{Tr}\Big[(-\partial_i\partial_k\tilde{G}_0^{-1})\frac{\partial\tilde{G}_0}{\partial\omega}(-\partial_j\partial_l\tilde{G}_0^{-1})\tilde{G}_0\Big]\\ &= J_{ij}J_{kl}\int\frac{d\omega}{2\pi}\ \mathrm{Tr}\Big[(-\partial_j\partial_l\tilde{G}_0^{-1})\tilde{G}_0(-\partial_i\partial_k\tilde{G}_0^{-1})\frac{\partial\tilde{G}_0}{\partial\omega}\Big],\end{aligned}\tag{C13}$$

where all Green's functions are evaluated at frequency $i\omega$ and we used cyclicity of the trace in the first and the last equality, the relation $\partial\tilde{G}_0/\partial\omega = -\hbar(\tilde{G}_0)^2$ for a noninteracting system in the first equality, and integration by parts and the fact that $(\partial_i\tilde{G}_0^{-1})$ is independent of ω in the second equality. By relabeling indices and using $J_{ij} = -J_{ji}$ one sees that the last line of Eq. (C13) is the negative of the second line and hence vanishes. Thus, the last term in Eq. (C7) does not contribute to the charge density in insulators,

$\int\frac{d\omega}{2\pi}\ \mathrm{Tr}\Big[$  $\Big] = 0.$ (C14)

In a similar way, one can show the relations

$\int\frac{d\omega}{2\pi}\ \mathrm{Tr}\Big[$  $+$  $-$  $\Big] = 0$ (C15)

and

$\int\frac{d\omega}{2\pi}\ \mathrm{Tr}\Big[$  $-$  $\Big] = 0.$ (C16)

Combining Eqs. (C12), (C7), and (C14)–(C16) leads to a simplified expression for the second-order correction to the charge density in insulators,

$\delta\rho^{(2)}(\mathbf{R}) = e\hbar\int\frac{d^3k}{(2\pi)^3}\int\frac{d\omega}{2\pi}\ \mathrm{Tr}\Big[$ 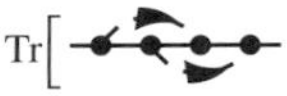
$+2\cdot\Big($  $+$  $\Big)\Big].$ (C17)

To make further progress, we now focus on the contribution of $\delta\rho^{(2)}$ to the total charge of a skyrmion, given by $\delta Q^{(2)} = \int d^3R\ \delta\rho^{(2)}(\mathbf{R})$. The expression for $\delta Q^{(2)}$ can be simplified by means of integration by parts in phase space. The last diagram on the right-hand side of Eq. (C17) represents the term $(\frac{i}{2})^2 J_{ij}J_{kl}\tilde{G}_0(-\partial_i\tilde{G}_0^{-1})\tilde{G}_0(-\partial_j\partial_k\tilde{G}_0^{-1})\tilde{G}_0(-\partial_l\tilde{G}_0^{-1})\tilde{G}_0$. It is structurally different from the other two diagrams in that it contains a second-order derivative in phase space. Integration by parts over the phase-space direction x_j or x_k, respectively, leads to the relations

$\int\frac{d^6x}{(2\pi)^3}$ 
$= -\int\frac{d^6x}{(2\pi)^3}\Big[$  $+$ 
$+$ 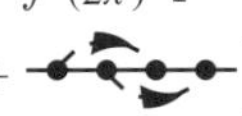 $\Big]$
$= -\int\frac{d^6x}{(2\pi)^3}\Big[$  $+$ 
$+$ 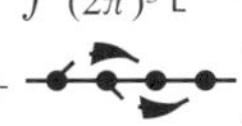 $\Big].$ (C18)

Combining Eqs. (C15)–(C18) leads to

$\delta Q^{(2)} = \frac{e\hbar}{3}\int\frac{d^6x}{(2\pi)^3}\int\frac{d\omega}{2\pi}\ \mathrm{Tr}\Big[$ 
$+$ 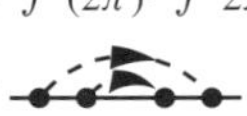 $-$  $\Big]$

$$= \frac{e\hbar}{3}\left(\frac{i}{2}\right)^2\int\frac{d^6x}{(2\pi)^3}\int\frac{d\omega}{2\pi}\ \mathcal{A}_{ijkl}\,\mathcal{B}_{ijkl}(i\omega,\mathbf{x})\tag{C19}$$

with

$$\begin{aligned}\mathcal{A}_{ijkl} &= J_{ij}J_{kl} + J_{il}J_{jk} - J_{ik}J_{jl} = -\epsilon_{ijkl},\\ \mathcal{B}_{ijkl}(i\omega,\mathbf{x}) &= \mathrm{Tr}\big[\tilde{G}_0\big(-\partial_i\tilde{G}_0^{-1}\big)\tilde{G}_0\big(-\partial_j\tilde{G}_0^{-1}\big)\tilde{G}_0\\ &\quad\times\big(-\partial_k\tilde{G}_0^{-1}\big)\tilde{G}_0\big(-\partial_l\tilde{G}_0^{-1}\big)\tilde{G}_0\big].\end{aligned} \tag{C20}$$

The right-hand side of Eq. (C19) is the second Chern number, which is a topological invariant [cf., e.g., Eq. (53) in Ref. 33]. A similar calculation as in Appendix C of Ref. 33 leads to Eq. (14) where the (in general) non-Abelian Berry curvature Ω is now a matrix in the space of occupied bands with elements given by

$$\begin{aligned}&\langle\mathbf{x},n|\Omega_{ij}|\mathbf{x},m\rangle\\ &= i\sum_{n'}\frac{\langle\mathbf{x},n|\frac{\partial H}{\partial x_i}|\mathbf{x},n'\rangle\langle\mathbf{x},n'|\frac{\partial H}{\partial x_j}|\mathbf{x},m\rangle - (i\leftrightarrow j)}{\big[\epsilon_n^{(0)}(\mathbf{x})-\epsilon_{n'}^{(0)}(\mathbf{x})\big]\big[\epsilon_m^{(0)}(\mathbf{x})-\epsilon_{n'}^{(0)}(\mathbf{x})\big]},\end{aligned} \tag{C21}$$

where $(i\leftrightarrow j)$ denotes the term with i and j exchanged.

4. Factorization of the skyrmion charge in insulators with Abelian Berry curvature

In this section, we show that the Berry curvature contribution to the charge in a two-dimensional insulator with Abelian Berry curvature is given by the product of the quantized Hall conductivity σ_{xy} and the skyrmion number Φ_0. An Abelian Berry curvature arises, e.g., if only a single band is occupied. We then derive an expression for the charge per length of a skyrmion line in a three-dimensional insulator.

For a two-dimensional system with only one occupied band, Eq. (14) reduces to

$$\begin{aligned}\delta Q^{(2)} &= -\frac{e\,\epsilon_{ijkl}}{8}\int\frac{d^4x}{(2\pi)^2}\Omega_{ij}\Omega_{kl}\\ &= -\frac{e\,\epsilon_{ijkl}}{4}\left(\int\frac{dx_j\,dx_k\,dx_l}{(2\pi)^2}A_j\Omega_{kl}\right)_{x_i=-\infty}^{x_i=+\infty},\end{aligned} \tag{C22}$$

where the symbols $\pm\infty$ denote either positions far away from the skyrmion or the boundaries of the Brillouin zone for a space or momentum direction x_i, respectively. In the second equality of Eq. (C22), we used the relation

$$\begin{aligned}\frac{\epsilon_{ijkl}}{8}\Omega_{ij}\Omega_{kl} &= \frac{\epsilon_{ijkl}}{2}\frac{\partial A_j}{\partial x_i}\frac{\partial A_l}{\partial x_k} = \frac{\epsilon_{ijkl}}{2}\frac{\partial}{\partial x_i}\left(A_j\frac{\partial A_l}{\partial x_k}\right)\\ &= \frac{\epsilon_{ijkl}}{4}\frac{\partial}{\partial x_i}(A_j\Omega_{kl}).\end{aligned} \tag{C23}$$

In Eq. (C22), Ω_{kl} only enters at the boundary of the x_i coordinate. At the boundary in spatial direction (i.e., far away from the skyrmion), the magnetization is collinear and therefore $\Omega^{\mathrm{RR}} = 0 = \Omega^{\mathrm{Rk}}$. Thus, if x_i is a spatial coordinate, only terms of the form $A_j^{\mathrm{R}}\Omega_{kl}^{\mathrm{kk}}$ contribute to the integral kernel in Eq. (C22). If x_i is a momentum coordinate, Ω_{kl} is evaluated at the boundary of the Brillouin zone. In an insulator, the charge must be quantized and we can adiabatically deform the Bloch functions such that they are independent of momentum in a narrow stripe around the Brillouin zone boundary. This is always possible since, in the absence of further symmetries, all noninteracting Hamiltonians of one-dimensional insulators are adiabatically connected.[48] Therefore, only terms of the form $A_j^{\mathrm{k}}\Omega_{kl}^{\mathrm{RR}}$ contribute if x_i is a momentum coordinate. In total, Eq. (C22) can be written as $\delta Q^{(2)} = \delta Q^{(2),\mathrm{R}} + \delta Q^{(2),\mathrm{k}}$ where

$$\delta Q^{(2),\mathrm{R}} = -\frac{e\,\epsilon_{ij}}{2}\left(\int\frac{dR_j\,d^2k}{(2\pi)^2}A_j^R\Omega_{xy}^{kk}\right)_{R_i=-\infty}^{R_i=+\infty} \tag{C24}$$

and $\delta Q^{(2),\mathrm{k}}$ is defined by formally exchanging all R and k. As the Berry curvature $\Omega_{xy}^{\mathrm{kk}}$ in Eq. (C24) is gauge independent, it can not depend on R_j for a collinear magnetization at $R_i = \pm\infty$. This implies

$$\begin{aligned}\delta Q^{(2),\mathrm{R}} &= -\frac{e\,\epsilon_{ij}}{2}\int\frac{d^2k}{(2\pi)^2}\left(\Omega_{xy}^{\mathrm{kk}}\int dR_j\big(A_j^{\mathrm{R}}\big)_{R_i=-\infty}^{R_i=+\infty}\right)\\ &= -\frac{e}{2}\int\frac{d^2k}{(2\pi)^2}\left(\Omega_{xy}^{\mathrm{kk}}\int d^2R\,\epsilon_{ij}\frac{\partial A_j^{\mathrm{R}}}{\partial R_i}\right)\\ &= -\frac{e}{2}\int\frac{d^2k}{(2\pi)^2}\left(\Omega_{xy}^{\mathrm{kk}}\int d^2R\,\Omega_{xy}^{\mathrm{RR}}\right)\\ &= +\frac{1}{2}\sigma_{xy}\Phi_0,\end{aligned} \tag{C25}$$

where σ_{xy} (Φ_0) is the quantized integral over $\frac{e^2}{\hbar}\Omega_{xy}^{\mathrm{kk}}$ ($\frac{\hbar}{|e|}\Omega_{xy}^{\mathrm{RR}}$) as defined in the main text (recall that, in our convention, $e<0$ is the electron charge). An analogous calculation leads to the same value for $\delta Q^{(2),k}$. Thus, we conclude that the Berry curvature contribution to the skyrmion charge in a two-dimensional insulator with Abelian Berry curvature is given by

$$\delta Q^{(2)} = \sigma_{xy}\Phi_0. \tag{C26}$$

In three-dimensional systems, skyrmions form line defects. From Eqs. (B5) and (B7), the Berry curvature contribution to the charge in a three-dimensional insulator is given by

$$\delta Q^{(2)} = \frac{3e}{3!(2\pi)^3}\int\hat{\omega}_0\wedge\hat{\Omega}\wedge\hat{\Omega}. \tag{C27}$$

We introduce dimensionless coordinates $\tilde{\mathbf{x}} \equiv (\tilde{\mathbf{R}},\tilde{\mathbf{k}})$ such that $\mathbf{R} = \tilde{R}_\alpha\mathbf{a}_\alpha$ and $\mathbf{k} = \tilde{k}_\alpha\mathbf{g}_\alpha/(2\pi)$. Here, the vectors $\mathbf{a}_\alpha$ are lattice vectors of the atomic lattice and $\mathbf{g}_\alpha$ are the corresponding reciprocal lattice vectors. The coordinates $\tilde{\mathbf{x}}$ are chosen such that momentum space is periodic in the three coordinate directions $\tilde{k}_\alpha$ and the Jacobian of the transformation is one. In the dimensionless coordinates, Eq. (C27) reads as

$$\begin{aligned}\delta Q^{(2)} &= -\frac{e\,\epsilon_{\alpha\beta\gamma\delta\mu\nu}}{16}\int\frac{d^6\tilde{x}}{(2\pi)^3}J_{\alpha\beta}\tilde{\Omega}_{\gamma\delta}\tilde{\Omega}_{\mu\nu}\\ &= -e\sum_{\alpha=1}^{3}\int d\tilde{R}_\alpha\int\frac{d\tilde{k}_\alpha}{2\pi}\frac{\epsilon_{\alpha(\alpha+3)\gamma\delta\mu\nu}}{8}\int\frac{d^4\tilde{x}}{(2\pi)^2}\tilde{\Omega}_{\gamma\delta}\tilde{\Omega}_{\mu\nu}\\ &= -e\sum_{\alpha=1}^{3}\int d\tilde{R}_\alpha\int\frac{d\tilde{k}_\alpha}{2\pi}n_\alpha^{\mathrm{R}}n_\alpha^{\mathrm{k}}\end{aligned} \tag{C28}$$

with

$$n_\alpha^{\mathrm{R}} = \frac{\epsilon_{\alpha\gamma\delta}}{2}\int d\tilde{R}_\gamma d\tilde{R}_\delta\, \tilde{\Omega}^{\mathrm{RR}}_{\gamma\delta};$$
$$n_\alpha^{\mathrm{k}} = \frac{\epsilon_{\alpha\mu\nu}}{2}\int \frac{d\tilde{k}_\mu d\tilde{k}_\nu}{(2\pi)^2}\, \tilde{\Omega}^{\mathrm{kk}}_{\mu\nu}. \quad \text{(C29)}$$

In the first line of Eq. (C28), $\tilde{\Omega}_{\gamma\delta}$ denotes the elements of the Berry-curvature tensor in coordinates $\tilde{\mathbf{x}}$. The components of the symplectic tensor $J_{\alpha\beta}$ are invariant under the transformation from coordinates $\mathbf{x}$ to $\tilde{\mathbf{x}}$ due to the relation $\mathbf{a}_\alpha \cdot \mathbf{g}_\beta = 2\pi\delta_{\alpha\beta}$. This allowed us to set $\beta = \alpha + 3$ in the second line of Eq. (C28). The last integral in the second line of Eq. (C28) runs over the four-dimensional subspace of phase space perpendicular to $(\mathbf{0},\mathbf{a}_\alpha)$ and $(\mathbf{g}_\alpha,\mathbf{0})$. Its value is given by the product of the real-space and the momentum-space winding numbers n_α^{R} and n_α^{k} by the same arguments that lead from Eq. (C22) to (C26).

The remaining integral over $\tilde{k}_\alpha$ in Eq. (C28) equates to a factor of 1 and the integral over $\tilde{R}_\alpha$ gives

$$\int d\tilde{R}_\alpha = \frac{\partial \tilde{R}_\alpha}{\partial R_i}\int dR_i = \frac{(\mathbf{g}_\alpha)_i}{2\pi}L_i = \frac{\hat{\mathbf{s}}\cdot\mathbf{g}_\alpha}{2\pi}L, \quad \text{(C30)}$$

where the unit vector $\hat{\mathbf{s}}$ points along along the skyrmion line and $L_i = L\hat{s}_i$ is the projection of the length L of the skyrmion line onto the coordinate direction R_i. For a generic skyrmion line that pierces all three position-space coordinate planes, the skyrmion number is $n_\alpha^{\mathrm{R}} = 1$ for all α. Combining Eqs. (C28) and (C30), we thus arrive at an expression for the charge per length of a skyrmion line in a three-dimensional insulator with Abelian Berry curvature

$$\frac{\delta Q^{(2)}}{L} = -e\sum_{\alpha=1}^{3}\frac{\hat{\mathbf{s}}\cdot\mathbf{g}_\alpha}{2\pi}n_\alpha^{\mathrm{k}}, \quad \text{(C31)}$$

where $\hat{\mathbf{s}}$ is the direction of the skyrmion line, $\mathbf{g}_\alpha$ are reciprocal lattice vectors, and $n_\alpha^{\mathrm{k}} \in \mathbb{Z}$ is defined in Eq. (C29).

*f.freimuth@fz-juelich.de
[1]I. Dzyaloshinsky, J. Phys. Chem. Solids **4**, 241 (1958).
[2]T. Moriya, Phys. Rev. **120**, 91 (1960).
[3]S. Mühlbauer, B. Binz, F. Jonietz, C. Pfleiderer, A. Rosch, A. Neubauer, R. Georgii, and P. Böni, Science **323**, 915 (2009).
[4]X. Z. Yu, Y. Onose, N. Kanazawa, J. H. Park, J. H. Han, Y. Matsui, N. Nagaosa, and Y. Tokura, Nature(London) **465**, 901 (2010).
[5]T. Schulz, R. Ritz, A. Bauer, M. Halder, M. Wagner, C. Franz, C. Pfleiderer, K. Everschor, M. Garst, and A. Rosch, Nat. Phys. **8**, 301 (2012).
[6]K. Everschor, M. Garst, B. Binz, F. Jonietz, S. Mühlbauer, C. Pfleiderer, and A. Rosch, Phys. Rev. B **86**, 054432 (2012).
[7]X. Z. Yu, N. Kanazawa, W. Z. Zhang, T. Nagai, T. Hara, K. Kimoto, Y. Matsui, Y. Onose, and Y. Tokura, Nat. Commun. **3**, 988 (2012).
[8]S. E. Barrett, G. Dabbagh, L. N. Pfeiffer, K. W. West, and R. Tycko, Phys. Rev. Lett. **74**, 5112 (1995).
[9]H. A. Fertig, L. Brey, R. Cote, and A. H. MacDonald, Phys. Rev. B **50**, 11018 (1994).
[10]S. L. Sondhi, A. Karlhede, S. A. Kivelson, and E. H. Rezayi, Phys. Rev. B **47**, 16419 (1993).
[11]D.-H. Lee and C. L. Kane, Phys. Rev. Lett. **64**, 1313 (1990).
[12]L. Brey, H. A. Fertig, R. Cote, and A. H. MacDonald, Phys. Rev. Lett. **75**, 2562 (1995).
[13]M. V. Berry, Proc. R. Soc. London, Ser. A **392**, 45 (1984).
[14]G. Sundaram and Q. Niu, Phys. Rev. B **59**, 14915 (1999).
[15]D. Xiao, M.-C. Chang, and Q. Niu, Rev. Mod. Phys. **82**, 1959 (2010).
[16]N. Nagaosa, J. Sinova, S. Onoda, A. H. MacDonald, and N. P. Ong, Rev. Mod. Phys. **82**, 1539 (2010).
[17]M. Lee, Y. Onose, Y. Tokura, and N. P. Ong, Phys. Rev. B **75**, 172403 (2007).
[18]Y. Yao, L. Kleinman, A. H. MacDonald, J. Sinova, T. Jungwirth, D.-s. Wang, E. Wang, and Q. Niu, Phys. Rev. Lett. **92**, 037204 (2004).
[19]X. Wang, J. R. Yates, I. Souza, and D. Vanderbilt, Phys. Rev. B **74**, 195118 (2006).
[20]S. Lowitzer, D. Ködderitzsch, and H. Ebert, Phys. Rev. Lett. **105**, 266604 (2010).
[21]J. Weischenberg, F. Freimuth, J. Sinova, S. Blügel, and Y. Mokrousov, Phys. Rev. Lett. **107**, 106601 (2011).
[22]I. Turek, J. Kudrnovský, and V. Drchal, Phys. Rev. B **86**, 014405 (2012).
[23]A. Neubauer, C. Pfleiderer, B. Binz, A. Rosch, R. Ritz, P. G. Niklowitz, and P. Böni, Phys. Rev. Lett. **102**, 186602 (2009).
[24]N. Kanazawa, Y. Onose, T. Arima, D. Okuyama, K. Ohoyama, S. Wakimoto, K. Kakurai, S. Ishiwata, and Y. Tokura, Phys. Rev. Lett. **106**, 156603 (2011).
[25]S. X. Huang and C. L. Chien, Phys. Rev. Lett. **108**, 267201 (2012).
[26]R. Cheng and Q. Niu, Phys. Rev. B **86**, 245118 (2012).
[27]F. Freimuth, S. Blügel, and Y. Mokrousov, arXiv:1308.5983.
[28]D. Xiao, J. Shi, and Q. Niu, Phys. Rev. Lett. **95**, 137204 (2005).
[29]P. J. Morrison, Rev. Mod. Phys. **70**, 467 (1998).
[30]A. Liechtenstein, M. Katsnelson, V. Antropov, and V. Gubanov, J. Magn. Magn. Mater. **67**, 65 (1987).
[31]F. Freimuth, S. Blügel, and Y. Mokrousov, arXiv:1305.4873.
[32]B.-J. Yang and N. Nagaosa, Phys. Rev. B **84**, 245123 (2011).
[33]X.-L. Qi, T. L. Hughes, and S.-C. Zhang, Phys. Rev. B **78**, 195424 (2008).
[34]T. Adams, S. Mühlbauer, C. Pfleiderer, F. Jonietz, A. Bauer, A. Neubauer, R. Georgii, P. Böni, U. Keiderling, K. Everschor, M. Garst, and A. Rosch, Phys. Rev. Lett. **107**, 217206 (2011).
[35]A. A. Mostofi, J. R. Yates, Y.-S. Lee, I. Souza, D. Vanderbilt, and N. Marzari, Comput. Phys. Commun. **178**, 685 (2008).
[36]F. Freimuth, Y. Mokrousov, D. Wortmann, S. Heinze, and S. Blügel, Phys. Rev. B **78**, 035120 (2008).
[37]Y. Ishikawa, G. Shirane, J. A. Tarvin, and M. Kohgi, Phys. Rev. B **16**, 4956 (1977).
[38]See http://www.flapw.de
[39]V. L. Moruzzi, J. F. Janak, and A. R. Williams, *Calculated Electronic Properties of Metals* (Pergamon, New York, 1978).
[40]T. Jeong and W. E. Pickett, Phys. Rev. B **70**, 075114 (2004).
[41]A. A. Mostofi *et al.*, Comput. Phys. Commun. **178**, 685 (2008).
[42]J. R. Yates, X. Wang, D. Vanderbilt, and I. Souza, Phys. Rev. B **75**, 195121 (2007).
[43]N. Marzari, A. A. Mostofi, J. R. Yates, I. Souza, and D. Vanderbilt, Rev. Mod. Phys. **84**, 1419 (2012).

[44]F. W. Warner, *Foundations of Differentiable Manifolds and Lie Groups* (Springer, New York, 1971).

[45]M. Göckeler and T. Schücker, *Differential Geometry, Gauge Theories, and Gravity* (Cambridge University Press, Cambridge, UK, 1989).

[46]J. Milnor and J. D. Stasheff, *Characteristic Classes*. (Princeton University Press, Princeton, NJ, 1974).

[47]J. Rammer and H. Smith, Rev. Mod. Phys. **58**, 323 (1986).

[48]S. Ryu, A. P. Schnyder, A. Furusaki, and A. W. W. Ludwig, New J. Phys. **12**, 065010 (2010).

Light-induced effective magnetic fields for ultracold atoms in planar geometries

G. Juzeliūnas,[1] J. Ruseckas,[1,2] P. Öhberg,[3] and M. Fleischhauer[2]
[1]*Institute of Theoretical Physics and Astronomy of Vilnius University, A. Goštauto 12, 01108 Vilnius, Lithuania*
[2]*Fachbereich Physik, Technische Universität Kaiserslautern, D-67663 Kaiserslautern, Germany*
[3]*Department of Physics, University of Strathclyde, Glasgow G4 0NG, Scotland*
(Received 21 November 2005; published 14 February 2006)

We propose a scheme to create an effective magnetic field for ultracold atoms in a planar geometry. The setup allows the experimental study of classical and quantum Hall effects in close analogy to solid-state systems including the possibility of finite currents. The present scheme is an extention of the proposal in Phys. Rev. Lett. **93**, 033602 (2004), where the effective magnetic field is now induced for three-level Λ-type atoms by two counterpropagating laser beams with shifted spatial profiles. Under conditions of electromagnetically induced transparency the atom-light interaction has a space-dependent dark state, and the adiabatic center-of-mass motion of atoms in this state experiences effective vector and scalar potentials. The associated magnetic field is oriented perpendicular to the propagation direction of the laser beams. The field strength achievable is one flux quantum over an area given by the transverse beam separation and the laser wavelength. For a sufficiently dilute gas the field is strong enough to reach the lowest Landau level regime.

DOI: 10.1103/PhysRevA.73.025602 PACS number(s): 03.75.Ss, 42.50.Gy, 42.50.Fx

One of the most fascinating subjects at the interface between ultracold atoms and solid-state systems is the possibility to experimentally study strong-correlation phenomena with the precision and the large degree of variability provided by atomic physics. For example, interacting Bose-Einstein condensates (BEC) or degenerate Fermi gases in rotating two-dimensional traps are studied in several laboratories with the goal to observe quantum-Hall-like effects [1–3]. The trap rotation provides an effective magnetic field for the electrically neutral atoms. However, in order to reach the fractional quantum-Hall regime it is necessary to rotate the trap close to the critical frequency. Furthermore, the atom density needs to be low enough such that the number of magnetic flux quanta approaches the number of atoms, which is an experimental challenge. Besides experimental difficulties this approach has some conceptual drawbacks: It is limited to rotational symmetric setups and does not allow to study transport phenomena, i.e., the effect of magnetic fields to a finite particle current.

In [4–6] we have suggested an alternative method based on light-induced gauge potentials for atoms with a space-dependent dark state. A dark state is created if three-level Λ-type atoms interact with two laser fields under conditions of electromagnetically induced transparency (EIT) [7–11]. If the dark state is space dependent, a vector gauge potential arises for the adiabatic center-of-mass motion [12]. As shown in [4,5] the vector potential is associated with a nonvanishing magnetic field, if at least one of the two light beams has a vortex, i.e., an orbital angular momentum (OAM). Yet the use of vortex light beams has similar drawbacks as the trap rotation regarding the spatial symmetry and transport phenomena.

We here propose a variation of this scheme which is free of the above-mentioned limitations. The scheme, shown in Fig. 1, once again involves two laser beams interacting with three-level atoms in the EIT configuration. Yet we are no longer dealing with light beams posessing an OAM with respect to their propagation axis. As we will show later on a nonvanishing magnetic field requires only a *relative* OAM between the two light beams. This can be achieved by two counter-propagating and overlapping laser beams with shifted spatial profiles. In this case an effective magnetic field appears perpendicular to the propagation direction and to the gradient of the relative intensity of the light beams. This configuration allows a planar geometry and a nonvanishing flow of atoms, e.g., an atomic BEC moving along an atomic waveguide [13].

Let us consider an ensemble of cold three-level atoms with lower levels $|1\rangle$ and $|2\rangle$ and electronically excited state $|3\rangle$. The atoms interact with two resonant laser beams in the EIT configuration, see Fig. 1. The first beam (to be referred to as the control beam) has a frequency ω_c, a wave-vector $\mathbf{k}_c$, and induces the atomic transitions $|2\rangle \to |3\rangle$ with Rabi frequency $\Omega_c \equiv \mu_{32}E_c/2$, where E_c is the electric field strength and μ_{32} is the transition dipole moment. The second (probe) beam with frequency ω_p, wave-vector $\mathbf{k}_p$ causes the transition $|1\rangle \to |3\rangle$ with a Rabi frequency $\Omega_p \equiv \mu_{31}E_p/2$. The two laser beams keep the atoms in their dark state [7–11]:

$$|D\rangle = |1\rangle\cos\theta - |2\rangle\sin\theta\exp(iS) \sim |1\rangle - \zeta|2\rangle, \qquad (1)$$

where $\zeta = \Omega_p/\Omega_c = |\zeta|e^{iS} \equiv \tan\theta\, e^{iS}$ is the ratio between the Rabi frequencies of the probe and control fields, S is their relative phase, and θ is the mixing angle between the states $|1\rangle$ and $|2\rangle$ in the atomic dark state $|D\rangle$.

The dark state depends on the atomic position through the $\mathbf{r}$ dependence of the Rabi frequencies $\Omega_p(\mathbf{r})$ and $\Omega_c(\mathbf{r})$, so an effective vector potential (generally known as the Berry connection [14,15]) appears in the adiabatic equation of motion for the atomic center of mass. The effective vector and trapping potentials governing the translational motion of the dark-state atoms read [5,6]

$$\mathbf{A}_{\text{eff}} = -\hbar\frac{|\zeta|^2}{1+|\zeta^2|}\nabla S = -\hbar\sin^2\theta\nabla S \qquad (2)$$

and

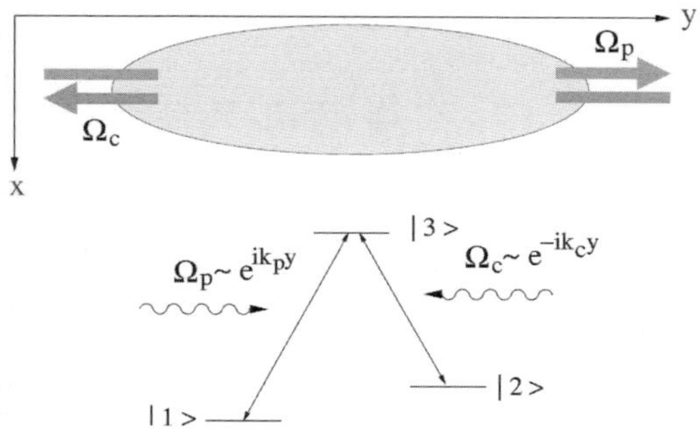

FIG. 1. (Color online) (top) Schematic representation of setup for light-induced effective magnetic fields: Two counterpropagating and overlapping laser beams interact with a cloud of cold atoms. (bottom) The level scheme for the Λ-type atoms interacting with the resonant probe and control beams characterized by Rabi frequencies Ω_p and Ω_c.

$$V_{\rm eff}(\mathbf{r}) = V_{\rm ext}(\mathbf{r}) + \frac{\hbar^2}{2m}\frac{|\zeta|^2(\boldsymbol{\nabla}S)^2 + (\boldsymbol{\nabla}|\zeta|)^2}{(1+|\zeta|^2)^2}, \tag{3}$$

where

$$V_{\rm ext}(\mathbf{r}) = \frac{V_1(\mathbf{r}) + |\zeta|^2(V_2(\mathbf{r}) + \hbar\omega_{21})}{1+|\zeta|^2} \tag{4}$$

is the external potential for the dark-state atoms, $V_j(\mathbf{r})$ is the trapping potential for an atom in the internal state j, and $\omega_{21}=\omega_2-\omega_1+\omega_c-\omega_p$ is the frequency of the two-photon detuning.

One easily recognizes that the vector gauge potential $\mathbf{A}_{\rm eff}$ yields a nonvanishing magnetic field only if the gradients of the relative intensity and the relative phase are both nonzero and not parallel to each other:

$$\mathbf{B}_{\rm eff} \equiv \boldsymbol{\nabla}\times\mathbf{A}_{\rm eff} = -\hbar\,\boldsymbol{\nabla}(\sin^2\theta)\times\boldsymbol{\nabla}S. \tag{5}$$

This equation has a very intuitive interpretation: $\boldsymbol{\nabla}(\sin^2\theta)$ is a vector that connects the "center of mass" of the two light beams; $\boldsymbol{\nabla}S$ is proportional to the vector of their relative momentum. Thus a nonvanishing $\mathbf{B}_{\rm eff}$ requires a *relative orbital angular momentum* of the two light beams. As discussed in [4–6] this is the case, e.g., for light beams with a vortex.

Here we consider, however, a different scenario. We suggest to use two counter-propagating light beams of finite diameter with an axis offset: $\Omega_p=\Omega_p^{(0)}e^{ik_p y}$ and $\Omega_c=\Omega_c^{(0)}e^{-ik_c y}$, where $\Omega_p^{(0)}$ and $\Omega_c^{(0)}$ are real amplitudes with shifted transverse profiles. The beams possess a relative orbital angular momentum similarly to two point particles with constant momenta passing each other at some finite distance. In such a situation the phase of the ratio $\zeta=\Omega_p/\Omega_c$ is given by

$$S = ky, \quad k = k_p + k_c, \tag{6}$$

so that $\boldsymbol{\nabla}S=k\hat{\mathbf{e}}_y$ where $\hat{\mathbf{e}}_y$ is a unit Cartesian vector.

The spatial dependence of the intensity ratio $|\zeta|^2=|\Omega_p/\Omega_c|^2$ is determined by the spatial profiles of both $|\Omega_p|^2$ and $|\Omega_p|^2$. Since the control and probe beams counter propagate along the y axis, their intensities depend weakly on y. Furthermore we shall disregard the z dependence of the intensity ratio $|\zeta|^2$. This is legitimate, for instance, if the atomic motion is confined to the xy plane due to a steep trapping potential in the z direction. Hence one finds

$$\mathbf{B}_{\rm eff} = \hat{\mathbf{e}}_z\hbar k\frac{\partial}{\partial x}\sin^2\theta. \tag{7}$$

The field strength $B_{\rm eff}$ depends generally on the x coordinate and has a weak y dependence as long as the paraxial approximation holds.

If we are interested in fractional quantum-Hall physics and thus in the possibility to enter the lowest Landau level (LLL) regime we have to estimate the maximum strength of the magnetic field. For this we determine the minimum area needed for a magnetic flux corresponding to a single flux quantum $2\pi\hbar$. From Eq. (7) we recognize that this area is given by the product $\lambda x_{\rm eff}$, where $x_{\rm eff}$ is the effective separation between the two beam centers. To reach the LLL in a two-dimensional gas the atomic density has thus to be smaller than one atom per $\lambda x_{\rm eff}$.

The above analysis holds as long as the atoms move sufficiently slow to remain in their dark states. This is the case if the adiabatic condition [5] holds: $\Omega \gg F$, where $F=|\nabla\zeta\cdot\mathbf{v}|/(1+|\zeta|^2)$ reflects the two-photon Doppler detuning. In the present situation we have

$$F^2 = \cos^2\theta\left[\left(v_x\frac{\partial}{\partial x}|\zeta|\right)^2 + (|\zeta|kv_y)^2\right] \ll \Omega^2. \tag{8}$$

where $\Omega=(|\Omega_c|^2+|\Omega_p|^2)^{1/2}$ is the rms Rabi frequency. The adiabatic condition implies that the quantity Ω^{-1} should be much less than the time an atom travels a characteristic length over which the amplitude or the phase of the ratio $\zeta=\Omega_p/\Omega_c$ changes considerably. For atoms moving along the y axis, such a length is $1/k \approx 1/2k_p \sim 10^{-7}$ m. On the other hand, the Rabi frequency can be of the order of 10^7 to 10^8 s^{-1} [16]. Therefore, the adiabatic condition should hold for atomic velocities up to meters per second.

The above estimation does not take into account a finite lifetime of the excited atoms, typically $\tau_3 \sim 10^{-7}$ s. If this is included, the atomic dark state acquires a finite lifetime $\tau_D \sim \tau_3\Omega^2/F^2$ due to nonadiabatic coupling [5]: For instance, if the atomic velocities are of the order of a centimeter per second, the atoms should survive in their dark states up to a second.

Much larger atomic velocities are possible, however, as long as the velocity spread $\Delta\mathbf{v}$ is much smaller than the central velocity $\mathbf{v}_0$. For atoms moving along the y axis, one can set a two-photon detuning $\omega_{21}=-(k_p+k_c)v_0$ to compensate the Doppler shift associated with $\mathbf{v}_0$. In that case it is the velocity spread $\Delta\mathbf{v}$ rather than the whole atomic velocity $\mathbf{v}$ that determines the nonadiabatic term F. For instance, in a recent experiment [17] on propagation of a BEC in a waveguide, the central atomic velocity is 5 cm/s, whereas the velocity spread is only 1.4 mm/s. Note that the two-photon detuning will also lead to a transversal slope in the trapping potential represented by the term with ω_{21} in Eq. (4).

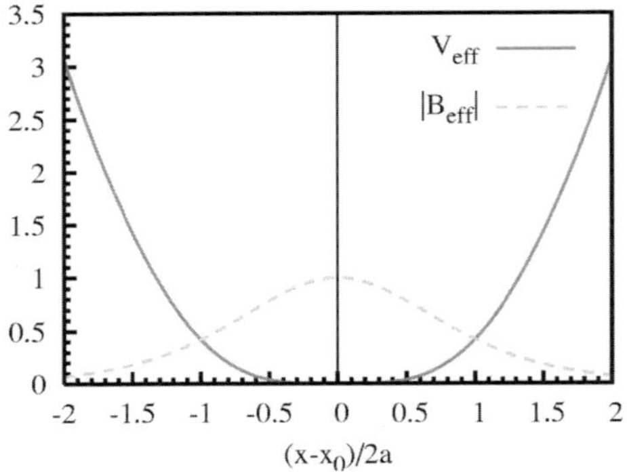

FIG. 2. (Color online) Effective trapping potential $V_{\rm eff}$ and an effective magnetic field $B_{\rm eff}$ produced by counter-propagating Gaussian beams. The external harmonic potential $V_{\rm ext}$ cancels the quadratic term in the overall potential $V_{\rm eff}$. The effective magnetic field is plotted in the units of $B_{\rm eff}(0)\equiv\hbar k/4a$, whereas the effective trapping potential is plotted in the units of $\hbar\omega_{\rm rec}(1+1/4a^2k^2)$, with $\omega_{\rm rec}=\hbar k^2/2m$.

Let us assume that both the control and probe beams are characterized by Gaussian profiles with the same amplitude Ω_0 and width σ:

$$|\Omega_j| = \Omega_0 \exp\left(-\frac{(x-x_j)^2}{\sigma^2}\right), \quad j=p,c. \tag{9}$$

In the paraxial approximation, the Gaussian beams have the width $\sigma\equiv\sigma(y)=\sigma_0[1+(\lambda y/\pi\sigma_0^2)]^{1/2}$, where $\sigma_0\equiv\sigma(0)$ is the beam waist and λ is the wavelength. Since $k_p\approx k_c\approx k/2$, we have $\lambda\approx 4\pi/k$ both for the control and probe beams. We are interested mostly in distances $|y|$ much less than the confocal parameter of the beams $b=2\pi\sigma_0^2/\lambda\approx k\sigma_0^2/2$. For such distances, $|y|\ll b$, the width $\sigma(y)$ is close to the beam waist: $\sigma(y)\approx\sigma_0$.

Suppose the beams are centered at $x_p=x_0+\Delta x/2$ and $x_c=x_0-\Delta x/2$, The intensity ratio reads then $|\zeta|^2\equiv|\Omega_p/\Omega_c|^2 =\exp[(x-x_0)/a]$, where $a\equiv a(y)=\sigma^2/4\Delta x$ is the relative width of the two beams. Thus we have

$$\mathbf{B}_{\rm eff} = -\hbar k\frac{1}{4a\cosh^2[(x-x_0)/2a]}\mathbf{e}_z, \tag{10}$$

$$V_{\rm eff}(\mathbf{r}) = V_{\rm ext}(\mathbf{r}) + \frac{\hbar^2k^2}{2m}\frac{(1+1/4a^2k^2)}{4\cosh^2[(x-x_0)/2a]}. \tag{11}$$

It is evident that both $\mathbf{B}_{\rm eff}$ and $V_{\rm eff}(\mathbf{r})$ are maximum at the central point $x=x_0$ and decrease quadratically for $|x-x_0|\ll a$. Similar to Ref. [5], the term quadratic in the displacement $x-x_0$ can be canceled in the effective trapping potential (11) by taking an external potential $V_{\rm ext}$ with the appropriate quadratic term. The frequency of the external potential fulfilling such a condition is

$$\omega_{\rm ext} = \frac{\hbar k}{4am}\sqrt{1+1/4a^2k^2}. \tag{12}$$

With this the overall effective trapping potential becomes constant up to terms of the fourth order in $x-x_0$. In the vicinity of the central point ($|x-x_0|\ll a$) the magnetic field strength is: $B_{\rm eff}\approx\hbar k/4a$. The corresponding magnetic length and cyclotron frequency are: $\ell_B\approx\sqrt{\hbar/B_{\rm eff}}=2\sqrt{a/k}$ and $\omega_c=B/m\approx\hbar k/4am$. The magnetic length ℓ_B is much smaller than the relative width of the two beams $\ell_B\ll a$ provided the latter is much larger than the optical wavelength: $ak\gg 1$. In that case many magnetic lengths fit within the interval $|x-x_0|<a$ across the beams. Furthermore the cyclotron frequency equals then approximately to the frequency of the external trap: $\omega_c\approx\omega_{\rm ext}$, both of them being much less than the recoil frequency.

Figure 2 shows the effective trapping potential and effective magnetic field calculated using Eqs. (10) and (11), with the external harmonic potential $V_{\rm ext}$ of frequency $\omega_{\rm ext}$ [Eq. (12)] added to cancel the quadratic term in the overall potential $V_{\rm eff}$. The magnetic field is seen to be close to its maximum value in the area of a constant potential where $|x-x_0|\ll a$. For larger distances the effective trapping potential forms a barrier, so the atoms can be trapped in the region of a large magnetic field.

In summary, we have shown how to create an effective magnetic field in ultracold gases with a planar geometry using two counter-propagating laser beams acting on three-level atoms in the EIT configuration. If the amplitude ratio of the two beams changes substantially in the transverse direction, an effective magnetic field appears in the plane perpendicular to the propagation direction of the beams. This can be achieved if the beams are shifted relative to each other (see Fig. 1), such that they have a relative OAM.

The suggested method provides a possibility to create an effective magnetic field over an extended area along the propagation direction. This allows for a geometrical setup similar to that used in solid-state systems for classical and quantum Hall measurements. In particular, finite currents perpendicular to the magnetic field are possible and Hall "voltages" can be detected by observing changes in the chemical potential perpendicular to both the current and magnetic field. Finally the suggested method is much more robust than that of Refs. [4,5], as it does not require vortex light beams.

This work was supported by the Marie-Curie Training-site at the Technical University of Kaiserslautern, the Royal Society of Edinburgh, and the Alexander von Humboldt Foundation through the collaborative grant between the TU Kaiserslautern and the Institute of Theoretical Physics and Astronomy of Vilnius University.

[1] V. Bretin, S. Stock, Y. Seurin, and J. Dalibard, Phys. Rev. Lett. **92**, 050403 (2004).
[2] V. Schweikhard, I. Coddington, P. Engels, V. P. Mogendorff, and E. A. Cornell, Phys. Rev. Lett. **92**, 040404 (2004).
[3] M. A. Baranov, K. Osterloh, and M. Lewenstein, Phys. Rev. Lett. **94**, 070404 (2005).
[4] G. Juzeliūnas and P. Öhberg, Phys. Rev. Lett. **93**, 033602 (2004).
[5] G. Juzeliūnas, P. Öhberg, J. Ruseckas, and A. Klein, Phys. Rev. A **71**, 053614 (2005).
[6] G. Juzeliūnas, J. Ruseckas, and P. Öhberg, J. Phys. B **38**, 4171 (2005).
[7] E. Arimondo, Prog. Opt. **35**, 259 (1996).
[8] S. E. Harris, Phys. Today **50**(7), 36 (1997).
[9] A. B. Matsko, O. Kocharovskaja, Y. Rostovtsev, G. R. Welch, A. S. Zibrov, and M. O. Scully, Adv. At., Mol., Opt. Phys. **46**, 191 (2001).
[10] M. D. Lukin, Rev. Mod. Phys. **75**, 457 (2003).
[11] M. Fleischhauer, A. Imamoglu, and J. P. Marangos, Rev. Mod. Phys. **77**, 633 (2005).
[12] R. Dum and M. Olshanii, Phys. Rev. Lett. **76**, 1788 (1996).
[13] R. Folman, P. Kruger, J. Schmiedmayer, J. Denschlag, and C. Henkel, Adv. At., Mol., Opt. Phys. **48**, 263 (2002).
[14] R. Jackiw, Comments At. Mol. Phys. **21**, 71 (1988).
[15] C.-P. Sun and M.-L. Ge, Phys. Rev. D **41**, 1349 (1990).
[16] L. V. Hau, S. E. Harris, Z. Dutton, and C. Behrooz, Nature (London) **397**, 594 (1999).
[17] S. Gupta, K. W. Murch, K. L. Moore, T. P. Purdy, and D. M. Stamper-Kurn, Phys. Rev. Lett. **95**, 143201 (2005).

Effect of Induced Spin-Orbit Coupling for Atoms via Laser Fields

Xiong-Jun Liu, Mario F. Borunda, Xin Liu, and Jairo Sinova
Department of Physics, Texas A&M University, College Station, Texas 77843-4242, USA
(Received 5 September 2008; published 26 January 2009)

We propose an experimental scheme to observe spin-orbit coupling effects of a two-dimensional Fermi atomic gas cloud by coupling its internal electronic states (pseudospins) to radiation in a Lambda configuration. The induced spin-orbit coupling can be of the Dresselhaus and Rashba type with a Zeeman term. We show that the optically induced spin-orbit coupling can lead to a spin-dependent effective mass under appropriate conditions with one of them able to be tuned between positive and negative effective masses. As a direct observable we show that in the expansion dynamics of the atomic cloud the initial atomic cloud splits into two clouds for the positive effective mass case regime, and into four clouds for the negative effective mass regime.

DOI: 10.1103/PhysRevLett.102.046402 PACS numbers: 71.70.Ej

Spin-orbit (SO) coupling effects in semiconductors have emerged in the solid-state community as a very active field of research, fueled in part by the field of spintronics [1], e.g., the engineering of devices where the spin degree of freedom of the electron is exploited for improved functionality. This has led to new developments in the anomalous Hall effect [2] and the spin Hall effect (SHE) [3,4]. In correspondence to the spin of an electron, the internal degree of freedom of an atom (pseudospin) is represented by the superposition of its electronic states (hyperfine levels). SO coupling can be equivalently depicted as the interaction between an effective non-Abelian gauge potential and a particle with (pseudo)spin. In quantum systems, the idea generating a gauge field adiabatically was proposed by Wilczek and Zee more than 20 years ago [5]. Recently, such an idea was applied to atomic systems, where the motion of atoms in a position-dependent laser configuration gives rise to an effective non-Abelian gauge potential [6–10], which can lead to an effective SO interaction in an ultracold atomic gas [11–13].

Realization of SO interaction in atomic gases opens new possibilities of studying spintronic effects, e.g., spin relaxation [11], *Zitterbewegung* [12], and SHE, in atomic systems which provide an extremely clean environment, allowing in a controllable fashion unique access to the study of complex physics. However, experimental detection of such SO effects in atoms requires one to measure the pseudospins (not just hyperfine levels) that are usually not directly observable for atomic systems. In this Letter, we propose an experimental scheme to study SO coupling effects, based on a trapped two-dimensional (2D) Fermi atomic gas with a simple internal three-level Λ-type setup. We demonstrate that an effective SO interaction, e.g., Rashba and linear Dresselhaus terms, can be obtained by coupling atoms with a three-level configuration to spatially varying laser fields. The optically induced SO coupling can lead to a spin-dependent effective mass under proper condition. A direct observable of this effect is in the expansion dynamics for each of the effective mass cases after the external trap is switched off, and we predict that the initial atomic cloud splits into two or four clouds.

We consider a cloud of quasi-2D (y-z plane) Fermi atomic gas with internal three-level Λ-type configuration [see Fig. 1(a)] coupled to radiation. The transition $|b\rangle \rightarrow |a\rangle$ is coupled by the laser field with Rabi frequency $\Omega_1 = \Omega_{10}\exp[i\phi_1(\mathbf{r})]$, and the transition $|c\rangle \rightarrow |a\rangle$ is coupled by another laser field $\Omega_2 = \Omega_{20}\exp[i\phi_2(\mathbf{r})]$, where $\phi_{1,2}(\mathbf{r})$ are position-dependent phases. The Hamiltonian of a single particle reads as $H = H_0 + V_{\text{trap}}(\mathbf{r}) + H_I$, where $V_{\text{trap}}(\mathbf{r})$ is the 2D trap, and

$$H_I = \hbar\Delta|a\rangle\langle a| - (\hbar\Omega_1|a\rangle\langle b| + \hbar\Omega_2|a\rangle\langle c| + \text{H.c.}). \quad (1)$$

Diagonalizing the interacting Hamiltonian H_I yields three eigenstates: $|\chi_D\rangle = (\sin\theta)|b\rangle - (\cos\theta)e^{i\phi}|c\rangle$, $|\chi_{B_1}\rangle = (\cos\alpha)(\cos\theta)|b\rangle + (\cos\alpha)(\sin\theta)e^{i\phi}|c\rangle + (\sin\alpha)e^{i\phi_1}|a\rangle$, and $|\chi_{B_2}\rangle = (\sin\alpha)(\cos\theta)|b\rangle + (\sin\alpha)(\sin\theta)e^{i\phi}|c\rangle - (\cos\alpha)e^{i\phi_1}|a\rangle$. Here $\phi = \phi_1 - \phi_2$, the mixing angles $\tan\theta = \Omega_{20}/\Omega_{10}$ and $\tan\alpha = \sqrt{\Omega_{10}^2 + \Omega_{20}^2}/\Delta \equiv \Omega_0/\Delta$. The corresponding eigenvalues are $E_D = 0$ and $E_{B_{1,2}} = \hbar(\Delta \mp \sqrt{\Delta^2 + 4\Omega_0^2})/2$. Since spatially varying lasers are employed, diagonalization of the interacting Hamiltonian H_I leads to a SU(3) gauge potential [6–9]. We consider the large detuning case, $\Delta^2 \gg \Omega_0^2$, where $|E_D - E_{B_1}| \ll \Omega_0$

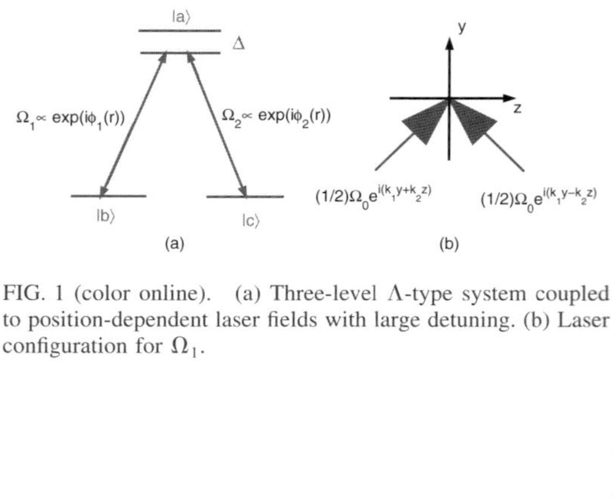

FIG. 1 (color online). (a) Three-level Λ-type system coupled to position-dependent laser fields with large detuning. (b) Laser configuration for Ω_1.

and $|\chi_D\rangle, |\chi_{B_1}\rangle$ spans a near-degenerate subspace, with their eigenvalues far separated from that of E_{B_2}. We can then apply the adiabatic condition by neglecting the state $|\chi_{B_2}\rangle$, which leads to a U(2) non-Abelian adiabatic gauge potential based on the near-degenerate subspace spanned by $|\chi_{D,B_1}\rangle$ [14]. This situation is different from the cases in Refs. [9,11] where the adiabatic condition is also assumed between the states $|\chi_D\rangle$ and $|\chi_{B_1}\rangle$, and thus the spin-dependent gauge potential is still Abelian in their case [9]. The adiabatic non-Ablelian gauge potential for the present near-degenerate subspace is obtained via

$$\mathbf{A}(\mathbf{r}) = i\frac{\hbar c}{e}\langle\chi_D| \otimes \langle\chi_{B_1}|\nabla|\chi_{B_1}\rangle \otimes |\chi_D\rangle. \tag{2}$$

For our purpose, we shall set the parameters $\phi_1 = \phi_2 = k_1 y$ and $\theta = k_2 z$, which means $\Omega_{1,2}$ are standing waves in the z direction but plane waves in the y direction. Such configuration can be achieved by applying two laser fields for each atomic transition. For Ω_1, for instance, we can set two laser fields with the same strength, where one travels with the wave vector $k_1\hat{\mathbf{e}}_y + k_2\hat{\mathbf{e}}_z$ and the other travels with $k_1\hat{\mathbf{e}}_y - k_2\hat{\mathbf{e}}_z$ [see Fig. 1(b)]. The Rabi frequencies are then given by $\Omega_1(\mathbf{r}) = \Omega_0\cos(k_2 z)e^{ik_1 y}$ and $\Omega_2(\mathbf{r}) = \Omega_0\sin(k_2 z)e^{ik_1 y}$. For simplicity, in what follows, we use the spin language and denote by $|\uparrow\rangle = |\chi_D\rangle$, $|\downarrow\rangle = |\chi_{B_1}\rangle$. Under this configuration the gauge potential (2) can be recast into $\mathbf{A} = m\lambda_1\sigma_z\hat{\mathbf{e}}_y - m\lambda_2\sigma_y\hat{\mathbf{e}}_z - m\lambda_1 I\hat{\mathbf{e}}_y$ with the coefficients $\lambda_1 = \hbar k_1\Omega_0^2/(2m\Delta_0^2)$ and $\lambda_2 = \hbar k_2/m$. In addition, the scalar potentials are given by $\varphi_{\uparrow\uparrow} = \frac{\hbar^2}{2m}\frac{\Omega_0^2}{\Delta^2}k_2^2$, $\varphi_{\downarrow\downarrow} = \frac{\hbar^2}{2m}\frac{\Omega_0^2}{\Delta^2}k_1^2$, and $\varphi_{\downarrow\uparrow} = 0$.

To this step we can obtain the effective Hamiltonian for the near-degenerate subspace with SO coupling in the form (neglecting constant terms):

$$H = H_0 + H_{\mathrm{SO}} + H_z + V_{\mathrm{trap}}, \tag{3}$$

where $H_0 = \frac{P_y^2}{2m} + \frac{P_z^2}{2m}$, $H_{\mathrm{SO}} = -\lambda_1\sigma_z P_y + \lambda_2\sigma_y P_z$, $H_z = M_0\sigma_z$ with $M_0 = \frac{\hbar^2}{4m}\frac{\Omega_0^2}{\Delta^2}k_2 + \hbar\frac{\Omega_0^2}{2\Delta}$, and the 2D harmonic trap $V_{\mathrm{trap}} = \frac{1}{2}m\omega^2(y^2 + z^2)$. A Hamiltonian of this form is predicted to give SHE [3] for $M_0 = 0$ and planar Hall effect [15] for $M_0 \neq 0$ in solid-state systems.

The term $H_{\mathrm{SO}} + H_z$ can be readily diagonalized in the momentum space, $(H_{\mathrm{SO}} + H_z)^{\mathrm{diag}} = \hbar\sqrt{[(M_0/\hbar) - \lambda_1 k_y]^2 + \lambda_2^2 k_z^2}\sigma_z$, where $k_{y,z}$ are wave vectors of atoms in the y and z directions. The associated eigenstates are given by $|+\rangle = [\cos\vartheta/2, i\sin\vartheta/2]^T$, $|-\rangle = [i\sin\vartheta/2, \cos\vartheta/2]^T$ with $\tan\vartheta = \lambda_2\hbar k_z/(M_0 - \lambda_1\hbar k_y)$. However, even in this situation, the full Hamiltonian H is not diagonalizable. Practically, we can consider the case where $M_0^2 \gg \lambda_2^2\hbar^2(k_F^z)^2 \gg \lambda_1^2\hbar^2(k_F^y)^2$ (the validity in realizable experimental setups will be discussed below), $k_F^{y,z}$ are the y/z components of the Fermi momenta, and expand the term $(H_{\mathrm{SO}} + H_z)^{\mathrm{diag}}$ to the k^2 order: $(H_{\mathrm{SO}} + H_z)^{\mathrm{diag}} \approx (M_0 + \frac{\lambda_2^2 P_z^2}{2M_0} - \lambda_1 P_y)\sigma_z$. Substituting this result into Eq. (3) yields

$$H^{(1)} = \frac{1}{2m}(P_y - \lambda_1 m\sigma_z)^2 + \frac{1}{2\tilde{m}}P_z^2 + M_0\sigma_z + V_{\mathrm{trap}}, \tag{4}$$

where $\frac{1}{\tilde{m}} = \frac{1}{m} + \frac{\lambda_2^2}{M_0}\sigma_z$. Equation (4) shows that the SO coupling leads to a spin-dependent effective mass of atoms moving in the z direction, with $\tilde{m}_\pm = (1 \pm \delta m/m)^{-1}m$ and $\delta m = \hbar^2 k_2^2/M_0$. The eigenfunction of Eq. (4) reads

$$\Psi^\eta_{n_y,n_z}(y,z) = \psi_{n_y}(\omega_y, y)\tilde{\psi}^\eta_{n_z}(\omega_z^\eta, z)e^{i\eta\lambda_1 my/\hbar}|\eta\rangle, \tag{5}$$

with $|\eta\rangle = |\pm\rangle$. Here ψ_{n_y}, $\tilde{\psi}^\eta_{n_z}$ are harmonic oscillator wave functions. The eigenvalues are given by $E^\eta_{n_y,n_z} = (n_y + \frac{1}{2})\hbar\omega_y + (n_z + \frac{1}{2})\hbar\omega_z^\eta + \eta M_0 - \frac{1}{2}m\lambda_1^2$, where $\omega_y = \omega$, $\omega_z^\eta = \omega\sqrt{1 + \eta\delta m/m}$ are spin-dependent effective trap frequencies due to SO coupling, and n_y, n_z are integers.

The results (4) and (5) are valid for the situation $\delta m < m$, where the effective mass of particles in state $|-\rangle$ is positive, i.e., $\tilde{m}_- > 0$. In that case the expansion of the Hamiltonian up to k^2 order suffices and its validity can be verified by comparing with the exact expression. On the other hand, the negative effective mass regime for the $|-\rangle$ state can be reached when $\delta m > m$. In this case one can check the dispersion relation $\varepsilon_k^- = H_0 + H_{\mathrm{SO}}^{(-)} + H_z^{(-)}$ (for atoms in $|-\rangle$) represents a double-well potential in the k_z axis; therefore a higher-order expansion with momentum k_z is required to derive the effective Hamiltonian. Similarly, we consider that $\lambda_2^4\hbar^4(k_F^z)^4 \ll M_0^4$ and can then expand ε_k^- up to the k_z^4 order, which gives a double well ϕ^4-type potential form in k_z momentum space ($\tilde{m}_- < 0$), i.e., $\varepsilon_{k_z}^- = \frac{\lambda_2^4}{8M_0^3}P_z^4 + \frac{1}{2\tilde{m}_-}P_z^2$ [16]. Higher-order terms in the expansion can equivalently lead to small corrections to the coefficients of k_z^2 and k_z^4 terms. Finally we get an effective Hamiltonian for atoms at state $|-\rangle$:

$$\begin{aligned} H_-^{(2)} = {} & \frac{1}{2m}(P_y + \lambda_1 m)^2 + \frac{\lambda_2^4}{8M_0^3}(1 + \gamma_1)P_z^4 \\ & + \frac{1}{2\tilde{m}_-}(1 + \gamma_2)P_z^2 - M_0 + \frac{1}{2}m\omega^2\mathbf{r}^2 - \frac{1}{2}m\lambda_1^2. \end{aligned} \tag{6}$$

Here the coefficient correction $\gamma_1 = \frac{4}{(\delta m/m)[1 + (\delta m/m)]^2} - 1$, $\gamma_2 = \frac{2}{1 + (\delta m/m)} - 1$ are small under the condition $\lambda_2^4\hbar^4(k_F^z)^4 \ll M_0^4$, say, $\gamma_{1,2}^2 \ll 1$ in the present case.

Unlike the positive mass regime, calculation of the Fermi energy in the negative mass case cannot be done analytically. Nevertheless, we are interested in systems with a large number of atoms, where the Thomas-Fermi (TF) approximation is suitable [17–19]. Then the Fermi energy can be calculated by solving the equation

$$N = \sum_{\eta=+,-}\int_{k<k_F^\eta, r<R_F^\eta} d^2\mathbf{r}\,d^2\mathbf{k}\,\rho_\eta^{(2)}(\mathbf{r}, \mathbf{k}, T = 0), \tag{7}$$

where the atomic distribution function in phase space is given by $\rho_\pm^{(2)} = (2\pi)^{-2}(e^{\beta(H_\pm^{(2)}(\mathbf{r},\mathbf{k}) - \mu_F)} + 1)^{-1}$ with

$\beta = 1/k_B T$, and the initial size of the atomic cloud is given by $R_F^{\mp} = (2\varepsilon_F \mp 2M_0 - 2\min\{\varepsilon_{k_z}^{\pm}\})^{1/2}/(m^{1/2}\omega)$. From Eq. (7) the relation of Fermi energy to the number of atoms N, trap frequency, and SO coupling strength can be obtained numerically as shown in Fig. 2.

There are specific observables of this spin-dependent effective mass induced by the spin-orbit coupling. First, the anisotropy of the effective mass can lead to the anisotropic momentum distribution. Considering the Thomas-Fermi approximation, at zero temperature and for the positive mass case, the momentum distribution is given by $n_{\pm}(\mathbf{k}) = \frac{1}{2\pi m\omega^2}[\varepsilon_F \mp M_0 - \frac{\hbar^2}{2m}(k_y^2 + \frac{m}{\tilde{m}_{\pm}}k_z^2)]$. This contrasts with the result for the usual isotropic mass, where the atomic distribution in the momentum space is always isotropic irrespective of the shape of the trap [17]. For the negative effective mass case, $n_+(\mathbf{k})$ has the same form, while $n_-(\mathbf{k}) = \frac{1}{2\pi m\omega^2}\{\varepsilon_F + M_0 - \frac{\hbar^2}{2m}[k_y^2 + \frac{m}{\tilde{m}_-}(1 + \gamma_2)k_z^2 + \frac{m\lambda_2^4\hbar^2}{4M_0^3}(1 + \gamma_1)k_z^4]\}$. Also, a fully polarized Fermi gas is obtained when $\varepsilon_F < M_0$, and $n_+(\mathbf{k}) = 0$. The anisotropy of the momentum distribution can lead to anisotropy of the Fermi velocity in the y-z plane. For instance, when $\delta m/m = 3/4$ and $\varepsilon_F < M_0$, we have $v_F^z \approx v_F^y/2$. The anisotropy of Fermi velocity can be directly detected by time-of-flight absorption [18,19].

However, the dramatic signature of SO effects in the present Fermi atomic gas resides in the expansion dynamics of the atomic cloud after the 2D external trap and the laser fields are switched off. The evolution of the atomic distribution in phase space $\rho(\mathbf{r}, \mathbf{k}, T, t)$ can be calculated by the Boltzmann transport equation assuming that at $t = 0$, $V_{\text{trap}} \to 0$ and $M_0 \to 0$. For the present cold dilute noninteracting Fermi gas, the evolution of $\rho_{b,c}(\mathbf{r}, \mathbf{k}, T, t)$ is followed by the ballistic law $\rho_{b,c}(\mathbf{r}, \mathbf{k}, T, t > 0) = \rho_{b,c}(\mathbf{r} - \hbar\mathbf{k}t/m, \mathbf{k}, T, 0)$ [here $|\mathbf{k}, \nu\rangle$, $(\nu = b, c)$ are the eigenstates after the trap and laser fields are turned off]. The temporal atomic spatial density can be calculated by $n_{b,c}(\mathbf{r}, T, t) = \int d^2\mathbf{k}\rho_{b,c}(\mathbf{r} - \hbar\mathbf{k}t/m, \mathbf{k}, T, 0)$. The eigenstates $|\mathbf{k}, \alpha\rangle$, $(\alpha = b, c)$ are related to the initial pseudospin basis $|\pm\rangle$ by $|\mathbf{k}, b\rangle \approx \frac{1}{2}|\mathbf{k} + k_2\hat{\mathbf{e}}_z, +\rangle + \frac{1}{2}|\mathbf{k} - k_2\hat{\mathbf{e}}_z, +\rangle + \frac{1}{2i}|\mathbf{k} + k_2\hat{\mathbf{e}}_z, -\rangle - \frac{1}{2i}|\mathbf{k} - k_2\hat{\mathbf{e}}_z, -\rangle$ and $|\mathbf{k}, c\rangle \approx \frac{1}{2}|\mathbf{k} + k_2\hat{\mathbf{e}}_z, -\rangle + \frac{1}{2}|\mathbf{k} - k_2\hat{\mathbf{e}}_z, -\rangle - \frac{1}{2i}|\mathbf{k} + k_2\hat{\mathbf{e}}_z, +\rangle + \frac{1}{2i}|\mathbf{k} - k_2\hat{\mathbf{e}}_z, +\rangle$. We find that the atomic density is

$$n_b(\mathbf{r}, T, t) = \sum_{\eta=\pm}\int\frac{d^2\mathbf{k}}{4}\left\{\rho_{\eta}^{(j)}\left[\mathbf{r} - \frac{\hbar t}{m}(\mathbf{k} + \eta k_2\hat{\mathbf{e}}_z), \mathbf{k}, T, 0\right] + \rho_{\eta}^{(j)}\left[\mathbf{r} - \frac{\hbar t}{m}(\mathbf{k} - \eta k_2\hat{\mathbf{e}}_z), \mathbf{k}, T, 0\right]\right\}, \tag{8}$$

where $\rho_{\pm}^{(j)}(\mathbf{r}, \mathbf{k}, T, 0)$ denote the distribution functions for atoms in the state $|\pm\rangle$ for the positive mass ($j = 1$) and negative mass ($j = 2$) cases. The function $n_c(\mathbf{r}, T, t)$ can be calculated in the same way. Practically, we can assume that before the expansion begins $\varepsilon_F < M_0$, and then $\rho_+^{(j)}(r, \mathbf{k}, T, 0) = 0$. For the positive mass case, the evolution of the atomic density can be calculated exactly:

$$n_b(\mathbf{r}, T, t) = \frac{\sqrt{m\tilde{m}_-}}{8(1 + \omega^2 t^2)\beta\pi\hbar^2}\left(\frac{1 + \omega^2 t^2}{1 + \frac{\tilde{m}_-}{m}\omega^2 t^2}\right)^{1/2} \times \left(\ln\frac{1 + e^{-\beta\tilde{E}_+(\mathbf{r},t)}}{e^{-\beta\tilde{E}_+(\mathbf{r},t)}} + \ln\frac{1 + e^{-\beta\tilde{E}_-(\mathbf{r},t)}}{e^{-\beta\tilde{E}_-(\mathbf{r},t)}}\right), \tag{9}$$

where $\tilde{E}_{\pm}(\mathbf{r}, t) = \mu_F + M_0 - \frac{m\omega^2 y^2}{2(1+\omega^2 t^2)} - \frac{m\omega^2(z \pm \hbar k_2 t/m)^2}{2[1 + (\tilde{m}_-/m)\omega^2 t^2]}$. One can verify that at $t = 0$ the maximum point of n_b is obtained at $\mathbf{r} = 0$, while after a sufficiently long time there are two maximum points at $y = 0$, $z = \pm\hbar k_2 t/m$. As a result, Eq. (9) represents an initial atomic cloud that splits into two clouds each moving in opposite directions with group velocities $v_g = \pm\hbar k_2\hat{e}_z/m$. Using typical parameters, $M_0 = 10^6\hbar \cdot \text{s}^{-1}$, $m \approx 0.963 \times 10^{-26}$ kg (^{6}Li atoms), and $k_2 = 0.87 \times 10^7\ \text{m}^{-2}$, one finds $\tilde{m}_- \approx 4$ m and $v_g \approx \pm 9.0$ cm/s.

The evolution of the atomic density in the negative effective mass regime is more complicated. Again, assuming $\varepsilon_F < M_0$, we obtain from Eq. (8) the temporal atomic density as (denoted by $\tilde{z}_\eta = z \pm \hbar k_2 t/m$):

$$n_b(\mathbf{r}, T, t) = \sum_{\eta=+,-}\int\frac{d^2\mathbf{k}}{16\pi^2}\left\{1 + \exp\left[\beta\left(\frac{1 + \omega^2 t^2}{2m}\hbar^2 k_y^2 + \frac{\Gamma}{2m\hbar^2}\left(\hbar k_z + \frac{\tilde{m}_-\omega^2 t\tilde{z}_\eta}{1 + \frac{\tilde{m}_-}{m}\omega^2 t^2}\right)^4 + \frac{m + \tilde{m}_-\omega^2 t^2}{2m\tilde{m}_-}\hbar^2 k_z^2 + \frac{m\omega^2 y^2}{2 + 2\omega^2 t^2} + \frac{m^2\omega^2\tilde{z}_\eta^2}{2m + 2\tilde{m}_-\omega^2 t^2} - M_0 - \mu_F\right)\right]\right\}^{-1}, \tag{10}$$

where $\Gamma = m\lambda_2^4\hbar^2(1 + \gamma_1)/(4M_0^3)$. To make a qualitative analysis, we first calculate the time independent momentum distribution: $n_b(\mathbf{k}, T, t) = \frac{1}{8\pi m\omega^2}(\ln\frac{1+e^{-\beta\tilde{\mathcal{E}}_+(k)}}{e^{-\beta\tilde{\mathcal{E}}_+(k)}} + \ln\frac{1+e^{-\beta\tilde{\mathcal{E}}_-(k)}}{e^{-\beta\tilde{\mathcal{E}}_-(k)}})$. Here $\tilde{\mathcal{E}}_{\pm}(\mathbf{k}) = \mu_F + M_0 + \frac{\hbar^2\Gamma\Upsilon^2}{2m} - \frac{\hbar^2}{2m} \times \{k_y^2 + \Gamma[(k_z \pm k_2)^2 - \Upsilon]^2\}$ with $\Upsilon = -m(1 + \gamma_2)/(2\tilde{m}_-\Gamma)$. Note that if $k_2 \neq \sqrt{\Upsilon}$, $n_b(\mathbf{k}, T, t)$ has four distinct maximums at $k_z = -k_2 - \sqrt{\Upsilon}$, $-k_2 + \sqrt{\Upsilon}$, $k_2 - \sqrt{\Upsilon}$, $k_2 + \sqrt{\Upsilon}$, which indicates that the initial atomic cloud is com-

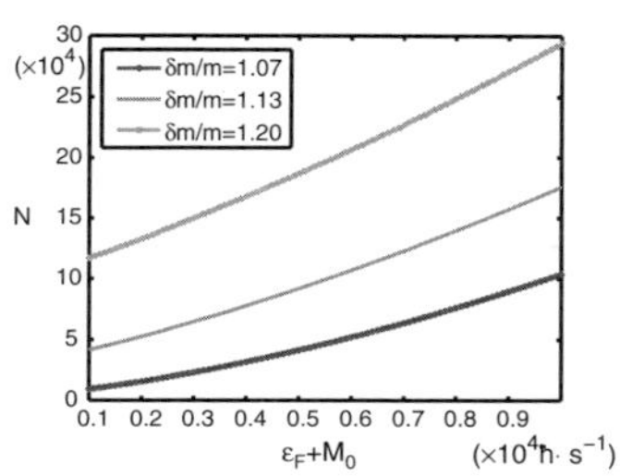

FIG. 2 (color online). Number of atoms versus Fermi energy at $\delta m/m = 1.07, 1.13, 1.20$, corresponding to the effective mass $\tilde{m}_- = -15m, -7.7m, -5m$.

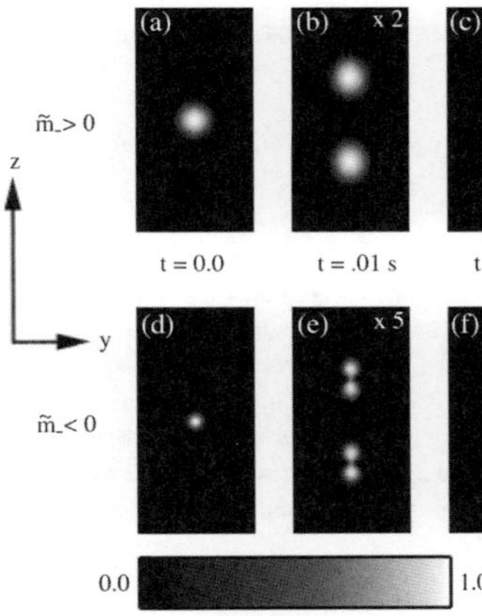

FIG. 3 (color online). Splitting of atomic cloud for the positive effective mass case (a)–(c) and negative mass case (d)–(f) after turning off the trap and laser fields. The $\times n$ label represents the value each panel has been multiplied by to keep figure to scale.

posed of four overlapping clouds that will travel in the z axis at different speeds. Using parameters similar to the previous system, $m \approx 0.963 \times 10^{-26}$ kg (^{6}Li atoms), $k_2 = 1.0 \times 10^7\ \mathrm{m}^{-1}$, and $M_0 = 0.90 \times 10^6 \hbar \cdot \mathrm{s}^{-1}$, we find that $\Gamma \approx 4.43 \times 10^{-15}\ \mathrm{m}^2$ and $\Upsilon = 8.9 \times 10^{12}\ \mathrm{m}^{-2}$. Thus the four maximums correspond to $v_g = \pm 13.4, \pm 7.4$ cm/s.

Figure 3 displays numerical estimates of the splitting. The time at which such splitting can be observed is found by estimating the initial size of the atomic cloud R_F. For a system with $N \approx 10^{4-5}$ atoms and a trap frequency $\omega \sim 35$ Hz, $R_F \approx 230\ \mu$m in the positive effective mass case and $R_F \approx 127.8\ \mu$m when the effective mass is negative, assuming $T < T_F$. As a result, for the positive effective mass regime, one can verify that the two atomic clouds will be fully separated after the system evolves by $\Delta t \sim 5$ ms. For the negative mass regime, the time is about $\Delta t \sim 9$ ms. It should be emphasized that measurement of the present expansion dynamics needs to detect the density of atoms in the hyperfine level $|b\rangle$ (or $|c\rangle$), rather than to detect the pseudospin states $|\pm\rangle$ that are not differentiable for the atomic system, thus the SO effect obtained here is directly observable in experiments by direct imaging of the separated atomic clouds.

Now we confirm the validity of the approximation $M_0^2 \gg \lambda_2^2 \hbar^2 (k_F^z)^2 \gg \lambda_1^2 \hbar^2 (k_F^y)^2$. The second inequality is always valid, since $\lambda_2 \gg \lambda_1$ in the large detuning case. Note that k_F^z can be calculated by $k_F^z = \sqrt{\Upsilon + 2m(\tilde{\varepsilon}_F + \hbar^2 \Gamma \Upsilon^2/2m)/(\hbar^2 \Gamma)}$ for $\tilde{m}_- < 0$ and $k_F^z = \sqrt{2\tilde{m}_- \tilde{\varepsilon}_F / \hbar^2}$ for $\tilde{m}_- > 0$. For the system composed of $N \sim 10^4$ atoms and a trap frequency $\omega = 35$ Hz, and using the previously employed parameters, we find $k_F^z \approx 1.41 \times 10^6\ \mathrm{m}^{-1}$ with $\lambda_2^2 \hbar^2 (k_F^z)^2 / M_0^2 \approx 0.019 \ll 1$ for $\tilde{m}_- > 0$, and $k_F^z \approx 3.55 \times 10^6\ \mathrm{m}^{-1}$ with $\lambda_2^4 \hbar^4 (k_F^z)^4 / M_0^4 \approx 0.028 \ll 1$ for $\tilde{m}_- < 0$ (note for the case $\tilde{m}_- < 0$ the approximation is up to the k_z^4 order). Thus, the first inequality is also valid. Finally we estimate the TF approximation that is considered in the calculation. TF approximation fails in a small periphery region $R_F - \delta R < r < R_F$ of the atomic cloud [17]. For the 2D Fermi atom gas, one can find that the ratio of δR to the atomic cloud size R_F satisfies $\delta R / R_F \sim N^{-1/2}$. Therefore such a small region can be safely neglected for the case with a large number of atoms, say $N > 10^4$. Note that this model can be readily extended to Bose-Einstein condensate systems [20], where, together with the atom-atom interaction, the SO coupling may lead to intriguing new physics.

In conclusion, we have proposed an experimental scheme to study SO coupling effects for a cloud of a 2D trapped Fermi gas. Under certain conditions, the optically induced SO coupling in atoms leads to a spin-dependent effective mass which can be positive or negative. In the expansion dynamics of the atomic cloud after switching off the trap, it is shown that the initial atomic cloud splits into two or four clouds moving oppositely depending on tunable spin-orbit coupling parameters. The present scheme provides an applicable way to directly observe the SO coupling in cold Fermi atoms.

We gratefully acknowledge discussions with V. Galastki. This work was supported by ONR under Grant No. ONR-N000140610122, by NSF under Grant No. DMR-0547875, and by SWAN-NRI. J. S. is a Cottrell Scholar of the Research Corporation.

[1] S. A. Wolf *et al.*, Science **294**, 1488 (2001).
[2] T. Jungwirth *et al.*, Phys. Rev. Lett. **88**, 207208 (2002).
[3] S. Murakami *et al.*, Science **301**, 1348 (2003); J. Sinova *et al.*, Phys. Rev. Lett. **92**, 126603 (2004).
[4] Y. K. Kato *et al.*, Science **306**, 1910 (2004); J. Wunderlich *et al.*, Phys. Rev. Lett. **94**, 047204 (2005).
[5] F. Wilczek and A. Zee, Phys. Rev. Lett. **52**, 2111 (1984).
[6] J. Ruseckas *et al.*, Phys. Rev. Lett. **95**, 010404 (2005).
[7] K. Osterloh *et al.*, Phys. Rev. Lett. **95**, 010403 (2005).
[8] X. -J. Liu *et al.*, Phys. Rev. Lett. **98**, 026602 (2007).
[9] S.-L. Zhu *et al.*, Phys. Rev. Lett. **97**, 240401 (2006).
[10] Yong Li *et al.*, Phys. Rev. Lett. **99**, 130403 (2007).
[11] T. D. Stanescu, C. W. Zhang, and V. Galitski, Phys. Rev. Lett. **99**, 110403 (2007).
[12] J. Y. Vaishnav *et al.*, Phys. Rev. Lett. **100**, 153002 (2008).
[13] A. M. Dudarev *et al.*, Phys. Rev. Lett. **92**, 153005 (2004).
[14] Note that in the large detuning case both the states $|\chi_{D,B_1}\rangle$ decouple from the excited state $|a\rangle$ since $\sin\alpha \ll 1$, and then the decay of them is negligible.
[15] H. X. Tang *et al.*, Phys. Rev. Lett. **90**, 107201 (2003).
[16] Y. Shin, Ph.D. thesis, Massachusetts Institute of Technology, 2005.
[17] D. A. Butts and D. S. Rokhsar, Phys. Rev. A **55**, 4346 (1997).
[18] B. DeMarco and D. S. Jin, Science **285**, 1703 (1999).
[19] S. R. Granade *et al.*, Phys. Rev. Lett. **88**, 120405 (2002).
[20] T. Stanescu *et al.*, Phys. Rev. A **78**, 023616 (2008).

Bose-Einstein Condensate in a Uniform Light-Induced Vector Potential

Y.-J. Lin, R. L. Compton, A. R. Perry, W. D. Phillips, J. V. Porto, and I. B. Spielman*
Joint Quantum Institute, National Institute of Standards and Technology, and University of Maryland, Gaithersburg, Maryland, 20899, USA
(Received 17 September 2008; published 30 March 2009)

We use a two-photon dressing field to create an effective vector gauge potential for Bose-Einstein-condensed ^{87}Rb atoms in the $F = 1$ hyperfine ground state. These Raman-dressed states are spin and momentum superpositions, and we adiabatically load the atoms into the lowest energy dressed state. The effective Hamiltonian of these neutral atoms is like that of charged particles in a uniform magnetic vector potential whose magnitude is set by the strength and detuning of the Raman coupling. The spin and momentum decomposition of the dressed states reveals the strength of the effective vector potential, and our measurements agree quantitatively with a simple single-particle model. While the uniform effective vector potential described here corresponds to zero magnetic field, our technique can be extended to nonuniform vector potentials, giving nonzero effective magnetic fields.

DOI: 10.1103/PhysRevLett.102.130401 PACS numbers: 03.75.Lm, 67.85.Hj

Ultracold atoms are an appealing system for the study of many-body correlated states relevant to condensed matter physics. These widely tunable, and nearly disorder-free systems have already realized one-dimensional Tonks-Girardeau gases [1], and the superfluid to Mott-insulator transition of a Bose-Einstein condensate (BEC) in an optical lattice [2]. The implementation of these simple iconic condensed matter systems paves the way for more interesting systems with exotic correlations and excitations, as in the fractional quantum Hall effect (FQHE) of a two-dimensional electron system in a strong magnetic field [3]. Simulating these systems with neutral atoms requires an effective Lorentz force, associated with a vector gauge potential. Current experimental approaches involve rotation of trapped BECs [4,5], where low field effects, such as the formation of an Abrikosov vortex lattice, have been observed. For technical reasons, this approach is limited to modest effective fields, too small for FQHE physics [5,6]. Most recent proposals to create significantly larger effective magnetic fields without rotation [7–10] involve optical coupling between internal atomic states, where the atoms are dressed in a spatially dependent manner. The effective magnetic field in such light-induced gauge potentials can be understood as a consequence of changing into a spatially varying basis of internal states. Even absent spatial gradients, light-induced gauge potentials can lead to remarkable physical effects such as an anisotropic Bogoliubov excitation spectrum [10], double well structures in the free-particle dispersion relation [11] (which we observe), and the spin-Hall effect in ultracold atoms [9].

Here we report the first observation of light-induced vector gauge potentials, produced by dressing a BEC with two counterpropagating Raman laser beams. The Raman beams couple internal (spin) states with linear momenta differing by twice the photon momentum. This gives rise to a spatial gradient of the phase difference between spin components of the dressed state. As we will show, this spatially varying state leads to a nonzero vector potential when the coupling is detuned from Raman resonance. We adiabatically load the BEC into the dressed state, and measure properties of the dressed-state dispersion relation by probing its spin and momentum decomposition. While the atoms are stationary in the lab frame (i.e., zero group velocity), the momenta of the individual spin components of the dressed state show a nonzero phase velocity depending on the strength and detuning of the Raman coupling. Our measurements agree with a single-particle model, and demonstrate the presence of an effective vector potential.

We dress a ^{87}Rb BEC in the $F = 1$ ground state with two Raman laser beams counterpropagating along $\hat{x}$. These beams couple states $|m_F, k_x\rangle$ differing in internal angular momentum by $\hbar$ ($\Delta m_F = \pm 1$), and differing in linear momentum $\hbar k_x$ by $2\hbar k_r$. Here, $\hbar k_r = h/\lambda$ is the single-photon recoil momentum, and λ is the wavelength of the Raman beams. We define $E_r = \hbar^2 k_r^2/2m$ as the recoil energy. The family of three states coupled by the Raman field, $\Psi(\tilde{k}_x) = \{|-1, \tilde{k}_x + 2k_r\rangle, |0, \tilde{k}_x\rangle, |+1, \tilde{k}_x - 2k_r\rangle\}$, is labeled by the wave vector of the quasimomentum, $\tilde{k}_x$. The Raman beams have frequencies ω_L and $\omega_L + \Delta\omega_L$, and a bias field $B_0\hat{y}$ produces a Zeeman shift $\hbar\omega_Z = g\mu_B B_0 \simeq \hbar\Delta\omega_L$ [see Figs. 1(a) and 1(b)]. Since the momentum transfer is along $\hat{x}$, the single-particle Hamiltonian can be written as $\mathcal{H} = \mathcal{H}_1(k_x) + [\hbar^2(k_y^2 + k_z^2)/2m + V(\vec{r})] \otimes \mathbf{1}$, where $\mathcal{H}_1$ is the Hamiltonian for the Raman coupling, the Zeeman energies and the motion along $\hat{x}$, and $\mathbf{1}$ is the 3×3 unit matrix acting on the spins. $V(\vec{r})$ is the state-independent trapping potential (arising from a far-off-resonance dipole trap and the scalar light shift of the Raman beams), and m is the atomic mass. In the rotating wave approximation for the frame rotating at $\Delta\omega_L$, $\mathcal{H}_1/\hbar$ expressed in the state basis of the family $\Psi(\tilde{k}_x)$ is

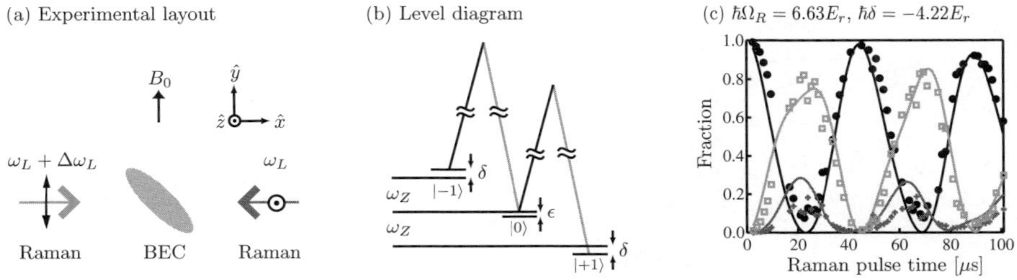

FIG. 1 (color online). (a) The ^{87}Rb BEC in a dipole trap created by two 1550 nm crossed beams in a bias field $B_0\hat{y}$ (gravity is along $-\hat{z}$). The two Raman laser beams are counterpropagating along $\hat{x}$, with frequencies ω_L and $(\omega_L + \Delta\omega_L)$, linearly polarized along $\hat{z}$ and $\hat{y}$, respectively. (b) Level diagram of Raman coupling within the $F = 1$ ground state. The linear and quadratic Zeeman shifts are ω_Z and ϵ, and δ is the Raman detuning. (c) As a function of Raman pulse time, we show the fraction of atoms in $|m_F = -1, k_x = 0\rangle$ (solid circles), $|0, -2k_r\rangle$ (open squares), and $|+1, -4k_r\rangle$ (crosses), the states comprising the $\Psi(\tilde{k}_x = -2k_r)$ family. The atoms start in $|-1, k_x = 0\rangle$, and are nearly resonant for the $|-1, 0\rangle \rightarrow |0, -2k_r\rangle$ transition at $\hbar\delta = -4.22E_r$. We determine $\hbar\Omega_R = 6.63(4)E_r$ by a global fit (solid lines) to the populations in $\Psi(-2k_r)$.

$$\begin{pmatrix} \frac{\hbar}{2m}(\tilde{k}_x + 2k_r)^2 - \delta & \Omega_R/2 & 0 \\ \Omega_R/2 & \frac{\hbar}{2m}\tilde{k}_x^2 - \epsilon & \Omega_R/2 \\ 0 & \Omega_R/2 & \frac{\hbar}{2m}(\tilde{k}_x - 2k_r)^2 + \delta \end{pmatrix}.$$

Here $\delta = (\Delta\omega_L - \omega_Z)$ is the detuning from Raman resonance, Ω_R is the resonant Raman Rabi frequency, and ϵ accounts for a small quadratic Zeeman shift [Fig. 1(b)]. For each $\tilde{k}_x$, diagonalizing $\mathcal{H}_1$ gives three energy eigenvalues $E_j(\tilde{k}_x)$ $(j = 1, 2, 3)$. For dressed atoms in state j, $E_j(\tilde{k}_x)$ is the effective dispersion relation, which depends on experimental parameters, δ, Ω_R, and ϵ (left panels of Fig. 2). The number of energy minima (from one to three) and their positions $\tilde{k}_{\rm min}$ are thus experimentally tunable. Around each $\tilde{k}_{\rm min}$, the dispersion can be expanded as $E(\tilde{k}_x) \approx \hbar^2(\tilde{k}_x - \tilde{k}_{\rm min})^2/2m^*$, where m^* is an effective mass. In this expansion, we identify $\tilde{k}_{\rm min}$ with the light-induced vector gauge potential, in analogy to the Hamiltonian for a particle of charge q in the usual magnetic vector potential $\vec{A}$: $(\vec{p} - q\vec{A})^2/2m$. In our experiment, we load a trapped BEC into the lowest energy, $j = 1$, dressed state, and measure its quasimomentum, equal to $\hbar\tilde{k}_{\rm min}$ for adiabatic loading.

Our experiment starts with a 3D ^{87}Rb BEC in a combined magnetic-quadrupole plus optical trap [12]. We transfer the atoms to a crossed dipole trap, formed by two 1550 nm beams, which are aligned along $\hat{x}$-$\hat{y}$ (horizontal beam) and $\sim 10°$ from $\hat{z}$ (vertical beam). A uniform bias field along $\hat{y}$ gives a linear Zeeman shift $\omega_Z/2\pi \simeq 3.25$ MHz and a quadratic shift $\epsilon/2\pi = 1.55$ kHz. The BEC has $N \approx 2.5 \times 10^5$ atoms in $|m_F = -1, k_x = 0\rangle$, with trap frequencies of $\approx$30 Hz parallel to, and $\approx$95 Hz perpendicular to the horizontal beam.

To Raman couple states differing in m_F by ± 1, the $\lambda = 804.3$ nm Raman beams are linearly polarized along $\hat{y}$ and $\hat{z}$, corresponding to π and σ relative to the quantization axis $\hat{y}$. The beams have $1/e^2$ radii of 180(20) μm [13], larger than the 20 μm BEC. These beams give a scalar light shift up to $60E_r$, where $E_r = h \times 3.55$ kHz, and contribute an additional harmonic potential with frequency up to 50 Hz along $\hat{y}$ and $\hat{z}$. The differential light shift between adjacent m_F states arising from the combination of misalignment and imperfect polarization is estimated to be smaller than $0.2E_r$. We determine Ω_R by observing population oscillations driven by the Raman beams and fitting to the expected behavior [Fig. 1(c)].

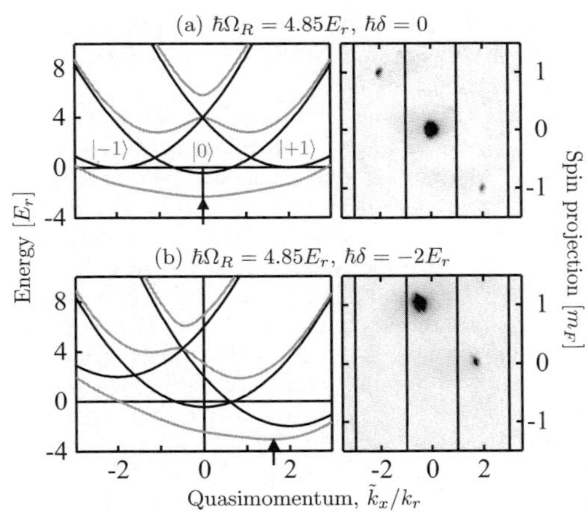

FIG. 2 (color online). Left panels: Energy-momentum dispersion curves $E(\tilde{k}_x)$ for $\hbar\epsilon = 0.44E_r$ and detuning $\hbar\delta = 0$ in (a) and $\hbar\delta = -2E_r$ in (b). The thin solid curves denote the states $|-1, \tilde{k}_x + 2k_r\rangle$, $|0, \tilde{k}_x\rangle$, $|+1, \tilde{k}_x - 2k_r\rangle$ absent Raman coupling; the thick solid, dotted and dash-dotted curves indicate dressed states at Raman coupling $\hbar\Omega_R = 4.85E_r$. The arrows indicate $\tilde{k}_x = \tilde{k}_{\rm min}$ in the $j = 1$ dressed state. Right panels: Time-of-flight images of the Raman-dressed state at $\hbar\Omega_R = 4.85(35)E_r$, for $\hbar\delta = 0$ in (a) and $\hbar\delta = -2E_r$ in (b). The Raman beams are along $\hat{x}$, and the three spin and momentum components, $|-1, \tilde{k}_{\rm min} + 2k_r\rangle$, $|0, \tilde{k}_{\rm min}\rangle$, and $|+1, \tilde{k}_{\rm min} - 2k_r\rangle$, are separated along $\hat{y}$ (after a small shear in the image realigning the Stern-Gerlach gradient direction along $\hat{y}$).

We developed a procedure to adiabatically load the BEC initially in $|-1, k_x = 0\rangle$ to the lowest energy, $j = 1$, Raman-dressed state. For a BEC initially in $|0, k_x = 0\rangle$, this could be achieved simply by slowly turning on the Raman beams at $\delta = 0$, resulting in the $j = 1$, $\tilde{k}_x = 0$ dressed state, located at the minimum of $E_{j=1}(\tilde{k}_x)$. However, our initial state $|-1, k_x = 0\rangle$, for which $k_x = \tilde{k}_x + 2k_r$, is a member of the family $\Psi(\tilde{k}_x = -2k_r)$. This loading procedure would therefore result in the $j = 1$, $\tilde{k}_x = -2k_r$ dressed state. To transfer our $\tilde{k}_x = -2k_r$ BEC to the desired $\tilde{k}_x = 0$ dressed state, we use an additional rf coupling which mixes families with $\tilde{k}_x$ differing by $2k_r$. The rf frequency is equal to the frequency difference between two Raman beams, $\Delta\omega_L/2\pi = 3.250$ MHz. The loading sequence is as follows: (i) We turn on the rf coupling in 1 ms to a resonant Rabi frequency $\Omega_{\rm rf}/2\pi = 12$ kHz, with an initial detuning $\hbar\delta = 15E_r$. We then sweep the detuning to resonance [14] in 9 ms (δ is varied by ramping the bias field B_0, leaving $\Delta\omega_L$ constant), loading the atoms into the lowest energy rf-dressed state, a spin superposition still at $k_x = 0$. (ii) We ramp on the Raman coupling in $t_{\rm on} = 20$ ms to a variable Rabi frequency Ω_R, and then turn off the rf in 2 ms. This loads the BEC into the $j = 1$, $\tilde{k}_x = 0$ dressed state, where it heats at $\sim$0.3 nK/ms, likely due to technical noise in B_0 or $\Delta\omega_L$. Therefore, in step (ii) we further evaporatively cool by decreasing the intensity of the horizontal trapping beam by 20% in 20 ms, after which it remains constant. The evaporation reduces the BEC number to $\approx 7 \times 10^4$.

In the final step (iii), we load the $\tilde{k}_x = 0$ dressed state into a nonzero $\tilde{k}_x$ state by sweeping the detuning from resonance to a variable δ in $t_{\rm sw}$ and holding for t_h. As we will show, this adiabatically loads the BEC into the $j = 1$ Raman-dressed state at $\tilde{k}_x = \tilde{k}_{\rm min}(\delta)$ for $t_{\rm sw} \gtrsim 20$ ms.

We characterize the Raman-dressed state by abruptly turning off the dipole trap and the Raman beams in less than 1 μs, projecting the atomic state onto its individual spin and momentum components. The atoms then expand in a magnetic field gradient applied during time-of-flight (TOF) approximately along $\hat{y}$, and the three spin states spatially separate due to the Stern-Gerlach effect. Imaging the atoms after a 20 ms TOF gives the momentum and spin composition of the dressed state (see right panels of Fig. 2). The quasimomentum $\hbar\tilde{k}_x$ is given by the shift along $\hat{x}$ of all three spin components from their positions when $\tilde{k}_x = 0$. We determine $\tilde{k}_x = 0$ to within 0.03 k_r by using the loading procedure above, except that in (ii), the Raman frequencies are sufficiently far-detuned that the atoms experience the scalar light shift but are not Raman-coupled; then the rf and Raman fields are snapped off concurrently before TOF.

It is easy to adiabatically load into the rf-dressed state in step (i) because the smallest relevant energy gap is the $\hbar\Omega_{\rm rf}/\sqrt{2}$ separation between rf-dressed states. The associated time scale of ≈ 20 μs is much less than the 1 ms turn-on time and the 9 ms ramp time and step (i) is easily adiabatic. To confirm subsequent adiabatic loading from the rf-dressed state into the Raman-dressed state in step (ii), we ramped up the Raman power in a time $t_{\rm on}$, then ramped it off in $t_{\rm on}$. Here the energy gap between dressed states at $\tilde{k}_x = 0$ is greater than $\hbar\Omega_{\rm rf}/\sqrt{2}$ (the added Raman coupling increases the gap). We observe no discernible excitation into other dressed states for $t_{\rm on} \geq 2$ ms, justifying the loading time $t_{\rm on} = 20$ ms in step (ii). To determine $t_{\rm sw}$ in step (iii), we measure $\tilde{k}_x$ as a function of $t_{\rm sw}$ for negligible t_h. Figure 3 shows $\tilde{k}_x$ versus $t_{\rm sw}$ at $\hbar\delta = 0$, 1, $2E_r$ and $\hbar\Omega_R = 5E_r$, for $t_h = 0.1$ ms. We use $t_{\rm sw} = 20$ ms, the time at which $\tilde{k}_x$ has nearly reached its equilibrium value. This adiabatic following is enabled by the small $\approx h \times 70$ Hz energy gap provided by the external trap, and the time scale observed for adiabatic loading is comparable to our trapping periods. We also modeled this using a 2D finite-temperature mean-field theory [15] and found the same time scale. After $t_{\rm sw}$, we add a hold time $t_h = 20$ ms during which residual excitations damp.

Figure 2 shows spin-resolved TOF images of adiabatically loaded Raman-dressed states at $\hbar\Omega_R = 4.85(35)E_r$ for $\hbar\delta = -2$ and $0E_r$. The resonance condition, $\delta = 0$, is determined from the symmetry of the rf-dressed state [16], with an uncertainty of $h \times 1.5$ kHz $= 0.4E_r$, limited by the stability of B_0. The quasimomentum $\hbar\tilde{k}_x = \hbar\tilde{k}_{\rm min}$, measured as a function of δ, is shown in Fig. 4(a), along with the calculated $\tilde{k}_{\rm min}(\delta)$. Each different $\tilde{k}_{\rm min}$ represents a spatially uniform vector gauge potential, where $\tilde{k}_{\rm min} = q\tilde{A}/\hbar$ is analogous to the magnetic vector potential with a magnetic field $\vec{B} = \vec{\nabla} \times \vec{A} = \vec{0}$.

For Rabi frequency $\Omega_R \leq 4.47E_r/\hbar$, $E_{j=1}(\tilde{k}_x)$ has multiple minima for some detunings δ. Figure 4(b) shows such multiple $\tilde{k}_{\rm min}$, expected for $\delta_a < |\delta| < \delta_b$, where $\hbar\delta_a = 0.35(20)E_r$ and $\hbar\delta_b = 1.2(1)E_r$, at $\hbar\Omega_R = 2.95(35)E_r$. At

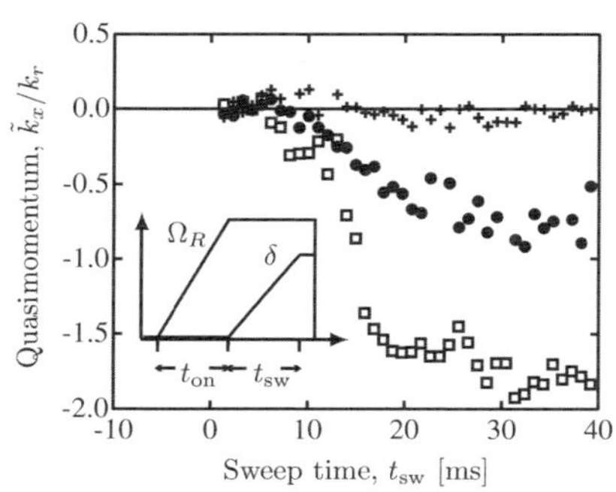

FIG. 3. Quasimomentum $\hbar\tilde{k}_x$ of the Raman-dressed state versus sweeping time $t_{\rm sw}$ of Raman detuning from $\hbar\delta = 0$ to $\hbar\delta = 1E_r$ (solid circles) and $\hbar\delta = 2E_r$ (open squares), at $\hbar\Omega_R = 5E_r$. The crosses denote $\tilde{k}_x = 0$ without sweeping ($\hbar\delta = 0$). The inset shows the time sequence of Ω_R and δ for the loading into a detuned Raman-dressed state.

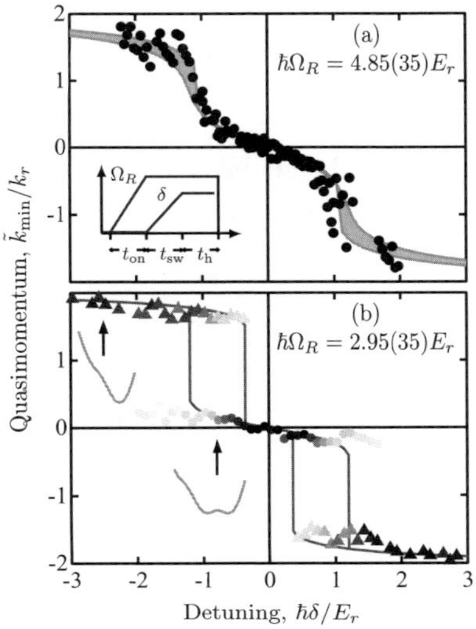

FIG. 4 (color online). (a) Measured quasimomentum $\hbar\tilde{k}_{\min}$ versus detuning δ, and the calculated $\hbar\tilde{k}_{\min}$ at Raman coupling $\hbar\Omega_R = 4.85(35)E_r$ (with the uncertainty indicated by the shaded region). The inset shows the time sequence of Ω_R and δ. (b) Quasimomentum $\hbar\tilde{k}_{\min 1}$ (circles) and $\hbar\tilde{k}_{\min 2}$ (triangles) versus detuning δ, and the calculated $\hbar\tilde{k}_{\min}$ (solid lines) at $\hbar\Omega_R = 2.95(35)E_r$. The intensity of the circles indicates the population of $|0, \tilde{k}_{\min 1}\rangle$ relative to its value at $\delta = 0$. The intensity of the triangles for $\delta < 0$ ($\delta > 0$) indicates the population of $|+1, \tilde{k}_{\min 2} - 2k_r\rangle$ ($|-1, \tilde{k}_{\min 2} + 2k_r\rangle$) relative to its value at $\hbar\delta = -3E_r$ ($\hbar\delta = 3E_r$). The insets picture $E(\tilde{k}_x)$ at $\hbar\delta = -0.8$ and $-2.5E_r$ as indicated by the arrows.

small and large detuning, we observe only one $\tilde{k}_{\min}$. However, at intermediate detuning, we observe two $\tilde{k}_{\min}$ at $\tilde{k}_{\min 1}$ and $\tilde{k}_{\min 2}$ [11]. In Fig. 4(b), the relative population is indicated by the gray scale, and the solid line is the calculated value(s). The observed $\tilde{k}_{\min}(\delta)$ agrees well with the theory, and a simultaneous fit of $\tilde{k}_{\min 1}$ and $\tilde{k}_{\min 2}$ with Ω_R as a free parameter gives $\hbar\Omega_R = 2.9(3)E_r$, which agrees well with the independently measured $\hbar\Omega_R = 2.95(35)E_r$. As expected, we observe population in only one $\tilde{k}_x$ for $|\delta| < \delta_a$, however, for some $|\delta| > \delta_b$, there is discernible population in two $\tilde{k}_x$ where we expect only one $\tilde{k}_{\min}$. We attribute this to nonadiabaticity when $\tilde{k}_{\min 1}$ disappears during the detuning sweep in loading step (iii).

In conclusion, we have prepared a Bose-Einstein condensate in a Raman-dressed state with a nonzero quasimomentum, controlled by the Rabi frequency Ω_R and the detuning δ. This technique, in conjunction with a spatial gradient of the bias field (and therefore the detuning) along $\hat{y}$, would give a spatial gradient of light-induced momentum $\hbar\tilde{k}_{\min}(y)\hat{x}$ and would create an effective magnetic field along $\hat{z}$ [17]. Using an easily tunable magnetic field gradient to create the effective magnetic field is likely a technical simplification compared with proposals which rely on spatial variations in the light-fields [8]. The analog of the magnetic length $l_B = \sqrt{\hbar/(qB)}$ is $(\partial\tilde{k}_{\min}/\partial y)^{-1/2}$, and is about the spacing between vortices in a vortex lattice like that formed in a rotating BEC [5]. For a gradient in $\tilde{k}_{\min}$ of k_r across a condensate of radius R, this gives an analog $l_B \approx 1.4$ μm for a $R \approx 10$ μm 3D BEC, and we expect a ~25-vortex lattice to form in the condensate. For dilute 2D systems, this is sufficient to reach the quantum Hall regime.

This work was partially supported by ONR, ODNI, ARO with funds from the DARPA OLE program, and the NSF through the JQI Physics Frontier Center. R. L. C. acknowledges support from NIST-NRC.

*ian.spielman@nist.gov

[1] B. Paredes *et al.*, Nature (London) **429**, 277 (2004); T. Kinoshita, T. Wenger, and D. Weiss, Science **305**, 1125 (2004).
[2] M. Greiner *et al.*, Nature (London) **415**, 39 (2002); T. Stöferle *et al.*, Phys. Rev. Lett. **92**, 130403 (2004); I. B. Spielman, W. D. Phillips, and J. V. Porto, Phys. Rev. Lett. **98**, 080404 (2007).
[3] D. C. Tsui, H. L. Stormer, and A. C. Gossard, Phys. Rev. Lett. **48**, 1559 (1982); R. B. Laughlin, Phys. Rev. Lett. **50**, 1395 (1983).
[4] J. R. Abo-Shaeer, C. Raman, J. M. Vogels, and W. Ketterle, Science **292**, 476 (2001); V. Bretin, S. Stock, Y. Seurin, and J. Dalibard, Phys. Rev. Lett. **92**, 050403 (2004).
[5] V. Schweikhard *et al.*, Phys. Rev. Lett. **92**, 040404 (2004).
[6] I. Bloch, J. Dalibard, and W. Zwerger, Rev. Mod. Phys. **80**, 885 (2008).
[7] D. Jaksch and P. Zoller, New J. Phys. **5**, 56 (2003); A. S. Sørensen, E. Demler, and M. D. Lukin, Phys. Rev. Lett. **94**, 086803 (2005); M. Hafezi, A. S. Sørensen, E. Demler, and M. D. Lukin, Phys. Rev. A **76**, 023613 (2007).
[8] G. Juzeliūnas and P. Öhberg, Phys. Rev. Lett. **93**, 033602 (2004); G. Juzeliūnas, J. Ruseckas, P. Öhberg, and M. Fleischhauer, Phys. Rev. A **73**, 025602 (2006).
[9] S.-L. Zhu *et al.*, Phys. Rev. Lett. **97**, 240401 (2006).
[10] J. Higbie and D. M. Stamper-Kurn, Phys. Rev. Lett. **88**, 090401 (2002); F. Papoff, F. Mauri, and E. Arimondo, J. Opt. Soc. Am. B **9**, 321 (1992).
[11] T. D. Stanescu, B. Anderson, and V. Galitski, Phys. Rev. A **78**, 023616 (2008).
[12] Y.-J. Lin *et al.* (to be published).
[13] Uncertainties reflect the uncorrelated combination of 1σ statistical and systematic uncertainties.
[14] All the detuning ramps are linear versus time, and the Raman beam power turns on nearly quadratically.
[15] P. B. Blakie *et al.*, Adv. Phys. **57**, 363 (2008).
[16] On resonance the rf-dressed state has equal population in $|-1, k_x = 0\rangle$ and $|+1, k_x = 0\rangle$, even for a nonzero ϵ. $\delta = 0$ determined from the symmetry of the Raman dressing is within $0.25E_r$ of that from the rf-dressed states.
[17] I. B. Spielman (to be published).

7

Non-canonical phase space and quantization

On the Dirac Theory of Spin 1/2 Particles and Its Non-Relativistic Limit*

LESLIE L. FOLDY
Case Institute of Technology, Cleveland, Ohio

AND

SIEGFRIED A. WOUTHUYSEN†
University of Rochester, Rochester, New York

(Received November 25, 1949)

By a canonical transformation on the Dirac Hamiltonian for a free particle, a representation of the Dirac theory is obtained in which positive and negative energy states are separately represented by two-component wave functions. Playing an important role in the new representation are new operators for position and spin of the particle which are physically distinct from these operators in the conventional representation. The components of the time derivative of the new position operator all commute and have for eigenvalues all values between $-c$ and c. The new spin operator is a constant of the motion unlike the spin operator in the conventional representation. By a comparison of the new Hamiltonian with the non-relativistic Pauli-Hamiltonian for particles of spin $\frac{1}{2}$, one finds that it is these new operators rather than the conventional ones which pass over into the position and spin operators in the Pauli theory in the non-relativistic limit.

The transformation of the new representation is also made in the case of interaction of the particle with an external electromagnetic field. In this way the proper non-relativistic Hamiltonian (essentially the Pauli-Hamiltonian) is obtained in the non-relativistic limit. The same methods may be applied to a Dirac particle interacting with any type of external field (various meson fields, for example) and this allows one to find the proper non-relativistic Hamiltonian in each such case. Some light is cast on the question of why a Dirac electron shows some properties characteristic of a particle of finite extension by an examination of the relationship between the new and the conventional position operators.

INTRODUCTION

A FREE Dirac particle of mass m is described by a four-component wave function Ψ satisfying the Dirac equation,

$$H\Psi=(\beta m+\boldsymbol{\alpha}\cdot\mathbf{p})\Psi=i(\partial\Psi/\partial t), \tag{1}$$

where $\mathbf{p}$ is the momentum operator for the particle, $\boldsymbol{\alpha}$ and β are the well known Dirac matrices (assumed here to be in their usual representation with β diagonal), and units have been used in which $\hbar$ and c are unity. The eigenfunctions of the Hamiltonian operator satisfy the equation:

$$(\beta m+\boldsymbol{\alpha}\cdot\mathbf{p})\psi=E\psi. \tag{2}$$

The eigenfunctions are of the form $u(\mathbf{p})e^{-\mathbf{p}\cdot\mathbf{x}}$. For each value of the momentum $\mathbf{p}$ there are four linearly independent spinors[1] $u(\mathbf{p})$ corresponding to the two eigenvalues $\pm(m^2+p^2)^{\frac{1}{2}}$ of the energy and the two eigenvalues ± 1 of the z component of an operator $\boldsymbol{\Sigma}$ (defined later in Eq. (26)) related to the spin operator for the particle.

Except in certain trivial cases, for a given sign of the energy at least three of the components of $u(\mathbf{p})$ are different from zero. However, two of the four components go to zero as the momentum goes to zero, while at least one of the other two components remains finite. In the above representation, for positive-energy eigenfunctions, the last two (or *lower*) components vanish with vanishing momentum, while for negative-energy eigenfunctions, the first two (or *upper*) components vanish with vanishing momentum.

Of course any spinor Ψ may be split into the sum of two spinors Φ and X, having only upper and lower components respectively, in the following manner:

$$\Psi=\Phi+X,\quad \Phi=(1+\beta/2)\Psi,\quad X=(1-\beta/2)\Psi,$$

but the spinors Φ and X do not represent states of definite energy in the above representation.

In the non-relativistic limit, where the momentum of the particle is small compared to m, it is well known that a Dirac particle (that is, one with spin $\frac{1}{2}$) can be described by a two-component wave function in the Pauli theory. The usual method of demonstrating that the Dirac theory goes into the Pauli theory in this limit makes use of the fact noted above that two of the four Dirac-function components become small when the momentum is small. One then writes out the equations satisfied by the four components and solves, approximately, two of the equations for the small components. By substituting these solutions in the remaining two equations, one obtains a pair of equations for the large components which are essentially the Pauli spin equations.[2]

The above method of demonstrating the equivalence of the Dirac and Pauli theories encounters difficulties, however, when one wishes to go beyond the lowest order approximation. One then finds that the "Hamiltonian" associated with the large components is no longer Hermitian in the presence of external fields because of the appearance of an "imaginary electric moment" for the particle. Furthermore, all four components must again be used in calculating expectation values of operators to

* Supported by the AEC and the ONR.

† Now at the Zeeman Laboratory, University of Amsterdam, Amsterdam, Holland.

[1] See, for example, W. Heitler, *The Quantum Theory of Radiation* (Oxford University Press, New York, 1944), Chapter 3. We shall use the term spinor for any four-component column (or row) matrix occurring in the Dirac theory.

[2] See, for example, W. Pauli, *Handbuch der Physik*, 2. Aufl., Bd. 24, Teil 1.

order v^2/c^2. Apart from these difficulties, however, there are some serious questions concerning the properties of corresponding operators in the Dirac theory and the Pauli theory even in lowest order. Thus, in the Dirac theory the operator representing the velocity of the particle is the operator $\boldsymbol{\alpha}$ whose components have only the eigenvalues ± 1. On the other hand, in the Pauli theory the operator representing the velocity of the particle is taken to be $\mathbf{p}/m$ whose components have eigenvalues embracing all real numbers. Furthermore, different components of the velocity operator in the Dirac theory do not commute and therefore are not simultaneously measurable with arbitrary precision, while different components of the velocity operator in the Pauli theory do commute and can therefore be measured simultaneously with arbitrary precision. One can well ask how the operators which purportedly represent the same physical variable in the two theories can have such completely different properties.

One can conclude from the above discussion that the relation between the Dirac theory and the Pauli theory is by no means clear from the usual method of descending from four- to two-component wave functions, and that further clarification of the connection between the theories would be desirable. In what follows we present an alternative method for passing from four- to two-component wave functions in the Dirac theory which alternative method clarifies many of the questions left open by the usual treatment. The method consists of a transformation to a new representation for the Dirac theory, putting the theory in a form closely analogous to the Pauli theory and permitting a direct comparison of the two. Specifically, we shall show that:

(1) For a free Dirac particle there exists a representation of the Dirac theory in which, for both relativistic and non-relativistic energies, positive-energy states and negative-energy states are separately described by two-component wave functions.

(2) There exists in the Dirac theory another position operator than the usual one; this operator has the property that its time derivative is the operator $\mathbf{p}/(m^2+p^2)^{\frac{1}{2}}$ for positive-energy states and $-\mathbf{p}/(m^2+p^2)^{\frac{1}{2}}$ for negative-energy states corresponding to the conventional concept of the velocity of the particle. It is this new position operator (which we call the *mean-position operator*) that in the non-relativistic limit is interpreted as the position operator of the Pauli theory.

(3) While the z component of the spin operator, $\boldsymbol{\sigma}=(1/2i)[\boldsymbol{\alpha}\times\boldsymbol{\alpha}]$, in the Dirac theory is not a constant of the motion, there exists in the Dirac theory another spin operator $\boldsymbol{\Sigma}$ (which we call the *mean-spin operator*) whose z component is a constant of the motion. In the non-relativistic limit, the operator $\boldsymbol{\Sigma}$ is the one which is interpreted as the spin operator in the Pauli theory.

(4) From a study of the transformation to the new representation further insight can be obtained into the question of why the Dirac particle has a magnetic moment and why it appears to show a behavior characteristic of a particle with finite extension of the order of its Compton wave-length.

(5) In the presence of interaction, such as with an external electromagnetic field, one can still make a transformation which leads to a representation involving two-component wave functions. The transformation, which previously could be made exactly, must now be made by an infinite sequence of transformations which process leads to a Hamiltonian which is an infinite series in powers of $(1/m)$, where m is the mass of the particle. This series is presumably semi-convergent in the sense that for given external fields, a finite number of terms of the series is a better-and-better approximation to the exact Hamiltonian, the larger the value of m. For strong interactions which strongly couple free-particle states of positive and negative energy, this representation is of little value; however, for sufficiently weak interactions, a finite number of terms of the series may be employed to obtain relativistic corrections to any order in $(1/m)$. In this way one can obtain the proper non-relativistic limit for the Hamiltonian representing a Dirac particle interacting with any type of external field.

THE FREE DIRAC PARTICLE

The essential reason why four components are in general necessary to describe a state of positive or negative energy in the representation of the Dirac theory corresponding to Eq. (1) is that the Hamiltonian in this equation contains odd operators,[3] specifically the components of the operator $\boldsymbol{\alpha}$. If it is possible to perform a canonical transformation on Eq. (1) which brings it into a form which is free of odd operators, then it will be possible to represent positive- and negative-energy states by wave functions having only two components in each case, the other pair of components being identically zero. We shall now show that there exists a canonical transformation which accomplishes just this end.

If S is an Hermitian operator, then the transformation,

$$\Psi' = e^{iS}\Psi, \tag{3}$$

$$H' = e^{iS}He^{-iS} - ie^{iS}(\partial e^{-iS}/\partial t), \tag{4}$$

[3] An odd operator in the Dirac theory is a Dirac matrix which has only matrix elements connecting upper and lower components of the wave function, while an even operator is one having no such matrix elements. Of the sixteen linearly independent matrices in the Dirac theory, the matrices 1, β, $\boldsymbol{\sigma}=1/2i[\boldsymbol{\alpha}\times\boldsymbol{\alpha}]$ and $\beta\boldsymbol{\sigma}$ are even, while the matrices $\boldsymbol{\alpha}$, $\beta\boldsymbol{\alpha}$, $\gamma^5=-i\alpha^1\alpha^2\alpha^3$ and $\beta\gamma^5$ are odd. The product of two even matrices or of two odd matrices is an even matrix, while the product of an odd matrix and an even matrix is an odd matrix. The matrix β commutes with all even matrices and anticommutes with all odd matrices in the Dirac theory. This last fact allows one to write any matrix as the sum of an odd and an even matrix in a simple way, namely:

$$m = \tfrac{1}{2}\{m+\beta m\beta\} + \tfrac{1}{2}\{m-\beta m\beta\},$$

where the first term on the right is the even part of the matrix and the second term is the odd part of the matrix m.

leaves Eq. (1) in the Hamiltonian form:

$$H'\Psi'=i(\partial\Psi'/\partial t). \tag{5}$$

(It must be remembered that in a canonical transformation of this type a physical variable whose operator-representative in the old representation is T has for its operator-representative in the new representation the operator:

$$T'=e^{iS}Te^{-iS}.) \tag{6}$$

Let us make a transformation of this type on Eq. (1) with S the non-explicitly time-dependent operator:

$$S=-(i/2m)\beta\boldsymbol{\alpha}\cdot\mathbf{p}w(p/m), \tag{7}$$

where w is a function of the operator[4] (p/m) whose form is to be determined such that H' is free of odd operators. With this choice of S one may readily show that

$$\begin{aligned}H'&=e^{iS}He^{-iS}=e^{2iS}H\\&=[\cos(pw/m)+(\beta\boldsymbol{\alpha}\cdot\mathbf{p}/p)\sin(pw/m)]H\\&=\beta[m\cos(pw/m)+p\sin(pw/m)]\\&\quad+\boldsymbol{\alpha}\cdot\mathbf{p}/p[p\cos(pw/m)-m\sin(pw/m)].\end{aligned} \tag{8}$$

We see from this that H' will be free of odd operators if we choose

$$w(p/m)=(m/p)\tan^{-1}(p/m), \tag{9}$$

and with this choice, we obtain

$$H'=\beta(m^2+p^2)^{\frac{1}{2}}=\beta E_p, \tag{10}$$

where E_p represents the *operator* $(m^2+p^2)^{\frac{1}{2}}$.

Equation (5), with H' given by (10), now has solutions such that the upper components represent positive energies and the lower components negative energies, for splitting Ψ' into its upper and lower components:

$$\Psi'=\Phi'+X',$$
$$\Phi'=(1+\beta/2)\Psi',\quad X'=(1-\beta/2)\Psi',$$

reduces Eq. (5) to the two uncoupled equations:

$$E_p\Phi'=i(\partial\Phi'/\partial t), \tag{11}$$

$$-E_pX'=i(\partial X'/\partial t). \tag{12}$$

In order for one to understand completely the nature of the transformation which we have performed, however, it is necessary to investigate in greater detail the manner in which the wave function and certain operators transform. A general wave function can be expressed in the form:

$$\Psi(\mathbf{x})=\int u(\mathbf{p}')\exp(i\mathbf{p}'\mathbf{x}')d\mathbf{p}'=\Psi_+(\mathbf{x})+\Psi_-(\mathbf{x}), \tag{13}$$

$$\Psi_+(\mathbf{x})=\int\frac{1}{2}\left[1+\frac{\beta m+\boldsymbol{\alpha}\cdot\mathbf{p}'}{E_{p'}}\right]u(\mathbf{p}')\exp(i\mathbf{p}'\cdot\mathbf{x})d\mathbf{p}', \tag{14}$$

$$\Psi_-(\mathbf{x})=\int\frac{1}{2}\left[1-\frac{\beta m+\boldsymbol{\alpha}\cdot\mathbf{p}'}{E_{p'}}\right]u(\mathbf{p}')\exp(i\mathbf{p}'\cdot\mathbf{x})d\mathbf{p}', \tag{15}$$

where Ψ_+ represents the positive energy part and Ψ_- the negative energy part of Ψ. Now

$$\begin{aligned}e^{iS}&=\exp\left(\frac{1}{2m}\frac{\beta\boldsymbol{\alpha}\cdot\mathbf{p}}{p}\tan^{-1}\frac{p}{m}\right)\\&=\cos[\tfrac{1}{2}\tan^{-1}(p/m)]+(\beta\boldsymbol{\alpha}\cdot\mathbf{p}/p)\sin[\tfrac{1}{2}\tan^{-1}(p/m)]\\&=\left[\frac{E_p+m}{2E_p}\right]^{\frac{1}{2}}+\frac{\beta\boldsymbol{\alpha}\cdot\mathbf{p}}{p}\left[\frac{E_p-m}{2E_p}\right]^{\frac{1}{2}}\\&=\frac{\beta\{\beta m+\boldsymbol{\alpha}\cdot\mathbf{p}+\beta E_p\}}{[2E_p(E_p+m)]^{\frac{1}{2}}},\end{aligned} \tag{16}$$

and hence

$$\Psi_+'=e^{iS}\Psi_+=\left(\frac{1+\beta}{2}\right)\int\left[\frac{2E_{p'}}{E_{p'}+m}\right]^{\frac{1}{2}}\times\left[1+\frac{\beta m+\boldsymbol{\alpha}\cdot\mathbf{p}'}{E_{p'}}\right]u(\mathbf{p}')\exp(i\mathbf{p}'\cdot\mathbf{x}')d\mathbf{p}'=\Phi', \tag{17}$$

$$\Psi_-'=e^{iS}\Psi_-=\left(\frac{1-\beta}{2}\right)\int\left[\frac{2E_p'}{E_{p'}+m}\right]^{\frac{1}{2}}\times\left[1-\frac{\beta m+\boldsymbol{\alpha}\cdot\mathbf{p}'}{E_{p'}}\right]u(\mathbf{p}')\exp(i\mathbf{p}'\cdot\mathbf{x}')d\mathbf{p}'=X', \tag{18}$$

$$\Psi'=e^{iS}\Psi=\Psi_+'+\Psi_-'. \tag{19}$$

We may legitimately identify Ψ_+' with Φ' and Ψ_-' with X' since the presence of the factors $(1+\beta)/2$ and $(1-\beta)/2$ shows that these functions have only upper and lower components, respectively. Thus we see that indeed the transformation is such as to lead to a representation in which upper components correspond to positive energies and lower components to negative energies.

Furthermore, since

$$u(\mathbf{p}')=\frac{1}{(2\pi)^3}\int\Psi(\mathbf{x}')\exp(-i\mathbf{p}'\cdot\mathbf{x}')d\mathbf{x}',$$

we have

$$\Psi'(\mathbf{x})=\int K(\mathbf{x},\mathbf{x}')\Psi(\mathbf{x}')d\mathbf{x}', \tag{20}$$

where

$$K(\mathbf{x},\mathbf{x}')=\frac{1}{(2\pi)^3}\int\left[\frac{2E_{p'}}{E_{p'}+m}\right]^{\frac{1}{2}}\times\left[1+\frac{\beta(\beta m+\boldsymbol{\alpha}\cdot\mathbf{p}')}{E_{p'}}\right]\exp(i\mathbf{p}'\cdot(\mathbf{x}-\mathbf{x}'))d\mathbf{p}'. \tag{21}$$

[4] A function $T(\mathbf{p})$ of the operator $\mathbf{p}$ is to be interpreted (in the coordinate representation) as defined by its Taylor expansion in powers of $(\mathbf{p}-\mathbf{p}_0)$, where $\mathbf{p}_0$ is any constant vector (c number), wherever the expansion converges, and by the analytic continuation of this series elsewhere; or, alternatively, by its integral-operator representation:

$$T(\mathbf{p})\Psi(\mathbf{x})=\frac{1}{(2\pi)^3}\int\int T(\mathbf{p}')\exp[i\mathbf{p}'\cdot(\mathbf{x}-\mathbf{x}')]\Psi(\mathbf{x}')d\mathbf{p}'d\mathbf{x}'.$$

Since $K(\mathbf{x}, \mathbf{x}')$ is not a Dirac delta-function in its space dependence, the transformation of Ψ is an integral rather than a point transformation. In general, Ψ' at a given point is constituted from contributions depending on Ψ over a neighborhood of dimensions of the order of a Compton wave-length of the particle about the point. Thus a wave function which in the old representation corresponded to a state in which the particle was definitely located at one point, passes over in the new representation into a wave function which apparently corresponds to the particle being spread out over a finite region.

The key to understanding this rather unusual statement lies in the fact that in the new representation, the operator-representative for the position of the particle is no longer the operator $\mathbf{x}$, but the rather complicated operator:

$$\mathbf{x}'=e^{iS}\mathbf{x}e^{-iS}=\mathbf{x}-\frac{i\beta\boldsymbol{\alpha}}{2E_p}+\frac{i\beta(\boldsymbol{\alpha}\cdot\mathbf{p})\mathbf{p}-[\boldsymbol{\sigma}\times\mathbf{p}]p}{2E_p(E_p+m)p}. \quad (22)$$

But if this is the case, then what is the operator-representative $\mathbf{X}$ in the old representation of the physical variable whose operator-representative in the new representation is $\mathbf{X}'=\mathbf{x}$? We find

$$\mathbf{X}=e^{-iS}\mathbf{x}e^{iS}=\mathbf{x}+\frac{i\beta\boldsymbol{\alpha}}{2E_p}-\frac{i\beta(\boldsymbol{\alpha}\cdot\mathbf{p})\mathbf{p}+[\boldsymbol{\sigma}\times\mathbf{p}]p}{2E_p(E_p+m)p}. \quad (23)$$

In order to interpret this new "position" operator, we calculate its time derivative:

$$\frac{d\mathbf{X}}{dt}=i[H,\mathbf{X}]=\frac{\mathbf{p}}{E_p}\cdot\frac{\{\beta m+\boldsymbol{\alpha}\cdot\mathbf{p}\}}{E_p}. \quad (24)$$

But since $\Lambda\mathbf{p}=(\beta m+\boldsymbol{\alpha}\cdot\mathbf{p})/E_p$ has the value $+1$ when applied to a positive energy wave function and the value -1 when applied to a negative energy wave function,[5] we see that the time derivative of $\mathbf{X}$ is just the "conventional" velocity operator $+\mathbf{p}/E_p$ for positive energy states and $-\mathbf{p}/E_p$ for negative energy states. The operator-representative of $d\mathbf{X}/dt$ in the new representation is

$$d\mathbf{X}'/dt=d\mathbf{x}/dt=i[H',\mathbf{x}]=\beta\mathbf{p}/E_p \quad (25)$$

with the same properties. We note also that since the operator $\mathbf{p}$ commutes with S, the operator-representative for the momentum in the new representation is still $\mathbf{p}$.

By comparing these results with certain well-known facts about the Dirac electron, we may understand the significance of our change in representation. In discussions of the Dirac theory one finds an analysis of the motion of a free Dirac electron which shows that the electron performs a very complicated motion, indeed.[6]

[5] E. Schrödinger, Berl. Ber. 419 (1930); 63 (1931).

[6] See, for example, P. A. M. Dirac, *The Principles of Quantum Mechanics* (Oxford University Press, New York, 1935), Second Edition, Chapter XII. See also in this connection, reference 2, pp. 230–231.

Its velocity can be written as the sum of two parts: the first is essentially the conventional velocity operator

$$\frac{\mathbf{p}}{E_p}\frac{\beta m+\boldsymbol{\alpha}\cdot\mathbf{p}}{E_p},$$

but to this is added a second term representing a rapidly oscillating motion ("Zitterbewegung") which ensures that a measurement of the instantaneous value of any velocity component shall yield the velocity of light. Our results above show that a corresponding division of the position operator for the particle is also possible, the first part $\mathbf{X}$ (in the old representation) representing a sort of mean position of the particle, and the second part $\mathbf{X}-\mathbf{x}$, oscillating rapidly about zero with an amplitude of the order of the Compton wave-length of the particle. While in the old representation the position operator $\mathbf{x}$ played the dominant role, in the new representation it is the position operator $\mathbf{X}'$, which we shall call the *mean-position* operator,[7] which plays the dominant role. Also, as will become obvious later when we consider the interaction of the particle with an external field, it is the *mean-position* operator which is identified with the position operator in the non-relativistic Pauli theory.

The modification in the interpretation of operators involved in our transformation does not end here, however, but new angular momentum operators also appear. In the old representation, the orbital angular momentum of the particle whose operator-representative is $[\mathbf{x}\times\mathbf{p}]$ and the spin angular momentum of the particle whose operator-representative is $1/2\boldsymbol{\sigma}$ are not separately constants of the motion,[6] although their sum is a constant of the motion. However, as one may readily verify, the operators $[\mathbf{X}\times\mathbf{p}]$ and

$$\boldsymbol{\Sigma}=\boldsymbol{\sigma}-\frac{i\beta[\boldsymbol{\alpha}\times\mathbf{p}]}{E_p}-\frac{[\mathbf{p}\times[\boldsymbol{\sigma}\times\mathbf{p}]]}{E_p(E_p+m)}, \quad (26)$$

whose analogues in the new representation are, respectively,

$$[\mathbf{X}'\times\mathbf{p}]=e^{iS}[\mathbf{X}\times\mathbf{p}]e^{-iS}=[\mathbf{x}\times p], \quad (27)$$

and

$$\boldsymbol{\Sigma}'=e^{iS}\boldsymbol{\Sigma}e^{-iS}=\boldsymbol{\sigma}, \quad (28)$$

[7] The operator, which we have designated the mean-position operator, has been discovered independently in other connections by several authors. M. H. L. Pryce (Proc. Roy. Soc. **150A**, 166 (1935)) found it useful to introduce this operator (as well as the operator which we later define as the mean spin angular momentum) in connection with the definition of coordinate and intrinsic angular momentum operators in the Born-Infeld theory, and again (Proc. Roy. Soc. **195A**, 62 (1948)) in a discussion of relativistic definitions of center of mass for systems of particles. In the latter paper, he also noted that this operator was connected with the usual position operator in the Dirac theory by a canonical transformation which is identical with the one performed above in this paper. T. D. Newton and E. P. Wigner (Rev. Mod. Phys. **21**, 400 (1949)) also found this operator in an investigation of localized states for elementary systems. The latter authors have also shown that this is the only position operator, with commuting components, in the Dirac theory which has localized eigenfunctions in the manifold of positive energy wave functions.

TABLE I. Table of operator-representatives of dynamical variables in old and new representations.

Dynamical variable	Operator-representative in old representation	Operator-representative in new representation
Position	$\mathbf{x}$	$\mathbf{x}'=\mathbf{x}-\frac{i\beta\boldsymbol{\alpha}}{2E_p}+\frac{i\beta(\boldsymbol{\alpha}\cdot\mathbf{p})\mathbf{p}-[\boldsymbol{\sigma}\times\mathbf{p}]p}{2E_p(E_p+m)p}$
Momentum	$\mathbf{p}\equiv(\hbar/i)\boldsymbol{\nabla}$	$\mathbf{p}'=\mathbf{p}$
Hamiltonian	$H=\beta m+\boldsymbol{\alpha}\cdot\mathbf{p}$	$H'=\beta(m^2+p^2)^{\frac{1}{2}}\equiv\beta E_p$
Velocity	$\boldsymbol{\alpha}=\dot{\mathbf{x}}$	$\boldsymbol{\alpha}'=\boldsymbol{\alpha}+\frac{\beta\mathbf{p}}{E_p}-\frac{(\boldsymbol{\alpha}\cdot\mathbf{p})\mathbf{p}}{E_p(E_p+m)}$
	β	$\beta'=(m\beta-\boldsymbol{\alpha}\cdot\mathbf{p})/E_p$
Orbital angular momentum	$[\mathbf{x}\times\mathbf{p}]$	$[\mathbf{x}'\times\mathbf{p}]$
Spin angular momentum	$\boldsymbol{\sigma}\equiv(1/2i)[\boldsymbol{\alpha}\times\boldsymbol{\alpha}]$	$\boldsymbol{\sigma}'=\boldsymbol{\sigma}+\frac{i\beta[\boldsymbol{\alpha}\times\mathbf{p}]}{E_p}-\frac{\mathbf{p}\times[\boldsymbol{\sigma}\times\mathbf{p}]}{E_p(E_p+m)}$
Mean position	$\mathbf{X}=\mathbf{x}+\frac{i\beta\boldsymbol{\alpha}}{2E_p}-\frac{i\beta(\boldsymbol{\alpha}\cdot\mathbf{p})\mathbf{p}+[\boldsymbol{\sigma}\times\mathbf{p}]p}{2E_p(E_p+m)p}$	$\mathbf{X}'=\mathbf{x}$
Mean velocity	$\dot{\mathbf{X}}=\frac{\mathbf{p}}{E_p}\cdot\left\{\frac{\beta m+\boldsymbol{\alpha}\cdot\mathbf{p}}{E_p}\right\}$	$\dot{\mathbf{X}}'=\frac{\beta\mathbf{p}}{E_p}$
Mean orbital angular momentum	$[\mathbf{X}\times\mathbf{p}]$	$[\mathbf{X}'\times\mathbf{p}]=[\mathbf{x}\times\mathbf{p}]$
Mean spin angular momentum	$\boldsymbol{\Sigma}=\boldsymbol{\sigma}-\frac{i\beta[\boldsymbol{\alpha}\times\mathbf{p}]}{E_p}-\frac{[\mathbf{p}\times[\boldsymbol{\sigma}\times\mathbf{p}]]}{E_p(E_p+m)}$	$\boldsymbol{\Sigma}'=\boldsymbol{\sigma}$
Λ_p	$\Lambda_p\equiv\frac{\beta m+\boldsymbol{\alpha}\cdot\mathbf{p}}{E_p}$	$\Lambda_p'=\beta$

are separately constants of the motion. We shall denote the physical variables of which these operators are the operator-representatives as the *mean-orbital angular momentum* and *mean spin angular momentum*, respectively, of the particle. It is again these variables which are conventionally identified with the orbital angular momentum and the spin angular momentum of the particle in the non-relativistic Pauli theory. For convenience, we have listed in Table I the operator-representatives in the old representation and in the new representation of the physical variables of principal interest in the Dirac theory.

DIRAC PARTICLE IN AN EXTERNAL ELECTROMAGNETIC FIELD

We consider now the case where a Dirac particle interacts with an external field such as the electromagnetic field. Before proceeding to the general case, however, it is instructive to consider an elementary special case in order to clarify some points regarding the classification of states as either of positive or negative energy. Let us consider first a Dirac particle moving in a weak static electric field derivable from a scalar potential φ. The Hamiltonian is then

$$H=\beta m+\boldsymbol{\alpha}\cdot\mathbf{p}-e\varphi.$$

For the case of a free particle, states were classified as of positive or negative energy according as they corresponded to values of $+1$ or -1, respectively, for the operator $(\beta m+\boldsymbol{\alpha}\cdot\mathbf{p})/E_p$ which was a constant of the motion. In the present case, however, this operator does not commute with the Hamiltonian and is therefore not a constant of the motion. If we regard the electric field as a perturbation, then one can say that the electric field induces transitions of the particle between the positive- and negative-energy states of a free particle. This is one way of viewing the physical situation.

On the other hand, one knows that for sufficiently weak fields the Hamiltonian above possesses a complete set of eigenfunctions with energy eigenvalues which may be classified according to whether they are positive or negative. There exists for these weak fields a clear-cut distinction between these two sets of stationary states[8] since they are separated by a relatively large energy gap of order $2m$. Furthermore, the wave functions corresponding to positive energies show a behavior of the particle appropriate to a particle of positive mass,[9] in that the particle tends to be localized in regions of low potential energy; while the negative-energy solutions show a behavior of the particle appropriate to a particle of negative mass, in that the particle tends to be localized in regions of high potential energy.

Either of the two descriptions of the behavior of the particle in a weak field given above is of course correct, although the distinction between what are called positive- and negative-energy states is different in the two descriptions. However, the question of terminology for positive- and negative-energy states being left to our own choice, we are free to choose our definitions in such a way as to give the more graphic (and perhaps more

[8] It is assumed that the constant in the energy is chosen so that zero energy occurs approximately midway in the energy gap between the two sets of states.

[9] It is perhaps better to classify the states as states of positive or negative *mass* rather than energy, since the addition of a constant to the energy (by adding a constant to φ, for example) may upset energy classification but not the mass classification.

intuitively satisfying) description of the actual physical events which are being described. In this spirit we feel that the second description is to be preferred since it has a perfectly reasonable classical limit. It would be difficult indeed to picture classically the motion of a particle in a weak field in terms of transitions between free-particle motions with positive and negative mass.

Consider now what happens when the particle interacts with strong rather than weak fields. Under such circumstances, the division of states into those of positive and negative mass is no longer clear-cut, since the energy, separation of the two sets of states is reduced to a relatively small amount. Furthermore, the wave functions describing these states no longer appropriately describe the motion of a particle of fixed sign of mass according to our customary notions. In fact, if we try to interpret the wave function in these terms, we encounter certain well-known paradoxes—the Klein paradox, for example. While the energy of any stationary state will still have a definite sign, the statement that the particle is in a state of positive energy will no longer carry with it the validity of any intuitive conceptions as to the behavior of a classical particle with positive energy, and there will be little qualitative difference between certain states of positive energy and certain states of negative energy. Hence, in the presence of strong fields, the usefulness of a description in terms of positive and negative-energy states will be lost.

These same ideas may be carried over to the case of more general interactions. If there are any advantages to be accrued by the employment of two-component wave functions to describe states of positive and negative energy for a free particle, we may expect these advantages to be present still in the case of weak interactions, but they can scarcely survive in the case of strong interactions. To define *weak interactions* in a more quantitative fashion, we prescribe that the interaction terms have no time Fourier components comparable with or greater than m, so that no transitions between free-particle states with energy differences of this latter order of magnitude are possible; and that the interaction terms have no space Fourier components comparable with or greater than m, so that no transitions between free-particle states with momentum differences of this order of magnitude are possible. Under these conditions, if the initial state is one in which no high momentum states occur, then the momentum of the particle will remain small compared to m under the influence of the interactions. In these circumstances we have essentially a non-relativistic problem, and hence it will be in the domain of *non-relativistic problems* that a representation by means of two-component wave functions may be expected to be of value.

In the presence of interaction it no longer appears possible to make a single, simple, canonical transformation to a representation in which the Hamiltonian is free of odd operators, but instead one can always make a sequence of transformations, each of which eliminates odd operators from the Hamiltonian to one higher order in the expansion parameter[10] $1/m$. In this way one obtains in the new representation a Hamiltonian free of odd operators which is an infinite power series in powers of $1/m$. While it can hardly be expected that this series is convergent, the series is presumably an asymptotic or semi-convergent series in the sense that the sum of a finite number of terms of the series is a better-and-better approximation to the true Hamiltonian, the larger the value of m, provided the magnitude of the interaction remains fixed. The usefulness of the series is then contingent on the interactions being sufficiently weak compared to m in just the sense described above for non-relativistic problems.

Turning our attention now to the explicit calculation in the case where a Dirac particle is subject to interactions, we note first that the Hamiltonian operator can always be put in the form[3]

$$H=\beta m+\mathcal{E}+\mathcal{O}, \tag{29}$$

where $\mathcal{E}$ is an even operator and $\mathcal{O}$ is an odd operator, both of which may be explicitly time-dependent. We assume (and this is ordinarily the case) that $\mathcal{E}$ and $\mathcal{O}$ are of no lower order in $(1/m)$ than $(1/m)^0$. Consider now the canonical transformation generated by the Hermitian operator

$$S=-(i/2m)\beta\mathcal{O}. \tag{30}$$

We note that if the operator S may be regarded as small, then we can make an expansion in powers of $(1/m)$ of the Hamiltonian in the new representation:

$$\begin{aligned} H' &= e^{iS}He^{-iS}-ie^{iS}\left(\frac{\partial e^{-iS}}{\partial t}\right) \\ &= H+\frac{\partial S}{\partial t}+i\left[S, H+\frac{1}{2}\frac{\partial S}{\partial t}\right] \\ &\quad +\frac{i^2}{2!}\left[S,\left[S, H+\frac{1}{3}\frac{\partial S}{\partial t}\right]\right]+\cdots. \end{aligned} \tag{31}$$

Carrying this out explicitly for the Hamiltonian above and retaining terms only to order $(1/m)^2$, we obtain

$$\begin{aligned} H' &= \beta m+\mathcal{E}+\frac{\beta}{2m}\mathcal{O}^2-\frac{1}{8m^2}\left[\mathcal{O},[\mathcal{O},\mathcal{E}]+\frac{\partial\mathcal{O}}{\partial t}\right] \\ &\quad -\frac{1}{2m}\beta\frac{\partial\mathcal{O}}{\partial t}+\frac{\beta}{2m}[\mathcal{O},\mathcal{E}]-\frac{1}{3m^2}\mathcal{O}^3+\cdots, \end{aligned} \tag{32}$$

[10] The expansion parameter may equally well be regarded as $1/c$ where c is the velocity of light. The actual dimensionless expansion parameters are the operators $(\hbar/mc)\nabla$ and $(\hbar/mc^2)(\partial/\partial t)$. From this, one sees that the successive terms of the series will decrease rapidly in magnitude if the interaction potentials do not vary appreciably in space over a Compton wave-length of the particle and in time over the period required for light to travel a Compton wave-length. This is equivalent to the restrictions imposed above on the space and time Fourier components of the interaction.

where we have made use of the fact that β commutes with all even operators and anticommutes with all odd operators in the Dirac theory. Remembering that the product of two even or two odd operators is an even operator and that the product of an odd and an even operator is an odd operator, we see that we have removed by this transformation all odd operators of order $(1/m)^0$ from the Hamiltonian although odd operators of order $(1/m)$ and higher remain and are given in the second line.

From this we see that by a sequence of further canonical transformations, the generator of the transformation at each step being chosen to be

$$S=-(i/2m)\beta\ \text{(odd terms in Hamiltonian of lowest order in } 1/m), \quad (33)$$

we may successively remove odd terms from the Hamiltonian to any desired order in $(1/m)$. If we continue this process indefinitely, we obtain, as mentioned earlier, a Hamiltonian which is an infinite power series in $(1/m)$ completely free of odd operators.

Let us now apply this method to the case where the Dirac particle interacts with an external electromagnetic field. In this case the Hamiltonian is given by

$$H=\beta m-e\varphi+\boldsymbol{\alpha}\cdot(\mathbf{p}-e\mathbf{A}), \quad (34)$$

where φ and $\mathbf{A}$ are the scalar and vector potentials of the electromagnetic field evaluated at the *position* of the particle. In accordance with the procedure outlined above, we then first make the canonical transformation generated by

$$S_1=-\frac{i}{2m}\beta\boldsymbol{\alpha}\cdot(\mathbf{p}-e\mathbf{A}),$$

obtaining then for the new Hamiltonian,

$$H_1=\beta m-e\varphi+\frac{\beta}{2m}(\mathbf{p}-e\mathbf{A})^2-\frac{e}{2m}\beta\boldsymbol{\sigma}\cdot\mathbf{H}$$
$$-\frac{e}{4m^2}\boldsymbol{\sigma}\cdot\mathbf{E}\times(\mathbf{p}-e\mathbf{A})+\frac{e}{8m^2}\operatorname{div}\mathbf{E}-\frac{ie}{2m}\beta\boldsymbol{\alpha}\cdot\mathbf{E}$$
$$-\frac{1}{3m^2}[(\mathbf{p}-e\mathbf{A})^2-e\boldsymbol{\sigma}\cdot\mathbf{H}]\boldsymbol{\alpha}\cdot(\mathbf{p}-e\mathbf{H})+\cdots,$$

to terms of order $(1/m)^2$. Following with the two canonical transformations generated by

$$S_2=-\frac{i}{2m}\beta\left\{-\frac{ie}{2m}\beta\boldsymbol{\alpha}\cdot\mathbf{E}\right\},$$

and

$$S_3=-\frac{i}{2m}\beta\left\{-\frac{e}{4m^2}\beta\boldsymbol{\alpha}\cdot\frac{\partial\mathbf{E}}{\partial t}\right.$$
$$\left.-\frac{1}{3m^2}[(\mathbf{p}-e\mathbf{A})^2-e\boldsymbol{\sigma}\cdot\mathbf{H}]\boldsymbol{\alpha}\cdot(\mathbf{p}-e\mathbf{A})\right\},$$

we eliminate odd operators from the Hamiltonian of order $(1/m)$ and $(1/m)^2$ respectively and obtain the Hamiltonian

$$H_3=\beta m-e\varphi+\frac{\beta}{2m}(\mathbf{p}-e\mathbf{A})^2-\frac{e}{2m}\beta\boldsymbol{\sigma}\cdot\mathbf{H}$$
$$-\frac{e}{4m^2}\boldsymbol{\sigma}\cdot\mathbf{E}\times(\mathbf{p}-e\mathbf{A})+\frac{e}{8m^2}\operatorname{div}\mathbf{E}+\cdots, \quad (35)$$

free of odd operators to order $(1/m)^2$. In the above $\mathbf{H}=\operatorname{curl}\mathbf{A}$ and $\mathbf{E}=-\nabla\varphi-\partial\mathbf{A}/\partial t$ are the magnetic and electric field strengths evaluated at what is now essentially the *mean-position* of the particle. By further canonical transformations we could remove odd operators of still higher order, but we shall limit our discussion of the Hamiltonian in the new representation to terms of order $(1/m)^2$.

Since the Hamiltonian (36) is free of odd operators, its eigenfunctions are two-component functions corresponding again to positive and negative energies. For positive energies, the Schrödinger equation is

$$\left\{m-e\varphi+\frac{1}{2m}(\mathbf{p}-e\mathbf{A})^2-\frac{e}{2m}\boldsymbol{\sigma}\cdot\mathbf{H}-\frac{e}{4m^2}\boldsymbol{\sigma}\cdot\mathbf{E}\times(\mathbf{p}-e\mathbf{A})\right.$$
$$\left.+\frac{e}{8m^2}\operatorname{div}\mathbf{E}\right\}\Phi=i\frac{\partial\Phi}{\partial t}, \quad (36)$$

which will be recognized as essentially the Pauli equations for a non-relativistic particle of spin $\frac{1}{2}$ interacting with the electromagnetic field. The presence of the terms corresponding to the interaction of the anomalous magnetic moment of the particle with the magnetic field and the spin-orbit interaction is evident. The term proportional to div $\mathbf{E}$ is a well-known correction[11] to the Pauli theory arising from the Dirac theory, and it is responsible for a relativistic shift of the S levels in the hydrogen atom (not to be confused with the Lamb-Retherford line shift).

The reason for the explicit appearance of these additional interaction terms (as well as further terms of higher order in $1/m$) in the new representation can now be understood in the light of our physical interpretation of the transformation to the new representation. In the old representation the Dirac particle interacted with the electromagnetic field only at its *position*. However, a particle which in the old representation was located at a point is in the new representation spread out over a region of dimensions of the order of a Compton wavelength in the space of its *mean-position variable*, $\mathbf{X}'=\mathbf{x}$. But in the new representation the interaction between the particle and the electromagnetic field is expressed in terms of the values of the electromagnetic-field quantities at its *mean-position*. Hence, one must expect a series of correction terms of the nature of a multipole

[11] C. G. Darwin, Proc. Roy. Soc. **118A**, 654 (1928).

expansion of the field, since the particle is actually spread out over a finite region in the space of its *mean-position variable*. From this point of view the explicit appearance of the magnetic-moment interaction and the accompanying spin-orbit interaction in the Hamiltonian is to be expected. In fact, the term in div**E**, which has previously been regarded as of rather mysterious origin, can now also easily be understood since it comes from the fact that the electric charge in the new representation is also spread out over a finite region. In a static potential, the particle then moves according to a suitable average of the potential over this region. But in lowest order such an averaging process is known to lead to a term proportional to the Laplacian of the potential and this is just the character of the term in div**E**.

(In employing a finite number of terms in a Hamiltonian such as (35) it must be remembered that wave functions, transition matrix elements, and expectation values of operators computed from the Hamiltonian are only correct to terms of the order in $(1/m)$ to which terms in the Hamiltonian are retained. Thus in the case of (36) it would not be consistent to retain only terms of order $(1/m)$ in the Hamiltonian and then to employ terms of this order in second-order perturbation theory where they generate terms of order $(1/m)^2$.)

The method employed above for reducing the Hamiltonian to non-relativistic two-component form for the case of interaction with an external electromagnetic field can be employed generally for interaction of a Dirac particle with any type of external field, such as various types of meson fields. With meson fields one obtains, in different cases, various types of spin interaction with the meson field and also various types of spin-orbit coupling terms such as have been employed in discussions of spin-orbit coupling in nuclei. The discussion of the reduction to non-relativistic form in the case of the many-particle theory is left for a later communication.

ACKNOWLEDGMENTS

This investigation arose out of a conversation between the authors and Dr. A. S. Wightman to whom we are indebted for subsequent valuable discussions on this subject. One of the authors (L.L.F.) wishes to express his gratitude to the University of Rochester for the hospitality extended to him in the period in which this work was initiated. This work was supported at Case Institute of Technology by the AEC and at the University of Rochester by the joint program of the ONR and the AEC.

Theory of Bloch Electrons in a Magnetic Field: The Effective Hamiltonian*

Walter Kohn
Department of Physics, Carnegie Institute of Technology, Pittsburgh, Pennsylvania
(Received February 26, 1959)

The Hamiltonian of a Bloch electron in a static magnetic field is $H=\frac{1}{2}\mathbf{P}^2+V(\mathbf{r})$, where $V(\mathbf{r})$ is the periodic potential, $\mathbf{P}=\mathbf{p}+\mathbf{A}/c$, and $\mathbf{A}$ is the vector potential giving rise to the magnetic field $\mathfrak{H}$. We consider the case of a nondegenerate band m. It is then shown that, with an error vanishing with $\mathcal{H}$ like $\mathcal{H}^{N+1}$ (N arbitrary), the eigenstates of H can be calculated from an equivalent Hamiltonian $\bar{H}_m(\mathbf{P})$ with the following properties: (1) It is a one-band Hamiltonian, obtained by transforming away all relevant interband matrix elements. (2) It depends only on the gauge-covariant operators P^α. (3) It has the periodicity property $\bar{H}_m(\mathbf{P}+\mathbf{K})=\bar{H}_m(\mathbf{P})$, where $\mathbf{K}$ is an arbitrary reciprocal lattice vector. (4) It can be written as a series $\bar{H}_m(\mathbf{P})=\Sigma_{i=0}{}^N s^i\bar{H}_{m;i}(\mathbf{P})$ where $s\equiv\mathcal{H}/c$ and the functions $\bar{H}_{m;i}(\mathbf{P})$ are completely symmetrized in the noncommuting operators P^α. Properties (3) and (4) can also be summarized in the equations $\bar{H}_m(\mathbf{P})=\Sigma_l a^{(l)}\times\exp[i\mathbf{R}^{(l)}\cdot\mathbf{P}]$, where the $\mathbf{R}^{(l)}$ are lattice vectors and the $a^{(l)}$ can be expanded as $a^{(l)}=\Sigma_{i=0}{}^N s^i a_i{}^{(l)}$. An algorithm is given for the construction of the $\bar{H}_{m;i}$ and carried through for $i=0, 1, 2$. The formalism is not restricted to the neighborhood of the bottom and top of the band. We believe that the equivalent Hamiltonian $\bar{H}_m(\mathbf{P})$ provides a sound basis for a discussion of wave functions and energy levels of Bloch electrons in a magnetic field.

1. Introduction

The study of the physical properties of metals and semiconductors in external magnetic fields has been among the most fruitful methods for obtaining insight into their electronic structure. In many cases the experiments give us information about the electronic energy levels in a magnetic field. Examples are the diamagnetic susceptibility, De Haas-Van Alphen effect, cyclotron resonance, and magneto-optic effects.

It is therefore not surprising that the theory of the motion of Bloch electrons in a uniform magnetic field has received a good deal of attention. Following the classic work of Landau[1] on the quantum theory of free electrons in a magnetic field, the first analysis of Bloch electrons in a magnetic field was carried through by Peierls.[2] This latter work was based on the tight-binding approximation and therefore its results have only qualitative validity. Since then a great many contributions to this problem have been made, some of them dealing with the individual energy levels,[3–10] others emphasizing the free energy of the entire system.[11–14] However, because of the great mathematical complexity of the problem, many authors have found it necessary to use one or the other uncontrolled approximation, so that the reliability of their results remains often in doubt. Thus much work is based on the so-called single-band Hamiltonian

$$H\equiv\epsilon_m(\mathbf{P}),\tag{1.1}$$

where $\epsilon_m(\mathbf{k})$ is the energy band in question,[15a]

$$\mathbf{P}=\mathbf{p}+\mathbf{A}/c,\tag{1.2}$$

and $\mathbf{A}$ is the vector potential giving rise to the uniform magnetic field. At least some interband matrix elements are left out of account in such theories and in certain cases it has been shown that this may lead to very serious errors.

Two relatively recent developments have been of particular interest. One is an expression due to Onsager,[4] which relates the levels in a magnetic field to simple geometrical properties of the energy bands and applies to bands of arbitrary shape. This expression has been extremely helpful in analyzing De Haas-Van Alphen experiments. However it is derived by means of a semi-classical argument whose range of validity is, as the author points out, not entirely clear.

The other question which has been recently discussed by several authors[6,8,9] concerns the extent to which magnetic levels, highly degenerate in the absence of a periodic potential, are spread into bands. That such banding occurs is beyond any doubt, and may be demonstrated in some simple examples. But the width of the bands is a matter which is not settled at present and there is substantial disagreement between different authors.

The present work was undertaken in the hope of clarifying some of the existing uncertainties. It is in a sense an outgrowth of some earlier work,[5,7] but unlike it is not restricted to the vicinity of the bottom or top

* A preliminary account was published in Proc. Phys. Soc. (London) **72**, 1147 (1958).
1 L. Landau, Z. Physik **64**, 629 (1930).
2 R. Peierls, Z. Physik **80**, 763 (1933).
3 J. M. Luttinger, Phys. Rev. **84**, 814 (1951).
4 E. N. Adams II, Phys. Rev. **85**, 41 (1952); L. Onsager, Phil. Mag. **43**, 1006 (1952).
5 J. M. Luttinger and W. Kohn, Phys. Rev. **97**, 869 (1955).
6 P. J. Harper, Proc. Phys. Soc. (London) **A68**, 874, 879 (1955).
7 T. Kjeldaas and W. Kohn, Phys. Rev. **105**, 806 (1957).
8 A. D. Brailsford, Proc. Phys. Soc. (London) **A70**, 275 (1957).
9 G. E. Zil'berman, J. Exptl. Theoret. Phys. U.S.S.R. **32**, 296 (1957) [translation: Soviet Phys. JETP **5**, 208 (1957)].
10 G. E. Zil'berman, J. Exptl. Theoret. Phys. U.S.S.R. **33**, 387 (1957) [translation: Soviet Phys. JETP **6**, 299 (1958)].
11 L. D. Landau and D. Shoenberg, Proc. Roy. Soc. (London) **A170**, 341 (1939), Appendix.
12 A. H. Wilson, Proc. Cambridge Phil. Soc. **49**, 292 (1953).
13 E. N. Adams II, Phys. Rev. **89**, 633 (1953).
14 I. M. Lifschitz and A. M. Kosevich, J. Exptl. Theoret. Phys. U.S.S.R. **29**, 730 (1955) [translation: Soviet Phys. JETP **2**, 636 (1956)].

15a Atomic units are used in this paper.

of an energy band. In this paper we derive an effective single-band Hamiltonian as well as the corresponding basis functions. The diagonalization of this Hamiltonian and the resulting energy levels, wave functions, and free energies will be discussed in a second paper.

Following is a summary of the results obtained in this paper. We are concerned with the solution of the Schrödinger equation

$$H\psi \equiv (\tfrac{1}{2}\mathbf{P}^2+V)\psi = E\psi, \tag{1.3}$$

where $\mathbf{P}$ is defined in Eq. (1.2) and V is a periodic potential. For $\mathbf{A}=0$, the eigenfunctions of H are the Bloch waves

$$\varphi_{n\mathbf{k}} = u_{n\mathbf{k}}(r)e^{i\mathbf{k}\cdot\mathbf{r}}, \tag{1.4}$$

and the corresponding eigenvalues are the energy bands $\epsilon_n(\mathbf{k})$. We fix our attention on solutions of (1.3) which in the limit of vanishing $\mathbf{A}$ go over into Bloch waves belonging to a nondegenerate band denoted by m.

In the presence of a magnetic field, the Hamiltonian H in (1.3) has nonvanishing matrix elements between Bloch waves associated with the band m and those belonging to other bands. We shall explicitly construct a new set of functions $\bar{\varphi}_{n\mathbf{k}}$ such that, *in a certain sense*, $H\bar{\varphi}_{m\mathbf{k}}$ can be expressed as a linear combination of the $\bar{\varphi}_{m\mathbf{k}'}$ with the *same* m. In fact we shall show that

$$H\,\bar{\varphi}_{m\mathbf{k}} \approx \textstyle\sum_{k'} \bar{\varphi}_{m\mathbf{k}'}(\mathbf{k}'|\bar{H}_m(\mathbf{P})|\mathbf{k}), \tag{1.5}$$

where $\bar{H}_m(\mathbf{P})$ is an explicitly constructed function of the operator $\mathbf{P}$, and the notation $(\mathbf{k}'|\ \ |\mathbf{k})$ denotes a matrix element between *plane*-wave states $\mathbf{k}'$ and $\mathbf{k}$.[15b]

The meaning of the phrase "in a certain sense" and of the $\approx$ sign in (1.5) is the following. We construct $\bar{\varphi}_{n\mathbf{k}}$ and $\bar{H}_m(\mathbf{P})$ by a step-by-step process, which can be carried to arbitrary order N in the magnetic field $\mathcal{H}$. If we stop at the Nth order, the equation (1.5) is correct apart from terms which vanish with $\mathcal{H}$ like $\mathcal{H}^{N+1}$. We do not yet know if for small enough $\mathcal{H}$ our process converges strictly or only in an asymptotic sense. To simplify our langugage, we shall assume the former, but if in fact the convergence is only asymptotic our results may have errors which vanish however more rapidly than $\mathcal{H}^N$, where N is arbitrary.[16]

The new basis functions $\bar{\varphi}_{m\mathbf{k}}$ have a "periodicity" property similar to that of ordinary Bloch waves. That is, if $\mathbf{K}$ is a reciprocal lattice vector,

$$\bar{\varphi}_{m\mathbf{k}+\mathbf{K}} = \pm\,\bar{\varphi}_{m\mathbf{k}}. \tag{1.6}$$

The sign here depends on the details of $V(\mathbf{r})$ as well as on $\mathbf{K}$. It is extremely convenient not to restrict $\mathbf{k}$ to a single Brillouin zone but to allow it to run over a very large volume in $\mathbf{k}$ space. Because of (1.6) the $\bar{\varphi}_{m\mathbf{k}}$ are then of course not linearly independent, but the difficulties associated with this *redundancy* are more than balanced by its advantages.

The effective one-band Hamiltonian, $\bar{H}_m(\mathbf{P})$ *includes* all interband effects. If we define

$$s \equiv \mathcal{H}/c, \tag{1.7}$$

it can be expanded in the form

$$\bar{H}_m(\mathbf{P}) = \bar{H}_{m;0}(\mathbf{P}) + s\bar{H}_{m;1}(\mathbf{P}) + s^2\bar{H}_{m;2}(\mathbf{P}) + \cdots, \tag{1.8}$$

where each $H_{m;i}$ is a completely symmetrized function of the noncommuting operators P^α. Furthermore, if we denote the lattice vectors of the crystal by $\mathbf{R}^{(l)}$, each $H_{m;i}$ has a Fourier series expansion of the form

$$\bar{H}_{m;i}(\mathbf{P}) = \textstyle\sum_l a_i^{(l)} \exp(i\mathbf{R}^{(l)}\cdot\mathbf{P}). \tag{1.9}$$

Explicit constructions for the functions

$$\bar{H}_{m;i}(\mathbf{k}) = \textstyle\sum_l a_i^{(l)} \exp(i\mathbf{R}^{(l)}\cdot\mathbf{k}) \tag{1.10}$$

(and hence for the a_i^l) are given. For $i=0$, we find of course

$$\bar{H}_{m;0}(\mathbf{k}) = \epsilon_m(\mathbf{k}), \tag{1.11}$$

which gives rise to the simple one-band Hamiltonian (1.1).

In view of (1.8) and (1.9), $\bar{H}_m(\mathbf{P})$ can also be expressed as a Fourier series,

$$\bar{H}_m(\mathbf{P}) = \textstyle\sum_l a^{(l)} \exp(i\mathbf{R}^{(l)}\cdot\mathbf{P}), \tag{1.12}$$

where

$$a^{(l)} = a_0^{(l)} + sa_1^{(l)} + s^2a_2^{(l)} + \cdots. \tag{1.13}$$

From this form the important periodicity property

$$\bar{H}_m(\mathbf{P}+\mathbf{K}) = \bar{H}_m(\mathbf{P}) \tag{1.14}$$

is apparent.

The Hamiltonian $\bar{H}_m(\mathbf{P})$ of Eq. (1.8) is far from unique. Any unitary transformation $U(\mathbf{P})$ with the appropriate periodicity properties will lead to an equivalent but different Hamiltonian. The leading term $\bar{H}_{m;0}(\mathbf{P})$ is of course common to all these forms.

When the crystal in question has a center of inversion, only even powers of s occur in our expansion (1.8). In this paper the expansion is explicitly carried out up to order s^2. When a center of inversion is absent, also odd powers of s occur. For this case the expansion is explicitly carried through up to order s.

This is the point where the present paper stops. It remains to find the solutions of Eq. (1.3) corresponding to the band m. These can be written in the form

$$\psi = \textstyle\sum_{\mathbf{k}} \bar{A}_m(\mathbf{k})\,\bar{\varphi}_{m\mathbf{k}}, \tag{1.15}$$

and by (1.5) will satisfy the Schrödinger equation, if $\bar{A}_m(\mathbf{k})$ satisfies the equation

$$\textstyle\sum_{\mathbf{k}'}(\mathbf{k}|\bar{H}_m(\mathbf{P})|\mathbf{k}')\bar{A}_m(\mathbf{k}') = E\bar{A}_m(\mathbf{k}). \tag{1.16}$$

The solution of this equation and related questions will be discussed in a subsequent paper.

The results derived in this paper have a very simple structure. It is therefore a pity that the methods by

[15b] For earlier discussions of an effective Hamiltonian, see especially references 3 and 4.

[16] This possibility is of importance in the question of the "banding" of the magnetic levels.

which they are derived here are shockingly complicated. Also the particular expressions we obtain for the higher-order Hamiltonians in Eq. (1.8) are very involved and depend rather surprisingly on such unexpected quantities as the values of the normalized Bloch waves at the arbitrarily chosen origin of coordinates. We have already mentioned the lack of uniqueness of the Hamiltonian. Perhaps much simpler expressions exist and much simpler derivations. If so, we hope that they will in time be found.

PART I

2. Formulation of the Mathematical Problem

Our problem is the solution of the Schrödinger equation

$$H\psi \equiv (\tfrac{1}{2}\mathbf{P}^2+V)\psi = E\psi, \tag{2.1}$$

where V is a periodic potential, $\mathbf{P}$ is the velocity operator

$$\mathbf{P}=\mathbf{p}+\mathbf{A}/c, \tag{2.2}$$

and $\mathbf{A}$ is the vector potential giving rise to a uniform magnetic field $\mathfrak{H}$:

$$\mathfrak{H}=\operatorname{curl}\mathbf{A}. \tag{2.3}$$

The boundary conditions are the usual periodicity conditions on the surface of a large box of volume $\bar{\Omega}$.

We shall be interested in the solutions of (2.1) which correspond to a simple band whose band index we denote by m. By "correspond" we mean that in the limit of vanishing $\mathbf{A}$ the solutions go over into Bloch waves with the band index m; and by a "simple" band we mean a band which has only one wave function for a given $\mathbf{k}$ and which does not touch or intersect another band.

Of course, except in some very special cases, such as $V\equiv 0$, an exact solution of (2.1) is not possible. We shall seek approximate solutions which are valid if the magnetic field is sufficiently weak. However, what constitutes a weak field for one part of the spectrum may not for another. Loosely speaking, we may say that a weak field is one in which the magnetic energy of the electron is small compared to some characteristic energy when the magnetic field is switched off. In atomic units, if we call

$$s\equiv\mathfrak{H}/c, \tag{2.4}$$

this means that

$$s\ll\gamma, \tag{2.5}$$

where γ is a pure number of the general order of magnitude 1. However in some cases, for example when dealing with very small effective masses, γ may be as small as 10^{-1} or 10^{-2}. We shall by a sequence of unitary transformations construct new Hamiltonian, equivalent to H, in the form of a series whose nth term contains a factor s^n. The validity of the expansion can then in each case be checked by inspection of successive terms. In most practical cases a few terms are sufficient.

It is perhaps worth mentioning that the validity of our procedure is quite independent of the temperature. For example, the observation of the De Haas-Van Alphen effect requires a "strong" field, but in the sense

$$s\gtrsim kT. \tag{2.6}$$

Since kT at 1°K equals 10^{-5} in atomic units, the condition (2.5) need by no means rule out the inequality (2.6).

3. Initial Basis Functions

Returning now to Eq. (2.1), it is of course well known that it cannot be solved by simple Schrödinger perturbation theory in powers of s. The effect of even a very weak magnetic field on the eigenfunctions is too profound. Nevertheless it seems natural to begin by writing Eq. (2.1) in the representation of the unperturbed eigenfunctions of H, the Bloch waves, and carry on from there. However, this approach led to technical difficulties which we could not overcome.

The procedure which we could carry through begins with the following basis, introduced in reference 5:

$$\chi_{n\mathbf{k}}\equiv u_{n0}e^{i\mathbf{k}\cdot\mathbf{r}}. \tag{3.1}$$

Here u_{n0} is the periodic part of the Bloch wave

$$\varphi_{n\mathbf{k}}=u_{n\mathbf{k}}e^{i\mathbf{k}\cdot\mathbf{r}}, \tag{3.2}$$

for $\mathbf{k}=0$. The normalization is fixed by the equation

$$\frac{(2\pi)^3}{\Omega}\int_{\text{cell}} u_{n\mathbf{k}}^* u_{n'\mathbf{k}}d\mathbf{r}=\delta_{nn'}, \tag{3.3}$$

where Ω is the volume of the unit crystal cell. Finally it follows from time reversal symmetry that we may fix the phase of $u_{n0}(\mathbf{r})$ such that

$$\operatorname{Im} u_{n0}(\mathbf{r})=0. \tag{3.4}$$

When $\mathbf{k}$ is restricted to the fundamental Brillouin zone, the set $\chi_{n\mathbf{k}}$ forms a complete orthogonal basis. This was the set used in the papers by Luttinger and Kohn[5] and Kjeldaas and Kohn[7] that were concerned with levels near the bottom or top of a band, which involved only small values of $\mathbf{k}$.[17] In the present paper we shall not limit ourselves to levels near the band edges. In particular we want to be able to describe levels which, in a semiclassical description, correspond to Bloch wave packets circulating under the influence of the magnetic field through *several* Brillouin zones. Two courses of action suggest themselves: One is to hold fast to the basis of Luttinger and Kohn and consider very carefully what happens when $\mathbf{k}$ is *on* the Brillouin zone boundary. This procedure we found not tractable. The other course, which we adopt, is to extend the basis (3.1) by letting $\mathbf{k}$ run over a very large volume $\bar{\tau}$ in $\mathbf{k}$ space. Now since the $\chi_{n\mathbf{k}}$ form a complete set when $\mathbf{k}$ runs over any one Brillouin zone, of volume τ, this extended basis has a redundancy of $(\bar{\tau}/\tau)$. The

[17] The results of the present work are in agreement with the conclusions of these earlier papers.

magnitude of $\bar{\tau}$ will of course drop out of our final results just as does the coordinate volume $\bar{\Omega}$ of the crystal.

When we come to count states, the finiteness of both $\bar{\tau}$ and $\bar{\Omega}$ will be made use of. But during most of the following developments it is convenient to regard them first as infinite. Then both **r** and **k** become continuous variables ranging over their entire respective spaces.

4. Schrödinger Equation in the Initial Representation

Since the functions χ_{nk} of our extended basis are more than complete, the solutions ψ of the Schrödinger equation can certainly be expanded in terms of them:

$$\psi=\sum A_n(\mathbf{k})\chi_{n\mathbf{k}}, \tag{4.1}$$

where the symbol $\sum$ includes both summation over n and integration over **k**. However, it is obvious that the coefficients $A_n(\mathbf{k})$ are not unique.

When we substitute (4.1) into the Schrödinger equation (2.3), we are led to consider the effect of $(H-E)$ operating of $\chi_{n\mathbf{k}}$. The resulting function has of course again a highly ambiguous expansion in terms of the $\chi_{n\mathbf{k}}$, but we are entirely at liberty to choose one which is particularly simple.

We follow in this largely the procedure of Kjeldaas and Kohn.[7] We wish first to compute the result of $\frac{1}{2}(\mathbf{p}+\mathbf{A}/c)^2$ acting on $\chi_{n\mathbf{k}}$. We make the following preliminary definitions. If Q is an operator which is a function of x^α and p^α, we define

$$(\mathbf{k}'|Q|\mathbf{k})\equiv\frac{1}{(2\pi)^3}\int e^{-i\mathbf{k}'\cdot\mathbf{r}}Qe^{i\mathbf{k}\cdot\mathbf{r}}d\mathbf{r}. \tag{4.2}$$

Further we call

$$p_{n'n}{}^{\alpha}\equiv\frac{(2\pi)^3}{\Omega}\int_{\text{cell}} u_{n'0}{}^{*}p^{\alpha}u_{n0}d\mathbf{r}. \tag{4.3}$$

With these notations we *may* write

$$\begin{aligned}p^{\alpha}\chi_{n\mathbf{k}}&=p^{\alpha}u_{n0}e^{i\mathbf{k}\cdot\mathbf{r}}\\&=\sum\chi_{n''\mathbf{k}''}[\delta_{n''n}k^{\alpha}(\mathbf{k}''-\mathbf{k})+p_{n''n}{}^{\alpha}\delta(\mathbf{k}''-\mathbf{k})]\\&=\sum\chi_{n''\mathbf{k}''}[\delta_{n''n}(\mathbf{k}''|p^{\alpha}|\mathbf{k})+p_{n''n}{}^{\alpha}\delta(\mathbf{k}''-\mathbf{k})],\end{aligned} \tag{4.4}$$

and

$$\begin{aligned}x^{\alpha}\chi_{n\mathbf{k}}&=x^{\alpha}u_{n0}e^{i\mathbf{k}\cdot\mathbf{r}}\\&=(1/i)(\partial/\partial k^{\alpha})u_{n0}e^{i\mathbf{k}\cdot\mathbf{r}}\\&=\sum\chi_{n''\mathbf{k}''}\delta_{n''n}(1/i)(\partial/\partial k^{\alpha})\delta(\mathbf{k}''-\mathbf{k})\\&=\sum\chi_{n''\mathbf{k}''}[\delta_{n''n}(\mathbf{k}''|x^{\alpha}|\mathbf{k})].\end{aligned} \tag{4.5}$$

Now in a uniform magnetic field we can write

$$P^{\alpha}=p^{\alpha}+s^{\alpha\alpha'}x^{\alpha'} \tag{4.6}$$

(summation over repeated indices implied), where

$$\mathcal{H}_1=c(s^{32}-s^{23}),\ \text{etc.} \tag{4.7}$$

Hence by (4.4) and (4.5) we can write

$$\begin{aligned}P^{\alpha}\chi_{n\mathbf{k}}=\sum&\chi_{n''\mathbf{k}''}\\&\times[\delta_{n''n}(\mathbf{k}''|P^{\alpha}|\mathbf{k})+p_{n''n}{}^{\alpha}\delta(\mathbf{k}''-\mathbf{k})].\end{aligned} \tag{4.8}$$

A second application of P^α results in

$$\begin{aligned}&\tfrac{1}{2}P^{\alpha}P^{\alpha}\chi_{n\mathbf{k}}\\&=\sum\chi_{n'k'}[\delta_{n'n}(\mathbf{k}'|\tfrac{1}{2}P^{\alpha}P^{\alpha}|\mathbf{k})+p_{n'n}{}^{\alpha}(\mathbf{k}'|P^{\alpha}|\mathbf{k})\\&\qquad+\tfrac{1}{2}\textstyle\sum_{n''}p_{n'n''}{}^{\alpha}p_{n''n}{}^{\alpha}\delta(\mathbf{k}'-\mathbf{k})].\end{aligned} \tag{4.9}$$

To this must be added

$$[V(\mathbf{r})-E]\chi_{n\mathbf{k}}=\sum\chi_{n'\mathbf{k}'}(V_{n'n}-E\delta_{n'n})\delta(\mathbf{k}'-\mathbf{k}), \tag{4.10}$$

giving

$$\begin{aligned}&(H-E)\chi_{n\mathbf{k}}\\&=\sum\chi_{n'\mathbf{k}'}[\delta_{n'n}(\mathbf{k}'|\tfrac{1}{2}P^{\alpha}P^{\alpha}|\mathbf{k})+p_{n'n}{}^{\alpha}(\mathbf{k}'|P^{\alpha}|\mathbf{k})\\&+(\tfrac{1}{2}\textstyle\sum_{n''}p_{n'n''}{}^{\alpha}p_{n''n}{}^{\alpha}+V_{n'n}-E\delta_{n'n})\delta(\mathbf{k}'-\mathbf{k})].\end{aligned} \tag{4.11}$$

This may be further simplified by noting that the special case $\mathbf{A}=0$, $\mathbf{k}=0$ gives

$$\tfrac{1}{2}\textstyle\sum_{n''}p_{n'n''}{}^{\alpha}p_{n''n}{}^{\alpha}+V_{n'n}=\epsilon_n\delta_{n'n}, \tag{4.12}$$

where ϵ_n is the energy of the nth band at $\mathbf{k}=0$. With this relation (4.11) now becomes

$$(H-E)\chi_{n\mathbf{k}}=\sum\chi_{n'\mathbf{k}'}(n'\mathbf{k}'|H-E|n\mathbf{k}), \tag{4.13}$$

where

$$\begin{aligned}(n'\mathbf{k}'|H|n\mathbf{k})\equiv\delta_{n'n}[\epsilon_n\delta(\mathbf{k}'-\mathbf{k})+\tfrac{1}{2}(\mathbf{k}'|P^{\alpha}P^{\alpha}|\mathbf{k})]\\+p_{n'n}{}^{\alpha}(\mathbf{k}|P^{\alpha}|\mathbf{k}'),\end{aligned} \tag{4.14}$$

and

$$(n'\mathbf{k}'|E|n\mathbf{k})=E\delta_{n'n}\delta(\mathbf{k}'-\mathbf{k}). \tag{4.15}$$

It should be noted that the matrix $(n'\mathbf{k}'|H|n\mathbf{k})$ is Hermitian.

When we substitute ψ, in the form of Eq. (4.1), into the Schrödinger equation (2.1) we obtain with the help of (4.13) the following *necessary and sufficient* condition on the coefficients $A_n(\mathbf{k})$:

$$\sum_{n'\mathbf{k}'}\sum_{n\mathbf{k}}\chi_{n'\mathbf{k}'}(n'\mathbf{k}'|H-E|n\mathbf{k})A_n(\mathbf{k})=0. \tag{4.16}$$

Now in the customary representation theory one infers from the orthogonality of the basis functions that (4.16) is equivalent to

$$\sum_{n\mathbf{k}}(n'\mathbf{k}'|H-E|n\mathbf{k})A_n(\mathbf{k})=0. \tag{4.17}$$

In the present case, since the $\chi_{n'\mathbf{k}'}$ are not linearly independent, the situation is different:

(a) Any set of coefficients $A_n(\mathbf{k})$, satisfying (4.17) will also satisfy (4.16) and hence the corresponding wave function,

$$\psi=\sum A_n(\mathbf{k})\chi_{n\mathbf{k}}, \tag{4.1}$$

will satisfy the Schrödinger equation (2.1). However the possibility exists (and is in fact realized) that a nonvanishing solution $A_n(\mathbf{k})$ of (4.17) may give rise to an identically vanishing wave function ψ. Equivalently, two different solutions of (4.17) can give rise to the same ψ.

(b) The totality of solutions of (4.17) give rise—via (4.1)—to the totality of solutions of the original Schrödinger equation (2.1). This result is demonstrated in Appendix A. It is not entirely trivial, since solutions of (4.16) may (and do) exist which are not solutions of (4.17). However, these solutions are equivalent (in the sense of giving rise to the same ψ) to other solutions of (4.17).

In view of this situation, the problem of finding all solutions of (2.1) can be divided into two parts: We first determine all solutions of (4.17). Then we construct the corresponding ψ's by means of (4.1) (some of which may vanish identically) and select from these a linearly independent set in terms of which they all can be expanded.

5. Elimination of Interband Matrix Elements

We now turn to the problem of solving Eq. (4.17), which on interchange of the primed and unprimed variables becomes

$$\sum_{n'\mathbf{k}'} (n\mathbf{k}|H|n'\mathbf{k}')A_{n'}(\mathbf{k}')=EA_n(\mathbf{k}); \tag{5.1}$$

here the matrix $(n\mathbf{k}|H|n'\mathbf{k}')$ is given by Eq. (4.14). This equation is an integral equation in $\mathbf{k}$ space and a matrix equation in the band index. Our first aim is to eliminate by a series of canonical transformations the interband matrix elements in (5.1) which connect the band in question, m, with other bands and thus to transform (5.1) into an equation of the form

$$\sum_{\mathbf{k}'}(\mathbf{k}|\bar{H}_m|\mathbf{k}')\bar{A}_m(\mathbf{k}')=E\bar{A}_m(\mathbf{k}). \tag{5.2}$$

First some preliminaries. As we have chosen the functions $u_{n0}(\mathbf{r})$ to be entirely real [see Eq. (3.4)] it follows that all matrix elements $p_{n'n}{}^{\alpha}$, Eq. (4.3), are purely imaginary and that

$$p_{nn'}{}^{\alpha}=-p_{n'n}{}^{\alpha}=p_{n'n}{}^{\alpha *}; \tag{5.3}$$

in particular,

$$p_{nn}{}^{\alpha}=0. \tag{5.4}$$

Further, it will be convenient to define the matrix operator $H(\mathbf{P})$ whose elements with respect to the band indices n are given by the operators

$$H_{nn'}(\mathbf{P})\equiv\delta_{nn'}[\epsilon_n+\tfrac{1}{2}\delta^{\alpha\beta}P^{\alpha}P^{\beta}]+(1-\delta_{nn'})p_{nn'}{}^{\alpha}P^{\alpha}. \tag{5.5}$$

By comparison with (4.14) we have

$$(n\mathbf{k}|H|n'\mathbf{k}')=(\mathbf{k}|H_{nn'}(\mathbf{P})|\mathbf{k}'). \tag{5.6}$$

Now we see from (5.5) that $H(\mathbf{P})$ consists of a part diagonal in n, of zeroth and second order in the operators P^{α}, and an off-diagonal part linear in P^{α}. Our next program is to regard P^{α} as formally small and, by successive unitary transformations, to remove, to higher and higher order in P^{α}, the off-diagonal elements of $H(\mathbf{P})$ which involve the band m.

As a first step we write

$$\bar{H}^{(1)}(\mathbf{P})\equiv\exp[-S^{(1)}(\mathbf{P})]H(\mathbf{P})\exp[S^{(1)}(\mathbf{P})], \tag{5.7}$$

where $S^{(1)}(\mathbf{P})$ is an anti-Hermitian matrix operator, which is taken to be linear in P^{α}:

$$S_{nn'}{}^{(1)}(\mathbf{P})=D_{nn'}{}^{\alpha}P^{\alpha}. \tag{5.8}$$

Expanding the right-hand side of (5.7) in powers of $S^{(1)}(\mathbf{P})$ gives

$$\bar{H}^{(1)}(\mathbf{P})=H(\mathbf{P})+[H(\mathbf{P}),S^{(1)}(\mathbf{P})] \\ +\tfrac{1}{2}[[H(\mathbf{P}),S^{(1)}(\mathbf{P})],S^{(1)}(\mathbf{P})]+\cdots. \tag{5.9}$$

In this expansion the terms linear in P^{α} have the matrix elements

$$(p_{nn'}{}^{\alpha}+\omega_{nn'}D_{nn'}{}^{\alpha})P^{\alpha}, \tag{5.10}$$

where

$$\omega_{nn'}\equiv\epsilon_n-\epsilon_{n'}. \tag{5.11}$$

In order that $\bar{H}^{(1)}(\mathbf{P})$ have no linear interband matrix elements involving the band m we must choose the coefficients $D_{nn'}{}^{\alpha}$ so that

$$p_{mn}{}^{\alpha}+\omega_{mn}D_{mn}{}^{\alpha}=0,\quad n\neq m, \tag{5.12}$$

and, because of the anti-Hermitian property of $S^{(1)}$,

$$D_{nm}{}^{\alpha}=-(D_{mn}{}^{\alpha})^{*}. \tag{5.13}$$

Obviously these requirements still leave us considerable freedom in the choice of the constants $D_{nn'}{}^{\alpha}$. The simplest choice would be $D_{mn}{}^{\alpha}=-(D_{nm}{}^{\alpha})^{*}=-p_{mn}{}^{\alpha}/\omega_{mn}$, for $n\neq m$, and all other $D_{nn'}{}^{\alpha}=0$. However for reasons which will become apparent later on we make a slightly more general choice, namely

$$D_{mn}{}^{\alpha}=\delta_{mn}iC_m{}^{\alpha}-(1-\delta_{mn})p_{mn}{}^{\alpha}/\omega_{mn}; \tag{5.14}$$

$$D_{nm}{}^{\alpha}=-(D_{mn}{}^{\alpha})^{*}; \tag{5.15}$$

and

$$D_{nn'}{}^{\alpha}=0,\quad n\neq m \text{ and } n'\neq m. \tag{5.16}$$

Here the $C_m{}^{\alpha}$ are real numbers, for the time being quite arbitrary, so that $iC_m{}^{\alpha}P^{\alpha}$ is properly anti-Hermitian.

The matrix elements of $\bar{H}^{(1)}(\mathbf{P})$ involving the band m are now, by Eq. (5.9),

$$\bar{H}_{mn}{}^{(1)}(\mathbf{P})=\delta_{mn}\left\{\epsilon_m+\left[\tfrac{1}{2}\delta^{\alpha\beta}+\sum_{n'}\frac{p_{mn'}{}^{\alpha}p_{n'm}{}^{\beta}}{\omega_{mn'}}\right]P^{\alpha}P^{\beta}+\cdots\right\} \\ +(1-\delta_{mn})\left\{\left[-iC_m{}^{\alpha}p_{mn}{}^{\beta}+\sum_{n'}\frac{p_{mn'}{}^{\alpha}p_{n'n}{}^{\beta}}{\omega_{mn}}\right]P^{\alpha}P^{\beta}+\cdots\right\}, \tag{5.17}$$

$$\bar{H}_{nm}{}^{(1)}(\mathbf{P})=[\bar{H}_{nm}{}^{(1)}(\mathbf{P})]^{\dagger}. \tag{5.18}$$

Thus by our construction $\bar{H}_{mn}{}^{(1)}(\mathbf{P})$ contains, for $n\neq m$, only terms of second order in P^{α}. To emphasize this feature we write $\bar{H}_{mn}^{(1)}(\mathbf{P})$ in the form

$$\bar{H}_{mn}^{(1)}(\mathbf{P})=\delta_{mn}\bar{H}_m^{(1)}(\mathbf{P}) + (1-\delta_{mn})\{Q_{mn}^{\alpha\beta}P^\alpha P^\beta+\cdots\}. \quad (5.19)$$

We now make a second unitary transformation and define

$$\bar{H}^{(2)}(\mathbf{P})\equiv\exp[-S^{(2)}(\mathbf{P})]\bar{H}^{(1)}(\mathbf{P})\exp[S^{(2)}(\mathbf{P})], \quad (5.20)$$

where $S^{(2)}$ is of second order in P^α and so chosen that $\bar{H}_{mn}^{(2)}$ has, for $n\neq m$, no terms of lower order than the third in P^α. Proceeding as before, this may be achieved by taking

$$S_{nn'}^{(2)}(\mathbf{P})=D_{nn'}^{\alpha\beta}P^\alpha P^\beta, \quad (5.21)$$

where

$$D_{mn}^{\alpha\beta}=\delta_{mn}iC_m^{\alpha\beta}-(1-\delta_{mn})Q_{mn}^{\alpha\beta}/\omega_{mn}; \quad (5.22)$$

$$D_{nm}^{\alpha\beta}=-(D_{mn}^{\alpha\beta})^*; \quad (5.23)$$

$$D_{nn'}^{\alpha\beta}=0, \quad n\neq m \text{ and } n'\neq m. \quad (5.24)$$

Here the constants $C_m^{\alpha\beta}$ are for the time being largely arbitrary, except that in view of the anti-Hermitian nature of $S^{(2)}$, $C_m^{\alpha\beta}P^\alpha P^\beta$ must be Hermitian:

$$(C_m^{\beta\alpha})^*P^\alpha P^\beta=C_m^{\alpha\beta}P^\alpha P^\beta. \quad (5.25)$$

Clearly, this procedure may be continued so that after the tth transformation we have

$$\bar{H}_{mn}^{(t)}(\mathbf{P})=\delta_{mn}\bar{H}_m^{(t)}(\mathbf{P}) + (1-\delta_{mn})[Q_{mn}^{\alpha_1\cdots\alpha_{t+1}}P^{\alpha_1}\cdots P^{\alpha_{t+1}}+\cdots]. \quad (5.26)$$

The subsequent transformation is then given by

$$S_{nn'}^{(t+1)}(\mathbf{P})=D_{nn'}^{\alpha_1\cdots\alpha_{t+1}}P^{\alpha_1}\cdots P^{\alpha_{t+1}}, \quad (5.27)$$

where

$$D_{mn}^{\alpha_1\cdots\alpha_{t+1}}=\delta_{mn}iC_m^{\alpha_1\cdots\alpha_{t+1}} -(1-\delta_{mn})Q_{mn}^{\alpha_1\cdots\alpha_{t+1}}/\omega_{mn}; \quad (5.28)$$

$$D_{nm}^{\alpha_1\cdots\alpha_{t+1}}=-D_{mn}^{\alpha_1\cdots\alpha_{t+1}}; \quad (5.29)$$

$$D_{nn'}^{\alpha_1\cdots\alpha_{t+1}}=0, \quad n\neq m \text{ and } n'\neq m; \quad (5.30)$$

and the numbers $C_m^{\alpha_1\cdots\alpha_{t+1}}$ are arbitrary, apart from the requirement that

$$C_m^{\alpha_1\cdots\alpha_{t+1}}P^{\alpha_1}\cdots P^{\alpha_{t+1}}=C_m^{\alpha_{t+1}\cdots\alpha_1}P^{\alpha_1}\cdots P^{\alpha_{t+1}}. \quad (5.31)$$

From the power series of $\exp(-S^{(t+1)}\bar{H}^{(t)}\exp(S^{(t+1)})$ we see that

$$\bar{H}_m^{(t+1)}(\mathbf{P})=\bar{H}_m^{(t)}(\mathbf{P})+O(P^{t+2}). \quad (5.32)$$

Thus all $\bar{H}_m^{(t)}$ for $t\geqslant t_0$ are identical up to order $(P^\alpha)^{t_0+1}$ inclusive. We now define $\bar{H}_m(\mathbf{P})$ as the formal limit of $\bar{H}_m^{(t)}$ as $t\to\infty$. It can be written in the form

$$\bar{H}_m(\mathbf{P})=\epsilon_m+E_m^{\alpha_1\alpha_2}P^{\alpha_1}P^{\alpha_2} +E_m^{\alpha_1\alpha_2\alpha_3}P^{\alpha_1}P^{\alpha_2}P^{\alpha_3}+\cdots, \quad (5.33)$$

where in view of (5.32) $E_m^{\alpha_1\cdots\alpha_t}$ can be obtained by calculating $H_m^{(t-1)}$.

We shall not discuss here the question of convergence of this procedure. However, in Sec. 10 we shall estimate the error of the solutions obtained.

6. Commutation Expansion of $H_m(P)$

We now wish to write $\bar{H}_m(\mathbf{P})$ in a series of terms each of which is "effectively" smaller by a factor $s(=\mathcal{H}/c)$ than the preceding one. Since, by (4.6),

$$P^\alpha=p^\alpha+s^{\alpha\alpha'}x^{\alpha'},$$

one's first inclination might be to order the series (5.33) in powers of s. However, in such a series successive terms are not effectively smaller by a factor of s. The reason for this can be most easily seen in the free electron case. Here

$$H(\mathbf{P})=\tfrac{1}{2}P^\alpha P^\alpha=\tfrac{1}{2}p^\alpha p^\alpha+s^{\alpha\alpha'}(p^\alpha x^{\alpha'}+x^\alpha p^{\alpha'}) +s^{\alpha\alpha'}s^{\alpha\alpha''}x^{\alpha'}x^{\alpha''}. \quad (6.1)$$

If we now consider solutions of (approximately) fixed energy, then as $s\to 0$, the dimensions over which the orbits extend behave as s^{-1}, so that terms of the type sx occurring in (6.1) are in fact independent of s.

A more appropriate procedure is to make what we call a commutator expansion of Eq. (5.33). Let us begin with some preliminaries. First we define

$$\langle P^{\alpha_1}\cdots P^{\alpha_t}\rangle_{\mathrm{Av}}, \quad (6.2)$$

as the average of $P^{\alpha_1}\cdots P^{\alpha_t}$ over all possible permutations of the factors. Since for $s=0$ the P^α's commute, it is clear that the difference between $P^{\alpha_1}\cdots P^{\alpha_t}$ and its average must be at least linear in s. For example,

$$P^{\alpha_1}P^{\alpha_2}-\langle P^{\alpha_1}P^{\alpha_2}\rangle_{\mathrm{Av}}=\tfrac{1}{2}(P^{\alpha_1}P^{\alpha_2}-P^{\alpha_2}P^{\alpha_1}) =\tfrac{1}{2}i(s^{\alpha_1\alpha_2}-s^{\alpha_2\alpha_1}) =\frac{is}{2}\left(\frac{s^{\alpha_1\alpha_2}-s^{\alpha_2\alpha_1}}{s}\right), \quad (6.3)$$

where the factor in parentheses is independent of the magnitude of s. In general, we can write

$$P^{\alpha_1}\cdots P^{\alpha_t}=\langle P^{\alpha_1}\cdots P^{\alpha_t}\rangle_{\mathrm{Av}}+s\Delta_1, \quad (6.4)$$

where Δ_1 is a linear combination of products of P^α, each of order $(P^\alpha)^{t-2}$. Now we define Δ_{r+1} by the equation

$$s\Delta_{r+1}=\Delta_r-\langle\Delta_r\rangle_{\mathrm{Av}}. \quad (6.5)$$

Then clearly we can develop $P^{\alpha_1}\cdots P^{\alpha_t}$ in a series of the form

$$P^{\alpha_1}\cdots P^{\alpha_t}=\langle P^{\alpha_1}\cdots P^{\alpha_t}\rangle_{\mathrm{Av}} +s\langle\Delta_1\rangle_{\mathrm{Av}}+\cdots s^{[t/2]}\langle\Delta_{[t/2]}\rangle_{\mathrm{Av}}, \quad (6.6)$$

where

$$[t/2]=t/2 \quad \text{for } t \text{ even}, \qquad =(t-1)/2 \quad \text{for } t \text{ odd}. \quad (6.7)$$

Such an expansion we call a commutator expansion.

We now develop each term of the power series (5.33) of $\bar{H}_m(\mathbf{P})$ in a commutator expansion and then collect all terms corresponding to a given power of s. This results in the commutator expansion of $\bar{H}_m(\mathbf{P})$ which has the form

$$\bar{H}_m(\mathbf{P})=\bar{H}_{m;0}(\mathbf{P})+s\bar{H}_{m;1}(\mathbf{P})+s^2\bar{H}_{m;2}(\mathbf{P})+\cdots. \quad (6.8)$$

Each term $\bar{H}_{m;i}(P)$ in this expansion is a linear combination of completely symmetrized products of the P^α. We shall see later that in general all $\bar{H}_{m;i}(\mathbf{P})$ are of similar magnitude so that (6.8) is a suitable expansion for sufficiently weak fields.

Let us now recall that any completely symmetrized *operator* $M(\mathbf{P})$ is completely defined by the *function* $M(\mathbf{k})$ to which it reduces when the operators P^α are replaced by the c-numbers k^α.[18] Hence each of the Hamiltonians $\bar{H}_{m;i}(\mathbf{P})$ is completely characterized by the corresponding function $\bar{H}_{m;i}(\mathbf{k})$. The procedure which we have followed gives us power series expansions for these functions. For example, from (5.17) and (6.8) we see that[19]

$$\bar{H}_{m;0}(\mathbf{k})=\epsilon_m+\left[\tfrac{1}{2}\delta^{\alpha\beta}+\sum_{n'}\frac{p_{mn'}{}^{\alpha}p_{n'm}{}^{\beta}}{\omega_{mn'}}\right]k^\alpha k^\beta+\cdots, \quad (6.9)$$

$$\bar{H}_{m;1}(\mathbf{k})=\left[\frac{i}{2}\frac{s^{\alpha\beta}}{s}\sum_{n'}\frac{p_{mn'}{}^{\alpha}p_{n'm}{}^{\beta}-p_{mn'}{}^{\beta}p_{n'm}{}^{\alpha}}{\omega_{mn'}}\right]+\cdots. \quad (6.10)$$

The deficiency of this formalism, as developed so far, is then that it is restricted to sufficiently small $\mathbf{k}$-vectors for which these series converge adequately. One of the main objectives of Part II of this paper is to obtain expressions for the functions $\bar{H}_{m;i}(\mathbf{k})$ without recourse to power series expansions.

PART II

To obtain explicit expressions for the functions $\bar{H}_{m;i}(\mathbf{k})$ we shall be following a rather complicated procedure. It may therefore be helpful if we give here a brief outline.

In Sec. 7 we show very easily that $\bar{H}_{m;0}(\mathbf{k})$ is just the energy band function $\epsilon_m(\mathbf{k})$. This has the well-known periodicity property

$$\epsilon_m(\mathbf{k}+\mathbf{K})=\epsilon_m(\mathbf{k}),$$

where $\mathbf{K}$ is a reciprocal lattice vector.

This suggests that also the functions $\bar{H}_{m;i}(\mathbf{k})$ for $i>0$ might have this property. However, these functions depend on the as yet largely arbitrary coefficients $C_m{}^{\alpha_1\cdots\alpha_t}$ occurring in the canonical transformations $S^{(t)}$ [see Eqs. (5.28), (5.31)]. In Secs. 8–14 it is shown how these constants can be chosen so that (a) the $\bar{H}_{m;i}(\mathbf{k})$ can be evaluated in terms of the Bloch waves $\varphi_{n\mathbf{k}}$ and without recourse to power series expansions in k^α, (b) they have the periodicity property

$$\bar{H}_{m;i}(\mathbf{k}+\mathbf{K})=\bar{H}_{m;i}(\mathbf{k}).$$

This is accomplished in several steps. In Sec. 8 it is shown that, when $s=0$, the $C_m{}^{\alpha_1\cdots\alpha_t}$ can be so chosen that the new basis functions generated by the unitary transformations $S^{(t)}$,

$$\varphi_{m\mathbf{k}}=\sum\chi_{n'\mathbf{k}'}(n'\mathbf{k}'|\exp(S^{(1)})(\exp S^{(2)})\cdots|m\mathbf{k}),$$

have the property of being *real* at the arbitrarily chosen origin $\mathbf{r}=0$. This fixes the phase of these functions—which are just Bloch waves—so as to assure their periodicity in the sense that

$$\varphi_{m\mathbf{k}+\mathbf{K}}=\pm\varphi_{m\mathbf{k}}.$$

In Secs. 9, 10, and 11 the same results are obtained when $s\neq 0$. The new basis functions,

$$\bar{\varphi}_{m\mathbf{k}}=\sum\chi_{n'\mathbf{k}'}(n'\mathbf{k}'|\exp(S^{(1)})\exp(S^{(2)})\cdots|m\mathbf{k}),$$

are constructed so as to be real at $\mathbf{r}=0$, and they are periodic in the same sense as $\varphi_{m\mathbf{k}}$. Furthermore the unitary transformation matrix connecting $\bar{\varphi}_{m\mathbf{k}}$ with $\chi_{n'\mathbf{k}'}$ is constructed without recourse to a power series expansion in k^α.

In Sec. 12 the algorithm for constructing the functions $\bar{H}_{m;i}(\mathbf{k})$, having the properties (a) and (b) above is developed; and in Secs. 13 and 14 it is applied to construct the first few $\bar{H}_{m;i}(\mathbf{k})$ $(i=0,1,\cdots)$ for crystals with and without a center of inversion, respectively.

7. Discussion of the Leading Term $\bar{H}_{m;0}(P)$

In this section we shall derive very easily an explicit expression for $\bar{H}_{m;0}(\mathbf{P})$ which makes evident the important periodicity property

$$\bar{H}_{m;0}(\mathbf{P}+\mathbf{K})=\bar{H}_{m;0}(\mathbf{P}), \quad (7.1)$$

where $\mathbf{K}$ is any reciprocal lattice vector. In subsequent sections, we shall discuss the higher order Hamiltonians, $\bar{H}_{m;i}(\mathbf{P})$ with $i>0$.

When the magnetic field is switched off, clearly

$$\bar{H}_m(\mathbf{P})\rightarrow\bar{H}_{m;0}(\mathbf{p}). \quad (7.2)$$

Now the eigenfunctions of $\bar{H}_{m;0}(\mathbf{p})$—as those of $\mathbf{p}$ itself—are plane waves $\exp(i\mathbf{k}\cdot\mathbf{r})$ and the corresponding eigenvalues are $\bar{H}_{m;0}(\mathbf{k})$. But these must be just the energies of the Bloch waves $\varphi_{m;\mathbf{k}}$ so that

$$\bar{H}_{m;0}(\mathbf{k})=\epsilon_m(\mathbf{k}). \quad (7.3)$$

We recall again that a completely symmetrized operator $M(\mathbf{P})$ is uniquely determined by the function $M(\mathbf{k})$ to which it reduces when $\mathbf{P}$ is replaced by $\mathbf{k}$. In particular, if

$$M(\mathbf{k})=\sum_{\mathbf{R}}A(\mathbf{R})e^{i\mathbf{R}\cdot\mathbf{k}}, \quad (7.4)$$

then

$$M(\mathbf{P})=\sum_{\mathbf{R}}A(\mathbf{R})e^{i\mathbf{R}\cdot\mathbf{P}}. \quad (7.5)$$

[18] The function $k^{\alpha_1}\cdots k^{\alpha_t}$ defines uniquely the operator $\langle P^{\alpha_1}\cdots P^{\alpha_t}\rangle_{\mathrm{Av}}$.

[19] It so happens that the leading term of order k^0, which is here given, vanishes.

This fact may be applied in the present situation. For $\epsilon_m(\mathbf{k})$ is a periodic function of $\mathbf{k}$ and hence may be written as a Fourier series of the form

$$\epsilon_m(\mathbf{k})=\sum_l a_0^{(l)} \exp(i\mathbf{R}^{(l)}\cdot\mathbf{k}), \tag{7.6}$$

where the $\mathbf{R}^{(l)}$ are the lattice vectors of the crystal. Therefore we have

$$\bar{H}_{m;0}(\mathbf{P})=\sum_l a_0^{(l)} \exp(i\mathbf{R}^{(l)}\cdot\mathbf{P}). \tag{7.7}$$

From this form we see at once that $\bar{H}_{m;0}(\mathbf{P})$ has the important periodicity property (7.1), analogous to the periodicity of $\epsilon_m(k)$.

It is sometimes a convenient notation to write

$$\bar{H}_{m;0}(\mathbf{P})=\langle\epsilon_m(\mathbf{P})\rangle_{\mathrm{Av}}, \tag{7.8}$$

where $\epsilon_m(\mathbf{P})$ is any function of the operator $\mathbf{P}$ which reduces to $\epsilon_m(\mathbf{k})$ when $\mathbf{P}$ is replaced by $\mathbf{k}$.

We wish next to draw attention to the *uniqueness* of $\bar{H}_{m;0}(\mathbf{P})$, in spite of the fact that the sequence of unitary transformations $S^{(t)}$ leading to $\bar{H}_m(\mathbf{P})$ contained a great deal of arbitrariness [see comments after Eqs. (5.13) and (5.30)]. How this uniqueness comes about may be understood as follows. Suppose we consider a different sequence of canonical transformations $S'^{(t)}$ which also uncouple the band m from the rest. The resulting effective Hamiltonian $\bar{H}_m'(\mathbf{P})$ must then be related to $\bar{H}_m(\mathbf{P})$ by a unitary transformation, $\exp T_m(\mathbf{P})$. Thus

$$\begin{aligned}\bar{H}_m'(\mathbf{P})&=\exp[-T_m(\mathbf{P})]\bar{H}_m(\mathbf{P})\exp[T_m(\mathbf{P})]\\&=\bar{H}_m(\mathbf{P})+[\bar{H}_m(\mathbf{P}),T_m(\mathbf{P})]+\cdots.\end{aligned} \tag{7.9}$$

When we now make a commutator expansion of $\bar{H}_m'(\mathbf{P})$ only $\bar{H}_m(\mathbf{P})$ in (7.9) can contribute to the leading term $\bar{H}_{m;0}'(\mathbf{P})$. For all the following terms, because of their form as commutators, contain one or more factors of s. The identity of $\bar{H}_{m;0}'(\mathbf{P})$ and $\bar{H}_{m;0}(\mathbf{P})$ is now obvious. On the other hand, the higher order terms in the commutator expansion (6.8) do depend on the particular choice of the transformations $S^{(t)}$.

8. Periodicity in k of Bloch Waves

We have just seen that for $i>0$, the functions $\bar{H}_{m;i}(\mathbf{k})$ depend on the choice of the coefficients $C_m^{\alpha_1\cdots\alpha_t}$. We shall find it extremely useful that they may be chosen in such a way that, in a certain gauge, the new basis functions are real at $\mathbf{r}=0$. These functions can then be shown to be "periodic" in the variable $\mathbf{k}$. We begin by demonstrating this in the absence of a magnetic field.

For $\mathbf{A}=0$ we have of course $\mathbf{P}=\mathbf{p}$, and therefore the Hamiltonian (4.14) is diagonal in $\mathbf{k}$:

$$(n\mathbf{k}|H|n'\mathbf{k}')=\delta(\mathbf{k}-\mathbf{k}')H_{nn'}(\mathbf{k}). \tag{8.1}$$

Consequently the solutions of the Schrödinger equation (5.1) are of the form

$$A_{n'}^{(n)}(\mathbf{k})=B_{n'}^{(n)}(\mathbf{k}^0)\delta(\mathbf{k}-\mathbf{k}^0), \tag{8.2}$$

and the corresponding eigenfunctions of H are just the Bloch waves $\varphi_{n\mathbf{k}^0}$ associated with the vector $\mathbf{k}^0$:

$$\begin{aligned}\varphi_{n\mathbf{k}^0}&=\sum_{n'} A_{n'}^{(n)}(\mathbf{k})\chi_{n\mathbf{k}}\\&=\exp(i\mathbf{k}^0\cdot\mathbf{r})\sum_{n'} B_{n'}^{(n)}(\mathbf{k}^0)u_{n'0}(\mathbf{r}).\end{aligned} \tag{8.3}$$

In this case, where the Hamiltonian is already initially diagonal in $\mathbf{k}$, the removal of the interband matrix elements leads to a complete diagonalization. Therefore

$$\varphi_{m\mathbf{k}}=\{\exp[S^{(1)}(\mathbf{p})]\exp[S^{(2)}(\mathbf{p})]\cdots\}\chi_{m\mathbf{k}}, \tag{8.4}$$

where $\varphi_{m\mathbf{k}}$ is a Bloch wave associated with quantum numbers m and $\mathbf{k}$. We say *a* Bloch wave, because its *phase* will depend on the disposition of the arbitrary constants in the transformations $S^{(t)}$.

We shall now show how the $C_m^{\alpha_1\cdots\alpha_t}$ can be chosen to make

$$\mathrm{Im}[\varphi_{m\mathbf{k}}(0)]\equiv 0. \tag{8.5}$$

The procedure will be to assure the property (8.5) up to any power of k^α, from which the reality for all $\mathbf{k}$ follows by analytic continuation in the variables $k^{(1)}$, $k^{(2)}$, and $k^{(3)}$.[20]

We have previously chosen the functions $u_{n0}(\mathbf{r})$ as real. Let us now further assume that $u_{m0}(0)>0$.[21] Next we write

$$\begin{aligned}\varphi_{m\mathbf{k}}^{(t)}&\equiv\sum\chi_{n\mathbf{k}'}\\&\times(n\mathbf{k}'|\{\exp[S^{(1)}(\mathbf{p})]\exp[S^{(2)}(\mathbf{p})]\cdots\}_t|m\mathbf{k}),\end{aligned} \tag{8.6}$$

where the notation $\{\ \}_t$ implies a power series expansion of $\exp[S^{(1)}]\exp[S^{(2)}]\cdots$ in powers of p^α up to order $(p^\alpha)^t$ inclusive. Clearly $\varphi_{m\mathbf{k}}^{(t)}$ has the form

$$\begin{aligned}\varphi_{m\mathbf{k}}^{(t)}(\mathbf{r})=e^{i\mathbf{k}\cdot\mathbf{r}}[&u_{m0}(\mathbf{r})+k^\alpha u_m^\alpha(\mathbf{r})\\&+\cdots k^{\alpha_1}\cdots k^{\alpha_t}u_m^{\alpha_1\cdots\alpha_t}(\mathbf{r})],\end{aligned} \tag{8.7}$$

where the functions $u_m^{\alpha_1\cdots\alpha_t}$ are periodic in $\mathbf{r}$. We must now show that $\varphi_{m\mathbf{k}}^{(t)}(0)$ can be made real for all t.

To lowest order we have

$$\varphi_{m\mathbf{k}}^{(0)}(0)=[u_{m0}(\mathbf{r})e^{i\mathbf{k}\cdot\mathbf{r}}]_{\mathbf{r}=0}, \tag{8.8}$$

and this is clearly real.

To first order we obtain, with the help of Eqs. (5.8), (5.14), and (5.15),

$$\begin{aligned}\varphi_{m\mathbf{k}}^{(1)}(0)&\\&=\{\textstyle\sum_n u_{n0}(\mathbf{r})[\delta_{nm}+S_{nm}^{(1)}(\mathbf{p})]e^{i\mathbf{k}\cdot\mathbf{r}}\}_{\mathbf{r}=0}\\&=\left\{e^{i\mathbf{k}\cdot\mathbf{r}}\left[u_{m0}(\mathbf{r})+k^\alpha\left(iC_m^\alpha u_{m0}-\sum_{n\neq m}u_{n0}\frac{p_{nm}^\alpha}{\omega_{nm}}\right)\right]\right\}_{\mathbf{r}=0}\\&=u_{m0}(0)+k^\alpha\left(iC_m^\alpha u_{m0}(0)-\sum_{n\neq m}u_{n0}(0)\frac{p_{nm}^\alpha}{\omega_{nm}}\right).\end{aligned} \tag{8.9}$$

[20] This assumes of course, without proof, the analyticity of $\varphi_{m\mathbf{k}}$ as a function of $\mathbf{k}$. For one-dimensional Bloch waves this analyticity has been established under mild restrictions by W. Kohn, Proc. Phys. Soc. (London) **72**, 301 (1958), and to be published.

[21] If $u_{m0}(0)$ should accidentally vanish, we choose another origin.

As $u_{m0}(0)$ and $u_{n0}(0)$ are real, and the $p_{nm}{}^{\alpha}$ are imaginary, this can be made real by choosing

$$C_m{}^{\alpha}=\frac{1}{i}\sum_{n\neq m} r_n \frac{p_{nm}{}^{\alpha}}{\omega_{nm}}, \tag{8.10}$$

where

$$r_n \equiv u_{n0}(0)/u_{m0}(0). \tag{8.11}$$

It will be seen that $C_m{}^{\alpha}$, defined by (8.10), is real as required. [See comment after Eq. (5.16).]

Proceeding in this manner it is easily seen that $\varphi_{m\mathbf{k}}{}^{(t)}(0)$ can be made real by an appropriate choice of real coefficients $C_m{}^{\alpha_1\cdots\alpha_t}$, which may be taken as invariant under permutation of the indices $\alpha_1\cdots\alpha_t$.

It may be mentioned in passing that if the crystal has a center of inversion at $\mathbf{r}=0$, the $C_m{}^{\alpha_1\cdots\alpha_t}$ can be taken as zero.

This procedure defines $\varphi_{m\mathbf{k}}(\mathbf{r})$ in terms of a power series in k^{α} of the form (8.7). By analytic continuation this function is then defined for all $\mathbf{k}$. It clearly satisfies the following conditions: It satisfies the wave equation; it is normalized; it is real at $\mathbf{r}=0$; and it has the quasi-periodicity property

$$\varphi_{m\mathbf{k}}(\mathbf{r}+\mathbf{R}^{(l)})=\exp[i\mathbf{k}\cdot\mathbf{R}^{(l)}]\varphi_{m\mathbf{k}}(\mathbf{r}). \tag{8.12}$$

These conditions define $\varphi_{m\mathbf{k}}$ uniquely, apart from sign. Consequently it must have the following "periodicity" property, as a function of $\mathbf{k}$:

$$\varphi_{m\mathbf{k}+\mathbf{K}}(\mathbf{r})=\theta\varphi_{m\mathbf{k}}(\mathbf{r}), \tag{8.13}$$

where for a given crystal and band index m, $\theta=\pm 1$ depending on the vector $\mathbf{K}$. Examples show that both signs do in fact occur. However for simplicity of writing we shall assume in the following that $\theta\equiv 1$, so that

$$\varphi_{m\mathbf{k}+\mathbf{K}}(\mathbf{r})=\varphi_{m\mathbf{k}}(\mathbf{r}). \tag{8.14}$$

Where the case $\theta=-1$ leads to significant differences, they will be explicitly mentioned.

9. Expansion of the New Basis Functions in Powers of P^{α}

The canonical transformations $S^{(1)}, S^{(2)}, \cdots$ discussed in Sec. 5 define a new set of basis functions which we shall denote by $\bar{\varphi}_{n\mathbf{k}}$:

$$\bar{\varphi}_{n\mathbf{k}}=\sum \chi_{n'\mathbf{k}'} \times(n'\mathbf{k}'|\exp[S^{(1)}(\mathbf{P})]\exp[S^{(2)}(\mathbf{P})]\cdots|n\mathbf{k}). \tag{9.1}$$

For $\mathbf{A}=0$ these were just the Bloch waves $\varphi_{n\mathbf{k}}$ discussed in the preceding section. In this section we begin a study of the $\bar{\varphi}_{n\mathbf{k}}$ in the presence of a magnetic field. We shall first show that if we define

$$\bar{\varphi}_{n\mathbf{k}}{}^{(t)}\equiv\sum \chi_{n'\mathbf{k}'} \times(n'\mathbf{k}'|\{\exp[S^{(1)}(\mathbf{P})]\exp[S^{(2)}(\mathbf{P})]\cdots\}_t|m\mathbf{k}), \tag{9.2}$$

where the symbol $\{\ \}_t$ denotes the truncated power series of the argument up to $(P^{\alpha})^t$, the constants $C_m{}^{\alpha_1\cdots\alpha_{t'}}$ can be so chosen as to make

$$\mathrm{Im}[\bar{\varphi}_{m\mathbf{k}}{}^{(t)}(0)]=0, \tag{9.3}$$

for all values of t. Since the phases of the $\bar{\varphi}_{n\mathbf{k}}$ depend on the gauge of the vector potential, we must settle on a particular choice. It is convenient to take

$$\mathbf{A}=\tfrac{1}{2}(\mathcal{H}\times\mathbf{r}); \tag{9.4}$$

in this gauge we write $s^{\alpha\alpha'}=\sigma^{\alpha\alpha'}$, so that

$$P^{\alpha}=p^{\alpha}+\sigma^{\alpha\alpha'}x^{\alpha'}, \tag{9.5}$$

and

$$\sigma^{32}=-\sigma^{23}=\mathcal{H}_1/c, \quad \text{etc.}, \tag{9.6}$$

$$\sigma^{11}=\sigma^{22}=\sigma^{33}=0. \tag{9.7}$$

Our final results will, however, be gauge invariant.

The procedure of removing off-diagonal elements to higher and higher order in P^{α}, which was described in Sec. 5, gives us $\bar{\varphi}_{n\mathbf{k}}{}^{(t)}$ in the form of a series

$$\bar{\varphi}_{n\mathbf{k}}{}^{(t)}=\sum_{t'=0}^{t} w_{n\mathbf{k}}{}^{(t')}, \tag{9.8}$$

where $w_{n\mathbf{k}}{}^{(t')}$ is of order $(P^{\alpha})^{t'}$. Let us begin by studying the first two of these, for $n=m$.

Clearly for $t'=0$ we have

$$w_{m\mathbf{k}}{}^{(0)}(\mathbf{r})=\chi_{m\mathbf{k}}(\mathbf{r})=u_{m0}(\mathbf{r})e^{i\mathbf{k}\cdot\mathbf{r}}, \tag{9.9}$$

and since all $u_{n0}(0)$ are real, so is $w_{m\mathbf{k}}{}^{(0)}(0)$. The next term is, in view of Eqs. (5.8), (5.14), and (5.15),

$$w_{m\mathbf{k}}{}^{(1)}(\mathbf{r})=\sum u_{n0}(\mathbf{r}) \times[iC_m{}^{\alpha}\delta_{nm}-(1-\delta_{nm})p_{nm}{}^{\alpha}/\omega_{nm}]P^{\alpha}e^{i\mathbf{k}\cdot\mathbf{r}}, \tag{9.10}$$

whose imaginary part at $\mathbf{r}=0$ is

$$\mathrm{Im}w_{m\mathbf{k}}{}^{(1)}(0)=u_{m0}(0)\left(iC_m{}^{\alpha}-\sum_{n\neq m} r_n\frac{p_{nm}{}^{\alpha}}{\omega_{nm}}\right)k^{\alpha}; \tag{9.11}$$

where r_n is defined by Eq. (8.11). This may be made to vanish by the same choice of $C_m{}^{\alpha}$, as in the absence of the magnetic field [see Eq. (8.10)].

So far there has been no significant difference between the cases of nonvanishing and vanishing magnetic fields. But in higher orders ($t'\geqslant 2$), the noncommutativity of the components of $\mathbf{P}$ does introduce such differences. We shall now show that nevertheless there exists a set of constants $C_m{}^{\alpha_1}, \cdots C_m{}^{\alpha_1\cdots\alpha_t}$ which will make $\varphi_{m\mathbf{k}}{}^{(t)}(0)$ real for arbitrary t, that these constants satisfy the requirement that the quantity

$$\sum_{\alpha_1\cdots\alpha_t} C_m{}^{\alpha_1\cdots\alpha_t}P^{\alpha_1}\cdots P^{\alpha_t} \tag{9.12}$$

is Hermitian [see Eq. (5.31)], and are real for odd t and imaginary for even t.

We give a proof by induction. Suppose we have chosen appropriate constants $C_m{}^{\alpha_1}, \cdots C_m{}^{\alpha_1\cdots\alpha_{t-1}}$. Then

$w_{m\mathbf{k}}^{(t)}$ has the form

$$w_{m\mathbf{k}}^{(t)}=\sum_n u_{n0} \sum_{\alpha_1\cdots\alpha_t} (i\delta_{nm}C_m^{\alpha_1\cdots\alpha_t}+Q_{nm}^{\alpha_1\cdots\alpha_t}) \times P^{\alpha_1}\cdots P^{\alpha_t}e^{i\mathbf{k}\cdot\mathbf{r}}, \quad (9.13)$$

where the constants $Q_{nm}^{\alpha_1\cdots\alpha_t}$ are real if t is even and imaginary if t is odd. Let us take the case of even t. From (9.13) we have, at $\mathbf{r}=0$,

$$w_{m\mathbf{k}}^{(t)}(0)=u_{m0}(0) \sum_{\alpha_1\cdots\alpha_t} (iC_m^{\alpha_1\cdots\alpha_t}+R_m^{\alpha_1\cdots\alpha_t}) \times[P^{\alpha_1}\cdots P^{\alpha_t}e^{i\mathbf{k}\cdot\mathbf{r}}]_{\mathbf{r}=0}, \quad (9.14)$$

where

$$R_m^{\alpha_1\cdots\alpha_t}\equiv\sum_n r_n Q_{nm}^{\alpha_1\cdots\alpha_t} \quad (9.15)$$

is real. We want to choose the $C_m^{\alpha_1\cdots\alpha_t}$ to make

$$\mathrm{Im}(w_{m\mathbf{k}}^{(t)})=0. \quad (9.16)$$

Now the product $P^{\alpha_1}\cdots P^{\alpha_t}$ may be divided, in a gauge-invariant manner, into a Hermitian and anti-Hermitian operator. One way of effecting this division is by a commutator expansion. For example,

$$\begin{aligned}P^{\alpha_1}P^{\alpha_2}&=\langle P^{\alpha_1}P^{\alpha_2}\rangle_{\mathrm{Av}}+\tfrac{1}{2}(P^{\alpha_1}P^{\alpha_2}-P^{\alpha_2}P^{\alpha_1})\\&=\langle P^{\alpha_1}P^{\alpha_2}\rangle_{\mathrm{Av}}+(i/2)\sigma^{\alpha_1\alpha_2}\\&=(P^{\alpha_1}P^{\alpha_2})_H+(P^{\alpha_1}P^{\alpha_2})_A, \quad (9.17)\end{aligned}$$

since

$$\sigma^{\alpha_1\alpha_2}=\tfrac{1}{2}(s^{\alpha_1\alpha_2}-s^{\alpha_2\alpha_1})=(1/i)[P^{\alpha_1},P^{\alpha_2}], \quad (9.18)$$

and is gauge invariant; the subscripts H and A denote, respectively, the Hermitian and anti-Hermitian parts. Similarly

$$\begin{aligned}&P^{\alpha_1}P^{\alpha_2}P^{\alpha_3}P^{\alpha_4}\\&=\langle P^{\alpha_1}P^{\alpha_2}P^{\alpha_3}P^{\alpha_4}\rangle_{\mathrm{Av}}\\&+i(\sigma^{\alpha_1\alpha_2}\langle P^{\alpha_3}P^{\alpha_4}\rangle_{\mathrm{Av}}+\sigma^{\alpha_1\alpha_3}\langle P^{\alpha_2}P^{\alpha_4}\rangle_{\mathrm{Av}}+\sigma^{\alpha_1\alpha_4}\langle P^{\alpha_2}P^{\alpha_3}\rangle_{\mathrm{Av}}\\&+\sigma^{\alpha_2\alpha_3}\langle P^{\alpha_1}P^{\alpha_4}\rangle_{\mathrm{Av}}+\sigma^{\alpha_2\alpha_4}\langle P^{\alpha_1}P^{\alpha_3}\rangle_{\mathrm{Av}}+\sigma^{\alpha_3\alpha_4}\langle P^{\alpha_1}P^{\alpha_2}\rangle_{\mathrm{Av}})\\&-(\sigma^{\alpha_1\alpha_2}\sigma^{\alpha_3\alpha_4}+\sigma^{\alpha_1\alpha_3}\sigma^{\alpha_2\alpha_4}+\sigma^{\alpha_1\alpha_4}\sigma^{\alpha_2\alpha_3})\\&=(P^{\alpha_1}P^{\alpha_2}P^{\alpha_3}P^{\alpha_4})_H+(P^{\alpha_1}P^{\alpha_2}P^{\alpha_3}P^{\alpha_4})_A, \quad (9.19)\end{aligned}$$

where the Hermitian part contains all terms even in σ, and the anti-Hermitian part all terms which are odd in σ. The general product $P^{\alpha_1}\cdots P^{\alpha_t}$ can be similarly decomposed.

We can therefore write $R_m^{\alpha_1\cdots\alpha_t}$ as the sum of two terms,

$$R_m^{\alpha_1\cdots\alpha_t}=H_m^{\alpha_1\cdots\alpha_t}+A_m^{\alpha_1\cdots\alpha_t}, \quad (9.20)$$

where

$$\sum_{\alpha_1\cdots\alpha_t} H_m^{\alpha_1\cdots\alpha_t}P^{\alpha_1}\cdots P^{\alpha_t} = \sum_{\alpha_1\cdots\alpha_t} R_m^{\alpha_1\cdots\alpha_t}(P^{\alpha_1}\cdots P^{\alpha_t})_H \quad (9.21)$$

and

$$\sum_{\alpha_1\cdots\alpha_t} A_m^{\alpha_1\cdots\alpha_t}P^{\alpha_1}\cdots P^{\alpha_t} = \sum_{\alpha_1\cdots\alpha_t} R_m^{\alpha_1\cdots\alpha_t}(P^{\alpha_1}\cdots P^{\alpha_t})_A; \quad (9.22)$$

both $H_m^{\alpha_1\cdots\alpha_t}$ and $A_m^{\alpha_1\cdots\alpha_t}$ are real.

We now choose

$$C_m^{\alpha_1\cdots\alpha_t}=iA_m^{\alpha_1\cdots\alpha_t}. \quad (9.23)$$

By (9.22) we have

$$\sum_{\alpha_1\cdots\alpha_t} C_m^{\alpha_1\cdots\alpha_t}P^{\alpha_1}\cdots P^{\alpha_t} =i \sum_{\alpha_1\cdots\alpha_t} R_m^{\alpha_1\cdots\alpha_t}(P^{\alpha_1}\cdots P^{\alpha_t})_A, \quad (9.24)$$

and this is properly Hermitian. Furthermore this choice gives, by (9.14),

$$w_{m\mathbf{k}}^{(t)}(0)=u_{m0}(0) \sum_{\alpha_1\cdots\alpha_t} R_m^{\alpha_1\cdots\alpha_t} \times[(P^{\alpha_1}\cdots P^{\alpha_t})_H e^{i\mathbf{k}\cdot\mathbf{r}}]_{\mathbf{r}=0}. \quad (9.25)$$

It remains to verify that, *in the gauge* (9.4), the imaginary part of (9.25) vanishes. Since the $R_m^{\alpha_1\cdots\alpha_t}$ are real, inspection of (9.17) and (9.19) shows that this will be so provided that for arbitrary t

$$\mathrm{Im}[\langle P^{\alpha_1}\cdots P^{\alpha_t}\rangle_{\mathrm{Av}}e^{i\mathbf{k}\cdot\mathbf{r}}]_{\mathbf{r}=0}=0. \quad (9.26)$$

To establish this fact substitute in the product $P^{\alpha_1}\cdots P^{\alpha_t}$ the expression (9.5) and pull all factors x to the right of all factors p. This results in

$$[(P^{\alpha_1}\cdots P^{\alpha_t})e^{i\mathbf{k}\cdot\mathbf{r}}]_{\mathbf{r}=0} =k^{\alpha_1}\cdots k^{\alpha_t}+i\sigma^{\alpha_1\alpha_2}k^{\alpha_3}\cdots k^{\alpha_t}+\cdots. \quad (9.27)$$

When we now perform the operation $\langle\ \rangle_{\mathrm{Av}}$ on (9.27), all terms containing one or more factors $\sigma^{\alpha\beta}$ vanish because of the antisymmetry of $\sigma^{\alpha\beta}$. Hence we have

$$[\langle P^{\alpha_1}\cdots P^{\alpha_t}\rangle_{\mathrm{Av}}e^{i\mathbf{k}\cdot\mathbf{r}}]_{\mathbf{r}=0}=k^{\alpha_1}\cdots k^{\alpha_t}, \quad (9.28)$$

which proves (9.26).

When t is odd an analogous demonstration can be given.

We have therefore succeeded in constructing functions of the form (9.2), or equivalently of the form

$$\begin{aligned}\bar{\varphi}_{n\mathbf{k}}^{(t)}&=\sum_{n'} u_{n'0}[\delta_{n'n}+\sum_{\alpha_1} G_{n'n}^{\alpha_1}P^{\alpha_1}+\cdots\\&\quad+\sum_{\alpha_1\cdots\alpha_t} G_{n'n}^{\alpha_1\cdots\alpha_t}P^{\alpha_1}\cdots P^{\alpha_t}]e^{i\mathbf{k}\cdot\mathbf{r}}\\&=\sum \chi_{n'\mathbf{k}'}(\mathbf{k}'|\delta_{n'n}+\sum_{\alpha_1} G_{n'n}^{\alpha_1}P^{\alpha_1}+\cdots\\&\quad+\sum_{\alpha_1\cdots\alpha_t} G_{n'n}^{\alpha_1\cdots\alpha_t}P^{\alpha_1}\cdots P^{\alpha_t}|\mathbf{k}), \quad (9.29)\end{aligned}$$

which for $n=m$ have the property (9.3).

10. Commutator Expansion of the New Basis Functions; Degree of Decoupling

The expansion (9.29) has the drawback of converging only for small values of k. Our previous experience suggests rearranging the terms in (9.29) in a commutator expansion. We therefore write formally

$$\delta_{n'n}+\sum_{\alpha_1} G_{n'n}{}^{\alpha_1}P^{\alpha_1}+\sum_{\alpha_1\alpha_2} G_{n'n}{}^{\alpha_1\alpha_2}P^{\alpha_1}P^{\alpha_2}+\cdots$$
$$=U_{n'n;0}(\mathbf{P})+sU_{n'n;1}(\mathbf{P})+s^2U_{n'n;2}(\mathbf{P})+\cdots, \quad (10.1)$$

where the operators $U_{n'n;i}(\mathbf{P})$ are completely symmetrized in the P^α. Because of this property they can be completely characterized by the functions $U_{n'n;i}(\mathbf{k})$ of the c-numbers k^α. For example, we see from (10.1) that

$$U_{n'n;0}(\mathbf{k})=\delta_{n'n}+\sum_{\alpha_1} G_{n'n}{}^{\alpha_1}k^{\alpha_1}+\sum_{\alpha_1\alpha_2} G_{n'n}{}^{\alpha_1\alpha_2}k^{\alpha_1}k^{\alpha_2}+\cdots,$$
$$U_{n'n;1}(\mathbf{k})=\sum_{\alpha_1\alpha_2} (i\sigma^{\alpha_1\alpha_2}/2s)G_{nn'}{}^{\alpha_1\alpha_2}+\cdots, \quad (10.2)$$

etc. The functions $U_{n'n;i}(\mathbf{k})$ are defined by their power-series in k^α where these converge, and by analytic continuation[22] elsewhere.

Combining (9.29) and (10.1) gives the following formal expansion for $\bar{\varphi}_{n\mathbf{k}}$:

$$\bar{\varphi}_{n\mathbf{k}}(\mathbf{r})=\sum u_{n'0}[U_{n'n;0}(\mathbf{P})+sU_{n'n;1}(\mathbf{P})+s^2U_{n'n;2}(\mathbf{P})+\cdots]e^{i\mathbf{k}\cdot\mathbf{r}} \quad (10.3)$$
$$=\sum \chi_{n'\mathbf{k}'}(\mathbf{k}'\,|\,U_{n'n;0}(\mathbf{P})+sU_{n'n;1}(\mathbf{P})+\cdots|\,\mathbf{k}).$$

For some purposes it is convenient to define

$$F_{n;i}(\mathbf{r};\,\mathbf{P})\equiv\sum_{n'} u_{n'0}(\mathbf{r})U_{n'n;i}(\mathbf{P}). \quad (10.4)$$

They are functions of $\mathbf{r}$ and completely symmetrized functions of the operators P^α. As the definition (10.4) shows, the P^α are to be thought of as on the *right* of the $\mathbf{r}$, i.e., not acting on it. With the help of these objects, we can rewrite (10.3) also in the compact form

$$\bar{\varphi}_{n\mathbf{k}}(\mathbf{r})=[F_{n;0}(\mathbf{r};\,\mathbf{P})+sF_{n;1}(\mathbf{r};\,\mathbf{P})+s^2F_{n;2}(\mathbf{r};\,\mathbf{P})+\cdots]e^{i\mathbf{k}\cdot\mathbf{r}}. \quad (10.5)$$

So far our entire development has been purely formal. We are now in a position to examine to what extent the functions $\bar{\varphi}_{m\mathbf{k}}$ have in fact been "decoupled" from the $\bar{\varphi}_{n\mathbf{k}}$ with $n\neq m$.

Let us first note that the transformation functions $S^{(t)}(\mathbf{P})$ have been so constructed that the equation

$$H_{nn'}(\mathbf{P})(n'\,|\exp[S^{(1)}(\mathbf{P})]\exp[S^{(2)}(\mathbf{P})]\cdots|\,m)$$
$$-(n\,|\exp[S^{(1)}(\mathbf{P})]\exp[S^{(2)}(\mathbf{P})]\cdots|\,m)\times\bar{H}_m(\mathbf{P})=0 \quad (10.6)$$

is a formal identity when both terms are formally expanded in powers of P^α. If now these power series are regrouped in a commutator expansion, (10.6) is also clearly a formal identity to all orders of s of this expansion. Let us now denote by $[Q(\mathbf{P})]_i$ the commutator expansion of any function Q of the operators P^α up to order s^i. Then clearly we must have

$$H_{nn'}(\mathbf{P})(n'\,|[\exp[S^{(1)}(\mathbf{P})]\exp[S^{(2)}(\mathbf{P})]\cdots]_i|\,m)$$
$$-(n\,|[\exp[S^{(1)}(P)]\exp[S^{(2)}(P)]\cdots]_i|\,m)\times[H_m(P)]_i=O(s^{i+1}). \quad (10.7)$$

Now let us denote by $\bar{\varphi}_{n\mathbf{k};i}$ the series (10.5) truncated after the power s^i:

$$\bar{\varphi}_{n\mathbf{k};i}\equiv[F_{n;0}(\mathbf{r};\,\mathbf{P})+\cdots s^iF_{n;i}(\mathbf{r};\,\mathbf{P})]e^{i\mathbf{k}\cdot\mathbf{r}}$$
$$=\sum u_{n'0}(n'\,|[\exp[S^{(1)}(\mathbf{P})]\exp[S^{(2)}(\mathbf{P})]\cdots]_i|\,n)\times e^{i\mathbf{k}\cdot\mathbf{r}}. \quad (10.8)$$

Then (10.7) can be rewritten as

$$H\bar{\varphi}_{m\mathbf{k};i}=\sum_{\mathbf{k}'} \bar{\varphi}_{m\mathbf{k}';i}(\mathbf{k}'\,|\,\bar{H}_{m;0}(\mathbf{P})+\cdots+s^i\bar{H}_{m;i}(\mathbf{P})\,|\,\mathbf{k})+O(s^{i+1}). \quad (10.9)$$

This answers our question concerning the degree of decoupling: $\bar{\varphi}_{m\mathbf{k};i}$ is decoupled up to order s^i inclusive. Consequently, if we solve the one-band equation

$$\sum_{\mathbf{k}'}(\mathbf{k}\,|[\bar{H}_m(\mathbf{P})]_i|\,\mathbf{k}')\bar{A}_m(\mathbf{k}')=E\bar{A}_m(\mathbf{k}), \quad (10.10)$$

where

$$[\bar{H}_m(\mathbf{P})]_i=\bar{H}_{m;0}(\mathbf{P})+s\bar{H}_{m;1}(\mathbf{P})+\cdots s^i\bar{H}_{m;i}(\mathbf{P}), \quad (10.11)$$

the energy E and the corresponding wave function

$$\psi=\sum \bar{A}_m(\mathbf{k})\,\bar{\varphi}_{m\mathbf{k};i} \quad (10.12)$$

will have errors which vanish with s like s^{i+1}.

It may be remarked that if we define

$$F_m(\mathbf{r})\equiv\sum A_m(\mathbf{k})e^{i\mathbf{k}\cdot\mathbf{r}}, \quad (10.13)$$

then (10.10) is equivalent to the equation

$$[\bar{H}_m(\mathbf{P})]_iF_m(\mathbf{r})=EF_m(\mathbf{r}). \quad (10.14)$$

11. Analytic Continuation of New Basis Functions

In the preceding section we have seen that the new basis functions $\bar{\varphi}_{n\mathbf{k}}(\mathbf{r})$ can be developed in a commutator expansion of the form (10.5). As the operators $F_{n;i}(\mathbf{r};\,\mathbf{P})$ are completely symmetrized in the components of P^α, they are uniquely determined by the functions $F_{n;i}(\mathbf{r};\,\mathbf{k})$. The procedure of the previous section gave us a series development of these functions in powers of k^α. In the present section we shall show how they can be obtained for arbitrary $\mathbf{k}$ without recourse to a power series.

To determine $F_{m;0}(\mathbf{r};\,\mathbf{k})$ we first consider the case $s=0$ which has already been treated in Sec. 8. Here one finds

$$\bar{\varphi}_{m\mathbf{k}}=\varphi_{m\mathbf{k}}=F_{m;0}(\mathbf{r};\,\mathbf{k})e^{i\mathbf{k}\cdot\mathbf{r}}, \quad (11.1)$$

so that

$$F_{m;0}(\mathbf{r};\,\mathbf{k})=u_{m\mathbf{k}}(\mathbf{r}), \quad (11.2)$$

where, by our construction, $\mathrm{Im}[u_{m\mathbf{k}}(0)]=0$.

[22] Again the analyticity is assumed without proof; see reference 20.

This gives us at once the result that in the presence of a magnetic field

$$\bar{\varphi}_{m\mathbf{k}}(\mathbf{r}) = u_{m\mathbf{P}}(\mathbf{r})e^{i\mathbf{k}\cdot\mathbf{r}} + O(s), \tag{11.3}$$

where $\mathbf{P}$ does not act on the argument $\mathbf{r}$ of $u_{m\mathbf{P}}$, and the operator $u_{m\mathbf{P}}$ is considered completely symmetrized in the components P^α. Now since

$$u_{m\mathbf{P}}(\mathbf{r})e^{i\mathbf{k}\cdot\mathbf{r}} = u_{m,\mathbf{k}+(1/c)\mathbf{A}(\mathbf{r})}(\mathbf{r})e^{i\mathbf{k}\cdot\mathbf{r}} + O(s), \tag{11.4}$$

we can also write

$$\begin{aligned}\bar{\varphi}_{m\mathbf{k}(\mathbf{r})} &= u_{m,\mathbf{k}+(1/c)\mathbf{A}(\mathbf{r})}(\mathbf{r})e^{i\mathbf{k}\cdot\mathbf{r}} + O(s) \\ &= \varphi_{m,\mathbf{k}+(1/c)\mathbf{A}(\mathbf{r})}(\mathbf{r}) \exp[-i(1/c)\mathbf{A}(\mathbf{r})\cdot\mathbf{r}] \\ &\quad + O(s).\end{aligned} \tag{11.5}$$

We now turn to the higher-order terms in the commutator expansion of $\bar{\varphi}_{m\mathbf{k}}$. As we are interested in the case of small s but arbitrary values of $\mathbf{k}$, it is natural to try and discuss the whole problem in a new representation based on the Bloch waves at an arbitrary point $\mathbf{g}$ in $\mathbf{k}$ space, rather than the $\chi_{n\mathbf{k}}$ representation which was based on the Bloch waves u_{n0} at $\mathbf{k}=0$. In analogy with (3.1) we therefore define the following new basis functions:

$$\chi_{n\mathbf{h}}(\mathbf{r};\mathbf{g}) \equiv \varphi_{n\mathbf{g}}(\mathbf{r}) \exp(i\mathbf{h}\cdot\mathbf{r}), \tag{11.6}$$

where the phases of the $\varphi_{n\mathbf{g}}$ are so chosen that

$$\operatorname{Im}\varphi_{n\mathbf{g}}(0) = 0, \tag{11.7}$$

and the sign of $\operatorname{Re}\varphi_{m\mathbf{g}}(0)$ is that which one obtains by analytic continuation of $\varphi_{m\mathbf{k}}$ from $\mathbf{k}=0$, under the condition (11.7). For $n\neq m$ the sign of $\varphi_{n\mathbf{g}}$ may be arbitrarily chosen. In complete analogy with Eqs. (5.5) and (5.6), we find that in this representation the Hamiltonian matrix is

$$(n\mathbf{h}|H|n'\mathbf{h}') = (\mathbf{h}|H_{nn'}(\mathbf{P};\mathbf{g})|\mathbf{h}'), \tag{11.8}$$

where

$$\begin{aligned}H_{nn'}(\mathbf{P};\mathbf{g}) &= \delta_{nn'}[\epsilon_n(\mathbf{g}) + p_{nn}{}^\alpha(\mathbf{g})P^\alpha + \tfrac{1}{2}\delta^{\alpha\beta}P^\alpha P^\beta] \\ &\quad + (1-\delta_{nn'})p_{nn'}{}^\alpha(\mathbf{g})P^\alpha;\end{aligned} \tag{11.9}$$

here $\epsilon_n(\mathbf{g})$ is the energy of band n at wave vector $\mathbf{g}$ while

$$p_{nn'}{}^\alpha(\mathbf{g}) \equiv \frac{(2\pi)^3}{\Omega}\int_{\text{cell}} \varphi_{n\mathbf{g}}{}^* p^\alpha \varphi_{n\mathbf{g}} d\mathbf{r}. \tag{11.10}$$

As before, we shall now eliminate the interband matrix elements of (11.9) and in this way be led to new basis functions $\bar{\varphi}_{n\mathbf{h}}(\mathbf{r};\mathbf{g})$. For $m=n$ it will be shown that these are related as follows to the functions $\bar{\varphi}_{m\mathbf{k}}(\mathbf{r})$ of Sec. 10:

$$\bar{\varphi}_{m\mathbf{h}}(\mathbf{r};\mathbf{g}) = \bar{\varphi}_{m\mathbf{g}+\mathbf{h}}(\mathbf{r}). \tag{11.11}$$

This connection will enable us to construct the function $F_{m;i}(\mathbf{r};\mathbf{k})$ of Eq. (10.5) as well as the functions $\bar{H}_{m;i}(\mathbf{k})$ of Eq. (6.8) without recourse to power series expansions in k^α.

We begin by eliminating the first-order off-diagonal matrix elements of $H_{nn'}(\mathbf{P};\mathbf{g})$ by means of the unitary transformation $\exp[S^{(1)}(\mathbf{P};\mathbf{g})]$:

$$\begin{aligned}\bar{H}^{(1)}(\mathbf{P};\mathbf{g}) &\equiv \exp[-S^{(1)}(\mathbf{P};\mathbf{g})]H(\mathbf{P};\mathbf{g})\exp[S^{(1)}(\mathbf{P};\mathbf{g})] \\ &= H(\mathbf{P};\mathbf{g}) + [H(\mathbf{P};\mathbf{g}), S^{(1)}(\mathbf{P};\mathbf{g})] \\ &\quad + \cdots.\end{aligned} \tag{11.12}$$

We take $S^{(1)}$ of the form

$$S_{nn'}{}^{(1)}(P;g) = D_{nn'}{}^\alpha(\mathbf{g})P^\alpha, \tag{11.13}$$

where the $D_{nn'}{}^\alpha(\mathbf{g})$ must satisfy the condition

$$p_{mn}{}^\alpha(\mathbf{g}) + \omega_{mn}(\mathbf{g})D_{mn}{}^\alpha(\mathbf{g}) = 0 \quad n\neq m, \tag{11.14}$$

and

$$D_{nn'}{}^\alpha(\mathbf{g}) = -[D_{n'n}{}^\alpha(\mathbf{g})]^*; \tag{11.15}$$

here

$$\omega_{nn'}(\mathbf{g}) \equiv \epsilon_n(\mathbf{g}) - \epsilon_{n'}(\mathbf{g}). \tag{11.16}$$

We choose

$$\begin{aligned}D_{mn}{}^\alpha(\mathbf{g}) &= \delta_{mn} iC_m{}^\alpha(\mathbf{g}) \\ &\quad - (1-\delta_{mn})p_{mn}{}^\alpha(\mathbf{g})/\omega_{mn}(\mathbf{g}),\end{aligned} \tag{11.17}$$

$$D_{nm}{}^\alpha = -(D_{mn}{}^\alpha)^*, \tag{11.18}$$

$$D_{nn'}{}^\alpha = 0, \quad n\neq m \text{ and } n'\neq m \tag{11.19}$$

[see Eq. (5.14) ff]. As before, we choose the real constants $C_m{}^\alpha(\mathbf{g})$ so as to make the new basis functions

$$\begin{aligned}\bar{\varphi}_{m\mathbf{h}}{}^{(1)}(\mathbf{r};\mathbf{g}) &= \sum_{n,\mathbf{h}'} \chi_{n\mathbf{h}'}(\mathbf{r};\mathbf{g}) \\ &\quad \times (n\mathbf{h}'|1+S^{(1)}(\mathbf{P};\mathbf{g})|m\mathbf{h}),\end{aligned} \tag{11.20}$$

real at $\mathbf{r}=0$, to first order in P^α. This requires

$$iC_m{}^\alpha(\mathbf{g}) = \frac{1}{\varphi_{m\mathbf{g}}(0)} \operatorname{Im} \sum_{n\neq m} \varphi_{n\mathbf{g}}(0)\frac{p_{nm}{}^\alpha(\mathbf{g})}{\omega_{nm}(\mathbf{g})}. \tag{11.21}$$

Continuing in this manner we construct, in analogy with Eq. (9.29), the functions

$$\begin{aligned}\bar{\varphi}_{n\mathbf{h}}{}^{(t)}(\mathbf{r};\mathbf{g}) &= \sum_{n'} \varphi_{n'\mathbf{g}}[\delta_{n'n} + \sum_{\alpha_1} G_{n'n}{}^{\alpha_1}(\mathbf{g})P^{\alpha_1} \\ &\quad + \cdots + \sum_{\alpha_1\cdots\alpha_t} G_{n'n}{}^{\alpha_1\cdots\alpha_t}(\mathbf{g})P^{\alpha_1}\cdots P^{\alpha_t}] \\ &\quad \times \exp(i\mathbf{h}\cdot\mathbf{r}).\end{aligned} \tag{11.22}$$

These series may be rearranged in a commutator series leading to an expansion analogous to Eq. (10.3)

$$\begin{aligned}\bar{\varphi}_{n\mathbf{h}}(\mathbf{r};\mathbf{g}) &= \textstyle\sum_{n'} \varphi_{n'\mathbf{g}}[U_{n'n;0}(\mathbf{P};\mathbf{g}) \\ &\quad + sU_{n'n;1}(\mathbf{P};\mathbf{g}) + \cdots] \exp(i\mathbf{h}\cdot\mathbf{r}),\end{aligned} \tag{11.23}$$

where the functions $U_{n'n;i}(\mathbf{P};\mathbf{g})$ are completely symmetrized in the P^α.

Now let us set

$$\mathbf{h} = \mathbf{k} - \mathbf{g}, \tag{11.24}$$

and use the following identity, valid for any function $Z(\mathbf{P})$:

$$\begin{aligned}Z(\mathbf{P}) &\exp[i(\mathbf{k}-\mathbf{g})\cdot\mathbf{r}] \\ &= \exp[-i\mathbf{g}\cdot\mathbf{r}]Z(\mathbf{P}-\mathbf{g})\exp[i\mathbf{k}\cdot\mathbf{r}].\end{aligned} \tag{11.25}$$

This allows us to re-write (11.23) as

$$\bar{\varphi}_{n\mathbf{k}-\mathbf{g}}(\mathbf{r};\,\mathbf{g})=\sum_{n'}\varphi_{n'\mathbf{g}}\exp(-i\mathbf{g}\cdot\mathbf{r})[U_{n'n;0}(\mathbf{P}-\mathbf{g};\,\mathbf{g})+sU_{n'n;1}(\mathbf{P};\,\mathbf{g})+\cdots]e^{i\mathbf{k}\cdot\mathbf{r}}. \quad (11.26)$$

Next we observe that the functions $\varphi_{n'\mathbf{g}}\exp(-i\mathbf{g}\cdot\mathbf{r})=u_{n'\mathbf{g}}(\mathbf{r})$ are completely periodic in $\mathbf{r}$, so that we can write

$$\varphi_{n'\mathbf{g}}\exp(-i\mathbf{g}\cdot\mathbf{r})=\sum_{n''}u_{n''0}(\mathbf{r})C_{n''n'}(\mathbf{g}),$$

$$C_{n''n'}(\mathbf{g})=\frac{(2\pi)^3}{\Omega}\int_{\text{cell}}u_{n''0}{}^*u_{n'\mathbf{g}}d\mathbf{r}, \quad (11.27)$$

where $C_{n''n'}$ is a unitary matrix. If we now define

$$\bar{U}_{n''n;i}(\mathbf{P};\,\mathbf{g})\equiv\sum C_{n''n'}(\mathbf{g})U_{n'n;i}(\mathbf{P}-\mathbf{g};\,\mathbf{g}), \quad (11.28)$$

Eq. (11.26) assumes the form

$$\begin{aligned}\bar{\varphi}_{n\mathbf{k}-\mathbf{g}}(\mathbf{r};\,\mathbf{g})&=\sum u_{n'0}[\bar{U}_{n'n;0}(\mathbf{P};\,\mathbf{g})+s\bar{U}_{n'n;1}(\mathbf{P};\,\mathbf{g})+\cdots]e^{i\mathbf{k}\cdot\mathbf{r}}\\&=\sum\chi_{n'\mathbf{k}'}(\mathbf{k}'|\bar{U}_{n'n;0}(\mathbf{P};\,\mathbf{g})+s\bar{U}_{n'n;1}(\mathbf{P};\,\mathbf{g})+\cdots|\mathbf{k}).\end{aligned} \quad (11.29)$$

These expressions have the same form as those occurring in Eq. (10.3). We shall now show that, for $n=m$, $\bar{\varphi}_{m\mathbf{k}-\mathbf{g}}(\mathbf{r};\,\mathbf{g})$ and $\bar{\varphi}_{m\mathbf{k}}(\mathbf{r})$ are in fact identical.

We begin by inverting the power series expansion (9.29) and rearranging the inverted series in a commutator expansion of the following form[23]:

$$\chi_{n'\mathbf{k}'}=\sum\bar{\varphi}_{n''\mathbf{k}''}\times(\mathbf{k}''|X_{n''n';0}(\mathbf{P})+sX_{n''n';1}(\mathbf{P})+\cdots|\mathbf{k}). \quad (11.30)$$

Substituting $\chi_{n'\mathbf{k}'}$ from (11.30) into (11.29) and regrouping in a commutator expansion gives an equation of the form

$$\bar{\varphi}_{n\mathbf{k}-\mathbf{g}}(\mathbf{r};\,\mathbf{g})=\sum\bar{\varphi}_{n'\mathbf{k}'}\times(\mathbf{k}'|W_{n'n;0}(\mathbf{P})+sW_{n'n;1}(\mathbf{P})+\cdots|\mathbf{k}). \quad (11.31)$$

Now for $s=0$, and $n=m$, we have by construction

$$\bar{\varphi}_{m\mathbf{k}-\mathbf{g}}(\mathbf{r};\,\mathbf{g})=\bar{\varphi}_{m\mathbf{k}}(\mathbf{r})=\varphi_{m\mathbf{k}}(\mathbf{r}), \quad (11.32)$$

where the phase of the Bloch wave is in accordance with (11.7) ff. Hence

$$W_{n'm;0}(\mathbf{P})\equiv\delta_{n'm}. \quad (11.33)$$

Furthermore the transformation matrix

$$W_{nn'}(\mathbf{P})=W_{nn';0}(\mathbf{P})+sW_{nn';1}(\mathbf{P})+\cdots \quad (11.34)$$

is formally unitary, i.e.,

$$\sum_{n''}W_{n'n''}{}^{\dagger}(\mathbf{P})W_{n''n'''}(\mathbf{P})=\delta_{n'n'''}, \quad (11.35)$$

where † denotes the Hermitian conjugate.

[23] This is of course again a formal expansion which has the following property: If $\bar{\varphi}_{n''\mathbf{k}''}$ is expressed as a formal commutator expansion, according to (10.3), substituted into (11.30) and the resulting terms regrouped in a commutator expansion, (11.30) is an identity to all orders in s.

Finally, since the original Hamiltonian H has no interband matrix elements connecting $\bar{\varphi}_{m\mathbf{k}}(\mathbf{r})$ with $\bar{\varphi}_{n\mathbf{k}}(\mathbf{r})$, $n\neq m$, or connecting $\bar{\varphi}_{m\mathbf{k}-\mathbf{g}}(\mathbf{r};\,\mathbf{g})$ with $\bar{\varphi}_{n\mathbf{k}-\mathbf{g}}(\mathbf{r};\,\mathbf{g})$, $n\neq m$, it follows that $\bar{\varphi}_{m\mathbf{k}-\mathbf{g}}(\mathbf{r};\,\mathbf{g})$ must be a linear combination of $\bar{\varphi}_{m\mathbf{k}'}(\mathbf{r})$ with the same m. Thus

$$W_{n'm;i}(\mathbf{P})=\delta_{n'm}W_i(\mathbf{P}), \quad (11.36)$$

and we have

$$\bar{\varphi}_{m\mathbf{k}-\mathbf{g}}(\mathbf{r};\,\mathbf{g})=\sum\bar{\varphi}_{m\mathbf{k}'}\times(\mathbf{k}'|1+sW_1(\mathbf{P})+s^2W_2(\mathbf{P})+\cdots|\mathbf{k}). \quad (11.37)$$

Now let us turn to $W_1(\mathbf{P})$. From the unitary property (11.35), it follows that

$$W_1(\mathbf{P})+W_1{}^*(\mathbf{P})=0. \quad (11.38)$$

Setting $s=0$ gives

$$W_1(\mathbf{k})+W_1{}^*(\mathbf{k})=0, \quad (11.39)$$

so that $W_1(\mathbf{k})$ is purely imaginary. Next let us use the reality of $\varphi_{m\mathbf{k}-\mathbf{g}}(\mathbf{r};\,\mathbf{g})$ and $\bar{\varphi}_{m\mathbf{k}}(\mathbf{r})$ at $\mathbf{r}=0$. This gives

$$\mathrm{Im}[\sum\bar{\varphi}_{m\mathbf{k}'}(0)(\mathbf{k}'|W_1(\mathbf{P})|\mathbf{k})]=0. \quad (11.40)$$

Now letting $s\to 0$ gives

$$\mathrm{Im}[W_1(\mathbf{k})]=0, \quad (11.41)$$

so that $W_1(\mathbf{k})$ is purely real. Hence

$$W_1(\mathbf{k})\equiv 0. \quad (11.42)$$

In the same way one shows next that $W_2(\mathbf{k})$ vanishes, etc. This gives finally the required result

$$\bar{\varphi}_{m\mathbf{k}-\mathbf{g}}(\mathbf{r};\,\mathbf{g})=\bar{\varphi}_{m\mathbf{k}}(\mathbf{r}), \quad (11.43)$$

which is equivalent to (11.11).

One important consequence of this result is the following. Since by construction

$$\bar{\varphi}_{m\mathbf{h}}(\mathbf{r};\,\mathbf{K})=\theta\,\bar{\varphi}_{m\mathbf{h}}(\mathbf{r}) \quad (11.44)$$

[see Eq. (8.13)], it follows from (11.43) that

$$\bar{\varphi}_{m\mathbf{K}+\mathbf{k}}(\mathbf{r})=\theta\,\bar{\varphi}_{m\mathbf{k}}(\mathbf{r}). \quad (11.45)$$

Thus, also in the presence of a magnetic field, the new basis functions have the quasi-periodic property (11.45).

It is obvious how (11.43) can be used to obtain the analytic continuation of $\bar{\varphi}_{m\mathbf{k}}(\mathbf{r})$ for arbitrary $\mathbf{k}$. Setting $\mathbf{g}=\mathbf{k}$, (11.43) gives

$$\bar{\varphi}_{m\mathbf{k}}(\mathbf{r})=\bar{\varphi}_{m0}(\mathbf{r};\,\mathbf{k}). \quad (11.46)$$

Now by Eq. (11.29) we have

$$\bar{\varphi}_{m0}(\mathbf{r};\,\mathbf{k})=\sum\chi_{n'\mathbf{k}'}(\mathbf{k}'|\bar{U}_{n'm;0}(\mathbf{P};\,\mathbf{k})+s\bar{U}_{n'm;1}(\mathbf{P};\,\mathbf{k})+\cdots|\mathbf{k}), \quad (11.47)$$

while by (10.3)

$$\bar{\varphi}_{m\mathbf{k}}(\mathbf{r})=\sum\chi_{n'\mathbf{k}'}(\mathbf{k}'|U_{n'm;0}(\mathbf{P})+2U_{n'm;1}(\mathbf{P})+\cdots|\mathbf{k}). \quad (11.48)$$

On subtracting and using (11.46), we get

$$0=\sum \chi_{n'\mathbf{k}'}(\mathbf{k}'|[U_{n'm;0}(\mathbf{P})-\bar{U}_{n'm;0}(\mathbf{P};\mathbf{k})] \\ +s[U_{n'm;1}(\mathbf{P})-\bar{U}_{n'm;1}(\mathbf{P};\mathbf{k})]+\cdots|\mathbf{k}). \quad (11.49)$$

As the $\chi_{n'\mathbf{k}'}$ are not orthonormal, we cannot immediately conclude that the coefficients of $\chi_{n'\mathbf{k}'}$ vanish. However, this fact may be established as follows. Set $s=0$ in (11.49), which then reads

$$0=\sum_{n'}\chi_{n'\mathbf{k}}[U_{n'm;0}(\mathbf{k})-\bar{U}_{n'm;0}(\mathbf{k};\mathbf{k})]. \quad (11.50)$$

Hence

$$U_{n'm;0}(\mathbf{k})=\bar{U}_{n'm;0}(\mathbf{k};\mathbf{k}), \quad (11.51)$$

and

$$U_{n'm;0}(\mathbf{P})=\bar{U}_{n'm;0}(\mathbf{P};\mathbf{k}). \quad (11.52)$$

Now, by using (11.52) in (11.49) and dividing the latter equation by s, we get

$$0=\sum \chi_{n'\mathbf{k}'}(k'|[U_{n'm;1}(\mathbf{P})-\bar{U}_{n'm;1}(\mathbf{P};\mathbf{k})] \\ +s[U_{n'm;2}(\mathbf{P})-\bar{U}_{n'n;2}(\mathbf{P};\mathbf{k})]+\cdots|k). \quad (11.53)$$

Now set $s=0$ and obtain

$$U_{n'm;1}(\mathbf{P})=\bar{U}_{n'm;1}(\mathbf{P};\mathbf{k}), \quad (11.54)$$

etc. Thus we see that for all i,

$$U_{n'm;i}(\mathbf{k})=\bar{U}_{n'm;i}(\mathbf{k};\mathbf{k}). \quad (11.55)$$

Finally, by (11.28), we have

$$\bar{U}_{n'm;i}(\mathbf{k};\mathbf{k})=\sum C_{n'n''}(\mathbf{k})U_{n''m;i}(0;\mathbf{k}). \quad (11.56)$$

But at $\mathbf{h}=0$, each coefficient $U_{n''m;i}(\mathbf{h};\mathbf{k})$ can be evaluated explicitly in terms of a finite number of the coefficients of the power series (11.22). Thus

$$U_{n''m;0}(0;\mathbf{k})=\delta_{n''m}, \quad (11.57)$$

$$U_{n''m;1}(0;\mathbf{k})=\sum_{\alpha_1\alpha_2}\frac{i\sigma^{\alpha_1\alpha_2}}{2s}G_{n''m}{}^{\alpha_1\alpha_2}(\mathbf{k}), \quad (11.58)$$

etc. We see then that for example to obtain $U_{n''m;1}(\mathbf{k})$ we require a knowledge of $G_{n''m}{}^{\alpha_1\alpha_2}(\mathbf{k})$ which may be obtained by two canonical transformations. This should be compared to Eq. (10.2) which required the summation of an infinite power series in k^α, i.e., the performance of an infinite sequence of canonical transformations.

12. Explicit Construction of the Effective Hamiltonian

In Sec. 6 we saw that the effective Hamiltonian $\bar{H}_m(\mathbf{P})$ could be expanded in a series of the form

$$\bar{H}_m(\mathbf{P})=H_{m;0}(\mathbf{P})+s\bar{H}_{m;1}(\mathbf{P})+s^2\bar{H}_{m;2}(\mathbf{P})+\cdots. \quad (12.1)$$

We have already remarked that since the $\bar{H}_{m;i}(\mathbf{P})$ are completely symmetrized functions of the P^α, they are determined by the functions $\bar{H}_{m;i}(\mathbf{k})$. In fact, if

$$\bar{H}_{m;i}(\mathbf{k})=\sum_{\mathbf{R}}B(\mathbf{R})e^{i\mathbf{R}\cdot\mathbf{k}}, \quad (12.2)$$

then

$$\bar{H}_{m;i}(\mathbf{P})=\sum_{\mathbf{R}}B(\mathbf{R})e^{i\mathbf{R}\cdot\mathbf{P}}. \quad (12.3)$$

However at the stage of Sec. 6, we had only expansions of $\bar{H}_{m;i}(\mathbf{k})$ in powers of k^α. We are now in a position to calculate these functions explicitly for arbitrary $\mathbf{k}$.

Let us begin by writing Eq. (10.3) for $n=m$ in the form

$$\bar{\varphi}_{m\mathbf{k}}(\mathbf{r})=\sum \chi_{n'\mathbf{k}'}(n'k'|U(\mathbf{P})|m\mathbf{k}), \quad (12.4)$$

where

$$U_{n'n}(\mathbf{P})=U_{n'n;0}(\mathbf{P})+sU_{n'n;1}(\mathbf{P})+\cdots \\ =(n'|\exp[S^{(1)}(\mathbf{P})]\exp[S^{(2)}(\mathbf{P})]\cdots|n). \quad (12.5)$$

Similarly, (11.23) can for $n=m$ be written as

$$\bar{\varphi}_{m\mathbf{h}}(\mathbf{r};\mathbf{g})=\sum_{n'}\chi_{n'\mathbf{h}'}(\mathbf{r};\mathbf{g})(n'\mathbf{h}'|U(\mathbf{P};\mathbf{g})|m\mathbf{h}), \quad (12.6)$$

where

$$U_{n'n'}(\mathbf{P};\mathbf{g}) \\ =U_{n'n;0}(\mathbf{P};\mathbf{g})+sU_{n'n;1}(\mathbf{P};\mathbf{g})+\cdots \\ =(n'|\exp[S^{(1)}(\mathbf{P};\mathbf{g})]\exp[S^{(2)}(\mathbf{P};\mathbf{g})]\cdots|n). \quad (12.7)$$

By (11.5) and (11.28), $U(\mathbf{P})$ and $U(\mathbf{P};\mathbf{g})$ are connected by the relation

$$U_{n''n}(\mathbf{P})=\sum C_{n''n'}(\mathbf{g})U_{n'n}(\mathbf{P}-\mathbf{g};\mathbf{g}). \quad (12.8)$$

We shall now prove the following identity. Let

$$\bar{H}_m(\mathbf{P})=U_{mn}^{-1}(\mathbf{P})H_{nn'}(\mathbf{P})U_{n'm}(\mathbf{P}), \quad (12.9)$$

where $U_{n'n}(\mathbf{P})$ is defined in (12.5) and

$$H_{nn'}(\mathbf{P})=\delta_{nn'}[\epsilon_n+\tfrac{1}{2}P^\alpha P^\alpha]+p_{nn'}{}^\alpha P^\alpha; \quad (12.10)$$

and let

$$\bar{H}_m(\mathbf{P};\mathbf{g})=U_{mn}^{-1}(\mathbf{P};\mathbf{g})H_{nn'}(\mathbf{P};\mathbf{g})U_{n'm}(\mathbf{P};\mathbf{g}), \quad (12.11)$$

where $U_{n'n}(\mathbf{P};\mathbf{g})$ is defined by (12.7) and

$$H_{nn'}(\mathbf{P};\mathbf{g})=\delta_{nn'}[\epsilon_n(\mathbf{g})+\tfrac{1}{2}P^\alpha P^\alpha]+p_{nn'}{}^\alpha(\mathbf{g})P^\alpha. \quad (12.12)$$

Then

$$\bar{H}_m(\mathbf{P})\equiv\bar{H}_m(\mathbf{P}-\mathbf{g};\mathbf{g}). \quad (12.13)$$

We first transform (12.9) with the help of (12.8):

$$\bar{H}_m(\mathbf{P})=U_{mn}^{-1}(\mathbf{P}-\mathbf{g};\mathbf{g})C_{nn''}{}^{-1}(\mathbf{g}) \\ \times H_{n''n'''}(\mathbf{P})C_{n'''n'}(\mathbf{g})U_{n'm}(\mathbf{P}-\mathbf{g};\mathbf{g}). \quad (12.14)$$

Next we note the identity

$$\delta_{n''n'''}\epsilon_{n''}=(u_{n''0},(p^2/2m+V)u_{n'''0}). \quad (12.15)$$

This allows us to write

$$C_{nn''}{}^{-1}(\mathbf{g})\delta_{n''n'''}\epsilon_{n''}C_{n'''n'}(\mathbf{g}) \\ =(u_{ng},u_{n''0})(u_{n''0},(p^2/2m+V)u_{n'''0})(u_{n'''0},u_{n'g}) \\ =(\varphi_{n\mathbf{g}}\exp(-i\mathbf{g}\cdot\mathbf{r}),(p^2/2m+V)\varphi_{n'\mathbf{g}}\exp(-i\mathbf{g}\cdot\mathbf{r})) \\ =(\epsilon_{n\mathbf{g}}+\tfrac{1}{2}\mathbf{g}^2)\delta_{nn'}-p_{nn'}{}^\alpha(\mathbf{g})g^\alpha. \quad (12.16)$$

Similarly

$$C_{nn''}^{-1}(\mathbf{g})p_{n''n'''}{}^{\alpha}C_{n'''n'}(\mathbf{g})=p_{nn'}{}^{\alpha}(\mathbf{g})-g^{\alpha}\delta_{nn'}. \quad (12.17)$$

Therefore (12.14) gives

$$\bar{H}_m(\mathbf{P})=U_{mn}^{-1}(\mathbf{P}-\mathbf{g};\mathbf{g})\times\{\delta_{nn'}[\epsilon_n(\mathbf{g})+\tfrac{1}{2}(P^{\alpha}-g^{\alpha})(P^{\alpha}-g^{\alpha})+p_{nn'}{}^{\alpha}(g)(P^{\alpha}-g^{\alpha})]\}U_{n'm}(\mathbf{P}-\mathbf{g};\mathbf{g})=\bar{H}_m(\mathbf{P}-\mathbf{g};\mathbf{g}), \quad (12.18)$$

as was to be shown.

This relation allows one to calculate explicitly the functions $\bar{H}_{m;i}(\mathbf{k})$ which determine the successive terms of the commutator expansion of $\bar{H}_m(\mathbf{P})$. We make a commutator expansion of both sides of (12.18),

$$\bar{H}_{m;0}(\mathbf{P})+s\bar{H}_{m;1}(\mathbf{P})+\cdots=\bar{H}_{m;0}(\mathbf{P}-\mathbf{g};\mathbf{g})+s\bar{H}_{m;1}(\mathbf{P}-\mathbf{g};\mathbf{g})+\cdots. \quad (12.19)$$

We equate corresponding terms, setting $\mathbf{P}=\mathbf{k}$ and $\mathbf{g}=\mathbf{k}$. This gives

$$\bar{H}_{m;i}(\mathbf{k})=\bar{H}_{m;i}(0;\mathbf{k}). \quad (12.20)$$

Now suppose that $\bar{H}_m(\mathbf{P};\mathbf{k})$ has the following *power* series expansion in P^{α}

$$\bar{H}_m(\mathbf{P};\mathbf{k})=\epsilon_m(\mathbf{k})+E_m{}^{\alpha_1}(\mathbf{k})P^{\alpha_1}+E_m{}^{\alpha_1\alpha_2}(\mathbf{k})P^{\alpha_1}P^{\alpha_2}+\cdots+E_m{}^{\alpha_1\cdots\alpha_t}(\mathbf{k})P^{\alpha_1}\cdots P^{\alpha_t}+\cdots. \quad (12.21)$$

This may be rearranged in a commutator expansion

$$\bar{H}_m(\mathbf{P};\mathbf{k})=\bar{H}_{m;0}(\mathbf{P};\mathbf{k})+s\bar{H}_{m;1}(\mathbf{P};\mathbf{k})+\cdots, \quad (12.22)$$

where

$$\bar{H}_{m;0}(\mathbf{P};\mathbf{k})=\epsilon_m(\mathbf{k})+E_m{}^{\alpha_1}(\mathbf{k})\langle P^{\alpha_1}\rangle_{\text{Av}}+E_m{}^{\alpha_1\alpha_2}(\mathbf{k})\langle P^{\alpha_1}P^{\alpha_2}\rangle_{\text{Av}}+\cdots, \quad (12.23)$$

$$\bar{H}_{m;1}(\mathbf{P};\mathbf{k})=i\sum_{\alpha_1\alpha_2}\frac{\sigma^{\alpha_1\alpha_2}}{s}E_m{}^{\alpha_1\alpha_2}(\mathbf{k})+O(P^{\alpha}), \quad (12.24)$$

$$\bar{H}_{m;2}(\mathbf{P};\mathbf{k})=-\sum_{\alpha_1\alpha_2\alpha_3\alpha_4}\frac{\sigma^{\alpha_1\alpha_2}\sigma^{\alpha_3\alpha_4}+\sigma^{\alpha_1\alpha_3}\sigma^{\alpha_2\alpha_4}+\sigma^{\alpha_1\alpha_4}\sigma^{\alpha_2\alpha_3}}{s^2}\times E_m{}^{\alpha_1\alpha_2\alpha_3\alpha_4}(\mathbf{k})+O(P^{\alpha}). \quad (12.25)$$

Hence by (12.20)

$$\bar{H}_{m;0}(\mathbf{k})=\epsilon_m(\mathbf{k}),$$

$$\bar{H}_{m;1}(\mathbf{k})=i\sum_{\alpha_1\alpha_2}\frac{\sigma^{\alpha_1\alpha_2}}{s}E_m{}^{\alpha_1\alpha_2}(\mathbf{k}),$$

$$\bar{H}_{m;2}(\mathbf{k})=-\sum_{\alpha_1\alpha_2\alpha_3\alpha_4}\frac{\sigma^{\alpha_1\alpha_2}\sigma^{\alpha_3\alpha_4}+\sigma^{\alpha_1\alpha_3}\sigma^{\alpha_2\alpha_4}+\sigma^{\alpha_1\alpha_4}\sigma^{\alpha_2\alpha_3}}{s^2}E_m{}^{\alpha_1\alpha_2\alpha_3\alpha_4}(\mathbf{k}), \quad (12.26)$$

etc. These functions can all be explicitly calculated.

Now by construction of $\bar{H}_m(\mathbf{P};\mathbf{k})$ the coefficients $E_m{}^{\alpha_1\alpha_2\cdots}(\mathbf{k})$ have the periodicity of the reciprocal lattice, so that we may write

$$\bar{H}_{m;i}(\mathbf{k})=\sum a_i{}^{(l)}\exp(i\mathbf{R}^{(l)}\cdot\mathbf{P}), \quad (12.27)$$

where the $\mathbf{R}^{(l)}$ are the lattice vectors of the crystal. Hence the effective Hamiltonian can be written in the form

$$\bar{H}_m(\mathbf{P})=\textstyle\sum_l a^{(l)}(s)\exp(i\mathbf{R}^{(l)}\cdot\mathbf{P}), \quad (12.28)$$

where

$$a^{(l)}(s)=a_0{}^{(l)}+sa_1{}^{(l)}+s^2a_2{}^{(l)}+\cdots. \quad (12.29)$$

Denoting by τ the volume of the fundamental Brillouin zone, we have

$$a_0{}^{(l)}=\frac{1}{\tau}\int_{\text{zone}}\epsilon_m(\mathbf{k})\exp(-i\mathbf{k}\cdot\mathbf{R}^{(l)})d\mathbf{k}, \quad (12.30)$$

$$a_1{}^{(l)}=\frac{1}{\tau}\int_{\text{zone}}(is^{\alpha_1\alpha_2}/s)E_m{}^{\alpha_1\alpha_2}(\mathbf{k})\times\exp(-i\mathbf{k}\cdot\mathbf{R}^{(l)})d\mathbf{k}, \quad (12.31)$$

etc.

Equation (12.28) has as a consequence the important periodicity property

$$\bar{H}_m(\mathbf{P}+\mathbf{K})=\bar{H}_m(\mathbf{P}), \quad (12.32)$$

where $\mathbf{K}$ is an arbitrary reciprocal lattice vector.

13. Application to Crystals with a Center of Inversion

Using the methods of the preceding sections, we shall now work out the first few terms of the commutator expansion of the effective Hamiltonian $\bar{H}_m(\mathbf{P})$ for crystals with a center of inversion.

For such crystals it is easily shown (Appendix B) that with our convention of

$$\text{Im}[\varphi_{n\mathbf{g}}(0)]=0, \quad (13.1)$$

one has the result[24]

$$\text{Im}[p_{n'n''}{}^{\alpha}(\mathbf{g})]=0. \quad (13.2)$$

This results in the vanishing of the terms of odd degree in s in the commutator expansion of $\bar{H}_m(\mathbf{P})$.

Working in the representation of the $\chi_{n\mathbf{h}}(\mathbf{r};\mathbf{g})$, we begin with the Hamiltonian

$$H(\mathbf{P};\mathbf{g})=H_0+H_1+H_2, \quad (13.3)$$

where

$$(H_0)_{nn'}=\delta_{nn'}\epsilon_n(\mathbf{g}), \quad (13.4)$$

$$(H_1)_{nn'}=p_{nn'}{}^{\alpha}(\mathbf{g})P^{\alpha}, \quad (13.5)$$

$$(H_2)_{nn'}=\tfrac{1}{2}\delta_{nn'}P^{\alpha}P^{\alpha}. \quad (13.6)$$

To this H we apply a sequence of unitary transformations $\exp[S^{(i)}(\mathbf{P};\mathbf{g})]$ to eliminate off-diagonal elements to higher and higher order in P^{α}. We take $S^{(i)}$ to be of

[24] For isolated values of $\mathbf{g}$, for which $\varphi_{n\mathbf{g}}(0)=0$, (13.1) does not fix the phase of $\varphi_{n\mathbf{g}}$. In such cases we define $\varphi_{n\mathbf{g}}=\lim_{\mathbf{g}'\to\mathbf{g}}\varphi_{n\mathbf{g}'}$. With this understanding, (13.2) holds also at such points.

order $(P^\alpha)^i$. Then, up to order $(P^\alpha)^4$ we have

$$
\begin{aligned}
\bar{H}(\mathbf{P};\mathbf{g})
&=\cdots\exp(-S^{(2)})\exp(-S^{(1)})H\exp(S^{(1)})\exp(S^{(2)})\cdots\\
&=H_0\\
&+\{H_1+[H_0,S^{(1)}]\}\\
&+\{H_2+[H_1,S^{(1)}]+\tfrac{1}{2}[[H_0,S^{(1)}],S^{(1)}]+[H_0,S^{(2)}]\}\\
&+\{[H_1,S^{(2)}]+[H_2,S^{(1)}]+[[H_0,S^{(1)}],S^{(2)}]\\
&+\tfrac{1}{2}[[H_1,S^{(1)}],S^{(1)}]+\tfrac{1}{6}[[[H_0,S^{(1)}],S^{(1)}],S^{(1)}]\\
&+[H_0,S^{(3)}]\}+\{[H_2,S^{(2)}]+\tfrac{1}{2}[[H_0,S^{(2)}],S^{(2)}]\\
&+[[H_1,S^{(1)}],S^{(2)}]+\tfrac{1}{2}[[H_2,S^{(1)}],S^{(1)}]\\
&+\tfrac{1}{2}[[[H_0,S^{(1)}],S^{(1)}],S^{(2)}]+\tfrac{1}{6}[[[H_1,S^{(1)}],S^{(1)}],S^{(1)}]\\
&+(1/24)[[[[H_0,S^{(1)}],S^{(1)}],S^{(1)}],S^{(1)}]\\
&+[[H_0,S^{(1)}],S^{(3)}]+[H_1,S^{(3)}]+[H_0,S^{(4)}]\}\\
&+\cdots. \qquad (13.7)
\end{aligned}
$$

The first-order off-diagonal elements are eliminated by the choice

$$
\begin{aligned}
&S_{mi}{}^{(1)}=-(S^{(1)})_{im}{}^{\dagger}=-(p_{mi}{}^{\alpha}/\omega_{mi})P^{\alpha}, \qquad i\neq m\\
&S_{n'n''}{}^{(1)}=0, \qquad \text{all other } n',\, n''.
\end{aligned} \qquad (13.8)
$$

Here we write $p_{nn'}{}^{\alpha}$ and $\omega_{nn'}$ short for $p_{nn'}{}^{\alpha}(\mathbf{g})$ and $\omega_{nn'}(\mathbf{g})$. Because of (13.1) and (13.2) we see that the corresponding new basis functions,

$$
\begin{aligned}
\bar{\varphi}_{m\mathbf{g}+\mathbf{h}}&=\exp(S^{(1)})\exp(S^{(2)})\cdots\chi_{m\mathbf{h}}=(1+S^{(1)}+\cdots)\chi_{m\mathbf{h}}\\
&=\varphi_{m\mathbf{g}}\exp(i\mathbf{h}\cdot\mathbf{r})-\textstyle\sum_i \varphi_{i\mathbf{g}}(p_{im}{}^{\alpha}/\omega_{im})P^{\alpha}\exp(i\mathbf{h}\cdot\mathbf{r})\\
&\quad+\cdots, \qquad (13.9)
\end{aligned}
$$

have to first order in P^α the required property

$$\mathrm{Im}[\bar{\varphi}_{m\mathbf{g}+\mathbf{h}}(0)]=0. \qquad (13.10)$$

The second transformation matrix $S^{(2)}$ must be chosen to eliminate the second-order off-diagonal elements from $\bar{H}(\mathbf{P};\mathbf{g})$. By (13.7), this requires that

$$\{H_2+[H_1,S^{(1)}]+\tfrac{1}{2}[[H_0,S^{(1)}],S^{(1)}]+[H_0,S^{(2)}]\}_{mi}=0, \qquad i\neq m, \qquad (13.11)$$

or

$$
\begin{aligned}
&\{p_{mm}{}^{\alpha}(-p_{mi}{}^{\alpha}/\omega_{mi})+(p_{mi}{}^{\alpha}/\omega_{mj})p_{ji}{}^{\beta}\}P^{\alpha}P^{\beta}\\
&\qquad+\omega_{mi}S_{mi}{}^{(2)}(\mathbf{P})=0, \quad i\neq m. \qquad (13.12)
\end{aligned}
$$

Here the band index j runs over all values except m. This equation can now be solved for $S_{mi}{}^{(2)}$. Recalling the anti-Hermitian property of $S^{(2)}$, we find

$$S_{mi}{}^{(2)}=-(S_{im}{}^{(2)})^{\dagger}=D_{mi}{}^{\alpha\beta}P^{\alpha}P^{\beta}, \qquad (13.13)$$

where

$$D_{mi}{}^{\alpha\beta}=(1/\omega_{mi})\{(1/\omega_{mi})p_{mm}{}^{\alpha}p_{mi}{}^{\beta}-(1/\omega_{mj})p_{mj}{}^{\alpha}p_{ji}{}^{\beta}\}, \qquad i\neq m. \qquad (13.14)$$

All other off-diagonal matrix elements of $S^{(2)}$ can be taken as zero. $S_{mm}{}^{(2)}$ is chosen so as to satisfy the condition (13.10) up to second order in P^α. This requires evidently that

$$\mathrm{Im}\{[\tfrac{1}{2}(S^{(1)})^2+S^{(2)}]\chi_{m\mathbf{h}}\}_{\mathbf{r}=0}=0. \qquad (13.15)$$

Now,

$$
\begin{aligned}
&[\tfrac{1}{2}(S^{(1)})_j\chi_{m\mathbf{h}}]_{\mathbf{r}=0}\\
&=\tfrac{1}{2}\varphi_{m\mathbf{g}}(0)\sum_{i\neq m}\frac{p_{mi}{}^{\alpha}}{\omega_{mi}}\frac{p_{im}{}^{\beta}}{\omega_{im}}[P^{\alpha}P^{\beta}\exp(i\mathbf{h}\cdot\mathbf{r})]_{\mathbf{r}=0}\\
&=\tfrac{1}{2}\varphi_{m\mathbf{g}}(0)\sum_{i\neq m}\frac{p_{mi}{}^{\alpha}}{\omega_{mi}}\frac{p_{im}{}^{\beta}}{\omega_{im}}[\langle P^{\alpha}P^{\beta}\rangle_{\mathrm{Av}}\exp(i\mathbf{h}\cdot\mathbf{r})]_{\mathbf{r}=0}\\
&=\tfrac{1}{2}\varphi_{m\mathbf{g}}(0)\sum_{i\neq m}\frac{p_{mi}{}^{\alpha}}{\omega_{mi}}\frac{p_{im}{}^{\beta}}{\omega_{im}}\mathbf{h}^{\alpha}\mathbf{h}_{\beta}, \qquad (13.16)
\end{aligned}
$$

and this is clearly real. The second term in (13.15) gives

$$
\begin{aligned}
&[S^{(2)}\chi_{m\mathbf{h}}]_{\mathbf{r}=0}=\varphi_{m\mathbf{g}}(0)[S_{mm}{}^{(2)}(\mathbf{P})\exp(i\mathbf{h}\cdot\mathbf{r})]_{\mathbf{r}=0}\\
&+\sum_{i\neq m}\varphi_{i\mathbf{g}}(0)\frac{1}{\omega_{im}}\left\{\frac{1}{\omega_{im}}(p_{ii}{}^{\alpha}p_{im}{}^{\beta}-p_{im}{}^{\alpha}p_{mm}{}^{\beta})+\frac{1}{\omega_{km}}p_{ik}{}^{\alpha}p_{km}{}^{\beta}\right\}\\
&\qquad\times[P^{\alpha}P^{\beta}\exp(i\mathbf{h}\cdot\mathbf{r})]_{\mathbf{r}=0}. \qquad (13.17)
\end{aligned}
$$

But

$$[P^{\alpha}P^{\beta}\exp(i\mathbf{h}\cdot\mathbf{r})]_{\mathbf{r}=0}=h^{\alpha}h^{\beta}+is^{\alpha\beta}. \qquad (13.18)$$

Therefore, if we call

$$r_{i\mathbf{g}}\equiv\varphi_{i\mathbf{g}}(0)/\varphi_{m\mathbf{g}}(0), \qquad (13.19)$$

then condition (13.15) requires that we choose

$$
\begin{aligned}
S_{mm}{}^{(2)}(\mathbf{P})=&-\sum_{i\neq m}r_{i\mathbf{g}}(0)\frac{1}{\omega_{im}}\frac{1}{\omega_{im}}\Big\{(p_{ii}{}^{\alpha}p_{im}{}^{\beta}-p_{im}{}^{\alpha}p_{mm}{}^{\beta})\\
&+\frac{1}{\omega_{km}}p_{ik}{}^{\alpha}p_{km}{}^{\beta}\Big\}\tfrac{1}{2}[P^{\alpha},P^{\beta}]. \qquad (13.20)
\end{aligned}
$$

Similarly we construct $S^{(3)}$. The result is

$$S_{im}{}^{(3)}=D_{im}{}^{\alpha\beta\gamma}P^{\alpha}P^{\beta}P^{\gamma}, \qquad (13.21)$$

where

$$
\begin{aligned}
D_{im}{}^{\alpha\beta\gamma}=&\frac{1}{\omega_{im}}\\
&\times\left\{-p_{ij}{}^{\alpha}\left[\frac{1}{\omega_{jm}{}^{2}}(p_{ij}{}^{\beta}p_{jm}{}^{\gamma}-p_{jm}{}^{\beta}p_{mm}{}^{\gamma})+\frac{1}{\omega_{jm}\omega_{km}}p_{jk}{}^{\beta}p_{km}{}^{\gamma}\right]\right.\\
&+\left[\frac{1}{\omega_{im}{}^{2}}(p_{ii}{}^{\alpha}p_{im}{}^{\beta}-p_{im}{}^{\alpha}p_{mm}{}^{\beta})+\frac{1}{\omega_{im}\omega_{km}}p_{ik}{}^{\alpha}p_{km}{}^{\beta}\right]p_{mm}{}^{\gamma}\\
&+\delta^{\alpha\beta}\frac{p_{im}{}^{\gamma}}{\omega_{im}}-\frac{p_{im}{}^{\alpha}}{\omega_{im}}\delta^{\beta\gamma}\\
&\left.+\frac{1}{3}\left(\frac{3}{\omega_{im}\omega_{jm}}+\frac{1}{\omega_{jm}{}^{2}}\right)p_{im}{}^{\alpha}p_{mj}{}^{\beta}p_{jm}{}^{\gamma}\right\}, \quad i\neq m \qquad (13.22)
\end{aligned}
$$

while

$S_{mm}^{(3)}(\mathbf{P})$

$$=\left[\sum_{i\neq m} r_{ig}\left(\frac{1}{6}\frac{p_{im}{}^{\alpha}}{\omega_{im}}\frac{p_{mj}{}^{\beta}}{\omega_{mj}}\frac{p_{jm}{}^{\gamma}}{\omega_{jm}}-D_{im}{}^{\alpha\beta\gamma}\right)+D_{mi}{}^{\alpha\beta}\frac{p_{im}{}^{\gamma}}{\omega_{im}}\right]$$
$$\times(\tfrac{1}{2}[P^{\beta},P^{\gamma}]P^{\alpha}+\tfrac{1}{2}[P^{\alpha},P^{\gamma}]P^{\beta}+\tfrac{1}{2}[P^{\alpha},P^{\beta}]P^{\gamma}). \quad (13.23)$$

The transformation $S^{(4)}$ is not required for computing the m-m block of $\bar{H}(\mathbf{P};\,\mathbf{g})$ up to fourth order in P^{α} [see Eq. (13.7)].

Substituting our results for $S^{(1)}$, $S^{(2)}$, and $S^{(3)}$ in Eq. (13.7) gives the following result:

$$\bar{H}_m(\mathbf{P};\,\mathbf{g})=\bar{H}_{m;0}(\mathbf{P};\,\mathbf{g})+s\bar{H}_{m;1}(\mathbf{P};\,\mathbf{g})+s^2\bar{H}_{m;2}(\mathbf{P};\,\mathbf{g})+\cdots, \quad (13.24)$$

where

$$\bar{H}_{m;0}(0;\,\mathbf{g})=\epsilon_{m\mathbf{g}},$$
$$\bar{H}_{m;1}(0;\,\mathbf{g})=0, \quad (13.25)$$
$$\bar{H}_{m;2}(0;\,\mathbf{g})=-\sum_{\alpha\beta\gamma\delta}\frac{\sigma^{\alpha\beta}\sigma^{\gamma\delta}+\sigma^{\alpha\gamma}\sigma^{\beta\delta}+\sigma^{\alpha\delta}\sigma^{\beta\gamma}}{s^2}E_m{}^{\alpha\beta\gamma\delta}(\mathbf{g}),$$

and

$$E_m{}^{\alpha\beta\gamma\delta}(\mathbf{g})=\frac{1}{\omega_{mi}}\left\{\frac{1}{\omega_{mj}}p_{mm}{}^{\alpha}p_{mj}{}^{\beta}-\frac{1}{\omega_{mi}}p_{mi}{}^{\alpha}p_{im}{}^{\beta}\right\}$$
$$\times\left\{\frac{1}{\omega_{mj}}p_{jm}{}^{\gamma}p_{mm}{}^{\delta}-\frac{1}{\omega_{mk}}p_{jk}{}^{\gamma}p_{km}{}^{\delta}\right\}$$
$$-\frac{1}{4}\frac{1}{\omega_{mi}{}^{2}}\{p_{mi}{}^{\alpha}p_{im}{}^{\beta}\delta^{\gamma\delta}-2p_{mi}{}^{\alpha}\delta^{\beta\gamma}p_{im}{}^{\delta}+\delta^{\alpha\beta}p_{mi}{}^{\gamma}p_{im}{}^{\delta}\}$$
$$-\frac{1}{2}\left(\frac{1}{\omega_{mi}{}^{2}\omega_{mk}}+\frac{1}{\omega_{mi}\omega_{mk}{}^{2}}\right)p_{mi}{}^{\alpha}p_{im}{}^{\beta}p_{mk}{}^{\gamma}p_{km}{}^{\delta}$$
$$-\sum_{i\neq m}\left[r_{ig}\left(\frac{1}{6}\frac{p_{im}{}^{\alpha}}{\omega_{im}}\frac{p_{mj}{}^{\beta}}{\omega_{mj}}\frac{p_{jm}{}^{\gamma}}{\omega_{jm}}-D_{im}{}^{\alpha\beta\gamma}\right)+D_{mi}{}^{\alpha\beta}\frac{p_{im}{}^{\gamma}}{\omega_{im}}\right]p_{mm}{}^{\delta}. \quad (13.26)$$

The coefficients $D_{im}{}^{\alpha\beta}$ and $D_{im}{}^{\alpha\beta\gamma}$ are defined in Eqs. (13.14) and (13.22).

Finally, using (13.25) and (13.26) in (12.26), we have now explicit expressions for $\bar{H}_{m;i}(\mathbf{k})$ for $i=0$, 1, and 2.

The absence of a term linear in s will be noted. It is easy to see in general that for the case of a crystal with a center of inversion, our construction leads to an expansion in even powers of s.

14. Application to Crystals without a Center of Inversion

In this section we construct the effective Hamiltonian for a crystal without a center of inversion. We again choose the phases of the $\varphi_{n\mathbf{g}}$ in accordance with the reality condition (13.1). But now, in contrast to the case of an inversion center, we have

$$p_{nn'}{}^{\alpha}(\mathbf{g})=q_{nn'}{}^{\alpha}(\mathbf{g})+it_{nn'}{}^{\alpha}(\mathbf{g}), \quad (14.1)$$

where q and t are real and in general neither vanishes.

We make the transformation (13.7) and find, as before, that

$$S_{mi}{}^{(1)}=-(S^{(1)})_{im}{}^{\dagger}=-(p_{mi}{}^{\alpha}/\omega_{mi})P^{\alpha}, \quad i\neq m. \quad (14.2)$$

However $S_{mm}{}^{(1)}$ does not now vanish. Instead, since

$$\bar{\varphi}_{m\mathbf{h}}(0;\,\mathbf{g})=\varphi_{m\mathbf{g}}(0)+\left[\left(-\sum_{i\neq m}\varphi_{i\mathbf{g}}(0)\frac{p_{im}{}^{\alpha}}{\omega_{im}}P^{\alpha}+\varphi_{m\mathbf{g}}(0)S_{mm}{}^{(1)}(\mathbf{P})\right)\exp(i\mathbf{h}\cdot\mathbf{r})\right]_{\mathbf{r}=0}+\cdots, \quad (14.3)$$

the reality requirement leads to

$$S_{mm}{}^{(1)}(P)=D_{mm}{}^{\alpha}P^{\alpha}, \quad (14.4)$$

where

$$D_{mm}{}^{\alpha}=-i\sum_{i\neq m}r_{ig}\frac{t_{im}{}^{\alpha}}{\omega_{im}}. \quad (14.5)$$

This is enough to determine $\bar{H}_m(\mathbf{P};\,\mathbf{g})$ up to second order in P^{α}, which is as far as we shall go. By (14.7) we find

$$\bar{H}_m(\mathbf{P};\,\mathbf{g})=\epsilon_m(\mathbf{g})+p_{mm}{}^{\alpha}P^{\alpha}+\{\tfrac{1}{2}\delta^{\alpha\beta}+p_{mi}{}^{\alpha}p_{im}{}^{\beta}/\omega_{mi}-i(D_{mm}{}^{\alpha}p_{mm}{}^{\beta}-p_{mm}{}^{\alpha}D_{mm}{}^{\beta})\}P^{\alpha}P^{\beta}+\cdots. \quad (14.6)$$

$$\bar{H}_{m;0}(\mathbf{k})=\bar{H}_{m;0}(0;\,\mathbf{k})=\epsilon_m(\mathbf{k}),$$
$$\bar{H}_{m;1}(\mathbf{k})=\bar{H}_{m;1}(0;\,\mathbf{k})=\frac{\sigma^{\alpha\beta}}{s}\left\{\frac{1}{2\omega_{mi}}(p_{mi}{}^{\alpha}p_{im}{}^{\beta}-p_{mi}{}^{\beta}p_{im}{}^{\alpha})-i(D_{mm}{}^{\alpha}p_{mm}{}^{\beta}-D_{mm}{}^{\beta}p_{mm}{}^{\alpha})\right\}. \quad (14.7)$$

It will be noted that $H_{m;1}$ does now *not* vanish. In general it cannot be removed by any unitary transformation. This fact is of importance for the location of the energy levels in the presence of a magnetic field.

ACKNOWLEDGMENTS

Most of this work was carried out while the author was a Senior Postdoctoral Fellow of the National Science Foundation at the Imperial College of Science and Technology in London. The Office of Naval Research also supported parts of this research. It is a great pleasure to thank these organizations for their assistance and to express my appreciation to the Imperial College and in particular to Professor H. Jones for their hospitality.

APPENDIX A. DEMONSTRATION THAT NO SOLUTIONS ARE OMITTED

In Sec. 4 it was stated that the solutions of Eq. (4.17) give rise, through Eq. (4.1), to all solutions of the original Schrödinger equation $H\psi = E\psi$. As $\mathbf{k}$ was taken to be a continuous variable, the dimensionality of the Hilbert space spanned by the $\chi_{n\mathbf{k}}$ is nondenumerable. To avoid the associated mathematical difficulties, we shall here prove the corresponding result for a vector space of finite dimensionality and assume that it can be generalized to a Hilbert space of the above type.

Consider then a vector space of dimensionality D, spanned by the orthonormal basis vectors

$$\mathbf{e}_\alpha, \qquad \alpha = 1, 2, \cdots D \tag{A.1}$$

$$\mathbf{e}_\alpha^* \mathbf{e}_{\alpha'} = \delta_{\alpha\alpha'}, \quad \alpha, \alpha' = 1, \cdots D. \tag{A.2}$$

Let H be a Hermitian Hamiltonian operator in this space, with matrix elements $(\alpha'|H|\alpha)$ defined by

$$H\mathbf{e}_\alpha = \sum_{\alpha'=1}^{D} \mathbf{e}_{\alpha'}(\alpha'|H|\alpha), \quad \alpha = 1, 2, \cdots D. \tag{A.3}$$

This operator has D orthonormal eigenvectors $\psi^{(i)}$:

$$H\psi^{(i)} = E^{(i)}\psi^{(i)}, \quad i = 1, \cdots D \tag{A.4}$$

$$\psi^{(i)*}\psi^{(i')} = \delta_{ii'} \quad i, i' = 1, \cdots D. \tag{A.5}$$

Now consider a *redundant* set of normalized vectors

$$\mathbf{e}_\alpha, \quad \alpha = 1, 2, \cdots ND \tag{A.6}$$

the first D of which are identical with the basis (A.1). Let the effect of operating on these $\mathbf{e}_\alpha$ with H be described by a matrix $(\alpha'|H|\alpha)$, such that

$$H\mathbf{e}_\alpha = \sum_{\alpha'=1}^{ND} \mathbf{e}_{\alpha'}(\alpha'|H|\alpha), \quad \alpha = 1, \cdots ND. \tag{A 7}$$

Because of the redundancy of the $\mathbf{e}_\alpha$, the matrix $(\alpha'|H|\alpha)$ is not completely defined by (A.7). But, in view of the completeness of the $\mathbf{e}_\alpha$, *some* matrix $(\alpha'|H|\alpha)$ certainly exists for which (A.7) holds. Finally assume that the matrix $(\alpha'|H|\alpha)$ is *Hermitian:*

$$(\alpha'|H|\alpha)^* = (\alpha|H|\alpha'). \tag{A.8}$$

This corresponds exactly to the situation of Sec. 4.

Now consider ND solutions of the eigenvalue problem

$$\sum_{\alpha'=1}^{ND} (\alpha|H|\alpha')A_{\alpha'}^{(l)} = E^{(l)}A_\alpha^{(l)}, \quad \alpha = 1, \cdots ND. \tag{A.9}$$

These may be taken as orthonormal in the sense

$$\sum_{\alpha=1}^{ND} A_\alpha^{(l)*}A_\alpha^{(l')*} = \delta_{ll'}; \quad l, l' = 1, \cdots ND. \tag{A.10}$$

In view of (A.7), the corresponding vectors

$$\boldsymbol{\varphi}^{(l)} = \sum_{\alpha'=1}^{ND} A_{\alpha'}^{(l)}\mathbf{e}_{\alpha'}, \quad l = 1, \cdots ND \tag{A.11}$$

satisfy the equation

$$H\boldsymbol{\varphi}^{(l)} = E\boldsymbol{\varphi}^{(l)}, \quad l = 1, \cdots ND. \tag{A.12}$$

Clearly, because of (A.1), *at most* D of these can be linearly independent.

We next show that also *at least* D of these $\psi^{(l)}$ are linearly independent. For, from (A.10) follow the reciprocal relationships

$$\sum_{l=1}^{ND} A_\alpha^{(l)*}A_{\alpha'}^{(l)} = \delta_{\alpha\alpha'}; \quad \alpha, \alpha' = 1, \cdots ND. \tag{A.13}$$

Now multiplying (A.11) by $A_\alpha^{(l)*}$ and summing over l from 1 to ND gives, by (A.13),

$$\mathbf{e}_\alpha = \sum_{l=1}^{ND} A_\alpha^{(l)*}\boldsymbol{\varphi}^{(l)}, \quad \alpha = 1, \cdots ND. \tag{A.14}$$

In particular therefore the D independent vectors $\mathbf{e}_1, \cdots, \mathbf{e}_D$ can be expressed linearly in terms of the $\boldsymbol{\varphi}^{(l)}$. Hence the latter must contain a linearly independent subset of D independent vectors, say

$$\boldsymbol{\varphi}^{(l_1)}, \cdots \boldsymbol{\varphi}^{(l_D)}. \tag{A.15}$$

Since the $\boldsymbol{\varphi}^{(l_j)}$ are D independent solutions of the same eigenvalue problem (A.12) as the $\psi^{(i)}$, each $\psi^{(i)}$ must be a linear combination of the $\boldsymbol{\varphi}^{(l_j)}$. (If there is no degeneracy there must in fact be a one-to-one correspondence.) It follows *a fortiori* that each $\psi^{(i)}$ is a linear combination of *all* the $\boldsymbol{\varphi}^{(l)}$ $(l = 1, \cdots ND)$ obtained from the solutions $A_\alpha^{(l)}$ of Eq. (A.9). Thus by solving (A.9) no solutions of the original eigenvalue problem (A.4) are lost.

APPENDIX B. REALITY OF MOMENTUM MATRIX ELEMENTS

Consider a crystal with a center of symmetry at $\mathbf{r} = 0$. Let $\mathbf{g}$ be a general point in $\mathbf{k}$ space and choose the phases of the Bloch waves $\varphi_{n\mathbf{g}}$ such that

$$\mathrm{Im}[\varphi_{n\mathbf{g}}(0)] = 0. \tag{B.1}$$

We shall then show that the momentum matrix elements are all real:

$$\mathrm{Im}[p_{nn'}{}^\alpha(\mathbf{g})] = 0. \tag{B.2}$$

Let us first assume that at $\mathbf{g}$ none of the set of Bloch energies $\epsilon_n(\mathbf{g})$ coincide and that none of the $\varphi_{n\mathbf{g}}(0)$ vanish. Then if the function

$$\varphi_{n\mathbf{g}}(\mathbf{r}) = u_{n\mathbf{g}}(\mathbf{r}) \exp(i\mathbf{g}\cdot\mathbf{r}) \tag{B.3}$$

satisfies the Schrödinger equation and periodicity condition described by $\mathbf{g}$, so does

$$\varphi_{n\mathbf{g}}(-\mathbf{r})^* = u_{n\mathbf{g}}^*(-\mathbf{r}) \exp(i\mathbf{g}\cdot\mathbf{r}). \tag{B.4}$$

From the assumed nondegeneracy of the $\epsilon_n(\mathbf{g})$ we conclude that

$$u_{n\mathbf{g}}^*(-\mathbf{r}) = C u_{n\mathbf{g}}(\mathbf{r}), \tag{B.5}$$

where C is a numerical constant. Setting $\mathbf{r}=0$, we see that $C=1$, so that

$$u_{n\mathbf{g}}(-\mathbf{r}) = u_{n\mathbf{g}}^*(\mathbf{r}). \tag{B.6}$$

Now let us divide $u_{n\mathbf{g}}$ into a real and imaginary part:

$$u_{n\mathbf{g}}(\mathbf{r}) = v_{n\mathbf{g}}(\mathbf{r}) + i w_{n\mathbf{g}}(\mathbf{r}). \tag{B.7}$$

Then, by (B.6) we have

$$v_{n\mathbf{g}}(-\mathbf{r}) = v_{n\mathbf{g}}(\mathbf{r}), \tag{B.8}$$

$$w_{n\mathbf{g}}(-\mathbf{r}) = -w_{n\mathbf{g}}(\mathbf{r}). \tag{B.9}$$

But as the operator

$$p_\alpha \equiv (1/i)(\partial/\partial x_\alpha) \tag{B.10}$$

is odd under inversion we have

$$\begin{aligned} p_{nn'}{}^\alpha(\mathbf{g}) &= \frac{(2\pi)^3}{\Omega}\int \varphi_{n\mathbf{g}}{}^* p^\alpha \varphi_{n'\mathbf{g}} d\mathbf{r} \\ &= \delta_{nn'} g^\alpha + \frac{(2\pi)^3}{\Omega}\int u_{n\mathbf{g}}{}^* p^\alpha u_{n'\mathbf{g}} d\mathbf{r} \\ &= \delta_{nn'} g^\alpha + \frac{(2\pi)^3}{\Omega}\int (v_{n\mathbf{g}} p^\alpha i w_{n'\mathbf{g}} - i w_{n\mathbf{g}} p^\alpha v_{n'\mathbf{g}}) d\mathbf{r}, \end{aligned} \tag{B.11}$$

and in view of (B.10) this is clearly real.

At special points $\mathbf{g}$, where either of the conditions mentioned after Eq. (B.2) fails, the property (B.2) will still hold if one chooses $\varphi_{n\mathbf{g}}$ as the limit of $\varphi_{n\mathbf{g}'}$ where $\mathbf{g}' \to \mathbf{g}$.

Bloch Electrons in a Magnetic Field

E. I. Blount
Bell Telephone Laboratories, Murray Hill, New Jersey
(Received August 14, 1961)

A formalism recently developed by the author is used to simplify Kohn's derivation of an effective Hamiltonian. The new derivation is valid for arbitrary symmetry and with the inclusion of spin-orbit effects and for bands with degeneracies. The actual form is obtained to second order in the magnetic field. It is explicitly shown that the procedure is asymptotically convergent. This Hamiltonian is then used to obtain the susceptibility in a convenient form. The nature of the actual energy levels is discussed with particular attention to the broadening, gaps in the continuous region, and magnetic breakdown.

Since the early days of solid-state physics it has been customary to treat the motion of Bloch electrons in a magnetic field by a Hamiltonian $H(\hbar\mathbf{k}+(e/c)\mathbf{a})$ obtained by substituting $\hbar\mathbf{k}+(e/c)\mathbf{a}$ for $\hbar\mathbf{k}$ in the energy function for a band. This was first justified by Peierls[1] within the limitations of the tight-binding approximation. Since then a number of further efforts have been made to justify this procedure culminating in the work of Kohn,[2] who, by methods which he called "shockingly complicated," was able to establish that the Hamiltonian can indeed be written as a function of $H(\hbar\mathbf{k}+(e/c)\mathbf{a})$, where $H(\mathbf{k})$ is itself a power series in the magnetic field $\mathbf{B}$.

A major purpose of the present paper is to reduce the complication of Kohn's method. We shall also demonstrate explicitly that it is asymptotically convergent. Our procedure is applicable also to bands which do have degeneracies with other bands and to electrons obeying more general wave equations than the nonrelativistic Schrödinger equation. These results will be applied to the steady susceptibility of a crystal and to the determination of the energy levels.

I. THE EFFECTIVE HAMILTONIAN

In this section we shall derive the effective diagonal-in-bands Hamiltonian for electrons described by any of the following one-electron Hamiltonians.

(a) *Schrödinger:*

$$\mathfrak{H}=\frac{1}{2m}\left(\mathbf{p}+\frac{e}{c}\mathbf{a}\right)^2+U, \tag{I.1a}$$

(b) *Pauli:*

$$\mathfrak{H}=\frac{1}{2m}\left(\mathbf{p}+\frac{e}{c}\mathbf{a}\right)^2+\frac{e^2}{8m^2c^2}[\mathbf{p}\cdot(\boldsymbol{\sigma}\times\nabla U)]+U, \tag{I.1b}$$

(c) *Dirac:*

$$\mathfrak{H}=c\boldsymbol{\alpha}\cdot[\mathbf{p}+(e/c)\mathbf{a}]+\beta mc^2+U, \tag{I.1c}$$

where U is the periodic crystalline potential. In the absence of the vector potential $\mathbf{a}$, each of the Hamiltonians has a velocity operator $\mathfrak{V}$, such that

$$\mathfrak{V}=-(i/\hbar)[\mathbf{x},H], \tag{I.2}$$

where H is the Hamiltonian with $\mathbf{a}=0$, whose eigenvalues are $E(\mathbf{k})$. All the above Hamiltonians can be written in the form

$$\mathfrak{H}=H+(e/2c)\{\mathfrak{V}\cdot\mathbf{a}\}+e^2a^2/2mc^2, \tag{I.3}$$

except that for the Dirac case we must set $1/m=0$ and the last term disappears. We now transform this equation to the crystal momentum representation (CMR). [A detailed discussion of this representation in the form used in this paper has been given by the author in another paper[3] which will hereafter be referred to as (F).] In the CMR, $\mathbf{x}$ has the representation $\mathbf{R}+\mathfrak{X}$ where $\mathbf{R}=i\partial/\partial\mathbf{k}$ and $\mathfrak{X}$ is diagonal in the wave vector $\mathbf{k}$, but not in the band index n. Correspondingly, we shall write $\mathbf{a}=A+\mathfrak{A}$, where A depends on $\mathbf{R}$ and $\mathfrak{A}$ on $\mathfrak{X}$. H and $\mathfrak{V}$ are diagonal in $\mathbf{k}$, and H is diagonal in n, while $\mathfrak{V}$ is a matrix in n. Finally,

$$\mathfrak{H}=H+\tfrac{1}{2}\{\mathfrak{V}\cdot(A+\mathfrak{A})\}+(A+\mathfrak{A})^2/2m,$$

where we have also chosen units such that $e/c=1$, or alternatively we have absorbed e/c into $\mathbf{a}$.

We will make much use of another object from the CMR,

$$S_{nn'}(\mathbf{k},\mathbf{k}')=\int u_{n\mathbf{k}}{}^*u_{n'\mathbf{k}'}d\tau,$$

where the u's are the periodic parts of the Bloch functions $\psi_{n\mathbf{k}}(\mathbf{r})=e^{i\mathbf{k}\cdot\mathbf{r}}u_{n\mathbf{k}}(\mathbf{r})$. S is related to $\mathfrak{X}$:

$$\mathfrak{X}_{nn'}(k)=\delta(\mathbf{k}-\mathbf{k}')i(\partial/\partial\mathbf{k}')S_{nn'}(\mathbf{k},\mathbf{k}'). \tag{I.5}$$

The whole difficulty in treating the Hamiltonian (I.4) is that $\mathbf{a}$ and in particular A, can be very large. Thus normal perturbation theory cannot be expected to give good results. The author has developed a procedure for treating perturbations which may be large, if they vary slowly from cell to cell [(F), Sec. III]. In this formalism one transforms from the CMR to a mixed representation to which we shall refer as pseudoclassical (PCR), in which an operator is written as a function of two variables $\mathbf{R}$ and $\mathbf{k}$, which can be treated as numbers. The quantum mechanics then rests in the multiplication rules, which are nonlocal. A typical operator then becomes $O_{nn'}(\mathbf{k},\mathbf{R})$. $\mathbf{x}$ is still written as $\mathbf{R}+\mathfrak{X}$, but $\mathbf{R}$ can now be considered a number not an operator, while $\mathfrak{X}$ is still a matrix and depends only on $\mathbf{k}$.

Alternatively, one can continue to regard $\mathbf{k}$ and $\mathbf{R}$ as operators, and our formalism provides a recipe for

[1] R. E. Peierls, Z. Physik **80**, 763 (1933).
[2] W. Kohn, Phys. Rev. **115**, 1460 (1959). Additional references will be found in this paper.
[3] E. I. Blount, in *Solid-State Physics*, edited by F. Seitz and D. Turnbull (Academic Press Inc., New York, 1961), Vol. 13.

obtaining other operators in the form of matrices each of whose elements is an operator function of **k** and **R**. Our multiplication rules then are equivalent to treating these functions as fully symmetrized Taylor series in **k** and **R** and multiplying them accordingly. Our formalism is valid asymptotically, however, in cases where the Taylor series diverge, as shown in (F).

We write

$$\mathfrak{H}_{nn'}(\mathbf{k},\mathbf{R})=H(\mathbf{k})+\tfrac{1}{2}\{\mathfrak{B}(\mathbf{k})\cdot[A(R)+\mathfrak{A}(\mathbf{k})]\}+[A(\mathbf{R})+\mathfrak{A}(\mathbf{k})]^2/2m. \quad \text{(I.6)}$$

(F) gives a procedure for removing interband matrix elements to arbitrary order, and we could proceed immediately in this direction, but we also wish to assure that our final expression is a function only of $\mathfrak{k}=(\mathbf{k}+A)$ in analogy with the free-electron problem. To this end we make the transformation $S_{nn'}(0,\mathbf{k})$. This has the effect of expressing the Hamiltonian in terms of functions whose periodic parts are the Bloch functions at $\mathbf{k}=0$ or Kohn-Luttinger functions. H, $\mathfrak{B}$, and $\mathbf{x}$ are altered as follows:

$$H(k)\rightarrow H(0)+\mathfrak{B}(0)\cdot\mathbf{k}+k^2/2m. \quad \text{(I.7)}$$

(We now choose $\hbar=1$.)

$$\mathfrak{B}\rightarrow\mathfrak{B}(0)+\mathbf{k}/m, \quad \text{(I.8)}$$

$$\mathbf{x}\rightarrow\mathbf{R}. \quad \text{(I.9)}$$

As a consequence of the last of these, $A+\mathfrak{A}$ is replaced by A and $\mathfrak{H}$ becomes

$$\mathfrak{H}'(k,R)=H(0)+\mathfrak{B}(0)\cdot(\mathbf{k}+A)+(\mathbf{k}+A)^2/2m =H(\mathbf{k}+A). \quad \text{(I.10)}$$

Thus we already have $\mathfrak{H}'$ as a function only of $\mathfrak{k}$, and our procedure as discussed in (F) will be to diagonalize this by successive transformations, each of which will depend only on $\mathfrak{k}$; thus our final result will be of this form. This follows from the fact that commutators of the components of $\mathfrak{k}$ obey the following rule:

$$[\mathfrak{k}_i,\mathfrak{k}_j]_0=i\epsilon_{ijk}B_k,$$

and are thus constant.

We now observe that (I.10) is of the same form as (I.7) with **k** replaced by $\mathfrak{k}$. Thus we can diagonalize it to lowest order, in the sense of (F), by applying the transformation $S_{nn'}(\mathfrak{k},0)$. This transformation itself is not unitary for it depends on both **k** and **R**, and the Poisson bracket $\langle S,S^\dagger\rangle$ does not vanish. It will thus be necessary to use the perturbation procedure of (F).

Before doing so, however, it will be desirable to obtain some simplification of certain formulas in (F) which arise from the fact that we will be using functions only of ($\mathfrak{k}$). In (F), we used the Poisson bracket of two operators N and O,

$$\langle N,O\rangle=\frac{\partial N}{\partial R_i}\frac{\partial O}{\partial k_i}-\frac{\partial N}{\partial k_i}\frac{\partial O}{\partial R_i}, \quad \text{(I.11)}$$

where we use the summation convention. If N and O are functions of $\mathfrak{k}$, we obtain

$$\langle N,O\rangle=\left(\frac{\partial N}{\partial \mathfrak{k}_j}\frac{\partial O}{\partial \mathfrak{k}_i}-\frac{\partial N}{\partial \mathfrak{k}_i}\frac{\partial O}{\partial \mathfrak{k}_j}\right)\frac{\partial a_j}{\partial k_i} =-\mathbf{B}\cdot\frac{\partial N}{\partial \mathfrak{k}}\times\frac{\partial O}{\partial \mathfrak{k}}, \quad \text{(I.12)}$$

where $\mathbf{B}=\nabla\times\mathbf{a}$ is the magnetic field.

This also makes it clear that in higher-order terms, where we also have such terms as

$$\left(\frac{\partial N}{\partial R_i}O\frac{\partial P}{\partial k_i}-\frac{\partial N}{\partial k_i}O\frac{\partial P}{\partial R_i}\right), \quad \text{(I.13)}$$

where N, O, P can themselves be derivatives, we can replace this sum by

$$\sum_i\left(\frac{\partial N}{\partial R_i}O\frac{\partial P}{\partial k_i}-\frac{\partial N}{\partial k_i}O\frac{\partial P}{\partial R_i}\right)=-B_l\epsilon_{ijl}\frac{\partial N}{\partial k_i}O\frac{\partial P}{\partial k_j}. \quad \text{(I.14)}$$

We now write from (F) the formula for the operator product $(SHS^\dagger)_O$ to second order in **B**. [In (F), we used "order" to indicate the number of differentiations with respect to **R** in a particular term. Here, this coincides with order in **B** since **R** appears only in A.]

$$\begin{aligned}(SHS^\dagger)_O=SHS^\dagger&+\tfrac{1}{2}i\left(\langle S,H\rangle S^\dagger+S\langle H,S^\dagger\rangle+\frac{\partial S}{\partial R_i}H\frac{\partial S^\dagger}{\partial k_i}-\frac{\partial S}{\partial k_i}H\frac{\partial S^\dagger}{\partial R_i}\right)-\frac{1}{8}\left(\frac{\partial^2 S}{\partial k_i\partial k_j}H\frac{\partial^2 S^\dagger}{\partial R_i\partial R_j}+2\frac{\partial S}{\partial k_i}\frac{\partial H}{\partial k_j}\frac{\partial^2 S^\dagger}{\partial R_i\partial R_j}\right.\\
&+S\frac{\partial^2 H}{\partial k_i\partial k_j}\frac{\partial^2 S^\dagger}{\partial R_i\partial R_j}-2S\frac{\partial^2 H}{\partial k_i\partial R_j}\frac{\partial^2 S^\dagger}{\partial k_j\partial R_i}+S\frac{\partial^2 H}{\partial R_i\partial R_j}\frac{\partial^2 S^\dagger}{\partial k_i\partial k_j}+\frac{\partial^2 S}{\partial R_i\partial R_j}\frac{\partial^2 H}{\partial k_i\partial k_j}S^\dagger+\frac{\partial^2 S}{\partial R_i\partial R_j}H\frac{\partial^2 S^\dagger}{\partial k_i\partial k_j}\\
&+2\frac{\partial^2 S}{\partial R_i\partial R_j}\frac{\partial H}{\partial k_i}\frac{\partial S^\dagger}{\partial k_j}-2\frac{\partial^2 S}{\partial k_i\partial R_j}\frac{\partial H}{\partial R_i}\frac{\partial S^\dagger}{\partial k_j}-2\frac{\partial^2 S}{\partial k_i\partial R_j}\frac{\partial^2 H}{\partial R_i\partial k_j}S^\dagger+\frac{\partial^2 S}{\partial k_i\partial k_j}\frac{\partial^2 H}{\partial R_i\partial R_j}S^\dagger-2\frac{\partial^2 S}{\partial k_i\partial R_j}H\frac{\partial^2 S^\dagger}{\partial k_j\partial R_i}\\
&-2\frac{\partial^2 S}{\partial k_i\partial R_j}\frac{\partial H}{\partial k_j}\frac{\partial S^\dagger}{\partial R_i}-2\frac{\partial S}{\partial k_i}\frac{\partial H}{\partial R_j}\frac{\partial^2 S^\dagger}{\partial R_i\partial k_j}+2\frac{\partial S}{\partial R_i}\frac{\partial H}{\partial R_j}\frac{\partial^2 S^\dagger}{\partial k_i\partial k_j}+2\frac{\partial S}{\partial k_j}\frac{\partial^2 H}{\partial k_i\partial R_j}\frac{\partial S^\dagger}{\partial R_i}-2\frac{\partial S}{\partial k_j}\frac{\partial^2 H}{\partial R_i\partial R_j}\frac{\partial S^\dagger}{\partial k_i}\\
&\left.-2\frac{\partial S}{\partial R_j}\frac{\partial^2 H}{\partial k_i\partial k_j}\frac{\partial S^\dagger}{\partial R_i}-2\frac{\partial S}{\partial R_j}\frac{\partial H}{\partial k_i}\frac{\partial^2 S^\dagger}{\partial k_j\partial R_i}+2\frac{\partial^2 S}{\partial k_i\partial k_j}\frac{\partial H}{\partial R_i}\frac{\partial S^\dagger}{\partial k_j}+2\frac{\partial S}{\partial R_j}\frac{\partial^2 H}{\partial k_j\partial R_i}\frac{\partial S^\dagger}{\partial k_i}\right),\end{aligned} \quad \text{(I.15)}$$

where all products on the right are matrix products. Using (I.14), this equation can be considerably simplified:

$$(SHS^\dagger)_O=SHS^\dagger-\tfrac{1}{2}iB_l\epsilon_{lij}(S_iH_jS^\dagger+SH_iS_j^\dagger+S_iHS_j^\dagger)$$
$$-\tfrac{1}{8}B_lB_m\epsilon_{lij}\epsilon_{mnp}(S_{in}HS_{jp}^\dagger+SH_{in}S_{jp}^\dagger+S_{in}H_{jp}S^\dagger+2S_iH_nS_{jp}^\dagger+2S_{in}H_jS_p^\dagger+2S_iH_{nj}S_p^\dagger), \quad \text{(I.16)}$$

where we have used subscripts to indicate differentiation with respect to $\mathfrak{k}$. To obtain $(SS^\dagger)_O$, we merely set $H=1$ in (I.16):

$$(SS^\dagger)_O=1-\tfrac{1}{2}iB_l\epsilon_{lij}S_iS_j^\dagger-\tfrac{1}{8}B_lB_m\epsilon_{lij}\epsilon_{mnp}S_{in}S_{jp}^\dagger. \quad \text{(I.17)}$$

We can now remove the first-order term in $(SS^\dagger)_O$ by multiplying S on the left by $(1+g^{(1)})$, where $g^{(1)}=\tfrac{1}{4}iB_l\epsilon_{lij}S_iS_j^\dagger$, which is Hermitian. Applying this to (I.16), we obtain to second order

$$\mathfrak{H}=E(\mathfrak{k})+\tfrac{1}{4}iB_l\epsilon_{lij}\{S_i(S_j^\dagger SHS^\dagger-H_jS^\dagger-HS_j^\dagger)+(SHS^\dagger S_i-S_iH-SH_i)S_j^\dagger-S_iH_jS^\dagger-SH_iS_j^\dagger\}$$
$$-\tfrac{1}{8}B_lB_m\epsilon_{lij}\epsilon_{mnp}\{(S_{in}HS_{jp}^\dagger+SH_{in}S_{jp}^\dagger+S_{in}H_{jp}S^\dagger+2S_iH_nS_{jp}^\dagger+2S_{in}H_jS_p^\dagger+2S_iH_{jn}S_p^\dagger)$$
$$-[S_iS_j^\dagger,(S_nH_pS^\dagger+S_nHS_p^\dagger+SH_nS_p^\dagger)]_{(+)}-[(S_{in}S_j^\dagger+S_iS_{jn}^\dagger),(SHS^\dagger)_p]_{(-)}$$
$$+\tfrac{1}{2}S_iS_j^\dagger SHS^\dagger S_nS_p^\dagger-\tfrac{1}{2}[S_{in}S_{jp}^\dagger,SHS^\dagger]_{(+)}+\tfrac{3}{4}[S_iS_j^\dagger S_nS_p^\dagger,SHS^\dagger]_{(+)}\}, \quad \text{(I.18)}$$

where the subscripts + and − indicate anticommutators and commutators, respectively. Here the first group of terms in second order comes from (I.16), the second group from $[g^{(1)},\langle SHS^\dagger\rangle]_{(+)}$, the third from $\langle g^{(1)},SHS^\dagger\rangle+\langle SHS^\dagger,g^{(1)}\rangle$, the fourth from $g^{(1)}SHS^\dagger g^{(1)}$, and the last two from $[g^{(2)},SHS^\dagger]_{(+)}$, where $g^{(2)}$ is chosen so that if $H=1$, the second-order terms vanish. That is, $g^{(2)}$ is chosen to make $(SS^\dagger)_O$ unitary to second order:

$$g^{(2)}=\tfrac{1}{16}B_lB_m\epsilon_{lij}\epsilon_{mnp}(S_{in}S_{jp}^\dagger-\tfrac{3}{32}S_iS_j^\dagger S_nS_p^\dagger).$$

The addition of $g^{(1)}$ and $g^{(2)}$ to S does not remove interband terms to second order, but before proceeding we rewrite (I.18) in terms of the $\mathfrak{X}$'s and $\mathfrak{V}$'s, using (I.2) and (I.5).

$$\mathfrak{H}=E(\mathfrak{k})+\tfrac{1}{4}B_l\epsilon_{lij}\{[\mathfrak{X}_i,v_j]_++[\mathfrak{X}_i,\mathfrak{V}_j]_+\}+B_lB_m\epsilon_{lij}\epsilon_{mnp}\{\tfrac{1}{32}[\mathfrak{X}_i[\mathfrak{X}_n,\alpha_{jp}]_+]_++\tfrac{1}{16}[v_j[\mathfrak{X}_n,\partial\mathfrak{X}_i/\partial\mathfrak{k}_p]_+]_+$$
$$+\tfrac{1}{16}[\mathfrak{X}_i,(\partial/\partial\mathfrak{k}_j)[\mathfrak{X}_n,\mathfrak{V}_p]_+]_++(\delta_{jn}/8m)\mathfrak{X}_i\mathfrak{X}_p-\tfrac{1}{32}[\partial\mathfrak{X}_i/\partial\mathfrak{k}_n,[\partial\mathfrak{X}_p/\partial\mathfrak{k}_j,E]_-]_-$$
$$+\tfrac{1}{16}i[((\partial\mathfrak{X}_i/\partial\mathfrak{k}_n)\mathfrak{X}_j\mathfrak{X}_p+\mathfrak{X}_i\mathfrak{X}_n\partial\mathfrak{X}_p/\partial\mathfrak{k}_j),\ SHS^\dagger]_-\}, \quad \text{(I.19)}$$

where v_j is the intraband velocity operator $v_j=\partial E/\partial\mathfrak{k}_j$. The first-order term is obtained straightforwardly from (I.18), while the second-order term requires complicated rearrangements. The last term has only interband elements and can be removed immediately.

Equation (I.19) can now be put in a very convenient form if we introduce the following notation:

$$\tfrac{1}{2}B_l\epsilon_{lij}\mathfrak{X}_i=\mathfrak{A}_j^s, \quad \text{(I.20)}$$

where the superscript indicates that this is the part of the vector potential associated with $\mathfrak{X}$ in the symmetric gauge, regardless of the gauge we started with;

$$\tfrac{1}{4}B_l\epsilon_{lij}[\mathfrak{X}_i,\mathfrak{V}_j]_+=-\mathbf{B}\cdot\mathfrak{M}', \quad \text{(I.21)}$$

where $\mathfrak{M}'$ is the nonsingular part of the magnetic moment operator. Equation (I.19) becomes to second order

$$\mathfrak{H}=E(\mathfrak{k}+\mathfrak{A}^s(\mathfrak{k}+\mathfrak{A}^s))-\mathbf{B}\cdot\mathfrak{M}'(\mathfrak{k}+\mathfrak{A}^s)+(\mathfrak{A}^s)^2/2m$$
$$+\tfrac{1}{8}[(\partial\mathfrak{A}_j^s/\partial k_n)[\partial\mathfrak{A}_n^s/\partial k_j,E]_-]_-, \quad \text{(I.22)}$$

where $E(\mathfrak{k}+\mathfrak{A}^s)$ and $\mathfrak{M}'(\mathfrak{k}+\mathfrak{A}^s)$ are defined by the power-series expansion of E and $\mathfrak{M}'$ about $\mathfrak{k}$. We have left out the last term of (I.19), which can be easily removed by a transformation. This expression except for the last term is exactly what the Hamiltonian of an ion in an external magnetic field looks like, the first term corresponding to $(1/2m)[\mathbf{p}+(e/c)\mathbf{a}]^2$ for the center-of-charge-and-mass motion, the second term being the magnetic moment, and the third the molecular diamagnetism. The fourth term is a specifically crystalline effect as is the appearance of $\mathbf{A}^s$ in the first two.

While (I.22) is a particularly concise expression it is not very convenient for computation, and we now show a different rewriting of (I.19) using a different magnetic moment operator $\mathfrak{M}$.

$$\mathfrak{M}_l=-\epsilon_{lij}\tfrac{1}{4}[\mathfrak{X}_i,V_j]_+, \quad \text{(I.23)}$$

where V_j has only interband matrix elements. This leads us to the form

$$\mathfrak{H}=E(\mathbf{k})-\tfrac{1}{2}[(\mathbf{B}\times\mathbf{v})\cdot\mathfrak{X}]_+-\mathbf{B}\cdot\mathfrak{M}+\tfrac{3}{8}[\mathfrak{A}_i^s,[\mathfrak{A}_j^s,\alpha_{ij}]_+]_++(\mathfrak{A}^s)^2/2m+\tfrac{1}{2}[v_j,[\mathfrak{A}_p^s,\partial\mathfrak{A}_j^s/\partial k_p]_+]_+$$
$$+\tfrac{1}{4}[[\mathfrak{A}_j,v_p]_-,\partial\mathfrak{A}_p/\partial k_j]_-+\tfrac{1}{8}[\partial\mathfrak{A}_j/\partial k_p,[\partial\mathfrak{A}_p^s/\partial k_j,E]_-]_--\tfrac{1}{2}[\mathfrak{A}_j^s,(\partial/\partial\mathfrak{k}_j)\mathbf{B}\cdot\mathfrak{M}]_+. \quad \text{(I.24)}$$

In this form the various orders of $\mathbf{B}$ are kept separate and terms involving $\mathbf{v}$ are separated from those involving V, which occurs only in $\mathfrak{M}$.

It is interesting to note that the second term can be written

$$-\tfrac{1}{2}[(\mathbf{B}\times\mathbf{v})\cdot\mathfrak{X}]_+=-\tfrac{1}{2}[\mathbf{F}_L\cdot\mathfrak{X}]_+. \quad \text{(I.25)}$$

This is the energy of a dipole $\mathfrak{X}$ in the Lorentz force $\mathbf{F}_L$, without the factor $\frac{1}{2}$ which occurs in the magnetic moment term. It means that if we add an electric field to the problem, the Hamiltonian to first order in both fields is

$$E(\mathfrak{k})-\mathbf{E}\cdot\mathbf{R}-\tfrac{1}{2}[(\mathbf{E}+\mathbf{F}_L),\cdot\mathfrak{X}]_+-\mathbf{B}\cdot\mathfrak{M}, \qquad (I.26)$$

that is, the electric and Lorentz forces appear on an equal footing. When the procedure we have used is applied to the Dirac equation, treated as a band theory, it is found that the intraband part of the first-order term is

$$-\tfrac{1}{2}[\mathbf{v}\times\mathfrak{X}]_+-\mathfrak{M}=\hbar\boldsymbol{\sigma}/m^*c,$$

where m^* is the transverse mass, this equation being valid for all $\mathbf{k}$. Thus, here it is the rather peculiar combination $\frac{1}{2}[\mathbf{v}\times\mathfrak{X}]_+ + \frac{1}{4}[\mathbf{V}\times\mathfrak{X}]_+$ which is simply related to $\boldsymbol{\sigma}$ the spin.

It is now appropriate to remove the first-order interband matrix elements, and for this purpose we simply use second-order perturbation theory, obtaining

$$\begin{aligned}\mathfrak{H}=E(\mathfrak{k})-(\mathbf{B}\times\mathbf{v})\cdot\mathfrak{x}-\mathbf{B}\cdot\mathbf{m}+\tfrac{3}{8}[\mathfrak{A}_i{}^s[\mathfrak{A}_j{}^s,\alpha_{ij}]_+]_+ +(\mathfrak{A}^s)^2/2m+\tfrac{1}{2}[v_j,[\mathfrak{A}_p{}^s,\partial\mathfrak{A}_j{}^s/\partial k_p]_+]_+\\ +E_{VD}+\tfrac{1}{4}[[A_i{}^s,v_p]_-,\partial A_p{}^s/\partial k_j]_- -\tfrac{1}{2}[\mathfrak{A}_i{}^s,(\partial/\partial k_i)\mathbf{B}\cdot\mathbf{M}]_+ +\tfrac{1}{8}[\partial A_j{}^s/\partial\mathfrak{k}_p,[\partial A_p{}^s/\partial\mathfrak{k}_j,E]_-]_-, \qquad (I.27)\end{aligned}$$

where the lower case $\mathfrak{x}$ and $\mathbf{m}$ designate intraband matrix elements of $\mathfrak{X}$ and $\mathfrak{M}$ (which for bands degenerate throughout $\mathbf{k}$ space is taken to mean a two×two matrix in the degenerate bands[3]), and E_{VD} is defined as follows:

$$E_{VD,nn}=-\tfrac{1}{2}\sum_{n'\neq n}\frac{|\mathbf{B}\cdot\mathbf{M}_{nn'}+[(\mathbf{B}\times\mathbf{v});\mathbf{X}_{nn'}]_+|^2}{E_n-E_{n'}}. \qquad (I.28)$$

The term in E_{VD} containing $|B\cdot M_{nn'}|^2$ is the Van Vleck paramagnetism, the term in $|X^2|$ is like a dielectric effect of the Lorentz force, and the cross term is an interference between these effects. The symbols $\mathbf{M}$, $\mathbf{X}$, $\mathbf{A}$ refer to *interband* matrix elements. We have left out some terms involving intraband matrix elements of $\mathfrak{A}$ because they vanish in the diagonal terms.

This is still not the most useful form for the second-order terms. As a first step, we eliminate all factors of the form $\partial A_\mu/\partial k_\nu$ for $\mu\neq\nu$. Then the second-order terms have the form

$$\begin{aligned}\mathfrak{h}_2=-\tfrac{1}{8}[\alpha_{\mu\nu}[A_\mu{}^sA_\nu{}^s]_+]_+ +\frac{(\mathfrak{A}^s)^2}{2m}-\tfrac{1}{2}[\nabla_\mathbf{k}\cdot\mathbf{A}^s[\mathbf{A}^s\cdot\mathbf{v}]_+]_+ +E_{VD}+\tfrac{1}{2}[\nabla_\mathbf{k}\cdot\mathbf{A}^s,\mathbf{B}\cdot\mathbf{M}]_+ +\tfrac{1}{8}[\nabla_\mathbf{k}\cdot\mathbf{A}^s[\nabla_\mathbf{k}\cdot\mathbf{A}^s,E]]\\ +\tfrac{3}{8}[\alpha_{\mu\nu}[\mathfrak{a}_\mu{}^s\mathfrak{a}_\nu{}^s]_+]_+ +\tfrac{1}{2}[v_\mu,[\mathfrak{a}_\nu{}^s\partial\mathfrak{a}_\mu{}^s/\partial k_\nu]_+]_+ -\tfrac{1}{2}[\mathfrak{a}_\mu{}^s,\partial\mathbf{B}\cdot\mathbf{m}/\partial k_\mu]_+\\ +\frac{1}{2}\frac{\partial}{\partial k_\nu}([A_\nu{}^s,[A_\mu{}^s,v_\mu]_+]_+ -\tfrac{1}{2}[A_\nu{}^s,[\nabla_\mathbf{k}\cdot\mathbf{A}^s,E]_-]_- -[A_\nu{}^s,\mathbf{B}\cdot\mathbf{M}]_+), \qquad (I.29)\end{aligned}$$

where $\mathfrak{a}^s$ is the intraband part of $\mathfrak{A}^s$.

$\nabla_\mathbf{k}\cdot\mathbf{A}^s$ and $\mathbf{B}\cdot\mathbf{M}$ have very similar structure which can be exploited to obtain a very simple form for the total Hamiltonian.

$$\begin{aligned}\mathfrak{H}=H(\mathfrak{k}+2\mathfrak{a}^s(\mathfrak{k}+2\mathfrak{a}^s))-\mathbf{B}\cdot\mathbf{m}(\mathfrak{k}+2\mathfrak{a}^s)+\mathfrak{a}^s\cdot\mathbf{v}_A+E_{VD}'+A^2/2m-\tfrac{1}{4}\alpha_{\mu\nu}[A_\mu{}^sA_\nu{}^s]_+\\ -\tfrac{1}{2}i[(R_\mu+\mathfrak{x}_\mu),(A_\mu{}^s\mathfrak{V}_\nu A_\nu{}^s+A_\nu{}^s\mathfrak{V}_\nu A_\mu{}^s+v_\nu[A_\nu{}^sA_\mu{}^s]_+)]_-. \qquad (I.30)\end{aligned}$$

In the first term, the argument is to be interpreted as $\mathfrak{k}$, plus $2\mathfrak{a}$ evaluated at $\mathfrak{k}+2\mathfrak{a}$. Furthermore, E_{VD}' is defined as follows:

$$E_{VD,n}'=\sum_{n'=n}\frac{|(\mathbf{A}^s\cdot\mathbf{V}+\{\mathbf{v}\cdot\mathbf{A}^s\})_{nn'}|^2}{E_n-E_{n'}}. \qquad (I.31)$$

We note that the operator $\mathbf{A}^s\cdot\mathbf{V}$ is *not* Hermitian. $\mathbf{v}_A$ is defined as

$$\mathbf{v}_A=-(i/\hbar)[\mathfrak{x},H-\mathbf{B}\cdot\mathbf{m}]_0,$$

the operator commutator of $\mathfrak{x}$ with $\mathfrak{H}$ evaluated to first order.

$$\mathbf{v}_A=+(i/\hbar)[\mathfrak{x},\mathbf{B}\cdot\mathbf{m}]+(i/\hbar)[\mathfrak{x},(\mathbf{B}\times\mathbf{v})\cdot(\mathfrak{x})].$$

We have used the form $\mathfrak{a}\cdot\mathbf{v}_A$ to emphasize that $\mathbf{v}_A$ is a part of the total velocity operator and this term thus has a natural place in the Hamiltonian. It was pointed out at the beginning that $1/m$ must be set equal to zero for Dirac electrons. We, nevertheless, expect a term like $\mathbf{A}^2/2m$ in the Hamiltonian. We now see that it comes from E_{VD}'. For the sum over n' must include negative energy states, for which $E_n-E_{n'}\sim 2mc^2$. The sum then becomes approximately $(\mathbf{A}\cdot\mathbf{VV}\cdot\mathbf{A})/2mc^2$. For the Dirac electron $\mathbf{VV}\sim 1c^2$ and the total contribution of the sum over negative energy states is approximately $A^2/2m$.

In (I.30), the first two terms are the only ones which exist for free electrons, while $\mathbf{A}^2/2m$ and E_{VD}' are the only ones for molecules and atoms. A different categorization is that the first three terms are determined by simple one-band parameters discussed in (F). These parameters have their analogs in the Dirac theory (F). The next two terms are explicitly interband or "molecular" in origin. The sixth and seventh terms are specifically connected with the interaction between intraband and interband effects.

The form (I.30) is valid for any choice of phases of the Bloch functions, and is indeed insensitive to the generalized phase transformation discussed in (F). The establishment of this fact serves only as a check on our result and will not be performed here.

One final rearrangement of (I.30) will be given which will be useful in the next section.

$$\begin{aligned}\mathfrak{H}=H(\mathfrak{k})-\mathbf{B}\cdot\mathbf{m}-\mathbf{B}\times\mathbf{v}\cdot\mathfrak{a}+(A^s)^2/2m+E_{VD}'-\tfrac{1}{4}\alpha_{\mu\nu}[A_\mu{}^sA_\nu{}^s]_+\\ -\tfrac{1}{2}i[(R_\mu+\mathfrak{x}_\mu),\,(A_\mu{}^s\mathfrak{B}_\nu A_\nu{}^s+A_\nu{}^s\mathfrak{B}_\nu A_\mu{}^s+v_\nu[A_\nu{}^sA_\mu{}^s]_+)]_-+(\partial/\partial k_\nu)(v_\mu[\mathfrak{a}_\mu{}^s\mathfrak{a}_\nu{}^s])\\ -\tfrac{1}{2}iv_\mu[\mathfrak{a}_\mu{}^s,[\mathfrak{x}_\nu,\mathfrak{a}_\nu]_-]_++\tfrac{1}{2}i[2R_\mu+\mathfrak{x}_\mu,\,[\mathfrak{a}^s,\mathbf{B}\cdot\mathbf{m}]_+]_--\tfrac{3}{8}[\mathbf{B}\cdot\mathbf{\Omega},\mathbf{B}\cdot\mathbf{m}]_+,\end{aligned}\qquad\text{(I.32)}$$

where

$$\mathbf{\Omega}=-i(\mathbf{R}+\mathfrak{x})\times(\mathbf{R}+\mathfrak{x})$$

and has been discussed in (F).

At this point we have accomplished our first objective, to show that an effective Hamiltonian can be found without interband matrix elements, to any desired order, and we have carried it out to second order.

Some final remarks are in order concerning the first-order terms in (I.27). In a crystal with inversion symmetry and no spin-orbit coupling, $\mathbf{m}$ vanishes and $\mathfrak{x}$ is a gradient $\nabla_{\mathbf{k}}\varphi$. For such a band, which in addition intersects no other bands, φ and $\mathfrak{x}$ can be made zero. In a band which has intersections, this cannot usually be done, though $\nabla_k\cdot\mathfrak{x}$ can be made zero. In crystals with inversion symmetry and spin-orbit coupling, all bands occur in degenerate pairs throughout $\mathbf{k}$ space, and we consider such pairs of bands to be one band of two-component wave functions. Then, in general $\mathbf{m}$ will not vanish nor will $\mathfrak{x}$, but the trace of $\mathbf{m}$ for each band vanishes everywhere, while the trace of $\mathfrak{x}$ behaves like $\mathfrak{x}$ in the absence of spin-orbit coupling. (In such bands $\mathbf{m}$ and $\mathfrak{x}$, like all other operators, are treated as 2×2 matrices, just as in the Pauli theory of spin.)

So far we have simply taken over the procedure of (F) without concerning ourselves with matters of convergence, either of the multiplication procedure (I.14), or the perturbation procedure. We now consider these in order.

In (F) it is pointed out that the multiplication formula (I.14) cannot be expected to converge if one factor has Fourier components $O(\mathbf{k};\mathbf{q})$ for $\mathbf{q}$ outside the range of convergence of another factor $N(\mathbf{k};\mathbf{q})$ as a function of $\mathbf{k}$. In our present problem let us consider the first step when we take the product $S(\mathbf{k}+A,0)H(\mathbf{k}+A)S^\dagger(\mathbf{k}+A,0)$. $H(\mathbf{k}+A)$ is quadratic in $\mathbf{k}$ and A and thus gives no trouble, but $S(\mathbf{k}+A,0)$ and its adjoint have Fourier components over the whole Brillouin zone (BZ), whereas as a function of $\mathbf{k}$ it does not in general have such a large range of convergence as a power series. We can, however, explicitly see that it is asymptotically convergent at small field. This is done in Appendix A.

Each step of the perturbation procedure involves the introduction of a new pair of operators, which must be multiplied by the expressions already obtained. These new operators are all functionals of S and the polynomials H and $\mathfrak{B}$. Consequently any finite number of them will not disturb the asymptotic character of the expressions obtained. Likewise, the removal of interband elements is asymptotic in that the error in the nth order is of order B^{n+1} or higher at small B. This establishes the asymptotic convergence of our total procedure.

One exception must be made to this statement. At degeneracies we have established the asymptotic convergence of the multiplication rules, but of course the perturbation theory will break down. In such cases, we usually use degenerate perturbation theory. Thus in the present problem, we might try to use degenerate perturbation theory at degeneracies. Unfortunately it does us no particular good, for in such neighborhoods the derivatives with respect to $\mathfrak{k}$ of the transformations become very large and there is no possibility of a development in power series. Indeed, the interband matrix elements have an important meaning in such places as we shall see. Of course, (I.24), which gives a Hamiltonian depending only on $\mathfrak{k}$, is valid and useful near degeneracies.

Finally, it is desirable to show what the basis function in the transformed representation looks like. We shall content ourselves with the zeroth order approximation since this contains the basic physical idea. From $S(\mathbf{k},\mathbf{R})=S(\mathbf{k}+A,\mathbf{k})$ we must calculate the matrix elements $S(\mathbf{k}',\mathbf{k}'')$ and apply them to the Bloch function. We find

$$\begin{aligned}\psi_{n\mathbf{k}}&=\sum_{n'}\int d^3k'\,S_{nn'}(\mathbf{k},\mathbf{k}')\psi_{n'\mathbf{k}'}(\mathbf{r})\\ &=\sum_{n',R}\int d^3k'\,S_{nn'}\left(\frac{\mathbf{k}+\mathbf{k}'}{2}+A,\,\frac{\mathbf{k}+\mathbf{k}'}{2}\right)e^{i\mathbf{R}\cdot(\mathbf{k}-\mathbf{k}')}e^{i\mathbf{k}'\cdot\mathbf{r}}u_{n'\mathbf{k}'}(\mathbf{r})\\ &=e^{ik\cdot r}\sum_{n',R}\int d^3k'\,S_{nn'}\left(\mathbf{k}+A+\frac{(\mathbf{k}'-\mathbf{k})}{2}\right)e^{i(\mathbf{k}'-\mathbf{k})\cdot(\mathbf{r}-\mathbf{R})}u_{n\mathbf{k}'}(\mathbf{r}).\end{aligned}\qquad\text{(I.33)}$$

This can be expanded as a series in B just as the multiplication rules can, and the result is

$$\psi_{n\mathbf{k}}' = e^{i\mathbf{k}\cdot\mathbf{r}} u_{n,\mathbf{k}+A}(\mathbf{r}) + \cdots. \tag{I.34}$$

This first term shows that the basic transformation is to change the periodic function $u_{n\mathbf{k}}$ to the nearly periodic function $u_{n,\mathbf{k}+A}$ characteristic of the crystal kinetic momentum $\mathbf{k}+A$. This is entirely in accord with one's intuitive feelings, and some authors have used this as a basis for attacking the problem of this paper.

Note added in proof. In the case of a band whose energy range does not overlap that of any other band, it is easy to see that the matrix elements with other bands can in principle be removed exactly. This follows from the fact that for fields below a certain B_c, the exact energy levels will be so arranged that a number of states equal to the number of states in the band will have energies well separated from the remaining states. Linear combinations of these wave functions can then be taken which will be infinite in extent and similar to our modified Bloch functions. Since these functions have no strict requirements to obey such as being of the form $e^{i\mathbf{k}\cdot\mathbf{r}}\times$periodic function, there is no difficulty. For this case, then, our proof gives an asymptotic form of the Hamiltonian, but we know that the removal of interband elements is exact.

If the bands do overlap, on the other hand, we cannot guarantee the exact removal of interband elements and this part too is only asymptotic. This distinction is important in connection with magnetic breakdown (see Sec. III).

II. THE STEADY SUSCEPTIBILITY

The first use we shall make of (I.30) is to calculate the susceptibility. This is determined from the free energy by the relation

$$\chi = \frac{M}{B} = -\frac{1}{B}\frac{\partial F}{\partial B}\bigg|_{\zeta}, \tag{II.1}$$

while F is given by

$$F = N\zeta + \mathrm{Tr} f(\mathfrak{H}), \tag{II.2}$$

$$f(\mathfrak{H}) = -kT \ln\{1+\exp[(\zeta-\mathfrak{H})/kT]\}. \tag{II.3}$$

To obtain f correct to order B^2, we would like to expand it as a power series in $\mathfrak{H}$ and use our multiplication rule [(I.16), with $\mathfrak{H}$ substituted for S and $S^\dagger$]. f has only a small finite radius of convergence however, which indeed vanishes at $T=0$. Therefore, we make use of the Laplace transform of f, since $\exp(s\mathfrak{H})$ has an infinite radius of convergence. Thus we write

$$f = \int_{c-i\infty}^{c+i\infty} \varphi(s)\exp(s\mathfrak{H})ds, \tag{II.4}$$

$$\exp(s\mathfrak{H}) = \sum_n \frac{s^n}{n!}\mathfrak{H}^n. \tag{II.5}$$

As shown in (F), the operator product $\mathfrak{H}^n$ evaluated at $(\mathbf{k},\mathbf{R})$ is not equal to the nth power of $\mathfrak{H}(\mathbf{k},\mathbf{R})$. Treating each term by the expansion procedure of (F), we obtain

$$\begin{aligned}\exp(s\mathfrak{H}) &= \sum_n \frac{s^n}{(n-2)!2}[\mathfrak{H}(k,R)]^{n-2}\{\tfrac{1}{8}i\langle\mathfrak{H},\mathfrak{H}\rangle - \tfrac{1}{8}\langle\langle\mathfrak{H},\mathfrak{H}\rangle\rangle\} \\ &+ \sum_n \frac{s^n}{(n-3)!}\frac{[\mathfrak{H}(k,R)]^{n-3}}{6}(-\tfrac{1}{8})\langle\langle\mathfrak{H}\mathfrak{H}\mathfrak{H}\rangle\rangle' \\ &+ \sum_n \frac{s^n}{(n-4)!}\frac{[\mathfrak{H}(k,R)]^{n-4}}{4!}(-\tfrac{1}{4})\langle\mathfrak{H},\mathfrak{H}\rangle^2 + \cdots,\end{aligned} \tag{II.6}$$

where the prime indicates that only terms in which each $\mathfrak{H}$ is differentiated at least once are to be included, the others being in the previous term. We note that $\langle\mathfrak{H},\mathfrak{H}\rangle=0$; summing over n, and inserting the result in (II.4), we find

$$\begin{aligned}f(\mathbf{k},\mathbf{R}) &= f(\mathfrak{H}(k,R)) \\ &\quad -\tfrac{1}{8}B_\lambda B_\mu \epsilon_{\lambda\iota\kappa}\epsilon_{\mu\nu\rho}(\tfrac{1}{2}f''\alpha^{\iota\nu}\alpha^{\kappa\rho} + \tfrac{1}{3}f'''\alpha^{\iota\nu}v^\kappa v^\rho).\end{aligned} \tag{II.7}$$

The integration over $\mathbf{k}$ and sum over bands yields

$$F(R) = F_0 - \frac{1}{48}B_\lambda B_\mu \epsilon_{\lambda\iota\kappa}\epsilon_{\mu\nu\rho}\left(\frac{1}{2\pi}\right)^3 \int d^3k\, \mathrm{Tr} f_0'\alpha^{\iota\nu}\alpha^{\kappa\rho},$$

where Tr signifies the trace of the matrix at fixed $\mathbf{k}$, $\mathbf{R}$; f_0 is f' the Fermi function, and F_0 is the trace of the first term in (II.7):

$$F_0 = F(B=0) + \left(\frac{1}{2\pi}\right)^3 \int d^3k\, \mathrm{Tr}\left\{f_0\mathfrak{h}^{(2)} + f_0'\frac{(\mathfrak{h}^{(1)})^2}{2}\right\}, \tag{II.9}$$

$\mathfrak{h}^{(1)}$ and $\mathfrak{h}^{(2)}$ are, respectively, the first- and second-order terms in (I.30). The first-order contribution of $\mathfrak{h}^{(1)}$ vanishes by time-reversal for any nonferromagnetic material.

The procedure outlined could in principle be extended to higher orders, but would remain asymptotic at best. In particular it would not be able to pick up the de Haas-van Alphen effect. This is most readily found by making use of the actual spectrum and has been done by Lifshitz and Kosevch[4] for arbitrary $\mathfrak{H}(\mathfrak{k})$. There is no need to reproduce it here.

The asymptotic character of our power series in B differs from that in the free electron case, however, for in that problem $\exp(-\beta\mathfrak{H})$ has a convergent power series in $\mathfrak{H}$ and the asymptotic character arises only on taking the inverse Laplace transform. In the present case $\exp(-\beta\mathfrak{H})$ itself is given only asymptotically by (II.6). The reason for this difference is that for free electrons each term in (II.5) is a polynomial and is thus exactly evaluable by the series (II.6) which terminates for each order n.

---

[4] I. M. Lifshitz and A. M. Kosevch, Doklady Akad. Nauk S.S.S.R. **96**, 963 (1954).

The total susceptibility can now be written in detail:

(1) The first term is χ_{LP}, the Landau diamagnetism with an effective mass tensor, given by the second term in (II.8). As a term in the free energy, it is written

$$F_{LP}=-\frac{1}{48}B_\lambda B_\mu \epsilon_{\lambda\iota\kappa}\epsilon_{\mu\nu\rho}\left(\frac{1}{2\pi}\right)^3\int \mathrm{Tr} f_0'\alpha_{\iota\nu}\alpha_{\kappa\rho}d^3k. \quad \text{(II.10)}$$

(2) The contribution of the last term in (II.9) is partially cancelled by some of the second term, namely, that part coming from the eighth through the eleventh terms of (I.30). (This requires the nonobvious result that $\int f_0 v_\mu \mathrm{Tr}\{\mathfrak{a}_\mu,[\mathfrak{x}_\nu,\mathfrak{a}_\nu]\}d^3k=0$.) The remainder will be called the Pauli paramagnetism, and it involves an effective magnetic moment

$$F_p=\left(\frac{1}{2\pi}\right)^3\int \mathrm{Tr} f_0'\frac{(\mathbf{B}\cdot\mathbf{m})^2}{2}d^3k. \quad \text{(II.11)}$$

(3) The Van Vleck-dielectric paramagnetism comes from the fifth term in (I.30):

$$F_{VD}=\left(\frac{1}{2\pi}\right)^3\int \mathrm{Tr} f_0 E_{VD}'d^3k.$$

(4) The atomic diamagnetism, from the fourth term of (I.30), strictly speaking from the Dirac point of view, arises from the previous term, as discussed following (I.30). The split is nevertheless useful since in solids, electrons are far from the relativistic region.

$$F_{at}=\left(\frac{1}{2\pi}\right)^3\int \mathrm{Tr} f_0\frac{(\mathbf{A}^s)^2}{2m}d^3k.$$

(5) The next term is similar in appearance to the last but involves an effective mass so that its sign is not definite, a characteristic common to the remaining terms, which distinguishes them from the previous terms. We shall call this contribution pseudoatomic:

$$F_{pa}=\left(\frac{1}{2\pi}\right)^3\int \mathrm{Tr} f_0(-\alpha_{\mu\nu})\left\{\frac{A_\mu{}^s A_\nu{}^s}{2}\right\}d^3k.$$

(6) The last term in (I.30) has no simple description. We shall denote its contribution by F_Ω:

$$F_\Omega=\left(\frac{1}{2\pi}\right)^3\int \mathrm{Tr} f_0(-\tfrac{3}{8})[\mathbf{B}\cdot\mathbf{\Omega},\mathbf{B}\cdot\mathbf{m}]_+d^3k.$$

(7) Finally we have another term whose existence depends on that of a Fermi surface for we can integrate by parts, obtaining

$$F_7=\left(\frac{1}{2\pi}\right)^3\int \mathrm{Tr} f_0'\left(\frac{v_\mu}{2}\right)(A_\mu{}^s\mathfrak{B}_\nu A_\nu{}^s + A_\nu{}^s\mathfrak{B}_\nu A_\mu{}^s + v_\nu\{A_\nu{}^s A_\mu{}^s\}d^3k.)$$

The relative importance of the various terms will differ considerably from one type of substance to another. In particular, of course, in insulators and intrinsic semiconductors the terms (1), (2), (7) are zero. When the band gaps are large in such materials, the atomic diamagnetism (4) will predominate. Narrow band gaps will increase the sizes of (3) and (5), which will tend to cancel, however, as (3) is definitely paramagnetic while the largest contribution to (5) will come from regions of large negative mass. Term (6) will be large only if, in addition to a small band gap, the material has strong spin-orbit coupling.[5] Otherwise, $\mathbf{m}$ will be about the free electron value and $\mathbf{\Omega}$ will be small. The contribution when large should be predominantly diamagnetic.

In a metal on the other hand the terms (1) and (2) may be comparable to or larger than the atomic diamagnetism. A discussion of the comparison between calculations and theory has been given by Wilson.[6] In the good metals the remaining terms are expected to be smaller, as the energy difference between energy levels at given $\mathbf{k}$ is generally large, while these terms, having effectively three such factors in denominators are increased by small energy differences.

In the case of doped semiconductors one has the opportunity to study the contribution of small numbers of added carriers, unmasked by the atomic and Van Vleck terms. This has been done in particular by Yafet and Bowers.[7] Any of the terms can be large in this case, though at small concentrations, the first two are especially enhanced by a factor of approximately the energy gap over the Fermi level.

In the semimetals one has effectively a doped semiconductor in which the intrinsic susceptibility is unknown. In Bi, for instance, it has long been supposed that the susceptibility is predominantly due to the carriers, but since it is now known that the electrons have a large magnetic moment,[5] these would appear to give a net paramagnetic contribution. On the other hand, the term (5) above could give rise to a very large diamagnetism from electrons in the filled bands near the band edges. But this is not the only possibility, since the atomic term should also be increased by small energy gaps, this contribution possibly coming from a larger region of $\mathbf{k}$ space. Term (6) F_Ω will also be diamagnetic in this case and comparable in size to (5). Whichever of these terms is the primary source of the diamagnetism, it should remain large if the electrons and holes are removed. This can apparently be achieved by making Bi-Sb alloys,[8] and the experiments of Shoenberg and Zaki Uddin[9] on such alloys provide some

[5] M. H. Cohen and E. I. Blount, Phil. Mag. **5**, 115 (1960); G. E. Smith, J. K. Galt, and F. R. Merritt, Phys. Rev. Letters 4, 276 (1960).

[6] A. H. Wilson, *The Theory of Metals* (Cambridge University Press, New York, 1953), 2nd ed.

[7] Y. Yafet and R. Bowers, Phys. Rev. **115**, 1165 (1959).

[8] A. L. Jain, Phys. Rev. **114**, 1518 (1959).

[9] D. Shoenberg and Zaki Uddin, Proc. Roy. Soc. (London) **A156**, 687 (1936).

confirmation of this prediction, though they did not examine specimens in the range which later proved to have well-defined semiconducting properties. Further experiments on such alloys are strongly indicated. In the semimetals in general, more information can be obtained about specific group of carriers from more specialized experiments such as de Haas-van Alphen effect, cyclotron resonance,[10] ultrasonic attenuation,[11] and anomalous skin effect,[12] but the steady susceptibility then serves as a check on the completeness of this information.

III. EIGENFUNCTIONS AND EIGENVALUES

Having now determined the effective one-band Hamiltonian, it is of interest to look at the eigenfunctions and eigenvalues as well. As a first step in this process, it is convenient to consider what constants of the motion can be determined in advance. At this point the simplicity of the problem depends on the choice of gauge. In particular, if we choose a doubly-singular gauge (see Appendix A), the effective Hamiltonian depends only on one component of **R**. This is not quite the same in general, as to say that we have two constants of the motion as we shall now see.

Consider first the case where the magnetic field is along a lattice vector of the crystal which we shall call the z axis. We choose the gauge so that $a_x=0$, $a_y=Bx$, $a_z=0$. Then our effective Hamiltonian is $H(k_x, k_y+Bx, k_z)$. k_y and k_z are constants of the motion, and the CMR wave equation is one dimensional. We could have come to this conclusion by considering the original Schrödinger Hamiltonian, so it does not depend on any of our approximations. If we now suppose that **B** is not along a short lattice vector but along a long one, so that its direction cosines have large denominators, nothing is changed in principal, but the simple statements can be made only in reference to a different unit cell. For we can obtain crystal momentum components as constants of the motion only if the Hamiltonian is periodic in the corresponding space directions. Thus we would proceed as follows for such a **B** as we have just described. We choose the z axis parallel to **B** and the shortest lattice vector in that direction as one edge of the unit cell. This determines a plane in reciprocal space which is perpendicular to **B**. If the lattice vector is large, the unit cell in this plane will also be large and the spacing of the planes small. Thus the BZ will be thin in the z direction and of great area, becoming thinner and more extensive as the z-lattice vector increases in length, that is, as the denominators of the direction cosines of **B** become larger. Finally, if the direction cosines are irrational, the zone has no finite thickness, and the BZ becomes in effect an infinite plane. An intermediate case exists if **B** is perpendicular to one reciprocal lattice vector, but has some irrational direction cosines. In this case, the BZ becomes a strip of infinite length but finite width. If the BZ has a finite thickness and a_z is chosen zero, and the other components of **a** are independent of z, $\mathbf{k}\cdot\mathbf{B}/|\mathbf{B}|$ is a constant of the motion. As the BZ becomes thinner, the range of values of $\mathbf{k}\cdot\mathbf{B}/|\mathbf{B}|$ decreases and the variation of wave functions and energies with $\mathbf{k}\cdot\mathbf{B}/|\mathbf{B}|$ becomes unimportant, until in the case of irrational orientations, it can be neglected. The eigenfunctions in all cases do not fall off at large z. For rational orientation their z dependence is like Bloch functions, albeit with very large periods. For irrational orientations, the eigenfunctions remain large at large z, even though not periodic. This reflects the fact that as the BZ becomes a pancake, the unit cell becomes a long thin rod, so that the periodic part of the wave function becomes a function only of z, assuming reasonably decent behavior; conversely the z dependence relates only to the periodic part. Thus, the Wannier wave function $f(\mathbf{R})$ should show no z dependence, but only depend on x and y. The preceding discussion will probably require some getting used to and is most easily seen by imagining the irrational orientations as a limiting case of rational orientations.

We have then that $\mathfrak{H}$ can be written as $\mathfrak{H}(\mathfrak{k}_x,\mathfrak{k}_y,``k_z")$, where the quotation marks are intended to mean that there is a k_z dependence only for rational field directions. At this point, we have reduced the problem to two dimensions, without in general, however, having any constants of the motion. We can now choose our gauge in such a way as to reduce the equation to one dimension, namely by picking the vector potential to have only one component and to vary in only one direction. We will choose $a_z=a_x=0$, $a_y=Bx$, so that k_y is a constant of the motion. In general k_y will have an infinite range, but when the BZ is a strip or a finite rectangle, the range of k_y is finite.

Thus, the situation is not unlike that for free electrons. In the latter case, there is only one other quantum number left n, which is discrete, and on which alone the energy depends. n labels the "harmonic oscillator states." In our present problem, we have yet no guarantee that n is a discrete quantum number, and it must clearly do the work of both the harmonic oscillator number and the band index. To see that n cannot be so simple, suppose we have a state $\psi_1=\psi(n,k_y,k_z;\mathbf{x})$. If we now make a gauge transformation, adding Bx_0 to the vector potential, the original Hamiltonian $\mathfrak{H}(p_x,p_y,p_z;\mathbf{x})$ becomes $\mathfrak{H}(p_x, p_y+Bx_0, p_z;\mathbf{x})$ and the wave function $\psi(n, k_y+Bx_0, k_z;\mathbf{x})$ has the same energy as ψ_1. If we now translate the system by a lattice vector **R**, the Hamiltonian is $\mathfrak{H}(p_x, p_y+B(x_0-R_x), p_y;\mathbf{x})$ and the wave function is $\psi(n, k_y+Bx_0, k_z;\mathbf{x}-\mathbf{R})$. Thus, if we choose x_0 a lattice vector **R**, we obtain from $\psi(n,k_y,k_z;\mathbf{x})$ a different eigenfunction $\psi(n, k_y+BR_x, k_z;\mathbf{x}-\mathbf{R})$ with the same energy. The spectrum considered as a function of k_y is then periodic

[10] J. K. Galt, W. A. Yager, F. R. Merritt, and B. B. Cetlin, Phys. Rev. **114**, 1396 (1959).
[11] D. H. Reneker, Phys. Rev. **115**, 303 (1959).
[12] G. E. Smith, Phys. Rev. **115**, 1561 (1959).

with a periodicity Br_x where r_x is the minimum R_x (which is zero for irrational orientations). The spectrum also has the periodicity K_y, where K_y is the minimum reciprocal lattice vector in the y direction. We now distinguish three cases:

(1) K_y and K_x are infinite; r_x and r_y are zero. The spectrum is independent of k_y (or of k_x had we chosen the "opposite" gauge $a_x=-By$, $a_y=0$).

(2) K_y and K_x are finite; r_x and r_y are also finite. Then the occurrence of two periodicities K_y and Br_x means that unless they are commensurate the spectrum is independent of k_y, and likewise of k_x in the "opposite" gauge. The periods are commensurate if Br_x/K_y is rational, that is, if the flux through a unit cell in the plane perpendicular to **B** is a rational multiple of hc/e. The spectrum is not independent of k_z.

(3) K_y is finite, K_x infinite; r_x is zero, r_y finite. Br_x is zero so the spectrum is independent of k_y, but in the opposite gauge; Br_y is finite and the spectrum is not constant, but may have a periodicity Br_y as a function of k_x.

Since we have only been able to show that a level at k_y is related to one at k_y+Br_x by translation symmetry and gauge invariance, we must suppose that for rational orientations, the corresponding levels at other (k_y+Bx_0) have different energies, though for small x_0 they must be arbitrarily close. This will soon be verified. We have seen, however, that except for particular values of B, the spectrum is independent of k_y. This leads to the conclusion that the spectrum for given k_y and k_z is not discrete.

The reader may have gotten the impression that there is a remarkable difference between rational and irrational orientation which seems physically unreasonable. This difference is only apparent, and is one of mathematical simplicity alone. If we did not attempt, at rational orientations, to make the most convenient choice of gauge, we would not notice the discontinuity. As we have seen, however, it is possible at rational orientations to make a choice of gauges, such that we greatly simplify the mathematics. In both cases, there is a continuous spectrum when all quantum numbers are considered. Because of the degeneracy of eigenvalues, it is possible to use very different-looking sets of eigenfunctions. A somewhat analogous situation would arise in a crystal, which we subject to a constant sinusoidal perturbation, whose repetition vector is **T**. If **T** is a lattice vector, the problem is still periodic, and the eigenfunctions can be written as Bloch functions. For other values of **T** this is not so, but we expect no discontinuities at the preferred values.

The above reasoning has not used the effective Hamiltonian, but rather has proceeded from the Hamiltonian (I.1). It will soon appear that the results from the effective Hamiltonian are in agreement. They will also give a more detailed picture of the spectrum and eigenfunctions than can be deduced from invariance arguments alone.

We have seen that the effective Hamiltonian can be written in the form $\mathfrak{H}(k_x, k_y+Bx, "k_z")$. For purposes of solving the equation of motion we can now forget about k_z immediately. We also wish to get rid of k_y. This can be done by making the transformation $e^{-ik_yk_x/B}$ which replaces the above Hamiltonian with $H(k_x,Bx)$. This is generally permissible, but not when K_x is finite, for in that case the wave function must be periodic in k_x. When k_y is removed from the Hamiltonian, we require that this boundary condition be replaced with one, that the wave function change by the factor $e^{-ik_yK_x/B}$ when k_x is increased by K_x.

Thus, we have replaced a set of different Hamiltonians with a set of different boundary conditions. When K_x is infinite no such action is required. The effect is that in either case we consider all the energy levels of $H(k_x,B_x)$ not merely those corresponding to eigenfunction periodic in k_x.

We now have a Hamiltonian $H(k_x,Bx)$, which is defined over a plane or set of planes, and which may be periodic in one or both arguments. Since the Hamiltonian is not simple, but involves only one pair of conjugate variables, the WKB procedure is indicated as a method of solution.

It is well known that the straightforward application of the WKB method leads to the familiar semiclassical quantization rules for the energy levels, and thus to the de Haas-van Alphen and related effects. It is also well known that some orbits called open extend indefinitely in k space and that the spectrum is continuous in the region of these orbits. It has been pointed out by Zil'berman and others[13] that the quantized levels are in fact subject to a slight broadening, while Kohn[2] has found that there are gaps in the open orbit region. Finally, it has recently been noted by Cohen and Falicov[14] that in strong fields the energy levels and wave functions can change in a qualitative way. These subjects are closely related and a unified discussion is desirable and appropriate at this point, because they are also closely related to the problem of the effective Hamiltonian.

We start with our two-dimensional Hamiltonian $\mathfrak{H}(k,Bx)$ which we rewrite as $\mathfrak{H}(p,q)$ where $q=k$ and $p=Bx$ and p, q satisfy the commutation relation $[p,q]=iB$.

In applying the WKB method, we suppose in effect that we have a relief map of energy as a function of p and q. We then make a two-dimensional plot of the contours for a particular energy E. The contour will consist in general of a number of curves which we shall call orbits, some closed, some open, that is infinite

[13] G. E. Zil'berman, Soviet Phys. (JETP), **3** (**30**), 835 (1957); **5** (**32**), 208 (1957); **6** (**33**), 299 (1958). P. G. Harper, Proc. Phys. Soc. (London) **A68**, 874, 879 (1955); A. D. Brailsford, *ibid.* **A70**, 275 (1957).

[14] M. H. Cohen and L. M. Falicov, Phys. Rev. Letters **7**, 231 (1961).

in extent. For the moment, we consider only closed orbits. If any orbit encloses an area of $(n+\frac{1}{2})B$ where n is an integer it corresponds to an eigenfunction. By going through this procedure for all energies, we obtain the quasi-classical energy levels. The procedure is only approximate, of course. The Hamiltonian has off-diagonal matrix elements; there are also overlap integrals since the wave functions are not exactly orthogonal. If we consider two neighboring orbits with n's differing by δn, then as $B \to 0$, the overlap and off-diagonal Hamiltonian matrix elements between them decrease as B^2 in the usual WKB approximation; by proceeding to the indicated higher approximations, they can be made to decrease more rapidly than any given power of B. For orbits differing by a given energy ϵ, or for orbits about different centers, on the other hand, these matrix elements fall off as $e^{-c/B}$, where c depends on the contours, but not on B. Its nature is discussed in Appendix B. This cannot be significantly improved by going to higher approximations in the WKB procedure.

On the other hand, the situation we have described is very similar to one encountered in certain idealized problems in solid-state physics, which are discussed by the tight-binding approximation. We have a number of approximate wave functions on different centers, which overlap very slightly, the matrix elements connecting them being small compared to the energy splittings between states on a given center. Two situations must be distinguished. In the first, the field is in an irrational direction and there is no periodicity. Neighboring orbits will have different energies in the WKB approximation and the spectrum is a continuum in that approximation. Then each level will be slightly shifted by interaction with its neighbors, somewhat as a nuclear spin may be thought of as having a slightly different local field from that of its neighbors. Since the spectrum is already continuous, no change in this characteristic will occur. When the field is in a rational direction, the WKB approximation will make orbits displaced from one another by a reciprocal lattice vector degenerate and although the over-all spectrum is continuous, the spectrum for a given k_z is discrete. In this case, discussed by Zil'berman, Harper, and Brailsford[13] interaction between neighboring orbits will produce a broadening into a band just as in the tight-binding treatment of bands. The width of the band will essentially be determined by the matrix element between neighboring orbits. In complex cases where not all the orbitals are identical, the band structure will also be more complicated, but can easily be determined in any given case. The physics of this broadening can be described as follows. The WKB levels do not take account of the atomic structure, but levels corresponding to different values of k_y have their centers at different positions relative to the unit cell (this point is meaningful only when the field is perpendicular to at least one reciprocal lattice vector), and thus have different energies. An additional broadening arises because levels whose k_y's differ by K_y have finite matrix elements, even though they are far apart. There will then be an interaction between them, and they will be somewhat mixed. These two sources of broadening are not actually different and appear so only in a given gauge. If the opposite gauge is chosen, the two types of broadening are interchanged. Our analysis in particular makes clear that at small field the bands are sharply defined with regions densely filled with states, alternating with energy gaps. The width of the levels is evidently of the order of $e^{-c/B}$, like the interorbit matrix elements.

We now turn to the case of open orbits. These are of two types: (1) Periodic open orbits which have periodic orbits in $\mathbf{k}$ space; these can occur only for field orientations such that the BZ in the plane is finite or a strip. (2) Nonperiodic open orbits, which can occur for irrational field directions. The latter are in no sense quantized. The former, with a proper choice of gauge, have k_y as a good quantum number; periodic boundary conditions on k_x result in a discrete spectrum for fixed k_y, though in the WKB approximation there is a continuous spectrum when all k_y's are considered.[15] As in the case of closed orbits, the open orbits will have matrix elements with each other, and with any closed orbits which may coexist with them, which decrease like $e^{-c/B}$. For the nonperiodic orbits, this results in shifts but does not change the character of the spectrum. For the periodic orbits on the other hand, the interactions introduce gaps in the spectrum.[2] This can be seen easily as follows. Just as we pictured the closed orbits as analogous to tightly bound electrons, we can imagine the open orbits to be analogous to nearly free electrons. Their orbits in (p,q) space are essentially identical to the orbits in phase space of nearly-free electrons in a weak one-dimensional periodic potential. The open orbits can be considered to belong to groups. These groups are separated by regions of closed orbits, and alternate groups represent electrons whose net velocities have opposite directions. We may suppose that the interactions between orbits within a group have been diagonalized to any desired degree and consider only interactions between orbits in different groups. So far there will be no energy gaps. But matrix elements will exist only between orbits for which k_y differs by BR_x. They are so small that we need only consider the interaction of orbits of practically the same energy. Two such orbits will now be split in energy by the amount of the matrix element between them. Thus just as in the one-dimensional nearly-free electron case, the energy levels will be practically unaffected, except for those orbits which have matrix elements with degenerate levels. At these points gaps will be introduced into the spectrum. For this purpose only neighboring groups of orbits need be considered.

[15] E. I. Blount, Phys. Rev. Letters 4, 114 (1960).

So far we have seen that for closed orbits the energy levels are broadened by an amount which decreases exponentially in $1/H$, and that similar gaps are opened in the open orbit spectrum for fixed k_z. Actually of course, when k_z is taken into account—or the infinite extent of the BZ in the plane when there is no k_z—the over-all spectrum is continuous. Finally, we must consider what happens in the transition region between closed and open orbits. This corresponds to the region which neither the tight-binding nor nearly-free electron approximations are very good, and requires more careful analysis. This can be done by an extension of the WKB method to cover the case where the energy is nearly equal to the height of a barrier. The necessary connection formulas can be found in Landau and Lifshitz.[16] Essentially this method has been used by Kohn, who found that there is a symmetry between energies above and below the barrier height, such that if an energy ϵ measured from the barrier height is permitted, the energy $-\epsilon$ is forbidden and conversely. This applies to periodic open orbits and the corresponding closed orbits, and is strictly true only for energies so close to the saddle point that the classical period is proportional to $\ln|\epsilon|$. It also considers interaction between only two adjacent groups of open orbits and the closed orbits between them. (The derivation of this result has not been published by Kohn, and the author is indebted to him for a discussion of it.) For low fields, the regions of validity of this approximation and our tight-binding and nearly-free-electron procedures overlap and they give equivalent results.

Since the broadening is, at low fields, negligible compared to the Landau splittings, it can for most purposes be neglected and the orbits treated as isolated. It is then much easier to map them back into a compact Brillouin zone and reintroduce k_z, for purposes of calculating densities of states. Only for open or very extended orbits is the mapping into a plane or flat zone necessary or useful in this approximation.

The final topic of this section is magnetic breakdown, which also represents the breakdown of the effective diagonal-in-bands Hamiltonian. For this discussion we revert to the Hamiltonian of (I.24) which includes interband matrix elements of order B, and consider the case where orbits of nearly the same energy in different bands also approach very closely in $\mathbf{k}$ space. To first order the Hamiltonian is

$$\mathfrak{H}(\mathfrak{k})-\tfrac{1}{2}\{\mathbf{B}\times\mathbf{v}\cdot\mathbf{X}\}-\mathbf{B}\cdot\mathbf{M}, \qquad \text{(III.1)}$$

where we have absorbed the diagonal first-order terms of (I.24) into the zero-order term. The last two terms give rise to matrix elements between two orbits. We now choose directions so that the line of closest approach between the orbits is in the y direction and set $k_y=0$ as before. Then the normalized WKB wave function for one of the orbits will be

$$\psi_1(k)=\left(\frac{1}{m_1 v_{y1}}\right)^{\frac{1}{2}}\exp\left[\frac{i}{B}\int_0^k R_1(k)dk\right], \qquad \text{(III.2)}$$

where k stands for k_x, R for R_x a function of k_x for a given orbit, m_1 is the effective mass of the orbit, $m_1=\oint dk/v_{1y}$, and v_{y1} the y component of velocity.

The matrix element between two such orbits in different bands is then

$$Q_{12}=\frac{B}{(m_1m_2)^{\frac{1}{2}}}\int\frac{\frac{1}{2}(v_{1y}+v_{2y})X-M}{(v_{1y}v_{2y})^{\frac{1}{2}}}\times\exp\left[\frac{i}{B}\int_0^k(R_1-R_2)dk'\right]dk, \qquad \text{(III.3)}$$

where X is the x component of $\mathbf{X}_{12}$ and M is the z component of $\mathbf{M}_{12}$. It is shown in Appendix B that if the orbits do not overlap Q_{12} decreases at small field as $e^{-c'/B}$. Here we consider the important special case of two bands, which are nearly degenerate. In fact we further suppose that they can be obtained from two bands which "cross" each other, by the addition of a small perturbation Δ. To a sufficient approximation, the Hamiltonian then looks like this in the unperturbed representation:

$$\begin{pmatrix}\mathbf{k}\cdot\mathbf{v}_1 & \Delta\\ \Delta & \mathbf{k}\cdot\mathbf{v}_2\end{pmatrix}, \qquad \text{(III.4)}$$

with eigenvalues

$$E_{\pm}=\tfrac{1}{2}\mathbf{k}\cdot(\mathbf{v}_1+\mathbf{v}_2)\pm\{\Delta^2+\tfrac{1}{4}[\mathbf{k}\cdot(\mathbf{v}_1-\mathbf{v}_2)]^2\}^{\frac{1}{2}}. \qquad \text{(III.5)}$$

Then for any field direction not in the direction of $\mathbf{v}_1-\mathbf{v}_2$ the orbits are hyperbolas and the bands have equal values of v_y. Furthermore the X term in the Hamiltonian is much larger than the M term. Thus for values of R near the turning point, our Hamiltonian becomes

$$H(k_x)-F_LR-F_LX, \qquad \text{(III.6)}$$

where $F_L=Bv_y$. This is identical in form to the Hamiltonian which produces Zener breakdown in insulators. Thus magnetic breakdown is essentially Zener breakdown, where the electric force is supplanted by the Lorentz force. We can then obtain a good approximation for the asymptotic behavior at low field by substituting in the formula for Zener breakdown as given by Kane[17]; the important factor is the transmission amplitude which we write

$$\alpha=\exp\left(-\frac{\pi}{4}\frac{|\Delta|}{F_L|X|}\right). \qquad \text{(III.7)}$$

We further have $X=v_x/2\Delta$ in our simple model, where $|v_x|$ is $|[\frac{1}{2}(\mathbf{v}_1-\mathbf{v}_2)\times\mathbf{B}/|\mathbf{B}|]|$ and $F_L=Bv_y$, so

$$\alpha=\exp\left(-\frac{\pi}{2}\frac{\Delta^2}{B|v_xv_y|}\right). \qquad \text{(III.8)}$$

[16] L. Landau and E. Lifshitz, *Quantum Mechanics—Nonrelativistic Theory* (Addison-Wesley Publishing Company, Reading, Massachusetts, 1959).

[17] E. O. Kane, J. Phys. Chem. Solids **12**, 181 (1960).

This result is obtained by approaching from low fields. Alternatively, we may approach from high fields and consider the effect of Δ on the orbits which are present when it is zero. We can indeed calculate the matrix elements for any perturbation $\Delta_{mn'}(\mathbf{k}',\mathbf{k}'')$ if we show how to do it for the case when $\mathbf{k}'$, $\mathbf{k}''$ differ by $\boldsymbol{\kappa}$. We then write $\Delta_{nn'}(\mathbf{k}',\mathbf{k}'')=\Delta_{nn'}(\mathbf{k})$ where $\mathbf{k}=(\mathbf{k}'+\mathbf{k}'')/2$. We also choose our gauge so that the x direction is parallel to $\mathbf{k}'-\mathbf{k}''=\boldsymbol{\kappa}$. Then between two wave functions $f_1(q)$ and $f_2(q)$, the matrix element will be

$$Q_{12}=\int f_{1m}(q-\tfrac{1}{2}\kappa)\Delta_{nn'}(q)f_{2n'}(q+\tfrac{1}{2}\kappa)dq. \quad \text{(III.9)}$$

If we suppose that the orbits for $\Delta=0$ belong to a particular band we have $n=n'$ and drop the index.

Now assuming that $\Delta(q)$ is slowly varying, we use WKB wave functions and the saddle point method to find that the matrix element consists of a contribution from each intersection of the orbits, with the orbit 2 displaced by κ. The absolute value of each contribution is

$$\pi^{\frac{1}{2}}\Delta B^{\frac{1}{2}}/[m^*|\tfrac{1}{2}(v_{1x}v_{2y}-v_{1y}v_{2x})|^{\frac{1}{2}}], \quad \text{(III.10)}$$

where m^* is the geometric mean of the cyclotron masses of the orbits and the v's are evaluated at the points of intersection. (This formula breaks down, of course, if one of the velocities gets very small.) This quantity in particular is quite insensitive to whether 1 and 2 are the same orbit. Provided 2 is a neighboring concentric orbit, the matrix element is nearly constant and m^* is simply the cyclotron mass of either. The ratio of this expression and the separation between neighboring levels is then

$$\pi^{\frac{1}{2}}\Delta/[\tfrac{1}{2}B(v_{1x}v_{2y}-v_{2x}v_{1y})]^{\frac{1}{2}}. \quad \text{(III.11)}$$

In our previous model, then this is just the square root of twice the exponent in (III.7). This corresponds to choosing κ to be a reciprocal lattice vector in the final structure.

The significance of the parameter we have found is more easily understood when we note that the matrix elements we have evaluated have had the form B/m^* multiplied by a function of the parameter, where B/m^* is the geometric mean of the frequencies of the two orbits involved. Now for degenerate orbits, the matrix element is the frequency of transfer from one orbit to the other in the simple-minded resonance picture, provided it is small compared to either orbit frequency. Thus, when we divide by either frequency, we obtain a sort of switching coefficient, the probability that the electron will jump from one orbit to the other on a given passage through the intersection or point of closest approach. When we divide, instead, by the geometric mean of the frequencies, we obtain a mean switching coefficient. Thus, in the low-field limit the switching coefficient between bands is given by (III.7), while in the high-field limit, the switching coefficient between "unperturbed" orbits is given by (III.11). We have derived the latter as applying between two "oscillator" levels of the same k_y, k_z. The same value would have been obtained had we considered the periodic potential to vary in the y direction and calculated the matrix element between equal oscillators at values of k_y differing by κ. This latter quantity is more closely related to the orbit picture as drawn in phase-space. It was the observation by Priestley of an orbit corresponding to the free-electron orbits in Mg that led Cohen and Falicov to postulate breakdown. These authors used the parameter $\Delta/\hbar\omega_c$ as a criterion. We now see that the correct parameter is much more favorable to breakdown. For some cases we can give a simpler form of our parameter. When the "unperturbed" orbit has an energy that can be written $E=m^*v^2/2$, and v_x and v_y are comparable, we have

$$Q\sim/\hbar\omega_c\sim\Delta/(E\hbar\omega_c)^{\frac{1}{2}},$$

where in experimentally attainable cases, E will usually be the Fermi energy which is much larger than $\hbar\omega_c$. Our derivation shows that this weakening applies not only to the case of breakdown, but to any interaction with long wavelength perturbations such as acoustical or optical vibrations.

From the emphasis we have put on magnetic breakdown it should not be inferred that this is the only case where the perturbation theory of (F) and Sec. I breaks down, though it is very likely unique in being a qualitative effect. Whenever $\frac{1}{2}\{\mathbf{B}\times\mathbf{v})\cdot\mathbf{X}_{nn'}\}$ or $\mathbf{B}\cdot\mathbf{m}$ is comparable to $E_n-E_{n'}$, care must be exercised. In such cases it will probably frequently be possible to devise methods better suited than that used here.

IV. SUMMARY

In this paper we have presented what we feel is a simple, even intuitive approach to the establishment of an effective one-band Hamiltonian for a Bloch electron in a magnetic field. Some of our manipulations are complicated, but we would emphasize that this complication arises only in an effort to simplify the actual expressions for the second-order terms. The actual proof that interband matrix elements can be removed to arbitrary order leaving a Hamiltonian $\mathfrak{H}(\mathfrak{k})$ is completed in the paragraph following (I.10).

Of previous attempts to accomplish this objective, we single out two for particular discussion. Zil'berman[13] has used a procedure equivalent to our zeroth-order transformation and showed that in the cases he considered, inversion symmetry and no spin-orbit coupling, the correction terms are of second order. Kohn[1] has used series of transformations in powers of $\mathfrak{k}$ and rearranged them to a series in powers of B to remove interband terms formally to all orders. There is a close relation between his method and ours. The simplification in our method arises from connecting the series in $\mathfrak{k}$ to one in B in the general formalism of (F), namely in the derivation of our basic multiplication rule. This has also made possible the systematic discussion of the validity of the rearrangement of the series, and in particular has made it possible to demonstrate the asymptotic character of its "convergence." This is of some importance both positively and negatively. For on the one hand, Zil'berman showed that for a particular model the eigenstates are broadened exponentially in $-1/B$, while working with a Hamiltonian accurate only to order B at best, whereas we have now established that the effective Hamiltonian can also be given with a similar asymptotic behavior. Also, the de Haas-van Alphen effect is associated with nonanalytic behavior at small B. On the other hand, if Kohn's procedure actually converged, the phenomenon of breakdown would be seriously modified, if not eliminated.

The difficulty in trying to treat the magnetic problem previously has been that one would like to "get one's hands" on both $\mathbf{k}$ and $\mathbf{R}$ simultaneously, because one has intuitively felt that the basis functions should be chosen differently at different positions in the lattice. With the usual techniques this has been difficult to do, whereas this is just the great merit of the PCR. The bands are defined as functions of $\mathbf{k}$ and $\mathbf{R}$ from the start, and our transformations are essentially a perturbation-theoretic approach to the diagonalization of the Hamiltonian for each point in "phase space." The noncommutativity leads to some specifically quantum-mechanical effects because one cannot completely localize in both $\mathbf{k}$ and $\mathbf{R}$, but basically one is "locally" diagonalizing the bands. We personally find it convenient to work with objects for which $\mathbf{k}$ and $\mathbf{R}$ can both be treated as c numbers, but this point is not essential and they can be thought of exclusively as operators if preferred.

Note added in proof. Recently, Roth [J. Phys. Chem. Solids (to be published)] and Wannier and Fredkin [Phys. Rev. **125**, 1910 (1962)] have written papers on the subject of Sec. I, which also greatly simplify the proof. Basically, both of these methods amount to the same thing as ours, though approached from quite different points of view; in both cases the method is applicable only to the present problem, rather than being a special case of a more general formalism.

The expression for the susceptibility in Sec. II is so far as we know, the first one given which is valid for electrons with spin-orbit coupling. It has also been possible to put the expression in what seems to be the most concise form yet given. We have also used simple one-band parameters whenever possible to facilitate comparison with experiments. It will also be noted that the PCR permits a relatively simple derivation of the Landau-Peierls formula.

Essentially equivalent formulas for the susceptibility have been given by Adams,[18] Hebborn and Sondheimer,[19] and Enz.[20]

[18] E. N. Adams, Phys. Rev. **89**, 633 (1953).
[19] J. E. Hebborn and E. H. Sondheimer, J. Phys. Chem. Solids **13**, 105 (1960).
[20] C. P. Enz, Helv. Phys. Acta **33**, 89 (1960).

APPENDIX A

In order to discuss the convergence of the multiplication rule (I.15), we start with a result of reference 3. Suppose we have a product $P=NM$ where N and M can be considered as belonging to families of functions $N(\mathbf{k},\mathbf{R};\lambda)$ or $N(\mathbf{k};\mathbf{q};\lambda)$ where λ is some parameter. Then consider the integrals

$$I_1=\int_0^{k_c(M)-\epsilon} |N(\mathbf{k};2\mathbf{q};\lambda)|\,|q|^n d^3q,$$
$$I_2=\int_{\mathrm{BZ}}' |N(\mathbf{k};\mathbf{q};\lambda)|\,d^3q, \tag{A1}$$

when the prime on the integral in I_2 means that the region of integration of I_1 is to be excluded. This region, in turn, is to be interpreted as one which extends to within some finite distance ϵ of the range of convergence $k_c(M)$ of the power series expansion of M as a function of $\mathbf{k}$ (*sic*). If both I_1 and I_2 are $O(\lambda^n)$ for all n at small λ, and the same is true of analogous integrals of M the series expansion of the product is asymptotically convergent in λ. In particular, if $N(\mathbf{k};\mathbf{q};\lambda)\sim e^{-\epsilon|\mathbf{q}|/\lambda}$ where ϵ is a positive number, and M is similar, the series is asymptotic. We are concerned with establishing such inequalities for the $S(\mathbf{k};\mathbf{q})$ corresponding to $S(\mathbf{k},\mathbf{R})=S(\mathbf{k}+A,\,\mathbf{k})$ since (1) they are the only nonpolynomial functions of $\mathbf{k}$ and $\mathbf{R}$ used in the first steps of our derivations, and (2) each later step in the perturbation theory involves beside polynomials such as $H_0(\mathfrak{k})$ and $V_0(\mathfrak{k})$, only a finite number of S's and derivatives thereof.

From $S(\mathbf{k},\mathbf{R})=S(\mathbf{k}+A,\,\mathbf{k})$ we obtain

$$S(\mathbf{k};\mathbf{q})=\sum_{R_i} S(\mathbf{k}+A_i,\,\mathbf{k})e^{i\mathbf{R}_i\cdot\mathbf{q}}, \tag{A2}$$

while $S(\mathbf{k}+A_i,\,\mathbf{k})$ can be expressed as

$$S(\mathbf{k}+A_i,\,\mathbf{k})=\int \mathcal{S}(\mathbf{k};\mathbf{Q})\exp(i\mathbf{Q}\cdot A_i)d^3Q, \tag{A3}$$

where

$$\mathcal{S}(\mathbf{k};\mathbf{Q})=\int S(\mathbf{k}+\mathbf{l},\,\mathbf{k})\exp(-i\mathbf{Q}\cdot\mathbf{l})d^3l, \tag{A4}$$

the last two integrals being carried out over all space and all $\mathbf{k}$ space, respectively. These equations yield

$$S(\mathbf{k};\mathbf{q})=\sum_{R_i}\int d^3Q\,\mathcal{S}(\mathbf{k};\mathbf{Q})\exp[i(\mathbf{Q}\cdot A_i+\mathbf{q}\cdot\mathbf{R}_i)]. \tag{A5}$$

Now for a linear vector potential we can write

$$A_i=G\cdot\mathbf{R}_i, \tag{A6}$$

where G will be called the gauge tensor. We further

define $\mathbf{T}(\mathbf{Q})=\mathbf{Q}\cdot G$ and write

$$S(\mathbf{k};\mathbf{q})=\sum_R \int d^3Q\, \mathcal{S}(\mathbf{k};\mathbf{Q}) \exp[i(\mathbf{q}+\mathbf{T}(\mathbf{Q}))\cdot\mathbf{R}_i] \quad \text{(A7)}$$

$$=\sum_{\mathbf{K}} \int \mathcal{S}(\mathbf{k};\mathbf{Q})\delta(\mathbf{q}+\mathbf{K}+\mathbf{T}(\mathbf{Q}))d^3Q. \quad \text{(A8)}$$

If $\mathbf{T}(\mathbf{Q})=\mathbf{q}+\mathbf{K}$ had a unique solution, this would take a simple form. In fact, however, the tensor G may be and almost always is a singular matrix, with one or two zero eigenvalues. In such cases the solutions $\mathbf{Q}$ of this equation fill a one- or two-dimensional manifold, respectively. Before discussing these complications however, we want to know what $\mathcal{S}(\mathbf{k};\mathbf{Q})$ is.

Starting from its definition we have

$$\mathcal{S}_{nn'}(\mathbf{k};\mathbf{Q})=\int d^3l\, S_{nn'}(\mathbf{k}+\mathbf{l},\mathbf{k}) \exp(-i\mathbf{l}\cdot\mathbf{Q})$$

$$=\int d^3l \int d\tau\, u_{n,\mathbf{k}+\mathbf{l}}^* u_{n'\mathbf{k}} \exp(-i\mathbf{l}\cdot\mathbf{Q}), \quad \text{(A9)}$$

where the spatial integration extends over one unit cell. This can now be rewritten as

$$\mathcal{S}_{nn'}(\mathbf{k};\mathbf{Q})$$

$$=\int d^3l \int d\tau\, \Psi_{n,\mathbf{k}+\mathbf{l}}^* \Psi_{n'\mathbf{k}} \exp[i\mathbf{l}\cdot(\mathbf{x}-\mathbf{Q})] \quad \text{(A10)}$$

$$=\int d^3l \int d^3x \int d^3k'\, \Psi_{n\mathbf{k}'}^* \Psi_{n'\mathbf{k}} \exp[i\mathbf{l}\cdot(\mathbf{x}-\mathbf{Q})], \quad \text{(A11)}$$

where the space integral is now over all space, as is the integral over l, while $\mathbf{k}'$ is integrated over one BZ. This last integral is now performed, yielding

$$\mathcal{S}_{nn'}(\mathbf{k};\mathbf{Q})=\int d^3l \int d^3x\, a_n^* \Psi_{n'\mathbf{k}} \exp[i\mathbf{l}\cdot(\mathbf{x}-\mathbf{Q})], \quad \text{(A12)}$$

where $a_n(\mathbf{x})$ is the Wannier function for the nth band. Finally, we obtain

$$\mathcal{S}_{nn'}(\mathbf{k};\mathbf{Q})=a_n^*(\mathbf{Q})\Psi_{n'\mathbf{k}}(\mathbf{Q}). \quad \text{(A13)}$$

We now have the basis for asymptotic convergence. For, as shown in reference 3, $a_n(\mathbf{Q})$ for a band which has no degeneracies with other bands, falls off exponentially in Q. Since G is linear in $|\mathbf{B}|$ we can write its inverse as $|\mathbf{B}|^{-1}g^{-1}$, for nonsingular gauges (that is, those with nonsingular gauge tensor). Then for this case, we have

$$S(\mathbf{k};\mathbf{q})$$
$$=\sum_K \mathcal{S}(\mathbf{k};|\mathbf{B}|^{-1}g^{-1}\cdot(\mathbf{q}+\mathbf{K}))$$
$$=\sum_K a_n^*(|\mathbf{B}|^{-1}g^{-1}\cdot(\mathbf{q}+\mathbf{K}))\Psi_{n'\mathbf{k}}(|\mathbf{B}|^{-1}g^{-1}\cdot(\mathbf{q}+\mathbf{K}))$$
$$\leq\sum_K |a_n(|\mathbf{B}|^{-1}g^{-1}\cdot(\mathbf{q}+\mathbf{K}))|). \quad \text{(A14)}$$

Each term in this sum decreases exponentially in $|\mathbf{B}|^{-1}$. At small $\mathbf{B}$, the sum will be dominated by its largest term, and it is clear that $|\mathbf{B}|$ plays the role of λ and that the conditions of I_1 and I_2 are satisfied. If we now consider singular gauges, there are no serious differences.

If however, we consider bands with degeneracies, there are serious difficulties because for them the Wannier functions will not fall off rapidly enough.[3] At this point the singular gauges play an important role, for with them we can demonstrate the asymptotic convergence.

To see how this comes about, we substitute (A10) in (A5), obtaining

$$S(\mathbf{k};\mathbf{q})=\sum_{R_i} \int d\tau\, \Psi_{n,\mathbf{k}+A_i}^* \Psi_{n'\mathbf{k}}$$
$$\times\exp[i(\mathbf{q}\cdot\mathbf{R}_i+A_i\cdot\mathbf{x})]. \quad \text{(A15)}$$

As in the passage from (A10) to (A11), we now introduce $\mathbf{k}'$ and extend the spatial integration over all space.

$$S(\mathbf{k};\mathbf{q})=[\delta(0)]^{l-3}\sum_{R_i} \int d^3x \int d^lk'$$
$$\times\Psi_{n\mathbf{k}'}^* \Psi_{n'\mathbf{k}} \exp[i(\mathbf{q}\cdot\mathbf{R}_i+A_i\cdot\mathbf{x})]. \quad \text{(A16)}$$

In this equation, however, $\mathbf{k}'$ need not extend over a whole BZ, but only over points $\mathbf{k}+\mathbf{m}$ where $\mathbf{m}$ is the general vector of the form $\mathbf{m}=G\cdot\mathbf{r}$ for all $\mathbf{r}$ or equivalently we simply integrate $\mathbf{m}$ over a fundamental region in extended k space. The factor $[\delta(0)]^{l-3}$ is required to assure that

$$S(\mathbf{k}+0,\mathbf{k})=\int S(\mathbf{k};\mathbf{q})d^3q=1.$$

To clarify this point, a digression is necessary. In a singular gauge, the vector potential cannot take on any value but is restricted to a line or plane in extended $\mathbf{k}$ space. In the case of a plane, when the points outside a BZ are mapped into it, they may lie on a family of discrete planes, if the direction perpendicular to the plane is parallel to any lattice vector (not reciprocal lattice vector). Otherwise, the points will be dense in the BZ. In the case of a line, three alternatives are possible; the points may lie on a discrete family of lines (this requires that the line be parallel to a reciprocal lattice vector), may be dense on a discrete family of planes (this requires that the line lie in a plane determined by two reciprocal lattice vectors), or may be dense in the BZ.

In the case of planes, the first possibility corresponds to a finite fundamental region, which is adopted because all points on the plane outside it are equivalent to a point inside. This means that the fundamental region is a section of a special BZ chosen to have faces parallel to the plane. The second possibility corresponds to the fundamental region being the whole plane. Likewise in the case of lines, the first possibility corresponds to a finite fundamental region, the second and third possibilities to the fundamental region being the whole line.

If the line or plane of **m** maps into discrete lines or planes in the BZ, we can easily see that there is a possibility of improving convergence. For if for given **k** we can choose the gauge so that the line or plane passing through **k** does not pass through a degeneracy, then the arguments used in (F) to prove the exponential fall-off of Wannier functions will work for these functions of the form

$$a_n(\mathbf{r};\mathbf{k},G)=\int d^l m\,\Psi_{n,\mathbf{k}+\mathbf{m}}(\mathbf{r}). \tag{A17}$$

These functions behave like Wannier functions in the direction corresponding to the manifold of **m** (which depends on G) and like Bloch functions perpendicular to it. This failure to fall off in the perpendicular directions does not hurt us, as the integral over that component of **x** simply kills the offensive $\delta(0)$'s.

The trouble with this procedure is that it is not generally applicable. To be sure, we can for any direction of magnetic field require the vector potential to lie in any desired plane, but it is not always possible to choose a plane which will avoid all degeneracies, still less to find a family of planes nearly all of which will avoid degeneracies. On the other hand, if we could confine **a** to a line, nearly all of the resulting family of lines would avoid degeneracies unless there were a plane of degeneracies, for which special measures are available. However, one cannot freely choose the line on which **a** must lie, for it is required to be perpendicular to the field direction. If as is true except for a set of field directions of measure zero, the ratios of the field components along crystal axes are irrational, there is no reciprocal lattice vector perpendicular to the field. Thus, if we wished to utilize the advantages of a doubly singular gauge, we would have points distributed densely in a BZ and thus passing through or arbitrarily close to degeneracies.

Fortunately, however, the doubly singular gauges have an additional advantage which offsets these difficulties. Namely, it permits us to use a one-dimensional integration, albeit over an infinite domain. Thus we have in effect a one-dimensional band structure over this domain. In reference 3 it is shown that in a one-dimensional band structure the Bloch functions are analytic functions of k even in the neighborhood of degeneracies on the real axis, though not in the neighborhood of those in the complex k space. The bands can be so labeled that this is true in our problem also. Thus the integration over **m** passes through no singularities. On the other hand it passes arbitrarily close to branch points, since the branch surfaces come right up to the real axis near degeneracies in real k space.[3] Thus, we cannot expect an exponential fall-off but we expect one which is faster than any power of x.

At this point we return to (A16) and rewrite it as

$$S_{nn'}(\mathbf{k};\mathbf{q})=[\delta(0)]^{-2}\sum_{R_i}\int d^3x\,a_n(\mathbf{x};G)\Psi_{n'\mathbf{k}}(\mathbf{x})\exp[i(\mathbf{q}\cdot\mathbf{R}_i+\mathbf{T}(\mathbf{x})\cdot\mathbf{R}_i)]$$
$$=[\delta(0)]^{-2}\sum_{K}\int d^3x\,a_n(\mathbf{x};G)\Psi_{n'\mathbf{k}}(\mathbf{x})\delta(\mathbf{q}+\mathbf{K}+\mathbf{T}(\mathbf{x})). \tag{A18}$$

It will now be easier if we choose a set of axes and a gauge. We pick **B** in the z direction and **a** in the y direction depending only on x. The crystal axes now have no particular relation to x, y, z. We have also specialized to the case of an irrational field orientation, so that there is no **k** dependence of **a**. Then

$$S_{nn'}(\mathbf{k};\mathbf{q})=[\delta(0)]^{-2}\sum_{K}\int dxdz\,a_n\!\left(x,\frac{q_x+K_x}{B},z;G\right)\Psi_{n',\mathbf{k}}\!\left(x,\frac{q_x+K_x}{B},z\right)\delta(q_y+K_y)\delta(q_z+K_z). \tag{A19}$$

Consider now one term of this sum. By the Schwarz inequality we have

$$|S_{nn'}(\mathbf{k};\mathbf{q};K)|^2\equiv\left|[\delta(0)]^{-2}\int dxdz\,a_n\!\left(x,\frac{q_x+K_x}{B},z;G\right)\Psi_{n'\mathbf{k}}\!\left(x,\frac{q_x+K_x}{B},z\right)\right|^2$$
$$\le[\delta(0)]^{-2}\int dxdz\left|a_n\!\left(x,\frac{q_x+K_x}{B},z;G\right)\right|^2[\delta(0)]^{-2}\int\left|u_{n'\mathbf{k}}\!\left(x,\frac{q_x+K_x}{B},z\right)\right|^2dxdz. \tag{A20}$$

For irrational orientations the second integral is independent of $(q_x+K_x)/B$ and simply cancels $[\delta(0)]^{-2}$. We now require the explicit form of a more general pseudo-Wannier function

$$a_n(x,y,z;G;Y)=\int_{-\infty}^{\infty}dk'\,\Psi_{n\mathbf{k}'}e^{-ik'Y}, \tag{A21}$$

where Y is a continuous variable. These functions are orthogonal for different values of Y. Furthermore, consider the integral

$$P(y;Y)=[\delta(0)]^{-2}\int|a_n(x,y,z;G;Y)|^2dxdz$$

$$=[\delta(0)]^{-2}\int dkdk'e^{i(k'-k)(y-Y)}\left(\int u_{nk}^*(x,y,z)u_{nk'}(x,y,z)dxdz\right). \quad \text{(A22)}$$

As in (A20), the integral over x and z is independent of y and cancels $[\delta(0)]^{-2}$ so that $P(y;Y)$ depends only on $(y-Y)$, and $P(y)$ is simply the Fourier transform of

$$S_{nn}(\kappa_y)=\int S_{nn}(k_y+\kappa_y,\,k_y)dk_y. \quad \text{(A23)}$$

Since our one-dimensional bands are analytic for real k, this is an analytic function of κ_y for all real κ_y, which, however, cannot be continued into the complex plane, because branch points are arbitrarily close to the real axis. This $P(y)$ falls off more rapidly than y^{-n} for any n at large y, but more slowly than $e^{-|\lambda y|}$ for any λ. (Of course, if the band has no degeneracies, the branch surfaces are not close to the real axis and the fall-off is exponential.) This is sufficient to show that $S(\mathbf{k};\mathbf{q};\mathbf{K})$ and $S(\mathbf{k};\mathbf{q})$ vanish faster than $|q/B|^{-n}$ and that the integrals I_1 and I_2 with S substituted for n satisfy the conditions for an asymptotic expansion. Thus the multiplication rules are asymptotically convergent at small B for all products used in this paper and for all k and R, provided we have chosen a doubly singular gauge. This is true even for points of degeneracy.

Although our proof has used a special gauge, it is clear that after obtaining the expansion in that gauge one can make the transformation

$$U(\mathbf{R})=e^{i\chi(\mathbf{R})}, \quad \text{(A24)}$$

which will change the vector potential by $\nabla\chi$ for arbitrary χ. Thus the product expansion will be valid for any gauge; presumably a guage-invariant proof exists. However, we have not seen it and do not consider it necessary to look further.

Our proof for degenerate bands has assumed an irrational field orientation, since any other has zero probability. The proof is no more difficult for rational orientations, but we leave it to the interested reader.

Finally, we should note that our differentiation of $S(k,R)$ with respect to R has been somewhat cavalier for we have differentiated $S(\mathfrak{k})$ with respect to $\mathfrak{k}$ and then $\mathfrak{k}$ with respect to R, thus treating $S(\mathfrak{k})$ as a continuous function of $\mathfrak{k}$. In fact, however, in reference 3 it was shown that the fact that S is defined only on lattice points is vital in giving it an infinite radius of convergence. In fact, we should use instead of $i^l\partial^lS/\partial R^l$,

$$\int_{BZ}\mathbf{q}^lS(k;q)e^{-i\mathbf{q}\cdot\mathbf{R}}d^3q=\sum_{R_i}\int d^3Q\int_{BZ}d^3q\,q^l\mathcal{S}(k;Q)\exp[i(\mathbf{Q}\cdot\mathbf{A}_i+\mathbf{q}\cdot(\mathbf{R}_i-\mathbf{R}))] \quad \text{(A25)}$$

$$=\sum_K\int d^3Q\int d^3q\;\mathcal{S}(k;Q)\delta(T(Q)+K+q)q^le^{-i\mathbf{q}\cdot\mathbf{R}}. \quad \text{(A26)}$$

The true derivative would be given by

$$\frac{i^l\partial^lS}{\partial k^l}=\int d^3Q\,[\mathbf{T}(\mathbf{Q})]^l\exp(i\mathbf{T}\cdot\mathbf{R})\mathcal{S}(\mathbf{k};\mathbf{Q})=\sum_K\int d^3q\;\mathcal{S}(\mathbf{k};\mathbf{Q})\delta(\mathbf{T}(\mathbf{Q})+\mathbf{K}+\mathbf{q})e^{-i\mathbf{q}\cdot\mathbf{R}}(\mathbf{q}+\mathbf{K})^l. \quad \text{(A27)}$$

These expressions differ by the presence of K in the term with $K\neq0$. Since all such terms get small very rapidly, however, the error is less than any power of B, and thus quite acceptable to us.

APPENDIX B

In this Appendix, we are concerned with the matrix elements between states represented by single orbits in the (p,q) plane. It is convenient to work in the PCR which also yields an intuitive appreciation of what is going on. We define the object

$$P_{nn'}(q,q')=\psi_n(q)\psi_{n'}^*(q'), \quad \text{(B1)}$$

and transform it into a PCR operator $P_{nn'}(p,q)$ as in Sec. III of (F). The matrix element $O_{nn'}$ of an operator O is simply $\mathrm{Tr}\{P_{nn'}O\}$ and it is easily shown that

$$O_{nn'}=\int\int dpdq\,P_{nn'}(p,q)O(p,q), \quad \text{(B2)}$$

a result obtained by Moyal[21] for the continuum case. Since most O of interest are slowly varying functions of p and q, one is encouraged to attempt the integration by the method of steepest descents, since $P_{nn'}$ may be

[21] J. E. Moyal, Proc. Cambridge Phil. Soc. **45**, 99 (1949).

expected to vary rapidly with p and q at small B. Thus in the WKB approximation a wave function for an orbit n is written

$$\psi_n(q)=\frac{1}{m_n^{*\frac{1}{2}}}\sum_\alpha \frac{e^{iS_{n\alpha}(q)/B}}{[v_{n\alpha}(q)]^{\frac{1}{2}}}, \quad \text{(B3)}$$

where $S_{n\alpha}=\int p_{n\alpha}(q)dq$, $p_{n\alpha}$ is a branch of the function $p(q)$ defined by $H(p,q)=E_n$, and m_n^* is given by

$$m_n^*=\sum_\alpha \int \frac{dq}{|v_{n\alpha}|}, \quad \text{(B4)}$$

where the sum is taken over all real branches and the integral is limited to the values of q for which p_α is real. This can be converted to

$$m_n^*=\oint \frac{dq}{v_n(q)}, \quad \text{(B5)}$$

v is $\partial H/\partial p$. Equation (B3) as it stands, is valid only for values of q such that all $p_\alpha(q)$ are real. When some are complex, more care is required in order to assure that the wave function can be said to belong to only one orbit, but Eq. (B3) will give valid results for our purposes. We then obtain for $P_{nn'}(p,q)$

$$P_{nn'}(p,q)=\sum_{\alpha\beta}\int dr \times\frac{\exp[(i/B)(S_{n\alpha}(q+\frac{1}{2}r)-S_{n'\beta}(q-\frac{1}{2}r)-pr)]}{(v_{n\alpha}v_{n'\beta})^{\frac{1}{2}}}. \quad \text{(B6)}$$

We will restrict ourselves to pairs of orbits such that a straight line can be drawn between them, and we choose our axes so that p is perpendicular to such a line. The integral over r can be performed by steepest descents, yielding

$$P_{nn'}(p,q)=\sqrt{\pi}\sum_{\alpha\beta}\frac{\exp[(i/B)(S_{n\alpha}(q+\frac{1}{2}r)-S_{n\beta}(q-\frac{1}{2}r)-pr)]}{(v_{n\alpha}v_{n'\beta})^{\frac{1}{2}}(|p_{n\alpha}'-p_{n'\beta}'|)^{\frac{1}{2}}}\Bigg|_{r=r_{\alpha\beta}(p,q)}, \quad \text{(B7)}$$

where $r_{\alpha\beta}$ satisfies the equation.

$$p_{n\alpha}(q+\tfrac{1}{2}r)+p_{n'\beta}(q-\tfrac{1}{2}r)-2p=0. \quad \text{(B8)}$$

In an integral involving $P_{nn'}$ the integration over p can also be performed by steepest descents and the major contribution will come from the region where

$$r_{\alpha\beta}=0,$$

and is given by

$$\bar{P}_{nn'}(q)=\frac{\pi}{2}\sum_{\alpha\beta}\frac{\exp[(i/B)(S_{n\alpha}(q)-S_{n\beta}(q))]}{(v_{n\alpha}v_{n'\beta})^{\frac{1}{2}}}. \quad \text{(B9)}$$

In this form it is clear that only the two branches of each orbit which are closest to each other will be important; for the exponential factor in the contributions from other pairs of branches will be much more rapidly varying and will give negligible integrals over q at small B.

Finally the main contribution of (B9) on integrating q can be made to come, on deforming the contour, from the point q_0 where

$$P_{n\alpha}(q_0)=P_{n'\beta}(q_0). \quad \text{(B10)}$$

We distinguish two cases.

(1) Orbits n and n' are in the same band. Then (B10) can be satisfied only if they have the same energy. In this case, the important factor is

$$\exp[(i/B)(S_n(q_0)-S_{n'}(q_0))]. \quad \text{(B11)}$$

If the orbits have different energies but are kept within a few level splittings of each other as $B\to 0$, the error in using this factor is small. If their energy difference is large compared to level splittings, more detailed treatment is required, but the matrix elements will then be small compared to the energy difference, so that such matrix elements will act only as small perturbations, shifting levels, but not capable of producing banding.

If the orbits have essentially the same energy the factor $\exp[(i/B)(S_n(q_0)-S_{n'}(q_0))]$ has a simple geometrical interpretation, for

$$S_n(q_0)=\int_a^{q_0} pdq,$$

where a is some point on orbit n. $S_n(q_0)-S_{n'}(q_0)$ is then $\int_a^b pdq$, where b is a point on n', and the path is a surface of constant energy. If we choose the path in imaginary q space, the integral is the area of the projection of the path on the real p—imaginary q plane. The imaginary part which determines the exponential fall-off of $O_{nn'}$ is invariant to our choice of end points on the orbits, but can be most easily visualized if we choose the points with the shortest distance between them.

It should be noted the main contribution to the Hamiltonian matrix elements will be $H(p_0,q_0)=E\times$the integral of (B9) which is just the overlap integral between the two orbits. Additional terms will be of order B smaller. Thus the principal effect of the matrix element is removed by orthogonalizing the orbitals without changing their energy.

(2) If the orbits are in different bands it should usually be possible to find solutions of (B10) even if their energies are different. In this case then, the

saddle point method is immediately applicable, and if the energies are nearly equal gives the same result (B11), with the same interpretation. In the special case considered in Sec. III, it gives the same exponential factor as was found there. For the M term in (III.3) this analysis is sufficient. The X term is somewhat more complicated, however, because X has the energy difference between the bands in its denominator, and the integral thus has a pole where the band energies are equal namely at q_0. This complicates the evaluation procedure, but only modifies the factor by which (B11) is multiplied. The actual evaluation for our model (III.4) is identical to that of Kane for Zener breakdown. If the orbit energies are unequal, the orbits may even overlap so that the integrals become larger, but this has no serious effects, because the energy difference required is so much greater than the matrix element that the orbits cannot be appreciably mixed.

Semiclassical diagonalization of quantum Hamiltonian and equations of motion with Berry phase corrections

P. Gosselin[1], A. Bérard[2], and H. Mohrbach[2,a]

[1] Institut Fourier, UMR 5582 CNRS-UJF, UFR de Mathématiques, Université Grenoble I, BP 74, 38402 Saint Martin d'Hères Cedex, France
[2] Laboratoire de Physique Moléculaire et des Collisions, ICPMB-FR CNRS 2843, Université Paul Verlaine-Metz, 57078 Metz Cedex 3, France

Received 16 August 2006 / Received in final form 26 June 2007
Published online 10 August 2007 –

Abstract. It has been recently found that the equations of motion of several semiclassical systems must take into account terms arising from Berry phases contributions. Those terms are responsible for the spin Hall effect in semiconductor as well as the Magnus effect of light propagating in inhomogeneous media. Intensive ongoing research on this subject seems to indicate that a broad class of quantum systems may be affected by Berry phase terms. It is therefore important to find a general procedure allowing for the determination of semiclassical Hamiltonian with Berry Phase corrections. This article presents a general diagonalization method at order $\hbar$ for a large class of quantum Hamiltonians directly inducing Berry phase corrections. As a consequence, Berry phase terms on both coordinates and momentum operators naturally arise during the diagonalization procedure. This leads to new equations of motion for a wide class of semiclassical system. As physical applications we consider here a Dirac particle in an electromagnetic or static gravitational field, and the propagation of a Bloch electrons in an external electromagnetic field.

PACS. 03.65.-w Quantum mechanics – 03.65.Sq Semiclassical theories and applications – 03.65.Vf Phases: geometric; dynamic or topological – 04.20.Cv Fundamental problems and general formalism

1 Introduction

Since the seminal work of Berry [1], the notion of Berry phase has found several applications in branches of quantum physics such as atomic and molecular physics, optics and gauge theories. Most studies consider the geometric phase that a wave function acquires when a quantum mechanical system has an adiabatic evolution. Yet, the Berry phase in momentum space has recently found unexpected applications in the topic of spintronics. Such a term may indeed be responsible for a transverse dissipationless spin-current in semiconductors in the presence of electric fields [2]. This effect is a particular case of the Spin-Hall effect which is now predicted and observed in many different physical situations and can be interpreted at the semiclassical level as due to the influence of Berry connections on semiclassical equations of motion of spinning particles, like electrons in electric [3] or magnetic fields [4]. In the above cited examples, the semiclassical equations of motion where derived from an approximate semiclassical Hamiltonian in a representation where this latter is diagonal. It was then shown that a noncommutative geometry, originating from the presence of a Berry phase which turns out to be a spin-orbit coupling, underlies the semiclassical dynamics. Spin-orbit contributions on the propagation of light have also been the focus of several other works [3, 5,6] and have led to a generalization of geometric optics called geometric spinoptics [7].

Semiclassical methods play a very important role in solid state physics too, in studying the dynamics of electrons to account for the various properties of metals, semiconductors and insulators [8]. In a series of papers [9] (see also [10]), a new set of semiclassical equations with a Berry phase correction was proposed to account for the semiclassical dynamics of electrons in magnetic Bloch bands (in the usual one band approximation). These equations were derived by considering a wave packet in a band and using a time-dependent variational principle in a Lagrangian formulation. The derivation of a semiclassical Hamiltonian was shown to lead to difficulties in the presence of Berry phase terms [9]. The apparent non-canonical character of the equations of motion with Berry phase corrections led the authors of [11] to conclude that the naive phase space volume is not conserved in the presence of a Berry phase and a magnetic field. There is nevertheless an invariant

[a] e-mail: mohrbach@univ-metz.fr

measure as found by themselves, so the Liouville theorem is not really violated in the end. This invariant measure is actually a general result of the well-established theory of non-canonical Hamiltonian dynamics, as pointed out by a number of authors [12–14]. However it is only in [15] that the non-canonical Hamiltonian formulation of a semiclassical electron in magnetic Bloch bands has been fully derived. This Hamiltonian approach allows deriving rigorously the semiclassical equations of motion, including explicitly the role of the Berry curvature and showed many similarities with the description of a Dirac electron in an electromagnetic field as in [4]. The common feature of the Hamiltonian formulations discussed above is that a noncommutative geometry underlies the algebraic structure of both coordinates and momenta. Actually, a Berry phase contribution to the dynamical operators stems from the representation where the kinetic energy is diagonal (for instance Foldy-Wouthuysen or Bloch representation). In this representation the physical coordinate and momentum become noncommutative operators.

The previous discussion shows that Berry phase terms could be present in semiclassical equations of motion of several physical systems ranging from electrons in vacuum, in solid or in semiconductor to photons in inhomogeneous media, with potential application in the field of spintronics and spinoptics. This in turn called for a general semiclassical Hamiltonian formalism from which semiclassical dynamics of a quantum system can be derived. This paper presents a general method of diagonalization at order $\hbar$ for a quantum mechanical Hamiltonian presenting bands structure, like for electrons in a periodic potential or for a Dirac (massive or massless) like-Hamiltonian. Starting with a Hamiltonian depending only on an invariant momentum $\mathbf{P}$ and whose diagonalization is known, we can introduce a dependence in the variable $\mathbf{R}$ and diagonalize the Hamiltonian in four steps (discussed in the text). During this process of diagonalization, we show that both position and momentum operators acquire a Berry-phase contribution making both the coordinate and momentum algebra noncommutative. As physical applications and to check to validity of our method we further consider the case of Dirac particle in an electromagnetic field and compare with the semiclassical diagonalization given in [4]. We also consider the case of a Dirac particle in a symmetric static gravitational field (the asymmetric case is studied in Ref. [16]) and compare with the articles [17,18]. As a last application, the reader will find the details of the diagonalization sketched in [15] for the propagation of a Bloch electron (spinless) in an external electromagnetic field. These various physical applications show that our semiclassical Hamiltonian diagonalization approach is potentially promising since it unifies several apparently unrelated problematics in one formalism.

The paper is organized as follows. In Section 2 we develop our formalism in the case of a general Hamiltonian which has an energy bands structure. We then derive the very general equations of motion in this case. Section 3 is devoted to the application of our method to the case of the Dirac Hamiltonian in an electromagnetic field but in a flat space, and to the diagonalization of the Dirac Hamiltonian in a symmetric static gravitational field, allowing us to check the validity of our method. Section 4 retrieves the equations of motion for an electron in a periodic potential within our general set up. Section 5 is for the conclusion.

2 A general process of semiclassical diagonalization

In this section we present a method to diagonalize at the semiclassical order ($\hbar$) a quantum mechanical system whose state space is a tensor product $L^2(\mathcal{R}^3) \otimes V$ with V some internal space. In other words, the Hamiltonian of this system can be written as a matrix $H_0(\mathbf{P}, \mathbf{R})$ of size $\dim V$ whose elements are operators depending on a couple of canonical variables $\mathbf{P}$ and $\mathbf{R}$. The archetype example is the Dirac Hamiltonian with $V = C^4$, but in Appendix B we show how a spinless electron in a periodic potential fits also in this set up. By diagonalization, we always mean here a unitary transformation setting the Hamiltonian in a diagonal matrix form, the diagonal elements being operators depending on $\mathbf{P}$ and $\mathbf{R}$. That is, we do not aim at finding the eigenvalues, but rather to derive the Band Hamiltonians, that are usually relevant for the semiclassical dynamics.

More precisely, our strategy will be to solve the exact diagonalization for $H_0(\mathbf{P}, \mathbf{R})$ at order $\hbar$, when the diagonalization of a fictitious Hamiltonian $H_0(\mathbf{P}, \widetilde{\mathbf{r}})$ is known for a parameter $\widetilde{\mathbf{r}}$ (replacing $\mathbf{R}$) which is supposed to commute with $\mathbf{P}$ (for instance this may be the Foldy-Wouthuysen transformation [19] for a free Dirac particle). We show how to compute the quantum corrections (at leading order in $\hbar$) that were neglected during this formal diagonalization (where position and momenta where considered as commuting quantities). The idea behind this procedure is that it is much easier to solve the diagonalization for $H_0(\mathbf{P}, \widetilde{\mathbf{r}})$, as seen in our applications, and only then turn to $H_0(\mathbf{P}, \mathbf{R})$.

2.1 Preliminary: products of operators series

To develop our process of diagonalization, the semiclassical expression of products of symmetrized expressions $S(\mathbf{P}, \mathbf{R})$ depending on the canonical couple of variables $\mathbf{P}$ and $\mathbf{R}$ is required. These expressions are also assumed to have series expansions in $\mathbf{P}$ and $\mathbf{R}$ whose coefficients can be of a matrix form (this last assumption allowing to deal with Dirac Hamiltonians). Let us consider two such expressions $S_1(\mathbf{P}, \mathbf{R})$ and $S_2(\mathbf{P}, \mathbf{R})$, supposed to be symmetrized in $\mathbf{P}$ and $\mathbf{R}$. By symmetrization, we mean that each expression has been written in a form where all the powers of $\mathbf{P}$ have been put half on the left and half on the right of the expression. Our aim is now to write the product $S_1(\mathbf{P}, \mathbf{R})\, S_2(\mathbf{P}, \mathbf{R})$ as a symmetric expression in terms of $\mathbf{P}$ and $\mathbf{R}$. This is easy to realize at order $\hbar$, since in that case, pushing half of the powers of $\mathbf{P}$ in $S_2(\mathbf{P}, \mathbf{R})$

on the left and half the powers of $S_1(\mathbf{P},\mathbf{R})$ on the right is equivalent to the computation of some commutators. One can easily see that at order $\hbar$

$$S_1(\mathbf{P},\mathbf{R})\,S_2(\mathbf{P},\mathbf{R}) = Sym(S_1(\mathbf{P},\mathbf{R})\,S_2(\mathbf{P},\mathbf{R})) + \frac{i}{2}\hbar Asym\nabla_{R_l}\nabla_{P^l}\left[S_1(\mathbf{P},\mathbf{R})\,S_2(\mathbf{P},\mathbf{R})\right] \quad (1)$$

where $Sym(S_1(\mathbf{P},\mathbf{R})\,S_2(\mathbf{P},\mathbf{R}))$ is the symmetrized expression for the product, and $Asym$ is defined through

$$Asym\nabla_{R_l}\nabla_{P^l}\left[S_1(\mathbf{P},\mathbf{R})\,S_2(\mathbf{P},\mathbf{R})\right] = \nabla_{R_l}S_1(\mathbf{P},\mathbf{R}) \times \nabla_{P^l}S_2(\mathbf{P},\mathbf{R}) - \nabla_{P^l}S_1(\mathbf{P},\mathbf{R})\,\nabla_{R_l}S_2(\mathbf{P},\mathbf{R}). \quad (2)$$

This formula can be easily generalized to an arbitrary product of n terms, but for the sequel of this paper, only three will be needed

$$\begin{aligned} &S_1(\mathbf{P},\mathbf{R})S_2(\mathbf{P},\mathbf{R})S_3(\mathbf{P},\mathbf{R}) \\ &= Sym(S_1(\mathbf{P},\mathbf{R})\,S_2(\mathbf{P},\mathbf{R})\,S_3(\mathbf{P},\mathbf{R})) \\ &+\frac{i}{2}\hbar Asym\nabla_{R_l}\nabla_{P^l}[S_1(\mathbf{P},\mathbf{R})\,S_2(\mathbf{P},\mathbf{R})]\,S_3(\mathbf{P},\mathbf{R}) \\ &+\frac{i}{2}\hbar S_1(\mathbf{P},\mathbf{R})\,Asym\nabla_{R_l}\nabla_{P^l}\left[S_2(\mathbf{P},\mathbf{R})\,S_3(\mathbf{P},\mathbf{R})\right] \\ &+\frac{i}{2}\hbar Asym\nabla_{R_l}\nabla_{P^l}\left[S_1(\mathbf{P},\mathbf{R})_{S_2(\mathbf{P},\mathbf{R})}S_3(\mathbf{P},\mathbf{R})\right] \end{aligned} \quad (3)$$

where

$$\begin{aligned} &Asym\nabla_{R_l}\nabla_{P^l}\left[S_1(\mathbf{P},\mathbf{R})_{S_2(\mathbf{P},\mathbf{R})}S_3(\mathbf{P},\mathbf{R})\right] = \\ &\quad[\nabla_{R_l}S_1(\mathbf{P},\mathbf{R})]\,S_2(\mathbf{P},\mathbf{R})\nabla_{P^l}S_3(\mathbf{P},\mathbf{R}) \\ &\quad-[\nabla_{P^l}S_1(\mathbf{P},\mathbf{R})]\,S_2(\mathbf{P},\mathbf{R})\,\nabla_{R_l}S_3(\mathbf{P},\mathbf{R}). \end{aligned} \quad (4)$$

Let us stress again that all these identities are valid only at order $\hbar$ and that considering higher order corrections would of course induce more corrections.

2.2 Diagonalization with a parameter $\widetilde{r}$

Let us consider a general Hamiltonian $H_0(\mathbf{P},\mathbf{R})$ which admits a series expansion in $\mathbf{P}$ and $\mathbf{R}$ written here for convenience in a symmetrical form. To perform the semiclassical diagonalization of this operator, we first consider in this subsection a fictitious Hamiltonian $H_0(\mathbf{P},\widetilde{\mathbf{r}})$ where a parameter $\widetilde{\mathbf{r}}$ commuting with $\mathbf{P}$ has replaced the operator $\mathbf{R}$. We further assume that $H_0(\mathbf{P},\widetilde{\mathbf{r}})$ is known to be exactly diagonalized through a matrix $U(\mathbf{P},\widetilde{\mathbf{r}}) \equiv U$. We assume that U and $U(\mathbf{P},\widetilde{\mathbf{r}})\,H_0(\mathbf{P},\widetilde{\mathbf{r}})\,U^+(\mathbf{P},\widetilde{\mathbf{r}})$ can be expanded in series of P (with positive or negative powers) as it will be the case in our applications, but in fact this assumption which is practical for our proofs could probably be relaxed.

As an typical example we can consider the following kind of Dirac Hamiltonian

$$H_0 = \alpha\cdot(\mathbf{P}-\mathbf{A}(\widetilde{\mathbf{r}}))+\beta m \quad (5)$$

where $\mathbf{A}(\widetilde{\mathbf{r}})$ mimics a formal magnetic potential. The potential $\mathbf{A}(\widetilde{\mathbf{r}})$ being $\mathbf{R}$ independent, it only shifts the momentum for each value of $\widetilde{\mathbf{r}}$. As a consequence, the usual Foldy Wouthuysen [19] transformation expressed in terms of $\mathbf{P}-\mathbf{A}$ instead of $\mathbf{P}$, diagonalizes the Dirac Hamiltonian exactly.

Going back to the general case, the diagonalization of $H_0(\mathbf{P},\widetilde{\mathbf{r}})$ will be written as

$$\varepsilon(\mathbf{P},\widetilde{\mathbf{r}}) = U(\mathbf{P},\widetilde{\mathbf{r}})\,H_0(\mathbf{P},\widetilde{\mathbf{r}})\,U^+(\mathbf{P},\widetilde{\mathbf{r}}) \quad (6)$$

where $\varepsilon(\mathbf{P},\widetilde{\mathbf{r}})$ is a diagonal matrix. For later use, let us notice that we have $U(\mathbf{P},\widetilde{\mathbf{r}})\,\mathbf{P}U^+(\mathbf{P},\widetilde{\mathbf{r}}) = \mathbf{P}$.

To gain some hints from our initial diagonalization, recall we assumed that the product $\varepsilon(\mathbf{P},\widetilde{\mathbf{r}}) = U(\mathbf{P},\widetilde{\mathbf{r}})\,H_0(\mathbf{P},\widetilde{\mathbf{r}})\,U^+(\mathbf{P},\widetilde{\mathbf{r}})$ can be expanded in series of monomial terms of the form: $\left(\varepsilon_1^{i_1}(\widetilde{\mathbf{r}})P_{i_1}^{k_{i_1}}\right)\ldots\left(\varepsilon_n^{i_n}(\widetilde{\mathbf{r}})P_{i_n}^{k_{i_n}}\right)$ where the $\varepsilon_l^{i_l}(\widetilde{\mathbf{r}})$ are some matrices commuting with $\mathbf{P}$, and $P_{i_l}^{k_{i_l}}$ is the i_lth component of $\mathbf{P}$ ($i_l = 1,2,3$) at some power k_{i_l} (this power should be written $k_{i_l,l}$ but we will avoid an excess of notation here). As usual, the sums over the i_l are implicit. Rearranging the series in powers of P_1, P_2, P_3 and given that $\widetilde{\mathbf{r}}$ is a parameter, we can write the energy in a symmetrical form:

$$\varepsilon(\mathbf{P},\widetilde{\mathbf{r}}) = \frac{1}{2}\sum_{\mathbf{X}}\left[A_{\mathbf{X}}(\widetilde{\mathbf{r}})\left(\prod_{i=1,2,3}P_i^{X_i}\right)+\left(\prod_{i=1,2,3}P_i^{X_i}\right)A_{\mathbf{X}}(\widetilde{\mathbf{r}})\right] \quad (7)$$

where $\left(\prod_{i=1,2,3}P_i^{X_i}\right)$ is a given momentum power and $A_{\mathbf{X}}(\widetilde{\mathbf{r}})$ is a combination of the $\varepsilon_l^{i_l}(\widetilde{\mathbf{r}})$. The $\mathbf{X}$ labels the multi index (X_i, $i = 1,2,3$).

The important consequence here is that the matrix $A_{\mathbf{X}}(\widetilde{\mathbf{r}})$ is diagonal. As an example, think of the Dirac Hamiltonian diagonalization, which involves some products $\alpha\cdot\mathbf{P}\alpha\cdot\mathbf{P}$, rearranged as

$$\frac{1}{2}\left[\alpha_i P_i \alpha_j P_j + \alpha_j P_j \alpha_i P_i\right] = \frac{1}{2}\left[\alpha_i\alpha_j + \alpha_j\alpha_i\right]P_iP_j \quad (8)$$

and $\frac{1}{2}\left[\alpha_i\alpha_j + \alpha_j\alpha_i\right]$ is diagonal, by the usual rules for Dirac matrices.

Let us conclude this subsection by noting that the symmetrizations we performed is unnecessary here, but will be of a practical interest when dealing with the exact diagonalization.

2.3 Introducing the R dependence. The transformation Ansatz

We are now going to reintroduce $\mathbf{R}$ into H_0 in order to diagonalize exactly $H_0(\mathbf{P},\mathbf{R})$ at the $\hbar$ order. In the following symmetrization in $\mathbf{R}$ and $\mathbf{P}$ will be assumed in all expressions. To find the diagonalization transformation for $H_0(\mathbf{P},\mathbf{R})$, we will use the following method. First notice that the Hamiltonian $H_0(\mathbf{P},\mathbf{R})$ is "almost" diagonalized, that is diagonalized at zeroth order in $\hbar$, through the transformation $U(\mathbf{P},\mathbf{R})$ (which is also symmetrized given our convention).

However, through the symmetrization process, the matrix $U(\mathbf{P},\mathbf{R})$ does not remain unitary. As a consequence, we will rather consider a matrix $U(\mathbf{P},\mathbf{R})+XU(\mathbf{P},\mathbf{R})$, where X is a contribution of order $\hbar$ that ensures the unitarity of the transformation at order $\hbar$ (the factor U is a normalization that simplifies the subsequent expressions).

The matrix X can be explicitly computed. Actually, from the unitary conditions

$$\left(U(\mathbf{P},\mathbf{R})+XU\right)\left(U^{+}(\mathbf{P},\mathbf{R})+U^{+}X^{+}\right)=1 \qquad (9)$$

and

$$\left(U^{+}(\mathbf{P},\mathbf{R})+U^{+}X^{+}\right)\left(U(\mathbf{P},\mathbf{R})+XU\right)=1 \qquad (10)$$

or equivalently

$$\begin{aligned} U(\mathbf{P},\mathbf{R})U^{+}(\mathbf{P},\mathbf{R})+X+X^{+}&=1\\ U^{+}(\mathbf{P},\mathbf{R})U(\mathbf{P},\mathbf{R})+U^{+}\left(X+X^{+}\right)U&=1. \end{aligned} \qquad (11)$$

To solve this equation, let us first notice that $U(\mathbf{P},\mathbf{R})U^{+}(\mathbf{P},\mathbf{R})\neq 1$ since $U(\mathbf{P},\mathbf{R})$ is not unitary. The crucial point here and in the sequel of this paper is the computation of a product of expressions in which the $\mathbf{R}$ dependence has been introduced. To do so let us use the initial relation for the parameter $\widetilde{\mathbf{r}}$

$$U(\mathbf{P},\widetilde{\mathbf{r}})U^{+}(\mathbf{P},\widetilde{\mathbf{r}})=1 \qquad (12)$$

expanded in a symmetric series (as we did for the Hamiltonian)

$$U(\mathbf{P},\widetilde{\mathbf{r}})U^{+}(\mathbf{P},\widetilde{\mathbf{r}})=\frac{1}{2}\sum_{\mathbf{X}}\left[B_{\mathbf{X}}(\widetilde{\mathbf{r}})\left(\prod_{i=1,2,3}P_i^{X_i}\right)+\left(\prod_{i=1,2,3}P_i^{X_i}\right)B_{\mathbf{X}}(\widetilde{\mathbf{r}})\right]=1. \qquad (13)$$

Therefore the series expansion $\sum_{\mathbf{X}}B_{\mathbf{X}}(\widetilde{\mathbf{r}})\left(\prod_{i=1,2,3}P_i^{X_i}\right)$ reduces to one constant term: the identity matrix.

Now, going back to $U(\mathbf{P},\mathbf{R})$, we use the symmetrization formula (1)

$$\begin{aligned} U(\mathbf{P},\mathbf{R})U^{+}(\mathbf{P},\mathbf{R})&=\\ \frac{1}{2}\sum_{\mathbf{X}}&\left[B_{\mathbf{X}}(\mathbf{R})\left(\prod_{i=1,2,3}P_i^{X_i}\right)+\left(\prod_{i=1,2,3}P_i^{X_i}\right)B_{\mathbf{X}}(\mathbf{R})\right]\\ &+\frac{i}{2}\hbar Asym\nabla_{R_l}\nabla_{P^l}\left[U(\mathbf{P},\mathbf{R})U^{+}(\mathbf{P},\mathbf{R})\right]. \end{aligned} \qquad (14)$$

Since our result about the $B_{\mathbf{X}}(\widetilde{\mathbf{r}})$ applies also to $B_{\mathbf{X}}(\mathbf{R})$ after replacing $\widetilde{\mathbf{r}}$ by $\mathbf{R}$, one has:

$$\begin{aligned} U(\mathbf{P},\mathbf{R})U^{+}(\mathbf{P},\mathbf{R})&=\\ 1+\frac{i}{2}\hbar Asym\nabla_{R_l}\nabla_{P^l}&\left[U(\mathbf{P},\mathbf{R})U^{+}(\mathbf{P},\mathbf{R})\right]\\ &=1-\frac{i}{2\hbar}\left[\mathcal{A}_{P_l},\mathcal{A}_{R_l}\right] \end{aligned} \qquad (15)$$

where we have defined the ("non projected", see below) Berry connections (we use the terminology of Ref. [20]) as

$$\begin{aligned} \mathcal{A}_{\mathbf{R}}&=i\hbar U(\mathbf{P},\mathbf{R})\nabla_{\mathbf{P}}U^{+}(\mathbf{P},\mathbf{R})\\ \mathcal{A}_{\mathbf{P}}&=-i\hbar U(\mathbf{P},\mathbf{R})\nabla_{\mathbf{R}}U^{+}(\mathbf{P},\mathbf{R}). \end{aligned} \qquad (16)$$

that are non diagonal Hermitian matrices of order $\hbar$. Similarly, we have also:

$$\begin{aligned} U^{+}(\mathbf{P},\mathbf{R})U(\mathbf{P},\mathbf{R})&=\\ 1+\frac{i}{2}\hbar Asym\nabla_{R_l}\nabla_{P^l}&\left[U^{+}(\mathbf{P},\mathbf{R})U(\mathbf{P},\mathbf{R})\right]\\ &=1-U^{+}(\mathbf{P},\mathbf{R})\frac{i}{2\hbar}\left[\mathcal{A}_{P^l},\mathcal{A}_{R_l}\right]U(\mathbf{P},\mathbf{R}). \end{aligned} \qquad (17)$$

Let us note that, as can be checked easily, the Berry connections are Hermitian. So is $\frac{i}{2\hbar}\left[\mathcal{A}_{P^l},\mathcal{A}_{R_l}\right]$. Therefore we can solve our problem with

$$X=\frac{i}{4\hbar}\left[\mathcal{A}_{P^l},\mathcal{A}_{R_l}\right]. \qquad (18)$$

Let us make an important remark at this point. Our choice for X is obviously not unique. Actually, it has been chosen to ensure the unitarity of the transformation and to obtain a transformation that reduces to the initial one when $\mathbf{P}$ and $\mathbf{R}$ do commute. We could thus add to X an expression like δXU where δX is anti-Hermitian. It is easy to see that the operator $U(\mathbf{P},\mathbf{R})+XU+\delta XU$ is still unitary. However, we will soon see that this non unicity is irrelevant and that our choice is sufficient to perform the diagonalization at order $\hbar$.

2.4 The quasidiagonalization

We will now consider the following quasi-diagonalization transformation

$$\left[\left(U(\mathbf{P},\mathbf{R})+XU\right)H_0(\mathbf{P},\mathbf{R})\left(U^{+}(\mathbf{P},\mathbf{R})+U^{+}X^{+}\right)\right]. \qquad (19)$$

To compute this last expression, decompose it at the first order in $\hbar$ as

$$\begin{aligned} U(\mathbf{P},\mathbf{R})H_0(\mathbf{P},\mathbf{R})U^{+}(\mathbf{P},\mathbf{R})&+XUH_0(\mathbf{P},\mathbf{R})U^{+}(\mathbf{P},\mathbf{R})\\ &+U(\mathbf{P},\mathbf{R})H_0(\mathbf{P},\mathbf{R})U^{+}X^{+}\simeq\\ U(\mathbf{P},\mathbf{R})H_0(\mathbf{P},\mathbf{R})U^{+}(\mathbf{P},\mathbf{R})&+X\varepsilon(\mathbf{P},\mathbf{R})+\varepsilon(\mathbf{P},\mathbf{R})X^{+}. \end{aligned} \qquad (20)$$

Let us first have a look to $U(\mathbf{P},\mathbf{R})H_0(\mathbf{P},\mathbf{R})U^{+}(\mathbf{P},\mathbf{R})$ and consider it, as before, as a series of products of operators in $\mathbf{P}$ and $\mathbf{R}$. If this two variables where commuting, we would recover the expansion $\varepsilon(\mathbf{P},\mathbf{R})$ given in Section 1.2. But now, since, $\mathbf{R}$ does not commute with $\mathbf{P}$, one has rather

$$\begin{aligned} U(\mathbf{P},\mathbf{R})H_0(\mathbf{P},\mathbf{R})U^{+}(\mathbf{P},\mathbf{R})&=\\ \frac{1}{2}\sum_{\mathbf{X}}&\left[A_{\mathbf{X}}(\mathbf{R})\left(\prod_{i=1,2,3}P_i^{X_i}\right)+\left(\prod_{i=1,2,3}P_i^{X_i}\right)A_{\mathbf{X}}(\mathbf{R})\right]\\ &+[\text{commutators}] \end{aligned} \qquad (21)$$

the commutators appearing while pushing the momentum powers on the left or on the right.

We can compute the first two terms of the right hand side by the same trick as before. Actually, by construction, the coefficients of the series expansion of $A_{\mathbf{X}}(\mathbf{R})$ in the variable $\mathbf{R}$, are the same as the coefficients (which are diagonal) of the expansion of $A_{\mathbf{X}}(\mathbf{r})$ in the parameter $\mathbf{r}$. As a consequence,

$$\frac{1}{2}\sum_{\mathbf{X}}\left[A_{\mathbf{X}}(\mathbf{R})\left(\prod_{i=1,2,3}P_i^{X_i}\right)+\left(\prod_{i=1,2,3}P_i^{X_i}\right)A_{\mathbf{X}}(\mathbf{R})\right] \tag{22}$$

is the series expansion of $\varepsilon(\mathbf{P},\mathbf{R})$, the powers of $\mathbf{P}$ being rejected symmetrically to the left and to the right.

As an example, consider again the case of the Dirac Hamiltonian with an electromagnetic field. The free "Benchmark" case is $\varepsilon^2(\mathbf{P}) = \mathbf{P}^2$, and given our conventions, replacing $\mathbf{P}$ by $\mathbf{P}-\mathbf{A}(\mathbf{R})$ leads us to define $\varepsilon^2(\mathbf{P}-\mathbf{A}(\mathbf{R})) = \mathbf{P}^2 - \mathbf{A}(\mathbf{R})\cdot\mathbf{P} - \mathbf{P}\cdot\mathbf{A}(\mathbf{R}) + \mathbf{A}^2(\mathbf{R})$, which is simply the usual operator $(\mathbf{P}-\mathbf{A}(\mathbf{R}))^2$. By the same way, we obtain as a series expansion

$$\begin{aligned}\varepsilon(\mathbf{P}-\mathbf{A}(\mathbf{R})) &= \varepsilon(\mathbf{P}) - \frac{1}{2}\left(\mathbf{A}(\mathbf{R})\cdot\frac{\mathbf{P}}{\mathbf{P}^2}+\frac{\mathbf{P}}{\mathbf{P}^2}\cdot\mathbf{A}(\mathbf{R})\right)\\ &\quad+\frac{1}{4}\left[\frac{1}{\mathbf{P}^2}\mathbf{A}^2(\mathbf{R})+\mathbf{A}^2(\mathbf{R})\frac{1}{\mathbf{P}^2}\right]+\ldots\end{aligned} \tag{23}$$

The last term in the right hand side of (21) involves just half the commutators obtained in pushing the momentum operators to the left or the right. As explained in Section 2.1 equation (3), they are simply given by

$$\begin{aligned}&[\text{commutators}] =\\ &\frac{i}{2}\hbar Asym\nabla_{R_l}\nabla_{P^l}\left[U(\mathbf{P},\mathbf{R})H_0(\mathbf{P},\mathbf{R})U^+(\mathbf{P},\mathbf{R})\right].\end{aligned} \tag{24}$$

As a consequence, we can write

$$\begin{aligned}&U(\mathbf{P},\mathbf{R})H_0(\mathbf{P},\mathbf{R})U^+(\mathbf{P},\mathbf{R}) = \varepsilon(\mathbf{P},\mathbf{R})\\ &\quad+\frac{i}{2}\hbar Asym\nabla_{R_l}\nabla_{P^l}\left[U(\mathbf{P},\mathbf{R})H_0(\mathbf{P},\mathbf{R})U^+(\mathbf{P},\mathbf{R})\right].\end{aligned} \tag{25}$$

A lengthy but straightforward computation presented in Appendix A leads to

$$\begin{aligned}&U(\mathbf{P},\mathbf{R})H_0(\mathbf{P},\mathbf{R})U^+(\mathbf{P},\mathbf{R}) =\\ &\varepsilon(\mathbf{P},\mathbf{R}) + \frac{1}{2}\left[\mathcal{A}_{R_l}\nabla_{R^l}\varepsilon(\mathbf{P},\mathbf{R}) + \nabla_{R^l}\varepsilon(\mathbf{P},\mathbf{R})\mathcal{A}_{R_l}\right]\\ &+\frac{1}{2}\left[\mathcal{A}_{P^l}\nabla_{P_l}\varepsilon(\mathbf{P},\mathbf{R}) + \nabla_{P_l}\varepsilon(\mathbf{P},\mathbf{R})\mathcal{A}_{P^l}\right]\\ &-\frac{i}{2\hbar}\left[\varepsilon(\mathbf{P},\mathbf{R}),\mathcal{A}_{P^l}\right]\mathcal{A}_{R_l} + \frac{i}{2\hbar}\left[\varepsilon(\mathbf{P},\mathbf{R}),\mathcal{A}_{R_l}\right]\mathcal{A}_{P^l}\\ &+\frac{i}{2\hbar}\left[\mathcal{A}_{R_l},\mathcal{A}_{P^l}\right]\varepsilon(\mathbf{P},\mathbf{R}).\end{aligned} \tag{26}$$

To end up with the quasi-diagonalization, we have to add the expression

$$X\varepsilon(\mathbf{P},\mathbf{R}) + \varepsilon(\mathbf{P},\mathbf{R})X^+. \tag{27}$$

Given the expression obtained previously for X, we have thus

$$\begin{aligned}&X\varepsilon(\mathbf{P},\mathbf{R}) + \varepsilon(\mathbf{P},\mathbf{R})X^+ =\\ &-\frac{i}{4\hbar}\left[\mathcal{A}_{R_l},\mathcal{A}_{P^l}\right]\varepsilon(\mathbf{P},\mathbf{R}) - \frac{i}{4\hbar}\varepsilon(\mathbf{P},\mathbf{R})\left[\mathcal{A}_{R_l},\mathcal{A}_{P^l}\right].\end{aligned} \tag{28}$$

We can thus ultimately write the diagonalization process as:

$$\begin{aligned}&\left[(U(\mathbf{P},\mathbf{R})+XU)H_0(\mathbf{P},\mathbf{R})\left(U^+(\mathbf{P},\mathbf{R})+U^+X^+\right)\right] =\\ &\varepsilon(\mathbf{P},\mathbf{R}) + \frac{1}{2}\left[\mathcal{A}_{R_l}\nabla_{R^l}\varepsilon(\mathbf{P},\mathbf{R}) + \nabla_{R^l}\varepsilon(\mathbf{P},\mathbf{R})\mathcal{A}_{R_l}\right]\\ &+\frac{1}{2}\left[\mathcal{A}_{P^l}\nabla_{P_l}\varepsilon(\mathbf{P},\mathbf{R}) + \nabla_{P_l}\varepsilon(\mathbf{P},\mathbf{R})\mathcal{A}_{P^l}\right]\\ &-\frac{i}{2\hbar}\left[\varepsilon(\mathbf{P},\mathbf{R}),\mathcal{A}_{P^l}\right]\mathcal{A}_{R_l} + \frac{i}{2\hbar}\left[\varepsilon(\mathbf{P},\mathbf{R}),\mathcal{A}_{R_l}\right]\mathcal{A}_{P^l}\\ &+\frac{i}{4\hbar}\left[\left[\mathcal{A}_{R_l},\mathcal{A}_{P^l}\right],\varepsilon(\mathbf{P},\mathbf{R})\right].\end{aligned} \tag{29}$$

Let us conclude this section by noting that our transformation is not a diagonalization at order $\hbar$, since it includes non diagonal contributions of order $\hbar$ through the Berry connections $\mathcal{A}_{R_l}$ and $\mathcal{A}_{P^l}$, justifying the name quasi-diagonalization. However, the next paragraph will show that these non diagonal terms are only an artifact. Actually, projecting our transformed Hamiltonian on the diagonal will in fact yield the genuine diagonalization.

2.5 The "exact" semiclassical diagonalization

As mentioned in Section 2.2, our choice of transformation $U(\mathbf{P},\mathbf{R})+XU$ is somewhat arbitrary, however it is sufficient to perform the exact diagonalization as shown in the present paragraph. Actually, since $U(\mathbf{P},\mathbf{R})$ diagonalizes H_0 at zeroth order in $\hbar$, it reasonable to consider an-unknown- true diagonalization unitary operator $U_1(\mathbf{P},\mathbf{R})$ reducing to $U(\mathbf{P},\mathbf{R})$ at zeroth order in $\hbar$ of the following form $U_1(\mathbf{P},\mathbf{R}) = U(\mathbf{P},\mathbf{R}) + \hbar(\ldots)$. As a consequence $U_1(\mathbf{P},\mathbf{R})$ and $U(\mathbf{P},\mathbf{R})+XU$ are equal at the zeroth order in $\hbar$ and the difference

$$\delta XU \equiv U_1(\mathbf{P},\mathbf{R}) - U(\mathbf{P},\mathbf{R}) - XU \tag{30}$$

is of order $\hbar$. Moreover $U_1(\mathbf{P},\mathbf{R})$ and $U(\mathbf{P},\mathbf{R})+XU$ being both unitary, δX is easily seen to be antihermitian. As a direct consequence, one can check that the difference between the exact diagonalization and our approximate one

$$\begin{aligned}&\left[(U(\mathbf{P},\mathbf{R})+XU)H_0(\mathbf{P},\mathbf{R})\left(U^+(\mathbf{P},\mathbf{R})+U^+X^+\right)\right]\\ &\quad-\left[U_1(\mathbf{P},\mathbf{R})H_0(\mathbf{P},\mathbf{R})U_1^+(\mathbf{P},\mathbf{R})\right]\end{aligned} \tag{31}$$

is equal to

$$\left[\delta X,\varepsilon(\mathbf{P},\mathbf{R})\right]. \tag{32}$$

Given that $\varepsilon(\mathbf{P},\mathbf{R})$ is diagonal, this last term is always non diagonal. As a consequence, if we project on the diagonal the difference between our transformations, one gets 0

$$P\left[\left(U\left(\mathbf{P},\mathbf{R}\right)+XU\right)H_0\left(\mathbf{P},\mathbf{R}\right)\left(U^+\left(\mathbf{P},\mathbf{R}\right)\right.\right.$$
$$\left.\left.+U^+X^+\right)\right]-P\left[U_1\left(\mathbf{P},\mathbf{R}\right)H_0\left(\mathbf{P},\mathbf{R}\right)U_1^+\left(\mathbf{P},\mathbf{R}\right)\right]=0. \quad (33)$$

Here we have denoted $P[...]$ the projection on the diagonal.

Now, given that $U_1\left(\mathbf{P},\mathbf{R}\right)H_0\left(\mathbf{P},\mathbf{R}\right)U_1^+\left(\mathbf{P},\mathbf{R}\right)$ is truly diagonal, one has

$$P\left[U_1\left(\mathbf{P},\mathbf{R}\right)H_0\left(\mathbf{P},\mathbf{R}\right)U_1^+\left(\mathbf{P},\mathbf{R}\right)\right]=$$
$$U_1\left(\mathbf{P},\mathbf{R}\right)H_0\left(\mathbf{P},\mathbf{R}\right)U_1^+\left(\mathbf{P},\mathbf{R}\right) \quad (34)$$

so that ultimately

$$P\left[\left(U\left(\mathbf{P},\mathbf{R}\right)+XU\right)H_0\left(\mathbf{P},\mathbf{R}\right)\left(U^+\left(\mathbf{P},\mathbf{R}\right)+U^+X^+\right)\right]=$$
$$U_1\left(\mathbf{P},\mathbf{R}\right)H_0\left(\mathbf{P},\mathbf{R}\right)U_1^+\left(\mathbf{P},\mathbf{R}\right). \quad (35)$$

We can therefore conclude, that the projection of our quasi-diagonalized Hamiltonian, by eliminating thus the non diagonal parts, is in fact the genuine diagonalized Hamiltonian at order $\hbar$.

2.6 The diagonal Hamiltonian

From the previous discussion we understand that the genuine semiclassical diagonal Hamiltonian H_D is simply given by the projection on the diagonal of the Hamiltonian equation (29):

$$\begin{aligned}H_D=&P\left[\left(U\left(\mathbf{P},\mathbf{R}\right)+XU\right)H_0(\mathbf{P},\mathbf{R})\left(U^+(\mathbf{P},\mathbf{R})+U^+X^+\right)\right]\\=&\varepsilon\left(\mathbf{P},\mathbf{R}\right)+\frac{1}{2}\left[A_{R_l}\nabla_{R_l}\varepsilon\left(\mathbf{P},\mathbf{R}\right)+\nabla_{R_l}\varepsilon\left(\mathbf{P},\mathbf{R}\right)A_{R_l}\right]\\&+\frac{1}{2}\left[A_{P_l}\nabla_{P_l}\varepsilon\left(\mathbf{P},\mathbf{R}\right)+\nabla_{P_l}\varepsilon\left(\mathbf{P},\mathbf{R}\right)A_{P_l}\right]\\&+P\left[-\frac{i}{2\hbar}\left[\varepsilon(\mathbf{P},\mathbf{R}),\mathcal{A}_{P_l}\right]\mathcal{A}_{R_l}+\frac{i}{2\hbar}\left[\varepsilon\left(\mathbf{P},\mathbf{R}\right),\mathcal{A}_{R_l}\right]\mathcal{A}_{P_l}\right]\end{aligned} \quad (36)$$

where we introduced the notation $A=P[\mathcal{A}]$. This Hamiltonian can be rewritten

$$\begin{aligned}H_D=&\varepsilon\left(\mathbf{P}+A_P,\mathbf{R}+A_R\right)+\frac{i}{2\hbar}P\left[\left[\varepsilon\left(\mathbf{P},\mathbf{R}\right),\mathcal{A}_{R_l}\right]\mathcal{A}_{P_l}\right.\\&\left.-\left[\varepsilon\left(\mathbf{P},\mathbf{R}\right),\mathcal{A}_{P_l}\right]\mathcal{A}_{R_l}\right]\\\simeq&\varepsilon\left(\mathbf{p},\mathbf{r}\right)+\frac{i}{2\hbar}P\left[\left[\varepsilon\left(\mathbf{p},\mathbf{r}\right),\mathcal{A}_{R_l}\right]\mathcal{A}_{P_l}-\left[\varepsilon\left(\mathbf{p},\mathbf{r}\right),\mathcal{A}_{P_l}\right]\mathcal{A}_{R_l}\right]\end{aligned} \quad (37)$$

where we have defined the projected dynamical operators

$$\begin{aligned}\mathbf{r}&=P\left[\left(U\left(\mathbf{P},\mathbf{R}\right)\right)\mathbf{R}U^+(\mathbf{P},\mathbf{R})\right]=\mathbf{R}+A_{\mathbf{R}}\\\mathbf{p}&=P\left[U\left(\mathbf{P},\mathbf{R}\right)\mathbf{P}U^+(\mathbf{P},\mathbf{R})\right]=\mathbf{P}+A_{\mathbf{P}}.\end{aligned} \quad (38)$$

The non-canonical dynamical variables $(\mathbf{p},\mathbf{r})$ have corrections of order $\hbar$ through the presence of the Berry connections.

2.7 The equations of motion

Given the Hamiltonian derived in the previous subsection, the equations of motion can now be easily derived. As usual [3,15] the dynamics has to be considered, not for the usual position $\mathbf{R}$ and momentum $\mathbf{P}$, but rather for the projected variables $\mathbf{r}$ and $\mathbf{p}$. These new dynamical operators which naturally appear in our diagonalization process at the $\hbar$ order have components which do not commute any more. Actually

$$\begin{aligned}\left[r_i,r_j\right]&=i\Theta_{ij}^{rr}=i\hbar\left(\nabla_{P_i}A_{R_j}-\nabla_{P_j}A_{R_i}\right)+\left[A_{R_j},A_{R_i}\right]\\\left[p_i,p_j\right]&=i\Theta_{ij}^{pp}=-i\hbar\left(\nabla_{R_i}A_{P_j}-\nabla_{R_j}A_{P_i}\right)+\left[A_{P_i},A_{P_j}\right]\\\left[r_i,p_j\right]&=i\hbar\delta_{ij}+i\Theta_{ij}^{rp}=i\hbar\delta_{ij}+i\hbar\left(\nabla_{R_j}A_{R_i}+\nabla_{P_i}A_{P_j}\right)\\&\quad+\left[A_{R_i},A_{P_j}\right]\end{aligned} \quad (39)$$

the Θ_{ij} being the so called non-Abelian Berry curvatures. They are of order $\hbar^2$ as lowest order corrections to the commutators, but will actually induce semiclassical corrections of order $\hbar$ to the equations of motion. Indeed the one band-Hamiltonian equation (37) yields directly to general equations of motion for $\mathbf{r}$, $\mathbf{p}$:

$$\begin{aligned}\dot{\mathbf{r}}=&\frac{i}{\hbar}\left[\mathbf{r},\varepsilon\left(\mathbf{p},\mathbf{r}\right)\right]+\frac{i}{\hbar}\left[\mathbf{r},\frac{i}{2\hbar}P\left[\left[\varepsilon\left(\mathbf{p},\mathbf{r}\right),\mathcal{A}_{R_l}\right]\mathcal{A}_{P^l}\right.\right.\\&\left.\left.-\left[\varepsilon\left(\mathbf{p},\mathbf{r}\right),\mathcal{A}_{P^l}\right]\mathcal{A}_{R_l}\right]\right]\\\dot{\mathbf{p}}=&\frac{i}{\hbar}\left[\mathbf{P},\varepsilon\left(\mathbf{p},\mathbf{r}\right)\right]+\frac{i}{\hbar}\left[\mathbf{p},\frac{i}{2\hbar}P\left[\left[\varepsilon\left(\mathbf{p},\mathbf{r}\right),\mathcal{A}_{R_l}\right]\mathcal{A}_{P^l}\right.\right.\\&\left.\left.-\left[\varepsilon\left(\mathbf{p},\mathbf{r}\right),\mathcal{A}_{P^l}\right]\mathcal{A}_{R_l}\right]\right]\end{aligned} \quad (40)$$

where the commutators can be computed through the previous commutation rules equation (39). The last term in each equation represents a contribution of "magnetization" type (see the following applications) and has the advantage to present this general form whatever the system initially considered. However, to put some flesh on these equations, we now turn to several examples covered by our formalism.

3 Application 1: the Dirac electron

To apply our previous formalism, we will consider two cases of Dirac Hamiltonians: the electromagnetic field and the static symmetrical gravitational field. These two cases have already been treated by different methods ([4,17, 18]), but in the second case (gravitational field) references to Berry phases was made for the first time in [16].

3.1 The Dirac electron in an electromagnetic field

The diagonalization of the Dirac Hamiltonian in the presence of an electromagnetic field is a difficult problem which was solved only approximately in the nonrelativistic limit in an m^{-1} expansion. Another approach consists in diagonalizing the Hamiltonian at the semiclassical order as was done in [4] using an approximate Foldy-Wouthuysen transformation [19]. From the semiclassical

$$\mathcal{A}_{R_i} = i\hbar U\nabla_{P_i}U^+ = \hbar\frac{i\alpha\cdot(\mathbf{P}-\mathbf{A}(\mathbf{R}))\,P_i\beta + i\beta E(E+m)\alpha_i - E\left(\mathbf{\Sigma}\times(\mathbf{P}-\mathbf{A}(\mathbf{R}))\right)_i}{2E^2(E+m)} \tag{45}$$

Hamiltonian the equations of motion were derived showing a topological spin-transport effect due to the presence of the Berry phases. Here we propose to apply our general formalism for the semiclassical diagonalization of the Dirac Hamiltonian to show the effectiveness of our general method. It is worth noticing that the method developed above has now to be adapted as we will transform the Dirac Hamiltonian into a (2×2) block-diagonal matrix (due to the spin degree of freedom). However, it is easy to check that, since the non diagonal components for the energy blocks are of order $\hbar$, these corrections do not impair our general formulas for the Berry phases contributions.

We thus start with the following Dirac Hamiltonian:

$$H_0(\mathbf{P},\mathbf{R}) = \alpha\cdot(\mathbf{P}-\mathbf{A}(\mathbf{R})) + \beta m + V(\mathbf{R}) \tag{41}$$

where the matrix α and β are the usual (4×4) Dirac matrices:

$$\alpha_i = \begin{pmatrix} \begin{matrix}0 & 0\\ 0 & 0\end{matrix} & \sigma_i \\ \sigma_i & \begin{matrix}0 & 0\\ 0 & 0\end{matrix}\end{pmatrix} \tag{42}$$

with σ_i the usual (2×2) Pauli matrices $(i=1,2,3)$ and

$$\beta = \begin{pmatrix} 1 & 0 & 0 & 0\\ 0 & 1 & 0 & 0\\ 0 & 0 & -1 & 0\\ 0 & 0 & 0 & -1\end{pmatrix}. \tag{43}$$

Replacing $\mathbf{R}$ by the parameter $\widetilde{\mathbf{r}}$, $\mathbf{A}(\widetilde{\mathbf{r}})$ just shifts the momentum, so that we can diagonalize $H_0(\mathbf{P},\widetilde{\mathbf{r}})$ through the well known Foldy-Wouthuysen transformation

$$U(\mathbf{P},\widetilde{\mathbf{r}}) = \frac{E+m+\beta\alpha\cdot(\mathbf{P}-\mathbf{A}(\widetilde{\mathbf{r}}))}{\sqrt{2E(E+m)}} \tag{44}$$

where $E=\sqrt{(\mathbf{P}-\mathbf{A}(\widetilde{\mathbf{r}}))^2+m^2}$. In this context, introducing the dependence in $\mathbf{R}$ we define the (non projected) Berry connections at first order in $\hbar$ as a (4×4) matrix

see equation (45) above

and

$$\mathcal{A}_{P_l} = -i\hbar U\nabla_{R_l}U^+ = \nabla_{R_l}A_k(\mathbf{R})\mathcal{A}_{R_k} \tag{46}$$

where E will now denote the symmetrized form of $E=\sqrt{(\mathbf{P}-A(\mathbf{R}))^2+m^2}$ (here at zeroth order in $\hbar$ we can consider that $\mathbf{P}$ and $\mathbf{R}$ commute). The spin matrices Σ_i $(i=1,2,3)$ are given by

$$\mathbf{\Sigma}_i = \begin{pmatrix} \sigma_i & \begin{matrix}0 & 0\\ 0 & 0\end{matrix} \\ \begin{matrix}0 & 0\\ 0 & 0\end{matrix} & \sigma_i\end{pmatrix}.$$

The general method developed in the previous section allows us to write the diagonal Hamiltonian H_D as a matrix generalization of equation (37)

$$\begin{aligned} H_D =& P\left[UH_0U^+\right] = \varepsilon(\mathbf{p},\mathbf{r}) + \frac{i}{2\hbar}P[[\varepsilon(\mathbf{p},\mathbf{r}),\mathcal{A}_{R_l}]\,\mathcal{A}_{P^l} \\ &- [\varepsilon(\mathbf{p},\mathbf{r}),\mathcal{A}_{P^l}]\,\mathcal{A}_{R_l}] + V(\mathbf{r}) \end{aligned} \tag{47}$$

where $\varepsilon(\mathbf{p},\mathbf{r})$ is the matrix $\beta\sqrt{(\mathbf{p}-\mathbf{A}(\mathbf{r}))^2+m^2}$ and $P[...]$ now projects on the diagonal blocks.

In equation (47) the operators $(\mathbf{p},\mathbf{r})$ are the physical dynamical variables satisfying the non-canonical commutations relations equation (39). Using the expressions for the Berry connections, H_D can be rewritten as:

$$\begin{aligned} H_D = \varepsilon(\mathbf{p},\mathbf{r}) + P\left[-\frac{i}{2}\left[\varepsilon(\mathbf{p},\mathbf{r}),U\nabla_{P^i}U^+\right]\varepsilon^{ijk}U\nabla_{P^j}U^+\right] \\ \times\frac{B^k(\mathbf{r})}{\hbar} + V(\mathbf{r}). \end{aligned} \tag{48}$$

Moreover, a straightforward computation shows that one can write $P\left[-\frac{i}{2}\left[\varepsilon(\mathbf{p},\mathbf{r}),U\nabla_{P^i}U^+\right]\cdot\varepsilon^{ijk}U\nabla_{P_j}U^+\right]\frac{B^k(\mathbf{r})}{\hbar} = \beta\frac{\hbar\mathbf{\Sigma}\cdot\mathbf{B}}{2E} - \beta\frac{\mathbf{L}\cdot\mathbf{B}}{E}$ at the first order in $\hbar$, where we have introduced the intrinsic angular momentum of semiclassical particles $\mathbf{L}=\mathbf{P}\times A_{\mathbf{R}}$ with $A_{\mathbf{R}}=P[\mathcal{A}_R]=\hbar\frac{(\mathbf{P}-\mathbf{A}(\mathbf{R}))\times\mathbf{\Sigma}}{2E(E+m)}$ the projection of the Berry connection on the diagonal. As a consequence, the Hamiltonian to be considered is given by

$$H_D = \varepsilon(\mathbf{p},\mathbf{r}) + \beta\frac{\hbar\mathbf{\Sigma}\cdot\mathbf{B}}{2E} - \beta\frac{\mathbf{L}\cdot\mathbf{B}}{E} + V(\mathbf{r}) \tag{49}$$

which is the Hamiltonian deduced in [4] from a different approach and which leads of course to the dynamics described in that paper.

3.2 The electron in a static gravitational field

The behavior of Dirac particles in static gravitational field is an important issue, at the crossroad of particle physics and cosmology. Different approaches for the diagonalization of the Hamiltonian lead to contradictory results in particular with regard to the existence of a dipole spin-gravity coupling [17,18]. It is not our goal to discuss this specific point but we study the semiclassical diagonalization of the Hamiltonian to get the velocity and momentum evolution. We can in particular compare our results with the article [18] where a Foldy-Wouthuysen transformation instead of a semiclassical approximation is used.

The interaction of a Dirac particle with a symmetric static gravitational field ($g_{00}=V(\mathbf{R}), g_{i0}=0, g_{ij}=\delta_{ij}F(\mathbf{R})$) is described by the Hamiltonian [18]

$$H_0 = \frac{1}{2}(\alpha\cdot\mathbf{P}F(\mathbf{R}) + F(\mathbf{R})\alpha\cdot\mathbf{P}) + \beta mV(\mathbf{r}). \tag{50}$$

$$\mathcal{A}_{R_i} = i\hbar U\nabla_{P_i}U^+ = \hbar\frac{iF^3(\mathbf{R})\alpha\cdot\mathbf{P}P_i\beta + i\beta F(\mathbf{R})E(E+mV(\mathbf{R}))\alpha_i - EF^2(\mathbf{R})\left(\mathbf{\Sigma}\times\mathbf{P}\right)_i}{2E^2\left(E+mV(\mathbf{R})\right)}$$
$$\mathcal{A}_{P_i} = -i\hbar U\nabla_{R_i}U^+ = -\hbar i\frac{m\left(\nabla_{R_i}\phi\right)\beta F^2(\mathbf{R})\alpha\cdot\mathbf{P}}{2E^2} \tag{52}$$

The Foldy Whouthuysen transformation when $\mathbf{R}$ is replaced by a parameter $\widetilde{\mathbf{r}}$ is given by

$$U\left(\mathbf{P},\widetilde{\mathbf{r}}\right) = \frac{E+mV(\widetilde{\mathbf{r}})+\beta F(\widetilde{\mathbf{r}})\alpha\cdot\mathbf{P}}{\sqrt{2E(E+mV(\widetilde{\mathbf{r}}))}} \tag{51}$$

with $E\left(\mathbf{P},\widetilde{\mathbf{r}}\right) = \sqrt{F^2(\widetilde{\mathbf{r}})\mathbf{P}^2+m^2V^2(\widetilde{\mathbf{r}})}$. This is quite the same as the free particle transformation. As a consequence introducing again the $\mathbf{R}$ dependence yields the non projected Berry connections for the position and the momentum operators:

see equation (52) above

with $\phi = V/F$, $\mathbf{\Sigma}$ the spin of the electron and the energy E is now given by $E\left(\mathbf{P},\mathbf{R}\right) = \sqrt{\left(F^2(\mathbf{R})\mathbf{P}^2+\mathbf{P}^2F^2(\mathbf{R})\right)/2+m^2V^2(\mathbf{R})}$.

The expressions for $\mathcal{A}_{R_i}$ and $\mathcal{A}_{P_i}$ allow us ultimately to define the semiclassical transformation: $U\left(\mathbf{P},\mathbf{R}\right)+\frac{i}{4\hbar}\left[\mathcal{A}_{P^l},\mathcal{A}_{R_l}\right]U\left(\mathbf{P},\mathbf{R}\right)$ and to compute the diagonal Hamiltonian

$$H_D = P\left[UH_0U^+\right] = \varepsilon\left(\mathbf{p},\mathbf{r}\right) - \beta\frac{F^3(\mathbf{r})}{2E^2}m\hbar\nabla\phi(\mathbf{r})\cdot\left(\mathbf{p}\times\mathbf{\Sigma}\right) \tag{53}$$

with $\varepsilon\left(\mathbf{p},\mathbf{r}\right) = \beta\sqrt{\left(F^2(\mathbf{r})\mathbf{p}^2+\mathbf{p}^2F^2(\mathbf{r})\right)/2+m^2V^2(\mathbf{r})}$ where the dynamical variables which are deduced from the projections $A_{\mathbf{R}} = P\left[\mathcal{A}_{\mathbf{R}}\right] = \hbar\frac{F^2\mathbf{\Sigma}\times\mathbf{P}}{2E(E+mV)}$ and $A_{\mathbf{P}} = P\left[\mathcal{A}_{\mathbf{P}}\right] = 0$ are given by

$$\mathbf{r} = \mathbf{R}+A_{\mathbf{R}} = \mathbf{R}-\hbar\frac{F^2(\mathbf{R})\mathbf{\Sigma}\times\mathbf{P}}{2E(E+mV(\mathbf{R}))}$$
$$\mathbf{p} = \mathbf{P}+A_{\mathbf{P}} = \mathbf{P}. \tag{54}$$

The commutators between these variables are thus

$$\left[r_i,r_j\right] = i\Theta_{ij}^{rr}$$
$$\left[r_i,p_j\right] = i\hbar\delta_{ij}+i\Theta_{ij}^{rp}$$
$$\left[p_i,p_j\right] = 0 \tag{55}$$

where the Berry curvatures expressed in terms of the operators $(\mathbf{p},\mathbf{r})$ are

$$\Theta_{ij}^{rr} = -\frac{\hbar^2F^3(\mathbf{r})\varepsilon^{ijk}}{2E^3\left(\mathbf{p},\mathbf{r}\right)}\left(m\phi(\mathbf{r})\mathbf{\Sigma}_k+\frac{F(\mathbf{r})\left(\mathbf{\Sigma}\cdot\mathbf{p}\right)\mathbf{p}_k}{E\left(\mathbf{p},\mathbf{r}\right)+mV(\mathbf{r})}\right)$$
$$\Theta_{ij}^{rp} = \frac{\hbar^2F^3(\mathbf{r})}{2E^3\left(\mathbf{p},\mathbf{r}\right)}m\nabla_i\phi(\mathbf{r})\left(\mathbf{\Sigma}\times\mathbf{p}\right)_j$$
$$\Theta_{ij}^{pp} = 0. \tag{56}$$

One can check, after developing $\mathbf{r}$ as a function of $\mathbf{R}$ and the Berry connection, that the Hamiltonian equation (53) coincides with the one given in [18] at order $\hbar$. This confirms also the validity of the Foldy Wouthuysen approach asserted in [18] in opposition with the transformation proposed in [17]. However our approach is more general since it does not require an expansion in V and F as done in [18]. Of course, we retrieve the result of [18] if we expand expression (56) at the leading order in F and V. Note also that when $m=0$ one recovers the Hamiltonian for the Neutrino or the photon proposed in [3,16].

To conclude this paragraph, we can derive the equations of motion with the help the noncanonical commutators between the coordinates and the spin

$$\Theta_{ij}^{r\Sigma} = \left[r_i,\Sigma_j\right] = i\hbar c^2\frac{-p_j\mathbf{\Sigma}_i+\mathbf{p}\cdot\mathbf{\Sigma}\delta_{ij}}{E\left(\mathbf{p},\mathbf{r}\right)\left(E\left(\mathbf{p},\mathbf{r}\right)+mV(\mathbf{r})\right)}$$
$$\Theta_{ij}^{p\Sigma} = \left[p_i,\Sigma_j\right] = 0. \tag{57}$$

We then can deduce the semiclassical equations of motion for the electron in a symmetric static gravitational field (by projecting on the positive energy subspace)

$$\dot{\mathbf{r}} = \left(1-\frac{\Theta^{rp}}{\hbar}\right)\nabla_{\mathbf{p}}E - \frac{1}{\hbar}\dot{\mathbf{p}}\times\Theta^{rr} + \frac{i}{\hbar}\Theta^{r\Sigma}\cdot\nabla_{\mathbf{\Sigma}}E$$
$$\dot{\mathbf{p}} = -\left(1-\frac{\Theta^{rp}}{\hbar}\right)\nabla_{\mathbf{r}}E \tag{58}$$

where we have defined the vectors Θ^{rr} through $\Theta_{ij}^{rr} = \varepsilon^{ijk}\Theta_k^{rr}$. We defined also in equation (58) the vector components $\left(\Theta^{rp}\nabla_{\mathbf{p}}E\right)_i = \Theta_{ij}^{rp}\nabla_{p^j}E$ and $\left(\Theta^{r\Sigma}\cdot\nabla_{\mathbf{\Sigma}}E\right)_i = \Theta_{ij}^{r\Sigma}\nabla_{\Sigma^j}E$. The velocity equation in equation (58) contains in particular an anomalous velocity term $\dot{\mathbf{p}}\times\Theta^{rr}/\hbar$ of order $\hbar$ which causes an additional displacement of the electrons orthogonally to the momentum $\mathbf{p}$. The dynamics of the system must be completed by the spin dynamics which is

$$\hbar\dot{\mathbf{\Sigma}} = \frac{mV(\mathbf{r})\hbar}{\varepsilon(\varepsilon+mV(\mathbf{r}))}\mathbf{\Sigma}\times\left(\nabla V(\mathbf{r})\times\mathbf{P}\right) - \frac{\hbar}{\varepsilon}\mathbf{\Sigma}\times\left(\nabla F(\mathbf{r})\times\mathbf{P}\right). \tag{59}$$

Although the position dynamics differs from the one of obtained in [18], due to our choice for the physical position operator ($\mathbf{r}$ instead of $\mathbf{R}$), one can show that the equations for $\dot{\mathbf{p}}$ and $\hbar\dot{\mathbf{\Sigma}}$ reduce to [18] in the case of weak fields.

4 Application 2: the electron in a periodic potential

This application has already been independently studied in [15], and is easily recovered by the present general setup. The purpose is to find the semiclassical Hamiltonian for an electron in a periodic potential facing an electromagnetic field. This topic was also already dealt with in [9] in the context of wave packets dynamics. We will show that the semiclassical equations of motions which are very essential in the solid state physics context must be corrected by Berry phases terms. To apply our formalism, consider an electron in an crystal lattice perturbated by the presence of an external electromagnetic field. As is usual, we express the total magnetic field as the sum of a constant field $\mathbf{B}$ and small nonuniform part $\delta\mathbf{B}(\mathbf{R})$. The Schrödinger equation reads $(H_0 - e\phi(\mathbf{R}))\,\Psi(\mathbf{R}) = E\Psi(\mathbf{R})$ with H_0 the magnetic contribution (ϕ being the electric potential) which reads

$$H_0 = \left(\frac{\mathbf{P}}{2m} + e\mathbf{A}(\mathbf{R}) + e\delta\mathbf{A}(\mathbf{R})\right)^2 + V(\mathbf{R}),\ \mathbf{P} = -i\hbar\nabla \tag{60}$$

where $\mathbf{A}(\mathbf{R})$ and $\delta\mathbf{A}(\mathbf{R})$ are the vectors potential of the homogeneous and inhomogeneous magnetic field, respectively, and $V(\mathbf{R})$ the periodic potential. The large constant part $\mathbf{B}$ is chosen such that the magnetic flux through a unit cell is a rational fraction of the flux quantum h/e. The advantage of such a decomposition is that for $\delta\mathbf{A}(\mathbf{R}) = 0$ the magnetic translation operators $\mathbf{T}(\mathbf{b}) = \exp(i\mathbf{K}\cdot\mathbf{b})$ defined in Appendix B, with $\mathbf{K}$ the generator of translation, are commuting quantities allowing to exactly diagonalize the Hamiltonian and to treat $\delta\mathbf{A}(\mathbf{R})$ as a small perturbation. The state space of the Bloch electron in the periodic zone scheme [21] is spanned by the basis vectors of plane waves $|n,\mathbf{k}\rangle = |\mathbf{k}\rangle\otimes|n\rangle$ with n corresponding to a band index and $\mathbf{k}$ vary in R^3. The state $|n\rangle$ can be seen as a canonical base vector $|n\rangle = (0...010...0...)$ (with 1 at the nth position) such that $U^+(\mathbf{k})\,|n\rangle = |u_n(\mathbf{k})\rangle$ with $|u_n(\mathbf{k})\rangle$ the periodic part (in space) of the magnetic Bloch waves [9,22]. In this representation $\mathbf{K}\,|n,\mathbf{k}\rangle = \mathbf{k}\,|n,\mathbf{k}\rangle$ and consequently the position operator is $\mathbf{R} = i\partial/\partial\mathbf{k}$, implying the canonical commutation relations $\left[\mathbf{R}_i, \mathbf{K}_j\right] = i\delta_{ij}$.

We first perform the diagonalization of the Hamiltonian in equation (60) for $\delta\mathbf{A} = 0$ by diagonalizing simultaneously H_0 and the magnetic translation operators $\mathbf{T}$. The diagonalization is performed as follows: start with an arbitrary basis of eigenvectors of $\mathbf{T}$. As explained in Appendix B, in this basis H_0 can be seen as a square matrix with operators entries. H_0 is diagonalized through a unitary matrix $U(\mathbf{K})$ which should depend only on $\mathbf{K}$ (since U should leave $\mathbf{K}$ invariant, i.e., $\mathbf{U}\mathbf{K}U^+ = \mathbf{K}$) and whose precise expression is not necessary for the derivation of the equations of motion, such that $UHU^+ = \mathcal{E}(\mathbf{K}) - e\phi(U\mathbf{R}U^+)$, where $\mathcal{E}(\mathbf{K})$ is the diagonal energy matrix made of elements $\mathcal{E}_n(\mathbf{K})$ with n the band index (i.e. the diagonal representation of H_0).

Now, to add a perturbation $\delta A(\mathbf{R})$ as in [15], that breaks the translational symmetry, we have to replace $\mathbf{K}$ in all expressions by

$$\tilde{\mathbf{K}} = \mathbf{K} + e\frac{\delta A(\mathbf{R})}{\hbar} \tag{61}$$

and as the flux $\delta\mathbf{B}$ on a plaquette is not a rational multiple of the flux quantum, we cannot diagonalize simultaneously its components $\tilde{K}_i$ since they do not commute anymore. Actually

$$\hbar[\tilde{K}^i, \tilde{K}^j] = -ie\varepsilon^{ijk}\delta B_k(\mathbf{R}). \tag{62}$$

As a consequence of this non-commutativity, we just aim at quasi-diagonalizing our Hamiltonian at the semiclassical order (with accuracy $\hbar$). To do that we replace $U(\mathbf{K})$ by $U\left(\tilde{\mathbf{K}}\right)$, so that the non projected Berry connections are $\mathcal{A}_{R_i} = iU\nabla_{\tilde{K}_i}U^+$ and $\mathcal{A}_{K_l} = \nabla_{R_l}\delta A_k(\mathbf{R})\mathcal{A}_{R_k}$. From these we can define the nth intraband position and momentum operators $\mathbf{r}_n = \mathbf{R} + A_n$ and $\tilde{\mathbf{k}}_n \simeq \tilde{\mathbf{K}} - eA_n(\tilde{\mathbf{k}}_n) \times \delta\mathbf{B}(\mathbf{r}_n)/\hbar + O(\hbar)$ with $A_n = P_n(U\nabla_{\tilde{\mathbf{K}}}U^+)$ the projection of the Berry connection on the chosen nth Band [15]. It can be readily seen that the matrix elements of A_n can be written $A_n(\mathbf{k}) = i\langle u_n(\mathbf{k})|\,\nabla_{\mathbf{k}}\,|u_n(\mathbf{k})\rangle$ (see also Ref. [22] for the derivation of the position operator in the diagonal representation). What is totally new here is the transformation on the momentum operator $\tilde{k}_n$ which get also a Berry connection correction.

Using our results of Section 2, the full Hamiltonian equation (60) can thus be diagonalized through the transformation $U(\tilde{\mathbf{K}}) + \frac{i}{4\hbar}\left[\mathcal{A}_{R_l}, \mathcal{A}_{P^l}\right]U(\tilde{\mathbf{K}})$ plus a projection on the chosen nth Band as it is usual in solid state physics (the so-called one band approximation)

$$\begin{aligned}
&P_n\left[U\left(\tilde{\mathbf{K}}\right)HU^+\left(\tilde{\mathbf{K}}\right)\right] = \\
&\quad P_n\left[\mathcal{E}\left(\tilde{\mathbf{k}}\right) - \frac{i}{4}\left[\mathcal{E}(\mathbf{K}), U\nabla_{K_i}U^+\right]\varepsilon^{ijk}\frac{\delta B^k(\mathbf{r})}{\hbar}U\nabla_{K_j}U^+\right. \\
&\qquad\left. - \frac{i}{4}U\nabla_{K_j}U^+\left[\mathcal{E}(\mathbf{K}), U\nabla_{K_i}U^+\right]\varepsilon^{ijk}\frac{\delta B^k(\mathbf{r})}{\hbar}\right] \\
&\qquad = \mathcal{E}_n\left(\tilde{\mathbf{k}}_n\right) - \mathcal{M}(\tilde{\mathbf{K}}).\delta\mathbf{B}(\mathbf{r}_n) + O(\hbar^2).
\end{aligned} \tag{63}$$

where the energy levels $\mathcal{E}_n\left(\tilde{\mathbf{k}}_n\right)$ are the same as $\mathcal{E}_n(\mathbf{K})$ with $\tilde{\mathbf{k}}_n$ replacing $\mathbf{K}$. The magnetization $\mathcal{M}(\tilde{\mathbf{K}}) = P_n(\frac{ie}{2\hbar}\left[\mathcal{E}(\tilde{\mathbf{K}}), \mathcal{A}(\tilde{\mathbf{K}})\right] \times \mathcal{A}(\tilde{\mathbf{K}}))$ can be written under the usual form [22] in the $(\mathbf{k}, n)$ representation

$$\mathcal{M}^i_{nn} = \frac{ie}{2\hbar}\varepsilon^{ijk}\sum_{n'\neq n}(\mathcal{E}_n - \mathcal{E}_{n'})(\mathcal{A}_j)_{nn'}(\mathcal{A}_k)_{n'n}.$$

We mention that this magnetization (the orbital magnetic moment of Bloch electrons), has been obtained previously in the context of electron wave packets dynamics [9].

From the expression of the energy equation (63) we can deduce the equations of motion (with the band index n now omitted)

$$\begin{aligned}
\dot{\mathbf{r}} &= \partial E(\tilde{\mathbf{k}})/\hbar\partial\tilde{\mathbf{k}} - \dot{\tilde{\mathbf{k}}} \times \Theta(\tilde{\mathbf{k}}) \\
\hbar\dot{\tilde{\mathbf{k}}} &= -e\mathbf{E} - e\dot{\mathbf{r}} \times \delta\mathbf{B}(\mathbf{r}) - \mathbf{M}\partial\delta\mathbf{B}/\partial\mathbf{r}
\end{aligned} \tag{64}$$

where $\left[r^i, r^j\right] = i\Theta^{ij}(\tilde{\mathbf{k}})$ with $\Theta^{ij}(\tilde{\mathbf{k}}) = \partial^i\mathcal{A}^j(\tilde{\mathbf{k}}) - \partial^j\mathcal{A}^i(\tilde{\mathbf{k}})$ the Berry curvature.

As explained in [15] these equations are the same as the one derived in [9] from a completely different formalism.

$$
\begin{aligned}
\frac{i}{2}\hbar Asym\nabla_{R_l}\nabla_{P^l}\left[U(\mathbf{P},\mathbf{R})H_0(\mathbf{P},\mathbf{R})U^+(\mathbf{P},\mathbf{R})\right] = {} & \frac{i}{2}\hbar Asym\nabla_{R_l}\nabla_{P^l}\left[U(\mathbf{P},\mathbf{R})\right]H_0(\mathbf{P},\mathbf{R})U^+(\mathbf{P},\mathbf{R}) \\
& + \frac{i}{2}\hbar U(\mathbf{P},\mathbf{R})Asym\nabla_{R_l}\nabla_{P^l}\left[H_0(\mathbf{P},\mathbf{R})\right]U^+(\mathbf{P},\mathbf{R}) \\
& + \frac{i}{2}\hbar U(\mathbf{P},\mathbf{R})H_0(\mathbf{P},\mathbf{R})Asym\nabla_{R_l}\nabla_{P^l}\left[U^+(\mathbf{P},\mathbf{R})\right] \\
& + \frac{i}{2}\hbar\nabla_{R_l}U(\mathbf{P},\mathbf{R})\nabla_{P^l}H_0(\mathbf{P},\mathbf{R})U^+(\mathbf{P},\mathbf{R}) \\
& - \frac{i}{2}\hbar\nabla_{P^l}U(\mathbf{P},\mathbf{R})\nabla_{R_l}H_0(\mathbf{P},\mathbf{R})U^+(\mathbf{P},\mathbf{R}) \\
& - \frac{i}{2}\hbar U(\mathbf{P},\mathbf{R})\nabla_{P^l}H_0(\mathbf{P},\mathbf{R})\nabla_{R_l}U^+(\mathbf{P},\mathbf{R}) \\
& + \frac{i}{2}\hbar\nabla_{P^l}U(\mathbf{P},\mathbf{R})\nabla_{R_l}H_0(\mathbf{P},\mathbf{R})\nabla_{P^l}U^+(\mathbf{P},\mathbf{R}) \\
& + \frac{i}{2}\hbar\nabla_{R_l}U(\mathbf{P},\mathbf{R})H_0(\mathbf{P},\mathbf{R})\nabla_{P^l}U^+(\mathbf{P},\mathbf{R}) \\
& - \frac{i}{2}\hbar\nabla_{P^l}U(\mathbf{P},\mathbf{R})H_0(\mathbf{P},\mathbf{R})\nabla_{R_l}U^+(\mathbf{P},\mathbf{R}) \qquad \text{(A.1)}
\end{aligned}
$$

5 Conclusion

Some recent applications of semiclassical methods to several branches of Physics, such as spintronics or solid state physics have shown the relevance of Berry Phases contributions to the dynamics of a system. However, these progresses called for a rigorous Hamiltonian treatment that would allow for deriving naturally the role of the Berry phase.

This paper has been devoted to derive a semiclassical diagonalization method for a broad class of quantum systems, including the electron in a periodic potential and the Dirac Hamiltonian. Doing so, we have exhibited a general pattern for this class of systems implying the role of the Berry phases both for the position and the momentum. In such a context, the coordinates and momenta algebra are no longer commutative, and the dynamical equations for these variables directly include the influence of Berry phases through the parameters of noncommutativity (Berry curvatures) and through an abstract magnetization term. Applications of our formalism consider the Dirac electron in an electromagnetic field, or in a particular case of static gravitational field, as well as the electron in a periodic potential. Our results are promising and indicate that our method will probably apply to several other systems.

The authors wish to thank Aileen Lotz for a critical reading of the manuscript.

Appendix A

To start our computation, we will use the formula given in the preliminary:

see equation (A.1) above

Let us first remark that $H_0(\mathbf{P},\mathbf{R})$ is already symmetrized in $\mathbf{R}$ and $\mathbf{P}$. As a consequence $Asym\nabla_{R_l}\nabla_{P^l}\left[H_0(\mathbf{P},\mathbf{R})\right] = 0$. Actually, remember that the asymmetrization term was the sum of commutators obtained by pushing the momentum half on the left and half on the right. For the same reason

$$
\begin{aligned}
Asym\nabla_{R_l}\nabla_{P^l}[U(\mathbf{P},\mathbf{R})] &= Asym\nabla_{R_l}\nabla_{P^l}\left[U^+(\mathbf{P},\mathbf{R})\right] \\
&= 0. \qquad \text{(A.2)}
\end{aligned}
$$

Now, we introduce the transformed variables and the non projected Berry Phases at order $\hbar$:

$$
\begin{aligned}
\mathbf{r} &= (U(\mathbf{P},\mathbf{R})+X)\,\mathbf{R}\left(U^+(\mathbf{P},\mathbf{R})+X^+\right) \\
&\simeq \mathbf{R} + \left[i\hbar U(\mathbf{P},\mathbf{R})\nabla_P U^+(\mathbf{P},\mathbf{R})\right] = \mathbf{R} + \mathcal{A}_R \\
\mathbf{p} &= U(\mathbf{P},\mathbf{R})\,\mathbf{P}U^+(\mathbf{P},\mathbf{R}) \\
&\simeq \mathbf{P} - \left[i\hbar U(\mathbf{P},\mathbf{R})\nabla_R U^+(\mathbf{P},\mathbf{R})\right] = \mathbf{P} + \mathcal{A}_P. \qquad \text{(A.3)}
\end{aligned}
$$

Before going further, we can find some relations on the Berry connections. Given that:

$$
U(\mathbf{P},\mathbf{R})\,U^+(\mathbf{P},\mathbf{R}) = 1 \qquad \text{(A.4)}
$$

at the zeroth order in $\hbar$ we have the following relations at the first order in $\hbar$:

$$
\begin{aligned}
\mathcal{A}_R &= i\hbar U(\mathbf{P},\mathbf{R})\nabla_P U^+(\mathbf{P},\mathbf{R}) \\
&= -i\nabla_P U(\mathbf{P},\mathbf{R})\,U^+(\mathbf{P},\mathbf{R}) \\
\mathcal{A}_P &= -i\hbar U(\mathbf{P},\mathbf{R})\nabla_R U^+(\mathbf{P},\mathbf{R}) \\
&= i\nabla_R U(\mathbf{P},\mathbf{R})\,U^+(\mathbf{P},\mathbf{R}). \qquad \text{(A.5)}
\end{aligned}
$$

Using these results as well as $\nabla_{P^l} = \frac{-i}{\hbar}[R_l,]$, $\nabla_{R_l} = \frac{i}{\hbar}\left[P^l,\right]$ and inserting the operators $U(\mathbf{P},\mathbf{R})$ and

$$\begin{aligned}
\frac{i}{2}\hbar Asym\nabla_{R_l}\nabla_{P^l}\left[U(\mathbf{P},\mathbf{R})H_0(\mathbf{P},\mathbf{R})U^+(\mathbf{P},\mathbf{R})\right] &= \frac{-i}{2\hbar}\mathcal{A}_{P^l}U(\mathbf{P},\mathbf{R})\left[R_l,H_0(\mathbf{P},\mathbf{R})\right]U^+(\mathbf{P},\mathbf{R})\\
&+\frac{i}{2\hbar}\mathcal{A}_{R_l}U(\mathbf{P},\mathbf{R})\left[P_l,H_0(\mathbf{P},\mathbf{R})\right]U^+(\mathbf{P},\mathbf{R})\\
&-\frac{i}{2\hbar}U(\mathbf{P},\mathbf{R})\left[R_l,H_0(\mathbf{P},\mathbf{R})\right]U^+(\mathbf{P},\mathbf{R})\mathcal{A}_{P^l}\\
&+\frac{i}{2\hbar}U(\mathbf{P},\mathbf{R})\left[P^l,H_0(\mathbf{P},\mathbf{R})\right]U^+(\mathbf{P},\mathbf{R})\mathcal{A}_{R_l}\\
&-\frac{i}{2\hbar}\mathcal{A}_{P^l}\varepsilon(\mathbf{P},\mathbf{R})\mathcal{A}_{R_l}+\frac{i}{2\hbar}\mathcal{A}_{R_l}\varepsilon(\mathbf{P},\mathbf{R})\mathcal{A}_{P^l}\\
&=\frac{-i}{2\hbar}\mathcal{A}_{P^l}\left[r_l,\varepsilon(\mathbf{P},\mathbf{R})\right]+\frac{i}{2\hbar}\mathcal{A}_{R_l}\left[p^l,\varepsilon(\mathbf{P},\mathbf{R})\right]-\frac{i}{2\hbar}\left[r_l,\varepsilon(\mathbf{P},\mathbf{R})\right]\mathcal{A}_{P^l}\\
&+\frac{i}{2\hbar}\left[p^l,\varepsilon(\mathbf{P},\mathbf{R})\right]\mathcal{A}_{R_l}-\frac{i}{2\hbar}\mathcal{A}_{P^l}\varepsilon(\mathbf{P},\mathbf{R})\mathcal{A}_{R_l}+\frac{i}{2\hbar}\mathcal{A}_{R_l}\varepsilon(\mathbf{P},\mathbf{R})\mathcal{A}_{P^l}\\
&=\frac{1}{2}\left[\mathcal{A}_{R_l}\nabla_{R^l}\varepsilon(\mathbf{P},\mathbf{R})+\nabla_{R^l}\varepsilon(\mathbf{P},\mathbf{R})\mathcal{A}_{R_l}\right]\\
&+\frac{1}{2}\left[\mathcal{A}_{P^l}\nabla_{P_l}\varepsilon(\mathbf{P},\mathbf{R})+\nabla_{P_l}\varepsilon(\mathbf{P},\mathbf{R})\mathcal{A}_{P^l}\right]-\frac{i}{2\hbar}\mathcal{A}_{P^l}\left[\mathcal{A}_{R_l},\varepsilon(\mathbf{P},\mathbf{R})\right]\\
&-\frac{i}{2\hbar}\left[\mathcal{A}_{R_l},\varepsilon(\mathbf{P},\mathbf{R})\right]\mathcal{A}_{P^l}+\frac{i}{2\hbar}\mathcal{A}_{R_l}\left[\mathcal{A}_{P^l},\varepsilon(\mathbf{P},\mathbf{R})\right]+\frac{i}{2\hbar}\left[\mathcal{A}_{P^l},\varepsilon(\mathbf{P},\mathbf{R})\right]\mathcal{A}_{R_l}\\
&-\frac{i}{2\hbar}\mathcal{A}_{P^l}\varepsilon(\mathbf{P},\mathbf{R})\mathcal{A}_{R_l}+\frac{i}{2\hbar}\mathcal{A}_{R_l}\varepsilon(\mathbf{P},\mathbf{R})\mathcal{A}_{P^l}. \qquad \text{(A.6)}
\end{aligned}$$

$U^+(\mathbf{P},\mathbf{R})$ when needed, we get:

see equation (A.6) above

Rearranging the commutators leads to:

$$\begin{aligned}
&\frac{i}{2}\hbar Asym\nabla_{R_l}\nabla_{P^l}\left[U(\mathbf{P},\mathbf{R})H_0(\mathbf{P},\mathbf{R})U^+(\mathbf{P},\mathbf{R})\right]=\\
&\frac{1}{2}\left[\mathcal{A}_{R_l}\nabla_{R_l}\varepsilon(\mathbf{P},\mathbf{R})+\nabla_{R_l}\varepsilon(\mathbf{P},\mathbf{R})\mathcal{A}_{R_l}\right]\\
&+\frac{1}{2}\left[\mathcal{A}_{P_l}\nabla_{P_l}\varepsilon(\mathbf{P},\mathbf{R})+\nabla_{P_l}\varepsilon(\mathbf{P},\mathbf{R})\mathcal{A}_{P_l}\right]\\
&-\frac{i}{2\hbar}\mathcal{A}_{R_l}\left[\varepsilon(\mathbf{P},\mathbf{R}),\mathcal{A}_{P_l}\right]-\frac{i}{2\hbar}\left[\varepsilon(\mathbf{P},\mathbf{R}),\mathcal{A}_{P_l}\right]\mathcal{A}_{R_l}\\
&+\frac{i}{2\hbar}\mathcal{A}_{P^l}\left[\varepsilon(\mathbf{P},\mathbf{R}),\mathcal{A}_{R_l}\right]+\frac{i}{2\hbar}\left[\varepsilon(\mathbf{P},\mathbf{R}),\mathcal{A}_{R_l}\right]\mathcal{A}_{P^l}\\
&-\frac{i}{2\hbar}\mathcal{A}_{P^l}\varepsilon(\mathbf{P},\mathbf{R})\mathcal{A}_{R_l}+\frac{i}{2\hbar}\mathcal{A}_{R_l}\varepsilon(\mathbf{P},\mathbf{R})\mathcal{A}_{P^l} \qquad \text{(A.7)}
\end{aligned}$$

rewrite $\frac{-i}{2\hbar}\mathcal{A}_{P^l}\varepsilon(\mathbf{P},\mathbf{R})\mathcal{A}_{R_l}+\frac{i}{2\hbar}\mathcal{A}_{R_l}\varepsilon(\mathbf{P},\mathbf{R})\mathcal{A}_{P^l}$ as $-\frac{i}{2\hbar}\mathcal{A}_{P^l}\left[\varepsilon(\mathbf{P},\mathbf{R}),\mathcal{A}_{R_l}\right]+\frac{i}{2\hbar}\mathcal{A}_{R_l}\left[\varepsilon(\mathbf{P},\mathbf{R}),\mathcal{A}_{P^l}\right]+\frac{i}{2\hbar}\left[\mathcal{A}_{R_l},\mathcal{A}_{P^l}\right]\varepsilon(\mathbf{P},\mathbf{R})$ so that:

$$\begin{aligned}
&\frac{i}{2}\hbar Asym\nabla_{R_l}\nabla_{P^l}\left[U(\mathbf{P},\mathbf{R})H_0(\mathbf{P},\mathbf{R})U^+(\mathbf{P},\mathbf{R})\right]=\\
&\frac{1}{2}\left[\mathcal{A}_{R_l}\nabla_{R^l}\varepsilon(\mathbf{P},\mathbf{R})+\nabla_{R^l}\varepsilon(\mathbf{P},\mathbf{R})\mathcal{A}_{R_l}\right]\\
&+\frac{1}{2}\left[\mathcal{A}_{P^l}\nabla_{P_l}\varepsilon(\mathbf{P},\mathbf{R})+\nabla_{P_l}\varepsilon(\mathbf{P},\mathbf{R})\mathcal{A}_{P^l}\right]\\
&-\frac{i}{2\hbar}\left[\varepsilon(\mathbf{P},\mathbf{R}),\mathcal{A}_{P^l}\right]\mathcal{A}_{R_l}\\
&+\frac{i}{2\hbar}\left[\varepsilon(\mathbf{P},\mathbf{R}),\mathcal{A}_{R_l}\right]\mathcal{A}_{P^l}+\frac{i}{2\hbar}\left[\mathcal{A}_{R_l},\mathcal{A}_{P^l}\right]\varepsilon(\mathbf{P},\mathbf{R}). \qquad \text{(A.8)}
\end{aligned}$$

And we thus have:

$$\begin{aligned}
&U(\mathbf{P},\mathbf{R})H_0(\mathbf{P},\mathbf{R})U^+(\mathbf{P},\mathbf{R})=\\
&\varepsilon(\mathbf{P},\mathbf{R})+\frac{1}{2}\left[\mathcal{A}_{R_l}\nabla_{R^l}\varepsilon(\mathbf{P},\mathbf{R})+\nabla_{R^l}\varepsilon(\mathbf{P},\mathbf{R})\mathcal{A}_{R_l}\right]\\
&+\frac{1}{2}\left[\mathcal{A}_{P^l}\nabla_{P_l}\varepsilon(\mathbf{P},\mathbf{R})+\nabla_{P_l}\varepsilon(\mathbf{P},\mathbf{R})\mathcal{A}_{P^l}\right]\\
&-\frac{i}{2\hbar}\left[\varepsilon(\mathbf{P},\mathbf{R}),\mathcal{A}_{P^l}\right]\mathcal{A}_{R_l}\\
&+\frac{i}{2\hbar}\left[\varepsilon(\mathbf{P},\mathbf{R}),\mathcal{A}_{R_l}\right]\mathcal{A}_{P^l}+\frac{i}{2\hbar}\left[\mathcal{A}_{R_l},\mathcal{A}_{P^l}\right]\varepsilon(\mathbf{P},\mathbf{R})
\end{aligned} \qquad \text{(A.9)}$$

as claimed in the text.

Appendix B: How solid states physics fits in our framework

In solid state Physics, we assume that the Hamiltonian is invariant through a discrete group of translations, for example a group of lattice translations, whose elements have the form

$$T(\mathbf{b})=\exp\left(-\frac{i}{\hbar}\int_0^{\mathbf{b}}A_i(\mathbf{R}+\mathbf{r})\mathbf{dr}\right)\exp\left(\frac{i}{\hbar}\mathbf{P}\cdot\mathbf{b}\right) \qquad \text{(B.1)}$$

where $\mathbf{b}$ is an arbitrary lattice vector. The eigenvalues of this operator are degenerated and have the form

$$\exp(i\mathbf{k}\cdot\mathbf{b}) \qquad \text{(B.2)}$$

where $\mathbf{k}$ belongs to some reduced dual lattice (a fraction of the dual lattice, i.e. a plaquette in solid state physics).

We aim at defining the generators $\mathbf{K}$ of these transformations as

$$\mathbf{K} \cdot \mathbf{b} = \log(T(\mathbf{b})) \quad \text{(B.3)}$$

so that we can define

$$K_i = \partial_{b_i} \mathbf{K} \cdot \mathbf{b} \quad \text{(B.4)}$$

However, this logarithm cannot be defined uniquely, and we will build an explicit choice in the sequel.

To do so, we work with the extended representation, so that the considered state space is defined, similarly to the Dirac case, by: $L^2(R^3) \otimes E$, where $L^2(R^3)$ is seen as the set of functions of the variable $\mathbf{k}$, running on R^3. E is a vector space of infinite size representing the bands. For each value of $\mathbf{k}$, $T(\mathbf{b})$ is diagonal in E, with eigenvalues $\exp(i\mathbf{k} \cdot \mathbf{b})$. We define K_i as acting diagonally as the multiplication by k_i. In other words, we have defined the momentum through an extension: $K \equiv K \otimes Id_E$. Consequently the position operator $\mathbf{R}$ is acting as $i\nabla_{\mathbf{k}}$. Therefore the state space of the Bloch electron is spanned by the basis vector of plane waves $|n, \mathbf{k}\rangle = |\mathbf{k}\rangle \otimes |n\rangle$ with n corresponding to a band index. The state $|n\rangle$ can be seen as a canonical base vector $|n\rangle = (0...010...0...)$ (with 1 at the nth position).

Turning now to the diagonalization process for the Hamiltonian, this last one can be performed independently for each value of $\mathbf{k}$, since the Hamiltonian commutes with the translations. We can thus see the Hamiltonian as a set of square matrices indexed by $\mathbf{k}$, each of them acting on each copy of E. As a consequence the diagonalization matrix is a Block acting on E for each value of $\mathbf{k}$.

Note that this diagonalization matrix can of course be seen as an operator $U(\mathbf{K}, \mathbf{R})$ or, and this is the point of view we adopt here, as a matrix acting on each copy of E, that is, a matrix $U(\mathbf{K})$, whose entries depend on $\mathbf{K}$ only. Actually, the dependence in $\mathbf{R}$ appears in the non diagonal elements, and we can discard them if we consider this "half matrix, half operator" version. This mixed representation has the advantage to do the connection with the Dirac Hamiltonian.

In this set up, $\mathbf{K}$ being diagonal and proportional to the identity, it commutes with every matrix $U(\mathbf{K}, \mathbf{R})$ preserving the Blocks. When considering the diagonalized Hamiltonian $\varepsilon(\mathbf{R}, \mathbf{K})$, it can also be seen as a diagonal matrix (implicitly denoted $\varepsilon(\mathbf{K})$) whose components are diagonal and denoted $\varepsilon_n(\mathbf{K})$, the n th band energy. The commutator

$$[\mathbf{R}, \varepsilon(\mathbf{K})] = \nabla\varepsilon(\mathbf{K}) \quad \text{(B.5)}$$

is again a diagonal matrix whose entries are $\nabla\varepsilon_n(\mathbf{K})$ (sketch of proof: $\varepsilon(\mathbf{K})$ is a series whose elements are products of powers of $\mathbf{K}$ and $\mathbf{R}$. For each power of $\mathbf{K}$, the fact that $\varepsilon(\mathbf{K})$ is diagonal implies that the dependence in $\mathbf{R}$ is a diagonal matrix. The gradient in $\mathbf{K}$ acting only on the power of $\mathbf{K}$, this diagonality is preserved). Turning now to the perturbation $\delta A(\mathbf{R})$, let us remark that if the operator $\delta A(\mathbf{R})$ preserves the bands, all operators $\varepsilon(\mathbf{K})\delta A(\mathbf{R})$, $[\varepsilon(\mathbf{K}), \delta A(\mathbf{R})]$ are diagonal, and given our previous remarks, the same is true for $\nabla\varepsilon(\mathbf{K})\delta A(\mathbf{R})$, $[\nabla\varepsilon(\mathbf{K}), \delta A(\mathbf{R})]$.

All this remarks that are obviously true for the Dirac case, appear to be useful in the solid state physics case (application 2), since it shows that the electron in an periodic potential fits in our framework. Actually, we can formally consider the Hamiltonian of such a problem as given by a matrix depending on the momenta and the coordinates.

References

1. M.V. Berry, Proc. R. Soc. A **392**, 45 (1984)
2. S. Murakami, N. Nagaosa, S.C. Zhang, Science **301**, 1348 (2003)
3. A. Bérard, H. Mohrbach, Phys. Rev. D **69**, 127701 (2004); A. Bérard, H. Mohrbach, Phys. lett. A **352**, 190 (2006)
4. K.Y. Bliokh, Europhys. Lett. **72**, 7 (2005)
5. K.Y. Bliokh, Y.P. Bliokh, Phys. Lett. A **333**, 181 (2004); K.Y. Bliokh, Y.P. Bliokh, Phys. Rev. E **70**, 026605 (2004); K.Y. Bliokh, Y.P. Bliokh, Phys. Rev. Lett. **96**, 073903 (2006)
6. M. Onoda, S. Murakami, N. Nagasoa, Phys. Rev. Lett. **93**, 083901 (2004)
7. C. Duval, Z. Horváth, P.A. Horváthy, Phys. Rev. D **74**, 021701 (2006), and J. Geom. Phys. **57**, 925 (2007)
8. N.W. Ashcroft, N.D. Mermin, *Solid State Physics* (Saunders, Philadelphia, 1976)
9. M.C. Chang, Q. Niu, Phys. Rev. Lett. **75**, 1348 (1995); M.C. Chang, Q. Niu, Phys. Rev. B **53**, 7010 (1996); G. Sundaram, Q. Niu, Phys. Rev. B **59**, 14915 (1999)
10. R. Shindou, K.I. Imura, Nucl. Phys. B **720**, 399 (2005); D. Culcer, Y. Yao, Q. Niu, Phys. Rev. B **72**, 085110 (2005)
11. D. Xiao, J. Shi, Q. Niu, Phys. Rev. Lett. **95**, 137204 (2005)
12. C. Duval, Z. Horvath, P. Horvathy, L. Martina, P. Stichel, Phys. Rev. Lett. **96**, 099701 (2006); D. Xiao, J. Shi, Q. Niu, Phys. Rev. Lett. **96**, 099702 (2006)
13. K.Y. Bliokh, Phys. Lett. A **351**, 123 (2006)
14. S. Ghosh, Phys. Lett. B **638**, 350 (2006); B. Basu, S. Ghosh, S. Dhar, Europhys. Lett. **76**, 395 (2006)
15. P. Gosselin, F. Ménas, A. Bérard, H. Mohrbach, Europhys. Lett. **76**, 651 (2006)
16. P. Gosselin, A. Bérard, H. Mohrbach, Phys. Rev. D **75**, 084035 (2007); P. Gosselin, A. Bérard, H. Mohrbach, Phys. Lett. A, in press.
17. Y.N. Obukhov, Phys. Rev. Lett. **86**, 192 (2001)
18. A.J. Silenko, O.V. Teryaev, Phys. Rev. D **71**, 064016 (2005)
19. L.L. Foldy, S.A. Wouthuysen, Phys. Rev. **78**, 29 (1949)
20. K.Y. Bliokh, Y.P. Bliokh, Ann. Phys. (NY) **319**, 13 (2005)
21. C. Kittel, *Quantum Therory of Solids* (Wiley, New York, 1963)
22. E.M. Lifshitz, L.P. Pitaevskii, *Statistical Physics* (Pergamon Press, 1981), Vol 9

TOPICAL REVIEW

Berry curvature, orbital moment, and effective quantum theory of electrons in electromagnetic fields

Ming-Che Chang[1] and Qian Niu[2]

[1] Department of Physics, National Taiwan Normal University, Taipei, Taiwan
[2] Department of Physics, University of Texas at Austin, Austin, TX, USA

E-mail: changmc@phy.ntnu.edu.tw and niu@physics.utexas.edu

Received 9 January 2008, in final form 20 March 2008
Published 11 April 2008
Online at stacks.iop.org/JPhysCM/20/193202

Abstract
Berry curvature and orbital moment of the Bloch state are two basic ingredients, in addition to the band energy, that must be included in the formulation of semiclassical dynamics of electrons in crystals, in order to give proper account of thermodynamic and transport properties to first order in the electromagnetic field. These quantities are gauge invariant and have direct physical significance as demonstrated by numerous applications in recent years. Generalization to the case of degenerate bands has also been achieved recently, with important applications in spin-dependent transport. The reader is assured that a knowledge of these ingredients of the semiclassical dynamics is also sufficient for the construction of an effective quantum theory, valid to the same order of the field, using a new quantization procedure that generalizes the venerable Peierls substitution rule. We cite the relativistic Dirac electron and the carrier in semiconductors as two prime examples to demonstrate our theory and compare with previous work on such systems. We also establish general relations between different levels of effective theories in a hierarchy.

(Some figures in this article are in colour only in the electronic version)

Contents

1. Introduction

Berry's phase is a unifying concept in physics [1, 2] and the semiclassical dynamics with Berry curvature included has found wide applications in solid state physics. The Berry curvature appeared quite early in history in the one-band effective dynamics of a Bloch electron [3, 4]. It generates an anomalous velocity that is transverse to the applied electric field. When applied to the quantum Hall system, the semiclassical theory can explain the Hall current and the quantization of the Hall conductivity [5–7]. In recent years, it has helped in solving the mystery of the anomalous Hall effect in ferromagnetic materials [8–10]. It is also a primary notion motivating the proposal of the intrinsic spin Hall effect [11, 12] and the optical Hall effect [13, 14].

Despite various attempts, the effective dynamics remains largely semiclassical, and a systematic quantum version is still lacking. Such a quantum theory would be needed if one wants to know, for example, the full intricate structure of a band under a magnetic field (the Hofstadter spectrum) [15]. Here we report our recent finding of a quantization procedure by generalizing the Peierls substitution rule [16].

For an effective quantum theory without the Berry phase, it is a common practice to use the Peierls substitution [17, 18]. This is clearly inadequate when the band has a Berry curvature. When this fails, one typically blames the neglect of interband couplings, and expands the range of the Hilbert space by including more bands. One message that we would like to convey is that there is sufficient information in the semiclassical formalism for constructing the correct quantum theory. The Berry curvature and orbital moment in the one-band semiclassical formulation contain all the information of interband coupling that is needed for an effective quantum theory for the band (or a subset of bands) under consideration. This finding may have an enormous impact on research based on first-principle band structure calculations. One is now assured that, by calculating the three essential physical quantities, the Bloch energy $E(\mathbf{q})$, the Berry curvature $\mathbf{F}(\mathbf{q})$, and the self-rotating angular momentum of the wavepacket $\mathbf{L}(\mathbf{q})$, all electronic properties to first order in external fields can be obtained.

On the other hand, in the 1950s and 1960s, systematic studies were taken to develop effective Hamiltonians for a Bloch electron in an electromagnetic field [19–21]. They are exact up to a certain order in the external fields. It is not immediately obvious how the Berry curvature emerges from such an effective quantum theory. The effective Hamiltonian involves quantities that are gauge dependent (depending on the phase choice of the basis Bloch functions), and the position operator conjugate to the momentum is not really the physical position operator. However, miraculously, when one considers the Ehrenfest equations of motion for the physical position and crystal momentum, one obtains the semiclassical theory with the gauge-invariant Berry curvature and the orbital moment properly included.

In this review, we start with the gauge-invariant semiclassical theory and quantize it using the generalized Peierls substitution [16]. Naturally, our effective Hamiltonians are in agreement with earlier results obtained by canonical transformation [22, 23]. Although this route toward an effective quantum theory is less direct compared with earlier works, our approach is physically more transparent. The wavepacket point of view is certainly intuitive, and the gauge-invariant Berry curvature and orbital moment in the semiclassical theory are of direct physical significance. Furthermore, our method of quantization can also be applied to more general situations, such as the crystal deformation described in [24].

We do not attempt to give a comprehensive review on the topics related to the Berry phase or semiclassical dynamics in solids. They can be found in the following reviews in [25–28]. Instead, we focus on the semiclassical transport in solid that involves spin (or quasi-spin) degrees of freedom. The development is mainly based on the multi-band generalization of the semiclassical dynamics [29], which will be reviewed in section 2. This formalism applies to carriers with spins and has a richer dynamics than the one-band theory.

Throughout this review, we will use the Dirac theory of relativistic electron to illustrate the formalism. Dirac theory is one of the simplest that admits a non-Abelian gauge structure. In section 3, the three basic quantities mentioned above are calculated exactly for the Dirac electron. They are used to obtain the equations of motion and the equation for spin precession.

Section 4 is devoted to the quantization of the semiclassical theory. The Landau levels in a semiconductor are not usually obtained by solving the Schrödinger equation treating the lattice potential and the vector potential on an equal footing. Such an approach is difficult and one ends up with a complex Hofstadter spectrum [15]. Instead, one first solves the band energy, then uses the Peierls substitution to obtain the effective quantum Hamiltonian. The Landau levels are the eigenenergies of this effective Hamiltonian [17, 18]. In section 4, we derive the generalized Peierls substitution and the effective Hamiltonians for systems with Berry curvature [16]. The Dirac theory is again used as an example.

Compared to the original quantum theory, the re-quantized version is an effective one applying to a smaller Hilbert space (of only one energy band, for example). The re-quantized theory can have its own semiclassical approximation. One can in turn re-quantize this semiclassical theory, and such a process can go on and on until the resulting theory is trivial. This way, one can generate a whole series of hierarchy theories with different levels of approximation. In the extreme case, in a system with self-similarity in energy scales, such as the Hofstadter spectrum, such a series is essentially infinite. The general relations between the theories with different levels of approximation will be explored in section 5.

In section 6, equipped with the tools developed and confirmed in preceding sections, we turn our attention to the carrier dynamics in semiconductors within the framework of the eight-band Kane model. This investigation offers fresh perspectives on familiar concept in semiconductors, such as the origin of the anomalous g-factor and the relation between Berry connection and spin–orbit interaction. The review is summarized and concluded in section 7.

2. Semiclassical theory

In this section, we introduce the basic physical quantities essential to the semiclassical theory. These include the Bloch energy, the Berry curvature, and the orbital moment. We will then briefly summarize the multi-band wavepacket formulation of Culcer *et al* [29, 30]. Such a formulation serves as a basic tool in this review. This summary is intended to be self-contained. However, readers not familiar with the wavepacket formulation are advised to consult earlier literature on the simpler one-band theory first [7, 24]. In the one-band version, the wavefunction has only one component and the gauge structure is Abelian; in the multi-band version, the wavefunction is a multi-component spinor and the gauge structure becomes non-Abelian.

2.1. Bloch band and Berry curvature

Consider an electron in a periodic crystal with the following Schrödinger equation:

$$H_0(\mathbf{r},\mathbf{p})\psi_{n\mathbf{q}} = E_{0,n}(\mathbf{q})\psi_{n\mathbf{q}}, \tag{1}$$

where n and $\mathbf{q}$ are band index and crystal momentum, and $E_{0,n}(\mathbf{q})$ and $\psi_{n\mathbf{q}}$ are the Bloch energies and Bloch states. The Bloch states can always be written in the form

$$\psi_{n\mathbf{q}}(\mathbf{r}) = \mathrm{e}^{\mathrm{i}\mathbf{q}\cdot\mathbf{r}}u_{n\mathbf{q}}(\mathbf{r}), \tag{2}$$

where $u_{n\mathbf{q}}(\mathbf{r})$ is a cell-periodic function. It is useful to define $\tilde{H} = \mathrm{e}^{-\mathrm{i}\mathbf{q}\cdot\mathbf{r}}H\mathrm{e}^{\mathrm{i}\mathbf{q}\cdot\mathbf{r}}$ and write the Schrödinger equation in an alternative form,

$$\tilde{H}_0(\mathbf{r},\mathbf{p};\mathbf{q})u_{n\mathbf{q}} = E_{0,n}(\mathbf{q})u_{n\mathbf{q}}. \tag{3}$$

In the semiclassical formulation, one considers a wavepacket moving within a subset $\{E_{0,n}(\mathbf{q}), n = 1\cdots D\}$ of the full spectrum. Such a subset is degenerate in energy and is gapped from the rest of the energy bands. Within such a subspace, one can define two geometric quantities: a metric that defines the quantum distance between two states [31], which is not a concern of this review, and a *connection* defined as

$$\mathbf{R}_{mn}(\mathbf{q}) = \mathrm{i}\left\langle u_{m\mathbf{q}}\middle|\frac{\partial u_{n\mathbf{q}}}{\partial \mathbf{q}}\right\rangle. \tag{4}$$

This is the non-Abelian Berry connection, which is a vector-valued $D\times D$ matrix [32].

Following the rule of parallel transport, a state circling a closed loop C in the Brillouin zone may acquire a unitary (Berry) rotation. The Berry rotation for an infinitesimal loop surrounding a point $\mathbf{q}$ is proportional to the curvature $\mathcal{F}$ at this point,

$$\mathcal{F}(\mathbf{q}) = \nabla_{\mathbf{q}}\times\boldsymbol{\mathcal{R}} - \mathrm{i}\boldsymbol{\mathcal{R}}\times\boldsymbol{\mathcal{R}}. \tag{5}$$

Throughout this review, symbols in calligraphic fonts represent matrices (boldfaced calligraphic fonts are vector-valued matrices). The Berry curvature matrix is gauge covariant under a gauge rotation.

If the subset comprises only of one band with $D = 1$, then $\boldsymbol{\mathcal{R}}$ and $\mathcal{F}$ become vectors $\mathbf{R}$ and $\mathbf{F}$, and the Berry rotation is simply a $U(1)$ Berry phase, which is invariant under a gauge transformation. One can prove that, if the system has time-reversal symmetry, then $\mathbf{F}(-\mathbf{q}) = -\mathbf{F}(\mathbf{q})$. Also, if there is space-inversion symmetry, then $\mathbf{F}(-\mathbf{q}) = \mathbf{F}(\mathbf{q})$. Therefore, when both symmetries exist, there can be no Berry curvature. However, in a multi-band theory, the Berry curvature can exist even if both symmetries exist.

2.2. Self-rotating angular momentum of a wavepacket

One can construct a wavepacket by linear superposition of the Bloch states. In order for the wavepacket to remain intact with a meaningful center of mass, we require it to reside on a set of bands closely packed in energy so that it will not split into separate packets. Before turning on an external field, the unperturbed Hamiltonian $H_0(\mathbf{r},\mathbf{p})$ is assumed to be solvable with degenerate Bloch states $|\psi_{n,\mathbf{q}}\rangle$, $n = 1\cdots D$. If necessary, a small degeneracy-breaking interaction can also be included. But it has to be treated as a perturbation[3].

The wavepacket is expanded as

$$|W\rangle = \sum_{n=1}^{D}\int \mathrm{d}^3 q\, a(\mathbf{q},t)\eta_n(\mathbf{q},t)|\psi_{n,\mathbf{q}}\rangle, \tag{6}$$

where $\eta_n(\mathbf{q},t)$ is normalized as a unit vector at each $\mathbf{q}$, $a(\mathbf{q},t)$ is a narrow distribution that centers at $\mathbf{q}_c(t)$ in momentum space and is normalized as $\int \mathrm{d}^3 q|a|^2 = 1$. The wavepacket is also centered at $\mathbf{r}_c$ in the real space. Therefore, one requires $\langle W|\mathbf{r}|W\rangle = \mathbf{r}_c$. Obviously, one also has $\langle W|H_0|W\rangle = E_0(\mathbf{q}_c)$.

The wavepacket is often found to be spinning with respect to its own center of mass. The self-rotating angular momentum of a wavepacket is $\mathbf{L} = \langle W|(\mathbf{r}-\mathbf{r}_c)\times\mathbf{p}|W\rangle$. It is difficult to calculate $\mathbf{L}$ under the present form since the exact shape of the wavepacket is not specified. Fortunately, for a wavepacket narrowly distributed around $\mathbf{q}_c$, it is possible to rewrite it in another form using the Bloch states only [7],

$$\mathbf{L}(\mathbf{q}_c) = \mathrm{i}\frac{m}{\hbar}\left\langle\frac{\partial u}{\partial \mathbf{q}_c}\right|\times\left[\tilde{H}_0 - E_0(\mathbf{q}_c)\right]\left|\frac{\partial u}{\partial \mathbf{q}_c}\right\rangle, \tag{7}$$

where the cell-periodic function without a subscript is defined as $|u\rangle = \sum_{n=1}^{D}\eta_n|u_n\rangle$. In the discussions below, we will also encounter an angular momentum matrix with the following matrix elements:

$$\mathbf{L}_{nl}(\mathbf{q}_c) = \mathrm{i}\frac{m}{\hbar}\left\langle\frac{\partial u_n}{\partial \mathbf{q}_c}\right|\times\left[\tilde{H}_0 - E_0(\mathbf{q}_c)\right]\left|\frac{\partial u_l}{\partial \mathbf{q}_c}\right\rangle. \tag{8}$$

Obviously, after spinor average, one obtains the angular momentum in equation (7): $\mathbf{L} = \eta^\dagger\boldsymbol{\mathcal{L}}\eta = \sum_{nl}\eta_n^*\mathbf{L}_{nl}\eta_l$.

2.3. Wavepacket energy in electromagnetic fields

An external electromagnetic field can be added to H_0 using a scalar potential $\phi(\mathbf{r})$ and a vector potential $\mathbf{A}(\mathbf{r})$, giving the full Hamiltonian H. The wavepacket energy $E(\mathbf{r}_c,\mathbf{q}_c) = \langle W|H|W\rangle$ differs from $E_0(\mathbf{q}_c)$ as a result of these external fields. Such an energy shift can be calculated using a gradient

[3] Such an addition is referred to as H_n in Culcer's paper [29].

expansion [24, 29]: since the wavepacket is localized, we only need to know the values of $\phi(\mathbf{r})$ and $\mathbf{A}(\mathbf{r})$ near $\mathbf{r}_c$. Therefore, they are expanded as

$$\phi(\mathbf{r}) \simeq \phi(\mathbf{r}_c) + (\mathbf{r} - \mathbf{r}_c) \cdot \frac{\partial \phi}{\partial \mathbf{r}_c}, \tag{9}$$

$$\mathbf{A}(\mathbf{r}) \simeq \mathbf{A}(\mathbf{r}_c) + (\mathbf{r} - \mathbf{r}_c) \cdot \frac{\partial \mathbf{A}}{\partial \mathbf{r}_c}. \tag{10}$$

Such an approximation is valid only if the potentials vary slowly across the wavepacket.

The values of $\phi_c = \phi(\mathbf{r}_c)$ and $\mathbf{A}_c = \mathbf{A}(\mathbf{r}_c)$ do not need to be small compared to the unperturbed part since they only contribute to a shift in energy by $-e\phi_c$ and a shift in crystal momentum by $e\mathbf{A}_c$ (the charge of the carrier is $-e$). However, the field strengths are limited by the adiabatic condition: the total probability in these D bands should remain conserved. There can be no tunneling out of this subspace.

Assuming the electric and magnetic fields are uniform, choosing the symmetric gauge for the vector potential, and neglecting nonlinear terms in $\mathbf{A}$, one finds

$$\begin{aligned} E(\mathbf{r}_c, \mathbf{q}_c) &= \langle W|H_c|W\rangle + e\langle W|(\mathbf{r} - \mathbf{r}_c)|W\rangle \cdot \mathbf{E} \\ &\quad + \frac{e}{2m}\langle W|(\mathbf{r} - \mathbf{r}_c) \times (\mathbf{p} + e\mathbf{A}_c)|W\rangle \cdot \mathbf{B} \\ &= E_0(\mathbf{k}_c) - e\phi(\mathbf{r}_c) + \frac{e}{2m}\mathbf{L}(\mathbf{k}_c) \cdot \mathbf{B}, \end{aligned} \tag{11}$$

where

$$H_c \equiv H_0(\mathbf{r}, \mathbf{p} + e\mathbf{A}_c) - e\phi_c. \tag{12}$$

Notice that $\langle W|(\mathbf{r} - \mathbf{r}_c)|W\rangle$ is zero and $\mathbf{k}_c \equiv \mathbf{q}_c + (e/\hbar)\mathbf{A}_c$. If a charged wavepacket is self-rotating, then it has an orbital magnetic moment $\mathbf{m} = -e/(2m)\mathbf{L}$ coupling with the magnetic field and gives the Zeeman energy.

Since nonlinear terms have been neglected in the calculation above, the following result is accurate only to linear order of the fields. The nonlinear electromagnetic effect is thus outside the scope of this review.

2.4. Wavepacket dynamics: Lagrangian and equations of motion

Instead of solving the Schrödinger equation, we use the time-dependent variational method to study the wavepacket evolution in a crystal. Recall that in the usual time-independent variational method one proposes a sensible wavefunction with a few adjustable parameters. These parameters are then determined by minimizing the energy $\langle H\rangle$ of the proposed state. In the time-dependent variational method, the wavepacket is parametrized by the center of mass, $\mathbf{r}_c(t)$ and $\mathbf{q}_c(t)$, and the spinor amplitude $\boldsymbol{\eta} = (\eta_1, \eta_2, \ldots, \eta_D)^T$. Any possible change of shape of the wavepacket will be ignored. One first calculates the effective Lagrangian,

$$L_{\text{eff}}(\mathbf{r}_c, \mathbf{q}_c, \boldsymbol{\eta}; \dot{\mathbf{r}}_c, \dot{\mathbf{q}}_c, \dot{\boldsymbol{\eta}}) = \langle W|\left(\mathrm{i}\hbar\frac{\mathrm{d}}{\mathrm{d}t} - H\right)|W\rangle, \tag{13}$$

then obtains the effective action $S_{\text{eff}} = \int \mathrm{d}t\, L_{\text{eff}}$ for a given trajectory. Unlike the energy $\langle H\rangle$ in the time-independent variational method that depends on only a few parameters, this effective action is a *functional* of the trajectory and the time evolution of the spinor amplitude: $(\mathbf{r}_c(t), \mathbf{q}_c(t), \boldsymbol{\eta}(t))$. Therefore, one needs to use the method of variation to determine the one that extremizes this action. The Euler–Lagrange equations for this effective Lagrangian give us equations of motion of the wavepacket.

By carefully analyzing the time-derivative term in equation (13), one obtains

$$\mathrm{i}\hbar\langle W|\frac{\mathrm{d}}{\mathrm{d}t}|W\rangle = \hbar\left\langle u\middle|\mathrm{i}\frac{\mathrm{d}u}{\mathrm{d}t}\right\rangle + \hbar\mathbf{q}_c \cdot \dot{\mathbf{r}}_c. \tag{14}$$

The first term can further be decomposed as

$$\hbar\left\langle u\middle|\mathrm{i}\frac{\mathrm{d}u}{\mathrm{d}t}\right\rangle = \mathrm{i}\hbar\boldsymbol{\eta}^\dagger\frac{\partial\boldsymbol{\eta}}{\partial t} + \hbar\dot{\mathbf{k}}_c \cdot \mathbf{R}(\mathbf{k}_c), \tag{15}$$

where we have used the gauge-invariant crystal momentum $\mathbf{k}_c = \mathbf{q}_c + (e/\hbar)\mathbf{A}_c$ in place of $\mathbf{q}_c$ as a basic variable. Therefore, the effective Lagrangian is

$$\begin{aligned} L_{\text{eff}}(\mathbf{r}_c, \mathbf{k}_c, \boldsymbol{\eta}; \dot{\mathbf{r}}_c, \dot{\mathbf{k}}_c, \dot{\boldsymbol{\eta}}) &= \mathrm{i}\hbar\boldsymbol{\eta}^\dagger\frac{\partial\boldsymbol{\eta}}{\partial t} \\ &\quad + \hbar\dot{\mathbf{k}}_c \cdot \mathbf{R} + (\hbar\mathbf{k}_c - e\mathbf{A}_c) \cdot \dot{\mathbf{r}}_c - E(\mathbf{r}_c, \mathbf{k}_c), \end{aligned} \tag{16}$$

in which $E = \langle W|H|W\rangle$ is the wavepacket energy calculated in equation (11) using the gradient expansion. Notice that there are three pairs of independent variables, as indicated in the argument of L_{eff}.

After lengthy derivations, the Euler–Lagrange equations of this effective Lagrangian give the following (coupled) dynamical equations of the wavepacket [29]:

$$\hbar\dot{\mathbf{k}}_c = -e\mathbf{E} - e\dot{\mathbf{r}}_c \times \mathbf{B}, \tag{17}$$

$$\hbar\dot{\mathbf{r}}_c = \frac{1}{\mathrm{i}}\boldsymbol{\eta}^\dagger\left[\mathrm{i}\frac{\partial}{\partial\mathbf{k}_c} + \boldsymbol{\mathcal{R}}, \mathcal{H}\right]\boldsymbol{\eta} - \hbar\dot{\mathbf{k}}_c \times \boldsymbol{\eta}^\dagger\boldsymbol{\mathcal{F}}\boldsymbol{\eta}, \tag{18}$$

$$\mathrm{i}\hbar\dot{\boldsymbol{\eta}} = \left(\frac{e}{2m}\boldsymbol{\mathcal{L}} \cdot \mathbf{B} - \hbar\dot{\mathbf{k}}_c \cdot \boldsymbol{\mathcal{R}}\right)\boldsymbol{\eta}, \tag{19}$$

where $\mathbf{E}$ and $\mathbf{B}$ are the external fields, and $\boldsymbol{\mathcal{R}}$, $\boldsymbol{\mathcal{F}}$, and $\boldsymbol{\mathcal{L}}$ are all evaluated at the center of the wavepacket. The Hamiltonian matrix within the commutator in equation (18) has the form

$$\mathcal{H}(\mathbf{r}_c, \mathbf{k}_c) = E_0(\mathbf{k}_c) - e\phi(\mathbf{r}_c) + \frac{e}{2m}\boldsymbol{\mathcal{L}}(\mathbf{k}_c) \cdot \mathbf{B}, \tag{20}$$

in which $E_0 - e\phi$ is multiplied by an identity matrix. Obviously, the spinor average of $\mathcal{H}$ is the wavepacket energy in equation (11), $E = \boldsymbol{\eta}^\dagger\mathcal{H}\boldsymbol{\eta} = \sum_{nl}\eta_n^* H_{nl}\eta_l$.

From the spinor equation (19), one can immediately obtain the equation for the spin vector, $\mathbf{J} = \boldsymbol{\eta}^\dagger\boldsymbol{\mathcal{J}}\boldsymbol{\eta}$, where $\boldsymbol{\mathcal{J}}$ is the usual angular momentum matrix in quantum mechanics,

$$\mathrm{i}\hbar\dot{\mathbf{J}} = \boldsymbol{\eta}^\dagger\left[\boldsymbol{\mathcal{J}}, \mathcal{H} - \hbar\dot{\mathbf{k}}_c \cdot \boldsymbol{\mathcal{R}}\right]\boldsymbol{\eta}. \tag{21}$$

The only term in $\mathcal{H}$ that would contribute to the commutator above is the Zeeman coupling. The correction to the Hamiltonian that is proportional to the Berry connection $\boldsymbol{\mathcal{R}}$ is in fact a spin–orbit coupling. This will become clear in later discussions.

3. Semiclassical theory for a relativistic electron

To have a better grasp of the theoretical formulation in the previous section, let us take the relativistic particle as an illustrative example [33, 34]. The wavepacket is living in the electron subspace. Since the electron states are separated from the positron states by a huge energy gap mc^2, the adiabatic condition is hardly violated. It fails only if electron–positron creation is not negligible. Furthermore, unlike the Bloch states in a solid, the energy eigenstates of a free Dirac particle are analytically known. Therefore, the calculations of **R**, **F**, and **L** are clean and exact. These quantities are essential to the dynamics of a Dirac wavepacket in an electromagnetic field.

3.1. Basics of the Dirac theory

The Dirac Hamiltonian in the presence of an electromagnetic field is

$$H = c\boldsymbol{\alpha} \cdot (\mathbf{p} + e\mathbf{A}) + \beta mc^2 - e\phi(\mathbf{r}), \tag{22}$$

where $\boldsymbol{\alpha}$ and β are the Dirac matrices,

$$\boldsymbol{\alpha} = \begin{pmatrix} 0 & \boldsymbol{\sigma} \\ \boldsymbol{\sigma} & 0 \end{pmatrix}, \qquad \beta = \begin{pmatrix} I & 0 \\ 0 & -I \end{pmatrix}, \tag{23}$$

in which $\boldsymbol{\sigma}$ are the Pauli matrices and I is the 2×2 identity matrix. The off-diagonal matrix $c\boldsymbol{\alpha}$ is the velocity operator for the Dirac electron.

In the absence of an external field, one can perform the following unitary transformation to diagonalize the Hamiltonian, $H_0 = e^{iS} H_{0d} e^{-iS}$, where $H_{0d} = \sqrt{c^2|\mathbf{p}|^2 + m^2c^4}\beta$ has positive-energy and negative-energy branches (each branch has two degenerate levels). The unitary operator $e^{iS} = e^{\frac{\omega}{2}\beta\boldsymbol{\alpha}\cdot\mathbf{p}/|\mathbf{p}|}$, where $\tan\omega = p/mc$ [35]. The four independent plane-wave solutions of H_0 are of the form $\psi_l(\mathbf{r}) = e^{i\mathbf{q}\cdot\mathbf{r}}u_l(\mathbf{r})$, where the spinor u_l can be obtained from the four-component *unit spinor* e_l via the unitary rotation, $u_l = e^{iS}e_l$.

We emphasize that these solutions remain normalized at any velocity. The equally valid solutions obtained from the unit spinors via a Lorentz boost are not normalized, since the particle density is not invariant under the Lorentz transformation. These two sets of solutions have slightly different normalization constants, but otherwise are of the same form. However, the Berry connection $\mathcal{R}$ in equation (4) can be Hermitian *only if* we use the normalized eigenstates, which are obtained from the unitary rotation.

3.2. Basics of the Dirac wavepacket

The Dirac wavepacket is constructed from the free-particle states with positive energy $E_0 = \sqrt{c^2\hbar^2q^2 + m^2c^4}$ (see figure 1),

$$|W\rangle = \sum_{n=1}^{2} \int d^3q\, a(\mathbf{q}, t)\eta_n(\mathbf{q}, t)|\psi_n\rangle. \tag{24}$$

In such a subspace with twofold spin degeneracy, the Berry connection and curvature are 2×2 matrices. Using the definitions in equations (4) and (5), after some calculations, one can obtain

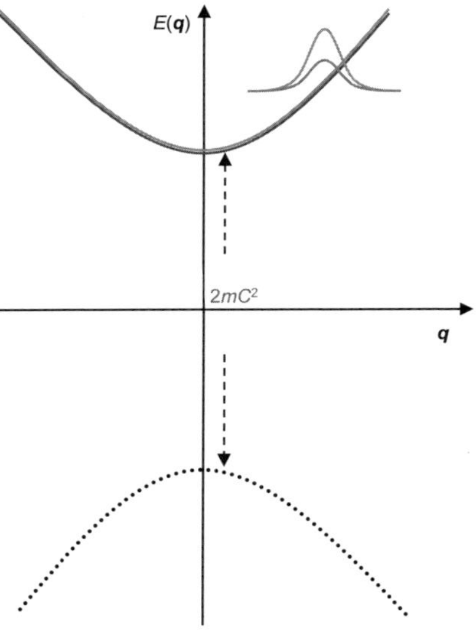

Figure 1. The Dirac spectrum has two branches separated by a huge mass gap. Each branch is twofold degenerate. Therefore, the wavepacket on the positive-energy branch has two localized probability amplitudes.

$$\mathcal{R} = \frac{\lambda_c^2}{2\gamma(\gamma+1)}\mathbf{q} \times \boldsymbol{\sigma}, \tag{25}$$

$$\mathcal{F} = -\frac{\lambda_c^2}{2\gamma^3}\left(\boldsymbol{\sigma} + \lambda_c^2 \frac{\mathbf{q}\cdot\boldsymbol{\sigma}}{\gamma+1}\mathbf{q}\right), \tag{26}$$

where $\lambda_c = \hbar/mc$ is the Compton wavelength, and $\gamma(q) \equiv \sqrt{1 + (\hbar q/mc)^2}$. The dilation factor γ reduces to one when the electron is at rest ($q = 0$).

It is remarkable that even a free electron would have a non-trivial gauge structure. The Berry connection vanishes at zero velocity but the Berry curvature reduces to a non-zero quantity proportional to the Pauli matrix, $-(\lambda_c^2/2)\boldsymbol{\sigma}$. The Berry curvature is not originated from level crossing and therefore is nothing like a monopole field. A similar result has been reported in [36, 37].

The other truly remarkable result is that such a free-electron wavepacket is in fact self-rotating with an orbital angular momentum [34]. A straightforward application of equation (8) gives[4]

$$\mathcal{L} = \frac{\hbar}{\gamma^2}\left(\boldsymbol{\sigma} + \lambda_c^2 \frac{\mathbf{q}\cdot\boldsymbol{\sigma}}{\gamma+1}\mathbf{q}\right). \tag{27}$$

[4] It has a functional form similar to $\mathcal{F}$. This is related to the fact that, when $\mathcal{L}$ is written in the form of equation (51) (shown later), it differs from an alternative form of the Berry curvature in equation (50) only by a factor of energy (apart from some constants). In the two-branch model we used, this only brings about a difference of a factor of $2E_0$.

After spinor average, this self-rotating wavepacket generates a magnetic moment $\mathbf{m} = -e/(2m)\mathbf{L}$ that couples with the magnetic field and gives rise to the Zeeman energy in equation (11). The coefficient $-e/(2m)$ finds its origin in the fact that $\mathbf{L} = \int d^3r(\mathbf{r} - \mathbf{r}_c) \times m\mathbf{j}(\mathbf{r})$, where $\mathbf{j}$ is the current density, while $\mathbf{m} = \frac{1}{2}\int d^3r(\mathbf{r} - \mathbf{r}_c) \times (-e)\mathbf{j}(\mathbf{r})$.

When the momentum q is zero, the orbital angular momentum $\mathcal{L}$ reduces to $\hbar\boldsymbol{\sigma}$, twice the spin angular momentum. It produces a Zeeman energy identical to the usual one in which the spin is $(\hbar/2)\boldsymbol{\sigma}$ and the g-factor is 2. However, we emphasize that the g-factor in our formulation equals one, not two! That is, in the wavepacket description, there is no need to assign a special g-factor to the electron spin.

Neither in the Dirac equation nor in its four-component wave functions can one find explicit trace of the spin. The established mathematical procedures, such as the Foldy and Wouthuysen transformation [22], for extracting the Zeeman interaction are usually complicated, and the appearance of the intrinsic magnetic moment for the electron has always been mysterious. The wavepacket approach offers an alternative and very intuitive picture explaining the origin of the electron spin: it is indeed a self-rotating motion that produces the angular momentum in equation (27). Such a delightful result is a natural consequence of the wavepacket theory and requires no forced assumption. For more discussions on the spinning Dirac wavepacket, one can also see [38].

3.3. Dynamics of the Dirac wavepacket

Once the quantities $\mathcal{H}$, $\mathcal{R}$, $\mathcal{F}$, and $\mathcal{L}$ are known, one can proceed to derive the equations of motion in equations (17)–(19). Correct to linear order in fields, the equations of motion for the center of mass are

$$\hbar\dot{\mathbf{k}}_c = -e\mathbf{E} - e\frac{\hbar\mathbf{k}_c}{\gamma m} \times \mathbf{B}, \tag{28}$$

$$\dot{\mathbf{r}}_c = \frac{\hbar\mathbf{k}_c}{\gamma m} + \frac{e}{\hbar}\left(\mathbf{E} \times \mathbf{F} + \mathbf{B} \cdot \mathbf{F}\frac{\hbar\mathbf{k}_c}{\gamma m}\right), \tag{29}$$

where $\gamma = \gamma(\mathbf{k}_c)$ and $\mathbf{F} = \eta^\dagger\mathcal{F}\eta$ is the spinor-averaged Berry curvature. The equation for spin precession in equation (21) for the Dirac electron has the following form:

$$\langle\dot{\boldsymbol{\sigma}}\rangle = \frac{e}{\gamma m}\left[\mathbf{B} + \mathbf{E} \times \frac{\hbar\mathbf{k}_c}{(\gamma + 1)mc^2}\right] \times \langle\boldsymbol{\sigma}\rangle. \tag{30}$$

This is the Bargmann–Michel–Telegdi (BMT) equation for a relativistic electron [39]. The second term proportional to the electric field couples the orbital motion with the spin. It is originated from the $-\hbar\dot{\mathbf{k}} \cdot \mathcal{R}$ term in equation (21) and thus is closely related to the Berry connection.

If we turn off the magnetic field in equations (28) and (29), then the trajectory is much simpler but still non-trivial. There is a linear acceleration along the direction of the electric field, $\hbar\dot{\mathbf{k}}_c = -e\mathbf{E}$, which is reasonable. However, there is a transverse velocity as a result of the Berry curvature in equation (29), even though there is no external force along this direction! The magnitude of the transverse particle momentum is $m(e/\hbar)\mathbf{E} \times \mathbf{F}$, which reduces to $\mathbf{E} \times \mathbf{m}/c^2$ at low velocity, where $\mathbf{m} = -e/(2m)\mathbf{L}$ (see figure 2).[5]

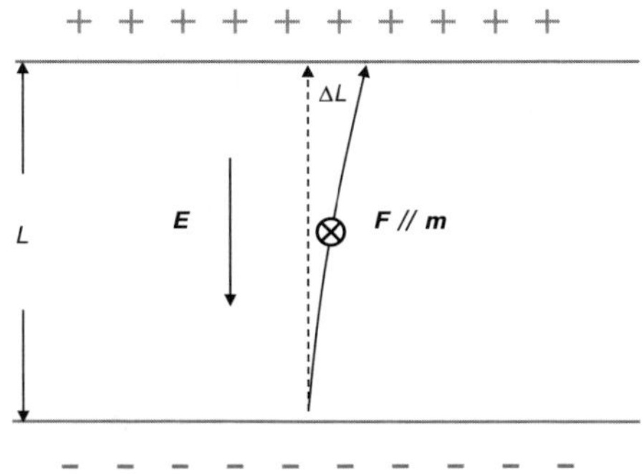

Figure 2. An electron is accelerated by a static electric field in the capacitor. The Berry curvature and the magnetic moment point into the page, giving rise to an anomalous velocity pointing to the right.

From a historical perspective, one can also understand such a curved trajectory in the following way: the theory of relativity tells us that a moving magnetic dipole $\mathbf{m}$ generates an effective electric dipole $\mathbf{p}_{\text{eff}} = \mathbf{v} \times \mathbf{m}/c^2$ [41]. Being in a uniform field, the electron feels no gradient dipole force. However, it acquires a dipole energy $-\mathbf{p}_{\text{eff}} \cdot \mathbf{E}$. Such a term contributes to the classical Hamiltonian. Therefore, from the Hamiltonian equation $\dot{\mathbf{r}} = \partial H/\partial\mathbf{p}$, one would get an extra velocity $\mathbf{v} = \mathbf{E} \times \mathbf{m}/(mc^2)$, the same as the particle velocity above[6].

It would be very impressive to confirm such a trajectory shift in free space. The electron can be accelerated by a parallel-plate capacitor with a potential difference V_0. Aligning the electron spin perpendicular to the electric field maximizes the effect. When moving from one plate to another, the electron trajectory will be shifted sideways by a distance $\lambda_c\sqrt{eV_0/2mc^2}$ (assuming the electron is initially at rest). That is, for a potential difference of 1 MeV, the shift is of the order of the Compton wavelength, which is two orders of magnitude smaller than the size of a hydrogen atom. One can alternate the polarity of the electrodes to synchronize with the electron motion and amplify the shift. Nevertheless, a huge potential difference is required and such an observation will not be easy.

4. Canonical quantization

The wavepacket approach for electron transport, even though approximate in general, can be accurate at a certain limit. One

[5] Whenever a magnetic dipole $\mathbf{m}$ produced by a current loop (such as the wavepacket here) is put in an electric field $\mathbf{E}$, it is accompanied by a hidden momentum $\mathbf{p}_{\text{hidden}} = \mathbf{m} \times \mathbf{E}/c^2$ [40, 41]. When this is taken into account, one finds that the part of $\hbar\mathbf{k}_c$ that is perpendicular to $\mathbf{E}$ remains conserved, consistent with its equation of motion.

[6] In addition to the velocity mentioned above, the effective electric dipole is closely related to the electric polarization generated by a spin current; see [42]. In the free space, a spin current can be modeled by counter-propagating beams of bare spin-up and spin-down electrons. It has been known for a long time that the charge center of the beams does not coincide with its mass center [43]. This is a direct result of the fact that the effective polarizations of both beams point in the same direction.

can take advantage of the simplicity of the semiclassical theory and re-quantize it to study various quantum phenomena. Under this certain limit, the prediction of the simplified (effective) quantum theory could be as accurate as the original quantum theory.

In an earlier work, we used the Bohr–Sommerfeld condition to re-quantize the cyclotron orbit and successfully predicted the positions of Landau levels in magnetic Bloch bands [7]. This approach, however, is limited to integrable systems with closed trajectories. Here we use the method of canonical quantization that does not have these restrictions and, moreover, can be applied to the multi-band cases. We demonstrate that the knowledge of $\mathcal{H}$, $\mathcal{R}$, $\mathcal{F}$, and $\mathcal{L}$ in the semiclassical formulation is sufficient for us to obtain the correct quantum theory.

In this section, we first find out the classical canonical variables, then promote the Poisson bracket to a commutator. As a result, the dynamical variables become operators acting on a Hilbert space and the theory is thus 'quantized'. That is, we can turn the wavepacket energy into an effective quantum Hamiltonian. The Dirac wavepacket will again be used as an example and its effective quantum Hamiltonian turns out to be the well known Pauli Hamiltonian.

4.1. Deriving canonical variables

The complementary pair $(\mathbf{r}, \mathbf{p})$ are canonical variables if their equations of motion have the following canonical form:

$$\dot{\mathbf{r}} = \frac{\partial E}{\partial \mathbf{p}}; \qquad \dot{\mathbf{p}} = -\frac{\partial E}{\partial \mathbf{r}}. \tag{31}$$

This also means that their Poisson brackets have the canonical form

$$\begin{aligned} \{r_\alpha, r_\beta\} &= 0, \\ \{p_\alpha, p_\beta\} &= 0, \\ \{r_\alpha, p_\beta\} &= \delta_{\alpha\beta}. \end{aligned} \tag{32}$$

The variables $\mathbf{r}_c$ and $\mathbf{k}_c$ that depict the trajectory of the wavepacket are not canonical variables because their equations of motion are not of the form above. This is due to the gauge potentials, $\mathbf{A}(\mathbf{r}_c)$ and $\mathbf{R}(\mathbf{k}_c)$, that generate rich dynamics but are otherwise harmless. However, in order to re-quantize the theory, one needs to find canonical variables with a simple Poisson bracket. In principle, this can always be done, as assured by the Darboux theorem [44]. Investigation along this path offers a powerful tool for the quantization.

In general, there is no easy way to exactly construct the canonical variables for systems with non-Abelian gauge potentials. However, for the wavepacket theory in this review, which is accurate only to linear fields, there is a simple way of finding them even for the multi-band version. It is crucial to observe that the source of non-canonicity is the presence of the gauge potentials, $\mathbf{A}(\mathbf{r}_c)$ and $\mathbf{R}(\mathbf{k}_c)$, in the Lagrangian. They need to be removed when the Lagrangian is written in the canonical variables $\mathbf{r}$ and $\mathbf{p}$. That is, once the Lagrangian in equation (16) can be written in the standard form (up to a total time derivative),

$$L_{\text{eff}} = \mathrm{i}\hbar\eta^\dagger \frac{\partial \eta}{\partial t} + \mathbf{p} \cdot \dot{\mathbf{r}} - E(\mathbf{r}, \mathbf{p}), \tag{33}$$

then the new variables $\mathbf{r}$ and $\mathbf{p}$ will automatically be canonical[7].

It is reasonable to make the initial guess

$$\hbar\mathbf{k}_c = \hbar\mathbf{k} + e\mathbf{A}(\mathbf{r}_c); \qquad \mathbf{r}_c = \mathbf{x} + \mathbf{R}(\mathbf{k}_c), \tag{35}$$

and rewrite the effective Lagrangian in equation (16) in the new variables, $\mathbf{x}$ and $\mathbf{k}$, as

$$L_{\text{eff}} = \mathrm{i}\hbar\eta^\dagger \frac{\partial \eta}{\partial t} + e\dot{\mathbf{A}} \cdot \mathbf{R} + \hbar\mathbf{k} \cdot \dot{\mathbf{x}} - E(\mathbf{x}, \mathbf{k}) + \frac{\mathrm{d}}{\mathrm{d}t}(\hbar\mathbf{k} \cdot \mathbf{R}), \tag{36}$$

in which the total time derivative will be ignored. This is still not of the standard form. To fix it, we choose a gauge $\mathbf{A} = \frac{1}{2}\mathbf{B} \times \mathbf{r}_c$ and write $e\dot{\mathbf{A}} \cdot \mathbf{R} = \frac{e}{2}\mathbf{R} \times \mathbf{B} \cdot (\dot{\mathbf{x}} + \dot{\mathbf{R}})$. After some rearrangement, one will get

$$e\dot{\mathbf{A}} \cdot \mathbf{R} + \hbar\mathbf{k} \cdot \dot{\mathbf{x}} = \left[\hbar\mathbf{k} + \frac{e}{2}\mathbf{R}(\mathbf{k}_c) \times \mathbf{B}\right] \cdot \dot{\mathbf{x}} + \mathbf{G}(\mathbf{k}_c) \cdot \hbar\dot{\mathbf{k}}_c, \tag{37}$$

where $G_\alpha(\mathbf{k}_c) \equiv (e/\hbar)(\mathbf{R} \times \mathbf{B}) \cdot \partial\mathbf{R}/\partial k_{c\alpha}$.

The term proportional to $\mathbf{G}$ can be removed by further shifting the position variable, $\mathbf{x} = \mathbf{r} + \mathbf{G}$. After throwing away a total time derivative $\mathrm{d}(\mathbf{G} \cdot \hbar\mathbf{k}_c)/\mathrm{d}t$ and a term explicitly nonlinear in B, we finally have

$$L_{\text{eff}} = \mathrm{i}\hbar\eta^\dagger \frac{\partial \eta}{\partial t} + \left[\hbar\mathbf{k} + \frac{e}{2}\mathbf{R}(\mathbf{k}_c) \times \mathbf{B}\right] \cdot \dot{\mathbf{r}} - E. \tag{38}$$

Obviously, the quantity in the square bracket should be identified as the canonical momentum $\mathbf{p}$ conjugate to the variable $\mathbf{r}$ (see equation (33)). Therefore, the connections between new and old variables are [16]

$$\begin{aligned} \mathbf{r} &= \mathbf{r}_c - \mathbf{R}(\mathbf{k}_c) - \mathbf{G}(\mathbf{k}_c), \\ \mathbf{p} &= \hbar\mathbf{k}_c - e\mathbf{A}(\mathbf{r}_c) - \frac{e}{2}\mathbf{B} \times \mathbf{R}(\mathbf{k}_c). \end{aligned} \tag{39}$$

The inverses of the relations are (again accurate to linear fields)

$$\begin{aligned} \mathbf{r}_c &= \mathbf{r} + \mathbf{R}(\boldsymbol{\pi}) + \mathbf{G}(\boldsymbol{\pi}), \\ \hbar\mathbf{k}_c &= \boldsymbol{\pi} + e\mathbf{B} \times \mathbf{R}(\boldsymbol{\pi}), \end{aligned} \tag{40}$$

where $\boldsymbol{\pi} = \mathbf{p} + e\mathbf{A}(\mathbf{r})$. This is the correct generalization of the Peierls substitution for systems with Berry connection $\mathbf{R}$. It reduces to the usual Peierls substitution when $\mathbf{R} = 0$. Similar equations can be found via a route slightly more complicated than the one reported here [46].

Two comments are in order. First, the center-of-mass variables are gauge invariant, but the canonical variables are not. Therefore, only the center-of-mass variables can be considered as physical variables. Second, the G-term would further shift the position operator, but does not influence the velocity to linear order of fields. Similarly, the $\mathbf{B} \times \mathbf{R}$-term would shift the momentum, but does not alter the force to the same order.

[7] It is possible to rewrite the first term in L_{eff} as the Wess–Zumino term,

$$\mathrm{i}\eta^\dagger \frac{\mathrm{d}\eta}{\mathrm{d}t} = -S \int_0^1 \mathrm{d}\lambda \hat{\mathbf{S}}_\lambda \cdot \frac{\partial \hat{\mathbf{S}}_\lambda}{\partial \lambda} \times \frac{\partial \hat{\mathbf{S}}_\lambda}{\partial t}, \tag{34}$$

where $\mathbf{S}$ is the spin vector. See [45] for more details. However, we use the spinor amplitude η as a basic variable in the Lagrangian, instead of the spin vector $\mathbf{S}$.

4.2. Effective quantum Hamiltonian

When written in the canonical variables, the wavepacket energy in equation (11) becomes

$$E(\mathbf{r},\mathbf{p}) = E_0(\boldsymbol{\pi}) - e\phi(\mathbf{r}) + e\mathbf{E}\cdot\mathbf{R}(\boldsymbol{\pi}) + \frac{e}{2m}\mathbf{B}\cdot\left[\mathbf{L}(\boldsymbol{\pi}) + 2\mathbf{R}\times m\frac{\partial E_0}{\partial \boldsymbol{\pi}}\right]. \tag{41}$$

The third term on the RHS comes from expanding the electrostatic potential $\phi(\mathbf{r}_c)$ about the canonical position $\mathbf{r}$ to first order, with nonlinear terms neglected. It has the form of a dipole energy, with $\mathbf{R}$ being the displacement of the charge center. Later we will see that this term represents the usual spin–orbit interaction. The last term has the form of the Zeeman energy, but in addition to the usual magnetic moment generated by $\mathbf{L}$ there is an extra term from expanding the band energy $E_0(\mathbf{k}_c)$ about $\boldsymbol{\pi}$ to first order. We will call it a Yafet term [21, 47, 48].

The effective quantum Hamiltonian can be obtained from the wavepacket energy by the standard procedure of promoting the variables to operators,

$$[r_\alpha, r_\beta] = 0, \qquad [p_\alpha, p_\beta] = 0, \qquad [r_\alpha, p_\beta] = \mathrm{i}\hbar\delta_{\alpha\beta}.$$

In addition, the basic quantities, $\mathbf{R}$, $\mathbf{F}$, and $\mathbf{L}$, also need to be promoted to matrices, $\mathcal{R}$, $\mathcal{F}$, and $\mathcal{L}$.

The effective Hamiltonian, being a function of $\boldsymbol{\pi}$, is invariant under the electromagnetic gauge rotation. However, the terms explicit in the Berry connection $\mathbf{R}$, such as the dipole energy and the Yafet term, are not invariant under the gauge rotation that re-shuffles the spinor basis. Therefore, it would be tricky to give a clear physical interpretation to each individual term. The original semiclassical theory based on the physical center-of-mass variables, on the other hand, is free of any gauge ambiguity.

4.3. Relativistic Pauli Hamiltonian

In this subsection, we will again take the Dirac electron as an example to illustrate the power of this re-quantization procedure. Combining equations (11) and (27), one has the energy of a Dirac wavepacket ($E_0 = \gamma mc^2$),

$$\begin{aligned} E(\mathbf{r}_c,\mathbf{k}_c) &= E_0(\mathbf{r}_c) - e\phi_c + \frac{e}{2m}\mathbf{L}(\mathbf{k}_c)\cdot\mathbf{B} \\ &= \gamma mc^2 - e\phi(\mathbf{r}_c) + \frac{\mu_B}{\gamma^2}\langle\boldsymbol{\sigma}\rangle\cdot\mathbf{B} \\ &\quad + \frac{\mu_B}{\gamma^2(\gamma+1)}\frac{\hbar\mathbf{k}_c}{mc}\cdot\langle\boldsymbol{\sigma}\rangle\frac{\hbar\mathbf{k}_c}{mc}\cdot\mathbf{B}, \end{aligned} \tag{42}$$

in which all the γ are functions of $\mathbf{k}_c$ and $\mu_B = e\hbar/2m$. In the next step, the generalized Peierls substitution in equation (40) is used to write the energy in canonical variables. One needs, for example,

$$\begin{aligned} \gamma(\mathbf{k}_c)mc^2 &= mc^2\sqrt{1+\left(\frac{\boldsymbol{\pi}+e\mathbf{B}\times\mathbf{R}}{mc}\right)^2} \\ &\simeq \gamma(\boldsymbol{\pi})mc^2 + \mu_B\langle\boldsymbol{\sigma}\rangle\cdot\mathbf{B}\left(\frac{\gamma-1}{\gamma^2}\right) \\ &\quad - \frac{\mu_B}{m^2c^2}\frac{\boldsymbol{\pi}\cdot\langle\boldsymbol{\sigma}\rangle\boldsymbol{\pi}\cdot\mathbf{B}}{\gamma^2(\gamma+1)}, \end{aligned} \tag{43}$$

in which all the γ are functions of $\boldsymbol{\pi}$. After some more calculations, one obtains an energy slightly simpler in form,

$$E(\mathbf{r},\mathbf{p}) = \gamma(\boldsymbol{\pi})mc^2 - e\phi(\mathbf{r}) + \frac{\mu_B}{\gamma}\langle\boldsymbol{\sigma}\rangle\cdot\mathbf{B} + \frac{\mu_B}{\gamma(\gamma+1)}\frac{\boldsymbol{\pi}}{mc^2}\times\langle\boldsymbol{\sigma}\rangle\cdot\mathbf{E}. \tag{44}$$

In the final step, one promotes the classical canonical variables $\mathbf{r}$ and $\mathbf{p}$ to quantum canonical variables, and $\langle\boldsymbol{\sigma}\rangle$ to Pauli matrices. The result is the effective quantum Hamiltonian for an electron living on the positive-energy branch. It allows no inter-branch transition (i.e. pair creation), but otherwise is valid for all velocities of the electron. At low velocity, it reduces to the familiar Pauli Hamiltonian.

This is by no means a trivial result since the gauge-invariant momentum operator $\boldsymbol{\pi}$ appears inside the square root in the denominator. In the usual textbook derivation, one performs successive Foldy–Wouthuysen (FW) transformations [49], treating the electron velocity as an expansion parameter. Such a unitary transformation quickly becomes very complicated at high velocity. One has to ingeniously come up with the correct unitary operator directly, assisted by some guessing, in order to reach the closed form in equation (44) [50]. By comparison, the approach presented here offers a simple and systematic way to derive an effective Hamiltonian when one has difficulty implementing an FW-like transformation.

4.4. Discussion

In addition to the re-quantization approach reported above, we would like to mention two other methods for obtaining an effective Hamiltonian. One method uses the unitary transformation to block-diagonalize a Hamiltonian. This is the method used by Foldy and Wouthuysen to obtain the Pauli Hamiltonian from the Dirac Hamiltonian [22]. Another method uses projection operators to restrict the extent of a Hilbert space [51]. This method has been implemented in single-particle as well as many-body quantum systems. More discussion of the effective Hamiltonian can also be found in Nenciu's review [27].

The zero-field Dirac Hamiltonian can be diagonalized by the unitary rotation given below equation (23). However, an external electric or magnetic field generates off-diagonal terms in the Hamiltonian matrix. One can perform the second unitary rotation U_{BD} to eliminate such couplings between positive- and negative-energy branches. This is the block-diagonalization process used in the FW transformation [22]. The Dirac Hamiltonian then consists of two disjoined pieces. Each piece is the effective 2×2 Hamiltonian for the corresponding energy branch. That is, $H_{\mathrm{eff},\pm} = P_\pm U_{BD}^\dagger H U_{BD} P_\pm$, where $P_\pm$ is the projection operator onto the positive-energy branch or the negative-energy branch.

This diagonalization method was used extensively in the 1950s and 1960s to obtain the effective Hamiltonian of a Bloch electron [4, 20, 21, 52]. In principle, one can systematically go beyond the linear order and obtain effective interactions nonlinear in field, but the derivation could be (in Kohn's words) 'shockingly complicated' [53].

One thing to keep in mind is that, as a result of the unitary transformation, the basis states are changed. Therefore, the explicit representation of an observable, such as the position operator, would also change. This fact is stated most clearly in the original paper by Foldy and Wouthuysen [22, 54]. For example, in the one-band effective theory, $\mathbf{r}_{\text{eff},n} = P_n U_{\text{BD}}^{\dagger} \mathbf{r} U_{\text{BD}} P_n$, where P_n projects to the nth band. In the simplest case when $U_{\text{BD}} = 1$, the effective position operator has a simple correction from the Berry connection [4],

$$\mathbf{r}_{\text{eff},n} = P_n \mathbf{r} P_n = \left\langle \psi_{n\mathbf{q}'} \left| \mathrm{i}\hbar \frac{\partial}{\partial \mathbf{p}} \right| \psi_{n\mathbf{q}} \right\rangle$$
$$= \left[\mathrm{i} \frac{\partial}{\partial \mathbf{q}} + \mathbf{R}_n(\mathbf{q}) \right] \delta(\mathbf{q} - \mathbf{q}'), \tag{45}$$

where $\mathbf{R}_n(\mathbf{q})$ is the Berry connection of the nth band. It is not unusual to apply only the projection and ignore the unitary transformation: the off-diagonal matrix elements in the Hamiltonian are often linear in field. After the unitary rotation, they would usually translate to quadratic terms in the block-diagonalized Hamiltonian. Therefore, if one is only interested in an effective Hamiltonian linear in field, then it may not be necessary to perform the rotation U_{BD}. In such cases, the effective position operator is obtained simply by the one-band projection.

It is also possible to obtain an effective Hamiltonian using nothing but the projection operator. One divides the complete Hilbert space into two orthogonal parts, one relevant and one irrelevant. Consider a time-independent Schrödinger equation,

$$(H_0 + V)|\Psi\rangle = E|\Psi\rangle, \tag{46}$$

where V is a perturbation, and $H_0|\Psi^{(0)}\rangle = E_0|\Psi^{(0)}\rangle$. It can be shown that the projected (relevant) state $P|\Psi\rangle$ satisfies the following equation [51, 55]:

$$H_{\text{eff}}(P|\Psi\rangle) = E(P|\Psi\rangle). \tag{47}$$

The effective Hamiltonian has the following form:

$$H_{\text{eff}} = H_0 + PVP + PV \frac{1}{E_0 - QHQ} VP$$
$$+ PV \frac{1}{E_0 - QHQ} V \frac{1}{E_0 - QHQ} VP + \cdots, \tag{48}$$

where $Q = 1 - P$ projects to irrelevant subspace. This projection method has been used, for example, in deriving the Fano–Feschbach resonance [56], the 'poor man's method' of renormalization [57], and the one-band effective Hamiltonians of quantum Hall systems [58, 59].

5. Hierarchy of effective theories

The process of constructing a semiclassical theory out of a given quantum theory, then using the re-quantization method to turn it back to an effective quantum theory, can be executed iteratively to generate a hierarchy of effective theories.

The Dirac theory in equation (22) for lattice electrons can be considered as the theory at the top of the solid-state-theory hierarchy. Instead of solving the Dirac–Bloch spectrum directly, which is more difficult, one can first treat the periodic potential $\phi(\mathbf{r})$ as a *smooth* external potential. This is justified since the Dirac wavepacket can be as small as the Compton wavelength, two orders of magnitude smaller than the size of an atom. Therefore, earlier discussion on the Dirac wavepacket in section 3 is fully applicable in such a periodic environment. One first has the semiclassical Dirac theory, then the re-quantization leads to the two-component Pauli theory. We emphasize again that, from the semiclassical point of view, the electron spin is a direct result of the self-rotation of the Dirac wavepacket.

Starting from the quantum Pauli theory, one can solve for the Bloch energy bands with the relativistic and the spin–orbit effects. By focusing on closely packed multiple bands of interest, one can climb down the hierarchy a notch further. For example, in semiconductor physics, one is often interested in only the conduction and the valence bands above and below the chemical potential. Therefore, depending on the desired accuracy, one can construct an effective theory having only four bands (Luttinger model), eight bands (Kane model), or 12 bands (extended Kane model) near the fundamental gap. This will be the subject of the next section. In this section, the hierarchy structure is studied from a more general perspective.

5.1. Gauge-invariant ingredients of an effective theory

In order for the re-quantization to work correctly, it is essential that the three gauge-invariant quantities, E_0, $\mathbf{F}$, and $\mathbf{L}$ (or their matrix version for multiple bands, which are gauge covariant), are encoded in the semiclassical description. Since the Bloch energy E_0 is a rather well known concept, we focus only on the latter two quantities here.

In the multi-band version, the Berry curvature and self-rotating angular momentum defined in equations (5), (8) can be rewritten in alternative forms more convenient for calculation [16]. First we write equation (5) in the component form,

$$\mathbf{F}_{mn} = \frac{\partial}{\partial \mathbf{q}} \times \left\langle u_m \left| \mathrm{i} \frac{\partial}{\partial \mathbf{q}} u_n \right.\right\rangle - \mathrm{i} \sum_{l \in \text{in}} \left\langle u_m \left| \mathrm{i} \frac{\partial}{\partial \mathbf{q}} u_l \right.\right\rangle \times \left\langle u_l \left| \mathrm{i} \frac{\partial}{\partial \mathbf{q}} u_n \right.\right\rangle$$
$$= \mathrm{i} \left\langle \frac{\partial}{\partial \mathbf{q}} u_m \right| \times \left| \frac{\partial}{\partial \mathbf{q}} u_n \right\rangle - \mathrm{i} \sum_{l \in \text{in}} \left\langle \frac{\partial}{\partial \mathbf{q}} u_m \middle| u_l \right\rangle \times \left\langle u_l \middle| \frac{\partial}{\partial \mathbf{q}} u_n \right\rangle, \tag{49}$$

where l sums over the states *inside* the wavepacket space. One can insert a complete set $\sum_l$ to the first cross-product term, where l sums over the states in the whole Hilbert space. After canceling with the $\sum_{l \in \text{in}}$ in the second term, one has

$$\mathbf{F}_{mn} = \mathrm{i} \sum_{l \in \text{out}} \mathbf{R}_{ml} \times \mathbf{R}_{ln}, \tag{50}$$

where $\mathbf{R}_{ml}$ is the Berry connection, and l sums over the states *outside* of the wavepacket space. Similarly, by inserting a complete set to the bracket in equation (8), and using $\tilde{H}_0|u_{l(\in \text{in})}\rangle = E_0|u_{l(\in \text{in})}\rangle$, one has

$$\mathbf{L}_{mn} = \frac{m_0}{\mathrm{i}\hbar} \sum_{l \in \text{out}} (E_{0,m} - E_{0,l}) \mathbf{R}_{ml} \times \mathbf{R}_{ln}. \tag{51}$$

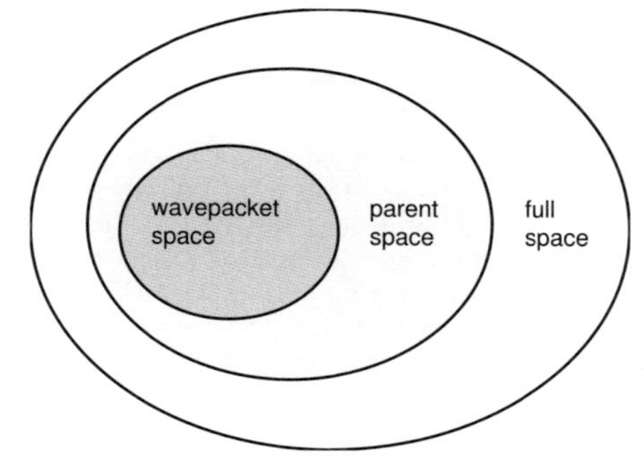

Figure 3. The relations between wavepacket space, parent space, and full space.

In the following, these two formulas are used to calculate the Berry curvature and the orbital angular momentum of a subspace.

In principle, equations (50) and (51) are exact if the summation is over the entire Hilbert space outside the wavepacket space. In practice, one often starts with a limited parent space, thereby neglecting terms from remote states outside this parent space (see figure 3). Let us call the approximate Berry curvature and orbital angular momentum obtained this way $\mathcal{F}'$ and $\mathcal{L}'$. In addition, let us assume that the neglected remote states generate a Berry curvature $\mathcal{F}_{\mathrm{p}}$ and an orbital angular momentum $\mathcal{L}_{\mathrm{p}}$ in the parent space, which is a subspace of the full space (see figure 3). Then from equations (50) and (51) they give corrections to the approximate $\mathcal{F}'$ and $\mathcal{L}'$ as follows:

$$\begin{aligned} \mathcal{F} &= \mathcal{F}' + P\mathcal{F}_{\mathrm{p}}P, \\ \mathcal{L} &= \mathcal{L}' + P\mathcal{L}_{\mathrm{p}}P, \end{aligned} \tag{52}$$

where P is the projection from the parent space to the wavepacket subspace. These relations describe precisely how the Berry curvature and orbital angular momentum should be renormalized. There is no need to refer back to the remote bands, once the Berry curvature and orbital angular momentum of the parent space are known.

Two comments are in order: first, the Berry curvature of the full Hilbert space for a given quantum model is always zero, since there is no 'outside' state in equation (50) [60]. The same is true for the self-rotating angular momentum. However, whether a Hilbert space is full or not is a relative concept. From the perspective of a more detailed quantum model with more degrees of freedom, the original Hilbert space is no longer full, but is only a subspace of a larger Hilbert space. Therefore, this subspace now can have its own Berry curvature. For example, from the viewpoint of the Dirac theory in equation (22), the positive-energy branch has a non-zero Berry curvature, whereas the four-component state itself is 'flat'. However, this latter fact is no longer true if the Dirac theory is considered as an effective theory of an even more complete theory (see section 5.3 for more discussions).

Second, the Berry curvature often can be attributed to a topological object in the parameter space. For example, it could be due to a $U(1)$ monopole at the location of level crossing [1]. The monopole charge is related to the first Chern number in the mathematical theory of fiber bundles [61]. The non-Abelian Berry curvature in this review is a result of projection, not level crossing. It has been shown that the Berry curvature in the Dirac theory is generated from a meron [37], an instanton with a topological charge $1/2$. In the four-band Luttinger model (see next section), the Berry curvature is generated from an instanton with integer topological charge [62, 63].

5.2. Position and velocity

As we mentioned earlier in section 4.4, rotating the basis states of a Hilbert space amounts to a rotation of all operators, not only the Hamiltonian operator. Therefore, the effective Hamiltonian is only part of the story. From the 'parent theory' point of view, one also needs to know how the physical variables (e.g. $\mathbf{r}_{\mathrm{c}}$) are related to the canonical variables (e.g. $\mathbf{r}$, which is unphysical due to its gauge dependence) in the Hamiltonian to avoid mistakes.

In the generalized Peierls substitution (neglecting the G-term), $\mathbf{r}_{\mathrm{c}} = \mathbf{r} + \mathbf{R}$, $\mathbf{r}$ is the position operator of the re-quantized theory. The re-quantized theory can have its own semiclassical theory, with its own wavepacket centered at $\mathbf{r}_a \equiv \langle \mathbf{r} \rangle$. On the other hand, there is the original wavepacket centered at $\mathbf{r}_{\mathrm{c}}$. How would the motions of these two wavepackets relate to each other if they are living in the *same* Hilbert space?

To illustrate the difference, we consider only the simpler case with $\mathbf{B} = 0$. A more detailed discussion using the Dirac electron as an example can be found in appendix A. For the original wavepacket, with the help of equations (17), (18), and (20), one has

$$\hbar \dot{\mathbf{r}}_{\mathrm{c}} = \left\langle \frac{\partial \mathcal{H}_0}{\partial \mathbf{k}_{\mathrm{c}}} \right\rangle + e\mathbf{E} \times \langle \mathcal{F} \rangle, \tag{53}$$

where the angular bracket performs the spinor average. On the other hand, the other wavepacket is assumed to live in the *full* Hilbert space of the *effective* theory with $H_{\mathrm{eff}}(\mathbf{r}) = H_0 - e\phi(\mathbf{r}) + e\mathbf{E} \cdot \mathcal{R}$. As we have just explained in the previous subsection, the Berry curvature of a full Hilbert space is always zero. Therefore, one can choose a trivial gauge for the Berry connection inside the bracket in equation (18) and obtain

$$\hbar \dot{\mathbf{r}}_a = \left\langle \frac{\partial \mathcal{H}_0}{\partial \mathbf{k}_a} \right\rangle + e \left\langle \frac{\partial}{\partial \mathbf{k}_a} \mathbf{E} \cdot \mathcal{R} \right\rangle, \tag{54}$$

in which $\mathbf{k}_a = \mathbf{k}_{\mathrm{c}}$ when $\mathbf{B} = 0$ (so the subscript will be dropped). The transverse velocities in equation (53) and equation (54) are obviously different.

One can improve the result in equation (54) by including the missing piece, $\langle \dot{\mathbf{R}} \rangle = \langle [\mathcal{R}, \mathcal{H}]/i\hbar \rangle + \langle \partial \mathcal{R}/\partial t \rangle$. It gives the following correction:

$$\hbar \langle \dot{\mathcal{R}} \rangle = -ie \langle [\mathcal{R}, \mathbf{E} \cdot \mathcal{R}] \rangle - e \left\langle \mathbf{E} \cdot \frac{\partial}{\partial \mathbf{k}} \mathcal{R} \right\rangle. \tag{55}$$

When combined with $\hbar\dot{\mathbf{r}}_a$, it is not difficult to check that one would indeed recover the correct result in equation (53).

For example, the Pauli theory is a valid effective theory of the Dirac theory. They both produce the same positive-energy spectrum. However, starting from the Pauli theory alone, *in the non-relativistic limit* and with $\mathbf{B} = 0$ for simplicity, the velocity of a Pauli wavepacket reads

$$\dot{\mathbf{r}}_a = \frac{\hbar\mathbf{k}}{m} + \frac{e\lambda_c^2}{4\hbar}\langle\boldsymbol{\sigma}\rangle \times \mathbf{E}. \tag{56}$$

On the other hand, for a Dirac wavepacket, one has

$$\dot{\mathbf{r}}_c = \frac{\hbar\mathbf{k}}{m} + \frac{e\lambda_c^2}{2\hbar}\langle\boldsymbol{\sigma}\rangle \times \mathbf{E}. \tag{57}$$

The transverse velocity of the Pauli wavepacket is smaller by a factor of two. That is, the Pauli theory does not always give a correct dynamical prediction *even in the low-velocity limit*. The lesson is that the effective Hamiltonian alone does not specify the effective theory completely. To avoid mistakes, one also needs to know how the physical variables are related to the canonical variables of the theory.

5.3. Dirac theory as an effective theory

At finer length scale or higher energy scale, even the Dirac theory itself is not complete. It needs to be embedded in a larger Hilbert space with more degrees of freedom. One example is the theory of electrons interacting with a quantized electromagnetic field. Another example is the theory for nucleons with inner structures. In these cases, the Dirac theory itself is an effective theory projected from a larger Hilbert space including photons or quarks.

By now, one should be familiar with the fact that such a projection can often generate a gauge structure (due to $\mathcal{R}$ and $\mathcal{F}$) and a magnetic moment (due to $\mathcal{L}$) in the effective theory. Indeed, the anomalous magnetic moment (AMM) of a Dirac particle originates either from the coupling with a quantized electromagnetic field, or from the internal motion of the constituent quarks in a nucleon. It contributes an energy in addition to those in equation (22) [49],

$$H' = a\mu_B\beta\left(\boldsymbol{\Sigma}\cdot\mathbf{B} - \mathrm{i}\boldsymbol{\alpha}\cdot\frac{\mathbf{E}}{c}\right), \tag{58}$$

where $a = g/2 - 1$ and $\boldsymbol{\Sigma}$ is the 4×4 spin matrix. Such an interaction will be treated as a perturbation to H_0.

The perturbation energy is calculated as $E' = \langle W|H'|W\rangle = \sum_{mn}\eta_m^*\langle u_m|H'|u_n\rangle\eta_n$ [29], where $|u_m\rangle$ ($m = 1, 2$) are the positive-energy eigenstates of H_0 (being evaluated at $\mathbf{k}_c$). With the help of the two identities,

$$\langle u_m|\beta\boldsymbol{\Sigma}|u_n\rangle = \left[\boldsymbol{\sigma} - \frac{\lambda_c^2}{\gamma(\gamma+1)}\mathbf{k}_c(\mathbf{k}_c\cdot\boldsymbol{\sigma})\right]_{mn}, \tag{59}$$

$$\mathrm{i}\langle u_m|\beta\boldsymbol{\alpha}|u_n\rangle = \left[\frac{\lambda_c}{\gamma}\boldsymbol{\sigma}\times\mathbf{k}_c\right]_{mn}, \tag{60}$$

one has

$$E' = \frac{a\mu_B}{\gamma mc^2}\hbar\mathbf{k}_c \times \langle\boldsymbol{\sigma}\rangle\cdot\mathbf{E} + a\mu_B\left[\langle\boldsymbol{\sigma}\rangle - \frac{\hbar\mathbf{k}_c\cdot\langle\boldsymbol{\sigma}\rangle\hbar\mathbf{k}_c}{\gamma(\gamma+1)m^2c^2}\right]\cdot\mathbf{B}. \tag{61}$$

Unlike the wavepacket energy in equation (11), the spin–orbit coupling is now explicit from the very beginning (before the Peierls substitution). It indicates that, when the particle has an AMM, the true center of charge $\mathbf{r}_c'$ has been displaced further from the $\mathbf{r}_c$ of a Dirac wavepacket. Such a displacement $\mathbf{R}'$ generates a spin–orbit interaction $e\mathbf{E}\cdot\mathbf{R}'$. By comparing with the spin–orbit energy in equation (61), we obtain $\mathbf{R}' = a(\lambda_c^2/2\gamma)\mathbf{k}_c \times \langle\boldsymbol{\sigma}\rangle$. Indeed, if one simply replaces the $\mathbf{r}_c$ and $e/2m$ in equation (11) with $\mathbf{r}_c' = \mathbf{r}_c + \mathbf{R}'$ and $eg/4m$ respectively, then the energy E' above can be exactly reproduced.

The equations of motion for a particle with AMM will not be reported in detail. A short discussion on its spin precession can be found in appendix B.

Following the same re-quantization scheme using the generalized Peierls substitution, we can again obtain the relativistic Pauli Hamiltonian with the AMM correction. First, by combining the energies in equations (42) and (61), the total energy becomes

$$\begin{aligned} E(\mathbf{r}_c, \mathbf{k}_c) &= \gamma mc^2 - e\phi(\mathbf{r}_c) + \mu_B\langle\boldsymbol{\sigma}\rangle\cdot\mathbf{B}\left(\frac{1}{\gamma^2} + a\right) \\ &\quad + \frac{a\mu_B}{\gamma mc^2}\hbar\mathbf{k}_c \times \langle\boldsymbol{\sigma}\rangle\cdot\mathbf{E} \\ &\quad + \frac{\mu_B}{m^2c^2}\hbar\mathbf{k}_c\cdot\langle\boldsymbol{\sigma}\rangle\hbar\mathbf{k}_c\cdot\mathbf{B}\left[\frac{1}{\gamma^2(\gamma+1)} - \frac{a}{\gamma(\gamma+1)}\right], \end{aligned} \tag{62}$$

in which $\gamma = \gamma(\mathbf{k}_c)$. Then one can use the generalized Peierls substitution to write the energy in canonical variables. Such a result is a generalization of equation (44). Promoting the canonical variables to quantum conjugate variables, we finally obtain the following relativistic Pauli Hamiltonian:

$$\begin{aligned} \mathcal{H}(\mathbf{r}, \mathbf{p}) &= \gamma mc^2 - e\phi(\mathbf{r}) + \mu_B\boldsymbol{\sigma}\cdot\mathbf{B}\left(\frac{1}{\gamma} + a\right) \\ &\quad + \frac{\mu_B}{\gamma mc^2}\boldsymbol{\pi}\times\boldsymbol{\sigma}\cdot\mathbf{E}\left(\frac{1}{1+\gamma} + a\right) \\ &\quad - \frac{a\mu_B}{m^2c^2}\frac{\boldsymbol{\pi}\cdot\boldsymbol{\sigma}\boldsymbol{\pi}\cdot\mathbf{B}}{\gamma(\gamma+1)}, \end{aligned} \tag{63}$$

in which $\gamma = \gamma(\boldsymbol{\pi})$. This result is an extension of the low-velocity one reported in [64].

We emphasize that the presence of the electric dipole energy indicates that the theory is only an effective one. The position of the particle in the effective theory should differ from the true position by a Berry connection. Of course, in predicting the trajectory of the particle, one always has to refer to its true position. Therefore, in the Dirac theory with $H' = a\mu_B\beta(\boldsymbol{\Sigma}\cdot\mathbf{B} - \mathrm{i}\boldsymbol{\alpha}\cdot\mathbf{E}/c)$, the true position is conjectured to be $\mathbf{r}_D + \mathbf{R}_D$, where $\mathbf{r}_D$ is the usual position operator in Dirac theory, and $\mathbf{R}_D = (a\mu_B/e)(-\mathrm{i}\beta\boldsymbol{\alpha})$ is the displacement required to generate the electric field term in H'. Thus, the true velocity operator of the particle should be

$$\dot{\mathbf{r}}_D + \dot{\mathbf{R}}_D = (1+a)c\boldsymbol{\alpha} - a\beta\frac{\boldsymbol{\pi}}{m} + a^2\frac{\mu_B}{mc}\left(\frac{\mathbf{E}}{c}\times\boldsymbol{\Sigma} + \gamma_5\mathbf{B}\right). \tag{64}$$

$$
H = \begin{pmatrix}
E_g + \frac{p^2}{2m_0} & 0 & -\frac{1}{\sqrt{2}}Vp_+ & \sqrt{\frac{2}{3}}Vp_z & \frac{1}{\sqrt{6}}Vp_- & 0 & -\frac{1}{\sqrt{3}}Vp_z & -\frac{1}{\sqrt{3}}Vp_- \\
0 & E_g + \frac{p^2}{2m_0} & 0 & -\frac{1}{\sqrt{6}}Vp_+ & \sqrt{\frac{2}{3}}Vp_z & \frac{1}{\sqrt{2}}Vp_- & -\frac{1}{\sqrt{3}}Vp_+ & \frac{1}{\sqrt{3}}Vp_z \\
-\frac{1}{\sqrt{2}}Vp_- & 0 & -P-Q & -L & -M & 0 & \frac{1}{\sqrt{2}}L & \sqrt{2}M \\
\sqrt{\frac{2}{3}}Vp_z & -\frac{1}{\sqrt{6}}Vp_- & -L^* & -P+Q & 0 & -M & -\sqrt{2}Q & -\sqrt{\frac{3}{2}}L \\
\frac{1}{\sqrt{6}}Vp_+ & \sqrt{\frac{2}{3}}Vp_z & -M^* & 0 & -P+Q & L & -\sqrt{\frac{2}{3}}L^* & \sqrt{2}Q \\
0 & \frac{1}{\sqrt{2}}Vp_+ & 0 & -M^* & L^* & -P-Q & -\sqrt{2}M^* & \frac{1}{\sqrt{2}}L^* \\
-\frac{1}{\sqrt{3}}Vp_z & -\frac{1}{\sqrt{3}}Vp_- & \frac{1}{\sqrt{2}}L^* & -\sqrt{2}Q & -\sqrt{\frac{3}{2}}L & -\sqrt{2}M & -\Delta - \frac{\gamma_1 p^2}{2m_0} & 0 \\
-\frac{1}{\sqrt{3}}Vp_+ & \frac{1}{\sqrt{3}}Vp_z & \sqrt{2}M^* & -\sqrt{\frac{3}{2}}L^* & \sqrt{2}Q & \frac{1}{\sqrt{2}}L & 0 & -\Delta - \frac{\gamma_1 p^2}{2m_0}
\end{pmatrix}, \qquad (65)
$$

The magnitude of the anomaly a for an electron is of the order of 10^{-3} and might be too small for such a difference to be observed. However, the anomaly for a nucleon is of order one. Therefore, it is possible to observe and verify such a deviation in experiments using relativistic nucleon beams[8].

6. Applications to semiconductor physics

We now turn our attention to an important application of the semiclassical approach. Using the eight-band Kane model as the parent theory, different gauge structures in different subspaces of the Kane model will be studied [16]. It is shown that the Berry curvature in the conduction band gives the wavepacket a spin-dependent transverse velocity *even in bulk materials*. A similar situation exists in valence bands and produces the spin Hall effect predicted by Murakami *et al* [11, 12]. We have also calculated the self-rotating angular momenta of the wavepackets in the conduction band and the valence band, from which one can obtain the anomalous g-factors. Finally, the effective Hamiltonians for different subspaces will be studied.

6.1. Berry connection in the Kane model

Starting with the eight-band Kane model, we study the wavepacket dynamics in its subspaces, such as the conduction band, the heavy hole–light hole (HH–LH) complex, and the spin–orbit split-off (SO) band (see figure 4). Each band has a twofold spin degeneracy. Assuming the fundamental gap is located at $\mathbf{k} = 0$, then for a small k one can expand the states as $u(\mathbf{k}) = \sum_j c_j u_j(0)$. The Schrödinger equation, $Hu(\mathbf{k}) = Eu(\mathbf{k})$, has the following Hamiltonian [23]:

(see equation (65) given above)

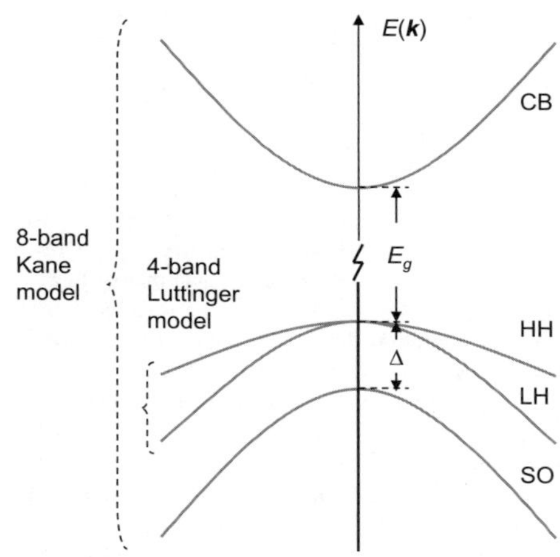

Figure 4. Schematic plot of the semiconductor band structure near the fundamental gap.

in which E_g is the fundamental gap, Δ is the spin–orbit split-off gap, $\mathbf{p} = \hbar\mathbf{k}$, and

$$
P = \frac{\gamma_1}{2m_0}p^2, \qquad Q = \frac{\gamma}{2m_0}(p_\parallel^2 - 2p_z^2),
$$
$$
L = -\frac{\sqrt{3}\gamma}{m_0}p_z p_-, \qquad M = -\frac{\sqrt{3}\gamma}{2m_0}p_-^2, \qquad (66)
$$

$$
V = \frac{\hbar}{m_0}\langle S|\hat{p}_x|X\rangle. \qquad (67)
$$

The matrix elements are ordered from higher energy to lower energy from the upper-left corner to the lower-right corner. That is, the upper-left 2×2 matrix is for the conduction electron. Simple truncation (projection) to the 4×4 matrix in the middle gives us the four-band Luttinger model for the HH–LH complex; the lower-right 6×6 matrix is for the six-band Luttinger model including the SO band. For demonstration, a simpler Hamiltonian with $\gamma_2 = \gamma_3 \equiv \gamma$ is chosen. Also, we do not distinguish between the usual band parameters and the reduced band parameters.

[8] However, when the velocity is projected to the subspace of a particle with positive energy, one has $P c\alpha P = \pi/(\gamma m)$. It would cancel with the term $P\beta\pi/mP = \pi/m$ to leading order of the velocity. Therefore, the leading (field-free) correction is of the order of $a(v/c)^2$ only.

One first needs to obtain the basic quantities $\mathcal{R}$, $\mathcal{F}$, and $\mathcal{L}$. Because of the anomalous velocity (the second term on the RHS) in equation (18), a non-zero Berry curvature in the conduction band or the valence band arising from the projection would immediately imply a transverse motion of the carrier that resembles the Hall effect. In this subsection, we focus first on the Berry connection $\mathcal{R}$ defined in equation (4), in which $|u_n\rangle$ are the eigenstates of the above Hamiltonian. It is unlikely for one to obtain these eigenstates by analytic diagonalization. Therefore, an alternative approach given in appendix C is used. It allows one to obtain the small-k limit of the Berry connection.

Instead of going through the details, we just show the result for the Berry connection of conduction electron. The result correct to the order of k^1 is[9]

$$\mathcal{R} = \frac{V^2}{3}\left[\frac{1}{E_g^2} - \frac{1}{(E_g + \Delta)^2}\right]\boldsymbol{\sigma} \times \mathbf{k}. \tag{68}$$

Therefore, the dipole term $e\mathbf{E} \cdot \mathcal{R}$ becomes

$$H_{so} = e\mathbf{E} \cdot \mathcal{R} = \alpha \mathbf{E} \cdot \boldsymbol{\sigma} \times \mathbf{k}, \tag{69}$$

where $\alpha \equiv (eV^2/3)[1/E_g^2 - 1/(E_g + \Delta)^2]$. This accounts for the spin–orbit coupling of a conduction electron. It requires neither bulk-inversion asymmetry, as in Dresselhaus coupling [65], nor structure-inversion asymmetry, as in Rashba coupling [66]. However, unlike the Rashba coupling in asymmetric quantum wells, in which the interface electric field plays a crucial role, here one needs to supply an *external* electric field E. When there is no spin–orbit split-off ($\Delta = 0$), the coefficient α vanishes as expected.

In addition, the Berry connections for the HH–LH complex and the split-off band are shown in table 1. They have the same functional form as the one in equation (68), but differ in coefficients and the dimensions of spin matrices. Notice that the Berry connection for the HH–LH complex in table 1 looks very different from the one in [11] (see equation (D.1) in appendix D). The parent theory there is the four-band Luttinger model, instead of the eight-band Kane model. The Berry curvature generated from the $\mathcal{R}$ in equation (D.1) in fact is zero, indicating a 'flat' space (see the discussion near the end of section 5.1). On the other hand, the Berry connection for the HH–LH complex in table 1 yields a non-zero Berry curvature. Because of equation (52), this gives a correction to the result in [11] (see appendix D).

6.2. *Berry curvature, orbital moment, and equations of motion*

The Berry curvature can be calculated from equation (5) once the Berry connection is known, while the angular momentum is most easily calculated from equation (51). Notice that, in order to calculate $\mathcal{L}$, the knowledge of off-diagonal $\mathcal{R}$ is required (see equation (C.5) in appendix C). Here we simply show the result. For the conduction band, to the lowest non-trivial order, one has

$$\mathcal{F} = \frac{2V^2}{3}\left[\frac{1}{E_g^2} - \frac{1}{(E_g + \Delta)^2}\right]\boldsymbol{\sigma}, \tag{70}$$

[9] This result remains valid up to a $SU(2)$ gauge rotation.

Table 1. Berry connection, Berry curvature, and orbital angular momentum of the *electron* wavepacket in three disjoint subspaces of the eight-band Kane model. Only the leading order (in k) terms are shown. E_g and Δ are the energy gaps in figure 4, $V = \hbar\langle S|p_x|X\rangle/m_0$, and $\mathcal{J}$ is the spin-3/2 matrix operator.

	Conduction band	HH–LH complex	Split-off band
$\mathcal{R}$	$\frac{V^2}{3}[\frac{1}{E_g^2} - \frac{1}{(E_g+\Delta)^2}]\boldsymbol{\sigma} \times \mathbf{k}$	$-\frac{V^2}{3E_g^2}\mathcal{J} \times \mathbf{k}$	$-\frac{V^2}{3}\frac{1}{(E_g+\Delta)^2}\boldsymbol{\sigma} \times \mathbf{k}$
$\mathcal{F}$	$\frac{2V^2}{3}[\frac{1}{E_g^2} - \frac{1}{(E_g+\Delta)^2}]\boldsymbol{\sigma}$	$-\frac{2V^2}{3E_g^2}\mathcal{J}$	$-\frac{2V^2}{3}\frac{1}{(E_g+\Delta)^2}\boldsymbol{\sigma}$
$\mathcal{L}$	$-\frac{2m_0V^2}{3\hbar}(\frac{1}{E_g} - \frac{1}{E_g+\Delta})\boldsymbol{\sigma}$	$-\frac{2m_0}{3\hbar}\frac{V^2}{E_g}\mathcal{J}$	$-\frac{2m_0}{3\hbar}\frac{V^2}{E_g+\Delta}\boldsymbol{\sigma}$

and

$$\mathcal{L} = -\frac{2m_0V^2}{3\hbar}\left(\frac{1}{E_g} - \frac{1}{E_g + \Delta}\right)\boldsymbol{\sigma}. \tag{71}$$

For a semiconductor electron, the E_0 in equation (11) should have already included the Zeeman energy from the *spin* magnetic moment. Through the Zeeman term in equation (11), the orbital magnetic moment generated from equation (71) contributes an extra g-factor,

$$\delta g = -\frac{4}{3}\frac{m_0V^2}{\hbar^2}\left(\frac{1}{E_g} - \frac{1}{E_g + \Delta}\right). \tag{72}$$

This is of course the familiar anomalous g-factor of the conduction electron. This result gives the clearest possible interpretation of the textbook formula [67]: the anomalous g-factor in solid is indeed a result of the self-rotating motion of the electron wavepacket.

The Berry curvatures and orbital angular momenta for the other two wavepacket subspaces can be found in table 1.

Finally, we selectively comment on the *electron* dynamics in the HH–LH complex using the semiclassical equations of motion. For brevity, the subscripts c in $\mathbf{r}_c$ and $\mathbf{k}_c$ are dropped. In order to focus on the effect introduced by the applied field, we set $\gamma = 0$ to avoid the complication from internal spin–orbit coupling. Therefore, this is only a toy-model calculation. Whenever necessary, one can always numerically solve the coupled dynamical equations with realistic parameters without difficulty.

In the simplest case, one applies only a constant electric field ($\mathbf{B} = 0$). The anomalous velocity in equation (18) due to the Berry curvature is $-2eV^2/(3E_g^2)\mathbf{E} \times \mathbf{J}$, where $\mathbf{J} = \langle\mathcal{J}\rangle$. Being transverse to the electric field and spin dependent, it is in fact a spin Hall current.

From the equation of spin precession in equation (21), one obtains

$$\dot{\mathbf{J}} = \frac{\kappa\mu_B/\hbar}{E_g/m_0c^2}\mathbf{B}_{\text{eff}} \times \mathbf{J}, \tag{73}$$

where $\kappa\mu_B \equiv eV^2/(3\hbar E_g)$, and $\mathbf{B}_{\text{eff}} = \mathbf{E} \times \hbar\mathbf{k}/(m_0c^2)$. The spin simply precesses around the effective magnetic field in its own reference frame.

In the presence of a constant magnetic field ($\mathbf{E} = 0$), the dynamics of the HH–LH wavepacket is slightly more complicated [68]. To linear order of the field, the quasi-momentum decouples from the other degrees of freedom, $\hbar\dot{\mathbf{k}} = (-e/\hbar)\partial E_0/\partial\mathbf{k} \times \mathbf{B}$, where $E_0 = -\gamma_1\hbar^2k^2/2m_0$. The orbit thus lies at the intersection of the Fermi surface and a plane perpendicular to the magnetic field. The component $\mathbf{k}_\parallel$ parallel to the magnetic field is a constant of motion. For a circular orbit, the frequency of rotation is the usual cyclotron frequency $\omega_c = eB\gamma_1/m_0$. The result so far is not different from the usual cyclotron motion without a Berry curvature [69].

The equation of spin precession is

$$\dot{\mathbf{J}} = \omega_V \left[\hat{\mathbf{B}} + \delta\mathbf{B}_\parallel(\mathbf{k}) + \delta\mathbf{B}_\perp(\mathbf{k})\right] \times \mathbf{J}, \tag{74}$$

where $\omega_V \equiv 2\kappa\mu_B B/\hbar$, $\delta\mathbf{B}_\parallel = -(E_0/E_g)\hat{\mathbf{k}}_\perp \times (\hat{\mathbf{k}}_\perp \times \hat{\mathbf{B}})$, and $\delta\mathbf{B}_\perp = -(E_0/E_g)\hat{\mathbf{k}}_\parallel \times (\hat{\mathbf{k}}_\perp \times \hat{\mathbf{B}})$, in which $\hat{\mathbf{B}}$ and $\hat{\mathbf{k}}_{\parallel,\perp}$ are unit vectors. The correction terms are negligible when the carrier is near the band edge (where $E_0 \simeq 0$).

After some inspection, it can be found that the correction $\delta\mathbf{B}_\perp$ rotates at the cyclotron frequency ω_c. On the other hand, $\delta\mathbf{B}_\parallel$ is proportional to $\mathbf{B}\sin^2\theta(t)$, where $\theta(t)$ is the angle between $\mathbf{B}$ and $\mathbf{k}(t)$. It can be steady or oscillating, depending on whether the orbit is circular or non-circular. In the simplest case of a circular orbit, the effective magnetic field is rotating around $\mathbf{B}$ with a fixed magnitude, resembling that in the electron spin resonance. Therefore, the spin would flip back and forth with a frequency ω_c. This is a natural consequence of combining the circular motion with the spin–orbit coupling.

6.3. Effective Hamiltonians

We can use the generalized Peierls substitution in equation (40) to obtain the effective Hamiltonians. The canonical Hamiltonian in equation (41) then has the following form for the conduction band:

$$H(\mathbf{r}, \mathbf{p}) = E_0(\boldsymbol{\pi}) - e\phi(\mathbf{r}) + \alpha\mathbf{E}\cdot\boldsymbol{\sigma}\times\boldsymbol{\pi} + \delta g\mu_B\mathbf{B}\cdot\frac{\hbar\boldsymbol{\sigma}}{2}, \tag{75}$$

where E_0 includes the Zeeman energy from the bare spin, α is given below equation (69), and δg is given in equation (72). The effective Hamiltonian for the spin–orbit split-off band has a similar form, but with $\alpha = -eV^2/[3(E_g + \Delta)^2]$ and $\delta g = -4m_0V^2/[3\hbar^2(E_g + \Delta)]$.

For a wavepacket in the HH–LH complex, one has

$$\begin{aligned} H(\mathbf{r}, \mathbf{p}) &= E_0(\boldsymbol{\pi}, \mathcal{J}) - e\phi(\mathbf{r}) + \alpha\mathbf{E}\cdot\mathcal{J}\times\boldsymbol{\pi} \\ &\quad + \delta g\mu_B\mathbf{B}\cdot\hbar\mathcal{J}, \end{aligned} \tag{76}$$

where $E_0(\boldsymbol{\pi}, \mathcal{J})$ is the four-band Luttinger Hamiltonian, $\alpha = -eV^2/(3E_g^2)$, and $\delta g = -2m_0V^2/(3\hbar^2 E_g)$. These results are in agreement with those in [23] obtained using the Löwdin partition.

The semiclassical approach is more than an alternative derivation of the effective Hamiltonians. It tells an intelligible story behind the formalism. The various interaction terms in the effective Hamiltonian have clear interpretations because of the direct connection with $\mathcal{R}$ and $\mathcal{L}$. Furthermore, these quantities can immediately be integrated with the equations of motion to study transport properties. Comparing with other approaches, such as the theory of linear response, this formulation offers a particularly useful visual guidance when the dynamics of charge and spin are coupled.

7. Conclusions

The wavepacket dynamics in a solid can successfully explain various phenomena, such as the Bloch oscillation and the de Haas–van Alphen effect [69]. In the early primitive version, the particle velocity is simply the group velocity of the wavepacket. Over the years, people gradually realized that the Berry curvature of Bloch states can have important effect on the electron transport. It contributes to an anomalous velocity that is crucial to the anomalous Hall effect [3, 8]. Subsequently, the (one-band) semiclassical formalism, with the Berry curvature included, was developed and applied successfully to the quantum Hall effect [6, 7]. A more general framework, allowing, e.g., adiabatic crystal deformation was then advanced [24]. Generalization of this formalism to the multi-band case makes it capable of investigating spintronic transport involving spin or quasi-spin degrees of freedom [29, 30].

Based on the semiclassical theory, one can use either the canonical quantization or other methods of quantization to obtain an effective quantum theory. This review focuses on the recent progress regarding the canonical quantization of the multi-band semiclassical theory. Because of this progress, the two-way passage between semiclassical and quantum theories is now achieved. We explain in great detail how to use this new recipe of quantization with the generalized Peierls substitution. It is applied successfully to the theories of Dirac electrons and semiconductor carriers.

It is necessary to include the Berry curvature and orbital moment in order to account for physical effects to first order in external fields. Furthermore, these gauge-covariant quantities are also sufficient for building a correct quantum theory. Although this route for an effective quantum theory is less direct, it is physically more transparent than other approaches. Most importantly, an architecture that is rich in concepts is revealed along the way. For example, we now understand the deep connection between the Berry connection and the spin–orbit interaction and the origin of the anomalous g-factor, to name only two. The wavepacket point of view is certainly both intuitive and productive.

The re-quantized effective theory can have its own semiclassical theory, which in turn can be re-quantized. Therefore, a whole series of effective theories can be generated. We study the connections between these hierarchical theories, and explain how the gauge-covariant quantities, $\mathcal{F}$ and $\mathcal{L}$, can be renormalized from one level to another. We now have a clear overview of these hierarchical theories. Such a fundamental understanding certainly is not restricted to solid state physics. It can be valuable to physical theories in other disciplines.

Most of the topics covered in this review have close connections with the field of spintronics, which has been vigorously pursued in recent years. In the past, the Bloch energy has played a key role in *ab initio* calculations. The values are calculated and tabulated for all kinds of crystalline materials. These results are useful for many purposes, but are incomplete in light of this work. The other two essential quantities, $\mathbf{F}$ and $\mathbf{L}$, should also be calculated and tabulated systematically. Such an endeavor would be very helpful for researchers to study more intricate electronic properties [70]. This subject is still at its dawning stage. We expect to see more exciting discoveries ahead of us.

Acknowledgments

The authors would like to thank D Culcer, C P Chuu, R Winkler, D Xiao, and W Yao for valuable discussions. They also thank the referee for pointing out the connection between this work and the electric polarization induced by spin current, which is added near the end of section 3.3. MCC was supported by the NSC of Taiwan (NSC 96-2112-M-003-010-MY3); QN was supported by DOE (DE-FG03-02ER45958), NSF (DMR-0404252/0606485), the Welch Foundation, and NSFC (No 10740420252).

Appendix A. The Pauli wavepacket

Starting from the non-relativistic Pauli Hamiltonian instead of the Dirac Hamiltonian, one can study the energy and dynamics of the Pauli wavepacket. The Hamiltonian can be expanded as

$$H = H_0 + \Delta H + H' = \frac{1}{2m}(\mathbf{p} + e\mathbf{A}_a)^2 - e\phi_a$$
$$+ \frac{e}{2m}\mathbf{L}\cdot\mathbf{B} + e\mathbf{E}\cdot(\mathbf{r} - \mathbf{r}_a) \tag{A.1}$$
$$+ \frac{e\hbar}{2m}\boldsymbol{\sigma}\cdot\mathbf{B} + \frac{e\hbar}{4m^2c^2}\boldsymbol{\sigma}\cdot\mathbf{E}\times(\mathbf{p} + e\mathbf{A}_a). \tag{A.2}$$

It consists of the unperturbed Hamiltonian H_0 (with two degenerate bands), the gradient correction ΔH, and the degeneracy-lifting perturbation H'. The third part H' has been mentioned in section 2 (footnote 3). Its influence on the equations of motion can be found in [29].

The Pauli Hamiltonian looks lengthier than the Dirac Hamiltonian. However, the eigenstates $|u_i\rangle$ of H_0 are simply the unit spinors, $(1, 0)^{\mathrm{T}}$ and $(0, 1)^{\mathrm{T}}$. Therefore, based on equations (4), (5), and (8), one has the trivial result with $\mathcal{R} = 0$, $\mathcal{F} = 0$, and $\mathcal{L} = 0$. This simplifies the equations of motion significantly [29],

$$\hbar\dot{\mathbf{k}}_a = -e\mathbf{E} - e\dot{\mathbf{r}}_a\times\mathbf{B}, \tag{A.3}$$

$$\hbar\dot{\mathbf{r}}_a = \frac{\partial E_0}{\partial\mathbf{k}_a} + \eta^\dagger\frac{\partial\mathcal{H}'}{\partial\mathbf{k}_a}\eta, \tag{A.4}$$

$$i\hbar\dot{\eta} = \mathcal{H}'\eta, \tag{A.5}$$

where $E_0 = \hbar^2k_a^2/2m$.

Substituting $\mathcal{H}' = \frac{e\hbar}{2m}\boldsymbol{\sigma}\cdot\mathbf{B} + \frac{e\hbar^2}{4m^2c^2}\boldsymbol{\sigma}\cdot\mathbf{E}\times\mathbf{k}_a$ into the equations above, one immediately has the following equations of motion:

$$\hbar\dot{\mathbf{k}}_a = -e\mathbf{E} - e\dot{\mathbf{r}}_a\times\mathbf{B}, \tag{A.6}$$

$$\dot{\mathbf{r}}_a = \frac{\hbar\mathbf{k}_a}{m} + \frac{e\hbar}{4m^2c^2}\langle\boldsymbol{\sigma}\rangle\times\mathbf{E}, \tag{A.7}$$

$$i\hbar\dot{\eta} = \frac{e\hbar}{2m}\left(\boldsymbol{\sigma}\cdot\mathbf{B} + \frac{\hbar}{2mc^2}\boldsymbol{\sigma}\cdot\mathbf{E}\times\mathbf{k}_a\right)\eta. \tag{A.8}$$

To compare these equations with those of the Dirac wavepacket in equations (28), (29), and (30), one needs the following two relations (see equation (40); $\mathbf{G}$ is neglected):

$$\begin{aligned}\mathbf{r}_c &= \mathbf{r}_a + \mathbf{R},\\ \hbar\mathbf{k}_c &= \hbar\mathbf{k}_a + e\mathbf{B}\times\mathbf{R}.\end{aligned} \tag{A.9}$$

One can check that the Lorentz force equation in equation (A.6) remains valid in either set of variables, $(\mathbf{r}_a, \mathbf{k}_a)$ or $(\mathbf{r}_c, \mathbf{k}_c)$. To linear order of fields, one can replace the $\mathbf{k}_a$ in equation (A.8) by $\mathbf{k}_c$. Therefore, it also gives the same (non-relativistic) BMT equation as given in equation (30).

There are two main differences between equation (29) and equation (A.7). First, the velocity transverse to the electric field in equation (A.7) is smaller by a factor of two if one does not distinguish between $\mathbf{r}_c$ and $\mathbf{r}_a$. Second, there is no magnetic-field-related velocity in equation (A.7). If the shift of position due to the Berry connection is taken into account, then in the non-relativistic limit one has

$$\dot{\mathbf{R}} = \frac{e\lambda_c^2}{4\hbar}\langle\boldsymbol{\sigma}\rangle\times\mathbf{E} + \frac{e\lambda_c^2}{4m}\mathbf{B}\times(\mathbf{k}_a\times\langle\boldsymbol{\sigma}\rangle). \tag{A.10}$$

The first term compensates the transverse velocity missing in equation (A.7) (see section 5.2). The second term would cancel with an extra term from the $e\mathbf{B}\times\mathbf{R}$ shift in momentum. Therefore, equation (A.7) gives

$$\dot{\mathbf{r}}_c = \frac{\hbar\mathbf{k}_c}{m} + \frac{e\hbar}{2m^2c^2}\langle\boldsymbol{\sigma}\rangle\times\mathbf{E}. \tag{A.11}$$

The $\mathbf{B}\cdot\mathbf{F}$-correction in equation (29) cannot be reproduced even if the R-shift has been considered. Such a correction amounts to a spin-dependent mass renormalization, which is not included in the Pauli equation from the beginning.

Appendix B. Spin precession for a particle with AMM

As a reference and without going into details, we show the spin precession equation for a particle with $a \neq 0$,

$$\langle\dot{\boldsymbol{\sigma}}\rangle = \frac{e}{\gamma mc}\left[(1+\gamma a)\mathbf{B} - a\frac{(\hbar\mathbf{k}_c\cdot\mathbf{B})\hbar\mathbf{k}_c}{(\gamma+1)m^2c^2}\right.$$
$$\left. + \frac{1}{mc^2}\left(\frac{1}{\gamma+1} + a\right)\mathbf{E}\times\hbar\mathbf{k}_c\right]\times\langle\boldsymbol{\sigma}\rangle. \tag{B.1}$$

If the particle is moving with the constraint $\boldsymbol{\beta}\cdot\mathbf{B} = \boldsymbol{\beta}\cdot\mathbf{E} = 0$, where $\boldsymbol{\beta} = \mathbf{v}/c = \hbar\mathbf{k}_c/\gamma mc$, then the spin precession follows $\langle\dot{\boldsymbol{\sigma}}\rangle = \boldsymbol{\omega}_s\times\langle\boldsymbol{\sigma}\rangle$, in which

$$\boldsymbol{\omega}_s = \frac{e}{m}\left[\frac{\mathbf{B}}{\gamma} - \frac{1}{\gamma+1}\frac{\boldsymbol{\beta}}{c}\times\mathbf{E} + a\left(\mathbf{B} - \frac{\boldsymbol{\beta}}{c}\times\mathbf{E}\right)\right]. \tag{B.2}$$

On the other hand, the velocity vector orbits at a slightly different frequency, $\dot{\boldsymbol{\beta}} = \boldsymbol{\omega}_c\times\boldsymbol{\beta}$, in which

$$\boldsymbol{\omega}_c = \frac{e}{m}\left[\frac{\mathbf{B}}{\gamma} - \frac{\gamma}{\gamma^2-1}\frac{\boldsymbol{\beta}}{c}\times\mathbf{E}\right]. \tag{B.3}$$

Unlike the frequency of spin precession, such a frequency is independent of a. It is not difficult to find that their difference, $\boldsymbol{\omega}_a = \boldsymbol{\omega}_s - \boldsymbol{\omega}_c$, is independent of the electric field if $\gamma_{\mathrm{magic}}^2 - 1 = 1/a$. This fact can be used on a beam line to eliminate the unwanted effect from the confining electric field [71].

Appendix C. Calculation of the Berry connection

We explain briefly the approach used in calculating the Berry connection of the conduction band. The connections for other subspaces can be calculated in a similar manner. For the Kane model in equation (65), we can separate the Hamiltonian into the following three parts:

$$\begin{aligned} H(\mathbf{k}) &= H_0 + H_1^D(\mathbf{k}) + H_1^{\mathrm{OD}}(\mathbf{k}) \\ &= \begin{pmatrix} h_0 & 0 \\ 0 & 0 \end{pmatrix} + \begin{pmatrix} h_1 & 0 \\ 0 & h_1' \end{pmatrix} + \begin{pmatrix} 0 & M \\ M^\dagger & 0 \end{pmatrix}, \end{aligned} \quad \text{(C.1)}$$

where h_0 is the part of the conduction-band Hamiltonian when $\mathbf{k} = 0$. The rest of the Hamiltonian is separated into H_1^D with only diagonal blocks, plus H_1^{OD} with only off-diagonal blocks.

The Hamiltonian can be diagonalized using a unitary transformation,

$$\begin{aligned} H^D &= \mathrm{e}^{-\mathrm{i}S} H \mathrm{e}^{\mathrm{i}S} \\ &= H + \mathrm{i}[H, S] - \tfrac{1}{2}[S, [S, H]] + \cdots \\ &= H_0 + H_1^D + H_1^{\mathrm{OD}} + \mathrm{i}[H_0, S] + \mathrm{i}[H_1^D, S] \\ &\quad + [H_1^{\mathrm{OD}}, S] - \tfrac{1}{2}[S, [S, H_0]] + \cdots. \end{aligned} \quad \text{(C.2)}$$

One divides the hermitian matrix S into the following form:

$$S = \begin{pmatrix} S_{00} & S_{01} \\ S_{01}^\dagger & S_{11} \end{pmatrix}, \quad \text{(C.3)}$$

where S_{00} has the same rank as h_0. We then demand the off-diagonal blocks in equation (C.2) to vanish to the leading power of k. This enables us to solve the off-diagonal part S_{01}, which has the same power of k as H_1^{OD}. To linear order of S, as far as the Berry curvature is concerned, one need not solve the block-diagonal part S_{00} since it merely re-shuffles the basis states of the wavepacket subspace. Such a gauge rotation would not change the Berry curvature. Once the unitary matrix $U = \mathrm{e}^{\mathrm{i}S}$ is known (to this order), the Berry connection can be calculated using $\mathcal{R} = \mathrm{i}U^\dagger \partial U/\partial \mathbf{k}$.

For example, the upper block-diagonal part of the Berry connection is

$$P_{\mathrm{U}} \mathcal{R} P_{\mathrm{U}} = \frac{\mathrm{i}}{2} S_{01}^\dagger \frac{\partial S_{10}}{\partial \mathbf{k}} + \mathrm{H.c.}, \quad \text{(C.4)}$$

where P_{U} projects to the upper diagonal block. The leading off-diagonal block is ($P_L + P_{\mathrm{U}} = 1$)

$$P_{\mathrm{U}} \mathcal{R} P_L = -\frac{\partial S_{01}}{\partial \mathbf{k}}. \quad \text{(C.5)}$$

Off-diagonal Berry connection is required in the calculation of the orbital angular momentum in equation (51).

Appendix D. Berry connection in the Luttinger model

In an earlier study, the Berry connection in the valence band was investigated based on the four-band Luttinger model [11]. For reference, the result is briefly reviewed here. The Berry connection calculated using the Luttinger model is (θ and ϕ are the angles of $\mathbf{k}$ in the spherical coordinate)

$$\mathcal{R} = \frac{1}{k} \mathcal{J}_y \hat{\theta} - \frac{1}{k} (\mathcal{J}_x - \operatorname{cosec}\theta \mathcal{J}_z) \hat{\phi}, \quad \text{(D.1)}$$

which generates zero Berry curvature, $\mathcal{F} = 0$.

After being projected to the HH (LH) subspace, one has

$$\begin{aligned} P_{\mathrm{HH}} \mathcal{R} P_{\mathrm{HH}} &= \frac{3}{2k} \operatorname{cosec}\theta \sigma_z \hat{\phi}, \\ P_{\mathrm{LH}} \mathcal{R} P_{\mathrm{LH}} &= \frac{1}{2k} \operatorname{cosec}\theta \sigma_z \hat{\phi} - \frac{\sigma_x}{k} \hat{\phi} + \frac{\sigma_y}{k} \hat{\theta}. \end{aligned} \quad \text{(D.2)}$$

It has been emphasized [11] that the Berry connection for the HH states (with $m_j = \pm 3/2$) is diagonal (Abelian) because the vector operator $\mathrm{i}\partial/\partial \mathbf{k}$ cannot connect states differing in angular momentum by more than one. On the other hand, the connection of the LH states is non-Abelian. The *projected* Berry connection generates a non-zero Berry curvature in each of their own subspaces, $\mathcal{F}'_{\mathrm{HH/LH}} = \mp 3/(2k^2)\sigma_z \hat{r}$. They have the same form as a magnetic monopole field in the k-space with a charge $\mp 3/2\sigma_z$.

According to the renormalization relations in equation (52), since the HH–LH complex inherited a Berry connection from the four remote bands in the Kane model, the Berry curvature of the HH (or LH) subspace would obtain a correction,

$$\mathcal{F}_{\mathrm{HH/LH}} = \mathcal{F}'_{\mathrm{HH/LH}} + P_{\mathrm{HH/LH}} \mathcal{F}_{\mathrm{p}} P_{\mathrm{HH/LH}}, \quad \text{(D.3)}$$

where $\mathcal{F}_{\mathrm{p}}$ is the curvature of the HH–LH complex listed in table 1 and $P_{\mathrm{HH/LH}}$ is the projection from four bands to two bands.

References

[1] Berry M V 1984 *Proc. R. Soc.* A **392** 45
[2] Shapere A and Wilczek F (ed) 1989 *Geometric Phases in Physics* (Singapore: World Scientific)
[3] Adams E N and Blount E I 1959 *J. Phys. Chem. Solids* **10** 286
[4] Blount E I 1962 *Solid State Physics* vol 13, ed F Seitz and D Turnbull (New York: Academic)
[5] Thouless D J, Kohmoto M, Nightingale P and den Nijs M 1982 *Phys. Rev. Lett.* **49** 405
Kohmoto M 1993 *J. Phys. Soc. Japan* **62** 659
[6] Chang M C and Niu Q 1995 *Phys. Rev. Lett.* **75** 1348
[7] Chang M C and Niu Q 1996 *Phys. Rev.* B **53** 7010
[8] Karplus R and Luttinger J M 1954 *Phys. Rev.* **95** 1154
[9] Taguchi Y, Oohara Y, Yoshizawa H, Nagaosa N and Tokura Y 2001 *Science* **291** 2573
Jungwirth T, Niu Q and MacDonald A H 2002 *Phys. Rev. Lett.* **88** 207208
Fang Z *et al* 2003 *Science* **302** 92
[10] Zeng C, Yao Y, Niu Q and Weitering H 2006 *Phys. Rev. Lett.* **96** 37204
[11] Murakami S, Nagaosa N and Zhang S C 2003 *Science* **301** 1348
[12] Murakami S, Nagaosa N and Zhang S C 2004 *Phys. Rev.* B **69** 235206
[13] Onoda M, Murakami S and Nagaosa N 2004 *Phys. Rev. Lett.* **93** 83901
Sawada K and Nagaosa N 2005 *Phys. Rev. Lett.* **95** 237402
Onoda M, Murakami S and Nagaosa N 2006 *Phys. Rev.* E **74** 66610
[14] Bliokh K Y *et al* 2006 *Phys. Rev. Lett.* **96** 073903
Bliokh K Y 2006 *Phys. Rev. Lett.* **97** 043901
Duval C *et al* 2006 *Phys. Rev.* D **74** 021701 (R)
[15] Hofstadter D 1976 *Phys. Rev.* B **14** 2239
[16] Chang M C and Niu Q 2008 submitted
[17] Peierls R 1933 *Z. Phys.* **80** 763
[18] Luttinger J M 1951 *Phys. Rev.* **84** 814

[19] Luttinger J M and Kohn W 1955 *Phys. Rev.* **97** 869
[20] Kohn W 1959 *Phys. Rev.* **115** 1460
[21] Roth L M 1962 *J. Phys. Chem. Solids* **23** 433
[22] Foldy L and Wouthuysen S 1950 *Phys. Rev.* **78** 29
[23] Winkler R 2003 *Spin–Orbit Coupling Effects in Two-Dimensional Electron and Hole Systems* (Berlin: Springer)
[24] Sundaram G and Niu Q 1999 *Phys. Rev.* B **59** 14 915
[25] Resta R 1994 *Rev. Mod. Phys.* **66** 899
Resta R 2000 *J. Phys.: Condens. Matter* **12** R107
[26] Bohm A, Mostafazadeh A, Koizumi H, Niu Q and Zwanziger J 2003 *The Geometric Phase in Quantum Systems* (Heidelberg: Springer)
[27] Nenciu G 1991 *Rev. Mod. Phys.* **63** 91
[28] Teufel S 2003 *Adiabatic Perturbation Theory in Quantum Dynamics* (Berlin: Springer)
[29] Culcer D, Yao Y and Niu Q 2005 *Phys. Rev.* B **72** 85110
[30] Shindou R and Imura K-I 2005 *Nucl. Phys.* B **720** 399
[31] Marzari N and Vanderbilt D 1997 *Phys. Rev.* B **56** 12 847
[32] Wilczek F and Zee A 1984 *Phys. Rev. Lett.* **52** 2111
[33] Bliokh K Y 2005 *Europhys. Lett.* **72** 7
[34] Chuu C P, Chang M C and Niu Q 2008 submitted
[35] Baym G 1990 *Lectures on Quantum Mechanics* (Reading, MA: Addison-Wesley)
[36] Mathur H 1991 *Phys. Rev. Lett.* **67** 3325
[37] Shankar R and Mathur H 1994 *Phys. Rev. Lett.* **73** 1565
[38] Huang K 1952 *Am. J. Phys.* **20** 479
[39] Bargmann V, Michel L and Telegdi V L 1959 *Phys. Rev. Lett.* **2** 435
[40] Griffiths D J 1999 *Introduction to Electrodynamics* 3rd edn (Englewood Cliffs, NJ: Prentice-Hall)
Jackson J D 1999 *Classical Electrodynamics* 3rd edn (New York: Wiley)
[41] Penfield P Jr and Haus H A 1967 *Electrodynamics of Moving Media* (Cambridge, MA: MIT Press)
[42] Katsura H, Nagaosa N and Balatsky A V 2005 *Phys. Rev. Lett.* **95** 057205
[43] Frenkel J 1926 *Z. Phys.* **37** 243
also see Bruno P and Dugaev V K 2005 *Phys. Rev.* B **72** 241302(R)
[44] Arnold V I 1989 *Mathematical Methods of Classical Mechanics* 2nd edn (New York: Springer)
Xiao D, Shi J and Niu Q 2005 *Phys. Rev. Lett.* **95** 137204
[45] Sachdev S 1999 *Quantum Phase Transition* (Cambridge: Cambridge University Press) chapter 13
[46] Gosselin P, Bérard A and Mohrbach H 2006 *Preprint* hep-th/0603192
Gosselin P, Hanssen J and Mohrbach H 2006 *Preprint* cond-mat/0611628
[47] Blount E I 1962 *Phys. Rev.* **126** 1636
[48] Yafet Y 1962 *Solid State Physics* vol 13, ed F Seitz and D Turnbull (New York: Academic)
[49] Bjorken J D and Drell S D 1964 *Relativistic Quantum Mechanics* (New York: McGraw-Hill) chapter 4
[50] See equation (33) in Silenko A J 2003 *J. Math. Phys.* **44** 2952
[51] Feschbach H 1958 *Ann. Phys.* **5** 363
[52] Kane E O 1957 *J. Phys. Chem. Solids* **1** 249
[53] For a recent generalization to many-body systems, see Kita T and Arai M 2005 *J. Phys. Soc. Japan* **74** 2813
[54] Blount E I 1962 *Phys. Rev.* **128** 2454
[55] Löwdin P-O 1951 *J. Chem. Phys.* **19** 1396
[56] See, for example Pethick C J and Smith H 2002 *Bose–Einstein Condensation in Dilute Gases* (Cambridge: Cambridge University Press)
[57] Anderson P W 1984 *Basic Notions of Condensed Matter Physics* (Reading, MA: Addison-Wesley) p 194
[58] Rammal R and Bellissard J 1990 *J. Physique* **51** 1803
[59] Chang M C, Yang M F and Hong T M 1997 *Phys. Rev.* B **56** 3602
[60] Avron J E 1995 *Les Houches Proc.* ed E Akkermans, G Montambaux and J L Pichard (Amsterdam: North-Holland)
[61] Simon B 1983 *Phys. Rev. Lett.* **51** 2167
[62] Samuel J 1996 *Phys. Rev. Lett.* **76** 717
[63] Avron J E, Sadun L, Segert J and Simon B 1989 *Commun. Math. Phys.* **124** 595
[64] See p 288 in Brown L S and Gabrielse G 1986 *Rev. Mod. Phys.* **58** 233
[65] Dresselhaus G 1955 *Phys. Rev.* **100** 580
[66] Rashba E I 1960 *Sov. Phys.—Solid State* **2** 1224
Bychkov Y A and Rashba E I 1984 *JETP Lett.* **39** 78
also see Chazalviel J-N 1975 *Phys. Rev.* B **11** 3918
and the nice review by Zutic I, Fabian J and Das Sarma S 2004 *Rev. Mod. Phys.* **76** 323–410
[67] See problem (9.16) in Yu P Y and Cardona M 2003 *Fundamentals of Semiconductors* 3rd edn (Berlin: Springer)
[68] Winkler R 2004 *Phys. Rev.* B **70** 125301
[69] Ashcroft N W and Mermin N D 1976 *Solid State Physics* (Philadelphia, PA: Saunders)
[70] Yao Y, Kleinman L, MacDonald A H, Sinova J, Jungwirth T, Wang D-s, Wang E and Niu Q 2004 *Phys. Rev. Lett* **92** 037204
Guo G Y, Yao Y and Niu Q 2005 *Phys. Rev. Lett.* **94** 226601
Wang X, Yates J R, Souza I and Vanderbilt D 2006 *Phys. Rev.* B **74** 195118
[71] Carey R M *et al* 1999 *Phys. Rev. Lett.* **82** 1632

Polarization and Adiabatic Pumping in Inhomogeneous Crystals

Di Xiao,[1,2,*] Junren Shi,[3] Dennis P. Clougherty,[4] and Qian Niu[1]

[1]*Department of Physics, The University of Texas at Austin, Austin, Texas 78712, USA*
[2]*Materials Science and Technology Division, Oak Ridge National Laboratory, Oak Ridge, Tennessee 37831, USA*
[3]*Institute of Physics and ICQS, Chinese Academy of Sciences, Beijing 100080, China*
[4]*Department of Physics, University of Vermont, Burlington, Vermont 05405, USA*
(Received 12 November 2007; published 24 February 2009)

We develop a general theory of electric polarization in crystals with inhomogeneous order. We show that the inhomogeneity-induced polarization can be classified into two parts: a perturbative contribution stemming from a correction to the basis functions and a topological contribution described in terms of the Chern-Simons form of the Berry gauge fields. The latter is determined up to an uncertainty quantum, which is the second Chern number in appropriate units. Our theory provides an exhaustive link between microscopic models and the macroscopic polarization.

DOI: 10.1103/PhysRevLett.102.087602 PACS numbers: 77.22.−d, 05.30.Pr, 71.10.Fd, 75.30.−m

In the study of dielectric and ferroelectric materials, it often happens that the system is in an inhomogeneous state, with the electric polarization strongly depending on the inhomogeneity. Classic examples include ferroelectric domain walls with induced polarization charge density [1] and flexoelectric polarization produced by a strain gradient [2,3]. Recently, another example of this type of problem has been found in a novel class of multiferroics, where the magnetic order varies in space and induces a spontaneous polarization [4]. A central quantity in the description of these phenomena is the inhomogeneity-induced polarization. However, despite its broad applications, the theory of polarization in inhomogeneous crystals remains primarily based on phenomenology [5].

Theoretically, it has been recognized that in an extended system only the change in polarization has physical meaning, and it can be quantified by using the Berry phase of the electronic wave functions [6–8]. For periodic insulators, the Berry-phase formula is conveniently expressed in terms of the Bloch orbitals, allowing one to compute the polarization from first principles. However, in inhomogeneous crystals the translational symmetry is broken, rendering previous methods based on Bloch's theorem inapplicable. To circumvent this, a supercell approach has been applied in numerical calculations [9]. But this approach has two shortcomings: If the inhomogeneity is incommensurate with the lattice, an enormous supercell is required to render an accurate result; furthermore, by its nature this approach is blind to the spatial variation within the supercell, making it difficult to investigate the role of inhomogeneity in the appearance of the polarization.

In this Letter, we present a general theory of electric polarization in inhomogeneous crystals that is exact up to first order in the spatial gradient. Our theory is based on the elementary relation between the change in polarization and the adiabatic current [7,8]. The latter can be evaluated using the semiclassical formalism of Bloch electron dynamics [10], a powerful framework for investigating the influence of slowly varying perturbations. We find that the inhomogeneity-induced polarization has two different origins: One results from a perturbative correction to the basis functions and is a direct generalization of previous work on insulators in finite electric fields [11,12]; the other is topological and is given in terms of the Chern-Simons form of the Berry gauge fields derived from the Bloch functions. Because of its topological nature, this contribution can be determined only up to an uncertainty quantum, which we identify as the second Chern number in appropriate units [13]. Our theory provides an exhaustive link between microscopic models and the macroscopic polarization, and it allows for a straightforward implementation in first-principles codes.

Let us consider an insulating crystal with an order parameter that varies slowly in space. We assume that, at least at the mean-field level, the system can be described as a perfect crystal under the influence of an external field $\boldsymbol{h}(\boldsymbol{r})$. Under this assumption, the Hamiltonian can be written as $\mathcal{H}[\boldsymbol{h}(\boldsymbol{r})]$, with $\mathcal{H}(\boldsymbol{h}=0)$ describing the unperturbed periodic crystal. If, for example, the order parameter is the magnetization, then $\boldsymbol{h}(\boldsymbol{r})$ can be chosen as the exchange field that yields the corresponding spin configuration $\boldsymbol{s}(\boldsymbol{r})$. Since we are interested in the inhomogeneity-induced polarization, we choose the initial state with a uniform $\boldsymbol{h}=\boldsymbol{h}_0$ and the final state with $\boldsymbol{h}(\boldsymbol{r})$ being fully turned on [14]. Our goal is to calculate the polarization difference between these two states.

As a first step, we make a local approximation, in which the system is described by a class of local Hamiltonians $\mathcal{H}_c[\boldsymbol{h}(\boldsymbol{r}_c)]$ which assume a constant value of the external field $\boldsymbol{h}(\boldsymbol{r}_c)$ at each position $\boldsymbol{r}_c$. Since $\mathcal{H}_c[\boldsymbol{h}(\boldsymbol{r}_c)]$ maintains the periodicity of the unperturbed crystal, its eigenstates still have the Bloch form: $|\psi_n(\boldsymbol{k},\boldsymbol{r}_c)\rangle = e^{i\boldsymbol{k}\cdot\boldsymbol{r}}|u_n(\boldsymbol{k},\boldsymbol{r}_c)\rangle$. Inserting $|u_n(\boldsymbol{k},\boldsymbol{r}_c)\rangle$ into the King-Smith and Vanderbilt formula [6] for periodic insulators, we obtain the zeroth-order contribution to the polarization:

$$\boldsymbol{P}^{\text{KS-V}} = -e\sum_n \int_{\text{BZ}} \frac{d\boldsymbol{k}}{(2\pi)^d} \langle u_{nk}|i\boldsymbol{\nabla}_{\boldsymbol{k}}|u_{nk}\rangle|_0^1, \tag{1}$$

where the sum is over occupied bands and 0 and 1 label the initial and final states, respectively. If the crystal structure is centrosymmetric and the uniform $\boldsymbol{h}$ does not break the inversion symmetry, then $\boldsymbol{P}^{\text{KS-V}}$ vanishes.

Next we look for first-order contributions. A natural step is to expand the Hamiltonian around $\boldsymbol{r}_c$:

$$\mathcal{H} = \mathcal{H}_c + (\nabla_\alpha^r h_\beta)(r_\alpha - r_{c,\alpha})\hat{s}_\beta, \tag{2}$$

where $\nabla_\alpha^r \equiv \partial/\partial r_\alpha$ and $\hat{s}_\beta \equiv \partial\mathcal{H}/\partial h_\beta$. (Summation over repeated indices is implied throughout our derivation.) Obviously, the gradient term in the Hamiltonian (2) will introduce a correction to the basis functions, i.e., $|\tilde{u}_{nk}\rangle = |u_{nk}\rangle + |\delta u_{nk}\rangle$, where $|\tilde{u}_{nk}\rangle$ is the modified basis and $|\delta u_{nk}\rangle$ is proportional to the gradient. If one assumes that the polarization is still given by Eq. (1) but with $|u_{nk}\rangle$ replaced by $|\tilde{u}_{nk}\rangle$, then a first-order contribution can be obtained as

$$\delta\boldsymbol{P}^{\text{KS-V}} = 2e\sum_n \int_{\text{BZ}} \frac{d\boldsymbol{k}}{(2\pi)^d} \,\text{Im}\langle \delta u_{nk}|\boldsymbol{\nabla}_k|u_{nk}\rangle|_0^1. \tag{3}$$

The explicit form of $|\delta u_{nk}\rangle$ can be derived using linear-response theory by taking the long wavelength limit [7,15]. In the special case of an applied electric field, the above formula reduces to the one obtained in Ref. [11].

However, the above derivation misses an important contribution due to the inhomogeneity-induced electron dynamics. As was emphasized in previous work [6–8], the proper calculation of the polarization relies on the elementary relation between the polarization and the adiabatic current [16], i.e.,

$$\boldsymbol{P} = \int_0^T dt\boldsymbol{j}(\boldsymbol{r}, t), \tag{4}$$

where $\boldsymbol{j}(\boldsymbol{r}, t)$ is the bulk current density as the system adiabatically evolves from the initial state ($t = 0$) to the final state ($t = T$). This process can be described by a smooth transformation of the Hamiltonian $\mathcal{H}[\boldsymbol{h}(\boldsymbol{r}; \lambda)]$, parametrized by a scalar λ. Following convention, we set $\lambda(0) = 0$ and $\lambda(T) = 1$. As we shall show below, the specific form of $\boldsymbol{h}(\lambda)$ is not important as long as the system remains an insulator during the process.

In order to find the current density $\boldsymbol{j}(\boldsymbol{r}, t)$, we adopt the formalism of wave packet dynamics of Bloch electrons [10]. In the following derivation, we shall consider only the case of nondegenerate bands and hence omit the band index n. It has been shown that the wave packet center satisfies the following equations of motion [10,17]:

$$\dot{r}_\alpha = \nabla_\alpha^k \varepsilon - \Omega_{\alpha\beta}^{kr}\dot{r}_\beta - \Omega_{\alpha\beta}^{kk}\dot{k}_\beta - \dot{\lambda}\Omega_\alpha^{k\lambda}, \tag{5a}$$

$$\dot{k}_\alpha = -\nabla_\alpha^r \varepsilon + \Omega_{\alpha\beta}^{rr}\dot{r}_\beta + \Omega_{\alpha\beta}^{rk}\dot{k}_\beta + \dot{\lambda}\Omega_\alpha^{r\lambda}, \tag{5b}$$

where ε is the electron energy. Here $\boldsymbol{\Omega}$ is the Berry curvature obtained from the gauge field $\mathcal{A}$ derived from $|\tilde{u}(\boldsymbol{k}, \boldsymbol{r}; \lambda)\rangle$. For example,

$$\mathcal{A}_\alpha^k = \langle\tilde{u}|i\nabla_\alpha^k|\tilde{u}\rangle, \qquad \mathcal{A}_\alpha^r = \langle\tilde{u}|i\nabla_\alpha^r|\tilde{u}\rangle, \tag{6}$$

$$\Omega_{\alpha\beta}^{kr} = \nabla_\alpha^k \mathcal{A}_\beta^r - \nabla_\beta^r \mathcal{A}_\alpha^k. \tag{7}$$

The Berry curvatures are antisymmetric tensors ($\Omega_{\alpha\beta}^{kr} = -\Omega_{\beta\alpha}^{rk}$).

The appearance of the modified basis $|\tilde{u}_k\rangle$ in the equations of motion (5) via the Berry curvature is undesirable and, in fact, is unnecessary in most terms. Recall that we are aiming at a first-order calculation in the spatial gradient. A quick look of Eq. (5) reveals that there are two zeroth-order terms: $\nabla_\alpha^k \varepsilon$ and $\dot{\lambda}\Omega_\alpha^{k\lambda}$; the rest are at least first-order in the gradient. The term $\nabla_\alpha^k\varepsilon$ has no effect on our final result. Therefore only $\Omega_\alpha^{k\lambda}$ needs to be evaluated using $|\tilde{u}_k\rangle$; for other Berry curvatures, it is sufficient to evaluate them using the unperturbed basis $|u_k\rangle$. For the same reason, the term $\dot{\lambda}\Omega_\alpha^{r\lambda}$ should be kept in the derivation. In previous work [18,19], this term was dropped based on the reasoning that the time and spatial variations are treated at the same order. However, this is not true in the calculation of the polarization because of the explicit time integral in Eq. (4).

Now we can calculate the induced adiabatic current given by

$$j = -e\int_{\text{BZ}} d\boldsymbol{k} D(\boldsymbol{k}, \boldsymbol{r})\dot{\boldsymbol{r}}, \tag{8}$$

where $D(\boldsymbol{r}, \boldsymbol{k})$ is the electron density of states, modified from its usual value of $1/(2\pi)^d$ in the presence of the Berry curvature [20]: $D(\boldsymbol{k}, \boldsymbol{r}) = (1 + \Omega_{\alpha\alpha}^{kr})/(2\pi)^d$. We solve for $\dot{r}_\alpha$ from Eq. (5) and then insert it into Eq. (8). It can be shown that only terms proportional to $\dot{\lambda}$ survive [21]. Physically, this result is required by the adiabaticity of the transformation. We find the total current is $\boldsymbol{j} = \boldsymbol{j}^{(1)} + \boldsymbol{j}^{(2)}$, where

$$j_\alpha^{(1)} = e\dot{\lambda}\int_{\text{BZ}} \frac{d\boldsymbol{k}}{(2\pi)^d}\tilde{\Omega}_\alpha^{k\lambda} \tag{9}$$

and

$$j_\alpha^{(2)} = e\dot{\lambda}\int_{\text{BZ}} \frac{d\boldsymbol{k}}{(2\pi)^d}(\Omega_{\alpha\beta}^{kk}\Omega_\beta^{r\lambda} + \Omega_{\beta\beta}^{kr}\Omega_\alpha^{k\lambda} - \Omega_{\alpha\beta}^{kr}\Omega_\beta^{k\lambda}). \tag{10}$$

We have added a tilde to $\tilde{\Omega}_\alpha^{k\lambda}$ to emphasize that it is evaluated using the modified basis $|\tilde{u}_k\rangle$. The current $\boldsymbol{j}^{(1)}$ is given by the first Chern form of the Berry curvature, hence the superscript (1). It was first derived by Thouless in his study of charge pumping [22]. The current $\boldsymbol{j}^{(2)}$ is given by the second Chern form of the Berry curvature. It originates from the competition between the intrinsic lattice period and incommensurate external field, and it will vanish in a uniform system. $\boldsymbol{j}^{(2)}$ provides another route for adiabatic charge pumping through spatial inhomogeneity [13].

To understand the origin of the inhomogeneity-induced current $\boldsymbol{j}^{(2)}$, we examine the physical meaning of the individual terms in Eq. (10). The first term is a Hall current: The integration of $\Omega^{kk}_{\alpha\beta}$ gives the Hall conductivity [23], and $\Omega^{r\lambda}_{\beta}$ is the driving force due to the spatial variation. The second term is a spatially modulated current: $\Omega^{k\lambda}_{\alpha}$ is the electron velocity along the α direction, and $\Omega^{kr}_{\beta\beta}$, coming from the modified density of states, represents the change in the unit length along the β direction. The third term does not have an obvious interpretation, and it has to be obtained through the above derivation or its equivalent [17].

Once we obtain the expression for the adiabatic current $\boldsymbol{j}$, the polarization can be found by integrating $\boldsymbol{j}$ [Eq. (4)]. The integration of $\boldsymbol{j}^{(1)}$ gives $\boldsymbol{P}^{(1)} = \boldsymbol{P}^{\text{KS-V}} + \delta\boldsymbol{P}^{\text{KS-V}}$. This prompts us to name $\delta\boldsymbol{P}^{\text{KS-V}}$ the perturbative contribution since it is a correction to the original King-Smith and Vanderbilt formula. On the other hand, the integration of $\boldsymbol{j}^{(2)}$ gives a purely topological contribution $\boldsymbol{P}^{(2)}$. The difference between $\delta\boldsymbol{P}^{\text{KS-V}}$ and $\boldsymbol{P}^{(2)}$ can be seen by considering a cyclic change of the system. During this process, both $\boldsymbol{P}^{(1)}$ and $\boldsymbol{P}^{\text{KS-V}}$ can change only by an integer (in appropriate units). Since $\delta\boldsymbol{P}^{\text{KS-V}}$ is a perturbative contribution, it must vanish. In contrast, $\boldsymbol{P}^{(2)}$ can be nonzero and takes an integer value as discussed below. Since the general properties of $\boldsymbol{P}^{(1)}$ have been discussed thoroughly in previous work [6–8], we shall focus on the properties of $\boldsymbol{P}^{(2)}$ in what follows.

We first show that $\boldsymbol{P}^{(2)}$ has the desired property that it depends only on the initial and final states. The gauge invariance of Eq. (10) allows us to calculate $\boldsymbol{P}^{(2)}$ with any gauge choice. In order to carry out the integration over λ, we choose the path-independent gauge by requiring that the phase difference between $|u^{(\lambda)}(\boldsymbol{k},\boldsymbol{r})\rangle$ and $|u^{(\lambda)}(\boldsymbol{k}+\boldsymbol{G},\boldsymbol{r})\rangle$ does not depend on λ, where $\boldsymbol{G}$ is a reciprocal lattice vector [8]. Under this gauge, $\boldsymbol{P}^{(2)}$ is found as

$$P^{(2)}_{\alpha} = e\int_{\text{BZ}}\frac{d\boldsymbol{k}}{(2\pi)^d}(\mathcal{A}^k_\alpha\nabla^r_\beta\mathcal{A}^k_\beta + \mathcal{A}^k_\beta\nabla^k_\alpha\mathcal{A}^r_\beta + \mathcal{A}^r_\beta\nabla^k_\beta\mathcal{A}^k_\alpha)|^1_0. \tag{11}$$

We recognize that the integrand in the above equation is of Chern-Simons form.

However, we have paid a price for carrying out the λ integration; namely, the spatially averaged polarization $\langle P^{(2)}_\alpha\rangle = (1/V)\int d\boldsymbol{r} P^{(2)}_\alpha$ resulting from this two-point formula (11) can be determined only modulo a quantum. Consider a cyclic change of the system with $\lambda(0) = \lambda(T)$, and assume $\boldsymbol{h}(\boldsymbol{r})$ is periodic in $\boldsymbol{r}$. The integral $\int d\boldsymbol{r}\int dt j^{(2)}_\alpha$ over a closed manifold spanned by $(k_\alpha, k_\beta, r_\beta, t)$ is an integer called the second Chern number [13]. Since Eq. (11) does not track the evolution of $\lambda(t)$, there is no information on how many cycles λ has gone through; hence $\langle P^{(2)}_\alpha\rangle$ can be determined only modulo a quantum. Assuming $\boldsymbol{h}(\boldsymbol{r})$ depends on y, we obtain the quantum for $P^{(2)}_x$ in a three-dimensional system:

$$\Delta\langle P^{(2)}_x\rangle = \frac{e}{l_y a_z}, \tag{12}$$

where l_y is the period of $\boldsymbol{h}(y)$ and a_z is the lattice constant along $\hat{z}$.

Next we discuss the general conditions on microscopic models [24,25] that display a nonzero $\boldsymbol{P}^{(2)}$. Since the Hamiltonian is assumed to have the form $\mathcal{H}[\boldsymbol{h}(\boldsymbol{r},\lambda)]$, both λ and $\boldsymbol{r}$ dependence enter through $\boldsymbol{h}$. Using the relation $\nabla^r_\alpha = \nabla^r_\alpha h_\mu \nabla^h_\mu$ and $\nabla^\lambda = \partial_\lambda h_\mu\nabla^h_\mu$, we find

$$P^{(2)}_\alpha = e\int_{\text{BZ}}\frac{d\boldsymbol{k}}{(2\pi)^d}\int_0^1 d\lambda\nabla^\lambda h_\mu\nabla^r_\beta h_\nu(\Omega^{kh}_{\alpha\mu}\Omega^{kh}_{\beta\nu} - \Omega^{kh}_{\alpha\nu}\Omega^{kh}_{\beta\mu} + \Omega^{kk}_{\alpha\beta}\Omega^{hh}_{\mu\nu}). \tag{13}$$

Besides having the crystal be inhomogeneous, there are three general conditions for $\boldsymbol{P}^{(2)}$ to be nonzero according to Eq. (13): (i) The system must be two-dimensional or higher; (ii) the order parameter $\boldsymbol{h}(\boldsymbol{r})$ must have two or more components; and (iii) the wave function must depend on four or more *independent* parameters. Conditions (i) and (ii) are obtained by realizing that the integrand in Eq. (13) is antisymmetric in k_α and k_β and in h_μ and h_ν. Furthermore, since the integrand is actually the second Chern 4-form $\Omega\wedge\Omega$, it would vanish identically in three or fewer dimensions; hence condition (iii) follows. Based on condition (iii), we can further deduce that $\dim(\mathcal{H}) > 2$. If $\dim(\mathcal{H}) = 2$, $\mathcal{H}$ has four components. However, since shifting and scaling energy has no effect on wave functions, the wave function can depend on only two *independent* parameters (for example, the spherical coordinates on a 2-sphere S^2), and $\boldsymbol{P}^{(2)}$ vanishes in this case.

The Chern-Simons form of our theory also allows us to deduce the general form of $\boldsymbol{P}^{(2)}$. We first consider a two-dimensional "minimal" model with $\boldsymbol{h}(\boldsymbol{r})$ having two components. Because of its antisymmetric properties, we can write the integrand of Eq. (13) as $\epsilon_{\alpha\beta}\epsilon_{\mu\nu}\chi$. Then, by assuming $\boldsymbol{h}(\boldsymbol{r},\lambda) = \lambda\boldsymbol{h}(\boldsymbol{r})$, we find Eq. (13) takes the following form:

$$\boldsymbol{P}^{(2)} = e\chi[(\boldsymbol{\nabla}\cdot\boldsymbol{h})\boldsymbol{h} - (\boldsymbol{h}\cdot\boldsymbol{\nabla})\boldsymbol{h}]. \tag{14}$$

Here χ, as a function of $\boldsymbol{h}(\boldsymbol{r})$, can be spatially dependent. Interestingly, if we identify $\boldsymbol{h}(\boldsymbol{r})$ with the magnetization order parameter $\boldsymbol{M}(\boldsymbol{r})$, the above result has the same form as that obtained from the Landau-Ginzburg approach [26–28]. However, unlike the Landau theories, Eq. (14) is a direct consequence of the minimal dimensionality, and we did not invoke any symmetry argument; in particular, the time-reversal invariance is irrelevant.

Next we consider three-dimensional systems with cubic symmetry. We start from Eq. (11) by making the variable substitution $\nabla^r_\beta = \nabla^r_\beta h_\gamma\nabla^h_\gamma$. By applying the antisymmetric properties of the Chern-Simons form, we can write $P^{(2)}_\alpha = \epsilon_{\beta\gamma\sigma}\nabla^r_\beta h_\gamma C_{\alpha\sigma}$, where $C_{\alpha\sigma}$ is a coefficient. Since the crystal has cubic symmetry, we have either (i) $C_{\alpha\sigma} = \delta_{\alpha\sigma}\chi$ or (ii) $C_{\alpha\delta} = \epsilon_{\eta\alpha\sigma}D_\eta$. In the former case, $P^{(2)}_\alpha = \epsilon_{\beta\gamma\alpha}\chi\nabla^r_\beta h_\gamma$. This is allowed only when $\boldsymbol{h}$ is invariant under time reversal. The antisymmetric tensor $\epsilon_{\beta\gamma\alpha}$ also prohibits its existence in two-dimensional systems. If $C_{\alpha\sigma} = \epsilon_{\eta\alpha\sigma}D_\eta$, then $\boldsymbol{D}$ must be pointing in the direction of $\boldsymbol{h}$, i.e., $D_\eta = \chi h_\eta$, and $P^{(2)}_\alpha = \chi(h_\beta\nabla^r_\beta h_\alpha - h_\alpha\nabla^r_\beta h_\beta)$, which again gives Eq. (14). For more complicated cases, one will have to carry out a careful symmetry analysis of the crystal structure to determine the general form of $\boldsymbol{P}^{(2)}$.

Finally, we discuss how to calculate $\boldsymbol{P}^{(2)}$ from a computational point of view. For simplicity, we again consider a two-dimensional system and assume $\boldsymbol{h}(\boldsymbol{r})$ is periodic in the y direction. After taking the spatial average, the two-point formula (11) for $P^{(2)}_x$ involves three nontrivial variables, namely, k_x, k_y, and y. We then make a tetrahedral mesh in the three-dimensional parameter space spanned by these variables. At each mesh point $\boldsymbol{R}_i = (k_{x_i}, k_{y_i}, y_i)$, one can solve the Kohn-Sham equation self-consistently with the constraint that the ground state energy is minimized for a constant $\boldsymbol{h}(y_i)$. In the case of multiferroics, this is equivalent to solving the Kohn-Sham equation with a fixed and uniform spin magnetization [29]. According to Ref. [30], the discretized version of the Chern-Simons action within each tetrahedron is given by

$$\tfrac{1}{6}(\phi_{12}\phi_{34} - \phi_{13}\phi_{24} + \phi_{14}\phi_{23}), \tag{15}$$

where ϕ_{ij} is the discretized Berry phase between two vertices labeled by i and j [31]:

$$\phi_{ij} = \operatorname{Im}\log\langle u_i \mid u_j\rangle. \tag{16}$$

The total polarization is then given by the sum over all tetrahedrons. The above process sketches the basic recipe for the calculation of $\boldsymbol{P}^{(2)}$, although for real crystals a more careful treatment is needed.

D.X. acknowledges useful discussions with S. Onoda and D. Culcer. We thank M. Stone for pointing out Ref. [30]. D.X. was supported by the NSF (No. DMR-0404252/0606485), J.R.S. by the NSF of China (No. 10604063), D.P.C. by the NSF (No. DMR-0814377), and Q.N. by the Welch Foundation, the DOE (No. DE-FG03-02ER45958), and the NSF of China (No. 10740420252). Work at ORNL was supported in part by the Division of Materials Sciences and Engineering, Office of Basic Energy Sciences, Department of Energy (No. DE-AC05-00OR22725 and No. DE-AC05-00OR22750).

Note added.—The formalism detailed in this paper was recently successfully applied by another group to magnetic field-induced polarization in topological insulators [32].

*xiaod@ornl.gov

[1] L.D. Landau, E.M. Lifshitz, and L.P. Pitaevskii, *Electrodynamics of Continuous Media* (Butterworth-Heinemann, Oxford, UK, 1984), 2nd ed.
[2] A.K. Tagantsev, Phys. Rev. B **34**, 5883 (1986).
[3] A.K. Tagantsev, Phase Transit. **35**, 119 (1991).
[4] For a review, see S.-W. Cheong and M. Mostovoy, Nature Mater. **6**, 13 (2007).
[5] P. Chandra and P.B. Littlewood, in *Physics of Ferroelectrics: A Modern Perspective*, edited by K.M. Rabe, C.H. Ahn, and J.-M. Triscone (Springer, Berlin, 2007).
[6] R.D. King-Smith *et al.*, Phys. Rev. B **47**, 1651 (1993).
[7] R. Resta, Rev. Mod. Phys. **66**, 899 (1994).
[8] G. Ortiz *et al.*, Phys. Rev. B **49**, 14 202 (1994).
[9] C. Wang *et al.*, Phys. Rev. Lett. **99**, 177202 (2007); A. Malashevich and D. Vanderbilt, *ibid.* **101**, 037210 (2008); S. Picozzi *et al.*, *ibid.* **99**, 227201 (2007).
[10] G. Sundaram and Q. Niu, Phys. Rev. B **59**, 14 915 (1999).
[11] R.W. Nunes *et al.*, Phys. Rev. B **63**, 155107 (2001).
[12] I. Souza *et al.*, Phys. Rev. Lett. **89**, 117602 (2002).
[13] H. Kunz, Phys. Rev. Lett. **57**, 1095 (1986).
[14] There is an arbitrariness in choosing the initial state. However, the polarization difference between two states with different but uniform $\boldsymbol{h}$ can be always obtained by using Eq. (1) for periodic insulators. Therefore this difference is on the zeroth order of the spatial gradient and is not our primary concern here.
[15] J. Shi *et al.*, Phys. Rev. Lett. **99**, 197202 (2007).
[16] Strictly speaking, there is an ambiguity in the definition of $\boldsymbol{P}$ given here because $\boldsymbol{j} = \partial\boldsymbol{P}/\partial t + \boldsymbol{\nabla}\times\boldsymbol{M}$ also contains a contribution from the magnetization current. As a result, $\boldsymbol{P}$ is defined only up to a divergence-free field. One can, of course, fix the gauge by imposing that $\boldsymbol{P}$ only have a longitudinal component. However, this is not a critical issue because the ambiguity can be removed by spatial averaging.
[17] G. Panati *et al.*, arXiv:0712.4365.
[18] R. Shindou *et al.*, Nucl. Phys. **B720**, 399 (2005).
[19] M. Lein, Ph.D. thesis, Technische Universität München, 2005.
[20] D. Xiao *et al.*, Phys. Rev. Lett. **95**, 137204 (2005).
[21] The integral of terms that do not contain λ is given by $-e\int_{\mathrm{BZ}} d\boldsymbol{k}[\nabla^k_\alpha\varepsilon + (\Omega^{kr}_{\beta\beta}\nabla^k_\alpha\varepsilon - \Omega^{kr}_{\alpha\beta}\nabla^k_\beta\varepsilon + \Omega^{kk}_{\alpha\beta}\nabla^r_\beta\varepsilon)]$. The integral of $\nabla^k_\alpha\varepsilon$ over the entire Brillouin zone obviously vanishes. After integration by parts and making use of the Bianchi identity $\nabla^k_\alpha\Omega^{rk}_{\beta\beta} + \nabla^k_\beta\Omega^{kr}_{\alpha\beta} + \nabla^r_\beta\Omega^{kk}_{\beta\alpha} = 0$, one can show that the integral of the last three terms contributes a divergence-free part, which can be discarded.
[22] D.J. Thouless, Phys. Rev. B **27**, 6083 (1983).
[23] D.J. Thouless *et al.*, Phys. Rev. Lett. **49**, 405 (1982).
[24] H. Katsura *et al.*, Phys. Rev. Lett. **95**, 057205 (2005).
[25] J. Hu, Phys. Rev. Lett. **100**, 077202 (2008).
[26] I.E. Dzyaloshinskii, JETP **19**, 960 (1964).
[27] M. Mostovoy, Phys. Rev. Lett. **96**, 067601 (2006).
[28] V.G. Bar'yakhtar *et al.*, Sov. Phys. JETP **37**, 673 (1983).
[29] D.M. Bylander *et al.*, Phys. Rev. B **61**, R11 875 (2000).
[30] P.R. Kotiuga, IEEE Trans. Magn. **25**, 2813 (1989).
[31] The vertices are labeled in such a way that the path connecting vertices 1, 2, and 3 follows the right-hand rule with respect to vertex 4.
[32] A.M. Essin *et al.*, arXiv:0810.2998.

Field Induced Positional Shift of Bloch Electrons and Its Dynamical Implications

Yang Gao,[1] Shengyuan A. Yang,[2] and Qian Niu[1,3]
[1]Department of Physics, The University of Texas at Austin, Austin, Texas 78712, USA
[2]Engineering Product Development, Singapore University of Technology and Design, Singapore 138682, Singapore
[3]International Center for Quantum Materials and Collaborative Innovation Center of Quantum Matter, Peking University, Beijing 100871, China
(Received 24 July 2013; published 25 April 2014)

We derive the field correction to the Berry curvature of Bloch electrons, which can be traced back to a positional shift due to the interband mixing induced by external electromagnetic fields. The resulting semiclassical dynamics is accurate to second order in the fields, in the same form as before, provided that the wave packet energy is derived up to the same order. As applications, we discuss the orbital magnetoelectric polarizability and predict nonlinear anomalous Hall effects.

DOI: 10.1103/PhysRevLett.112.166601 PACS numbers: 72.15.-v, 72.10.Bg, 73.43.-f, 85.75.-d

The response of Bloch electrons to external fields has been a central topic in solid state physics. Because of the Berry curvature of Bloch states, the semiclassical dynamics acquires a noncanonical structure [1–3]. This is manifested as an anomalous velocity and a modification of the phase-space density of states, with important consequences on the thermodynamic and transport properties [2,4–6]. Together with a first order correction to the band energy due to the orbital magnetic moment, the Berry curvature provides the essential ingredient for a full theory of the electron response to first order in external fields.

However, response functions such as electric polarizability, magnetic susceptibility, and magnetoelectric polarizability would require a theory that is accurate up to second order in external fields. The difficulty in establishing this type of theory originates from the unboundedness of the perturbative Hamiltonian. Blount pioneered the work of systematically extending semiclassical theory up to second order by using phase-space quantum mechanics [7]. However, his method uses variables which are not fully gauge invariant with respect to the phase choice in the basis Bloch states, rendering it difficult to understand the physical meaning of his results, especially so because of some unresolved gauge issues [8].

In this Letter, we present a second order semiclassical theory for Bloch electrons under uniform electromagnetic fields in terms of physical position and crystal momentum which are fully gauge invariant. A central concept is a gauge-invariant positional shift due to field induced interband mixing. It leads to a field correction to the Berry curvature and modifies the relationship between the physical position and crystal momentum with the canonical ones. Remarkably, the resulting equations of motion up to second order still retain the same form as in the first order theory, provided that the band energy is also corrected to second order in the fields.

The field induced positional shift of Bloch electrons has profound implications. It is solely responsible for the cross-gap part of the orbital magnetoelectric polarizability [9–14]. Moreover, its resulting field correction to the Berry curvature also leads to nonlinear anomalous Hall effects, with a Hall conductivity proportional to an external electric or magnetic field. The electric nonlinear anomalous Hall conductivity is intimately related to the orbital magnetoelectric polarizability and requires the system to have both time reversal and spatial inversion symmetry breaking. The magnetononlinear anomalous Hall effect does not have such symmetry restrictions, and it competes with the ordinary Hall effect in relatively dirty samples. Besides these two well-analyzed applications, our complete second order semiclassical theory also provides straightforward methods to evaluate magnetic susceptibility, electric polarizability, magnetoresistance, intrinsic thermoelectric current, etc. All results from our theory can be easily evaluated in the first principles calculations to be compared with experiments.

Positional shift.—The basic idea of the semiclassical theory is to study the evolution of a wave packet constructed from a single Bloch band (labeled by the subscript 0, and we focus on the Abelian case for simplicity). One starts from the local Hamiltonian obtained from the full quantum Hamiltonian by evaluating the gauge potentials at the center of mass position $\boldsymbol{r}_c$ of the wave packet $\hat{H}_c(\hat{\boldsymbol{p}},\hat{\boldsymbol{q}}) = \hat{H}_0(\hat{\boldsymbol{p}} + \frac{1}{2}\boldsymbol{B}\times\boldsymbol{r}_c, \hat{\boldsymbol{q}}) + \boldsymbol{E}\cdot\boldsymbol{r}_c$, where $\hat{H}_0(\hat{\boldsymbol{p}},\hat{\boldsymbol{q}})$ is the unperturbed Hamiltonian, $\hat{\boldsymbol{p}}$ and $\hat{\boldsymbol{q}}$ are momentum and position operators, and we have set $e = \hbar = 1$ to simplify notations. In the first order semiclassical theory, the wave packet is constructed by superposing the Bloch eigenstates $e^{i\boldsymbol{q}\cdot\boldsymbol{p}}|u_0(\boldsymbol{p} + \frac{1}{2}\boldsymbol{B}\times\boldsymbol{r}_c)\rangle$ of this local Hamiltonian with the crystal momentum $\boldsymbol{p}$ centered around some point $\boldsymbol{p}_c$, satisfying the self-consistency requirement that the position expectation value of the wave packet coincides with the presumed value $\boldsymbol{r}_c$. A noncanonical geometry of the semiclassical dynamics emerges in the equations of motion [5] for the physical position $\boldsymbol{r}_c$ and the gauge-invariant crystal momentum $\boldsymbol{k}_c = \boldsymbol{p}_c + \frac{1}{2}\boldsymbol{B}\times\boldsymbol{r}_c$. This involves the Berry curvature and the orbital magnetic moment in the unperturbed Bloch band, which are necessary and sufficient to make the dynamics and its quantum extension accurate to first order in external fields [1].

In the second order theory, we need the first order correction $|u_0'\rangle$ to the unperturbed band $|u_0\rangle$ by the gradient perturbation $\hat{H}' = (1/4m)\boldsymbol{B}\cdot[(\hat{\boldsymbol{q}}-\boldsymbol{r}_c)\times\hat{\boldsymbol{V}} - \hat{\boldsymbol{V}}\times(\hat{\boldsymbol{q}}-\boldsymbol{r}_c)] + \boldsymbol{E}\cdot(\hat{\boldsymbol{q}}-\boldsymbol{r}_c)$ to the local Hamiltonian $\hat{H}_c$, where $\hat{\boldsymbol{V}} = -i[\hat{\boldsymbol{q}}, \hat{H}_0]$ is the velocity operator. Based on the perturbed band, the wave packet now acquires a shift in its center of mass position given by $\boldsymbol{a}' = \langle u_0|i\partial_{\boldsymbol{p}}|u_0'\rangle + \text{c.c.}$ It corresponds to a first order correction to the Berry connection $\boldsymbol{a} = \langle u_0|i\partial_{\boldsymbol{p}}|u_0\rangle$ of the unperturbed band but is gauge invariant, as can be easily checked by using the orthogonality between $|u_0'\rangle$ and $|u_0\rangle$. Therefore, it is a physical quantity and represents the shift of the wave packet center due to external fields. Indeed, $\boldsymbol{a}'$ transforms as a spatial vector under symmetry operations; e.g., it is odd under spatial inversion and even under time reversal. Furthermore, it should be noted that $\boldsymbol{a}'$ is a periodic function of the lattice, which does not cause any macroscopic charge density gradient; hence, it will not affect the electron's chemical potential profile. This positional shift is the central concept of our theory.

Following the standard perturbation scheme [15], we can derive the following explicit formula for the positional shift (of band 0) in terms of the unperturbed Bloch bands and external fields:

$$a_i' = F_{ij}B_j + G_{ij}E_j, \tag{1}$$

where

$$\begin{aligned} F_{ij} &= \text{Im}\sum_{n\neq 0}\frac{(V_i)_{0n}(\omega_j)_{n0}}{(\varepsilon_0-\varepsilon_n)^2}, \\ G_{ij} &= 2\text{Re}\sum_{n\neq 0}\frac{(V_i)_{0n}(V_j)_{n0}}{(\varepsilon_0-\varepsilon_n)^3}, \end{aligned} \tag{2}$$

with $\boldsymbol{\omega}_{n0}$ defined as

$$(\omega_j)_{n0} = -i\epsilon_{jk\ell}\sum_{m\neq 0}\frac{[(V_k)_{nm} + (v_k)\delta_{nm}](V_\ell)_{m0}}{\varepsilon_m - \varepsilon_0}. \tag{3}$$

Here, ε_0, ε_n, and ε_m are band dispersions for bands 0, n, and m, respectively; i, j, k, and ℓ refer to the spatial components; $\epsilon_{jk\ell}$ is the antisymmetric tensor; $v_\ell = \partial_{p_\ell}\varepsilon_0$ is the group velocity of band 0; and $(V_k)_{nm}$ is the matrix element of the velocity operator. Here and hereafter, summation is implied over repeated spatial indices. Since $\boldsymbol{a}'$ contains the interband velocity, it is not a single-band property. All the quantities in Eq. (2) can be readily evaluated in first principles calculations.

To illustrate the positional shift, we consider a generic two-band model Hamiltonian with

$$\hat{H}_0 = h_0 + \boldsymbol{h}\cdot\boldsymbol{\sigma}, \tag{4}$$

where $\boldsymbol{\sigma}$ is the vector of Pauli matrices and h's have arbitrary dependence on the crystal momentum. The energy band dispersion is $\varepsilon_\pm = h_0 \pm h$. Assume the two bands are fully gapped with $h \neq 0$. The positional shift for the lower band can be calculated from Eqs. (1) and (2) with

$$F_{ij} = -\frac{g_{ik}\epsilon_{k\ell j}\partial_{p_\ell}h_0}{4h} - \frac{1}{8}\epsilon_{jk\ell}\Gamma_{\ell ki}, \qquad G_{ij} = -\frac{1}{4h}g_{ij}, \tag{5}$$

where $g_{ik} = \partial_{p_i}\boldsymbol{n}\cdot\partial_{p_k}\boldsymbol{n}$ (with $\boldsymbol{n} = \boldsymbol{h}/h$) is the quantum metric of the band and $\Gamma_{\ell ki} = \frac{1}{2}(\partial_{p_i}g_{\ell k} + \partial_{p_k}g_{\ell i} - \partial_{p_\ell}g_{ki})$ is the corresponding Christoffel symbol [16]. Like the Berry curvature, the quantum metric is also a geometric physical quantity, which defines the infinitesimal distance in the Hilbert space on the Brillouin zone. Meanwhile, the Christoffel symbol defines the affine geometry of the Brillouin zone [16]. They together make the Brillouin zone a Riemannian manifold. It has been proposed that the quantum metric could be probed by measuring the current noise spectrum [17]. Our result shows that g and Γ are also closely connected with the positional shift and hence might be probed in second order effects.

Second order semiclassical theory.—To see how the positional shift enters the second order semiclassical dynamics, we derive the effective Lagrangian for the wave packet dynamics [15]:

$$L = -(\boldsymbol{r}_c - \boldsymbol{a} - \boldsymbol{a}')\cdot\dot{\boldsymbol{k}}_c - \frac{1}{2}\boldsymbol{B}\times\boldsymbol{r}_c\cdot\dot{\boldsymbol{r}}_c - \tilde{\varepsilon}, \tag{6}$$

where $\boldsymbol{k}_c = \boldsymbol{p}_c + \frac{1}{2}\boldsymbol{B}\times\boldsymbol{r}_c$ is the gauge-invariant crystal momentum and $\tilde{\varepsilon}$ is the semiclassical energy accurate to second order (see a discussion of $\tilde{\varepsilon}$ in the Supplemental Material [15]). In deriving L, the crystal momentum $\boldsymbol{p}$ for each Bloch component of the wave packet has been integrated out, so quantities such as $\boldsymbol{a}$ and $\boldsymbol{a}'$ here are now functions of $\boldsymbol{k}_c$, instead of $(\boldsymbol{p} + \frac{1}{2}\boldsymbol{B}\times\boldsymbol{r}_c)$.

One direct consequence of the positional shift is that the Berry curvature Ω now acquires a field correction given by $\Omega' = \boldsymbol{\partial}\times\boldsymbol{a}'$. (Here and hereafter, the partial derivative $\boldsymbol{\partial}$ is with respect to $\boldsymbol{k}_c$ unless being explicitly pointed out otherwise.) Surprisingly, with this modified Berry curvature $\tilde{\Omega} = \Omega + \Omega'$ and the second order wave packet energy $\tilde{\varepsilon}$, the Euler-Lagrange equations of motion have the same form as in the first order theory [5,15]:

$$\dot{\boldsymbol{r}}_c = \frac{\partial\tilde{\varepsilon}}{\partial\boldsymbol{k}_c} - \dot{\boldsymbol{k}}_c\times\tilde{\Omega}, \tag{7}$$

$$\dot{\boldsymbol{k}}_c = -\boldsymbol{E} - \dot{\boldsymbol{r}}_c\times\boldsymbol{B}. \tag{8}$$

The force equation remains the same as before, and the velocity equation now involves the modified quantities with field corrections. Similar to the first order theory [2], the phase-space density of states has a correction factor $\mathcal{D} = 1 + \boldsymbol{B}\cdot\tilde{\Omega}$, which is now accurate to second order in the fields. It is important to note that even though the Berry curvature is corrected, the Chern number, which is the integral of Berry curvature over the Brillouin zone, is not affected. This is because $\boldsymbol{a}'$ is well defined and periodic in the Brillouin zone; hence the integral of its curl over the entire zone necessarily vanishes.

The positional shift also modifies the relationship between the physical variables ($\boldsymbol{r}_c$, $\boldsymbol{k}_c$) and the canonical variables ($\boldsymbol{q}$, $\boldsymbol{p}$) [1]. These relations are important for requantizing the semiclassical theory and are directly related to physical quantities such as charge polarization discussed in the following. As detailed in the Supplemental Material [15], we follow the procedure in Ref. [1] and obtain that

$$\boldsymbol{r}_c = \boldsymbol{q} + \boldsymbol{a} + \frac{1}{2}(\boldsymbol{B}\times\boldsymbol{a}\cdot\partial_{\boldsymbol{p}})\boldsymbol{a} + \frac{1}{2}\Omega\times(\boldsymbol{B}\times\boldsymbol{a}) + \boldsymbol{a}', \quad (9)$$

$$\boldsymbol{k}_c = \boldsymbol{p} + \frac{1}{2}\boldsymbol{B}\times\boldsymbol{q} + \boldsymbol{B}\times(\boldsymbol{r}_c - \boldsymbol{q}). \quad (10)$$

Note that in Eqs. (9) and (10), the argument for $\boldsymbol{a}$ and $\boldsymbol{a}'$ is $\boldsymbol{p} + \frac{1}{2}\boldsymbol{B}\times\boldsymbol{q}$. It was previously thought that the first four terms on the right-hand side of Eq. (9) would be sufficient to first order in the fields [1], but now one observes that this is incomplete without the positional shift $\boldsymbol{a}'$.

Orbital magnetoelectric polarizability.—In the absence of external fields, the polarization from electrons in a filled band is given by the integral of the Berry connection $\boldsymbol{a}$ [18,19]. From a semiclassical point of view, a magnetic field $\boldsymbol{B}$ modifies this formula in two ways: (1) the density of states in the integral should contain the factor $\mathcal{D} = 1 + \boldsymbol{B}\cdot\tilde{\Omega}$ and (2) the Berry connection $\boldsymbol{a}$ should be replaced by $\boldsymbol{a} + (\boldsymbol{B}\times\boldsymbol{a}\cdot\partial_{\boldsymbol{p}})\boldsymbol{a}/2 + \Omega\times(\boldsymbol{B}\times\boldsymbol{a})/2 + \boldsymbol{a}'$ according to the relationship between the physical position and the canonical position in Eq. (9). Combining these modifications up to first order in the magnetic field and rewriting them in terms of the gauge-invariant crystal momentum $\boldsymbol{k}_c$, we obtain the polarization $\boldsymbol{P}'$ that is first order in the $\boldsymbol{B}$ field:

$$\boldsymbol{P}' = -\int\frac{d^3k}{(2\pi)^3}\left[\frac{1}{2}(\Omega\cdot\boldsymbol{a})\boldsymbol{B} + \boldsymbol{a}'\right], \quad (11)$$

where the integral is over the Brillouin zone and we drop the subscript c of momentum k for simple notations.

The first term in Eq. (11) is the Abelian Chern-Simons form, which plays a central role in the classification of three-dimensional topological insulators [10,20–25]. It is also derived in a more general setting where the crystal periodicity is not required [26]. It only involves the Berry connection and Berry curvature of the unperturbed band and can be derived within the framework of the first order semiclassical theory [13,27]. The additional term from the field induced band mixing was envisioned in Ref. [27], but its validation had to wait for a full quantum perturbation treatment in Ref. [14]. We now see that this additional term actually comes in the nice form of the positional shift integrated over the Brillouin zone. Our result agrees exactly with the full quantum result, confirming the reliability of our semiclassical theory.

Since the topological part [the first term in Eq. (11)] is quantized and well understood [10,20–26], we focus on the magnetoelectric polarization due to the positional shift, which requires broken time reversal and spatial inversion symmetry [14]. To show its connection with the nonlinear anomalous Hall effect discussed later, we consider the two-band model in Eq. (4), in which the second term in Eq. (11) for the lower band gives $P_i' = \int G_i B d^3k/(2\pi)^3$, where $\boldsymbol{G} = (\hat{z}\cdot\partial h_0\times\partial n_j)\partial n_j/(4h)$, with $\hat{z}$ being the direction of the magnetic field. We note that, if h_0 is a constant, $\boldsymbol{G}$ would vanish. This is consistent with previous observations that a nonzero orbital magnetoelectric polarization must require particle-hole symmetry breaking of the system [14]. A minimal lattice model that realizes this effect can be constructed in 2D. Notice that for the model equation (4) in 2D, the topological part of magnetoelectric polarization, i.e., the first term in Eq. (11) vanishes [27]; hence, only the contribution from $\boldsymbol{a}'$ exists. Moreover, since $\boldsymbol{a}'$ transforms as a spatial vector and it must lie in the plane, in general, it must vanish if the system has in-plane rotational symmetry. And, if in-plane mirror symmetry exists, $\boldsymbol{P}'$ would be restricted to be along the normal direction of the mirror line [see Fig. 1(a)]. These symmetry constraints provide guidance for the construction of the lattice model, as discussed in the Supplemental Material [15].

Nonlinear anomalous Hall effect.—In the semiclassical approach, the transport current is given by $\boldsymbol{j} = -\int\dot{\boldsymbol{r}}_c f(\boldsymbol{k})\mathcal{D}d^3k/(2\pi)^3$, where $f(\boldsymbol{k})$ is the distribution function. Because our theory is accurate up to second order in external fields, it allows us to evaluate the nonlinear current response. Here, we focus on the intrinsic

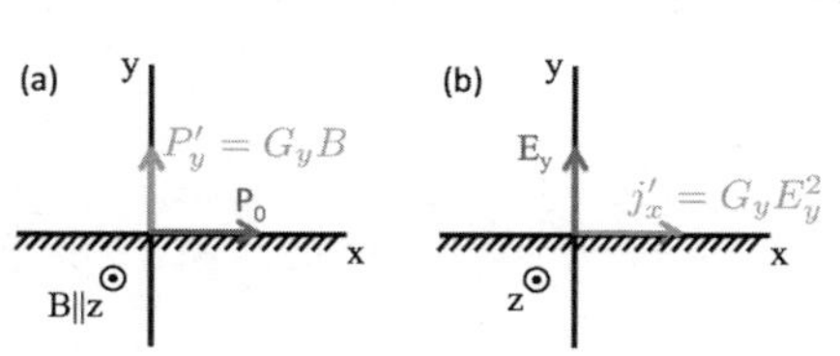

FIG. 1 (color online). (a) Magnetoelectric polarization and (b) the electric nonlinear anomalous Hall effect in a 2D system with a mirror line along the x axis. In (a), the mirror symmetry requires the zeroth order polarization $\boldsymbol{P}_0$ to be along the mirror line and requires the first order $\boldsymbol{P}'$ to lie in the perpendicular direction. In (b), the linear anomalous Hall current vanishes due to the mirror symmetry, but the nonlinear anomalous Hall current can exist along the mirror line if the electric field is applied along the perpendicular direction.

contribution to the Hall current, which is purely from the band structure effects without disorder scattering [28–34]. Under fixed temperature and uniform electromagnetic fields, we obtain the intrinsic current $\boldsymbol{j}'$ that is second order in external fields:

$$\boldsymbol{j}' = \boldsymbol{E} \times \int [\boldsymbol{v} \times \boldsymbol{a}' + \Omega(\boldsymbol{B} \cdot \boldsymbol{m})] \frac{\partial f_0}{\partial \varepsilon_0} \frac{d^3k}{(2\pi)^3}. \quad (12)$$

Or, more explicitly, we can write

$$\frac{\partial^2 j'_i}{\partial E_j \partial E_\ell} = \int (v_i G_{j\ell} - v_j G_{i\ell}) \frac{\partial f_0}{\partial \varepsilon_0} \frac{d^3k}{(2\pi)^3}, \quad (13)$$

$$\frac{\partial^2 j'_i}{\partial E_j \partial B_\ell} = \int (v_i F_{j\ell} - v_j F_{i\ell} + \epsilon_{ijk} \Omega_k m_\ell) \frac{\partial f_0}{\partial \varepsilon_0} \frac{d^3k}{(2\pi)^3}, \quad (14)$$

where f_0 is the equilibrium Fermi-Dirac distribution function and $\boldsymbol{m} = -\frac{1}{2}\mathrm{Im}\langle \boldsymbol{\partial} u_0 | \times (\varepsilon_0 - \hat{H}_c) | \boldsymbol{\partial} u_0 \rangle$ is the orbital magnetic moment [4]. The effects associated with the two response functions in Eqs. (13) and (14) shall be termed as the electric nonlinear anomalous Hall effect and the magnetononlinear anomalous Hall effect, respectively. We call such currents anomalous because they are not caused by the Lorentz force, as in the case of the ordinary Hall effect. These response functions can be directly evaluated by first principles methods. From Eq. (12), we see that the intrinsic nonlinear current is purely of Hall type, and the appearance of $\partial f_0/\partial \varepsilon_0$ shows that it is a Fermi surface property. The second term in the square brackets of Eq. (12) comes from a correction of the band energy which could be envisioned from an naive extension of the first order theory and has been discussed in the study of anomalous Hall transport in multivalley systems [35]. The first term comes from the correction of the Berry curvature due to the positional shift found in this work, which is quite nontrivial. Moreover, we note that for in-phase oscillating $\boldsymbol{E}$ and $\boldsymbol{B}$ fields, the first order intrinsic anomalous Hall response vanishes upon the time average; hence, the dc intrinsic anomalous Hall current would be dominated by the nonlinear response $\boldsymbol{j}'$.

First, let us consider the electric nonlinear anomalous Hall effect with $\boldsymbol{B} = 0$. Then, the intrinsic nonlinear Hall conductivity $\sigma'_{xy} = \partial j_x / \partial E_y$ is proportional to the electric field and only the term with positional shift in Eq. (12) contributes. For the generic two-band model [Eq. (4)], the result is $\sigma'_{xy} = -\int (\partial f_0/\partial \varepsilon_0) \boldsymbol{G} \cdot \boldsymbol{E} d^3k/(2\pi)^3$. Interestingly, it also involves the $\boldsymbol{G}$ vector found for the orbital magnetoelectric polarizability. In fact, the two effects have the same symmetry properties, requiring both time reversal and spatial inversion symmetries to be broken in the system. This nonlinear anomalous Hall current will dominate if the corresponding linear current vanishes due to symmetry constraints. For example, if a 2D system has a mirror line perpendicular to the electric field, then the linear intrinsic current vanishes because the (unperturbed) Berry curvature has a sign change under mirror operation, while the nonlinear current could be finite [see Fig. 1(b)].

In comparison, the magnetononlinear anomalous Hall effect does not have such a stringent symmetry constraint. In fact, since the current transforms in the same way as the product of electric and magnetic fields under both time reversal and spatial inversion, this is much easier to realize in real systems. Furthermore, if the system itself has time reversal symmetry (neglect the small Zeeman splitting due to the external magnetic field), both the linear anomalous Hall effect and the electric nonlinear anomalous Hall effect vanish, and the magnetononlinear anomalous Hall effect dominates. For the two-band model, we find that

$$\sigma'_{xy} = \int \frac{d^3k}{(2\pi)^3} \frac{\partial f_0}{\partial \varepsilon_0} \left[\frac{g_{ij}}{4h} (\hat{z} \times \boldsymbol{v})_i (\boldsymbol{B} \times \boldsymbol{\partial} h_0)_j - \frac{1}{8} (\hat{z} \times \boldsymbol{v})_i \epsilon_{k\ell j} B_k \Gamma_{j\ell i} + h(\Omega \cdot \hat{z})(\Omega \cdot \boldsymbol{B}) \right], \quad (15)$$

where $\boldsymbol{v} = \boldsymbol{\partial}(h_0 - h)$ and $\Omega = \frac{1}{2}\epsilon_{ijk} n_i \boldsymbol{\partial} n_j \times \boldsymbol{\partial} n_k$ is the unperturbed Berry curvature. As a concrete example, let us consider a 2D gapped Dirac model with $h_0 = 0$ and $\boldsymbol{h} = (vk_x, vk_y, \Delta)$, which is widely used to study systems such as symmetry-breaking graphene, MoS_2, topological insulator surfaces with time reversal symmetry breaking, and topological insulator thin films [20,21,36,37]. Here, v is the Fermi velocity and Δ is the gap parameter. Considering an in-plane electric field and an out-of-plane magnetic field, we obtain that (at zero temperature)

$$\sigma'_{xy} = -e^3 \frac{v^2(v^2 p_F^2 + 2\Delta^2)}{16\pi(v^2 p_F^2 + \Delta^2)^2} B, \quad (16)$$

where $p_F = \hbar k_F$ is the Fermi momentum, and we assume the Fermi level is in the upper band. We have recovered the factors e and $\hbar$ in Eq. (16). We point out that for MoS_2 or graphene (with inversion symmetry breaking) with two inequivalent valleys K and K' connected by time reversal symmetry, the contributions to this magnetononlinear anomalous Hall effect from the two valleys in fact add together rather than cancel each other, as in the first order response [38].

Besides this effect, note that due to the magnetic field, there is also the ordinary Hall response due to Lorentz force. Both effects have linear B dependence in the transport coefficient. However, there is an important difference. The ordinary Hall conductivity is proportional to the square of relaxation time (or longitudinal conductivity), while our intrinsic nonlinear conductivity does not have such extrinsic dependence. On the other hand, in terms of the Hall resistivity, the ordinary effect has an intrinsic looking form $\rho_{xy}^{\mathrm{ord}} = -B/ne$, where n is the carrier density, while intrinsic nonlinear Hall resistivity acquires a dependence on the square of the longitudinal resistivity. This is well understood as a result of matrix inversion between the conductivity and resistivity tensors with the usual condition that longitudinal coefficients are much bigger than the transverse ones. Therefore, from this discussion, like the linear anomalous Hall effect, our nonlinear effect would become important for more resistive samples.

The argument above can be quantified for the gapped Dirac model by calculating the ratio between the two contributions explicitly. From Eq. (16), we have

$$\frac{\rho'_{xy}}{\rho^{\mathrm{ord}}_{xy}} = \left(\rho_{xx}\frac{e^2}{4h}\right)^2\left[1-\left(\frac{\Delta}{\varepsilon_F}\right)^4\right], \quad (17)$$

where $\varepsilon_F = \sqrt{v^2 p_F^2 + \Delta^2}$ is the Fermi energy, $\rho'_{xy} \simeq \sigma'_{xy}\rho^2_{xx}$ is obtained by inverting σ_{xy}', and ρ_{xx} is the longitudinal resistivity. In Eq. (17), the first factor is proportional to ρ^2_{xx}, which is universal (model independent). The second factor shows a simple dependence on Fermi energy for the gapped Dirac model: it vanishes at the band bottom and quickly saturates with increasing Fermi energies. Our predictions can be tested by the standard Hall bar measurement on single layer MoS_2 or Bi_2Se_3 thin films with longitudinal resistivity tuned by temperature, film thickness, or doping. The difference in the scaling in terms of the longitudinal resistivity, as shown in Eq. (17), can be used to disentangle the ordinary Hall effect and the nonlinear anomalous Hall effect.

We acknowledge useful discussions with Z. Qiao, H. Chen, J. Zhou, X. Li, R. Cheng, and L. Zhang. Q. N. is supported by NBRPC (No. 2012CB921300 and No. 2013CB921900), and NSFC (No. 91121004). S. A. Y. is supported by SUTD-SRG-EPD-2013062. Y. G. is supported by DOE (DE-FG03-02ER45958, Division of Materials Science and Engineering) and Welch Foundation (F-1255).

[1] M.-C. Chang and Q. Niu, J. Phys. Condens. Matter **20**, 193202 (2008).
[2] D. Xiao, J. Shi, and Q. Niu, Phys. Rev. Lett. **95**, 137204 (2005).
[3] R. G. Littlejohn and W. G. Flynn, Phys. Rev. A **44**, 5239 (1991).
[4] D. Xiao, M.-C. Chang, and Q. Niu, Rev. Mod. Phys. **82**, 1959 (2010).
[5] G. Sundaram and Q. Niu, Phys. Rev. B **59**, 14915 (1999).
[6] M.-C. Chang and Q. Niu, Phys. Rev. Lett. **75**, 1348 (1995).
[7] E. I. Blount, Phys. Rev. **126**, 1636 (1962).
[8] Blount systematically classified the contribution to the effective Hamiltonian Eq. [I.(30)] in Ref. [7]. However, the effective Hamiltonian is gauge dependent since the third term $\boldsymbol{a}^s \cdot \boldsymbol{v}_A$ depends on the gauge choice, while all the other terms are gauge invariant.
[9] L. Fu, C. L. Kane, and E. J. Mele, Phys. Rev. Lett. **98**, 106803 (2007).
[10] X.-L. Qi, T. L. Hughes, and S.-C. Zhang, Phys. Rev. B **78**, 195424 (2008).
[11] J. E. Moore and L. Balents, Phys. Rev. B **75**, 121306(R) (2007).
[12] R. Roy, Phys. Rev. B **79**, 195322 (2009).
[13] A. M. Essin, J. E. Moore, and D. Vanderbilt, Phys. Rev. Lett. **102**, 146805 (2009).
[14] A. M. Essin, A. M. Turner, J. E. Moore, and D. Vanderbilt, Phys. Rev. B **81**, 205104 (2010).
[15] See Supplemental Material at http://link.aps.org/supplemental/10.1103/PhysRevLett.112.166601 for construction of a proper wave packet state, the derivation of the semiclassical dynamics, the realization of the cross-gap magneto electric polarization in a 2D system, and the derivation of the nonlinear anomalous Hall current.
[16] J. Anandan and Y. Aharonov, Phys. Rev. Lett. **65**, 1697 (1990).
[17] T. Neupert, C. Chamon, and C. Mudry, Phys. Rev. B **87**, 245103 (2013).
[18] R. D. King-Smith and D. Vanderbilt, Phys. Rev. B **47**, 1651 (1993).
[19] R. Resta, Rev. Mod. Phys. **66**, 899 (1994).
[20] M. Z. Hasan and C. L. Kane, Rev. Mod. Phys. **82**, 3045 (2010).
[21] X.-L. Qi and S.-C. Zhang, Rev. Mod. Phys. **83**, 1057 (2011).
[22] P. Hosur, S. Ryu, and A. Vishwanath, Phys. Rev. B **81**, 045120 (2010).
[23] T. L. Hughes, E. Prodan, and B. A. Bernevig, Phys. Rev. B **83**, 245132 (2011).
[24] L. Fu, C. L. Kane, and E. J. Mele, Phys. Rev. Lett. **98**, 106803 (2007).
[25] C. Fang, M. J. Gilbert, and B. A. Bernevig, Phys. Rev. B **86**, 115112 (2012).
[26] B. Leung and E. Prodan, J. Phys. A **46**, 085205 (2013).
[27] D. Xiao, J. Shi, D. P. Clougherty, and Q. Niu, Phys. Rev. Lett. **102**, 087602 (2009).
[28] N. Nagaosa, J. Sinova, S. Onoda, A. H. MacDonald, and N. P. Ong, Rev. Mod. Phys. **82**, 1539 (2010).
[29] P. Středa and T. Jonckheere, Phys. Rev. B **82**, 113303 (2010).
[30] P. Středa, Phys. Rev. B **82**, 045115 (2010).
[31] T. Jungwirth, Q. Niu, and A. H. MacDonald, Phys. Rev. Lett. **88**, 207208 (2002).
[32] M. Onoda and N. Nagaosa, J. Phys. Soc. Jpn. **71**, 19 (2002).
[33] R. C. Fivaz, Phys. Rev. **183**, 586 (1969).
[34] R. Karplus and J. M. Luttinger, Phys. Rev. **95**, 1154 (1954).
[35] T. Cai, S. A. Yang, X. Li, F. Zhang, J. Shi, W. Yao, and Q. Niu, Phys. Rev. B **88**, 115140 (2013).
[36] A. H. Castro Neto, F. Guinea, N. M. R. Peres, K. S. Novoselov, and A. K. Geim, Rev. Mod. Phys. **81**, 109 (2009).
[37] D. Xiao, G.-B. Liu, W. Feng, X. Xu, and W. Yao, Phys. Rev. Lett. **108**, 196802 (2012).
[38] D. Xiao, W. Yao, and Q. Niu, Phys. Rev. Lett. **99**, 236809 (2007).

8

Nonlinear quantum systems

Nonlinear Evolution of Quantum States in the Adiabatic Regime

Jie Liu,[1] Biao Wu,[1,2] and Qian Niu[1]
[1]Department of Physics, The University of Texas, Austin, Texas 78712
[2]Condensed Matter Sciences Division, Oak Ridge National Laboratory, Oak Ridge, Tennessee 37831-6032
(Received 2 August 2002; published 1 May 2003)

We investigate adiabatic evolution of quantum states as governed by the nonlinear Schrödinger equation and provide examples of applications with a nonlinear tunneling model for Bose-Einstein condensates. Our analysis not only spells out conditions for adiabatic evolution of eigenstates but also characterizes the motion of noneigenstates which cannot be obtained from the former in the absence of the superposition principle. We find that Aharonov-Anandan phases play the role of classical canonical actions and are conserved in the adiabatic evolution of noneigenstates.

DOI: 10.1103/PhysRevLett.90.170404 PACS numbers: 03.65.Ge, 03.65.Vf, 03.75.Lm, 71.35.Lk

Adiabatic evolution has been an important method of preparation and control of quantum states [1,2]. The main guidance comes from the adiabatic theorem of quantum mechanics [3], which dictates that an initial nondegenerate eigenstate remains to be an instantaneous eigenstate when the Hamiltonian changes slowly compared to the level spacings. More precisely, the quantum eigenstate evolves only in its phase, given by the time integral of the eigenenergy (dynamical phase) and a quantity independent of the time duration (geometric phase). The linearity of quantum mechanics then immediately allows a precise statement on the adiabatic evolution of noneigenstates through the superposition principle.

Our concern here is how the adiabatic theorem gets modified in nonlinear evolution of quantum states. Nonlinearity has been introduced as possible modifications of quantum mechanics on the fundamental level [4]. Our motivation, however, derives from practical applications in current pursuits of adiabatic control of Bose-Einstein condensates (BECs) [5], which can often be accurately described by the nonlinear Schrödinger equation. Here the nonlinearity stems from a mean field treatment of the interactions between atoms. Difficulties in theoretical study of adiabatic control of the condensate arise not only from the lack of unitarity [6] but also from the absence of the superposition principle [7,8].

In this Letter we attempt to overcome these difficulties by combining ideas from classical adiabatic dynamics and quantum geometric phases. Noticing that the eigenstates correspond to extremum points of the system energy, we find that their adiabatic condition depends on the Bogoliubov excitation spectrum about such points and has nothing to do with level spacings between the eigenstates. Also, because of nonlinearity, the adiabatic evolution of noneigenstates cannot be expressed as a superposition of such eigenstates with conserved probabilities as in the linear case. We find that the Aharonov-Anandan (AA) phases [9], which can be defined for cyclic or quasicyclic quantum states (at fixed control parameters), can serve as adiabatic invariants analogous to canonical actions in classical systems [10]. As an illustration and application, we study a simple two-mode BEC system.

General formalism.—The Schrödinger equation, linear or nonlinear, can be written in the form [11] ($\hbar = 1$)

$$i\frac{d\psi_j}{dt} = \frac{\partial}{\partial\psi_j^*}\mathcal{H}(\psi, \psi^*, \boldsymbol{R}), \qquad (1)$$

where ψ_j is the jth amplitude of the wave function ψ over an orthonormal basis set, $\mathcal{H}$ is the energy of the system, and $\boldsymbol{R}$ denotes all the system parameters subject to adiabatic change. It is known that the Schrödinger equation has a canonical structure of classical dynamics. One can show, for example, that for each j the probability $|\psi_j|^2$ and phase $\arg\psi_j$ form a canonical pair, which satisfies the Hamilton's equations of motion with the energy $\mathcal{H}$ serving as the classical Hamiltonian. In this sense, the nonlinear Schrödinger equation is often termed classical [12], although the linear Schrödinger equation has not been called so.

However, our original system is quantum mechanical, which should impose additional structures in the corresponding canonical dynamics. Specifically, the system has the gauge symmetry that $\mathcal{H}$ is invariant under shifts in the overall phase of the wave function. This implies that the total probability must be conserved and the dynamics of the overall phase can be separated from the rest of the degrees of freedom. For example, in a finite dimensional problem, with $j = 1, 2, \ldots, N$, we can choose a new set of canonical variables: $p_j = |\psi_j|^2$, $q_j = \arg\psi_j - \arg\psi_N$, $j = 1, 2, \ldots, N-1$, and $p_N = \sum_{j=1}^N |\psi_j|^2$, $q_N = \arg\psi_N$. Because the Hamiltonian is independent of q_N, so that p_N is conserved and can be set to unity, the other variables form a closed set of Hamiltonian dynamics,

$$\frac{dq_j}{dt} = \frac{\partial\mathcal{H}}{\partial p_j}, \quad \frac{dp_j}{dt} = -\frac{\partial\mathcal{H}}{\partial q_j}, \quad (j = 1, \ldots, N-1). \qquad (2)$$

The gauge symmetry also allows the introduction of a geometric phase for the quantum state. Let λ be the overall phase of the wave function; we took it to be

$\arg\psi_N$ in the above discussion. We split off this overall phase by writing $\psi_j = e^{i\lambda}\phi_j$; then ϕ_j belongs to the so-called projective Hilbert space. For many physical problems, the new wave function varies cyclically in time, for which one can introduce a geometric phase in generalization of Berry's phase. From Eq. (1), we obtain

$$\frac{d\lambda(t)}{dt} = \sum_{j=1}^{N} \phi_j^* i \frac{\partial}{\partial t}\phi_j - \sum_{j=1}^{N} \phi_j^* \frac{\partial}{\partial \phi_j^*}\mathcal{H}. \tag{3}$$

For a linear quantum system, the second term is the same as the energy [11], whose time integral gives the so-called dynamical phase. The time integral of the first term, $\int_0^\tau dt \sum_{j=1}^N \phi_j^* i \frac{\partial}{\partial t}\phi_j$, gives an additional contribution to the overall phase and is called the Aharonov-Anandan phase [9]. For the nonlinear case, we continue to call it the Aharonov-Anandan phase [13]. We show that this geometric phase plays the role of classical action in the canonical dynamics and is thus an adiabatic invariant.

Eigenstates.—We first consider adiabatic evolution of eigenstates, which is defined by $\frac{\partial}{\partial\psi_j^*}\mathcal{H}(\psi,\psi^*,\boldsymbol{R}) = E(\boldsymbol{R})\psi_j$ $(j = 1,\ldots,N)$. Besides the ground state which has obvious physical significance, higher eigenstates (also called nonlinear coherent modes) can be prepared experimentally [14]. The eigenstates correspond to extremum energy or fixed points of the classical dynamics (2) at a given $\boldsymbol{R}$ [15]. For an elliptic fixed point, we expect it to be able to follow adiabatically the control parameter provided the latter changes slowly compared with the fundamental frequencies of periodic orbits around the fixed point. These frequencies can be evaluated by linearizing Eq. (2) about the fixed point [10] and are identical to the Bogoliubov excitation spectrum of the corresponding eigenstate. In linear quantum mechanics, these frequencies are just the level spacings, so that breakdown of adiabaticity occurs by level crossing. In the nonlinear quantum problem, the fundamental frequencies are generally different from the level spacings, so that adiabaticity can often be maintained even if the energy levels cross as demonstrated below with a two-level model.

Nonlinearity in our quantum problem not only makes different eigenstates nonorthogonal but also can produce more eigenstates than the dimension N of the Hilbert space. Some of these additional eigenstates correspond to hyperbolic points in the classical dynamics, characterized by complex fundamental frequencies and strong sensitivity to small perturbations. We thus expect that such eigenstates are not able to follow adiabatically the control parameter. The complex fundamental frequencies correspond to a complex Bogoliubov spectrum on top of the mean field solution, signifying spontaneous production of quasiparticles. In practice, we need to traverse the dynamical instability regime fast enough to avoid the quasiparticle production and slow enough to avoid non-adiabatic excitations [16].

As an illustration, we consider a nonlinear two-level model,

$$\begin{aligned} i\frac{\partial\psi_1}{\partial t} &= \left(\frac{R}{2} + \frac{c}{2}(\psi_2^*\psi_2 - \psi_1^*\psi_1)\right)\psi_1 + \frac{v}{2}\psi_2, \\ i\frac{\partial\psi_2}{\partial t} &= \frac{v}{2}\psi_1 - \left(\frac{R}{2} + \frac{c}{2}(\psi_2^*\psi_2 - \psi_1^*\psi_1)\right)\psi_2. \end{aligned} \tag{4}$$

This model was proposed to describe the tunneling of BEC in an optical lattice [17] or in a double-well potential [18]. The parameter c characterizes the interaction strength between atoms; v is the coupling strength between the two modes. The parameter R can be the Bloch wave number or energy difference between the two wells. We are interested in the tunneling between the energy levels shown in the top panels of Fig. 1 when R is increased slowly from the far negative end to the far positive end.

Following our general formalism, we choose the total phase as $\lambda = \arg\psi_2$ and introduce a pair of canonical variables, $q = \arg\psi_1 - \arg\psi_2$ and $p = |\psi_1|^2$. The total energy of the system is $\mathcal{H} = \frac{v}{2}(\psi_1^*\psi_2 + \psi_1\psi_2^*) + \frac{R}{2}(\psi_1^*\psi_1 - \psi_2^*\psi_2) - \frac{c}{4}(\psi_1^*\psi_1 - \psi_2^*\psi_2)^2$. The wave function without the total phase takes the form $\phi_1 = \sqrt{p}e^{iq}$, $\phi_2 = \sqrt{1-p}$. Then we have the equivalent classical Hamiltonian as in Eq. (2),

$$\mathcal{H} = v\sqrt{p(1-p)}\cos q + \frac{R}{2}(2p-1) - \frac{c}{4}(2p-1)^2. \tag{5}$$

In Fig. 1, the structure of eigenenergy levels of Eq. (4) is shown in the top panels while the phase space orbits of the corresponding classical system (5) are shown in the bottom panels. When $c < v$, there are only two eigenstates and two fixed points. Since both f_1 and f_2 are elliptic with finite fundamental frequencies, the corresponding quantum states are expected to be able to follow adiabatically with R. This has been corroborated by our numerical simulations.

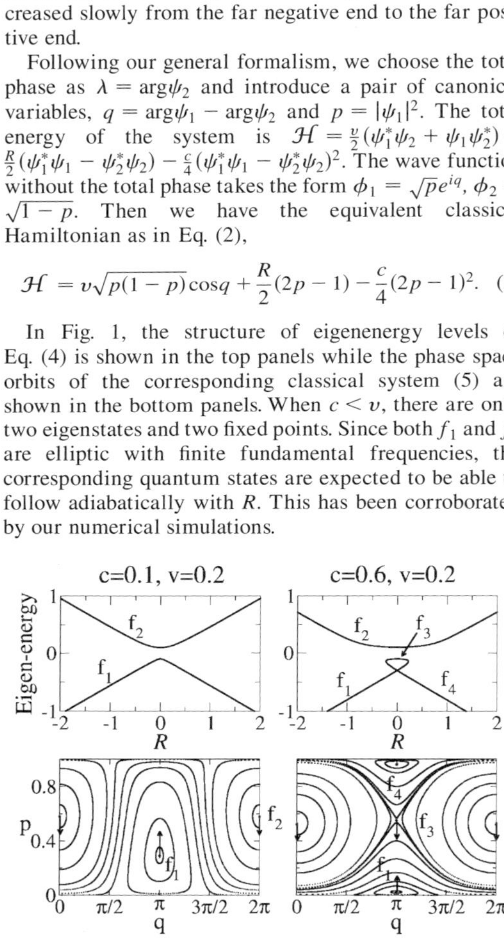

FIG. 1. The two top panels show eigenenergies as a function of R for two typical cases $c < v$ and $c > v$. The two bottom panels show the corresponding phase space portraits at a given value of R ($= -0.05$). The arrows on the fixed points indicate the directions of their movements as R increases.

When $c > v$, there are two more eigenstates, forming a loop structure; in the phase space portrait, there appear two more fixed points with one of them, f_3, being hyperbolic. Because of this structure change, the adiabatic evolution becomes very different here. First, the eigenstate corresponding to f_3 will not be able to follow the adiabatic change of R since f_3 is hyperbolic. This has been checked by our numerical integration of Eq. (4). Second, the fixed point f_1 can annihilate itself by colliding with f_3 as R changes slowly, leading to the breakdown of adiabaticity of the tunneling as reported numerically in Ref. [17]. Finally we notice that there is a level crossing between f_1 and f_4 at $R = 0$; however, our calculation shows that their fundamental frequencies are $v[(\frac{c}{v})^2 - 1]^{1/2} \neq 0$. This clearly illustrates our statement in the general formalism that the fundamental frequencies are not related to the level spacing in the nonlinear case.

Cyclic and quasicyclic states.—Compared with eigenstates, adiabatic evolution of noneigenstates is in general very complicated as the motions given by Eq. (2) may be chaotic [15]. We choose to focus on the states around an elliptic point, where the classical orbits are regular. Here the motions are confined on the $(N-1)$-dimensional torus, and we may introduce a set of action-angle variables, $\boldsymbol{I} = (I_1, I_2, \ldots, I_{N-1}), \boldsymbol{\Theta} = (\Theta_1, \Theta_2, \ldots, \Theta_{N-1})$ [10]. The angular variables change with time at frequencies $\boldsymbol{\omega} = (\omega_1, \omega_2, \ldots, \omega_{N-1})$ while the actions $\boldsymbol{I}$ are constants. More importantly, according to the classical adiabatic theorem [10], the actions $\boldsymbol{I}$ are adiabatic invariants in the sense that they remain constant even if the control parameter $\boldsymbol{R}$ changes (slowly) in time. The existence of these adiabatic invariants presents strong constraint on the motion and guarantees a state initially close to an eigenstate (elliptic point) to stay near it as the system is changed slowly.

Furthermore, we can attach a physical meaning to these adiabatic invariants in the effective classical description by making connection to the Aharonov-Anandan phase of the states. The AA phase is defined as the time integral of the first term in Eq. (3) for a periodic orbit or a quasiperiodic state [19],

$$\gamma_{AA}(\boldsymbol{R}) = \int_0^T dt \sum_{j=1}^{N} \phi_j^* i \frac{\partial}{\partial t} \phi_j. \tag{6}$$

We can rewrite it with the canonical variables (q_j, p_j) and further with the action-angle variables,

$$\gamma_{AA} = \sum_{j=1}^{N-1} \int_0^T p_j dq_j = \boldsymbol{I} \cdot \boldsymbol{\Omega}, \tag{7}$$

where $\boldsymbol{\Omega} = (\omega_1 T, \omega_2 T, \ldots, \omega_{N-1} T)$ and T is a time period. Therefore, the actions are related to the AA phase γ_{AA}, which is an observable physical quantity [20]. In the special case of $N = 2$, there is only one independent action, so the AA phase is simply $\gamma_{AA} = 2\pi I$. This simple connection can be expanded to the general case of $N > 2$, where one can single out a particular cyclic state that involves only one action I_j. For this cyclic state, we again have the simple relation $\gamma_{AA} = 2\pi I_j$.

How do the above adiabatic invariants connect to the familiar notions in the standard linear quantum mechanics? Consider the time evolution of a general state in a linear quantum system for a given $\boldsymbol{R}$, $\psi_j(t) = \psi_j(0)e^{-iE_j t}$, $j = 1, \ldots, N$, where E_j's are the eigenenergies. This is a (quasi-)cyclic state with the projective wave functions given by $\Phi_j(t) = \psi_j(0)e^{-i(E_j - E_N)t}$, $j = 1, \ldots, (N-1)$ and $\Phi_N(t) = \psi_N(0)$. Its AA phase can be computed with Eq. (6); after comparing with Eq. (7), we immediately find that $I_j = |\psi_j(0)|^2$. Therefore, in linear quantum mechanics, these adiabatically invariant actions I_j are nothing but the probabilities on the energy levels. In this way, we have rederived the adiabatic theorem of linear quantum mechanics [21].

The conservation of probabilities on the energy levels can be generalized to the case where the system is nonlinear in an intermediate range of the parameter $\boldsymbol{R}$ (Fig. 2). When the actions (AA phases) are conserved during the entire process, the initial and final probabilities on the energy levels must remain the same, $I_j = I'_j$, because the system is linear at the beginning and end of the process so the actions are just the occupation probabilities. In the intermediate range where the nonlinearity is present, the probabilities will change because the conserved actions are not probabilities on the energy levels.

We now illustrate this important result using our two-level model, where, at the two ends with $|R| \gg c$, the nonlinear term can be ignored and the system is effectively linear. For $c < v$, where all the fixed points are elliptic, the fundamental frequency ω for the periodic orbit remains finite and the AA phase (action) is conserved [see lines (*a*) in the right panels of Fig. 3]. The initial and final probabilities on each level are indeed the same [Fig. 3(a)], although they oscillate in the intermediate range of the parameter where the system is nonlinear. As the nonlinearity gets strong, the occurrence of tunneling begins to depend on the choice of the initial state. In Fig. 3(b), where one starts with probability $I = 0.1$ on level two, tunneling happens; however, in Fig. 3(c) where one starts with probability $I = 0.8$ on level two, there is no tunneling. The difference is whether there is collision

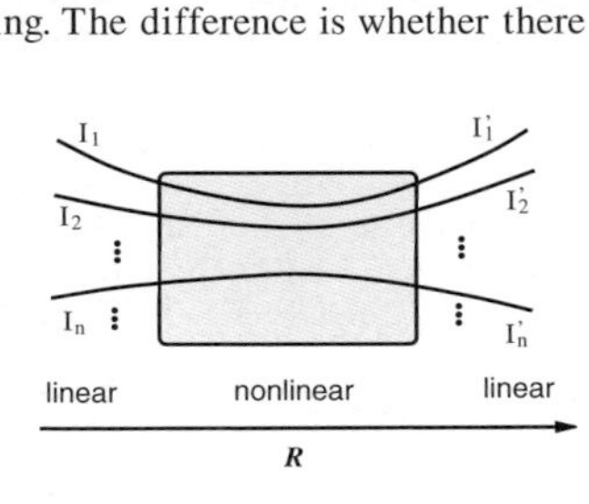

FIG. 2. Nonlinear tunneling of a system that is nonlinear in an intermediate range of the parameter $\boldsymbol{R}$. I's and I''s are the occupation probabilities on different eigenstates, at the beginning and the end of the tunneling process, respectively.

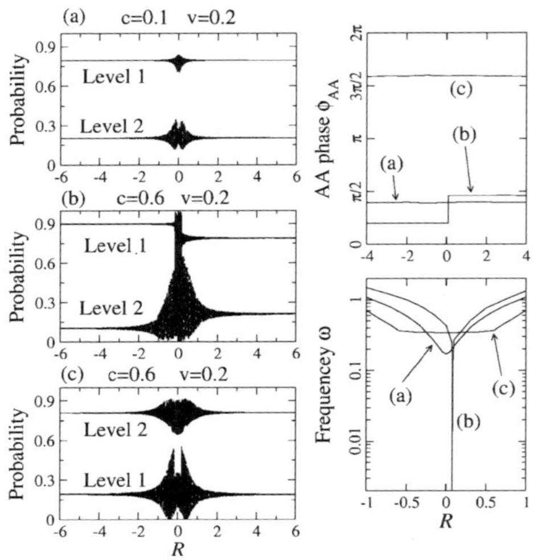

FIG. 3. Left panels: Change of probabilities on the two levels with R, which changes with rate $\alpha = 0.0001$, for three different cases. The right two panels show how the AA phases and the fundamental frequencies change with R in these three cases, respectively.

with the hyperbolic point f_3. In Fig. 3(b), the initial noneigenstate falls on a periodic orbit surrounding the fixed point f_1, which will later collide with the hyperbolic point f_3, where the fundamental frequency drops to zero and the AA phase has a finite jump [see lines (*b*) in the right panels]. The jump height is proportional to the tunneling probability. In Fig. 3(c), the initial state falls on a periodic orbit around the fixed point f_2, which will not collide with f_3.

In summary, we have generalized the quantum adiabatic theorem to systems governed by the nonlinear Schrödinger equation. Adiabaticity of an eigenstate requires that the control parameters vary slowly with respect to the Bogoliubov excitation frequencies. We further show that the Aharonov-Anandan phase can serve as the general adiabatic invariant, in analogy to the canonical actions in classical mechanics. This finding provides an important application of the Aharonov-Anandan phase and is expected to be useful in guiding adiabatic manipulation of the condensated atoms.

We acknowledge the support of NSF (DMR-0071893), R. A. Welch Foundation, and ORNL, managed by UT-Battelle, LLC for the USDOE (DE-AC0500OR22725).

[1] K. Bergmann, H. Theuer, and B.W. Shore, Rev. Mod. Phys. **70**, 1003 (1998).

[2] M. B. Dahan *et al.*, Phys. Rev. Lett. **76**, 4508 (1996); S. R. Wilkinson *et al.*, Phys. Rev. Lett. **76**, 4512 (1996).

[3] L. D. Landau and E. M. Lifshitz, *Quantum Mechanics* (Pergamon Press, New York, 1977); F. Wilczek and A. Shapere, *Geometric Phases in Physics* (World Scientific, Singapore, 1989).

[4] S. Weinberg, Phys. Rev. Lett. **62**, 485 (1989).

[5] A. J. Leggett, Rev. Mod. Phys. **73**, 307 (2001); T. L. Gustavson *et al.*, Phys. Rev. Lett. **88**, 020401 (2002); M. Mackie, R. Kowalski, and J. Javanainen, *ibid.* **84**, 3803 (2000); R. B. Diener *et al.*, *ibid.* **89**, 070401 (2002).

[6] Although the norm of a state is conserved, the inner product of two states changes with time.

[7] Y. S. Kivshar and B. A. Malomed, Rev. Mod. Phys. **61**, 763 (1989).

[8] Y. B. Band, B. Malomed, and M. Trippenbach, Phys. Rev. A **65**, 033607 (2002); Y. B. Band and M. Trippenbach, Phys. Rev. A **65**, 053602 (2002); G. J. de Valcárcel, cond-mat/0204406.

[9] Y. Aharonov and J. Anandan, Phys. Rev. Lett. **58**, 1593 (1987).

[10] V. I. Arnol'd, *Mathematical Methods of Classical Mechanics* (Springer-Verlag, Berlin, 1978); A. J. Lichtenberg and M. A. Lieberman, *Regular and Stochastic Motion* (Springer-Verlag, Berlin, 1983).

[11] One can often write $\partial\mathcal{H}/\partial\psi^* = H\psi$, with H being an operator (matrix) depending on the wave function in the nonlinear case. The simple relation, $\mathcal{H} = \langle\psi|H|\psi\rangle$, is valid only in the linear case.

[12] One thus often regards the mean field equation as the classical analog of the original many-body quantum problem. However, we do not address this correspondence in the present work. See R. Franzosi and V. Penna, Phys. Rev. A **63**, 043609 (2001).

[13] J. C. Garrison and R. Y. Chiao, Phys. Rev. Lett. **60**, 165 (1988); J. Anandan, Phys. Rev. Lett. **60**, 2555 (1988).

[14] V. I. Yukalov, E. P. Yukalova, and V. S. Bagnato, Phys. Rev. A **66**, 043601 (2002).

[15] For an eigenstate ψ, its time evolution involves only a phase, $\psi(t) = e^{-iEt}\psi(0)$. With the erasion of the overall phase, the projected state $\phi(t)$ does not change with time and is thus a fixed point. This connection may have been known to many; e.g., see P. Coullet and N. Vandenberghe, J. Phys. B **35**, 1593 (2002); R. Franzosi and V. Penna, Phys. Rev. A **65**, 013601 (2002).

[16] V. A. Yurovsky and A. Ben-Reuven, Phys. Rev. A **63**, 043404 (2001); A. Vardi, V. A. Yurovsky, and J. R. Anglin, *ibid.* **64**, 063611 (2001); B. Wu and Q. Niu, *ibid.* **64**, 061603 (2001); J. R. Anglin, cond-mat/0210504.

[17] B. Wu and Q. Niu, Phys. Rev. A **61**, 023402 (2000); J. Liu *et al.*, Phys. Rev. A **66**, 023404 (2002).

[18] G. J. Milburn *et al.*, Phys. Rev. A **55**, 4318 (1997); I. Zapata, F. Sols, and A. J. Leggett, *ibid.* **57**, R28 (1998); A. Smerzi *et al.*, Phys. Rev. Lett. **79**, 4950 (1997); A. Vardi and J. R. Anglin, *ibid.* **86**, 568 (2001); S. Kohler and F. Sols, *ibid.* **89**, 060403 (2002).

[19] For quasiperiodic motions, where the frequencies ω_i are not commensurate, one can use rational numbers n/m to approach the ratios ω_i/ω_j as closely as possible. In this spirit, Eq. (7) also holds for quasicyclic motions.

[20] D. Suter, K. T. Mueller, and A. Pines, Phys. Rev. Lett. **60**, 1218 (1988); H. Weinfurter, *ibid.* **64**, 1318 (1990).

[21] This is different from the semiclassical relation discussed in M. V. Berry, J. Phys. A **17**, 1225 (1984).

Geometric Phase for Adiabatic Evolutions of General Quantum States

Biao Wu,[1,2] Jie Liu,[3] and Qian Niu[4]

[1]*Institute of Physics, Chinese Academy of Sciences, P.O. Box 603, Beijing 100080, China*
[2]*Condensed Matter Sciences Division, Oak Ridge National Laboratory, Oak Ridge, Tennessee 37831, USA*
[3]*Institute of Applied Physics and Computational Mathematics, P.O. Box 8009, Beijing 100088, China*
[4]*Department of Physics, The University of Texas at Austin, Austin, Texas 78712, USA*
(Received 29 September 2004; published 13 April 2005)

The concept of a geometric phase (Berry's phase) is generalized to the case of noneigenstates, which is applicable to both linear and nonlinear quantum systems. This is particularly important to nonlinear quantum systems, where, due to the lack of the superposition principle, the adiabatic evolution of a general state cannot be described in terms of eigenstates. For linear quantum systems, our new geometric phase reduces to a statistical average of Berry's phases. Our results are demonstrated with a nonlinear two-level model.

DOI: 10.1103/PhysRevLett.94.140402 PACS numbers: 03.65.Vf, 03.65.Ca

A complete theory for the adiabatic evolution of quantum systems rests on three pillars. First, the adiabatic theorem dictates that the probability on each instantaneous eigenstate (nondegenerate) remains constant when the external condition, described by a set of control parameters, changes slowly in time [1]. Second, besides the usual dynamical phase, given by the time integral of the eigenenergy, the phase of an evolving eigenstate has a geometric part, called Berry's phase, which depends only on the geometric path in the parameter space [2]. Third, the linearity of the Schrödinger equation allows one to write the evolution of an arbitrary state as a linear superposition of the eigenstates with amplitudes whose magnitudes and phases are as prescribed above [3].

For a nonlinear quantum system, such as that described by the nonlinear Schrödinger equation, the third pillar is missing. The superposition principle no longer applies. While it is still possible to define eigenstates and describe their adiabatic evolution, one is clueless on the evolution of a general state when the control parameters are changed slowly. Nonlinear quantum systems have become increasingly important in physics. They often arise in the mean field treatment of many-body quantum systems, such as Bose-Einstein condensates (BECs) of dilute atomic gases [4,5], and as a possible fundamental nonlinear modification of quantum mechanics [6].

In this Letter we introduce the concept of a geometric phase for the adiabatic evolution of general quantum states including noneigenstates. It applies to both linear and nonlinear quantum systems. In the linear case, our geometric phase reduces to a statistical average of Berry's phases for the eigenstates, weighted by the probabilities that the system finds itself in the eigenstates.

Our effort to define a geometric phase for general quantum states is aided by the fact that quantum systems (linear or nonlinear) have an exact canonical structure of Hamiltonian dynamics, which has been known since long ago but not as widely as it deserves [6,7]. We can thus introduce classical concepts such as actions and angles to describe the motion of the quantum states without any approximation. The actions remain invariant during an adiabatic process, and the change in the angles has a geometric part called Hannay's angles [8]. Our new geometric phase is found to be related to these Hannay's angles in a derivative form. In defining the new phase, an average is taken over all possible initial states of the same actions. Therefore, our geometric phase can be viewed as a phase for a manifold of fixed actions, which is an invariant subspace of states in the adiabatic process (see Fig. 1).

The physical meaning of our phase is discussed in the context of a heavy particle interacting with a nonlinear quantum system. Our phase appears in the effective Lagrangian for the heavy particle. Similar to the original Berry's phase [2], we show that our phase for a general quantum state can be regarded as a geometric part of a generalized total phase. In the end, a nonlinear two-level model is used to illustrate our phase.

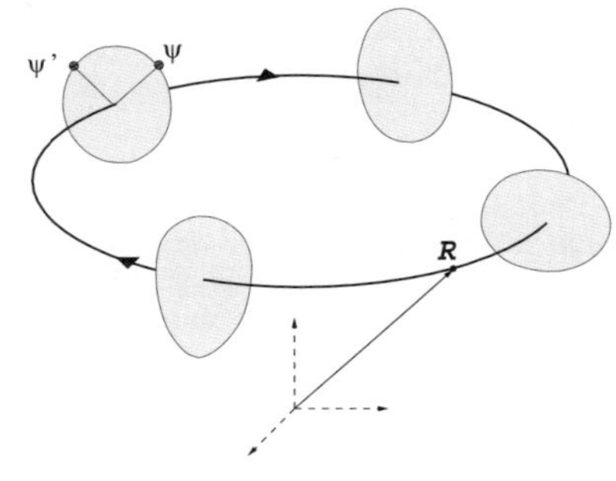

FIG. 1. Evolution of quantum states as the control parameters $\boldsymbol{R}$ traverse slowly along a closed path $\mathcal{O}$. At the end of the loop, an individual state (dot) may not come back to itself, but a collection of states (disk) does. This collection is characterized by a set of adiabatically invariant actions (illustrated by the disk area). Different disk shapes represent symbolically distinct dynamics at different values of $\boldsymbol{R}$.

We consider an N-level quantum system governed by a general nonlinear Schrödinger equation ($\hbar = 1$),

$$i\frac{d}{dt}|\psi\rangle = H(\psi^*, \psi; \boldsymbol{R})|\psi\rangle. \tag{1}$$

Here $|\psi\rangle = (\psi_1, \psi_2, \ldots, \psi_N)$ with ψ_j being its jth component over an orthonormal basis; the vector $\boldsymbol{R}$ represents all the control parameters subject to adiabatic change. For all problems of physical interest, we can assume that the system is globally gauge invariant. Therefore, the norm of the wave function is conserved. When H is independent of ψ^* and ψ, Eq. (1) is the usual linear Schrödinger equation.

Eigenstates for Eq. (1) can be defined as usual,

$$H(\boldsymbol{R})|\varphi_n(\boldsymbol{R})\rangle = \mu_n(\boldsymbol{R})|\varphi_n(\boldsymbol{R})\rangle. \tag{2}$$

Then for fixed values of $\boldsymbol{R}$, $e^{-i\mu_n(\boldsymbol{R})t}|\varphi_n(\boldsymbol{R})\rangle$ is a solution of the nonlinear Schrödinger equation (1). In the context of BEC physics, μ_n plays the role of a chemical potential. When $\boldsymbol{R}$ varies slowly in time, such a state remains to be a solution of Eq. (1) with the phase replaced by $\int dt\mu_n(\boldsymbol{R})$ (dynamical phase), minus Berry's phase defined by

$$\gamma_n(\mathcal{O}) = \int_{\mathcal{O}} d\boldsymbol{R}\langle\varphi_n(\boldsymbol{R})|i\frac{\partial}{\partial \boldsymbol{R}}|\varphi_n(\boldsymbol{R})\rangle. \tag{3}$$

This phase depends on the path $\mathcal{O}$ in the parameter space, but does not depend on the rate of change in $\boldsymbol{R}$.

However, it is not clear how a similar geometric phase can be defined for noneigenstates, which is imperative for nonlinear quantum systems due to their lack of the superposition principle as mentioned in the introduction. One can check that various existing generalizations of Berry's phase do not help here [9]. To overcome this difficulty, we notice a long-known fact [6,7] that the quantum system Eq. (1) is mathematically equivalent to Hamilton's equations of motion with a Hamiltonian energy functional $\mathcal{H}(\psi^*, \psi; \boldsymbol{R})$ and Poisson brackets $\{\psi_j^*, \psi_k\} = i\delta_{jk}$. That is, Eq. (1) can be reformulated as

$$i\frac{d}{dt}\psi_j = \frac{\partial}{\partial \psi_j^*}\mathcal{H}(\psi^*, \psi; \boldsymbol{R}). \tag{4}$$

In this formulation, the eigenstate defined in Eq. (2) is variationally an extremum solution of the energy functional $\mathcal{H}$, with μ_n viewed as the value of the Lagrange multiplier enforcing the conserved norm of a wave function. We emphasize that this canonical Hamiltonian structure is exact and applies to both the linear and the nonlinear Schrödinger equations. It is not related to the usual classical-quantum correspondence in the limit of $\hbar \to 0$. Using the language of Hamiltonian dynamics, the system of Eq. (1) can be classified into integrable or nonintegrable as for any classical systems. In this sense, all linear quantum systems are integrable.

We focus on the case that Eq. (1) is locally integrable even in the presence of nonlinearity. As we point out later, generalization to chaotic systems is possible. For fixed control parameters $\boldsymbol{R}$, in the integrable regions, the system can be described by a set of action and angle variables, $\boldsymbol{I} = \{I_1, I_2, \ldots, I_N\}$, $\boldsymbol{\theta} = \{\theta_1, \theta_2, \ldots, \theta_N\}$; the actions are constant while the angles vary in time with frequencies $\omega_\ell = \partial\mathcal{H}/\partial I_\ell$. In other words, the wave function can be expressed in terms of these action-angle variables, $|\psi\rangle = |\psi(\boldsymbol{I}, \boldsymbol{\theta}; \boldsymbol{R})\rangle$, with a ($\boldsymbol{R}$-dependent) canonical transformation parametrizing the amplitudes ψ_j with these new variables. When $\boldsymbol{R}$ are time dependent, we can still use the same expression of the wave function, $|\psi\rangle = |\psi(\boldsymbol{I}, \boldsymbol{\theta}; \boldsymbol{R})\rangle$, although the action-angle variables will lose the above simple properties.

However, if $\boldsymbol{R}$ vary slowly in time, the actions continue to be conserved [10]. In this case, $|\psi\rangle$ changes with time only through the rotating $\boldsymbol{\theta}$ and the slowly varying $\boldsymbol{R}$. This allows us to do the following expansion:

$$\langle\psi|d|\psi\rangle = \sum_{\ell=1}^{N}\langle\psi|\frac{\partial}{\partial\theta_\ell}|\psi\rangle d\theta_\ell + \langle\psi|\frac{\partial}{\partial \boldsymbol{R}}|\psi\rangle\cdot d\boldsymbol{R}. \tag{5}$$

The second term is clearly related to the geometry of the parameter space $\boldsymbol{R}$. It motivates us to define our geometric phase as

$$\gamma(\mathcal{O}) = \overline{\int_{\mathcal{O}} d\boldsymbol{R}\langle\psi(\boldsymbol{I}, \boldsymbol{\theta}; \boldsymbol{R})|i\frac{\partial}{\partial \boldsymbol{R}}|\psi(\boldsymbol{I}, \boldsymbol{\theta}; \boldsymbol{R})\rangle}, \tag{6}$$

where the overbar indicates an average after the integration along $\mathcal{O}$ over all initial angles with the same actions. This averaging wipes out the dependence of our phase γ on the dynamical angles $\boldsymbol{\theta}$ and makes it geometric. Such an averaging technique has been used in defining Hannay's angles [8].

The phase γ in Eq. (6) is the geometric phase that we have sought for general quantum states. Two remarks are in order: (i) One can check that our phase γ reduces to γ_n for an eigenstate $|\varphi_n\rangle$. (ii) Because of the averaging, the phase γ is a geometric characterization of the manifold of all states with a given set of actions. Such a manifold is an invariant subspace in an adiabatic process.

The physical meaning of our phase γ can be seen by considering a coupled system described by the following Lagrangian:

$$\mathcal{L} = \langle\psi|i\frac{d}{dt}|\psi\rangle - \mathcal{H}(\psi^*, \psi; \boldsymbol{R}) + \mathcal{L}'(\boldsymbol{R}, \dot{\boldsymbol{R}}), \tag{7}$$

where the first two terms are the Lagrangian for the nonlinear Schrödinger Eq. (1) and $\mathcal{L}'$ is a Lagrangian for a heavy particle moving slowly. In terms of action and angle variables, the Lagrangian becomes

$$\mathcal{L} = \boldsymbol{I}\cdot\dot{\boldsymbol{\theta}} + \langle\psi|i\frac{\partial}{\partial \boldsymbol{R}}|\psi\rangle\cdot\dot{\boldsymbol{R}} - \mathcal{H}(\boldsymbol{I}; \boldsymbol{R}) + \mathcal{L}'(\boldsymbol{R}, \dot{\boldsymbol{R}}). \tag{8}$$

Since the angles $\boldsymbol{\theta}$ change much faster than the particle's coordinates $\boldsymbol{R}$, we can take an average over the angles and obtain an effective Lagrangian for the dynamics of the heavy particle,

$$\mathcal{L}_{\text{eff}} = \overline{\langle\psi|i\frac{\partial}{\partial \boldsymbol{R}}|\psi\rangle} \cdot \dot{\boldsymbol{R}} - \mathcal{H}(\boldsymbol{I};\boldsymbol{R}) + \mathcal{L}'(\boldsymbol{R},\dot{\boldsymbol{R}}). \quad (9)$$

Our phase γ shows how the fast dynamics of a nonlinear quantum system influences the heavy particle.

Let us go back to Eq. (5) and examine the other two terms. After integration, the term on the left becomes

$$\beta = i\int_{\mathcal{O}} \langle\psi|d|\psi\rangle = \int_{\mathcal{O}} \langle\psi|H|\psi\rangle dt, \quad (10)$$

which is apparently a dynamical phase. For the first term on the right of Eq. (5), we have

$$\mathcal{T} = \int_{\mathcal{O}} \sum_{\ell=1}^{N} d\theta_\ell \langle\psi(\boldsymbol{I},\boldsymbol{\theta};\boldsymbol{R})|i\frac{\partial}{\partial\theta_\ell}|\psi(\boldsymbol{I},\boldsymbol{\theta};\boldsymbol{R})\rangle. \quad (11)$$

We similarly average $\mathcal{T}$ over all initial angles. Because of such an averaging, the integrand in Eq. (11) can be replaced with I_ℓ [11]. Letting $\Delta\theta_\ell$ be the total change in angle θ_ℓ for an adiabatic process along $\mathcal{O}$, we find

$$\overline{\mathcal{T}} = \sum_{\ell=1}^{N} I_\ell \Delta\theta_\ell, \quad (12)$$

which is the total action of the quantum state on average. Putting all these together, we arrive at

$$\sum_{\ell=1}^{N} I_\ell \Delta\theta_\ell = \overline{\beta} - \gamma(\mathcal{O}), \quad (13)$$

where the overbar stands for the averaging over all the initial angles as before. For an adiabatically evolving eigenstate $|\psi\rangle = e^{-i\lambda(t)}|\varphi_n(\boldsymbol{R})\rangle$, there is only one pair of nonzero action and angle, the norm $\langle\psi|\psi\rangle$ and the phase λ, respectively. In this special case, our total action $\overline{\mathcal{T}} = \mathcal{T}$ becomes the total phase $\Delta\lambda$, accumulated during the adiabatic evolution of the eigenstate. Therefore, the total action is also a generalization of the total phase, and relation (13) says that our phase γ is the geometric part of this generalized total phase. Finally, we note that this relation (13) can also be regarded as another expression for our phase γ, which is easier to be implemented numerically and is used to compute the results in Fig. 2.

According to Hannay [8], the rotating angles $\boldsymbol{\theta}$ pick up geometric angles $\boldsymbol{\alpha} = \{\alpha_1, \alpha_2, \ldots, \alpha_N\}$ in addition to the dynamical part as $\boldsymbol{R}$ traverse slowly round a loop $\mathcal{O}$. These angles $\boldsymbol{\alpha}$, known as Hannay's angles and regarded as a classical concept, have now shown up naturally in quantum systems. In light of this recognition, the geometric part of the total action $\mathcal{T}$ is apparently $\sum_\ell I_\ell \alpha_\ell$, which should equal our phase γ. Taking the derivative of Eq. (13) with respect to I_ℓ, we then obtain

$$\alpha_\ell(\mathcal{O}) = -\frac{\partial}{\partial I_\ell}\gamma(\mathcal{O}), \quad (14)$$

which relates our geometric phase and Hannay's angle in a derivative form. This relation can be viewed from another angle. The quantum system (linear or nonlinear) of Eq. (1) may be regarded as the classical limit of a second-quantized system $\hat{\mathcal{H}}$, where ψ_j becomes an operator $\hat{\psi}_j$. According to Ref. [12], our phase γ is the usual Berry's phase for this second-quantized system in the semiclassical limit. In other words, the relation (14) is also the semiclassical relation between Hannay's angles and Berry's phases, which has been known for a while [12–14].

So far, we have focused on completely integrable systems (or regions). Our phase γ can be generalized to other nonintegrable systems, including the class of chaotic systems studied in Ref. [15]. One can achieve this by expanding the averaging in Eq. (6). Namely, if the system has only $K(<N)$ adiabatic invariants, the averaging in Eq. (6) is then over all initial states with the same K invariants.

We now illustrate the new phase γ with two examples. First, we consider a general quantum state in linear systems, where the evolving quantum state can be expanded in terms of the normalized eigenstates,

$$|\psi(t)\rangle = \sum_{n=1}^{N} a_n(t)|\varphi_n(\boldsymbol{R})\rangle. \quad (15)$$

According to the quantum adiabatic theorem [1], the occupation probabilities of different eigenstates $|a_n|^2$ are adiabatic constants. In fact, they are actions $I_n = |a_n|^2$ when the system is regarded mathematically as a classical Hamiltonian system [16]; their corresponding angle variables θ_n's are the phases of a_n's. With these in mind, the computation of Eq. (6) with the state (15) is straightforward. We find that the off-diagonal terms become zero after the averaging, and the geometric phase γ is

$$\gamma(\mathcal{O}) = \sum_{n=1}^{N} |a_n|^2 \gamma_n(\mathcal{O}). \quad (16)$$

Therefore, in linear quantum systems, the phase γ is just a weighted summation of Berry's phases of all the eigenstates involved. Interestingly, this kind of weighted summation of Berry's phases has already been applied in real physical systems [17,18]. Interestingly, if we apply Eq. (14) to Eq. (16), we find that these Hannay's angles differ from Berry's phases only by a sign, $\alpha_n = -\gamma_n$.

Second, we study a nonlinear two-level model given by

$$i\frac{d\psi_1}{dt} = \left[c|\psi_2|^2 + \frac{Z}{2}\right]\psi_1 + \frac{X - iY}{2}\psi_2, \quad (17)$$

$$i\frac{d\psi_2}{dt} = \left[c|\psi_1|^2 - \frac{Z}{2}\right]\psi_2 + \frac{X + iY}{2}\psi_1. \quad (18)$$

This simple model can be used to describe the Josephson effect of Bose-Einstein condensates residing in a double-well potential [5,19]. The complex coupling constant, as denoted by X and Y, can be realized in experiment through phase imprinting on one of the two wells [20].

We first look at the geometric phase γ for eigenstates of this nonlinear quantum system. This is to find all the eigenstates $|\varphi_n(\boldsymbol{R})\rangle$, and then use them to calculate the

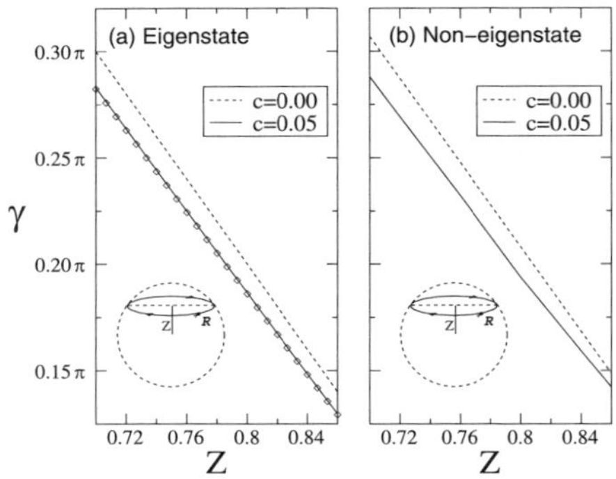

FIG. 2. Geometrical phases γ of the nonlinear two-level model (17) and (18): (a) the lower eigenstate and (b) the noneigenstate near the lower eigenstate. The noneigenstate has an action (or *AA* phase) of $I = 0.005$. The inset shows the closed path $\mathcal{O}$ used that is a circle on a unit sphere with Z fixed. The circle is traversed with a rate of 0.001. The results for eigenstates are compared to the analytical expression of Eq. (19) (denoted by diamonds): a very good match is found.

phase with Eq. (3). They could be done numerically; fortunately for this simple case, analytical results can be obtained. When the path is restricted on the unit sphere $X^2 + Y^2 + Z^2 = 1$, we find that the phases for these nonlinear eigenstates are

$$\gamma(\mathcal{O}) = \int_{\partial S=\mathcal{O}} \frac{\eta^3(\boldsymbol{R} + c\eta\hat{z}) \cdot d\boldsymbol{S}}{(c\eta + Z)^2(c\eta^3 + Z)}, \tag{19}$$

where $\hat{z}$ is the unit vector along the z axis and η is one of the real roots of

$$c^2\eta^4 + 2cZ\eta^3 + (1 - c^2)\eta^2 - 2cZ\eta - Z^2 = 0. \tag{20}$$

Different real roots η correspond to different eigenstates. Equation (20) can have more than two real roots, indicating that there can be more than two eigenstates [21]. Here we limit ourselves to the situations where Eq. (20) has only two real roots. For the path $\mathcal{O}$ that is a circle with a fixed Z, the geometric phase in Eq. (19) becomes $\gamma = (1 - \eta)\pi$. The diamonds in Fig. 2 show how γ for the lower eigenstate changes with Z.

For a general quantum state, we have to resort to numerical means. The path $\mathcal{O}$ is picked to be a circle with fixed Z. We then solve Eqs. (17) and (18) numerically after choosing a changing rate $v = 0.001$ for the parameters $\boldsymbol{R} = \{X, Y, Z\}$. The evolving states are recorded and used to compute the phase γ with Eq. (13), where the averaging is done for different initial states with the same action (or *AA* phase) I. Results for $c = 0.05$ are plotted in Fig. 2, showing how the phase changes with Z. Computation is done for both eigenstate and noneigenstate, and the results (solid lines) are compared to the phases for the linear case $c = 0.0$ (dashed lines). The changing rate of $\boldsymbol{R}$ ($v = 0.001$) is slow enough to be considered as adiabatic. This is witnessed by the good fit between the solid line and the diamonds in Fig. 2(a) as the diamonds are the analytical results of Eq. (19).

In conclusion, a geometric phase for the adiabatic evolution of general quantum states is introduced. It applies to noneigenstates as well as eigenstates in both linear and nonlinear quantum systems. Our geometric phase is related to Hannay's angles, which arise naturally in a canonical description of adiabatic dynamics of quantum states. Like the usual Berry's phase, it also affects the effective dynamics of slow variables. A nonlinear two-level system is used to demonstrate our results.

B. W. acknowledges the support of "BaiRen" program of Chinese Academy of Sciences, NSF (DMR-0306239), and LDRD of ORNL, managed by UT-Battelle; J. L. is supported by NNSF of China (10474008), 973 project on strong field physics, and NSF (PHY-0244688); Q. N. is supported by DOE (DE-FG03-02ER45958) and the Welch Foundation.

[1] A. Messiah, *Quantum Mechanics* (Dover, New York, 2000).
[2] M. V. Berry, Proc. R. Soc. London A **392**, 45 (1984).
[3] J. Anandan and L. Stodolsky, Phys. Rev. D **35**, 2597 (1987).
[4] F. Dalfovo, S. Giorgini, L. P. Pitaevskii, and S. Stringari, Rev. Mod. Phys. **71**, 463 (1999).
[5] A. J. Leggett, Rev. Mod. Phys. **73**, 307 (2001).
[6] S. Weinberg, Ann. Phys. (N.Y.) **194**, 336 (1989).
[7] A. Heslot, Phys. Rev. D **31**, 1341 (1985).
[8] J. H. Hannay, J. Phys. A **18**, 221 (1985).
[9] Y. Aharonov and J. Anandan, Phys. Rev. Lett. **58**, 1593 (1987); J. C. Garrison and R. Y. Chiao, *ibid.* **60**, 165 (1988); F. Wilczek and A. Zee, *ibid.* **52**, 2111 (1984); J. Samuel and R. Bhandari, *ibid.* **60**, 2339 (1988).
[10] V. I. Arnol'd, *Mathematical Methods of Classical Mechanics* (Springer-Verlag, Berlin, 1978).
[11] One can exchange the order of the averaging and the integration along $\mathcal{O}$.
[12] M. V. Berry, J. Phys. A **18**, 15 (1985).
[13] J. Anandan, Phys. Lett. A **129**, 201 (1988).
[14] J. Liu, B. Hu, and B. Li, Phys. Rev. Lett. **81**, 1749 (1998); Phys. Rev. A **58**, 3448 (1998).
[15] C. Jarzynski, Phys. Rev. Lett. **74**, 1732 (1995); J. M. Robbins and M. V. Berry, Proc. R. Soc. London A **436**, 631 (1992).
[16] Jie Liu, Biao Wu, and Qian Niu, Phys. Rev. Lett. **90**, 170404 (2003).
[17] D. J. Thouless, Ping Ao, and Qian Niu, Phys. Rev. Lett. **76**, 3758 (1996).
[18] Yugui Yao *et al.*, Phys. Rev. Lett. **92**, 037204 (2004).
[19] G. J. Milburn *et al.*, Phys. Rev. A **55**, 4318 (1997).
[20] J. Denschlag *et al.*, Science **287**, 97 (2000).
[21] Biao Wu and Qian Niu, Phys. Rev. A **61**, 023402 (2000).

Berry phase in nonlinear systems

J. Liu and L. B. Fu
Institute of Applied Physics and Computational Mathematics, Post Office Box 8009, Beijing 100088, China and
Center for Applied Physics and Technology, Peking University, Beijing 100084, China
(Received 13 April 2009; published 17 May 2010)

The Berry phase acquired by an eigenstate that experienced a nonlinear adiabatic evolution is investigated thoroughly. The circuit integral of the Berry connection of the instantaneous eigenstate cannot account for the adiabatic geometric phase, while the Bogoliubov excitations around the eigenstates are found to be accumulated during the nonlinear adiabatic evolution and contribute a finite phase of geometric nature. A two-mode model is used to illustrate our theory. Our theory is applicable to Bose-Einstein condensate, nonlinear light propagation, and Ginzburg-Landau equations for complex order parameters in condensed-matter physics.

DOI: 10.1103/PhysRevA.81.052112 PACS number(s): 03.65.Vf, 03.75.Kk, 03.65.Ge

I. INTRODUCTION

Adiabatic theory, as a fundamental issue of quantum mechanics, involves two aspects: (i) when the Hamiltonian changes slowly compared to the level spacings, an initial nondegenerate eigenstate remains to be an instantaneous eigenstate [1]; (ii) the phase acquired by the eigenstate is the sum of the time integral of the eigenenergy (dynamical phase) and a quantity independent of the time duration and related to the geometric property of the closed path in parameter space (adiabatic geometric phase or Berry phase) [2,3]. Adiabatic theory has played a crucial role in the preparation and control of quantum states [4]. Recently, the Berry phase and related geometric phases [5,6] have received renewed interest due to their important use in the implementation of quantum-computing gates [7] and applications in condensed-matter physics [8].

For a nonlinear quantum system, such as that described by the nonlinear Schrödinger equations, how does the adiabatic theory get modified? Nonlinear quantum systems have become increasingly important in physics. They often arise in the mean-field treatment of many-body quantum systems, such as Bose-Einstein condensates (BECs) of dilute atomic gases [9]. Other applications include nonlinear light propagation [10]. Recently, extending the first aspect of adiabatic theory to the nonlinear systems, i.e., investigating the adiabatic condition and adiabaticity for the nonlinear quantum evolution, has been done [11,12]. Interestingly, it was found that the adiabaticity of an eigenstate only requires that the control parameters vary slowly with respect to the Bogoliubov excitation frequencies that in general are not equivalent to the level spacings. Nevertheless, the Berry phase issue in such nonlinear systems is far from understood.

On the other hand, it should be mentioned that the geometric phase for the cyclic evolution with a finite time duration T (i.e., nonadiabatical motion) in nonlinear systems was studied many years ago [13]. It was found that the nonadiabatic geometric phase takes the form of $-\int_0^T i\langle\phi|\frac{\partial}{\partial t}|\phi\rangle dt$, analogous to its linear counterpart [6]. Here, ϕ is the wave function in the projective Hilbert space satisfying the cyclic requirement $\phi(t=0)=\phi(t=T)$. A similar formula was deduced recently in investigating the geometric phase in a Bose-Einstein-Josephson junction [14]. However, the Berry phase associated with the adiabatic evolution of an eigenstate in nonlinear systems is still unresolved. Along the considerations of [13], one might imagine that when the parameter $\mathbf{R}$ vector moves in a circuit adiabatically, the adiabatic geometric phase acquired by an eigenstate would take the usual form of $-\oint i\langle\phi(\mathbf{R})|\frac{\partial}{\partial\mathbf{R}}|\phi(\mathbf{R})\rangle d\mathbf{R}$ [15]. Here, $\phi(\mathbf{R})$ denotes one eigenstate of the nonlinear system and $\langle\phi(\mathbf{R})|\frac{\partial}{\partial\mathbf{R}}|\phi(\mathbf{R})\rangle$ is called the Berry connection. The expression indicates that the Berry phase still equals the circuit integral of the Berry connection. In a recent study of the Berry phase for a specific BEC system described by a nonlocal Gross-Pitaevskii (nonlinear) equation with a quadratic potential, the exact solutions were constructed and the Berry phase was calculated in explicit form [16]. Their obtained adiabatic geometric phase, however, does not equal the above expression; an additional term emerges that is directly provoked by the presence of nonlinearity. The reason was not determined [16]. This controversy is not properly resolved yet, calling for further investigation; it may indicate that some subtle and important aspects were missed in previous theoretical considerations.

In the present paper, we have made a thorough analysis of the Berry phase issue for nonlinear systems. Our analytic deduction clearly indicates that the Berry phase is dramatically modified by the nonlinearity. The underlying mechanism has been revealed: for a nonlinear system because the Hamiltonian is a functional of the instantaneous wave functions, the Bogoliubov excitations induced by the slow change of the system are allowed to feed back to the Hamiltonian. They are accumulated during an adiabatic evolution and eventually contribute a finite phase of geometric nature.

Our paper is organized as follows. Sec.II is our general formalism. In Sec. III a two-mode BEC model is used to illustrate our theory and the geometric meaning of the new phase is discussed accordingly. Section IV presents our discussion and conclusion.

II. GENERAL FORMALISM

Without losing generality, our discussion is follows the Schrödinger equation with a quadratic nonlinear term, and our deduction is readily extended to other forms of nonlinearity with the invariance of gauge transformation,

$$i\frac{\partial\psi}{\partial t} = H_0\psi + g|\psi|^2\psi, \tag{1}$$

where $H_0 = -\frac{1}{2}\nabla^2 + V(\mathbf{R}; r)$. Physically, the above system can be considered as a BEC trapped in a potential $V(\mathbf{R}; r)$, with $\hbar = m = 1$ [17], where $\mathbf{R}$ is the parameter vector that varies in time slowly, and g is the nonlinear parameter representing the interaction between the coherent atoms. The total energy of the system $E_T = \int dr E(\psi^*, \psi)$, where the energy density $E(\psi, \psi^*) = \psi^* H_0 \psi + \frac{1}{2} g |\psi|^4$. The above system is invariant under gauge transformations of the first kind, $\psi(r,t) \rightarrow \exp(i\eta)\psi(r,t)$ with constant η. The gauge symmetry implies that the total atom number is conserved, i.e., $\int dr |\psi|^2 = 1$.

Let λ be the overall phase of the wave function, we may take it to be the phase of the wave function at a fixed position r_0, for example, $\lambda = -\arg(\psi(r_0,t))$. We split off this overall phase by writing $\psi = e^{-i\lambda}\phi$, and ϕ belongs to the so-called projective Hilbert space. From (1) we obtain

$$\frac{d\lambda}{dt} = -i\langle\phi|\frac{\partial}{\partial t}|\phi\rangle + \int dr E(\phi^*, \phi) + \frac{g}{2}\langle\phi|\phi^*\phi|\phi\rangle. \quad (2)$$

The eigenequation of the system is

$$H_0\overline{\psi} + g|\overline{\psi}|^2\overline{\psi} = \mu\overline{\psi}, \quad (3)$$

where $\overline{\psi}$ is the eigenfunction and μ is the eigenvalue (or chemical potential).

Now we assume the parameter vector $\mathbf{R}$ varies slowly in time and introduce the dimensionless adiabatic parameter of $\varepsilon \sim |\frac{d\mathbf{R}}{dt}| \sim \frac{1}{T}$ as the measure of how slow the parameters change. The adiabatic parameter tends to zero, i.e., $\varepsilon \rightarrow 0$, indicating the adiabatic limit. T is the time duration.

Consequently, the expression of the total phase can be expanded in a perturbation series in the adiabatic parameter, i.e.,

$$\frac{d\lambda}{dt} = \alpha_0(\varepsilon^0) + \alpha_1(\varepsilon^1) + o(\varepsilon^2). \quad (4)$$

When the parameters move in a circuit, the eigenstate evolves for an infinitely long time duration in the adiabatic limit. The time integral of the zero-order term gives the so-called dynamic phase because it is closely related to the temporal process of the evolution. The time integral of the first-order term makes an additional contribution to the overall phase, which will be shown later to be of a geometric nature, that is, it only depends on the geometry of the closed path in the parameter space. The contribution of the higher-order term vanishes in the adiabatic limit.

In the quantum evolution with slowly changing parameters, we assume $\phi = \overline{\phi}(\mathbf{R}) + \delta\phi(\mathbf{R})$, where $\overline{\phi}(\mathbf{R})$ is the wave function of the instantaneous eigenstate corresponding to the local minimum energy. $\delta\phi$ denotes the secular part of the Bogoliubov excitations induced by the system's slow change, while the rapid oscillations in the excitations are ignored because they vanish after a long-term average. $\delta\phi$ depends on the adiabatic parameter and is of order ε, then from Eq. (2) and with the help of relation Eq. (3), we have the explicit expressions as follows:

$$\alpha_0(\varepsilon^0) = \mu(\mathbf{R}), \quad (5)$$

$$\alpha_1(\varepsilon^1) = -i\int dr\left(\overline{\phi}^*\frac{\partial}{\partial t}\overline{\phi}\right) + g\int dr(\overline{\phi}^2\overline{\phi}^*\delta\phi^* + \overline{\phi}^{*2}\overline{\phi}\delta\phi). \quad (6)$$

From the above expressions, we see that the dynamical phase has been modified to be the time integral of the chemical potential rather than the energy. This is because the instantaneous eigenstates are feedback to the Hamiltonian. More interestingly, the first-order term, i.e., the Berry phase term has been modified due to the feedback of the Bogoliubov excitations to the Hamiltonian. To evaluate it qualitatively and express the modified geometric phase explicitly, let us introduce a set of orthogonal basis $|k\rangle$ ($k = 1, 2, \ldots, N$) and the variable ψ_j, which is the jth component, i.e., $\psi_j = \langle j|\psi\rangle$. Without losing generality, the projective Hilbert space is set to be of a specific gauge such that the phase of the Nth component is zero. In the projective Hilbert space, the new variables (n_j, θ_j) are introduced through $\phi_j = \sqrt{n_j}e^{i\theta_j}$. Substituting the expression of $\psi_j = \sqrt{n_j}e^{i\theta_j}e^{-i\int^t \beta dt}$ into the nonlinear equation, and separating real and imaginary parts, we have the following differential equations for the density n_j and phase θ_j, respectively:

$$\frac{dn_j}{dt} = f_j, \quad \frac{d\theta_j}{dt} = h_j, \quad j = 1, 2, \ldots, N-1, \quad (7)$$

where f_j and h_j are functions of the amplitudes, relative phases, matrix elements $C_{jk}(\mathbf{R}) = \langle j|H_0(\mathbf{R})|k\rangle$, and the overlap integral $D_{j,k,l,m} = \langle j|\langle k|l\rangle|m\rangle$. Their explicit expressions can be readily deduced but are not given here.

The norm conservation condition $n_N = 1 - \sum_{k=1}^{N-1} n_k$ has been used to remove the variable n_N in the above equations. In the representation of new variables, $(\overline{n}_j, \overline{\theta}_j)$ satisfy equations of the equilibrium state, i.e.,

$$\left(\frac{dn_j}{dt}, \frac{d\theta_j}{dt}\right)\Bigg|_{(\overline{n}_j, \overline{\theta}_j)} = 0. \quad (8)$$

$(\overline{n}_j, \overline{\theta}_j)$ are functions of the parameter $\mathbf{R}$ corresponding to the eigenstates of the system. Let us make a perturbation expansion around the eigenstate with

$$n_j = \overline{n}_j(\mathbf{R}) + \delta n_j(\mathbf{R}), \quad \theta_j = \overline{\theta}_j(\mathbf{R}) + \delta\theta_j(\mathbf{R}). \quad (9)$$

Here, $\overline{\phi}_j(\mathbf{R}) = \sqrt{\overline{n}_j(\mathbf{R})}e^{i\overline{\theta}_j(\mathbf{R})}$, and $\delta n_j(\mathbf{R}), \delta\theta_j(\mathbf{R})$ are the excitations of order ε. Then, inserting the above expansion into equations (7) and, ignoring the higher-order terms such as $\frac{\partial\delta n_j}{\partial t}, \frac{\partial\delta\theta_j}{\partial t}$, and denoting $\upsilon = (n_1, \theta_1; \ldots; n_{N-1}, \theta_{N-1})$, we obtain

$$\frac{d\overline{\upsilon}}{d\mathbf{R}}\frac{d\mathbf{R}}{dt} = \mathcal{L}\delta\upsilon, \quad (10)$$

where the matrix takes the form

$$\mathcal{L} = \{\mathcal{L}_{jk}\}_{(N-1,N-1)}, \quad \mathcal{L}_{jk} = \begin{pmatrix} \frac{\partial f_j}{\partial n_k} & \frac{\partial f_j}{\partial \theta_k} \\ \frac{\partial h_j}{\partial n_k} & \frac{\partial h_j}{\partial \theta_k} \end{pmatrix}_{\upsilon=\overline{\upsilon}}. \quad (11)$$

Then, inversely, we have

$$\delta\upsilon = \mathcal{L}^{-1}\frac{d\overline{\upsilon}}{d\mathbf{R}}\frac{d\mathbf{R}}{dt}. \quad (12)$$

The differential relation between the new variables and old ones takes the form

$$\begin{pmatrix} \delta\phi_j \\ \delta\phi_j^* \end{pmatrix} = \Pi_j\begin{pmatrix} \delta n_j \\ \delta\theta_j \end{pmatrix}, \quad (13)$$

in which

$$\Pi_j = \begin{pmatrix} \frac{1}{2}\overline{n}_j^{-1/2} e^{i\overline{\theta}_j} & i\sqrt{\overline{n}_j} e^{i\overline{\theta}_j} \\ \frac{1}{2}\overline{n}_j^{-1/2} e^{-i\overline{\theta}_j} & -i\sqrt{\overline{n}_j} e^{-i\overline{\theta}_j} \end{pmatrix}. \tag{14}$$

Substituting (12) and (13) into (6), we finally obtain the explicit expression of the adiabatic geometric phase that contains two terms,

$$\gamma_g = \gamma_B + \gamma_{NL}, \tag{15}$$

where the first term is the usual Berry phase formula,

$$\gamma_B = -i \oint \langle \overline{\phi} | \nabla_{\mathbf{R}} | \overline{\phi} \rangle \, d\mathbf{R} = \oint \sum_{j=1}^{N-1} \overline{n}_j \frac{\partial \overline{\theta}_j}{\partial \mathbf{R}} d\mathbf{R}, \tag{16}$$

and the additional term is from the nonlinearity, taking the form

$$\gamma_{NL} = g \oint \left\langle \Lambda | \Pi \circ \mathcal{L}^{-1} \left| \frac{d\overline{v}}{d\mathbf{R}} \right. \right\rangle d\mathbf{R}. \tag{17}$$

Here,

$$\Lambda = \left(\left(\overline{n}_1 + \sum_{j=1}^{N-1} \overline{n}_j - 1 \right) \overline{n}_1^{1/2} e^{i\overline{\theta}_1}, \left(\overline{n}_1 + \sum_{j=1}^{N-1} \overline{n}_j - 1 \right) \overline{n}_1^{1/2} e^{-i\overline{\theta}_1}, \ldots \right),$$

$$\frac{d\overline{v}}{d\mathbf{R}} = \left(\frac{d\overline{n}_1}{d\mathbf{R}}, \frac{d\overline{\theta}_1}{d\mathbf{R}}, \ldots, \frac{d\overline{n}_{N-1}}{d\mathbf{R}}, \frac{d\overline{\theta}_{N-1}}{d\mathbf{R}} \right)^T,$$

and diagonal matrix $\Pi = \mathrm{diag}(\Pi_1, \Pi_2, \ldots, \Pi_{N-1})$. Notice that to simplify the expression of Λ, we use the approximation that the overlap intergral $D_{j,k,l,m} \simeq 0$ when the subscripts are not all identical.

Both γ_B and γ_{NL} have the geometric property of parameter space. The novel second term indicates that the Bogoliubov excitations induced by the slow change of the system, which is negligible in the linear case, could be accumulated in the nonlinear adiabatic evolution and could contribute to the finite phase of a geometric nature.

III. TWO-LEVEL MODEL

As an illustration of our theoretical formalism, we consider the simple two-mode BEC model described by the nonlinear equation [18]

$$i\frac{d}{dt}\begin{pmatrix} \Psi_1 \\ \Psi_2 \end{pmatrix} = H(\Psi_1, \Psi_2) \begin{pmatrix} \Psi_1 \\ \Psi_2 \end{pmatrix}, \tag{18}$$

with

$$H(\Psi_1, \Psi_2) = \begin{pmatrix} Z|\Psi_1|^2 & \frac{\rho}{2} e^{-i\varphi} \\ \frac{\rho}{2} e^{i\varphi} & Z|\Psi_2|^2 \end{pmatrix}, \tag{19}$$

and the $\mathbf{R} = (X = \rho\cos\varphi,\ Y = \rho\sin\varphi, Z)$ are parameters. For simplicity, we fix ρ and Z and change the parameter φ from 0 to 2π adiabatically.

The eigenequations read

$$H \circ \begin{pmatrix} \overline{\Phi}_1 \\ \overline{\Phi}_2 \end{pmatrix} = \mu \begin{pmatrix} \overline{\Phi}_1 \\ \overline{\Phi}_2 \end{pmatrix}. \tag{20}$$

For the nonlinear system, the number of eigenstates may be larger than the dimension of the Hilbert space and the eigenstate could be unstable [11]. We have obtained four eigenstates for the case $Z > \rho$ by solving the above eigenequation. Three of them are stable and one is unstable. We choose the following stable eigenstate to illustrate our theory, i.e.,

$$\begin{aligned} \overline{\Phi}_1 &= \left[\frac{1}{2} \left(1 - \sqrt{1 - \frac{\rho^2}{Z^2}} \right) \right]^{1/2}, \\ \overline{\Phi}_2 &= \left[\frac{1}{2} \left(1 + \sqrt{1 - \frac{\rho^2}{Z^2}} \right) \right]^{1/2} e^{i\varphi}, \end{aligned} \tag{21}$$

with eigenvalue of $\mu = Z$.

The Berry term of the geometric phase is readily deduced,

$$\gamma_B = -i \int_0^{2\pi} \left(\overline{\Phi}_2 \frac{\partial}{\partial\varphi} \overline{\Phi}_2 \right) d\varphi = \pi \left(1 + \sqrt{1 - \frac{\rho^2}{Z^2}} \right). \tag{22}$$

Now we are going to derive the additional term γ_{NL} of the geometric phase. Let us introduce the new variables (n, θ) through $(\Phi_1, \Phi_2) = (\sqrt{1-n}, \sqrt{n} e^{i\theta})$. Substituting $(\Psi_1, \Psi_2) = e^{-i\int^t \beta dt}(\Phi_1, \Phi_2)$ into Eq. (18), and separating the real and imaginary parts, we have four differential equations, two of which are identical due to the norm conservation:

$$\frac{d}{dt} n = -\rho\sqrt{n - n^2} \sin(\theta - \varphi), \tag{23}$$

$$\frac{d}{dt}\theta = -\frac{\rho\sqrt{1-n}}{2\sqrt{n}} \cos(\theta - \varphi) - Zn + \beta, \tag{24}$$

$$\beta = Z(1-n) + \frac{\rho}{2}\sqrt{\frac{n}{1-n}} \cos(\theta - \varphi). \tag{25}$$

The eigenstate makes up the fixed point of Eqs. (23) and (24), i.e.,

$$\overline{n} = \frac{1}{2}\left(1 + \sqrt{1 - \frac{\rho^2}{Z^2}} \right), \quad \overline{\theta} = \varphi. \tag{26}$$

Let us make a perturbation expansion around the eigenstate with $n = \overline{n}(\varphi) + \delta n$ and $\theta = \overline{\theta}(\varphi) + \delta\theta$. Then inserting the above expansion into Eqs. (23) and (24), we get

$$\begin{pmatrix} \frac{\partial \overline{n}}{\partial \varphi} \\ \frac{\partial \overline{\theta}}{\partial \varphi} \end{pmatrix} \frac{d\varphi}{dt} = \mathcal{L} \begin{pmatrix} \delta n \\ \delta\theta \end{pmatrix}, \tag{27}$$

with

$$\mathcal{L} = \begin{pmatrix} 0 & -\rho\sqrt{\overline{n} - \overline{n}^2} \\ -2Z + \frac{\rho}{4(\overline{n} - \overline{n}^2)^{3/2}} & 0 \end{pmatrix}. \tag{28}$$

Then, we have

$$\gamma_{NL} = Z \int_0^{2\pi} \left\langle \Lambda | \Pi \mathcal{L}^{-1} \circ \left(\frac{\partial \overline{n}}{\partial\varphi}, \frac{\partial \overline{\theta}}{\partial\varphi} \right)^T \right\rangle d\varphi, \tag{29}$$

where $\Lambda = \left((2\overline{n} - 1)\sqrt{\overline{n}} e^{-i\overline{\theta}}, (2\overline{n} - 1)\sqrt{\overline{n}} e^{i\overline{\theta}} \right)$. Finally, we get

$$\gamma_{NL} = \frac{\pi\rho^2}{Z\sqrt{Z^2 - \rho^2}}. \tag{30}$$

Combining formulas (22) and (29), we arrive at the explicit expression of the adiabatic geometric phase:

$$\gamma_g = \pi \left(1 + \sqrt{1 - \frac{\rho^2}{Z^2}} + \frac{\rho^2}{Z\sqrt{Z^2 - \rho^2}}\right). \qquad (31)$$

This new geometric phase has been verified by numerically solving the nonlinear Schödinger equation (18). This phase can also be interpreted as the flux of a virtual magnetic field M through the surface enclosed by the close circuit in parameter space. The virtual field has been deduced to take form of

$$M = \frac{\mathbf{R}}{2(Z^2 - \rho^2)^{3/2}}, \qquad (32)$$

in contrast to the Dirac monopole field $\frac{\mathbf{R}}{2R^3}$ claimed by Berry for the linear case [2].

Now we discuss the geometric meaning of the new phase. The state vector ψ can be parametrized in a Bloch sphere according to

$$|\psi\rangle\langle\psi| = \tfrac{1}{2}(I + s \cdot \sigma). \qquad (33)$$

Here, $\psi = (\cos\frac{\alpha}{2}e^{-i\delta/2}, \sin\frac{\alpha}{2}e^{i\delta/2})$, $s = (\sin\alpha\cos\delta,\ \sin\alpha \sin\delta, \cos\alpha)$, and I and σ are unit and Pauli matrix, respectively. In the new variables, the nonlinear phase can be expressed as a function of the solid angle in a Bloch sphere [i.e., $\Omega_B = 2\pi(1 - \cos\alpha)$] and the solid angle in the parameter space [i.e., $\Omega_P = 2\pi(1 - Z/\sqrt{\rho^2 + Z^2})$]. The two angles usually are not identical in the nonlinear case. To understand it, let us first consider the linear case $Z = 0$. The system can be viewed as a spin s driven by an external magnetic field $B = (\rho\cos\varphi, \rho\sin\varphi, Z)$. When the magnetic field varies in time adiabatically, from the dynamical equation

$$\frac{ds}{dt} = B \times s, \qquad (34)$$

we know that the spin will stay in parallel with the field. Obviously, in this case, $\Omega_B = \Omega_P$. While in the presence of the nonlinearity, the magnetic field will be modulated self-consistently by the spin, the effective field $B^* = (\rho\cos\varphi, \rho\sin\varphi, Z\cos\alpha)$. This can be seen from Eq. (19) where parameter Z has been renormalized by $|\Psi_1|^2 - |\Psi_2|^2 = \cos\alpha$. The adiabatic evolution of the spin is expected to parallel the modulated field B^* rather than B. Thus, for a cyclic adiabatic evolution of the spin, the solid angles in the Bloch sphere and parameter space are not identical in general. On the other hand, from Eqs. (22) and (29) and using the relations $\overline{n} = \frac{\Omega_B}{4\pi}$ and $\rho/Z = \tan\arccos(1 - \Omega_P/2\pi)$, we can express our nonlinear geometric phase in terms of these solid angles, i.e.,

$$\gamma_g = \frac{\Omega_B}{2} + \frac{\pi\left(1 - \frac{\Omega_B}{2\pi}\right)\left(1 - \frac{\Omega_P}{2\pi}\right)}{\left(1 - \frac{\Omega_P}{2\pi}\right) - \sqrt{\left(2 - \frac{\Omega_P}{2\pi}\right)\left(\frac{\Omega_P}{2\pi}\right)} \Big/ \left[\left(2 - \frac{\Omega_B}{2\pi}\right)\left(\frac{\Omega_B}{2\pi}\right)\right]^{\frac{1}{2}}}. \qquad (35)$$

For the linear case $Z = 0$, $\Omega_P = 2\pi$, the above expression reduces to the well-known relation $\gamma_g = \frac{\Omega_B}{2}$, i.e., the Berry phase equals half the solid angle.

IV. DISCUSSION AND CONCLUSION

Before concluding, we present some discussion. Let us first recall the nonadiabatic geometric phase for a cyclic motion. By inserting $\psi = e^{-i\lambda}\phi$ into the Schrödinger equation $i\frac{\partial\psi}{\partial t} = H\psi$, one can readily obtain that $\frac{d\lambda}{dt} = -i\langle\phi|\frac{\partial}{\partial t}|\phi\rangle + \langle\phi|H|\phi\rangle$. Here, ϕ is the wave function in the projective Hilbert space satisfying the cyclic requirement $\phi(t = 0) = \phi(t = T)$. The total phase acquired during the cyclic evolution contains two parts, i.e., $-\int_0^T i\langle\phi|\frac{\partial}{\partial t}|\phi\rangle dt$ and $\int_0^T \langle\phi|H|\phi\rangle dt$. The former has no relation to Hamiltonian H and therefore can be regarded as the geometric phase for the cyclic state [13]. The above deduction is correct regardless of nonlinearity, just requiring that the system is invariant under gauge transformation (obviously the $H = H_0 + g|\psi|^2$ discussed in the present paper has this symmetry). A question arises as to why the above simple deduction does not apply to our adiabatic case. The main reason is that the adiabatic limit is a process, i.e., one can approach it but one cannot reach it. That is, for any small adiabatic parameter ϵ, i.e., $\mathbf{R}$ sweeps at any small rate, the dynamical solution obtained by solving the Schrödinger equation deviates from the adiabatic solution by a small quantity of order ϵ. This point has been clarified by the discussion in previous sections and expressed explicitly by $\phi = \overline{\phi}(\mathbf{R}) + \delta\phi(\mathbf{R})$. Moreover, as revealed by our previous discussion, in the presence of nonlinearity, this infinitesimal deviation could be accumulated to contribute a finite phase during an infinitely long time evolution $T \sim \frac{1}{\epsilon} \to \infty$. Whereas, for the cyclic evolution with a finite time duration, this kind of accumulation cannot emerge. Therefore, the adiabaticity is crucial to the emergence of our nonlinear correction.

In conclusion, we have investigated the phase acquired by an eigenstate during the nonlinear evolution in which the parameter vector moves adiabatically in a circuit. The acquired phase consists of two parts: the dynamical phase and the adiabatic geometric phase. Compared to the linear case, the dynamical phase is found to be modified as the time integral of the chemical potential rather than the energy. This is because the instantaneous eigenstates are feedback to the nonlinear Hamiltonian. More interestingly, the adiabatic geometric phase or Berry phase is found also to be modified by the nonlinearity. The underlying mechanism has been revealed. The above nonlinear corrections are of significance, for instance, they could affect the interferences of BEC matter waves, and therefore are expected to be observable in future experiments. Since some of the theories covered by our results are mean-field limits of quantum many-body theories, the possibility of generalizing these considerations to quantized field theories from the correspondence principle is of great interest for future study.

ACKNOWLEDGMENTS

This work is supported by the National Natural Science Foundation of China (Nos. 10725521, 10604009), and the 973 project of China under Grant Nos. 2006CB921400 and 2007CB814800.

[1] P. Ehrenfest, Ann. Phys. (Berlin) **51**, 327 (1916); M. Born and V. Fock, Z. Phys. **51**, 165 (1928); L. D. Landau, Zeitschrift **2**, 46 (1932); C. Zener, Proc. R. Soc. A **137**, 696 (1932); J. Schwinger, Phys. Rev. **51**, 648 (1937).

[2] M. V. Berry, Proc. R. Soc. Lond. A **392**, 45 (1984).

[3] *Geometric Phase in Physics*, edited by A. Shapere and F. Wilczek (World Scientific, Singapore, 1989); A. Bohm, A. Mostafazadeh, H. Koizumi, Q. Niu, and J. Zwanziger, *The Geometric Phase in Quantum Systems* (Springer, New York, 2003).

[4] K. Bergmann, H. Theuer, and B. W. Shore, Rev. Mod. Phys. **70**, 1003 (1998).

[5] F. Wilczek and A. Zee, Phys. Rev. Lett. **52**, 2111 (1984).

[6] Y. Aharonov and J. S. Anandan, Phys. Rev. Lett. **58**, 1593 (1987).

[7] *Quantum Computation and Quantum Information Theory*, edited by C. Macchiavello, G. M. Palma, and A. Zeilinger (World Scientific, Singapore, 2000); J. Jones, V. Vedral, A. K. Ekert, and C. Castagnoli, Nature (London) **403**, 869 (2000); G. Falci, R. Fazio, G. M. Palma, J. Siewert, and V. Vedral, *ibid.* **407**, 355 (2000); L.-M. Duan, I. Cirac, and P. Zoller, Science **292**, 1695 (2001).

[8] C. Zhang, A. M. Dudarev, and Q. Niu, Phys. Rev. Lett. **97**, 040401 (2006); D. Xiao, Y. Yao, Z. Fang, and Q. Niu, *ibid.* **97**, 026603 (2006); W. Yao and Q. Niu, *ibid.* **101**, 106401 (2008).

[9] C. J. Pethick and H. Smith, *Bose-Einstein Condensation in Dilute Gases* (Cambridge University Press, London, 2002); *Bose-Einstein Condensation in Atomic Gases*, edited by M. Inguscio, S. Stringari, and C. E. Wieman (IOS Press, Amsterdam, 1999).

[10] Y. Lahini, F. Pozzi, M. Sorel, R. Morandotti, D. N. Christodoulides, and Y. Silberberg, Phys. Rev. Lett. **101**, 193901 (2008), and references therein.

[11] J. Liu, B. Wu, and Q. Niu, Phys. Rev. Lett. **90**, 170404 (2003).

[12] H. Pu, P. Maenner, W. Zhang, and H. Y. Ling, Phys. Rev. Lett. **98**, 050406 (2007).

[13] J. C. Garrison and R. Y. Chiao, Phys. Rev. Lett. **60**, 165 (1988).

[14] R. Balakrishman and M. Metha, Euro. Phys. J. D **33**, 437 (2005).

[15] B. Wu, J. Liu, and Q. Niu, Phys. Rev. Lett. **94**, 140402 (2005); X. X. Yi, X. L. Huang, and W. Wang, Phys. Rev. A **77**, 052115 (2008).

[16] F. N. Litvinets *et al.*, J. Phys. A. **39**, 1191 (2006).

[17] E. P. Gross, Nuovo Cimento **20**, 454 (1961); L. P. Pitaevskii, Zh. Eksp. Teor. Fiz. **40**, 646 (1961) L. P. Pitaevskii, [Sov. Phys. JETP **13**, 451 (1961)].

[18] A. Smerzi, S. Fantoni, S. Giovanazzi, and S. R. Shenoy, Phys. Rev. Lett. **79**, 4950 (1997); A. J. Leggett, Rev. Mod. Phys. **73**, 307 (2001).

Berry Phase Effects on the Dynamics of Quasiparticles in a Superfluid with a Vortex

Chuanwei Zhang,[1,2] Artem M. Dudarev,[1,3] and Qian Niu[1]
[1]*Department of Physics, The University of Texas, Austin, Texas 78712-1081 USA*
[2]*Condensed Matter Theory Center, Department of Physics, University of Maryland, College Park, Maryland 20742 USA*
[3]*Max-Planck-Institut für Physik Komplexer Systeme, Dresden, 01187 Germany*
(Received 5 July 2005; published 28 July 2006)

We study quasiparticle dynamics in a Bose-Einstein condensate with a vortex by following the center of mass motion of a Bogoliubov wave packet, and find important Berry-phase effects due to the background flow. We show that the Berry phase invalidates the usual canonical relation between the mechanical momentum and position variables, leading to important modifications of quasiparticle statistics and thermodynamic properties of the condensates. Applying these results to a vortex in an infinite uniform superfluid, we find that the total transverse force acting on the vortex is proportional to the superfluid density. We propose an experimental setup to directly observe Berry phase effects through measuring local thermal atoms' momentum distribution around a vortex.

DOI: 10.1103/PhysRevLett.97.040401 PACS numbers: 03.75.−b, 03.65.Sq, 03.65.Vf

Quasiparticles in a superfluid play an essential role in its thermodynamic as well as dynamic properties [1]. Owing to the realization of Bose-Einstein condensates (BEC) with ultracold atoms, much detailed investigation of the quasiparticles becomes possible [2,3]. In the past several years, extensive experimental and theoretical work has been devoted to the study of quasiparticle dynamics in static BECs [3–6], while quasiparticles in a flowing condensate have not received much attention [7].

In this Letter, we study the quasiparticle dynamics in a BEC superfluid with a vortex, a typical flowing condensate. We adopt a semiclassical approach by following the center of mass motion of a Bogoliubov wave packet, finding important Berry phase effects [8] due to the background flow. Unlike the case with a static BEC, the quasiparticle position and momentum are no longer canonical variables, and there are also important modifications to the quasiparticle statistics and thermodynamic properties of the condensates. We use these results to calculate the momentum circulation around a vortex in an infinite superfluid and find the total transverse force acting on the vortex is proportional to the superfluid density [9], which rules out the existence of the Iordanskii force that is supposed to originate in the asymmetric scattering of quasiparticles by the vortex [10–12]. Finally, we propose an experimental setup to directly observe Berry phase effects through measuring local thermal atoms' momentum distribution around a vortex.

Wave packet dynamics.—For simplicity, we present the theory here for a single vortex, although it is applicable to a general flowing condensate such as vortex lattices. The dynamics of the condensate are described by the Gross-Pitaveskii equation [13]

$$i\frac{\partial \psi_0}{\partial t} = \left(\frac{\mathbf{p}^2}{2} + g|\psi_0|^2 + V(\mathbf{r}) - \mathbf{\Omega}\cdot(\mathbf{r}\times\mathbf{p})\right)\psi_0, \quad (1)$$

where, for simpler notation, we have used units such that $\hbar$ and the atomic mass m are both unity, g is the interatom coupling constant, and $V(\mathbf{r})$ is the trapping potential. We have also included an angular momentum term to take into account the possibility of using a rotating reference frame, with $\mathbf{\Omega}$ being its angular frequency. The condensate wave function may be written as $\psi_0 = \sqrt{n(\mathbf{r})}e^{i\alpha(\mathbf{r})}e^{-i\mu t}$, where μ is the chemical potential, $n(\mathbf{r})$ and $\alpha(\mathbf{r})$ are the vortex density and phase, respectively. The gradient of phase yields local velocity of the vortex $\mathbf{v}_s = \nabla\alpha(\mathbf{r})$.

A small deviation $\delta\psi e^{-i\mu t}$ (elementary excitations) from the condensate wave function ψ_0 is described by the Bogoliubov equation [13]

$$i\sigma_z\frac{\partial}{\partial t}|\Phi\rangle = \mathcal{Q}|\Phi\rangle, \qquad \mathcal{Q} = \begin{pmatrix} H_+ & H_2e^{2i\alpha(\mathbf{r})} \\ H_2e^{-2i\alpha(\mathbf{r})} & H_- \end{pmatrix}. \quad (2)$$

The wave function Φ has two components, u and v, which are related to $\delta\psi$ through $\delta\psi = ue^{-i\omega t} + v^*e^{i\omega t}$, where ω is the quasiparticle energy. The entries of the matrix operator are given by $H_\pm = \mathbf{p}^2/2 + 2gn(\mathbf{r}) + V(\mathbf{r}) - \mu \mp \mathbf{\Omega}\cdot(\mathbf{r}\times\mathbf{p})$, and $H_2(\mathbf{r}) = gn(\mathbf{r})$. Because of the Pauli matrix σ_z on the left-hand side, the Bogoliubov equation is non-Hermitian.

We now consider a quasiparticle wave packet centered at $\mathbf{r}_c$ with its spread small compared to the length scale of the slowly varying potentials [including trap potential $V(\mathbf{r})$, condensate wave function ψ_0, and terms related to $\mathbf{\Omega}$]. The dynamics of the wave packet is approximately governed by the local Bogoliubov operator $\mathcal{Q}_c \equiv \mathcal{Q}(\mathbf{p}, \mathbf{r} = \mathbf{r}_c)$ plus its linear gradient correction. The local Bogoliubov operator has plane wave eigenstates $\mathcal{Q}_c e^{i\mathbf{q}\cdot\mathbf{r}}|\phi(\mathbf{q}, \mathbf{r}_c)\rangle = \omega_c\sigma_z e^{i\mathbf{q}\cdot\mathbf{r}}|\phi(\mathbf{q}, \mathbf{r}_c)\rangle$, where $\mathbf{q}$ is the wave vector. The amplitude satisfies a simple 2×2 matrix equation, which can be solved easily to yield

$$\omega_c(\mathbf{q}, \mathbf{r}_c) = [H_1^2(\mathbf{q}) - H_2^2]^{1/2} - \mathbf{q}\cdot(\mathbf{\Omega}\times\mathbf{r}_c) \quad (3)$$

for the local quasiparticle energy, where $H_1(\mathbf{q}) = \mathbf{q}^2/2 + 2gn(\mathbf{r}_c) + V(\mathbf{r}_c) - \mu$, and

$$|\phi(\mathbf{q},\mathbf{r}_c)\rangle = \frac{1}{2}\begin{pmatrix} \zeta+\zeta^{-1} \\ (\zeta-\zeta^{-1})e^{-2i\alpha(\mathbf{r}_c)} \end{pmatrix}$$

for the two-component amplitude of the local eigenstate, where $\zeta = (\frac{H_1-H_2}{H_1+H_2})^{1/4}$. We note that the wave vector $\mathbf{q}$ and the wave packet center $\mathbf{r}_c$ enter the local quasiparticle energy and wave function parametrically. The wave function is normalized in the sense that $\langle\phi(\mathbf{q},\mathbf{r}_c)|\sigma_z|\phi(\mathbf{q},\mathbf{r}_c)\rangle = 1$. We have also chosen the phase of the wave function such that it is smooth and single valued in the parameters $(\mathbf{q},\mathbf{r}_c)$. We shall see that the parametric dependence of the eigenstates on the center position of the wave packet will manifest as Berry phase terms in the equations of motion.

We now turn our attention to the wave packet itself, which is to be constructed out of these eigenstates as

$$|\Phi\rangle = \int d^3q\, a(\mathbf{q},t)e^{i\mathbf{q}\cdot\mathbf{r}}|\phi(\mathbf{q},\mathbf{r}_c)\rangle, \tag{4}$$

where the superposition amplitude $a(\mathbf{q},t)$ may be taken as a Gaussian in $\mathbf{q}$. The normalization is taken to be $\langle\Phi|\sigma_z|\Phi\rangle = \int d^3q|a(\mathbf{q},t)|^2 = 1$. We assume that the Gaussian is centered at $\mathbf{q}_c = \int d^3q|a(\mathbf{q},t)|^2\mathbf{q}$ and has a width narrow compared to momentum scales of the energy dispersion and of the eigenstates. Microscopic calculation shows that $\mathbf{q}_c = \langle\Phi|\sigma_z\mathbf{p}|\Phi\rangle = \int \delta\psi^*\mathbf{p}\delta\psi d\mathbf{r}$, implying that it represents the mechanical momentum of the quasiparticle. To be self consistent, the wave packet must yield the preassigned center position $\mathbf{r}_c = \langle\Phi|\sigma_z\mathbf{r}|\Phi\rangle$.

The dynamics of the quasiparticle can be derived from a time-dependent variational principle for the Bogoliubov equation, where the action (defined as time integral of the Lagragian) is extremized with respect to the quasiparticle wave function. Here the Lagrangian is given by

$$L = -\langle\Phi|\mathcal{Q}|\Phi\rangle + \left\langle\Phi|i\sigma_z\frac{d}{dt}|\Phi\right\rangle, \tag{5}$$

and the wave packet (4) is chosen as the variational wave function with time-dependent parameters $\mathbf{r}_c$ and $\mathbf{q}_c$. We use d/dt to mean the derivative with respect to the time dependence of the wave function explicitly or implicitly through $\mathbf{r}_c$ and $\mathbf{q}_c$. Under the previously discussed conditions that the wave packet is narrow both in position and momentum spaces, the Lagrangian (5) can be evaluated as a function of variational parameters $\mathbf{r}_c$ and $\mathbf{q}_c$, and their time derivatives, independent of the width and shape of the wave packet in position or momentum.

Quasiparticle energy and the Berry phase.—The first term in Lagrangian (5) corresponds to the total energy of the quasiparticle wave packet, which is found to be

$$\omega = [H_1^2(\mathbf{q}_c) - H_2^2]^{1/2} - \mathbf{q}_c\cdot(\mathbf{\Omega}\times\mathbf{r}_c) + (1-\rho^2)\mathbf{q}_c\cdot\mathbf{v}_s, \tag{6}$$

where $\mathbf{v}_s = \nabla\alpha(\mathbf{r}_c)$ is the local velocity of the vortex. The first two terms are the local quasiparticle energy Eq. (3) with momentum $\mathbf{q}_c$, stemming from the expectation value $\langle\Phi|\mathcal{Q}_c|\Phi\rangle$ of the local Bogoliubov operator. The last term is the correction of the wave packet energy originating from the linear gradient expansion of the Bogoliubov operator $\Delta\mathcal{Q} = \frac{1}{2}((\mathbf{r}-\mathbf{r}_c)\cdot\frac{\partial\mathcal{Q}_c}{\partial\mathbf{r}_c} + \text{c.c})$, and $\rho = \langle\Phi|\Phi\rangle$ is the total atomic mass contained in the quasiparticle wave packet.

To understand the quasiparticle energy expression, it will be instructive to consider the simpler situation of a uniformly flowing superfluid of velocity $\mathbf{v}_s$. If $\mathbf{p}_0$ is the quasiparticle momentum in the reference frame where the superfluid is static, then we have the relationship $\mathbf{q}_c = \mathbf{p}_0 + \rho\mathbf{v}_s$. In terms of $\mathbf{p}_0$, the quasiparticle energy (6) can be expressed (in the limit of small $\mathbf{v}_s$) as $\omega = \varepsilon(\mathbf{p}_0) + \mathbf{p}_0\cdot\mathbf{v}_s$, where $\varepsilon(\mathbf{p}_0) = [H_1^2(\mathbf{p}_0) - H_2^2]^{1/2} = [p_0^2(p_0^2/4 + gn)]^{1/2}$ is the energy dispersion in a static superfluid. This is just the standard Landau formula obtained using Galilean transformation [1].

The evaluation of the second term of the Lagrangian (5) is similar to that in Ref. [14], which yields

$$L = -\omega + (\mathbf{q}_c + \mathbf{A})\cdot\dot{\mathbf{r}}_c. \tag{7}$$

Here the term $\mathbf{q}_c\cdot\dot{\mathbf{r}}_c$ comes from the time dependence of the superposition amplitude $a(\mathbf{q},t)$ in the wave packet (4), while the vector potential $\mathbf{A} = i\langle\phi|\sigma_z|\partial\phi/\partial\mathbf{r}_c\rangle = -(\rho-1)\nabla\alpha(\mathbf{r}_c)$ is obtained from the position dependence of the two-component local eigenvector, which is similar to the case of a spin in a position-dependent Zeeman field [8]. The vector potential, also called Berry connection, arises for a vortex because of the nonzero local velocity. Its line integral over a path $\mathcal{C}$ gives a Berry phase $\Gamma(\mathcal{C}) = \oint_{\mathcal{C}} d\mathbf{r}_c\cdot\mathbf{A} = -\oint_{\mathcal{C}}(\rho-1)d\alpha(\mathbf{r}_c)$ of the eigenvector [15]. For instance, the accumulated Berry phase for a quasiparticle moving around a vortex in an infinite uniform superfluid is $-2\pi(\rho-1)$, where ρ represents the total atom number (notice that the atom mass m has been taken as unity) in the quasiparticle wave packet. Apart from a nonessential 2π, it means that the phase acquired by moving each atom around a vortex is 2π, which agrees with Feynman's argument [16].

We note that the vector potential vanishes at very high momenta, where $\rho\to 1$. Likewise, the energy correction due to superfluid flow [the last term of Eq. (6)] vanishes in this limit. This is quite reasonable, because the quasiparticle becomes a free particle and is decoupled from the condensate at high momenta. The condensate has influence on the quasiparticle only at low and intermediate momenta.

Semiclassical dynamics.—Following the standard procedure of analytical mechanics, we may introduce the canonical momentum $\mathbf{k}_c = \partial L/\partial\dot{\mathbf{r}}_c = \mathbf{q}_c + \mathbf{A}$ conjugate to the coordinate vector $\mathbf{r}_c$. We see that the vector potential makes the canonical and mechanical momenta different. The energy $\omega = \mathbf{k}_c\cdot\dot{\mathbf{r}}_c - L$ becomes the Hamiltonian and the equations of motion are

$$\dot{\mathbf{r}}_c = \frac{\partial\omega}{\partial\mathbf{k}_c}, \qquad \dot{\mathbf{k}}_c = -\frac{\partial\omega}{\partial\mathbf{r}_c}, \tag{8}$$

where the quasiparticle energy ω is now recasted as a function of $\mathbf{k}_c$, $\omega=[H_1^2(\mathbf{k}_c)-H_2^2]^{1/2}-(\mathbf{k}_c-\mathbf{A})\cdot(\mathbf{\Omega}\times\mathbf{r}_c)+\mathbf{k}_c\cdot\mathbf{A}$. We see that the vector potential $\mathbf{A}$ modifies the energy expression in this canonical formulation.

In terms of the physical quantity of mechanical momentum $\mathbf{q}_c$, the quasiparticle equations of motion are dramatically altered:

$$\dot{\mathbf{r}}_c=\frac{\partial\omega}{\partial\mathbf{q}_c}+\frac{\partial\rho}{\partial\mathbf{q}_c}(\dot{\mathbf{r}}_c\cdot\mathbf{v}_s),$$
$$\dot{\mathbf{q}}_c=-\frac{\partial\omega}{\partial\mathbf{r}_c}-\dot{\mathbf{r}}_c\times\left(\frac{\partial\rho}{\partial\mathbf{r}_c}\times\mathbf{v}_s\right)+\left(\dot{\mathbf{q}}_c\cdot\frac{\partial\rho}{\partial\mathbf{q}_c}\right)\mathbf{v}_s. \tag{9}$$

They follow directly from (8) by the transformation $\mathbf{k}_c=\mathbf{q}_c+(1-\rho)\mathbf{v}_s$. Here ω is the original energy expression (6). We see that additional terms depending on superfluid velocity and total atom mass in the quasiparticle wave packet appear in both equations, therefore the equations of motion are no longer of canonical form. This is a result very different from the case of static superfluid studied in [5].

In the statistical treatment of quasiparticles, an important semiclassical quantity is the density of states of phase space. For the canonical variables $\mathbf{k}_c$ and $\mathbf{r}_c$, this should be taken as the constant $1/(2\pi)^3$. Because of the vector potential, the position $\mathbf{r}_c$ and mechanical momentum $\mathbf{q}_c$ are no longer canonical variables, which leads to a modification of the density of states [17]

$$D(\mathbf{r}_c,\mathbf{q}_c)=\frac{1}{(2\pi)^3}\det\left(\mathbf{I}-\frac{\partial\rho}{\partial\mathbf{q}_c}\mathbf{v}_s\right), \tag{10}$$

where $\mathbf{I}$ is the 3×3 unit matrix.

The thermodynamics properties of condensates are determined by the physical quantities of quasiparticles such as energy and density of states, therefore they are strongly affected by the Berry phase. For instance, the thermal depletion of the condensate density is given by

$$\delta n(\mathbf{r}_c,T)=\int d^3\mathbf{q}_c D(\mathbf{r}_c,\mathbf{q}_c)\rho/(e^{\omega/k_BT}-1) \tag{11}$$

according to our theory, which is different from that for a static condensate because of the modifications of energy ω and density of state $D(\mathbf{r}_c,\mathbf{q}_c)$. Notice that the integration of this distribution over position space yields the total number of thermal atoms.

Transverse force on a vortex.—The general form of the transverse force per unit length acting on a moving vortex with velocity $\mathbf{v}_L$ in an infinite uniform superfluid can be written as $\mathbf{F}=\kappa(\mathcal{C})\hat{\mathbf{z}}\times\mathbf{v}_L$, where $\kappa(\mathcal{C})$ is the momentum circulation along a path far away from the core of the vortex, and we assume the normal fluid velocity is zero [12]. In the past several decades, there has been some controversy about the expression of $\kappa(\mathcal{C})$, which was argued to be either $2\pi\hbar n_{\rm tot}/m$ or $2\pi\hbar n_s/m$ through different approaches [9–11]. Here $2\pi\hbar/m$ is the quantum of circulation, $n_{\rm tot}$ is the total mass density, and n_s is the superfluid density.

According to our theory, the total momentum circulation $\kappa(\mathcal{C})$ contains two contributions: $2\pi\hbar[n_{\rm tot}-\delta n(\mathbf{r}_c,T)]/m$ from the condensate and $\kappa_1(\mathcal{C})=\oint_{\mathcal{C}}\mathbf{W}\cdot d\mathbf{r}_c$ from quasiparticles. $\mathbf{W}$ is the momentum density of quasiparticles under equilibrium Bose-Einstein distribution and is obtained by summing over the mechanical momenta from all states,

$$\mathbf{W}=\int d^3\mathbf{q}_c D(\mathbf{r}_c,\mathbf{q}_c)\frac{\mathbf{q}_c}{e^{\omega(\mathbf{r}_c,\mathbf{q}_c)/k_BT}-1}$$
$$=\int\frac{d^3\mathbf{k}_c}{(2\pi)^3}\frac{\mathbf{k}_c+(\rho-1)\nabla\theta_c}{e^{\omega(\mathbf{r}_c,\mathbf{k}_c)/k_BT}-1}, \tag{12}$$

where T is the temperature, k_B is the Boltzman constant, $D(\mathbf{r}_c,\mathbf{q}_c)=\frac{1}{(2\pi)^3}[1-\frac{1-\rho^2}{\omega_c r_c^2}(\mathbf{r}_c\times\mathbf{q}_c)_z]$, and $1/(2\pi)^3$ are the density of states for the noncanonical momentum $\mathbf{q}_c$ and canonical momentum $\mathbf{k}_c$, respectively.

Since we choose the circular contour $\mathcal{C}$ to be far away from the core of the vortex, we can expand the Bose-Einstein distribution $f(\mathbf{k}_c)\approx f_0(\mathbf{k}_c)-\Delta f$ with $f_0(\mathbf{k}_c)=\frac{1}{e^{\omega_c/k_BT}-1}$, and $\Delta f=(1-\rho)\mathbf{k}_c\cdot\nabla\theta_c e^{\omega_c/k_BT}/[k_BT(e^{\omega_c/k_BT}-1)^2]$. Considering a closed orbit with fixed radius R and taking the limit of $R\to\infty$, we find the thermal depletion of the condensate $\delta n(\mathbf{r}_c,T)=\frac{1}{2\pi^2}\times\int_0^\infty f_0(\mathbf{k}_c)\rho k_c^2 dk_c$, and quasiparticle momentum circulation

$$\kappa_1(\mathcal{C})=\frac{1}{\pi}\int_0^\infty f_0(\mathbf{k}_c)\rho k_c^2 dk_c-\frac{1}{3\pi k_BT}\times\int_0^\infty\frac{e^{\omega_c/k_BT}k_c^4 dk_c}{(e^{\omega_c/k_BT}-1)^2}. \tag{13}$$

The first term of $\kappa_1(\mathcal{C})$ stems from the difference between the canonical momentum $\mathbf{k}_c$ and mechanical momentum $\mathbf{q}_c$, and cancels with $-2\pi\hbar\delta n(\mathbf{r}_c,T)/m$. The second term of $\kappa_1(\mathcal{C})$ comes from the deviation Δf in the distribution function, and is found to be $2\pi\hbar\rho_n/m$, where $\rho_n=2\pi^2k_B^4T^4/45\hbar^3s^5$ is the normal fluid density. Summing up the contributions from both condensate and quasiparticles, we find the total momentum circulation $\kappa(\mathcal{C})=2\pi\hbar(n_{\rm tot}-\rho_n)/m=2\pi\hbar n_s/m$, and the transverse force per unit length $\mathbf{F}=2\pi\hbar n_s\hat{\mathbf{z}}\times\mathbf{v}_L/m$. This form of transverse force agrees with that obtained by Thouless *et al.* using general properties of superfluid order [9], and rules out the existence of the Iordanskii force that is proportional to the normal fluid density ρ_n, and is supposed to originate in the asymmetric scattering of quasiparticles by the vortex [10–12].

Experimental observation.—The above discussions show that the modifications of mechanical momentum, energy dispersion, and density of states of quasiparticles due to the Berry phase affect the thermodynamics of atoms. Our prediction of a transverse force proportional T^4 can in principle be measured in experiments. In the following, we propose an experiment for a direct observation of the local

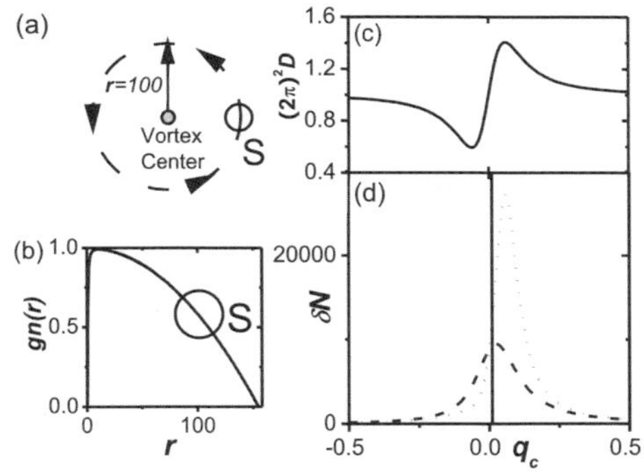

FIG. 1. Thermal-atom momentum distribution around a trapped vortex. Length scales are in units of $l = 0.73$ μm for a Rb^{87} BEC in a trap specified in the text. (a) A schematic plot of experimental geometry. Measurement of thermal-atom momentum distribution is to be made in a region S at 100 unit lengths away from the vortex center with a radius of 20 unit lengths. This region contains about 9740 condensate and 1650 thermal atoms. (b) Condensate atom density (multiplied by the interaction strength g) in the presence of a trapped vortex. (c) Density of states for quasiparticles in the measurement region S as a function of quasiparticle momentum q_c along the circulation flow. (d) Momentum distribution of thermal atoms in the measurement region S. Dotted and dashed lines correspond to our Eq. (14) and the naive expression $\int_S d\mathbf{r}_c\rho/[\exp(\frac{\omega_c+\mathbf{q}_c\cdot\nabla\theta}{k_BT})-1]$, respectively. For reference, the momentum of condensate atoms is indicated by the vertical solid line.

thermal-atom momentum distribution around a vortex, in which the effect of the Berry phase can be seen clearly.

Figure 1(a) is a schematic plot of experimental geometry. The length unit is $l = \sqrt{\frac{\hbar}{m(\mu+\hbar\Omega)}}$, chosen for convenience. It corresponds to $l \approx 0.73$ μm for a BEC of $N = 5\times 10^5$ Rb^{87} atoms in a quasi-two-dimensional magnetic trap [18] with axial and radial trapping frequencies $\omega_z = 2\pi\times 800$ Hz, $\omega_r = 2\pi\times 2$ Hz, in which a single vortex is created at a rotation frequency of $\Omega = 0.4\omega_r$ [19]. The condensate density profile is shown in Fig. 1(b), where one can see that the observation region S is chosen relatively far away from the vortex core, which was to make sure that the semiclassical approximation remains valid. We will consider a temperature of $T = 2\mu/k_B \approx 21.1$ nK which is about a third of the BEC transition temperature $T_c \approx 52.8$ nK for this system.

The atoms are supposed to be in a hyperfine state ($|1\rangle$). By focusing two copropagating Raman beams at region S one may drive both the thermal and condensate atoms there to another hyperfine state ($|2\rangle$). The remaining atoms in state $|1\rangle$ are to be rapidly expelled from the trap by applying radio-frequency radiation that flips them to an anti-trapped hyperfine state. Finally, the time of flight measurement [20] of atoms in state $|2\rangle$ can then determine their momentum distribution, which, according to our theory, is given by

$$\delta n(\mathbf{p}_c, T) = \int_S d^2\mathbf{r}_c D(\mathbf{r}_c, \mathbf{q}_c)\rho/(e^{\omega/k_BT}-1). \qquad (14)$$

In Fig. 1(c), we see that there are more quasiparticles moving along the circulating superfluid flow than in the opposite direction and the deviation from the naive value of the density of states $1/(2\pi)^2$ is substantial. In Fig. 1(d), we find that the momentum distribution is quite different from the commonly believed one that does not take into consideration the Berry phase effects.

We gratefully thank Junren Shi, Alexander L. Fetter, and Erich Mueller for stimulating discussions. This work is supported by the NSF and the R. A. Welch Foundation.

[1] L. D. Landau and E. M. Lifshitz, *Statistical Physics: Part II* (Butterworth-Heinemann, Oxford, 1980); C. J. Pethick and H. Smith, *Bose-Einstein Condensation in Dilute Gates* (Cambridge University Press, Cambridge, England, 2002).

[2] F. Dalfovo *et al.*, Rev. Mod. Phys. **71**, 463 (1999); A. J. Leggett, *ibid.* **73**, 307 (2001).

[3] R. Ozeri *et al.*, Rev. Mod. Phys. **77**, 187 (2005).

[4] M. Edwards *et al.*, Phys. Rev. Lett. **77**, 1671 (1996); P. A. Ruprecht *et al.*, Phys. Rev. A **54**, 4178 (1996); A. L. Fetter, Phys. Rev. A **53**, 4245 (1996); V. M. Pérez-García *et al.*, Phys. Rev. Lett. **77**, 5320 (1996); Y. Castin and R. Dum, Phys. Rev. Lett. **77**, 5315 (1996); Yu Kagan *et al.*, Phys. Rev. A **54**, R1753 (1996); S. Stringari, Phys. Rev. Lett. **77**, 2360 (1996); P. Öhberg *et al.*, Phys. Rev. A **56**, R3346 (1997).

[5] M. Fliesser *et al.*, Phys. Rev. A **56**, R2533 (1997); M. Fliesser *et al.*, *ibid.* **56**, 4879 (1997); A. Csordás *et al.*, *ibid.* **56**, 5179 (1997).

[6] B. Wu and Q. Niu, Phys. Rev. A **64**, 061603 (2001); L. Fallani *et al.*, Phys. Rev. Lett. **93**, 140406 (2004).

[7] S. J. Woo *et al.*, Phys. Rev. Lett. **92**, 170402 (2004).

[8] M. V. Berry, Proc. R. Soc. A **392**, 45 (1984).

[9] D. J. Thouless, P. Ao, and Q. Niu, Phys. Rev. Lett. **76**, 3758 (1996).

[10] R. J. Donnelly, *Quantized Vortices in Helium II* (Cambridge University Press, London, 1991).

[11] E. B. Sonin, Phys. Rev. B **55**, 485 (1997).

[12] M. Stone, Phys. Rev. B **61**, 11 780 (2000).

[13] S. Ghosh, Phase Transit. **77**, 625 (2004).

[14] G. Sundaram and Q. Niu, Phys. Rev. B **59**, 14 915 (1999).

[15] Our Berry phase differs by a nonessential 2π from the result of M. J. Bijlsma and H. T. C. Stoof, Phys. Rev. A **62**, 013605 (2000).

[16] R. P. Feynman, in *Progress in Low Temperature Physics*, edited by C. J. Gorter (New Holland, Amsterdam, 1955), Vol. 1.

[17] D. Xiao, J. Shi, and Q. Niu, Phys. Rev. Lett. **95**, 137204 (2005).

[18] A. Görlitz *et al.*, Phys. Rev. Lett. **87**, 130402 (2001).

[19] M. R. Matthews *et al.*, Phys. Rev. Lett. **83**, 2498 (1999).

[20] Momentum distributions of a small condensate may be measured using methods developed in T. P. Meyrath *et al.*, Phys. Rev. A **71**, 041604(R) (2005).

9

Geometric phase in collective coordinates

Transverse Force on a Quantized Vortex in a Superfluid

D. J. Thouless,[1] Ping Ao,[1,2] and Qian Niu[3]
[1]*Department of Physics, University of Washington, Box 351560, Seattle, Washington 98195*
[2]*Department of Theoretical Physics, Umeå University, Umeå, Sweden*
[3]*Department of Physics, University of Texas, Austin, Texas 78712*
(Received 31 October 1995)

We have derived an exact expression for the total nondissipative transverse force acting on a quantized vortex moving in a uniform background. The derivation is valid for neutral boson or fermion superfluids, provided the order parameter is a complex scalar quantity. This force is determined by the one-particle density matrix far away from the vortex core, and is found to be the Magnus force proportional to the superfluid density. We conclude that contributions of the localized core states do not change this force. [S0031-9007(96)00124-X]

PACS numbers: 67.40.Vs, 47.37.+q

For a vortex moving in a superfluid there is a force transverse to its velocity which is the counterpart of the Magnus force in classical fluid mechanics [1]. This was used in the experiment by which Vinen [2] established the quantization of circulation round a wire in ^{4}He. The obvious generalization of the form of the Magnus force to the situation in superfluid helium is to take the force per unit length to $\mathbf{F}_M = \rho_s \mathbf{K} \times (\mathbf{v}_V - \mathbf{v}_s)$, where $\mathbf{K}$ is the circulation, $\mathbf{v}_V$ is the vortex velocity, ρ_s is the superfluid density, and $\mathbf{v}_s$ is the superfluid velocity in the neighborhood of the vortex. Even for the case of ^{4}He there has been some controversy about the form of this transverse force, despite the experimental measurements of Vinen [2] and of Whitmore and Zimmermann [3] at temperatures where there is an appreciable difference between ρ and ρ_s. This controversy is partly based on the observation by Iordanskii [4] that excitations such as phonons are asymmetrically scattered by a vortex, and this should lead to a transverse force proportional to ρ_n and to the velocity $\mathbf{v}_n - \mathbf{v}_V$ of the normal fluid component relative to the vortex. The magnitude of this term is not immediately obvious, nor is it obvious whether this contribution should be added to the Magnus force as we have written it here, or to the expression for $\mathbf{F}_M$ with the superfluid density replaced by the total fluid density. It has been convincingly argued [5] that the correct form of the transverse force is

$$\mathbf{F} = \rho \mathbf{K} \times (\mathbf{v}_V - \mathbf{v}_s) + \rho_n \mathbf{K} \times (\mathbf{v}_n - \mathbf{v}_V), \qquad (1)$$

which can also be written as $\mathbf{F}_M$ plus a term proportional to $\mathbf{v}_n - \mathbf{v}_s$. The problem of determining the correct form of the force is discussed in detail by Sonin [6].

For a fermion superfluid there appear to be additional complications introduced by the spectrum of low-lying fermion excitations associated with the vortex core. Volovik [7,8] has identified an additional contribution proportional to $\mathbf{v}_n - \mathbf{v}_V$ associated with these excitations, which, in some estimates, cancels most of the Magnus force. However, measurements of the circulation quantization in the B phase of superfluid ^{3}He [9] show a Magnus force of the expected size. In superconductors there seems to be little unambiguous evidence about the existence or magnitude of this transverse force.

In earlier work we have shown the close connection between the Magnus force and a Berry phase [10], but our arguments were dependent on a detailed model of the vortex structure. In this Letter we consider a single, isolated vortex forced to move through an infinite superfluid with uniform velocity $\mathbf{v}_V$ by some moving pinning potential. We are able to make an exact evaluation of the coefficient of $\mathbf{v}_V$ in the force which has to be applied to the vortex in terms of the integral of the momentum density round a large loop that surrounds the vortex, and this is essentially the term proportional to $\mathbf{v}_V$ in Eq. (1). The first step in the argument involves a perturbative calculation of the effect of the motion of the vortex in terms of the instantaneous eigenstates of the Hamiltonian. In the case of a system with no inhomogeneous substrate this expression can be written in terms of a Berry phase [11]. This expression can be written in terms of the derivatives of one- and two-particle Dirac density matrices, or, alternatively, as the commutator of two components of the total momentum operator. This can be turned into an integral over a surface (or a loop for the two-dimensional case) surrounding the vortex and at a large distance from it. From this we can conclude that there are no corrections to the term proportional to $\mathbf{v}_V$ in Eq. (1).

For simplicity we work with a two-dimensional system, but the argument can readily be applied to a three-dimensional system. We assume that the system is homogeneous and infinite. The isolated vortex in the superfluid can be pinned to a position $\mathbf{r}_0$ by applying a potential $\sum_i V(\mathbf{r}_0 - \mathbf{r}_i)$ which acts on all the particles in the superfluid. For a static problem this potential can be arbitrarily weak, but when the pinning potential is made to move through the superfluid it must have a strength sufficient that the vortex does not get detached from the pinning center, either by tunneling with the aid of the Magnus force or by acquiring enough energy from the phonon system. The potential has to be strong enough to break the degeneracy of this broken symmetry vortex state and allow us to use a perturbative treatment of the velocity. The situation is somewhat reminiscent of the theory of the Stark effect in atoms, where a perturbative treatment of a weak uniform electric field is possible, despite the fact that the electric field connects the bound levels of the electrons in the atom to a continuum of ionized states of the same energy. The pinning potential may also be very strong, as it is in the Vinen experiment, where the circulation is pinned to a solid wire. Our results are quite independent of the details of the pinning potential, provided that it is strong enough to allow perturbative methods to be used, and does not break cylindrical symmetry.

In terms of the instantaneous eigenvalues $E_\alpha(t)$ and eigenstates $|\Psi_\alpha(t)\rangle$, for which we choose phases such that $\langle\Psi_\alpha \mid \dot{\Psi}_\alpha\rangle = 0$, the time-dependent solutions of the Schrödinger equation can be written as

$$|\Psi(t)\rangle = a_\alpha(t)e^{-i\int^t E_\alpha(t')\,dt'/\hbar}|\Psi_\alpha(t)\rangle + \sum_{\beta\neq\alpha} a_\beta(t)e^{-i\int^t E_\beta(t')\,dt'/\hbar}|\Psi_\beta(t)\rangle, \quad (2)$$

where, to first order in the velocity, $a_\alpha(t) = 1$, and

$$\begin{aligned} a_\beta(t) &= -\int^t dt'\langle\Psi_\beta \mid \dot{\Psi}_\alpha\rangle e^{-i\int^{t'}(E_\alpha - E_\beta)\,dt''/\hbar} \\ &= -\int^t dt'\langle\Psi_\beta|\mathbf{v}_V\cdot\nabla_0\Psi_\alpha\rangle e^{-i\int^{t'}(E_\alpha - E_\beta)\,dt''/\hbar}. \end{aligned} \quad (3)$$

Here ∇_0 denotes the partial derivative with respect to the position $\mathbf{r}_0$ of the pinning potential. This gives the expectation value of the force on the pinning potential as

$$\begin{aligned} \mathbf{F} = &-\sum_\alpha f_\alpha\langle\Psi_\alpha|\nabla_0 H|\Psi_\alpha\rangle \\ &+ \sum_\alpha f_\alpha\langle\Psi_\alpha|\nabla_0 H e^{i\int^t(E_\alpha - H)\,dt'/\hbar} \\ &\times \mathcal{P}_\alpha\int^t dt' e^{-i\int^{t'}(E_\alpha - H)\,dt''/\hbar}\mathbf{v}_V\cdot\nabla_0 + \text{H.c.}|\Psi_\alpha\rangle, \end{aligned} \quad (4)$$

where H is the total Hamiltonian of the system, which depends on $\mathbf{r}_0$ only through the pinning potential, f_α is the occupation probability of the state α, $\mathcal{P}_\alpha$ is the projection operator off the state α, and H.c. denotes the Hermitian conjugate term. The first term on the right side of this equation includes any forces on the vortex proportional to the normal and superfluid velocities, and we have not attempted to evaluate these. We concentrate on the second term, which gives those forces proportional to the vortex velocity. There are energy-conserving dissipative forces included in this expression, but, as with the Kubo formula for longitudinal conductivity, their evaluation requires a careful limiting process, and we have not done this here. We concentrate on the contributions from intermediate states whose energy is different from that of the initial state, and it is these that give rise to the transverse force perpendicular to the vortex velocity.

For the homogeneous system that we are considering the eigenvalues are independent of time, so Eq. (4) can be written in the simpler form

$$\begin{aligned} \mathbf{F} = &-\sum_\alpha f_\alpha\langle\Psi_\alpha|\nabla_0 H|\Psi_\alpha\rangle \\ &+ \sum_\alpha f_\alpha\langle\Psi_\alpha|\nabla_0 H\frac{i\hbar\mathcal{P}_\alpha}{E_\alpha - H}\mathbf{v}_V\cdot\nabla_0 + \text{H.c.}|\Psi_\alpha\rangle. \end{aligned} \quad (5)$$

Since $\nabla_0 H$ is the commutator of the operator ∇_0 with H, the commutator cancels the energy denominator, and the part linear in the vortex velocity can be written as

$$\mathbf{F}\times\hat{\mathbf{n}} = -i\hbar\mathbf{v}_V\sum_\alpha f_\alpha\left(\left\langle\frac{\partial\Psi_\alpha}{\partial x_0}\,\middle|\,\frac{\partial\Psi_\alpha}{\partial y_0}\right\rangle - \left\langle\frac{\partial\Psi_\alpha}{\partial y_0}\,\middle|\,\frac{\partial\Psi_\alpha}{\partial x_0}\right\rangle\right), \quad (6)$$

where $\hat{\mathbf{n}}$ is the unit vector normal to the plane. The Berry phase associated with a closed loop in $\mathbf{r}_0$ can be written as the integral over the area enclosed by the loop of $F_M/\hbar v_V$. Equation (6) corresponds to the familiar form for the Berry phase [11]. Since we can choose the wave functions in such a way that the dependence on $\mathbf{r}_0$ is entirely through $\mathbf{r} - \mathbf{r}_0$, the partial derivatives with respect to $\mathbf{r}_0$ can be replaced by a sum over partial derivatives with respect to the particle coordinates $\mathbf{r}_j$. Upon thermal average, this expression can now be written in terms of the Dirac density matrices as

$$\begin{aligned} F/v_V &= -i\hbar\sum_\alpha f_\alpha\sum_{ij}\left(\left\langle\frac{\partial\Psi_\alpha}{\partial x_i}\,\middle|\,\frac{\partial\Psi_\alpha}{\partial y_j}\right\rangle - \left\langle\frac{\partial\Psi_\alpha}{\partial y_j}\,\middle|\,\frac{\partial\Psi_\alpha}{\partial x_i}\right\rangle\right) \\ &= -i\hbar\hat{\mathbf{n}}\cdot\int\!\!\int d^2r[\nabla\times\nabla'\rho(\mathbf{r}',\mathbf{r})]_{r=r'} - i\hbar\hat{\mathbf{n}}\cdot\int\!\!\int d^2r_1\int\!\!\int d^2r_2[2\nabla_1\times\nabla'_2\Gamma(\mathbf{r}'_1,\mathbf{r}'_2;\mathbf{r}_1,\mathbf{r}_2)]_{r=r'}, \end{aligned} \quad (7)$$

where ρ and Γ are the one- and two-particle Dirac density matrices for the system, and the sum over i and j denotes a sum over all the particles in the system.

The integral over the two-particle density matrix Γ vanishes. Because of the symmetry between $\mathbf{r}_1$ and $\mathbf{r}_2$, the integrand in this integral can also be written as

$$[(\nabla_1 + \nabla_1') \times (\nabla_2 - \nabla_2')\Gamma(\mathbf{r}_1', \mathbf{r}_2'; \mathbf{r}_1, \mathbf{r}_2) - \rho(\mathbf{r}_1', \mathbf{r}_1)\rho(\mathbf{r}_2', \mathbf{r}_2)]_{r=r'}; \quad (8)$$

the replacement of Γ by $\Gamma - \rho\rho$ is also obvious, since the integral over $\mathbf{r}_1$ of $\mathrm{grad}_1 \rho(\mathbf{r}_1)$ vanishes. For any value of $\mathbf{r}_2$ this can be integrated over $\mathbf{r}_1$ to give a correlation function between distant points which should vanish if the range of integration is sufficiently large.

A more formal way of getting the same result is to say that the first line of Eq. (7) gives the transverse force in terms of the expectation value of the commutator of the x and y components of the total momentum of the particles. This commutator is a one-particle operator given by

$$[P_x, P_y] = \int\int dx\, dy \left(\frac{\partial \psi^\dagger}{\partial x} \frac{\partial \psi}{\partial y} - \frac{\partial \psi^\dagger}{\partial y} \frac{\partial \psi}{\partial x} \right), \quad (9)$$

where the $\psi^\dagger$, ψ are creation and annihilation operators for fermions or bosons. Since it is a one-particle operator, its expectation value is given by the one-particle Dirac density matrix.

The integrand of the first term in Eq. (7) can be written as half the curl of $(\nabla' - \nabla)\rho(\mathbf{r}', \mathbf{r})$ evaluated at $\mathbf{r} = \mathbf{r}'$. Now divide the integrals over $\mathbf{r}$ and $\mathbf{r}_1$ into a sum over finite areas labeled with an index σ; the first of these areas will be centered on the position of the vortex, and the others are all well away from the vortex. Stokes' theorem can be used to write the result as

$$F/v_V = \sum_\sigma \oint_\sigma \frac{i\hbar}{2} dx \cdot [(\nabla - \nabla')\rho(\mathbf{r}', \mathbf{r})]_{r=r'}. \quad (10)$$

This result is exact, and there are no contributions to the integrals from the neighborhood of the vortex core, since we have chosen the boundaries of the regions of integration to be well away from the core; any contributions from the core states or from the properties of the core must be reflected in the density matrices well away from the core. In particular, because localized states inside the vortex core, such as occur for an s-wave superconductor [12], do not influence the one-particle density matrix far away from the core, their contributions to the transverse force must be zero. It should be pointed out that this property is not transparent in Eq. (7), where both detailed forms of density matrices and explicit integrations are needed to recover it.

For a neutral superfluid the integrand in Eq. (10) is just the momentum density, equal, by definition, to $\rho_s \mathbf{u}_s + \rho_n \mathbf{u}_n$, where $\mathbf{u}_s$ and $\mathbf{u}_n$ are the local values of the superfluid and normal velocities. If the circulation of the normal fluid is zero, this term gives $\mathbf{F} = \rho_s \mathbf{K} \times \mathbf{v}_V$, the Magnus force $\mathbf{F}_M$ [2,9,10]. From the Galilean invariance of the problem we can deduce that $\mathbf{v}_V$ should be replaced by $\mathbf{v}_V - \mathbf{v}_s$, but we cannot determine the coefficient of the term proportional to $\mathbf{v}_n - \mathbf{v}_s$ without explicitly putting in different background velocities for the normal fluid and superfluid components.

At first sight it is surprising that we can obtain an exact result for this problem, since there are very few problems to which quantum many-body theory gives exact answers, but there is a similar argument that can be used in classical hydrodynamics. If fluid is streaming past a fixed vortex with velocity $\mathbf{v}$, one can calculate the momentum balance on a very large cylinder centered on the vortex. The net pressure force per unit length on the interior of this cylinder is $\rho_n \mathbf{K} \times \mathbf{v}/2$, the momentum flow out from the cylinder is $-\rho_n \mathbf{K} \times \mathbf{v}/2$, and these must be balanced by a force $-\rho_n \mathbf{K} \times \mathbf{v}$ acting on the vortex core. It is by an elaboration of this argument that Barenghi, Donnelly, and Vinen [5] obtain their expression for the contribution of the flow of the normal fluid past the vortex.

Our conclusion is that for any infinite, homogeneous, neutral superfluid with a scalar order parameter the transverse force has the form $\rho_s \mathbf{K} \times \mathbf{v}_V$, and those mechanisms which have been suggested as giving corrections must actually give alternative ways of looking at the same thing. Because our argument is, like the classical argument quoted in the previous paragraph, basically an argument that balances the forces on the vortex core with forces and momentum flow at large distances for the vortex, there is no obvious generalization that allows for the effect of the disordered background potential which is normally important for real supeconductors.

We acknowledge the hospitality of the Aspen Center for Physics, where this work was conceived and completed. We thank Daniel Fisher, Carlos Wexler, Boris Spivak, Xiao-Mei Zhu, and numerous other people for helpful comments on this problem. The work was supported by U.S. NSF Grant No. DMR-9220733 and Swedish Natural Science Research Council (P. A.).

[1] H. Lamb, *Hydrodynamics* (Dover, New York, 1945).
[2] W.F. Vinen, Proc. R. Soc. London A **260**, 218 (1961).
[3] S.C. Whitmore and W. Zimmermann, Phys. Rev. **161**, 181 (1968).
[4] S.V. Iordanskii, Sov. Phys. JETP **22**, 160 (1966).
[5] C.F. Barenghi, R.J. Donnelly, and W.F. Vinen, J. Low Temp. Phys. **52**, 189 (1983); see also R.J. Donnelly, *Quantized Vortices in Helium II* (Cambridge University Press, Cambridge, 1991), pp. 86–104.
[6] E.B. Sonin, Rev. Mod. Phys. **59**, 87 (1987).
[7] G.E. Volovik, Sov. Phys. JETP **77**, 435 (1993); N.B. Kopnin and M.M. Salomaa, Phys. Rev. B **44**, 9667 (1991).
[8] G.E. Volovik, JETP Lett. **62**, 66 (1995).

[9] R. J. Zieve, Yu. M. Mukharsky, J. D. Close, J. C. Davis, and R. E. Packard, J. Low Temp. Phys. **91**, 315 (1993).

[10] P. Ao and D. J. Thouless, Phys. Rev. Lett. **70**, 2158 (1993); D. J. Thouless, P. Ao, and Q. Niu, Physica (Amsterdam) **200A**, 42 (1993); P. Ao, Q. Niu, and D. J. Thouless, Physica (Amsterdam) **194B–196B**, 1453 (1994); Q. Niu, P. Ao, and D. J. Thouless, Phys. Rev. Lett. **72**, 1706 (1994).

[11] *Geometric Phases in Physics,* edited by A. Shapere and F. Wilczek (World Scientific, Singapore, 1989).

[12] C Caroli, P. G. de Gennes, and J. Matricon, Phys. Lett. **9**, 307 (1964).

Spin-Wave Dynamics in Real Crystals

Qian Niu and Leonard Kleinman

Department of Physics, The University of Texas, Austin, Texas 78712-1081

(Received 30 September 1997)

We derive the spin-wave dynamics in crystals from the time-dependent variational principle, which involves Berry phase terms as well as the energy of frozen spin waves. The general formulation is based on the adiabatic and harmonic approximations. Formulas for ferromagnets and antiferromagnets are derived in terms of Kohn-Sham wave functions, which are particularly suitable for practical *ab initio* calculations. [S0031-9007(98)05490-8]

PACS numbers: 75.30.Ds, 03.65.Bz, 75.30.Fv, 75.40.Gb

In spite of the fact that when applied to itinerant electrons, the Heisenberg Hamiltonian is at best only a parametrized model, all calculations of spin-wave dispersion curves have to this day been based on that venerable model [1]. This is exemplified in a recent monograph where Aharoni [2] chooses "to use the assumption of localized magnetic moments on lattice sites, and the Heisenberg Hamiltonian, because it is the only way to include the variation of the magnetization in space." Contrary to this widely held belief, in this Letter we derive spin-wave equations of motion which in the harmonic and adiabatic limit are exact everywhere between the localized moment and jellium limits, and are universally applicable to ferromagnets, antiferromagnets, and ferrimagnets. We then show how the quantities appearing in the equations of motion can be exactly evaluated within density functional theory [3], which is known to yield accurate values for the ground state magnetization of Fe, Co, and Ni. Interestingly, the formulas for the ferromagnetic spin-wave frequency, when written in terms of the frozen spin-wave energy, become identical in the jellium and localized limits.

General formulation.—Suppose we have a symmetry broken ground state $|\psi_0\rangle$ with a magnetization density $m(\mathbf{r})\hat{\mathbf{e}}(\mathbf{r}) = \langle\psi_0|\hat{\rho}(\mathbf{r})\boldsymbol{\sigma}|\psi_0\rangle$, where $\boldsymbol{\sigma}$ is the Pauli matrix vector and $\hat{\rho}(\mathbf{r})$ is the density operator. The direction $\hat{\mathbf{e}}(\mathbf{r})$ (constant for a ferromagnet) may be a function of position in general, but we assume it has a periodicity commensurate with the crystal lattice, while $|\psi_0\rangle$ has the periodicity of the magnetic lattice defined by $m(\mathbf{r})\hat{\mathbf{e}}(\mathbf{r})$. With the excitation of a spin wave, the magnetization density becomes $m(\mathbf{r})[\hat{\mathbf{e}}(\mathbf{r}) + \mathbf{S}(\mathbf{r})]$, where for a small amplitude of $\mathbf{S}(\mathbf{r})$ we can write the spin wave in the form

$$\mathbf{S}(\mathbf{r}) = \mathrm{Re}\left\{e^{i\mathbf{k}\cdot\mathbf{r}}\sum_{\mathbf{G}}\mathbf{S}_{\mathbf{G}}e^{i\mathbf{G}\cdot\mathbf{r}}\right\}, \tag{1}$$

where the sum is over the reciprocal lattice vectors of the magnetic unit cell. Our task is to establish the equations of motion governing the time evolution of the spin-wave excitations. When these equations are linearized for small amplitudes, we should be able to obtain the normal modes and their frequency dispersions.

Our derivation of the spin dynamics is based on the adiabaticity assumption that the electronic wave function can always follow the instantaneous spin-wave configuration [4]. We first introduce a frozen spin-wave state $|\psi\rangle$ defined as the lowest energy state with a given spin-wave configuration $m(\mathbf{r})[\hat{\mathbf{e}}(\mathbf{r}) + \mathbf{S}(\mathbf{r})] = \langle\psi|\hat{\rho}(\mathbf{r})\boldsymbol{\sigma}|\psi\rangle$. This makes $|\psi\rangle$ a functional of $\mathbf{S}(\mathbf{r})$ or a function of its Fourier components $\mathbf{S}_{\mathbf{G}}$. Then we demand that $\mathbf{S}_{\mathbf{G}}$ should evolve with time in such a way that the adiabatic wave function $e^{if(t)}|\psi[\mathbf{S}_{\mathbf{G}}(t)]\rangle$ satisfies the time-dependent Schrödinger equation for some phase $f(t)$.

This requirement can be conveniently implemented by employing the time-dependent variational principle that the action $\int L\,dt$ should be extremized by physical wave functions, where $L = \langle\Psi|i\hbar\frac{\partial}{\partial t}|\Psi\rangle - \langle\Psi|H|\Psi\rangle$. When we substitute the adiabatic wave function into this expression, we obtain a Lagrangian for the variables $\mathbf{S}_{\mathbf{G}}$,

$$L = \hbar\sum_{\mathbf{G}j}\dot{S}^j_{\mathbf{G}}\langle\psi|\frac{i\partial}{\partial S^j_{\mathbf{G}}}|\psi\rangle - E\,, \tag{2}$$

where the dot on $S^j_{\mathbf{G}}$ indicates a time derivative, and $E = \langle\psi|H|\psi\rangle$ is the energy of the system for the given spin configuration. The variable $S^j_{\mathbf{G}}$ is the real or imaginary part of a vector component of $\mathbf{S}_{\mathbf{G}}$, and j is an index labeling the six different possibilities. The phase $f(t)$ in the adiabatic wave function only contributes a term of total time derivative, which we have dropped from the Lagrangian without any physical consequences [5]. The terms in the sum are related to the Berry phase of the adiabatic wave function, and their appearance is universal in adiabatic Lagrangians of slow variables [6].

The Euler-Lagrange equations for extremizing the action then governs the dynamics of these variables,

$$-\sum_{j',\mathbf{G}'}\hbar\Omega^{jj'}_{\mathbf{G}\mathbf{G}'}\dot{S}^{j'}_{\mathbf{G}'} + \frac{\partial E}{\partial S^j_{\mathbf{G}}} = 0\,, \tag{3}$$

where the Ω matrix (the Berry curvature) is defined as

$$\Omega^{jj'}_{\mathbf{G}\mathbf{G}'} = \frac{\partial}{\partial S^j_{\mathbf{G}}}\langle\psi|\frac{i\partial}{\partial S^{j'}_{\mathbf{G}'}}|\psi\rangle - \frac{\partial}{\partial S^{j'}_{\mathbf{G}'}}\langle\psi|\frac{i\partial}{\partial S^j_{\mathbf{G}}}|\psi\rangle\,. \tag{4}$$

Consistent with our assumption of small amplitudes, we make the harmonic approximation by linearizing the force $\partial E/\partial S^j_{\mathbf{G}}$, yielding

$$\sum_{j',\mathbf{G}'}[-\hbar\Omega^{jj'}_{\mathbf{GG}'}\dot{S}^{j'}_{\mathbf{G}'} + K^{jj'}_{\mathbf{GG}'}S^{j'}_{\mathbf{G}'}] = 0\,, \tag{5}$$

where $K^{jj'}_{\mathbf{GG}'}$ is the second derivative of the energy E with respect to $S^j_{\mathbf{G}}$ and $S^{j'}_{\mathbf{G}'}$. Both Ω and K are evaluated at the ground state configuration, where the first derivatives of the energy vanish.

The above set of equations is a central result of this work, which in principle allows one to study the spin dynamics for a general many-body system. Since the spin variables are real, K is obviously real, and so is the Ω matrix [7]. Also, by definition, K is symmetric in the simultaneous interchange of $\mathbf{G}$ and j with $\mathbf{G}'$ and j', while Ω is antisymmetric. The secular equations for the normal modes are obtained by replacing the time derivative by $-i\omega$. The above properties of the matrices guarantee that the eigenfrequencies are real.

The elements of the K matrix may be obtained in a straightforward manner. Let $E(S^j_{\mathbf{G}})$ be the energy (relative to the ground state energy) in the state of the frozen spin configuration (1) with a single amplitude $S^j_{\mathbf{G}}$ present. Then we have $\frac{1}{2}K^{jj}_{\mathbf{GG}}(S^j_{\mathbf{G}})^2 = E(S^j_{\mathbf{G}})$, from which the diagonal elements of the K matrix are obtained. To obtain the off-diagonal elements, let $E(S^j_{\mathbf{G}}, S^{j'}_{\mathbf{G}'})$ be the energy in the state of the frozen spin configuration (1) with the two spin amplitudes present. Then, $K^{jj'}_{\mathbf{GG}'}S^j_{\mathbf{G}}S^{j'}_{\mathbf{G}'}$ is just the energy difference $E(S^j_{\mathbf{G}}, S^{j'}_{\mathbf{G}'}) - E(S^j_{\mathbf{G}}) - E(S^{j'}_{\mathbf{G}'})$.

The evaluation of the Ω matrix needs some consideration, but the final formula is straightforward to implement. For small amplitudes, the product $\frac{1}{2}\Omega^{jj'}_{\mathbf{GG}'}S^j_{\mathbf{G}}S^{j'}_{\mathbf{G}'}$ is equal to the phase, or the imaginary part of the logarithm of the following quantity:

$$\langle\psi(0)|\psi(S^j_{\mathbf{G}})\rangle\langle\psi(S^j_{\mathbf{G}})|\psi(S^{j'}_{\mathbf{G}'})\rangle\langle\psi(S^{j'}_{\mathbf{G}'})|\psi(0)\rangle\,, \tag{6}$$

where $|\psi(0)\rangle$ is the ground state, and $|\psi(S^j_{\mathbf{G}})\rangle$ is the state with a single spin-wave amplitude $S^j_{\mathbf{G}}$. This result can be shown by a brute force expansion to quadratic order in the spin-wave amplitudes, bearing in mind that $\langle\psi|(\partial/\partial S)|\psi\rangle$ is purely imaginary [7]. The above result was motivated from a relation between the Berry curvature and the Berry phase [8].

The quantity (6) is quite suitable for numerical calculations, because one needs to deal only with wave functions of a single spin-wave amplitude. Also, because each wave function and its complex conjugate appears in the expression, one need not worry about the arbitrary phases of the wave functions [9]. Finally, if we choose the phases of the wave functions such that their projections on the ground state are real and positive, then the first and last factors in (6) may be dropped, yielding

$$\Omega^{jj'}_{\mathbf{GG}'} = \frac{\text{phase of } \langle\psi(S^j_{\mathbf{G}})|\psi(S^{j'}_{\mathbf{G}'})\rangle}{\frac{1}{2}S^j_{\mathbf{G}}S^{j'}_{\mathbf{G}'}}\,. \tag{7}$$

Reduction by symmetries.—It has been a common practice to separate the dynamics of the transverse spin components from the longitudinal. This does not seem to be possible in the presence of spin-orbit coupling, unless the ground state magnetization and the Bloch wave vector of the spin wave lie along a crystal axis. In this Letter, we will consider the simpler situation where spin-orbit coupling is neglected. Then the ground state magnetization is along a fixed direction, say, the z direction ($+z$ and $-z$ for an antiferromagnet), and the matrices Ω and K are invariant under spin rotations about the z axis, with the following consequences: (i) All the matrix elements coupling the z with the x and y components must vanish; (ii) the xx and the yy elements of the matrices are equal; and (iii) the xy and yx elements of the matrices differ only by a sign. Because of these properties, the spin dynamics of $S^x + iS^y$, $S^x - iS^y$, and S^z are mutually decoupled. Moreover, the spectra of the two transverse polarizations are simply related by time reversal.

In the rest of this Letter, we will be concerned with one of the transverse modes, which is known to have a spectral branch (Goldstone mode) whose frequency vanishes as $\mathbf{k} \to 0$. Our spin wave (1) will be restricted to the form

$$\mathbf{S}(\mathbf{r}) = \mathrm{Re}\left\{e^{i(\mathbf{k}\cdot\mathbf{r}-\omega t)}\sum_{\mathbf{G}}(\hat{\mathbf{x}} + i\hat{\mathbf{y}})S_{\mathbf{G}}e^{i\mathbf{G}\cdot\mathbf{r}}\right\}. \tag{8}$$

Using the new independent variables $S^1_{\mathbf{G}} = \mathrm{Re}\, S_{\mathbf{G}}$ and $S^2_{\mathbf{G}} = \mathrm{Im}\, S_{\mathbf{G}}$, we can now repeat the calculations through Eq. (7) without changing any formulas, except that the superscript j now labels the two possibilities instead of the previous six.

Further simplifications are achieved by the observation that a spin rotation about the z axis is equivalent to a phase change of the amplitude $S_{\mathbf{G}}$, that is, a rotation between the real and imaginary parts. Therefore, invariance of the Ω and K matrices under the spin rotation implies that $\Omega^{11}_{\mathbf{GG}'} = \Omega^{22}_{\mathbf{GG}'}$ and $\Omega^{12}_{\mathbf{GG}'} = -\Omega^{21}_{\mathbf{GG}'}$, with the same relations for K. Because of these, the variables $S_{\mathbf{G}} = S^1_{\mathbf{G}} + iS^2_{\mathbf{G}}$ and its complex conjugate are decoupled, and the equations of motion (5) are reduced to

$$\sum_{\mathbf{G}'}[i\hbar\omega\Omega_{\mathbf{GG}'} + K_{\mathbf{GG}'}]S_{\mathbf{G}'} = 0\,, \tag{9}$$

where we have replaced the time derivative by $-i\omega$. The new matrices are defined as $\Omega_{\mathbf{GG}'} = \Omega^{11}_{\mathbf{GG}'} - i\Omega^{12}_{\mathbf{GG}'}$ and $K_{\mathbf{GG}'} = K^{11}_{\mathbf{GG}'} - iK^{12}_{\mathbf{GG}'}$. Because of the symmetry properties among the matrix elements mentioned above and earlier [the paragraph below that of Eq. (5)], the new K matrix is Hermitian, $K^*_{\mathbf{G}'\mathbf{G}} = K_{\mathbf{GG}'}$, while the new Ω matrix is anti-Hermitian, $\Omega^*_{\mathbf{G}'\mathbf{G}} = -\Omega_{\mathbf{GG}'}$. This implies that for $\mathbf{G}' = \mathbf{G}$, $K^{12}_{\mathbf{GG}}$ and $\Omega^{11}_{\mathbf{GG}}$ vanish.

Kohn-Sham.—We now consider how our formalism may be implemented in the Kohn-Sham density functional theory [3], in which the many-body wave function is represented by a determinant of single particle states. In the absence of spin-orbit coupling, the single particle states in the ground state of the system may be taken as either spin up, or spin down, with the spatial wave functions being Bloch waves. To fix ideas, we consider how a spin-up state mixes with a spin-down component, when the magnetization is tilted away from the z axis due to a spin wave of wave vector $\mathbf{k}$ and amplitude $S^j_{\mathbf{G}}$ [10]. The spin-density functional theory dictates that the Kohn-Sham potential becomes a position dependent matrix whose eigenspinors line up with the local magnetization direction [11,12]. Thanks to an insight by Herring [13], one can show that the potential matrix and thus the Kohn-Sham Hamiltonian are invariant under a lattice translation combined with a spin rotation. This leaves the wave function in the form $|u(S^j_{\mathbf{G}})\rangle|\uparrow\rangle + |v(S^j_{\mathbf{G}})\rangle|\downarrow\rangle$, where $|u\rangle$ and $e^{i(\mathbf{k}+\mathbf{G})\cdot\mathbf{r}}|v\rangle$ are Bloch waves [14] which can be calculated using standard band structure techniques.

The K matrix can be obtained from standard procedures for the energetics of the frozen spin-wave states, so we will focus on the calculation of the Ω matrix. It is straightforward to show that the Berry curvature of the many-body wave function is equal to the sum of Berry curvatures of the occupied states [15]. The same kind of reasoning leading to Eqs. (6) and (7) can be applied to a single particle state. Therefore, the contribution to $\Omega^{jj'}_{\mathbf{GG}'}$ from a spin-up state is the phase of

$$\langle u(S^j_{\mathbf{G}})|u(S^{j'}_{\mathbf{G}'})\rangle + \langle v(S^j_{\mathbf{G}})|v(S^{j'}_{\mathbf{G}'})\rangle \tag{10}$$

divided by $\frac{1}{2}S^j_{\mathbf{G}}S^{j'}_{\mathbf{G}'}$. It is understood that this formula should be used after choosing the phases of $|u(S)\rangle$, $\{S = S^j_{\mathbf{G}}, S^{j'}_{\mathbf{G}'}\}$, such that the projections $\langle u(0)|u(S)\rangle$ are real and positive. Consistent with this condition and to second order in the spin-wave amplitudes, the phase of (10) is equal to the imaginary part of $\langle v(S^j_{\mathbf{G}})|v(S^{j'}_{\mathbf{G}'})\rangle$ [16]. This shows that when the amplitudes $S^2_{\mathbf{G}'}$ and $S^1_{\mathbf{G}'}$ are given the same value, the contribution to $\Omega_{\mathbf{GG}'} \equiv \Omega^{11}_{\mathbf{GG}'} - i\Omega^{12}_{\mathbf{GG}'}$ from the spin-up state may be written as

$$\mathrm{Im}\langle v(S^1_{\mathbf{G}})|v(S^1_{\mathbf{G}'})\rangle - i\,\mathrm{Im}\langle v(S^1_{\mathbf{G}})|v(S^2_{\mathbf{G}'})\rangle \tag{11}$$

divided by $\frac{1}{2}S^1_{\mathbf{G}}S^1_{\mathbf{G}'}$. This is just $i\langle v(S^1_{\mathbf{G}})|v(S^1_{\mathbf{G}'})\rangle^* = i\langle v(S^1_{\mathbf{G}'})|v(S^1_{\mathbf{G}})\rangle$, because of the relation $|v(S^2_{\mathbf{G}'})\rangle = e^{-i(\pi/2)}|v(S^1_{\mathbf{G}'})\rangle$ [17]. Finally, after a similar consideration for spin-down states, we find that $\Omega_{\mathbf{GG}'}$ is given by

$$\Omega_{\mathbf{GG}'} = \frac{i\sum[\langle v(S^1_{\mathbf{G}'})|v(S^1_{\mathbf{G}})\rangle_\uparrow - \langle u(S^1_{\mathbf{G}'})|u(S^1_{\mathbf{G}})\rangle_\downarrow]}{\frac{1}{2}S^1_{\mathbf{G}}S^1_{\mathbf{G}'}}, \tag{12}$$

where the sum is over the occupied single particle states, and the arrows indicate the spin orientation of the state in the absence of the spin wave [18].

The almost-jellium regime.—Here, the lowest branch of the spin waves is dominated by the $\mathbf{G} = 0$ Fourier component except near the zone boundary. Then, we can immediately solve for the eigenfrequency from (9) as

$$\hbar\omega = \frac{K_{00}}{-i\Omega_{00}} = \frac{\frac{1}{2}K^{11}_{00}(S^1_0)^2}{\sum[\langle v|v\rangle_\uparrow - \langle u|u\rangle_\downarrow]}, \tag{13}$$

using (12) with $\mathbf{G} = \mathbf{G}' = 0$. As previously discussed, the numerator is just the energy $\Delta E(\mathbf{k})$ to create the spin wave. The denominator is simply the reduction of the total crystal spin S due to the spin wave, because $\sum\langle v|v\rangle_\uparrow$ represents the reduction of the up component of the spin-up electrons and $\langle u|u\rangle_\downarrow$ represents the reduction of the down component of the spin-down electrons. Therefore, we have the more transparent result

$$\hbar\omega = 2\Delta E(\mathbf{k})/S\theta^2, \tag{14}$$

where θ is the angle of tilt of the local magnetization from the z direction.

Several remarks are in order. First, $\Delta E(\mathbf{k})$ is proportional to θ^2, so that ω is independent of tilt angle. Second, although $\Delta E(\mathbf{k})$ and S are total crystal quantities, we may also take them to be the spin-wave energy and net spin per unit cell. Third, at long wavelengths, the energy $\Delta E(\mathbf{k})$ is proportional to the square of the spin-density gradient, so that $\omega \propto k^2$. Finally, for a Heisenberg Hamiltonian with spin S per unit cell, one also obtains (14) [19], although the energy $\Delta E(\mathbf{k})$ for the two cases is not the same.

Near the Brillouin zone boundary, coupling between different Fourier components becomes important. At $\mathbf{k} = -\mathbf{G}/2$, the spin waves of amplitude S_0 and $S_{\mathbf{G}}$ have equal but opposite wave vectors, and they are degenerate because the 00 and $\mathbf{GG}$ components are equal to each other for both the K and Ω matrices [20]. This degeneracy will be lifted to produce a gap at the zone edge when their coupling is taking into account

$$\begin{aligned}
[i\hbar\omega\Omega_{00} + K_{00}]S_0 + [i\hbar\omega\Omega_{0\mathbf{G}} + K_{0\mathbf{G}}]S_{\mathbf{G}} &= 0, \\
[i\hbar\omega\Omega_{\mathbf{G}0} + K_{\mathbf{G}0}]S_0 + [i\hbar\omega\Omega_{\mathbf{GG}} + K_{\mathbf{GG}}]S_{\mathbf{G}} &= 0.
\end{aligned} \tag{15}$$

In addition, if the ground state has space inversion symmetry, then the $0\mathbf{G}$ and $\mathbf{G}0$ components are also equal for both the K and Ω matrices [21]. Because of the Hermiticity of K and anti-Hermiticity of Ω, the above symmetry arguments imply that the elements K^{12}_{00}, $K^{12}_{0\mathbf{G}}$, Ω^{11}_{00}, and $\Omega^{11}_{0\mathbf{G}}$ all vanish. Thus,

$$\hbar\omega = \frac{K_{00} \pm K_{0\mathbf{G}}}{i(\Omega_{00} \pm \Omega_{0\mathbf{G}})} = \frac{K^{11}_{00} \pm K^{11}_{0\mathbf{G}}}{\Omega^{12}_{00} \pm \Omega^{12}_{0\mathbf{G}}}. \tag{16}$$

In the case of an antiferromagnetic ground state, the denominators in (13) and (14) vanish, so it is essential to include other Fourier components of the spin wave even for $\mathbf{k}$ away from the zone boundary. For this purpose, we consider the effect of coupling to a single $\mathbf{G}$ of minimal size for which the spin-wave changes sign under translations connecting the up and down spin sublattices. In this case, the matrix elements Ω_{00}, $\Omega_{\mathbf{GG}}$, and $K_{0\mathbf{G}}$ are zero [22], with which Eq. (15) yields

$$\hbar\omega = \frac{\sqrt{K_{00}K_{\mathbf{GG}}}}{|\Omega_{0\mathbf{G}}|}, \quad (17)$$

where we have used the relation $\Omega_{\mathbf{G}0} = -\Omega^*_{0\mathbf{G}}$.

The above result confirms the usual behavior for antiferromagnetic spin waves that $\omega \propto k$ as $\mathbf{k} \to 0$, because K_{00} is of order k^2, while $K_{\mathbf{GG}}$ and $\Omega_{0\mathbf{G}}$ approach finite constants. This qualitative result remains true even if the coupling to other Fourier components is included, although the frequency may be changed quantitatively.

In summary, within the adiabatic and harmonic approximations, we have obtained the exact spin-wave equations of motion for any ferromagnetic, ferrimagnetic, or antiferromagnetic crystal. We showed how to evaluate the parameters therein in both the many-body and the one electron pictures. We also demonstrated that the frequency has the correct $\mathbf{k}$ dependence at small k and at the zone boundary for both the ferromagnetic and antiferromagnetic crystals.

This work was supported by NSF (DMR 9614040 and PHY 9722610), and by the Welch Foundation.

[1] W. Heisenberg, Z. Phys. **49**, 619 (1928).
[2] A. Aharoni, *Introduction to the Theory of Ferromagnetism* (Clarendon Press, Oxford, 1996).
[3] W. Kohn and L. J. Sham, Phys. Rev. **140**, A1133 (1965).
[4] V. P. Antropov *et al.*, Phys. Rev. B **54**, 1019 (1996).
[5] A total time derivative contributes terms to the action depending only on the initial and final configurations, and therefore does not affect the Euler-Lagrange equations.
[6] J. Moody, A. Shapere, and F. Wilczek, in *Geometric Phases in Physics*, edited by A. Shapere and F. Wilczek (World Scientific, Singapore, 1989), p. 160.
[7] The quantity $\langle\psi|(\partial/\partial S)|\psi\rangle$ is purely imaginary, because the sum of its complex conjugate and itself is just the derivative of $\langle\psi|\psi\rangle = 1$.
[8] The quantity $\frac{1}{2}\Omega^{jj'}_{\mathbf{GG}'}S^j_{\mathbf{G}}S^{j'}_{\mathbf{G}'}$ may be regarded as an areal integral of $\Omega^{jj'}_{\mathbf{GG}'}$ inside the right triangle in the $S^j_{\mathbf{G}} - S^{j'}_{\mathbf{G}'}$ plane whose corners are at (0,0), $(S^j_{\mathbf{G}}, 0)$, and $(0, S^{j'}_{\mathbf{G}'})$. In view of Eq. (4) and Stokes's theorem, this may be turned into a line integral of the vector $\langle\psi|(i\partial/\partial\mathbf{S})|\psi\rangle$ along the sides of the triangle. This is just the Berry phase around the triangle, which coincides with the phase of the expression (6) when the triangle is small.
[9] R. Resta, Rev. Mod. Phys. **66**, 899 (1994).
[10] One must apply an external "spin-wave field" to prevent the system from collapsing back into its magnetic ground state [11]. When calculating the spin-wave energy the contribution of the external field is subtracted off.
[11] M. Uhl, L. M. Sandratski, and J. Kübler, Phys. Rev. B **50**, 291 (1994).
[12] M. Körling and J. Ergon, Phys. Rev. B **54**, 8293 (1996).
[13] Conyers Herring, *Magnetism*, edited by G. T. Rado and H. Suhl (Academic Press, New York, 1966), Vol. 4.
[14] In order to keep the wave function continuous at zero spin-wave amplitude which is necessary for the Berry curvature calculations later, the spin rotation operator about the z axis used in [11,12] is modified into the form $e^{i\phi(\sigma_z-1)/2}$, where σ_z is the z component Pauli matrix, and $\phi = (\mathbf{k} + \mathbf{G}) \cdot \mathbf{r}$ is the angle of rotation.
[15] The easiest way to prove this is to show the same statement for the quantity $\langle\psi|(\partial/\partial S^j_{\mathbf{G}})|\psi\rangle$. Since the derivative acts separately on the single particle wave functions, it may be regarded as a sum of single particle operators.
[16] In the presence of a spin wave, the potential matrix has off-diagonal elements first order in the spin-wave amplitude, while the diagonal elements differ from the unperturbed one only in second order. Moreover, we have required $\langle u(0)|u(S)\rangle$ to be real and positive. Therefore, $|v(S)\rangle$ is of first order in the spin-wave amplitude S, and $|u(S)\rangle - |u(0)\rangle$ is of second order.
[17] The spin wave of amplitude $S_{\mathbf{G}} = iS^2_{\mathbf{G}}$ can be obtained from a spin wave of amplitude $S_{\mathbf{G}} = S^1_{\mathbf{G}}$ by a spin rotation. Applying the spin rotation operator in Ref. [14] with $\phi = -\pi/2$, we obtain the relation in the text.
[18] In the atomic sphere approximation [11,12] where the direction of magnetization is taken to be constant over each Wigner-Seitz sphere, $\mathbf{k}$ and $\mathbf{k} + \mathbf{G}$ are equivalent, so $\mathbf{G}$'s other than $\mathbf{G} = 0$ are redundant. In a full potential calculation, one can force a spin wave with $S_{\mathbf{G}}$ as its main component, but self-consistency will cause it to be dressed with other Fourier components of smaller amplitude. Equation (12) still holds for the dressed spin-wave amplitudes with the understanding that $S_{\mathbf{G}}$ stands for the amplitude of the main component of a spin wave.
[19] Although it can be shown in general, here we show it for a nearest neighbor linear chain. Following C. Kittel [*Introduction to Solid State Physics* (Wiley, New York, 1996), 7th ed.], we have $\hbar\omega = 4JS(1 - \cos ka)$, whereas $E(k) = -2J\mathbf{S}_p \cdot \mathbf{S}_{p+1} = -2J(S_z^2 + S_\perp^2 \cos ka)$ and $\Delta E = E(k) - E(0) = 2JS_\perp^2(1 - \cos ka) \approx 2JS^2\theta^2(1 - \cos ka)$.
[20] The ground state is invariant under complex conjugation which is equivalent to time reversal followed by a spin rotation of 180° about the y axis. For a spin-wave state with $\mathbf{k} = -\mathbf{G}/2$, this operation reflects the spin y component, and thus replaces the roles of S_0 and $S_{\mathbf{G}}$ by $S^*_{\mathbf{G}}$ and S^*_0, respectively.
[21] Under space inversion, the roles played by S_0 and $S_{\mathbf{G}}$ are interchanged. Therefore, the matrix elements of K and Ω must be invariant under $0 \leftrightarrow \mathbf{G}$.
[22] In the presence of spin waves of amplitude S_0 and $S_{\mathbf{G}}$, the energy above the ground state equals half of $K_{00}|S_0|^2 + K_{0\mathbf{G}}S^*_0S_{\mathbf{G}} + K_{\mathbf{G}0}S^*_{\mathbf{G}}S_0 + K_{\mathbf{GG}}|S_{\mathbf{G}}|^2$. Under the symmetry operation of time reversal followed by the translation $\mathbf{r} \to \mathbf{r} + \mathbf{R}/2$ between the sublattices, the spin-wave amplitude S_0 changes a phase of $\mathbf{k} \cdot \mathbf{R}/2$, while $S_{\mathbf{G}}$ changes a phase of $(\mathbf{k} + \mathbf{G}) \cdot \mathbf{R}/2$. The extra phase in $S_{\mathbf{G}}$ makes the second and third terms of the energy switch their signs. In order for the energy to be invariant under the symmetry operation, these terms must vanish for arbitrary S_0 and $S_{\mathbf{G}}$, which implies the vanishing of $K_{0\mathbf{G}}$.

Adiabatic Dynamics of Local Spin Moments in Itinerant Magnets

Q. Niu,[1,2] Xindong Wang,[2] L. Kleinman,[1] Wu-Ming Liu,[1,2] D. M. C. Nicholson,[2] and G. M. Stocks[2]
[1]*Department of Physics, University of Texas, Austin, Texas 78712*
[2]*Solid State Division, Oak Ridge National Laboratory, Oak Ridge, Tennessee 37831*
(Received 8 December 1998)

Using the adiabatic approximation, we derive the equations of motion for local spin moments which are valid for itinerant as well as tight-binding spins. Material parameters in the equations of motion can be obtained using standard density functional methods, because they depend only on the energy and Berry phase of the constrained ground state of frozen spin configurations. For the calculation of spin waves in a collinear magnet, it is sufficient to know the quadratic forms of total energy and spin component along the symmetry axis as functions of the spin deviations from the ground state configuration.

PACS numbers: 75.30.Ds, 03.65.Bz, 75.30.Fv, 75.40.Gb

Historically, there have been two opposing points of view of magnetism [1]: The atomic moment picture of Heisenberg and the itinerant electron picture of Bloch and Stoner. Because of its conceptual simplicity, the Heisenberg model has enjoyed a dominant position in the theory of magnetism, despite the tremendous success of the itinerant picture for transition metals. There have also been steady and serious efforts trying to push the atomic moment picture beyond its obvious domain of validity [2,3]. Recently, it has been proposed to use such local moments to formulate large amplitude spin dynamics which are needed for finite temperature effects on magnetism [4].

In this Letter, we show that while it is still possible to define local moments for itinerant spins, the usual Landau-Lifshitz equations for the moments must be replaced by more accurate equations of motion. This we establish purely based on the adiabatic assumption of the local moments with respect to the system's other degrees of freedom. For collinear magnets, the Berry curvatures involved in the equations of motion are shown to describe how the total spin component along the symmetry axis changes due to spin deviations from the ground state configuration. A simple formula for the spin wave spectrum, previously derived for ferromagnets in either the free or tight-binding limits, is now shown to be valid everywhere between these limits and for all collinear magnets.

Adiabatic dynamics of the local moments.—Following the usual practice, we partition the space into spin cells of volume V_j [5]. There is some empirical knowledge as to how to choose the partition effectively, but one does not have to assign one cell for each atom. We concentrate first on how to derive the best set of equations of motion in the adiabatic limit for a given partition. The spin for a cell is defined as $\mathbf{s}_j = \langle\psi|\hat{\mathbf{s}}_j|\psi\rangle$, where $\hat{\mathbf{s}}_j = \int_{V_j} d^3r\,\hat{\mathbf{s}}(\mathbf{r})$, and $\hat{\mathbf{s}}(\mathbf{r})$ is the spin density operator. (In this Letter, we use the convention that spin angular momenta have an extra factor of $\hbar$ beside s.) Let $|\psi[\mathbf{s}]\rangle$ be the lowest energy state of the electronic system which yields a given spin configuration denoted by $[\mathbf{s}]$. Using a similar method as in [6], we find

$$-\sum_{\alpha' j'} \hbar\Omega_{jj'}^{\alpha\alpha'} \dot{s}_{j'}^{\alpha'} + \frac{\partial E}{\partial s_j^{\alpha}} = 0\,, \tag{1}$$

where α labels the spin components, E is the energy of the constrained ground state, and the Ω matrix elements (Berry curvatures [7]) are defined as

$$\Omega_{jj'}^{\alpha\alpha'} = \frac{\partial}{\partial s_j^{\alpha}}\left\langle \psi \,\middle|\, \frac{i\partial}{\partial s_{j'}^{\alpha'}} \,\middle|\, \psi \right\rangle - \frac{\partial}{\partial s_{j'}^{\alpha'}}\left\langle \psi \,\middle|\, \frac{i\partial}{\partial s_j^{\alpha}} \,\middle|\, \psi \right\rangle. \tag{2}$$

Therefore, the adiabatic dynamics of the spin moments is completely determined from the energy and the Berry curvatures of the constrained ground state. These quantities can be extracted using standard density functional methods [8], which are simpler and thus more practical than methods based on the linear response or time-dependent density functional theory [9]. There has already been much work on the calculation of the energy [5], and one also has a few tricks to extract the Berry curvatures [6]. For spin wave calculations, the Berry curvatures are needed only at the ground state, and we will show in this Letter that there is a powerful method to extract them when the spin ground state has an axial symmetry.

The equations of motion (1) are exact in the adiabatic limit [10] for a given partition of the space into cells, and are valid beyond the harmonic approximation or linear response. Based on the variational nature of the theory, the local moment description of the spins becomes better for finer partition of the space, simply because more degrees of freedom are included in the variational wave function [11]. There is nothing intrinsically wrong with the local moment description of itinerant spins; the specification of local moments for every point of space becomes equivalent to the description of spin configuration in terms of its Fourier components as used in [6].

The Landau-Lifshitz equations.—In the special situation of tight-binding or rigid spins, the wave function

$|\psi[\mathbf{s}_j]\rangle$ is obtained from the absolute ground state of spin configuration $\mathbf{s}_j^0$ by rigidly rotating the spins within each cell, i.e.,

$$|\psi[\mathbf{s}_j]\rangle = \prod_j e^{i\vec{\theta}_j\cdot\hat{\mathbf{s}}_j}|\psi_0\rangle, \tag{3}$$

where $\vec{\theta}_j$ is in the direction of $\mathbf{s}_j^0 \times \mathbf{s}_j$ and has a magnitude given by the angle between $\mathbf{s}_j$ and $\mathbf{s}_j^0$. The spin operators $\hat{\mathbf{s}}_j$ are for the total spin in each cell, and they satisfy the usual commutation relations for angular momenta, i.e., $[\hat{s}_j^\alpha, \hat{s}_{j'}^{\alpha'}] = i\delta_{jj'}\epsilon_{\alpha\alpha'\beta}\hat{s}_j^\beta$, where $\epsilon_{\alpha\beta\gamma}$ is the fundamental antisymmetric tensor. Then, the Berry curvatures can be calculated as

$$\Omega_{jj'}^{\alpha\alpha'} = \delta_{jj'}\epsilon^{\alpha\alpha'\beta}s_j^\beta/(s_j)^2, \tag{4}$$

which says that there is no Berry curvature between different cells, and that the Berry curvature within a single cell is given by that of a rigid spin [12]. With this form of the Berry curvature, our general equations of motion (1) reduce to the classical Landau Lifshitz equations

$$\hbar\dot{\mathbf{s}}_j = \mathbf{s}_j \times \frac{\partial E}{\partial \mathbf{s}_j}. \tag{5}$$

However, the wave function (3) is not the constrained ground state for the given spin configuration in general. The Landau-Lifshitz equations can thus fail for itinerant spins, or when the local moments interact with itinerant electrons [13]. There are at least two ways that inter-moment Berry curvatures arise and thus invalidate the Landau-Lifshitz equations. (i) In the presence of electron hopping, the moments at different sites may share the same spin and thus have Berry curvatures between them. (ii) Additional Berry phases may arise due to the adiabatic following of the neglected degrees of freedom to the local moments.

To illustrate point (i), we have solved exactly a $t - J - B$ Hamiltonian on two sites,

$$-\frac{t}{2}\sum_\sigma[a_{1\sigma}^\dagger a_{2\sigma} + a_{2\sigma}^\dagger a_{1\sigma}] - 4J\hat{\mathbf{s}}_1\cdot\hat{\mathbf{s}}_2 - \sum_j \mathbf{B}_j\cdot\hat{\mathbf{s}}_j,$$

where $a_{j\sigma}$ is the destruction operator on site j and spin state σ. When there are two electrons present in the system, there are six different states, four with one electron on each site and two with both electrons on either site. A uniform magnetic field, $\mathbf{B}_j = B\hat{z}$, is added to select a direction of magnetization for the ferromagnetic ground state, and our final results have a well defined limit as $B \to 0$. There are six eigenstates consisting of a triplet with energies $-J - B, -J$, and $-J + B$, and of three singlets with energies at 0 and $\frac{3}{2}J \pm [(\frac{3}{2}J)^2 + (2t)^2]^{1/2}$. If the hopping energy is not too large ($t^2 < J^2 + 5BJ + B^2$), the ground state is one of the triplet, in which the average spin on each site is in the z direction with the full magnitude of 1/2.

The average spins are tilted when small transverse fields are added to the system. The new ground state can be calculated perturbatively, from which the Berry curvatures at $\mathbf{B}_j = B\hat{z}$ are extracted exactly. The calculations are tedious but straightforward, and we have obtained the following simple results:

$$\Omega_{11}^{xy} = \Omega_{22}^{xy} = -2 - \frac{t^2}{(J+B)^2}, \tag{6}$$

$$\Omega_{12}^{xy} = \Omega_{21}^{xy} = \frac{t^2}{(J+B)^2}, \tag{7}$$

where the xx and yy components all vanish because of the rotational symmetry about the spin z axis. These results clearly demonstrate the effect of intersite hoppings. When t is zero, we go back to the original results of the tight-binding limit. When t is nonzero, not only the on-site Berry curvature is modified, but also an intersite Berry curvature is induced.

To illustrate point (ii), we consider the three spin model $H = -J(\mathbf{s}_1\cdot\mathbf{s}_2 + \mathbf{s}_2\cdot\mathbf{s}_3) - B(s_1^z + s_2^z + s_3^z)$. Two of these spins (1 and 3) will be taken into account in the definition of the local moments, and the other simulates a neglected degree of freedom such as an orbital moment or an interstitial spin pocket. The eigenstates consist of a quartet with energies $-2J \pm 2B$ and $-2J \pm B$, and of two doublets with energies $\pm B$ and $4J \pm B$, respectively. For ferromagnetic coupling ($J > 0$), the ground state is one of the quartet, with all the spins fully in the z direction ($B > 0$). We apply small transverse fields on spins 1 and 3 (leaving spin 2 free), causing transverse deviations of the spins in the new ground state. The Berry curvatures with respect to the transverse components of spins 1 and 3 are then calculated, with the results

$$\Omega_{11}^{xy} = \Omega_{33}^{xy} = -2\frac{B^2 + 4BJ + 5J^2}{(B+2J)^2}, \tag{8}$$

$$\Omega_{13}^{xy} = \Omega_{31}^{xy} = -2\frac{J^2}{(B+2J)^2}. \tag{9}$$

Significance of the Berry curvatures.—A transparent physical picture for the Berry curvatures can be obtained if we confine our attention to the case of ground states with rotational symmetry of the spins about an axis. This includes the usual ferromagnets, ferrimagnets, and antiferromagnets, but excludes canted or spiral magnetic structures. No translational symmetry will be assumed except in a special discussion. The Berry curvatures will be related to the reduction of the total spin z component due to the excitations of the spins, and a simple formula for the frequency dispersion will be obtained.

Let the symmetry axis of the ground state be the z direction, then the xy spin components due to a spin wave eigenmode can be written as [14]

$$\mathbf{s}_j^\perp = \mathrm{Re}[e^{-i\omega t}(\hat{x} + i\hat{y})s_j], \tag{10}$$

where s_j is the amplitude (complex) on the jth site. Using the complex amplitudes in Eq. (10) as the independent variables, we can obtain the same set of equations of motion as in Eq. (1), except that the upper indices now refer to the real and imaginary parts of the amplitudes. By linearizing the energy gradient and writing $K_{jj'}^{\alpha\alpha'}$ for the second derivative of the energy with respect to s_j^α and $s_{j'}^{\alpha'}$, we obtain the eigenmode equations as

$$\sum_{j'} [i\hbar\omega\Omega_{jj'} + K_{jj'}]s_{j'} = 0, \tag{11}$$

where $\Omega_{jj'} = \Omega_{jj'}^{11} - i\Omega_{jj'}^{12}$, and $K_{jj'} = K_{jj'}^{11} - iK_{jj'}^{12}$.

Therefore, once we know the K and Ω matrices, the above equations of motion determine the spin wave spectrum and the associated eigenmodes. It is easy to see that the quadratic form $\frac{1}{2}s^\dagger K s = \frac{1}{2}\sum_{jj'} s_j^* K_{jj'} s_{j'}$ represents the energy ΔE for creating the spin excitations $\{s_j\}$. The diagonal element K_{jj} can thus be extracted from the energy increase in the system when only a single spin s_j is excited. The off-diagonal elements $K_{jj'}$ and $K_{j'j} = K_{jj'}^*$ $(j \neq j')$ can be obtained from the energy increase when two spins are excited. How about the Ω matrix? In the following, we will show that $-i\frac{1}{2}s^\dagger\Omega s = -i\frac{1}{2}\sum_{jj'} s_j^*\Omega_{jj'}s_{j'}$ is, in fact, the quadratic form for the reduction, $\Delta S_z = S_z^0 - \langle \hat{S}_z \rangle$, of the total spin z component of the ground state due to the spin excitations, i.e.,

$$-i\frac{1}{2}s^\dagger\Omega s = \Delta S_z. \tag{12}$$

This relation then allows a simple evaluation of the Ω matrix in a similar manner as for the K matrix, and it is not necessary to invoke directly the constrained ground state wave functions, which are usually much harder to obtain than the spin expectation values.

That the relation (12) is valid for all collinear magnets can be easily seen in the tight-binding limit. The spin configuration of a collinear magnetic ground state may be written as $\mathbf{s}_j = s_j^0\hat{z}$, where s_j^0 can be positive (ferromagnet) or take both signs (ferrimagnets or antiferromagnets). The total reduction of the spin-z component due to a spin wave (10) is just $\Delta S_z = \frac{1}{2}\sum_j |s_j|^2/s_j^0$. The same result is obtained from the left-hand side of Eq. (12), if we use $\Omega_{jj'} = -i\Omega_{jj'}^{12} = i\Omega_{jj'}^{xy} = i\delta_{jj'}(s_j^0)^{-1}$, which is justified according to Eq. (4) and from the rotational invariance of the ground state about the z axis.

To establish the relation (12) beyond the tight-binding limit, we first show that $i\frac{1}{2}s^\dagger\Omega s$ is actually a Berry phase in the complex plane of the overall amplitude of the spin wave. Let us write $s_j = AR_j$, where $A = A_1 + iA_2$ is the overall amplitude, and $R_j = R_j^1 + iR_j^2$ are the relative amplitudes which are regarded as known and fixed. The Berry curvature at the origin of the complex A plane is given by

$$\Omega_A = \frac{\partial}{\partial A_1}\left\langle \psi \left| \frac{i\partial}{\partial A_2} \right| \psi \right\rangle - \frac{\partial}{\partial A_2}\left\langle \psi \left| \frac{i\partial}{\partial A_1} \right| \psi \right\rangle. \tag{13}$$

The derivatives can be transferred to those with respect to the s_j variables, with the result

$$\sum_{jj'} [\Omega_{jj'}^{11}(R_j^2R_{j'}^1 - R_j^1R_{j'}^2) + \Omega_{jj'}^{12}(R_j^1R_{j'}^1 + R_j^2R_{j'}^2)],$$

where we have used the relations $\Omega_{jj'}^{11} = \Omega_{jj'}^{22}$ and $\Omega_{jj'}^{12} = -\Omega_{jj'}^{21}$, due to the rotational symmetry of the ground state. On the other hand, by expanding in terms of the real and imaginary parts of A and R_j, we find that $i\frac{1}{2}s^\dagger\Omega s$ is just $\frac{1}{2}|A|^2$ times the expression (13), or the Berry phase around a small loop of area $\frac{1}{2}|A|^2$.

Next we show that this Berry phase is, in fact, equal to the reduction of the total spin-z component. Consider a circular wedge of radius $|A|$ and angle ϕ. The Berry phase around the sides of this wedge is just Ω_A (13) times the area $\frac{1}{2}\phi|A|^2$. It can also be calculated directly as the phase of $\langle\psi|\psi_\phi\rangle$, where $|\psi\rangle$ is the constrained ground state for the case of $A_1 = |A|, A_2 = 0$, while $|\psi_\phi\rangle$ is for $A_1 = |A|\cos\phi, A_2 = |A|\sin\phi$. Here we have taken the phase convention of wave functions such that their overlaps with the ground state are real and positive. Because a phase rotation of the amplitude A is equivalent to a rotation of the total spin about the z axis, we may write $|\psi_\phi\rangle = e^{i\phi(\hat{S}_z - S_z^0)}|\psi\rangle$, where, in order to be consistent with the above phase convention, we have subtracted from $\hat{S}_z$ its expectation value S_z^0 in the ground state. For sufficiently small ϕ, we may expand the exponential of the spin rotation operator, and find that the Berry phase along the wedge is equal to $\phi(\langle S_z\rangle - S_z^0)$ as desired.

A simple formula for the spinwave spectrum.—Multiplying s_j^* on the terms of Eq. (11), and summing over j, we find that the eigenfrequency can be expressed as

$$\hbar\omega = \frac{\frac{1}{2}s^\dagger K s}{-i\frac{1}{2}s^\dagger\Omega s} = \frac{\Delta E}{\Delta S_z}, \tag{14}$$

where we have used the relations of the quadratic forms of the K and Ω matrices to the energy increase and spin-z reduction of the ground state due to the spin excitations [15]. This was obtained by Niu and Kleinman [6] for the special cases of ferromagnets in the jellium and tight-binding limits. It is now shown to be valid everywhere between those limits, and to be valid for all collinear magnets including ferrimagnets and antiferromagnets. Because our derivation does not depend on how the space is partitioned into cells, this formula is an exact result of the adiabatic limit and the axial symmetry of the ground state.

There is, in fact, a deep connection between the above simple formula and the quantum theory of spin waves. According to the latter, the energy of a magnon (the spin wave quantum) is just $\hbar\omega$, and there is a single spin flip from the ground state per magnon in a ferromagnet. The above formula simply says that the magnon energy is given by the energy of the spin wave divided by the number of magnons in the spin wave. Interestingly, the connection also applies to ferrimagnetic and antiferromagnetic spin waves. In the latter case, the ground state magnetization S_0^z is zero, but $\langle\hat{S}_z\rangle$ is finite in a spin wave of the form (10) [14]. This can be verified directly from the classical spin wave solutions of the Heisenberg model [16]. The fact that the spectrum goes linearly with k at small k is explained by the observation that $\langle \hat{S}_z \rangle$, which enters the denominator of the formula (14), vanishes linearly as $k \longrightarrow 0$. Quantum mechanically [16], the z component of the total spin, which is a good quantum number, has eigenvalues of ± 1 in each degenerate subspace of one-magnon states, where the negative eigenvalue corresponds to a spin wave of the form (10), while the positive eigenvalue corresponds to a spin wave of the opposite chirality.

The formula (14) can also be used to evaluate the spin wave spectrum. Suppose, for instance, the cell spins in the ground state form a Bravais lattice; then one knows the spin wave eigenmode for each wave vector purely based on the translational symmetry of the system. All one needs to do is to calculate the energy increase due to the presence of the spin wave and the corresponding reduction of the total spin z component, and take the ratio of the two. This method has been used implicitly by Halilov *et al.* [5] and explicitly by Brown *et al.* [18] for the evaluation of spin wave spectra of the ferromagnets of crystalline Ni, Co, and Fe.

The only approximation in this method is the assumption of adiabaticity regarding the motion of the cell spins relative to the rest of the degrees of freedom in the system. One type of such degrees of freedom are spin fluctuations within each cell, and they become important when there are more than one atom per cell. In such a case, one should take a finer partition of space for accuracy, and pay the price of not being able to know the spin wave eigenmodes based on symmetry arguments alone. One may still use the formula (14) to calculate the spin wave spectrum variationally based on trial spin wave configurations. Of course, one can always use Eq. (11) or even (1) to obtain the spectrum systematically.

Q. N. is indebted to Walter Kohn for numerous discussions, and to S. Girvin for his suggestion of the three spin model. This work was supported in part by NSF (DMR9705406, DMR9614040, PHY9722610) and the Welch Foundation. Q. N. and W. M. L. acknowledge the hospitality of the Institute for Theoretical Physics at Santa Barbara, where the work was completed and supported by NSF Grant No. PHY9407194.

[1] C. Herring, *Exchange Interactions among Itinerant Electrons, in Magnetism,* edited by G. T. Rado and H. Suhl (Academic Press, New York, 1966), Vol. IV; J. F. Cooke, J. W. Lynn, and H. L. Davis, Phys. Rev. B **21**, 4118 (1980).

[2] J. Hubbard, Phys. Rev. B **19**, 2626 (1979); **20**, 4584 (1979); H. Hasegawa, J. Phys. Soc. Jpn. **49**, 963 (1980); M. You and V. Heine, J. Phys. F **12**, 177 (1983); **13**, 979 (1983); A. J. Pindor *et al.*, J. Phys. F **13**, 979 (1983).

[3] Toru Moriya, *Spin Fluctuations in Itinerant Electron Magnetism,* Springer Series in Solid State Sciences Vol. 56 (Springer-Verlag, Berlin, 1985).

[4] V. P. Antropov *et al.*, Phys. Rev. B **54**, 1019 (1996).

[5] S. V. Halilov, E. Eschrig, A. Y. Perlov, and P. M. Oppeneer, Phys. Rev. B **58**, 293 (1998); G. M. Stocks, B. Ujfalussy, Xindong Wang, D. M. C. Nicholson, W. A. Shelton, Yang Wang, A. Canning, and B. L. Gyorffy, Philos. Mag. B **78**, 665 (1998).

[6] Qian Niu and Leonard Kleinman, Phys. Rev. Lett. **80**, 2205 (1998).

[7] R. Resta, Rev. Mod. Phys. **66**, 899 (1994).

[8] In a density functional calculation, it is necessary to apply external forces to achieve a given spin configuration. The energy E is just the internal energy of the system, meaning that it does not include the expectation value of the external perturbation. Following practices in the Berry phase method for polarization calculations [7], the Berry curvature can be taken, as an approximation, as the sum of those from the Kohn-Sham orbitals.

[9] K. L. Liu and S. H. Vosko, Can. J. Phys. **67**, 1015 (1989).

[10] This condition is satisfied at the long wavelength limit of a spin wave, but is violated in the region of the so-called Stoner continuum where the spin waves are heavily damped.

[11] With finer partition, there will be more branches of the spin wave included. The higher branches may be erroneous because of the violation of the adiabatic assumption; the lower branch(es) should become more accurate nevertheless.

[12] *Geometric Phases in Physics,* edited by A. Shapere and F. Wilczek (World Scientific, Singapore, 1989).

[13] Ya. B. Bazaliy, B. A. Jones, and S.-C. Zhang, Phys. Rev. B **57**, R3213 (1998).

[14] Another choice of the chirality is with $\hat{x} - i\hat{y}$, but the discussion will be entirely parallel and is therefore neglected here. For antiferromagnets, it is also possible to choose the reference of chirality to be with respect to the local magnetization directions, but such a choice is not necessary despite our (wrong) intuition that the spins should precess in opposite senses on different sublattices. The fact is that there is a high degree of degeneracy in the spin wave spectrum, and different linear combinations of the eigenmodes can give different pictures of spin motion. It is even possible that the spin motion does not have to be chiral at all.

[15] Because the energy and the magnetization scale with the size of the system, the quantities ΔE and ΔS_z can also be regarded as those of a unit cell.

[16] C. Kittel, *Introduction to Solid State Physics* (Wiley, New York, 1998), 7th ed.

[17] C. Kittel, *Quantum Theory of Solids* (Wiley, New York, 1963).

[18] R. H. Brown, D. M. C. Nicholson, Xindong Wang, and T. C. Schulthess, J. Appl. Phys. **85**, 4830 (1999).

Fe magnon dispersion curve calculated with the frozen spin-wave method

D. M. Bylander, Qian Niu, and Leonard Kleinman
University of Texas, Austin, Texas 78712
(Received 8 February 2000)

Since Niu and Kleinman's recent derivation of the frozen spin-wave procedure for calculating magnon dispersion curves, two calculations have appeared which differ by about a factor of 2 in their Fe magnon frequencies at small wave vector. Because the frozen spin-wave state relaxes back to the ferromagnetic state upon iterating to self-consistency, those calculations included a constraining term in the Hamiltonian. Here we describe a quasi-self-consistent method of imposing the frozen spin wave and compare our computational results with those we obtain using a constraining term. We demonstrate that the constrained results converge to the quasi-self-consistent results in the limit that the coefficient of the constraining term approaches zero.

Within the adiabatic and harmonic approximation we[1] recently derived an exact method for calculating magnon dispersion curves which involves Berry phase terms as well as the energy of frozen spin waves. In those cases not involving an expansion in reciprocal-lattice vectors, such as the opposite limits of a tight-binding (i.e., Heisenberg) Hamiltonian and jellium, this exact result reduced to

$$\hbar\omega = \Delta E/\Delta S_z, \tag{1}$$

where ΔE is the increase in energy caused by impressing the frozen spin wave on the system and ΔS_z is the total reduction in the z component of spin. As long as one remains in the harmonic limit, ΔE and ΔS_z are both proportional to θ^2, where θ is the tilt angle of the frozen spin wave, and hence independent of the chosen value of θ. More recently we[2] showed that Eq. (1) is, within the harmonic and adiabatic limits, exact in all cases, *providing that one knows the exact spin-wave eigenstate*. Although the derivation of Eq. (1) was quite tedious, the result should have been immediately obvious. Each quantized spin wave that is excited results in the reversal of a single spin and thus reduces S_z by unity. Hence, Eq. (1) merely says that the spin-wave quantum is equal to the energy of the frozen spin wave divided by the number of quanta it represents.

Because the frozen spin-wave state is not the ground state or even a local minimum in the energy, any self-consistent calculation will converge to the ferromagnetic ground state unless a constraining term is added to the Hamiltonian. After the calculation is converged this term is discarded before ΔE is calculated. There have already been two calculations using our methods. The first[3] found Fe magnon frequencies about half those of the second.[4] It used the atomic sphere approximation and applied a constraining **B** field transverse to the magnetization of the frozen spin wave at an azimuthal angle fixed within a Wigner-Seitz (W-S) sphere but increasing with increasing x coordinate of the W-S sphere within a supercell. However, these workers used the force theorem to evaluate ΔE. Because their perturbing **B** is perpendicular to the magnetization of the spin wave, ΔE is second order in the change it induces in the spin-density matrix $\tilde{\rho}(\mathbf{r})$. But the force theorem, as shown in the Appendix, says that ΔE is given by the change in occupied one-electron eigenvalues and is valid only to first order in $\delta\tilde{\rho}(\mathbf{r})$, so it is unclear what their results means.

Before going on to describe the difference between our method of calculation and that of Gebauer and Baroni[4] (GB) and to compare the numerical results, we should note that for many years magnon dispersion curves have been obtained from poles in the dynamic susceptibility.[5] That method has the advantage that it includes damping effects but the disadvantage that it depends on quasiparticle excitation energies which cannot be obtained for transition metals with much accuracy from first-principles calculations, whereas the frozen spin-wave method depends only on ground-state properties which are obtained with fairly good accuracy from density-functional calculations. We also note that the spin stiffness correction[6] to the local spin-density approximation (LSDA) that we[7] found was necessary to obtain an fcc-Fe spiral spin-density wave (SSDW) energy vs wave-vector curve consistent with the double peak in the neutron-scattering data, need not be applied in these spin-wave calculations to obtain good agreement with experiment. This might indicate that different corrections to the LSDA are required for fcc and bcc iron; but it more likely indicates that the ground state of fcc Fe is not a simple one atom per unit cell SSDW.

COMPUTATIONAL METHODS

This spin-wave calculation in bcc iron closely parallels our full potential SSDW calculation[8] in fcc iron. We used the identical two projector ultrasoft pseudopotential[9] for the d electrons but in order to eliminate the rapidly oscillating parts of the reconstructed s and p components of the valence charge density,[10] we here used standard s and p pseudopotentials.[11] We expand in all plane waves with kinetic energy less than 40 Ry, at 4394 points in the first Brillouin zone (BZ) on a 26×26×26 mesh (which covers the first four BZ and use an eigenvalue Gaussian broadening of 3.77 mRy full width half maximum. The spin-density matrix is evaluated at 8000 points in the unit cell on a 20×20×20 mesh. The exchange-correlation energy and potential as well as the multipole components of the Coulomb potential arising from the compensation charge[10] are evaluated on a 36 $\times 36 \times 36$ mesh as in Ref. 8.

The calculation of GB (Ref. 4) would be expected to be less accurate than ours since they sample only 512 points on a $8\times 8\times 8$ mesh in the first BZ with a Gaussian smearing of 30 mRy. They also use a single projector norm conserving pseudopotential[11] which should be somewhat less accurate than the two projector pseudopotential[9] we used for the d electrons. Like us, and unlike Ref. 3, they use Herring's[12] observation to reduce the calculation to the primitive unit cell rather than using a supercell. There are two basic differences between their calculation and ours. Rather than use Eq. (1), GB work directly with the Berry phase. Although this method would allow one to approach the exact result within the adiabatic and harmonic approximations, it would require an expansion in a large number of frozen spin waves to do so.[1] Since GB use only a single frozen spin wave and since Eq. (1) follows directly from the Berry phase formulation in that case, this difference in our methods should have no effect on our results. What does have an effect is the difference in how we apply the constraint which determines the frozen spin wave. GB add a term to the many body Hamiltonian which corresponds to adding

$$E_{\rm const}[\tilde{\rho}(\mathbf{r})]=A\int[\mathbf{m}(\mathbf{r})-\mathbf{M}(\mathbf{r})]^2 d\mathbf{r} \tag{2}$$

to the density functional, leading to an additional term in the Kohn-Sham Hamiltonian

$$\tilde{V}_{\rm const}(\mathbf{r})=2A\tilde{\sigma}\cdot[\mathbf{m}(\mathbf{r})-\mathbf{M}(\mathbf{r})] \tag{3}$$

(we use tildes to indicate spinor matrices). Here $\mathbf{M}(\mathbf{r})$ is a somewhat arbitrarily chosen magnetization which GB took to be the magnetization of the ferromagnet, tilted by a small angle θ away from the z axis with a fixed azimuthal angle within the inscribed sphere. They took $A\to\infty$ so that $\mathbf{m}(\mathbf{r})$, the self-consistently calculated magnetization which is given by

$$\mathbf{m}(\mathbf{r})=\sqrt{(\rho_{\uparrow\uparrow}-\rho_{\downarrow\downarrow})^2+4\rho_{\uparrow\downarrow}\rho_{\downarrow\uparrow}} \tag{4}$$

will approach $\mathbf{M}(\mathbf{r})$ in the inscribed sphere where it is defined. The total energy of the ferromagnet with the spin-density wave imposed is

$$E=N^{-1}\sum_{n,\mathbf{k}}^{\rm occ}\epsilon_{n\mathbf{k}}-\int(\tilde{V}_{\rm Hxc}+\tilde{V}_{\rm const})\cdot\tilde{\rho}d\mathbf{r}+E_{\rm Hxc}[\tilde{\rho}], \tag{5}$$

where N is the number of points in the BZ at which the Kohn-Sham eigenvalues $\epsilon_{n\mathbf{k}}$ are evaluated and $\tilde{V}_{\rm Hxc}[\tilde{\rho}]$ and $E_{\rm Hxc}[\tilde{\rho}]$ are the Hartree plus exchange-correlation potential and energy density functional.

To avoid having to arbitrarily choose an $\mathbf{M}(\mathbf{r})$ we have developed what we call the quasi-self-consistent frozen spin wave. We choose a value of $v_{\uparrow\downarrow}(\mathbf{q})$, where $\mathbf{q}$ is the wave vector of the spin wave and $v_{\uparrow\downarrow}(\mathbf{q})=v^*_{\downarrow\uparrow}(\mathbf{q})$ is the $\mathbf{q}$th Fourier transform of the off-diagonal matrix element of $\tilde{V}_{\rm Hxc}(\mathbf{r})$. Then after each iteration we multiply $v_{\uparrow\downarrow}(\mathbf{q})$ and all other Fourier components $v_{\uparrow\downarrow}(\mathbf{q}+\mathbf{G})$ by that factor which returns $v_{\uparrow\downarrow}(\mathbf{q})$ to its chosen value. This factor converges to a fixed value and the output potential converges to be proportional to the input potential which is as close to self consistency as one can come. The choice of $v_{\uparrow\downarrow}(\mathbf{q})$ does not in principle effect the result because it is equivalent to choosing the tilt angle θ. The total energy is calculated from Eq. (5) without the $\tilde{V}_{\rm const}$ term. We note that $\tilde{V}_{\rm Hxc}$ is the input potential to the last iteration and therefore contains the constraint, i.e., the multiplicative factor.

There is in fact a theoretical basis for the quasi-self-consistent method, a variational principle for the spin-wave spectrum. Within the class of frozen spin waves of a given Bloch wave vector, the one that minimizes the ratio $\Delta E/\Delta S_z$ gives the lowest branch of the spin wave eigenmodes, and the minimal value of this ratio gives the corresponding spectrum. This naturally leads to the condition that the first-order variation of this ratio should be zero for the lowest eigenmode. This only requires a quasi-self-consistent convergence in the spinwave amplitude or $\rho_{\uparrow\downarrow}$, because a constant overall factor in the tilt of the spins drops out from this ratio. Also, to first order, the quasi-self-consistence in the spins is achieved when that in the potentials $v_{\uparrow\downarrow}$ is achieved.

To prove the variational principle, we note that, close to the ground state, the total reduction in z component of spin can be regarded as a quadratic form of the frozen spin-wave amplitudes (transverse spin deviations at each position). This quadratic form is moreover positive definite for a ferromagnet. It can thus be used to define a scalar product in the linear space of the amplitudes, and define the norm for each amplitude. The matrix for this quadratic form is in fact the Berry-curvature matrix Ω involved in the eigenmode equation[1,2]

$$\hbar\omega\Omega s=Ks, \tag{6}$$

where s is an abstract vector notation of the spin-wave amplitude, and K is the matrix whose quadratic form, $s^\dagger Ks$, is the energy ΔE. It can be easily shown that the eigenmodes form an orthogonal basis and can be chosen to be normalized with respect to the scalar product mentioned above. An arbitrary spin-wave amplitude can thus be expanded uniquely in the eigenmodes, and one can thus write

$$\frac{\Delta E}{\Delta S_z}=\frac{\Sigma_n|a_n|^2\hbar\omega_n}{\Sigma_n|a_n|^2}, \tag{7}$$

where a_n is the expansion coefficient on the nth eigenmode. This expression is clearly bounded from below by the lowest branch of the spin-wave spectrum.

RESULTS AND DISCUSSION

We chose $\mathbf{q}$'s to be commensurate with our $26\times 26\times 26$ BZ mesh, i.e., $\mathbf{q}=(2\pi/a)(n/13,\ 0,\ 0)$, up; to $n=7$. Our choice of θ was a compromise between it being large enough so that ΔS_z and ΔE were large compared to errors introduced by the finite sampling of the BZ and the unit cell [for $\mathbf{q}=(2\pi/a)(1/13,\ 0,\ 0)$, ΔE is only one part in 10^7 of the calculated energy[13] of the ferromagnet] and being small enough that ΔE and ΔS_z are still proportional to θ^2. Since θ varies throughout the unit cell, we estimate it by $\theta=\cos^{-1}(S_z^{sw}/S^f)$, where $2S^f=2.26246$ and $2S_z^{sw}$ are the excess number of up spins, $\rho_{\uparrow\uparrow}(\mathbf{r})-\rho_{\downarrow\downarrow}(\mathbf{r})$, integrated over the unit cell of the ferromagnet without and with the spin wave. Our largest $2\Delta S_z=2(S^f-S_z^{sw})$ occurs at $n=2$ in Table I and results in $\theta=0.29671$ rad. This introduces an error of 0.034%

TABLE I. ΔE in meV/atom, $2\Delta S_z$ in electrons/atoms, and magnon energy $\hbar\omega$ in meV for wave vectors $\mathbf{q}=(2\pi/a)(n/13,\ 0,\ 0)$ calculated with the quasi-self-consistent method and with the $A=2$ Ry a_0^3 constrained method results in parenthesis.

n	ΔE	$2\Delta S_z$	$\hbar\omega$
1	0.190 (0.191)	0.0984 (0.0955)	3.853 (4.006)
2	0.998 (1.054)	0.0989 (0.0997)	20.19 (21.14)
3	2.913 (2.933)	0.0930 (0.0925)	62.68 (63.43)
4	5.096 (5.061)	0.0898 (0.0888)	113.5 (114.0)
5	7.374 (7.268)	0.0894 (0.0869)	165.0 (167.3)
6	9.876 (9.642)	0.0897 (0.0853)	220.2 (225.5)
7	12.56 (12.07)	0.0921 (0.0865)	272.8 (278.9)

in approximating $\cos\theta$ by $1-\frac{1}{2}\theta^2$ and it seems unlikely that it would violate the harmonic approximation with an error more than order of magnitude larger than this.

In Table I we list ΔE, $2\Delta S_z$, and $\hbar\omega$ obtained from Eq. (1) calculated with both the quasi-self-consistent method with $v_{\uparrow\downarrow}(\mathbf{q})=6$ mRy and the constraint method of Eqs. (2) and (3) with $A=2$ Ry a_0^3. These $\hbar\omega$ are displayed in Fig. 1 along with the 10 K pure iron neutron-scattering data[14] and the four values calculated by GB. The smooth curve, drawn as a guide to the eye is

$$\hbar\omega=\alpha q^2(1-\beta q^2) \tag{8}$$

with α=275.5 meV a_0^2 and β=0.2155 a_0^2. The experimental data is anything but smooth, leading us to wonder whether this is an experimental artifact or a real band-structure effect. Room-temperature data[15] on a 12% Si alloy of Fe is smooth and lies below the 10 K pure iron data points. None of the calculations indicate a lack of smoothness but even our BZ

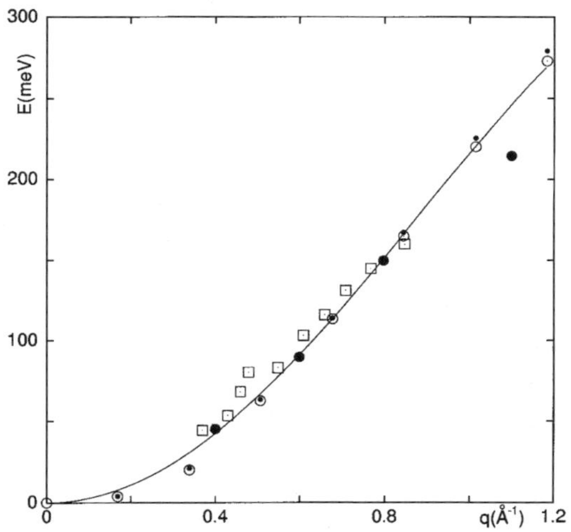

FIG. 1. Magnon energies $\hbar\omega$ in meV vs wave vector. The curve is a guide to the eye, the squares are experimental data of Ref. 14, the open circles are our quasi-self-consistent calculated results, the small dark circles are our calculated $A=2$ Ry a_0^3 constrained results, and the large dark circles are the calculated constrained results of Ref. 4.

TABLE II. Same as Table I except for different values in A in Ry a_0^3.

n	A	ΔE	$2\Delta S_z$	$\hbar\omega$
2	0.2	0.9609	0.09543	20.139
2	2.0	1.0543	0.09974	21.141
2	20	1.3803	0.09962	27.712
4	2.0	5.0613	0.08883	113.96
4	20	6.0778	0.09093	133.68

mesh is fairly gross and the Gaussian eigenvalue smearing it requires could smear out band-structure effects. There does appear to be one real crystal effect in the computational results. For fixed A or input $v_{\uparrow\downarrow}(\mathbf{q})$ one would expect ΔS_z to decrease smoothly and monotonically with increasing ΔE which it fails to do. The frequencies we obtained for the smallest two $\mathbf{q}$'s are obviously too small, which we attribute to taking small differences of large numbers in obtaining ΔE. Slightly improved values could be obtained with a larger input $v_{\uparrow\downarrow}(\mathbf{q})$. The only other point in the figure which seems out of line is the highest $\mathbf{q}$ point calculated by GB. Although they were able to draw a smooth curve through their four points which fit the experimental data somewhat better than our smooth curve, one expects a larger value of β in Eq. (8) for the experimental data because of spin-wave lifetime effects which are not included in the frozen spin-wave calculations. Thus one should expect the calculated points at large q to lie above the extrapolation of the best fit to the experimental data. The first three points of GB are consistent with

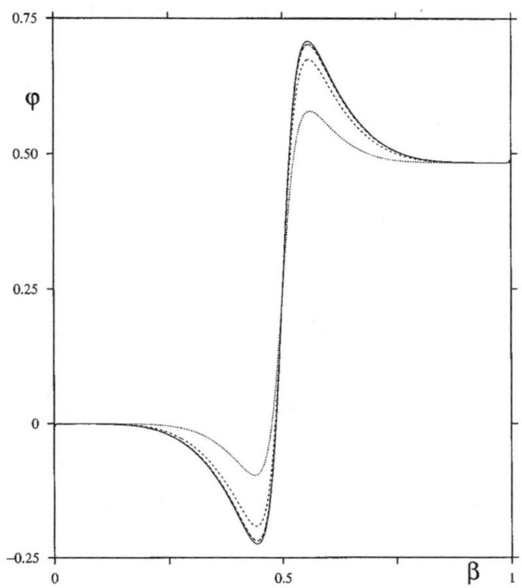

FIG. 2. The azimuthal angle of the $\mathbf{q}=(2\pi/a)(2/13,\ 0,\ 0)$ frozen spin wave in radians vs β, the fraction of the distance to a nearest-neighbor atom. The solid line is the quasi-self-consistent result and the longer dash, short dash, and dotted curves represent the constrained results with A=0.2, 2, and 20 Ry a_0^3, respectively.

our calculations; there is no apparent reason why the fourth should not be equally so.

Finally, in Table II we look at the effect of the constraint parameter A on the magnon frequency. As one would expect, the larger A, the harder $\mathbf{m}(\mathbf{r})$ is forced toward the arbitrarily chosen $\mathbf{M}(\mathbf{r})$ and the less variational freedom it has to approach its natural value throughout the unit cell. This raises the calculated value of $\hbar\omega$ well above that obtained from the natural $\mathbf{m}(\mathbf{r})$. In Fig. 2 we plot the azimuthal angle $\varphi(\mathbf{r})$ of $\mathbf{m}(\mathbf{r})$ between nearest-neighbor atoms for $\mathbf{q}=(2\pi/a)(2/13,\ 0,\ 0)$ and $A=0.2$, 2, and 20 Ry a_0^3 and compare it with the quasi-self-consistent result. We see that as $A\rightarrow\infty$, $\varphi(\mathbf{r})$ becomes a step function, indicating that better results would be obtained if the constraint sphere radius were chosen to be less than that of the inscribed sphere. We also see as $A\rightarrow 0$, $\varphi(\mathbf{r})$ approaches the quasi-self-consistent $\varphi(\mathbf{r})$, demonstrating that the quasi-self-consistent method yields the maximum variational freedom.

ACKNOWLEDGMENTS

This work was supported by the Texas Advanced Computing Center of the University of Texas at Austin, by the Welch Foundation, and the NSF under Grant Nos. DMR9705406 and DMR9614040.

APPENDIX

The force theorem was originally derived for displaced atoms in crystals and is fairly complicated.[16] However, when applied to a simple perturbation it is quite simple and merely says that within density-functional theory, the change in total energy of a perturbed system is given *to first order* by

$$\Delta E=\sum_i^{\mathrm{occ}} [\epsilon_i(v+\delta v,\ V_{\mathrm{Hxc}})-\epsilon_i(v,V_{\mathrm{Hcx}})], \qquad \text{(A1)}$$

where the Kohn-Sham eigenvalues of the perturbed system are to be evaluated using the self-consistent Hartree and exchange-correlation potentials of the unperturbed system. Since Ref. 3 is not unique in misinterpreting this, we point out that Eq. (A1) is easily obtained from the total-energy expression

$$E(v)=\sum \epsilon_i(v,V_{\mathrm{Hxc}})-\int V_{\mathrm{Hxc}}\rho d\mathbf{r}+E_{\mathrm{Hxc}}[\rho]. \qquad \text{(A2)}$$

Under a perturbation, this expression is changed by adding δv to the external potential plus corresponding shifts in the density $\delta\rho$ and in the Hartree-exchange-correlation potential δV_{Hcx}. Equation (A1) is then obtained if we also note that, to first order,

$$\epsilon_i(v+\delta v,V_{\mathrm{Hxc}}+\delta V_{\mathrm{Hxc}})=\epsilon_i(v+\delta v,V_{\mathrm{Hxc}})+\int \delta V_{\mathrm{Hxc}}\rho d\mathbf{r},$$

$$(V_{\mathrm{Hxc}}+\delta V_{\mathrm{Hxc}})(\rho+\delta\rho)=V_{\mathrm{Hxc}}\rho+V_{\mathrm{Hxc}}\delta\rho+\delta V_{\mathrm{Hxc}}\rho,$$

$$E_{\mathrm{Hxc}}[\rho+\delta\rho]=E_{\mathrm{Hxc}}[\rho]+\int V_{\mathrm{Hxc}}\delta\rho d\mathbf{r}. \qquad \text{(A3)}$$

The second-order terms can also be easily found for ΔE and they are given by

$$\sum_i^{\mathrm{occ}}\sum_j^{\mathrm{unocc}} |(\delta V_{\mathrm{Hxc}})_{ij}|^2/\Delta\epsilon_{ij}-\frac{1}{2}\int \delta V_{\mathrm{Hxc}}\delta\rho d\mathbf{r}. \qquad \text{(A4)}$$

There is no reason that these should cancel each other in general.

[1] Qian Niu and Leonard Kleinman, Phys. Rev. Lett. **80**, 2205 (1998).

[2] Q. Niu, X. Wang, L. Kleinman, W.-M. Liu, D. M. C. Nicholson, and G. M. Stocks, Phys. Rev. Lett. **83**, 207 (1999).

[3] R. H. Brown, D. M. C. Nicholson, Xingdong Wang, and T. C. Schulthess, J. Appl. Phys. **85**, 4830 (1999).

[4] Ralph Gebauer and Stephan Baroni, Phys. Rev. B **61**, 6459 (2000).

[5] J. F. Cooke, J. W. Lynn, and H. L. Davis, Phys. Rev. B **21**, 4118 (1980).

[6] Leonard Kleinman, Phys. Rev. B **59**, 3314 (1999).

[7] D. M. Bylander and Leonard Kleinman, Phys. Rev. B **60**, 9916 (1999).

[8] D. M. Bylander and Leonard Kleinman, Phys. Rev. B **58**, 9207 (1998).

[9] David Vanderbilt, Phys. Rev. B **41**, 7892 (1999).

[10] The ultrasoft pseudopotential is ultrasoft because the pseudofunctions are not norm conserving. The charge density which is reconstructed is a near perfect approximation to the actual, rather than pseudo, charge density.

[11] Leonard Kleinman and D. M. Bylander, Phys. Rev. Lett. **48**, 1425 (1982).

[12] Conyers Herring, in *Magnetism*, edited by George T. Rado and Harry Suhl (Academic, New York, 1966), Vol. IV.

[13] Because of the nonlinearity of the xc functional, our calculated total energy represents the entire xc energy plus all other contributions to the valence electron ionization energy.

[14] C. K. Loong, J. M. Carpenter, J. W. Lynn, R. A. Robinson, and H. A. Mook, J. Appl. Phys. **55**, 1895 (1984).

[15] J. W. Lynn, Phys. Rev. B **11**, 2624 (1975).

[16] M. Weinert, R. E. Watson, and J. W. Davenport, Phys. Rev. B **32**, 2115 (1985).

Index